ANNOTATED TEACHER'S EDITION

SCIENCEPLUS®
TECHNOLOGY AND SOCIETY

LEVEL RED

Project Directors

International: **Charles McFadden**
Professor of Science Education
The University of New Brunswick
Fredericton, New Brunswick

National: **Robert E. Yager**
Professor of Science Education
The University of Iowa
Iowa City, Iowa

Project Authors

Earl S. Morrison *(Author in Chief)*
Alan Moore *(Associate Author in Chief)*
Nan Armour
Allan Hammond
John Haysom
Elinor Nicoll
Muriel Smyth

This new United States edition has been adapted from prior work by the Atlantic Science Curriculum Project, an international project linking teaching, curriculum development, and research in science education.

HOLT, RINEHART AND WINSTON
Harcourt Brace & Company
Austin • New York • Orlando • Atlanta • San Francisco • Boston • Dallas • Toronto • London

Portions of this work were published in previous editions.

Acknowledgments: See page S192, which is an extension of this copyright page.

Printed in the United States of America

ISBN 0-03-095094-5

1 2 3 4 5 6 7 032 00 99 98 97 96

ACKNOWLEDGMENTS

Project Advisors

Herbert Brunkhorst
Director, Institute for Science Education
California State University,
San Bernardino
San Bernardino, California

David L. Cross
Science Consultant
Lansing School District
Lansing, Michigan

Jerry Hayes
Associate Director, Science Outreach
Teacher's Academy,
Mathematics and Science
Chicago, Illinois

William C. Kyle, Jr.
Director, School Mathematics and Science Center
Purdue University
West Lafayette, Indiana

Mozell Lang
Science Education Specialist
Michigan Department
of Education
Lansing, Michigan

Annotated Teacher's Edition Writers

Harry Dierdorf
New Brighton, Pennsylvania

Shirley Key, Ph.D.
Multicultural Consultant and Writer
Missouri City, Texas

Pamela Russ, Ph.D.
Multicultural Consultant and Writer
Turlock, California

Sandy Tauer
Derby, Kansas

Nancy Wesorick
Longmont, Colorado

Feature Writers

Judith Edgington
Austin, Texas

Bruce R. Mulkey
Austin, Texas

David Stienecker
Forest Hills, New York

Teacher Reviewers

Robert W. Avakian
Alamo Junior High
Odessa, Texas

Barry Lynne Bishop
San Rafael Junior High
Ferron, Utah

Carol Bornhorst
Bonita Vista Middle School
Chula Vista, California

Paul Boyle
Parry Heights Middle School
Evansville, Indiana

Renae Cartwright
Cedar Park Middle School
Cedar Park, Texas

Kenneth Creese
White Mountain Junior High
Rock Springs, Wyoming

Leila R. Dumas
LBJ Science Academy
Austin, Texas

Kenneth Horn
Fallston Middle School
Fallston, Maryland

Roberta Jacobowitz
C. W. Otto Middle School
Lansing, Michigan

Pamela Jones
Birmingham Covington School
Birmingham, Michigan

David Negrelli
Miami Lakes Middle School
Hialeah, Florida

Kevin Reel
Thacher School
Ojai, California

Steve Siegel
McCormick Junior High
Cheyenne, Wyoming

Patricia Soto
G. W. Carver Middle School
Miami, Florida

Larry Tackett
Andrew Jackson Middle School
Cross Lanes, West Virginia

Nancy Wesorick
Sunset Middle School
Longmont, Colorado

Donald Yost
Cordova High School
Rancho Cordova, California

Academic Reviewers

David Armstrong, Ph.D.
Department of EPO Biology
University of Colorado
Boulder, Colorado

Kenneth Brown, Ph.D.
Professor of Chemistry
Department of Chemistry
Northwestern Oklahoma State
University
Alva, Oklahoma

Patricia Buis, Ph.D.
Assistant Professor
Department of Geology
and Geological Engineering
University of Mississippi
University, Mississippi

Linda K. Butler, Ph.D.
Division of Biological Sciences
University of Texas
Austin, Texas

Mark Coyne, Ph.D.
Assistant Professor of Agronomy
University of Kentucky
Lexington, Kentucky

Albert B. Dickas, Ph.D.
Professor of Geology
University of Wisconsin
Superior, Wisconsin

David J. Froehlich
University of Texas
J. J. Pickle Research Center
Austin, Texas

Frederick R. Heck
Associate Professor of Geology
Department of Physical Sciences
Ferris State University
Big Rapids, Michigan

Arthur Huffman, Ph.D.
Department of Physics
University of California
at Los Angeles
Los Angeles, California

James Kaler, Ph.D.
Professor of Astronomy
University of Illinois
Urbana, Illinois

Doris I. Lewis, Ph.D.
Professor of Chemistry
Suffolk University
Boston, Massachusetts

R. Thomas Myers, Ph.D.
Professor of Chemistry Emeritus
Department of Chemistry
Kent State University
Kent, Ohio

Jeffrey Karl Ochsner,
Department of Architecture
University of Washington
Seattle, Washington

Thomas Troland, Ph.D.
Associate Professor
Department of Physics
and Astronomy
University of Kentucky
Lexington, Kentucky

Blue-Ribbon Committee

The following teachers constituted a special committee of current *SciencePlus* users who provided information and insights on how to improve the *SciencePlus* program. Their input was invaluable.

Patricia Barry
Milwaukee, Wisconsin

Carol Bornhorst
Chula Vista, California

Catherine Carlson
Plano, Texas

Harry Dierdorf
New Brighton, Pennsylvania

Jeff Felber
Spring Valley, New York

Barbara Francese
Madison, Connecticut

Kenneth Horn
Forest Hills, Maryland

Doug Leonard
Toledo, Ohio

Betsy Mabry
Enid, Oklahoma

Debbie Melphi
Syracuse, New York

Lynn Roudabush-Novak
Midlothian, Virginia

Donna Robinson
Rapid City, South Dakota

Sandy Moerke-Schaefer
Marysville, Washington

Marvin Selnes
Sioux Falls, South Dakota

Patricia Soto
Coral Gables, Florida

Margaret Steinheimer
St. Louis, Missouri

Sandy Tauer
Derby, Kansas

Joy Ward
Chicago, Illinois

Gary Weaver
Vacaville, California

Nancy Wesorick
Longmont, Colorado

Brenda West
Scott Depot, West Virginia

Project Associates

We wish to thank the thousands of science educators, teachers, and administrators from the scores of universities, high schools, junior high and middle schools who have contributed to the success of *SciencePlus.*

CONTENTS ANNOTATED TEACHER'S EDITION

OWNER'S MANUAL

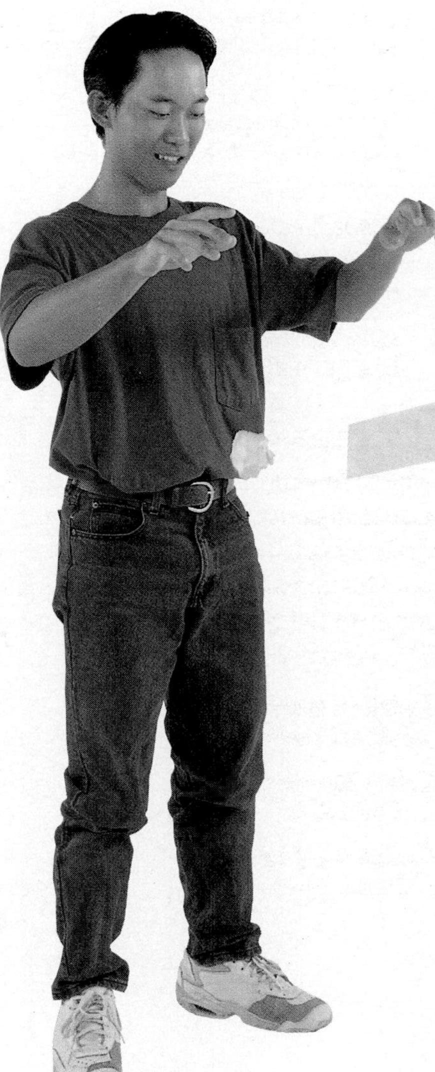

UNIT INTERLEAF INFORMATION

CONTENTS

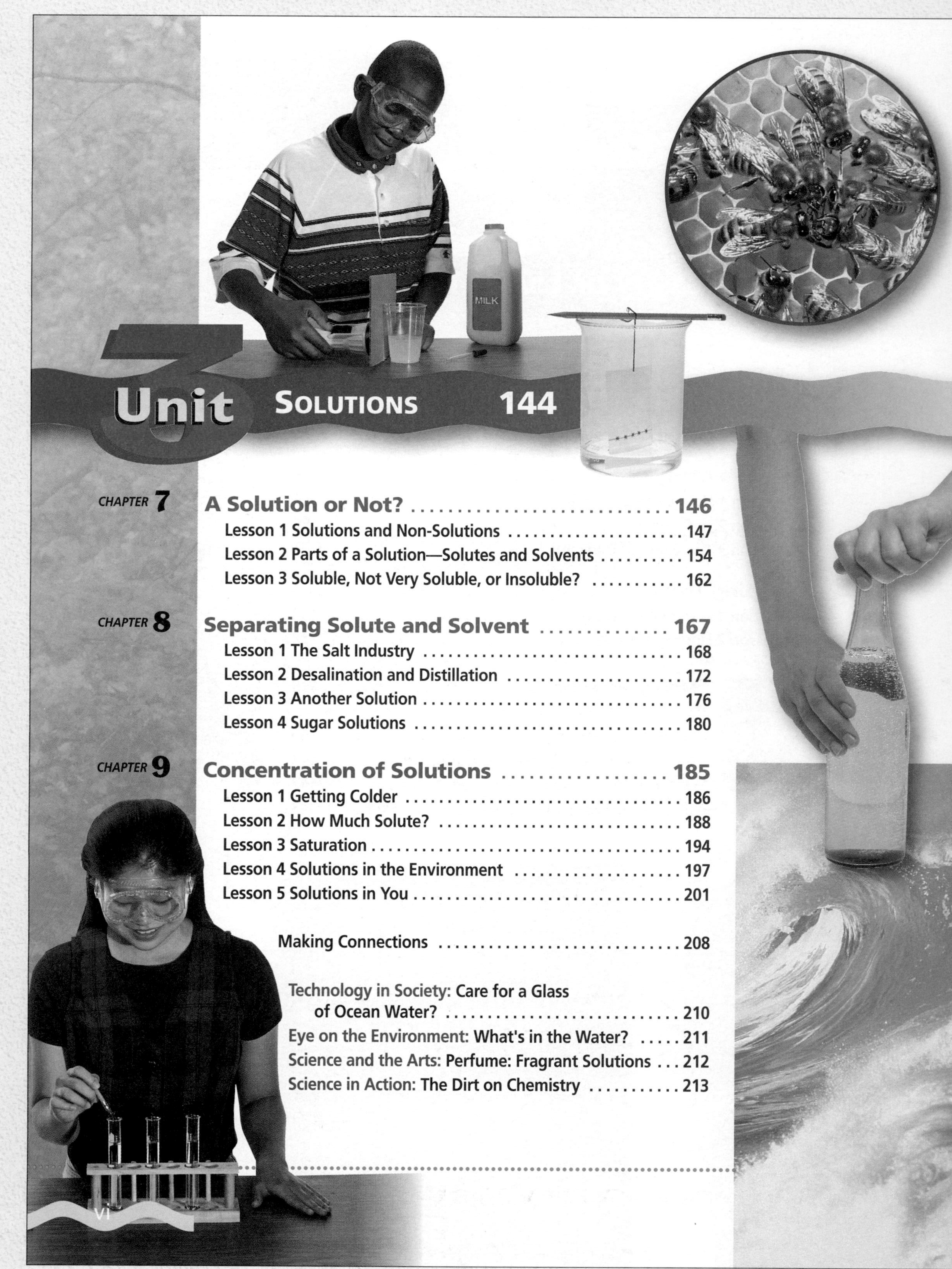

Unit 3 Solutions 144

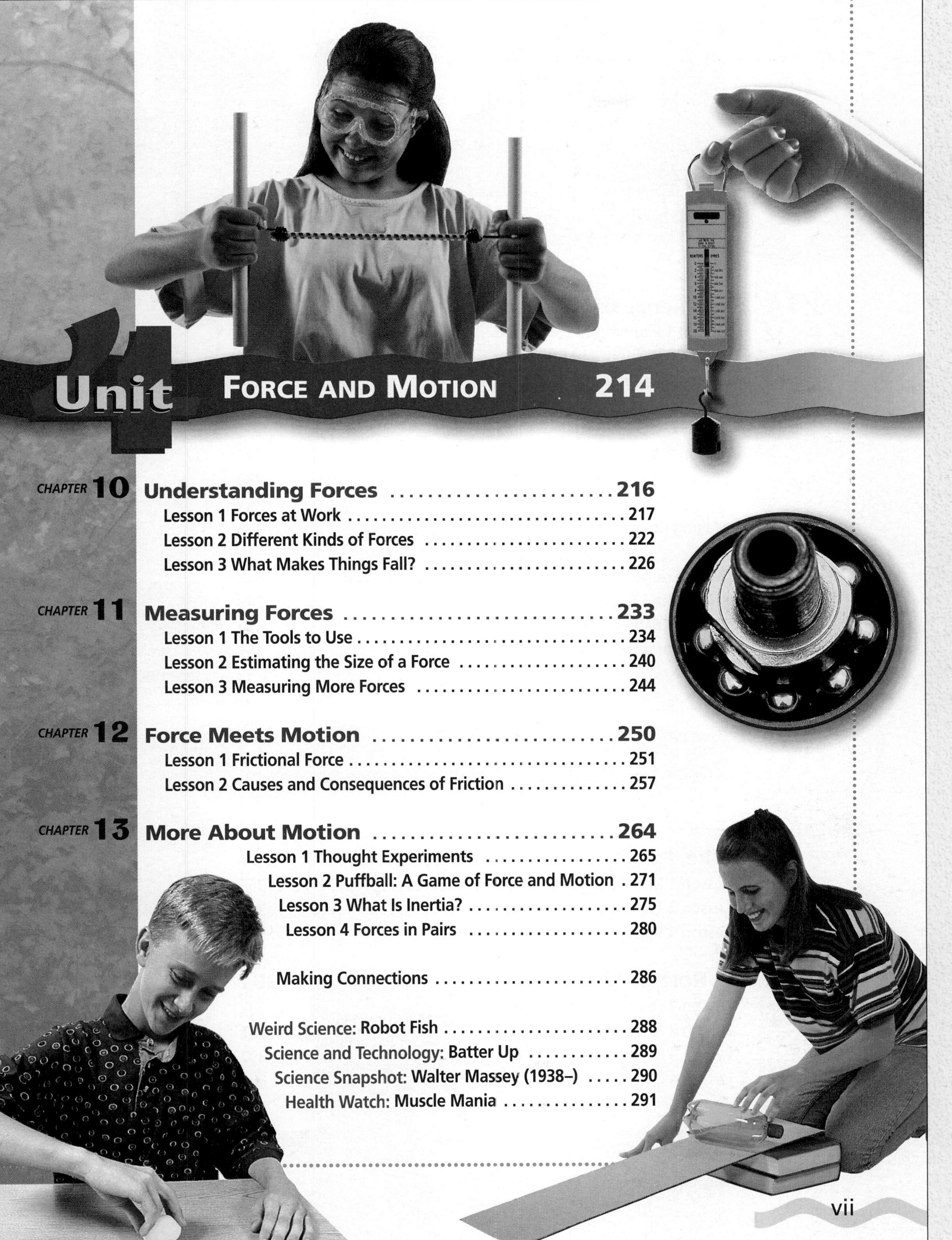

Unit 4 Force and Motion 214

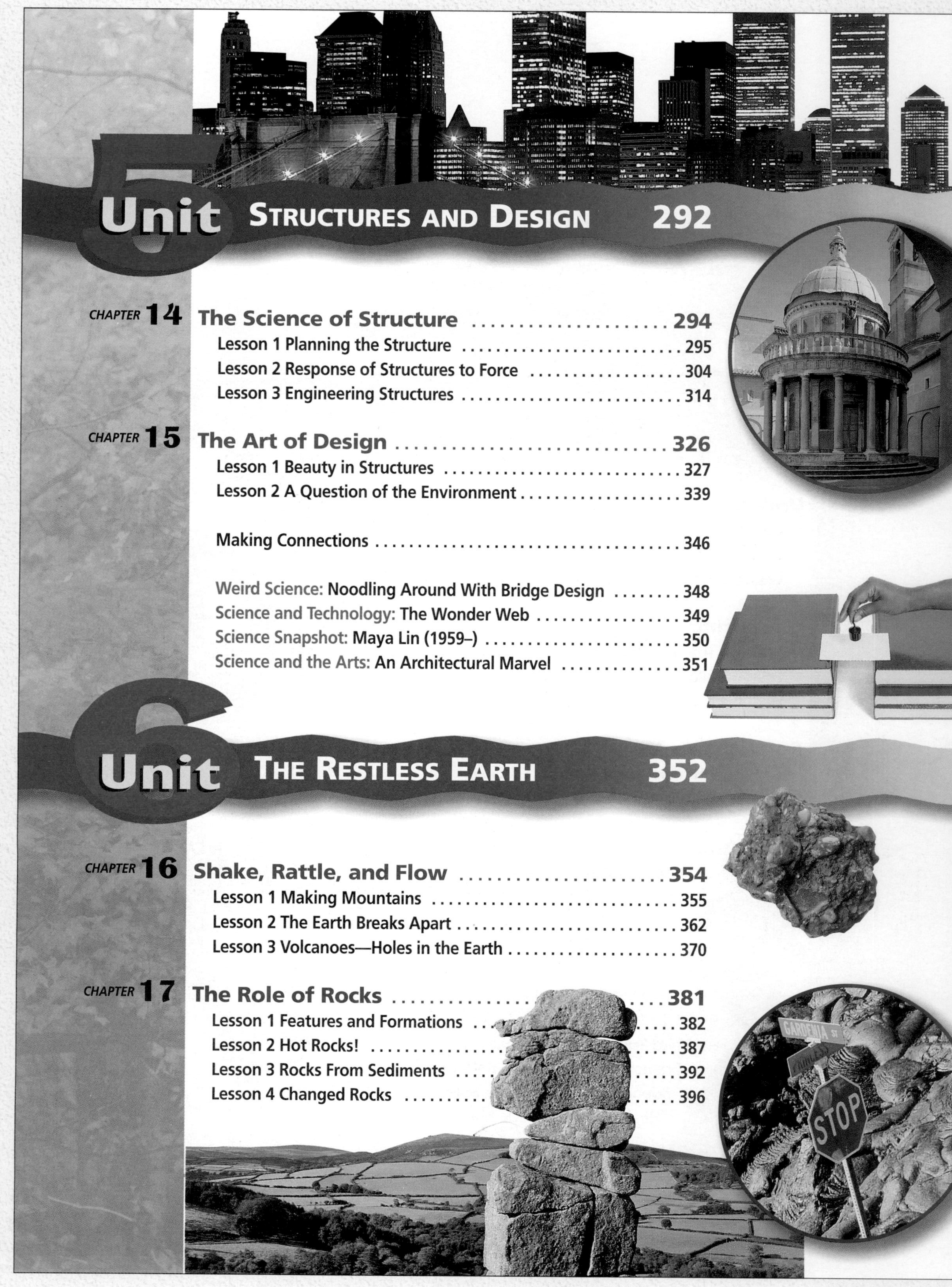

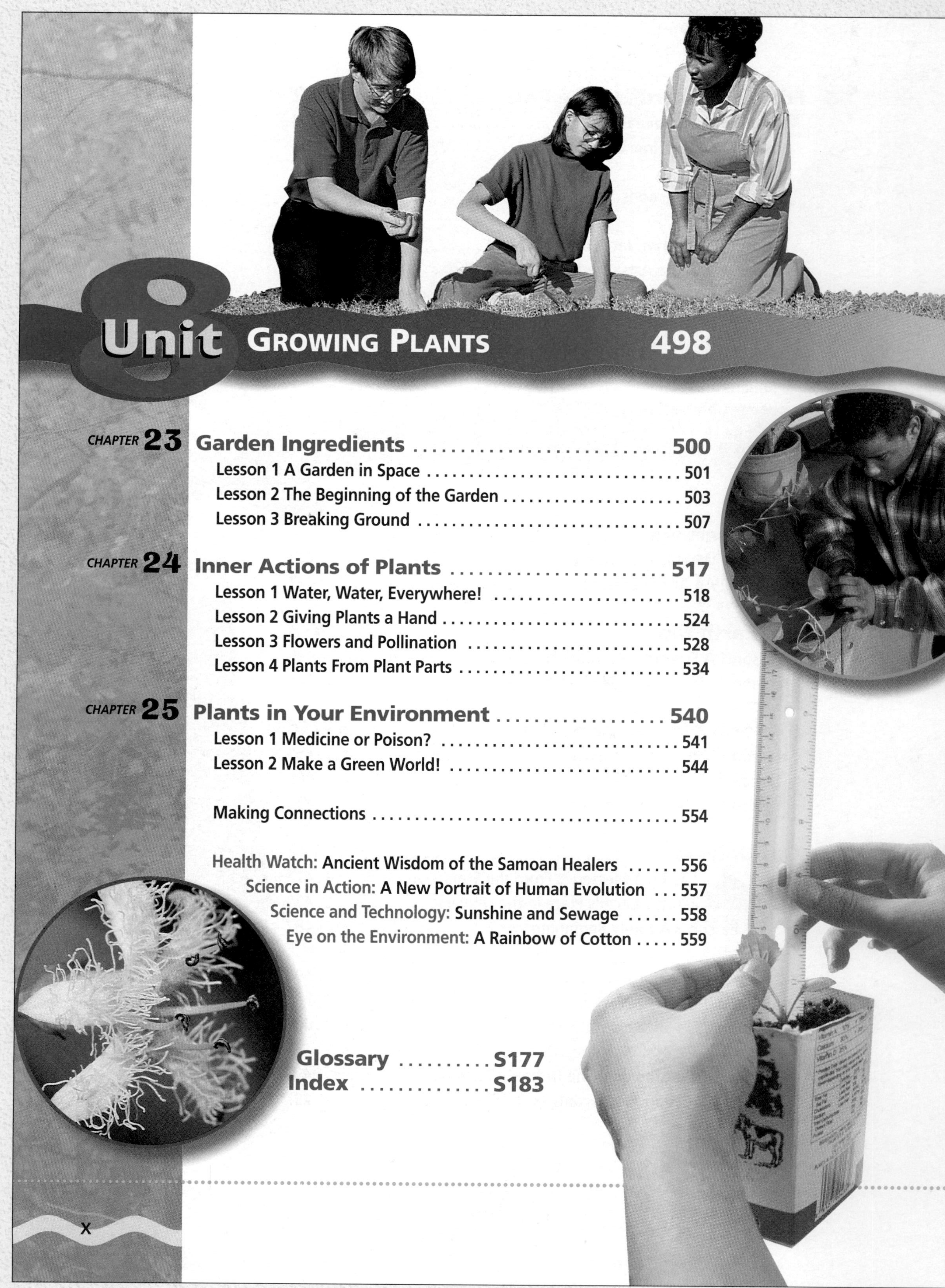

Unit 8 GROWING PLANTS 498

SOURCEBOOK

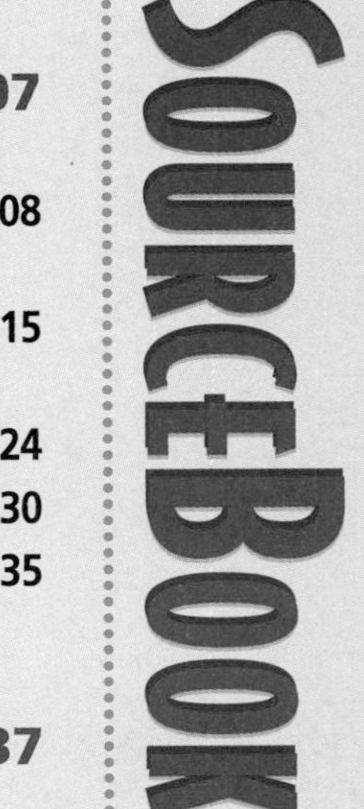

SciencePlus Owner's Manual

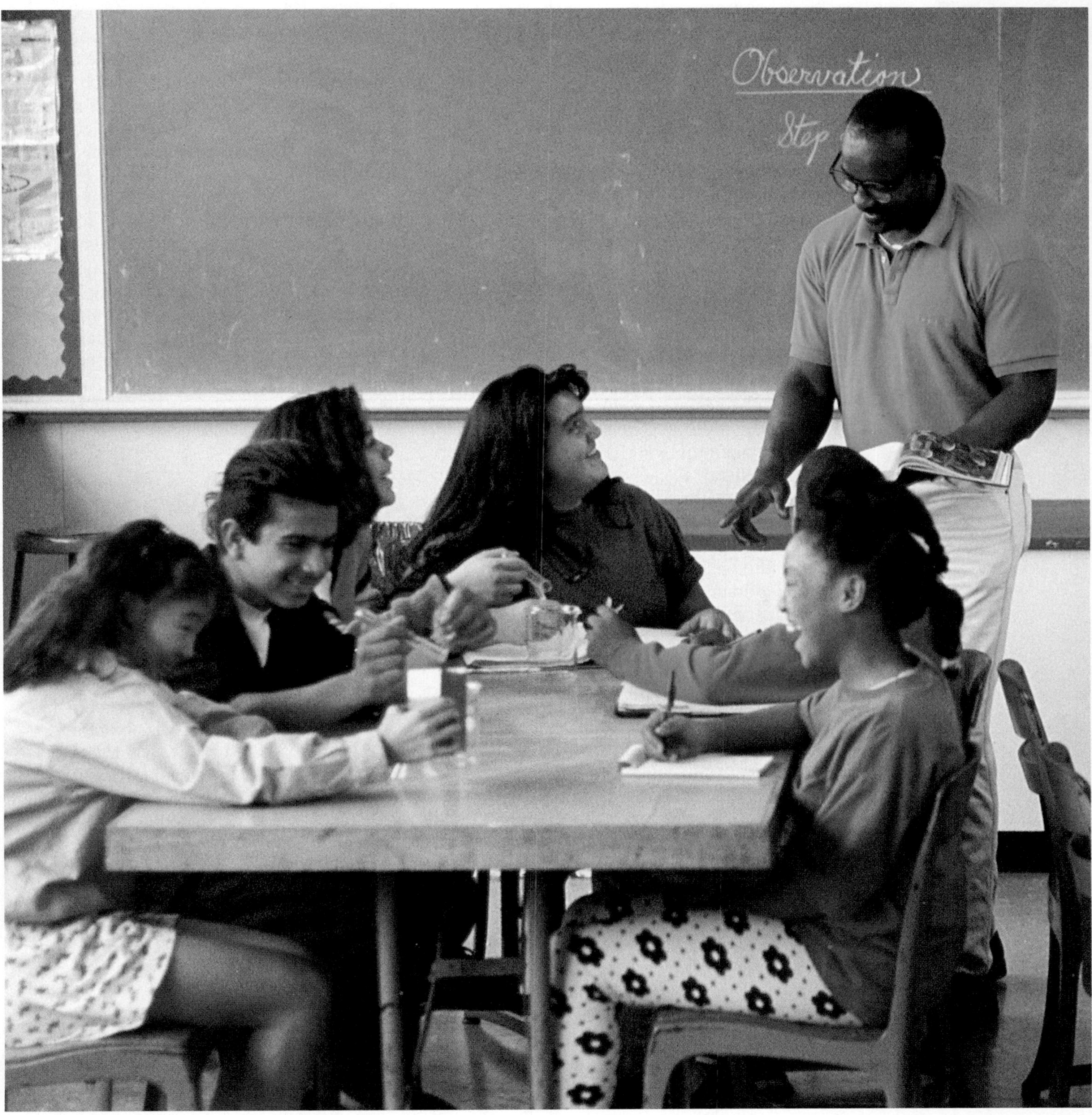

THE SCIENCEPLUS PHILOSOPHY

SciencePlus is lively, engaging, and relevant to the student's world.

Welcome to *SciencePlus*, an innovative approach to science education. *SciencePlus* is unlike any science program you have used before. It is designed from the ground up to teach science in precisely the way that students learn best—by thinking, talking, and writing about what they do and discover. *SciencePlus* is activity- and inquiry-based. In other words, it is both hands-on and minds-on. *SciencePlus* is lively, engaging, and relevant to the students' world. *SciencePlus* is loaded with thought-provoking activities designed to challenge students' thinking skills while introducing them to realistic methods of science.

SciencePlus works for students and teachers alike. Students enjoy and benefit from its varied and active approach, while teachers find it to be teachable under real conditions. Laboratory-type activities require about 30 to 40 percent of class time. The remainder of class time is taken up by a rich variety of learning activities.

SciencePlus is tailored to meet the needs of middle school–aged students. It accommodates every type of student you are likely to encounter: the basic student, who requires substantial guidance; the average student, who is not yet fully equipped with abstract-thinking skills; and the advanced student, who needs only to be pointed in the right direction to succeed.

At every stage, the *SciencePlus* program emphasizes concept and skill development over memorization of facts. By doing science activities and then thinking about the results, students learn the *whys* and *hows,* not just the *whats* and *whens,* of science. Ultimately, students come to see science as a system for making sense of the world.

Origins

The *SciencePlus* program was originally developed by the Atlantic Science Curriculum Project (ASCP) of Canada to replace the traditional recall-based curriculum, which had proven to be ineffective. *SciencePlus* represents a ground-breaking effort, the culmination of many years of labor by dozens of talented, dedicated science educators.

The *SciencePlus* development team was guided every step of the way by the latest insights into how children actually learn. The *SciencePlus* program has been thoroughly tested on real students in realistic settings, refined, and then retested. The result is a program that works! Teachers using *SciencePlus* have reported dramatic gains in student comprehension and retention of scientific concepts. Above all, students enjoy using *SciencePlus* and develop a heightened interest in science.

A Continuing Tradition

This edition of *SciencePlus*, now in its second edition as an American publication, continues the tradition of excellence begun in Canada many years ago. The program has undergone many changes in response to recommendations of curriculum reform movements, such as Project 2061 and the NSTA Scope, Sequence, and Coordination. In addition, the revision of this program was heavily influenced by *SciencePlus* teachers, whose combined and varied experiences helped us make the program even more relevant and easier to implement.

An Interactive, Effective Program

SciencePlus employs proven teaching strategies: guided and open-ended investigations, small-group discussions, exploratory writing and reflective-reading tasks, games, picture and word puzzles, and independent long-range projects. This variety motivates and helps maintain the interest of students and teachers alike.

SciencePlus develops scientific process skills as an essential goal. As the curriculum progresses, the students will master increasingly complex tasks. For example, students will move from directed to open-ended inquiry and from reading and completing tables and graphs to constructing them from experimental data they have collected on their own.

In general, each of the units in *SciencePlus* is self-contained and may be taught as a separate instructional module. *SciencePlus* contains a balance of physical, biological, earth/space, and environmental science topics.

Guiding Principles

The guiding principles of *SciencePlus* are simple and few:

- **Anyone can learn science**
The image of science as the private domain of the superintelligent is wrong and damaging. Science is for everyone. Children exposed to science for the first time take to it naturally. It is only later, after science explorations have been replaced by fill-in-the-blank worksheets and recall drills, that love for science is replaced by boredom and even dread. *SciencePlus* can rekindle the sense of wonder and fascination that lies dormant within your students.
- **Science is a natural endeavor**
Whether we realize it or not, each of us applies science nearly every day. Hardly a day passes that we don't ask ourselves, "How does this work?" or "Why does that happen?" or "What happens if . . . ?" Scientists differ from other people only in that it is their profession, rather than their avocation, to figure out "how," "why," and "what if."

Unfortunately, stereotypes about science and scientists abound. Many students feel that only "nerds" or "geeks" enjoy science. This falsehood may do as much to turn people away from science as any curricular shortcoming. *SciencePlus* actively refutes these stereotypes. It portrays science as a rewarding, quintessentially human undertaking. Scientists are portrayed as normal people, not aloof geniuses who talk in equations.

- **Science is its own reward**
There is no feeling quite like the thrill of discovery or the sense of accomplishment that comes from rising to a difficult challenge. Science can be thought of as a voyage into the unknown. This voyage can be exciting and rewarding for all.

Aims

SciencePlus is designed to help you further develop each of the following in your students:

- Understanding of the interrelationships among science, technology, and society
- Understanding of important science concepts, processes, and ideas
- Use of higher-order thinking skills
- Ability to solve problems and apply scientific principles
- Commitment to environmental protection
- Interest in independent study of scientific topics
- Social skills
- Communication skills

To accomplish these goals, a wide variety of teaching strategies is employed. The common denominator among these is their emphasis on doing. At all times, students are to be active and involved.

SciencePlus can rekindle the sense of wonder and fascination that lies dormant within your students.

> To prepare students for the challenges of the future, they must become science literate.

Science, Technology, and Society

Science and technology are flip sides of the same coin; each supports the other. Neither should be studied in isolation. *SciencePlus* explores the relationship between science and technology, and the effect of both on our society as a whole. Even people who never again set foot in a laboratory after leaving school can benefit from an understanding of science and technology and how both relate to each other and to society at large.

Our society, complex as it is, will become even more so in the years to come. Science and technology will play roles in every aspect of life. In the future, "high-tech" will be more than a catch phrase—it will permeate every aspect of life. To prepare students for the challenges of the future, they must become science literate. They must be given the tools that will enable them to become responsible and productive individuals in a highly technological world.

Science for All

SciencePlus is designed to put the "process" back into science education and, in so doing, to provide students with the mental skills they need to truly understand and apply science. Today, as never before, a thorough grounding in science is absolutely essential; without it, students—the citizens of tomorrow—cannot expect to be fully conversant in and responsive to the complex issues of the twenty-first century.

No program can teach itself. You, with your energy, enthusiasm, and ability, are the key to a successful outcome. *SciencePlus* will help you help your students develop all of the skills they need to learn independently. *SciencePlus* fosters a spirit of joint exploration. Let the journey begin.

THE SCIENCEPLUS METHOD

BUILDING A BETTER UNDERSTANDING OF SCIENCE

The *SciencePlus* program is based on the Constructivist Learning Model (CLM). Constructivism is based as much on common sense as on the results of research. With the CLM, students "construct" an understanding of concepts step by step. Students begin by identifying what they already know about a topic. Any misconceptions they may have about a topic are exposed at this point. Identifying these misconceptions is a critical part of the process. Next, students do hands-on activities to experience the subject matter directly. Their experiences cause them to amend, add to, or scrap altogether the mental model they already have of the subject in question.

Constructivism is based on a few key steps:

1. **Invitation**
   The Invitation stimulates students' curiosity and engages their interest. At this stage, students note the unexpected, pose questions, or define a problem.
2. **Exploration**
   Explorations engage students in the search for solutions or explanations. Students look for alternative sources of information, collect and evaluate data, and clarify their findings through discussion and debate.
3. **Proposing explanations and solutions**
   At the conclusion of the Explorations, students propose their response to the problem or question posed in the Invitation. The class is exposed to a variety of possible responses, and students have the opportunity to consider each.
4. **Taking action**
   Students make decisions about a course of action based on the various proposals offered. If the class reaches consensus, then this stage may bring about closure of the lesson. It may happen, though, that this stage identifies new questions to explore.

Another way to think of the Constructivist Learning Model is in terms of the five *E*s: Engage, Explore, Explain, Elaborate, and Evaluate.

For an in-depth discussion of Constructivism, see "The Constructivist Learning Model," by Bob Yager,* The Science Teacher, *September 1991.

With the Constructivist Learning Model, students "construct" an understanding of concepts step by step.

CONCEPTUAL FRAMEWORK

The information and chart that follow outline the conceptual framework for *SciencePlus*, Level Red. The information on this page will help you interpret the chart that follows on the next two pages.

Concept Focus

Each unit focuses on one major scientific concept. This concept is developed through a thematic approach. Although other concepts are covered in each unit, this column indicates the major focus.

Content Integration

Content Focus

The content of each unit integrates the traditional disciplines of life, earth, and physical science. This column lists the main discipline covered in each unit.

Supporting Content

The secondary content focus listed in this column supports the major content focus of each unit and helps achieve integration.

Thematic Focus

SciencePlus uses five themes as its organizational framework. These themes allow students to see content relationships and promote conceptual understanding. This column lists the main themes of each unit.

Science, Technology, and Society (STS)

This column lists three STS topics introduced in the unit and allows students to consider the development and impact of technology, to examine social issues, and to explore possible solutions.

Process Skills

All Major Skills Covered

SciencePlus places a strong emphasis on process skills. There are 11 major science process skills, and when a check mark appears in this column, all 11 skills have been used in the unit. The 11 process skills covered in *SciencePlus* can be found on page T47.

Process Skills Focus

Of the 11 skills used in *SciencePlus*, the 3 skills listed in this column are emphasized in the unit.

CONCEPTUAL FRAMEWORK

CONCEPT FOCUS	CONTENT INTEGRATION				PROCESS SKILLS	
	Content Focus	Supporting Content	Thematic Focus	STS	All Major Skills Covered	Process Skills Focus
Unit 1 **Interactions** Interactions that occur among plants, animals, and the environment	Life science	•Physical science •Earth science	•Changes Over Time •Energy •Cycles •Systems •Structures	•Recycling materials •Pollution caused by humans •Researching acid precipitation	X	•Observing •Classifying •Inferring
Unit 2 **Diversity of Living Things** The great diversity of living things, probable causes of diversity, and how scientists make sense of diversity	Life science	•Earth science •Physical science	•Changes Over Time •Systems •Structures	•The impact of the Industrial Revolution •The role of humans in extinction and population recovery •Medical research on cancer	X	•Observing •Comparing •Classifying
Unit 3 **Solutions** The characteristics, uses, and properties of solutions	Physical science	•Earth science •Life science	•Energy •Changes Over Time •Cycles •Structures	•Properties and uses of common solutions •The impact of oil spills •Methods of mining salt	X	•Observing •Measuring •Contrasting
Unit 4 **Force and Motion** The concepts of force and motion and methods of measuring forces	Physical science	•Earth science •Life science	•Energy •Cycles •Systems •Structures	•Constructing machines to measure force •Constructing machines that work easier by applying certain forces •Uses and problems associated with friction	X	•Measuring •Predicting •Inferring
Unit 5 **Structures and Design** The technological and societal impact of structures and their design	Physical science	•Life science •Earth science	•Changes Over Time •Systems •Structures	•Constructing architectural models •Testing shapes for strength •Interrelationship between architecture and culture	X	•Analyzing •Organizing •Comparing

CONCEPTUAL FRAMEWORK

CONCEPT FOCUS	CONTENT INTEGRATION				PROCESS SKILLS	
	Content Focus	Supporting Content	Thematic Focus	STS	All Major Skills Covered	Process Skills Focus
Unit 6 **The Restless Earth** The processes of geological change and the effects of these changes on society and the environment	Earth science	• Life science • Physical science	• Energy • Changes Over Time • Cycles • Systems • Structures	• Measuring earthquakes • Using rock and minerals • Radiometric dating	X	• Observing • Inferring • Hypothesizing
Unit 7 **Toward the Stars** The structure of the universe and the characteristics and movements of celestial bodies	Earth science	• Physical science • Life science	• Energy • Changes Over Time • Cycles • Systems • Structures	• Using space probes and radio telescopes to explore the universe • Planning future space stations • Colonizing other planets	X	• Observing • Comparing • Inferring
Unit 8 **Growing Plants** The growth and structure of plants, as well as their role in environmental processes	Life science	• Earth science • Physical science	• Changes Over Time • Cycles • Structures	• Planning self-contained biospheres • Using chemical fertilizers • Agricultural and genetic engineering • Landscape analysis and planning	X	• Analyzing • Predicting • Communicating

COMPONENTS OF SCIENCEPLUS

PUPIL'S EDITION

SciencePlus is no ordinary textbook. *SciencePlus* is a student-friendly text: lively; abundantly illustrated with clever, colorful illustrations; and loaded with engaging activities. Every effort has been taken to make this text the sort of book that students will actually want to use.

Units, Chapters, Lessons

SciencePlus contains eight units, which are further divided into chapters and lessons on closely related subject matter. Each lesson includes a wide variety of activities and explorations designed to develop the lesson content.

Each chapter begins with a ScienceLog page that invites students to express what they already know about the subject of the chapter. Highly graphic and playful questions help expose misconceptions students may have about the topics to be covered. This important process sets the stage for firsthand exploration of the subject area. The answers to the ScienceLog questions are the equivalent of hypotheses or predictions, which are either supported or disproved as the students explore the chapter content. Either way, the students become invested in what they are about to learn.

Explorations

Scattered throughout each unit are a series of Explorations—hands-on, inquiry-based activities. These Explorations allow students to see scientific principles in action. The Explorations are essential for inducing real learning in students. As students do the Explorations, they have the opportunity to compare their mental models of scientific principles with the real things. As weaknesses are exposed, students adjust their thinking to accommodate what they have learned.

Most of the Explorations are designed so they can be done cooperatively in small groups. In this way, the Explorations model real scientific experiences, in which scientists work together to solve problems. By working in cooperative groups, students develop important skills such as communicating ideas and sharing responsibility. And the cooperative groups not only make science more interactive and more fun, but also provide valuable opportunities to develop social skills. They are also more economical to implement.

For more information on cooperative learning, see pages T41 through T46 of this Annotated Teacher's Edition.

Most of the supplies needed for the Explorations consist of very common equipment and materials. In most cases, they can be gathered from the home. For help in gathering supplies, the *Teaching Resources* package contains a comprehensive *Materials Guide,* which contains both a master materials list and individual unit materials lists. In addition, the second page of the Home Connection parent letter in each *Teaching Resources* booklet provides a checklist of materials that you can use to invite parent donations. Using this portion of the parent letter will help you keep your budget low and at the same time get parents involved in the *SciencePlus* experience.

Assessment

The *SciencePlus* Pupil's Edition contains a variety of methods for checking student learning.

- **Chapter Review—Challenge Your Thinking**
  To make sure your students comprehend the new information, each chapter concludes with Challenge Your Thinking questions. These questions challenge students to apply newly learned material in a variety of ways. Many of the questions are like brain-teasers in that they are unusual and highly creative. Because the questions require more than simple recall, students find them fun to figure out.

  The chapter review pages conclude with an invitation to students to rewrite their answers to the ScienceLog questions located at the beginning of the chapter. This exercise gives students the opportunity to confront and discard any misconceptions they may have had at the outset of the chapter.

- **Unit Review—Making Connections**
  The Making Connections pages consist of two parts: The Big Ideas and Checking Your Understanding. The Big Ideas asks students to formulate their own summary of the unit, using a list of questions as a guide. The Checking Your Understanding poses a selection of comprehensive questions designed to gauge students' understanding of the unit's subject matter.

  Also included on the Making Connections pages is a short description and table of contents of the corresponding SourceBook unit. This reference directs students to the appropriate unit of the SourceBook, where they can extend their knowledge of the concepts they just explored.

All answers to the Challenge Your Thinking and Making Connections questions are located in the Wrap-Around Margins of this Annotated Teacher's Edition.

For more information about assessment, see pages T56 through T64 of this Annotated Teacher's Edition.

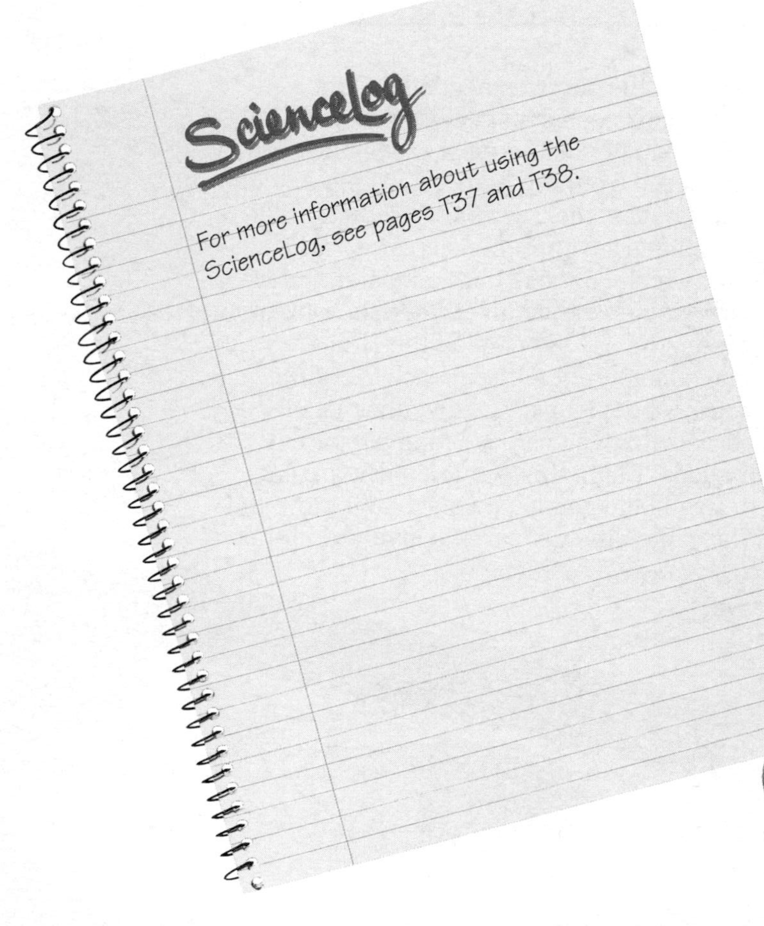

ScienceLog
For more information about using the ScienceLog, see pages T37 and T38.

Special Features

A common complaint among students is that the material they learn is not relevant. The end-of-unit features in *SciencePlus* help show students how science is an integral part of their lives.

In keeping with the spirit of *SciencePlus*, all of the features are interactive, with thought-provoking questions and research ideas that encourage students to explore the topic further. Each unit concludes with four features selected from the menu below.

- **Science in Action** features illustrate working scientists today. These features focus on the excitement and variety that characterize science careers. Students will learn about real experiences and what the scientists like most and least about their work. They may even learn about the scientists' personal career aspirations as well as their greatest disappointments.
- **Science Snapshots** are glimpses into the lives of famous scientists. Such scientists are featured to give students a sense of the colorful history of science. Present and future scientists really do stand on the shoulders of giants!
- **Health Watch** features provide insights into the variety of connections between science and health.
- **Eye on the Environment** features focus on the effects that science and technology have on our environment—both good and bad.
- **Science and Technology** features explore the science on which new technologies are based. In many cases, explorations in science open the doors to new technology. Highlighting the impact of this technology on our everyday lives shows students how relevant scientific research can be.
- **Technology in Society** features take the logical step beyond the technology that has been created through the use of science. These features explore the impact of technology on our society and put forth the inevitable question: Do the benefits outweigh the potential problems?
- **Science and the Arts** features show that science and art are not necessarily polar opposites. The connection between science and society is reinforced by showing students how artists have been inspired by nature and how scientific methods or principles have enhanced or empowered their work.
- **Weird Science** features showcase some of the odd, outlandish, and even unbelievable creatures and phenomena that make up our unique world.

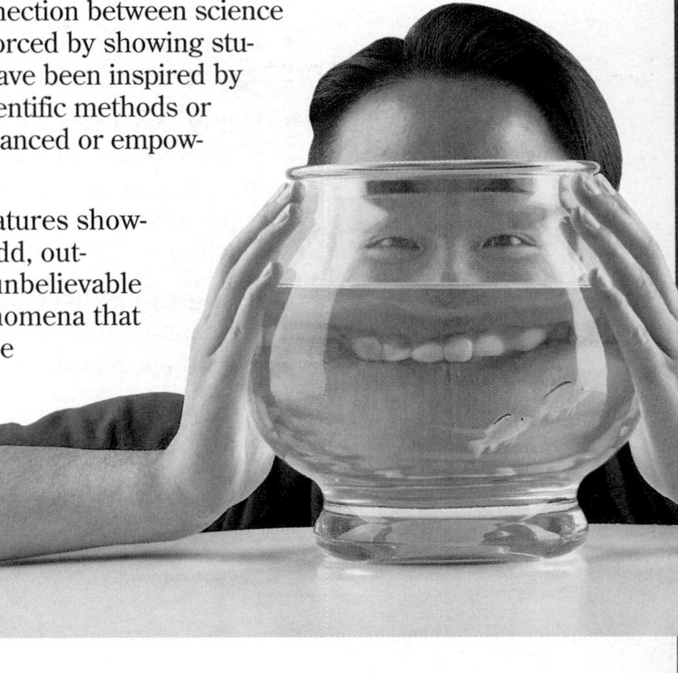

SOURCEBOOK

The SourceBook is an in-text science reference located at the back of the Pupil's Edition. This unique reference includes information that is organized to match the units of *SciencePlus*. Each SourceBook unit both reinforces and extends beyond the material presented in the Pupil's Edition, providing an excellent resource that students can use to add depth to their understanding of the topics presented in *SciencePlus*.

Because the SourceBook is not designed for hands-on exploration, it lends itself to reading at home or outside the lab. It's handy for homework assignments or for when you have a substitute teacher. Each unit of the SourceBook ends with a Unit CheckUp for checking students' comprehension. Worksheets are also available in the *Teaching Resources* booklets.

The SourceBook is referenced in many ways throughout the Pupil's Edition.

The *SciencePlus* Annotated Teacher's Edition will help you achieve the full potential of *SciencePlus*. The Annotated Teacher's Edition consists of two major parts: the Unit Interleaf and the Wrap-Around Margins. Each Unit Interleaf consists of a six-page insert preceding each unit. The Wrap-Around Margins provide on-page annotations and teaching suggestions.

Using the Unit Interleaf

Each Unit Interleaf consists of six pages of information to help you plan your lessons. Each interleaf has the following planning and preparation aids.

- **Unit Overview** is a quick overview of the concepts and content presented in the unit.
- **Using the Themes** provides descriptions of the themes that are relevant to the unit, including short explanations of how they apply and how to integrate them into your teaching.
- **Using the SourceBook** is a concise description of the content in the accompanying SourceBook unit.

- **Bibliography** consists of three subsections categorized for teachers; for students; and for films, videotapes, software, and other media.
- **Unit Organizer** is a comprehensive chart that identifies the objectives, time requirements, and teaching resources that are available for all of the chapters and lessons in the unit.
- **Materials Organizer** is a chart that identifies the materials required to carry out each Exploration and Activity in the unit.
- **Advance Preparation** identifies what you may need to do in advance to prepare for the Explorations and Activities.
- **Unit Compression** provides recommendations on how to compress or reduce the time spent teaching the unit without sacrificing the integrity of the unit.
- **Homework Options** is a chart that lists the various homework opportunities, suggestions, and worksheets available to support each of the chapters in the unit.
- **Assessment Planning Guide** is a chart that identifies the assessment opportunities and worksheets available at the lesson, chapter, and unit levels.
- **Using the *Science Discovery* Videodiscs** provides an overview and barcodes for the Science Sleuth mystery and Image and Activity Bank selections that correspond to the unit.

Using the Wrap-Around Margins

The entire Pupil's Edition of *SciencePlus* has been reduced in size and placed on the pages of this Annotated Teacher's Edition. The margins on the reduced pages have been filled with teaching suggestions and commentary to help you teach each lesson with maximum effectiveness.

UNIT OPENER

Each unit opener has the following teacher's information to help you quickly engage your students in the subject of the unit.

- **Unit Focus** provides interactive suggestions for introducing students to the unit.
- **Connecting to Other Units** is a table that shows at a glance the integration between that unit and other units in the program.
- **Using the Photograph** provides suggestions for using the unit photograph to start students thinking and asking questions about the subject of the unit topic.
- **Answer to In-Text Question** provides the answer to any question that is posed to the students on the unit opener pages.

CHAPTER OPENER

Each chapter has the following information to help you prepare to teach the chapter.

- **Connecting to Other Chapters** is a chart that identifies the basic topics explored in each chapter of the unit. This chart allows you to see the logical progression of concepts through the unit.
- **Prior Knowledge and Misconceptions** describes how to use the ScienceLog questions and suggests other ways to access students' prior knowledge or misconceptions about the chapter topics.

LESSON

Each lesson of the unit has extensive commentary and teaching notes to help you in teaching the material. The lesson information is divided into three parts: Focus, Teaching Strategies, and Follow-Up. In addition, a Lesson Organizer is provided at the beginning of each lesson.

Lesson Organizer

The Lesson Organizer is provided at the beginning of each lesson. The Lesson Organizer provides easy access to the following information about the lesson:

- Time Required
- Process Skills
- Theme Connections
- New Terms
- Materials
- Teaching Resources

FOCUS

The Focus consists of Getting Started and the Main Ideas. Getting Started provides suggestions for an activity and discussion to begin each lesson. The Main Ideas list the main concepts of the lesson.

TEACHING STRATEGIES

Teaching Strategies provide a variety of methods for successfully guiding your students through each lesson. The following categories of strategies are provided to serve your needs.

Meeting Individual Needs

Today's classrooms are places of diversity—populated by students with different ability levels and from a variety of ethnic backgrounds and cultures. The following categories of teaching suggestions are provided to help you tailor your instruction to meet the needs of all of your students.

- Gifted Learners
- Second-Language Learners
- Learners Having Difficulty

CROSS-DISCIPLINARY FOCUS

Suggestions and activities are provided that cross the boundaries between science and other disciplines, such as art, health, mathematics, social studies, music, language arts, foreign language, and industrial arts. By doing these activities, students will experience the interrelatedness of science with other areas of study and will come to understand that science is not just for scientists.

Multicultural Extension

Under this heading you will find activities and suggestions that serve to highlight cultural diversity and show its positive influence on science as well as other disciplines. These activities build on the multicultural elements already included in the Pupil's Edition to ensure ample opportunities to show students how culture and science are integrated to the benefit of us all.

Integrating the Sciences

These strategies help you explore concepts from the standpoint of more than one scientific discipline. In this way, you can help students to see how life, earth, and physical sciences work together to help us understand the world around us.

Theme Connection

Each Theme Connection consists of a Focus question and its answer. By asking the Focus question, you can help students organize their learning within an overarching theme to help them make better sense of the ever-increasing pool of scientific knowledge.

ENVIRONMENTAL FOCUS

Many lessons lend themselves to an environmental focus. In these cases, information is provided both to inform and to invite class discussion

Homework

Homework suggestions are provided as well as recommendations as to which activities and worksheets make good homework assignments.

Portfolio

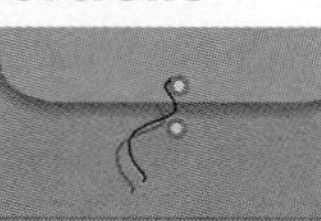

Students are encouraged to choose a portion of their work to include in a personal Portfolio. The suggestions provided in Teaching Strategies help you guide your students in developing this Portfolio. You may then choose to use the Portfolio as part of your final assessment.

Cooperative Learning

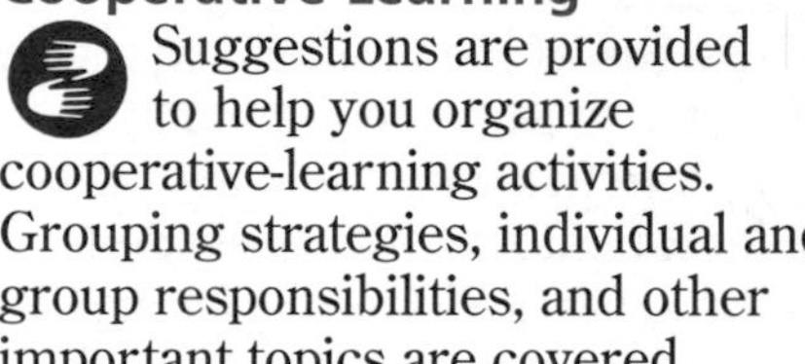

Suggestions are provided to help you organize cooperative-learning activities. Grouping strategies, individual and group responsibilities, and other important topics are covered.

Guided Practice

Questions are provided to help you lead group discussions related to the concepts presented in the lesson. The questions help invite and encourage group participation.

Independent Practice

Suggestions are provided for additional activities that students can do independently to help them solidify their understanding of difficult concepts.

Answers to Questions

All questions that have been posed in the course of the lessons, either within the running text or in the review materials, are answered in the Wrap-Around Margins for your convenience. In only a few instances, where the answer requires a graph or chart, will you have to turn pages to refer to the answer. In addition, all questions in the running text have lettered annotations on the reduced Pupil's Edition pages for quick and easy reference.

Did You Know. . .

Science facts are provided to add special insights and interest to your lessons.

Safety Alert/Waste Disposal

Additional information is provided when there are special safety and waste-disposal concerns.

FOLLOW-UP

The Follow-Up consists of four parts:

- Reteaching
- Assessment
- Extension
- Closure

This information provides effective methods to close out the lesson and to evaluate whether students have grasped the main concepts. If you find that your students need additional help, a reteaching strategy is provided. The extension provides an idea or activity for learning more about a concept covered in the lesson.

Review Pages

All questions appearing in each Challenge Your Thinking and Making Connections are fully answered in the Wrap-Around Margins. All answers for the SourceBook Unit CheckUps are provided in the back of this Annotated Teacher's Edition. Annotations in the Wrap-Around Margins reference the appropriate page numbers.

Features

All of the special-interest features at the end of each unit include background information, answers to questions, and a selection of teaching strategies, including ideas for extending, discussing, debating, and analyzing the feature.

TEACHING RESOURCES

All of the worksheets you will need or want are bound into eight convenient unit booklets. Now you do not have to spend precious time compiling worksheets from various sources. Each *Teaching Resources* booklet has all the worksheets you need to teach an entire unit.

Each *Teaching Resources* booklet contains the following types of worksheets:

Home Connection

The Home Connection begins with a two-page parent letter that introduces the unit of study to the parents and includes home activities to encourage parents' participation. The last page lists the supplies needed to do the Explorations and Activities in the unit. This page makes it easy for you to invite donations in order to keep your budget low.

Chapter Worksheets

- **Resource Worksheets** Worksheet versions of charts, graphs, puzzles, and activities from the Pupil's Edition
- **Exploration Worksheets** Worksheet versions of the Explorations in the Pupil's Edition
- **Discrepant Event Worksheets** Demonstrations and activities that spur students' curiosity and motivate further exploration
- **Theme Worksheets** Worksheets that draw thematic connections between concepts
- **Transparency Worksheets** Review, reinforcement, and assessment worksheets that correspond to the overhead transparencies
- **Math and Graphing Practice Worksheets** Worksheets that reinforce important math concepts and graphing skills
- **Review Worksheets** Worksheet versions of the Challenge Your Thinking review pages in the Pupil's Edition
- **Chapter Assessment** Tests consisting of a variety of questioning strategies to check students' comprehension. Challenge questions that require students to use higher-order thinking skills are included.

Unit Worksheets

- **Unit Activity Worksheets** Extension and review activities, usually in the form of a crossword puzzle or word search
- **Unit Review Worksheets** Worksheet versions of the Making Connections review pages from the Pupil's Edition
- **Self-Evaluation of Achievement** Checklist that helps students evaluate their progress throughout the unit
- **End-of-Unit Assessment** Test that consists of a variety of question types, including word usage, short response, correction/completion, short essay, numerical problems, and interpreting illustrations, graphs, and charts. At least two Challenge questions are also provided to check higher-order thinking skills.
- **Activity Assessment** Performance-based test that requires students to use materials and data to solve a problem
- **Spanish Resources** Spanish blackline masters consisting of Home Connection letters, Big Ideas, and unit glossaries

SourceBook Worksheets

- **SourceBook Activity Worksheets** Hands-on laboratory explorations and activities to illustrate and reinforce the content in the SourceBook
- **SourceBook Review Worksheets** Worksheet versions of the CheckUp review pages in the SourceBook
- **SourceBook Assessment** Tests consisting of multiple-choice, true-false, and short-answer questions

OTHER TEACHING AIDS

The *SciencePlus* program includes the following support materials to make your teaching both effective and efficient.

Teaching Transparencies

Full-color transparencies that highlight key points from each chapter and organize important content with charts, graphs, and puzzles from the Pupil's Edition and Teaching Resources

Getting Started Guide (for Levels Red and Blue)

Blackline master booklet designed to orient both teachers and students who are new to the *SciencePlus* method of teaching and learning science

Assessment Checklists and Rubrics

Blackline master booklet that provides a variety of assessment checklists and rubrics to make your assessment tasks efficient and timely. Electronic versions of these blackline masters are available on the *SnackDisc.*

Materials Guide

Complete listing of all the supplies and equipment needed to teach the entire level, as well as each individual unit, of *SciencePlus*

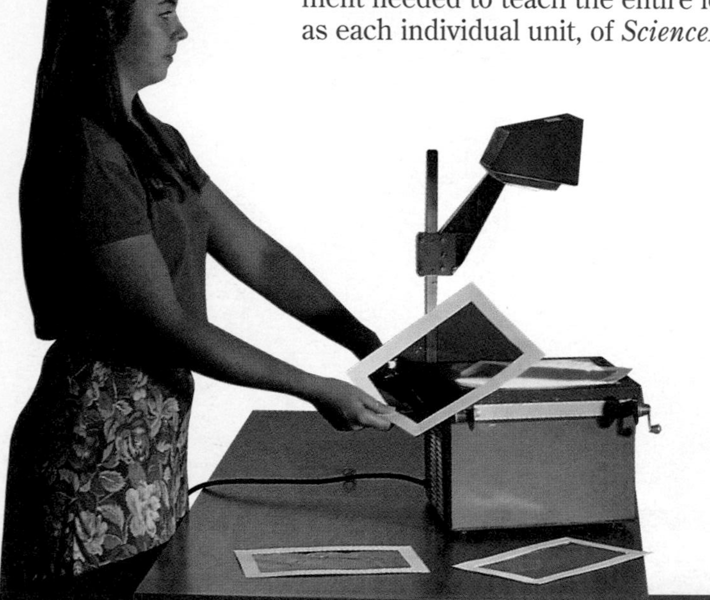

Test Generator

Test item software for Macintosh® and Windows® allows you to create your own tests from an extensive bank of tests and test items. In addition, you can edit the items and even add your own original items. A comprehensive *Test Item Listing* is provided to preview the test items in hard copy.

English/Spanish Audiocassettes

Audiocassette tapes that provide important preview information for each unit in both English and Spanish. Each unit begins with an attention-grabbing skit or story and then takes students on a visual tour of the unit.

Videodisc Resources

Instructions, barcodes, and worksheets for using the *Science Discovery* videodiscs with the *SciencePlus* program. For more information about the *Science Discovery* videodisc program, see page T31 of this Annotated Teacher's Edition.

SnackDisc

CD-ROM disc for Macintosh® and Windows® that contains a vast array of worksheets and information, most of which is customizable to fit your specific needs. The following is a partial listing of the items on the *SnackDisc.*

- Safety First! including safety review, test, and contracts
- Student Review Worksheets, providing review of math, graphing, and the metric system
- Science Sites, showing points of scientific interest on maps of the United States, Canada, and Mexico
- Home Connection letters in English and Spanish
- Assessment Checklists and Rubrics
- Lab Inventory Checklists
- District Requirements Reference Chart
- *SciencePlus* Communicator

SCIENCE DISCOVERY VIDEODISC PROGRAM

Science Discovery is a comprehensive videodisc program that blends imagination and state-of-the-art technology into an exciting program that will add a new dimension to your science teaching. The program includes the following:

- Science Sleuths videodisc
- Science Sleuths Teacher's Guide
- Science Sleuths Resource Directory
- Image and Activity Bank videodisc
- Image and Activity Bank Directory
- Image and Activity Bank Quick Reference Card

Science Sleuths

The Science Sleuths videodisc consists of 24 open-ended science mysteries, one for each unit of the *SciencePlus* textbook series. Each mystery begins with a dramatization in which a problem is presented and the students are asked to serve as consultants (sleuths) to figure it out. Problems include such mysteries as Exploding Lawn Mowers, Dead Fish on Union Lake, and The Misplaced Fossil. The videodisc then becomes a "videophone" that connects students to the laboratory, where they can request information such as the testing of samples, experiments, interviews, tables, newspaper clippings, news broadcasts, and many other bits of information from an extensive menu.

The students work as a class or in small groups to explore the mystery, develop hypotheses, and support their position using the data from the videodisc. They try to solve the mystery using as few of the video segments as possible. In this way they either challenge themselves or other groups to see who can solve the mystery most efficiently.

In working as Science Sleuths, students improve their problem-solving and reasoning skills while applying their science knowledge from the textbook. Students work in a realistic scientific mode in which information comes from a variety of sources. Students must also judge the accuracy of each source and separate raw data from interpretation and inference.

Science Sleuths Teacher's Guide

The Science Sleuths Teacher's Guide contains the teaching plans for using the videodisc.

Science Sleuths Resource Directory

The Science Sleuths Resource Directory contains the barcodes and frame numbers for using each mystery. There are five directory sets so that you can have five working cooperative groups at a time.

Image and Activity Bank

The Image and Activity Bank videodisc consists of a still and motion image database designed to reinforce and extend concepts presented in *SciencePlus*. The still images include hundreds of photographs and computer graphics related to life, earth, and physical sciences. The motion images include demonstrations, experiments, and selected motion footage.

Image and Activity Bank Directory

The Image and Activity Directory provides descriptions, frame numbers, and barcodes for the Image and Activity Bank videodisc. A separate Reference Card provides barcodes and frame numbers for selected topics.

For complete directions on how to use* Science Discovery *with* SciencePlus*, see the Videodisc Resources booklet described on page T30.

Using SciencePlus

Themes in Science

One subject can be addressed from the viewpoint of many different themes.

The traditional division of science into three branches—life, earth, and physical—often leads students to believe that nature is similarly arranged. Too often, students view science as a system of separate, unrelated abstractions or as a compilation of facts and difficult-sounding terms. But science is simply the study of nature, and there are certain underlying principles, or themes, that unite the study of all areas of science. The themes provided here are not meant to replace the traditional teaching of scientific disciplines, but rather to create a framework for the unification of these disciplines.

The following themes are emphasized in *SciencePlus*. These themes are intended to integrate facts and ideas to provide a context for discussing the textual matter in a meaningful way. You can employ these themes as an organizational tool to reinforce understanding of the subject matter.

Energy

Energy puts matter into motion and causes it to change. Energy is what makes the universe and everything in it dynamic. The study of dynamic systems in any field of science requires an understanding of energy: its origins, how it flows through systems, how it is converted from one form to another, and how it is conserved. Energy provides the basis for all interactions, whether biological, chemical, or physical. Thematically, energy connects all disciplines.

Systems

A system is any collection of objects that influence one another. The parts of a system can be almost anything—planets, organisms, or machines, for example. And a system may be very small, such as a cell nucleus; very large, such as a galaxy; or very complex, such as the human body. All of the science disciplines involve the study of some kind of system or systems. Understanding a system involves knowing what its important parts are and how those parts work together.

Structures

Structure provides a basis for studying all matter, from the most basic forms to the most complex. Structure is closely related to function, so scientists study the structures of things to learn how they work. For example, the structure of an eye reveals much about the process of vision. And the key to a diamond's strength lies in the tight lattice structure of carbon atoms within the diamond.

Changes Over Time

A change over time is not a single alteration, but a progression of alterations that occurs over the continuum of time. Biological evolution is an example of change over time. Evolution has gradually changed the characteristics of organisms ranging from the starfish to the giraffe. Changes over time occur in physical and earth science as well. For example, the chemical change of rust forming on an iron nail can be traced through time, as can the movements of the Earth's plates. An understanding of how changes occur over time allows students to appreciate not only the present state of the world around them but also what it may have looked like in the past and what it may look like in the future.

Cycles

A cycle is a pattern of events that recurs regularly over time or a circular flow of materials in a system. Cycles occur throughout nature and appear in all of the scientific disciplines. The water cycle, the rock cycle, and the life cycle of a particular organism are examples of cycles at work. A defining feature of a cycle is that it has no beginning or end; therefore, the study of any part of a cycle is incomplete without consideration of the other elements in the cycle.

Using the Themes

A major strength of the thematic approach is that seemingly different processes, structures, or systems can be shown to have underlying similarities. Although many thematic organizations are possible, each Unit Interleaf in this Annotated Teacher's Edition suggests at least three major themes that can be discussed in relation to the unit and chapter material. Focus questions are also provided in the Wrap-Around Margins to promote discussion and an understanding of how the themes relate to specific text material.

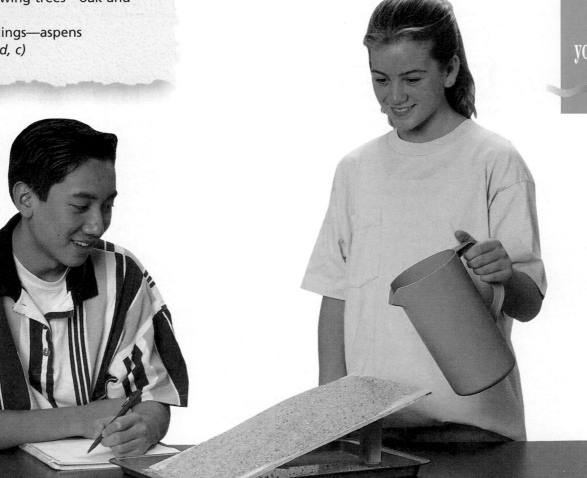

Theme Connection

Changes Over Time

Focus question: When succession occurs at an abandoned farm, in what order would the following plants appear?

a. taller plants—goldenrod and grasses
b. weeds that thrive in sunny locations—wild carrots and dandelions
c. slower growing trees—oak and hickory
d. young saplings—aspens
(Answer: b, a, d, c)

Although major themes have been identified in each Unit Interleaf and in the Wrap-Around Margins, it is up to you to decide which themes are most appropriate. The direction that your class discussion takes will most likely guide you in your choices.

In your class discussions, use the themes to provide a framework of understanding for your students. For example, whether you are studying photosynthesis or the way in which the forces of nature have shaped the physical appearance of the Earth, the theme of **Energy** can be discussed. Similarly, one subject can be addressed from the viewpoint of many different themes. For example, in discussing an organism such as a zebra, **Energy** can be applied in a discussion of how the zebra takes in food from the environment; **Changes over Time** can be discussed in relation to how the zebra's structures are adaptations to its environment; and **Cycles** can be introduced in a discussion of how the zebra's migration habits are based on the seasons.

In your class discussions, use the themes to provide a framework of understanding for your students.

INTEGRATING THE SCIENCES

Throughout *SciencePlus*, students are asked to look at things from many different viewpoints. Since *SciencePlus* is not organized according to life, earth, and physical sciences, these disciplines are integrated throughout the program. Each discipline comes into play as needed in covering a main concept or theme in *SciencePlus*. In this way, science disciplines are blended together so that the emphasis is on student comprehension of the "big picture" rather than on isolated components of different areas of science.

Additional integration suggestions are provided in the Wrap-Around Margins. Look for the Integrating the Sciences within the Teaching Strategies of the lessons. These suggestions will help you build connections between the science disciplines so that students can get a fuller, more in-depth view of the concept at hand.

Integrating the Sciences

Life, Earth, and Physical Sciences

Every field of science employs its own classification system in order to make sense of diversity. Students will learn later in this unit how biologists apply a classification system to living things. You might want to explain to the class how some other disciplines classify things in their own field.

CROSS-DISCIPLINARY CONNECTIONS

SciencePlus is also integrated in terms of nonscience disciplines. For example, students are asked to write a headline for a science story, to write advertising copy highlighting geological features for a travel agency, and even to make dioramas of animal habitats and ways of life. In doing these activities, students see firsthand the connections between science and a variety of other disciplines, including history, geography, social studies, and others. And the students even have fun making the connections.

Although the Pupil's Edition contains a wide variety of cross-disciplinary activities and Explorations, additional suggestions are included in the Wrap-Around Margins of this Annotated Teacher's Edition. Look for the Cross-Disciplinary Focus in the Teaching Strategies of the lessons. Each Cross-Disciplinary Focus provides a suggestion for an activity that highlights the interrelatedness of science and disciplines such as mathematics, social studies, art, music, health, language arts, foreign language, and industrial arts.

Cross-Disciplinary Focus

Foreign Languages

The naming system used by scientists makes use of many Latin words. Have students make a dictionary of some of the commonly used Latin words, examples of each word's use, and the common name for the animal. For example, *Homo* means "man" and is used in such species names as *Homo erectus* and *Homo sapiens*, the name given to modern humans.

SCIENCE, TECHNOLOGY, AND SOCIETY

STS is an approach to teaching science in which the impacts of scientific, technological, and social matters are explored. STS teaches science from the context of the human experience and in so doing leads students to think of science as a social endeavor. STS emphasizes personal involvement in science. Students become active participants in the scientific experience. For this reason, *SciencePlus* incorporates an STS approach.

STS contrasts with the traditional "basic science" approach to science education, in which students follow a carefully sequenced program of the basic models, concepts, laws, and theories of science. The fundamental flaw of the basic science approach is that the student is more or less a passive participant in the learning process.

There are three parts to STS:

Science concept and skill development, and knowledge of the nature of science

This component of STS introduces science as a system for learning about the natural world and gives students the foundation they need to actually practice science in and out of the classroom. The ultimate goal of STS, and of *SciencePlus*, is to turn students into scientists, at least for the duration of their science education. To accomplish this, students first learn the methods of science and the skills that scientists draw on. Whenever possible, the major ideas of science are introduced from the standpoint of those who developed them. In this way, students come to see the reasoning that went into the development of these ideas.

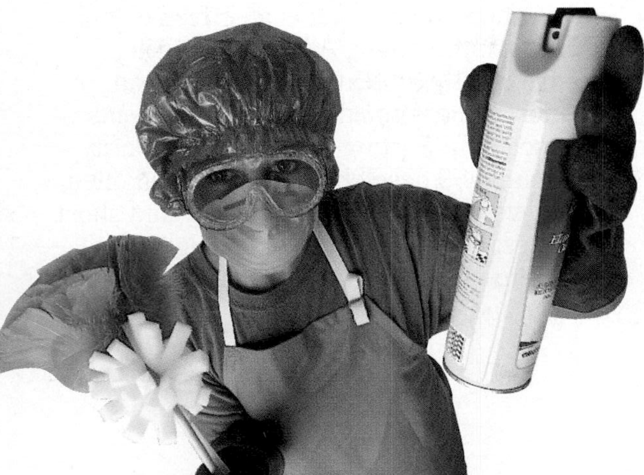

Knowledge of the relationship of science and technology, and engagement in science-based problem solving

Students who understand the real-world applications of science are better able to appreciate and enjoy it. This component of STS reinforces the practical value of science. Students see that science is a system for solving practical problems. Students themselves become practical scientists—first identifying problems and then developing solutions to them. Students learn to analyze, to plan, to organize, to design, and to refine models and designs.

Engagement in science-related social issues and attention to science as a social institution

This component of STS deals with the ways in which science serves human needs. The benefits may be tangible or intangible, but either way, they are real. To emphasize the social responsibility of science, scientists are shown to be concerned about the impact of their work on society as a whole. Advances in science and technology sometimes lead to thorny ethical issues. Such issues are often used as a focus for discussion and investigation. From these investigations, students draw conclusions and form reasoned opinions.

Studies have shown that students begin the study of science full of curiosity and enthusiasm. But after a few years of a traditional curriculum, the curiosity is squelched and the enthusiasm has all but vanished. By contrast, under the STS approach, student interest builds through the years and is maintained long after formal study ends. The key to this success is the active involvement in science that STS imparts. The end products of the STS approach are students who appreciate and understand science and who are equipped to deal sensibly with the complex issues that will become commonplace in the decades ahead.

For a fuller discussion of STS, see the NSTA position paper, July 1990.

STS emphasizes personal involvement in science. Students become active participants in the scientific experience.

SciencePlus challenges students to communicate ideas in novel ways.

One of the most important skills that students can acquire is the ability to communicate what they have learned, both orally and in writing. *SciencePlus* challenges students to develop and communicate their mastery of new ideas in novel ways—for example, by writing for one another or for some audience other than the teacher. Students' comprehension is enhanced when they are called upon to reformulate in their own words what they have learned.

Throughout *SciencePlus*, students are called on to communicate what they have learned in many different ways. Students may be called on to interpret a passage, illustrate a paragraph, write a headline, label a diagram, or write a caption for a photo or drawing. These strategies complement the inquiry approach followed throughout *SciencePlus.*

Reading and Writing in the Classroom

The time spent in helping students prepare to read is critical in fostering comprehension. Two strategies are particularly important in the pre-reading phase of instruction: building on prior knowledge and establishing a purpose for reading.

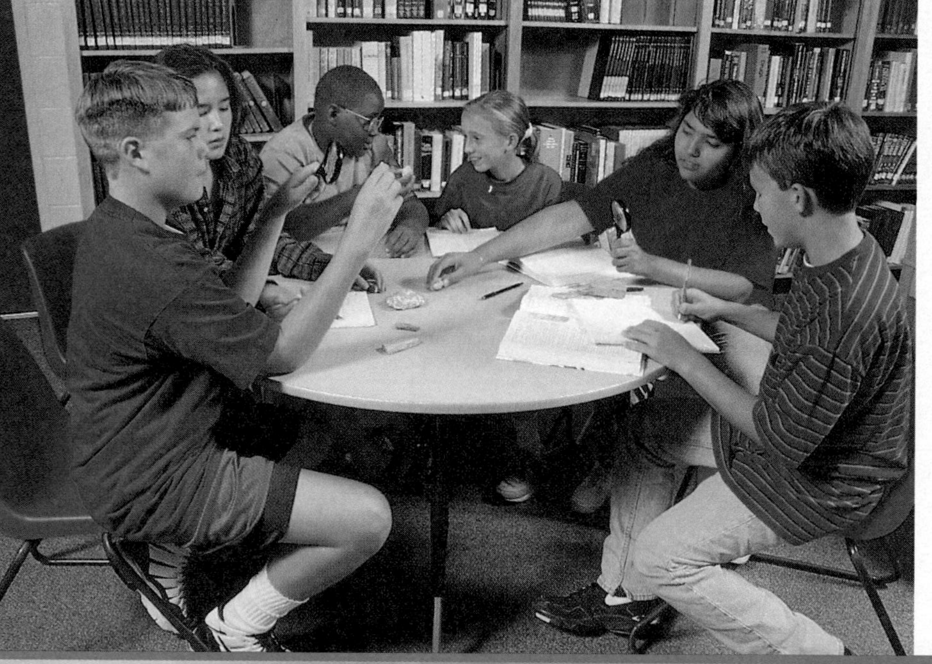

The Power of Prior Knowledge

The amount of prior knowledge that students have about a topic directly influences their comprehension of that topic. The more students know about something, the easier it is for them to grasp new information about it. Helping students identify the information they already know about a topic assists them in relating the new information to existing knowledge.

Research strongly suggests that students tend to retain misconceptions they may have about a topic. If the text information seems to conflict with their preconceptions, students may ignore or reject new information. It is therefore extremely important to identify these misconceptions so that they may be dispelled.

Reading to Understand

One way to establish a purpose for reading is to make a study guide with questions that students can answer as they read. You can also help students learn to make their own study guides. First teach students to preview a unit by looking at all the unit headings, illustrative material, and terms and phrases in boldface or italics. This technique helps students gain a feel for the unit and helps them build a basic structure for the new information. Then show them how to use the captions, headings, and highlighted words to devise study-guide questions to answer after reading the unit. Following this method, students will read with a purpose in mind, a purpose of their own devising. You may also wish to consider using the English/Spanish Audiocassettes, which are an excellent means for previewing the unit with students.

Writing to Understand

Studies show that writing is an effective tool for improving reading. As students write, they are creating a text for others to read. The most important advice to give students about scientific writing is to strive for clarity and accuracy. These characteristics can often be achieved with simple vocabulary and short sentences. One useful approach might be to have them imagine that they are writing for a younger audience.

Keeping a Journal

One highly successful tool for improving students' performance in science is the journal. The *SciencePlus* version of the journal is called the ScienceLog. The ScienceLog has many functions. First and foremost, it is an ongoing record of students' learning. Students begin the study of a new topic by recording prior knowledge of that topic. Any misconceptions that students may have are thus exposed. As the lesson progresses, students record any and all new findings. In many cases, students find that what they learned in the activities contradicts their preconceptions.

CHAPTER 7
A Solution or Not?
What is the difference between these two liquids?
Why can't you see sugar after it is stirred into iced tea?
Describe each of these substances in terms of its ability to dissolve in water.
ScienceLog
Think about these questions for a moment, and answer them in your ScienceLog. When you've finished this chapter, you'll have the opportunity to revise your answers based on what you've learned.
146

Much of the work that students do should be recorded in their ScienceLog. The ScienceLog is a constant reminder to students that learning is occurring. Students can look back and compare their early work with later work to see and take pride in the progress they have made. To supplement their other work, you may want to ask students to briefly summarize what they have learned each week. This makes a very handy capsule history of their work.

What makes a good ScienceLog? Insofar as is possible, the ScienceLog should be neat and easy to follow. Students may organize their ScienceLogs in any of a number of ways—chronologically, by unit, by chapter, or by lesson, to name a few. Some kind of heading should set off each major entry.

A spiral-bound notebook or hard-bound lab-type notebook makes a good ScienceLog. Or you may make copies of the sample ScienceLog pages in the *Teaching Resources* or *SnackDisc* and distribute them to your students to use if you wish. These ScienceLog pages are located at the beginning of the Unit Worksheets for Unit 1.

Portfolios: What, Why, and How

Definitions and descriptions vary as to what constitutes a portfolio. Very simply, a portfolio is a collection of work that is done by the student during the course of the year. Usually, the students themselves have a say as to what goes into the portfolio, but selections should represent the objectives outlined by the curriculum. The purpose of the portfolio is to represent the student's mastery of skills and knowledge within the subject area. This may sound simple enough, and, in fact, it can be very simple. On the other hand, portfolios can be elaborate collections of work that are limited only by the teacher's and students' imaginations.

Your initial decision to use portfolios in your assessment leads to a host of other decisions that must be made. However, before you launch into making these decisions, it is a good idea to seek input from other teachers as well as from your school administration. You may even want to discuss your ideas with parents and other members of your community. This will make it easier to get feedback later as to the impact portfolios are having on student learning and attitudes.

You will also need to plan a system for organizing and managing the portfolios. You will need to establish guidelines for what types of information will be admitted into the portfolios, how and by whom the selections will be made, and when materials can be added or revised. These decisions will be based on your individual preferences and on the level and attitudes of your students. The following guidelines may help you in developing your plans.

The most important advice to give students about scientific writing is to strive for clarity and accuracy.

- Allow students to select for themselves the sample materials that best represent their level of understanding and mastery. Although you may require that certain projects or materials be placed in the portfolios, it is advised that the students play an active role in selecting their best work. This gives students ownership and encourages them to take increasing responsibility for the quantity and quality of their work.
- Allow students to revise the selections in their portfolios at any time. The portfolios should evolve as your students' skills improve.
- Identify well in advance when you will be assessing the portfolios. Students should be given plenty of warning before the portfolios are collected for assessment.
- Determine in advance the criteria you will use for grading the portfolios. Share this criteria with the students up front. You may want to make a criteria checklist that each student can place inside his or her portfolio. This checklist would provide both you and your students with a ready reference to the grading criteria.
- You may want to keep the portfolios in the classroom or in some other area, allowing frequent but controlled access to them. This will decrease the chances of portfolios being lost. This may require a significant amount of room, depending on the nature of the portfolios.
- Make available examples of high-quality portfolios so that students can see examples of excellent work. This, of course, may be possible only after you have used portfolio assessment for at least one course.

Sciencelog as a Portfolio

The ScienceLog provides an excellent opportunity for portfolio assessment. By directing your students to include in their ScienceLogs representative samples of their work, as well as comments and reflections about their samples, you can add depth to your evaluation method. To help you utilize portfolios effectively in your teaching, specific strategies are provided in Portfolio boxes in the Wrap-Around Margins of this Annotated Teacher's Edition.

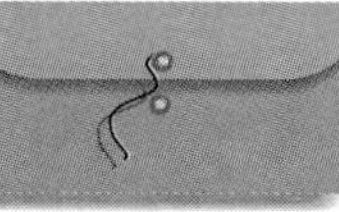

PORTFOLIO
Students may wish to include their written composition in their Portfolio.

CONCEPT MAPPING

Too often, students are able to master the individual elements of a topic without truly grasping the "big picture." If students fail to understand how the elements fit together or relate to one another, they cannot truly comprehend the topic. Concept mapping is a very effective method of helping students see how individual ideas or elements connect to form a larger whole. Concept maps are a highly effective tool for helping students make those logical connections.

The most effective concept maps are those that students construct on their own. Used in this way, concept maps are both a self-teaching system and a diagnostic tool. To construct a proper concept map, the student must first examine closely his or her mental model of the topic at hand. Any flaws or shortcomings in that model will be reflected in the concept map.

Concept maps are flexible. They can be simple or highly detailed, linear or branched, hierarchical or cross-linked, or they can contain all of these major elements. Students can construct their own maps from scratch or can finish incomplete maps. Concept maps can take almost any form as long as they are logically arranged.

Making Concept Maps

The steps involved in making a concept map are outlined below. To provide guidance to your students in making concept maps, direct them to page xvii in the Pupil's Edition.

1. Make a list of the concepts to be mapped. Concepts are signified by a noun or short phrase equivalent to a noun.
2. Choose the most general, or the main, idea. Write it down and circle it.
3. Select the concept most directly related to the main idea. Place it underneath the main idea and circle it. If two or more concepts bear the same relationship to the main idea, they should be placed at the same level.
4. Draw a line between the related concepts, leaving a space for a short action phrase that shows how the concepts are related. These are linkages.
5. Continue in this way until every concept in the list is accounted for.

The simple concept map below shows the relationship among the following terms: plants, photosynthesis, carbon dioxide, water, and sun's energy. More detailed maps are shown on the facing page.

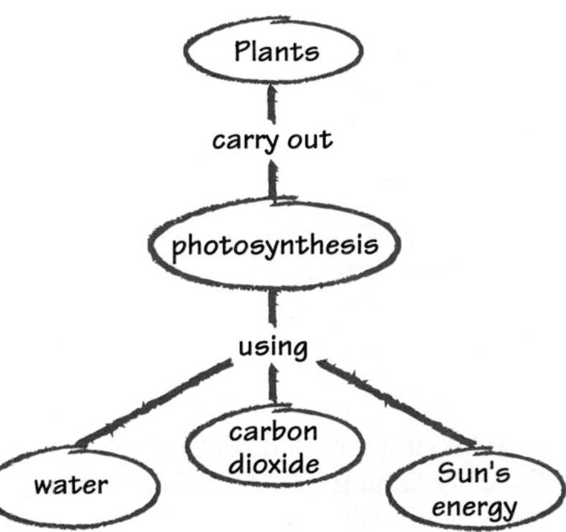

For any given topic, there is no single "correct" concept map. Not all maps are equally valid, however. Good concept maps have most or all of the following characteristics:

- start with a single, general concept—a big idea—and work down to more specific ideas
- represent each concept with a noun or short phrase, each of which appears only once
- link concepts with linkage words or short phrases
- show cross-linkages where appropriate
- consist of more than single path
- include examples where appropriate

Concept maps are a highly effective tool for helping students make logical connections.

> The most effective concept maps are those that students construct on their own.

Using Concept Maps

Concept maps can be applied in many ways.

- to gauge prior knowledge of a topic
- as end-of-lesson, chapter, or unit evaluation
- as pre-test review
- to help summarize special presentations, such as films, videos, or guest speakers
- as an aid to note taking
- for reteaching

You may also want to use partially completed concept maps as pop quizzes or as devices for summarizing particularly difficult class sessions. Also, be sure to use the concept map in each Making Connections review.

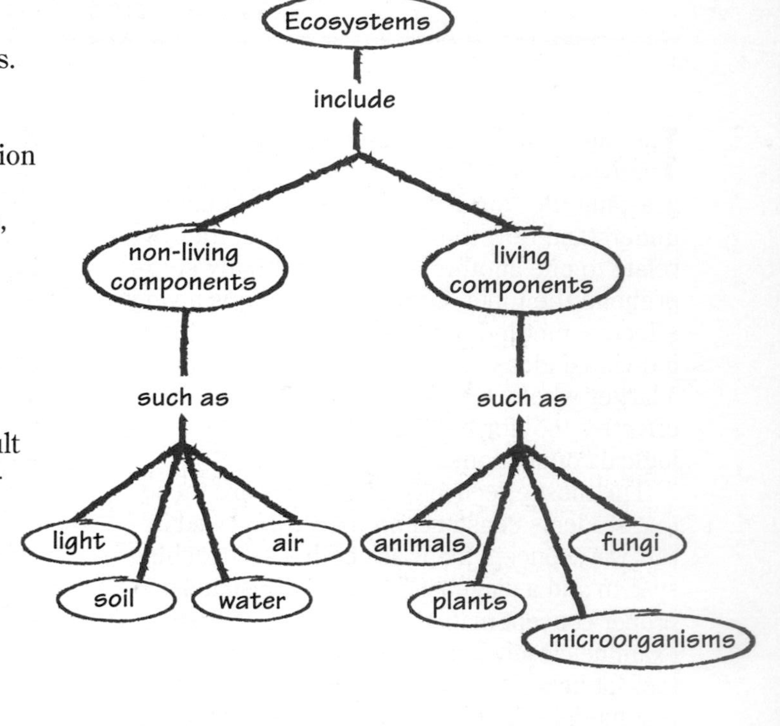

Evaluating Concept Maps

Again, there is no single correct concept map. However, you should consider the following criteria as you evaluate your students' concept maps.

- how comprehensive the map is (Are all relationships shown?)
- how clearly the concepts are linked (proper relationship between concepts, use of linkage terms between all concepts)
- overall clarity of presentation (Could the map be simpler? Is it redundant? Is it logically arranged? Are linkage terms used properly?)

Used properly, concept maps can increase comprehension, improve retention, and sharpen study skills in your students. They are a valuable addition to any student's arsenal of learning strategies.

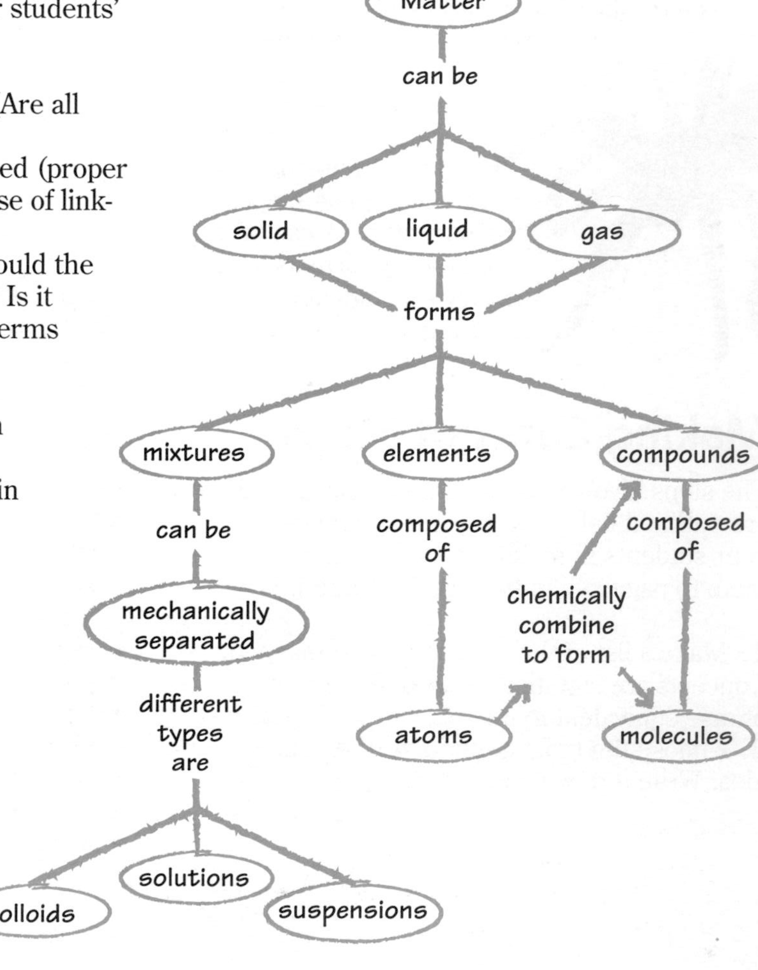

COOPERATIVE LEARNING

Cooperative learning is a teaching technique that brings students together to learn in small, heterogeneous groups. In these groups, students work interdependently without constant and direct supervision from the teacher. Assignments are structured so that everyone contributes. Challenges as well as rewards are shared. Brainstorming, lively discussion, and collaboration are the hallmarks of the cooperative-learning classroom.

Cooperative learning is an ideal complement to the *SciencePlus* approach. The discussions, explorations, research projects, games, and puzzles of *SciencePlus* can all be used in a cooperative-learning format.

What It's Not!

- Cooperative learning is **not** the same as ability grouping, where a teacher divides up the class in order to instruct students with similar skills.
- Cooperative learning is **not** having students sit side by side at the same table to talk while they complete individual assignments.
- Cooperative learning is **not** assigning a task to a group in which one student does the work and the others get equal credit.

Benefits of Cooperative Learning

Traditionally, teachers structure lessons so that students work individually to achieve learning goals or compete against one another to be "the best." While these formats can be useful, cooperative learning provides an important alternative.

- **Cooperative learning models the scientific experience.**
  Students working in groups learn about the joys as well as the frustrations involved in scientific inquiry. Cooperative learning models real scientific experience, in which scientists work together, not in isolation, to solve difficult problems. With cooperative learning, the classroom becomes a fertile environment for ideas and novel solutions.
- **Cooperative learning empowers and involves students.**
  Cooperative learning raises students' self-esteem because they are learning something on their own through cooperation, rather than being handed prepackaged knowledge. It helps students become self-sufficient, self-directed, lifelong learners. In a cooperative-learning environment, students are less dependent on you for knowledge.

Students working in groups learn about the joys as well as the frustrations involved in scientific inquiry.

- **Cooperative learning serves the heterogeneous classroom.**
With group work, everyone has the chance to participate, and everyone has a role to play. As students join forces to achieve a common goal, they come to recognize commonalities that cut across differences related to ethnicity, socioeconomic background, and gender. Likewise, cooperative learning provides an excellent vehicle for students of differing ability levels to work together in a positive way. Basic students can interact successfully with average and advanced students, and in so doing can learn that they, too, have something to offer.

- **Cooperative learning strengthens interpersonal skills.**
Group tasks are structured so that students must cooperate to succeed. Students quickly understand that they will "sink or swim" together by how constructively they interact. Consequently, students develop important interpersonal and social skills that help them function in a group setting and that will ultimately benefit them socially, at work, and in other situations.

- **Cooperative learning develops appropriate social skills.**
When doing cooperative group work, students channel their energies into constructive tasks while satisfying their fundamental need for social interaction.

- **Cooperative learning is an effective management tool.**
Establishing cooperative learning in the classroom requires you to relinquish some control, so the students themselves can become responsible for building their own knowledge. Working in groups to probe and investigate ideas, answer questions, and draw conclusions about observations allows students to discover and discuss concepts in their own language. When students learn through cooperation, the knowledge derived becomes their own, not just a loan of your ideas or those from the textbook.

- **Cooperative learning increases achievement.**
Since the 1920s, there has been extensive research on cooperative-learning techniques. Results clearly indicate that cooperative learning promotes higher achievement for all grade levels in all subject areas.[1]

[1] Johnson, Johnson, Holubec, and Roy. *Circles of Learning, Cooperation in the Classroom.* Association for Supervision and Curriculum Development. ©1984

Using Cooperative Learning With *SciencePlus*

Cooperative learning is an essential component of the *SciencePlus* philosophy. To help you take full advantage of this important component, several cooperative-learning activities have been highlighted in each unit of this Annotated Teacher's Edition. These activities are broken down to provide suggestions for group size, group goal, positive interdependence, and individual accountability.

This symbol designates cooperative-learning activities in the Wrap-Around Margins of this Annotated Teacher's Edition. Cooperative Learning worksheets are are also provided in the **Teaching Resources** booklets.

Group Size Although group size will vary depending upon the activity, the optimum size for cooperative learning is between three and four students. For students unaccustomed to this learning style, keep the group size to about two or three students.

Group Goal Students need to understand what is expected of them. Identify the group goal, whether it be to master specific objectives or to create a product such as a chart, a report, or an illustration. Identify and explain the specific cooperative skills required for each activity.

Positive Interdependence A learning activity becomes cooperative only when everyone realizes that no group member can be successful unless all group members are successful. The "we're all in this together" part of group work is the positive interdependence. Encourage positive interdependence by assigning each student some meaningful role, or allow students to do this themselves. You can also encourage positive interdependence by dividing materials, resources, or information among group members.

Individual Accountability Each group member should have some specific responsibility that contributes to the learning of all group members. At the same time, each group member should reach a certain minimum level of mastery.

Meeting Individual Needs With Cooperative Learning

Cooperative learning is an effective tool for meeting the individual needs of your students. Cooperative learning builds relationships among students where relationships might not have developed before. Students are forced to interact with each other as individuals with common goals. In so doing, students learn more about each other's personal characteristics, and as a result, many stereotypes are destroyed.

There is no single set of cooperative-learning strategies that will work with all students in all situations. However, the following strategies may provide you with insight and guidance in developing your own set of strategies that will work for your students.

Balance the needs of students of all levels and learning styles

Your ultimate goal will be to ensure that all students are able to work effectively in any group. Initially, however, you may wish to develop special grouping strategies to foster the growth of learners having difficulty and second-language learners, and to assure gifted students that their grade will not be affected by slower learners. More information is provided about grouping strategies on the next page.

Have a clearly defined goal

When you tell students what is expected of them, be sensitive to their special needs. Be sure that each student understands the group goal and his or her own personal responsibility.

It is also important that assignments be specifically appropriate for groups. In other words, simply having students fill in the blanks of a worksheet or answer the end-of-unit questions is not creating an adequate cooperative-learning assignment. Students need tasks that cannot be easily completed alone. Students should see that if they work together, the end product will be better and more complete than if they had worked alone.

Answer questions only when the whole group has the same question. It's a good idea to designate one person per group as the liaison between you and the group.

Praise success

If students seem unmotivated or feel that their individual tasks are unimportant to the success of the group, you may wish to consider offering group rewards. Reward the groups as they successfully complete each activity. Reward successful project results as well as positive interaction and effective group-process skills. However, rewards should not become automatic. They should be used only for the short term. For the long term, students should take pride in their group's achievements and should benefit from the knowledge that these achievements contribute to their success as individuals.

Encourage interpersonal problem solving within groups

Pulling a disruptive student out of a group is sometimes necessary, but be sure that the isolation is only temporary. Difficult students need the support of others. Build into each group the spirit of encouraging each other. Suggest that groups evaluate their own performance after an activity is finished. Encourage students to suggest solutions to problems without criticizing individuals.

Grouping Strategies

With cooperative learning, you can either place students in particular groups or assign students to groups at random. There are advantages to both approaches. At first, however, it is recommended that you assign students to particular groups.

The* SciencePlus SnackDisc *CD-ROM and the* SciencePlus Assessment Checklists and Rubrics *booklet contain several checklists to help you implement cooperative-learning groups.

Assigned Grouping

Composing groups yourself lets you create groups that are heterogeneous in terms of academic ability, gender, ethnicity, and cultural background. Heterogeneous groups are preferred because cooperation among diverse students not only teaches the widest range of interpersonal skills but also promotes frequent exchange of explanations and greater perspective in discussions. This increases depth of understanding and retention of concepts.

To create *effective* heterogeneous groups, balance each group with students who have different strengths. First decide who your resource students are. These are students you think will facilitate group work—either because of their academic ability or because of their interpersonal skills. Assign at least one resource student to each group. Distribute students who may be disruptive and students who lack academic skills evenly throughout the groups. Avoid putting close friends together to prevent cliques from disrupting teamwork. Put students who have limited English proficiency in groups with bilingual students who can act as translators.

Random Grouping

Random grouping can be especially effective with experienced cooperative learners or if you plan to change group membership often. To create random groups, you can simply have students count off from 1 to 4. All of the 1s form a group, all of the 2s form another group, and so on.

There are many other fun ways to assign groups randomly. For example, you can hold a lottery in which students pick numbers out of a hat. Numbers 1 to 4 form one group, numbers 5 to 8 form another, and so forth. You can also use cards naming sets of a particular type of item. All students who draw items belonging to the same set form a group. For example, all students whose cards name types of flowers belong in one group, students whose cards name farm animals belong in another group, those with cards naming heavy-metal bands form a third group, etc. Students have a fun and lively time discovering who belongs in the same group.

You can also combine lesson content with assigning groups. First decide how many students you want in each group. For each group, write a different scientific term or principle on a flashcard. Then for each group's term, list on separate cards the definition of the term, a synonym, or an example of what the term means. Mix up the cards and hand them out as students come into the room or once students are seated. Students use the cards as clues to find the others in their group.

Assigning Roles

Assign roles at first; students can choose their own later.

Assigning roles is very important, especially at first. Consider the behavior patterns of students, and assign roles that will complement those patterns. Group work needs to be structured so that everyone has a part to play. In other words, there needs to be positive interdependence. Just as members of a surgical team work together, with each person contributing his or her own special skill, students work effectively in teams when everyone has a unique role that is vital to the group's success.

Some examples of useful roles are listed here. Use as many as you need, modify them, combine them, or invent roles yourself.

- **Facilitator** The facilitator is a leadership role. The facilitator keeps an activity running smoothly by presiding over the work flow. He or she manages the group so that all members have a chance to talk, questions are answered, students listen to one another's ideas, and ideas are substantiated with reasons and explanations.
- **Recorder** The recorder records data and answers questions posed to the group.
- **Reporter** The reporter explains the group's findings to the teacher or the entire class.
- **Safety officer** The safety officer makes sure safety practices are followed and notifies the teacher of any unsafe situations.
- **Checker** The checker makes sure that everyone has finished his or her worksheet or other individual assignment.
- **Materials manager** The materials manager gathers activity materials at the outset, monitors their use during the activity, and organizes the cleanup and return of materials to their proper place after an activity.

Again, assign roles carefully, especially at first, taking into account students' behavior patterns. A shy student might be most comfortable as a recorder, while a student who likes attention might make the best reporter. The facilitator is a role that some students will always want and others will avoid. Be careful not to stereotype. Sometimes the most unlikely students will make the best leaders.

Assessment

Assessment within a cooperative-learning setting is not as difficult as it may seem. Like any other assessment, you must determine in advance what you would like to assess and to what degree. You will also need to develop some slightly different monitoring skills. In addition to the following information, the *SciencePlus* program includes several checklists and rubrics to make this task simple and efficient.

A variety of checklists designed for monitoring and assessing cooperative group work are provided on the* SciencePlus Snackdisc *CD-ROM and in the* SciencePlus Assessment and Rubrics *booklet. Both student and teacher checklists are provided.

Monitor Groups

Resist the temptation to get caught up on paperwork as the groups do their work—this is the time to observe, monitor, and coach. As you monitor the groups, you can reinforce cooperative behavior with a formal observation sheet. Appropriate checklists are available on the *SnackDisc* CD-ROM and in *Assessment Checklists and Rubrics* booklet. Record how many times you observe each student using a collaborative skill, such as contributing ideas or asking questions.

If a group seems hopelessly confused or "stuck," you can intervene to guide students to a solution. But make sure students have the opportunity to reason through problems themselves first. Consider the following differences between direct supervision and the kind of monitoring that supports cooperative learning.

DIRECT SUPERVISION	SUPPORTIVE MONITORING
Lecturing	Giving feedback
Disciplining	Encouraging problem solving
Telling Students what to do	Providing resources
Leading discussions	Observing

What to Assess

What should you assess in a cooperative-learning activity? Individual success? Group success? Cooperative skills? Actually, many teachers find it useful to evaluate all three. And there are many ways to assess each of these areas.

Individual success can be evaluated by asking students to fill out answers to a worksheet as they progress through an activity; by having them record, analyze, and submit data; or by having them take a quiz. Some activities are structured so that each student turns in a product, such as a report or a poster, that can be individually graded. The individual accountability portions of the cooperative-learning features in this Annotated Teacher's Edition also provide ways to assess individual performance.

Group success is evaluated according to how well the group accomplished its assigned task. Was the task completed? Were the results accurate? If not, were errors explained and accounted for? Criteria such as these provide a framework for group evaluation.

Cooperative skills are evaluated based on your observations of students' behavior in their groups. Evaluating students' use of cooperative skills will motivate students to use them. If you intend to grade cooperative skills, it is helpful to use a formal observation checklist as you monitor students at work. Log the frequency with which group members exhibit cooperative skills or disruptive behavior.

Weigh each of the three grades as you wish to compute a single overall grade. Stress the factors that you consider most important. Use cooperative learning to meet the needs of your students. Use it and enjoy!

PROCESS SKILLS

Process skills are a means for learning and are essential to the conduct of science. For this reason, *SciencePlus* is strongly process oriented.

Perhaps the best way to teach process skills is to let students carry out scientific investigations and then point out to them the process skills they used in the course of the investigations. The Lesson Organizer at the beginning of each lesson identifies the process skills that are emphasized in the corresponding text.

SciencePlus makes regular use of many different process skills, as highlighted below. These and other process skills are called upon in *SciencePlus* in virtually every lesson.

Observing
An observation is simply a record of a sensory experience. Observations are made using all five senses. Scientists use observation skills in collecting data.

Communicating
Communicating is the process of sharing information with others. Communication can take many different forms: oral, written, nonverbal, or symbolic. Communication is essential in science, given its collaborative nature.

Measuring
Measuring is the process of making observations that can be stated in numerical terms. In *SciencePlus*, all measurements are given in SI units.

Comparing
Comparing involves assessing different objects, events, or outcomes for similarities. This skill allows students to recognize any commonality that exists between seemingly different situations. A companion skill to comparing is contrasting, in which objects, events, or outcomes are evaluated according to their differences.

Contrasting
Contrasting involves evaluating the ways in which objects, events, or outcomes are different. Contrasting is a way of finding subtle differences between otherwise similar objects, events, or outcomes.

Organizing
Organizing is the process of arranging data into a logical order so it is easier to analyze and understand. The organizing process includes sequencing, grouping, and classifying data by making tables and charts, plotting graphs, and labeling diagrams.

Classifying
Classifying involves grouping items into like categories. Items can be classified at many different levels, from the very general to the very specific.

Analyzing
The ability to analyze is critical in science. Students use analysis to determine relationships between events, to identify the separate components of a system, to diagnose causes, and to determine the reliability of data.

Inferring
Inferring is the process of drawing conclusions based on reasoning or past experience.

Hypothesizing
Hypothesizing is the process of developing testable explanations for phenomena. Testing either supports a hypothesis or refutes it.

Predicting
Predicting is the process of stating in advance the expected result of a tested hypothesis. A prediction that is accurate tends to support the hypothesis.

CRITICAL THINKING

Requiring students to think critically as they learn improves their comprehension and increases their motivation.

Critical-thinking skills are essential for making sense of large amounts of information. Too often, science lessons leave students with a set of facts and little ability to integrate those facts into a comprehensible whole. Requiring students to think critically as they learn improves their comprehension and increases their motivation.

Loosely defined, critical thinking is the ability to make sense of new information based on a set of criteria. Critical-thinking skills draw on higher-order thinking processes, especially synthesis and evaluation skills. Critical thinking takes a number of different forms.

Validating Facts

This type of critical thinking involves judging the validity of information presented as fact. Too often, people will accept as valid almost any statement, no matter how outrageous, as long as it comes from a supposedly authoritative source. It is important for scientists to treat all untested data with suspicion, no matter how reasonable it may seem.

Students may validate facts in a number of ways: by observing, by testing, or by rigorously examining the logic of the so-called fact. *SciencePlus* presents students with many opportunities to critically evaluate facts and hypotheses.

Making Generalizations

A scientist must often be able to identify similarities among disparate events. Generalizations are drawn based on a limited set of observations that can be applied to an entire class of phenomena. One does not have to test every substance known to make the generalization that solid substances melt when heated. Generalizations allow scientists to make predictions. Once the rule is known, future outcomes can be forecast with a high degree of confidence.

It is important that students base their generalizations on an adequate amount of information. A generalization that is formed too quickly may be wrong or incomplete or may lead the student down a dead-end path.

Making Decisions

Many students would not regard science as a field requiring decision-making skills. But in fact, scientists must make decisions routinely in the course of their work. Any time a scientist works through a problem or develops a model, a whole series of decisions must be made. A single faulty decision can throw the entire process into disarray. Making informed decisions requires knowledge, experience, and good judgment.

Interpreting Information

Having all the information in the world is useless unless one also has the tools to interpret that information. Scientists must know how to separate the meaningful information from the "noise." Information can come in any form detectable by the five senses. It is important that scientists and students alike interpret information to determine its meaning, validity, and usefulness.

ENVIRONMENTAL AWARENESS

No species affects its surroundings as dramatically as does the human species. Thanks to recent highly publicized events—such as Chernobyl, destruction of the rain forests, the depletion of the ozone layer, and the greenhouse effect—people have come to realize the global impact that human actions can have. It is incumbent upon the educational system to promote environmental awareness among students. *SciencePlus* addresses environmental issues in a way that students can easily grasp.

Environmental issues run the gamut from global to local. While large-scale problems get headlines, they can be hard to grasp for many students, who may have never directly observed their impact. In most cases it is best to introduce your students to local issues to start building their awareness. Local issues not only are more relevant to their lives, but also are more likely to lead to direct involvement.

Environmental awareness serves two purposes: it promotes understanding of the living world and the place of humans within it, and it produces a positive change in students' behavior toward the environment. *SciencePlus* pursues both goals. You may involve students directly in environmental issues by using the suggested activities in the Pupil's Edition.

In addition, this Annotated Teacher's Edition contains teaching strategies and special Environmental Focus boxes throughout the Wrap-Around Margins. See the sample below. The Environmental Focus boxes contain relevant environmental information that you can use to add depth to the topic at hand and to encourage discussion or some other action.

Environmental awareness produces a positive change in students' behavior toward the environment.

ENVIRONMENTAL FOCUS

Many organisms produce poisonous chemicals called *biotoxins* that are used as a defense against predators or in capturing prey. The most-studied biotoxins are snake venoms because they are so easy to obtain. Have students do some research or contact a local government agency to find out about poisonous organisms in their area. Students can then create a booklet of information about biotoxins from poisonous plants and animals found in their area.

MULTICULTURAL INSTRUCTION

> Respect for your students' cultures will go a long way in helping each of your students achieve and maintain a positive self-concept.

The success of our nation would not be possible without the contributions of the many cultures and ethnic groups that make up our country. Multicultural instruction serves to ensure that all students have the chance to learn, to succeed, and to become whoever they would like to become, regardless of race, gender, socioeconomic background, or disability. Multicultural instruction affirms the positive nature of this country's diversity by helping students develop an open mind, a positive self-concept, and a realistic understanding of the world that surrounds them.

Meeting the needs of culturally diverse students is perhaps the most demanding challenge faced by today's teachers. We must constantly strive to arouse adolescent curiosity, minimize risks of failure, and be as relevant as possible to our individual students' needs. The more culturally relevant we make our science programs, the better we will be at serving our changing class populations. These challenges are especially difficult with middle-school students because they are also going through the physical and emotional changes associated with adolescence.

By their very nature, middle-school science programs have the potential for promoting the full development of individual learners—especially when science classrooms are perceived as places of inquiry and discovery. When such environments exist in science classrooms, students become more successful learners.

Relevance, Positive Self-Concept, and Multiculturalism

Let your students help you incorporate multicultural learning into the classroom by allowing them the freedom to express their feelings and attitudes during your classes. With a program tailored specifically to the personal experiences of your students, you will find that your students are more curious about the world around them. With an increased level of relevance, learning is more important to young thinkers.

Likewise, it is very important to be sensitive to the cultural identities of your students. You must view these cultural identities on an equal footing. Having a healthy respect for your students' cultures will go a long way in helping each of your students achieve and maintain a positive self-concept.

Using Multicultural Instruction

A strong program of multicultural instruction can begin by implementing a few of these basic strategies. While none of the strategies are exclusively multicultural, they can provide proper contexts and situations that capitalize on the cultural backgrounds of students.

- Recognize and convey to students that all languages are equally valid. The learning of English, however, increases the range of opportunities available to the students.
- Draw special attention to the diversity of role models in *SciencePlus*. At every opportunity, provide information about past and current scientists from diverse cultures.

  The power of such role models should not be underestimated. Role models may create interest and motivation and may even influence a student's pursuit of a career.
- Use cooperative learning to diversify student groups. You will find that students develop more of an open-minded awareness as well as more positive, accepting, and supportive relationships with peers. Labels concerning ethnicity, gender, ability, social class, and handicaps cease to exist. For more detailed information about cooperative learning, see pages T41–T46.
- Peer and cross-age tutoring is an excellent strategy for fostering better understanding among individuals. Peer tutoring involves students tutoring students their own age. Cross-age tutoring involves older students tutoring younger students. These strategies are beneficial both to the tutors and to the students being tutored.

  In using either of these strategies, be very careful when pairing students. Although this is an excellent opportunity to integrate students, both the tutor and the student must be willing participants.

- Take every opportunity to relate science to personal experiences. Invite your students to discuss any of their own experiences that may apply. You may discover some very relevant connections and analogies, and the learning process will become more interactive and personalized to the class. This process might also add to the richness of the class by highlighting the cultural differences among your students.
- Allow students to select independent projects relevant to their own world. These projects should permit students to create new, positive avenues of self-expression from their own experiences. Students find opportunities to select, design, and articulate their own interest within science programs while developing their creative-thinking and problem-solving skills. In addition, these activities promote the development of positive attitudes toward general academics, social interactions, and the study of science.

Multicultural Instruction and *SciencePlus*

Science is for everyone, and *SciencePlus* is designed to serve the multiethnic and multicultural classrooms of today. Students, regardless of their ethnic backgrounds, will not have to look hard to find positive role models in the pages of *SciencePlus*. In addition, content that shows events, concepts, and issues from diverse ethnic and cultural perspectives is provided. As students work through *SciencePlus*, they will come to understand that science is a human endeavor that has been advanced by the contributions of many cultures and ethnic groups.

To add depth to your multicultural instruction, the Wrap-Around Margins of this Annotated Teacher's Edition periodically include a feature called Multicultural Extension. This information provides activities to help you focus on cultural diversity, highlighting the individuality and contributions of different ethnic groups.

Meeting the needs of culturally diverse students is perhaps the most demanding challenge faced by teachers today.

Obviously, to teach effectively you must be able to reach every individual in your class. This is seldom easy, given the diverse nature of most of today's classrooms. In addition, certain students present special challenges. Dealing adequately with these students requires special preparation and strategies. In many cases a minimal amount of preparation is sufficient to make the classroom a place where all can learn. Some of the more common situations you are likely to encounter are discussed below.

Learners Having Difficulty

Learners having difficulty are those who, for any of a number of reasons, are liable to perform poorly and who have a high probability of dropping out. *SciencePlus* is engaging and interesting throughout, appealing to all students. Throughout *SciencePlus*, clear easy-to-read prose and straightforward, attractive graphics reduce the potential for students to become bored. The style of *SciencePlus* is intentionally friendly and unintimidating. Field-testing has shown that the performance of students considered at-risk in science increased substantially when using *SciencePlus*.

Additional activities and teaching suggestions for learners having difficulty are provided under the learners Meeting Individual Needs: Learners Having Difficulty heading in the Wrap-Around Margins of this Annotated Teacher's Edition.

Meeting Individual Needs

Learners Having Difficulty

Using a hot plate, heat a few spoonfuls of sugar in a nonstick frying pan, stirring frequently. Once the sugar has caramelized, pour it onto wax paper to cool. Provide students with a cube of sugar and a piece of the caramelized sugar for comparison. Students should describe the metamorphosis that has taken place, compare the appearances of the two substances, and relate the change to the formation of metamorphic rock.

Second-Language Learners

Because *SciencePlus* places so much emphasis on doing science rather than reading about it, the program is ideal for students who are not proficient in English. Science is a universal language—the language of curiosity and logical reasoning. Many *SciencePlus* activities are easy to follow and require a minimum of reading. Lengthy explanations are seldom called for. You need only to get students started in the right direction; thereafter their intuition and common sense take over. The cooperative approach emphasized in *SciencePlus* helps to give second-language learners the extra support they need.

Additional activities and teaching suggestions for second-language learners are found in the Wrap-Around Margins of this Annotated Teacher's Edition. Look for the heading Meeting Individual Needs: Second-Language Learners.

Meeting Individual Needs

Second-Language Learners

Have students tell the story of a volcanic eruption using a comic-strip format. Students can create or gather a series of drawings or photographs to illustrate the sequence of events. Then encourage them to write a simple dialogue or commentary to accompany the pictures.

The Spanish Resources section of the *Teaching Resources* booklets also contains useful information, including blackline masters of parent letters translated into Spanish, as well as unit summaries and unit glossaries in Spanish.

Also available are *English/Spanish Audiocassettes,* which provide important preview information to assist Spanish-speaking students and students who are auditory learners.

Gifted Learners

The difficulty of teaching gifted students lies in keeping them interested, motivated, and challenged. Gifted students who are inadequately challenged may become bored, withdrawn, or even openly disruptive. *SciencePlus* includes many activities suitable for even the most advanced student. Open-ended activities, in particular, are especially suited for gifted students.

The *SciencePlus* approach emphasizes creative problem solving. In many cases there is no single right answer to a problem or question, so students' answers can reflect their individual abilities. This approach is ideal for gifted students, as they may extend the activities to fit their interests and talents.

Additional activities and teaching suggestions for gifted students are found in the Wrap-Around Margins of this Annotated Teacher's Edition. Look for the heading Meeting Individual Needs: Gifted Students.

Meeting Individual Needs

Gifted Learners

Have students create their own guidebooks of local plants. Students could either sketch or photograph each plant. They can then list the common name of the plant along with its scientific name and some of its characteristics.

Physically Impaired Students

Make your classroom as easy to move about in as possible. Remove or bypass any obvious barriers. Encourage your students to assist physically impaired students. If the student uses a wheelchair, make the aisles wide enough to accommodate the chair. Make sure that the student can reach any equipment he or she needs. You may wish to enlist the aid of other students in the class to assist the disabled student as necessary.

As much as possible, adapt the classroom to make it possible for physically impaired students to engage in the same activities as other students. Use a mobile demonstration table so that it can be moved to different areas of the room for maximum visibility.

Visually Impaired Students

Seat students with marginal vision near the front of the room to maximize their view of both you and the chalkboard, or assign a student to make copies of what you write. You could also assign a student to explain all visual materials in detail as they are presented.

Students who are completely blind should be allowed to become familiar with the classroom layout before the first class begins. Promptly inform these students of any changes to your classroom layout. Whenever possible, provide blind students with Braille or taped versions of all printed materials. Blind students may also use hand-held devices for converting written text into speech.

Hearing-Impaired Students

If you have hearing-impaired students in your class, remember to always face the class while speaking. Minimize classroom noise, and arrange seating in a circle or semicircle so that hearing-impaired students can see others. This arrangement facilitates speech reading. Speak in simple, direct language and avoid digressions or sudden changes in topic. During class discussions, periodically summarize what students are saying and repeat students' questions before answering them. Use visual media such as filmstrips, overhead projectors, and close-captioned films when appropriate. You might arrange a buddy system in which another student provides copies of notes about activities and assignments.

A student who is completely deaf may require a sign-language interpreter. If so, let the student and the interpreter determine the most convenient seating arrangement. When asking the student a question, be sure to look at the student, not at the interpreter. If the student also has a speech impairment, group assignments for oral reports may be advisable.

In many cases a minimal amount of preparation is sufficient to make the classroom a place where all can learn.

Computer-assisted instruction is an extremely useful tool for some learning-disabled students.

Speech-Impaired Students

Mainstreaming speech-impaired students is generally not very difficult. Patience is essential when dealing with speech-impaired students, however. For example, resist the temptation to finish sentences for a student who stutters. At the same time, do not show impatience. Also pay attention to nonverbal cues, such as facial expression and body language.

Be supportive and encouraging. You need not leave the speech-impaired student out of normal classroom discussions. For example, you may call on a speech-impaired student to answer a question and then allow the student to write out his or her response on the chalkboard or overhead projector. Use multisensory materials whenever possible to create a more comfortable learning environment for the speech-impaired student.

Learning-Disabled Students

Learning disabilities are any disorders that obstruct a person's listening, reasoning, communication, or mathematical abilities, and they range from mild to severe. An estimated 2 percent of all adolescents have some type of learning disability. Learning disabilities are the most common type of disability. Provide a supportive and structured environment in which rules and assignments are clearly stated. Use familiar words and short, simple sentences. Repeat or rephrase your instructions as needed.

Students may require extra time to complete exams or assignments, with the amount of extra time being dependent on the severity of their disability. Some students may need to tape-record lectures and answers to exam questions. For those who have difficulty organizing materials, you might provide chapter or lecture outlines for them to fill in. Having peer tutors work with learning-disabled students on specific assignments and review materials can be effective.

Computer-assisted instruction is an extremely useful tool for some learning-disabled students. This mode of instruction can even help these students develop good learning skills. For learning-disabled students, computers serve as a tireless instructor with unlimited patience. In addition, students receive simplified directions; proceed in small, manageable steps; and receive immediate reinforcement and feedback with computerized instruction.

Students With Behavioral Disorders

Behavioral disorders are emotional or behavioral disturbances that hinder a student's overall functioning. The behaviorally impaired may exhibit any of a variety of behaviors, ranging from extreme aggression to complete passivity.

Obviously, no single teaching strategy can accommodate all behavioral disorders. In addition, behavioral psychologists disagree on the best way to deal with students who have behavioral disorders. As a general rule, try to be fair and consistent, yet flexible, in your dealings with behaviorally disabled students. Make sure to state rules and expectations clearly. Reinforce desirable behavior or even approximations of such behavior, and ignore or mildly admonish undesirable behavior.

Because learning disabilities often accompany behavioral disorders, you might also wish to refer to the guidelines for learning disabilities.

MATERIALS AND EQUIPMENT

SciencePlus is designed to be teachable even by those with a limited budget for materials. Most activities use common household items that can be brought to class by students or parents or that can otherwise be easily obtained. For a comprehensive listing of the materials and equipment you will need to carry out the activities of *SciencePlus*, see the *Materials Guide*. The *Materials Guide* includes both a master list of materials and a unit-by-unit list.

In addition, the second page of the parent letter contained in the *Teaching Resources* booklets lists the supplies needed for each unit. This page makes it easy for you to invite donations from parents to keep your budget as low as possible.

For teachers who would rather have the convenience of purchasing supplies through the mail, Science Kit is the official *SciencePlus* materials and equipment supplier. For your convenience, Science Kit offers kits that contain all of the materials and equipment you will need to teach each unit of *SciencePlus*. Or, if you prefer, you can order needed materials and equipment individually as necessary.

Science Kit ordering information is available from your local HRW representative.

SCIENCEPLUS TEACHER'S NETWORK

The *SciencePlus* Teacher's Network (SPTN) is part of the Atlantic Science Curriculum Project (ASCP), linking teaching, curriculum development, and research in science education. The ASCP, which produced the Canadian version of *SciencePlus*, has existed for over 15 years, beginning as a collaborative, grass-roots activity to improve science teaching in middle schools and junior high schools. The goal of the SPTN is to continue this collaborative effort, linking teachers with teachers through newsletters, electronic bulletin boards, conferences, and small-group and regional meetings.

The *SciencePlus Communicator* is the official U.S. publication of the SPTN. The communicator is written by and for *SciencePlus* teachers. It provides a forum for *SciencePlus* teachers to share their classroom experiences with other *SciencePlus* teachers. Each edition includes copy masters, activities, current research, assessment ideas, and much more that can help make your teaching more effective. The *SciencePlus Communicator* links you to a community of like-minded teachers whose breadth of experience can provide encouragement and support.

Assessing Student Performance

A Comprehensive Approach to Assessment

Developing strategies for assessing student progress is an important step in realizing the goals of *SciencePlus*. Students pay the most attention to those aspects of a lesson on which they know they will be graded. Teachers who want their students to be successful should therefore teach with continual assessment in mind.

In *SciencePlus*, there is no distinct boundary between teaching and assessing. You will find that most of the tests and assessment activities in this program are designed to teach as well as to evaluate comprehension and performance. This emphasis can help correct the preoccupation with measuring and sorting students. The suggestions here are intended to aid you in your primary task: teaching.

Assessment Aids in *SciencePlus*

SciencePlus includes a wide variety of assessment aids to help you measure your students' mastery of the concepts and processes covered in this course. In addition to the assessment materials contained in the Pupil's Edition, including the Challenge Your Thinking chapter reviews, Making Connections unit reviews, and SourceBook Unit CheckUp reviews, the following materials are available for comprehensive and convenient assessment of your students.

In the *Teaching Resources* booklets

Blackline-master tests are available in the *Teaching Resources* booklets in the following categories:

- **Chapter Assessment**
- **Activity Assessment**
- **End-of-Unit Assessment**
- **SourceBook Assessment**

In addition, a checklist is available to help you implement self-assessment into your classroom. The **Self-Evaluation of Achievement** checklist will allow you to add this facet of assessment to your other assessment methods. The checklists are tailored to each unit in the Pupil's Edition.

In the *Assessment Checklists and Rubrics* booklet

This booklet contains over 40 different checklists for student self-evaluation, peer evaluation, and teacher assessment. Checklists are also provided to aid ongoing assessment.

This booklet contains a variety of assessment rubrics that serve as models for grading writing assignments, portfolios, reports, presentations, experiments, and technology projects.

Several progress reports are also provided to help you keep track of your students' progress.

On the *SnackDisc* CD-ROM

All of the checklists and rubrics contained in the *Assessment Checklists and Rubrics* booklet are also available on the *SnackDisc.* This CD-ROM for Macintosh® and Windows® makes these assessment materials extremely easy to access and fully customizable. Now you can quickly and easily create assessment checklists and rubrics that fit your own criteria and class situation. You can add to the recommended criteria or replace the criteria with your own.

In the *Test Generator* and *Test Item Listing* booklet

The *Test Generator* for Macintosh® and Windows® contains five categories of tests for each unit of the Pupil's Edition: Chapter Assessment, End-of-Unit Assessment, SourceBook Assessment, Activity Assessment, and Extra Assessment Items. The Extra Assessment Items consist of questions found only in the *Test Generator* and *Test Item Listing.* All of the other tests have blackline masters in the *Teaching Resources* booklets.

The *Test Generator* is easy to use, includes graphics, and is fully customizable. You can easily change the questions or add questions of your own.

The *Test Item Listing* booklet is a handy way to preview the tests and questions contained in the *Test Generator.*

In the *Annotated Teacher's Edition*

Each Unit Interleaf in this Annotated Teacher's Edition contains a comprehensive **Assessment Planning Guide.** This guide identifies, in chart form, all of the assessment components available in the program. A sample Assessment Planning Guide is shown below.

The Wrap-Around Margins of this Annotated Teacher's Edition also contain valuable assessment options. Each lesson contains a Follow-Up section that includes an optional assessment strategy. See page T28 for an example of these strategies.

Assessment Planning Guide

Lesson, Chapter, and Unit Assessment	SourceBook Assessment	Ongoing and Activity Assessment	Portfolio and Student-Centered Assessment
Lesson Assessment Follow-Up: see Teacher's Edition margin, pp. 303, 313, 323, 338, and 343 **Chapter Assessment** Chapter 14 Review Worksheet, p. 35 Chapter 14 Assessment Worksheet, p. 38* Chapter 15 Review Worksheet, p. 46 Chapter 15 Assessment Worksheet, p. 48* **Unit Assessment** Unit 5 Review Worksheet, p. 52 End-of-Unit Assessment Worksheet, p. 54*	SourceBook Unit Review Worksheet, p. 64 SourceBook Assessment Worksheet, p. 69*	Activity Assessment Worksheet, p. 59* **SnackDisc** Ongoing Assessment Checklists ♦ Teacher Evaluation Checklists ♦ Progress Reports ♦	Portfolio: see Teacher's Edition margin, pp. 298, 309, 320, 323, and 334 **SnackDisc** Self-Evaluation Checklists ♦ Peer Evaluation Checklists ♦ Group Evaluation Checklists ♦ Portfolio Evaluation Checklists ♦

* Also available on the Test Generator software

♦ Also available in the Assessment Checklists and Rubrics booklet

Assessment should be ongoing and should measure performance on exams and quality of class work. Homework, lab work, and ScienceLog entries should all be factors in assigning grades.

The authors strongly discourage reliance on recall-based assessment strategies. Teachers who currently rely heavily on such assessment strategies may find it difficult at first to adopt new methods of assessment. However, once the transition is made, the reward—in the form of improved student performance and motivation—will more than offset the inconvenience.

ScienceLog Assessment

SciencePlus provides many opportunities for students to demonstrate their understanding of specific concepts. The first page of each chapter is devoted to getting students to think about what they already know about the concepts in the chapter. Students are asked several questions and are encouraged to write down what they already know or think they know. After students complete the chapter, they are given the opportunity to revise their ScienceLog entries in the Challenge Your Thinking chapter review. Here students will confront any misconceptions they may have had in the beginning. In this way students actually assess their own prior knowledge and make adjustments accordingly.

Although you should not grade these ScienceLog entries beyond checking that students have done them, viewing students' initial entries can give you a good idea of their understanding of the main concepts. In this way, these entries can provide you with an excellent diagnostic tool to determine where to start your teaching and what concepts will need the most emphaisis.

By checking your students revised ScienceLog entries, you can get a good idea of the progress that has been made as well as what topics may need to be revisited to ensure understanding.

In addition, the ScienceLog serves as a companion notebook for most of the written work that is assigned throughout the program. You may want to have students hand in their entire ScienceLog periodically to check their work and progress.

Portfolio Assessment

SciencePlus is ideally suited to the use of Portfolios as one method of assessing your students, performance and accomplishments. For more information about Portfolios and their use in *SciencePlus*, see page T38 of this Annotated Teacher's Edition.

For help in assessing student Portfolios, several checklists and rubrics have been provided in the *Assessment Checklists and Rubrics* booklet and on the *SnackDisc.*

Assessing Scientific, Psychomotor, and Communication Skills

In *SciencePlus*, knowledge and understanding are closely linked to the development of important process skills such as observing, measuring, graphing, writing, predicting, inferring, analyzing, and hypothesizing. All learning tasks are designed to help develop these skills. The teacher can assess such skill development by inspecting student work and by observing student performance.

The sample tables below are suitable models for evaluating student performance.

Assessing Environmental Awareness

SciencePlus was written with a commitment to environmental awareness. Many activities that promote such awareness are included. The teacher is provided with suggestions on extending this theme through creative projects, cleanup or recycling projects, and so on. Tasks such as these promote environmental consciousness. The care that students take in carrying out these activities is a measure of their awareness of environmental issues.

A successful science program is reflected in a student body with outside interests in science.

ASSESSING SCIENTIFIC BEHAVIOR

Behavior	Poor	Satisfactory	Good	Very Good	Excellent
• Cooperates with others in small groups					
• Observes and records observations					
• etc.					

ASSESSING TECHNICAL SKILLS

Task	Yes	No	Uncertain
• Is able to read thermometer correctly			
• Is able to use spring scale to measure force			
• etc.			

Assessing Scientific Attitudes

It can be useful to survey your students about the types of science-related hobbies and interests that they pursue outside of class. In a direct way, this provides feedback on the success of your school's science program. A successful science program is reflected in a student body with outside interests in science. Ask your students to keep a tally of any science-related activities they undertake outside of class. These could include reading or writing about science and technology, science-related projects, visits to museums, attending lectures on scientific and technological topics, and viewing science programs on television.

For developing and assessing individual students' interest in and attitudes toward science, the assignment of elective reading and independent projects is essential. In addition to the numerous project ideas and extension activities included in the units of *SciencePlus*, the Science in Action features provide a range of project ideas from which students may choose. Students may also want to read further about a topic in the SourceBook or in another science-related book or magazine.

Student work on elective projects should count as a significant part of overall assessment. This type of work provides the surest indication of a student's interest and proficiency in science, especially the student's ability to study, plan, and research independently.

A Balanced Assessment

The authors of *SciencePlus* recommend a balance between the different forms of assessment. As a general rule, the following proportions are suggested:

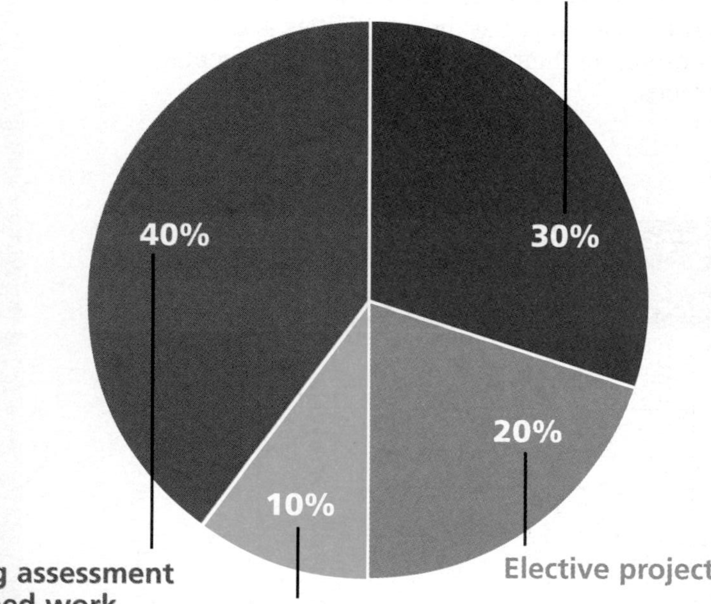

Assessing Science Projects

SciencePlus offers students abundant opportunities for independent investigation. Many open-ended, curiosity-stimulating questions are posed to students. Some of these questions are natural starting points for science projects. Students using *SciencePlus* have often developed successful science-fair projects based on questions in the text. The Wrap-Around margins of this Annotated Teacher's Edition also contain additional project ideas where appropriate.

Undertaking a project provides students with a host of positive experiences. Students learn to organize, plan, and piece together many separate ideas and pieces of information into a coherent whole. Undertaking a project also allows students to experience the sense of accomplishment that comes from tackling and completing a difficult task. It has even been argued that no science education is complete without having undertaken and completed a major project.

Many students will resist the idea of undertaking a major project because they feel that it is too much work or that they are simply not up to the task. The following suggestions may help you overcome students' reluctance:

- Allow students select their own project ideas.
- Encourage students to be creative.
- Provide a clear set of guidelines for developing and completing projects.
- Help students locate sources of information, including people in science-related fields who might advise students about their projects.
- Allow students the option of presenting their completed projects to the class.
- Emphasize the satisfaction students will derive from completing their projects.
- Inform students of the general areas on which assessment may be made, such as scientific thought, originality, and presentation.
- Do not emphasize the details of assessment. "Scoring points" should not be a major incentive.

Do not allow preconceived notions about how the project should be done to detract from the students' interest, enjoyment, and satisfaction in doing original work. Rather than forcing all students to fit their project work into a mold suited to scientific research, establish three sets of criteria, as described in the tables on pages T62 to T64.

DEVELOPING ASSESSMENT ITEMS

When developing tests, you should bear in mind the sort of skill required to answer each assessment item. The manner of testing determines what is learned: tests that require tick-mark responses teach tick-marking. Tests that require verbal, graphic, illustrative, and numeric responses develop writing, speaking, graphing, drawing, and mathematical skills. A superior test draws on as many skills as possible.

The assessment-item development model below is designed to help you construct comprehensive tests that will meet the educa-tional goals of *SciencePlus*. The model features four categories (verbal, graphic, illustrative, and numeric) and twelve types of items. The variety of tests and test items that accompany this program have been developed in accordance with this model.

The manner of testing determines what is learned: tests that require tick-mark responses teach tick-marking.

ASSESSMENT-ITEM DEVELOPMENT MODEL

VERBAL

Word Usage: The words given are to be used in a prescribed situation.

Correction/Completion: Incorrect or incomplete sentences and paragraphs are given for correction or completion.

Short Essays: Information is given or a question is posed for short-essay response.

Short Responses: Answers to these questions require a tick mark, a line, or a single word, phrase, or sentence.

ILLUSTRATIVE

Illustrations for Interpretation: Illustrations (drawings or photographs) are presented for interpretation.

Illustrations for Correction or Completion: An incorrect or incomplete illustration is given for correction or completion.

Answering by Illustration: A question requiring a drawing as the expected answer is asked.

GRAPHIC

Graphs for Interpretation: A graph of a relationship between two variables is given for interpretation.

Graphs for Correction or Completion: An incorrect or incomplete graph is given for correction or completion.

Graphing Data: Data are given to be graphed.

NUMERIC

Data for Interpretation: A data table is given for interpretation.

Numerical Problems: A problem requiring a numerical solution is given.

Rubric for Reports and Presentations

SCIENTIFIC THOUGHT (40 POINTS POSSIBLE)				
40–36	35–31	30–26	25–21	20–10
Complete understanding of topic; topic extensively researched; variety of primary and secondary sources used and cited; proper and effective use of scientific vocabulary and terminology	Good understanding of topic well-researched; a variety of sources used and cited; good use of scientific vocabulary and terminology	Acceptable understanding of topic; adequate research evident; sources cited; adequate use of scientific terms	Poor understanding of topic; inadequate research; little use of scientific terms	Lacks an understanding of topic; very little research, if any; incorrect use of scientific terms

ORAL PRESENTATION (30 POINTS POSSIBLE)				
30–27	26–23	22–19	18–16	15–5
Clear, concise, engaging presentation; well-supported by use of multisensory aids; scientific content effectively communicated to peer group	Well-organized, interesting, confident presentation supported by multisensory aids; scientific content communicated to peer group	Presentation acceptable; only modestly effective in communicating science content to peer group	Presentation lacks clarity and organization; ineffective in communicating science content to peer group	Poor presentation; does not communicate science content to peer group

EXHIBIT OR DISPLAY (30 POINTS POSSIBLE)				
30–27	26–23	22–19	18–16	15–5
Exhibit layout self-explanatory, and successfully incorporates a multisensory approach; creative use of materials	Layout logical, concise, and can be followed easily; materials used in exhibit appropriate and effective	Acceptable layout of exhibit; materials used appropriately	Organization of layout could be improved; better materials could have been chosen	Exhibit layout lacks organization and is difficult to understand; poor and ineffective use of materials

Rubric for Experiments

SCIENTIFIC THOUGHT (40 POINTS POSSIBLE)	
40–36	35–5
An attempt to design and conduct an experiment or project with all important variables controlled	An attempt to design an experiment or project, but with inadequate control of significant variables

ORIGINALITY (16 POINTS POSSIBLE)				
16–14	13–11	10–8	7–5	4–2
Original, resourceful, or novel approach; creative design and use of equipment	Imaginative extension of standard approach and use of equipment	Standard approach and good treatment of current topic	Incomplete and unimaginative use of resources	Lacks creativity in both topic and resources

PRESENTATION (24 POINTS POSSIBLE)				
24–21	20–17	16–13	12–9	8–5
Clear, concise, confident presentation; proper and effective use of vocabulary and terminology; complete understanding of topic; able to arrive at conclusions	Well-organized, clear presentation; good use of scientific vocabulary and terminology; good understanding of topic	Presentation acceptable; adequate use of scientific terms; acceptable understanding of topic	Presentation lacks clarity and organization; little use of scientific terms and vocabulary; poor understanding of topic	Poor presentation; cannot explain topic; scientific terminology lacking or confused; lacks understanding of topic

EXHIBIT (20 POINTS POSSIBLE)				
20-19	18-16	15-13	12-11	10-6
Exhibit layout self-explanatory, and successfully incorporates a multisensory approach; creative and very effective use of materials	Layout logical, concise, and can be followed easily; materials used appropriate and effective	Acceptable layout; materials used appropriately	Organization of layout could be improved; better materials could have been chosen	Layout lacks organization and is difficult to understand; poor and ineffective use of materials

Rubric for Technology Projects

SCIENTIFIC TECHNICAL THOUGHT (40 POINTS POSSIBLE)				
40–36	35–31	30–26	25–21	20–10
An attempted design solution to a technical problem; the problem is significant and stated clearly; the solution reveals creative thought and imagination; underlying technical and scientific principles are very well understood	An attempted design solution to a technical problem; the solution may be a standard one for similar problems; underlying technical and scientific principles are recognized and understood	A working model; underlying technical and scientific principles are well understood; model is built from a standard blueprint or design	Model is built from a standard blueprint or design or from a kit; underlying technical and scientific principles are recognized but not necessarily understood	Model is built from a kit; underlying technical and scientific principles are not recognized or understood

PRESENTATION (30 POINTS POSSIBLE)				
30–27	26–23	22–19	18–16	15–5
Clear, concise, confident presentation; proper and effective use of vocabulary and terminology; complete understanding of topic; able to extrapolate	Well-organized, clear presentation; good use of scientific vocabulary and terminology; good understanding of topic	Presentation acceptable; adequate use of scientific terms; acceptable understanding of topic	Presentation lacks clarity and organization; little use of scientific terms and vocabulary; poor understanding of topic	Poor presentation; cannot explain topic; scientific terminology lacking or confused; lacks understanding of topic

EXHIBIT (30 POINTS POSSIBLE)				
30–27	26–23	22–19	18–16	15–5
Exhibit layout self-explanatory, and successfully incorporates a good sensory approach; creative and very effective use of material	Layout logical, concise, and easy to follow; materials used in exhibit appropriate and effective	Acceptable layout of exhibit; materials used appropriately	Organization of layout could be improved; better materials could have been chosen	Layout lacks organization and is difficult to understand; poor and ineffective use of materials

SCIENCE PLUS
TECHNOLOGY AND SOCIETY

TO THE STUDENT

This book was written with you in mind!
There are *many things to try*, to *create*, and to *investigate*—both in and out of class.
There are *stories to read*, *articles to think about*, *puzzles to solve*, and even *games to play*.

GET INVOLVED!

The best way to learn is by doing. In the words of an old Chinese proverb:

Tell me—*I will forget*
Show me—*I may remember*
Involve me—*I will understand*

The activities in this book will allow you to make some basic and important scientific discoveries on your own. You will be acting much like the early investigators in science who, without expensive or complicated equipment, contributed so much to our knowledge.

What these early investigators had, and had in abundance, was curiosity and imagination. If you have these qualities, you are in good company! And if you develop sharp scientific skills, who knows?— you might make your own contributions to science someday.

Scientists are usually interested in understanding things that happen in nature. However, the discoveries that scientists make are often used by inventors and engineers. Using science in this way has resulted in our most sophisticated technology, including such things as computers, laser discs, nuclear reactors, and instant global communication.

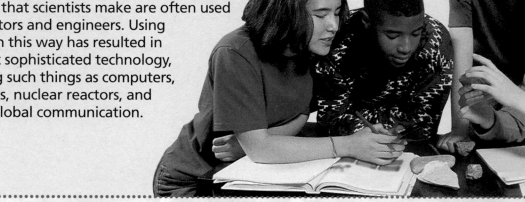

SCIENCE & TECHNOLOGY

There is an interaction between science and technology. Science makes technology possible. On the other hand, the products of technology are used to make further scientific discoveries. In fact, much of the scientific work that is done today has become so technically complicated and expensive that no one person can do it entirely alone. But make no mistake, the creative ideas for even the most highly technical and expensive scientific work still come from individuals.

A built-in reference section is located at the back of this book. It's called the SourceBook. **CHECK IT OUT!**

GO FOR IT!

Science is a process of discovery, a trek into the unknown. The skills you develop as you do the activities in this book—like observing, experimenting, and explaining observations and ideas—are the skills you will need in order to be a part of science in the future. There is a universe of scientific exploration and discovery awaiting those who take up the challenge.

Keep a ScienceLog

A journal is an important tool in creative work. In this book, you will be asked to keep a type of journal, called a ScienceLog, to record your thoughts, observations, experiments, and conclusions. As you develop your ScienceLog, you will see your own ideas taking shape over time. This is often the way scientists arrive at new discoveries. You too may log some discoveries as you develop your own journal.

xiii

SAFETY FIRST!

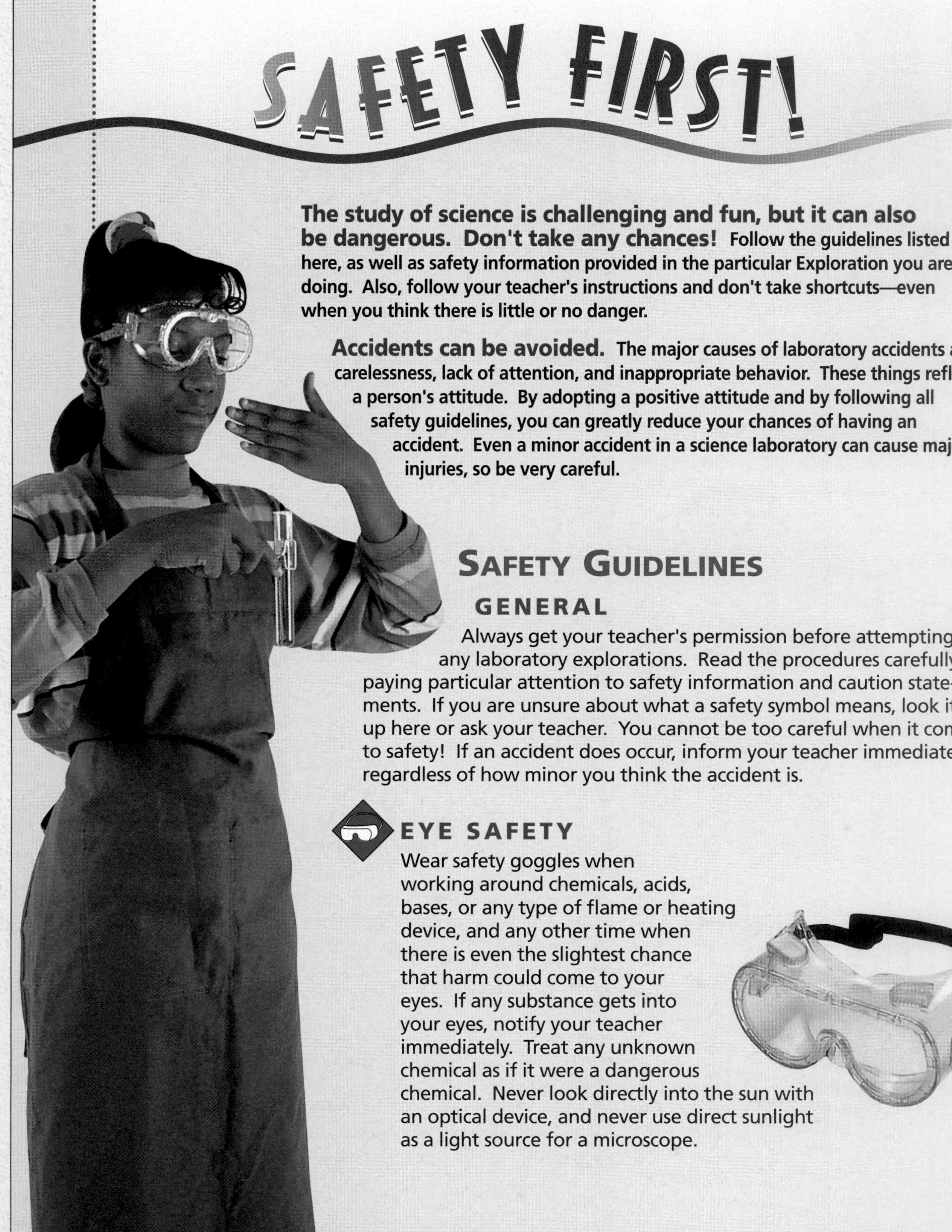

The study of science is challenging and fun, but it can also be dangerous. Don't take any chances! **Follow the guidelines listed here, as well as safety information provided in the particular Exploration you are doing. Also, follow your teacher's instructions and don't take shortcuts—even when you think there is little or no danger.**

Accidents can be avoided. **The major causes of laboratory accidents are carelessness, lack of attention, and inappropriate behavior. These things reflect a person's attitude. By adopting a positive attitude and by following all safety guidelines, you can greatly reduce your chances of having an accident. Even a minor accident in a science laboratory can cause major injuries, so be very careful.**

SAFETY GUIDELINES

GENERAL

Always get your teacher's permission before attempting any laboratory explorations. Read the procedures carefully, paying particular attention to safety information and caution statements. If you are unsure about what a safety symbol means, look it up here or ask your teacher. You cannot be too careful when it comes to safety! If an accident does occur, inform your teacher immediately, regardless of how minor you think the accident is.

EYE SAFETY

Wear safety goggles when working around chemicals, acids, bases, or any type of flame or heating device, and any other time when there is even the slightest chance that harm could come to your eyes. If any substance gets into your eyes, notify your teacher immediately. Treat any unknown chemical as if it were a dangerous chemical. Never look directly into the sun with an optical device, and never use direct sunlight as a light source for a microscope.

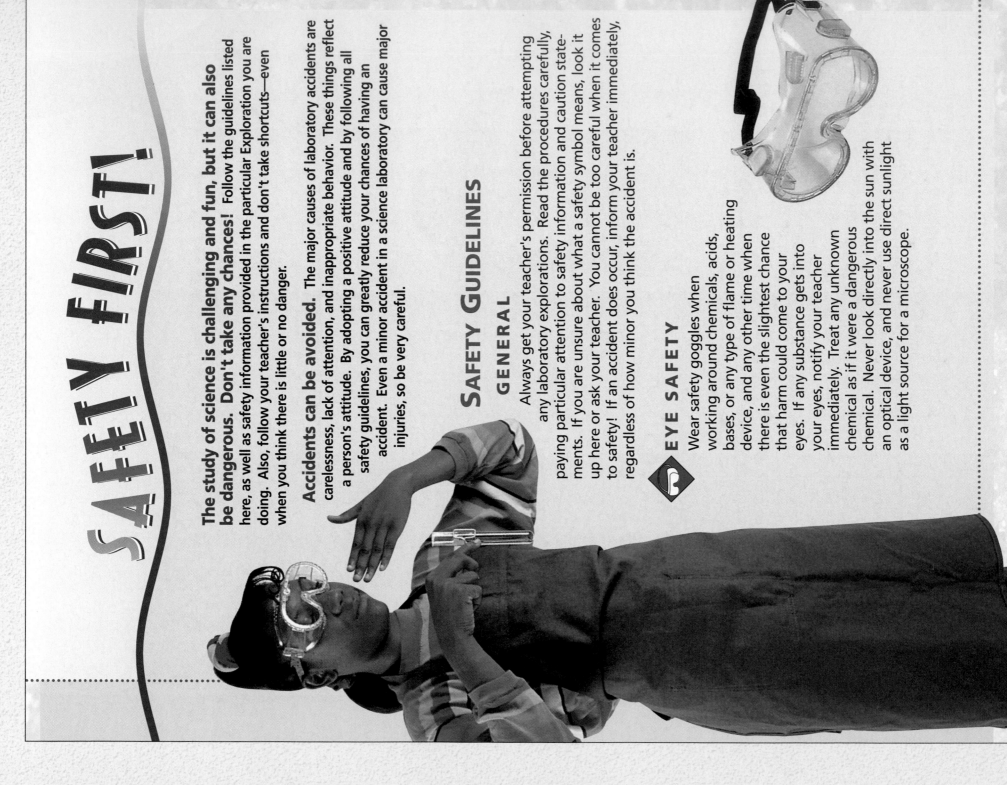

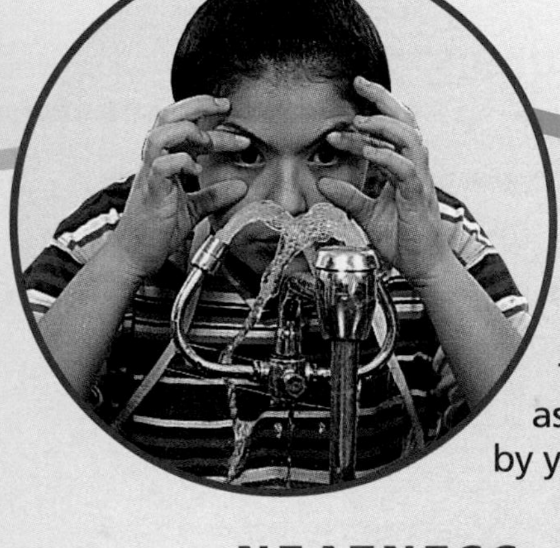

SAFETY EQUIPMENT

Know the location of and how to use the nearest fire alarms and any other safety equipment, such as fire blankets and eyewash fountains, as identified by your teacher.

NEATNESS

Keep your work area free of all unnecessary books and papers. Tie back long hair and secure loose sleeves or other loose articles of clothing such as ties and bows. Remove dangling jewelry. Don't wear open-toed shoes or sandals in laboratory situations. Never eat, drink, or apply cosmetics in a laboratory setting; food, drink, and cosmetics can easily become contaminated with dangerous materials.

SHARP/POINTED OBJECTS

Use knives and other sharp instruments with extreme care. Never cut objects while holding them in your hands. Place objects on a suitable work surface for cutting.

HEAT

Wear safety goggles when using a heating device or a flame. Whenever possible, use an electric hot plate instead of a flame as a heat source. When heating materials in a test tube, always slant the test tube away from yourself and others. Wear oven mitts, when instructed to do so, to avoid burns.

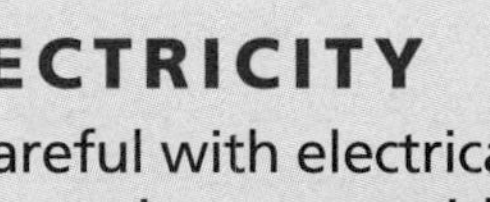

ELECTRICITY

Be careful with electrical wiring. When using a microscope with a lamp, do not place the cord where it could cause someone to trip. Do not let cords hang over a table edge in a way that could cause equipment to fall if the cord is accidentally pulled. Do not use equipment with damaged cords. Be sure your hands are dry and that the electrical equipment is in the "off" position before plugging it in. Turn off equipment when you are done.

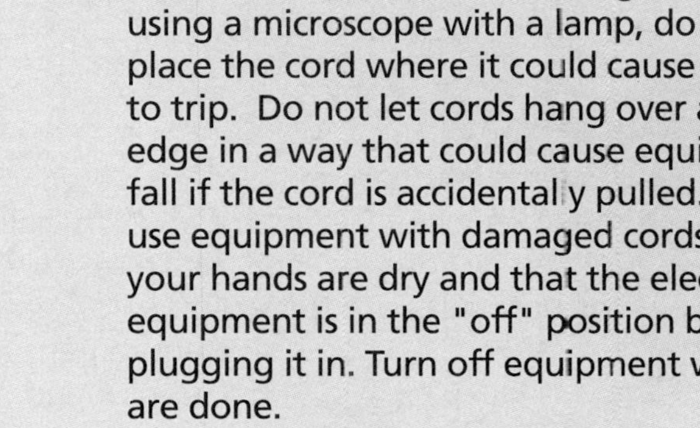

xv

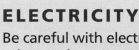

Never taste, touch, or smell chemicals unless you are specifically directed to do so by your teacher.

If you are instructed to note the odor of a substance, wave the fumes toward your nose with your hand. Never put your nose close to the source. Never mix chemicals unless you are told to do so by your teacher.

CHEMICALS

Wear safety goggles when handling any potentially dangerous chemicals, acids, or bases. If a chemical is unknown, handle it as you would a dangerous chemical. Wear an apron and latex gloves when working with acids or bases or when told to do so in the Exploration or Activity. If a spill gets on your skin or clothing, rinse it off immediately with water for at least 5 minutes while calling your teacher.

ANIMAL SAFETY

Always obtain your teacher's permission before bringing any animal into the school building. Handle animals only as your teacher directs. Always treat animals carefully and with respect. Wash your hands thoroughly after handling any animal.

PLANT SAFETY

Do not eat any part of a plant or plant seed used in the laboratory. Wash hands thoroughly after handling any part of a plant.

GLASSWARE

Examine all glassware before using. Be sure that it is clean and free of chips and cracks. Report damaged glassware to your teacher. Glass containers used for heating should be made of heat-resistant glass.

CLEANUP

Before leaving, clean up your work area. Put away all equipment and supplies. Dispose of all chemicals and other materials as directed by your teacher. Make sure water, gas, burners, and electric hot plates are turned off. Hot plates and other electrical equipment should also be unplugged. Wash hands with soap and water after working in a laboratory situation.

concept mapping

A Way to Bring Ideas Together

What Is a Concept Map?

Have you ever tried to tell someone about a book or a chapter you've just read, and you find that you can remember only a few isolated words and ideas? Or maybe you've memorized facts for a test, and then weeks later you're not even sure what topic those facts are related to.

In both cases, you may have understood the ideas or concepts by themselves, but not in relation to one another. If you could somehow link the ideas together, you would probably understand them better and remember them longer. This is something a concept map can help you do. A concept map is a visual way of choosing how ideas or concepts fit together. It can help you see the "big picture."

How to Make a Concept Map

1. **Make a list of the main ideas or concepts.**

   It might help to write each concept on its own slip of paper. This will make it easier to rearrange the concepts as many times as you need to before you've made sense of how the concepts are connected. After you've made a few concept maps this way, you can go directly from writing your list to actually making the map.

2. **Spread out the slips on a sheet of paper, and arrange the concepts in order from the most general to the most specific.**

   Put the most general concept at the top and circle it. Ask yourself, "How does this concept relate to the remaining concepts?" As you see the relationships, arrange the concepts in order from general to specific.

3. **Connect the related concepts with lines.**

4. **On each line, write an action word or short phrase that shows how the concepts are related.**

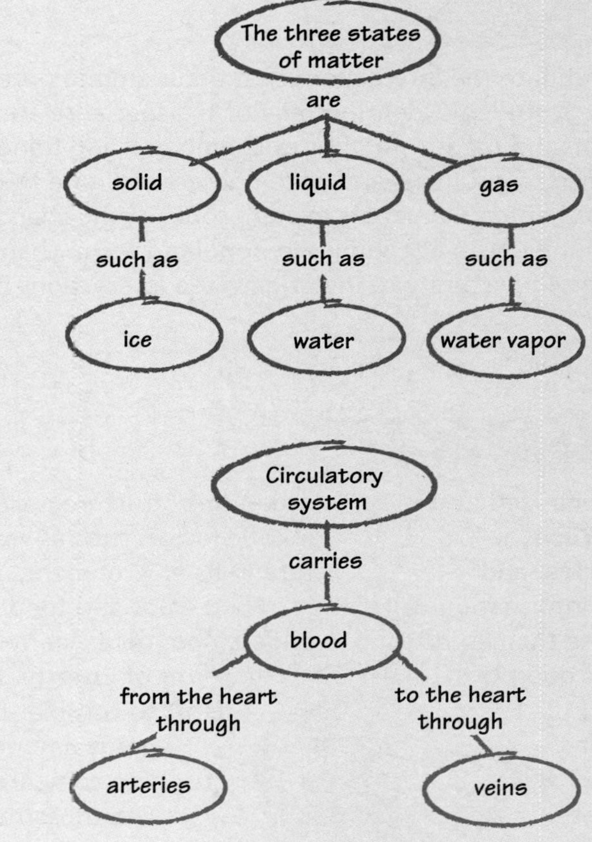

Look at the concept maps on this page and then see if you can make one for the following terms: plants, water, photosynthesis, carbon dioxide, and sun's energy.

The answer is provided below, but don't look at it until you try the concept map yourself.

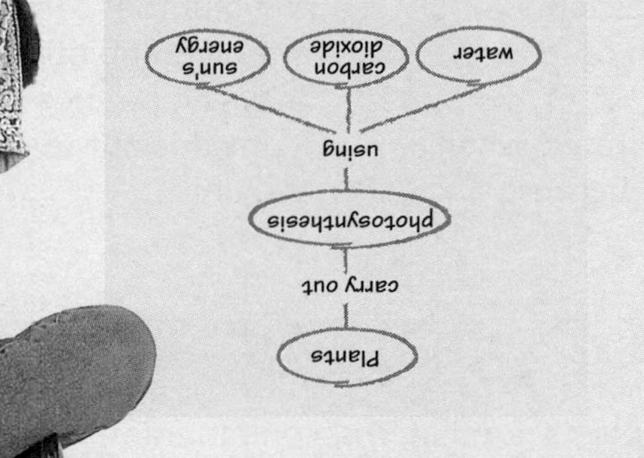

Unit 1 INTERACTIONS

Unit Overview

In this unit, students investigate the interactions that occur among organisms and their environments. In Chapter 1, students identify some of the special relationships, such as commensalism, mutualism, and parasitism, that exist among living things. In Chapter 2, they observe how producers, consumers, and decomposers facilitate the flow of energy through a community. In Chapter 3, students explore some of the changes that occur within a biological community, such as changes in population. The chapter concludes by providing students with an opportunity to investigate the interactions between humans and the environment.

Using the Themes

The unifying themes emphasized in this unit are **Changes Over Time, Structures, Energy, Cycles,** and **Systems.** The following information will help you incorporate these themes into your teaching plan. Focus questions that correspond to these themes appear in the margins of this Annotated Teacher's Edition on pages 3, 22, 31, 35, 45, 49, and 53.

Changes Over Time can be discussed in a variety of contexts throughout this unit. As students study natural interactions, they should always keep in mind that these systems are constantly changing. During discussions of succession, have students consider specifically succession as it occurs at an abandoned farm or in a rotting-log community.

Structures can be incorporated when discussing ranges of tolerance because an organism's structure is closely linked to abiotic factors such as water and temperature in its environment.

Food webs and food chains can be understood more clearly when thought of in terms of **Energy.** Invite your students to consider transfers of energy among organisms and to compare these transfers to those that occur in electrical systems.

Food webs are also conducive to incorporating the theme of **Cycles.** As energy is passed along a food chain, for instance, decomposers serve to return nutrients to the soil where they can be reused by producers.

Finally, to the extent that natural communities comprise a network of interacting biotic and abiotic factors, they can be thought of as **Systems.** Point this out to students throughout the unit.

Using the SourceBook

Unit 1 expands on the concepts of populations, communities, succession, and the role of humans in the environment. Interactions among organisms are explored in a variety of ways. Students study characteristics of populations, abiotic resources, biomes, pollution, and conservation.

Bibliography for Teachers

Asimov, Isaac, and Frederik Pohl. *Our Angry Earth.* New York City, NY: Tom Doherty Associates, Inc., 1991.

Burton, Robert, ed. *Nature's Last Strongholds.* New York City, NY: Oxford University Press, 1991.

Halliday, Tim, ed. *Animal Behavior.* Norman, OK: University of Oklahoma Press, 1994.

Quinn, John R. *Wildlife Survivors: The Flora and Fauna of Tomorrow.* Blue Ridge Summit, PA: TAB Books, 1994.

Bibliography for Students

Aldis, Rodney. *Rainforests.* New York City, NY: Dillon Press, 1991.

Pollock, Steve. *Ecology.* New York City, NY: Dorling Kindersley, Inc., 1993.

Reed, Catherine. *Environment.* Vero Beach, FL: Rourke Publications, Inc., 1992.

Rescue Mission Planet Earth: A Children's Edition of Agenda 21. New York City, NY: Kingfisher Books, 1994.

Films, Videotapes, Software, and Other Media

A Field Trip to the Rainforest
Software (Macintosh, Windows, or Apple II)
Sunburst
101 Castleton St.
P.O. Box 100
Pleasantville, NY 10570-9963

Animal Communities
Videodisc (English/Spanish bilingual)
AIMS Media
9710 DeSoto Ave.
Chatsworth, CA 91311-4409

Animal Populations: Nature's Checks and Balances
Film and videotape
Britannica
310 S. Michigan Ave.
Chicago, IL 60611

Tracing Cycles in the Environment: Ecology
Software (Macintosh, MS-DOS, or Apple II)
Queue, Inc.
338 Commerce Dr.
Fairfield, CT 06430

Unit Organizer

Unit/Chapter	Lesson	Time*	Objectives	Teaching Resources
Unit Opener, p. 2				Science Sleuths: Dead Fish on Union Lake English/Spanish Audiocassettes Home Connection, p. 1
Chapter 1, p. 4	**Lesson 1, The Parts Make Up the Whole, p. 5**	2	1. Identify interactions taking place in the environment. 2. Explain the dependence of living and nonliving things on one another. 3. Classify parts of the environment as either biotic or abiotic.	Exploration Worksheet, p. 3
	Lesson 2, The Living Players, p. 10	2 to 3	1. Describe the difference between a habitat and a niche. 2. Define and compare commensalism, mutualism, and parasitism. 3. Identify examples of commensalism, mutualism, and parasitism.	Image and Activity Bank 1-2 Resource Worksheet, p. 6
	Lesson 3, The Nonliving Players, p. 16	4	1. Identify water, light, and temperature as important abiotic factors of the environment. 2. Explain what is meant by the "range of tolerance" for a given abiotic factor in the environment. 3. Provide examples of how plants and animals respond to changing abiotic factors in the environment.	Image and Activity Bank 1-3 Exploration Worksheet, p. 7 ▼ Activity Worksheet, p. 9 Activity Worksheet, p. 11 ▼ Exploration Worksheet, p. 12 Resource Worksheet, p. 13 ▼ Transparency 4
End of Chapter, p. 25				Chapter 1 Review Worksheet, p. 17 Chapter 1 Assessment Worksheet, p. 20
Chapter 2, p. 27	**Lesson 1, Energy Flow, p. 28**	2	1. Describe the differences between producers, consumers, and decomposers. 2. Identify the differences between herbivores, carnivores, and omnivores. 3. Identify how predator-prey relationships facilitate the flow of energy through a community. 4. Explain the importance of scavengers in a community. 5. Explain how energy flows through a community of living things.	Image and Activity Bank 2-1 Transparency Worksheet, p. 23 ▼
	Lesson 2, Who Eats What? p. 32	2 to 3	1. Explain the difference between a food chain and a food web. 2. Diagram food chains and food webs. 3. Describe how a change in one part of a food web affects other parts of the web. 4. Identify the organisms in a food chain or a food web as either producers or consumers, and describe the role that they play in their community.	Image and Activity Bank 2-2 Theme Worksheet, p. 25 Transparency Worksheet, p. 27 ▼ Resource Worksheet, p. 29 ▼ Exploration Worksheet, p. 30 ▼ Exploration Worksheet, p. 32 Activity Worksheet, p. 39
End of Chapter, p. 44				Chapter 2 Review Worksheet, p. 41 Chapter 2 Assessment Worksheet, p. 44
Chapter 3, p. 46	**Lesson 1, Natural Changes, p. 47**	5 to 6	1. Explain what is meant by biological succession. 2. Describe changes that take place in biological communities. 3. Explain how a change in the population of one organism may affect the populations of other organisms. 4. Describe the characteristics of a specific biological community.	Image and Activity Bank 3-1 Exploration Worksheet, p. 48 Graphing Practice Worksheet, p. 51 Transparency 9
	Lesson 2, Humans: What Is Our Role? p. 56	4	1. Identify some positive and negative influences that people can have on the environment. 2. Explain the consequences of acid rain on the environment. 3. Identify some of the sources of acid rain. 4. Suggest ways of preventing acid rain and of reversing its effects on the environment. 5. Identify ways in which people can have a positive effect on the environment.	Image and Activity Bank 3-2 Discrepant Event Worksheet, p. 53 Transparency 10
End of Chapter, p. 64				Chapter 3 Review Worksheet, p. 54 Chapter 3 Assessment Worksheet, p. 58
End of Unit, p. 66				Unit 1 Activity Worksheet, p. 61 ▼ Unit 1 Review Worksheet, p. 63 ▼ Unit 1 End-of-Unit Assessment, p. 65 Unit 1 Activity Assessment, p. 71 Unit 1 Self-Evaluation of Achievement, p. 74

* Estimated time is given in number of 50-minute class periods. Actual time may vary depending on period length and individual class characteristics.

▼ Transparencies are available to accompany these worksheets. Please refer to the Teaching Transparencies Cross-Reference chart in the Unit 1 Teaching Resources booklet.

Materials Organizer

Chapter	Page	Activity and Materials per Student Group
1	6	**Exploration 1:** large ball of string or yarn; 1 index card per student; 1 small piece of tape per student
	17	**Exploration 2:** 30 radish seeds; 3 petri dishes; three 10 cm × 10 cm pieces of cardboard; about 200 mL of water; 12 paper towels; alcohol thermometer; scissors; 3 resealable plastic bags; refrigerator; radiator or other heat source
	20	**Exploration 3:** about 30 radish seeds; a few small flowerpots filled with soil; 1–2 L of water; 100 mL graduated cylinder; about 30 seeds of various types
2	38	**Exploration 2, A Woodland Community:** field guides for identifying organisms common to your area; metric ruler or meter stick; **A Forest-Floor Community:** field guides for identifying organisms; meter stick; newspaper or white plastic tray; magnifying glass; **A Pond Community:** field guides for identifying organisms; plastic bottles, containers, and bags; kitchen strainer or aquarium net; white plastic tray (additional optional teacher materials: glass jar or aquarium; aquarium pump)
3	50	**Exploration 1:** field guides for identifying or-ganisms; paring knife; alcohol thermometer; magnifying glass; trowel; white plastic tray or newspaper
	63	**Learn More About It!** colored markers; poster board

Advance Preparation

Exploration 2, page 17: Make sure that a refrigerator and a heat source are available for students to use.

Exploration 2, page 38, and Exploration 1, page 50: You will need to locate appropriate sites to investigate prior to carrying out the Explorations.

Unit Compression

In this unit, students get a glimpse of the full range of biological, physical, and environmental interactions that occur in their world. The material that students encounter in this unit is important for a full appreciation of the material in the units that follow. Consequently, substantial compression of this unit is discouraged. Nonetheless, there are many opportunities to reduce the amount of class time necessary to cover the unit.

First, you may wish to have students read certain sections ahead of time to speed in-class discussion. Possibilities in-clude Types of Interactions in Communities on pages 12–15, Lesson 1 of Chapter 2 on pages 28–31, and Population Changes on pages 53–55.

Second, you may find that several sections may be assigned as homework. These include Problems to Consider on pages 14–15, Other Projects on page 18, Exploration 1 on page 35, and Some Things to Think About on page 52.

If time constraints are particularly severe, you may find it necessary to skip some activities that are meant to reinforce student understanding of topics covered in previous sections. You may wish to consider leaving the following sections to be covered at a later time: Exploration 2 on pages 38–43 and Exploration 1 on pages 50–51.

Homework Options

Chapter 1	See Teacher's Edition margin, pp. 8, 10, 14, 18, 19, 20, 23, and 26 Activity Worksheet, p. 9 Activity Worksheet, p. 11 Resource Worksheet, p. 13 SourceBook, pp. S2, S5, and S8
Chapter 2	See Teacher's Edition margin, pp. 30, 33, 34, 35, 37, and 42 Theme Worksheet, p. 25 Resource Worksheet, p. 29 Exploration Worksheet, p. 30 Activity Worksheet, p. 39 SourceBook, pp. S3, S5, and S7
Chapter 3	See Teacher's Edition margin, pp. 47, 49, 52, 53, 57, 62, and 65 Graphing Practice Worksheet, p. 51 SourceBook, pp. S11 and S19
Unit 1	Activity Worksheet, p. 61 Activity Assessment, p. 71 SourceBook Activity Worksheet, p. 74

Assessment Planning Guide

Lesson, Chapter, and Unit Assessment	SourceBook Assessment	Ongoing and Activity Assessment	Portfolio and Student-Centered Assessment
Lesson Assessment Follow-Up: see Teacher's Edition margin, pp. 9, 15, 24, 31, 43, 55, and 63 **Chapter Assessment** Chapter 1 Review Worksheet, p. 17 Chapter 1 Assessment Worksheet, p. 20* Chapter 2 Review Worksheet, p. 41 Chapter 2 Assessment Worksheet, p. 44* Chapter 3 Review Worksheet, p. 54 Chapter 3 Assessment Worksheet, p. 58* **Unit Assessment** Unit 1 Review Worksheet, p. 63 End-of-Unit Assessment Worksheet, p. 65*	SourceBook Review Worksheet, p. 77 SourceBook Assessment Worksheet, p. 81*	Activity Assessment Worksheet, p. 71* **SnackDisc** Ongoing Assessment Checklists ♦ Teacher Evaluation Checklists ♦ Progress Reports ♦	Portfolio: see Teacher's Edition margin, pp. 8, 11, 17, 24, 31, 34, 39, and 47 **SnackDisc** Self-Evaluation Checklists ♦ Peer Evaluation Checklists ♦ Group Evaluation Checklists ♦ Portfolio Evaluation Checklists ♦

* Also available on the Test Generator software
♦ Also available in the Assessment Checklists and Rubrics booklet

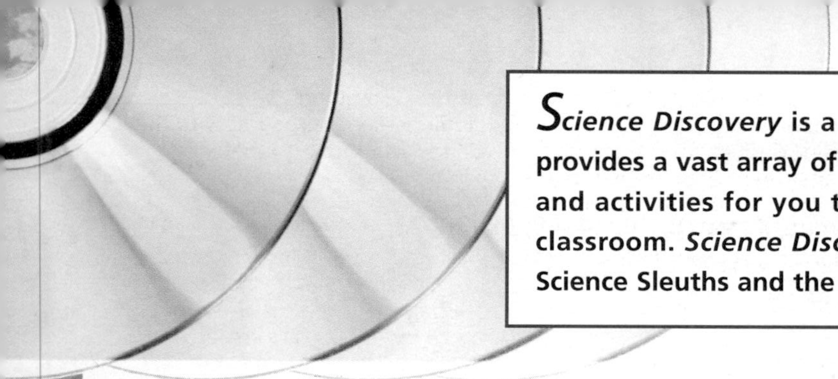

*S*cience *Discovery* is a versatile videodisc program that provides a vast array of photos, graphics, motion sequences, and activities for you to introduce into your *SciencePlus* classroom. *Science Discovery* consists of two videodiscs: Science Sleuths and the Image and Activity Bank.

Using the *Science Discovery* Videodiscs

Science Sleuths: Dead Fish on Union Lake
Side A

The 4Cs, a civic organization, coordinated the cleanup of a polluted urban lake and its surrounding parks. Suddenly, the fish in the lake are dying in large numbers. The 4Cs think someone is polluting the lake again. The Science Sleuths must analyze the evidence and determine why the fish are dying.

Interviews

1. Setting the scene: Spokesperson for the 4Cs **35975 (play ×2)**

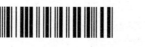

2. Parks Department employee **36772 (play)**

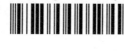

3. Elementary school students **37799 (play)**

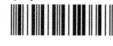

4. Fisherman **38223 (play)**

5. Sewage engineer **39120 (play)**

6. Limnologist **39882 (play)**

7. Fisheries biologist **40403 (play)**

Documents

8. Lake cleanup report **40949 (step ×2)**

9. Lake cleanup map **40953**

10. Fish habitats brochure **40955 (step ×2)**

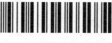

11. 4Cs brochure **40959 (step ×2)**

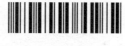

12. Parks Department fishing brochure **40963 (step ×4)**

Literature Search

13. Search on the words: ALGAE, FISH, GEESE, PARKS, POLLUTION, UNION LAKE **40969**

14. Article #1 ("4Cs Introduce New Initiative") **40971 (step)**

15. Article #2 ("Geese Get Going") **40974 (step)**

16. Article #3 ("Those Active Algae") **40977 (step)**

Sleuth Information Service

17. Temperature and rainfall graph **40980**

18. Blue-green algae **40982 (step)**

19. DO and BOD tests **40987 (step ×2)**

Sleuth Lab Tests

20. Lake temperature **40992**

21. Lake water pH **40994 (step)**

22. DO measurement **40997 (step)**

23. BOD analysis **41000 (step ×3)**

24. Trawl analysis **41005 (step ×3)**

25. Preliminary analysis of fish **41010 (step)**

26. Fish dissection **41013 (step ×6)**

27. Sediment analysis **41021 (step ×5)**

Still Photographs

28. Dead fish **41028**

Image and Activity Bank
Side A or B

A selection of still images, short videos, and activities is available for you to use as you teach this unit. For a larger selection and detailed instructions, see the Videodisc Resources booklet included with the Teaching Resources materials.

1-2 The Living Players, page 10

Bat taking a fig 48505–48981 (Side B only) (play ×2)
Most species of bats eat insects, but some eat fruit, nectar from flowers (these species also act as pollinators), or blood.

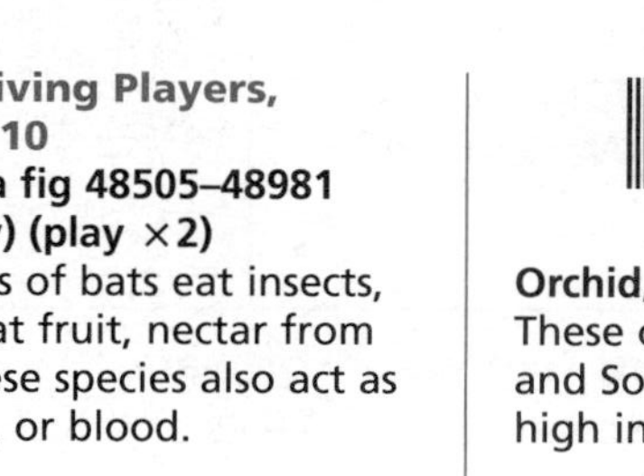

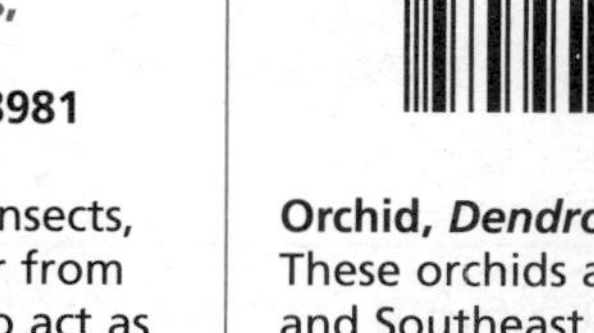

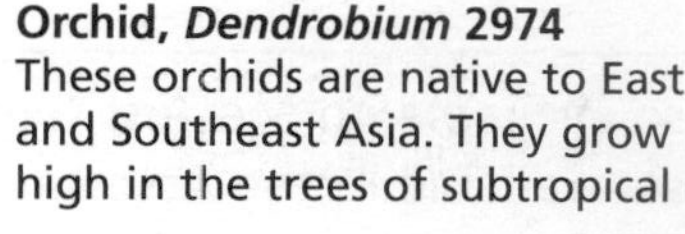

Orchid, *Dendrobium* 2974
These orchids are native to East and Southeast Asia. They grow high in the trees of subtropical or

Step Reverse **Play** **Pause** **Step Forward**

tropical jungles and get their nourishment from air, rain, and decaying vegetable matter without harming their host tree.

Shrimp, cleaner 3535
The cleaner shrimp picks off and eats parasites from any fish that needs cleaning. The fish benefits from the removal of the unwanted parasites, and the shrimp receives easily attained food. This is a *symbiotic* relationship.

Anemone, colonial 3108
Green anemones get their green color from symbiotic algae that live in their tissues. The algae use sunlight to make sugar, which the anemones use for food.

Tick 3346
Ticks wait on bushes for long periods of time until an animal that they can drop on walks by. These ticks are on the belly of a tapir. When ticks suck blood, they swell up like balloons. Female ticks lay their eggs after a meal.

Mistletoe; as parasite 3034
Mistletoe is a parasite that gets all of its nutrients from its host tree.

1-3 The Nonliving Players, page 16

Cyanobacteria; in geysers 2729
Cyanobacteria are among the most hardy organisms. Some grow in hot springs (shown here). Others grow under the ice in antarctic lakes, in lakes choked with salt, or on moist soil, tree bark, or snowbanks.

Camel, dromedary 3937
Arabian camels were bred to carry about 200 kg for 64 km or a person for almost 200 km in one day. Broad and soft feet, long legs, long eyelashes, nostrils that close, and wooly fur are adaptations that allow the camels to survive in sandy deserts. Camels do not store water in their humps.

Cactus, barrel 3003
This cactus is well adapted to its desert environment. Its fleshy trunk stores water, its thick skin reduces evaporation, and its spines discourage predators.

2-1 Energy Flow, page 28

Algae 2745
Algae are one of the primary producers in many communities of animals and microorganisms.

Robber fly 3462
Robber flies pounce on other insects and suck out their juices. Some are able to capture horseflies while flying. Robber flies are predators that help control growing insect populations, thereby contributing to the balance of nature.

Scavengers; hyenas, jackals, and vultures 4008
Lions have left the remains of a kill, and hyenas, jackals, and vultures have moved in for the scraps. Nothing is wasted. Although scavengers can kill their own prey, they prefer to let other animals do it for them.

Mold; on pumpkin pie 2767
This mold is a decomposer grown from a spore on a pumpkin pie. The spore matures and releases more spores, all of which can create more mold.

Fungus, shelf 2786
This fungus typically grows in shelflike groups on rotting wood or in wounds of living trees. It appears in the late winter or early spring.

2-2 Who Eats What? page 32

Food chain 4320
Plankton produce food through photosynthesis. They are eaten by small fish, which are in turn eaten by large fish. The original food molecules produced by the plankton end up in the cormorant (a type of water bird), which eats the large fish.

Pond ecosystem 4326
Algae produce basic sugars through photosynthesis; these nutrients are passed to the heron. When the organism dies, bacteria and molds recycle its nutrients.

Water boatman 3410
Water boatmen are aquatic insects that collect algae with their short, hairy front legs and swim with their back legs. They are an important food source for many fish, forming the second link in the food chain; the fish are the third link, the herons the fourth, and so on.

3-1 Natural Changes, page 47

Mount St. Helens; volcanic devastation 1025
Many tens of thousands of hectares of forest land were destroyed or damaged by the eruption of Mount St. Helens on May 18, 1980.

Mount St. Helens; plant succession 705–708 (step ×3)
A few months after the Mount St. Helens explosion, surviving plants and pioneer species began to thrive. (step ×3) This process, called *plant succession*, allows one type of plant to replace another type of plant until, eventually, the forest regrows.

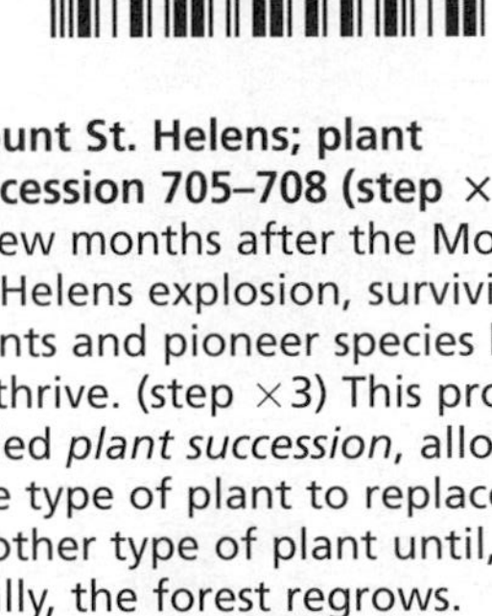

Grass, pampas 2968
Pampas grass was introduced to California as an ornamental plant, but its windborne seeds spread over the state and became a threat to native plants. The plant has a fountainlike base of razor-sharp leaves and central plumes of flowers.

Starling, common 3864
Starlings were introduced into the United States in the nineteenth century by a man who was a fan of Shakespeare. He imported all of the English birds, including starlings and house sparrows, that are mentioned in Shakespeare's plays. Starlings and house sparrows have since spread over the entire country, crowding out local birds.

3-2 Humans: What Is Our Role? page 56

DDT facts 2660–2662 (step ×2)
Facts on the use of DDT

Mining, open-pit copper 1122
Copper is mined layer by layer, terrace by terrace, deep into the ground.

Pollution, air 1458
Cars on the freeway are one source of air pollution.

Waste, hazardous 2604
Corroding metal drums of chemicals in a dumping area

Deforestation 24472–24950 (Side B only) (play ×2)
It is estimated that most tropical forests will be gone by the year 2000. This will result in the extinction of millions of species and could contribute to global warming.

Ozone plot 1492–1493 (step)
This concentration plot shows the ozone layer over the South Pole. The blue and pink areas near the center indicate lower ozone concentrations. (step) Upper left, 8/1/87; lower left, 9/1/87; upper right, 8/25/88; lower right, 9/8/88

◀❙ Step Reverse  Play ▶ Pause ❙❙ Step Forward ❙▶

Unit 1 INTERACTIONS

UNIT FOCUS

Ask students to identify the following statements as true or false:

- The nonliving things in the environment have no effect on living things. *(False)*
- Pollution is such a large problem that nothing can be done about it. *(False)*
- Most of the energy that humans use to live came originally from the sun. *(True)*
- Once an area has been destroyed by a fire, nothing can grow there again. *(False)*

Tally student responses on the board and use their answers to promote discussion.

A good motivating activity is to let students listen to the English/Spanish Audiocassettes as an introduction to the unit. Also, begin the unit by giving Spanish-speaking students a copy of the Spanish Glossary from the Unit 1 Teaching Resources booklet.

Unit 1

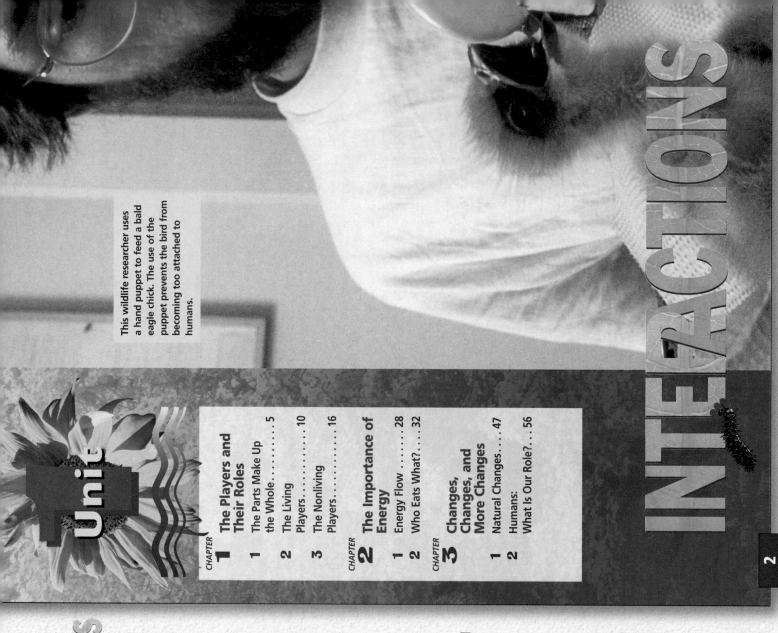

This wildlife researcher uses a hand puppet to feed a bald eagle chick. The use of the puppet prevents the bird from becoming too attached to humans.

INTERACTIONS

Connecting to Other Units

This table will help you integrate topics covered in this unit with topics covered in other units.

Unit	Connection
Unit 2 Diversity of Living Things	The process of natural selection is mediated through many biotic and abiotic interactions.
Unit 3 Solutions	Acid rain and salt water are just two solutions that play an important role in many communities.
Unit 5 Structures and Design	Carefully planned building designs can minimize negative human impacts on the environment.

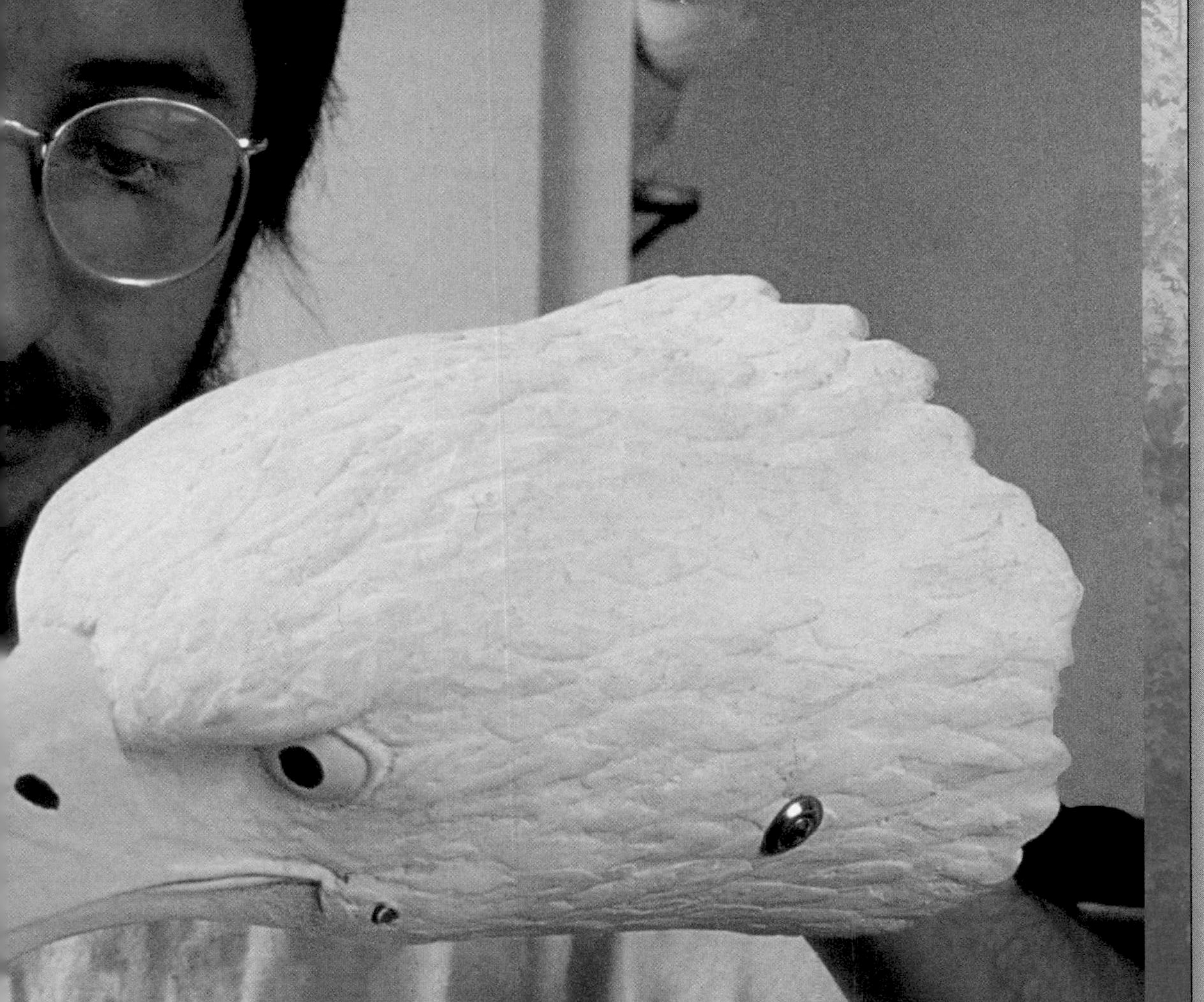

A puppet show for a baby bird? Not exactly, but roles are being played here. This 1-day-old chick has a big part in a species recovery program for bald eagles. In such programs, eagle chicks are raised in captivity until they are old enough to be released into the wild. This allows researchers to study the birds while ensuring that they reach maturity safely, away from predators and other dangers.

As many as 50,000 bald eagles once nested in the nation's wetlands. By 1973, water pollution, loss of habitat, and other threats reduced the number to about 1600. The bald eagle soared to the top of the endangered species list. But thanks to recovery programs and other factors, the number has risen in recent years. In fact, the status of the bald eagle was upgraded from endangered to threatened in June 1994. Scientists remain hopeful that the population will continue to grow.

In this story of survival, one player (the eagle) interacts with another (the human). Yet the story is more complicated. What other players have roles? In this unit, you will explore the complex world of interactions around you—you'll even examine your own role. Ⓐ

3

Connecting to Other Units, *continued*

Unit 6 The Restless Earth	Fossils help us understand how ancient life-forms interacted with each other and the environment.
Unit 7 Toward the Stars	The sun is the ultimate source of most energy that is transferred through the Earth's food chains.
Unit 8 Growing Plants	The interactions of plants with animals and the environment are defining factors of ecosystems.

Using the Photograph

Ask students to formulate an opinion about the photograph. They should consider what natural processes are at work in this example. Ask: What should be the extent of humanity's role in the environment? Is the activity of the man in the photograph destructive or helpful to the environment, or neither? Is the relationship between the man and the bird mutually beneficial, or is one of them gaining more from the interaction than the other? *(Accept all reasonable responses. Students may say that the man is helping both the bird and his own environment by restoring the population of bald eagles. The hatchling is also benefiting by being fed and cared for. This is an example of mutualism.)*

Theme Connection

Changes Over Time

You might want to bring a historical atlas to class and discuss some of the large-scale historical changes that have taken place around the world. Some examples are the human population explosion, the growing utilization of land for agriculture and urbanization, the destruction of the rain forest and other topographic changes, seasonal climatic changes, the migrations of animals, and the variability of many animal populations.

Answer to *In-Text Question*

Ⓐ The food the eagle eats, the plants with which it makes its nest, and even the air it breathes also have roles.

The Players and Their Roles

Connecting to Other Chapters

Chapter 1
introduces students to ecological interactions and the characteristics of these interactions.

Chapter 2
explores food chains, food webs, and the flow of energy within natural communities.

Chapter 3
examines the types of changes that occur in nature, including the role of humans in these changes.

CHAPTER 1

The Players and Their Roles

1 What are the important "players" in this environment? Are they all alive? Can they all be seen?

2 What are some roles these players play?

3 What are some of the interactions that can occur between the players?

ScienceLog

Think about these questions for a moment, and answer them in your ScienceLog. When you've finished this chapter, you'll have the opportunity to revise your answers based on what you've learned.

4

Prior Knowledge and Misconceptions

Your students' responses to the ScienceLog questions on this page will reveal the kind of information—and misinformation—they bring to this chapter. Use what you find out about your students' knowledge to choose which chapter concepts and activities to emphasize in your teaching.

In addition to having students answer the ScienceLog questions on this page, you might want to involve the class in the following discussion: Tell them that they are a group of investors interested in starting a shoe business. Ask: What must you do to prepare for opening day? Suggest that their list should include such activities as selecting a site for the business, raising money, advertising, learning about the competitors, appointing managers and other supervisory personnel, and hiring employees for tasks such as stocking the shelves and operating the cash register. Have a volunteer write the students' suggestions on the board.

Then point out to the class that establishing and running a business requires many different people and objects and that the existence of each of these elements influences the success of the entire business. Also point out that a business must adapt to change in order to be successful. Then ask: What types of roles do you think plants and animals have in nature? How do they depend on each other? How is the functioning of a community in nature similar to the functioning of a business? Emphasize that there are no right or wrong answers. Use the discussion to find out what students know about the roles of living and nonliving things in nature, what misconceptions they may have, and what interests them about these roles.

The Parts Make Up the Whole

Your **environment** is everything that surrounds you—trees, grass, soil, buildings, air, water, roads, animals, insects, and even other people. How important do you think the environment is to your life? Could you live without it? Are there some parts you could live without? Are different parts of your environment related to other parts? Give all of these questions some thought. The game on pages 6 and 7 will help you find some answers.

LESSON 1 ORGANIZER

Time Required
2 class periods

Process Skills
observing, predicting, classifying

New Terms
Abiotic factors—the parts of the environment that are nonliving and were never alive
Biotic factors—the parts of the environment that are living or were once alive
Environment—the physical surroundings of an organism, including all of the conditions and circumstances that affect its deve opment
Interaction—a relationship between parts of the environment

Materials (per student group)
Exploration 1: large ball of string or yarn; 1 index card per student; 1 small piece of tape per student

Teaching Resources
Exploration Worksheet, p. 3
SourceBook, p. S2

The Parts Make Up the Whole

FOCUS

Getting Started

Ask students to look around them and describe their environment. Encourage them to identify as many factors as they can. Then ask students what roles they play in their environment, and record their responses on the chalkboard. *(Student, athlete, cousin, brother, sister, friend, consumer, and so on)* When students have exhausted their ideas, point out that everyone plays more than one role in the environment.

Main Ideas

1. Every part of the environment interacts with other parts of the environment.
2. Organisms that are alive or were once alive are the biotic factors of an environment.
3. Those things that are not and never were alive are the abiotic factors of an environment.

TEACHING STRATEGIES

Invite students to identify factors in the environment depicted on page 5. Write their responses on the board. Then call on volunteers to draw some connective lines to show relationships that exist among factors in the environment.

INDEPENDENT PRACTICE Ask students to take out their ScienceLog and to write short responses to the questions posed in the introduction. Students should provide at least one example in support of each of their answers. Invite volunteers to share some of their responses.

Exploration 1

Explain to students that they will gain new insights by doing this "real-life simulation." Direct their attention to the photograph on page 6, and encourage them to predict some of the connections that will be made. Then have them read page 7 to see if their predictions were correct.

If possible, perform the activity outdoors where students have room to spread out. Before the game begins, have each student select one part of the environment that he or she will represent. To ensure that a variety of environmental factors are selected, list some factors on the chalkboard for students to choose from. Ask students to identify the part of the environment they represent by writing it on an index card. The index cards should be taped to their clothes, as shown in the photograph.

Divide the class into groups of 10 to 15 students. Be sure to have several skeins or balls of string or yarn ready for the activity.

You may wish to take a few minutes with one of the groups to demonstrate how the game should be played. Then have students begin. Offer helpful hints to keep the game moving.

GUIDED PRACTICE After each group has had time to form its "web," have one student let go of his or her string. Encourage students to describe what happens when this part of the environment is disconnected. *(Students should recognize that when one part of the environment is disconnected, it affects or even destroys other parts of the environment.)*

After students have had time to play the game, have them return to the classroom to discuss the questions on pages 6–7 that refer to the illustration.

★ **An Exploration Worksheet is available to accompany Exploration 1 (Teaching Resources, page 3).**

Exploration 1

Relationships in Nature

You Will Need

- one index card per student
- a ball of string or yarn
- a pen or marker
- tape

What to Do

Choose from the environment something that you'd like to be. You might be the wind, the sun, an ant, a daisy, the soil, or any one of many other things. Each person should play a different role. Use an index card to label what your role is.

One person should hold the end of the string and pass the ball of string to another student in the circle with whom she or he can be "related." The first student will then explain to the whole group what this relationship is.

For example, the "daisy" student holds the end of the string. He or she passes the ball to the "water" student and says, "I need water to grow." The "water" student takes hold of the string, passes the ball to the "fish" student, and says, "I am your home." This activity continues one move at a time, showing relationships in the circle. You may end up holding several portions of the string that are connected to many different things.

What would happen if one part of the environment were removed? Test your prediction. As a group, pick one person to let go of the string. Which part(s) of the environment would you *not* want to release the string? Which parts seem to be the most important for maintaining the relationships in the circle? Look at the connections to air and to water. Why are there so many? **A**

The illustration at right shows the connections a group of students made in playing the game. What does the circle look like? Does any living thing in the circle exist alone? Would removing mosquitoes from the circle make any difference? How? **B**

6

Cross-Disciplinary Focus

Social Studies

Human society requires individuals to play important roles just as nature does. To keep societies healthy, individuals choose to play a variety of supportive roles. You might wish to have students draw a large circle on a piece of paper and tell them to place the names of important people in their lives around the circle. Students should then draw lines among the people and describe how their interactions form a supportive network. After students have finished the assignment, you might wish to compare the results. Do students of different backgrounds emphasize different roles? How do ties among family members compare with those among friends?

Why does the string connect the earthworm and the soil? The answer is quite simple: the earthworm lives in the soil and gets its food there. The earthworm *interacts* with the soil. In the game, wherever one part of the environment is connected by the string to another part, there is an interaction. What interaction takes place between the earthworm and the bird? between the air and the soil? **C**

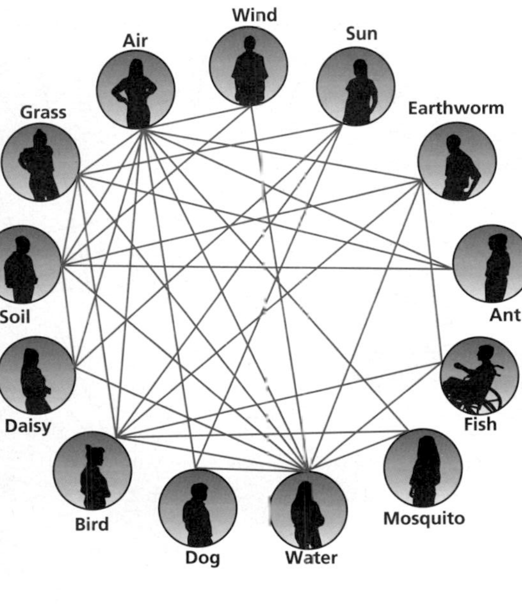

Think About It!

1. What are some of the different ways the connected pairs interact? In other words, describe the various ways in which one member of a pair depends on the other member.
2. Are all of the possible interactions shown in the diagram? If not, describe what other interactions could take place.

Answers to *In-Text Questions, pages 6–7*

A If one part of the environment were removed, many other parts of the environment would be affected. Most students will probably agree that the students representing air and water should not release the string because these are extremely important parts of the environment. Many living things require air and water; therefore, many connections show these interactions.

B The circle may look like a spider's web. No living thing in the circle should exist alone. Everything is connected directly or indirectly to everything else. If mosquitoes were removed from the circle, some birds would lose a source of food. There would be one less type of organism dependent on water and air.

C The bird uses the earthworm for food. Air is an important part of soil. It provides oxygen and other gases to plants and animals that live in the soil.

Answers to *Think About It!*

1. Answers will vary. Interactions can include the use of something as a food source, a place to live, a provider of nutrients, and so on.
2. Not all possible interactions are shown in the diagram. For example, the earthworm could be connected to the daisy because it helps to supply the roots of the daisy with food and air.

ENVIRONMENTAL FOCUS

Some experts have estimated that up to three species of animals are becoming extinct every day, and the rate is increasing. By the year 2000, up to 20 percent of all known species may have disappeared. Ask: How would the world be different without these animals? *(Delicate systems of interactions among animals would probably be badly damaged all over the world.)*

Biotic and Abiotic

This activity provides an opportunity for students to summarize the concept of environmental interaction. Have students read the page silently, and give them time to think about the questions. Then involve them in a discussion of their responses.

For the exercise at the top of page 9, you may wish to divide the class into small groups and ask them to make a list of the biotic and abiotic factors in their schoolyard. Then ask them to draw a diagram in their ScienceLog showing the various factors arranged in a circle. Students should draw lines among the factors that are connected. When this activity is completed, have the groups share and discuss their ideas as a class. This will provide students with the additional experience they need to add further connections to the diagram produced by Ms. Wilkie's class.

Possible definitions for *biotic* and *abiotic* include the following: *Biotic* means living or once living. *Abiotic* refers to things that are not and never were alive.

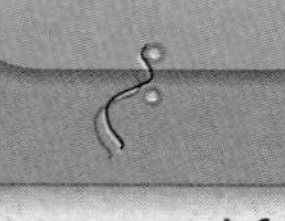

PORTFOLIO

Suggest that students include their list of environmental factors in the schoolyard in their Portfolio. Students may wish to illustrate their lists.

Homework

As a homework assignment, have students select pictures from old magazines to create an interactions diagram like those shown on pages 7–8.

Biotic and Abiotic

Ms. Wilkie's class played the interaction string game. Look at the parts they chose.

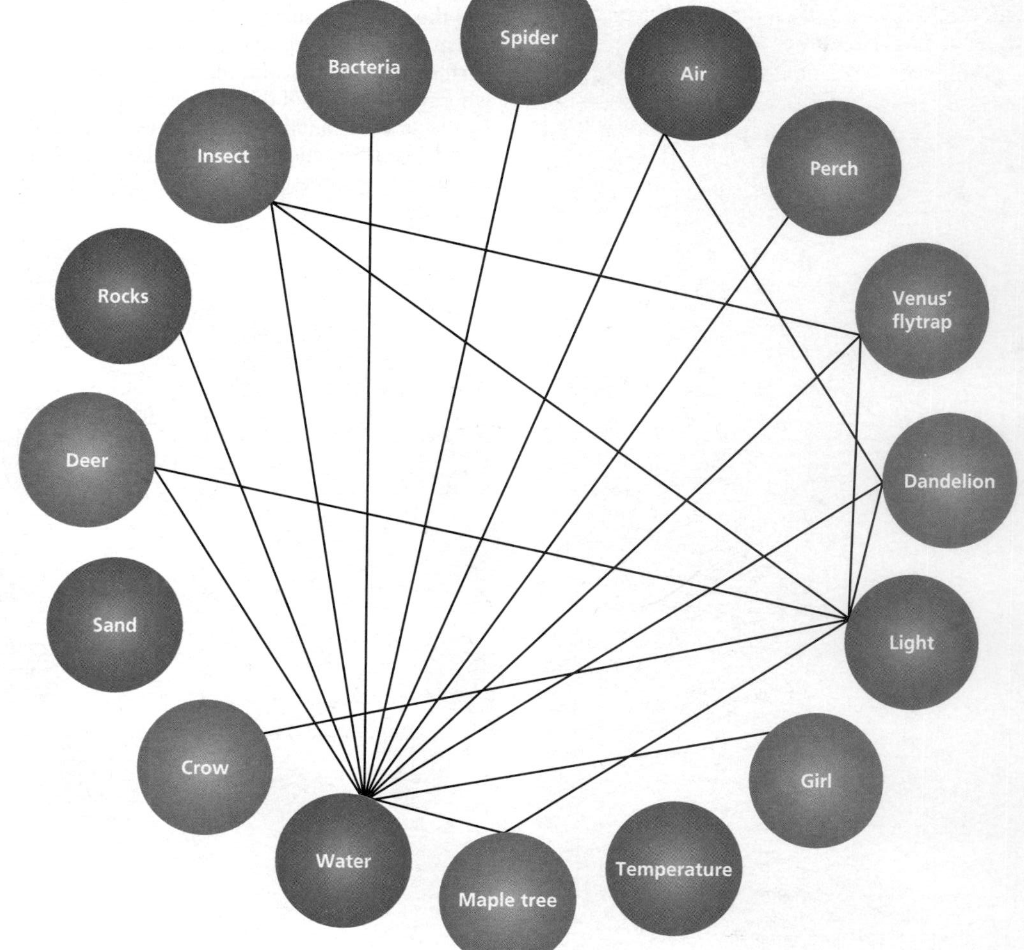

What do the green items have in common? the brown items? The green parts are considered to be **biotic,** and the brown parts are considered to be **abiotic.** Based on the common characteristics of the green items, what do you think *biotic* means? And based on the common characteristics of the brown items, what do you think *abiotic* means?

If you said that the biotic parts of the environment are living things, you are right. But here are a few more examples of biotic parts of the environment:

- a rotting log
- a dead animal
- a bone

Together with a classmate, devise new definitions of *biotic* and *abiotic*.

Look around your schoolyard. Which parts of the schoolyard environment are biotic? Which parts are abiotic? Arrange the various parts in a circle, and connect the ones that you think interact. Draw lines only for those interactions that you are sure of. (A)

Look back at the diagram of the string game played by Ms. Wilkie's class. Is the diagram complete? What parts of that environment (both biotic and abiotic) are not connected but could be? Give a reason for each new connection you propose. (B)

Some Detective Work

Take a close look at the illustration on page 4. How many interactions can you find? Classify each interaction under one of the following headings:

- Animal—Animal
- Animal—Plant
- Plant—Plant
- Living—Nonliving
- Living—Once Living

What abiotic parts of the environment did you include in your lists?

Which parts of this illustration are abiotic? Which parts are biotic? (C)

What do you call the environment in which interactions occur between abiotic parts and biotic parts? Find out on page S2 in the SourceBook.

Answers to *In-Text Questions and Caption*

(A) The biotic parts of the schoolyard could include grass, trees, ants, grasshoppers, students, and birds. The abiotic parts could include rocks, water, and air. Sample connection: *Students* use the *grass* as a place to run and play.

(B) Sample connection: The *girl* and the *air* could be connected because the girl (biotic) breathes the air (abiotic).

(C) The waves and rock formation represent abiotic parts, and the women, bird, plant, and insect represent biotic parts.

Answers to *Some Detective Work*

Possible responses include the following:

- Animal—Animal: mountain lion with deer, hawk with field mouse
- Animal—Plant: mountain lion with tree, deer with grass
- Plant—Plant: moss with tree, grass with tree
- Living—Nonliving: tree with water, deer with water, deer with land, flower with sunlight
- Living—Once Living: mushrooms with decaying plants, bacteria with dead leaves

In this list, water, land, and sunlight are abiotic parts of the environment. (You may wish to allow students to include mushrooms and other fungi as "plantlike" living things.)

FOLLOW-UP

Reteaching

Have students select pictures from old magazines and newspapers to make a poster that illustrates environmental interactions. Have students write a caption describing each one. Display the posters around the classroom.

Assessment

Display pictures of different environments, and have students write brief descriptions of the interactions taking place in each one. Ask students to classify their interactions as animal—animal, animal—plant, plant—plant, living—nonliving, or living—once living.

Extension

Suggest that students do research about ecosystems. Have them make a drawing or diagram of one and identify some of the interactions that take place there. Possible ecosystems include a desert, pond, or rain forest.

Closure

Some students may enjoy doing field research. Suggest that they select one area to observe for an hour or two. The area might be their backyard, a park, or the schoolyard. They should keep track of all the interactions that they observe or that they can infer. Have them share their observations with the class by presenting an oral summary.

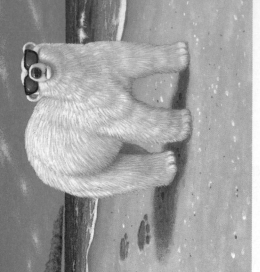

The Living Players

FOCUS

Getting Started

Ask students: Which environment has a greater number of living things: a tropical rain forest or a forest in a cooler area? *(A tropical rain forest)* Students may be interested to know that even though tropical rain forests occupy only 2 percent of the Earth's surface, at least 50 percent of the Earth's species reside there.

Main Ideas

1. An organism's habitat is the place where it lives.
2. An organism's niche is its way of life in its habitat, including its biotic and abiotic relationships.
3. Commensalism, mutualism, and parasitism are three kinds of biotic relationships.

TEACHING STRATEGIES

Communities of Living Things

GUIDED PRACTICE Ask students to think about the kinds of relationships they have with different people. Invite them to suggest words that describe these relationships, such as friendly, supportive, unpleasant, helpful, or useful. Explain that just as people have different kinds of relationships with each other, organisms in the natural world also interact with each other in many different ways. These interactions may either help or harm individual organisms.

Have students read page 10 silently. Involve them in a discussion of the difference between niche and habitat.

Communities of Living Things continued ►

Homework

Have students write five sentences using either *habitat* or *niche* at least once in each sentence. *(Student sentences should demonstrate an understanding of the meanings of these terms.)*

LESSON 2

The Living Players

Communities of Living Things

You would probably be surprised to see carrots attacking a rabbit or a mouse chasing a cat. It would also seem a bit odd to find orchids growing in the Arctic or a polar bear roaming the beaches of Florida!

Your experiences have shown you that every organism has its own way of living. Each exists in the environment that best meets its needs. Also, every living thing relates to other organisms in specific ways. The relationships between an organism and its physical surroundings and neighbors make up an organism's **niche.** Therefore, the niche is not the place where an organism lives, but is the organism's entire way of life—the organism's *role* in the community where it lives. The *place* where an organism lives is called its **habitat.**

For instance, your habitat might be San Antonio, Sacramento, Portland—any city or town. Your niche (right now) is student, son or daughter, nephew or niece, member of the school choir or basketball team, and so on.

Being a student in school, for example, reveals a great deal about your way of life, your environment, and those with whom you interact.

Do you understand the difference between habitat and niche? In your ScienceLog, complete this definition: A ___?___ is the way of life that an organism adopts to survive in a particular ___?___. (A)

Does your habitat look like this? (B)

Is your niche similar to this? (C)

LESSON 2 ORGANIZER

Time Required
2 to 3 class periods

Process Skills
observing, analyzing, inferring, classifying

New Terms

Commensalism—an association between two organisms in which one benefits and the other is unaffected by the relationship

Habitat—the place where an organism lives

Mutualism—an association between two organisms in which both benefit

Niche—an organism's way of life, including its relationship with other organisms and its physical surroundings

Parasitism—an association between two organisms in which one (the parasite) benefits and one (the host) is harmed

Materials (per student group)
none

Teaching Resources
Resource Worksheet, p. 6
SourceBook, p. S5

Copy this table into your ScienceLog. Complete the table by thoroughly describing the niches of the organisms shown on this page. **D**

Organism	Habitat	Description of niche	
		How it uses other things (living and nonliving)	How it is used by other things in its environment
robin	Open areas, lawns, bushes, trees, air, or ground	Uses bush or tree for a nesting site or for protection. Uses twigs and grass for nest. Eats worms and insects for food.	Eaten by larger birds, snakes, or cats. Eggs eaten by other birds. Parasites—mites and lice—live in its feathers.
turtle			
spruce tree			
beetle			

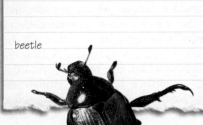

Communities of Living Things, *continued*

GUIDED PRACTICE Direct students' attention to the table on page 11. Review the sample entry. After students complete the table, involve them in a discussion of their ideas. You may wish to give students time to do research in order to fill out the table more thoroughly.

PORTFOLIO Suggest that students include their table from page 11 in their Portfolio. They may wish to illustrate their table by drawing a picture of each organism named and then creating an interactions diagram with the drawings.

★ **A Resource Worksheet is available to accompany the material on this page (Teaching Resources, page 6).**

Answers to *In-Text Questions and Caption, pages 10–11*

A A *niche* is the way of life that an organism adopts to survive in a particular *habitat*.

B Students should understand that a habitat is where you live; if they live in an urban area, their habitat may look like the one in this photograph.

C Since your niche consists of all of your biotic and abiotic interactions, most students should agree that their niche is similar to this one.

D Answers to the table may vary; possible responses include the following:

The turtle lives in and around water areas such as streams, ponds, lakes, swamps, and rivers. It eats insects, minnows, and tadpoles, and it suns itself on rocks and floating logs. It may be food for animals such as raccoons, bears, alligators, and otters, and its eggs are eaten by birds.

The spruce tree grows in northern, temperate climates with cool summers and cold winters. It gets nutrients from the soil and uses the sun's energy to make food. Birds and squirrels build nests in spruce trees and may use the seeds for food. People use spruce trees for lumber.

Beetles live on plants and under rocks. They use plants and other insects for food, and they bury their eggs in the soil or inside dying or dead trees. Some mammals, snakes, and birds eat beetles, and some plants use beetles to disperse pollen.

"One-Way" Benefits

Direct students' attention to the photograph of the robins, and have them identify what it shows. Encourage them to discuss the relationship between the robins and the tree. *(Accept all reasonable comments.)* Have students read the section silently, or call on volunteers to read it aloud. Involve students in a discussion of the example of the barnacles and the horseshoe crab.

"Two-Way" Benefits

Call on a volunteer to read the first two paragraphs aloud. This relationship is different from commensalism because both organisms benefit.

GUIDED PRACTICE Encourage students to articulate the difference between commensalism and mutualism. Have them read the example about the lichens and discuss how it demonstrates mutualism. You may wish to review the terms *fungus, algae,* and *photosynthesis* after students have read about them in the textbook.

Answers to *In-Text Questions*

A Other ways that the robin benefits from its relationship with the tree include the following: The tree also supports food for the robin, such as caterpillars and insects. The tree also provides the robin with shelter from predators.

B By riding on the back of crabs, barnacles avoid overcrowding and increase their food resources. Students should recognize that this is an example of commensalism because one organism, the barnacle, benefits from the relationship, while the other organism, the horseshoe crab, neither benefits nor is harmed. Other examples of commensalism include remoras riding on sharks and Spanish moss growing on trees.

C The bee is helped by the flower because the flower provides it with food (nectar). The plant (flower) is helped by the bee because the bee distributes the flower's pollen.

D The lichens' names probably came from their appearance.

Types of Interactions in Communities

"One-Way" Benefits

Is the scene in the photo at right familiar to you? Robins like these might even live in your backyard, where trees provide them with a convenient place to build their nests. Let's analyze the relationship between a robin and a tree.

In this situation, two different kinds of organisms live close together. One organism—the robin—benefits from the relationship. The other organism—the tree—is neither helped nor harmed by the nest. This relationship is called **commensalism.** How else does the robin benefit from this relationship? **A**

Now let's look at another example of commensalism. Adult barnacles cannot move from place to place on their own. But they often attach themselves to the shell of a horseshoe crab and travel with the crab. The crab is unaffected by the presence of the barnacles getting the free ride. How do the barnacles benefit from traveling with the crab? Why is this an example of commensalism? Think of other examples of commensalism. **B**

"Two-Way" Benefits

The bee and the flowering plant provide an example of another kind of relationship. How does this relationship differ from commensalism? Think about these facts:

- The bee gets nectar (a liquid food that it uses to make honey) from the flower.
- While the bee is feeding from the flower, some of the flower's pollen sticks to the bee's body.
- When the bee flies to another flower, it deposits some of the pollen on that flower. *Pollination* is part of the process by which a flowering plant produces seeds.

How is the bee helped in this relationship? How is the plant helped? A relationship in which both organisms benefit is called **mutualism.** **C**

Now consider *lichens,* which are plantlike organisms that can be found in almost every part of the world. Lichens can be seen on rocks and on the bark of trees. One kind, called old man's beard, hangs from tree branches. Another kind, called matchstick lichens or British soldiers, can be found on the ground in the shape of little red-tipped, vertical sticks or stems. Look at the photographs of lichens shown here. How do you think these organisms got their names? **D**

Matchstick lichens

Old man's beard

Look at the other examples of lichens shown below. Does it look like there is more than one type of organism in each picture? It might surprise you to find out that a lichen is actually two kinds of organisms that live together for mutual benefit. You would have to use a microscope to see both organisms.

Golden lichens

Reindeer-moss lichens

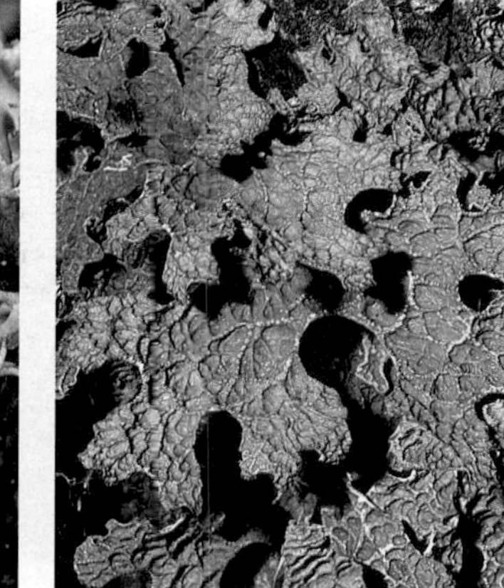
Foliose lichens

One organism in a lichen is a *fungus* (plural, fungi), which has many threadlike branches. These branches provide both protection and moisture for green, single-celled organisms called *algae* (singular, alga). The algae make food by a process called *photosynthesis*. The fungi in lichen do not make their own food. Instead, they use the food made by the algae.

Name as many other examples of mutualism as you can. (E)

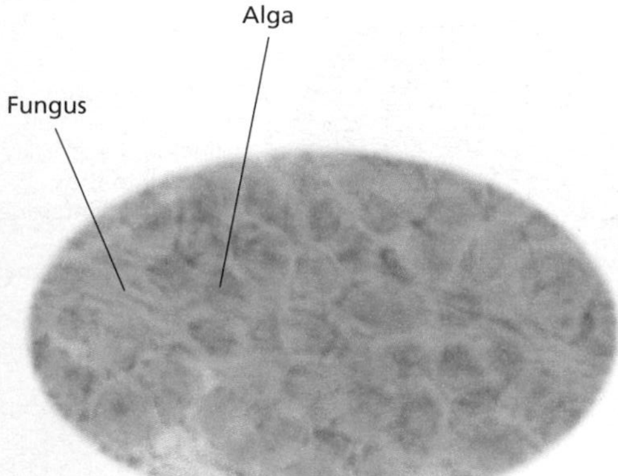

This microscopic view of a lichen (400×) shows clumps of single-celled algae among strands of fungus.

Benefit and Harm

Compare commensalism and mutualism with a relationship called **parasitism.** In parasitism, one organism—a *parasite*—lives in or on another organism—a *host*. The parasite obtains its food from the host. In this relationship, the host is harmed instead of helped. Sometimes the host is even killed by the parasite.

Find out more about the following pairs of living things that are sometimes linked together through parasitism:

- people and tapeworms
- trichinae (worms) and pigs
- rust fungus and wheat
- spruce trees and spruce budworms

Which organisms are the hosts? Which are the parasites? (F)

Benefit and Harm

Have students read the material silently, or call on a volunteer to read it aloud. Students should discuss the differences between commensalism, mutualism, and parasitism. Ask: In which of these biotic relationships is there a threat to one of the organisms involved? *(Parasitism)*

Direct students' attention to the list of parasitic relationships. You may wish to have students make predictions about which of the organisms is the parasite and which is the host. Then have them research to find out if their predictions were correct.

Answers to *In-Text Questions*

(E) Student examples of mutualism may include butterflies drinking nectar from flowers while dispersing the flowers' pollen and seed dispersal by fruit-eating animals.

(F) Tapeworms are parasites, and people are hosts; trichina worms are parasites, and pigs are hosts; rust fungus is the parasite, and wheat is the host; and spruce budworms are parasites, and spruce trees are hosts.

Multicultural Extension

Lichens

Share the following information with your students: Lichens have been used in a variety of cultures in a variety of products. Iceland moss, a lichen that grows in alpine areas of the Northern Hemisphere, has been used medicinally by Europeans to relieve chest ailments, to suppress appetite, and to make soap. Other lichens have been used for food, dyes, and even perfumes.

Problems to Consider

Read the two problems presented on page 14 to the class. Students could either call out answers as a part of class discussion, or they could record their responses in their ScienceLog. Either way, have students give reasons for their choices.

Provide students with time to complete their diagrams. Then call on several volunteers to share their diagrams and describe them to the class. Encourage students to research specific examples to illustrate their diagrams. To extend the exercise, have students use their diagrams to make bulletin-board displays, or have groups of students use their ideas to create murals. Allow students time to review all of the examples and record their ideas in their ScienceLog. As a class, have them share and discuss their responses.

Answers to *Problems to Consider*

1. Students' diagrams should include reasonable examples of each relationship.
2. Students' responses should be similar to the following:
   - **a.** Mutualism—The cow's digestion is aided by the bacteria. The bacteria benefit from the cow by having a place to live and food to eat.
   - **b.** Parasitism—The ichneumon larvae benefit from other insects' eggs or larvae. The host eggs or larvae are destroyed.
   - **c.** Commensalism—The burdock plant benefits from animals. The animals neither benefit from nor are harmed by the burdock plant.
   - **d.** Mutualism—The protozoa help the termites digest wood. The termites provide the protozoa with food and a place to live.
   - **e.** Parasitism—The sheep give the threadworms a place to live and food to eat. The sheep are harmed by the threadworms.

Answers to Problems to Consider continued ▶

Problems to Consider

1. Make three diagrams that will help you remember these three relationships: commensalism, mutualism, and parasitism.
2. Examine each of the pictures on these two pages, and read the accompanying facts. Then identify the type of relationship in each case.

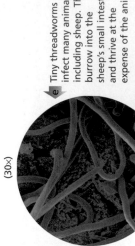

a The stomach of a cow is home to a type of bacteria that digests the grass eaten by the cows.

b Ichneumons (pronounced ik NEW muns), insects that resemble wasps, lay their eggs in or on the eggs or larvae of other insects. Once hatched, the ichneumon larvae feed on the other insects' eggs or larvae.

c Animals can pick up burdock seeds (burs) when walking through the brush. The seeds are often deposited in other areas.

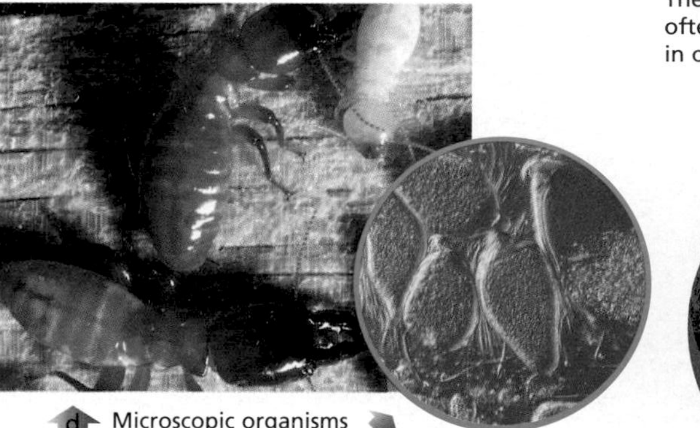

(65×)

d Microscopic organisms called protozoa live inside termites and digest wood for the termite and for themselves.

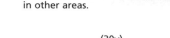

(30×)

e Tiny threadworms infect many animals, including sheep. They burrow into the sheep's small intestine and thrive at the expense of the animal.

Homework

You may wish to assign the first problem under Problems to Consider as an at-home activity and then use the second problem to assess the students' understanding the next day.

Did You Know. . .

Cattle and cattle egrets are often seen together in pastures. What's going on here? Commensalism! The egrets get a free ride from the cattle and take advantage of the grasshoppers, crickets, and other insects stirred up by the browsing cows. The cattle are neither helped nor harmed by this interaction.

f Dodder, a plant with no leaves, winds around wheat and other plants for support. Dodder then takes food from the other plant through specialized roots.

h A remora attaches itself to a shark by a sucker on the top of its head. It gets a free ride and shares some of the scraps of food left over from the shark's meal.

i Certain ants actually "herd" aphids. The ants move the aphids from place to place and protect them from predators. The ants collect the sweet liquid (honeydew) produced by the aphids.

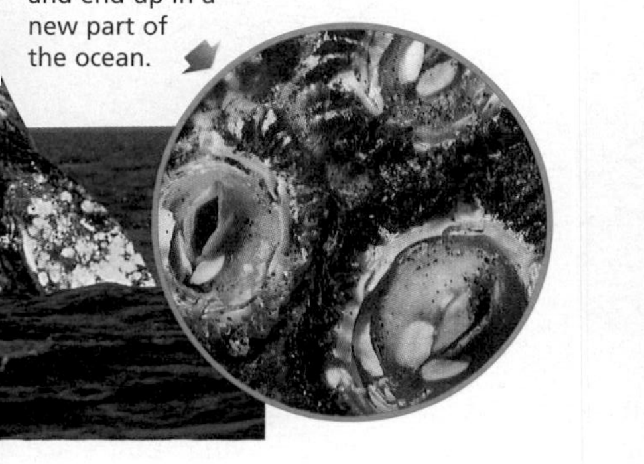

g Barnacles often attach to whales and end up in a new part of the ocean.

k Orchids grow on other plants for mechanical support only.

j Mistletoe grows on trees such as oak, obtaining both support and food while weakening the tree.

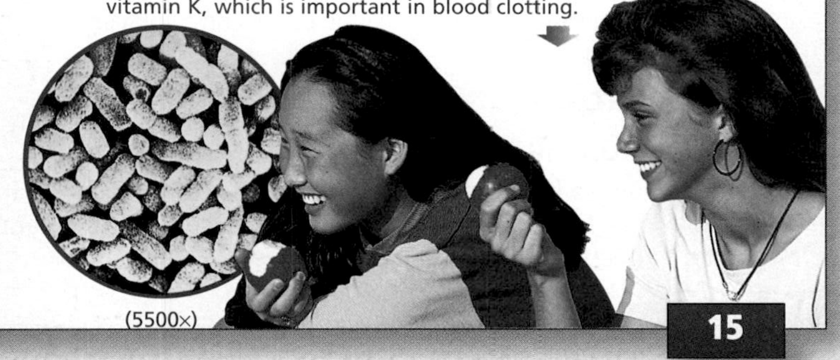

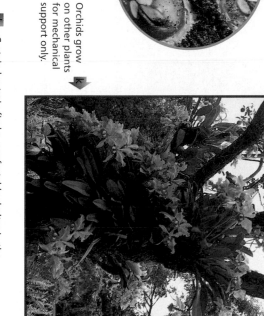

l Certain bacteria find a comfortable shelter in the human digestive tract. The bacteria help remove water from indigestible material. They also help produce vitamin K, which is important in blood clotting.

(5500×)

15

Answers to *Problems to Consider, continued*

f. Parasitism—The dodder plant benefits from the wheat. The wheat is destroyed by the dodder plant.

g. Commensalism—The barnacles use the whales to move from place to place. The whales neither benefit from nor are harmed by the barnacles.

h. Commensalism—The remora benefits from the shark. The shark neither benefits from nor is harmed by the remora.

i. Mutualism—The aphids receive protection, and the ants benefit by getting food from the aphids.

j. Parasitism—The tree provides the mistletoe with support and food. The mistletoe weakens the tree.

k. Commensalism—The orchids grow on other plants for support. The other plants neither benefit from nor are harmed by the orchids.

l. Mutualism—The people give the bacteria a place to live and food to eat. The bacteria aid people's digestion and produce vitamin K.

FOLLOW-UP

Reteaching

Students may enjoy drawing cartoons to illustrate commensalism, mutualism, and parasitism. Ask them to use the animal and plant examples from the lessons in their cartoons. Use the finished cartoons to create a bulletin-board display.

Assessment

Have students make a list of several organisms that live in the schoolyard or in a neighborhood park. Then have them make a table similar to the one on page 11 using the organisms from their list. When students finish, have them share the information with their classmates.

Extension

Point out to students that the terms *community, population*, and *habitat* have a scientific meaning in relation to groups of organisms living together. Suggest that students do research to discover what these terms mean in relation to the environment. Have them present their information to the class in an oral report.

Closure

Suggest that students make dioramas to show the habitat for a particular organism, including biotic and abiotic relationships. Display the dioramas in the classroom for others to enjoy.

LESSON 3

The Nonliving Players

FOCUS

Getting Started

Ask a volunteer to explain why people sweat when they get hot. *(When a person's body gets hot, the sweat glands release sweat. When the sweat evaporates, the body cools.)* Students may be interested to know that their skin contains about 100 sweat glands per square centimeter!

Main Ideas

1. Temperature, water, and light are significant abiotic factors for living things.
2. Every plant and animal has a specific range of tolerance for the abiotic factors in its environment.
3. Animals and plants have many adaptations that enable them to withstand the extremes in temperature that result from seasonal changes.
4. Humans can survive much longer without food than without water.
5. Light affects the growth, health, color, and structure of plants.

TEACHING STRATEGIES

Have students read page 16 silently. Then involve them in a discussion of the temperature range that they find most comfortable. Help students to recognize that a comfortable temperature often depends on the type of activity they are doing.

Answers to *In-Text Questions*

Ⓐ Answers will vary. Accept all reasonable responses.

Ⓑ A temperature range between 20°C and 25°C is comfortable for most people. A temperature of around 22°C is usually considered optimum.

LESSON 3

The Nonliving Players

If you were wearing a sweatshirt in your classroom and the temperature rose above 30°C, how would you feel? How would you feel if the temperature fell below 10°C? Ⓐ

There is a range of temperatures that you can tolerate comfortably. This is called your *range of tolerance for temperature.* Predict what your range of tolerance for temperature is for the place where you are now.

Within the range of tolerance, there is a temperature that you like best—one that is most comfortable for you. That temperature is called the *optimum temperature.* What would you guess the optimum temperature is for doing science in a classroom? Ⓑ

You have a range of tolerance for other conditions too. What might some of these conditions be? You and all living things are affected by conditions in the environment. At the beginning of this unit, you identified abiotic (nonliving) parts of the environment, such as temperature, moisture, light, and wind. How do these factors affect various living things? For each of the abiotic factors, how does the range of tolerance vary for different organisms? You will investigate these questions in the next few pages as you learn more about the nonliving players.

Temperature

Temperature Tolerance in Plants

Do you think plants can tolerate a wide range of temperatures? In the following Exploration, you'll see for yourself whether temperature affects the growth of radish seeds.

16

LESSON 3 ORGANIZER

Time Required
4 class periods

Process Skills
observing, comparing, contrasting, analyzing

Theme Connection
Structures

New Terms
none

Materials (per student group)
Exploration 2: 30 radish seeds; 3 petri dishes; three 10 cm × 10 cm pieces of cardboard; about 200 mL of water; 12 paper towels; alcohol thermometer; scissors; 3 resealable plastic bags; refrigerator; radiator or other heat source
Exploration 3: about 30 radish seeds; a few small flowerpots filled with soil; 1–2 L of water; 100 mL graduated cylinder; about 30 seeds of various types

Teaching Resources
Exploration Worksheets, pp. 7 and 12
Activity Worksheets, pp. 9 and 11
Resource Worksheet, p. 13
Transparencies 1, 2, 3, and 4
SourceBook, p. S8

EXPLORATION 2

Too Hot or Too Cold?

You Will Need
- 30 radish seeds
- 3 petri dishes
- 3 pieces of cardboard
- paper towels
- a thermometer
- water
- scissors
- plastic bags

What to Do

1. Cut four pieces of paper towel into the shape of a petri dish and layer them in the bottom of the dish. Do this for the other petri dishes as well.

2. Now pour water into each petri dish until the layers of paper towels are soaked. Pour off any extra water.
3. Put 10 radish seeds on the paper towels in each dish. Spread out the seeds so that they do not touch each other.
4. Cover each petri dish with a piece of cardboard.

5. Place each petri dish in a plastic bag. Put one petri dish in a refrigerator, place another in a warm location (such as near a radiator), and place the third dish in the classroom in a safe place where it won't be disturbed.
6. Measure the temperature at each place.

7. Over several days, count the number of seeds that sprout each day in each petri dish. Add water as needed. In your ScienceLog, make a table like the one shown below and record your observations.

Day	Number of seeds sprouted		
	In refrigerator _____°C	In warm place _____°C	In classroom _____°C
1			
2			
3			
4			

Judging from your results, at what temperature do radish seeds sprout best? How would you redesign this experiment to find the range of temperature tolerance shown by radish seeds? Ⓒ

17

★ **An Exploration Worksheet (Teaching Resources, page 7) and Transparency 1 are available to accompany Exploration 2.**

EXPLORATION 2

Before students begin this Exploration, be sure that a refrigerator and a heat source are available for their use. Have students work in small groups to complete the activity. Observations should be made daily over a period of four or five days. Remind students that in order to make this a fair test, they need to keep all of the variables constant except for temperature.

Answers to *Exploration 2*

Ⓒ Students should observe that the radish seeds germinate best in a warm place and worst in a cold place. Actual temperatures may vary.

Involve students in a discussion of how they would redesign the experiment. Help them to recognize that they would have to try germinating seeds at several different temperatures over a narrower range in order to find the actual range of tolerance.

PORTFOLIO

Suggest that students include their table from Exploration 2 in their Portfolio. Students may annotate their table by giving an explanation of what the table shows.

ENVIRONMENTAL FOCUS

Plants and animals depend on one another in terms of their abiotic interactions. Most plants require carbon dioxide and emit oxygen, while animals use oxygen and expel carbon dioxide. Also, plants can regulate local temperatures by providing shade. Ask the class to suggest other ways that animals benefit from plants. *(Sample answer: Trees provide shelter for animals such as birds and squirrels. Humans use wood for fuel and to make tools and furniture.)*

Other Projects

You may wish to divide the class into groups and assign a project to each group. Students will need to keep track of daily temperatures in order to make their charts. Help students adapt their activity to the climate in their area.

Suggest that students organize their responses to the first two questions in a chart and then write a short paragraph in response to the third question. Temperature tolerance of different plants varies greatly.

Provide seed catalogs for students, or suggest that they contact a local nursery to find out the information that they need. To extend the activity, have students provide additional information about the fruit trees that will grow in their area.

You may wish to suggest some types of areas—such as forested, mountainous, semitropical, temperate, or coastal—for students to use in their comparisons. Adaptations make plants well suited to their environment.

Temperature Tolerance in Animals

Suggest that students look in almanacs or contact the National Weather Service to find out the hottest and coldest temperatures ever recorded in their area.

Help students recognize that physiological responses to temperature, such as shivering and sweating, are ways that living things adjust to changes in temperature.

You may wish to have students work on the questions on page 19 in small groups. Each group should complete all five questions.

Other Projects

To find out more about temperature tolerance in plants, select one of these research projects. Keep in mind that some of them need to be investigated at a specific time of year.

1. During the fall season, observe which plants survive the cold nights. Which plants survive frost the longest? Which plants do you think will survive the winter? Make a chart showing temperature and plant survival.
2. A spring follow-up to the fall activity: Which plants are the first to grow in the spring? Where are these plants located? What have you discovered about the temperature tolerance of different plants?
3. Compare plants that are native to your area with plants that are native to an entirely different type of habitat. For example, if you live in a desert-like environment, you might want to compare native plants with the plants in a tropical rain forest. What makes plants well suited to their habitat?
4. You are interested in growing a fruit tree in a planter or in your yard. Look in some seed catalogs to determine what kinds of fruit trees will grow successfully in your area.

Temperature Tolerance in Animals

What is the hottest temperature ever recorded for your area? the coldest temperature ever recorded? Ⓐ

How well do you think these students are tolerating these temperature extremes? Ⓑ

What, then, is the range of temperatures your body has to endure? How do you prepare for such temperature extremes? The most obvious answer is "by the clothes I wear." Can you think of some body responses that help you tolerate different temperatures? Ⓒ

How do other living things tolerate temperature extremes? To help answer this question, form into small groups of three or four students and choose at least two of the following questions or activities. You may need to look in books on the subject and ask informed people, such as zookeepers or veterinarians.

Answers to *In-Text Questions*

Ⓐ Answers will vary. Accept all reasonable responses.

Ⓑ It does not appear that these students are tolerating these temperature extremes very well. The students in the first picture are wearing too much clothing, which holds the heat in and increases body temperature. The students in the second picture are not wearing enough clothing to protect themselves from the cold.

Ⓒ The body shivers when it is cold. Shivering is a series of small muscle contractions that produce heat. The body sweats when it is hot. When the sweat evaporates, the body cools.

Homework

You may wish to assign one or more of the research projects described in Other Projects as homework.

1. Discuss and list all the possible hardships experienced by living things (both animals and plants) in the winter. Check the ones that you think are the most stressful.
2. List ways in which living things cope with the difficulties you have identified.
3. In the winter, where would you expect to find the following animals: deer, bear, frog, robin, woodpecker, skunk, snake, and squirrel? How does each one survive the winter?
4. How can snow help some plants and animals survive cold temperatures? Name some specific examples.
5. Noreen lives in Bermuda, where the temperature never drops below 15°C. Write a letter telling her about the changes you might see in the animals and plants in your area as winter approaches.

Moisture

Moisture is also a vital abiotic factor for living things. Living things have both a range of tolerance for moisture and an optimum amount of moisture needed. Compare the ranges of moisture tolerance for the two plants in each of the pairs below.

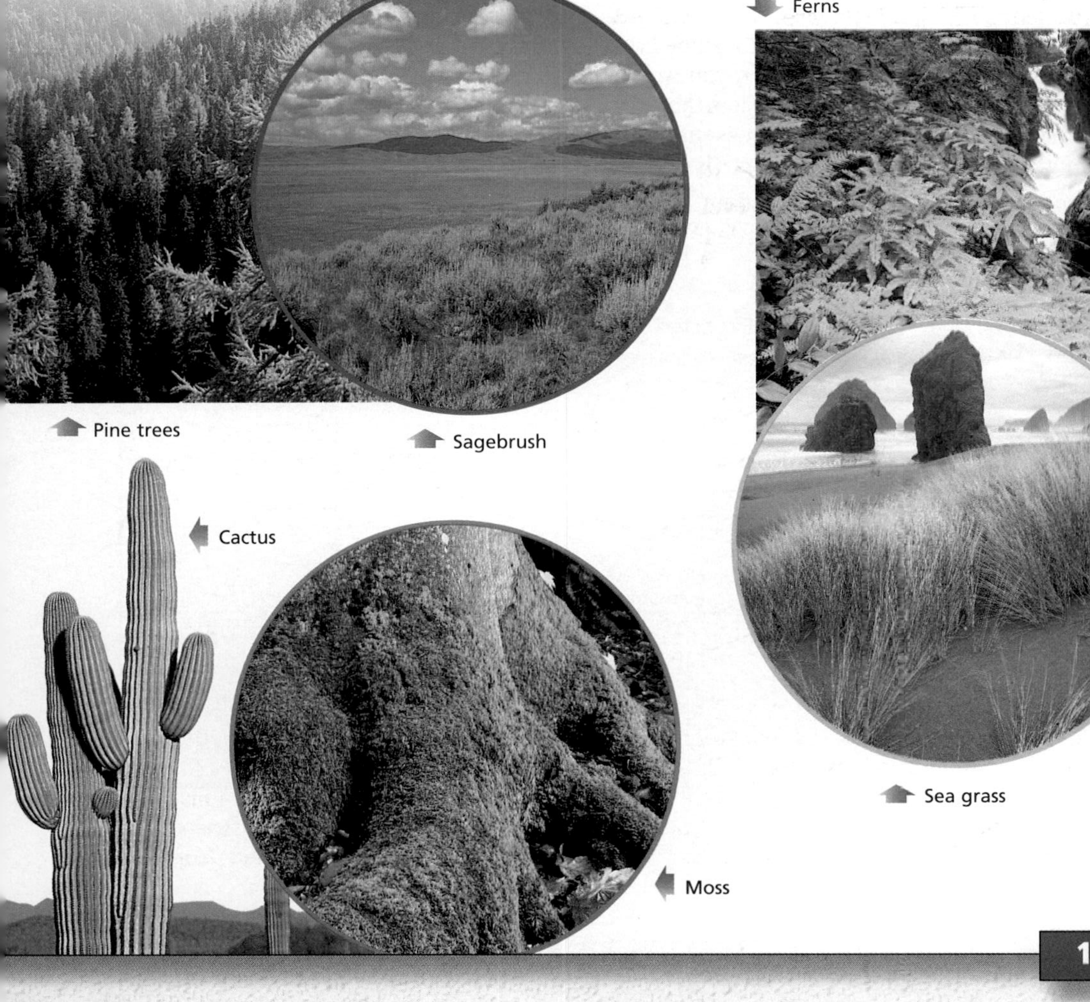

Moisture

Students will probably need to do some research to make their comparisons properly. You could assign groups of students to research one plant from each pair.

Homework

Two Activity Worksheets are available to be assigned as homework (Teaching Resources, pages 9 and 11). If you choose to use the second one in class, Transparency 2 is available to accompany it.

Answers to *Moisture*

Pine trees need a moist environment; sagebrush thrives in a dry environment.

Both sea grass and ferns need a moist environment.

Moss needs a very moist environment; cactus thrives in a very dry climate.

Answers to *In-Text Questions*

1. Answers may vary, but possible responses include finding food, keeping warm, finding shelter, and hiding from predators. Of these, finding food is probably the most stressful for animals. Protection from temperature extremes may be the most stressful for plants.
2. Answers may vary, but possible responses include the following: Animals may grow heavy coats of fur to protect themselves against cold weather; some animals add insulating layers of fat that their bodies use in the winter when food is scarce; some animals hibernate; some animals protect themselves from predators by changing color in the winter to blend in with snowy backgrounds; and some plants become dormant (alive, but not actively growing) to survive the winter.
3. Deer would be grazing in open fields and pastures. Bears would be sleeping in dens. Frogs would be hibernating in the bottoms of pools, streams, and lakes. Robins would have flown south for the winter. Woodpeckers would be looking for dormant insects in the bark of trees. Skunks would be hibernating in dens. Snakes would be hibernating underground. Squirrels would be looking for the food they stored in the fall. (Animal activities may vary depending on which part of the country is being discussed.)
4. Snow serves as insulation for many plants and animals and keeps their temperature above freezing. For example, mice live quite successfully under the snow. Many pond animals survive under the snow and ice, where the water is not frozen. Plants may be kept from freezing if they are covered with snow.
5. Students' letters will vary depending on the type of winter in their area. However, student responses should indicate an understanding of how plants and animals cope with the lower temperatures in winter.

EXPLORATION 3

This Exploration provides an excellent opportunity for students to design and perform their own experiments. This Exploration can be done at home or in the classroom. Make sure that students share their experimental designs with you before they begin. Helpful suggestions may be made at that time. If the experiment is to be done at home, write a note to the parents describing the activity so that they can cooperate and monitor the experiment for safety considerations.

Make sure that the seeds you use have not been treated in any way (such as with a fungicide or fertilizer). When all the experiments have been performed and conclusions have been drawn, have students discuss and display their findings. Students should conclude that although radish seeds may germinate with varying amounts of water, there is an optimum range of moisture for maximum germination. Also, students should conclude that different kinds of seeds have different optimum moisture levels for germination.

★ **An Exploration Worksheet is available to accompany Exploration 3 (Teaching Resources, page 12).**

Answers to *Water Needs—Some Questions to Discuss*

1. Water is essential to humans because our cells and tissues are composed largely of water. It serves as a vehicle to remove salts and wastes from the body, in the form of urine. Water is also the major component of blood.
2. **a.** With 4 L of water, a human can live about 8 days at 35°C. However, without water, a human can survive for only 5 or 6 days.
   b. With 10 L of water, a human can survive about 19 days at 25°C.
   c. The length of survival time is determined by the temperature of the environment and the amount of available water.

Answers to Water Needs—Some Questions to Discuss continued ►

EXPLORATION 3

Experimental Design for Moisture Tolerance

Here is a challenge to your experimental abilities! Plan experiments to answer the following questions:

- Is there an optimum amount of moisture needed for radish seeds to sprout?
- Do all seeds have the same optimum moisture level for sprouting?

Show your experimental design to your teacher so that he or she can make sure that you are on the right track and that your plan is safe. Then perform your experiments and find the answers to the questions.

Water Needs—Some Questions to Discuss

1. Humans can survive for several weeks without food, but not nearly so long without water. In fact, humans need almost 3 L of water every day! In a hot desert, you would need 11 L a day to replace water lost through perspiration. Why is water essential to humans?
2. Look at the graph below, and then answer the following questions. The amounts of water given are total starting quantities.
   a. How long can a human live at 35°C with 4 L of water? with no water?
   b. How long can a human survive at 25°C with 10 L of water?
   c. What are the two factors on the graph that determine length of survival time?

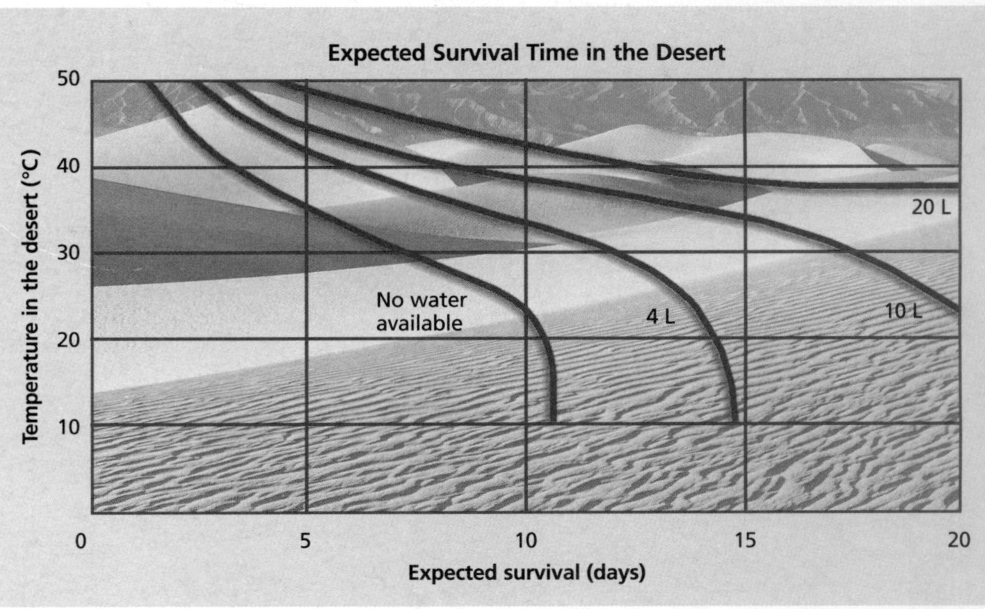

Homework

The Resource Worksheet on page 13 of the Unit 1 Teaching Resources booklet makes an excellent homework activity. If you choose to use this worksheet in class, Transparency 3 is available to accompany it.

3. Even when you eat solid food and drink liquids other than water, you are taking in part of the 3 L of water that you need daily. Suppose you have just finished this delicious dinner:

Quantity	Food	Mass of food	Percentage of water in food	Mass of water
1 small glass	orange juice	90 g	87%	78.3 g
1 slice	bread	30 g	35%	?
1 serving	corn kernels	85 g	79%	?
1 serving	roast beef	100 g	59%	?
1	baked potato	100 g	75%	?
1 serving	squash	110 g	96%	?
2 slices	tomato	30 g	95%	?
2 leaves	lettuce	50 g	94%	?
1 slice	watermelon	800 g	93%	?
1 glass	milk	250 g	87%	?

Can you calculate how many liters of water you ate and drank in this meal? This will require a little arithmetic. First, you will need to calculate the mass of water for each food listed above. Example: 90 g orange juice × .87 = 78.3 g water. (Helpful hint: 1 mL of water has a mass of about 1 g.)

More Problems to Ponder

1. Marine fish and sea gulls can drink salt water. Why can't you?
2. In the desert you find plants such as cacti and animals such as kangaroo rats and lizards. How do they survive long periods without drinking water?
3. Animals that live along coastlines are covered with water only half the time (when the tide is in). How do they survive those times when the tide is out?
4. Plants need water. What do they use it for?
5. Find out what the average monthly rainfall for your area was last year. Compare it with the rainfall in other parts of the country. On how many days did you have rain? Do the amount and frequency of rainfall in places near the ocean differ from those inland? near mountains? in prairie regions?

Who Needs Water Most?

Consider this scenario: The summer was particularly dry, and water levels were very low. Many animals had made their home in or around the creek bed—a mouse, a frog, some earthworms, a fish, and a snake. The drought period was a crisis for most of them, but it was especially dangerous for some.

Imagine that these animals could discuss their predicament. What might they say? Compose a conversation among the animals that shows the relative importance of water to each. Make sure their conversation includes the level of tolerance that each animal has to the water shortage.

Answers to *Water Needs—Some Questions to Discuss, continued*

3. The table below provides the mass of water for each food item. The total amount of water for the meal is 1432.6 g, which is about 1.4326 L.

Food	Mass of water
orange juice	78.3 g
bread	10.5 g
corn kernels	67.2 g
roast beef	59 g
baked potato	75 g
squash	105.6 g
tomato	28.5 g
lettuce	47 g
watermelon	744 g
milk	217.5 g

CROSS-DISCIPLINARY FOCUS

Mathematics

Ask students to calculate what percentage of the dinner was water. *(The total mass of the dinner = 1645 g. Therefore, 1432.6 g ÷ 1645 g = 0.87 = 87%.)*

Who Needs Water Most?

As students begin this writing exercise, make it clear that humor, wit, and knowledge are all acceptable in the conversations they create. Though students' compositions will vary, they should indicate that the fish would suffer the most by the water shortage, followed by the frog, earthworm, snake, and mouse.

Answers to *More Problems to Ponder*

1. People cannot tolerate high concentrations of salt. To get rid of the excess salt in salt water, humans would have to excrete so much water along with the salt that they would become dehydrated. Marine fish get rid of excess salt through special salt-secreting cells in their gills. Sea gulls excrete salt through glands above their eyes.
2. Cacti have tough, rubbery surfaces and tiny leaves that permit very little water loss. Cacti also store water in their fleshy stems. Many desert animals need little water and are able to survive on the water in the food they eat. Desert animals also tend to be nocturnal, which minimizes the water loss that would occur during the heat of the day.
3. Animals in a tidal region often have hard coverings or shells. They keep some water in their shell, which prevents them from drying out. Some also move down into burrows during low tide.
4. Plants use water in several ways. Minerals in soil are dissolved in water before they are absorbed into the roots and transported throughout the plant. Plants also use water during photosynthesis and transpire water as a cooling mechanism.
5. Student responses will vary depending on where they live. Students should discover that the amount of rainfall differs in each of these areas.

Light

Involve students in a discussion of what they already know about how plants respond to light. For example, most students have probably observed that plants bend in the direction of light. Most will know that plants need light in order to grow. Some students may remember from earlier studies that plants need sunlight in order to make food. Direct students' attention to the pictures on page 22. Call on a volunteer to identify each one, and involve the class in a discussion of the questions.

Answers to *Light*

Sunflowers (b) and orange trees (c) need the most light. Ferns (a) and mushrooms (d) thrive in the shade.

Answers to *In-Text Questions*

Ⓐ Students should conclude that the picture on the right (b) shows the plant at dawn because its leaves are folded in a protective fashion. On the left (a), the plant's leaves are open at noon to receive sunlight.

Light

Still another vital abiotic factor to living things is light. In your opinion, which of the organisms shown in (a) through (d) needs the most light? the least light?

a Ferns

b Sunflower

d Mushrooms

Which of the photographs below do you think was taken at dawn, and which was taken at noon? Explain your reasoning. Ⓐ

a

b

22

Homework

Propose the following scenario to students: You have just been hired in a flower shop. The manager tells you that light and water are equally important to a plant's survival. Make a short list of reasons why each of these factors is important. If you had to choose, which would you say is more important? Why? *(Accept all reasonable responses.)*

Theme Connection

Structures

Focus question: How does an organism's tolerance of temperature, moisture, and light relate to its structure? *(An organism's tolerance and structure are very closely related. For example, the structure of a cactus allows it to store water and survive long droughts; broad leaves allow a plant to catch more light and therefore tolerate lower levels of light; and a whale's thick layer of fat allows it to stay warm in cold water.)*

Some Light Thinking

Light, especially sunlight, clearly plays a major role in the growth of plants. Can you explain the following situations?

- Geranium plants stored in the basement over the winter develop very long stems.
- The grass under a garbage can becomes yellowish.
- Most woodland flowers grow quickly and bloom very early each spring.
- Plants that grow in a forest under the branches of trees are small and scarce compared with those in open spaces.
- Moss tends to grow more abundantly on the north side of trees than on the other sides.

What conclusions can you draw about plants and light from your explanations for these interesting facts?

Learn More About Light

The following activities can help you learn even more interesting facts about living things and light. Work in pairs or groups.

1. Observe the changes that occur in the leaves, stems, and flowers of plants at different times of the day, evening, and night. Record your observations.

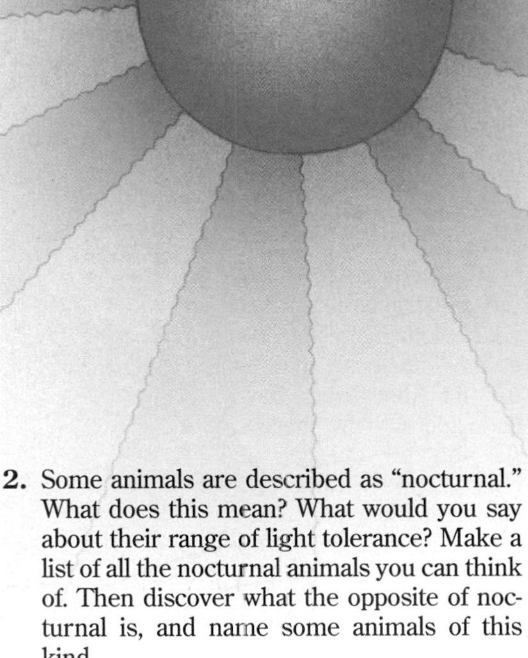

2. Some animals are described as "nocturnal." What does this mean? What would you say about their range of light tolerance? Make a list of all the nocturnal animals you can think of. Then discover what the opposite of nocturnal is, and name some animals of this kind.
3. Find out how the amount of light affects each of the following:
   a. bats
   b. houseplants
   c. migrating animals and birds
   d. seed germination (sprouting)
   e. poinsettias
   f. humans
4. Are there animals that can get along without any light at all? Research information about animals that live underground and in the depths of the sea.
5. How does the amount of light affect the activity of different kinds of insects? If possible, observe the activities of some insects at different times of the day.

23

Answers to *Some Light Thinking*

- The longer stems serve to gather the maximum amount of light.
- Grass requires light to remain green. Once the light is cut off, the leaves become yellowish due to a loss of chlorophyll.
- Woodland plants develop and bloom early because sunlight is readily available on the forest floor. Once the trees produce leaves, the forest floor becomes shaded, and the plants there grow more slowly.
- Very little light penetrates the branches of forest trees. As a result, very few types of plants can survive on the forest floor.
- A tree's north side receives less sunlight, so plants such as moss that require less light and more moisture grow there.

Students should conclude that light is very important to the way in which plants grow and respond to their environment.

Answers to *Learn More About Light*

1. Students' responses will depend on the plants they observe. Some plants fold their leaves at night and open them in the morning. Many plants close their blossoms at night and open them in the morning. Some, like the morning glory, close them during the day.
2. Nocturnal animals are active at night. They have a very limited tolerance for light. Some nocturnal animals are bats, skunks, raccoons, owls, and moles. Animals active during the day are called diurnal. Diurnal animals include squirrels, groundhogs, and most small birds.
3. **a.** Bats are nocturnal. They can see, but they do not have good eyesight. They rely on other senses for obtaining their food, so their behavior is not hindered by low light levels.
   b. Houseplants may develop pale, elongated stems if they are kept away from light. Also, they may turn their leaves toward light.
   c. It is possible, but not confirmed, that shorter days with fewer hours of daylight help stimulate animals to migrate.
   d. Light (or the absence of light) is a requirement for the sprouting of many kinds of seeds.
   e. When the hours of daylight are restricted, poinsettia leaves begin to turn red.
   f. Research indicates that extended periods of low light can cause depression in humans.
4. Students should discover that animals that live in the deepest parts of the oceans, in caves, and underground survive without any light at all. (Encourage students to discover how these animals are adapted to living in darkness.)
5. Students' responses will depend on the insects that they observe. Many insects, such as crickets, begin their calls at dusk. Fireflies come out after dark. Bees, on the other hand, are active during the day.

Abiotic Factors and the Seasons

GUIDED PRACTICE Involve students in a discussion of the seasonal changes that occur in their area. List key words and phrases on the chalkboard, such as cold, snowy winters, warm, rainy summers, and cool, dry falls. Help students to recognize that these phrases represent abiotic factors that affect living things in their environment. Have students read the page and review the chart. When students have completed their charts, involve them in a discussion of their ideas and responses. Help resolve any differences of opinion.

Encourage students to consult almanacs and encyclopedias to find out the necessary information.

Transparency 4 is available to accompany Abiotic Factors and the Seasons.

Answer to *Caption*

Ⓐ Each of the abiotic factors listed in the table is affected by the change in seasons. In the photograph, light is the most obviously affected factor.

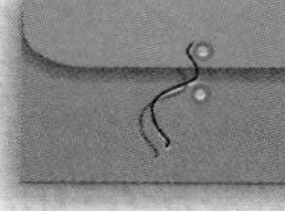

PORTFOLIO Suggest that students add their chart of abiotic factors to their Portfolio.

Abiotic Factors and the Seasons

Although you may not have realized it, you've been reading about some of the effects that seasons have on abiotic factors. For example, the temperature decreases in winter and increases in summer. There is more light in summer than in winter. The amount of moisture available also varies with the seasons.

To help you track seasonal changes in abiotic factors, create a chart like the one to the right and complete it in your ScienceLog. Note that two abiotic factors that you haven't discussed have been added to the chart. As you fill out the chart, also think about the biotic factors that change as the seasons change.

Place (your choice):

Abiotic factor	Spring	Summer	Autumn	Winter
temperature				
moisture				
light				
wind				
soil				

Can you identify any seasonal differences in abiotic factors? Ⓐ

24

FOLLOW-UP

Reteaching

Have students make cartoons to illustrate how temperature, moisture, and light affect plants and animals. Suggest that they refer to the illustration on page 16 as an example. Have students share their work by displaying it in the classroom or organizing it into a booklet for others to read and enjoy.

Assessment

Have students work in small groups to make bulletin-board displays to illustrate how abiotic factors can affect living things. Suggest that they use pictures from magazines, drawings of their own, and original poetry and writings to describe these abiotic factors.

Extension

Point out to students that naturally occurring phenomena, such as volcanoes, and events triggered by human actions, such as oil-well fires, can affect the amount of sunlight that reaches the Earth. Suggest that they do some research to find out what other kinds of events, both natural and human-made, can affect abiotic factors in the environment. Have them share what they learn.

Closure

Introduce students to the term *phototropism* (a plant's reaction to light). Suggest that they do some research to learn more about phototropism and other kinds of growth patterns that are exhibited by plants and animals.

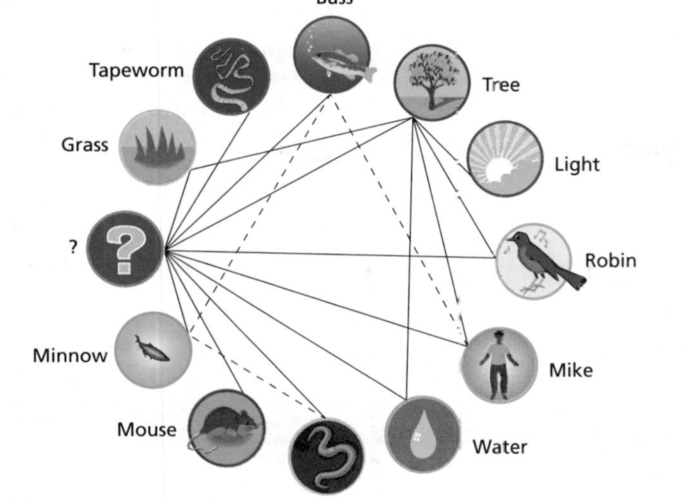

1. Going in Circles

Answer the following questions about the diagram shown above.

a. What abiotic or biotic part of the environment might "?" represent?
b. Tell the story represented by the dotted line.
c. Six lines originate at "tree." What might the interactions be?
d. "Grass" has two lines leaving it. How many more lines could you put in?
e. Suppose a line joined "Mike" and "tapeworm." How would you describe this interaction?

2. Interaction Haiku

A haiku is a Japanese poem that has three lines of 5, 7, and 5 syllables, respectively. Write a haiku describing any interactions that occur in nature. Ask a classmate to read your haiku to discover what type of interaction you've described.

Here are some examples of haiku:

Hummingbirds hover
To drink the orchids' nectar
And to pollinate.

(Do you recognize mutualism?)

Crows like it, owls don't.
Plants cannot live without it,
Though bats will shun it.

(What is it? Hint: It's an abiotic factor.)

Integrating the Sciences

Earth and Life Sciences

Explain to students that there are several different kinds of biological communities in the world and that these communities are called *biomes.* Biomes have particular climates and populations of plants and animals. Suggest that students make a map of the world and show where the different biomes are located. To extend the activity, individual students may wish to choose a biome and investigate its altitude, latitude, rainfall, and temperature. Other possible areas of interest could include geology of the area and the species living there.

Multicultural Extension

Animals in American Indian Cultures

Have students research the economic, cultural, and religious importance of animals in American Indian cultures. Examples include the wolf, bison, eagle, and bear. Have students also find out where these animals live today. Have them share their results with the class.

Answers to *Challenge Your Thinking*

1. **a.** The question mark might represent air. The large number of interactions that it has indicates that the missing part is critical to all other items on the chart.
   b. Sample answer: Mike ate a bass that had eaten a minnow that had consumed an earthworm.
   c. Sample answer: The tree casts a shadow on the grass and limits its exposure to sunlight. The water, along with dissolved minerals, is absorbed through the tree's roots. The tree's leaves use carbon dioxide in the air to produce food. Mike uses the tree for recreation, wood, or shade. The tree provides a habitat for the robin. The tree's leaves use sunlight to make food.
   d. Sample answer: Four. Grass helps produce oxygen in the air. The grass provides the mouse with protection from predators. The grass decays and provides food for the earthworm. The grass needs water to grow.
   e. Parasitism
2. Answers will vary. Students should emphasize interactions between organisms and their environment. (The answer to the second riddle is light.)

Answers to Challenge Your Thinking continued ►

★ **You may wish to provide students with the Chapter 1 Review Worksheet that is available to accompany this Challenge Your Thinking (Teaching Resources, page 17).**

Sciencelog

The following are sample revised answers:

1. Students should recognize that there are many important players in this scene, such as the mountain lion, the deer, and the hawk. Students should also recognize the other important players that are not so obvious—for example, the moss, the mushrooms, and the sunlight. Nonliving players include the sunlight and water. Some players, like air and bacteria, cannot be seen.
2. Student answers may include the following: All of the plants (including grass, flower, and trees) can be used as food by the animals. Some of the animals (deer) eat the plants, and some (mountain lion and hawk) eat the other animals. The toadstools break down dead materials. The water and sunlight make it possible for the other players to live.
3. Students may recognize several interactions, including mutualism (butterfly and flower) and commensalism (moss and tree). Students should also recognize the interactions occurring between the abiotic and biotic factors (sunlight and trees, and air and animals).

Homework

Have students draw a diagram to answer the following question: Why do plants with many leaves lose water faster than plants with few leaves? *(Diagrams should show that plants with many leaves have more surface area from which water is lost through transpiration.)*

3. Todd's Temperature Test

Todd placed 20 sunflower seeds on damp paper towels in each of five petri dishes. Each dish was kept at a different temperature. Todd recorded the number of seeds that sprouted in each dish in the table below. How would you describe the temperature tolerance of sunflower seeds?

	0°C	10°C	15°C	20°C	25°C
Day 5	0	2	4	1	0
Day 10	0	5	11	4	3
Day 15	0	8	18	12	7

4. A Day in the Life . . .

Write a short story describing one day in your life. In your story, use these words: habitat, niche, environment, abiotic, biotic.

5. Grasping the Concept

Copy this concept map into your ScienceLog, and complete it using the following words: mutualism, parasitism, no effect, benefit, harm.

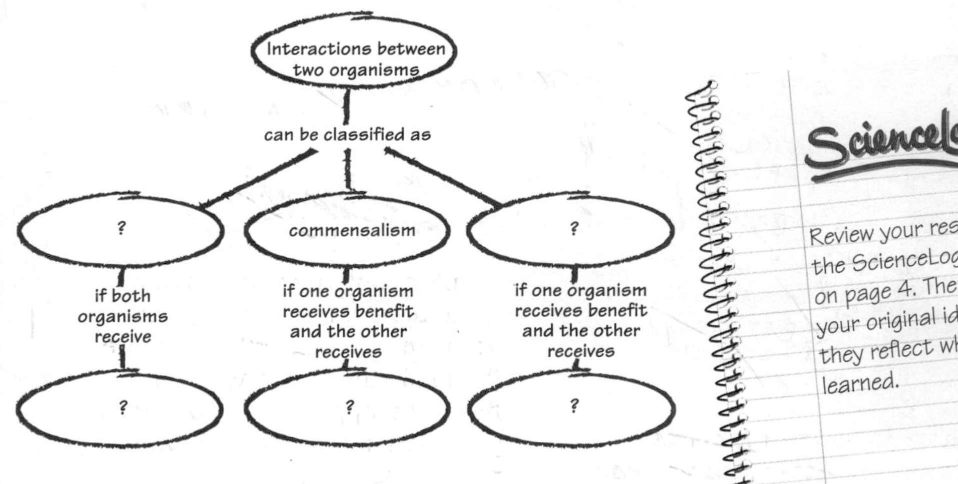

Sciencelog

Review your responses to the ScienceLog questions on page 4. Then revise your original ideas so that they reflect what you've learned.

Answers to *Challenge Your Thinking, continued*

3. The sunflower seeds appear to have a tolerance ranging from 10°C to 25°C. The optimum temperature for growth appears to be about 15°C.
4. Student answers will vary but should include the given vocabulary. For example, students will probably describe their hometown as a habitat. Their role as a student and family member would be their niche. Abiotic factors included in their story could be air, water, or sunlight, and biotic factors could be anything from park squirrels to a dinner salad. Students should identify all of these things as part of their environment.
5.

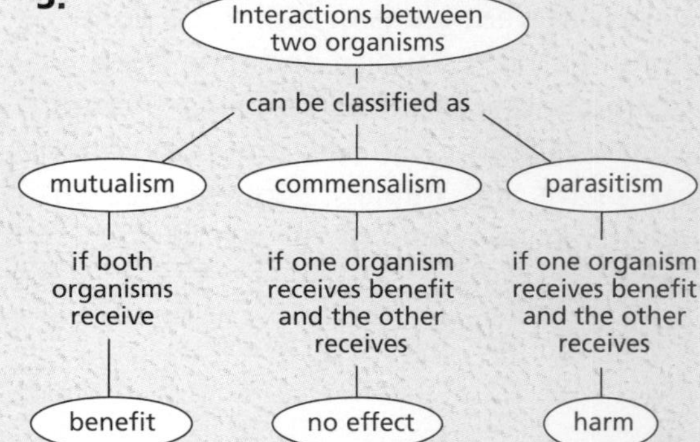

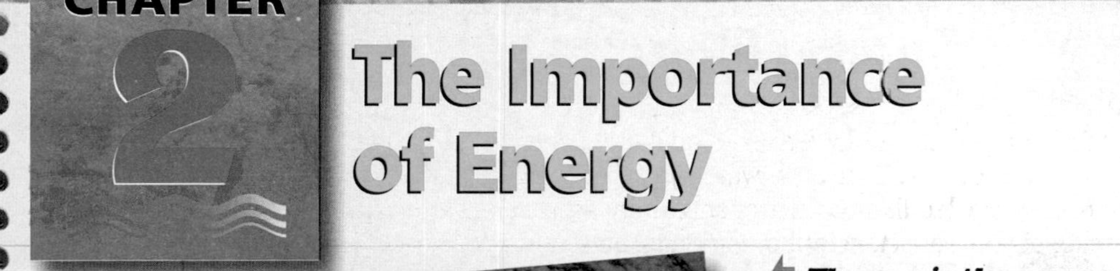

CHAPTER 2 The Importance of Energy

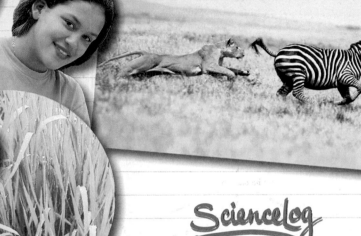

1 *The sun is the source of my food supply.* **Does this statement apply to you? Why or why not?**

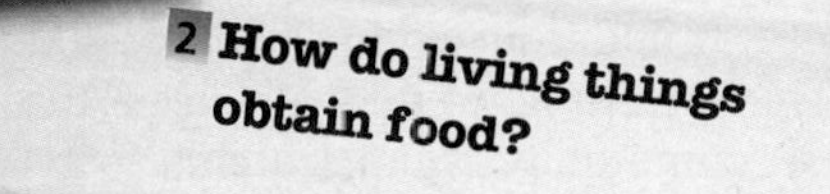

2 **How do living things obtain food?**

3 **How do these organisms interact? How could you show this with a diagram?**

ScienceLog

Think about these questions for a moment, and answer them in your ScienceLog. When you've finished this chapter, you'll have the opportunity to revise your answers based on what you've learned.

CHAPTER 2 The Importance of Energy

Connecting to Other Chapters

Chapter 1 ***introduces students to ecological interactions and the characteristics of these interactions.***

Chapter 2 ***explores food chains, food webs, and the flow of energy within natural communities.***

Chapter 3 ***examines the types of changes that occur in nature, including the role of humans in these changes.***

Prior Knowledge and Misconceptions

Your students' responses to the ScienceLog questions on this page will reveal the kind of information—and misinformation—they bring to this chapter. Use what you find out about your students' knowledge to choose which chapter concepts and activities to emphasize in your teaching.

In addition to having students answer the questions on this page, you might want to assign a "free-write" in order to assess their prior knowledge. To do this, instruct students to write for 3 to 5 minutes on the role of energy in nature, including what they think is meant by energy in nature and some examples of this energy. Tell them to keep their pens moving at all times, writing in a stream-of-consciousness fashion. Emphasize that there are no right or wrong answers in this exercise. It might be best to ask students not to put their names on their papers. Collect the papers, but do not grade them. Instead, read them to find out what students know about the existence and movement of energy in ecosystems, what misconceptions they may have, and what is interesting to them about this topic.

LESSON 1 Energy Flow

FOCUS

Getting Started

Ask several students to tell what they had for dinner the night before. Write their responses on the board. Trace the energy used to produce each food item back to its original source. Help students to see that all of the energy in the foods that they eat originates with the sun.

Main Ideas

1. Organisms interact in communities.
2. All living things depend on sunlight as their initial source of energy.
3. Green plants are called producers because they use the sun's energy to produce food.
4. Consumers are organisms that eat other organisms.
5. Decomposers are organisms that break down dead organisms into substances that enrich the soil.

TEACHING STRATEGIES

GUIDED PRACTICE Involve students in a discussion of their community. Have them name some members of their community. *(Possibilities include friends, neighbors, and relatives.)* Explain that in a biological sense the word *community* has a similar meaning—it refers to all of the organisms that inhabit an area and that influence each other through food chains or interactions with the environment.

Encourage students to identify some other kinds of communities that exist in the natural world. *(A pond community, a desert community, a mountain community)* Help students recognize that energy flows through every community in a similar way.

★ **A Transparency Worksheet (Teaching Resources, page 23) and Transparency 5 are available to accompany the material on this page.**

LESSON 1 Energy Flow

You live in a community. Your community is made up of your family, neighbors, friends, pets, and all the other living things in your neighborhood. Similarly, a forest contains many kinds of organisms. In a forest, some organisms cooperate, some compete, and some never directly interact with each other. We can call this a forest, or woodland, community.

Every community of living things, whether in a forest or city, needs energy to support life. You know that an automobile gets its energy from burning a fuel, usually gasoline. Well, a community of living things also gets its energy from "burning" a fuel, but it gets energy in a different way and from a different fuel—food.

Think about how energy "moves" through a forest community. Some animals get their energy (or food) by eating other animals. Some animals eat plants. But what do plants eat? What is their source of energy?

Energy → Plants → Animals → Animals

A few plants, such as the pitcher plant and the Venus' flytrap, eat animals (insects). But this is not the usual niche that plants occupy. So how do plants usually get their energy? During **photosynthesis**, green plants change carbon dioxide (a gas they get from the air) and water (from the soil) into food. This food supplies energy not only for the plants, but also for the animals that eat the plants. The process of photosynthesis also *requires* energy. Green plants use energy from sunlight to manufacture their food, as you can see from the following energy diagram:

Sun's energy → Plants → Animals → Animals

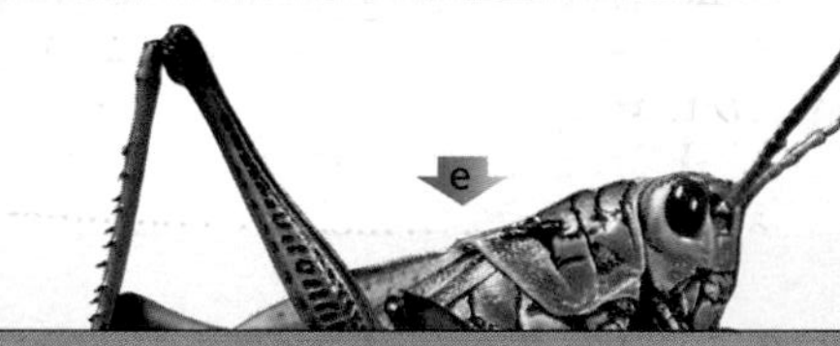

LESSON 1 ORGANIZER

Time Required
2 class periods

Process Skills
observing, analyzing, inferring, classifying

Theme Connection
Energy

New Terms

Bacteria—a group of microorganisms that includes the most common decomposers
Carnivores—consumers of animals only
Consumers—organisms that depend on other organisms for food
Decomposers—organisms that break down dead plants and animals into substances that enrich the soil
Herbivores—consumers of plants only
Omnivores—consumers of plants and animals
Photosynthesis—the process of using sunlight, water, and carbon dioxide to make food
Predator—a consumer that hunts or captures other organisms (prey) for food

continued ►

Types of Roles in a Community

Green plants play a particular and valuable role in a community. They make food energy, which they use themselves. But this food energy is also distributed throughout the community when plants are eaten by other organisms.

Because green plants produce food for themselves and others, they are called **producers**. This is their niche. The organisms that eat green plants and other organisms are called **consumers**. Why? (A)

Which of the organisms on these two pages are producers? Which are consumers? (B)

Types of Roles in a Community

Have students read the paragraphs on page 29. Discuss the differences between consumers and producers. Have students identify which of the organisms in the photographs are producers and which are consumers.

Organisms that eat green plants or other animals are called consumers because they obtain energy by consuming food; they do not produce their own food. Producers produce their own food.

Answers to *In-Text Questions*

(A) They are called consumers because they obtain energy by consuming food they did not produce.

(B) Students should recognize at least the following producers and consumers that are featured in the photographs:
- Producers: grass, apple tree, and tomato plant
- Consumers: deer, mice, mountain lion, chameleon, hawk, snake, shrew, fox, human, grasshopper, and rabbit

Some students may recognize that other organisms in the photographs may also be classified as producers or consumers. These include the following organisms:
- Producers: wildflowers, trees, and shrubs
- Consumers: organisms being eaten by each of the following: mountain lion, chameleon, shrew, bird, snake, and fox

ORGANIZER, continued

Prey—an organism that is hunted or captured and eaten by a consumer
Producers—green plants that make food for themselves and for others
Scavenger—a consumer of dead or decaying plants and animals

Materials (per student group)
none

Teaching Resources
Transparency Worksheet, p. 23
Transparency 5
SourceBook, p. S5

A deer and a fox are both called consumers. But because their diets are different, they are different types of consumers. The deer is called a **herbivore** (plant eater) because its main food is plants. The fox is called a **carnivore** (meat eater) because its main food is other animals. (A) Which name fits humans? If you eat a turkey sandwich with lettuce and tomato, you are eating parts of both plants and animals. That would make you an **omnivore**. What do you think the term means? Make an energy diagram, like the ones on page 28, that contains the following items: producer, sunlight, carnivore, herbivore. (B)

Consumer Relations

Some consumers, such as the fox, kill other animals for food. The fox is a **predator**, and its food is its **prey**. What other examples of the predator-prey relationship can you think of? (C)

Unlike the fox, some consumers do not simply eat and run. The tapeworm (a parasite) can live comfortably in a human (a host) for long periods of time, consuming food that the person eats. Certain parasites spend their entire lives in a single host. Other parasites spend part of their life in one host and the rest of their life in a second host. Still other organisms are parasites for part of their life but are free-living the rest of the time.

Some consumers wait until their food source is already dead instead of killing it. A vulture eating a dead bird is called a **scavenger**. An earthworm eating dead organic material in the soil is another example of a scavenger. Look back at the interaction diagram on page 7. What lines could you draw to show the earthworm's role as a scavenger? (Remember that all of the organisms in the diagram will eventually be dead.) (D)

Not all living things are killed and eaten by consumers. Nor do scavengers eat every last bit of the dead organisms they feed on. What happens to the remains? Something must happen; otherwise the Earth would be covered with the bodies of dead plants and animals. Can you imagine what that situation would be like?

Fortunately, some consumers clean up the Earth by breaking down plant and animal bodies into substances that enrich the soil, which then supports the growth of new plants. Such consumers are called **decomposers**. Mushrooms growing on tree stumps and the black fuzzy mold on tomatoes are examples of decomposers that you can easily see.

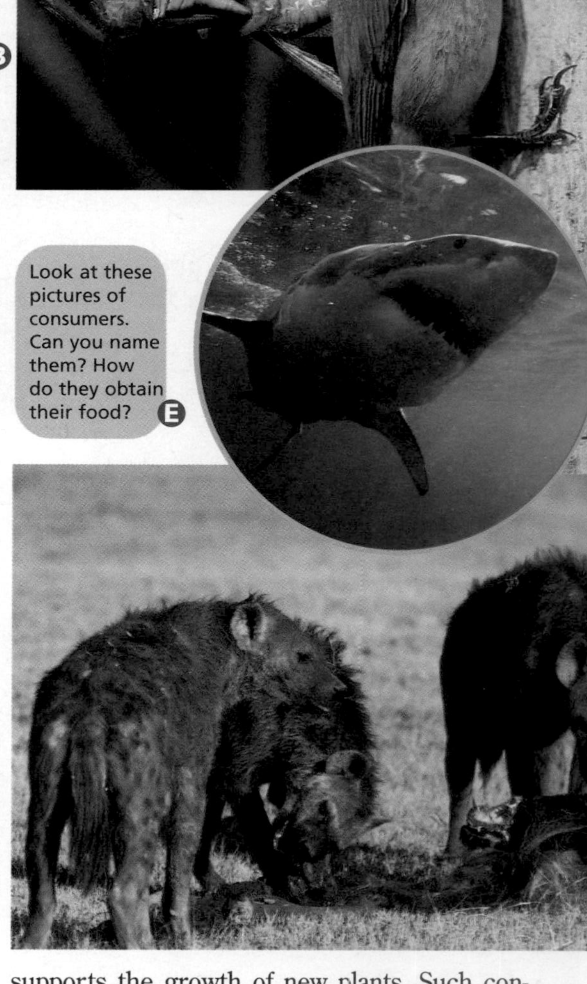

Look at these pictures of consumers. Can you name them? How do they obtain their food? (E)

Answers to
In-Text Questions and Caption

(A) Most humans are neither strictly herbivores nor strictly carnivores.

(B) Omnivores eat both plants and animals. The energy diagram is the following: sunlight → producer → herbivore → carnivore.

(C) Additional predator-prey relationships include cats and mice, lions and zebras, and owls and shrews.

(D) Lines could be drawn from the earthworm to all of the organisms in the diagram because these organisms will die and provide organic material for the earthworm to eat.

(E) The preying mantis hunts insects. Birds eat worms and insects; they obtain their food by hunting in trees and on the ground. Sharks hunt and scavenge for their food. Hyenas scavenge their food from animals that are already dead.

Consumer Relations

Call on a volunteer to read this section aloud. As the class reads along, encourage discussion of the in-text questions. Involve students in a discussion of decomposers. Help them realize how important decomposers are to the environment. Encourage students to offer scenarios of decomposers at work. *(Possible examples include a rotting log, spoiled food, sour milk, a decaying animal, a dead tree, a compost pile, and so on.)*

Homework

Propose the following scenario to students: You have been hired by a local magazine to write a mystery entitled "The Predator." The mystery should be about one page long and should contain the terms *predator* and *prey.* *(Student compositions should indicate an understanding of the meaning of these two terms.)*

The most common decomposers, however, can be seen only with a microscope. They are found in almost all environments on land and in water. These are **bacteria** (singular, bacterium).

Let's look at an example of many decomposers at work. Have you ever seen a compost pile—a mixture of leaves, grass, potato peelings, soil, and other biotic materials? If so, you may have noticed a matted layer of fine white threads on top of the pile. These threads are parts of fungi. Molds and mushrooms are also fungi. These organisms obtain their food from dead plants and animals. They also break down dead matter into soluble substances that can be absorbed by new plants. Deeper down in a compost pile is a whole world of bacteria that also feed on the dead material and make it usable by plants.

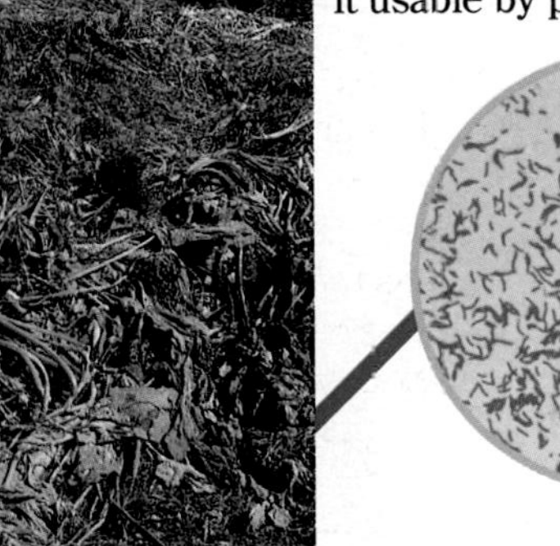

Fungi

Compost pile

Bacteria (400×)

Match Game

Below is a list of interactions. Can you match each numbered event on the left with the type of interaction it represents?

1. A butterfly drinks nectar from a flower.
2. A mosquito sucks blood.
3. Mold grows on oranges.
4. An aphid lives on the leaves of trees.
5. A toad eats a fly.
6. A millipede eats dead plants.
7. A dog has fleas.
8. A human kills and eats a deer.
9. Bacteria live on a dead sparrow.
10. A sow bug eats rotting wood and leaves.

a. predator-prey

b. parasite-host

c. scavenger and its food

d. decomposer and its food

e. none of the above

Tell Me About It!

You have learned many new terms that are used to describe an organism's niche. Take another look at the picture on page 4. Suppose you were asked to describe this scene to a friend who is visually impaired. In your ScienceLog, write down what you might say. Use your new scientific vocabulary to describe the picture. In your description, use words such as those in the box at right.

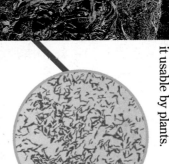

- producer
- consumer
- herbivore
- carnivore
- parasite
- host
- predator
- prey
- scavenger
- decomposer
- mutualism
- commensalism
- energy

Answers to *Match Game*

1. e, none of the above

2. b, parasite-host

3. d, decomposer and its food

4. b, parasite-host

5. a, predator-prey

6. c, scavenger and its food

7. b, parasite-host

8. a, predator-prey

9. d, decomposer and its food

10. c, scavenger and its food

Answers to *Tell Me About It!*

Student answers should include some or all of the following:

- producers: plants of all sizes
- consumers: animals
- herbivores: butterfly, field mouse, and deer
- carnivores: eagle and mountain lion
- parasite: possibly a parasitic worm inside one of the animals
- host: hawk, deer, mountain lion, or field mouse with parasitic worm
- predators and prey: hawk and field mouse, mountain lion and deer
- scavenger: possibly the hawk
- decomposers: mushrooms
- mutualism: butterfly pollinating a flower and getting nectar at the same time
- commensalism: moss on the tree
- energy: the sun and the flow of food from one organism to another

FOLLOW-UP

Reteaching

Have students write riddles that give clues to the identity of a herbivore, a carnivore, an omnivore, a predator, and a prey. Encourage them to exchange riddles with each other.

Assessment

Have students make poster diagrams showing the flow of energy through a community of living things. Display the posters around the room.

Extension

Take students on a field trip to identify producers, consumers, decomposers, and scavengers. Ask them to keep a list of their observations in their ScienceLog.

Closure

Students may enjoy working in groups to make murals of a forest, pond, or desert community. Their murals should depict the relationships and roles of the various organisms living in that community.

Theme Connection

Energy

Focus question: How could an asteroid hitting the Earth have caused dinosaurs and other animals to become extinct? *(Because the resulting cloud of dust might block the sun and cause the extinction of many plants, all of the organisms in the ecosystem could be affected.)* A Theme Worsheet is available (Teaching Resources, page 25).

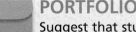

PORTFOLIO

Suggest that students include their descriptions from Tell Me About It! in their Portfolio.

Who Eats What?

FOCUS

Getting Started

Tell students to imagine that all of the insects on the planet are gone. Ask students whether they think they could get along without insects. Have students write their responses in their ScienceLog. *(Some students will probably answer yes.)* Explain to students that insects are an important part of many food chains and food webs. Insects are a food source for fish, birds, and many other animals that are in turn eaten by humans. Without insects, we would lose many food crops, such as oranges, apples, plums, strawberries, pears, grapes, peas, onions, carrots, and cabbages, that are pollinated by insects. Insects also aerate the soil so that plants can grow, and they help to recycle dead organisms.

Main Ideas

1. A food chain shows the flow of energy, in the form of food, from one organism to another.
2. A food web is a complex system of food chains.
3. When one organism in a food web is removed, many other organisms are affected.

TEACHING STRATEGIES

Ask students to identify where they get their energy. *(From the food they eat)* Then list some of their favorite foods on the chalkboard. Point to several of the items, and ask students to identify the organisms that the items come from and where those organisms got their energy. Help students recognize that the energy they receive from food has already passed through several other organisms before it reaches them. Then have students read the lesson introduction.

Food Chains

Direct students' attention to the photographs on page 32, and ask them to identify each one. Then have students read the first paragraph on page 33 and examine the diagram of the food chain.

Who Eats What?

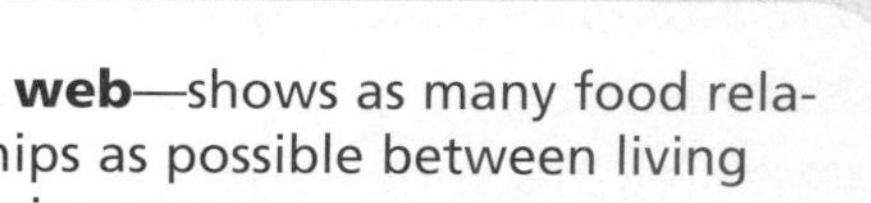

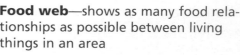

If you sit quietly in a field or in the woods and watch closely, you can usually spot a number of animals. What are you most likely to find them doing? Do you think they would be gathering or eating food? Most animals *do* spend a great deal of time gathering and eating food, which is their source of energy.

Each type of animal prefers certain kinds of foods. And in many cases, each animal becomes food for other animals. Patterns of relationships among living things are easy to find if you look for them.

Food Chains

Look at the pictures at right (which are in random order) and think about which organisms would complete the following statement:

Plants are eaten by ___?___, which are eaten by ___?___, which are eaten by ___?___, which are eaten by ___?___.

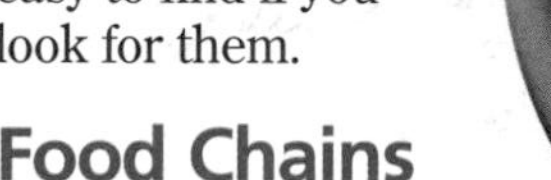

LESSON 2 ORGANIZER

Time Required
2 to 3 class periods

Process Skills
observing, classifying, inferring, analyzing

Theme Connections
Cycles, Systems

New Terms
Food chain—shows how certain living things depend on one another for food energy

Food web—shows as many food relationships as possible between living things in an area

Materials (per student group)
Exploration 2, A Woodland Community: field guides for identifying organisms common to your area; metric ruler or meter stick; **A Forest-Floor Community:** field guides for identifying organisms; meter stick; newspaper or white plastic tray; magnifying glass; **A Pond Community:** field

continued ►

These relationships can be shown in a linked diagram or can be written out as a "chain." Such a diagram or statement describes a **food chain**. It shows how living things depend on one another for energy. Each organism can be called a "link" in the chain. The food chain for the organisms shown on page 32 would have five links and would look like this:

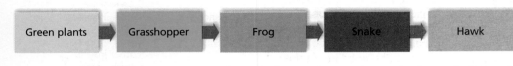

Arrows are used to show the direction in which energy moves. What would happen to the other organisms if the green plants were removed? Ⓐ

Green plants begin almost all food chains. Is it therefore true to say all members of food chains that begin with plants depend on the sun? Ⓑ

Study the food chain shown above and see if you can answer these questions:

1. Which links in the food chain are producers? Which are consumers?
2. The grasshopper is called a *primary consumer*. Why?
3. Animals that eat herbivores are called *secondary consumers*. Why? Which organisms in the food chain above are secondary consumers?
4. Which consumers are herbivores? Which consumers are carnivores?

At the end of the chain is the hawk. It apparently has no natural enemies in this situation. In this case, the hawk is the top carnivore.

Construct a food chain that shows what may happen or is happening in the illustration below. Identify the producer(s) and any primary or secondary consumers. Which organism is the top carnivore? Ⓒ

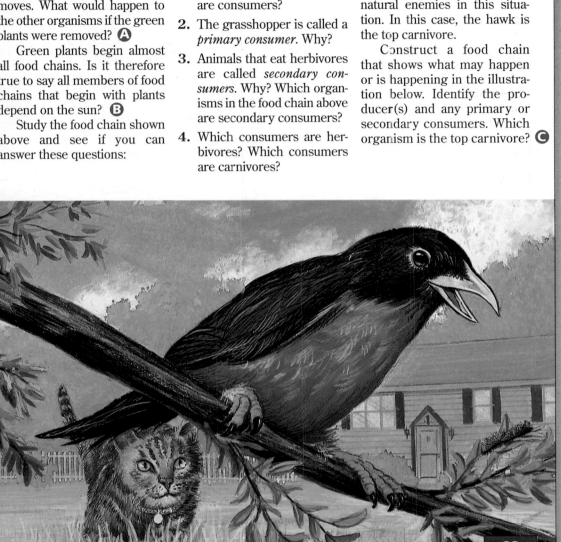

33

Answers to *In-Text Questions*

Ⓐ Students should realize that if the green plants were removed, all of the other organisms would die because the food producer for the food chain would no longer exist.

Ⓑ All members of a food chain depend on the sun's energy because it is the source of energy for green plants.

Ⓒ The food chain in the drawing can be depicted as follows: leaf → caterpillar → bird → cat. The producer is the tree. The primary consumer is the caterpillar. The secondary consumers are the bird and the cat. The cat is the top carnivore.

Answers to *Food Chains, pages 32–33*

Sample completed statement: Plants are eaten by *grasshoppers,* which are eaten by *frogs,* which are eaten by *snakes,* which are eaten by *hawks.*

1. The green plants are producers. The other organisms in the food chain are consumers.
2. The grasshopper is called a primary consumer because it is an organism that eats a producer.
3. The frog, snake, and hawk are secondary consumers because they eat other consumers.
4. The grasshopper is a herbivore. The frog, snake, and hawk are carnivores.

★ **A Transparency Worksheet (Teaching Resources, page 27) and Transparency 6 are available to accompany the material on page 32.**

Homework

The Theme Worksheet on page 25 of the Unit 1 Teaching Resources booklet may be assigned as homework in conjunction with pages 32–33 of the student textbook.

ORGANIZER, continued

guides for identifying organisms; plastic bottles, containers, and bags; kitchen strainer or aquarium net; white plastic tray (additional optional teacher materials: glass jar or aquarium; aquarium pump)

Teaching Resources
Theme Worksheet, p. 25
Transparency Worksheet, p. 27
Resource Worksheet, p. 29
Exploration Worksheets, pp. 30 and 32
Activity Worksheet, p. 39
Transparencies 6, 7, and 8
SourceBook, pp. S3 and S7

Complicating the Food Chain: Food Webs

Have students read the paragraphs at the top of the page. Then direct their attention to the list of organisms in the table.

GUIDED PRACTICE Ask students to identify which of the organisms in the list are producers *(grass, leaves)*, primary consumers *(earthworm, rabbit, grasshopper, and other insects)*, secondary consumers *(robin, snake, hawk, mouse)*, and the top carnivore *(hawk)*. Remind students that the arrows show the direction that energy flows.

Answers to *In-Text Questions*

Ⓐ Many connections are possible in this diagram. For instance, arrows may be drawn from the grasshoppers to the hawk, robin, and snake and from the leaves to the rabbit and insects. Students should be able to justify any connection that they make in terms of energy flow within the food web.

Ⓑ The animals named might be found in a forest community.

Multicultural Extension

Food Webs

Remind students that in this unit they learned how to construct a food web to show the interrelationships between plants and animals in a community. Suggest that they make a diagram of a food web to show the interrelationships in a biological community in a different part of the world. They should also include how humans interact with the organisms in their chosen location. Invite volunteers to share their finished diagrams with the class.

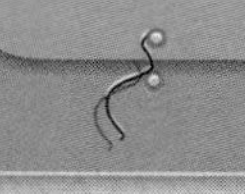

PORTFOLIO

Suggest that students include their food web in their Portfolio. Students may wish to describe any interesting things that they noticed while creating their food web.

Complicating the Food Chain: Food Webs

Real food chains are not as simple or straightforward as the ones you have studied so far. Frogs get energy from organisms other than grasshoppers; snakes eat animals other than frogs. If you were to include all of the food for every single animal in an area, your food chain would be very complicated.

Let's examine what it might look like. At the right is more information about the animals you have been discussing. (Keep in mind that even this expanded food list is far from complete.)

Circles containing the names of the various organisms are randomly arranged below. Sketch this arrangement in your ScienceLog. Draw arrows showing the movement of food energy from one organism to another. A few arrows have already been drawn. (Arrows can cross one another.) Ⓐ

Animal	Food
earthworm	leaves (dead)
grasshopper	grass, leaves
robin	grasshoppers, other insects, earthworms
snake	mice, insects, young rabbits
hawk	snakes, insects, mice, rabbits
rabbit	grass, leaves

Hawk
Snake
Mouse
Rabbit
Other insects
Robin
Grass
Grasshoppers
Leaves
Earthworm

This kind of diagram is called a **food web**. A food web shows as many food relationships as possible between living things in an area. In what kind of area (environment) might you find the animals named above? Since many more animals than those listed live in such an environment, you can see that a food web including all of them would be very large. It would show many links and a complex pattern of connections. To be complete, a food web must include all scavengers, parasites, and decomposers as well. Ⓑ

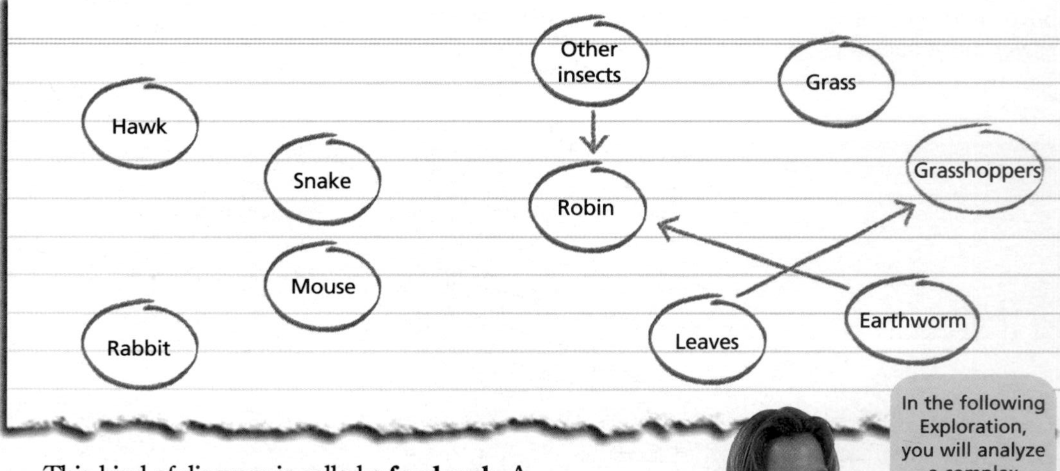

In the following Exploration, you will analyze a complex food web.

34

Meeting Individual Needs

Learners Having Difficulty

Have students take out their ScienceLog and list or draw several food chains that form part of the food web on this page. *(Answers may vary. Possible responses include the following: grass → rabbit → hawk; leaves → grasshopper → snake → hawk; leaves → earthworms → robin → hawk; grass → grasshopper → snake → hawk.)*

Homework

You may wish to assign the Resource Worksheet on page 29 of the Unit 1 Teaching Resources booklet as homework. If you choose to use this worksheet in class, Transparency 7 is available to accompany it.

EXPLORATION 1

Analysis of a Food Web

Organism	Food
hawk	squirrels, grasshoppers, mice
coyote	squirrels, grasshoppers, mice, deer
bobcat	squirrels, mice, deer
squirrel	seeds, tree buds
grasshopper	grass
mouse	seeds, grass
deer	tree buds, twigs, grass
fungi and bacteria	hawks, coyotes, bobcats, squirrels, grasshoppers, mice, deer, seeds, tree buds, twigs, grass

In your ScienceLog, draw a food web for the organisms (or parts of organisms) listed below.

- hawk
- bobcat
- coyote
- squirrel
- grasshopper
- mouse
- deer
- seeds
- twigs
- grass
- tree buds
- fungi and bacteria

Now look at the food web you drew and identify the producers, primary consumers (herbivores), and secondary consumers (carnivores).

Do you notice that there are a lot of arrows going to the fungi and bacteria? What role do fungi and bacteria play in this community?

Suppose the mice were eliminated from the community by disease. What effect would this have? The chart below will help you answer this question. Copy the chart into your ScienceLog, and complete the chart.

What If . . . ?

Answer the following questions in terms of the effects on other members of the food web:

1. What if the mice were secondary consumers instead of primary consumers?
2. What if the mice were both primary and secondary consumers (omnivores)?
3. What if the hawk, coyote, and bobcat didn't eat mice?

When mice are removed . . .

These organisms have lost a source of food:	These are the remaining food sources for each of the organisms in the first square:
These organisms (or parts of organisms) are less likely to be eaten:	These primary consumers are left:

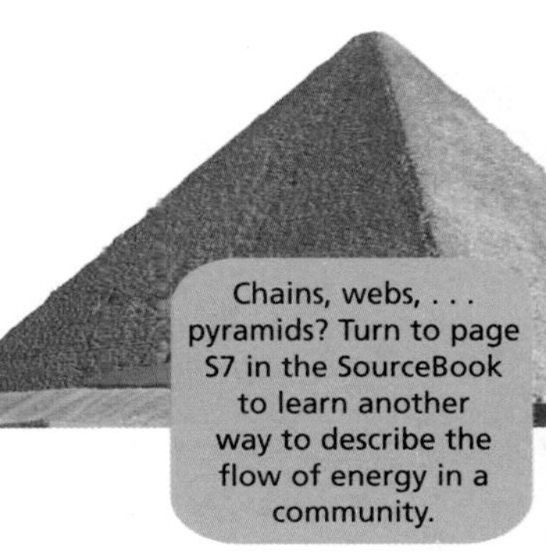

Chains, webs, . . . pyramids? Turn to page S7 in the SourceBook to learn another way to describe the flow of energy in a community.

EXPLORATION 1

Provide students with time to draw their food webs. Call on volunteers to draw their webs on the board.

Answers to *Exploration 1*

A sample food web is shown on page S202.

The producers are the grass, tree buds, seeds, and twigs; they receive their energy from the sun. The primary consumers (herbivores) are the squirrel, deer, mouse, and grasshopper; they get their energy by eating producers. The secondary consumers (carnivores) are the hawk, coyote, and bobcat; they get energy from eating other animals. (Note that some students may recognize that mice can be omnivores and therefore could be considered secondary consumers.)

As the organisms die, they become food sources for the fungi and bacteria, which are decomposers.

When mice are removed from a community, the following results could probably be observed:

- These organisms lose a source of food: coyote, bobcat, hawk, fungi, and bacteria.
- These organisms (or parts of organisms) are less likely to be eaten: seeds and grass.
- These are the remaining food sources for each of the organisms in the first square: grasshoppers, deer, and squirrels for the coyote; squirrels and deer for the bobcat; squirrels for the hawk; and all of the remaining organisms for the fungi and bacteria.
- These primary consumers are left: squirrels, deer, and grasshoppers.

Answers to *What If . . . ?*

1. If mice were secondary consumers instead of primary consumers, they would eat the grasshoppers and other herbivores such as other insects and earthworms. This would cause a decrease in the populations of these organisms.
2. If mice were primary and secondary consumers, they would eat grasshoppers and other insects along with seeds and grass.
3. If the hawk, coyote, or bobcat didn't eat mice, there would be an increase in the mouse population and a corresponding decrease in the seed and grass populations.

Homework

You may wish to have students complete the Exploration Worksheet that accompanies Exploration 1 as homework (Teaching Resources, page 30). If you choose to use this worksheet in class, Transparency 8 is available to accompany it.

Theme Connection

Cycles

Focus question: How could parts of a food web be described as a cycle? *(Energy is passed from organism to organism and may eventually be recycled. For instance, a mouse feeds on grass that absorbed nutrients from the soil. These nutrients were deposited by fungi and bacteria that fed on the remains of a dead mouse. Thus, the pattern of energy transfer has no beginning or end and can be described as a cycle.)*

A Visit to a National Park

You may wish to display a map of the United States and have a volunteer locate the Great Smoky Mountains National Park. Students may find it interesting to know that the first people known to live in the region were the Cherokee Indians. The area was made into a park by an act of the United States Congress in 1926.

Provide students with time to read the lists of organisms and their food sources and to identify the pictures. Then have them construct their food webs as instructed on page 37. When students finish, involve them in a discussion of their diagrams and ideas. You may want students to work in pairs to create poster-sized food webs, which could then be displayed around the classroom. To summarize the lesson, involve students in a discussion of the questions on page 37.

Multicultural Extension

World Cuisine

Remind students that animals, including people, get their energy from the food they eat. Assign each student a different country, and have them do research to find out what kinds of food people eat in different parts of the world. Encourage them to share unusual recipes with the class. If there is sufficient interest in the subject, suggest that students compile an international cookbook. Students can also construct food chains that include each of the ingredients in the foods they reported on.

Integrating the Sciences

Life and Physical Sciences

Suggest that students do research to discover how different kinds of pollution affect people's health. Some areas to investigate include air pollution, toxic waste, water pollution, and pesticide contamination. For example, there may be a link between the pesticide DDT and certain types of cancer. Students could find out the answers to the following questions: What is DDT? Is it still used in the world today? How does it make its way through food chains?

A Visit to a National Park

A wide variety of animals make their home in Great Smoky Mountains National Park in Tennessee and North Carolina. Read this brochure to learn about some of these animals and the foods they eat. Then do the activity and answer the questions on the last page of the brochure.

▲ **Deer** Wild apples, poplar twigs

▲ **Mouse** Grain, seeds of wild apples

▲ **Porcupine** Inner bark of spruce and pine trees

▶ **Snake** Frogs, salamanders, mice

▲ **Bear** Wild apples, blueberries, mice

▲ **Rabbit** Grass, twigs

36

Meeting Individual Needs

Gifted Learners

Explain that all of the living things in a biological community, or ecosystem, make up its *biomass.* Scientists know that an ecosystem can support only a certain amount of biomass. Challenge students to do some research on biomass and to design a biomass pyramid that shows how total biomass changes from the level of producers to the higher levels of consumers.

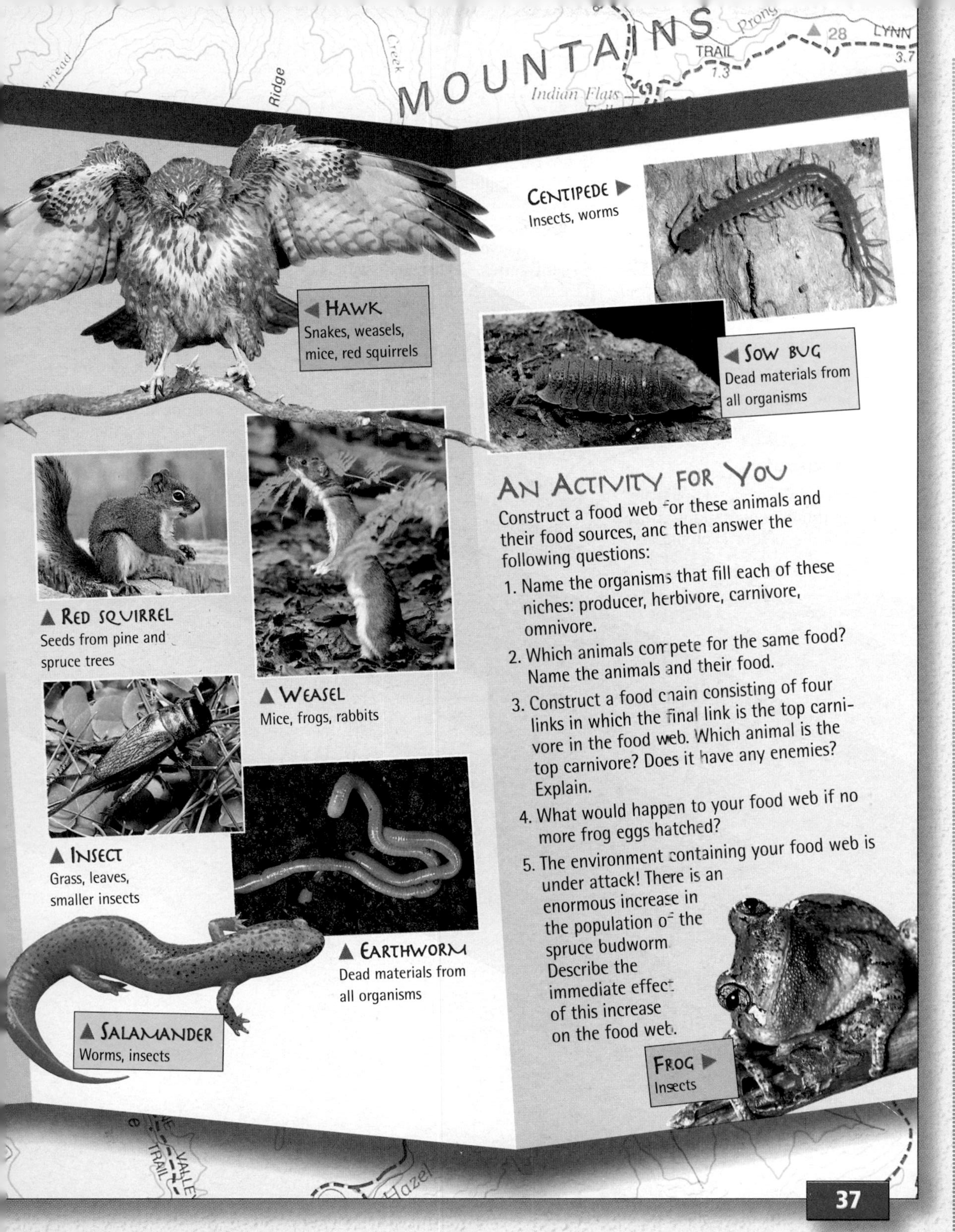

MOUNTAINS
Ridge
Creek
Indian Flats
TRAIL
7.3
Prong
28
LYNN
3.7
CENTIPEDE
Insects, worms
HAWK
Snakes, weasels, mice, red squirrels
SOW BUG
Dead materials from all organisms
RED SQUIRREL
Seeds from pine and spruce trees
WEASEL
Mice, frogs, rabbits
INSECT
Grass, leaves, smaller insects
EARTHWORM
Dead materials from all organisms
SALAMANDER
Worms, insects
FROG
Insects
AN ACTIVITY FOR YOU
Construct a food web for these animals and their food sources, and then answer the following questions:
1. Name the organisms that fill each of these niches: producer, herbivore, carnivore, omnivore.
2. Which animals compete for the same food? Name the animals and their food.
3. Construct a food chain consisting of four links in which the final link is the top carnivore in the food web. Which animal is the top carnivore? Does it have any enemies? Explain.
4. What would happen to your food web if no more frog eggs hatched?
5. The environment containing your food web is under attack! There is an enormous increase in the population of the spruce budworm. Describe the immediate effect of this increase on the food web.
Hazel
TRAIL
VALLEY
37

Homework

Have students create a brochure such as the one shown on pages 36–37 to advertise a natural area or park in their community.

Answers to *An Activity for You*

1. Producers: trees (apple, poplar, spruce, pine), leaves, grass, grain, blueberries
Herbivores: deer, red squirrels, porcupines, rabbits
Carnivores: weasels, snakes, frogs, hawks, centipedes, salamanders
Omnivores: bears, insects, mice

2. The deer, mice, and bears compete for wild apples. The deer and rabbits compete for twigs. Weasels, bears, snakes, and hawks compete for mice. Weasels and snakes compete for frogs. Insects, frogs, salamanders, and centipedes compete for insects. Frogs and salamanders compete for earthworms. Sow bugs and earthworms compete for dead materials from all organisms. (You may wish to point out that, even though the porcupines and red squirrels both depend on pine and spruce trees for food, they eat different parts of the trees and therefore are not in competition.)

3. Answers may vary. Possible food chains might include the following: grass → rabbit → weasel → hawk; grain → mouse → snake → hawk. In both cases, the hawk is the top carnivore.

Students should recognize that top carnivores do not have enemies because there are no predators that prey on them. However, humans are always a potential threat to top carnivores.

4. If no frog eggs hatched, the weasel and snake would lose a source of food. There would also be fewer animals eating insects, and the insect population would increase.

5. An enormous increase in the population of the parasitic spruce budworm could damage the entire food web. For example, the spruce trees would begin to die, and red squirrels and porcupines would have less food to eat and might die also.

EXPLORATION 2

Ideally, every student should have the opportunity to examine one of the three communities listed in the Exploration. One way to facilitate this is to divide the class into several groups. Assign each group one of the communities to study. After the field trip, provide each group with time to prepare an oral report for the class. Encourage class members to ask questions and to discuss the information that has been presented to them by each group. Instruct students to be prepared by having their ScienceLog and pencil with them to record their observations.

The three communities listed in the Exploration are only a representation of the many biological communities that exist. If you live in an area that does not support one of these communities, substitute others that are more appropriate. For example, you may be in an area where desert, beach, or grassland communities are more accessible than woodland or forest communities. If your school is in an urban area, suggest that students study a community in a park or vacant lot. Remember to get permission before entering any restricted areas.

The following guidelines will be helpful when students engage in fieldwork:

- Contact park rangers or site administrators ahead of time to find out what safety precautions should be observed when entering the area. Discuss these precautions with students and answer any questions they have.
- Students should dress in a manner that will keep them comfortable, warm, and dry. Long pants and sturdy shoes should be worn at all times. Leather gloves may be useful as well.
- Bring along a first-aid kit, a snake-bite kit, and insect repellent. Students should report any injury immediately.
- Caution students not to touch any wild plants or animals unless instructed to do so by you.
- Have students travel in pairs and keep their partner's safety, as well as their own, in mind.

EXPLORATION 2

A Community Study

What communities exist in your area? Is there a nearby forest, pond, or stream? These examples and others are interesting communities for study. With your classmates, choose a community that you wish to explore. The following are three possible communities for study:

- a woodland community
- a forest-floor community
- a pond community

A Woodland Community

Woodlands provide people with many things: quiet hiking trails, beautiful spots for camping and picnicking, lakes for fishing, and so on. But too often we overlook much of the life that a forest contains.

Of course, one visit is not enough to investigate all of a forest's mysteries. Perhaps you will have the opportunity to return with your classmates or family and continue to discover the wonders of the woodlands.

You Will Need

- field guides for identifying organisms
- your ScienceLog and a pen or pencil
- a metric ruler or meter stick for making measurements

Some Things to Keep in Mind

When you leave your study area, make sure it is clean and there are no signs of human disruption. (Carry out those aluminum cans you find among the bushes!) If you turn over stones, be sure to return them to their original positions. Instead of taking samples of organisms back with you, try sketching them in your ScienceLog. If you must have a sample, take one organism, not a handful. Try to obtain samples in a way that won't harm the organism. For example, instead of pulling a leaf from a tree, take a leaf that has already fallen from the tree.

38

A Woodland Community

Have students read the material and collect the items they will need before they take the field trip. Encourage each group to thoroughly discuss procedures ahead of time.

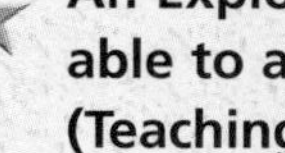

An Exploration Worksheet is available to accompany Exploration 2 (Teaching Resources, page 32).

A Tree Study

Decide which group of trees makes up the main population of the area. If the trees have needle-like leaves, produce cones, and keep their leaves all winter, they are *coniferous* trees. If they have flat, broad leaves, produce flowers, and shed their leaves in autumn, the trees are *deciduous*. Choose a tree that you would like to study.

Using a field guide, look up the group (coniferous or deciduous) and the name of the tree you chose. To help in the identification, pay particular attention to leaf size and shape, how the leaves are arranged on a stem, and how many leaves grow from a single stem. Sketch a leaf of your tree. If you can find fallen leaves, take one with you so that you can learn more about the tree later.

Each type of tree within the two main groups has distinctive bark color and markings. In your ScienceLog, describe and sketch the bark of the tree you chose to study.

Cherry blossoms

If you do your study in spring and are studying deciduous trees, check for tree flowers. Some are colorful and obvious, like cherry or apple blossoms. Others, such as the flowers of the birch and oak, are simple, small, and plain. If flowers are present on the tree, sketch them in your ScienceLog. Then look for seeds on the tree or on the ground beneath it. Sketch these too.

If you are studying a coniferous tree, check for cones on the tree or on the ground. Cones are made up of scales, and seeds are attached to these scales. Remove some scales and try to find the seeds or the scars showing where the seeds were once attached. Sketch a cone, a scale, and a seed or scar.

Bark of the papery birch tree

Bark of the yew tree

Observe the tree for signs of other organisms. Check the bark, the leaves, and the area at the base of the trunk. List the organisms you find. Try to discover the relationship between the tree and each organism. Then construct a food chain or food web that shows how the organisms interact.

What influence do you think your tree has on other plants and animals nearby? What other organisms and abiotic conditions does your tree depend on? Consider its range of tolerance for each of the **A** abiotic factors you studied earlier. Can you discover any examples of mutualism or commensalism between your tree and organisms in its immediate environment? **B**

In what way do you think squirrels are helpful to coniferous trees? **C**

Exploration 2 continued ►

Answers to *In-Text Questions and Caption*

A The tree affects the availability of moisture and light and provides food and shelter for various organisms. The tree needs water and light. Animals may pollinate the tree, and soil microorganisms provide nutrients.

B Students may observe examples of birds' nests in trees (commensalism), lichens on tree trunks (mutualism), spiders and cobwebs on bushes (commensalism), and bracket fungi on trees (parasitism).

C Squirrels may help in the dispersal of seeds.

A Tree Study

Students should easily recognize the difference between coniferous and deciduous trees. Reference books and field guides will help students identify trees by leaf shape and bark characteristics. Leaf or bark rubbings are an excellent way of recording information for later research. When examining seeds, have students consider how the seeds are adapted for dispersal. Encourage them to think about the role of animals, especially squirrels and birds, in the dispersal of seeds.

Point out to students that a forest can be divided into layers. The upper layer, or canopy, receives the most sunlight and moisture. The amount of growth in the middle layer and in the lowest layer, or undergrowth, depends on the thickness of the canopy. The undergrowth is also affected by the thickness of the middle layer, the amount of moisture that penetrates the middle layer, and the soil of the forest floor.

INDEPENDENT PRACTICE Have students examine areas of the forest floor where the amount of undergrowth varies. They should be able to suggest reasons for the differences they observe.

Students may have difficulty constructing food chains because of an absence of observable wildlife. However, they may be able to form food chains from the insects and other small animals that they observe on tree trunks and under rocks and logs. Encourage students to look not only for the actual animals, but also for the evidence of animals, such as nests, webs, tracks, or droppings. Based on these observations, students may be able to construct a large food web of the woodland area rather than individual food chains.

Exploration 2 continued ►

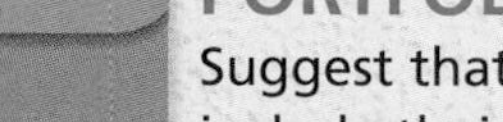

PORTFOLIO

Suggest that students include their food chains or food webs from Exploration 2 in their Portfolio.

EXPLORATION 2, *continued*

A Forest-Floor Community

Have students read the material and collect the items they will need ahead of time. Suggest that they make notes on what they expect to observe. The procedures for this study are somewhat more complicated than the previous one. It is important, therefore, that each group thoroughly discuss the procedures before the field trip.

Smaller Plants

Encourage students to be as specific as possible when they complete their maps. Have them begin by noting that the litter on the forest floor consists of leaves, needles, twigs, and animal droppings. All of this constitutes the beginning of the decomposition process that leads to the formation of new soil. Students will observe more in the litter if they use a magnifying glass. For the purposes of this activity, naming the plants is not as important as being aware of their presence and recognizing their importance as soil builders and providers of food and shelter.

Meeting Individual Needs

Second-Language Learners

Many forest animals serve as mascots for sports teams. Have students pick one of these animals to investigate. They should find out what it eats, what roles it plays in its community, and what characteristics make it a good mascot. Evaluate student work based on scientific content rather than use of language.

A Forest-Floor Community

A whole world exists on the forest floor. Small flowering plants, ferns, mosses, lichens, and fungi can be seen on the ground. And a multitude of organisms live under the litter of the forest floor.

You Will Need

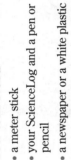

- a meter stick
- your ScienceLog and a pen or pencil
- a newspaper or a white plastic tray
- a magnifying glass
- field guides for identifying organisms

Smaller Plants

Use a meter stick to mark off a 1 m × 1 m plot. Examine the area closely, and make a map of the plant and animal life you find. Use a scale of 1:5 to draw your map. In other words, your map will be 20 cm × 20 cm. You can use symbols to indicate each plant species, but be sure to give a key for your symbols. Also note the soil conditions of your sample area.

If the area you examined did not contain any lichens, make a special effort to search for some nearby. Some lichens that are easy to recognize include matchstick lichens and old man's beard. Other lichens are scaly looking and gray or orange-red and grow on rocks. Sketch the lichens, and use the pictures on pages 12 and 13 or a field guide to identify them. Also look for interesting examples of mosses.

Mosses thrive on the damp forest floor

What's Under the Forest Litter?

Mark off a 50 cm × 50 cm plot on the forest floor. Carefully look under rocks, stones, twigs, leaves, and other debris in the area.

You will be surprised by what you find. Record all traces of plant and animal life. Try to construct food chains and food webs for the organisms you discover.

Gather some decomposing leaves, and spread them on a piece of newspaper or on a white plastic tray. You might find animal life such as millipedes, spiders, beetles, salamanders, and snails.

Look around you for examples of commensalism, mutualism, and parasitism. Can you see signs of different types of living things cooperating with each other? Keep a record of your observations.

Sit down and look up. What do the plants and animals that live in the top layers of a forest add to the forest floor? How do the large trees affect the abiotic conditions on the forest floor?

Describe the tolerance ranges of the organisms that live in the leaf litter. How do these compare with the tolerance ranges of the organisms that live in the upper levels of the forest? Explain your answer.

Exploration 2 continued ►

What's Under the Forest Litter?

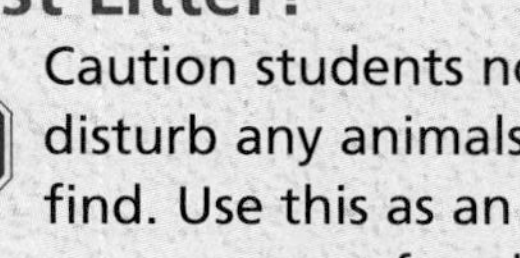

Caution students not to touch or disturb any animals that they find. Use this as an opportunity to encourage respect for all living things. Also, encourage students to exercise caution because some organisms, such as centipedes, might bite or sting. For this reason, it is a good idea for students to wear leather gloves for this part of the activity.

GUIDED PRACTICE When students have returned to the classroom, help them to construct food webs. In the forest litter and on the forest floor, the plants are represented mainly by decaying leaves and twigs. Consumers are primarily arthropods such as insects, millipedes, centipedes, sow bugs, and spiders. Millipedes and some insects are the main herbivores. Centipedes, most spiders, and some of the insects are the main carnivores. Larger animals such as salamanders are also carnivores.

Exploration 2 continued ►

Answers to *What's Under the Forest Litter?*

Many of the organisms that live in the forest litter and on the forest floor are scavengers, such as sow bugs, slugs, and snails. Most of the parasites are microorganisms such as bacteria. Some mites may attach to other animals in the litter and be parasitic, but these will be extremely difficult to observe, even with a magnifying glass.

Commensalism is easy to observe on the forest floor. For example, many of the animals that live in the litter use leaves and other debris for shelter and protection. Some animals find shelter in the roots of growing plants. Mutualism may be harder to observe. There are examples of insects—often beetles—that cooperate with one another in food gathering and storage, but the likelihood of observing such behavior is remote. Parasitism is also difficult to see in many cases. One such case is that of a tapeworm in the stomach of a mouse or other forest rodent.

Plants and animals from the top layers of the forest supply the forest-floor community with leaves, seeds, nuts, and berries, as well as decaying organisms. They also supply waste products that add nutrients to the soil. The trees block much of the light that might otherwise reach the forest floor. Sunlight, temperature, and moisture on the forest floor are therefore greatly affected by the surrounding trees.

The tolerance range for these abiotic factors varies widely among organisms in the upper levels of the forest. On the forest floor, light, moisture, and temperature tend to be more stable, so organisms require a smaller range of tolerance in order to live there.

EXPLORATION 2, continued

A Pond Community

Scout the pond community prior to the Exploration to be sure it is safe. Be aware of the conditions of the bank and pond bottom and of the location of deep water. If students are going to wade in the water, they should wear life jackets. Also, an adult skilled in water safety and resuscitation should accompany the group.

Have students read the page and collect the items they will need ahead of time. Suggest that they make some notes about what they expect to observe. Have each group discuss the procedures they will use to explore the pond community before the field trip begins. Have students make a list of the instructions and questions in the text to take with them to the pond.

If possible, have a field guide available for each group of students, or develop fact sheets illustrating some of the plants and animals commonly found in a pond community. This information will enable students to identify organisms quickly and to recognize some of their specific features. Have students try to classify the animals they find as either carnivores or herbivores.

Setting up a pond aquarium in the classroom allows students to observe the organisms over a long period of time. Be careful not to overload the aquarium with organisms—a small number of organisms is often more successful than a large number. Proper aeration is essential for the organisms living in the aquarium.

Be sure students recognize that the pond is visited by many animals, such as deer and hawks, that are searching for food and water. These should be considered part of the pond community and should be included in the food webs constructed by the students.

Homework

For Exploration 2, you may wish to have students concentrate on collecting data while they are in the field. Then they can use the data to answer the in-text questions as homework.

EXPLORATION 2, continued

A Pond Community

You Will Need

- field guides for identifying organisms
- plastic bottles, containers, and bags for taking samples
- your ScienceLog and a pen or pencil
- a long-handled "muck scoop" (kitchen strainer) or aquarium net
- a white plastic tray
- a glass jar or an aquarium (optional)
- an aquarium pump (optional)

Be Careful: Use caution around the pond. Wear suitable clothing—preferably rubber boots.

What to Do

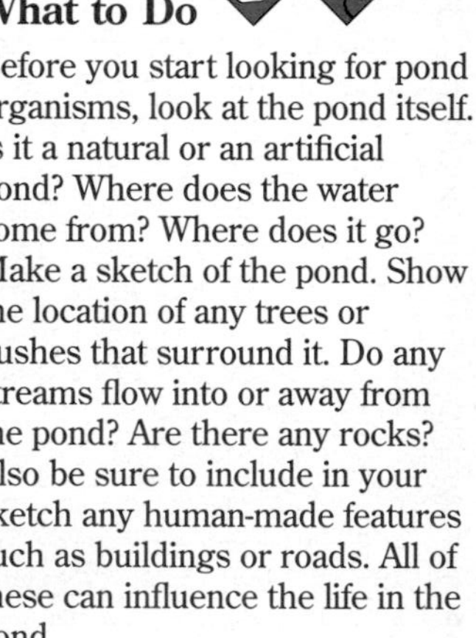

Before you start looking for pond organisms, look at the pond itself. Is it a natural or an artificial pond? Where does the water come from? Where does it go? Make a sketch of the pond. Show the location of any trees or bushes that surround it. Do any streams flow into or away from the pond? Are there any rocks? Also be sure to include in your sketch any human-made features such as buildings or roads. All of these can influence the life in the pond.

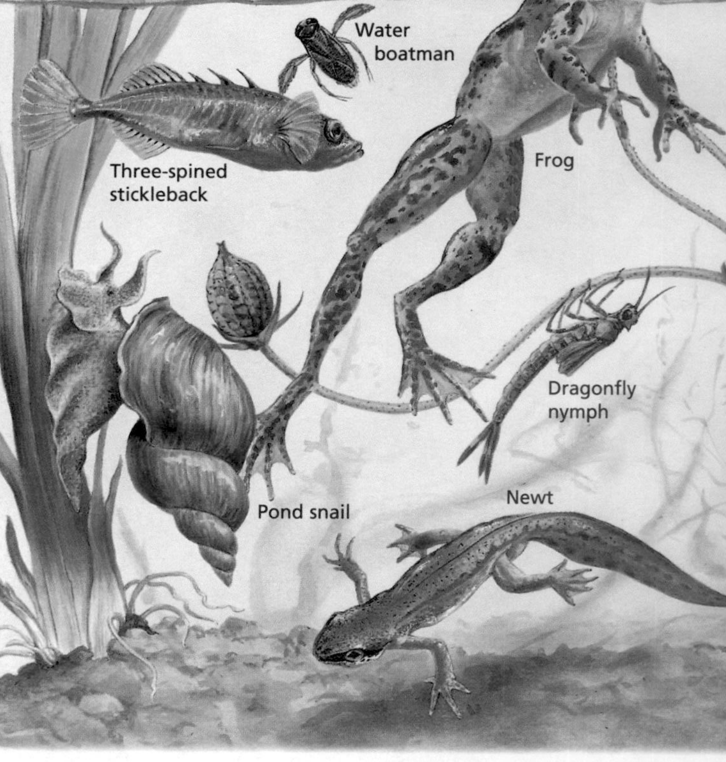

One of the most interesting features of a pond is its great variety of animal life. The illustration shows some of the pond life you might see.

The animals that are most difficult to find are the ones that live in the muck at the bottom. For these you will need a "muck scoop." A kitchen strainer with a long handle should work. Use the scoop to bring up some of the decaying material on the bottom. Place a scoopful on a white plastic tray and gently poke through it with the scoop, looking for anything that moves. You may find larvae of the dragonfly, damselfly, stonefly, and other insects. There may also be different types of worms.

As you find these organisms, carefully move them to a container that has only pond water in it. This will allow you to look at the organisms more closely. Using

42

CROSS-DISCIPLINARY FOCUS

Language Arts

Some students may enjoy setting up a reading center by gathering materials about the interrelationships of plants and animals in biological communities.

- Have students create their own anthology of poetry. They should pick a few topics from the unit and write poems about them. The poems could be organized into a booklet.
- Include newspaper and magazine articles on the ways that plants and animals live and on different kinds of biological communities.

Homework

The Activity Worksheet on page 39 of the Unit 1 Teaching Resources booklet is available to be assigned as homework once students have completed Lesson 2.

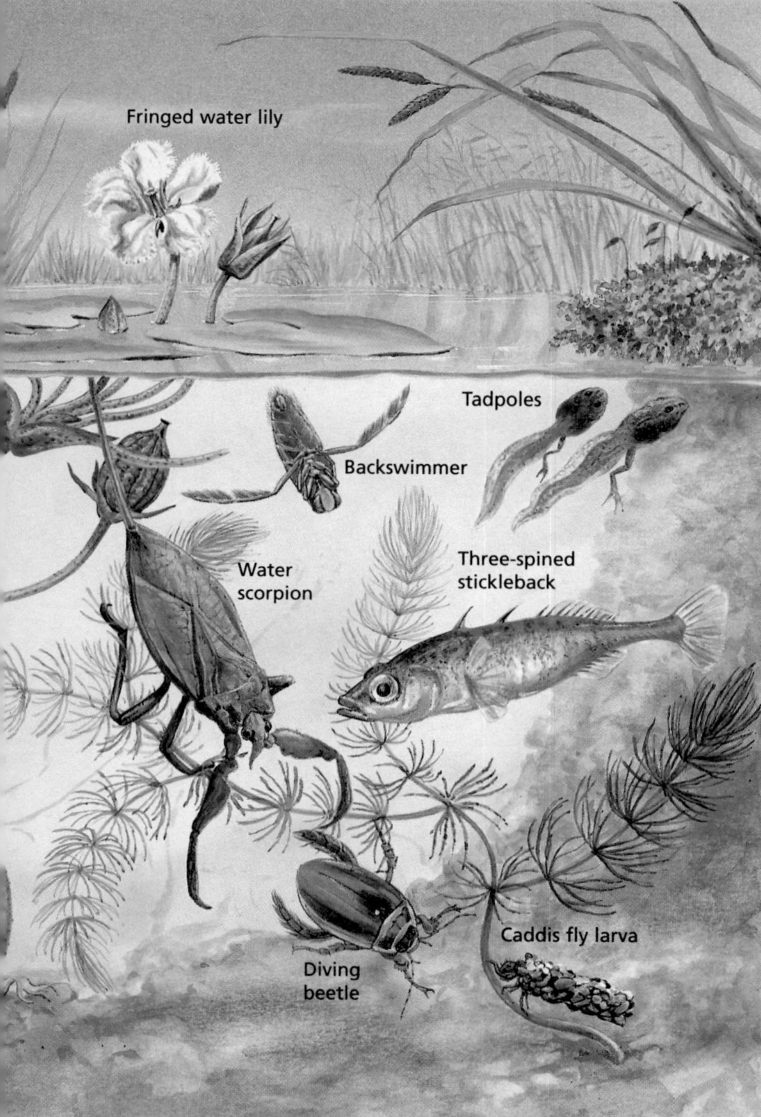

a field guide, identify as many of them as possible. Can you determine which of the animals are herbivores? carnivores? scavengers? What do you think the mouth of a carnivore looks like?

There may be some larger animals around the pond you visit. You may see the animals or perhaps just signs of them. In spring, frog eggs may be found. Can you see signs of muskrats or beavers? Make a list of any animals or birds that you see.

Look for examples of commensalism, mutualism, and parasitism. Can you find signs of different organisms living or working together?

Describe the tolerance ranges of the organisms that live in the bottom muck. How are the abiotic factors different for the organisms that live in the area surrounding the pond?

When you finish your study of the pond, leave as few human signs as possible behind. Your footprints will remain, but take your plastic bottles, containers, bags, and other equipment with you.

Before you leave the pond, collect some of the muck from the bottom, some of the plants, a variety of animals, and some pond water. You can use these materials to set up a small pond aquarium at home or in your classroom. Use a large glass jar or an aquarium. Whichever you use, try to make it as natural as possible. The water in your aquarium will have to be *aerated* (have oxygen added). You can do this with a small aquarium pump.

Watch the animals in the aquarium closely. Describe how each one moves. How does each feed? How do they react to each other?

Construct one large food web for the pond community, including all of the organisms that you either found or saw signs of. Are there some that you didn't see but that you think might visit the pond occasionally? How do they fit into your web? Classify the organisms as herbivores, carnivores, omnivores, or producers. Did you find any decomposers in or around the pond?

43

Answers to *A Pond Community, pages 42–43*

Carnivores tend to have mouthparts and other appendages that are adapted for catching and holding prey. The mouthparts of herbivores are less obvious.

As with a forest community, commensalism is easy to observe in a pond community. For example, fish and frogs use water and land plants for shelter and protection. Mutualism is harder to observe. There are examples of aquatic organisms that eat parasites and dead tissue from other organisms such as fish, but the likelihood of observing such behavior is remote. Leeches that live on the blood of other organisms are a good example of parasitism.

The tolerance ranges of organisms living in the bottom muck are quite small because the abiotic conditions in this area of a pond remain fairly constant throughout the year. Abiotic conditions, such as light level, availability of oxygen, amount of water, and temperature, will differ for organisms living around the pond.

Did You Know. . .

One of the factors that contributes energy to a pond community is humans! By actions such as feeding ducks and other wildlife, humans become significant members of natural food webs. (This is a good opportunity to stress the importance of maintaining a safe distance from wildlife.)

FOLLOW-UP

Reteaching

Suggest that students make mobiles of food chains. The top layer in the mobile should represent the top carnivore. The second layer should represent secondary consumers. The third layer should show primary consumers. The bottom layer should show producers.

Assessment

Have students identify the herbivores, predators, carnivores, producers, decomposers, scavengers, and omnivores in one of the biological communities studied in this lesson, such as a woodland, forest-floor, or pond community.

Extension

Suggest that students write and act out a skit that describes life in a forest community. The different characters should include producers, consumers, herbivores, carnivores, omnivores, scavengers, decomposers, predators, and prey found in the community. Have students perform a skit for the class.

Closure

Suggest that students visit a park or some other undeveloped area and take notes on the plants and animals that live there. They should use their notes to draw a food web of the community.

Answers to *Challenge Your Thinking*

1. **a.** The lion is a predator of the zebra, its prey.
   b. The lion, a carnivore, eats the zebra.
   c. The zebra, an herbivore, grazes on the savanna grasses.
   d. The vulture, a scavenger, feeds on dead animals, such as the remains of the zebra, after the lion feeds.
   Other interactions include the action of bacteria to decompose dead organisms and the grass producing food by using sunlight.
2. **a.** Plants use *energy* from the *sun* to make their own food from *water* and *carbon dioxide*. Because they produce their own food, plants are called *producers.*
   b. Student answers will vary. Answers should demonstrate an understanding of what constitutes a food chain. For example, a student might write the following paragraph: Grass, a *producer,* is eaten by grasshoppers, which are *primary consumers.* These grasshoppers are then eaten by *secondary consumers,* birds. This makes the grasshoppers *herbivores* and the birds *carnivores.*
3. Sample answer: A food chain describes the direct flow of energy as it moves from one organism to the next. It connects one organism at a time in a linear progression. A food web, on the other hand, shows the complex network of interactions between many organisms in a community. In a diagram of a food web, organisms are connected to many other organisms by lines that overlap each other.
4. Because plankton (which encompasses a wide variety of marine life, including plant and animal species) is such an important source of food in this food web, the removal of plankton might have dire consequences, such as the collapse of the entire web.

You may wish to provide students with the Chapter 2 Review Worksheet that is available to accompany this Challenge Your Thinking (Teaching Resources, page 41).

1. Name That Interaction

Identify the following interactions that might occur between the organisms shown below:

a. a predator-prey interaction
b. an interaction involving a carnivore
c. an interaction involving a herbivore
d. an interaction involving a scavenger

Can you identify interactions other than those listed above? If so, list them in your ScienceLog.

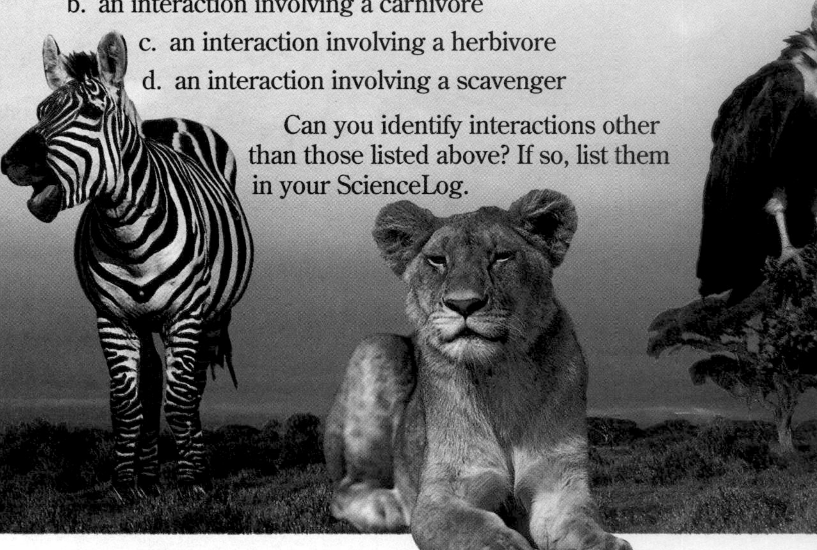

2. Putting It to the Test

a. Copy the sentences below into your ScienceLog. Fill in the blanks with these words:

- producers
- energy
- water
- carbon dioxide
- sun

Plants use ___?___ from the ___?___ to make their own food from ___?___ and ___?___. Because they produce their own food, plants are called ___?___.

b. Now create your own fill-in-the-blank paragraph. In your paragraph, describe a food chain that links at least four organisms. Leave blanks for the words *producer*, *primary consumer*, and *secondary consumer*. (You may also choose to include blanks for *carnivore*, *herbivore*, and *scavenger*.) When you are done, exchange exercises with a classmate. How did your test work? Could you complete your partner's exercise?

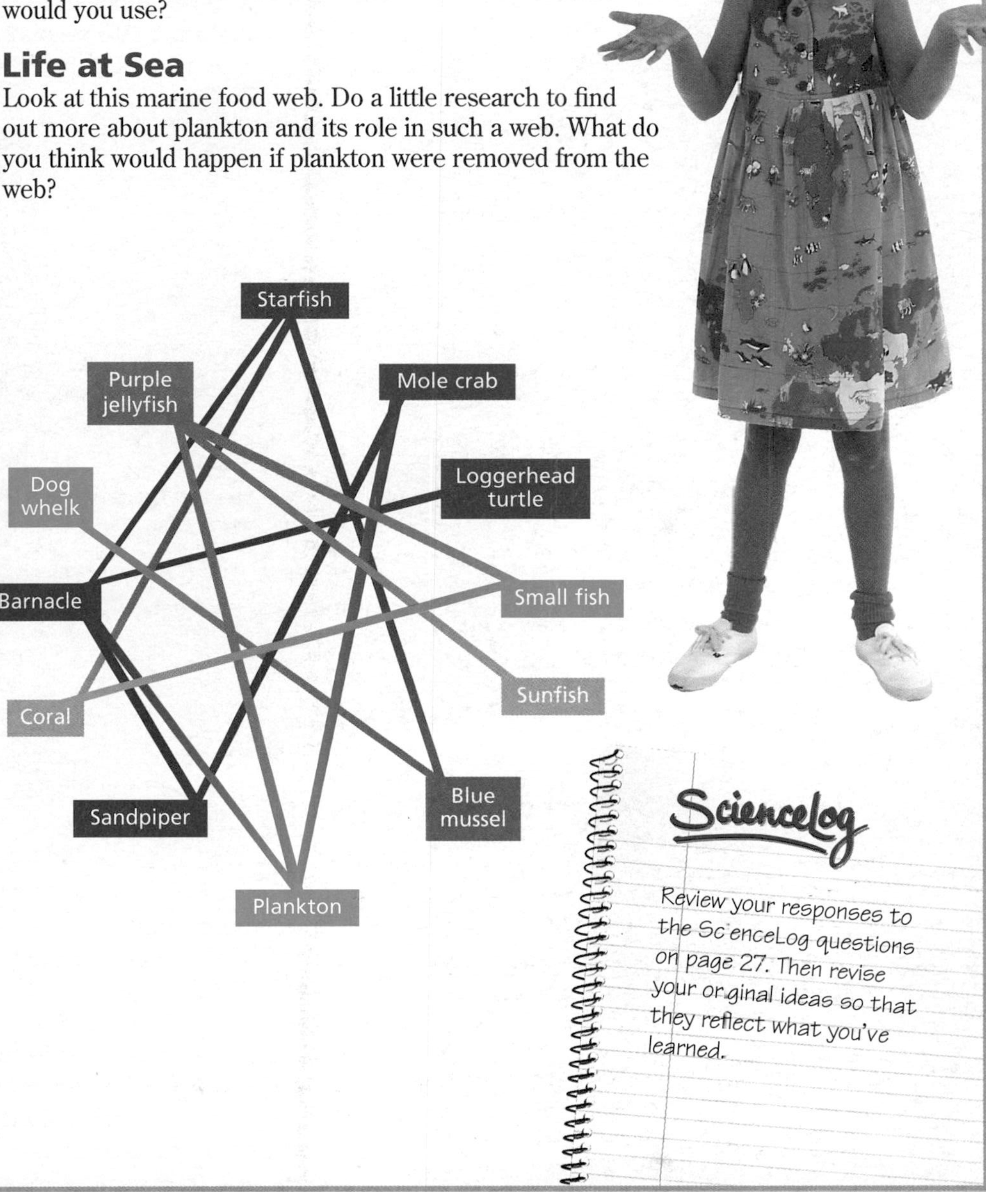
3. Clear Up the Confusion
"I don't understand," said Yasmin, a student who is a few years younger than you. "What is the difference between a food chain and a food web? Aren't they the same thing?" How would you answer her question? What examples would you use?
4. Life at Sea
Look at this marine food web. Do a little research to find out more about plankton and its role in such a web. What do you think would happen if plankton were removed from the web?
Starfish
Purple jellyfish
Mole crab
Dog whelk
Loggerhead turtle
Barnacle
Small fish
Sunfish
Coral
Sandpiper
Blue mussel
Plankton
ScienceLog
Review your responses to the ScienceLog questions on page 27. Then revise your original ideas so that they reflect what you've learned.
45

ScienceLog

The following are sample revised answers:

1. Yes. This statement applies to most living things because sunlight allows plants to produce food and oxygen, both of which are vital to all organisms. Consumers get energy from eating these plants. Like other omnivores, we can get energy by eating plants directly or by eating the consumers of these plants.
2. Living things obtain food in many ways. Producers, such as grass and trees, produce their own food. Primary consumers (including the zebra) get their food by eating producers. Secondary consumers (such as the lion) get their food by eating other consumers. Decomposers and scavengers get their food from dead organisms. Parasites feed off living organisms.
3. Grass makes food from sunlight and water, the cow eats grass, and the girl eats meat from the cow. To illustrate this in a diagram, you would draw a food chain linking these three in a direction that moves from grass to cow to girl.

Theme Connection

Systems

Focus question: How are the interactions within a natural system similar to the interactions among human body parts? How are they different? *(Accept all reasonable responses. Although body parts, like organisms, depend on each other to function, they cannot become part of a new system when changes occur.)*

Changes, Changes, and More Changes

Connecting to Other Chapters

Chapter 1
introduces students to ecological interactions and the characteristics of these interactions.

Chapter 2
explores food chains, food webs, and the flow of energy within natural communities.

Chapter 3
examines the types of changes that occur in nature, including the role of humans in these changes.

Prior Knowledge and Misconceptions

Your students' responses to the ScienceLog questions on this page will reveal the kind of information—and misinformation—they bring to this chapter. Use what you find out about your students' knowledge to choose which chapter concepts and activities to emphasize in your teaching.

In addition to having students answer the questions on this page, you might want to have students generate their own questions about environmental changes. Remind them that they have already confronted some environmental changes in the previous two chapters. For instance, organisms must continually adapt to changes in the environment. Also, the flow of energy in a community often changes over time.

Students' questions should address environmental changes with which they are already familiar or about which they want to learn more. Students should consider the role of humans in the environment as well. You might have them choose one of their questions to pose to the class. Emphasize that there are no right or wrong answers in this exercise. Collect the papers and read them to find out what students know about changes in the environment, what misconceptions they may have, and what about these changes are of interest to them.

CHAPTER 3

Changes, Changes, and More Changes

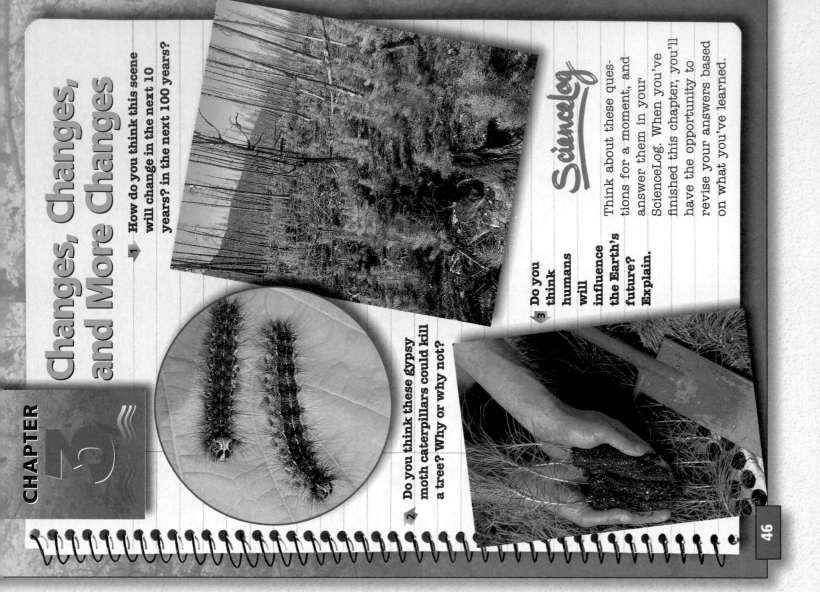

1 How do you think this scene will change in the next 10 years? in the next 100 years?

2 Do you think these gypsy moth caterpillars could kill a tree? Why or why not?

3 Do you think humans will influence the Earth's future? Explain.

ScienceLog

Think about these questions for a moment, and answer them in your ScienceLog. When you've finished this chapter, you'll have the opportunity to revise your answers based on what you've learned.

46

LESSON 1 Natural Changes

Succession—Environmental Change

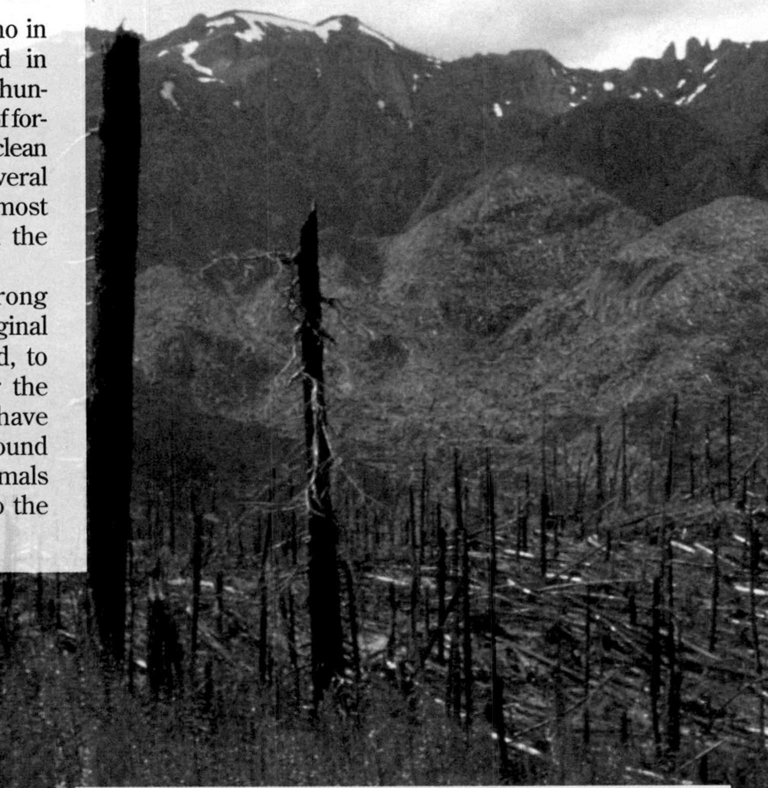

Mount St. Helens, a volcano in Washington state, erupted in 1980. The blast destroyed hundreds of square kilometers of forest. Hillsides were stripped clean and then buried under several meters of volcanic ash. Almost nothing remained alive in the blast area.

But nature has a strong ability to preserve its original state and, if it is disturbed, to return to that state. Over the years, many changes have taken place in the area around the volcano. Plants and animals have gradually returned to the charred land.

Have you ever noticed changes in the plant and animal communities where you live? Imagine visiting one of the following places, and then describe your experience in either poetry or prose.

- an abandoned farm
- a pond that you have not seen for many years
- a place where loggers cleared trees a few years earlier
- a garden left unattended for several weeks
- a field left unplowed for several years
- an abandoned lot

LESSON 1 ORGANIZER

Time Required
5 to 6 class periods

Process Skills
observing, predicting, analyzing, inferring

Theme Connection
Changes Over Time

New Term
Succession—changes in an area that cause one group of organisms to be replaced by another

Materials (per student group)
Exploration 1: field guides for identifying organisms; paring knife; alcohol thermometer; magnifying glass; trowel; white plastic tray or newspaper

Teaching Resources
Exploration Worksheet, p. 48
Graphing Practice Worksheet, p. 51
Transparency 9
SourceBook, p. S11

LESSON 1 Natural Changes

FOCUS

Getting Started

Promote discussion by asking the following questions: What changes have you seen in your neighborhood over the years? What caused these changes? Were these changes beneficial or harmful? If students are interested, have them write a narrative or story that reflects the changes they have witnessed in their environment.

Main Ideas

1. Changes are constantly occurring within biological communities.
2. Nature can restore itself through succession.
3. Human actions may adversely affect the organisms in biological communities.
4. Changes in the population size of one organism often affect the population sizes of other organisms.

TEACHING STRATEGIES

Succession—Environmental Change

Direct students' attention to the photograph on page 47 and have them identify what it shows. Involve students in a discussion of how a volcanic eruption would affect the communities of living things around it. Other photos appear on pages 370–373.

Succession—Environmental Change continued ►

Homework

You may wish to have students complete the writing assignment described on this page as homework.

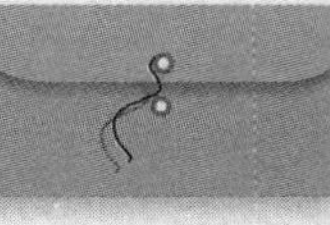

PORTFOLIO
Students may wish to include their written composition in their Portfolio.

Succession—Environmental Change, *continued*

If students have not had the opportunity to visit one of the places listed on page 47, involve them in a discussion of the changes they think might occur in each location. For example, an abandoned farm might become overgrown with weeds, shrubs, and fast-growing trees. Review the definition of the term *succession,* and have students name areas where they have seen succession occurring. As suggested on page 48, invite students to take a trip to an area where succession is taking place and to record the changes that they observe.

Answer to *In-Text Question*

Ⓐ In the sequence pictured here, the lily pads are growing and reproducing, and then grasses start growing and reproducing. Eventually, the area will no longer be covered by water, but will be completely filled in with soil and land plants.

Changes in which one group of organisms replaces another may take days, weeks, years, or centuries, depending on the type of environment that is changing. All of the changes that occur in an area are collectively known as **succession.** The illustration below shows succession in a forest community. During succession, one group of organisms follows, or *succeeds,* another. Succession is a slow, natural process. If possible, study an area where succession is taking place. Make regular trips to the area on your own, and record the changes over time that you find in the plant and animal life.

Look at the three photographs shown at right. Describe how succession is occurring in this pond community. Ⓐ

a

b

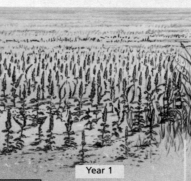

c

Succession in a forest community

Year 1

Year 2

Years 3–10

48

Theme Connection

Changes Over Time

Focus question: When succession occurs at an abandoned farm, in what order would the following plants appear?

a. taller plants—goldenrod and grasses
b. weeds that thrive in sunny locations—wild carrots and dandelions
c. slower growing trees—oak and hickory
d. young saplings—aspens

(Answer: b, a, d, c)

Homework

Using the illustration on pages 48–49 as a model, have students make an illustration that shows succession occurring in another type of community. They may wish to use the writing assignment that they completed earlier as a starting point. (See page 47.)

EXPLORATION 1

Review all safety procedures with students before they arrive at the Exploration site. Helpful guidelines can be found in the margin on page 38.

This activity provides students with an opportunity to examine the biotic and abiotic changes occurring in a rotting-log community, how these changes are brought about, and their effect on the organisms that live there. Visit the area before the field trip in order to locate suitable logs for students to study and to determine the length of time that should be allotted for the activity. If a field trip is not possible, you may wish to bring one or two small rotting logs into the classroom. Students can complete the activities for Zones A, B, and C.

The day before the field trip, provide some classroom time for students to discuss what they will be doing. Ask them to make some predictions about what they think they will observe. Then divide the class into four groups, and assign each group one zone to study. Provide time for each group to read and discuss the instructions. You may wish to have students copy the instructions and the questions from the textbook to use as a reference during the field trip.

Once students have been given a log to study, allow them to work on their own. Be available to answer any questions and to make procedural suggestions. For each zone, students should record abiotic factors such as temperature, light, and moisture. Students can then use these factors to determine the tolerance ranges of the organisms found in each zone. Remind students that they should keep careful notes and make sketches in their ScienceLog.

Caution students to avoid sitting on logs that may be unstable or cave in. Suggest that they determine whether the kinds of organisms living on the top of the log are the same as those living near the bottom of the log. Encourage students to identify how the characteristics of the bark change from place to place. For example, the bark on top of the log may be drier than the bark near the ground.

Have students wash their hands with soap and water once they have completed this Exploration.

Interactions = Change

Although succession is a slow process that can take years, small changes occur every day in natural communities. For example, many interactions occur as a fallen tree decomposes. What do you think some of these interactions might be? What changes would these interactions lead to? Think about these questions for a minute and write your thoughts down in your ScienceLog. In Exploration 1, you will get the chance to find out the answers to these questions firsthand.

EXPLORATION 1

Field Trip to a Changing Community

Where?

You will visit a rotting-log community in a nearby wooded area.

When?

Before the ground freezes. It may take quite a few hours to complete your visit.

Why?

You are going to discover what goes on in a fallen tree—who lives there, what work they do, and what they eat.

You Will Need

For each group:

- a field guide
- a paring knife
- your ScienceLog and a pen or pencil
- a thermometer
- a magnifying glass
- a trowel
- a white plastic tray or newspaper

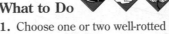

What to Do 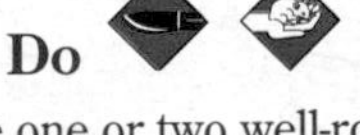

1. Choose one or two well-rotted logs to examine.
2. Approach the log slowly and quietly. You might not be the only visitor! Look carefully for signs of animals that may be using the log.
3. Divide into four groups, with each group examining just one zone of the rotting-log community.
   Zone A: Bark
   Zone B: Wood beneath bark
   Zone C: Bottom of log
   Zone D: The top 6 cm of soil beneath the log
4. For each zone, describe and record the abiotic factors of temperature, light intensity, and dampness.
5. For each zone, consider the tolerance ranges of the organisms for the abiotic factors listed in step 4.

50

Meeting Individual Needs

Gifted Learners

Have students imagine that a forest fire has destroyed an entire forest community. In another forest, toxic waste from a chemical plant has killed all of the organisms. Ask: In which community will the process of succession take longer? Explain. *(In the latter, because healthy soil still remains even after a forest fire)* Do you think that the destruction of an area is ever beneficial? *(Answers will vary.)*

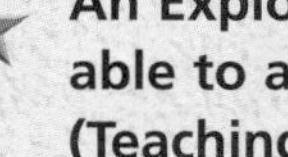

An Exploration Worksheet is available to accompany Exploration 1 (Teaching Resources, page 48).

ZONE A

Bark

Describe the size, shape, and color of any fungi on the log.

Describe any green plants growing on the bark. The plants could be mosses, ferns, or tree seedlings.

Examine the bark closely for lichens. Observe the lichens through a magnifying glass, and make a drawing of what you see. Remember, this organism has two parts: a fungus (consumer) and an alga (producer). Since these are not visible without a microscope, you may wish to review the photograph of the parts of lichens on page 13.

Look carefully for animal life or evidence of life (such as spider webs, animal droppings, and discarded seed shells or cones). Are there holes in the bark to suggest animal life beneath?

ZONE B

Wood Beneath Bark

Carefully remove some of the outer bark. Use a paring knife to break apart the wood underneath. With your magnifying glass, search it carefully for any small animals that make their home in the log. Try to identify as many as you can. Draw those that you cannot identify. Do plants grow in this zone? Why or why not?

ZONE C

Bottom of Log

Examine the bark and wood here carefully, as suggested for Zones A and B. Record all plant and animal life that you see.

Does life on the bottom of the log differ from that in Zones A and B? Why? Carefully roll the log away, if possible, and examine any life on the ground under it. You might find snakes, salamanders, toads, insects, or evidence of other animals (tunnels, pathways, food).

ZONE D

Soil Beneath Log

How does the vegetation growing next to the log differ from that growing on the log? Push a trowel down 5 cm into the soil beneath the log. Place some of this soil on a white plastic tray or newspaper and carefully observe any life. List the different kinds of organisms and types of movement you see.

Zone A

SAFETY TIP When exploring or investigating in an outdoor area, students should be aware of poisonous plants. Help them to become familiar with the appearance of plants such as poison ivy, poison oak, and poison sumac. Also instruct students never to eat any plants found in the wild. Wild plants that can be poisonous include mushrooms, holly, mistletoe, buttercups, and nightshade.

Zone B

Have students remove the bark very slowly so that they do not disturb any organisms living under it. Students should watch carefully as the bark is removed to glimpse any organisms that scurry away.

Green plants cannot live under the bark because of the absence of sunlight.

Zone C

When students move the log, they should be especially wary of organisms that may be lurking under it.

Students should carefully roll the log over in order to examine the bottom of it. Suggest that they move the log as little as possible. Instruct students to watch for any organisms that may quickly dart away.

Students should recognize that many of the organisms living in this zone are adapted to high levels of moisture and low levels of light. Students may find some organisms like those in Zones A and B, such as fungi and certain types of insects.

Zone D

Have students carefully examine the ground beneath the log before they disturb it in any way. Students should observe that most of the plants growing next to the log have roots that extend into the soil. Most of the plants growing on the log do not have similar root systems. They should conclude that the plants growing on the log get their nourishment from the log itself. The plants growing next to the log get their nourishment from the soil beneath the log.

Some Things to Think About

1. A tree that falls may have a mass of hundreds of kilograms. The tree will eventually disappear, leaving no visible trace. What will happen to the organisms that once lived on or in the tree? the log?
2. It takes many years for a tree to rot away completely. What changes in plant and animal life might occur during this time? Can you identify fallen trees in different stages of decay?
3. If snow falls and covers the log, what will happen to the life in and on the log? What effect do you think winter has on the decaying process?
4. Compare tolerance ranges of the organisms living in the different zones of a rotting log. Do all the organisms on the log need to tolerate the same range of conditions? Explain your reasoning.
5. In nature, destructive processes such as the rotting away of trees are balanced by constructive processes. In other words, forces that renew life are constantly at work in the rotting log. How many plants and animals can you identify that have benefited from the rotting away of the log?
6. Write an article for a class newspaper explaining the niche of an animal that is associated with the rotting-log community.

So tell me, what do you like about your community?
PRESS

In your study of the rotting-log community, you saw that the log was home to many organisms. These plants and animals used the log for shelter as well as for food. But organisms that you didn't see—decomposers—were also present. Think back to what you read on page 31 about compost piles. Decomposers there were using the compost for food and breaking it down so that other organisms could use it for food. The decomposers in the rotting log were doing the same thing.

When the remains of living things are used in this way to help new life develop, the materials are being *recycled*. People, too, are learning to recycle many of the things they need and use. You can see many examples of recycling going on constantly in the natural world.

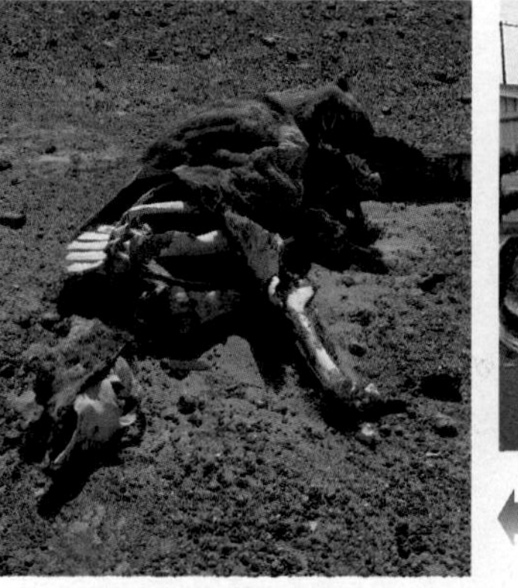

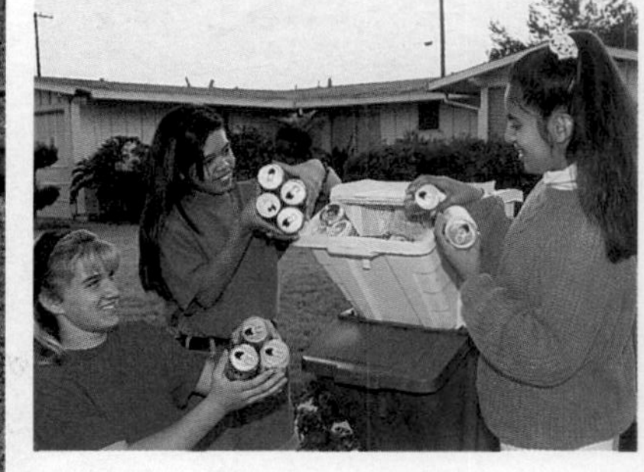

Two examples of recycling

Answers to
Some Things to Think About

1. Organisms that lived in or on the tree or log will find new places to live; some may not survive because of predators, lack of a safe location to reproduce, or lack of food.
2. As a tree rots, food supplies change. The bark and the wood may loosen, allowing beetles, ants, slugs, and worms to make their homes there. Plants may send roots down into the newly formed spaces. The following is an example of succession that could take place in a decaying tree:

   At first, the bark and wood remain solid. Lichens may grow on the bark, and small rodents may still live in the tree.

   Next, the bark loosens. Bacterial action begins to soften the wood under the bark. Small insects burrow into the bark and wood. Lichens and mosses cover much of the bark.

   Then the wood begins to break down and decay. The evidence of burrowing organisms becomes more apparent. The roots of grasses and small trees penetrate the wood, causing it to break apart.

   Finally, the log becomes noticeably smaller. The wood becomes very soft. The populations of insects, spiders, and centipedes increase. A variety of plant life grows on the log.
3. In winter, most plant life and fungi become dormant or produce seeds or spores that will germinate in the spring. Most animals either hibernate, migrate, or die. Many lay eggs that will hatch in the spring. The decaying process slows down in the winter due to cold temperatures, reduced moisture, and dormant plant life.
4. Ranges of tolerance for temperature and moisture are difficult to determine. Students might infer, however, that there are more extremes of temperature and moisture in Zone A than in the other zones. For example, the moisture and temperature levels in the soil under the log would not change as much as those on the log's surface.
5. Answers may vary, but possible responses include lichens, mosses, small plants, ferns, snails, ants, salamanders, slugs, rodents, snakes, beetles, earthworms, spiders, moles, and centipedes.
6. Content will vary depending on the animal selected by the student. However, articles should demonstrate an understanding of how the animal fills a niche in a rotting-log community. Consider having students publish their articles in a classroom or school newspaper.

Homework

Some Things to Think About makes an excellent homework activity.

Population Changes

You know that a natural community contains a large number of niches for producers and consumers. Environmental changes that disturb the producer-consumer relationship often change the natural balance within a community. Sometimes this can create serious problems.

Many settlers who came to the United States wanted to farm the land. Before they could plant crops, though, they had to cut down areas of forest. What do you think happened to the many animals and other living things that depended on the forest for food and shelter? What would have happened if all the forests had been cut down? Do you think the settlers' fields provided a home for any living things? Ⓐ

In many parts of the United States today, forests and fields are being replaced by houses and other buildings. As human communities grow larger, communities of other living things get smaller.

Some changes in a community, however, occur because of what seem to be unrelated events. Consider the following statement:

The number of mice in a community indirectly affects the number of rabbits in the same community.

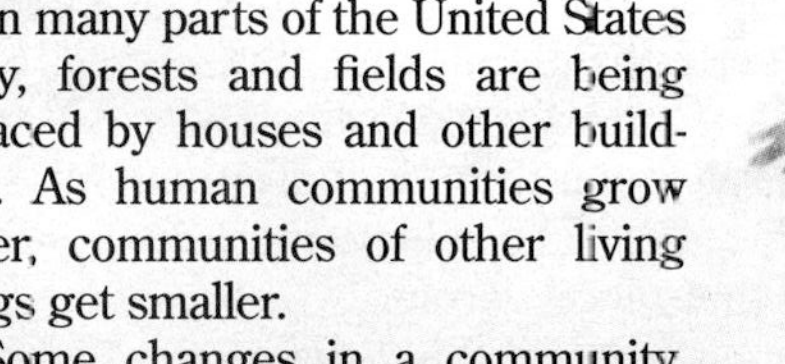

Answers to *In-Text Questions*

Ⓐ Students should realize that when the settlers cut down a forest, the plants and animals that were dependent on it died unless they could get to another forest.

If all of the forests had been cut down, all of the plants and animals that were dependent on them would have become extinct if they could not adapt to their new environment.

Settlers' fields were probably less diverse because many types of plants were replaced by a few types of plants. Therefore, fewer kinds of organisms could survive there. However, the settlers' fields provided a new environment for many animals, such as birds, snakes, rabbits, deer, mice, and shrews. Some of these animals attracted predators, such as owls and hawks. Help students to conclude that any human impact on the environment affects animal and plant populations.

Population Changes

GUIDED PRACTICE Before students begin reading, direct their attention to the illustration, and have them identify and discuss some of the roles and relationships that exist between the animals and plants shown. *(Possible responses include the following: the fox eats mice, which eat grasses and seeds. The hawk eats rabbits, which eat grasses and leaves. The mouse is prey for the fox, which is a predator. The rabbit is prey for the hawk, which is a predator. The fox and the hawk are carnivores. The rabbit and mouse are herbivores.)*

Introduce the term *population* to the students. Explain that a population is a group of organisms of the same species living in the same place. Have students read the first two paragraphs and pause to consider the questions. Then have students read the rest of the page and consider the statement that appears at the end of the text. Ask them to speculate about what the statement might mean.

Population Changes continued ►

Theme Connection

Changes Over Time

Focus question: Do the changes that occur in the rotting-log community fit the definition of succession? Why or why not? *(Most students will answer yes to this question. Bacterial action softens the wood under the bark. This paves the way for insects to burrow into the bark and into the wood under the bark. Plant roots penetrate the log, causing it to break apart. The spaces created allow more organisms to make their homes in the rotting log.)*

Homework

You may wish to assign the Graphing Practice Worksheet on page 51 of the Unit 1 Teaching Resources booklet as homework. This worksheet accompanies Population Changes.

Population Changes, *continued*

Have students read the paragraphs at the top of the page, pausing at the population graph. Involve them in a discussion of what they have just read in order to answer the in-text questions.

Answers to *In-Text Questions*

A Because of the increased food supply, the mice will thrive and reproduce. The immediate result will be an increase in their population. Because the conditions affecting the rabbits have not changed, there will be no initial effect on the rabbit population.

B The increased mouse population indirectly increases the rabbit population. This is because mice become easier to catch than rabbits. Therefore, hawks and foxes will hunt more mice and fewer rabbits, and the rabbit population will increase slightly.

C When the population of rabbits begins to increase, they are more easily seen and caught by predators than the mice. Now the pressure is off the mouse population, which had been in decline because of excessive predation and lack of food. The mouse population begins to increase to its original size.

D
- **a.** 7—the original population of rabbits
- **b.** 10—the largest population of rabbits
- **c.** 3—the largest population of mice
- **d.** 5—the smallest population of mice
- **e.** 2, 8—The mouse population begins to increase much more rapidly than the rabbit population does.
- **f.** 4—The mouse population begins to decline.
- **g.** 9—The rabbit population increases dramatically.
- **h.** 11—The rabbit population begins to decline.
- **i.** 6—The mouse population begins to return to normal.
- **j.** 6, 12—Both populations return to where they were at the beginning.

★ **Transparency 9 is available to accompany this page.**

How can this be? Do rabbits eat mice, or vice versa? Mice and rabbits are both herbivores—that is, they eat green plants. Suppose that the type of food mice like to eat starts growing faster than usual in the area. Soon there is more than enough food for the mice. There is no increase in the type of food that rabbits prefer.

What immediate effect will this have on the mouse and rabbit populations? **A**

Foxes and hawks eat both mice and rabbits. Mice are easier to catch, so with the increase in the mouse population, the foxes and hawks will eat more mice than rabbits. What is the reason for this change of diet? What will happen to the rabbit population now that the foxes and hawks have changed their diets? Judging from what you've learned, does the mouse population in this community affect the rabbit population? **B**

What do you think will eventually happen in this community? Why do you think this will happen? **C**

The changes in population of mice (—) and rabbits (- -) are shown in the graph below.

Each of the events listed at right is represented by a number on the graph. Can you match the number to the event? **D**

a. The original population of rabbits
b. The largest population of rabbits
c. The largest population of mice
d. The smallest population of mice
e. Mouse and rabbit populations when the increase in plants that mice prefer is first noticed
f. The mouse population when their food supply is exhausted and when foxes and hawks are still preying on them more than on rabbits
g. The rabbit population when the foxes and hawks begin preying on mice more than on rabbits
h. The rabbit population when the mice run out of food and the foxes and hawks are preying on rabbits more than on mice
i. The mouse population when foxes and hawks are preying more on the rabbits
j. A return to original population counts of mice and rabbits

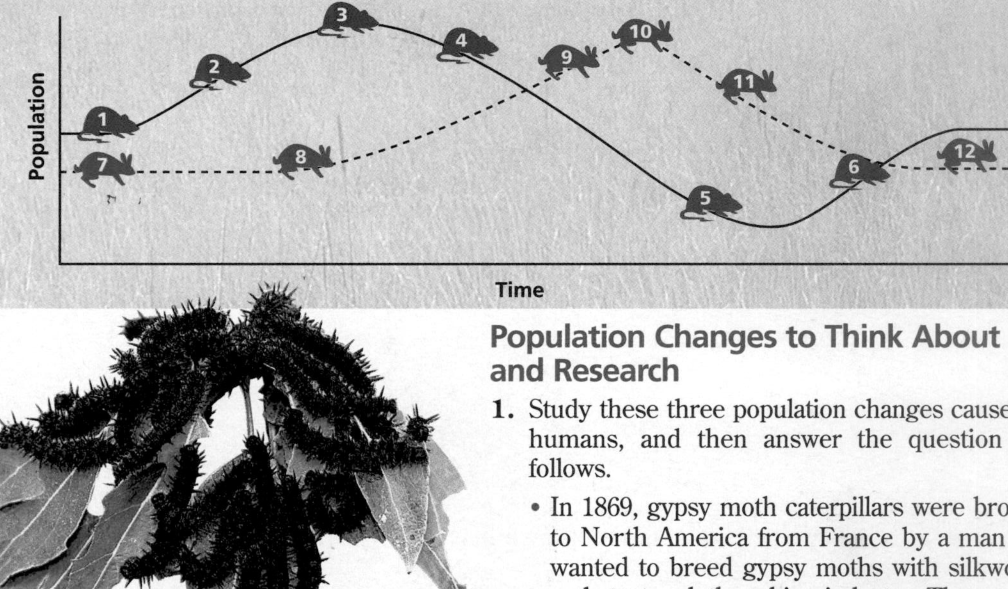

Gypsy moth caterpillars

Population Changes to Think About and Research

1. Study these three population changes caused by humans, and then answer the question that follows.
   - In 1869, gypsy moth caterpillars were brought to North America from France by a man who wanted to breed gypsy moths with silkworms and start a cloth-making industry. These caterpillars now kill shade trees by eating their leaves.

54

Population Changes to Think About and Research

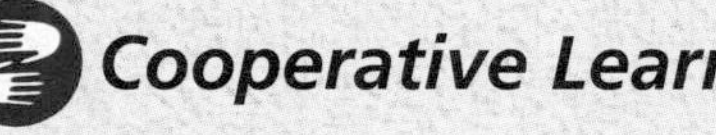

Cooperative Learning

Group size: 2 to 3 students
Group goal: to research and report on one of the topics listed in the text
Positive interdependence: Assign each student a role such as principal investigator, data compiler, or editor.
Individual accountability: Each member of the group should be able to give a short, oral description of the group's research and results.

Answers to *Population Changes to Think About and Research, pages 54–55*

1. When a new species is introduced into a location where it has no natural enemies, its population can increase unchecked. The resulting effect on native plants and animals may be disastrous. Other examples of introduced species that have caused problems include melaleuca trees in Florida and the Nile perch in Lake Victoria, Africa.

Answers to Population Changes to Think About and Research continued ►

- In 1859, 24 European rabbits were introduced into one area of Australia. These rabbits had no natural enemies in their new environment: there were no predators, parasites, or diseases to affect them. The rabbits quickly became pests, destroying all plant life in the area. Within just six years, more than 20,000 of the rabbits had to be killed in an effort to control their growth.
- In 1890, an Englishman brought starlings to North America from Europe. Now the starling, a noisy and destructive bird, is found all over North America. It has displaced many native songbirds from their niches.

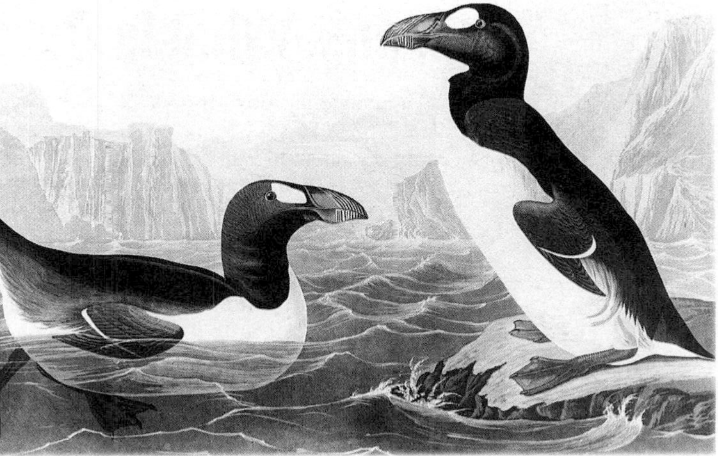
Painting of great auks by John James Audubon

How do some animals get out of control and become pests? Search for other examples of "out-of-place" animals or plants that have become pests.

Lemming

2. The lemming is a small rodent that lives in cold, northern parts of the world. Every few years, according to legend, lemmings leave their habitats in large numbers and drown themselves in the ocean. However, scientists now know that lemmings do not deliberately kill themselves. Find out more about lemmings and the reasons they leave their homes and die in large numbers.

3. In 1534, Jacques Cartier landed on the east coast of North America. He found thousands of flightless birds that were clumsy on land but were powerful swimmers. These birds, known as great auks, soon became regarded as a source of profit. They were used for food, their feathers were used for making featherbeds, and they were even boiled for their oil! As a result, no one has seen a great auk alive since about 1844. Find out more about why and how this happened.

   What other animals have become extinct in the past century? What animals are in danger of becoming extinct?

Are there also plants that are in danger of becoming extinct? E

Answers to *Population Changes to Think About and Research, continued*

2. Lemmings are small rodents that inhabit the mountains of Canada and Scandinavia. Every few years, their food supply increases. As a result, there is a rapid increase in the size and frequency of lemming litters. After several years, the population of lemmings is too great for the food supply. Thousands of lemmings leave in search of food. Many of them die due to exposure and predation.
3. Eventually, so many auks were killed that there were not enough left to maintain the population—that is, more auks died each year than were born. The great auk was hunted to extinction. Some other animals that have become extinct in the last century include the passenger pigeon, Carolina parakeet, Labrador duck, moa, and Steller's sea cow. Well-known endangered species include the grizzly bear, Florida panther, red wolf, grey wolf, bobcat, cheetah, Asian elephant, lowland gorilla, leopard, giant panda, black rhinoceros, tiger, hooded crane, California condor, whooping crane, American peregrine falcon, and American crocodile.

Answer to *Caption*

E Yes, the Santa Cruz cypress and the white bladderpod are two examples. Many plants in rain forests around the world are endangered due to habitat destruction.

FOLLOW-UP

Reteaching

Discuss with students how a small store can go out of business and be replaced by a different store that occupies the same location. Point out that this is an example of succession in a human context.

Assessment

Have students detail the succession process as it happens after a forest fire. Encourage them to include details of the levels of the forest, from the forest floor to the canopy.

Extension

Have students do research on what happens when forests are clear-cut for logging or for use as farmland. Students could focus their research on the rain forests of South America or the old-growth forests of the Pacific Northwest.

Closure

Have students imagine that they are preparing land to make a garden. The land is plowed, raked, and ready to be used. But then something happens, and the garden never gets planted. Ask students to draw three illustrations that show what happens to the garden by the end of the summer, after one year, and after several years. Display the finished drawings around the classroom.

Humans: What Is Our Role?

FOCUS

Getting Started

Have students discuss what is meant by Chief Seattle's statement. *(Sample answer: Anything humans do to the environment eventually affects humans themselves.)*

Main Ideas

1. The impact of humans on the environment has often upset the delicate balance of nature.
2. Acid rain is one example of the serious consequences that human activity can have on the environment.
3. People can have a positive influence on the environment and can correct mistakes that they made in the past.

TEACHING STRATEGIES

GUIDED PRACTICE Ask students to consider how people interact with the environment. Write some of their ideas on the chalkboard. *(People use natural resources, dispose of wastes, grow food, and create pollution.)* Use the list to point out that people can either help or harm natural communities.

Answers to In-Text Question, Caption, and Facing the Problems are on the next page.

Humans: What Is Our Role?

Man did not weave the web of life, he is merely a strand in it. Whatever he does to the web, he does to himself.
—attributed to Chief Seattle

Sometimes it's easy to see ourselves as separate from animals, plants, and other parts of nature. We know we influence our environment, but we often behave as if the environment has little influence on us. Sometimes we humans act as though we have complete authority over both living and nonliving things. How do the population changes you just studied illustrate this? (A)

It is true that humans can manipulate and use the environment in ways that no other living thing on Earth can. With this ability comes a responsibility to consider the effects of our actions not just on ourselves and future generations of humans but on every other living thing on Earth.

Facing the Problems

When we do not carefully consider our relationships with the environment, difficult situations sometimes result. Here are some issues to consider. Think about the issues, research them, and discuss them with your fellow students.

1. What harm has resulted from the careless use of pesticides? Is this a problem in the area where you live? What evidence of a problem is there? What is being done about it? What future problems might result from the careless use of pesticides?
2. What are the problems created by smog? Is there air pollution where you live? What is being done about it? What effect might cleaning up air pollution have on the environment 100 years from now?

What is the problem? What is the solution? (B)

LESSON 2 ORGANIZER

Time Required
4 class periods

Process Skills
observing, analyzing, inferring

New Terms
Acid rain—precipitation that occurs when pollution from burning gas, oil, and coal mixes with water vapor in the air to form acid; this acid then falls as snow or rain
pH scale—a scale ranging from 0 to 14 that is used to measure the acidic or alkaline content of a substance

Materials (per student group)
Learn More About It! colored markers; poster board

Teaching Resources
Discrepant Event Worksheet, p. 53
Transparency 10
SourceBook, p. S19

Did You Know . . .

In the early 1800s, Chief Seattle was the chief of several tribes in the Puget Sound area. He is remembered in part for his eloquent writings and speeches concerning the role of humans in the environment. The city of Seattle, Washington, is named after him.

3. What have industrial wastes done to many rivers and lakes? What are "hazardous wastes"? How do these kinds of wastes affect the way humans live today? How might it affect the way humans live in the future?
4. What happens to forests and farmland when human communities, including industries and highways, expand? Is this a problem in your area? What effect does this type of development have on animals that make their homes in these areas?

We have come to realize that when we damage living things or thoughtlessly change nonliving things, we may hurt ourselves. Because we are part of the web of life, we are affected by changes in our surroundings. How can we ensure that these changes are not destructive to the millions of species that live on this planet, including humans? Survival means thoughtful, careful interactions between us and all parts of the environment—living, once living, and nonliving.

A Sustainable Future

Careful thought about the needs of other organisms and of future generations of humans when we plan our own use of land, air, and water is known as *sustainable development.* In other words, if humans carefully consider all the players on Earth and conserve natural resources now, we sustain the ability of all living things to survive on Earth in the long run. Sustainable development will be most effective if it takes place at many levels—international, national, local, and individual.

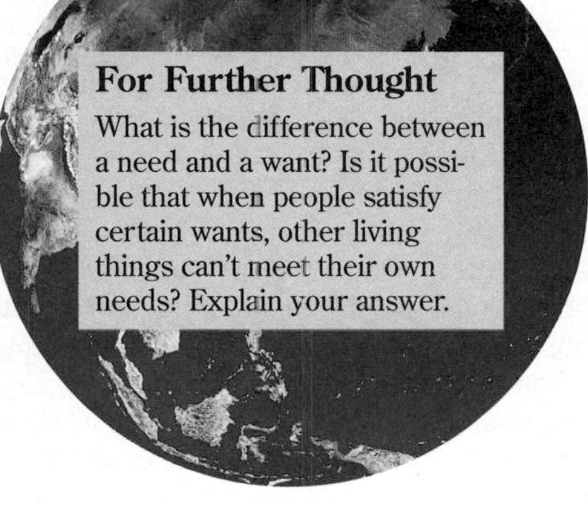

For Further Thought

What is the difference between a need and a want? Is it possible that when people satisfy certain wants, other living things can't meet their own needs? Explain your answer.

International Cooperation, National Responsibility

The quality of the environment affects living things all around the world. For example, air that is polluted in one part of the world will one day be breathed by humans and other organisms in another part of the world. Because we are so connected, the people of many nations have decided to cooperate in efforts to improve our global future.

In 1992, a subcommittee of the United Nations held a large meeting in Rio de Janeiro, Brazil, called the Earth Summit. A summit is a meeting of government officials or other high-ranking officials to talk about important issues. Representatives from 179 nations (including the United States) attended the Earth Summit. At the meeting, the representatives discussed ways to reduce pollution, save endangered species, and protect the world's forests.

57

Answers to *Facing the Problems, pages 56–57*

The following are sample answers:

1. Pesticides have contaminated the food we eat, killed wildlife, and polluted water. Answers will vary depending on where the students live. Warning labels are printed on pesticides. Future problems might include human illnesses and resistance in the organisms that the pesticides are supposed to kill.
2. Smog can cause serious illnesses, such as lung cancer, emphysema, and heart disease. Cleaning up air pollution might allow natural communities to remain healthy for the next 100 years.
3. Industrial wastes have polluted natural supplies of soil and water. Hazardous wastes include chemical and nuclear wastes. These wastes could interfere with human life by causing birth defects, disease, and hardship. People who live in an area where they might be exposed sometimes have to move. In the future, humans might have special filters on their water faucets and might have to wear special protective clothing.
4. Forests are cut down and are never replaced. Farmland can be lost to development. Answers will vary depending on where students live. Animals in these areas have their habitats destroyed and have to find new homes, or they will die.

A Sustainable Future

Discuss the term *sustainable development* with students. Ask students for examples of how to conserve natural resources. *(Possible student responses could include carpooling to save gasoline, limiting water use, and conserving electricity.)*

Answers to *In-Text Question and Caption, page 56*

A In the examples at the end of Lesson 1, humans failed to consider the long-term consequences of their actions.

B The problem is smog created by vehicle exhaust (left). The solution is commuting by bus, which helps to reduce traffic and pollutant emissions (right).

Answers to *For Further Thought*

A *need* is a necessity. Food, clothing, and shelter are examples of needs. A *want* is a wish for something. An example is, "I want a new fur coat." If the wants of humans are satisfied, it is possible that other living things will not be able to meet their needs. (In the above example, an animal's life is sacrificed so that a human's wish can be fulfilled.)

Homework

You may wish to have students answer the questions posed in For Further Thought as homework.

Teaching Strategies for International Cooperation, National Responsibility are on the next page. ►

International Cooperation, National Responsibility,

pages 57–58

Have students read the material silently. Then involve them in a discussion of the benefits of the Earth Summit. *(Students may conclude that this is a beginning to help the Earth. If people are made aware of the global problems, then they might cooperate to help save the Earth from destruction caused by mismanagement of resources.)*

Answer to *In-Text Question*

Ⓐ Student answers will vary. All nations may not have the same want or need to protect the environment. If a plant or animal species is important to the livelihood of a particular nation, then protecting it would be detrimental to that nation.

In our own nation, it is the job of our government, law enforcement agencies, and citizens to enforce various aspects of the agreement.

Answers to *Check It Out!*

One of the agreements that came out of the Earth Summit was the Biodiversity Treaty. This treaty encourages wealthier countries to give money to poorer countries for the protection of potentially valuable species.

Another result of the Earth Summit was an international agreement to curb global warming that went into effect on March 21, 1994.

Where Do You Fit In?

Have students read this section silently and copy the table into their ScienceLog. Brainstorm with the students ways that individuals, families, cities, and nations can contribute to sustainable development. Then instruct students to list three activities that could contribute to sustainable development in each category. The tables should be similar for each student.

The president of Brazil and a young girl plant a tree during the Earth Summit in 1992

Perhaps the major achievement of the Earth Summit was to launch a global partnership for sustainable development. Let us not forget that such a partnership should include and benefit the young, in whose hands lies the future of this planet.

—Boutros Boutros-Ghali, Secretary-General of the United Nations

One of the agreements made in the Earth Summit is called Agenda 21. Agenda 21 outlined a plan for nations to help each other protect the environment and practice sustainable development. While Agenda 21 suggests that nations help each other in this effort, each nation is responsible for interpreting and enforcing the agreement as it sees fit. This means that no nation can tell another nation how to protect the environment. Do you agree with this type of agreement? Why or why not? In our own nation, whose job is it to make sure we live up to the agreement? Ⓐ

Young people also held a summit at Rio. At the conference, these young people from around the world expressed their concerns about the global environment and shared their ideas about what all people—youth and adults—could do to protect the future of the Earth. Attendees of the youth summit even published a book so that they could share their concerns and hopes with people everywhere.

Check It Out!

Has the Earth Summit produced results? Have any other international agreements been made to solve environmental problems?

Where Do You Fit In?

How do you think individuals can contribute to sustainable development? Can you and your classmates make a difference at a local and individual level? What are some things that you and your family could do at home or in your community to contribute to sustainable development?

Copy this table into your ScienceLog. List at least three activities that could contribute to sustainable development in each category. (Remember, an important factor of sustainable development is ensuring that our activities today do not cause problems for the generations of tomorrow.) How does your table compare with those completed by your classmates?

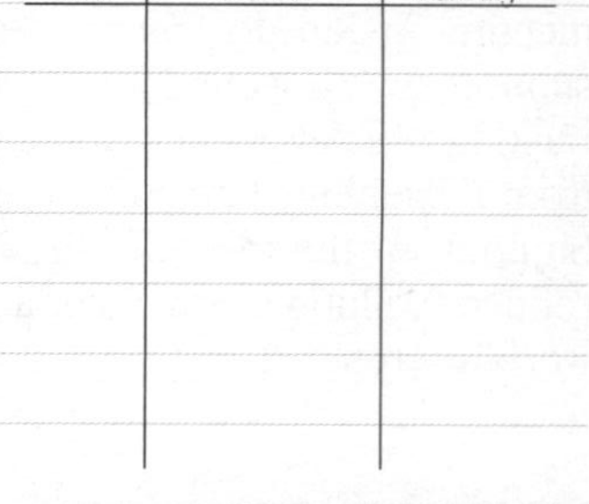

Contributions to Sustainable Development

Me	Class	Family

58

Answers to *Where Do You Fit In?*

Sample answers:

Individual students could turn off water when not using it, recycle trash, and turn off the television when not watching it.

The class could start a recycling program, buy and plant trees with money from recycled cans, and turn off lights when not in the classroom.

The students' families could walk instead of driving a car whenever possible, recycle newspapers and trash, and conserve water.

In the next few pages, you'll read about how a group of students decided to get involved with solving one environmental problem—acid rain.

The Story of Mr. Ripley's Class

Mr. Ripley's class in environmental science is no ordinary class. They study environmental problems that affect their community and beyond. They have expressed their concerns in such practical ways as writing a page on environmental issues in the local newspaper and participating in energy conservation activities. They even held an environmental fair at school to inform their classmates and teachers about ways to help the environment.

The students are quick to initiate action. So when the city council announced the formation of a committee to study acid rain, Mr. Ripley's class decided to get involved. Here are some of the questions they investigated:

- What is acid rain?
- What causes rain to become acidic?
- Is it a natural problem, or is it caused by people?
- What harm does it do?
- Is it a widespread problem?
- Does it do any harm locally?
- If acid rain is harmful, what can we do as a group of people and as a government to control the problem?

How well informed are you? How many of these questions can you answer? What other questions would you have asked? **B**

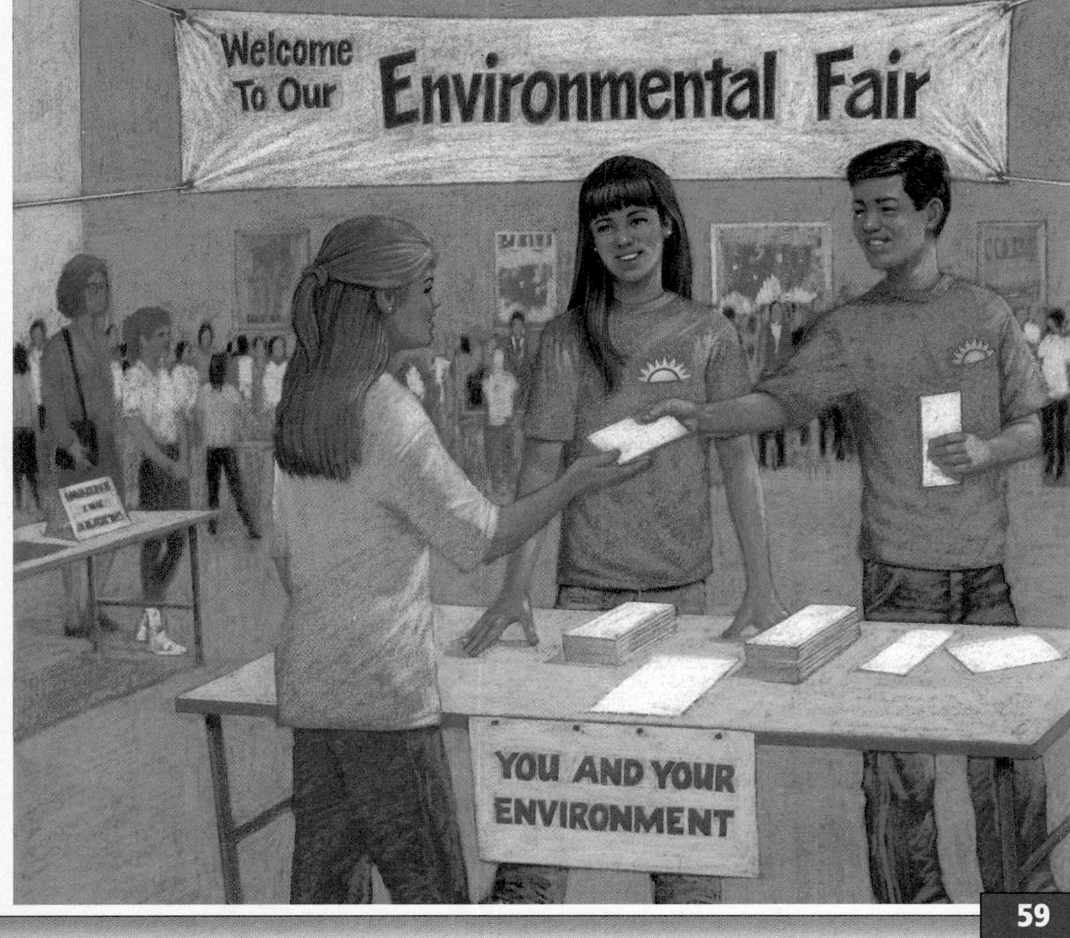

★ **The Discrepant Event Worksheet on page 53 of the Unit 1 Teaching Resources booklet decribes a teacher demonstration that can be used to introduce The Story of Mr. Ripley's Class.**

Answers to *In-Text Questions*

B Some other questions that students might have asked include the following: What are some long-term effects of acid rain? Where are the highest concentrations of acid rain in this country? in other countries?

The Story of Mr. Ripley's Class

The value of this lesson is that it examines a current environmental problem and illustrates how students can effectively contribute to its solution. The lesson also provides considerable information about acid rain, including its harmful effects, causes, and possible solutions. If possible, have a variety of resources about acid rain available for students. Remind students to seek parental guidance when getting involved in community activities.

GUIDED PRACTICE Have students read the page silently. Then involve them in a discussion of each of the questions. Accept all reasonable responses without comment. Suggest that students record their ideas in their ScienceLog. Point out that as they read more about Mr. Ripley's class, they will discover more about acid rain and the answers to these questions.

Meeting Individual Needs

Learners Having Difficulty

Students that have access to cameras might enjoy taking pictures for a photo essay about how animals and plants interrelate in biological communities. Allow students some flexibility in how they interpret the concepts. Arrange the finished photo essays around the classroom in a manner that suggests an art gallery.

Meeting Individual Needs

Second-Language Learners

Point out to students that in this unit they have learned many terms that relate to the interrelationships and roles of organisms in biological communities. Suggest that they make a list of the terms and use them to create a bilingual dictionary. Suggest that students add illustrations to clarify their definitions. Invite them to make a cover for their bilingual dictionary and to think of a title. When they are finished, review their work for science content, with minimal emphasis on language proficiency.

What the Students Found Out in Their Research

Have students read the page silently. Then call on a volunteer to describe how acid rain results from air pollution. If students have difficulty grasping the concept, draw a diagram on the chalkboard. Have students write explanations of how acid rain is formed. Point out that some natural events, such as volcanic eruptions, can also contribute to air pollution. In addition, you may wish to explain that some scientists prefer to use the term *acid precipitation* because this phenomenon includes forms of precipitation other than rain. Continue the class discussion until students are comfortable with these concepts.

Direct students' attention to the pH scale. Be sure that they understand which end of the scale represents bases and which represents acids. Point out that the number 7.0 on the scale is neutral because it lies exactly in the middle of the acid-base range.

Answers to *In-Text Questions*

1. Lemon juice
2. Milk of magnesia has a pH of about 10.5. It is a base and helps to neutralize acids in the stomach.
3. Yes
4. Soils are often acidic. Calcium oxide, commonly called lime, is a very basic compound, with a pH of about 12.4. Spreading lime on acidic soil neutralizes, or reduces, the acidic content of the soil. This is especially helpful in the spring after acidic snow melts into lawns.
5. 100 times stronger

What the Students Found Out in Their Research

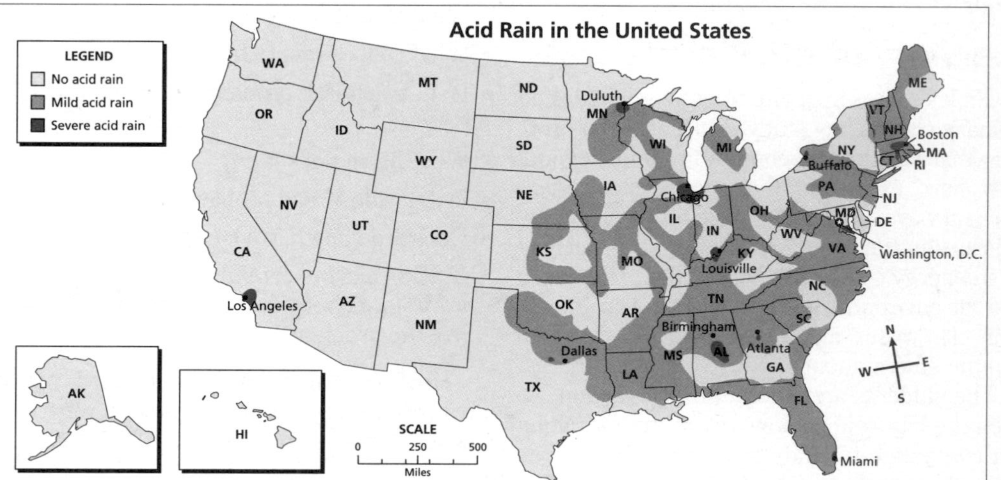

When gas, oil, and coal are burned, compounds that are released combine with water vapor in the air to form acid. The acid then falls to Earth with snow and rain. This **acid rain** can kill trees and make ponds and lakes unable to support life. Some lakes in the eastern United States have been so affected by acid rain that they can barely support life.

The acidity of a substance is measured on a **pH scale** of 0 to 14. Look at the scale below. It shows the acidity of many different liquids and foods. A pH of 0 indicates the most acidic solution, and a pH of 14 indicates the most basic. Anything that has a pH value below 7 is acidic. The lower the number, the greater the acidity. The middle point, pH = 7, is neutral.

A pH value greater than 7 indicates the presence of a *base*. Basic solutions are often referred to as *alkaline*. Alkaline substances can *counteract*, or neutralize, acids; that is, they reduce the acidity of a substance.

Look at the scale to answer these questions:

1. Which is more acidic—tomato juice or lemon juice?
2. Have you heard of people taking milk of magnesia for an acid stomach? Why would they do this?
3. Is ordinary rainwater acidic?
4. Homeowners often put lime on their lawns in the spring. Why?
5. When the pH drops by one point, the acidity increases by 10 times. For example, a solution with a pH of 4 is 10 times as acidic as a solution with a pH of 5. If the pH dropped by two points, what would the increase in acidity be?

60

Meeting Individual Needs

Learners Having Difficulty

Have students collect some rainwater and then test its pH using litmus paper. The pH of lakes and streams can also be measured. Other schools in different parts of the country can also be contacted to set up similar experiments. Compile the data collected to determine any patterns in acid-rain locations.

Integrating the Sciences

Earth and Physical Sciences

Ask students to research a chemical equation for acid-rain formation. *(There are several formulas that describe the formation of acid rain from different chemicals in the atmosphere. Sample answer: Sulfur oxides and water vapor combine and react in a series of steps to form sulfuric acid.)*

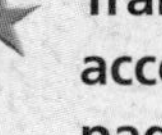

Transparency 10 is available to accompany the material on this page.

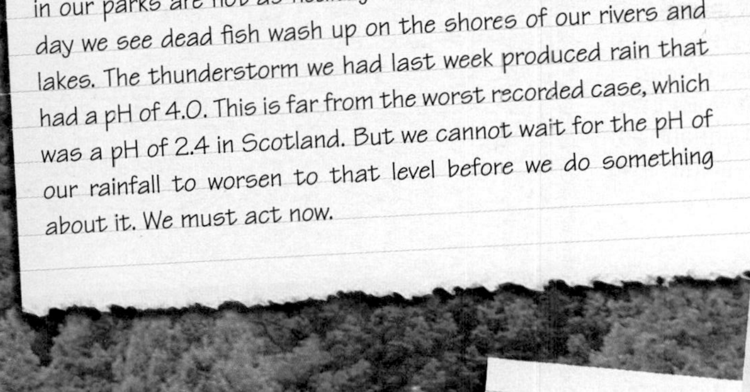

The Class Report

Now read what the students in Mr. Ripley's class wrote in their report to the city council concerning the pH of rainwater in their city:

Acid rain has become a major problem in our city. The trees in our parks are not as healthy as they used to be, and every day we see dead fish wash up on the shores of our rivers and lakes. The thunderstorm we had last week produced rain that had a pH of 4.0. This is far from the worst recorded case, which was a pH of 2.4 in Scotland. But we cannot wait for the pH of our rainfall to worsen to that level before we do something about it. We must act now.

Look at the pH chart again. What liquids have a pH near 2.4? Imagine rain made of any of these liquids! (A)

What do you think causes the pH of some rainwater to be low? Mr. Ripley's class easily discovered the answer. Industries, cars, and trucks release gases (sulfur oxides and nitrogen oxides) into the air. When rain or snow falls, these gases dissolve in the water vapor to form sulfuric acid and nitric acid. Below is more information from the students' report.

In North America, air pollution is a major concern. Every day, at least 165,000 metric tons of sulfur and nitrogen oxides are released into the atmosphere. This pollution remains in the air for up to 10 days and can travel thousands of kilometers from its source, depending on the wind currents. In other words, sulfuric acid clouds rising from factories in and around the Pittsburgh area might be over the Great Lakes region in two days. So our pollution can affect more than our community. Likewise, another region's pollution might be making our situation worse. If every community reduced the amount of sulfur and nitrogen oxides that are released by cars and industries, we would all be better off.

Cross-Disciplinary Focus

Language Arts

After students have read the report that Mr. Ripley's class presented to the city, ask the following questions:

1. How does sulfur and nitrogen oxide released in Pittsburgh cause acid rain at Lake Erie?
   - **a.** by the formation of clouds over Lake Erie
   - **b.** by thunderstorms that carry acid rain
   - **c.** from an increased amount of pollution from cars
   - **d.** by wind currents that carry pollutants from one region to another *(Correct)*
2. In the first paragraph, the students believe that the fish in the rivers and lakes are dying because of
   - **a.** increased acidity of the rivers and lakes. *(Correct)*
   - **b.** greater predation.
   - **c.** dumping of toxic wastes by industries.
   - **d.** diseases caused by sulfur and nitrogen oxides.

The Class Report

Have students begin reading the page, pausing after the first part of the report to the city council. Have students turn back to the pH scale on the previous page to find out what liquid has a pH near 2.4.

GUIDED PRACTICE Encourage students to speculate why acid rain kills fish and trees. Accept all reasonable responses. *(Acid rain finds its way into lakes, rivers, and streams and can lower the pH. This can upset the delicate balance of abiotic factors in those communities and cause many fish and other aquatic organisms to die. Acid rain falls on forests and can kill the plants living in the forest community that cannot tolerate the higher acidity.)* Help students recognize that acid rain upsets the existing balance in the abiotic factors of a biological community, which results in serious consequences for the organisms that live there.

Answer to *In-Text Question*

(A) The pH of vinegar, cola, and lemon juice is near 2.4.

Multicultural Extension

What's the Environment Like?

Point out to students that people inhabit many different kinds of environments. Suggest that they find out what the environment is like in a different part of the world. Students should include information on how different cultures respond to their environments by investigating such topics as clothing, shelter, and food. Have students try to determine whether the practices of the culture they investigate are in harmony with the culture's natural surroundings. Have students use this information to write travel brochures that describe different environments. Students should illustrate their travel brochures and include them in a bulletin-board display. If you have students from different countries, invite them to describe the environment of their native area.

The Effects of Acid Rain

Have students read the material silently. Point out that acid rain can kill not only fish, but also other aquatic organisms, such as algae, frogs, salamanders, and insects. Involve students in a discussion of how acid rain may affect the food webs of aquatic communities. Help them recognize that when one organism is affected by acid rain, chances are that many other organisms will be affected too. For example, if there are no more frogs, then wading birds, such as herons, cranes, and egrets, will stop coming to ponds and lakes to look for food. If the fish die, eagles will also stop coming in search of food.

Answers to
In-Text Question and Caption, pages 62–63

Ⓐ In response to the research suggestion, students should find that the high acidity of the water raises the acidity of the blood in the fish and poisons them. Lakes become more acidic over time because acid rain continues to fall on them. Also, acid rain falls on the soil and leaches out minerals, which may eventually kill much, if not all, of the life found in the lake.

Ⓑ Acid rain killed the trees in this forest. The forest can no longer provide the wood needed to make paper and build houses and other structures.

Ⓒ The United Nations Sofia Protocol was signed by 27 nations in 1988. It required a reduction in nitrogen oxide emissions. The United States agreed to this protocol in 1989. Reduction of these emissions has helped to control the acid-rain problem.

Some Controls

GUIDED PRACTICE Before students read the selection, involve them in a discussion of what might be done about the acid-rain problem. List their suggestions on the chalkboard. Then have students read the material to see if they included the recommendations made by Mr. Ripley's class. *(Additional suggestions include the following: encourage the use of nonpolluting automobiles, and provide more funding for research into nonpolluting energy sources.)*

The Effects of Acid Rain

Mr. Ripley's class found plenty of information about the effects of acid rain on the environment. They learned that acid rain can lower fish populations and can seriously harm aquatic life. Acid rain can also destroy soil and trees. The students realized that when soil is affected, crops may not grow as well as they would in soil with normal acidity. When acid rain kills trees, many animals lose their homes and sources of food.

Do some research to find out how acid rain actually affects the life processes of aquatic animals. Why do lakes become more acidic over time? Ⓐ

Even the economy may suffer when trees are harmed by acid rain. Why do you think this is so? Ⓑ

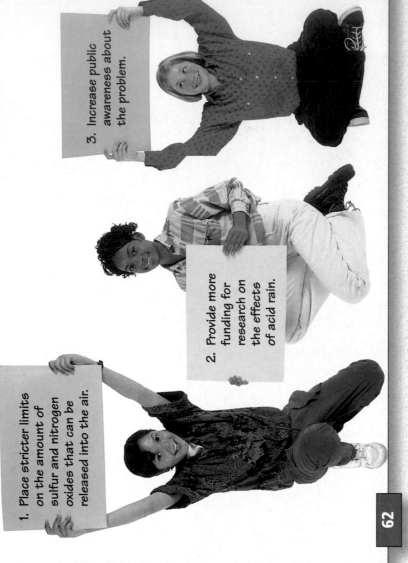

Some Controls

With your classmates, discuss what could be done about the acid rain problem. Make a class list of possible solutions.

Below are three of the eight specific recommendations that Mr. Ripley's class made in their report to the city council. Did you include any of these among your own suggestions?

1. Place stricter limits on the amount of sulfur and nitrogen oxides that can be released into the air.
2. Provide more funding for research on the effects of acid rain.
3. Increase public awareness about the problem.

62

ENVIRONMENTAL FOCUS

Ask: What are some changes or sacrifices that you would be willing to make in your own life to solve environmental problems? *(Answers will vary. Students may list recycling, using public transportation, carpooling, walking or riding bicycles, turning off appliances that are not in use, and educating others about conservation.)*

Homework

As homework, you may wish to have students answer one or more of the questions posed in Learn More About It! on page 63.

Learn More About It!

1. Prepare posters for the school and community alerting people to the acid rain problem. Use the information on acid rain provided in these pages.
2. Answer these questions: What occupations in your state are likely to be affected by acid rain? What effect do you think the increasing problem of acid rain will have on the general public? How does the public contribute to the problem? Compare your answers with those of your classmates.
3. Rank these factors in terms of how they contribute to the acid rain problem today: weather conditions, industry, cars and trucks, an uninformed public, a government that is dealing with other matters, and inadequate education.
4. Call your local weather service and ask if they monitor the pH level of local rain. If so, chart the pH values during the next couple of months to see how they fluctuate. Would you expect the same values in the fall as in the spring? Why or why not? What local factors might influence the pH values? Is acid rain a problem in your community?
5. Find out what programs have been initiated on a local, national, and international level to help deal with the problem of acid rain. Who started these programs? What are their goals? What progress have they made so far? Present your findings to the class.

Find out about the United Nations Sofia Protocol, signed in 1988. What effect has it had on acid rain?

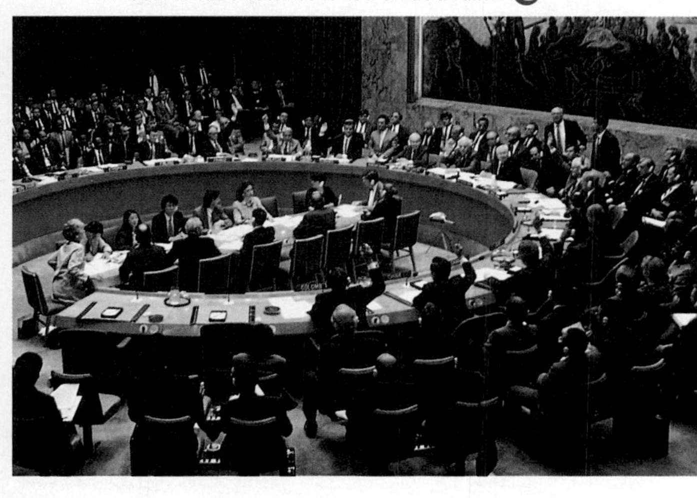

What are some things humans can do to make the future better for all organisms on Earth? Check it out on pages S22–S24 in the SourceBook.

Answers to *Learn More About It!*

1. Posters should reflect an understanding of the causes, effects, and possible solutions of the problem of acid rain.
2. Acid rain strongly affects the fishing, lumber, and tourist industries. As the public becomes more aware of the problem, more pressure may be exerted on public officials to find a solution. The public itself contributes to the problem by relying heavily on automobiles for transportation.
3. One possible ranking is the following: industry, cars and trucks, uninformed public, government preoccupied with other matters, inappropriate education, and weather conditions.
4. The pH levels in many locations may be higher in the spring because melting acidic snow can cause large amounts of acidic water to be released into natural water sources. Answers will depend on the students' local environment. Local factors might include a nearby factory, local weather patterns, or the amount of road traffic in the area.
5. Answers will vary. Sample answer: On an international level, several treaties have been enacted. For instance, the United Nations Helsinki Declaration, which required countries to cut sulfur oxide emissions by 30 percent over 10 years, was enacted in 1985. The declaration was signed by 18 nations. The United States did not sign this agreement.

FOLLOW-UP

Reteaching

Have students bring in articles about acid rain that they find in magazines and newspapers. Use the articles to set up a reading center. Have students add to the center during the weeks that follow.

Assessment

Provide students with local newspaper articles that describe an interaction between humans and the environment. Ask students to write a paragraph explaining how they feel about the interaction and whether it is harmful, is helpful, or has no effect on the environment.

Extension

Point out to students that in addition to acid rain, there are other environmental problems facing the world, such as the greenhouse effect, overpopulation, the deterioration of the ozone layer, and deforestation. Suggest that students choose one of these topics to research. Students should describe the nature of the problem as well as any debates about the problem's severity.

Closure

Point out to students that acid rain is a global problem. Suggest that they do some research to discover other countries that are affected by it and what these countries are doing to solve it. Have them share what they learn by presenting a "world news report" on acid rain to the class.

Answers to *Challenge Your Thinking*

1. Sarah sees humans as separate from other animals. She sees humans as controlling and taking responsibility for the welfare of other living things.
   Jim believes that humans influence the environment, but that humans are likewise influenced by the environment and all it contains, just as other animals are.
   Student opinions may differ. Students should be able to support their opinion with concrete examples.
2. The world's population grew steadily and then skyrocketed in the last millennium. This differs from other animal populations, which tend to fluctuate and not to experience such explosive growth. Students may identify a number of reasons for this difference. One example might be that advances in medicine have decreased the death rate and increased the birth rate. Students may predict problems with limited housing, food, and natural resources. They may also choose to discuss technological advances that might alleviate these problems.
3. The graph should show a dramatic decline in the frog population following the introduction of snakes. The snake population will grow because of the available food source. Eventually the population of frogs will drop so low that it cannot support the population of snakes, resulting in a decline in the snake population. The two populations will reach an equilibrium.
4. The student dialogues should describe the natural succession of a pond. This would include the growth of weeds and aquatic plants as well as the growth of trees and undergrowth surrounding the pond. Students may also include information on how acid rain has affected the pond.
5. Household spiders eat many household insects. To kill off these spiders would mean that more insects would survive to reproduce, resulting in an increase in the population of household insects.

CHALLENGE YOUR THINKING

1. Humans: Bosses or Co-workers?

Sarah and Jim used diagrams to show their understanding of the role of humans in their environment. What is Sarah's view? What does Jim think? Which view do you agree with? Explain your answer.

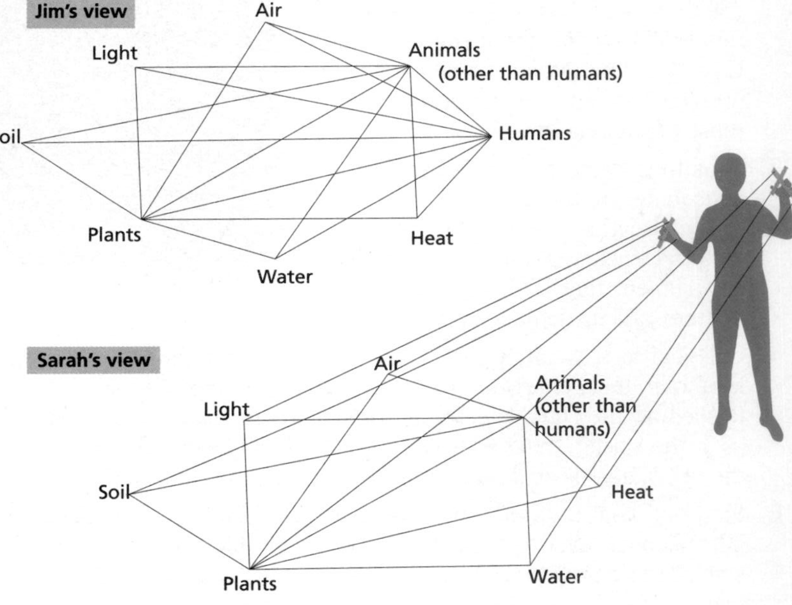

2. A Planet of People

Describe the changes in the world's population shown on the graph at left. How are these changes different from the changes in other populations you have studied? What do you think causes this difference? How do you think human population growth will affect your future?

★ You may wish to provide students with the Chapter 3 Review Worksheet that is available to accompany this Challenge Your Thinking (Teaching Resources, page 54).

3. Frogs Versus Snakes

The graph below shows the population of frogs in a pond over a period of many years. For a long time, the frogs had no predators. Then some snakes were introduced into the community. Draw this graph in your ScienceLog. Extend the graph to show how you think the frog population will change after the snakes arrive. What will happen to the snake population over the next few years?

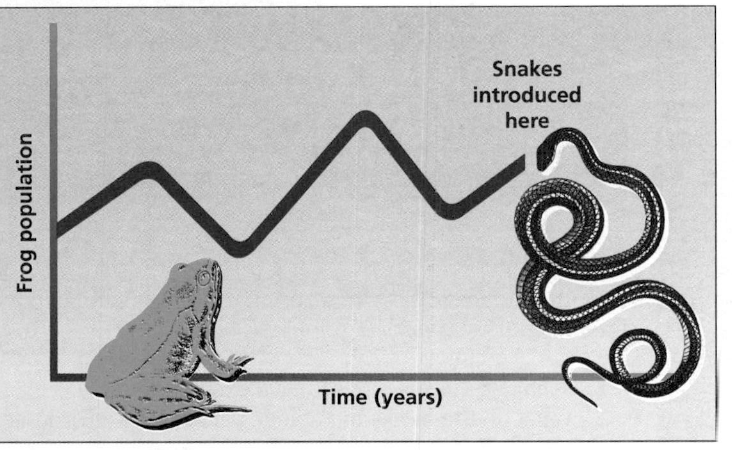

4. When I Was Young . . .

Imagine that you are the oldest catfish in a small pond. A young catfish swims up to you and asks, "Have things changed much since you were a hatchling?" What would you tell the young catfish about the natural changes in the pond during your lifetime?

5. Good Luck Charm

A traditional saying suggests that tearing down a spider's web in your home will bring bad luck. Based on what you've learned about interactions and populations, explain why tearing down a web might cause problems.

Review your responses to the ScienceLog questions on page 46. Then revise your original ideas so that they reflect what you've learned.

The following are sample revised answers:

1. In 10 years, the seedlings will have grown taller, and there will probably be some undergrowth in the new forest. In 100 years, the forest will be mature. There will be a larger variety of plant and animal life. The student may infer that undergrowth might not be as dense due to tree leaves that limit the amount of sunlight reaching the undergrowth.
2. The caterpillars eat tree leaves. Because they have no natural predators in this country, the caterpillars multiply uncontrolled. The numerous caterpillars can eat enough leaves to kill a tree.
3. Student answers will vary greatly. Accept all reasonable answers. Students may focus on negative or positive human influences.

Have students speculate on what the course of events will be if the human population rate continues to skyrocket. Have them consider the residual effects on forests, ponds, and grasslands.

Did You Know. . .

While the human population in the twentieth century has undergone tremendous growth, it doesn't compare with that of the *E. coli* bacterium. A population of these microorganisms can double in size in 20 minutes!

Making Connections

Unit 1

SourceBook

Turn to pages S1–S26 in the SourceBook to take a look at the role of organisms in ecosystems. You will learn more about how energy and materials are used in these ecosystems and how pollution affects our biosphere.

Here's what you'll find in the SourceBook:

The Big Ideas

In your ScienceLog, write a summary of this unit, using the following questions as a guide:

1. What are "interactions"? Give examples.
2. What do the biotic and abiotic parts of the environment include?
3. How is an organism's "niche" different from its "habitat"?
4. Can you illustrate the differences between commensalism, mutualism, and parasitism?
5. How would you describe the roles of producers, consumers, scavengers, and decomposers in a community?
6. What is the difference between a food chain and a food web?
7. How do each of the abiotic parts of the environment contribute to the welfare of a community?
8. What kinds of changes take place in a community over time?
9. Why are there continuous changes in plant and animal populations?

Making Connections

The Big Ideas

The following is a sample unit summary:

Interactions are relationships in which one part of the environment depends on, uses, or is used by another part of the environment. Interactions can take place between living things, (biotic-biotic) or between living and nonliving things (biotic-abiotic). Examples include a parasite invading a host's body or a plant interacting with sunlight to make food. (1)

The biotic parts include plants, animals, fungi, and bacteria—the living parts of the environment. The abiotic, or nonliving, parts include rocks, light, air, water, chemicals, and heat. (2)

The place where an organism lives is its habitat, and its way of life is its niche. (3) Commensalism is the dependence of one organism on another organism that is neither harmed nor helped by the relationship. In parasitism, one organism depends on another for habitat and food; the second organism is harmed or killed by the relationship. Mutualism is a sharing of benefits between organisms. (4)

Producers are plants that make food for the community from the sun's energy, carbon dioxide, and water. Consumers are generally animals. They do not make their food as plants do. Consumers that eat only plants are called herbivores. Consumers that eat only animals are called carnivores. Scavengers eat and dispose of the bodies of dead animals. Decomposers reduce the remains of plants and animals to basic substances by causing decay. (5)

A food chain shows how a series of organisms depends on one another for food. Food chains move from producer to primary consumer to secondary consumer to top carnivore. A food web consists of many interconnected food chains. It shows how a given organism is linked to many other members of the community. (6)

Plants and animals depend on an optimum temperature for growth, and they exhibit a range of tolerance for temperatures. Similarly, living organisms require moisture, and they vary in moisture tolerance. Light is required by all plants for the production of food. Plants tend to turn toward light, and animals have varied responses to light. (7)

Disrupted communities tend to return to their original state. After an environment is destroyed, one group of organisms can begin to grow again. This is followed by another group that succeeds the first group. In turn, more complex plants succeed the early plants until the original state is reached. This process is called succession. (8)

Changes occur in a community because of the organisms' dependence on each other for food. As a certain plant increases in number, so will the herbivores that use the plant as a primary food source. This allows for more predation on the herbivores, and the population of the predators increases. Eventually the predators reduce the herbivore population. Other events, such as natural disasters and human development, can also affect natural communities. (9)

Checking Your Understanding

1. Organisms get their food in many different ways. Use the clues to fill in the blanks. Then, identify the mystery word that tells what it's all about! (Do not write in this book.)

I am a vegetarian.

I'm not fussy. I eat anything!

I stalk my food.

I can make my own food.

I try to keep from being eaten.

I prefer animals for my food.

I change dead plants and animals into soil.

I am generous! I allow others to live on me.

I live on the source of my food!

I never make my own food.

I feed on dead animals.

2. Arrange these organisms into a food chain: grass, grasshopper, bear, fish, and frog.

3. Suppose that during the drought last summer, a forest fire burned a large portion of a forest near your town or city. Predict the kinds of changes that will occur in the plant and animal communities there over the next 100 years.

4. concept map Draw a concept map that shows the energy relationships between the following items: producer, sunlight, carnivore, and herbivore.

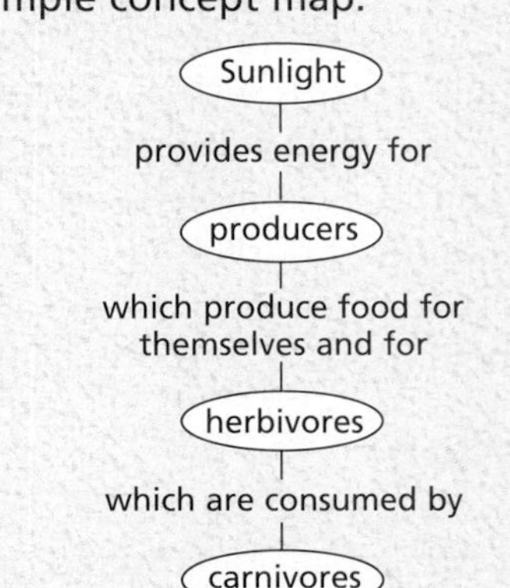

Answers to *Checking Your Understanding*

1. The answers, in order, are herbivore, omnivore, predator, producer, prey, carnivore, decomposer, host, parasite, consumer, and scavenger. The mystery word is *interacting*.
2. The order of the food chain is grass, grasshopper, frog, fish, and bear.
3. Over the years, the land where the forest stood will undergo succession. Succession is a natural series of changes in which one population is replaced by another. The first population to appear will be lichens, mosses, and grasses. Next, small seed plants like shrubs and bushes will appear. Finally, larger plants and trees will appear. When the community returns to its original state, succession will be complete.
4. Sample concept map:

Sunlight
provides energy for
producers
which produce food for themselves and for
herbivores
which are consumed by
carnivores

You may wish to provide students with the Unit 1 Review Worksheet that is available to accompany this Making Connections (Teaching Resources, page 63). Transparency 12 is also available to accompany this material.

Homework

The Activity Worksheet on page 61 of the Unit 1 Teaching Resources booklet makes an excellent homework activity.

Science Snapshot

Background

In 1976, Michael became the second gorilla to be taught sign language at the Gorilla Foundation. Since then, both he and Koko have been very busy. In 1984, the foundation sponsored an exhibition of art created by the gorillas. In 1985, Koko and All Ball were featured in a photograph chosen as *Time* magazine's Picture of the Year for 1985. In 1992 Koko was introduced to a second male gorilla, named Ndume, with the hope that they might produce offspring.

The declining population of gorillas in the wild has raised a great deal of concern. Of all the great apes, the gorilla is thought to be the most psychologically sensitive. A gorilla's reproductive pattern is easily disrupted by environmental stresses. This makes environmental problems such as human population growth and deforestation of gorilla habitats even more serious. Also, the population of gorillas in the world's zoos is just barely enough to ensure the survival of the species, even in captivity.

Discussion

You may wish to ask the class to suggest different traits that humans and gorillas share. Have a volunteer list them on the board. Then explain to the class that both gorillas and humans are classified by biologists as members of a mammal group called primates because they have so many similarities. *(Gorillas have the same number of legs, fingers, toes, and teeth as humans. They also use a variety of methods to communicate with one another, such as gestures, facial expressions, and vocal sounds. They are capable of understanding spoken language, express intelligence, and exhibit many emotions. A gorilla cries with sounds rather than with tears.)*

Multicultural Extension

Sign Language

Invite a member of the deaf community to teach the class about sign language and its strengths and weaknesses compared with spoken language. You might even have the students learn some interesting terms and phrases that are unique to sign language.

Science Snapshot

Francine Patterson (1947–)

Her favorite color is red, and her favorite foods are nuts and apples. She loves to play with dolls, draw pictures, and play chase with her friends. She's Koko, a 113 kg gorilla, and she can tell you all about it—in American Sign Language! How did Koko become so outspoken? She's the star pupil of Dr. Francine Patterson.

Koko Communicates

In 1972 Dr. Patterson began to study how gorillas could learn and use language. Koko, a gorilla born at the San Francisco Zoo, was only a year old. Dr. Patterson taught Koko how to sign words such as *eat, drink, love, good,* and *bad*. Four years later, Dr. Patterson and her research team established the Gorilla Foundation, an organization that studies intelligence, behavior, and physical development in gorillas.

▲ Koko makes the sign for sip.

By showing Koko signs and shaping the signs with the gorilla's hands, Dr. Patterson has taught Koko how to communicate. Koko currently uses over 500 signs, often speaking spontaneously with her human friends. Sometimes Koko even makes up signs of her own!

The Human Connection

Koko demonstrates many abilities that scientists used to believe only humans had. For example, Koko had a pet kitten called All Ball. She would cuddle and tickle the kitten much as a person might. When All Ball was killed by a car, Koko cried and signed "cry, sad, frown." Years later, Koko still recognized the cat in photographs. Although she has other pets, Koko still remembers All Ball.

◀ Koko and All Ball (Koko's kitten)

Dr. Patterson has worked with Koko for over 20 years. Much of what Dr. Patterson has learned about gorilla development can help protect these endangered animals. Dr. Patterson and the Gorilla Foundation are creating a new preserve for endangered gorillas. Located on the island of Maui in Hawaii, the preserve will allow gorillas to roam in a tropical setting and socialize with other gorillas. Scientists are hopeful that the gorillas will also have offspring.

Signs of Language

In an encyclopedia or other source, look up the alphabet for American Sign Language. Learn how to sign your name using the ASL alphabet.

Answers to *Signs of Language*

After students have learned to sign their names, invite volunteers to sign their names for the class. You may want to have students make a poster of the sign-language alphabet.

EYE ON THE ENVIRONMENT

Alien Invasion

A group of tiny aliens left their ship in Mobile, Alabama. Their bodies were red and shiny, and they walked on six legs. The aliens looked around and then quietly crawled off to make homes in their new land.

Westward Ho!

In 1918, fire ants were accidentally imported into the United States in a freighter ship from South America. In their new environment, the imported fire ants had no natural predators or competitors. In addition, these ants are extremely agressive, and their colonies can harbor many queens, instead of just one queen like many other ant species. With all these advantages, it is not surprising that the ants have spread like wildfire. By 1965, imported fire-ant mounds were popping up on the southeastern coast and as far west as Texas. Today they are found in portions of at least 10 southern states and may soon reach as far as California.

Jaws of Destruction

Imported fire ants have done a lot of damage as they have spread across the United States. Because they are attracted to electrical currents, these ants chew through wire insulation, causing shorts in electrical circuits. The invaders have also managed to disturb the natural balance of native ecosystems. In some areas, they have killed off 70 percent of the native ant species and 40 percent of other native insect species. Each year, about 25,000 people seek medical attention for painful fire-ant stings.

▼ **Range of the imported fire ant**

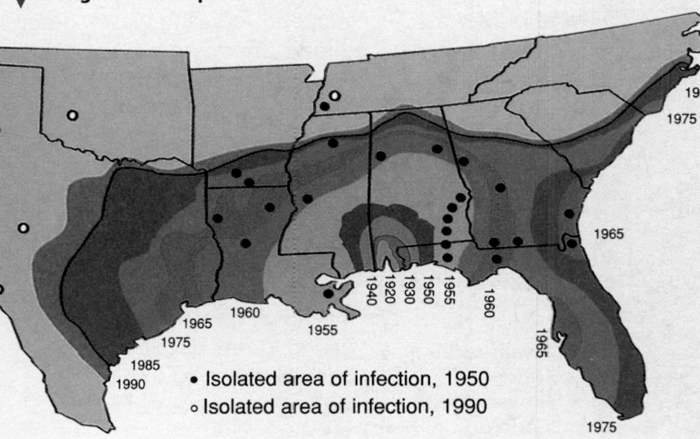

▲ **Three types of fire ants are found in a colony: the queen, workers, and males. Notice how the queen ant dwarfs the worker ants.**

Fighting Fire

Nearly 80 years after the fire ants' introduction into the United States, the destructive ants continue to multiply. Although about 157 chemicals, including ammonia, gasoline, extracts from manure, and harsh chemical pesticides, are registered for use against fire ants, most have little or no success. Unfortunately, many of these remedies harm the environment. By 1995 the government had approved only one fire-ant bait for large-scale use.

An Ant-Farm Census

How many total offspring does a single fire-ant queen produce if she lives for 5 years and produces 1000 eggs a day? If a mound contains 300,000 ants, how many mounds is this?

69

Answers to *An Ant Farm Census*

A typical queen may produce over 1.8 million eggs in her life: 1000 eggs/day × 365 days/year × 5 years = 1,825,000 eggs. Point out that, realistically, a queen may not produce eggs every day.

You might also want to explain to the students that by dividing by 300,000 ants per mound, this number is equal to just over six mounds, or more than one new mound every year!

EYE ON THE ENVIRONMENT

Background

In combating the spread of fire ants, people must be careful not to harm the environment. In the 1960s, ground-up corncobs and soybean oil were mixed with an ant poison and sprayed over huge areas of land. Studies have since shown that the poison used is toxic to many species.

In addition, many researchers speculate that this strategy may have backfired because fire ants can recover from disasters much quicker than other types of ants. After an area was treated with insecticides, the first ants that were able to reinhabit it were fire ants. Similar scenarios occurred in many different areas. As a result, fire ants have become a serious ecological problem in many areas.

Did You Know. . .

- *The tunnels of a fire-ant colony can extend up to 15 m away from the central mound.*
- *By weight, fire-ant venom is more powerful than cobra venom.*

Discussion

Discuss with students some safe and natural ways that they can eliminate ants around their homes. For instance, they can squeeze fresh lemon or lime juice into holes or cracks in their homes and then leave the peels where they have seen ants. They can also scatter aromatic substances, such as mint, red pepper, or paprika, around spots where ants enter homes. These substances repel ants.

Background

According to the Environmental Protection Agency, more than 240 million tires are discarded every year. However, because many landfills do not accept tires, the used tires are often simply stockpiled above ground, where they sit indefinitely. They not only are an eyesore but also can be a health hazard: water-filled tires are ideal breeding grounds for insects such as mosquitoes and flies.

Old tires can be recycled for use in "green" buildings. Sturdy walls can be made by stacking old tires like bricks. The tires are filled with dirt and covered with cement or clay. This technique recycles materials and is also energy efficient. These tire walls absorb heat when the weather is hot and release heat when the weather is cool, so they can act as huge temperature regulators, without setting a single thermostat control!

Discussion

You might want to lead a discussion about what activities the students could do to make the classroom into a "green" room. *(Examples might include installing low-wattage lights and putting insulating sheets of plastic over the windows in winter.)*

Extension

Have students write a letter to an imaginary prospective home builder. They should stress the importance of "green" building techniques by reiterating what they have learned from the class discussion and the text and by incorporating any ideas of their own.

Design It Yourself!

In their designs, students may wish to incorporate some of the materials discussed in the text. Encourage them to be innovative by thinking about what materials are available and what use these materials might serve. For example, they could use cardboard for walls, soda straws for thatched roofs, or old clothes for curtains or tablecloths.

Green Buildings

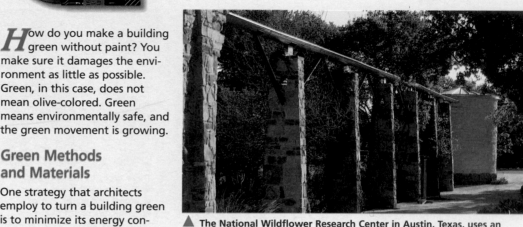

How do you make a building green without paint? You make sure it damages the environment as little as possible. Green, in this case, does not mean olive-colored. Green means environmentally safe, and the green movement is growing.

▲ The National Wildflower Research Center in Austin, Texas, uses an extensive rainwater-harvesting system to water its plants, reducing the need for additional water.

Green Methods and Materials

One strategy that architects employ to turn a building green is to minimize its energy consumption. They also reduce water use wherever possible, such as creating landscapes using only native plants that require little watering. Green builders also use fewer new materials in their projects by selecting recycled building materials. For example, crushed light bulbs can be recycled into floor tiles, and recycled cotton can replace fiberglass as insulation.

Seeing Green

Although green buildings cost more than conventional buildings to construct, they save money in the long run. For example, the Audubon building in Manhattan saves $100,000 in maintenance costs every year. It uses over 60 percent less energy and electricity, and inside, the workers enjoy natural lighting, cleaner air, and an environment that is free of unnecessary chemicals.

Some designers want to create buildings that are even greener than the Audubon building. Walls can be made of straw bales or packed dirt, and landscapes can be maintained with rainwater collected from rooftops. By conserving, recycling, and reducing waste, green builders are doing a great deal to help the environment.

◀ The walls of this building are being made out of worn-out tires packed with earth. The walls will later be covered with stucco.

Design It Yourself!

Design a building, a home, or even a doghouse that is made of only recycled materials. Be inventive! If you like your design, create a scale model. Describe how your green structure saves resources.

WeiRD Science

Cool Turtles

A frozen animal is a dead animal. Try telling that to the baby painted turtles, who spend their first winter exposed to temperatures several degrees below freezing. These cool turtles can survive being frozen for at least 11 days!

Frozen in Time

Painted turtle eggs are laid in a sandy underground nest. They hatch in the autumn, but to avoid predators, they stay in the nest all winter instead of digging themselves out. During the winter, temperatures in a painted turtle nest can reach as low as –8°C. That is 5°C below the normal freezing point of the painted turtle's body fluids.

Freezing an animal cell will damage it beyond repair. Therefore, most animals cannot survive extended exposure to below-freezing temperatures because their body's cells would also freeze. Baby painted turtles, however, have a secret weapon.

◀ **These two cells have been frozen. The freezing process damaged the one on top, while the one below it was protected by chemical antifreeze.**

A Natural Antifreeze

As the baby turtle begins to freeze, ice forms first on its skin. The cold gradually creeps inward until virtually all brain activity, breathing, blood flow, and movement stop. While fluids that are outside the cells freeze, such as blood plasma and urine, the fluid inside the cells stays liquid. This is because these young turtles have a special chemical that prevents the fluid inside their cells from freezing. The chemical acts much like the antifreeze that is added to car radiators each winter. Some snakes, lizards, frogs, and other turtles have similar adaptations, but few can survive at such low temperatures for such a long time as the painted turtle hatchlings.

▼ **Painted turtles can be found across the United States and in southern Canada.**

The Wonders Never Cease

As if freezing temperatures weren't enough, during their frosty first winter, baby painted turtles also face oxygen deprivation. Because oxygen supplies are limited underground, and because blood flow stops when the turtles freeze, the turtles must be able to survive without oxygen. Another adaptation, called metabolic arrest, allows the turtles to do just that. Metabolic arrest lowers the rate at which cells use energy, extending the time that the turtles can survive on the limited supply of oxygen already in the body. Without metabolic arrest, the turtle's special antifreeze would be useless. The painted turtle maintains the ability to undergo metabolic arrest, but strangely, after their first year of life, the turtles lose the ability to survive being frozen alive.

Cryogenic Creatures

What if scientists could create a nontoxic chemical that would prevent the fluid inside human cells from freezing? What are some ways this new technology could be put to use?

71

Discussion Questions

1. What other strategies do animals employ to survive the winter? *(Many animals, such as birds, migrate to warmer areas. Others, such as wolves, have thick fur or other features that help to keep them warm.)*
2. What mammals do you know of that hibernate? *(Hibernating mammals include bears, squirrels, raccoons, and bats.)*

WeiRD Science

Background

In addition to natural "antifreeze," baby painted turtles also produce substances called *ice-nucleating proteins.* When the turtles' extracellular fluid reaches freezing temperatures, these proteins actually accelerate the freezing process, acting as nuclei around which ice crystals can form. Although this would seem dangerous for body tissues, the proteins actually protect tissues by keeping the size of the ice crystals small.

Did You Know. . .

- Many animals prepare for hibernation by storing fats in their body.
- The arctic woolly bear caterpillar spends the winter in an ice-crystal cocoon.
- The arctic ground squirrel is the first mammal discovered that can survive temperatures below 0°C.

Answers to *Cryogenic Creatures*

(You might want to first explain to the class what the word *cryogenic* means. Cryogenics deals with low temperatures and their effects on matter.) Some potential uses for such a nontoxic chemical might include temporarily preserving a patient with a fatal disease or wound until a cure or treatment can be found. The chemical might also allow organs to be preserved for more than a few days so that they can be transplanted into other people. This technique is called *cryopreservation.*

Diversity OF LIVING THINGS

Unit Overview

This unit focuses on the diversity of living things, the possible reasons for diversity, and how scientists make sense of this diversity. In Chapter 4, students consider the great diversity of living things and how plants and animals are adapted to their habitats and niches. In Chapter 5, students are introduced to adaptations, Darwin's theory of natural selection, and the value of diversity. In Chapter 6, students develop their own classification systems and then learn how scientists classify organisms.

Using the Themes

The unifying themes emphasized in this unit are **Changes Over Time, Systems,** and **Structures.** The following information will help you incorporate these themes into your teaching plan. Focus questions that correspond to these themes appear in the margins of this Annotated Teacher's Edition on pages 76, 95, 97, 108, and 127.

Throughout the unit, discussions of **Changes Over Time** will be useful as students explore the processes of succession and natural selection. Specifically, focus questions in Chapters 4 and 5 encourage students to consider the changes in diversity that might follow a natural disaster, major changes in climate, or a disturbance caused by humans.

Another focus question in Chapter 5 addresses social **Systems** that are found in many types of organisms. Students are asked to consider what benefits might be gained by having such a system.

The theme of **Structures** also emerges in this unit. Because all organisms have acquired anatomical features that make them well-suited to their habitat, it is illuminating to consider the structural aspects of these features.

Using the SourceBook

In Unit 2, students are introduced to organisms from each of the five kingdoms and learn how the history of life on Earth relates to the development of complex organisms. The unit examines artificial selection, the theory of evolution by natural selection, and evidence for the common ancestry of all life on Earth.

Bibliography for Teachers

Martin, James. *Masters of Disguise: A Natural History of Chameleons.* New York City, NY: Facts On File, Inc., 1992.

McGowan, Chris. *Diatoms to Dinosaurs: The Size and Scale of Living Things.* Washington, DC: Island Press, 1994.

Morris, Desmond. *The World of Animals.* New York City, NY: Viking, 1993.

Bibliography for Students

Bousquet, Catherine. *Incredibly Hidden.* New York City, NY: New Discovery Books, 1993.

Penny, Malcolm. *Animal Adaptations.* New York City, NY: The Bookwright Press, 1989.

Tesar, Jenny. *Mammals.* Woodbridge, CT: Blackbirch Press, 1993.

Films, Videotapes, Software, and Other Media

Amazing Animals
Software (Macintosh or MS-DOS)
Sunburst Communications
101 Castleton St.
P.O. Box 100
Pleasantville, NY 10570-9963

***Animals With Backbones* (2nd ed.)**

***Animals Without Backbones* (2nd ed.)**
Films or videotapes
Coronet/MTI
108 Wilmot Rd.
Deerfield, IL 60015

***Camouflage in Nature: Form, Color, Pattern Matching* (2nd ed., rev.)**
Film or videotape
Coronet/MTI
108 Wilmot Rd.
Deerfield, IL 60015

Mammals: A Multimedia Encyclopedia
CD-ROM (Macintosh or MS-DOS)
National Geographic Society
Educational Services
1145 17th St. NW
Washington, DC 20036-4688

Unit Organizer

Unit/Chapter	Lesson	Time*	Objectives	Teaching Resources
Unit Opener, p. 72				Science Sleuths: Neo-Cassava: The Tropical Miracle English/Spanish Audiocassettes Home Connection, p. 1
Chapter 4, p. 74	**Lesson 1, A Cycle of Diversity, p. 75**	1	1. Examine the cycle of diversity before and after a volcanic explosion.	none
	Lesson 2, Diversity All Around, p. 78	2 to 3	1. Observe and describe the diversity of living things, including size, shape, and structure. 2. Compare groups of organisms in terms of their diversity.	Image and Activity Bank 4-2 Exploration Worksheet, p. 3 Exploration Worksheet, p. 9 ▼
End of Chapter, p. 86				Chapter 4 Review Worksheet, p. 12 Chapter 4 Assessment Worksheet, p. 14
Chapter 5, p. 88	**Lesson 1, How Does It Happen? p. 89**	2 to 3	1. Analyze and compare the theories of Lamarck and Darwin. 2. Predict how a species will adapt to changes in its habitat. 3. Observe Darwin's finches and use the theory of natural selection to analyze their differences.	Image and Activity Bank 5-1 Resource Worksheet, p. 17 Resource Worksheet, p. 19 ▼
	Lesson 2, To Survive, p. 96	2 to 3	1. Observe and describe animal and plant adaptations. 2. Infer how animal and plant adaptations help organisms to survive in their particular habitats.	Image and Activity Bank 5-2 Discrepant Event Worksheet, p. 22 Resource Worksheet, p. 23
	Lesson 3, The Value of Diversity, p. 106	2 to 3	1. Explain how the loss or addition of one species can affect other species in the community. 2. Explain some of the natural and human-made pressures that can cause extinction. 3. Describe how humans can cause or help prevent the extinction of an organism.	Image and Activity Bank 5-3 Theme Worksheet, p. 24 Exploration Worksheet, p. 26 Exploration Worksheet, p. 30 ▼
End of Chapter, p. 111				Graphing Practice Worksheet, p. 32 Chapter 5 Review Worksheet, p. 34 Chapter 5 Assessment Worksheet, p. 37
Chapter 6, p. 113	**Lesson 1, Ordering Diversity, p. 114**	5	1. Classify living things into two kingdoms—plants and animals. 2. Classify animals into two groups—vertebrates and invertebrates. 3. Classify plants into two groups—mosses and plants with conducting tubes.	Image and Activity Bank 6-1 Transparency Worksheet, p. 40 ▼ Transparency 17 Transparency 18 Resource Worksheet, p. 42 ▼ Resource Worksheet, p. 44 ▼ Transparency 21
	Lesson 2, Finding a Name, p. 132	2	1. Explain the system that biologists use to name living things. 2. Observe and describe diversity within the species *Homo sapiens*.	Image and Activity Bank 6-2 Transparency 22 Exploration Worksheet, p. 45 ▼ Activity Worksheet, p. 47 ▼
End of Chapter, p. 136				Chapter 6 Review Worksheet, p. 49 Chapter 6 Assessment Worksheet, p. 53
End of Unit, p. 138				Unit 2 Activity Worksheet, p. 56 Unit 2 Review Worksheet, p. 58 Unit 2 End-of-Unit Assessment Worksheet, p. 60 Unit 2 Activity Assessment Worksheet, p. 69 Unit 2 Self-Evaluation of Achievement, p. 72

* Estimated time is given in number of 50-minute class periods. Actual time may vary depending on period length and individual class characteristics.

▼ Transparencies are available to accompany these worksheets. Please refer to the Teaching Transparencies Cross-Reference chart in the Unit 2 Teaching Resources booklet.

Materials Organizer

Chapter	Page	Activity and Materials per Student Group
4	79	**Exploration 1, Activity 1:** four 15–20 cm long craft sticks; metric measuring tape; 5 m of heavy string; scissors; magnifying glass; field guide for plant identification
5	106	**The Web:** large ball of string or yarn; 2 index cards per student; 2 small pieces of tape per student
	108	**Exploration 1:** 1 die; 20 cards and 1 token for each animal; 20 blank cards; 2 copies of each extinction card (See Advance Preparation below. Additional teacher materials: Exploration Worksheet of animal and extinction cards, animal tokens, and the game board; scissors)
6	126	**Putting It All Together:** optional materials: a variety of preserved or live specimens of plants and animals

Advance Preparation

Exploration 1, The Extinction Game, page 108: You may wish to provide students with the animal and extinction cards, animal tokens, and game board necessary to play this game. These can be copied using the Exploration Worksheet on page 26 of the Unit 2 Teaching Resources booklet. They will need to be cut out before students can play the game.

Unit Compression

An understanding of the process of natural selection is essential in studying life science. Therefore, certain sections of Unit 2 should not be omitted. As you proceed through the unit, however, you may find that some of the supporting content may be left out without seriously disrupting the integrity of the unit's conceptual progression.

Chapter 4 explores the nature of diversity, both on a global scale and on a local scale. Exploration 1 on page 79, which invites students to examine diversity in a study site, should be considered essential because it serves to give students a concrete picture of diversity in their environment. You may wish to make decisions about omitting the other material in this chapter based on the time that is available to you.

Similarly, Lesson 2 of Chapter 5 provides many examples of adaptations, but you may find that students are able to grasp the concept of adaptation without covering the entire lesson.

In addition, keep in mind that several sections may be assigned as homework to reduce the amount of class time necessary to teach the unit. Examples include Exploration 2 on page 82, The Case of the Peppered Moth on page 93, and parts of Chapter 6, Lesson 1, which begins on page 114.

Homework Options

Chapter 4	See Teacher's Edition margin, pp. 79, 82, 84, and 87 Exploration Worksheet, p. 3 Exploration Worksheet, p. 9
Chapter 5	See Teacher's Edition margin, pp. 91, 93, 94, 97, 100, 103, 107, 108, 110, and 111 Resource Worksheet, p. 17 Resource Worksheet, p. 19 Resource Worksheet, p. 23 Theme Worksheet, p. 24 Exploration Worksheet, p. 30 Graphing Practice Worksheet, p. 32 SourceBook, pp. S39, S43, and S46
Chapter 6	See Teacher's Edition margin, pp. 115, 118, 120, 123, 127, 131, and 135 Activity Worksheet, p. 47 SourceBook, pp. S28 and S44
Unit 2	SourceBook Activity Worksheet, p. 73

Assessment Planning Guide

Lesson, Chapter, and Unit Assessment	SourceBook Assessment	Ongoing and Activity Assessment	Portfolio and Student-Centered Assessment
Lesson Assessment Follow-Up: see Teacher's Edition margin, pp. 77, 85, 95, 105, 110, 131, and 135 **Chapter Assessment** Chapter 4 Review Worksheet, p. 12 Chapter 4 Assessment Worksheet, p. 14* Chapter 5 Review Worksheet, p. 34 Chapter 5 Assessment Worksheet, p. 37* Chapter 6 Review Worksheet, p. 49 Chapter 6 Assessment Worksheet, p. 53* **Unit Assessment** Unit 2 Review Worksheet, p. 58 Unit 2 End-of-Unit Assessment Worksheet, p. 60*	SourceBook Review Worksheet, p. 75 SourceBook Assessment Worksheet, p. 79*	Unit 2 Activity Assessment Worksheet, p. 69* **SnackDisc** Ongoing Assessment Checklists ♦ Teacher Evaluation Checklists ♦ Progress Reports ♦	Portfolio: see Teacher's Edition margin, pp. 80, 93, 116, and 126 **SnackDisc** Self-Evaluation Checklists ♦ Peer Evaluation Checklists ♦ Group Evaluation Checklists ♦ Portfolio Evaluation Checklists ♦

* Also available on the Test Generator software

♦ Also available in the Assessment Checklists and Rubrics booklet

Science Discovery is a versatile videodisc program that provides a vast array of photos, graphics, motion sequences, and activities for you to introduce into your *SciencePlus* classroom. *Science Discovery* consists of two videodiscs: Science Sleuths and the Image and Activity Bank.

Using the *Science Discovery* Videodiscs

Science Sleuths: Neo-Cassava: The Tropical Miracle, Side A

A foundation has transplanted the neo-cassava from a tropical rain forest. The foundation hopes to grow this inexpensive, nutritionally balanced plant on experimental farms to serve as a remedy for world hunger. After a promising initial harvest, the crop is dying. The Science Sleuths must analyze the evidence for themselves to determine the cause of the problem.

Interviews

1. Setting the scene: President of the Neo-Cassava Foundation **41036 (play ×2)**

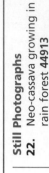

2. Neo-Cassava Foundation farmer **41932 (play)**

3. Ethnobotanist **42995 (play)**

4. Tropical entomologist **43863 (play)**

Documents

5. Press release from the Foundation **44864 (step ×2)**

6. Fax from plant pathologist **44868 (step)**

Literature Search

7. Search on the words: CROP FAILURE, DROUGHT, LOCUSTS, NEO-CASSAVA, PLANT DISEASE, RUSTY SMUT **44871**

8. Article #1 ("Locusts Swarm in Ros Dashen") **44873**

9. Article #2 ("Caring for Houseplants") **44875 (step ×2)**

Sleuth Information Service

10. Weather in the Foundation area for the past three years **44879 (step)**

11. Climate chart for tropical rain forest **44882**

12. Climate chart for Foundation area **44884**

13. Rain-forest arthropod sample **44886 (step ×2)**

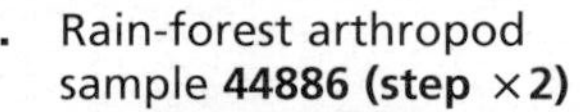

14. Foundation arthropod sample **44890**

15. Food web **44892 (step)**

16. Rusty smut **44895 (step)**

17. Monoculture **44898 (step)**

Sleuth Lab Tests

18. Nutritional analysis of neo-cassava **44901**

19. Rain-forest soil analysis **44903 (step ×2)**

20. Foundation soil analysis **44907 (step ×2)**

21. Chemical analysis of insects on farm **44911**

Still Photographs

22. Neo-cassava growing in rain forest **44913**

23. Neo-cassava growing on farm **44915**

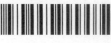

24. Neo-cassava leaves in rain forest **44917**

25. Neo-cassava leaves on farm **44919**

26. Neo-cassava roots in rain forest **44921**

27. Neo-cassava roots on farm **44923**

28. Neo-cassava roots in rain forest, cross section **44925**

29. Neo-cassava roots on farm, cross section **44927**

Step Reverse · Play · Pause · Step Forward

Image and Activity Bank
Side A or B

A selection of still images, short videos, and activities is available for you to use as you teach this unit. For a larger selection and detailed instructions, see the Videodisc Resources booklet included with the Teaching Resources materials.

4-2 Diversity All Around, page 78

Jellyfish; bioluminescent 44317–44527 (Side B only) (play ×2)
Bioluminescence occurs when certain proteins interact with oxygen and enzymes to create a chemical that releases light. Other organisms with this property are fireflies, some marine fish, crustaceans, worms, and some types of fungi.

Flowers, blooming 42408–43042 (Side B only) (play ×2)
Flowers are adapted to attract pollinators (insects, birds, bats, and other animals). When a flower opens, the reproductive parts of the plant are exposed. If pollination occurs, a seed develops that helps to ensure survival of the species.

Elephant; mud bath 52605–53146 (Side B only) (play ×2)
Elephants bathe daily, if possible. Because of their huge mass, they overheat easily; water cools them. Their skin is very sensitive and cracks without daily baths and dusting. Caked mud on their skin keeps insects from biting them.

Caterpillars feeding 48054–48504 (Side B only) (play ×2)
All of a caterpillar's body parts are soft except its jaws, which are hard in order to tear and eat leaves.

5-1 How Does It Happen? page 89

Moth, peppered 4252
Peppered moths come in pale and dark forms. Originally the moths were mostly pale colored. As pollution killed pale lichens on the trees, the dark moths were selected because they were camouflaged against the dark trees.

Darwin's finches 4253–4254 (step)
From one type of finch blown to the Galápagos Islands, many types of finches evolved. Each type of finch has adapted a different kind of beak that enables it to eat different foods, thus avoiding competition.

5-2 To Survive, page 96

Katydid 3367–3368 (step ×2)
The shape, color, and veins of this katydid's wings camouflage it. (step) Leaf mimicry by katydids can be very elaborate. Even dead leaves may be imitated.

Fox, arctic 3994
Arctic foxes have furry feet to help them walk on snow. Their legs and ears are short to retain heat. Every 4 years their population peaks, 1 year after the lemming population peaks. Lemmings, a kind of rodent, are their favorite food.

Zebra; herd 3981
Although zebras live in either brown or green savannas, their bold black and white stripes may serve as camouflage. Traveling in herds, they blend together, and predators can't single out a victim.

Frog, Tinctorius poison-dart 3620
The bright colors and bold patterns on this frog are warnings of danger.

Octopus, blue-ring 3217
An octopus can jet backward to travel or to escape predators. It can also walk forward or sideways on its arms. It has difficulty walking, however, because each arm literally has a mind of its own.

Pheasant, golden 3795
The male pheasant creates an elaborate display with the striped feathers on the back of its neck. It does this to bully other males and to attract females.

Caterpillar, two-tailed swallowtail 3452
What looks like this creature's head is really the tail end of a caterpillar. The spots on the tail, which look like eyes, bulge to look even more startling.

Plant adaptations 3046
This image shows four adaptations of plants: competing with other plants for light; living on water; being carnivorous; and storing water.

Puffball 2772
Puffballs have white interiors when immature. These later change to yellow or black as spores are produced. When a puffball matures, a hole forms on its top. Spores are expelled through the hole when the puffball is struck by a raindrop.

Palm, coconut 2944
The coconut palm has a unique method of reproduction. Wild trees grow near or lean over the edge of water. Their seeds (coconuts) fall into the water and float to other places to sprout.

5-3 The Value of Diversity, page 106

Crab, boxer 3292
Boxer crabs carry anemones on their arms, probably for self-defense. These crabs are one of many animals that haven't been satisfactorily studied. Marine habitats are being polluted, and many species are being lost before they can be studied.

Warbler, yellow 3866
The yellow warbler is one of the many tropical birds that fly north in the summer to nest in North America. They are becoming scarce because of the loss of habitat both in the North American woods and in the southern tropical jungles.

Rhinoceros, black 3984
All rhino species are endangered because of habitat loss or poaching for their horns, which are prized by some cultures.

Musk ox 3962
Musk oxen were once fairly abundant in the North American Arctic, despite their bleak habitat. Their powerful defense instinct of forming a solid wall of horns to fight wolves did not work against rifles, however. The number of musk oxen was reduced to just 500 in the 1800s.

Deforestation 2588
This deforested area is in South America. Deforestation leads to soil erosion, is a threat to species living in that ecosystem, and contributes to increased concentrations of carbon dioxide in the atmosphere.

Tropical plants 3029
Unlike agricultural plants, plants in the wild grow in diverse groups. This diversity limits the spread of diseases and pests that may affect only one species.

6-1 Ordering Diversity, page 114

Classification; the five kingdoms 2807
A five-kingdom system of living organisms

6-2 Finding a Name, page 132

Grouping; all living things 2806
Classification system created by Carolus Linnaeus

◀| Step Reverse   Play ▶   Pause || Step Forward |▶

UNIT FOCUS

Ask students to take out their ScienceLog and list as many living things as they can in one minute. Then have them share their lists with each other. Tell them that some biologists estimate that there may be as many as 100 million different kinds of living things on Earth. Fewer than 2 million have been named and described. Ask students if they have ever thought about the great diversity of living things in the world or why such diversity exists. Encourage discussion.

A good motivating activity is to let students listen to the English/Spanish Audiocassettes as an introduction to the unit. Also, begin the unit by giving Spanish-speaking students a copy of the Spanish Glossary from the Unit 2 Teaching Resources booklet.

Unit 2 Diversity of Living Things

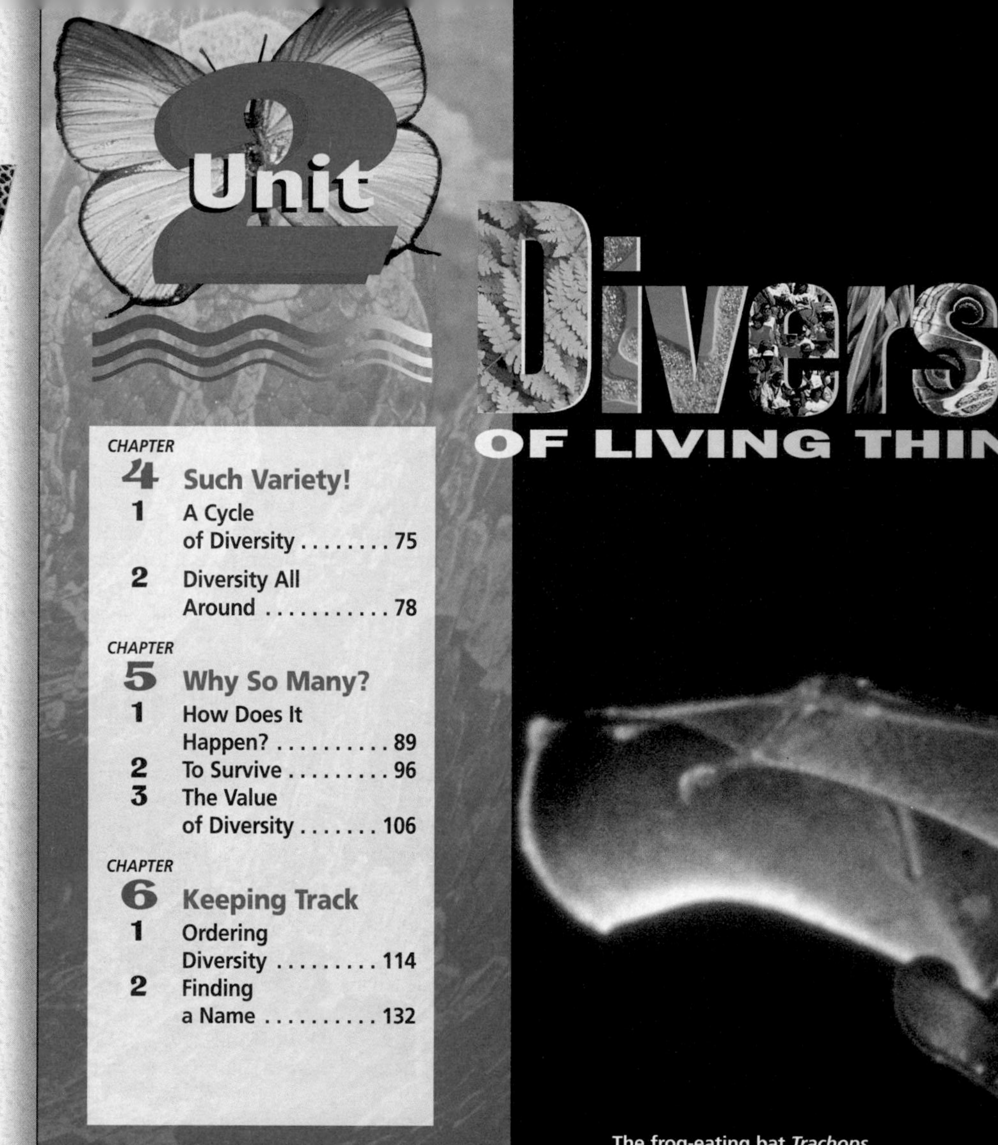

The frog-eating bat *Trachops cirrhosus* hunts its prey in the forests of southern Mexico and several South American countries.

Connecting to Other Units

This table will help you integrate topics covered in this unit with topics covered in other units.

Unit 1 Interactions	Predator-prey interactions are an important part of the natural-selection process.
Unit 4 Force and Motion	Over time, species have adapted to the force of gravity with appendages such as wings, feet, and fins.
Unit 5 Structures and Design	Structures such as thick leg bones are important to both the form and function of living things.
Unit 6 The Restless Earth	Geologic processes influence the environment and thus have an indirect effect on natural selection.
Unit 8 Growing Plants	Both living and extinct plants exhibit a wide array of diversification.

By looking at this photo, you would probably guess that this little frog became a hearty bat dinner. But believe it or not, the frog survived to croak again! The frog was minding its own business when a large frog-eating bat plunged toward it in a flurry of wings and jaws. But just as the bat clamped down on the frog with its jaws, it spat the frog back out! The little frog was poisonous. How did the bat know?

Look closely at the bat's mouth and you will see tiny bumps that look like a fringe. These are called skin teeth. With its skin teeth, this bat was able to sense the poison quickly enough to avoid a very bad meal. The skin teeth seem to be a specific defense against poisonous frogs.

This special feature helps the bat survive. What special feature does the frog have that helps it survive? In this (A) unit you will learn about such special features and how they come about. You will also explore how these features have led to the diversity of life on our planet. Finally, you will learn about the system that scientists all over the world use to name and classify living things.

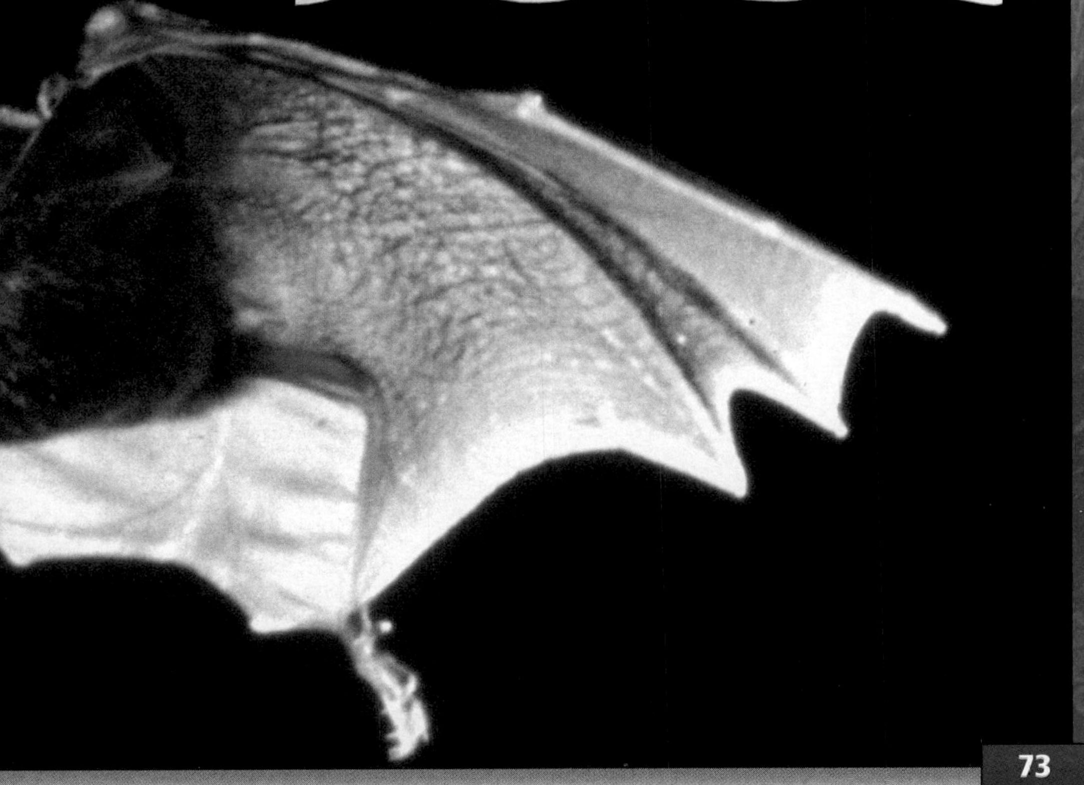

Answer to *In-Text Question*

(A) The poison prevents the frog from being eaten.

Using the Photograph

Students may find the following background information interesting. There are about 900 known species of bats, 40 of which are found in North America. (Only a few species of bats are predators of frogs.) About 2700 species of frogs are known, of which about 53 can be found in the United States. Both frogs and bats live everywhere except Antarctica and the Arctic. The great diversity of bats and frogs is a result of the diverse environments to which these animals have had to adapt.

Most bats are nocturnal predators and have developed a special method of finding prey, called *echolocation.* Bats also have a good sense of smell. Frogs, on the other hand, have developed a wide range of defense mechanisms. Some protect themselves with special colorations designed to help them blend in with their environment. Some display brilliant colors that warn predators that they are poisonous.

Bats and frogs play important roles in their communities. Have students think about what some of these roles might be. In addition, humans use bat guano as fertilizer and frogs for food and scientific research.

Critical Thinking

Have students look closely at the bat's claws. Ask: What do you think is the function of the bat's claws? Why do you think the bat is using its mouth rather than its claws to catch the frog? *(Students may say that they are used to catch prey, but a bat uses its claws primarily to hang upside down when it roosts.)* Tell students that bats' legs are very weak—some bats can't even walk. Ask students to suggest other land-dwelling creatures that experience difficulty walking. *(Answers include ducks and seals.)* You may want to tell the class that some birds, such as eagles, catch their prey with talons, while other birds, such as pelicans, use their beaks. Ask: Why do you think they have adapted in such different ways? *(Eagles hunt many land creatures, such as rodents and snakes. Pelicans use their beaks to scoop fish out of the water.)*

CHAPTER 4 Such Variety!

Connecting to Other Chapters

Chapter 4 *introduces the concept of diversity and how the diversity in a community can change.*

Chapter 5 *explores the theory of natural selection and the value of diversity to a community.*

Chapter 6 *challenges students to create classification systems for a wide range of living things.*

Prior Knowledge and Misconceptions

Your students' responses to the ScienceLog questions on this page will reveal the kind of information—and misinformation—they bring to this chapter. Use what you find out about your students' knowledge to choose which chapter concepts and activities to emphasize in your teaching.

In addition to having students answer the questions on this page, you may wish to have the students play the following game: Give one student in the class the name of an animal or plant and explain why you chose this organism. Have the student use the last letter of the name as the first letter for another animal or plant name. The student should then explain why he or she chose this organism and should give the name to a second student. Continue in this way around the room until every student has been given a chance to think of an animal or plant. For instance, you might give a student "poison ivy." The student might respond with "yellow jacket." The next student might then respond with "tree frog," and so on. If a student has trouble thinking of a name, you may ask for suggestions from the class or give the student another name yourself. Pay attention to student responses to find out what kinds of organisms are familiar to students, what misconceptions they may have, and what is interesting to them about plant and animal diversity. Afterward, explain to the class that in this chapter they will further explore the diversity of the natural world.

CHAPTER 4

Such Variety!

1 **How many different types of plants and animals do you think live on Earth?**

2 **How does the term *diversity* relate to living things?**

3 **What are some ways that organisms can be different from each other?**

ScienceLog

Think about these questions for a moment, and answer them in your ScienceLog. When you've finished, you'll have the opportunity to revise your answers based on what you've learned.

74

A Cycle of Diversity

An Island Explodes

On the morning of August 26, 1883, the sky was hazy and gray over the Indonesian island of Krakatau. Clouds of billowing ash spewing from the island's volcano had blotted the sun. The volcano had been active off and on that summer, and the eruptions had become increasingly violent in the previous days.

Suddenly, at 10:00 A.M., a massive eruption blew the top off the volcano, producing the loudest sound in recent history. The roar of the blast was so great that it was heard in Australia—3620 km away. About 4 km^3 of ash shot almost 30 km into the air, resulting in complete darkness for two days. Heavy layers of ashy dust fell on the decks of ships more than 2500 km away for several days. Particles of the fine dust drifted around the Earth several times, creating spectacular red sunsets for more than a year.

The force of the explosion was equivalent to almost 150 megatons of dynamite. This great force triggered giant tidal waves—some as far away as South America. The largest wave, created just after the explosion on Krakatau, rose to a height of more than 36 m and destroyed 295 towns on the nearby islands of Java and Sumatra. More than 35,000 people were killed by the tidal wave.

It had gradually been growing dark since 9 A.M. There were terrible noises from the volcano and the sky was filled with forked lightning. By the time the squall struck, it was like midnight at noon. There was a heavy shower of ashes and a strong smell of sulfur. It was difficult to breathe.
(From the logbook of the *W. H. Besse*, an American ship sailing nearby when the volcano erupted)

75

LESSON 1 ORGANIZER

Time Required
1 class period

Process Skills
observing, analyzing

Theme Connection
Changes Over Time

New Terms
none

Materials (per student group)
none

Teaching Resources
none

A Cycle of Diversity

FOCUS

Getting Started

In this lesson, students will learn about the natural cycle of regrowth and diversification that occurs in a region devasted by a natural or human-made disaster. Point out to students that the volcanic eruption described in this lesson is one example of a sudden natural change, but emphasize that diversification also occurs through slower changes in the environment. One example is the ice ages. Glacial movements over million-year cycles have influenced the migration patterns and adaptations of many animals, including horses, camels, bison, deer, and elephants.

Main Ideas

1. Natural disasters cause destruction and death.
2. A cycle of regrowth occurs after a natural disaster, and eventually the diversity of living organisms is restored.

TEACHING STRATEGIES

An Island Explodes

Explain to students that when a volcano erupts, it affects many things. The living organisms nearby are killed if they cannot escape quickly enough. If a volcano is near an ocean, enormous tidal waves can be created that also cause destruction. The ash also affects the atmosphere and can interfere with weather patterns.

Did You Know . . .

The diversity of plant and animal life in a region varies with the latitude of the region. Species diversity increases from the poles to the tropics. In tropical rain forests there can be up to 10 times more kinds of trees per hectare than in forests farther to the north or south.

What Was Left of Krakatau?

Ask students to visualize what Krakatau looked like before and after the destruction. Have them list plants and animals that might have lived on the island before and after the eruption.

Answer to *In-Text Question*

Ⓐ Students should understand that few living things on the island could have survived Krakatau's eruption.

Theme Connection

Changes Over Time

Focus question: Describe a typical succession of organisms after a natural disaster. *(The first organisms to grow or appear are grasses and small insects or spiders. Shrubs will gradually start to grow, and larger animals will begin to thrive. Eventually, many species of plants and animals will thrive because the environment can provide food and shelter.)*

Multicultural Extension

Human Diversity

Humans are also a very diverse group. Ask students to point out on a world map the origin of some ethnic groups that are familiar to them. Students may be interested to know that many scientists have abandoned attempts to divide the human species into subgroups or "races" because neighboring human groups tend to blend into one another instead of being distinctly divided. Therefore, no scientifically meaningful divisions can be made.

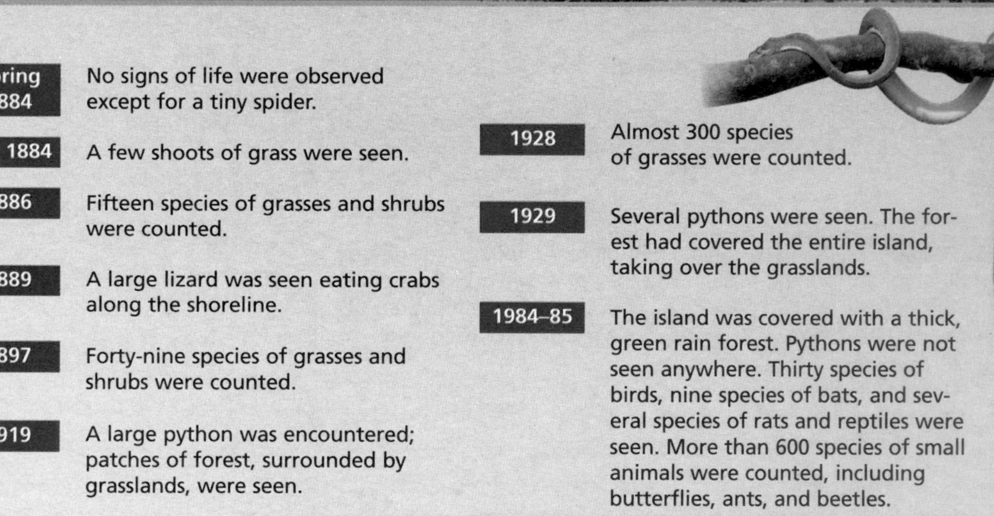

What Was Left of Krakatau?

Look at an atlas or a map of the world to see where Krakatau used to be. You'll have to look in the Sunda Strait of the Indian Ocean, between Java and Sumatra. There, the small Indonesian island collapsed into an underwater crater that was created by the volcanic explosion. The crater was 7 km long and 270 m deep. A mountain on the island, called Rakata, was all that remained above the sea. This tiny new island was covered by a layer of ash and lava that was 40 m thick. The temperature was between 300°C and 850°C following the eruption. Do you think any living things survived the explosion? Ⓐ

New Beginnings

In the century following Krakatau's eruption, several expeditions landed on the shores of Rakata. Notes and reports from these expeditions included the observations below.

Spring 1884	No signs of life were observed except for a tiny spider.
Fall 1884	A few shoots of grass were seen.
1886	Fifteen species of grasses and shrubs were counted.
1889	A large lizard was seen eating crabs along the shoreline.
1897	Forty-nine species of grasses and shrubs were counted.
1919	A large python was encountered; patches of forest, surrounded by grasslands, were seen.
1928	Almost 300 species of grasses were counted.
1929	Several pythons were seen. The forest had covered the entire island, taking over the grasslands.
1984–85	The island was covered with a thick, green rain forest. Pythons were not seen anywhere. Thirty species of birds, nine species of bats, and several species of rats and reptiles were seen. More than 600 species of small animals were counted, including butterflies, ants, and beetles.

Integrating the Sciences

Earth, Life, and Physical Sciences

Every field of science employs its own classification system in order to make sense of diversity. Students will learn later in this unit how biologists apply a classification system to living things. You might want to explain to the class how some other disciplines classify things in their own fields. For example, geologists classify rocks according to hardness, color, luster, density, and cleavage. Chemists classify elements into a periodic table. You might have students develop their own classification system for other kinds of objects. Some possibilities include types of tools, bridges, or clouds.

Rakata is now the home of many species, including the Tokay gecko.

Coming to Conclusions

What conclusions can you make from these observations? With a few of your classmates, write a conclusion under each of the following headings:

- Means of Animal Transportation
- Types of Living Things
- Interactions Among Living Things
- The Process of Change

For each conclusion, write down the supporting evidence you found in the observations.

The rain forest that covers the tiny Indonesian island of Rakata is now home to a diverse population of living things. Do you think there is diversity in the living things around you? In the next lesson, you'll find out!

Coming to Conclusions

GUIDED PRACTICE Have students use the observations on page 76 to discuss the succession of living organisms on the island of Krakatau. Point out that these changes occurred gradually. Ask them to suggest how these changes might have occurred. *(Students should conclude that the first organisms were not complex. Eventually, more complex and diverse organisms were able to survive on the island.)* Then discuss student answers.

Answers to *Coming to Conclusions*

Sample answers may include the following:

- Means of Animal Transportation—Some organisms utilize natural processes for transportation. The spider was probably carried to the island by the wind or a bird.
- Types of Living Things—As time went on, conditions were favorable for more types of organisms to survive. Ships may have transported seeds and animals to the island unknowingly. The ocean waves could have deposited organisms. Many different kinds of organisms began to flourish.
- Interactions Among Living Things—The diverse array of living organisms allowed for predator-prey relationships to develop. The pythons could have become prey to more complex animals on the island.
- The Process of Change—Change occurred as a result of processes such as wind and wave action, which may have deposited living organisms on the barren island.

FOLLOW-UP

Reteaching

Provide students with materials to make a "flip book" illustrating the natural changes that occurred on the island following the eruption. The book should include as many appropriate plants and animals as possible.

Assessment

Ask: Why did changes occur on Rakata in the years following the eruption? *(Students should understand that as the island's biological community gradually became diverse, it was able to support even more types of organisms. Once organisms found sufficient resources to survive, they were able to multiply and thrive on the island.)*

Extension

Have students write a story describing the changes that occurred on Rakata from the perspective of a python. Their story should include details of their arrival on the island as well as the ecological transformations that they witnessed.

Closure

Have students answer the following question and support their answer with examples: Why are biotic interactions important to consider when studying the changes that occurred on Rakata? *(Answers should include the idea that every consumer depends on other organisms in order to survive.)*

Diversity All Around

FOCUS

Getting Started

Before class, sketch a map of an imaginary zoo on a sheet of poster board. Label each zoo exhibit with the name of the organisms in the exhibit, such as bacteria, insects, birds, trees, mushrooms, flowers, or mammals. Provide each student with a nature or life-science magazine, and tell them that they are to find pictures of as many organisms as they can to populate the exhibits in this zoo. Ask: What are the natural habitats of these organisms? Are some more common than others in your local area? Afterward, you may wish to have the class vote on which organisms in their zoos are the most interesting.

Main Ideas

1. Living things exhibit diversity in size, shape, and physical structure.
2. A diversity of living things can be found in all environments.

TEACHING STRATEGIES

INDEPENDENT PRACTICE Before class, collect some specimens of plants, small insects, and other living things. Set up some workstations with a few different specimens at each station. Have students work in small groups to examine the diversity among the organisms. Ask them to compare the size, structure, and shape of the specimens. Students should make notes and illustrations in their ScienceLog about their observations.

Answer to *In-Text Question*

Ⓐ Student answers will vary. Look for accurate observations and descriptions. Student answers could describe physical characteristics or infer behavioral patterns from the picture.

Diversity All Around

There are more than 1.4 million *known* kinds of living things in the world (estimates of the *actual* number of kinds range from 10 to 100 million). These living things come in all shapes, sizes, and structures. **Diversity** is the term that biologists use to describe differences among living things. In this unit you will look at the diversity of living things. On this page and the next are illustrations of four living things, or **organisms**. Look at each one carefully. How do they differ from one another? Ⓐ

Sabelid worm Tube-dwelling worm that lives at the bottom of the ocean

Horsetail Small, flowerless plant related to the fern

LESSON 2 ORGANIZER

Time Required
2 to 3 class periods

Process Skills
observing, contrasting, communicating

New Terms
Diversity—a term biologists use to describe different characteristics in all living things
Organisms—living things

Materials (per student group)
Exploration 1, Activity 1: four 15–20 cm long craft sticks; metric measuring tape; 5 m of heavy string; scissors; magnifying glass; field guide for plant identification

Teaching Resources
Exploration Worksheets, pp. 3 and 9
Transparency 13

Gila monster
Poisonous lizard of the American Southwest

Fly agaric
Toxic wild mushroom

With a classmate, discuss the diversity you see. Here are some questions to help you begin.

1. What is the relative size of each organism?
2. What features does each organism have? Does it have roots, stems, leaves, flowers, eyes, legs, hair, scales, skin, or any other notable features?
3. What one characteristic does each organism have that distinguishes it from the other three?

You have just described four living things that illustrate the great diversity of organisms. Living things can differ in size, habitat, appearance, eating habits, and methods of self-protection, to name just a few characteristics. Now you will investigate diversity a little further. Do either Activity 1 or Activity 2 of Exploration 1.

EXPLORATION 1

Looking for Diversity

ACTIVITY 1

Diversity in a Lawn

While playing ball on the grass or stretching out in your lawn chair to enjoy the sunshine, have you ever wondered about what is happening beneath you? There may be a food-hunting expedition or a ferocious battle going on there. You can learn a lot by getting down on your hands and knees and carefully observing a small area of lawn. In this activity you will mark off 1 sq. m of ground, preferably the day before the Exploration. Then you will study the diversity of animal and plant life in this square meter of lawn.

You Will Need

- sticks (4 per group)
- a measuring tape
- heavy string and scissors
- a magnifying glass
- your ScienceLog and a pencil
- a field guide for plant identification

What to Do

Find a convenient grassy area. Using the measuring tape, measure out 1 m of ground and place a stick at both ends of the tape.

Place one end of the tape at a 90° angle from one of the sticks. Measure out 1 m, and push a stick into the ground. Repeat this step twice to make a square. Then tie string to the sticks to outline your square.

Exploration 1 continued

INDEPENDENT PRACTICE Have students examine the organisms pictured on pages 78 and 79. Ask students to work in pairs to read and answer the questions asked on these pages. Then reassemble the class and involve them in a discussion of their responses to the questions. You may wish to have them answer these same questions for the specimens they observed at their workstations. Encourage the use of the words *diversity* and *organism* in all class discussions.

EXPLORATION 1

ACTIVITY 1

Have students read pages 79–81 before doing this Activity. Tell them that on the first day, they will set up their lawn areas, observe the plant life there, and fill in the table for plants.

On the second day, students will explore animal life and fill in the table on page 81. Note that the tables require a count of the number of kinds, or *species,* of plants and animals. Students need only a general notion of species as a kind of organism to complete this Activity.

Before students go outside, remind them not to disturb the area they will be working in any more than is necessary. Prior to the Exploration, "scout out" areas for the students to avoid, such as those with poisonous plants or snakes, or areas where students might cut themselves on broken bottles. An area that has not been mowed for a while, such as an open field or abandoned lot, will probably provide the best results. You may wish to supply students with field guides for local plants and animals. Students should first get a general feel for the area by noting any large plants or animals that they see. Then they can take a closer look at the organisms in their marked-off area.

Exploration 1 continued

Homework

The Exploration Worksheet on page 3 of the Unit 2 Teaching Resources booklet makes an excellent homework activity to accompany Exploration 1.

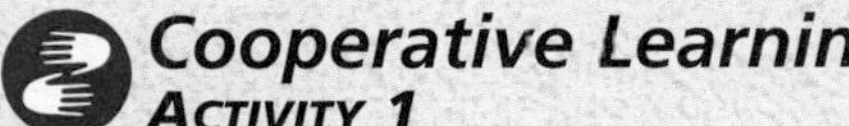

Cooperative Learning
Activity 1

Group size: 4 to 5 students
Group goal: to observe and record plant and animal life in a study site
Positive interdependence: Assign each student a role such as chief investigator, site preparer, materials coordinator, and data compiler.
Individual accountability: Students should include their completed tables from pages 80 and 81 in their ScienceLog.

EXPLORATION 1, *continued*

It may be impossible to count the exact number of certain kinds of plants or animals, so discuss some guidelines for students to use when recording their observations. Suggest that they estimate the number by sampling (counting the number in a representative 10 cm × 10 cm square and then multiplying that number by 100).

At first, students may think that all of the plants look the same. However, on closer examination with a magnifying glass, they should be able to see that there are differences in leaf shape and size.

Prior to working in the field, remind students never to touch an unknown plant and never to touch, tease, or annoy any animal.

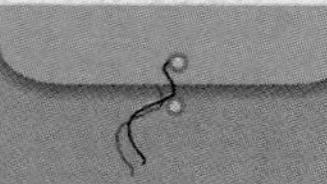

PORTFOLIO

Suggest that students include in their Portfolio their illustrations of plants observed during Activity 1. Students may evaluate their field experience by describing their most interesting observation.

Observe the plant life in your study site. In your ScienceLog, fill in a table like the one shown below. Then analyze the information you collected by considering the questions at right.

Quick sketch of each plant	Number of same kinds (species) of plants	Description of each plant (including size, appearance, color, and any other features observed)	Common name

Questions

1. How many different types of plants did you find?
2. Which is the smallest plant you found? the biggest?
3. Which plant did you find in the greatest numbers? the fewest?
4. Do all the plants you found have some feature in common? If so, what is it?
5. How does the plant life affect the animal life in your study site?

The next day, carefully approach your staked area to see whether any larger animals, such as butterflies, are present. If you come to the area too quickly or noisily, you may scare away some animals. Using a magnifying glass, look closely for small animals that might be hidden among the plants. In your ScienceLog, fill in a table similar to the one shown below. When you have completed the table, analyze this information by considering the questions at right. Then share your findings with your classmates.

Quick sketch of each animal	Number of same kinds (species) of animal	Description of each animal (including size, appearance, color, and any other features observed)	Common name

Further Questions

1. How many different types of animals did you find?
2. Did you hear animal sounds but not actually see the animal?
3. Did you find any evidence of animal life but not see the animals themselves?
4. Which is the smallest animal you found? the biggest?
5. Which animal did you find in the greatest numbers? the fewest?
6. How does animal life affect plant life in your study site?

ACTIVITY 2

Diversity Around You

What to Do  

In your ScienceLog, prepare two tables like those in Activity 1. Next, walk around your school, your neighborhood, or a park, and fill in your tables. Then answer the questions that followed each table in Activity 1.

Meeting Individual Needs

Second-Language Learners

Students with access to cameras may enjoy documenting the scientific methods used in Exploration 1. Instruct student photographers to take photographs of the different plants and animals they see. When the photos are developed, students can include their best pictures in a photo essay about the diversity of organisms they studied in the Exploration.

EXPLORATION 1, *continued*

Instruct students to look for evidence of animal life, even if the animals themselves are not present. For example, they may find plants that have been partially eaten, footprints, empty nests, or webs.

Students may know the common names for some of the organisms they see. If not, as they draw each organism, allow them to give the organism a descriptive name. For the purposes of this Exploration, it is not important for students to learn the formal names of the organisms.

GUIDED PRACTICE When the activity has been completed, involve the class in a discussion of the results. Students may use the Further Questions on this page to help them summarize their results. Ask: Did all of you find the same kinds of plants and animals? Did you find these plants and animals in the same numbers? Students may have different names for the same organism. This may be a good time to introduce the concept of standardization of names. Explain that scientists standardize plant and animal names so that they can communicate with each other about different organisms.

ACTIVITY 2

You may wish to have students complete this activity in class or as a homework assignment. Tell students that this activity is similar to Activity 1, except that they are asked to observe a much larger area from a greater distance. Explain that rather than observing small plants and animals found in a tiny patch of ground, they will be observing and describing larger organisms, such as trees, shrubs, mammals, and birds. After the activity has been completed, involve the class in a discussion of their results. It will be interesting for students to compare their tables. If they observed different areas, they are likely to have observed different kinds of plant and animal life.

Exploration 2

ACTIVITY 1

Have students work individually, at home or in class, to solve each of the riddles. There are enough details provided in each riddle to lead students to the realization that every species has unique features. Ask students to list in their ScienceLog all of the information given and to research any information that they do not understand. If they cannot solve the riddles from the clues given, the coded answers on page 84 should help.

Answers to *Activity 1, pages 82–84*

1. Dragonfly
2. Human
3. Toad
4. White pelican
5. Woodpecker
6. Onion
7. Rainbow trout
8. Apple
9. Brown bear
10. Elk

Multicultural Extension

Folk Tales About Animals

African American folk tales include stories that often describe natural phenomena, human psychology, or legendary origins of the world. Many of these stories offer explanations for some of the adaptations of different animals. Have students research how animals are portrayed in various folk tales, legends, or myths.

Homework

The Exploration Worksheet on page 9 of the Unit 2 Teaching Resources booklet makes an excellent homework activity. If you choose to use this worksheet in class, Transparency 13 is also available to accompany Exploration 2.

In Exploration 1 you discovered the diversity that can be found in the living things around you. Here are some riddles showing how much more diversity exists. See how many you can solve!

Exploration 2

The Puzzling Diversity of Living Things

ACTIVITY 1

Who Am I?

In the following riddles, you'll find information about the structures, habits, and habitats of some organisms. Read each riddle carefully, and think about what it tells you. Which living thing does each riddle describe? If you're having trouble figuring out what a riddle is describing, look at the pictures for some hints. (The pictures are not in order.) If you are really stuck, the answers are in code on page 84. All you have to do is break the code!

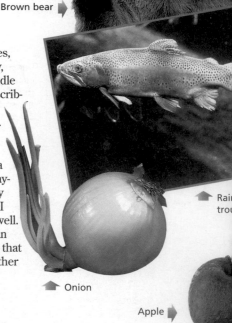

Brown bear

Rainbow trout

Onion

Apple

Woodpecker

Riddle 1

I move slowly when I am young but very quickly when I'm an adult. I eat flying insects, which I hunt near water. I have to be a strong flier to catch my food. When I stretch my four wings, I look like a helicopter. I have two more legs than a dog, and I have very large eyes. I am cold-blooded and have an external (outside) skeleton. Sometimes I'm very colorful. Who am I?

Riddle 2

I can walk, run, and swim. I can see well, but my sense of smell is not very sharp. I am warm-blooded. I am very adaptable and can live in many different environments. I really enjoy changing my environment. I care for my young for many years. I stand upright. Who am I?

Riddle 3

I must live in damp or wet places, avoiding the dry heat of summer and the cold of winter. If I am living in a cold climate, I become dormant in the winter. If I'm a female, I produce young by laying eggs in water. I survive by eating any moving thing that I can swallow. I can sing very well. Some of my close relatives can secrete a sticky, white poison that can kill or paralyze dogs or other enemies who may try to eat them. Who am I?

Riddle 4

I live in lakes, marshes, salt bays, and on beaches. I eat mostly fish and crustaceans. Although I can fly, I catch fish only by swimming. My great throat pouch is handy for scooping up fish. I fly by alternating several flaps of my wings with a glide. I always fly with my head hunched behind my shoulders. I nest on the ground in colonies. I have a wingspan of 2.5–3.0 m. My close cousins live only by the ocean, but I can venture inland. I am happy to report that these cousins are growing in number, even though they suffered from DDT poisoning a few years ago. Who am I?

White pelican

Toad

Elk

Dragonfly

Human

Riddle 5

I have a very high body temperature. My feet are well adapted for grasping things. I have four toes on each foot: two point forward, and two point backward. I have stiff, spiny tail feathers that act as a prop when I hunt food. I eat tree-boring insects, ants, acorns, flying insects, berries, and sap. My home, which I make myself, is a hole in a tree. I use my bill to chisel away the wood. Who am I?

Riddle 6

I have pointed, green stalks above the ground and a rounded, brown bulb below. People must pull me out of the soil before I can be useful to them. Cooks use me to improve the taste of food. If people bite me, I can bite back, making their eyes water. Who am I?

Riddle 7

I live in cold, well-oxygenated water, and I'm a fast, strong swimmer. I am slim, sleek, and colorful. I'm a carnivore; I eat mostly insects and smaller members of my own kind. I spawn my eggs during the spring in small, clear streams. I'm coldblooded. Who am I?

Riddle 8

I undergo wondrous changes during my life. At the beginning, I am a sweet-smelling, pink-and-white blossom. Later I'm a hard, green ball that makes your eyes water and your mouth pucker if you try to eat me. Finally, I become a sweet, juicy, red or yellow fruit. People say I keep physicians away. Who am I?

Riddle 9

I am a big animal. My mass is about 225 kg, but my tail is only about 15 cm long. I am dark in color. Generally, I live on forest floors and in thickets. When it starts to get really cold, I enter my shelter for the winter. I don't have very good sight, but my senses of hearing and smell are keen. Using these senses, I find lots of food—small animals, insects, garbage, leaves, grasses, berries, nuts, and fruits. Who am I?

Riddle 10

I am warmblooded and hairy. I feed milk to my young. I chew my cud, and I have a complex stomach. The males of my kind have huge, branching antlers. I have a heavily maned neck. Humans, wolves, and mountain lions are my only enemies, but mountain lions usually won't attack me when I'm fully grown. My young are not camouflaged from these enemies until their winter hair grows out. Sometimes you can hear the males of my kind give a high-pitched bugle call. If this call is answered by another male, a battle may follow. Who am I?

Exploration 2 continued

GUIDED PRACTICE As an extension to this Exploration, you may wish to read the following poem to your students. Before beginning, ask students to list several words that they associate with bats. After they have heard the poem, ask students if their ideas about bats have changed at all.

Bats

A bat is born
Naked and blind and pale.
His mother makes a pocket of her tail
And catches him. He clings to her long fur
By his thumbs and toes and teeth.
And then the mother dances through the night.
Doubling and looping, soaring, somersaulting—
Her baby hangs on underneath.
All night, in happiness, she hunts and flies.
Her high sharp cries
Like shining needlepoints of sound
Go out in the night and, echoing back,
Tell her what they have touched.
She hears how far it is, how big it is,
Which way it's going:
She lives by hearing.
The mother eats the moths and gnats she catches
In full flight; in full flight
The mother drinks the water of the pond
She skims across. Her baby hangs on tight.
Her baby drinks the milk she makes him
In moonlight or starlight, in mid-air.
Their single shadow, printed on the moon
Or fluttering across the stars,
Whirls on all night; at daybreak
The tired mother flaps home to her rafter.
The others all are there.
They hang themselves up by their toes.
They wrap themselves in their brown wings.
Bunched upside down, they sleep in air.
Their sharp ears, their sharp teeth, their quick sharp faces
Are dull and slow and mild.
All the bright day, as the mother sleeps,
She folds her wings about her sleeping child.

—Randall Jarrell

ACTIVITY 2

Have students work on this activity individually. Make sure that students have enough information to write their riddles. Instruct them to include the same amount of detail in their riddles as in those from the text.

To help students get started, you may wish to have them find answers to the following questions about their chosen organism: Where does it live? What does it eat? How does it move? What appendages does it have? (Tell students that an *appendage* is any part of an organism that is attached to the main body, such as a tail or finger.) What other structures does it have? How does it produce young? Is it coldblooded or warmblooded?

Students may be encouraged to select animals or plants that are unfamiliar to them so that they can practice their research skills as well as their writing skills. You may wish to supply students with the names of some organisms that they can use to write riddles about. You might even write the names of organisms on slips of paper and let students randomly choose the one they will write about. Include species of plants and animals that are common in your area. This will give students an opportunity to become more familiar with the living things around them.

CROSS-DISCIPLINARY FOCUS

Art

As a class, select a specific habitat, such as a desert sand dune, coral reef, tropical rain forest, or prairie. Then have students use arts and crafts materials to construct a model of the habitat in a section of the room. Allow some students to design the plant life, soil, and features of the environment. Other students should construct replicas of animals that are indigenous to the habitat using wire, cardboard, or papier-mâché. Encourage students to exhibit the diversity of the plants and animals in the model habitat they have built.

Here are the coded answers to the riddles:

1. CQZFNMEKX
2. GTLZM
3. SNZC
4. VGHSD ODKHBZM
5. VNNCODBJDQ
6. NMHNM
7. QZHMANV SQNTS
8. ZOOKD
9. AQNVM ADZQ
10. DKJ

Hint for decoding: ZMS = ANT

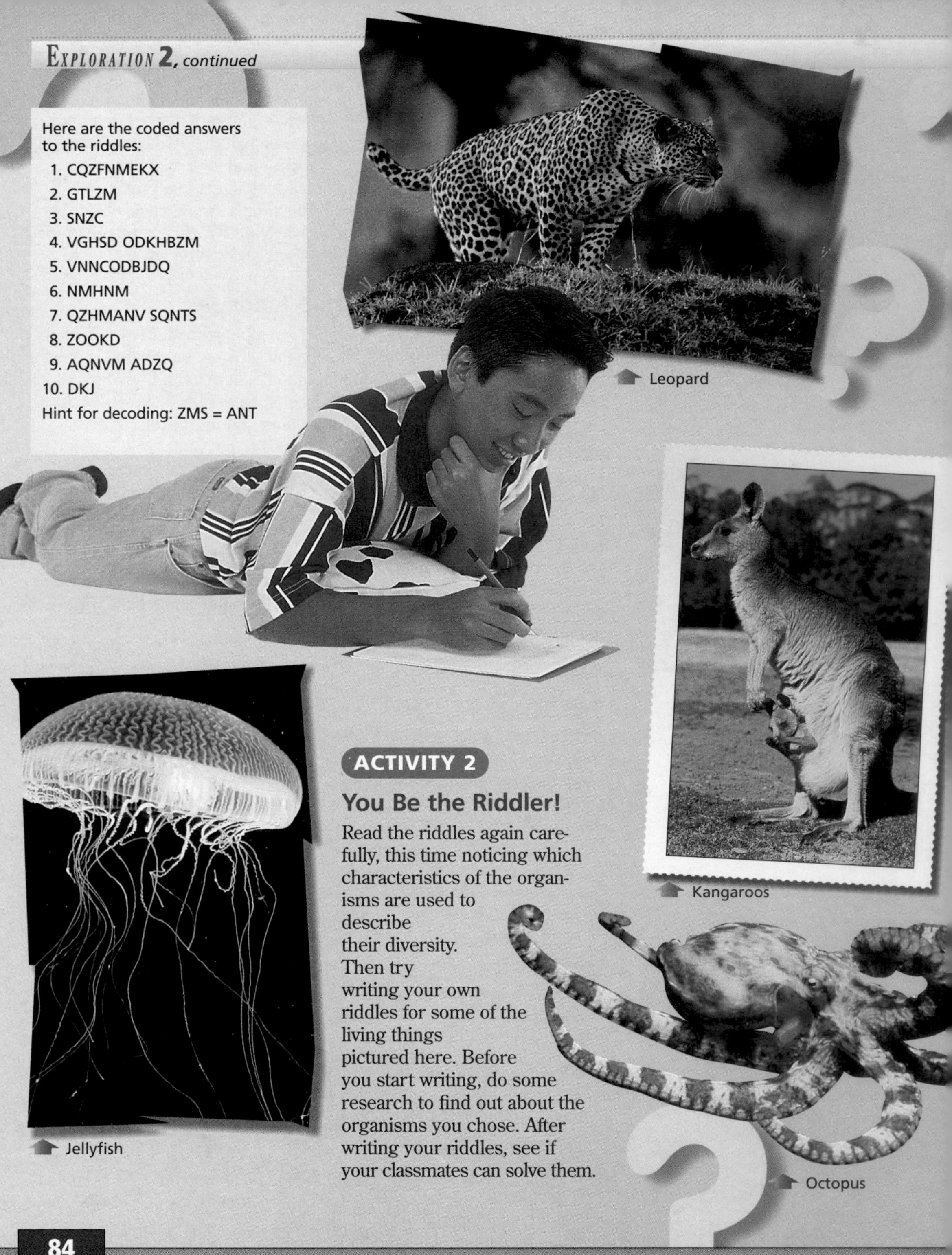

Leopard

Kangaroos

Jellyfish

Octopus

ACTIVITY 2

You Be the Riddler!

Read the riddles again carefully, this time noticing which characteristics of the organisms are used to describe their diversity. Then try writing your own riddles for some of the living things pictured here. Before you start writing, do some research to find out about the organisms you chose. After writing your riddles, see if your classmates can solve them.

Homework

Using the riddles on pages 82–83 as models, have students write a riddle that describes their favorite flower.

Fisherman bat

Narcissus flowers

Sea slug

Hammerhead shark

Great horned owl

Fungi

Duck-billed platypus

Pine branch with cones

85

FOLLOW-UP

Reteaching

Play a nature-sounds tape, such as a tape of pond or marsh sounds. Ask the students to listen to the tape quietly, imagining that they are sitting on the banks of this body of water. After a few minutes, ask: What living things can you hear? What living things live here that you cannot hear? Have students name as many different kinds of plants, animals, and microorganisms as possible.

Assessment

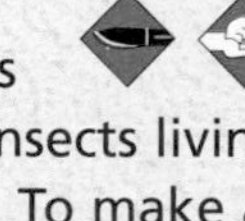

Have students use sweep nets to sample and compare the insects living in two different grassy areas. To make a sweep net, bend a wire coat hanger into the shape of a circle. Put an old pillow case on the wire. Use wire or strong tape to fasten the net to an old broom or mop handle. To take a sample, mark off a 1 m × 1 m area, and sweep the net back and forth across the grass. Hold the net closed until the contents are emptied into a jar. Students should try to identify the organisms using field guides that you have provided. They should then tally the number of each type of organism. Ask students to speculate about factors in the sample areas that account for the different organisms they encountered.

Extension

Take students on a trip to a zoo to observe the diversity of animals. Assign each student a type of animal to research in more depth. For example, students could study primates, snakes, fish, marsupials, cats, bears, or other groups of animals. They should find out about the native environment of the animal, what it eats, and how it produces its young. Reports and illustrations could be displayed in the classroom.

Closure

As an extension to Exploration 2, Activity 2, students could play twenty questions. A volunteer could begin by saying, "I'm thinking of an animal" or "I'm thinking of a plant." The remainder of the class would be allowed to ask 20 questions to guess the identity of the animal or plant. Each question should have a yes or no answer. Encourage students to think about the differences among various living organisms.

Answers to *Challenge Your Thinking*

1. Answers will depend on location of study and individual experience. Log entries should demonstrate thorough observations.
2. Students may highlight the variety of plants and animals in their area, focus on diversity of human life in their area, or both. Answers will vary according to region and location of the community.
3. Students should base the ordering and content of their time line on the information provided in Lesson 1. The drawings should show the variation of diversity among island plant and animal species. This includes the rich diversity before the explosion, the lack of diversity immediately after the explosion, and the increasing diversity in the years following the explosion. You may wish to display the time lines in the classroom.
4. Answers will vary. A student might answer that all three animals have elongated bodies and no legs. Visible differences include the type of skin, markings on their skin, and body size. Students should realize that all three animals share common internal characteristics (blood circulation and digestion) but probably also have internal differences because of differences in their environments and food sources.
5. Students should recognize the presence of many different plants and plant products all around. For example, we sit on chairs made of wood, wear clothes made of cotton, eat vegetables, write on paper made from trees, and sleep in beds that may have wooden frames and cotton sheets. Students should identify other ways that plants affect their lives, such as the production of the oxygen we need in order to survive. Each student's description of what life would be like without plant diversity should reflect an understanding of the many ways that plants are involved in his or her lifestyle.

1. Seven Days of Diversity

Write a series of log entries that describe the different plants and animals you see during one week. Where did you see them? What were they doing? Were there any changes over the week? Did you see anything that surprised you? Compare your log entries with those of your classmates.

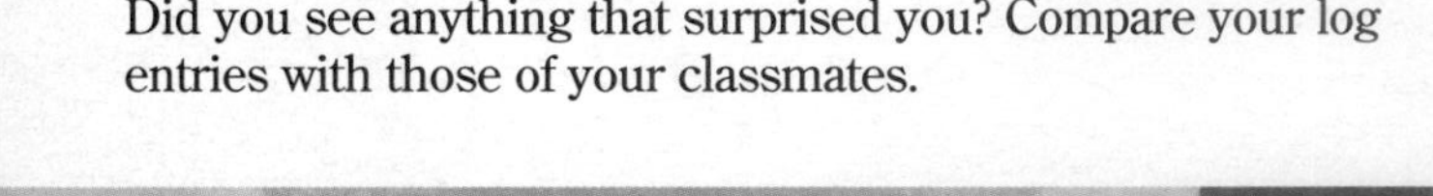

2. Tourist Attraction

Your city's tourist bureau has asked the members of your class to write a section in their new brochure called "Diversity in My Community." What would you write about the diversity of living things in your community in order to attract tourists to visit?

3. Now You See It, Now You Don't

Create a pictorial time line that shows Krakatau before the volcano erupted and that includes the changes on Rakata up to 1985.

★ **You may wish to provide students with the Chapter 4 Review Worksheet that is available to accompany this Challenge Your Thinking (Teaching Resources, page 12).**

4. Invisible Differences

Look at these three animals. What similarities do you see? What differences do you see? Do you think there are differences and similarities that are not visible? Explain your reasoning.

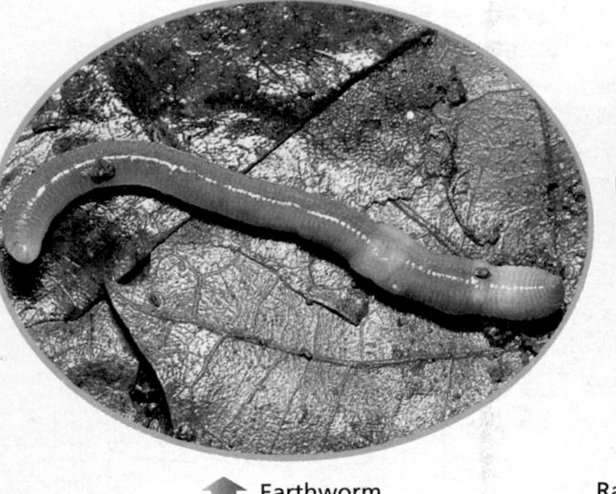

Earthworm

Rattlesnake

Moray eel

5. They're Everywhere!

What plants do you sit on? What plants do you wear, eat, write on, and sleep in? Think about the ways that plants affect your life, and list as many as you can. Describe what your life would be like if there were not so much diversity in plants.

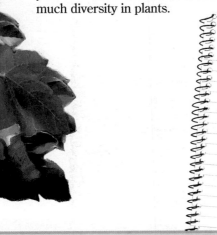

Sciencelog

Review your responses to the ScienceLog questions on page 74. Then revise your original ideas so that they reflect what you've learned.

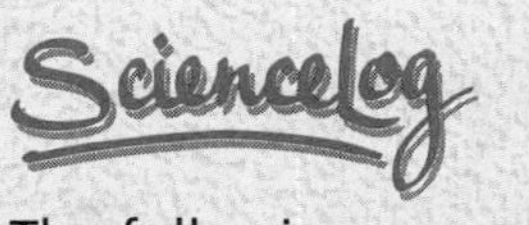

Sciencelog

The following are sample revised answers:

1. More than 1.4 million kinds of living things are known to exist on Earth. Estimates of the actual number range from 10 to 100 million.
2. *Diversity* is the term that biologists use to describe the many differences among living things.
3. Organisms can differ in countless ways, including appearance (such as shape and markings), external physiological function (such as locomotion), internal physiological function (such as respiration or reproduction), and behavior (such as methods of rearing young). The three birds shown differ in size, beak shape, and coloration. They also eat different things and behave differently.

Homework

Animals and animal products play important roles in our lives. Have students describe three animal products that are important to humans. Students should explain their choices. *(Sample answer: Leather is important because it is used to make many articles of clothing as well as upholstery and purses.)*

Why So Many?

Connecting to Other Chapters

Chapter 4 *introduces the concept of diversity and how the diversity in a community can change.*

Chapter 5 *explores the theory of natural selection and the value of diversity to a community.*

Chapter 6 *challenges students to create classification systems for a wide range of living things.*

Prior Knowledge and Misconceptions

Your students' responses to the ScienceLog questions on this page will reveal the kind of information—and misinformation—they bring to this chapter. Use what you find out about your students' knowledge to choose which chapter concepts and activities to emphasize in your teaching.

In addition to having students answer the questions on this page, you may wish to have them complete the following activity. Have students write a short skit that takes place in a world in which adjectives do not exist. Alternatively, you may want to eliminate a different element of grammar, such as nouns, verbs, adverbs, or prepositions. By appreciating the way that parts of speech work together to form an interacting system, students may begin to understand how organisms can work together to form an interacting system in a natural community. The skit should feature two students discussing how they get to school in the morning. You may wish to have volunteers act out a few of the skits.

After students finish their skits, have them write a paragraph about the difficulties they had in writing their skit and what role the missing element of grammar serves in communication. Ask students to explain why it is useful to have a variety of different elements of grammar when communicating. *(Explain to them that each element of grammar serves a purpose in language and that it is this diversity that allows meaningful communication to be possible.)*

Collect the paragraphs, but do not grade them. Instead, read them to find out whether students are aware of the importance of diversity in language, what misconceptions they have, and what is interesting to them about the need for diversity. In this chapter they will explore the need for diversity in the natural world and the process by which it occurs.

CHAPTER 5

Why So Many?

1 **Why are there no brown polar bears?**

2 **Why is there so much diversity among living things?**

3 **Can you see the chameleon in this picture? How does this animal benefit from blending so well into its environment?**

ScienceLog

Think about these questions for a moment, and answer them in your ScienceLog. When you've finished this chapter, you'll have the opportunity to revise your answers based on what you've learned.

88

LESSON 1

How Does It Happen?

You've seen that there are many different kinds of plants and animals—perhaps more than you realized. Think about the following questions:

- Why are there so many different types of living things? Why isn't there a single "all-purpose" organism or at least one "all-purpose" plant and one "all-purpose" animal?
- Have there always been so many different types of plants and animals?
- Why have some plants and animals survived while others have not?

Are there other questions you wonder about? Keep reading to find some answers! (A)

A cycad, a palmlike tropical plant (150 million years old)

American mastodon (10,000 years old)

Diversity Over Time

We learn a lot about plants and animals that lived a long time ago by examining fossils. Look closely at the pictures on this page. What evidence do you see that there has been diversity in organisms for a long time?

Leaves from a gingko tree (250 million years old)

Fish (50 million years old)

Fly trapped in amber (35 million years old)

LESSON 1 ORGANIZER

Time Required
2 to 3 class periods

Process Skills
observing, predicting, hypothesizing, analyzing

Theme Connection
Changes Over Time

New Term
Natural selection—theory developed by Charles Darwin that explains how the characteristics of a species can change over many generations to make the species better suited to the environment

Materials (per student group)
none

Teaching Resources
Resource Worksheets, pp. 17 and 19
Transparency 14
SourceBook, p. S39

LESSON 1

How Does It Happen?

FOCUS

Getting Started

Before class, cut out a moth shape from white paper and a moth shape from brightly colored gift paper. Glue the two shapes together and mount the moth on a dowel rod. Hold the moth, colored side up, against white poster board. Ask students to imagine that they are hungry birds. Would they have trouble finding this moth to eat? *(Probably not)* Turn the moth over. Ask students to imagine that the colorful moth had offspring with white wings. Would hungry birds find the white moths as easily? *(No)* Tell students that a feature that helps an organism survive is called an *adaptation.* Therefore, white color is an adaptation for the paper moth. In this lesson, students will learn about how species change over time to adapt to their surroundings.

Main Ideas

1. Natural selection explains how the different features of a species change over generations as a result of changing environmental conditions.
2. Darwin developed his theory of natural selection by observing the differences in the shapes and sizes of closely related organisms.

TEACHING STRATEGIES

Have students read the introduction silently and then study the pictures on page 89. Direct their attention to the diversity that exists among the fossils. Then discuss their ideas about the questions asked. Students will develop more complete answers to these questions as they proceed through the unit.

Answer to *In-Text Question*

(A) Accept all reasonable responses. Sample question: Why don't dinosaurs exist today?

Answer to
Diversity Over Time, page 89

The differences in shape, size, and composition of the fossil organisms are evidence that there was diversity among these organisms.

Survival of the Fittest

Ask student volunteers to read aloud this selection and to play the roles of the people interviewed. Discuss with students the similarities and differences of the beliefs of Lamarck and Darwin. *(Students should conclude that both scientists believed that living things survived because they were suited to their environment. However, Lamarck believed that organisms became suited to their environments by acquiring characteristics that enabled them to survive. Darwin believed that those organisms best suited to their environment survived and reproduced, passing their characteristics to their offspring.)*

Meeting Individual Needs

Learners Having Difficulty

Have students simulate survival of the fittest. Divide the class into groups of four or five students, and distribute a full sheet of black-and-white newspaper to each group. Then distribute to each group 100 circular cut-out shapes of newspaper and 100 circular cutouts of brightly colored paper. Instruct the groups to place their sheet of newspaper on a flat surface and randomly distribute the 200 circular shapes on the newspaper. The students are the predators and the circular shapes are the prey. On your signal, have students begin predation. They are to pick up one prey at a time as fast as they can and drop their catch in a predation pile 1 m away. They repeat this process for 1 minute. At the end of 1 minute, students should count and record the number of prey that they caught for each color. Encourage students to record their results in their ScienceLog. Compile class results. Students should find that most of the prey caught were brightly colored.

Survival of the Fittest

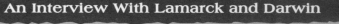

Why do some types of plants and animals survive while others do not? This question has been asked for many years by many people. Not everyone has come to the same conclusion.

Some of the earliest studies were conducted by two famous scientists—Jean Baptiste de Lamarck and Charles Darwin. Even though they didn't work together (Darwin was only 20 years old when Lamarck died in 1829), they had similar ideas. They both thought that living things survived because they were somehow suited to their environment. They also believed that these survivors passed on their "suitable" characteristics to their offspring. However, Lamarck and Darwin did not agree on how this all happened. Read on to find out what Lamarck and Darwin might tell Maria, a curious student, about their ideas today.

An Interview With Lamarck and Darwin

Did You Know . . .

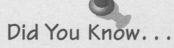

The outer ear of humans helps collect sound waves. Yet most birds have no outer ear because the ear would add air resistance and cause them to expend greater energy when flying. Bats, on the other hand, depend so much on hearing that they have very large outer ears despite the aerodynamic disadvantage that the ears create.

MARIA: Why do you think there is so much diversity in the color of butterflies?

LAMARCK: At one time there was only one color of butterfly, perhaps yellow. When yellow butterflies sat on yellow flowers, it was hard for predators to see them. This means that they were suited to their environment, so there was less chance they would be eaten and a greater chance they would survive. However, when yellow butterflies landed on red flowers, they gradually changed and acquired a new characteristic—a red color—so that they couldn't be seen easily. These changed butterflies were able to pass on this new characteristic to their offspring.

DARWIN: I agree with some of what Lamarck says. Yellow butterflies are more likely to survive if they sit on yellow flowers than if they sit on red flowers. However, I don't think that yellow butterflies became red just because they sat on red flowers. I think that over the years, there were many, many colors of butterflies. Butterflies of certain colors were better suited to their environment than were butterflies of other colors; not all types survived. Those that we see today had ancestors with colors that matched the flowers in their environment and therefore weren't seen by predators. They lived long enough to reproduce and pass on the suitable color to their offspring. Butterflies with unsuitable colors would have been seen by predators and would have been eaten, so they would not have reproduced and passed on the unsuitable color to their offspring. They didn't have any offspring.

Which explanation do you agree with? Why? (A)

For a time, Lamarck's idea received acceptance—even Darwin believed it for a while. However, we have a much greater understanding of inheritance today. It is now known that a characteristic that is acquired by a parent organism during its lifetime is *not* passed to the organism's offspring. If it were, a mother cat that had lost her tail in an unfortunate accident would have kittens without tails! We know this is not the case.

Lamarck's explanation is not accepted today, but Darwin's idea has become an important part of science.

Answers to *Survival of the Fittest*

(A) Accept all reasonable responses. Students should understand that Darwin's theory involves a more plausible mechanism of change.

Homework

Ask students to choose a local species of plant or animal and to observe a specimen of that species in its natural environment. Students should record observations about the organism's physical characteristics and behavior and should speculate about what adaptations enable this organism to survive.

Natural Selection

Have students read page 92 to learn about Darwin and his theory of natural selection. Ask volunteers to summarize Darwin's ideas and to think of examples that support his theory.

Make sure students understand that the theory requires that variations occur naturally in populations of organisms. Sometimes pressures from the environment, such as a reduced food supply, allow individuals with certain variations to better survive in that environment. Those organisms that survive are able to reproduce, and their characteristics are then passed on to the next generation.

Answers to *In-Text Questions and Caption*

- **A** Students should conclude that the dark bears probably went extinct in the arctic regions.
- **B** The white bears were better suited to their environment. Therefore, they were more successful than the dark bears in passing their coloration on to future generations.
- **C** Yes, because most of the dark-colored moths survived in polluted areas. Since more of these moths survived, they passed their dark coloration on to future generations.
- **D** The polar bear walking on ice and snow is harder to see and would have better luck sneaking up on its prey than the polar bear walking in the woods.

Natural Selection

Did you know that at one time, not all bears living in the Arctic regions of the world were white? Some were dark. However, the white bears were more successful in finding food. They blended in with their snow-covered environment and were able to sneak up on their prey. Therefore, they were well fed and healthy, enabling them to survive and reproduce. The white bears passed on their suitable color to their offspring. These bears are known today as polar bears. What do you think happened to the dark bears that lived in the Arctic regions? **A**

As you read earlier, Darwin had a theory about how organisms like butterflies had *evolved,* or changed, over many generations. He thought that organisms with traits or characteristics that are well suited to the environment have a greater chance of survival. They reproduce at a greater rate than organisms that are not as well suited to the environment. The well-suited organisms pass desirable traits on to their offspring. Darwin called this process **natural selection** because the environment favors, or selects, organisms that are best suited for survival. How does natural selection explain how the dark bears were replaced by white ones? **B**

Sometimes the habitat of a species will change in some way. This change may force the species to move to a new habitat, adapt over many generations to the changed habitat, or become extinct. One example of a species that adapted to a changed habitat is the peppered moth, which you are about to examine in detail. Study the research data given on page 93. Do these facts support the theory of natural selection? **C**

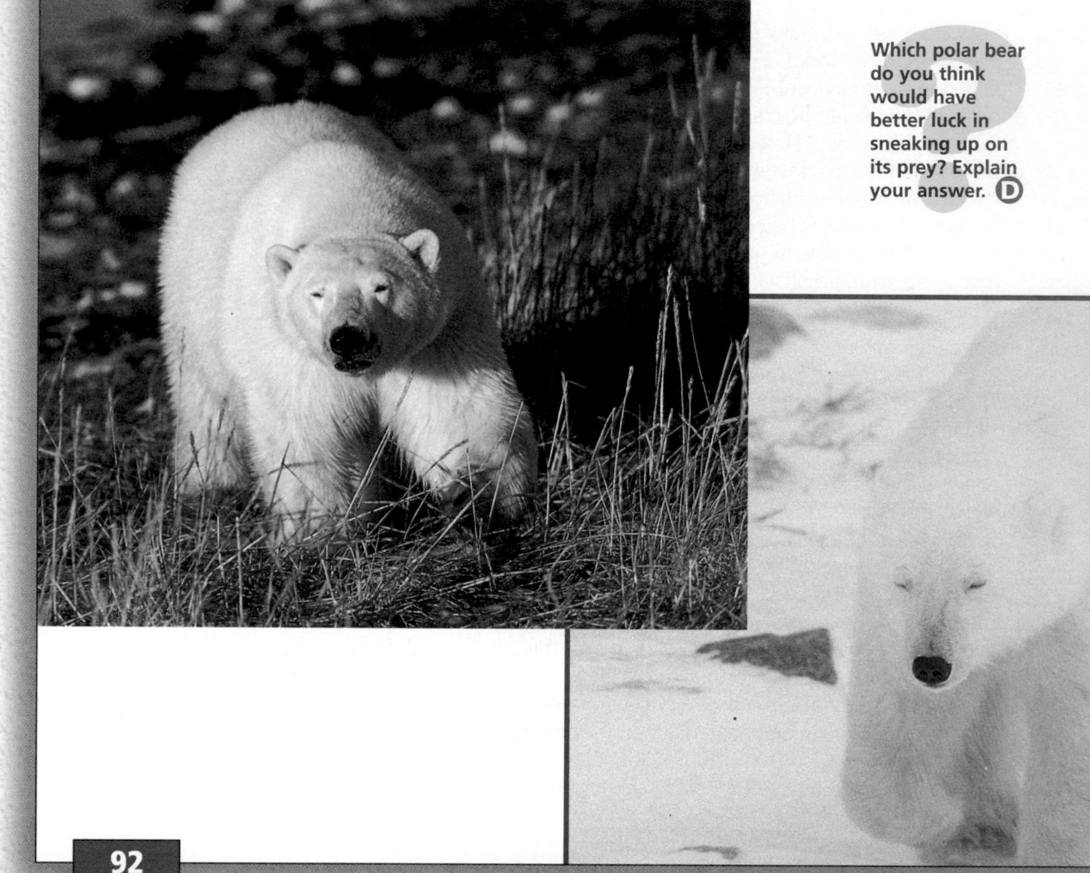

Which polar bear do you think would have better luck in sneaking up on its prey? Explain your answer. **D**

92

The Case of the Peppered Moth

Research Subject
Peppered moths have lived in the forests of England for thousands of years. They rest on the trunks of trees during the day and are a source of food for many birds. Peppered moths vary in color, from light-colored to dark-colored.

Research Problem
Did air pollution, which covered tree trunks with black soot from the many new factories during the Industrial Revolution of the 1800s, affect the survival of the peppered moth?

Conditions
Before the Industrial Revolution, the tree trunks were light-colored. The trunks and branches were also covered with silvery white lichens. As the Industrial Revolution progressed, pollution killed the lichens and blackened the tree bark.

Hypothesis
In the 1950s, Oxford University professor H.B.D. Kettlewell and his assistants formed a hypothesis about the peppered moth. Form your own hypothesis. It should include the effect that you think the Industrial Revolution had on the peppered moth's survival. (E)

Procedure
With his assistants, Kettlewell performed an experiment to test his hypothesis. They used the following procedure:

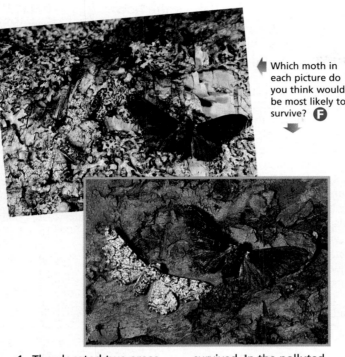

Which moth in each picture do you think would be most likely to survive? (F)

1. They located two areas. One was a wooded area with lichen-covered oak trees. The other was a wooded area that had been subjected to pollution for many years.
2. They released a known number of light-colored and dark-colored peppered moths into each area.
3. After a given amount of time, they recaptured as many moths as they could.

Results
In the unpolluted area, more light-colored moths survived. In the polluted area, more dark-colored moths survived.

Discussing the Results
1. Do these results support your hypothesis? (They supported Kettlewell's hypothesis.)
2. How can you explain the results?
3. Why didn't Kettlewell release the moths only into the polluted area?
4. Do Kettlewell's findings support the theory of natural selection? Explain your answer.

93

Homework

The Resource Worksheet that is available to accompany The Case of the Peppered Moth makes an excellent homework activity (Teaching Resources, page 17).

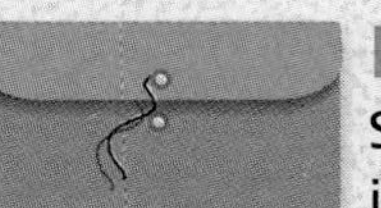

PORTFOLIO

Students may wish to include their answers to Discussing the Results in their Portfolio.

The Case of the Peppered Moth

This activity is a good illustration of natural selection. Students are asked to formulate their own hypotheses and to compare their ideas with historical facts.

Have students work in small discussion groups to read the page and answer the questions. Suggest that they stop reading just before they get to Kettlewell's procedure. Then ask each group to formulate their own hypothesis before continuing.

Answers to *In-Text Question and Caption*

(E) Sample hypothesis: The darker tree trunks that resulted from the Industrial Revolution allowed dark-colored moths to survive better than light-colored moths.

(F) In the top picture, the light-colored moth would be most likely to survive. In the bottom picture, the dark-colored moth would be most likely to survive.

Answers to *Discussing the Results*

1. Answers will depend on the hypothesis that was developed by the students.

2. When the light-colored bark was darkened, the light-colored moths were no longer camouflaged. They then became easy prey for predators such as birds. Under the same circumstances, however, the dark-colored moths had the advantage of camouflage and survived better.

3. The moths released into the unpolluted area were used as the control group in the experiment.

4. Kettlewell's findings support the theory of natural selection because more of the dark-colored moths survived in the polluted area and would therefore be able to pass their dark coloration on to later generations. By contrast, the light-colored moth population decreased in the polluted area.

Darwin's Finches

Point out to students that the Galápagos Islands are a living laboratory of natural selection. This is because they are volcanic islands isolated from the South American mainland by some 960 km of ocean. Ancestors of the plants and animals that inhabit the islands today were brought from the mainland by wind currents or floating debris. The organisms that arrived there first had no competition from other species because there were no species already there.

Answers to *In-Text Questions, page 95*

1. The beak of the small tree finch is broad and short, enabling it to eat aphids and small berries. The beak of the large tree finch is parrotlike, enabling it to obtain and grind larger fruit and insects than the small tree finch is able to eat. The beak of the small ground finch is long and pointed, enabling it to obtain tiny seeds from the ground and ticks from iguanas. The large ground finch has a strong conical beak that enables it to crush hard seeds. The cactus finch has a long, narrow beak that it uses to probe for nectar in cactus flowers.
2. Yes. The finches that had the appropriate beak for feeding on a particular food source had the greatest chance of surviving and passing that characteristic on to their offspring.
3. **a.** Accept all reasonable responses. (You may wish to tell students that scientists believe that the original finches probably ate seeds like those found on the South American mainland.)
   b. Possibly, but probably not, because the finches have so many similarities other than their beaks
   c. Yes. The birds' beaks adapted to the food sources available, which varied from island to island. Also, living on one large island would have allowed the finches to keep breeding with each other as a large group, and they might not have diversified.
   d. **i.** There would have been no competition from other species for the same food.
   ii. The finch population could feed and reproduce with no threat to its existence from predators.
   iii. The finch population could remain healthy and strong without any damaging effects from parasites.
4. Because of the diversity of their beaks, the finches were able to successfully inhabit many different habitats on the islands. Diversity also reduced competition for limited food sources.

Darwin's Finches

In 1835, Darwin spent five weeks visiting the Galápagos Islands as the naturalist on a ship called the H.M.S. *Beagle.* He observed, recorded, collected, and preserved everything he could of the islands' natural history. There were many strange and colorful animals, but what interested him most were drab little birds that made unmusical sounds—finches. The finches on all the islands resembled each other closely, except for one set of features—the size and shape of their beaks. Separated for thousands of years on the different islands of the Galápagos chain, the birds had adapted in unique ways to their environments. The differences among the finches of the various islands are shown in the table.

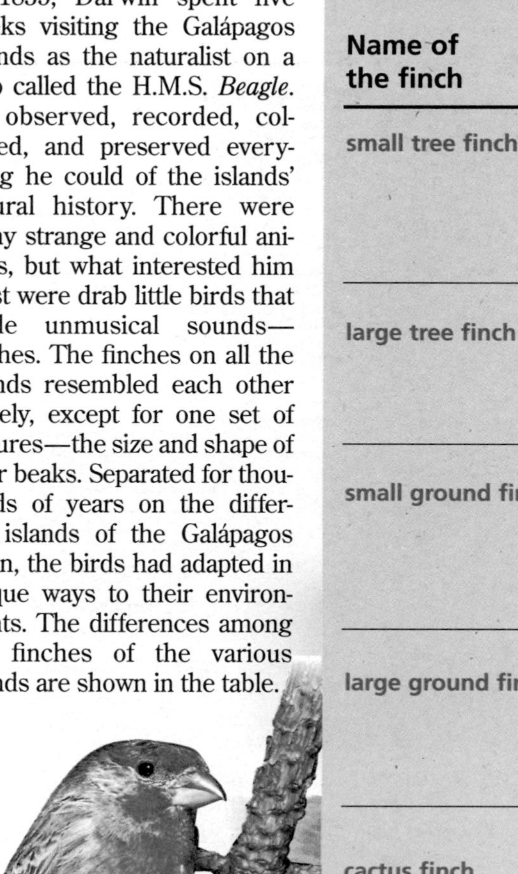

Differences Among Some of Darwin's Finches

Name of the finch	Feeding habit	Form of beak
small tree finch	Uses delicate bill to eat aphids and small berries.	
large tree finch	Grinds fruit and insects with parrotlike bill.	
small ground finch	Uses pointed bill to eat tiny seeds and to pick ticks from iguanas.	
large ground finch	Conical bill enables it to eat large, hard seeds.	
cactus finch	Long bill probes for nectar in cactus flowers.	

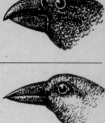

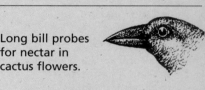

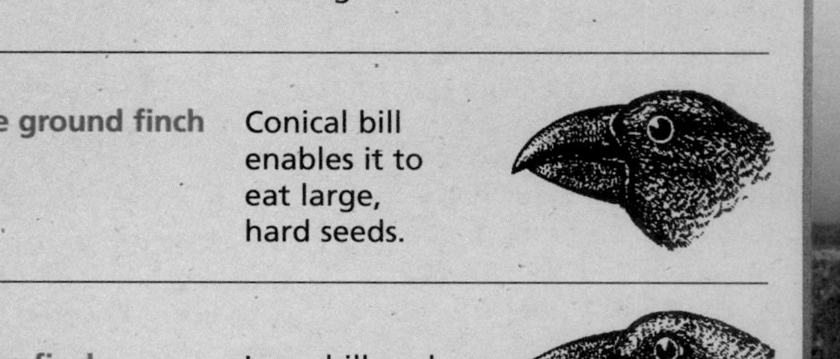

Homework

The Resource Worksheet that is available to accompany Darwin's Finches makes an excellent homework activity (Teaching Resources, page 19). If you choose to use this worksheet in class, Transparency 14 is also available.

1. How is the structure of the beak well suited to the diet of each group of finches?
2. Do the differences in Darwin's finches support the theory of natural selection? Give reasons for your answer.
3. Scientists have speculated that Darwin's finches reached the Galápagos Islands from the mainland of South America as a single flock perhaps a million or more years ago. Think about the following questions, and explain your answers:
   a. What do you think the original finches looked like? Why?
   b. Is it possible that the original birds were various species that arrived on the islands at different times?
   c. Assume that one flock of finches gave rise to the 14 different species now existing on the islands. If this occurred, would it be significant that the Galápagos chain consists of many small islands rather than one large one?
   d. What advantages would the finches have had in arriving on the islands under the following conditions?
      i. There were no other species with exactly the same diet.
      ii. There were no predators.
      iii. There were no parasites to live on the finches and weaken them.
4. How has diversity helped Darwin's finches survive?

Aerial view of one of the Galápagos Islands

To learn more about Darwin's findings and the theory of natural selection, turn to pages S41–S43 in the SourceBook.

Adventures in a New Environment

Would you like to live in a new, vastly different environment? How about on the moon? under water? on a planet in outer space? What would help you survive? Think about the technological adaptations that were required in order for the *Apollo 11* astronauts to live and travel on the moon for several days.

Create a tale about how you and a group of your friends would survive in a very different environment. Describe the conditions of the new environment, the difficulties you would encounter, and the adaptations you would need in order to survive. You might be able to develop technological solutions to the new problems quickly, but changes in the structure or function of organisms' bodies (natural adaptations) often take many, many generations to occur—or they may never occur at all. What might happen to your group if the necessary technology were not available?

Did You Know. . .

One of the Galápagos finches, the vampire finch, drinks the blood of other birds!

Theme Connection

Changes Over Time

Focus question: What changes might occur to organisms if the Earth were to undergo another ice age? *(As the glaciers pushed forward from the polar icecaps, organisms would have to either adapt to the changing climatic conditions or migrate to a more favorable environment. Those organisms that cannot migrate or that are unable to adapt will most likely become extinct.)*

Adventures in a New Environment

Encourage students to read the paragraphs and then write some creative tales about a strange new habitat and the adaptations they would need for survival there. Suggest to students that they make illustrations of the adaptations to accompany their stories.

Integrating the Sciences

Earth and Life Sciences

Darwin was greatly influenced by the theories of the geologists Charles Lyell and James Hutton. Have students find out what these theories were and how they influenced Darwin. The students can then present these theories to the class. You may also wish to tell students that they will learn more about these theories in Unit 6.

FOLLOW-UP

Reteaching

Ask students to describe in writing how different species of finches came to inhabit the Galápagos Islands and how their feeding habits became specialized.

Assessment

The fennec, a fox that lives in the hot deserts of North Africa, has very large ears. The red fox that lives in the United States has medium-sized ears. The arctic fox lives in a very cold climate and has very small ears. Ask students how the different ear sizes can be explained by the theory of natural selection.

Extension

Have students find out about finches that are native to your area. These may include closely related birds such as grosbeaks and goldfinches.

Closure

Have students imagine that they are a cabin boy or girl on the H.M.S. *Beagle*. The captain has assigned them to go to the Galápagos Islands with Charles Darwin and to be his assistant for a day. Have them write a letter home to a friend describing what their day as Darwin's helper was like.

To Survive

FOCUS

Getting Started

Obtain jaws or entire skulls of a herbivore (such as a cow or deer) and a carnivore (such as a cat or dog) from a veterinarian, university, or museum. Make sure that all bones are cleaned and sanitized. Display the jaws, and allow students to examine them. Ask: What do you think these animals ate? *(Accept all reasonable responses.)* Then point out the different tooth adaptations. The carnivores have large canines and sharp molars for cutting and tearing meat. The herbivores have scissorlike front teeth for cutting grass and leaves and large, flat molars for grinding tough plant material. Explain to the class that an organism's adaptations are the features that help it fit into its environment.

Main Ideas

1. There are many unique environments on Earth, with many habitats for living things in each of these environments.
2. Adaptations enable organisms to survive in their habitats.
3. Plants and animals have a variety of structures and behaviors that help them adapt to their habitats.

TEACHING STRATEGIES

Have students read page 96 and study the photographs on pages 96 and 97. Ask them to list the plant and animal adaptations that they observe. Have students categorize the adaptations in terms of whether they are used for obtaining food, protection, or locomotion.

The Discrepant Event Worksheet on page 22 of the Unit 2 Teaching Resources booklet describes a teacher demonstration that works well as an introduction to The Fine Art of Survival on page 97.

To Survive

There are many different environments on our planet. These include deserts and oceans, grasslands and mountains, forests and tundra. In each of these environments, there are many different habitats for living things.

Since there is such a diversity of habitats, no single organism can survive in all of them. Obviously, an organism that lives on the ocean floor would have great difficulty surviving in a desert!

Each organism plays a particular role in its habitat. Just as no single organism can survive in all habitats, no single organism can play all roles in one habitat. All this leads to another question: How is each organism suited to play a certain role in a certain habitat? Keep on reading!

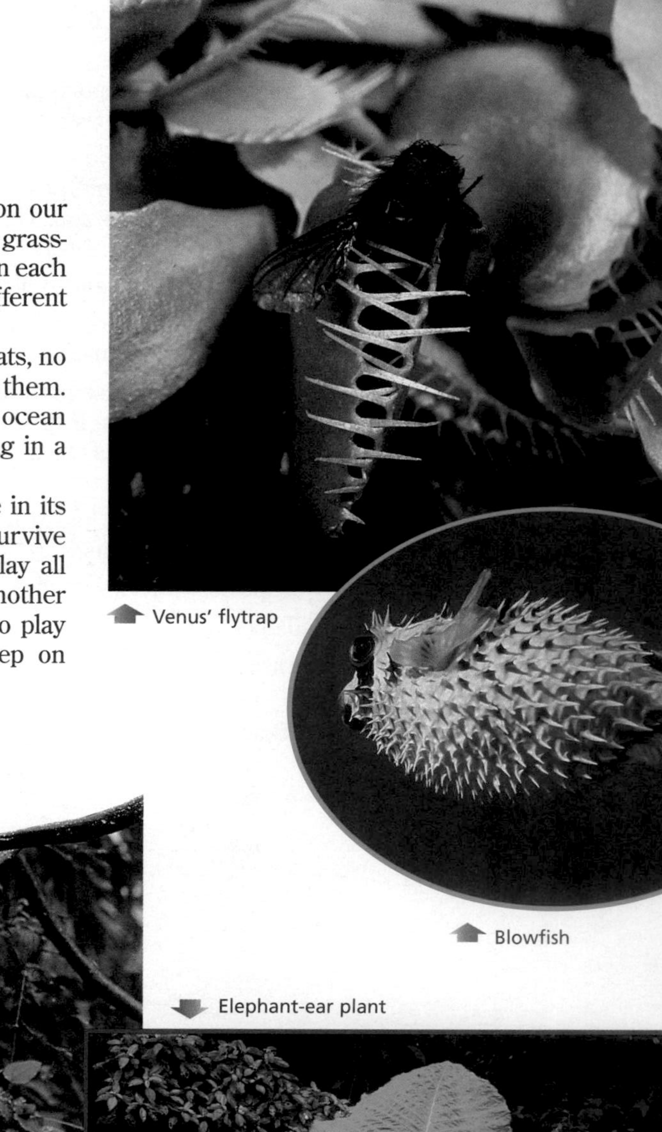

Venus' flytrap

Blowfish

Gibbon

Elephant-ear plant

LESSON 2 ORGANIZER

Time Required
2 to 3 class periods

Process Skills
observing, inferring, analyzing

Theme Connection
Systems

New Terms
Adaptations—inherited features that enable organisms to survive and produce young
Camouflage—adaptation that enables an organism to blend in with its environment
Mimicry—adaptation in which one organism gains protection by looking like another type of organism that predators avoid because of its undesirable smell, taste, or sting

Materials (per student group)
none

Teaching Resources
Discrepant Event Worksheet, p. 22
Resource Worksheet, p. 23
SourceBook, p. S43

Mosquito

Cactus

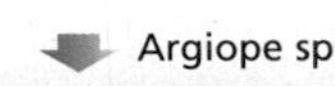

Argiope spider

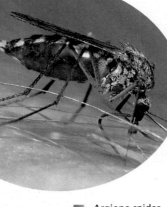

Lion

Hummingbird

The Fine Art of Survival

Adaptations are inherited features that help increase an organism's chances of surviving and reproducing. Both animals and plants have many different kinds of structures and behaviors that help them survive in their environments. Below are three categories of adaptations to discuss: adaptations for obtaining food, adaptations for protection, and adaptations for locomotion. The organisms on these pages will give you some hints.

1. Name some examples of the great variety of structures and behaviors (adaptations) that animals and plants have for obtaining their food.
2. Can you think of animal and plant adaptations that could be used for protection? Give some examples.
3. Consider an animal's locomotion—its movement from place to place. Name some adaptations that various animals have for locomotion. Is each adaptation you thought of related in some way to the organism's habitat? Explain.

97

The Fine Art of Survival

INDEPENDENT PRACTICE Have students select an animal of their choice and draw that organism in its correct habitat. Display the students' pictures in the classroom according to the habitat in which each animal lives.

Answers to *The Fine Art of Survival*

1. Structures and adaptations for obtaining food include the Venus' flytrap's leaves (for catching insects); the spider's glands that secrete silk (to build the webs that catch its prey); the hummingbird's long, thin bill (for reaching the nectar in flowers); the elephant-ear plant's large leaves (for catching sunlight to make food); the lion's sharp claws and teeth (for catching and eating its prey); and the mosquito's proboscis (a specialized sucking tube used to obtain blood from its victims).
2. Structures and adaptations for protection include the cactus's sharp spines that protect it from being eaten and the blowfish's spines and ability to puff itself up as a warning to potential predators.
3. Adaptations for locomotion include the long arms of the gibbon, which enable it to swing from tree branches, the wings of the hummingbird and mosquito, which enable them to fly, and the fins and tail of the blowfish, which enable it to swim.

Cross-Disciplinary Focus

Social Studies

As students explore diversity within a species, they may exhibit an interest in hereditary links and family trees. Invite students to trace the lineage of a renowned family such as the Tudor and Stuart royalty of England, the family tree of author Laura Ingalls Wilder, or the genealogy of Quanah, the Comanche chief. After students have developed a family tree, have them link historical events, scientific inventions, and social changes to each generation.

Homework

The Resource Worksheet on page 23 of the Unit 2 Teaching Resources booklet makes a good homework activity to accompany The Fine Art of Survival.

Theme Connection

Systems

Focus question: How can the division of labor in a species improve the species' chance of survival? *(For humans, the allocation of specific roles within an extended family allows the family members to complete tasks more efficiently. The group of related individuals is able to survive many adverse conditions, which contributes to the survival of the species.)* You might want to discuss with the class the similar social systems of ants and bees.

Disappearing Acts

The photographs on pages 98 and 99 show various animals that use camouflage to hide from either their predators or, like the polar bear or flower mantis, from their prey. Have students read the text and observe the photographs.

GUIDED PRACTICE If there is time, have students try another activity to demonstrate how camouflage can help an animal to survive. Scatter an equal number of differently colored toothpicks on some green grass. Tell students that the toothpicks represent insects. Allow each student 10 seconds to pick up as many toothpicks as possible. Count the number of toothpicks of each color found. Ask: Which colors were easiest to find? *(The colors most different from green)* Which colors were most difficult to find? *(The colors most similar to green)* Why? *(Green toothpicks are camouflaged better than the other toothpicks.)*

ENVIRONMENTAL FOCUS

Suppose an organism lives on the pine needles of evergreen trees. Most organisms of its species are normally brown, but this one has undergone a mutation that makes it green. Ask: How could this mutation become a permanent trait of the entire species? *(Because it is likely to blend in better with its environment, the organism stands a better chance of avoiding predation and producing offspring. It can then pass on its mutation to the next generation, who will also be more likely to survive. The new trait will be possessed by a larger part of the population in each subsequent generation until the entire population has the trait. Then the trait will be a permanent characteristic.)*

Disappearing Acts

Some animals display spectacular survival techniques, but you have to be a very good observer to see them. Like escape artists or magicians, some animal species have developed an amazing variety of illusions. To escape the ferocious jaws, beaks, and fangs of hungry predators, they simply "disappear" into their environment! Other animals do the same trick, not to avoid becoming prey, but to catch their own prey.

Living things pull off their "disappearing acts" by a method called **camouflage,** which allows them to blend in with their environment. Some organisms resemble twigs or leaves. Others resemble nonliving things like stones. Many organisms blend in perfectly with the color of their surroundings. The following examples illustrate how the shape and color of organisms can protect them in their environments.

Some types of katydids look like green leaves. If caterpillars nibble at the edges of the katydid's wings, brown spots surrounded by yellow rings appear. These markings are just like the spots of decay that would appear on a leaf.

Chameleons usually appear green or brown to blend in with their surroundings. They also have a flat, oval shape that helps them to look like leaves. However, when chameleons are attracting a mate, defending their territory, or responding to light or temperature changes, they may change to black, red, yellow, white, or orange.

Polar bears blend in with the snow and ice that are usually part of their surroundings.

Snowshoe hares are white in winter to blend in with the snow and are brown in summer to blend in with trees, grasses, and weeds.

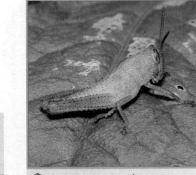

The flower mantis looks like a flower. Its appearance attracts insects to feed on what they think is flower nectar. The unsuspecting insects are then devoured by the deceptive mantis.

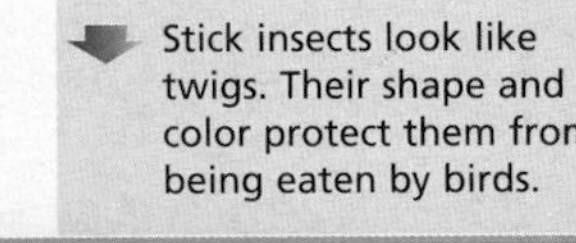

Stick insects look like twigs. Their shape and color protect them from being eaten by birds.

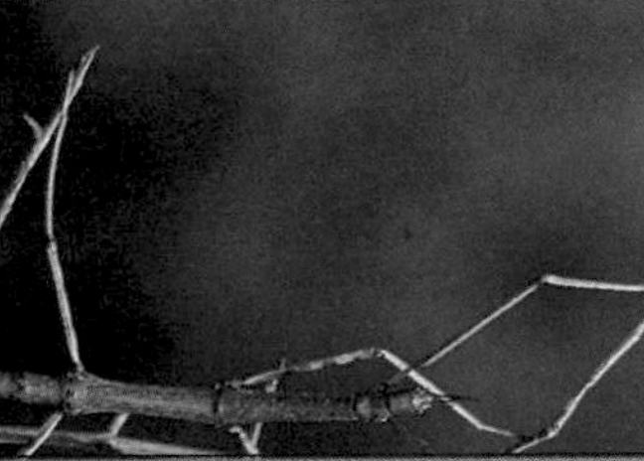

Some young grasshoppers, called nymphs, can change their color to match their environment and avoid being eaten by predators. For instance, nymphs that eat green leaves become green, those that eat pink flowers become pink, and those that eat burned grass turn black!

Who Am I?

Match each rhyme below to one of the animals shown on these pages.

I'm an extremely
changeable fellow;
Any color will do—
black, white, or yellow!

As for me,
I specialize in white—
So I can match
my habitat site.

I am white too,
for part of the year;
But when summer comes,
then brown will appear.

Colors—green and brown—
are important for me too;
But it's looking like a leaf
that really sees me through.

Yes, shape is important—
I look like a flower;
But that's to bring insects
for me to devour.

I could change color
when I was small;
Then I was a tasty treat
for all.

I have the best trick
that you've ever heard;
I look like a twig—
it'll fool any old bird.

Answers to *Who Am I?*

The answers are, from top to bottom, chameleon, polar bear, snowshoe hare, katydid, flower mantis, grasshopper, and stick insect.

Meeting Individual Needs

Learners Having Difficulty

Have interested students draw, color, and cut out pictures of objects or animals that they believe would be hard to see if placed in the school yard. Then have half the students go out into the school yard and place their animals in plain sight. Have the other half see how many of the animals they can find. Repeat this activity, having the two groups switch roles. Ask: Which animals were the easiest to find? Which were the most difficult to find? Why? *(Students should find out that the objects that blend in with their background the best are the hardest to find because they are better suited to their environment.)*

Attention, Please!

The photographs on page 100 show examples of animals that announce their presence with bright colors, rather than using their coloring as camouflage. The bright colors may attract a mate or warn predators to stay away. Often, brightly colored animals have another form of defense, such as a sting or a bad smell or taste. Have students work in groups of three or four to observe the pictures and answer the questions. Then reassemble the class and involve them in a discussion of their answers.

Answers to *Attention, Please!*

- The male peacock, tiger swallowtail butterfly, and cardinal use their bright colors to attract mates.
- The lionfish, skunk, and wasp have markings to warn predators away.

Homework

In addition to bright colors, some animals make distinct sounds in order to attract mates or communicate with others of their species. Have students list at least five types of animals besides birds that do this. *(Sample answer: Wolves, whales, frogs, monkeys, and hippopotamuses)*

Attention, Please!

Some animals have adaptations that help them attract attention. Which animals shown here have an adaptation that

- attracts members of their own kind for mating purposes?
- warns possible enemies (predators) of their bitter or smelly secretions or powerful stings?

Male peacock

Skunk

Wasp

Lionfish

Male cardinal

Tiger swallowtail butterfly

100

Mimicry: Looking Dangerous

Some animals, such as the king snake, have warning features that protect them from being eaten by predators. The king snake is not poisonous, but it looks like the coral snake, which predators avoid because the coral snake's bite is poisonous. The king snake *mimics* the coral snake. Through **mimicry,** the harmless king snake avoids being eaten.

King snake

Coral snake

Eyespots on the wings of an io moth

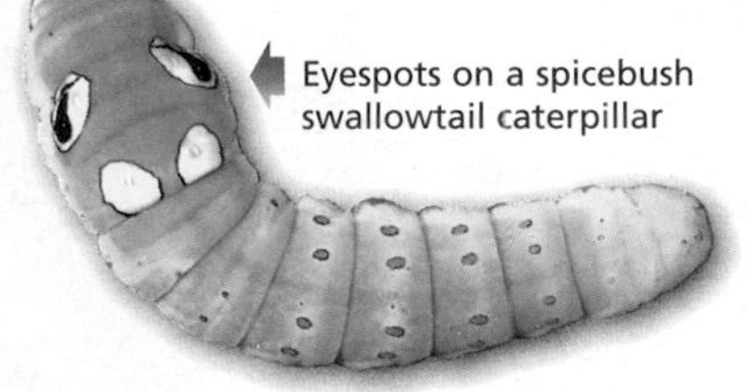

Eyespots on a spicebush swallowtail caterpillar

Something to Research

Find out about the mimicry of the drone fly. What does it mimic? From which animals is the drone fly protected by its mimicry?

There are some very interesting examples of mimicry in the animal kingdom. The larvae of the hawkmoth resemble certain poisonous snakes. Another type of moth mimics hornets; its wings and coloring look like a hornet's, and it even has what looks like a stinger on its abdomen. When threatened, the moth bends and twists like a hornet searching for a place to inject its venom.

Another interesting form of mimicry is found in insects that have large spots resembling eyes on their bodies. On winged insects, these eyespots are often on the upper side of the wings, and they are usually covered when the insect is at rest. When the insect is disturbed, however, it spreads its wings, exposing the eyespots. Through experiments, scientists have shown that many birds are frightened by these eyespots.

Mimicry: Looking Dangerous

GUIDED PRACTICE Have students read the text silently. Then ask: How does the king snake use mimicry to protect itself from predators? *(The king snake mimics the appearance of the coral snake. Potential predators stay away because it resembles the poisonous coral snake.)* How do some moths use mimicry to scare away birds? *(They have large eyespots on the upper surface of their lower wings. When a potential predator approaches, the moth spreads its wings to expose the eyespots and frighten away the predator.)*

Answers to *Something to Research*

The drone fly mimics stinging bees. This mimicry protects it from insect-eating birds and toads.

Plants Also Adapt to Their Habitats

Point out that plants, like animals, show a variety of adaptive structures. This variety is necessary for different plants to survive in different habitats.

You may wish to arrange a field trip so that students can observe plant adaptations. If there are different habitats near the school (for example, a grassy field, marsh, or forest), take students to each habitat to observe and compare the different plants in each area. Then ask students to explain how the plants are adapted to each different habitat.

SAFETY ALERT Warn students not to touch any plants and not to touch, tease, or annoy any animals.

While on the field trip, suggest that students participate in a scavenger hunt to find plant adaptations. Make a list of plant adaptations they could find. Possible examples include a plant that an animal might avoid (a plant with thorns, spines, or stinging hairs), a plant that can live in the shade, a plant that can climb, a plant that can live in water, a plant that has a strong odor, a plant with flowers or fruits, a plant growing on another plant, and a plant that can keep its leaves all winter. Have students look for these adaptations, record the names of the plants in their ScienceLog, and make a quick sketch of each one. To evaluate the hunt, ask volunteers to show their sketches to the rest of the class.

Ask students to read pages 102 and 103 and to look at the illustrations. There are 13 plants shown. Divide the class into small groups of two or three students. Ask each group to use the list of plant names on page 103 to identify the plants. Then have the groups match the 13 varieties of plants with the eight habitat adaptations listed on page 102.

Meeting Individual Needs

Gifted Learners

Have students create their own guidebooks of local plants. Students could either sketch or photograph each plant. They can then list the common name of the plant along with its scientific name and some of its characteristics.

Plants Also Adapt to Their Habitats

Like animals, plants are uniquely suited to their habitats. Although plants cannot move about like animals, they do have certain adaptations that allow them to survive and reproduce in particular habitats.

Look at the plants shown on these two pages. Then decide which plant best matches each of the following habitat adaptations:

1. a plant that is able to hold on to rocks in swift rivers and streams
2. a plant that can store water
3. a plant that can catch insects
4. a plant that can compete with other plants for sunlight
5. a lawnmower-proof plant
6. a tree that can withstand high winds
7. a plant that grazing animals would not eat because of the plant's protective surface features
8. a plant that is capable of living on the surface of a pond

Perhaps you were able to identify some of these plants because you have seen them before. Match the following common names to the appropriate pictures:

- barnyard grass
- barrel cactus
- birch tree
- broadleaf plantain
- coconut palm
- dandelion
- duckweed
- pitcher plant
- pondweed
- Scotch thistle
- sundew
- white waterlily
- wild rose

103

Answers to *Plants Also Adapt to Their Habitats, pages 102–103*

Some plants may be placed in more than one category. In general, the plants that correspond to the habitats described are as follows:

1. d (pond weed)
2. b (barrel cactus)
3. j (sundew), l (pitcher plant)
4. c (birch tree), i (coconut palm)
5. a (broadleaf plantain), e (barnyard grass), f (dandelion), g (Scotch thistle)
6. i (coconut palm)
7. b (barrel cactus), g (Scotch thistle), h (wild rose)
8. k (white waterlily), m (duckweed)

Homework

Have students make sketches of at least five plants that are found in their community. They should identify adaptations that make each plant uniquely suited to its habitat.

Adaptations for Seed Dispersal

Have students make sketches to show how the pea seed could be adapted to meet each of the needs listed. Drawing the seed adaptations will help students see the relationship between form and function. Encourage students to be creative and clear in their drawings.

Answers to *In-Text Questions*

A Sample answers:
- Add waterproof air sacs so that the seed could float in water.
- The seeds could grow in a pod that would pop open and eject its seeds, like a touch-me-not pod. Also, if the seeds grow high on a tall, flexible stalk, they could be thrown farther from the parent plant.
- If the seeds were surrounded by a tough coat that was resistant to digestion and then surrounded by tasty flesh, animals would eat the flesh and eliminate the seeds in different locations.
- Having a sticky substance or burrs on its surface would enable the seed to be picked up and carried by furry animals.
- Having wings or tufts would enable the seeds to be carried away by the wind.

A Picture Puzzle

This exercise enables students to check the ideas they thought of in the previous exercise. Have them make tables in their ScienceLog with the headings "Plant" and "Adaptation for travel." Students may not be familiar with the names of all the plants shown, but they should still be able to identify seed adaptations based on plant structure.

Adaptations for Seed Dispersal

Most plants produce seeds that grow into new plants. But seeds cannot always survive if they simply drop directly beneath the plant that produced them. They may need to be moved to another area. Many types of seeds, therefore, are adapted to travel. This helps ensure that new plants will grow and survive. Look at a pea seed. Do you think it could travel very far on its own? How might you change it so that it would have a greater chance to travel and survive? Working with some classmates, design changes in the pea seed that would enable it to do the following: A

- float on water
- be thrown a distance by the parent plant
- attract an animal that would carry it to another location
- hitchhike on an animal for some distance
- be carried by the wind

A Picture Puzzle

Examine each seed closely. Identify the adaptation that is used to help each seed travel away from the parent plant. Do these adaptations resemble the suggestions you made for the pea seed?

104

Answers to *A Picture Puzzle*

- **a.** acorn—attractive as a food source for animals; spread or buried by these animals
- **b.** squash—eaten by animals; seeds dispersed in feces
- **c.** cucumber—eaten by animals; seeds dispersed in feces
- **d.** pod—drying, bursting, and expelling seeds
- **e.** pod—drying, bursting, and expelling seeds
- **f.** burdock—seed has hooks that stick to clothing or fur
- **g.** blueberries—eaten by animals; seeds dispersed in feces
- **h.** maple seed—wings allow the seed to drift in the air
- **i.** milkweed—pod breaks open and silky hair on seed helps it to become airborne, even in a gentle breeze
- **j.** hay—small hooks on the seeds attach to clothing and fur
- **k.** cones—wash away during heavy rains and open in forest fires
- **l.** apple tree—fruit eaten by animals; seeds dispersed in feces
- **m.** coconut—floats in water
- **n.** poppy—pod throws out seeds
- **o.** mountain ash—berries eaten by birds; seeds dispersed in feces
- **p.** dandelion—airborne by silky tufts of hair

INDEPENDENT PRACTICE Have students go on a seed hunt. Encourage them to use their observational and deductive skills to find out how each kind of seed is dispersed.

- Parachute seeds (e.g., dandelion, sycamore, and milkweed) can be blown to see how they are carried away by the wind.
- Helicopter seeds (e.g., maple, ash, elm, tree of heaven, and linden seeds) can be tossed into the air to see how they descend.
- Slingshot seeds (e.g., jewelweed, also called touch-me-nots) can be touched to see how they burst and fling their seeds.
- Hitchhiker seeds (e.g., burdocks and beggar's ticks) will stick to clothing, hair, and fur.
- Indigestible seeds in fruits can be seen when the fruit is cut open.
- Boat seeds (e.g., coconuts, sea beans, and cranberries) can be seen floating in water.

FOLLOW-UP

Reteaching

Provide students with pictures of animals. Ask them to observe the animals' structures, to describe the animals' adaptations, and to deduce where the animals could live.

Assessment

Ask students to design a plant that could survive in a particular habitat, such as a shady spot, a pond, a swiftly flowing stream, or a desert. Then ask them to write a short paragraph explaining how its adaptations help it to survive.

Extension

Have students do research on insect diversity. Have students select a certain type of insect, such as beetles, ants, or wasps, and document the many different environments that they inhabit.

Closure

Take students on a field trip to do some bird-watching. Ask them to pay particular attention to the adaptations of the birds' beaks and feet. Have students use their observational and deductive skills to discover some facts about each bird's way of life. Use photographs of birds if a field trip is not possible.

LESSON 3

The Value of Diversity

FOCUS

Getting Started

Ask students to imagine an animal species that is adapted to eating the leaves and seeds of only one type of tree. Ask: What do you think would happen to this animal species if all of the trees that it feeds on were killed in a forest fire or cut down? *(Students should realize that this animal species would probably not survive.)*

Main Ideas

1. Extinction results from a number of natural and human-made causes.
2. In an environment with little diversity, there is a greater chance of living things becoming extinct.
3. Human actions can help prevent animals from becoming extinct.

TEACHING STRATEGIES

Answers to *In-Text Questions*

Ⓐ The first web is more complex because there are more organisms to form connections.

Ⓑ The first organisms to be affected by the loss of an animal are the predators of that animal. Because their food is gone, they will probably die. In addition, any organisms that the missing animal consumed would be able to multiply. If the plants were removed, the whole ecosystem might collapse because the producers would have been eliminated.

Ⓒ If a web had a lot of diversity (and thus more types of organisms), it would be more likely to recover from the loss of a species. Predators of a missing animal, for instance, might be able to find other animals to prey upon.

Ⓓ The rosy periwinkle grows in rain-forest habitats. Because these habitats are disappearing, it is threatened with extinction.

LESSON 3

The Value of Diversity

What is the value of diversity? What are the disadvantages of living in an environment where there is very little diversity of living things?

The Web

Did you do the Relationships in Nature activity on pages 6–7 in Unit 1? If not, try it now. All you need are your classmates and some yarn or string. What you end up with is a complex web connecting all of the different types of living things in the environment.

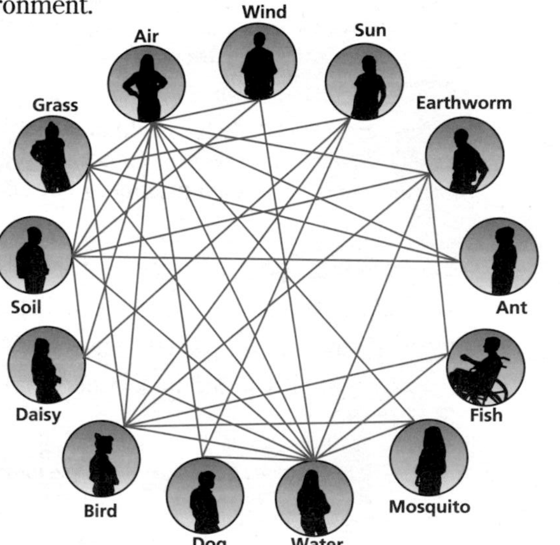

Do this activity again but with a change. This time, form your web with only one type of plant, one bird, one insect, one mammal, etc. After you have made all the connections, compare this new web with the first one you did. Which is more complex? Why? Ⓐ

Now remove one of the animals from your web. Which organisms are first affected by this loss? What do you think will happen to them? Which organisms will be the next to be affected? What would happen if you removed the plant from the web first? Why? Ⓑ

How would this new activity be different if there were more types of organisms in the web? Which web has a better chance of recovering from the loss of one or two species—a web that has a lot of diversity or one that has little diversity? Why? Ⓒ

Diversity and Medicine

Have you heard that some plants are important in making medicines that cure human diseases? As species of plants become extinct, are we losing possible cures? Many people feel that this is a serious concern and another reason why we cannot afford to allow so many species to disappear.

One famous example is the rosy periwinkle, a small plant that grows in Madagascar, off the coast of southeast Africa. It has been very important in the search for a cure for cancer, but it is being threatened with extinction. Why? Consult your local library to find out more about this important plant. Ⓓ

The rosy periwinkle

LESSON 3 ORGANIZER

Time Required
2 to 3 class periods

Process Skills
analyzing, predicting

Theme Connection
Changes Over Time

New Term
Extinction—the disappearance of a species

Materials (per student group)
The Web: large ball of string or yarn; 2 index cards per student; 2 small pieces of tape per student

Exploration 1: 1 die; 20 cards and 1 token for each animal; 20 blank cards; 2 copies of each extinction card (additional teacher materials: Exploration Worksheet of animal and extinction cards, animal tokens, and the game board; scissors)

Teaching Resources
Theme Worksheet, p. 24
Exploration Worksheets, pp. 26 and 30
Transparency 15
SourceBook, p. S46

The Loss of Diversity: An Example

Sometimes a loss of diversity occurs because one or a few species dominate all the others. A plant named purple loosestrife is an example of such a species. This plant was introduced into a habitat where it did not exist previously, with interesting results.

Purple Loosestrife: A Plant Profile

- perennial plant that grows to almost 2 m tall, with long, spiky clusters of pinkish purple flowers, each one capable of producing hundreds of thousands of seeds
- likes moisture, prefers damp soil, and is often found in marshy areas
- native to Europe and Asia
- eaten regularly in Europe by several insects, particularly a few types of beetles
- not native to North America, but carried here on ships
- used by many North Americans as an ornamental plant to decorate gardens and yards
- no natural predators in North America
- beginning to dominate wetland habitats in many regions of North America
- has choked out other plants, resulting in less diversity and wetlands that have only a single plant species

Purple loosestrife

Questions

- Why has purple loosestrife been able to dominate plant life in North American wetlands? Give at least three reasons.
- What effect does purple loosestrife have on other plants that live in a wetland?
- Lots of ducks live in wetlands. What effect does purple loosestrife have on them?
- What do you think a marsh would look like if purple loosestrife were the only plant species? What effect would this have on the diversity of the area?

You Be the Researcher

Proposed action: Import beetles that eat loosestrife, and introduce them into the habitats where purple loosestrife has become a problem.

With a group of your classmates, design a research project to determine whether this is a good solution. Your design should contain the following:

- the main purpose
- key questions that the research should answer
- types of data that should be collected
- reasons why this would be a good solution to the problem

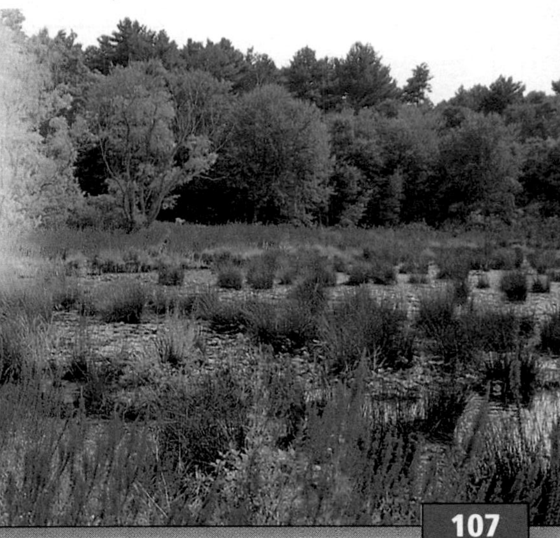

Wetland overtaken by purple loosestrife

The Loss of Diversity: An Example

Have a student volunteer read the selection aloud. Explain to students that if species not native to a region are introduced, they can threaten the native species because native species have no natural defenses against them. The introduced species are called exotic species. Other examples of exotic species include the melaleuca tree, kudzu, the Chinese tallow tree, the water hyacinth, and the nutria.

Answers to *Questions*

- It produces hundreds of thousands of seeds, has no natural predators, and grows well in damp soil.
- It chokes out the native plants.
- It destroys marshes and interferes with ducks' feeding grounds.
- It would not be a marsh anymore because the plant would grow everywhere the marsh had been. There would be less diversity because the only plant species would be purple loosestrife.

You Be the Researcher

A main purpose might be to halt the growth of purple loosestrife. Students could include the following questions: What type of beetle prefers purple loosestrife? If a beetle is introduced into the area, what natural predators exist to keep the beetle population in check? Data collection could include areas of known populations of purple loosestrife, numbers of plants, and numbers of beetles introduced. One reason why this would be a good solution to the problem would be new species growing in an area that had been inhabited by purple loosestrife.

ENVIRONMENTAL FOCUS

Some researchers have suggested that a large mirror should be put into orbit in order to simulate a full moon, allowing humans to see better at night. Yet many animals are only active at night. Ask: What effect might this light source have on nocturnal animals? When do the needs of animals outweigh the needs of humans? *(Accept all reasonable answers.)*

Homework

Have students write a short newspaper article that documents the spread of purple loosestrife in North American wetlands.

EXPLORATION 1

The Extinction Game

Divide the class into groups of two to eight students. If you wish, provide enough paper for each group to make 20 of each animal card and 2 of each extinction card. Invite each group to play the game. The Unit 2 Teaching Resources booklet contains animal and extinction cards that can be copied so that students do not have to make the cards themselves. There are five more extinction cards in the Teaching Resources booklet than in the Pupil's Edition.

You might want to make reference materials available to students who want to learn more about the animals on the extinction cards.

GUIDED PRACTICE The extinction cards contain realistic information about the dangers and possible causes of extinction. You may wish to pose the following questions to students to assess their understanding of the activity: What types of natural changes influence the survival of a species? What are some human-made changes that can influence the survival of a species?

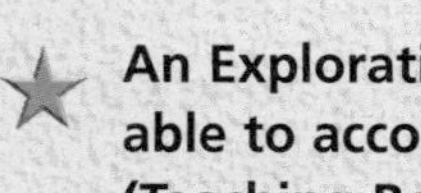

An Exploration Worksheet is available to accompany Exploration 1 (Teaching Resources, page 26).

EXPLORATION 1

The Extinction Game

Animals may become *extinct* when their habitat changes significantly. The Extinction Game will help you discover the kinds of changes that may affect the survival of certain animals. Two to eight people can play at one time.

To play the game, choose one of the animals shown below. Begin with a population of 20 of your chosen animal. To be a winner in the game, you need to finish before your animal becomes extinct. You will be using two sets of cards for this game: "animal cards," which represent each of the eight animals, and "extinction cards," which tell you how many members to add or subtract from your animal's population.

Place the extinction cards facedown in a pile on the table. (After drawing an extinction card, place it on the bottom of the pile.) Roll a die to determine who begins—the highest goes first. Then, in turn, each player throws the die and moves the number of spaces shown on the die. Remove animal cards from your pile or add them to your pile as required by the extinction cards. If your animal becomes extinct, you are out of the game.

Extinction is a process that usually happens over a long period of time. This period may be represented by traveling the path a second time with your surviving population.

After you've finished the game, discuss what you learned about extinction with your classmates.

Getting Started

Make 20 cards each for the following animals: whooping crane, bowhead whale, wood bison, sea otter, Peary caribou, eastern cougar, white pelican, and Eskimo curlew. Have 20 blank animal cards on hand in case some of the animal populations grow beyond 20 members. Make two copies of each extinction card shown on this page and the next. Also make one cardboard token for each animal, to be used on the gameboard path.

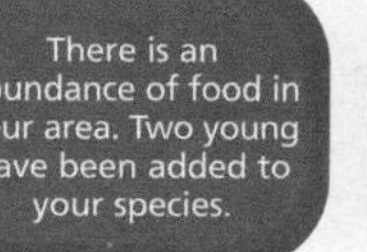

ANIMAL CARDS

White pelican
Peary caribou
Eastern cougar
Sea otter
Whooping crane
Wood bison
Bowhead whale
Eskimo curlew

EXTINCTION CARDS

Humans are in your territory. They have partially destroyed your habitat and killed one member of your species.

There is an abundance of food in your area. Two young have been added to your species.

Harsh winter weather has killed four members of your species.

A park has been built in the middle of your habitat. One member of your species has died as a result of this disturbance.

Humans are in your territory. They have been successful in killing two members of your species.

Theme Connection

Changes Over Time

Focus question: How can human activity affect the diversity of life on Earth? *(Accept all reasonable responses. Students should understand that any time human activities affect the environment, the diversity of affected ecosystems could decrease.)* The Theme Worksheet on page 24 of the Unit 2 Teaching Resources booklet makes an excellent homework activity to accompany this Theme Connection.

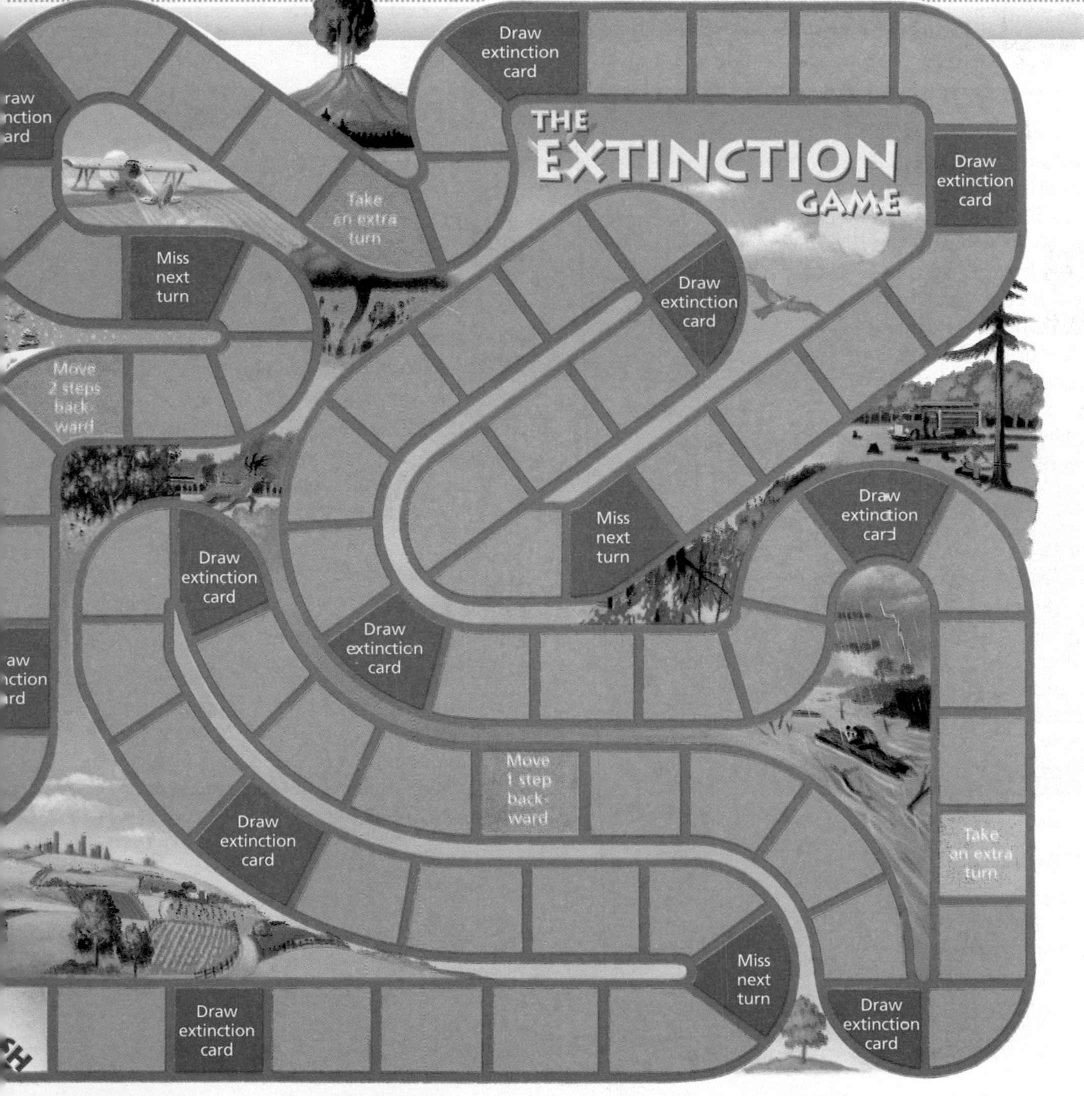

The weather has been exceptionally favorable, and your species has reproduced well. Your population has increased by five members.

Humans have cultivated crops in the habitat of your species. Four members of your species have died

A disease has killed all of the predators of your species. As a result, your species has begun to multiply. It has two new members.

Predators are plentiful in your area. They have killed three members of your species.

The mating of your species with a similar species has resulted in no young being born. In addition, one older member has died.

A drought has reduced the food supply in your species' habitat. One member has died.

Flooding in the habitat of your species has caused two deaths.

Pesticides have been sprayed on the food your species eats. Two members have died, and the rest are at risk.

A tornado has swept through the habitat of your species and killed three members.

Volcanic dust has reduced solar radiation and caused a food shortage in the habitat of your species. Two members have died.

109

ENVIRONMENTAL FOCUS

Ask: How might an increase in the extinction of species be related to the increase in the human population? *(Possible student answers could include the following: As humans become more numerous, they need more space to live, to work, for recreation, and to grow crops. Humans often change or even destroy habitats to meet their own needs. An increase in the number of people has also greatly increased the amount of pollution in the environment as well as the consumption of natural resources, which can have devastating effects on other species.)*

Did You Know. . .

Two hundred years ago, whooping cranes foraged for food from Canada to Mexico. But settlers destroyed their habitat, and the birds were frequently shot by farmers to protect their crops. In 1941, only 41 whooping cranes remained. Now, thanks to international efforts, the population is growing and has reached over 200 birds.

EXPLORATION 2

Back From the Brink

The brown pelican is a fish-eating coastal bird that nests along the Atlantic, Pacific, and Gulf Coasts of the Americas. At right are some facts about brown pelicans on West Anacapa Island, a major breeding colony for brown pelicans in California. The facts trace the brown pelican's history from the days when it lived in large numbers on West Anacapa Island, through its decline, and to its comeback from the brink of extinction. The facts are not in the right sequence, however. Your task is to put them in order.

Brown pelicans on West Anacapa Island

a. Investigators discovered that a chemical company had been dumping DDT into the Los Angeles sewer system for some time.
b. An average of 5000 pairs of brown pelicans nested on West Anacapa Island from 1985 to 1989.
c. In 1973 the United States placed the brown pelican on its list of endangered species—those species that may not survive in the wild unless they are protected.
d. When brown pelicans ate DDT-contaminated fish, the DDT accumulated in their bodies. This caused the shells of their eggs to be so thin and fragile that they often broke while the eggs were being laid or incubated.
e. In 1970, only one brown pelican hatched on West Anacapa Island.
f. Brown pelicans nested in large numbers on West Anacapa Island.
g. In 1972 the use of DDT was banned.
h. DDT worked well in killing mosquitoes and other insects, but biologists discovered that DDT had contaminated the Pacific Ocean and the fish in it. This caused considerable harm to the food chain.

Another Close Call

Examine the diagram below closely. What does it tell you about the survival of the brown pelican in Texas and Louisiana?

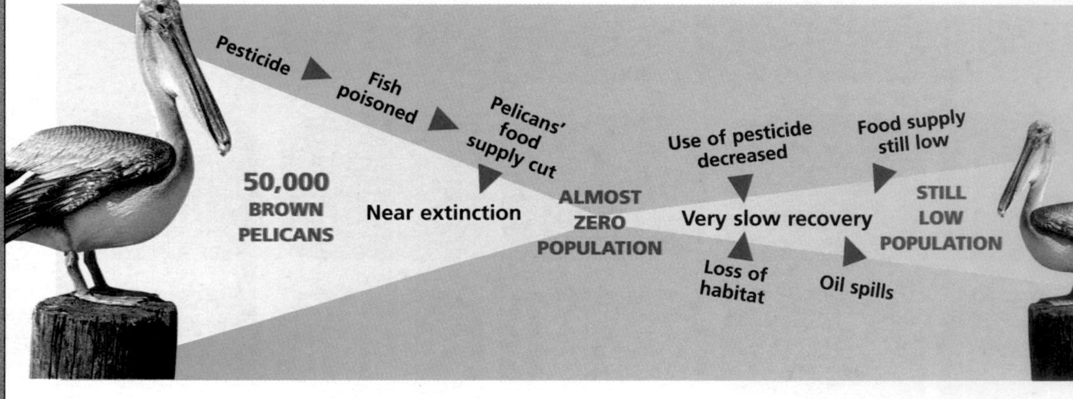

EXPLORATION 2

Have students read about the brown pelican on this page. Ask: Why should we be concerned about the brown pelican? What difference would it make if it became extinct? *(The arguments for saving the brown pelican, or any other species, can be categorized in terms of ethical values [respect for all forms of life], scientific values [contributions that a species can make to the study of science], and ecological values [an organism's role in maintaining an ecosystem].)*

Homework

The Exploration Worksheet that is available to accompany Exploration 2 makes an excellent homework activity (Teaching Resources, page 30). If you choose to use this worksheet in class, Transparency 15 is also available.

Answers to *Back From the Brink*

The sequence students choose should indicate their understanding that DDT caused the decline of the brown pelican on West Anacapa Island but that the pelican population there was restored to viable numbers in the 1980s. Suggested sequence: f, h, d, a, e, g, c, b

Answers to *Another Close Call*

Students' answers will vary but should indicate that in Texas and Louisiana, a pesticide poisoned the fish that brown pelicans ate. When the fish died, the pelicans' food supply was cut off, and the population of pelicans dropped to almost zero. The use of the pesticide has since decreased, but the numbers of pelicans in Texas and Louisiana will probably remain low because of decreased food supply, oil spills, and the loss of habitat.

FOLLOW-UP

Reteaching

Ask students to write a short story told from the point of view of an endangered species. The story should incorporate some of the natural and human-made changes that have threatened the species and some that have helped it.

Assessment

Have students select one or two important reasons for saving wildlife. Then have them write a newspaper editorial on the question, Why save wildlife?

Extension

Show students pictures of species that became extinct due to natural causes (e.g., dinosaurs and saber-toothed tigers). Then show them pictures of animals that are now extinct due to human interference (e.g., the dodo bird and passenger pigeon). Ask them to do some research on one of these animals to find out more details about what happened.

Closure

Obtain a list of local endangered species from your state wildlife agency, the United States Fish and Wildlife Service, or the Department of Agriculture. Ask students to select a species from one of the lists and to prepare a report about its problems. After the reports are finished, discuss how citizens might assist in the preservation of these species.

1. Tall Tales

Evidence in the fossil record suggests that the ancestors of modern giraffes had very short necks. They lived on grasslands in Africa where there was a lot of vegetation that they were able to use for food. There were short grasses, bushes, and tall trees. There is also evidence of long periods of drought in that region of Africa.

How do you think Lamarck would explain the development of giraffes with long necks? How would Darwin explain it?

2. To the Editor

An article in the local newspaper stated, "Everything that humans do to the environment causes animals to become extinct." Write a letter to the editor stating whether you agree or disagree with this statement. Give several reasons for your viewpoint.

3. A Different World

Imagine what the world would be like if

- all plants were 5 cm tall.
- all bears were black.
- all rabbits were white.
- no insects could fly.
- only seals lived in the ocean.

For one of these situations, list all of the ways you think the world would be different from the way it is now.

★ **You may wish to provide students with the Chapter 5 Review Worksheet that is available to accompany this Challenge Your Thinking (Teaching Resources, page 34).**

Homework

As a homework activity, you may wish to assign the Graphing Practice Worksheet on page 32 of the Unit 2 Teaching Resources booklet.

Answers to *Challenge Your Thinking*

1. Both scientists would agree that during the long periods of drought, larger plants with deep roots would survive longer than smaller plants with shallow roots. Both would agree that it would be advantageous for a herbivore to be able to reach the taller plants. Lamarck would suggest that when the short-necked giraffes encountered the drought, they stretched their necks in order to reach the trees. Darwin would say that some of these early giraffes had longer necks than others. Giraffes with longer necks had a better chance for survival. Over time, the giraffes with longer necks survived more often to reproduce, resulting in more giraffes with long necks. Eventually, all giraffes had long necks.
2. Answers will vary. Students should be able to support their opinions. Students in agreement with the statement may cite such evidence as land development and the use of pesticides. Students who disagree with the statement may cite such evidence as the protection that wildlife departments and zoos offer to endangered species. Encourage students to include in their letters their own innovative ways to prevent extinction.
3. Accept all reasonable answers. Sample responses include the following:
   - If all plants were 5 cm tall, overcrowding would result. Plants would not be able to take advantage of the resources available in the higher and lower layers of a forest ecosystem.
   - If all bears were black, those in the Arctic would have no camouflage.
   - If all rabbits were white, those that live in woods and fields would be seen easily by predators.
   - If no insects could fly, some flowers might not be pollinated.
   - If the only living things in the oceans were seals, then seals would have nothing to eat.

Answers to Challenge Your Thinking continued ►

Answers to *Challenge Your Thinking, continued*

4. Encourage students to be creative and to incorporate all of the appendages into their design. (Students will probably be interested to learn that all of the listed features are present on a lobster or a crayfish.)

   Lobsters have two pairs of feelers for testing food; appendages around the mouth for creating water currents; two large pincers for holding food, for breaking open shells, and for defense; four pairs of legs for walking; smaller appendages under the tail for swimming and holding eggs; and a tail appendage for swimming.

The following are sample revised answers:

1. Over time, arctic bears with white fur were more successful at sneaking up on their prey unnoticed than were arctic bears with dark fur because the white bears blended in with their ice-covered environment. As a result, the white bears were healthier and better fed than the dark bears, and they survived to reproduce more white bears. Eventually, there were no longer any dark bears in the Arctic. The white bears are known as polar bears.
2. The Earth supports a wide variety of environments. Organisms have different adaptations that help them survive conditions unique to their environment. Such adaptations include methods of protection, movement, dispersal, and feeding. The variety of adaptations displayed by organisms accounts for a great deal of the diversity among living things.
3. By blending into its environment, this chameleon avoids being detected by predators and being seen by its own prey. This protective coloration is an adaptation that helps the chameleon to survive in its environment. (You may wish to remind students that chameleons do not, as is commonly believed, change color to blend in with their surroundings. Like several other lizards, the chameleon changes color due to light variations and temperature changes and as a fear response.)

4. Critter Creation

Appendages are adaptations designed to help an animal perform various specialized tasks. Your thumb is an appendage; in fact, your entire hand and arm is an appendage. Appendages help an animal live successfully in its environment.

Design and draw an animal with appendages that will give it the following characteristics:

a. The animal lives in water.
b. Its heavy body needs a lot of support as the animal walks.
c. It can walk on the bottom of a body of water for many kilometers without stopping.
d. It can dart away suddenly from its enemies by swimming.
e. It can create water currents to bring food in the water to its mouth.
f. It tests its food before eating it.
g. It has appendages that enable it to hold large pieces of food.
h. It can break apart hard bits of food.
i. Its diet includes shelled animals.
j. It has appendages that enable it to hold its young.
k. It has formidable defensive weapons.

Does the animal you drew look like any animal you have seen before? Do you think an animal exists that has all of the above characteristics? Explain your reasoning.

Do you think the characteristics listed describe this imaginary animal?

ScienceLog

Review your responses to the ScienceLog questions on page 88. Then revise your original ideas so that they reflect what you've learned.

page 88

CHAPTER 6 Keeping Track

1 How would you sort the living things shown on this page into groups?

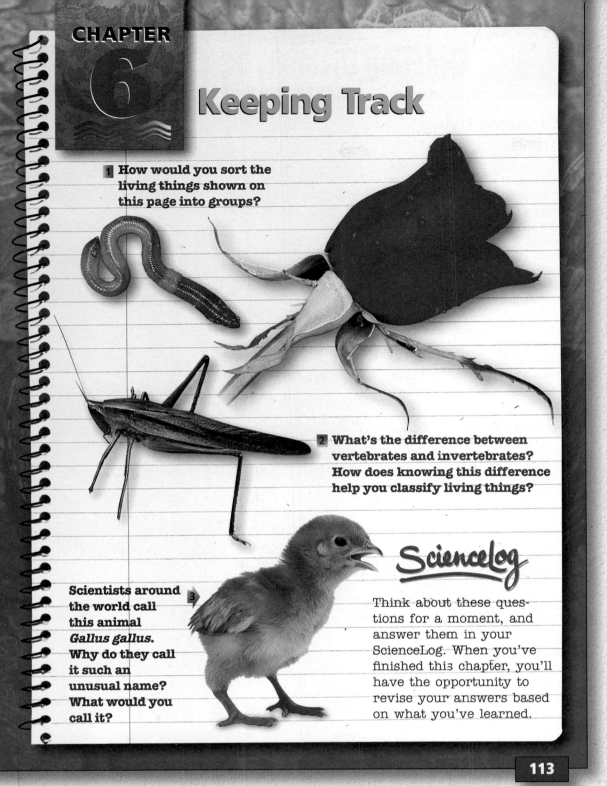

2 What's the difference between vertebrates and invertebrates? How does knowing this difference help you classify living things?

3 Scientists around the world call this animal *Gallus gallus*. Why do they call it such an unusual name? What would you call it?

ScienceLog

Think about these questions for a moment, and answer them in your ScienceLog. When you've finished this chapter, you'll have the opportunity to revise your answers based on what you've learned.

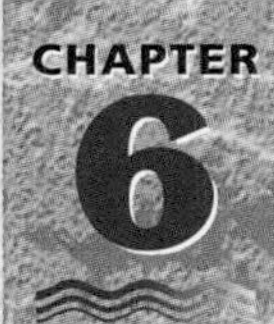

Connecting to Other Chapters

Chapter 4 introduces the concept of diversity and how the diversity in a community can change.
Chapter 5 explores the theory of natural selection and the value of diversity to a community.
Chapter 6 challenges students to create classification systems for a wide range of living things.

Prior Knowledge and Misconceptions

Your students' responses to the ScienceLog questions on this page will reveal the kind of information—and misinformation—they bring to this chapter. Use what you find out about your students' knowledge to choose which chapter concepts and activities to emphasize in your teaching.

In addition to having students answer the questions on this page, you may wish to have students perform the following activity: Tell them that the names of all of the plants and animals in the world have been forgotten and that their job is to create new ones. Divide the class into groups of two to five students. Each member of the group should take a turn describing a plant or animal. Then the other members of the group should decide on a new name for the animal, trying to be as descriptive as they can. Then have each group share their results with the rest of the class by writing their animal names on the board with a brief description of each one. Have each student write a paragraph discussing what scheme they would use to best classify these animals.

Emphasize that there are no right or wrong answers in this exercise. Collect the paragraphs, but do not grade them. Instead, read them to find out what students know about naming and classifying organisms, what misconceptions they may have, and what is interesting to them about how organisms are named. After the game, explain to the class that in this chapter they will explore how scientists order the diversity of life and how biologists give scientific names to organisms.

LESSON 1 Ordering Diversity

FOCUS

Getting Started

Ask students to estimate how many different items a supermarket sells. *(The correct answer is approximately 30,000, but accept all reasonable responses.)* Ask: How could you find a bottle of ketchup among so many thousands of things? *(Lead students to realize that a supermarket groups similar items together.)* Point out that scientists organize the millions of known plants, animals, and microorganisms in a similar way, based on how closely related they are.

Main Ideas

1. Diversity creates the need to classify living things into groups.
2. Most living things can be classified into two kingdoms—the plant kingdom and the animal kingdom.
3. The animal kingdom is divided into vertebrates and invertebrates.
4. The plant kingdom is divided into two groups—mosses and plants with tubes for conducting food and water.

TEACHING STRATEGIES

GUIDED PRACTICE Ask students to explain how they group things in their everyday lives. Ask: Do you arrange your clothes in a certain way? Do you arrange cassette or compact-disc collections in a particular way?

Discuss the problems that would arise if there were no classification systems in libraries, telephone books, or the mail system. Students should conclude that things are classified for the convenience of people who study or use them.

A Transparency Worksheet (Teaching Resources, page 40) and Transparencies 16 and 17 are available to accompany the material on this page.

LESSON 1 Ordering Diversity

Grouping Living Things

As you may well imagine, it's difficult to keep track of 1.4 million kinds of living things! The easiest way to do this is to group the organisms. You already have lots of experience with grouping things. For example, in your kitchen, how is the silverware grouped? Are the dishes kept in one place and the dish towels somewhere else? Are the pots and pans separated from the drinking glasses? Here are more situations to think about in which items are sorted, or *classified,* into groups:

- How are the books classified in the school library? (A)
- How does the telephone company use grouping to keep track of everyone with a phone? (B)
- There are over 240 million people in the United States. How does the postal service group all these people so that mail can be delivered to every person? (C)

Name some other situations in which items are divided into groups. Now think about some ways in which you might classify living things. (D)

114

LESSON 1 ORGANIZER

Time Required
5 class periods

Process Skills
comparing, contrasting, organizing, classifying, analyzing

Theme Connection
Structures

New Terms
Classify—to sort into groups
Invertebrates—animals without backbones
Vertebrates—animals that have backbones to support their body

Materials (per student group)
Putting It All Together: optional materials: a variety of preserved or live specimens of plants and animals

Teaching Resources
Transparency Worksheet, p. 40
Resource Worksheets, pp. 42 and 44
Transparencies 16–21
SourceBook, p. S28

Try to place all the living things in the illustration on page 114 into two groups. Then further divide each group into two subgroups. Invent a name for the living things in each group and subgroup. **E**

Katie devised the concept map shown here to classify the organisms. Katie's system is only one of many ways of classifying organisms. Compare your system with Katie's. Try classifying the living things in the picture according to Katie's system. How useful do you think her system is? **F**

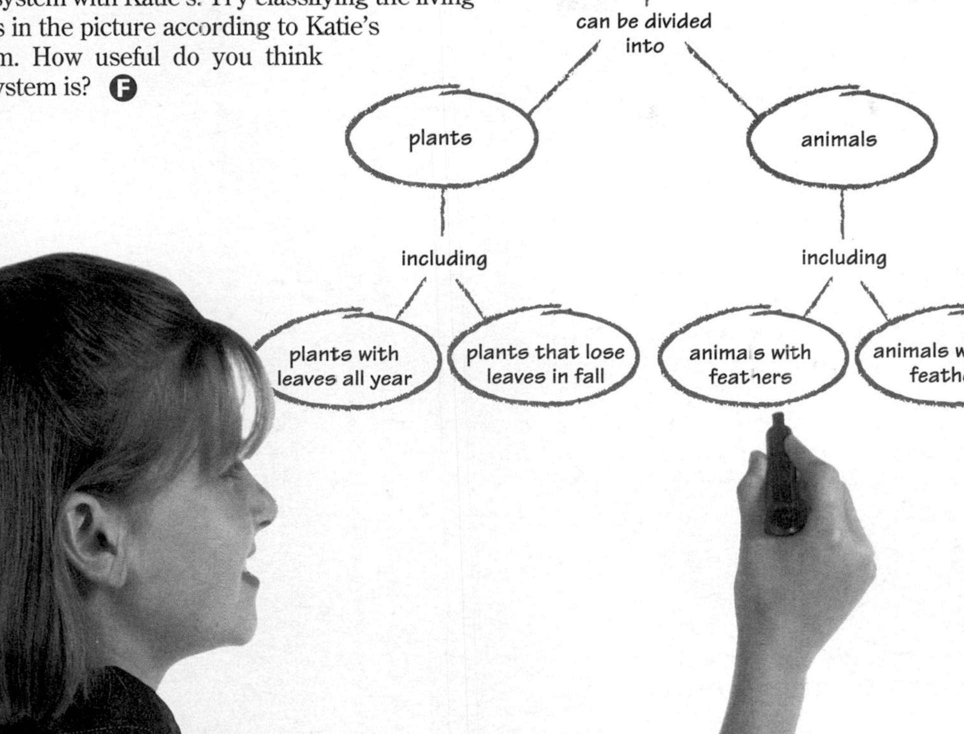

Now let's look more closely at the differences between organisms—differences that help you make a classification.

1. Choose one of the living things in the illustration on page 114. In your ScienceLog, list the distinguishing features that you could use to identify it. Read your list to some of your classmates. Ask them to suggest other organisms that would also match the features on your list. How many different organisms did your classmates suggest? Do you always need to talk about *each single living thing*, or can you talk about *a group of living things*?
2. Try doing the same exercise with another organism.

Integrating the Sciences

Earth and Physical Sciences

Geologists classify rocks as igneous, sedimentary, or metamorphic depending on how they were formed. Extreme temperatures and pressures can transform sandstone, a sedimentary rock, into quartzite, a metamorphic rock. Ask the class to find out what parent rocks *gneiss* and *marble* come from. *(Gneiss, a metamorphic rock, comes from granite, an igneous rock. Marble is also metamorphic, but it is derived from limestone, a sedimentary rock.)*

Homework

Ask students to design a system of classification for household items. Students should describe the characteristics of each grouping.

Grouping Living Things, *pages 114–115*

Divide the class into groups of three or four students, and have the groups work through the classification activities on pages 115–117. Tell them that there is more than one way to classify the organisms in the picture on page 114. Ask them to figure out their own classification system before looking at Katie's system in the middle of page 115.

Answers to *In-Text Questions, pages 114–115*

A Library books are classified by author, topic, and title.

B Area codes indicate a specific region of each state, and the first 3 digits of a telephone number indicate which part of the city the number will access. Telephone books are alphabetized.

C The mail system uses zip codes, state names, street names, and building or apartment numbers.

D Other situations include articles in a newspaper and food items in a cafeteria.

E Sample answer: Living things can be divided into air breathers (land plants and land animals) and water breathers (aquatic plants and aquatic animals).

F Students should conclude that her system works well for classifying the organisms in the picture.

Answers to *In-Text Questions*

1. Answers will vary, but the descriptions should include distinguishing features that can be used to identify the animal or plant. Students should be able to guess each other's organisms from the descriptions and should be able to name other related organisms that have the same features. Students will realize as they do the activity that a group of living things—not just a single living thing—may match the descriptions they have written.
2. See the answer for (1) above.

Answers to In-Text Questions continued ►

Answers to
In-Text Questions, continued

3. Concept maps may vary. One possible concept map could identify living things as the main group, with subgroups of plants and animals. Plants could be divided further into land plants and water plants. Animals could be divided further into those that live only in the water and those that can live out of the water.

4. Concept maps may vary. One possible concept map could divide all plants into two groups—trees, and shrubs and smaller plants. Trees could be further divided into trees that have leaves year-round and trees that lose their leaves in the fall. Shrubs and smaller plants could be divided into those with woody stems and those with soft, green stems.

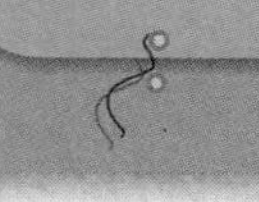

PORTFOLIO

Suggest that students include their classification concept maps from page 116 in their Portfolio.

ENVIRONMENTAL FOCUS

Many organisms produce poisonous chemicals called *biotoxins* that are used as a defense against predators or in capturing prey. The most-studied biotoxins are snake venoms because they are so easy to obtain. Have students do some research or contact a local government agency to find out about poisonous organisms in their area. Students can then create a booklet of information about biotoxins from poisonous plants and animals found in their area.

3. Look at the living things in this pond setting. Draw a concept map that could be used to identify them.

4. Try to develop a slightly different classification system for the four types of plants below.

116

Answer to
In-Text Question

A The eagle (a) and worm (e) can be classified as animals. The flowers (b), palm trees (d), ferns (g), and cactus (f) can be classified as plants. The mushrooms (h), spirogyra (c), and amoeba (i) cannot be classified into either group, since there are distinct differences between them and plants or animals. For instance, the mushroom has no leaves and does not produce its own food. Therefore, Katie's system is not inclusive enough for all of the organisms shown.

INDEPENDENT PRACTICE Have students read the first paragraph, which discusses the kingdom system of classification. Ask students to classify the organisms on page 117 again after the kingdom system has been discussed. *(Answers are given in the diagram at the bottom of page 118 in the Pupil's Edition.)*

Homework

For homework, you may wish to have students read the detailed information about each of the five kingdoms provided in the SourceBook on pages S28–S38.

Meeting Individual Needs

Gifted Learners

Have students create a computer program that classifies animals according to phylum, class, and order. Students should prepare a database of animal traits and the names of the different phyla, classes, and orders. The program should prompt the user with questions such as, "Does the animal, ______, give birth to live young or lay eggs?" When complete, the program could be used as a resource for classroom research. Suggest that students begin their project with no more than 9 or 10 different phyla. If interest in the project is maintained, students can expand the program to accommodate the remaining phyla in the animal kingdom as well as living organisms from other kingdoms. Encourage small groups of students to work together on the project.

ENVIRONMENTAL FOCUS

The American Indians of the Great Plains called them *wishtonwish.* The French called them *petits chiens,* or little dogs. Merriwether Lewis, of the explorers Lewis and Clark, first dubbed them barking squirrels. He then revised the name to burrowing squirrels. Lewis finally settled on a name that is still used by Americans—prairie dogs.

Biologists, the scientists who study living things, realized the need for a consistent system of classification. For this reason, they divided all living things into two kingdoms: the animal kingdom and the plant kingdom. However, throughout the past century, scientists have been gaining more knowledge about the differences among living things. This has created the need to devise more precise systems for grouping all organisms. Most biologists now group living things into five kingdoms, and many biologists have even proposed a six-kingdom system. Either way, there are many living things that are regarded as neither plants nor animals.

The diagram below shows how a modern biologist would group the living things shown in the photos on page 117. Notice that the organisms that are neither plants nor animals have been placed in the "Other kingdoms" category.

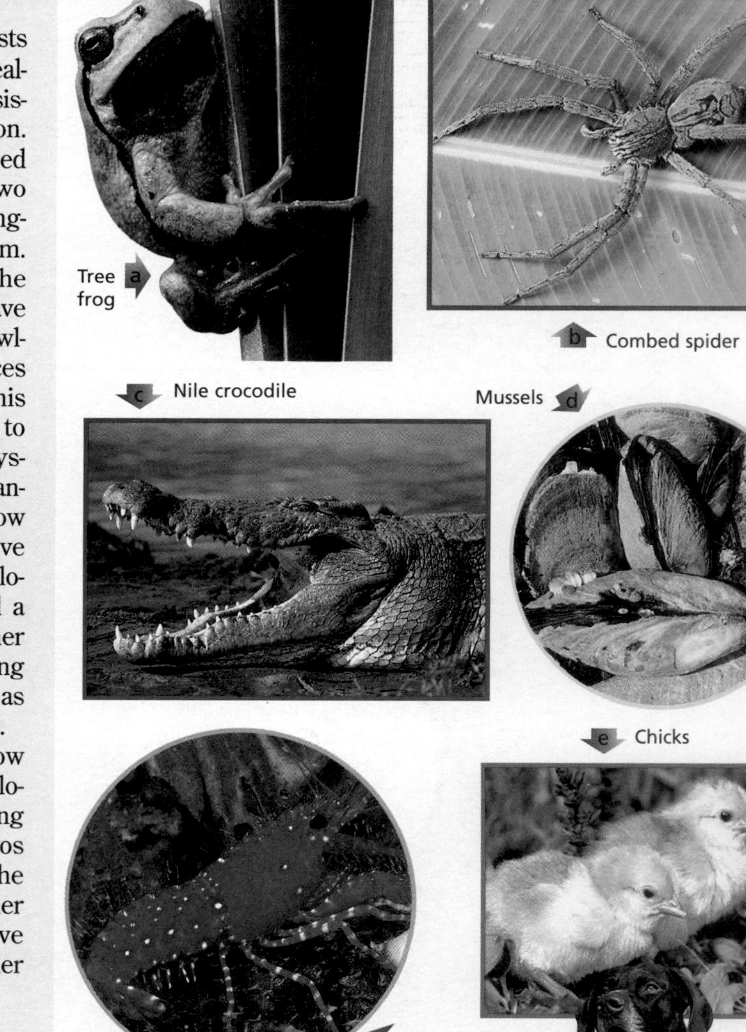

a Tree frog

b Combed spider

c Nile crocodile

d Mussels

e Chicks

f Scarlet reef lobster

g German short-haired pointer

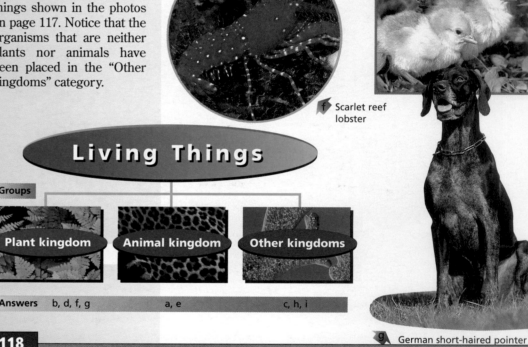

Living Things

Groups	Plant kingdom	Animal kingdom	Other kingdoms
Answers	b, d, f, g	a, e	c, h, i

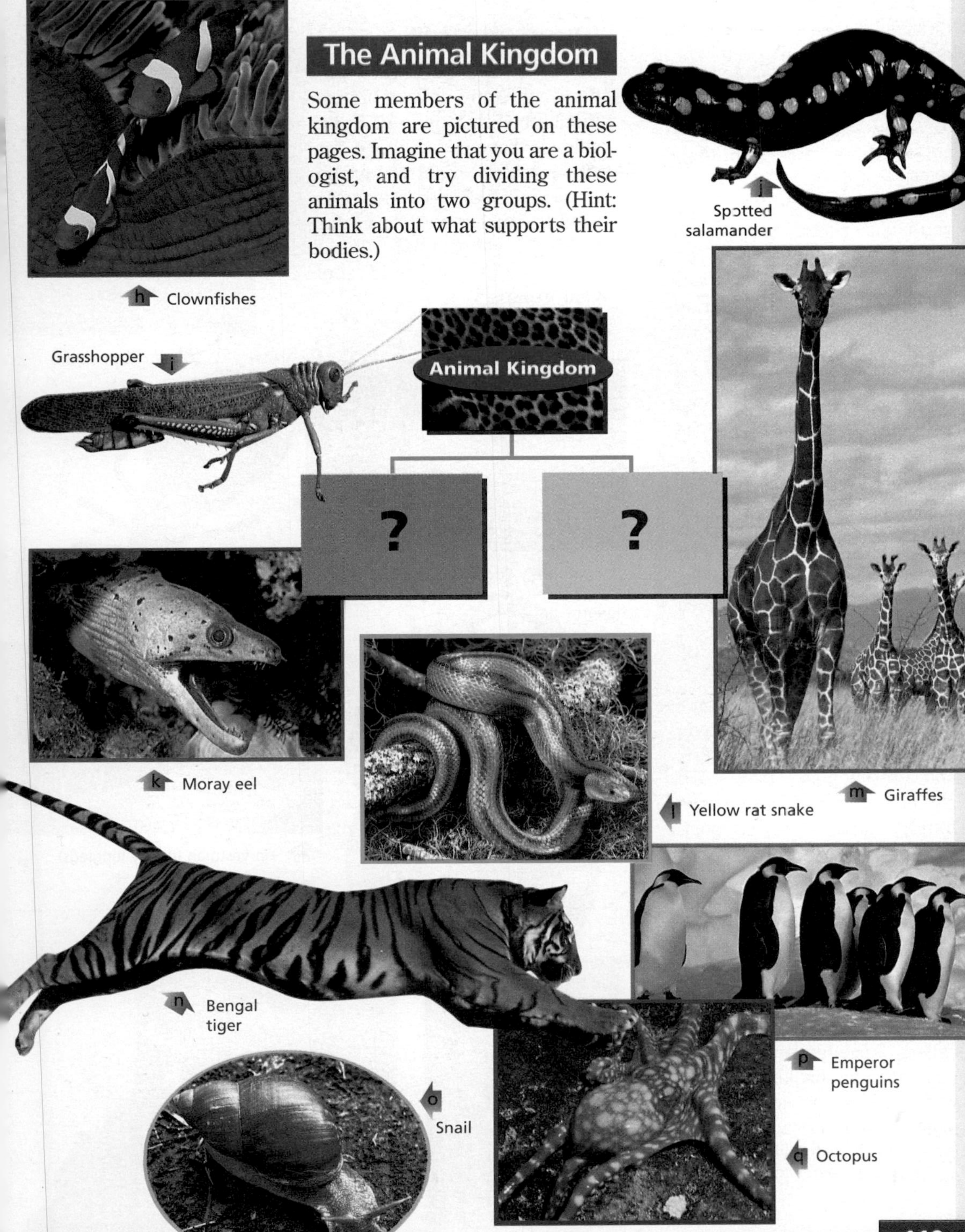

The Animal Kingdom

Some members of the animal kingdom are pictured on these pages. Imagine that you are a biologist, and try dividing these animals into two groups. (Hint: Think about what supports their bodies.)

119

The Animal Kingdom

As a class, devise several ways in which the organisms in the illustrations on these pages could be classified into two groups. Consider all possibilities (e.g., with or without legs, with or without shells, with or without fur, with or without tails, living in the water or on the land, do or do not fly, do or do not have tails). Have students classify all of the animals using each suggested classification system. After students have presented their own classification systems, have them read page 120, which presents the animal classification system that biologists use. Then ask students to classify the animals pictured on these pages as either vertebrates or invertebrates. Emphasize that the major feature distinguishing vertebrates from invertebrates is the presence of a backbone.

Answers to *The Animal Kingdom*

The organisms pictured can be divided into two groups: those with backbones (vertebrates) and those without backbones (invertebrates). The invertebrates in the pictures include the snail (o), mussel (d), grasshopper (i), lobster (f), octopus (q), and spider (b). The snail, mussel, and octopus are mollusks. Most mollusks have hard shells to protect their soft bodies, although some mollusks, such as the octopus, lack an external shell or even a shell at all. The grasshopper, lobster, and spider are arthropods. They have hard external skeletons that protect their bodies, and they have jointed appendages. Their muscles are attached to these exoskeletons. The vertebrates include the eel (k), penguins (p), snake (l), giraffes (m), frog (a), dog (g), fish (h), chicks (e), salamander (j), tiger (n), and crocodile (c).

Meeting Individual Needs

Learners Having Difficulty

To show how an animal's form is related to function, have students prepare permanent casts of footprints found in soft soil or mud along a stream. They will need some cardboard, paper clips, a small sack of plaster of Paris, water, a pan, and a mixing spoon. Make sure students follow the precautions as listed on the plaster of Paris container. To make a cast, students should form a cylinder with the cardboard, securing it with a paper clip. Place it on top of a track, and pour a mixture of plaster into the cardboard cylinder. As the plaster hardens, a cast of the track forms. Students should make casts of at least two different kinds of footprints. Then they can compare the structure and how the feet are used (e.g., for wading, walking, swimming, or carrying objects).

Answer to
In-Text Question

Ⓐ For a sample classification based on type of body support, see Answers to The Animal Kingdom on the previous page.

Answers to
Invertebrates

Features that could be used to classify the organisms include number of legs, presence of antennae, type of habitat (on land or in water), and existence of eyes. The names used to label the subgroups could be based on the characteristics chosen.

Homework

Have students research the habitat and behavior of one of the animals pictured on pages 120–121. They should present their findings in the form of a poster or pamphlet.

The more you know about different animals, the easier it is to make useful groupings. To classify, you need to observe the animals' features carefully—not only their outside features, but also their inside features.

What characteristics did you use to classify the animals on pages 118 and 119? As you probably realize, your job was difficult because you had to depend mostly on what you could observe in the pictures. Biologists, though, study the actual animals, examining them on both the outside and the inside. The type of body support that an animal has is one feature that biologists use to classify animals. Many animals have internal skeletons to support themselves. These animals are placed in a group called **vertebrates**. Vertebrates are animals with a *vertebral column*, or backbone.

Other animals depend on a hard, external support system. For example, clams, lobsters, and grasshoppers have hard shells. Earthworms have tough muscles for support. Other animals don't really need a means of support. Their surroundings are their support. This is true for many of the animals that live and drift about in the water. All the animals mentioned in this paragraph can be placed in a group called **invertebrates**—animals without backbones. Did you classify the animals on pages 118 and 119 according to type of body support? If not, do so now. Ⓐ

Invertebrates

On these two pages are pictures of some invertebrates. Remember that none of these animals have a backbone for support. How might you group these animals? What features would you use as a basis for making subgroups? What names might you use to label your groups?

a Argiope spider

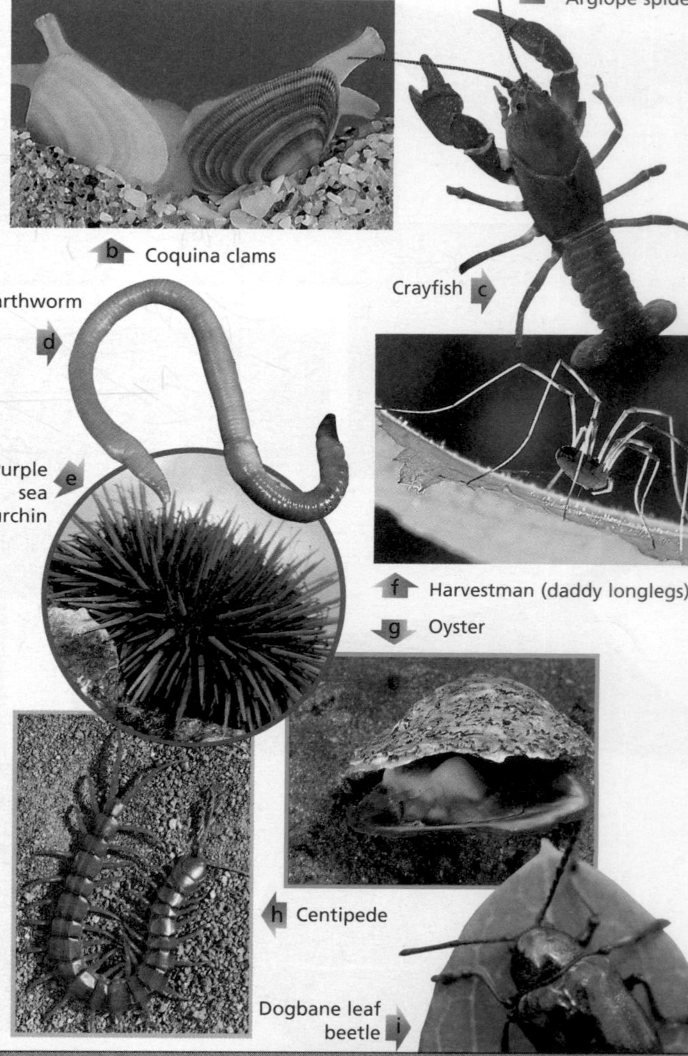

b Coquina clams

c Crayfish

d Earthworm

e Purple sea urchin

f Harvestman (daddy longlegs)

g Oyster

h Centipede

i Dogbane leaf beetle

Integrating the Sciences

Life and Physical Sciences

Students may be interested to learn the following information about animals with external skeletons: Insects and arthropods have skeletons on the outside of their bodies. These external structures are called exoskeletons. The main substance in an exoskeleton is a compound known as chitin. In most arthropods, chitin is a soft, leathery complex carbohydrate. However, in lobsters and other marine arthropods, the exoskeletons also contain calcium carbonate, which provides additional firmness. Most animals cannot digest chitin, so if they consume too many insects, they might get an upset stomach!

The Classification System Used by Biologists

Ask students to read the material on page 122. Call on volunteers to read aloud the material in the tables.

GUIDED PRACTICE If there is a beach or riverbank nearby, take students on a field trip to observe and classify the animals or animal remains that they find on the beach or in shallow water. (An adult skilled in water safety and resuscitation should accompany you.) Students should illustrate or photograph their observations. Set up a classroom display of the different phyla of invertebrates, to which students can add their specimens.

If there is no beach nearby, choose a seashore area to research as a class. Assign a specific organism to each student. Students should find out as much as possible about the organism and how its features allow it to survive in its habitat. Students can illustrate their organisms, and the illustrations can be organized into a display.

SAFETY ALERT Many invertebrates have poisons that are used in defense against predators or to catch prey. Do not handle any invertebrates unless a field guide specifically states that the invertebrate is safe to handle. Some poisonous invertebrates include scorpions, bees, wasps, jellyfish, certain spiders, ants, and sea anemones. Others, such as crayfish, can deliver painful wounds.

Transparency 18 is available to accompany The Classification System Used by Biologists.

The Classification System Used by Biologists

By studying invertebrates in detail, biologists have developed a classification system for them. To help you understand their system, study these two tables. The first table identifies the features that biologists use to classify invertebrates into different subgroups. The second table divides one subgroup, the arthropods, into even more subgroups.

After studying the tables, see whether you can place all the animals pictured on pages 120 and 121 into the classification system on page 123. Your task will not be easy; using pictures of animals to identify features is more difficult than using the actual animals.

Identifying Features of Invertebrates

Distinguishing features used to group animals	Subgroup	Examples of animals in the subgroups
wormlike, round bodies with many segments	annelids (means "arranged in rings")	earthworm
very soft bodies, which are protected by a shell in most cases	mollusks (means "soft-bodied")	clam (example with a shell) octopus (example without a shell)
covered with a spiny skin	echinoderms (means "spiny-skinned")	starfish
many jointed appendages (attachments such as legs, feelers, tail)	arthropods (means "jointed legs")	lobster

Identifying Features of Arthropods

Distinguishing features of arthropod subgroups	Subgroup	Examples of animals in the subgroups
six legs, three main body parts, wings	insects	grasshopper
eight legs, three main body parts	arachnids	spider
more than eight legs, many other kinds of appendages, hard covering	crustaceans	lobster
numerous legs	diplopods chilopods	millipede centipede

A Simplified Classification System for Invertebrates

Look at the animals pictured on pages 120 and 121. Next, copy the classification system shown below into your ScienceLog. Then fill in each blank with an animal's name.

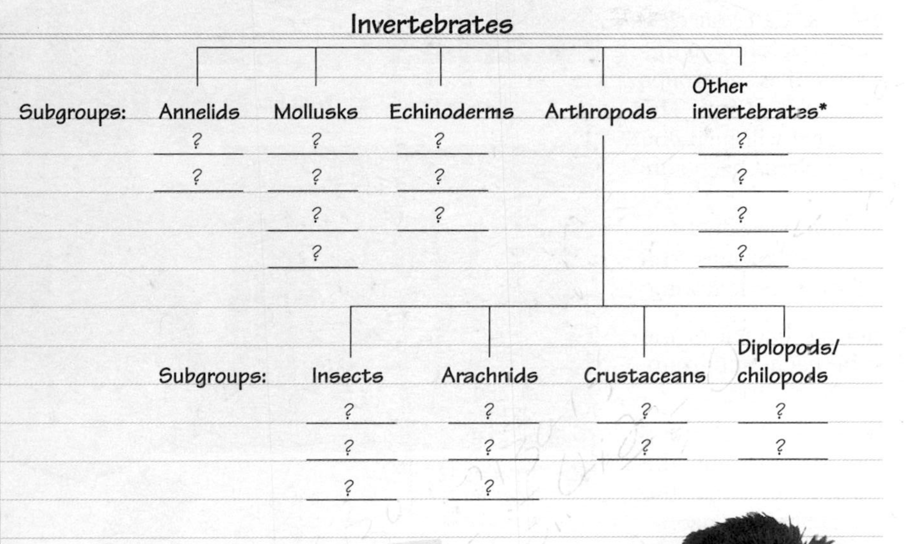

*"Other invertebrates" includes other worms, such as flatworms and unsegmented roundworms, along with many water animals, such as jellyfish, sea anemones, corals, and sponges.

If you need a hand in classifying invertebrates, turn to pages S34–S36 in the SourceBook.

Now think some more about the subgroups.

1. Did you have any problems deciding which subgroup each invertebrate belonged to?
2. It's interesting to think about where invertebrates live. How many are found in water? in moist places? on dry land?
3. Does it appear that the structure of invertebrates enables them to live successfully in various places? Why do you think this is the case?

Answers to *A Simplified Classification System for Invertebrates*

The invertebrates should be classified as follows:

Invertebrate Subgroups
- Annelids—earthworm (d) and leech (v)
- Mollusks—clam (b), oyster (g), scallop (n), and octopus (u)
- Echinoderms—sea star (t), sand dollar (l), and sea urchin (e)
- Other Invertebrates—jellyfish (m), flatworm (p), anemone (j), and rotifer (r)

Arthropod Subgroups
- Insects—beetle (i), butterfly (s), and grasshopper (k)
- Arachnids—harvestman (f) (daddy longlegs), argiope spider (a), and tarantula (q)
- Crustaceans—crayfish (c) and crab (w)
- Diplopods/chilopods—millipede (o) and centipede (h)

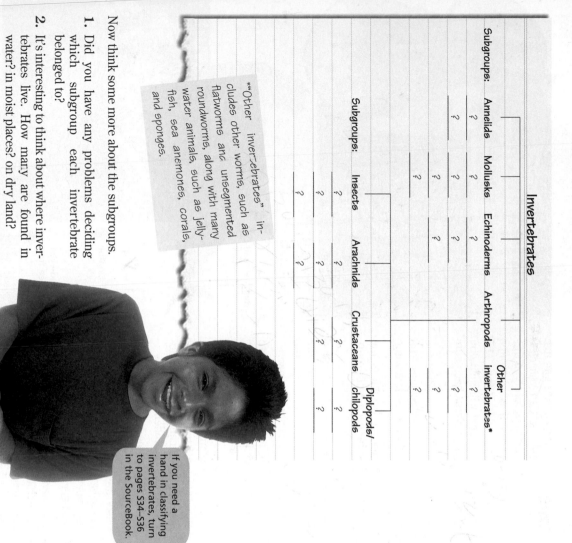

1. Students will probably have some problems classifying invertebrates. It is difficult to classify invertebrates using only photographs, especially if they are not familiar with the organisms' structures.
2. Most of these animals live in the water. Exceptions include insects and arachnids that can live on dry land and earthworms, snails, centipedes, and millipedes, which prefer moist land environments.
3. Yes. Because of the diversity of structures among invertebrates, they can live successfully in many different places.

★ **A Resource Worksheet (Teaching Resources, page 42) and Transparency 19 are available to accompany A Simplified Classification System for Invertebrates.**

Homework

Have students examine the photograph on this page. For homework, have them write a new speech bubble to accompany this photograph. Encourage students to be creative, but their submissions should pertain to the topic of invertebrates. *(Sample answer: There's nothing I like better than a hands-on activity that helps me learn about invertebrates!)*

Vertebrates

INDEPENDENT PRACTICE To help students remember how vertebrates are classified, have them summarize the characteristics of each subgroup in their ScienceLog. Encourage them to include simple drawings of animals that belong in each subgroup.

Answers to *Vertebrates, pages 124–125*

Ⓐ The five subgroups are as follows:

- Fish—scales and fins; includes (e) lionfish, (k) emperor angelfish, and (d) hammerhead shark
- Amphibians—no scales; includes (f) red-eyed tree frog, (o) changeable toad, and (l) salamander
- Reptiles—scales, no fins; includes (g) yellow Philippine snake, (i) American alligator, and (c) red-eared slider turtle
- Birds—feathers; includes (a) common loon, (j) wood duck, and (m) great horned owl
- Mammals—fur or hair; includes (b) fruit bat, (h) human, and (n) fur seal

Answers to Vertebrates continued ►

Vertebrates

Because you are so well acquainted with the vertebrates around you (including yourself!), you will probably be able to create vertebrate subgroups easily. All vertebrates have backbones. Biologists classify vertebrates into five well-known subgroups. Can you classify the 15 vertebrates pictured on this page and the next into five subgroups of three animals each? (Hint: The covering of each animal will help you find the five groups.) Name each subgroup. Ⓐ

Now that you have classified the vertebrates into five subgroups and named each one, discuss the following:

1. Identify the normal habitat of the living things in each subgroup. How do their habitats differ?
2. Do the members of each subgroup move in specific ways? Identify the method of locomotion that the members of each subgroup use.
3. Members of a subgroup share other features besides body covering, habitat, and means of locomotion. Keep in mind, though, that such features may vary from one subgroup to another. What are some of these features for the subgroups you have identified?
4. The ability of animals to survive in the place where they live is obviously important. Are any features of a subgroup specifically related to the survival of the subgroup's members? Explain.
5. The names of the subgroups of vertebrates commonly used by biologists are found on page 127. Perhaps you suggested a different way to group vertebrates. If so, describe your method of grouping and the reasons for your classification.

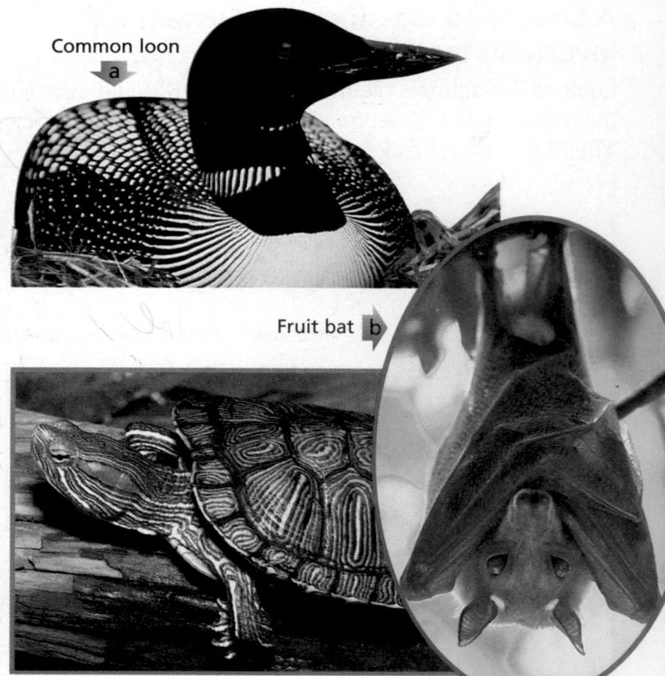

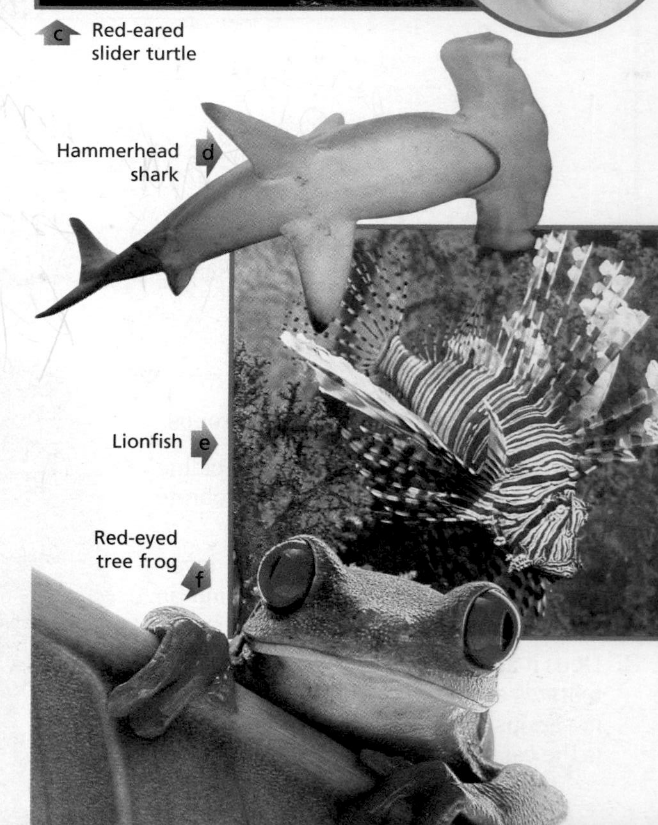

Common loon a

Fruit bat b

c Red-eared slider turtle

Hammerhead shark d

Lionfish e

Red-eyed tree frog f

Answers to *Vertebrates, continued*

1. Fish live in water. Amphibians live in water and on land. Reptiles live mostly on land. Birds live mostly on land and in the air. Mammals live mainly on land, although some live in the air or in the water.
2. Fish swim by using their fins, tails, and body movements. Amphibians walk or run by using their legs. Some also swim or jump using their large hind legs. Reptiles (except for snakes) walk or run on two pairs of legs. Snakes use their internal muscles to crawl. Birds use their wings to fly and their legs to walk, run, or hop. Most mammals use their two pairs of legs to walk or run.
3. Answers may vary. Students may suggest methods of reproduction and respiration that subgroups have in common. Fish and amphibians lay their eggs in water. Reptiles also lay eggs, but they are fertilized inside the female's body. The young of most mammals develop in the mother's body until they are born. Fish use gills to breathe. Amphibians breathe through their moist skin and lungs. Reptiles, birds, and mammals breathe with lungs.
4. The features described in answers (2) and (3) help the animals in each subgroup to survive in their specific habitats. For example, fish have fins, tails, and streamlined bodies that enable them to swim. Gills enable them to breathe in water.
5. Accept all reasonable responses. Respect every attempt that students make to classify according to consistent criteria. They might suggest classification in a number of ways, including habitat, size, color, locomotion, reproduction, or type of appendages.

Putting It All Together

This activity gives students an opportunity to consolidate all of the information they have learned so far about classification. Have them copy the tables on pages 126–127 into their ScienceLog and then fill them in using the illustrations on pages 126–129. They can also include any preserved or live specimens available in the classroom. Answers to Animal Kingdom Classification System can be found in the margin of this Annotated Teacher's Edition on page 129.

GUIDED PRACTICE To help students remember how to classify vertebrates and invertebrates, set up a chart on a bulletin board titled "The Animal Kingdom." The chart could have the following headings: Group, Subgroup, Examples, and Major Characteristics. Divide students into small groups, and have each group fill in part of the chart. Students could select pictures from old magazines or draw pictures of animals and place them around the bulletin board.

INDEPENDENT PRACTICE Find an area near the school that is an interesting place for students to search for animals. Remind students not to touch any plants and not to touch, tease, or annoy any animals. Have students take their ScienceLog, a pencil, and a spoon for digging. Have them look under rocks and logs and dig up some soil. They should record the names of any animals they recognize and make sketches of those that are unfamiliar. You may wish to take some field guides for insects or other animals so that students can use them to identify the animals they do not know. Instruct students not to disturb the animals they find. Remind them to put back any logs, rocks, or soil that they have moved. After the activity is completed, have students classify all of the animals they found into the correct groups. Arthropods can also be classified into the correct classes.

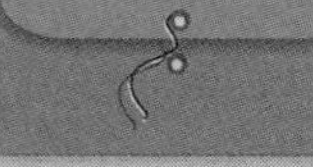

PORTFOLIO

Students may wish to include their Animal Kingdom Classification System in their Portfolio. Since this activity requires that students compile almost all of the information from the lesson, they may use their charts to review and reteach the concepts they have learned.

Putting It All Together

Many animals are shown on this page and the next three pages. Your task is to classify them in your ScienceLog using classification tables like the two on this page and the next. These tables are an overall classification system for animals. They bring together everything you have learned about classification for the animal kingdom.

If your classroom contains any living or preserved specimens that are not represented in the pictures, classify them as well. Remember to write only in your ScienceLog.

Animal Kingdom

Invertebrate subgroups	Examples
Annelids	
Mollusks	
Echinoderms	
Arthropods	
Insects	
Crustaceans	
Arachnids	
Diplopods/chilopods	
Other invertebrates	

a Scorpion

b Tree snail

c Manatees

d Sally lightfoot crab

e White-faced tree ducks

126

A Resource Worksheet (Teaching Resources, page 44) and Transparency 20 are available to accompany Putting It All Together.

Classification System

Vertebrate subgroups	Examples
Fishes	
Amphibians	
Reptiles	
Birds	
Mammals	

f Soldier beetle

g Millipede

h Clam

i Arabian horses

Theme Connection

Structures

Focus question: What structures allow the ostrich to adapt to its environment? *(The ostrich cannot fly, but its long, strong legs, two-toed feet, and flexible knees are adapted to running quickly across the varied terrain of Africa. The ostrich is the fastest creature on two legs—it can run at speeds of up to 70 km/h. Ostriches have long necks and large eyes—the largest eyes of any terrestrial vertebrate—that enable them to spot predators quickly.)*

Homework

You may wish to have students complete the Animal Kingdom Classification System tables on pages 126–127 as homework.

Answer to
In-Text Question

A Even though the turtle has a shell, it is not a mollusk. Mollusks have soft bodies covered by a hard shell made of calcium compounds. Turtles do not have soft bodies. They have all the characteristics of reptiles, including a backbone, a scale-covered body, and lungs. Their shells are made of a hard, hornlike material that lies over a deeper, bony layer. Students should understand that many characteristics determine the classification of an organism.

Sometimes an animal's outward appearance can fool you, so be sure to examine each animal carefully. For example, the turtle has a hard covering, but it is not a mollusk. Why not? **A**

j Bearded bristleworm

k Emerald tree boa

l Red squirrel

m Adélie penguins

n Green sea turtle

o Ostriches

128

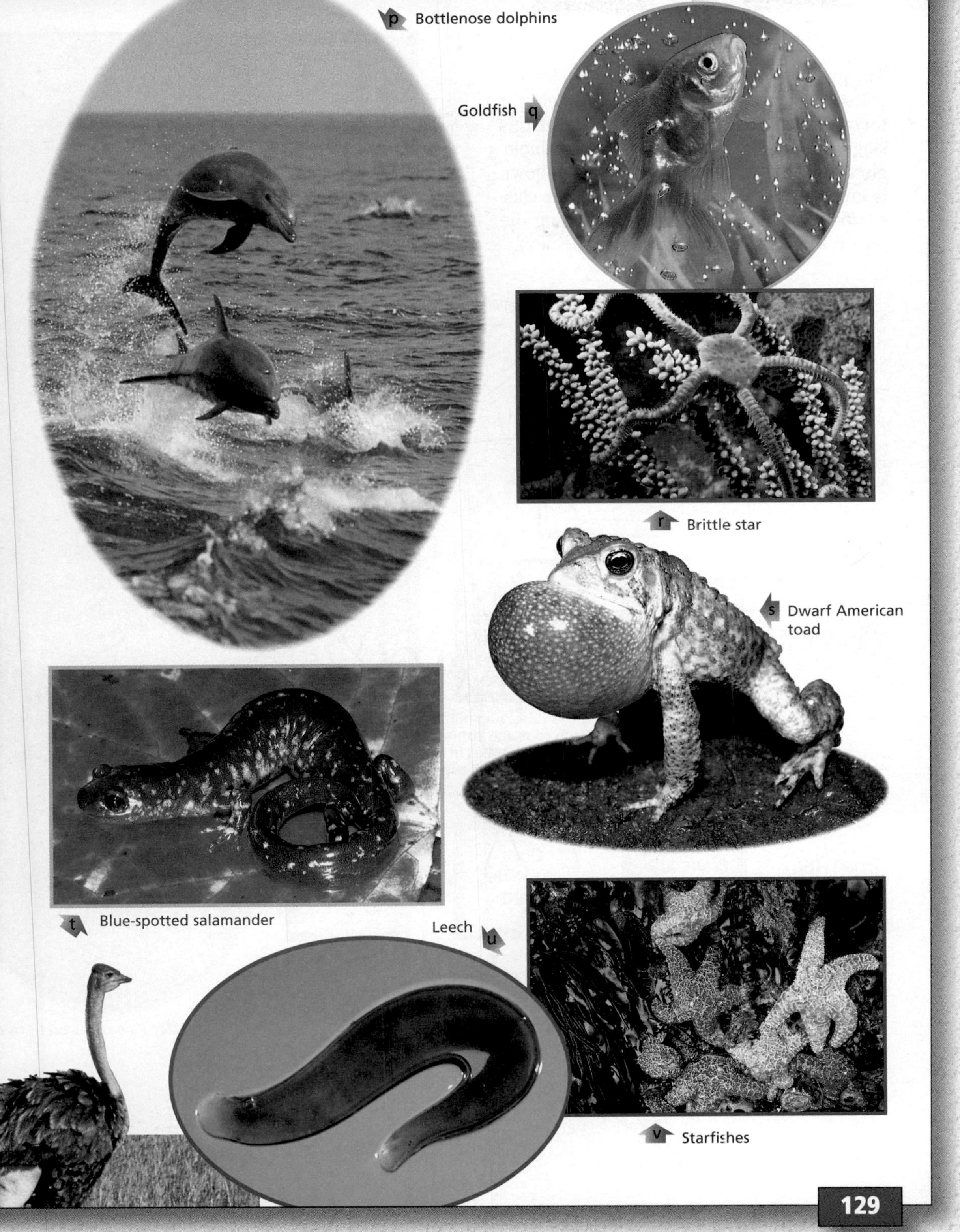

Answers to *Animal Kingdom Classification System*

The animals in the illustrations should be classified as shown below.

Invertebrate subgroups and examples:

- Annelids: (j) bearded bristleworm, (u) leech
- Mollusks: (b) tree snail, (h) clam
- Echinoderms: (v) starfishes, (r) brittle star
- Arthropods:
  - Insects—(f) soldier beetle
  - Crustaceans—(d) Sally lightfoot crab
  - Arachnids—(a) scorpion
  - Diplopods/chilopods—(g) millipede
- Other Invertebrates: There are no other invertebrates.

Vertebrate subgroups and examples:

- Fish: (q) goldfish
- Amphibians: (s) dwarf American toad, (t) blue-spotted salamander
- Reptiles: (n) green sea turtle, (k) emerald tree boa
- Birds: (e) white-faced tree ducks, (m) Adélie penguins, (o) ostriches
- Mammals: (c) manatees, (i) Arabian horses, (l) red squirrel, (p) bottlenose dolphins

The Plant Kingdom

Have students use the classification diagram on page 130 to classify the plants in the illustrations on pages 130–131. If possible, bring in some sample specimens of plants from each subgroup. You may wish to ask students to bring in small twig cuttings, leaves, flowers, and seeds or fruits from cone-bearing and flower-bearing plants.

Students may not realize that all trees other than conifers have flowers that develop into pods or fruits. Deciduous trees (those that lose their leaves in autumn or winter) usually flower in the spring before the leaves develop.

GUIDED PRACTICE Take students on a scavenger hunt to discover how many examples of both flowering and nonflowering plants they can find in the schoolyard or a park. (Algae, lichens, mosses, ferns, and conifers are examples of nonflowering plants.) Ask: Where do the plants grow? How are they adapted to their habitats? *(One example is a lichen, which is actually a fungus and an alga living together. Lichens can be found growing on rocks and on trees. They do not need the nutrients that soil provides. Ferns and mosses grow in damp and shady places. Conifers live in colder places and can be found on north-facing slopes.)*

Answers to *The Plant Kingdom*

The plants in the illustrations should be classified as follows:

- Mosses: moss
- Ferns and relatives: cinnamon fern
- Cone-bearing plants: fir tree
- Flower-bearing plants: black cherry tree, white oak tree

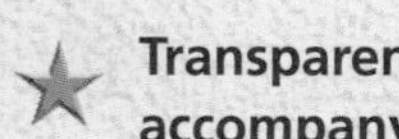

Transparency 21 is available to accompany The Plant Kingdom.

The Plant Kingdom

The great variety of plants in the world led scientists to develop a classification system for them, just as was done for animals. At this stage, it may not be clear to you why biologists use the particular groupings shown below. Nonetheless, you can use them to classify plants around you. Practice by using the pictures of plants on this page and the next.

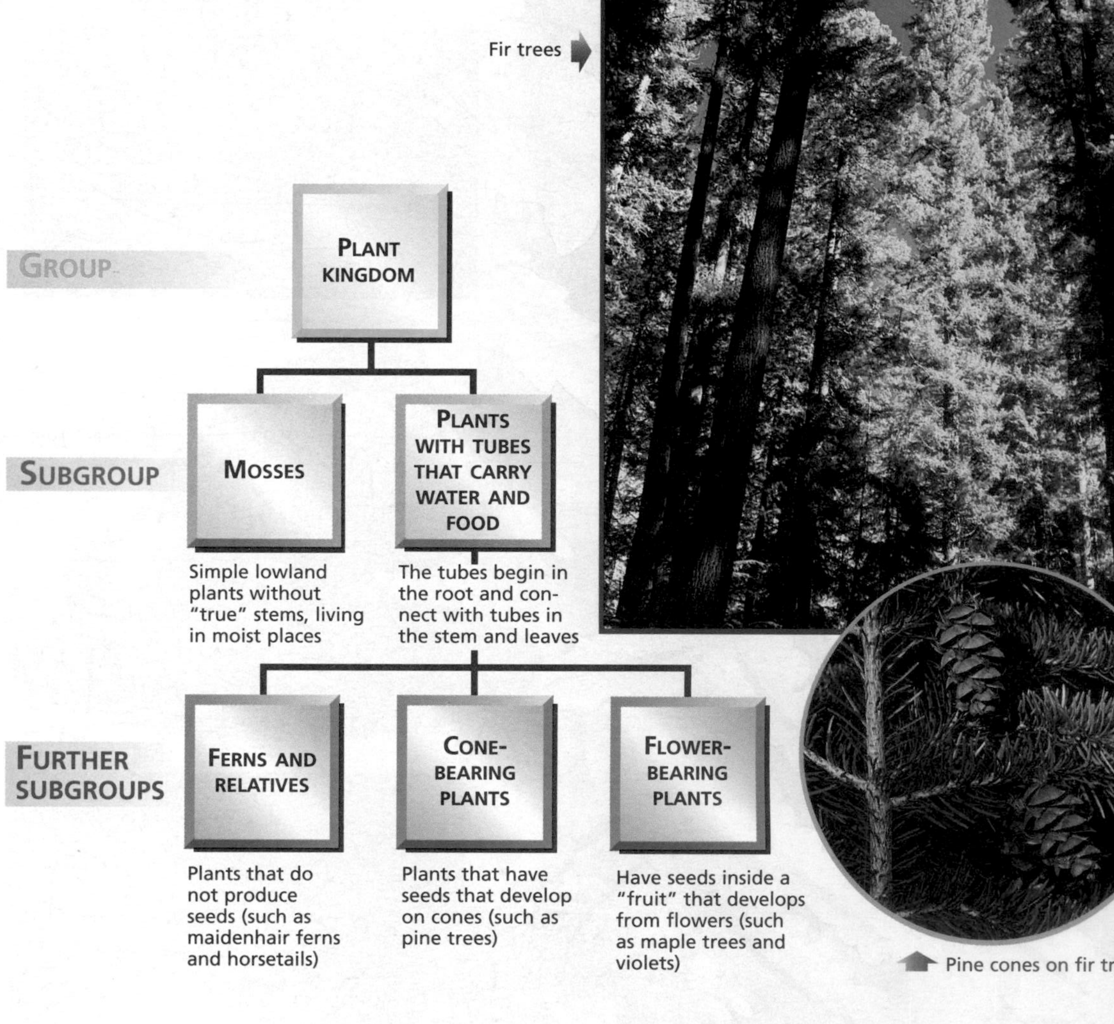

Fir trees

Pine cones on fir tree

130

Multicultural Extension

Plants From the Tropics

A great variety of plants can be found in tropical countries, including many fruits, such as citrus fruits and bananas. Many other fruits, such as mangoes, passion fruits, papayas, and guavas, are native to tropical countries. Have interested students find out more about tropical fruits or houseplants. If possible, bring in some unfamiliar tropical fruits for students to taste. If there are students in your class who come from a tropical country, have them discuss the fruits that are available in their native countries. Before tasting the fruit, make sure it is safe to eat and that the skins of the fruit have been cleaned. Have students research the effects that the plants grown in an area have on the culture of the people living there. For example, in the rain forests of South America, many plants are used for medicinal purposes, building materials, or food.

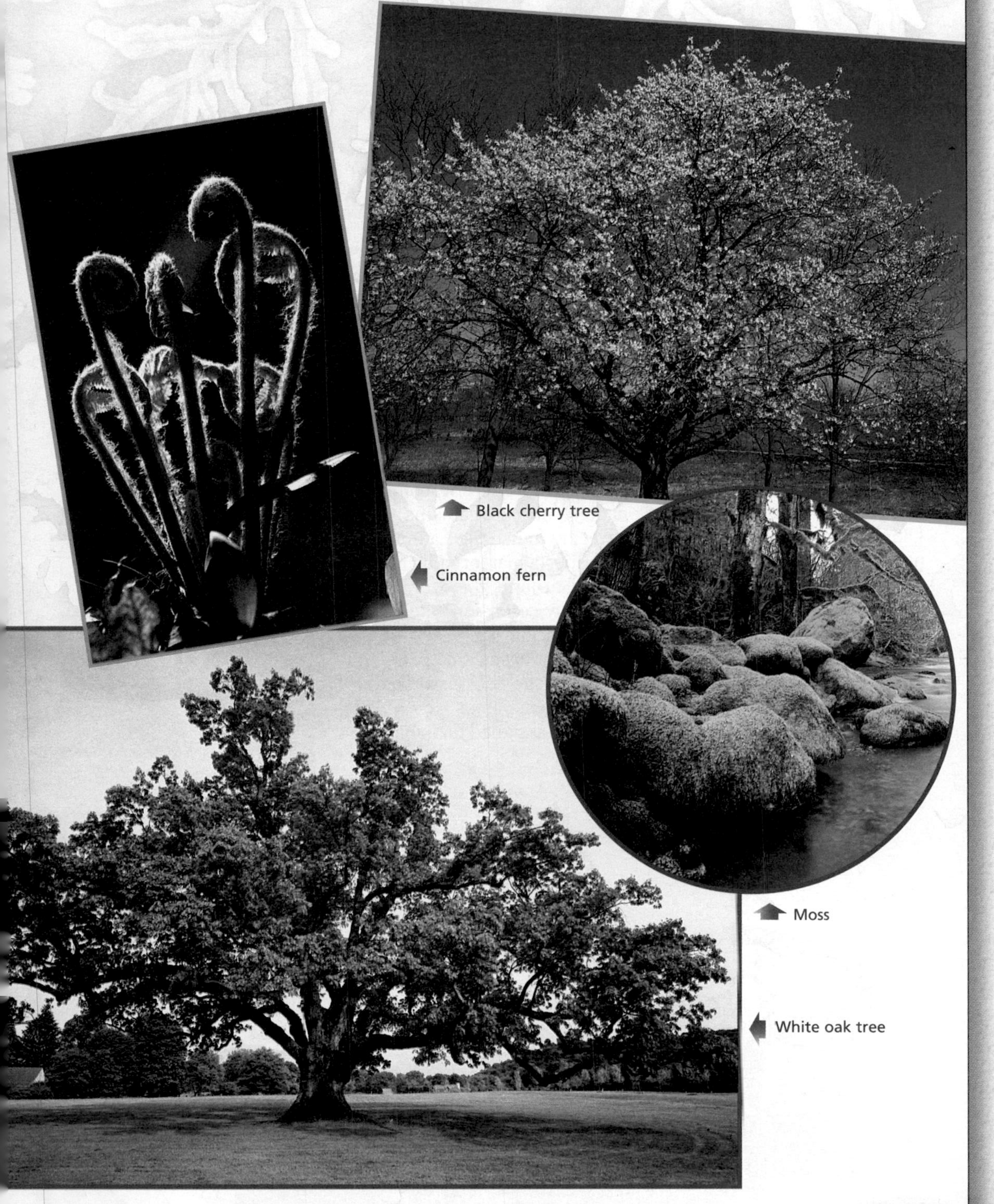
Black cherry tree

Cinnamon fern

Moss

White oak tree

131

FOLLOW-UP

Reteaching

Bring in pictures of vertebrates and invertebrates. Ask students to classify the animals into the correct groups and subgroups and to explain their reasoning.

Assessment

Bring in samples of several different kinds of plants, including mosses, fern leaves, branches from deciduous trees, cones and branches from conifers, and seeds and fruits from flowering plants. Ask students to classify these specimens into the accepted groups and subgroups. Have students explain their reasons for classifying the specimens the way that they did.

Extension

Remind students that microscopic organisms cannot be classified as plants or animals. For example, protozoans, which belong to the kingdom Protista, are one-celled, animal-like organisms that can be seen only with a microscope. Ask students to fill a jar half-full of mud and water that you have collected from the bottom of a pond or stream so that they can observe these one-celled creatures. Suggest that they add about a dozen grains of cooked rice to the jar as a source of nutrients and then leave it uncovered and undisturbed for a week, away from direct sunlight. Then ask them to use a microscope to observe the variety of protozoans in one drop of the water.

Closure

Ask students to look at the pictures on pages 124–125 and to make a list of the pictures in which more than one kind of living thing can be seen. *(For instance, the picture of the turtle on page 124 also shows a log.)* They should classify and, if possible, name the other organisms. Ask students why they think so many photos had more than one kind of living thing present. *(Most organisms are dependent on the presence of other organisms for survival.)*

Homework

You may wish to assign the Closure activity described on this page as homework.

LESSON 2 Finding a Name

FOCUS

Getting Started

Before class, make a poster of the word *dog* translated into other languages, such as Spanish, French, Italian, Swahili, German, Russian, or Chinese. In order for scientists to work together, it is necessary to have one name for each living thing. This name is called the *scientific name*. For a dog, it is *Canis familiaris.*

Main Ideas

1. Scientists classify living things into progressively more specific groups, according to their similarities.
2. The scientific name of a living thing is made up of two words: the genus name and the species name.
3. There is diversity even within a species.

TEACHING STRATEGIES

Have students read the introductory material on Linnaeus. Tell students that Aristotle developed a system for classifying plants and animals more than 2000 years ago. Aristotle's system divided all organisms into two groups, plants and animals. He grouped animals according to where they lived (in the air, on land, or in the water). He grouped plants according to their size (trees, shrubs, or herbs).

By contrast, Linnaeus's system classified organisms on the basis of structure. After Darwin's theory gained acceptance, biologists decided that classifying organisms according to their evolutionary relationships would be more useful. Structural characteristics usually reflect evolutionary relationships, so biologists have retained most of Linnaeus's groupings.

Transparency 22 is available to accompany the material on this page.

LESSON 2 Finding a Name

The Swedish scientist Carolus Linnaeus, who lived in the eighteenth century, can be called the "master of order." He devised a scientific system for classifying living things according to their structure and anatomy. It is his system that we still use today. You can see Linnaeus's classification system at the right.

You might be wondering why some of the groups in his classification system have Latin names. In Linnaeus's time, Latin was the written language used by scholars. In fact, "Carolus Linnaeus" is a latinized form of "Karl von Linne"—his given name!

It was Linnaeus's ambition to name all the living things in the world. He did not quite achieve this goal, but he did name more than 770 plants and 4400 animals. He gave each of these organisms a Latin name.

Latin Lingers On

Scientists continue to name newly discovered organisms every day. And even though no one communicates in Latin anymore, scientists still give Latin names to organisms. Why do you think this is so?

People in different countries have different common names for organisms. By agreeing to use the Latin name for a particular organism, scientists around the world can avoid any confusion when they communicate with each other about the organism. Also, no one has to go back and rename all the organisms to which Linnaeus assigned Latin names!

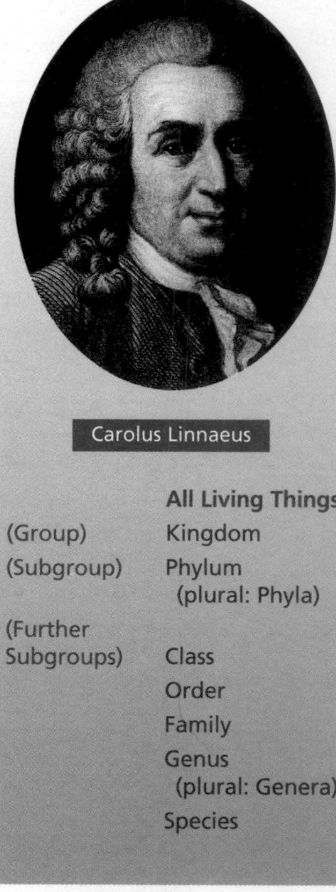

Carolus Linnaeus

	All Living Things
(Group)	Kingdom
(Subgroup)	Phylum (plural: Phyla)
(Further Subgroups)	Class
	Order
	Family
	Genus (plural: Genera)
	Species

How does the use of Latin lead to clear communication? Ⓐ

LESSON 2 ORGANIZER

Time Required
2 class periods

Process Skills
observing, contrasting, analyzing

New Term
Scientific name—the genus and species names of a living thing

Materials (per student group)
none

Teaching Resources
Exploration Worksheet, p. 45
Activity Worksheet, p. 47
Transparencies 22, 23, and 24
SourceBook, p. S44

A Name You Can Claim

Where do you fit into Linnaeus's classification system? First, you are clearly a member of the category of All Living Things. Next, you belong to the animal kingdom. Because you are a vertebrate (you have a backbone), you belong to the phylum Chordata. Then, because you are warmblooded, have hair (not feathers), and were born alive (instead of hatching from an egg), you fit into the class Mammalia. There are certain similarities between you and animals such as gorillas and chimpanzees, so the next grouping you belong to is the order Primates. You belong to the family Hominidae and the genus *Homo*. The most specific grouping is species. Your specific name is *sapiens*. Combine the names of our genus and our species, and you get *Homo sapiens*—the name for all human beings. (You might find it interesting to look in a dictionary to see what our name means.) B

You have more in common with other humans than you do with other primates, such as chimpanzees. But you have more in common with chimpanzees than you do with other mammals, such as cats. However, there are still more similarities between you and cats than between you and maple trees, which are also living things.

Every single type of living thing, from a fungus to a frog, has its own **scientific name**. A scientific name is a combination of an organism's genus and species names. It's important to remember that a scientific name contains both of these parts. You wouldn't refer to a wolf, for instance, simply as *Canis*. A wolf's scientific name is *Canis lupus*. Don't get him confused with *Canis familiaris*—your pet dog!

Now do some research. What are some other organisms that belong to the animal kingdom? to the phylum Chordata? to the class Mammalia? to the order Primates? to the family Hominidae? to the genus *Homo*? C

A member of the species *Homo sapiens* walking with her companion, a member of the species *Canis familiaris*

Answers to
Caption and In-Text Questions, pages 132–133

A Latin is a "dead" language because it is no longer used for communication, and therefore it will not change. People all over the world can avoid the confusion of using common names in different languages.

B The name *Homo sapiens* comes from the Latin words *homo*, meaning "man," and *sapiens*, meaning "wise."

C Other animals include worms, insects, lizards, and cats. Other members of the phylum Chordata include frogs, lizards, and cats. Other members of the class Mammalia include cats and cows. Other members of the order Primates include gorillas, monkeys, and aye-ayes. Other members of the family Hominidae include extinct humans. Other members of the genus *Homo* include *Homo habilis*, *Homo rudolfensis*, and *Homo erectus* (all extinct).

A Name You Can Claim

Ask students to do some research and list some animals that belong in each grouping. In doing this research, they will discover that each subgroup is more specific than the preceding one.

Point out to students that the first word in the scientific name, the genus, is usually a noun and is always capitalized. The species is generally an adjective and is always lowercase. The scientific name of an organism is always underlined or written in italics.

Did You Know. . .

In many cases, scientists wish to be especially precise when naming groups of organisms. A third name can define a smaller group within a species that is called a *subspecies*. For example, the subspecific name for the Siberian tiger is *Panthera tigris altaica* which distinguishes the Siberian tiger from the other tigers.

Meeting Individual Needs

Second-Language Learners

Often, the names given to animals reflect their outstanding features (e.g., crossbill, woodpecker, and grasshopper). Challenge the students to think of as many descriptive names for animals as they can. The names can be of actual organisms, or students can think up new names for organisms that reflect the organisms' characteristics.

CROSS-DISCIPLINARY FOCUS

Foreign Languages

The naming system used by scientists makes use of many Latin words. Have students find out the meanings of the following names for dinosaurs: *Tyrannosaurus rex (tyrant-lizard king), Apatosaurus (headless lizard),* and *Triceratops (three-horned face).* You may wish to have students research other dinosaur names. Also, point out that the word *dinosaur* is derived from Latin and means "terrible lizard."

Homo sapiens: Same Name but Diverse Characteristics

GUIDED PRACTICE Ask students to read the introductory material on page 134.

EXPLORATION 1

Ask students to copy the table on page 135 into their ScienceLog and fill in the appropriate information for group and family members. You may need to clarify some of the characteristics listed on the table. Point out the difference between attached and free ear lobes and between pointed and straight hairlines (i.e., widow's peak or not). Demonstrate the two thumb positions when the hands are folded together, and have a student demonstrate tongue rolling.

An Exploration Worksheet (Teaching Resources, page 45) and Transparency 23 are available to accompany Exploration 1.

Answer to *In-Text Question*

Ⓐ Answers will vary, but possible responses include eye, hair, and skin color; height and body build; shape of eyes, nose, and mouth; and hair texture.

Did You Know. . .

Many parasites have unusual adaptations that help them survive. The tapeworm, a parasite found in human intestines, is a good example. It has a special coating that protects it from digestive juices. Its head is equipped with suckers and hooks that allow it to attach itself to a host. It has no mouth or digestive system. Instead it absorbs predigested food through its skin.

Homo sapiens: Same Name but Diverse Characteristics

You're a member of the species *Homo sapiens,* and every person you know is also a member. All organisms within a species are alike. Thus, you share certain characteristics with all human beings—characteristics such as the general arrangement of your facial features, the number of your fingers and toes, and the ability to reason, to name just a few. People who are related tend to resemble each other even more closely. This is because every person inherits certain characteristics from his or her parents, grandparents, and so on.

Despite resemblances to ancestors, however, everyone is unique. You are "one of a kind." Not one of your friends or acquaintances is exactly the same as you. Nor are you a duplicate of your parents or grandparents or brothers or sisters (even if you are a twin).

EXPLORATION 1

Tracing Similarities and Differences

In this Exploration, you will investigate some of your inherited characteristics and those of some of your classmates.

What to Do

In the classroom, work in groups of four. Examine six inherited characteristics for each member of the group. In your ScienceLog, construct a table like the one on page 135, and record each member's characteristics. Repeat this procedure at home with family members or other relatives.

Every human is unique. How many examples of the diversity of features in *Homo sapiens* can you identify? Ⓐ

134

Multicultural Extension

The Language Tree

Hundreds of languages are spoken throughout the world. Most linguists believe that many of these languages were derived from a few progenitor languages. For instance, English is part of the Indo-European family of languages. Have students research the origin of some modern languages. Ask: What other languages are of Indo-European origin? What other families of languages are there? You may wish for students to make a poster showing how different languages are related. Ask: How is the classification of languages similar to the biological classification of animals? *(Both systems use similar traits to develop a hierarchy.)*

Characteristic	You	Student 2	Student 3	Student 4	Family 1	Family 2
Hand folding, thumb position						
1. Left over right						
2. Right over left						
Ear lobes						
3. Attached						
4. Free						
Hairline						
5. Pointed						
6. Straight						
Tongue rolling						
7. Can roll						
8. Can't roll						
Middle segment of fingers						
9. Hair						
10. No hair						
Toe next to big toe						
11. Same length or longer than big toe						
12. Shorter than big toe						

Analyze Your Data

1. Study your table and compare your results with those of the other groups in your class.
2. Calculate the percentage of individuals in your group who have each characteristic. What does this tell you?
3. Does anyone have all of characteristics 1, 3, 5, 7, 9, and 11? Does anyone have all of characteristics 2, 4, 6, 8, 10, and 12? Are there any two people in the class who have identical characteristics? Did you know that there are 64 possible combinations of these six characteristics?
4. You have looked at only six characteristics. There are hundreds of other characteristics you might have considered. What are some of these other characteristics?

135

Homework

The Activity Worksheet on page 47 of the Unit 2 Teaching Resources booklet makes an excellent homework activity. If you choose to use this worksheet in class, Transparency 24 is available to accompany it.

Answers to *Analyze Your Data*

As students complete the table and answer the questions, they will realize that there are hundreds of characteristics within the human species that can vary. Other examples include tooth size and shape, hair texture, and presence of freckles. Students should conclude that no two people are exactly alike.

FOLLOW-UP

Reteaching

Review the classification system that Linnaeus developed. Use mnemonic devices to help students remember the categories of the classification system—kingdom, phylum, class, order, family, genus, species. One possibility is the sentence, "King Philip came over for greasy soup." You may wish to have students develop their own mnemonic device.

Assessment

Display pictures of various vertebrates and invertebrates. Ask students to do the following:

a. Identify the animals that belong to the same kingdom as humans. *(All animals)*

b. Identify the ones that belong to the same phylum as humans. *(All those with backbones)*

c. Identify the ones that belong to the same class as humans. *(All mammals)*

d. Identify the ones that belong to the same order as humans. *(Primates)*

e. Explain how the animals in each group or subgroup are similar to humans. *(Answers will vary.)*

Extension

Ask students to imagine that they are explorers on a newly discovered planet. Have each student draw a picture of one plant or animal that they have observed and write a brief description of the organism on the back of their picture. Then ask students to trade papers and name each other's organisms. If time permits, have students try classifying the organisms into one of the five kingdoms.

Closure

Have students research the scientific names for various organisms that are familiar to them, such as plants and animals that live in their area. Have students find the derivations of these names in the dictionary. They will discover that most scientific names have a Latin origin.

Answers to *Challenge Your Thinking*

1. The whole pie represents all of the animal kingdom, three-quarters of which are arthropods. Of the arthropods, the insects represent the largest section. The missing portion contains all other invertebrate subgroups, but primarily it contains vertebrates, or animals with backbones.

2. Sample answer:

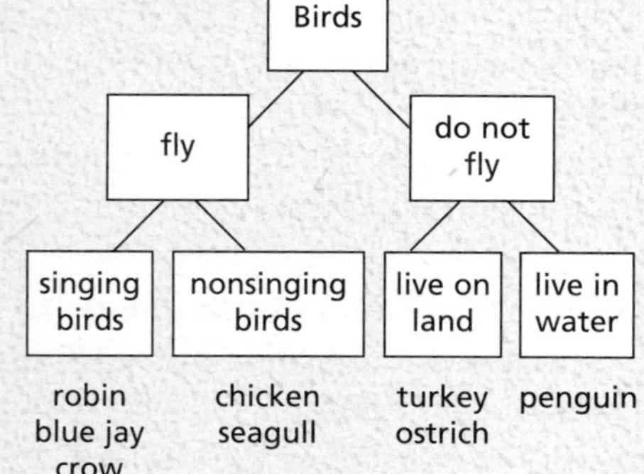

   Other possible classification categories students might consider include migratory and nonmigratory, nests in trees and nests on land, fast flyer and slow flyer, or fast runner and slow runner.

3. **a.** The green anole lizard's scientific name is *Anolis carolinensis.* The name *carolinensis* comes from the American states in which it is found—North Carolina and South Carolina.

   Viola tricolor is the scientific name for the three-colored pansy. The prefix *tri-* means three.

   Ursus arctos horribilis is the scientific name for grizzly bear. While *Ursus* is the Latin name for bear, *horribilis* refers to the bear's supposed ferocity.

   The frog is named *Rhinoderma darwinii.* It is named for the famous scientist Charles Darwin.

   b. Accept all reasonable answers. Students should show an understanding of the difference between genus and species and should use the proper capitalization rules for scientific names. For instance, a new type of wolf that drools a lot could be called *Canis slobberensis.*

4. **a.** A turtle is more closely related to an alligator than to an eel because both the turtle and the alligator are reptiles, while the eel is a fish.

   b. An earthworm is more closely related to a grasshopper than to a snake because both the earthworm and the grasshopper are invertebrates, while the snake is a vertebrate.

   c. An octopus is more closely related to a clam than to a lobster because both the octopus and the clam are mollusks, while the lobster is a crustacean.

   d. A whale is more closely related to a tiger than to a shark because both the whale and the tiger are mammals, while the shark is a fish.

   e. A maple tree is more closely related to a rosebush than to a pine tree because both the maple tree and the rosebush are flower-bearing plants, while the pine tree is a cone-bearing plant.

1. The Animal-Kingdom Pie

Can you explain the meaning of this pie chart? Which animals would you place in the missing piece of pie?

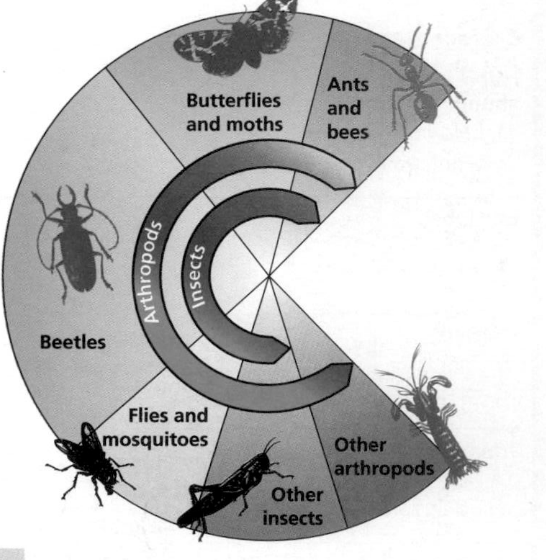

2. Birds of a Feather

Classify the birds shown according to their characteristics. Divide them into two groups, and write the characteristic shared by each group in the boxes of the first row. Then divide each of these groups into two subgroups, and write the characteristic shared by each subgroup in the boxes of the second row. Under each of the four boxes in the second row, write the names of the birds that fit into that subgroup.

3. The Name Game

When scientists discover an organism, they may choose a name for a number of reasons. The name may reflect a characteristic of the species or the location where it was found. The name might even incorporate the name of a well-known scientist.

a. Shown here are a list of scientific names and photographs of organisms with their common name. Try to match each organism with its scientific name. On what basis did you make each match?

b. Create a scientific name. Choose your favorite animal, and find out its scientific name. Now describe an imaginary animal that might be classified in the same genus. Complete the animal's scientific name by creating a species name. Give your reasoning for the name you chose.

Names:

- *Ursus arctos horribilis*
- *Rhinoderma darwinii*
- *Viola tricolor*
- *Anolis carolinensis*

This green anole lizard is found in two American states that share a name.

This pansy has three-colored flowers.

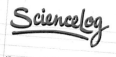

This frog was named in honor of a famous scientist.

The grizzly bear fiercely protects itself, its family, and its food.

4. Can You Relate?

According to the classification systems used by biologists, why is

a. a turtle more closely related to an alligator than to an eel?

b. an earthworm more closely related to a grasshopper than to a snake?

c. an octopus more closely related to a clam than to a lobster?

d. a whale more closely related to a tiger than to a shark?

e. a maple tree more closely related to a rosebush than to a pine tree?

ScienceLog

Review your responses to the ScienceLog questions on page 113. Then revise your original ideas so that they reflect what you've learned.

137

★ **You may wish to provide students with the Chapter 6 Review Worksheet that is available to accompany this Challenge Your Thinking (Teaching Resources, page 49).**

ScienceLog

The following are sample revised answers:

1. The chick, earthworm, and grasshopper are animals, while the rose is a plant. The animals could be further sorted into vertebrates (the chick) and invertebrates (the earthworm and grasshopper).
2. Vertebrate animals have backbones, while invertebrates do not. Knowing this difference allows us to classify animals into two different groups. These two groups can then be further classified into subgroups.
3. The name *Gallus gallus,* this animal's scientific name, identifies its genus and species. This Latin name is used by scientists across the globe to refer to the same animal. By using scientific names, even scientists that speak different languages can understand each other. Most American students would call this animal a chick.

Making Connections

The Big Ideas

The following is a sample unit summary:

Diversity describes the many differences in size, shape, and structure that exist among living things. (1) It arises as organisms adapt over time to a great variety of habitats and conditions. (2)

Animal adaptations used for protection include coloring, shape, mimicry, odors, stingers, claws, and teeth. Adaptations used in obtaining food include different types of teeth, beaks, and sensing organs. Adaptations used in attracting a mate include bright color patterns and communication signals. And adaptations for locomotion include structures such as wings, legs, and fins. (3)

Plants have root structures to anchor them and to obtain water and other nutrients. They have leaves that collect the sun's energy to make food. Some have sharp spikes to ward off animals. Some have flowers to attract pollinators. Plants can spread seeds by attaching them to passing animals, or by air, wind, water, or mechanical propulsion. (4)

Natural selection means that organisms with favorable traits are the most likely to survive, reproduce, and pass on those traits to their young. (5)

Darwin examined the differences between the beaks of finches on the Galápagos Islands. He concluded that after being separated for thousands of years on the different islands of the Galápagos, the finches of each island had adapted in a unique way to their environment. (6)

Species can become extinct when they cannot adapt to changes in environmental conditions. These changes can occur naturally or through human manipulation of the environment. (7)

We classify things because dividing things into organized groups makes it easier for us to locate items or information quickly and efficiently. Grouping things also helps us recognize similarities and differences in a systematic way. (8)

Living things are grouped and subgrouped according to similarities and differences in their evolutionary relationships. (9)

Our scientific name is *Homo sapiens.* More generally, we belong to the family Hominidae, the order Primates, the class Mammalia, the phylum Chordata, and the kingdom Animalia. (10, 11)

Making Connections

SourceBook

You have been introduced to the great diversity of life. You also have learned how to classify animals and plants. On pages S27–S50 in the SourceBook, take a closer look at classification, at the ways diversity came about, and at the origin of life.

UNIT 2

Here's what you'll find in the SourceBook:

The Diversity of Modern Life S28
The Evolution of Diversity S39

138

The Big Ideas

In your ScienceLog, write a summary of this unit, using the following questions as a guide:

1. What is diversity?
2. Why is there diversity among living things?
3. What are some adaptations animals have for protection? for obtaining food? for attracting a mate? for locomotion?
4. How are plants adapted to survive and reproduce?
5. What is meant by natural selection?
6. What evidence did Darwin use to develop his theory of how life evolved?
7. What kinds of conditions cause species to become extinct?
8. Why do we classify things?
9. How, in a general way, are living things classified?
10. How would you classify yourself, according to Linnaeus's system?
11. What is the scientific name of your species?

Checking Your Understanding

1. For each group of words below, write one or two sentences that show how the words are related to each other.
   a. adaptation, predator, peppered moths, habitat
   b. organisms, diversity, environment, survival
   c. diversity, species, inherited characteristics, unique

2. Classify the living things in these photos according to some consistent system. Explain the system you used.

Salamander

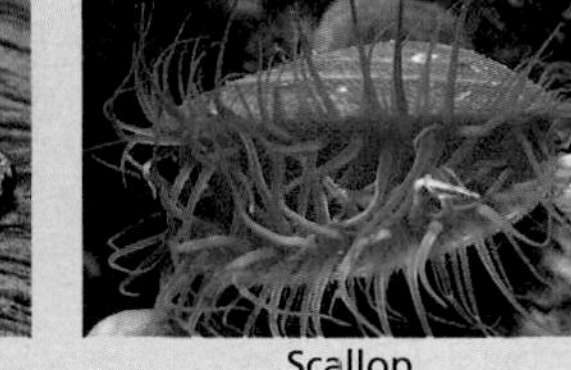
Scallop

Iguana

Spider

Peach tree

Pony

Asters

Sparrow

Spruce tree

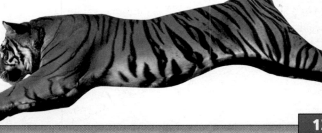

Sea urchin

Beetle

Centipede

Surgeonfish

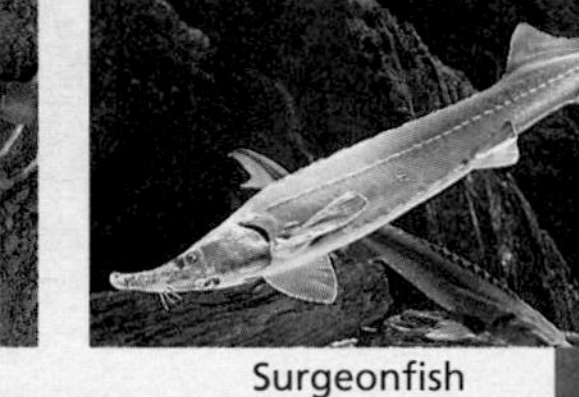
Whale

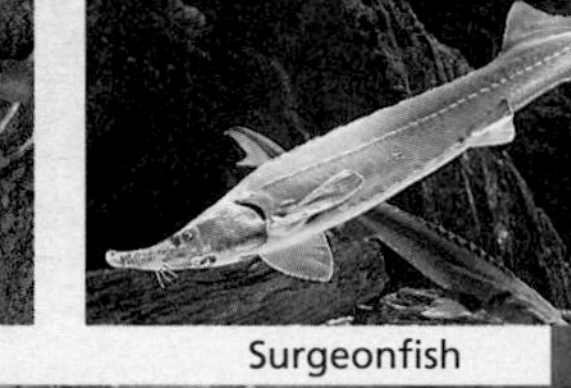
Swans

Crayfish

3. (concept map) Draw a concept map that shows how the following words are related to each other: coral snake; camouflage; mimicry; snowshoe hare; adaptations; snow; grasses, trees, and weeds; brown in summer; white in winter; and king snake.

Answers to *Checking Your Understanding*

1. Sample answers:
   - **a.** Peppered moths use camouflage as an adaptation that protects them from predators in their habitat.
   - **b.** Adaptations help organisms in their struggle for survival in a given environment. Diversity arises from variations in how each species adapts over time to its habitat.
   - **c.** While inherited characteristics make members of a species alike in many ways, there is also great diversity within a species. People belonging to the same family, for instance, may look similar, but despite family resemblance, every human is unique.
2. Accept all reasonable responses. See the sample classification diagram on page S202.
3. See the sample concept map on page S203.

★ **You may wish to provide students with the Unit 2 Review Worksheet that is available to accompany this Making Connections (Teaching Resources, page 58). The Unit 2 Activity Worksheet may also be used once students have completed Unit 2.**

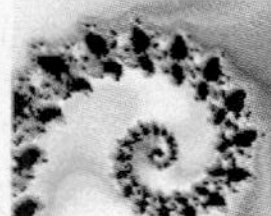

Background

The Iroquois *(EER uh kwoy)* was a league of tribes that lived in the present-day state of New York. The league, begun in the 1600s, consisted of the Mohawks, Oneida, Onandaga, Cayuga, and Seneca tribes, which were known as the Five Nations. The Tuscarora tribe later joined the Iroquois, after which the league was known as the Six Nations. The Iroquois participated in the French and Indian War, siding with the British, but they disbanded in the eighteenth century after their loyalties were divided between the British and the Americans during the American Revolution.

The Sioux, who lived on the Great Plains of North America, were also composed of a number of subgroups. Besides the Brule Sioux, there were the Santee in Minnesota, the Yankton in the eastern Dakotas, and the Lakota in the western Dakotas and Nebraska. Famous Sioux leaders include Crazy Horse and Sitting Bull.

Extension

Have a volunteer find and compare the definitions of *myth* and *legend. (The definitions are very similar, although myths tend to deal specifically with origins.)* Have students identify other types of stories. *(Folktales, epics, sagas, fables)* Ask: In human cultures, what might be the purpose of legends and myths? *(Sample answer: to understand nature, reflect cultural values, convey anxieties, or express religious beliefs)* Students may find out more by consulting an encyclopedia or another reference on oral traditions.

Answers to *Tell a Tale*

Accept all reasonable stories. Encourage students to be creative. Students should understand that many stories have specific authors but that most legends do not. Legends are passed from generation to generation and may be altered or supplemented in ways that reflect changing cultural values.

Animal Legends

Why do birds fly? Why do moles live underground? Why do beavers have flat tails? Although these are questions that scientists have studied, they are also the subjects of legends. The following two American Indian legends have likely been told and retold in one form or another for thousands of years.

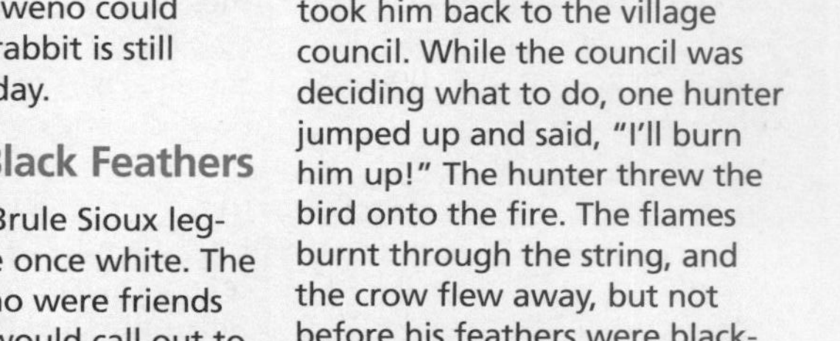

Why Rabbit Has No Fierce Claws

According to an Iroquois legend, when Rabbit was being created, he asked for long hind legs, long ears, sharp fangs, and fierce claws. Raweno, who made the animals, did not mind making Rabbit the way he wanted to be.

Raweno was forming Rabbit's legs when Owl interrupted and demanded a long neck, red feathers, a long beak, and a crown of plumes. Raweno, now shaping Rabbit's ears, scolded, "Be quiet. Close your eyes. Don't you know that no one is allowed to watch me work?"

But Owl disobeyed, saying, "I like watching you, and watch I will." That made Raweno so angry that he pulled Owl from his branch, stuffed his head into his body, pulled on his ears until they stuck straight up, and shook him so hard that his eyes grew big. Owl flew off, and his body stayed that way. Raweno returned to his work, but Rabbit was gone. He had run away in fright before Raweno could finish. And the rabbit is still nervous to this day.

The Crow's Black Feathers

According to a Brule Sioux legend, crows were once white. The white crows, who were friends of the buffalo, would call out to warn the buffalo when hunters were coming. The buffalo would stampede, and the people would go hungry.

To solve this problem, a wise chief came up with a plan. He stood on the plain dressed in buffalo skins. When the crow leader flew down to warn him, he caught the bird, tied him to a stone with a rawhide string, and took him back to the village council. While the council was deciding what to do, one hunter jumped up and said, "I'll burn him up!" The hunter threw the bird onto the fire. The flames burnt through the string, and the crow flew away, but not before his feathers were blackened by the fire.

Tell a Tale

Choose two or three animals and write a story to explain how the animals got their characteristics. What is the difference between a story and a legend?

Genies in Your Cells

Ashanthi DeSilva was born with a disease of the immune system called adenosine deaminase deficiency, or ADA deficiency. Her body was unable to fight off even the mildest infection. This meant that minor cuts and scrapes could easily become fatal. In 1990, Ashanthi, then four years old, was treated with gene therapy. With this revolutionary treatment, doctors helped Ashanthi conquer her disease by turning malfunctioning cells into healthy ones.

Instructions for Cells

Almost every cell in your body contains a set of 46 chromosomes. Chromosomes are microscopic structures that contain genes. Genes pass hereditary traits from parent to offspring and determine the function and characteristics of each cell. Sometimes these genes function improperly and produce harmful effects known as *genetic disorders*. Some examples of genetic disorders are muscular dystrophy and certain types of cancer. Ashanthi's genetic disorder came from the malfunctioning ADA gene. The ADA gene controls the production of chemicals that the immune system needs in order to work properly.

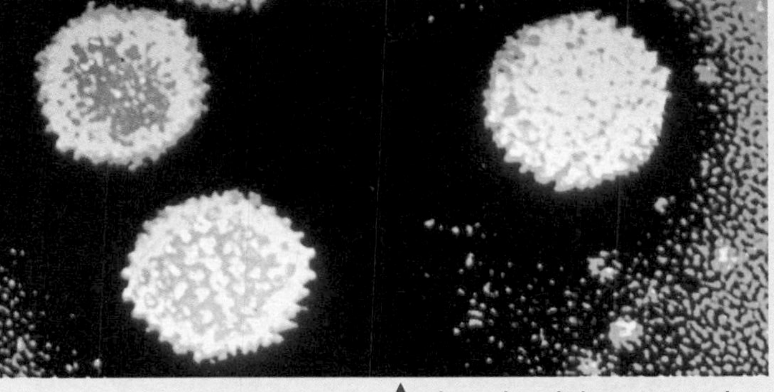

▲ These altered viruses are used to transport desired genes into unhealthy cells during gene therapy.

Delivering the Goods

Gene therapy helped Ashanthi overcome her genetic disorder. In gene therapy, healthy genes are injected into unhealthy cells. In most cases, doctors use viruses to transport the desired genes. These viruses normally take over cells by injecting their own genes into the cells. Doctors use these viruses to their advantage by altering the viruses so that the viruses carry and inject the desired genes.

In Ashanthi's case, doctors used a virus to insert properly functioning ADA genes into a few of Ashanthi's cells that contained the malfunctioning ADA gene. The new ADA gene disabled the malfunctioning ADA gene. With the healthy genes at work, Ashanthi's immune system began to function much better.

◀ A faulty gene on this chromosome may cause malignant melanoma, a type of skin cancer.

Curing the Incurable

Gene therapy may some day provide cures for diseases that were once considered incurable, such as cystic fibrosis. Treatments for viral infections such as AIDS are also being tested. As scientists refine the technology of gene therapy, they hope to help even more victims of genetic disorders.

The Cutting Edge

Gene therapy has provoked great debate. Some people believe that gene therapy could be used for unethical genetic alterations. Others think that gene therapy should be encouraged so that all genetic disorders can be eliminated. Think about these issues and write a paragraph or two that reflects your ideas about the ethics of gene therapy.

141

Debate

You might want to stimulate critical awareness by posing certain scenarios to the class and asking students to explain why gene therapy should or should not be used in each case. Divide the class into small groups and have each group discuss the subject. Then have each group elect a member to debate their position in front of the class. Some diseases whose faulty genes may have been identified include osteoporosis, psoriasis, kidney cancer, and dwarfism.

The Cutting Edge

Accept all thoughtful responses. Many students will probably say that gene therapy has value but should not be taken too far.

TECHNOLOGY in SOCIETY

Background

Gene therapy is a medical treatment that is still in its infancy. Genes are segments of chromosomes that instruct a cell how to produce proteins. Proteins are a key component of many structures in the body, from muscles to hemoglobin in the blood. For hereditary diseases, doctors can insert a normal gene into the patient's cells. For a nonhereditary disease, doctors can insert a gene that will make the cells more sensitive to other types of treatment.

However, there are problems with gene therapy. Currently, it is extremely expensive—up to $300,000 for a single treatment. Also, in some cases the injection technique is not efficient enough. AIDS patients, for example, require 20 to 50 percent of their cells to contain the antiviral gene to receive any benefit, but gene therapy typically affects only 5 to 10 percent of all cells. Cancer patients are also difficult to treat because if even one cancerous cell is left untreated, the cell could eventually grow into another tumor.

CROSS-DISCIPLINARY FOCUS

Art

Have students draw a comic strip showing the biological steps involved in gene therapy. The first frame could show a gene-carrying virus approaching a cell. Next the virus attaches to the cell wall and injects the good gene into the cell. Then the good gene makes its way to the nucleus of the cell, where it incorporates itself into one of the chromosomes. Encourage students to be creative in their pictures and to label the important features. You may wish to do this activity with the class, drawing a cartoon on an overhead transparency and soliciting input from students.

SCIENCE in ACTION

Background

It is unclear when the first zoo, or zoological garden, was established. However, pictures on Egyptian tombs dating back to 2500 B.C. depict a variety of wild animals wearing collars. Royalty in both China and the Middle East also established collections of exotic animals as early as 1500 B.C. Since that time, zoos have become an integral part of societies all over the world. The San Diego Zoo is one of the nation's most popular zoos, and it attracts many tourists.

In addition to educating and entertaining visitors, the San Diego Zoo is committed to animal conservation. As Lieras explains, "Zoos are involved in a lot of programs that the public is unaware of. One of these is the national Species Survival Plan, in which individuals in the zoo community and others create a survival plan for a specific species." Regardless of who keeps the animals, everyone cooperates to enhance the breeding pools and preserve the species.

Being a zookeeper requires a lot of persistence and hard work. Lieras's first job at the zoo was in food services. She moved to the animal care center after seven months and then to a position as a keeper. In 1992, she left the zoo to teach, but she soon returned and eventually became the team area leader at the children's zoo.

CROSS-DISCIPLINARY FOCUS

Art

If possible, take students on a field trip to a zoo or wildlife park in your community. Have students take notes on the types of exhibits and animals that they encounter. Encourage students to photograph or sketch interesting things that they see. Then have the class design a scrapbook illustrating the field trip to the zoo. You may wish to have students add to the scrapbook by doing research on the wildlife that they learned about on the trip.

SCIENCE in ACTION

Mary Lieras: Zookeeper

Mary Lieras caters to some pretty wild customers. Some are loud, some are dangerous, and some stay up all night. But it's all a part of the job for Lieras, who leads a team of workers at the San Diego Children's Zoo and Wildlife Park. The Children's Zoo allows people to get up close to a wide variety of exotic and common species of animals.

Hanging Out With Wildlife

Taking care of zoo animals is not easy. Of course, the animals must be fed and their cages must be cleaned, but the job doesn't stop there. "We also monitor the animal's health, maintain the exhibits, and keep detailed records," Lieras says. "One of the most tricky and dangerous parts of the job, however, is catching and moving the animals. Once, while trying to catch a spider monkey for a routine medical examination, the monkey attacked me. It used its strong prehensile tail to get out of my net and started climbing up my body, biting me at the same time. And believe me, spider monkeys have very sharp teeth! Luckily, a fellow zookeeper helped pull the monkey off me before I was seriously injured."

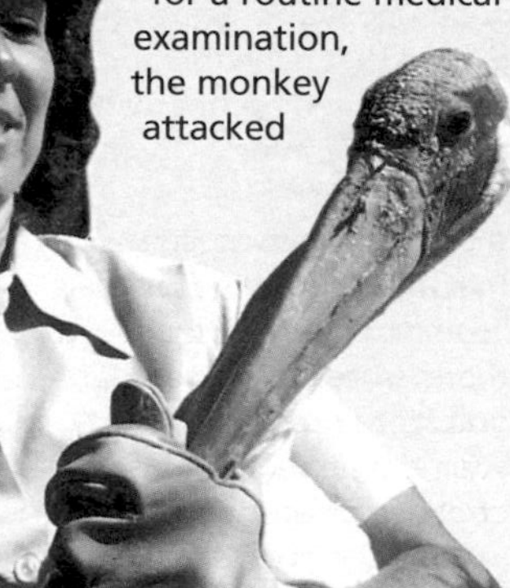

Mary Lieras moving a pelican at the San Diego Zoo

Designed for Diversity

In an attempt to provide realistic habitats for its animals, the San Diego Zoo has built some exhibit areas called bioclimatic zones. Bioclimatic zones approximate natural habitats and interactions by grouping animals that normally live together with many of the plants of their natural habitat. For example, gorillas are exhibited in an area surrounded by tropical plants of that region. Animals that might be found in the same region, such as crowned eagles and kikuyu colubus monkeys, are exhibited nearby.

Attempting to mimic nature's diversity presents several problems for animal-care workers. For example, special care is required for a large field exhibit where gazelles, wildebeests, deer, rhinos, and giraffes live together with several species of birds. In order to ensure the safety of each animal, these exhibits must be monitored very carefully. In this case, the birds seem especially particular. "The birds need to have areas where they can get away when they need to," Lieras explains.

"If you can affect people with what you're doing at the zoo, they may get more involved with conservation in their own communities."

Captive or Protected?

Some people question whether animals should be held in captivity at all. Lieras believes that zoos are a good thing. "As we see more and more of the animals' habitats being destroyed, it becomes pretty obvious that there may be a day when the only animals we have left are those in captivity. As we learn more about the animals we're taking care of, we may be able to reintroduce some of these animal species back into the wild."

An Exhibit Plan

Design your own wildlife exhibit using animals and plants that are found together in the wild. Draw a plan of the exhibit, and identify the types of living things that will inhabit the area. How will you keep the animals from escaping? How will you control predation?

142

Multicultural Extension

Worldly Plants and Animals

Divide the class into groups of three to five students. Have each group choose a country and prepare a wildlife profile for that country. The profile should include 10 native animal species and 5 native plant species. Have each group present their profiles to the class. You may wish to stage this event in the style of an academic conference. Explain that most countries have strict laws governing importation or exportation of native plants and animals.

Answers to *An Exhibit Plan*

Student designs are likely to vary considerably. Make sure students consider issues such as predator-prey interactions, necessary land area, solitary or herd behavior, sleeping patterns, and the toxicity of certain plants to certain animals. Also encourage students to consult an encyclopedia or similar resource.

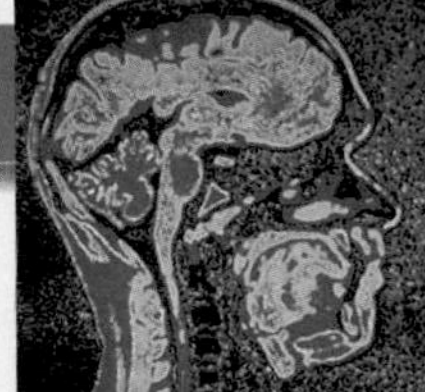

Worms That Get Under Your Skin

You are writing a horror story in which small creatures invade the bodies of human beings. These creatures grow quickly into large worms that live just under the skin. Is this a great start to a fictional story? No, this actually happens! The creatures are called Guinea worms, and they infect millions of people in Africa, India, and the Middle East.

The Life Cycle of the Guinea Worm

The Guinea worm spends most of its larval stage inside water fleas that live in pools, ponds, and open wells. When humans drink the water, they swallow the water fleas. Digestive juices in the stomach kill the fleas, but the larvae survive and burrow into the lining of the digestive tract. The larvae eventually migrate to tissues just under the surface of the skin, where they grow into adult worms that can reach lengths of more than 1 m!

In order to reproduce, the adult female worm makes a small blister in the host's skin. On contact with water, the blister breaks open, releasing up to 2 million larvae into the water. Then, over a period of many weeks, the worm exits the body through the skin, causing extreme pain to the human host. Some people try to pull the worms out, wrapping them around pieces of cloth. This must be done very slowly and carefully because the worm can cause a deadly infection if it breaks and slides back under the skin.

▲ A woman in Cameroon is using a filter to remove water fleas (top), which carry the Guinea worm larvae. If you look closely, you can see a worm larva inside the flea.

The Power of Prevention

The most effective treatment is prevention. Boiling the drinking water or filtering it through a muslin cloth will kill or remove the larvae. Keeping infected skin from touching public water supplies prevents the spread of larvae. Building closed wells also prevents contamination.

International efforts to educate people about prevention and to provide them with safe water supplies promise great success in eliminating the Guinea worm across the globe.

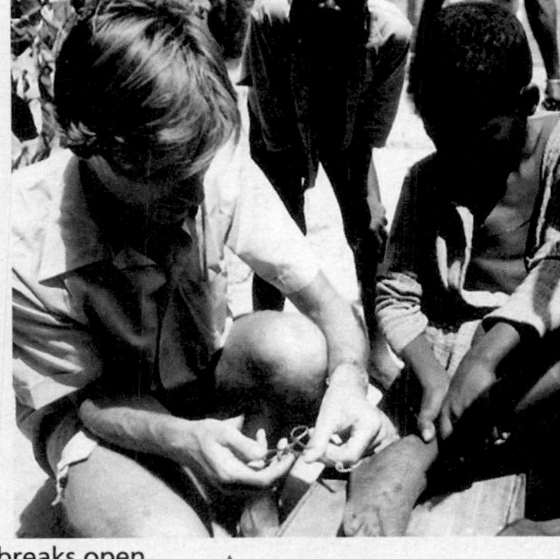

▲ Here a Guinea worm emerges through the skin and is being removed by a health-care worker.

Kill the Carrier?

Many people suggest using pesticides to kill the water fleas that transmit the Guinea worm larvae to humans. Can you identify any problems with this approach? What else could be done to prevent the spread of this disease?

Answers to *Kill the Carrier?*

The obvious problem of using pesticides is that it might kill not only the Guinea worms, but also other lifeforms. In addition, drinking water that contains toxic chemicals could be harmful, and if a pesticide seeped into the ground, it could be carried to other water sources.

Successful alternatives to chemical treatment could include stocking water reservoirs with fish that eat water fleas.

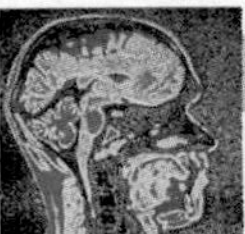

HEALTH WATCH

Background

The Guinea worm, or *Dracunuculus medinensis,* threatens the health of about 100 million people in Africa, India, and the Middle East. It is estimated that this disease has caused up to $1 billion in lost labor in a single year.

Some attempts to control the disease, such as teaching people about prevention techniques, using filters and safer wells, and treating contaminated water, have been successful. For instance, in 1994 no cases of Guinea worm disease were reported in Pakistan, compared with over 2000 cases in 1987. Similarly, the Ivory Coast reported 67,000 cases in 1966 but only 592 cases in 1985.

Although Guinea worm disease is very serious, the Guinea worm is not the most common parasitic worm in humans. Other worms, such as hookworm and tapeworm, are much more common.

Extension

Bring to class some water from a nearby stream or pond. Put a few drops on a glass slide, add a coverslip, and place the slide under a microscope. Have students comment on what they observe in the water. You might want them to compare this water with tap water. *(You will probably see some microorganisms in the pond water but very few, if any, in the tap water.)*

Debate

You might want to organize a class debate on public health issues versus the need for biodiversity. Specific examples include the Guinea worm and the smallpox virus. In 1980, the World Health Organization announced that the smallpox virus had been eliminated worldwide, even though samples of the virus were being kept alive in containment. Should the virus have been eliminated completely? What purpose could keeping live samples of it serve?

Unit 3 SOLUTIONS

Unit Overview

In this unit, students are introduced to the properties of solutions. In Chapter 7, students use their prior knowledge to formulate a definition of solutions, solutes, and solvents. Then students investigate the familiar characteristics of hard and soft water. In Chapter 8, students investigate distillation, evaporation, and chromatography as methods used to separate solutes from solvents. In Chapter 9, students explore the characteristics of solutions that have different concentrations as well as solutions in the environment and solutions in the body.

Using the Themes

The unifying themes emphasized in this unit are **Structures, Energy, Cycles,** and **Changes Over Time.** The following information will help you incorporate these themes into your teaching plan. Focus questions that correspond to these themes appear in the margins of this Annotated Teacher's Edition on pages 157, 169, 174, 183, and 198.

When learning about the nature of solutions in Chapter 7, a discussion of **Structures** will help students understand the size and orientation of the particles that make up solutions.

Later, in Chapter 8, focus questions invite students to consider the role of **Energy** both in the process of making Cibwa salt and in the atmospheric water cycle.

Just as the atmospheric water cycle describes a pattern of water movement in the air, the theme of **Cycles** appears again in a focus question that challenges students to consider the nature of underwater water cycles. Studying these cycles helps students relate the topic of solutions to both Earth and life sciences.

Finally, the concentration of solutes in natural bodies of water is affected by evaporation and other natural processes. Considering this fact from the perspective of **Changes Over Time** will help students to analyze it in greater depth.

Using the SourceBook

Unit 3 reviews the important terms discussed in the text and introduces the new term *colloid.* Special classes of substances known as acids, bases, and salts are then discussed, with emphasis on their properties and important practical uses.

Bibliography for Teachers

Barber, Jacqueline. *Crime Lab Chemistry.* Great Explorations in Math and Science (GEMS) Series, ed. Lincoln Bergman and Kay Fairwell. Berkeley, CA: Lawrence Hall of Science, 1990.

Center for Multisensory Learning. *Mixtures and Solutions.* Berkeley, CA: Lawrence Hall of Science, 1990.

Sumrall, William J., and Fred W. Brown. "Consumer Chemistry in the Classroom: Science From the Supermarket." *The Science Teacher,* 58 (April 1991): 29–31.

Bibliography for Students

Kramer, Alan. *How to Make a Chemical Volcano and Other Mysterious Experiments.* New York City, NY: Franklin Watts, 1991.

Lewis, James. *Measure, Pour, and Mix: Kitchen Science Tricks.* New York City, NY: Meadowbrook Press, 1990.

Morgan, Sally, and Adrian Morgan. *Water.* New York City, NY: Facts On File, Inc., 1994.

Films, Videotapes, Software, and Other Media

Alterations in the Atmosphere
Videotape
Films for the Humanities, Inc.
P.O. Box 2053
Princeton, NJ 08543

Mixtures and Solutions
Videotape
Eureka!
Lawrence Hall of Science
University of California
Berkeley, CA 94720

Solubility
Software (Apple II family)
Educational Materials and Equipment Co.
Old Mill Plain Rd.
P.O. Box 2805
Danbury, CT 06813-2805

Solutions
Software (Apple II family)
Prentice-Hall Allyn & Bacon
113 Sylvan Ave.
Englewood Cliffs, NJ 07632

Unit Organizer

Unit/Chapter	Lesson	Time*	Objectives	Teaching Resources
Unit Opener, p. 144				Science Sleuths: Green Thumb Plant Rentals #1 English/Spanish Audiocassettes Home Connection, p. 1
Chapter 7, p. 146	**Lesson 1, Solutions and Non-Solutions, p. 147**	2 to 3	1. Distinguish between mixtures that are solutions and those that are not. 2. Recognize and describe the Tyndall effect.	Image and Activity Bank 7-1 Exploration Worksheet, p. 3
	Lesson 2, Parts of a Solution—Solutes and Solvents, p. 154	4	1. Identify the parts of a solution as the solvent and solute. 2. Name solutions in which the solvent and solute are various combinations of gas, liquid, and solid. 3. Name useful and harmful solutions. 4. Identify and test for hard water.	Image and Activity Bank 7-2 Discrepant Event Worksheet, p. 8 Resource Worksheet, p. 9 ▼ Exploration Worksheet, p. 11 Exploration Worksheet, p. 13
	Lesson 3, Soluble, Not Very Soluble, or Insoluble? p. 162	2	1. Distinguish between solutes or substances that are soluble, not very soluble, or insoluble. 2. Identify variables that will increase the rate of dissolving.	Image and Activity Bank 7-3 Exploration Worksheet, p. 14 ▼ Exploration Worksheet, p. 17
End of Chapter, p. 165				Review Worksheet, p. 19 Chapter 7 Assessment Worksheet, p. 21
Chapter 8, p. 167	**Lesson 1, The Salt Industry, p. 168**	1 to 2	1. Define *evaporation* and *boiling*, and explain how these methods can be used to separate a solute from a solvent.	Image and Activity Bank 8-1 Exploration Worksheet, p. 24
	Lesson 2, Desalination and Distillation, p. 172	1 to 2	1. Describe the processes of desalination and distillation. 2. Describe the process of the water cycle.	Image and Activity Bank 8-2 Graphing Practice Worksheet, p. 25
	Lesson 3, Another Solution, p. 176	2	1. Identify and analyze the methods that scientists use for solving problems. 2. Explain the process of paper chromatography.	Image and Activity Bank 8-3 Exploration Worksheet, p. 27 ▼
	Lesson 4, Sugar Solutions, p. 180	1 to 2	1. Explain the effect of solutes on the boiling point of a solvent.	Image and Activity Bank 8-4 Exploration Worksheet, p. 30
End of Chapter, p. 183				Review Worksheet, p. 31 Chapter 8 Assessment Worksheet, p. 34
Chapter 9, p. 185	**Lesson 1, Getting Colder, p. 186**	2	1. Explain the effect of solutes on the freezing or melting point of a solvent.	Image and Activity Bank 9-1 Exploration Worksheet, p. 37 Exploration Worksheet, p. 38 Exploration Worksheet, p. 39
	Lesson 2, How Much Solute? p. 188	2	1. Correctly use the terms *concentrated* and *dilute*. 2. Compare the concentrations of two solutions made up of the same solvent and solute.	Image and Activity Bank 9-2 Exploration Worksheet, p. 40 Exploration Worksheet, p. 43
	Lesson 3, Saturation, p. 194	2	1. Explain saturation and solubility. 2. Analyze the effect of temperature on solubility.	Image and Activity Bank 9-3 Transparency Worksheet, p. 45 ▼ Exploration Worksheet, p. 47 Activity Worksheet, p. 48 ▼
	Lesson 4, Solutions in the Environment, p. 197	2	1. Recognize that the atmosphere and the ocean are solutions. 2. Name some solutes that, in excess, can pollute the atmosphere. 3. Correctly use the vocabulary of solutions.	Image and Activity Bank 9-4 Theme Worksheet, p. 49 Resource Worksheet, p. 51 ▼
	Lesson 5, Solutions in You, p. 201	1 to 2	1. Recognize the many roles of water solutions in the body. 2. Explain why blood is both a solution and a suspension.	Resource Worksheet, p. 52
End of Chapter, p. 206				Activity Worksheet, p. 53 Review Worksheet, p. 54 Chapter 9 Assessment Worksheet, p. 57
End of Unit, p. 208				Unit 3 Activity Worksheet, p. 60 ▼ Unit 3 Review Worksheet, p. 62 Unit 3 End-of-Unit Assessment, p. 65 Unit 3 Activity Assessment, p. 72 Unit 3 Self-Evaluation of Achievement, p. 74

* Estimated time is given in number of 50-minute class periods. Actual time may vary depending on period length and individual class characteristics.

▼ Transparencies are available to accompany these worksheets. Please refer to the Teaching Transparencies Cross-Reference chart in the Unit 3 Teaching Resources booklet.

Materials Organizer

Chapter	Page	Activity and Materials per Student Group
7	148	**Exploration 1:** 6 clear plastic cups or 250 mL beakers; 100 mL graduated cylinder; 10 mL graduated cylinder; about 1.5 L of water; stirring rod; 1 mL of powdered drink mix; 10 mL of rubbing alcohol; 5 mL of milk; eyedropper; 1 mL of sugar; 1 mL of vegetable oil; 10 mL of sand; 10 mL of salt; bright light, lamp, or flashlight; piece of cardboard, about 12 cm × 12 cm; sharp pencil or knife; metric ruler; various substances to test, such as 15 mL of instant coffee, bag of tea, 15 mL of chalk dust, 200 mL of a soft drink, 15 mL of salt; several index cards; safety goggles; lab aprons; latex gloves
	154	**What Is Dissolving?** 2 different colors of modeling clay, 1 stick each; 500 mL beaker
	160	**Exploration 2:** 50 mL graduated cylinder; 35 mL of distilled water or other soft water; small cup or 50 mL beaker; 10 mL of tap water; several flakes of basic, unscented hand soap; 2 test tubes with stoppers; test-tube rack; eyedropper; 10 mL of water from each of several local sources; safety goggles
	161	**Exploration 3:** a few crystals each of sugar, sodium chloride, magnesium chloride, calcium chloride, iron chloride, calcium nitrate, and magnesium nitrate; 7 flat toothpicks; 50 mL graduated cylinder; about 200 mL of distilled water; 7 test tubes with stoppers; test-tube rack; a handful of flakes of basic, unscented hand soap; eyedropper; stirring rod; safety goggles; latex gloves; lab aprons
	162	**Exploration 4:** 10 mL graduated cylinder; 20 test tubes or 50 mL beakers; test-tube racks to hold 20 test tubes; stirring rod; small spoon or flat toothpicks; 200 mL of water; a pinch or two each of flour, candle wax, cornstarch, chalk dust, pepper, salt, gelatin, sugar, ground coffee, tea, baking soda, Epsom salts, flavored drink-mix powder, instant-coffee crystals, instant-tea powder, chili powder, and solid watercolor paint; 2 bouillon cubes; 2 aspirin tablets; effervescent tablet; safety goggles
	164	**Exploration 5:** 2 sugar cubes; 250 mL graduated cylinder; 300 mL of water; two 500 mL beakers; stopwatch; optional materials: heat source, stirring rod, safety goggles
8	171	**Exploration 1:** about 20 g of mixture containing equal parts of sand and salt; heat source; beaker of any size; funnel with a piece of filter paper; about 200 mL of water; 1 or 2 stirring rods; several paper towels; tweezers; magnifying glass; metal ring; ring stand; safety goggles; oven mitts
	173	**Distillation:** various materials for distilling water, such as 2 ring stands, heat source, 250 mL flask with stopper, glass tubing, 250 mL beaker, safety goggles, and oven mitts (See Advance Preparation below.)
	177	**Exploration 2:** metric ruler; strip of filter paper or blank newspaper, about 10 cm × 6 cm; 1 L beaker or other wide-mouth container; about 150 mL of distilled water; equipment to hold paper upright in container, such as pencil and 20 cm of thread; several samples of water-soluble ink or felt-tip pens; pencil; clock or watch; safety goggles
	182	**Exploration 3:** 250 mL beaker; 65 g of white sugar; metric balance; 50 mL graduated cylinder; 35 mL of water; 3 drops of maple flavoring; eyedropper; hot plate; stirring rod; candy thermometer, or other thermometer that reads over 100°C; safety goggles; oven mitts
9	186	**Exploration 1:** 250 mL graduated cylinder; large plastic-foam cup; about 10 ice cubes; alcohol thermometer; 150 mL of water; stirring rod
	187	**Exploration 2:** optional materials: 15 mL measuring spoon (1 Tbsp.); 3 juice glasses, plastic cups, or 250 mL beakers; 45 mL of sugar; 15 mL of salt; stirring rod; 100 mL graduated cylinder; about 250 mL of water (additional optional teacher materials: freezer)
	190	***Exploration 4:** about 500 mL each of solutions *A* and *B;* two 500 mL beakers; metric balance; 10 mL graduated cylinder; 2 hard-boiled eggs; 2 test tubes; test-tube rack; several drops of food coloring; clear plastic drinking straw or glass tube; large plastic-foam cup; about 10 ice cubes; metric ruler; alcohol thermometer; stirring rod (additional teacher materials to prepare solutions *A* and *B:* 375 g of salt and 2 L of water for every two student groups)
	192	**Exploration 5:** 25 mL each of 3 salt solutions; 3 large test tubes; test-tube rack; 3 used pencils about 8 cm long and 6 thumbtacks, or 2 plastic drinking straws cut in half, 6 nails, and small ball of modeling clay; metric balance; 25 mL graduated cylinder; metric ruler; scissors (additional teacher materials: 375 g of salt and 3 L of water per class)
	194	**Exploration 6:** about 10 mL of salt; 10 mL of water; 10 mL graduated cylinder; test tube; stirring rod

* You may wish to set up this activity in stations at different locations around the classroom.

Advance Preparation

Exploration 1, page 148: Make sure that the powdered drink mix does not show the Tyndall effect (i.e., it should form a true solution).

Exploration 2, page 160, and Exploration 3, page 161: You may wish to use a cheese grater to make the soap flakes necessary for the soapy water ahead of time.

Distillation, page 173: Before beginning this activity, students will need to submit their procedure and equipment list to you so that you can review them for safety considerations and method.

Exploration 2, page 187: The materials in this Exploration are listed as optional because it is intended as an at-home activity. However, if a freezer is available for your use, you may wish to have students conduct this Exploration in class.

Exploration 4, page 190, and Exploration 5, page 192: To make a concentrated salt solution, dissolve 300 g of salt in 1 L of water. To make a dilute salt solution, dissolve 75 g of salt in 1 L of water.

Unit Compression

There are several ways that this unit can be shortened. First, several of the activities in this unit may be assigned as homework. These include Solutions From A to Z on page 156, A Project: Oil Spills on page 159, A Life-Saving Solution on page 173, and Detective Work at Home on page 179. You may find others that are suitable to be assigned as homework as well.

Some sections of this unit, although they are very interesting and help students to understand the relevance of the material, are not essential to the development of the principal concepts. Thus, you may find it necessary to omit some or all of these sections: A Project: Oil Spills on page 159; Exploration 2 on page 160; and Lesson 3 of Chapter 8, which begins on page 176. In addition, Lesson 5 of Chapter 9, which begins on page 201, may be omitted without affecting the integrity of the unit.

Homework Options

Chapter 7
See Teacher's Edition margin, pp. 150, 155, 157, 158, 160, and 165
Resource Worksheet, p. 9
SourceBook, pp. S52, S53, and S54

Chapter 8
See Teacher's Edition margin, pp. 169, 173, 175, 177, 179, and 183
Graphing Practice Worksheet, p. 25
SourceBook, pp. S56 and S67

Chapter 9
See Teacher's Edition margin, pp. 186, 192, 196, 198, 199, 204, and 206
Exploration Worksheet, p. 38
Activity Worksheet, p. 48
Theme Worksheet, p. 49
Resource Worksheet, p. 51
Resource Worksheet, p. 52
Activity Worksheet, p. 53
SourceBook, pp. S53, S54, S56, and S57

Unit 3
Unit 3 Activity Worksheet, p. 57

Assessment Planning Guide

Lesson, Chapter, and Unit Assessment	SourceBook Assessment	Ongoing and Activity Assessment	Portfolio and Student-Centered Assessment
Lesson Assessment Follow-Up: see Teacher's Edition margin, pp. 153, 161, 164, 171, 175, 179, 182, 187, 193, 196, 200, and 205 **Chapter Assessment** Chapter 7 Review Worksheet, p. 19 Chapter 7 Assessment Worksheet, p. 21* Chapter 8 Review Worksheet, p. 31 Chapter 8 Assessment Worksheet, p. 34* Chapter 9 Review Worksheet, p. 54 Chapter 9 Assessment Worksheet, p. 57* **Unit Assessment** Unit 3 Review Worksheet, p. 62 End-of-Unit Assessment Worksheet, p. 65*	SourceBook Review Worksheet, p. 77 SourceBook Assessment Worksheet, p. 81*	Activity Assessment Worksheet, p. 72* **SnackDisc** Ongoing Assessment Checklists ♦ Teacher Evaluation Checklists ♦ Progress Reports ♦	Portfolio: see Teacher's Edition margin, pp. 159, 170, 173, and 191 **SnackDisc** Self-Evaluation Checklists ♦ Peer Evaluation Checklists ♦ Group Evaluation Checklists ♦ Portfolio Evaluation Checklists ♦

* Also available on the Test Generator software
♦ Also available in the Assessment Checklists and Rubrics booklet

Science Discovery is a versatile videodisc program that provides a vast array of photos, graphics, motion sequences, and activities for you to introduce into your *SciencePlus* classroom. *Science Discovery* consists of two videodiscs: Science Sleuths and the Image and Activity Bank.

Using the *Science Discovery* Videodiscs

Science Sleuths: Green Thumb Plant Rentals #1 Side A

Plants rented by MicroDiscovery (MDY) from Green Thumb Incorporated (GTI) are dying. GTI suspects a rival company of sabotage. The Science Sleuths must analyze the evidence for themselves to determine the true cause of the plant failures.

Interviews

1. Setting the scene: Vice-president of GTI **44935 (play ×2)**

2. Employee of MDY **45710 (play)**

3. Marketing developer of rival plant company **46315 (play)**

4. Botanist **46714 (play)**

5. City water engineer **47294 (play)**

6. The Galloping Gardener **48061 (play)**

Documents

7. GTI technician's site report **48442 (step ×2)**

8. GTI brochure **48446 (step ×3)**

9. Evergreen brochure **48451 (step)**

10. List of all GTI customers **48454 (step ×8)**

11. List of all customers with dying plants **48464 (step ×2)**

12. Advertisement for full-spectrum lights **48468**

Literature Search

13. Search on the words: HERBICIDE, HOUSEPLANTS, MICRODISCOVERY, PLANT DISEASE, RESERVOIR, WELLS **48470 (step)**

14. Article #1 ("Farmers Protest in Favor of Insecticide") **48473 (step)**

15. Article #2 ("Water Hazards") **48476 (step)**

16. Article #3 ("Some Greens to Get the Blues Out") **48479 (step ×2)**

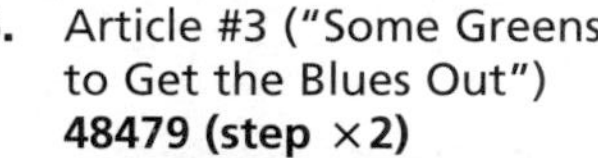

17. Article #4 ("Caring for Houseplants") **48483 (step ×2)**

Sleuth Information Service

18. Map of city **48487**

19. Map of city water supply **48489**

20. Temperature chart **48491**

21. Humidity chart **48493**

22. Fertilizer-composition chart **48495 (step)**

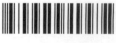

Sleuth Lab Tests

23. Soil-composition analysis **48498 (step ×2)**

24. Water-evaporation analysis **48502 (step)**

25. Redissolved-soil analysis **48505 (step)**

26. Simulation showing the effect of salt water on plants **48508 (step)**

Still Photographs

27. MDY plants **48511 (step)**

28. MDY lighting **48514**

29. Plants at GTI greenhouse **48516**

30. Close-up of MDY plant **48518**

31. MDY plant leaves **48522**

32. GTI plant leaves **48520**

33. MDY plant roots **48526**

34. GTI plant roots **48524**

35. Different kinds of light bulbs **48528 (step ×4)**

◀| Step Reverse   Play ▶      Pause || Step Forward |▶

Image and Activity Bank
Side A or B

A selection of still images, short videos, and activities is available for you to use as you teach this unit. For a larger selection and detailed instructions, see the Videodisc Resources booklet included with the Teaching Resources materials.

7-1 Solutions and Non-Solutions, page 147

Tyndall effect 46878–47389 (Side A only) (play ×2)
The Tyndall effect is the scattering of light by a colloid. Add milk to the water to form a colloid. Light is scattered in a colloid, but it is not scattered in a true solution.

Sunlight on the atmosphere 2248
The shorter wavelengths of visible light are scattered by gases and particles in the atmosphere, which have radii similar to the wavelength of blue light. For this reason the sky appears blue.

Brass horn 2333
Brass is used to make many musical instruments, such as the French horn. Brass is an alloy containing about 20 percent zinc and 80 percent copper by mass.

7-2 Parts of a Solution—Solutes and Solvents, page 154

Salt dissolving 47731–48190 (Side A only) (play ×2)
Salt dissolves in water to form a solution.

Salt dissolving; animation 47390–47730 (Side A only) (play ×2)
A cube of NaCl dissolves in water. Notice how Na dissociates from Cl.

What makes water hard? 2539–2540 (step)
Pure water, such as rain and snow, is known as *soft water*. (step) *Hard water* contains salts such as sodium chloride, calcium carbonate, or magnesium chloride.

7-3 Soluble, Not Very Soluble, or Insoluble, page 162

Drink mix; solution 2512–2516 (step ×4)
Drink powder and water ready for mixing (step) The powder is mixed with the water and stirred. (step ×3)

8-1 The Salt Industry, page 168

Salt solution; evaporated 2524
Salt and other materials are left on the glass after the water in the solution has evaporated.

8-2 Desalination and Distillation, page 172

Desalination plant 2517–2520 (step ×3)
This is a model of a desalination plant in Santa Barbara, California. Desalination removes salt from sea water, making it suitable for consumption. (step) Water is piped in from the ocean to be desalted. (step) The process used to desalt the water is called *reverse osmosis*. Powerful pumps drive the sea water at high pressure through reverse-osmosis membranes, which separate the salt from the water. (step) These water tanks are used for storage.

Oil, crude; distillation 2663
Several products derived from crude oil

Solar distillation 1521–1525 (step ×4)
Materials needed to make a solar distiller for water (step) The sun shines through the glass, heats the water in the pan, and allows it to evaporate. (step) The water condenses on the lower surface of the glass. (step) The water runs down the glass into the pan. (step) The water running into the pan is free of dirt and other material.

Water cycle 1527
Processes involved in the water cycle

8-3 Another Solution, page 176

Observation and inference 472
Statements of observation and inference

Scientific method 473–476 (step ×3)

Chromatography; dyes 34405–36762 (Side B only) (play ×2 then step)
Time-lapse photography shows pigment travel and separation in a chromatography experiment. (step) Test results

8-4 Sugar Solutions, page 180

Osmosis 40186–41568 (Side B only) (play ×2)
A membrane filled with red-dyed sugar solution is weighed and then placed in distilled water. After 3 hours, the bag weighs more. This is because distilled water passes through the membrane into the sugar solution, which has relatively fewer water molecules.

9-1 Getting Colder, page 186

Salt, rock 2521–2522 (step)
Rock salt is loaded into a truck for application onto icy highways. (step) Salt lowers the freezing point of water. Adding salt to icy roads speeds the melting process of the ice.

Solution; salt and freezing 2533
Four trial solutions of salt and water, each with a different concentration, are prepared. Then their freezing times are calculated.

9-2 How Much Solute? page 188

Solution, concentrated 2534
Preparing a concentrated solution

Buoyancy; egg in two solutions 48191–48429 (Side A only) (play ×2)
The egg sinks in water because it is denser than the water. The salt solution is denser than the egg, so the egg floats.

9-3 Saturation, page 194

Fog, ground 1478
Ground fog forms at night as nearly still air at the surface cools. It is most prevalent in valleys and depressions, because colder, denser air tends to sink to these areas at night.

Solubility; effects of temperature 2537
Solubility generally increases with an increase in temperature. The relationship, however, is nonlinear.

9-4 Solutions in the Environment, page 197

Atmospheric composition 1519
Pie chart showing the composition of gases in the Earth's atmosphere

Sea water, composition 1494
The ionic concentration of sea water

◀| Step Reverse  Play ▶  Pause ‖  Step Forward |▶

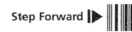

Unit 3 SOLUTIONS

UNIT FOCUS

To introduce the unit, pose the following questions. Accept all reasonable responses.

- What is a mixture? *(A substance made up of two or more identifiable components)*
- What is a solution? *(Students may suggest that a solution is something dissolved in water.)*
- Can there be a solution with no water? *(Yes)* What might such a solution be composed of? *(Students may suggest other liquid solvents.)*
- Can gases form solutions? If so, name one. *(Yes. Soda water is a solution of carbon dioxide gas dissolved in water.)*

A good motivating activity is to let students listen to the English/Spanish Audiocassettes as an introduction to the unit. Also, begin the unit by giving Spanish-speaking students a copy of the Spanish Glossary from the Unit 3 Teaching Resources booklet.

Unit 3

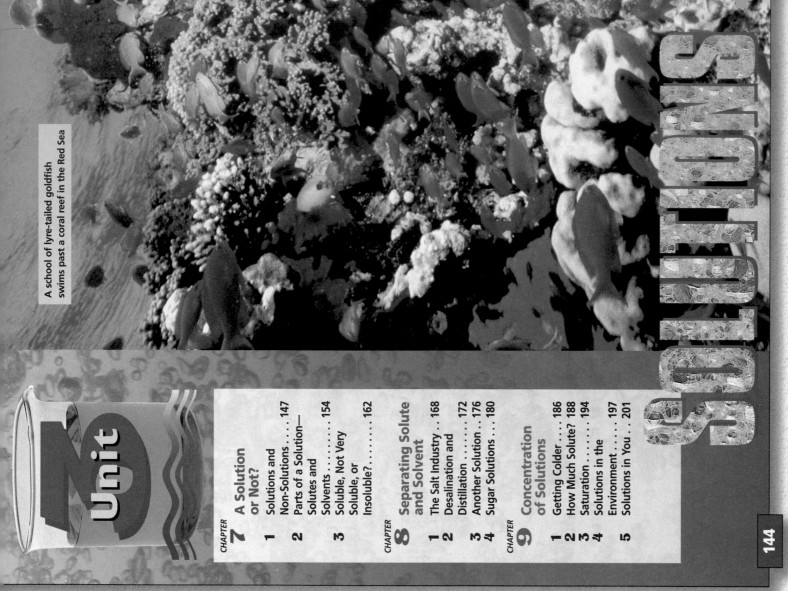

A school of lyre-tailed goldfish swims past a coral reef in the Red Sea

144

Connecting to Other Units

This table will help you integrate topics covered in this unit with topics covered in other units.

Unit 1 Interactions	The many nutrients and gases that are dissolved in the oceans support a complex marine food web.
Unit 2 Diversity of Living Things	The atmosphere is a complex solution that, if altered, can cause a loss of diversity in nature.
Unit 4 Force and Motion	An increase or decrease in solutes alters the density of a solution and thus the buoyancy force of that solution.

As impossible as it may seem, there are more than a million animals in this picture. Can you see them? Look behind the fish and you will notice what looks like a garden of brightly colored plants. Actually, this is a colony of tiny animals called coral polyps. These polyps build their skeletons using calcium absorbed from the sea. When the polyps die, new polyps build on top of their skeletons. Eventually the polyps form long coral reefs (up to 2010 km long). These reefs provide a barrier against powerful waves for fish, clams, lobsters, sea stars, and other ocean species.

Although coral reefs protect a great diversity of living things, the reefs themselves are fragile. Since coral polyps live in shallow water, a rainstorm at low tide can lower the salt content of the water, killing large numbers of polyps. Silt stirred up by a storm can smother them. And pollutants, such as those released by oil spills, can also threaten the coral polyps.

The health of the coral polyps—and therefore the ocean life protected by the coral reef—depends largely on the balance of many substances in the water. Why can't you see these substances, such as salt and calcium, in this photograph? What would it mean if you could see them? (A) Turn the page to dive into Solutions!

145

Using the Photograph

Ocean water is a very complex solution. Calcium is only one of at least 61 elements that are dissolved in ocean water in various forms.

The largest coral reef in the world is the Great Barrier Reef off the northeastern coast of Australia. It is the home of about 400 types of polyps and 1500 species of fish.

You might want to point out to students that rock formations in caves form in a way that is similar to how coral reefs form, although in caves the process does not depend on living organisms. Dripping water slowly deposits calcium on the ceiling (forming a *stalactite*) and on the floor (forming a *stalagmite*). Other minerals in the water contribute to the dramatic coloration of many cave formations. You may also wish to encourage students to observe a saltwater aquarium, if possible. What kinds of organisms are in the aquarium? How do they interact with one another?

Critical Thinking

Have students read the text on page 145. Explain to them that the salt and calcium particles in the water are too small to be seen. Ask: Can you explain how the fish breathe? *(Fish pass water across their gills to absorb oxygen that is also dissolved in the water.)* What other kinds of marine life-forms might live in coral reefs? *(Answers include sea anemones, crabs, and sea turtles.)*

Answer to *In-Text Question*

(A) The substances cannot be seen in the water because they are dissolved. If you could see them, it would mean that the ocean-water solution was saturated.

Connecting to Other Units, *continued*

Unit 5 Structures and Design	Many historical structures are deteriorating because of the action of acidic solutions in the atmosphere.
Unit 6 The Restless Earth	Sedimentary rocks form when water evaporates from a solution, leaving behind a layer of minerals that binds soil particles together.
Unit 8 Growing Plants	The concentrations of the solutions in soil determine the rate of osmosis in plants.

CHAPTER 7 A Solution or Not?

Connecting to Other Chapters

Chapter 7
introduces students to solutions and non-solutions, components of a solution, and the concept of solubility.

Chapter 8
explores methods used to separate solutions, focusing on the separation of sugar and salt solutions.

Chapter 9
introduces students to the concept of saturation and explores solutions in the environment and the human body.

Prior Knowledge and Misconceptions

Your students' responses to the ScienceLog questions on this page will reveal the kind of information—and misinformation—they bring to this chapter. Use what you find out about your students' knowledge to choose which chapter concepts and activities to emphasize in your teaching.

In addition to having students answer the questions on this page, you might want to assign a "free-write" in order to assess their prior knowledge. To do this, instruct students to write for three to five minutes on the subject of solutions, including what they think a solution is and what it is not, as well as some examples of solutions and non-solutions. Tell them to keep their pens moving at all times, writing in a stream-of-consciousness fashion. Emphasize that there are no right or wrong answers in this exercise. It might be best to ask students not to put their names on their papers. Collect the papers, but do not grade them. Instead, read them to find out what students know about solutions, what misconceptions they may have, and what about solutions is interesting to them.

CHAPTER 7 A Solution or Not?

1 What is the difference between these two liquids?

2 Why can't you see sugar after it is stirred into iced tea?

3 Describe each of these substances in terms of its ability to dissolve in water.

ScienceLog

Think about these questions for a moment, and answer them in your ScienceLog. When you've finished this chapter, you'll have the opportunity to revise your answers based on what you've learned.

146

Solutions and Non-Solutions

A Case of Foul Play

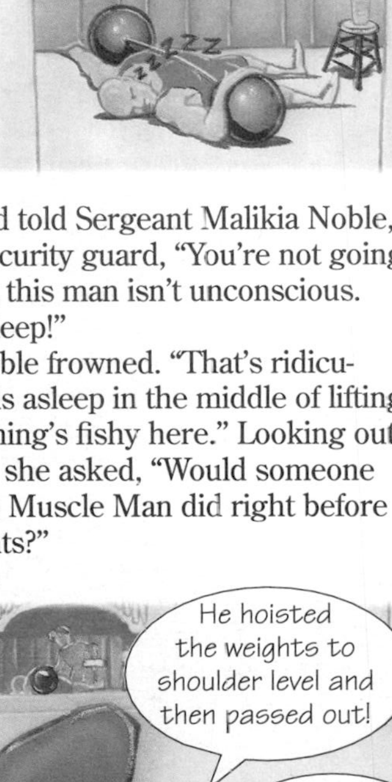

At first, it appeared that the carnival's Muscle Man had passed out from trying to lift too much weight or possibly from the heat.

Binti Sirba, a paramedic, examined the Muscle Man and told Sergeant Malikia Noble, the carnival's security guard, "You're not going to believe it, but this man isn't unconscious. He's just fast asleep!"

Sergeant Noble frowned. "That's ridiculous. No one falls asleep in the middle of lifting weights. Something's fishy here." Looking out at the audience, she asked, "Would someone tell me what the Muscle Man did right before lifting the weights?"

A boy spoke up. "Sure. He bragged that he was going to lift more weight than we'd ever seen anyone lift before. He gulped down a bunch of whatever's in that cup there, took a few deep breaths, lifted the weights to shoulder level, and then boom! Down he went." Others nodded. "Yep. Out like a light."

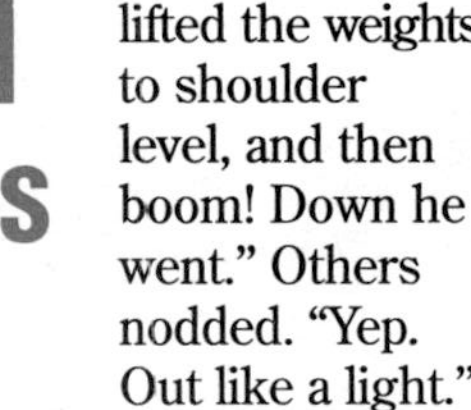

It worked! Now maybe I'll finally get to be the Muscle Man!
LEMONADE

After hearing this, Ms. Sirba carefully examined the Muscle Man's drink. It was yellowish and clear, and it smelled like lemons.

"That's lemonade from the stand across the way," Sergeant Noble told Ms. Sirba. "He drinks it all the time, although it's not very tasty. It's that powdered lemonade mix you add to water."

"Hmm," Ms. Sirba said. "That helps, but I still need to make one more test." She put some of the lemonade into a clear glass, shook it, and held it up to the light. "This isn't a pure lemonade solution," she concluded. She held it up for the security guard to see. "My guess is that someone's added sleeping powder to the lemonade in that cup."

Chemical tests at a forensic lab showed that a potent dose of a sleeping powder had been added to the Muscle Man's lemonade. Thanks to the paramedic's astute observations, further investigations were carried out, and the man who operated the lemonade stand was later convicted of the crime.

LESSON 1 ORGANIZER

Time Required
2 to 3 class periods

Process Skills
comparing, observing, predicting, classifying

New Terms
none

Materials (per student group)
Exploration 1: 6 clear plastic cups or 250 mL beakers; 100 mL graduated cylinder; 10 mL graduated cylinder; about 1.5 L of water; stirring rod; 1 mL of powdered drink mix; 10 mL of rubbing alcohol; 5 mL of milk; eyedropper; 1 mL of sugar; 1 mL of vegetable oil; 10 mL of sand; 10 mL of salt; bright light, lamp, or flashlight; piece of cardboard, about 12 cm × 12 cm; sharp pencil or knife; metric ruler; various substances to test, such as 15 mL of instant coffee, bag of tea, 15 mL of chalk dust, 200 mL of a soft drink, 15 mL of salt; several index cards; safety goggles; lab aprons; latex gloves

Teaching Resources
Exploration Worksheet, p. 3
SourceBook, p. S52

Solutions and Non-Solutions

FOCUS

Getting Started

Before class, set up two beakers to test for the Tyndall effect (as shown on page 150). Fill both beakers with water, and add 5 drops of food coloring to one and 5 drops of milk to the other. With students present, darken the room and turn on the flashlights. The beam of light should be visible only in the mixture of milk and water. Explain that failure to demonstrate this effect is one of the characteristics of solutions.

Main Ideas

1. Most solutions are clear, liquid mixtures.
2. The parts of a solution are spread uniformly throughout the mixture.
3. The Tyndall effect is a phenomenon that can be used to identify a solution; no true solution shows the Tyndall effect.

TEACHING STRATEGIES

A Case of Foul Play

The story on page 147 sets the stage for Exploration 1. Call on a volunteer to read it aloud. Ask students to make preliminary evaluations of the situation described, based on the facts given.

SAFETY TIP Explain to students that poison-control centers provide advice over the phone for treating emergencies that involve poisoning. Phone numbers for such centers are usually located inside the front covers of phone books. The caller will usually be asked to identify the poison, when the poisoning occurred, how much of the poison was ingested or inhaled, and the age of the victim.

EXPLORATION 1

Before beginning Exploration 1, make sure that the powdered drink mix does not show the Tyndall effect (i.e., it should form a true solution).

PART 1

Divide the class into small groups and distribute the materials. Encourage each group member to prepare at least one mixture. The graduated cylinder and stirring rod should be cleaned after the preparation of each mixture.

Be sure students understand that all examples in Part 1 are mixtures; however, only the mixtures of powdered drink mix and water, rubbing alcohol and water, and sugar and water are solutions. Emphasize that in contrast to solutions, the other mixtures are formed by two or more substances that remain identifiable when blended together. The most obvious example of this is the mixture of sand and salt.

Cooperative Learning
PART 1

Group size: 2 to 3 students
Group goal: to distinguish solutions from non-solutions
Positive interdependence: Assign each student a role such as group leader, materials organizer, or data recorder.
Individual accountability: Have each student describe in detail each mixture that his or her group prepared and explain whether it is a solution or non-solution and why.

Answers to
Part 1

Explanations of the differences between solutions and non-solutions will vary. Accept all reasonable responses. In general, the difference between the two groups of mixtures is that all of the solid particles in the solutions have dissolved. In the non-solution group, the particles have not dissolved and remain distinguishable.

The main characteristics of a liquid solution are that (1) it is a mixture, (2) light can pass through it, (3) the substances are distributed uniformly, and (4) the individual particles in the solution are indistinguishable.

EXPLORATION 1

Solution or Non-Solution?

How do you think I was able to distinguish the non-solution in the Muscle Man's cup from a pure solution?

PART 1

To tell a solution from a non-solution, try this experiment.

You Will Need
- 6 clear plastic cups or beakers
- a 100 mL graduated cylinder
- a 10 mL graduated cylinder
- the mixtures listed in the What to Do section
- a stirring rod

What to Do
In separate beakers, prepare these six mixtures. Stir each one well to ensure that it is completely mixed.

Mixture 1:
Add 1 mL of powdered drink mix to 100 mL of water.
Mixture 2:
Add 10 mL of rubbing alcohol to 100 mL of water.
Mixture 3:
Add 2 or 3 drops of milk to 100 mL of water.
Mixture 4:
Add 1 mL of sugar to 100 mL of water.
Mixture 5:
Add 1 mL of oil to 100 mL of water.
Mixture 6:
Add 10 mL of sand to 10 mL of salt.

Now separate the mixtures into two groups: those that are solutions (Mixtures 1, 2, and 4) and those that are not solutions (Mixtures 3, 5, and 6). Examine the two groups carefully. What differences do you see? Based on your observations, how will you know a solution when you see one?

Suppose that you have been asked to appear in court as the scientific expert in the trial of the man who operated the lemonade stand. In giving evidence before the court, you are asked how to tell a solution from a non-solution. To illustrate your answer, you decide to show the court the six mixtures and explain the differences between the three solutions and the three non-solutions. Prepare your explanation in writing.

On the next page is a memorandum from Sergeant Noble to the forensic laboratory. How does your definition of a solution compare with Sergeant Noble's?

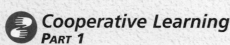

Meeting Individual Needs

Second-Language Learners

Make certain that students understand the new terms presented throughout this unit. For example, have students look up definitions for the words *mixture, clear,* and *insoluble*. Students can add these definitions to their ScienceLog. Discuss with students any definitions that they do not understand. Encourage students to add to their list of vocabulary words as they proceed through the unit.

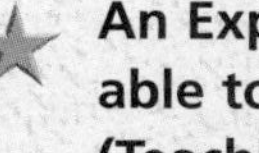

An Exploration Worksheet is available to accompany Exploration 1 (Teaching Resources, p. 3).

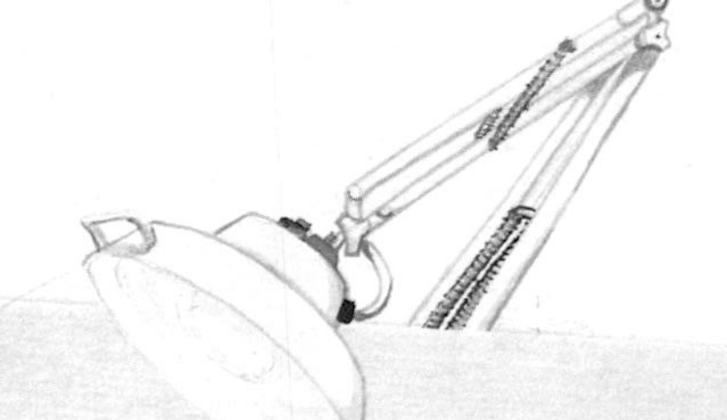

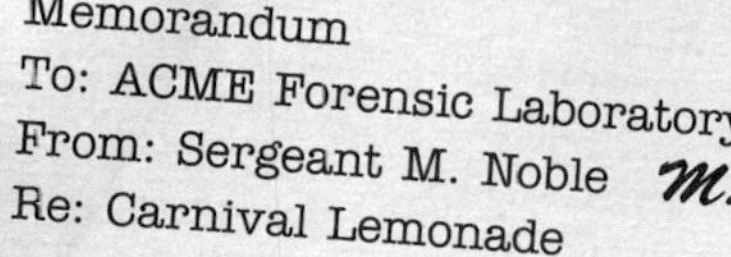

Memorandum
To: ACME Forensic Laboratory
From: Sergeant M. Noble *M. Noble*
Re: Carnival Lemonade

After a preliminary examination of the crime scene, it appeared that the victim had merely fallen asleep on the job. However, further investigation showed that the lemonade the carnival employee had consumed was not a pure solution of lemonade powder and water. This led us to suspect criminal activity.

Anyone with even a little background in chemistry knows that a solution is a mixture of two or more things but that not all mixtures are solutions. In a solution, one of the materials becomes hidden so that the mixture appears to be only one thing. You might say that one part becomes camouflaged. Solutions are always clear. Sometimes, as in the case of some kinds of food coloring and water, a solution can be colored and clear.

When the lemonade was held up to the light, it showed the Tyndall effect. This test provided evidence that the lemonade was not a pure solution. I suspect foul play and recommend that you test the enclosed sample to find out what was added to the lemonade consumed by the carnival employee.

Exploration 1 continued ►

EXPLORATION 1, *continued*

Since students will be appearing as scientific experts in the trial of the lemonade man, suggest that they review their written explanations on the differences between solutions and non-solutions and prepare to discuss them. Call on volunteers to share their explanations.

INDEPENDENT PRACTICE As an alternative, you may want to organize a classroom dramatization. Students could work in groups to study the expert testimony on page 149 and then develop a set of questions to ask the witnesses in courtroom-style proceedings. You may choose a student to serve as a judge or moderator. Other students could play the roles of Sergeant Noble, paramedic Binti Sirba, and attorneys for each side.

Exploration 1 continued ►

Meeting Individual Needs

Gifted Learners

The use of expert witnesses, especially scientists, is increasing in courts and legislatures throughout the United States. For example, scientific evidence is often used to help make decisions about environmental and health issues. Have students locate and discuss newspaper articles in which the testimony of expert witnesses has played a role. Students should write a summary of the information in the articles or act out a dramatization of the courtroom proceedings.

EXPLORATION 1, *continued*

Have students read the information about the Tyndall effect on page 150. This information describes and explains the phenomenon known as the Tyndall effect, in which a liquid that contains large, suspended particles will scatter or reflect the path of a beam of light passing through it. As a result, the boundaries of the beam of light become visible. An example of the Tyndall effect is shown in the illustration at the bottom of page 150. True solutions do not exhibit the Tyndall effect (shown in the illustration on the bottom of page 151) because the light travels through the mixture without being reflected or scattered.

Homework

Have students write a paragraph explaining in their own words why the Tyndall effect is a good test to determine whether a mixture is a true solution. *(If light passing through a mixture is scattered, then the particles that are scattering the light are too big to be truly dissolved.)*

One section of a reference book in the forensic lab deals with the Tyndall effect. Page 18 of the reference book describes and explains it, and page 19 tells how to test for it. Study these pages and then test Mixture 3—milk and water—with the solution tester shown on the bottom of the pages.

THE TYNDALL EFFECT

Description
Some mixtures that appear to be solutions may prove not to be solutions after all. If the path of a bright light shining through a mixture can easily be seen, then that mixture is not a solution. In other words, the path of light passing through a solution is not visible. No matter what color they are, true solutions made of liquids are clear.

Explanation
A non-solution contains particles large enough to scatter or reflect light, showing the path of the light as it passes through the mixture. This scattering is called the Tyndall effect. True solutions do not exhibit the Tyndall effect.

A non-solution

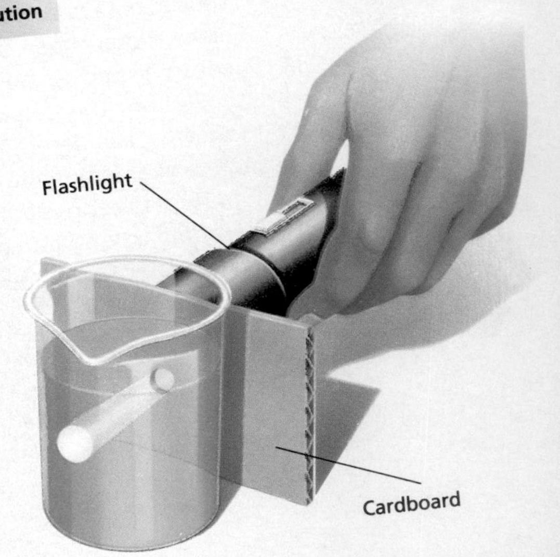

18

TESTING FOR THE TYNDALL EFFECT

Materials

- a bright light, lamp, or flashlight
- a piece of cardboard
- a pencil or knife
- the mixture to be tested, in a clear container
- a metric ruler

Making the Solution Tester

Using a pencil or knife, make a small, neat hole in the piece of cardboard. The hole should be no more than 0.5 cm in diameter. Place the cardboard over the light source so that the only light you can see comes through the hole.

The Test

Press the light and the cardboard tightly against the container holding the mixture. If you can see the path of the light through the mixture, then it is not a true solution. The test should be done in a darkened room.

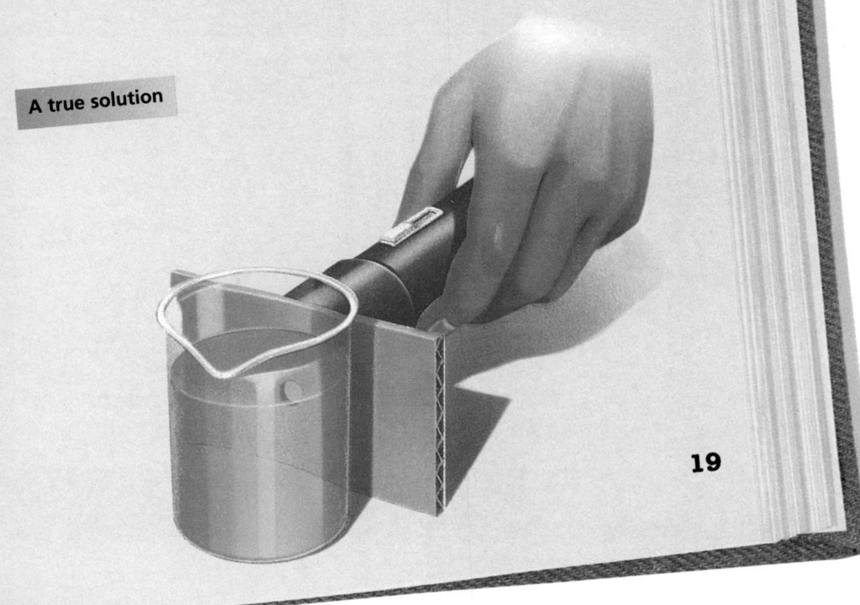

A true solution

19

Exploration 1 continued ►

EXPLORATION 1, continued

GUIDED PRACTICE Direct students' attention to the illustration at the bottom of page 151. It shows the results of testing a true solution. Ask: What does the illustration show? *(A light source passing through a clear liquid without any visible traces of its path)* Then have students look back at the illustration on page 150 that shows the results of testing a non-solution. Ask: Why is the path of the light visible in the non-solution? *(The non-solution has larger particles that scatter and reflect the light.)*

Before continuing to page 152, have students review their data from Part 1 and predict the results of testing the liquid mixtures from Part 1 for the Tyndall effect. *(The mixtures of powdered drink mix and water, rubbing alcohol and water, and sugar and water should show results similar to that for the true solution on page 151. The mixtures of milk and water and of oil and water should show results similar to that for the non-solution on page 150.)*

Relate the information on these pages to the memorandum from Sergeant Noble by asking the following questions:

- Why did Sergeant Noble test for the Tyndall effect? *(To find out whether the liquid in the cup was a pure solution of powdered lemonade)*
- If the results of the test for the Tyndall effect had been negative, would that have proven that the liquid was a pure powdered lemonade solution? Why or why not? *(No; the substance added to the lemonade could have been soluble and could have given a negative result for the Tyndall effect.)*

Exploration 1 continued ►

EXPLORATION 1, continued

PART 2

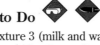

Students should be warned not to ingest the coffee, tea, soft drink, or salt solution.

Divide the class into the cooperative-learning groups that were formed for Part 1, and distribute the materials for this part of the Exploration. Have groups read page 152 and perform Part 2. The following are some hints for obtaining the best results:

- A darkened room is helpful when testing for the Tyndall effect.
- All mixtures should be diluted so that they are clear and nearly colorless.
- Use the purest water possible.
- Glassware should be as clean as possible.

Be sure that students understand that the Tyndall effect is the actual path of the light shown by the particles in the mixture. The reason that the path of light can be seen in a non-solution is that the particles are large enough to scatter the light; the particles of a true solution are too small to scatter light.

Answers to *Part 2*

Solutions: sugar and water, coffee and water, tea and water, salt and water, and the soft drinks (if they are flat or still in the sealed bottle)

Non-solutions: freshly opened soft drinks, milk and water, and chalk dust and water

Note that freshly opened soft drinks are not true solutions because the mixture is heterogeneous where the bubbles are formed. (The bubbles are pockets of pure gas.) Freshly opened soft drinks might display the Tyndall effect because the pockets of gas could scatter the light. Soda in a sealed bottle and flat soda have an equilibrium concentration of gas in solution. (This can be shown by heating the flat soda and forcing out more gas bubbles.)

Answer to *Caption*

Ⓐ Because milk is not a true solution, its particles will reflect light. Therefore, when Scott turns on the flashlight, the path of the light will be visible in the mixture.

PART 2

What to Do

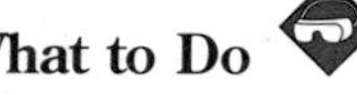

Test Mixture 3 (milk and water) with the solution tester. Can you see the light path through it? Now test Mixture 4 (sugar and water).

Next, repeat the test with other mixtures. Some that you can test include instant coffee and water, tea and water, chalk dust and water, soft drinks, and salt and water. To make your mixtures, stir about 1 g of a solid or several milliliters of a liquid into a beaker of water. Which mixtures are true solutions?

What will happen when Scott turns on the flashlight? Ⓐ

Detective Work

Are all of the mixtures pictured below solutions? Discuss your answers with a classmate. (Hint: Water is present in all of the beakers.)

Solution Facts

You will come across many interesting facts about solutions and non-solutions. Perhaps your class could set aside some space on a bulletin board for "Solution Facts." Use index cards to record any interesting facts about solutions or non-solutions, and post them on the bulletin board.

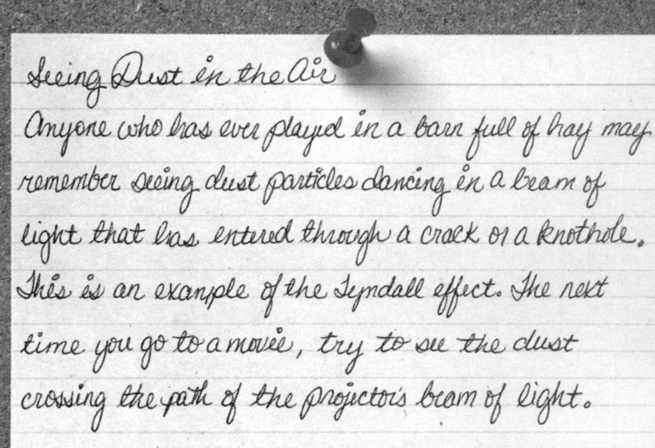

Seeing Dust in the Air
Anyone who has ever played in a barn full of hay may remember seeing dust particles dancing in a beam of light that has entered through a crack or a knothole. This is an example of the Tyndall effect. The next time you go to a movie, try to see the dust crossing the path of the projector's beam of light.

Detective Work

You may wish to set up the three beakers as shown in the illustration on page 152. Place water in one; food coloring and water in the second; and food coloring, water, and a drop of milk in the third. Ask: Are all of these solutions? How can you find out? *(You could use the solution tester.)*

Students should recognize that the third (red) substance cannot possibly be a solution because it is not clear. Point out that the first two substances, although clear, cannot automatically be classified as solutions—each could be a single substance, such as water or alcohol.

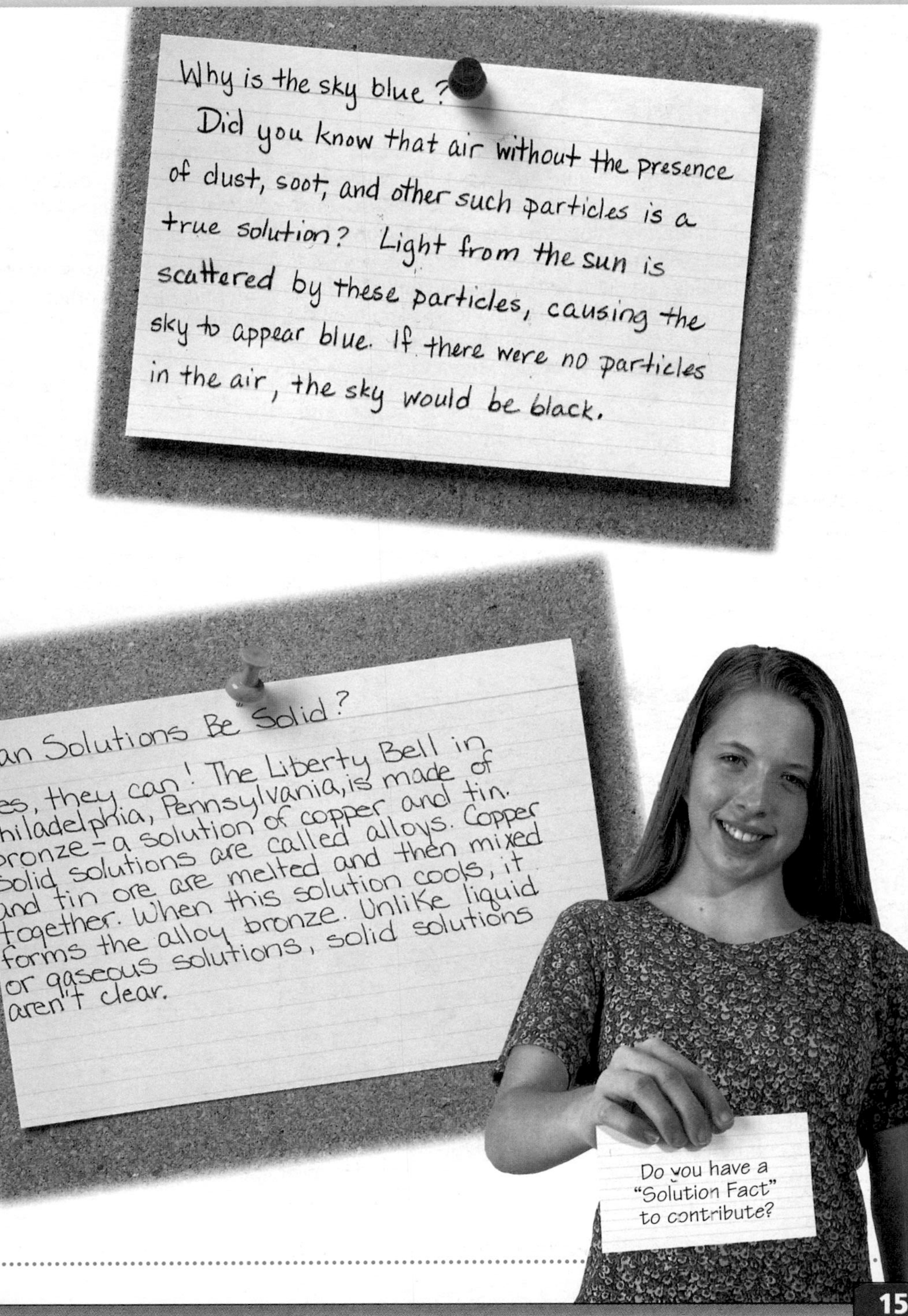

FOLLOW-UP

Reteaching

Many substances in the students' everyday lives are solutions. Ask them to make a list of five such substances and explain why they think each is a solution. *(Examples include cleaning solutions such as bleach and detergent, clear drinks such as juice, and cosmetic solutions such as perfume. Students should explain that these are all clear mixtures.)*

Assessment

Present students with the following scenario: You are describing to someone who is several years younger than you why vinegar is a solution. What would you say? *(Students should include a basic definition of a solution, indicate that vinegar is a clear mixture of several components, and describe how vinegar could be tested to prove that it is a solution and not just a mixture.)*

Extension

Have students investigate alloys—what they are, how they are made, and why they are useful. Examples include bronze, stainless steel, and sterling silver.

Closure

Have students form pairs. Give several index cards to one student from each pair. Each index card should have a word from the lesson printed on it. Have students take turns challenging each other to a game of twenty questions using the words on the cards.

Solution Facts, *page 152*

The purpose of this exercise is to introduce new information and expand students' understanding of what a solution is. Solutions can be combinations of gases and gases, liquids and gases, liquids and solids, liquids and liquids, or solids and solids. Examples of these combinations are given in the table on page 154. All solutions have a homogeneous composition, and all solutions except solids and solid mixtures are clear.

Have students create a bulletin-board display of their own solution facts. Encourage them to include drawings or magazine photographs showing examples of solutions to make their displays more interesting.

LESSON 2

Parts of a Solution—Solutes and Solvents

FOCUS

Getting Started

The Discrepant Event Worksheet on page 8 of the Unit 3 Teaching Resources booklet describes a teacher demonstration that makes an excellent introduction to this lesson.

Main Ideas

1. Solutions can be different combinations of gases, liquids, and solids.
2. In solutions, the substance that is in greater quantity is the solvent, and the substance that is in lesser quantity is the solute.
3. Hard water is a solution containing iron, calcium, and magnesium compounds.

TEACHING STRATEGIES

Involve students in a discussion of the table on this page. Ask them to suggest other examples of each combination. *(Examples of solutions include the following: solid and liquid—sugar and water; gas and liquid—air and water; liquid and liquid—antifreeze and water; solid and solid—iron and carbon [steel]; gas and gas—nitrous oxide and oxygen [laughing gas])*

Teaching Strategies for Naming the Parts of a Solution are on the next page. ▶

★ **Transparency 25 accompanies the material on this page.**

LESSON 2

Parts of a Solution—Solutes and Solvents

What Is Dissolving?

What happens when you mix substances to make a solution? According to the particle model of matter, all matter is made up of particles. When some solids and liquids are mixed together, the particles making up the solid separate and spread evenly throughout the liquid, occupying the spaces between the liquid particles. This process is called **dissolving**.

In the model below, clay balls are used to represent sugar and water before and after the sugar is dissolved in the water. You may wish to make this model yourself. How does this model explain the following observations about a solution of sugar and water? Ⓐ

- The volume of sugar and water before mixing is greater than the volume of the solution after mixing.
- The masses of sugar and water before mixing are equal to the mass of the solution after mixing.
- One part of the solution is identical to any other part of the solution.

In reality, we cannot see either the particles of water or the particles of sugar, even when the solution is tested for the Tyndall effect.

Naming the Parts of a Solution

Earlier you prepared a solution of sugar and water. Sugar is a solid, and water is a liquid. You also prepared a solution of rubbing alcohol and water—both liquids. Other possible combinations are listed in the table at right. For each combination, an example is given. Can you provide another example of each combination?

Some combinations	Example	Your example
solid and liquid	salt water	?
gas and liquid	soft drinks	?
liquid and liquid	rubbing alcohol and water	?
solid and solid	bronze (a solution of copper and tin)	?
gas and gas	air	?

LESSON 2 ORGANIZER

Time Required
4 class periods

Process Skills
analyzing, classifying, comparing, inferring

Theme Connection
Structures

New Terms
Dissolving—the process in which a solute separates into particles that spread evenly throughout a solvent
Hard water—water that contains certain types of solutes and will not form many suds when mixed with soap
Soft water—water that is mostly free of dissolved matter
Solute—a substance that is dissolved in a solvent to make a solution
Solvent—a substance that dissolves a solute to make a solution

Materials (per student group)
What Is Dissolving? 2 different colors of modeling clay, 1 stick each; 500 mL beaker

continued ▶

If a solid or a gas is dissolved in a liquid, the liquid is usually the **solvent**. The solid or gas is the **solute**. What happens when you have other combinations, such as two liquids, two gases, or two solids? In these cases, the substance present in a *greater* quantity is the solvent. The substance present in a *lesser* quantity is the solute.

The *solvent* dissolves the *solute*.
The *solute* dissolves *in* the *solvent*.

In these examples, which part is the *solute* and which part is the *solvent*? B

Stir
Water
Sol—?
Sol—?
Salt
SPARKLING
WATER
Gas and water
Oil
OIL
Shake after mixing
Gasoline
A Solution Fact
The particles in a solution can't be seen even with a microscope. If a spoonful of sugar is dissolved in a glass of water, there are approximately 20,000,000,000,000,000,000,000 particles of sugar and 6,000,000,000,000,000,000,000,000 particles of water. Amazing!

ORGANIZER, *continued*

Exploration 2: 50 mL graduated cylinder; 35 mL of distilled water or other soft water; small cup or 50 mL beaker; 10 mL of tap water; several flakes of basic, unscented hand soap; 2 test tubes with stoppers; test-tube rack; eyedropper; 10 mL of water from each of several local sources; safety goggles
Exploration 3: a few crystals each of sugar, sodium chloride, magnesium chloride, calcium chloride, iron chloride, calcium nitrate, and magnesium nitrate; 7 flat toothpicks; 50 mL graduated cylinder; about 200 mL of distilled water; 7 test tubes with stoppers; test-tube rack; a handful of flakes of basic, unscented hand soap; eyedropper; stirring rod; safety goggles; latex gloves; lab aprons

Teaching Resources
Discrepant Event Worksheet, p. 8
Resource Worksheet, p. 9
Exploration Worksheets, pp. 11 and 13
Transparencies 25–27
SourceBook, p. S53

Naming the Parts of a Solution, *pages 154–155*

GUIDED PRACTICE Students may need help with the idea that the particles in a solution are so small that they cannot be seen even with a microscope. To illustrate this, dissolve a single crystal of potassium permanganate in a beaker of water. The individual particles cannot be seen, but their presence is indicated by the color change.

Answers to *In-Text Question and Caption, pages 154–155*

A
- Because the dissolved sugar particles occupy spaces between the water molecules, the volume of the solution will be less than that of the unmixed sugar and water.
- The sugar and water have the same mass in solution as they do unmixed because just as much matter is present.
- Because the particles are spread evenly throughout the solution, any part of the solution has the same number of particles in it.

B In the beaker, the water is the solvent and the salt is the solute. In the bottle, the water is the solvent and the gas (carbon dioxide) is the solute. In the gasoline container, the gasoline is the solvent and the oil is the solute.

Meeting Individual Needs

Second-Language Learners
Have students create a scrapbook of examples of solutions and mixtures. Suggest that they include examples from publications in their native language, if available.

Homework

Have students explain how they could use a beaker, a bag of large marbles, and a bag of small marbles to explain how a solute dissolves. *(Students should arrange the small marbles between the large marbles to represent how particles of the solute fill the spaces between the particles of the solvent.)*

Solutions From A to Z

Here are some situations that involve real solutions. Read through them with a classmate. For each situation, do the following:

Type of solution	Situations (by letter)
solid and liquid	c, h, etc.
gas and liquid	
liquid and liquid	
solid and solid	
gas and gas	
other (specify)	

1. *Identify the type of solution* (solid in a liquid, liquid in a liquid, etc.). In your ScienceLog, enter this information in a table similar to the one at right.
2. *Identify the solvent and solute in each solution.* Enter this information in your ScienceLog in a table similar to the one on page 157. Again, remember to write only in your ScienceLog.
   a. Fish breathe oxygen, which is dissolved in water.
   b. A 100 g sample of vinegar is made up of 5 g of liquid acetic acid and 95 g of water.
   c. A clear mixture of baking soda and water often helps an upset stomach.
   d. Brass is an alloy that can be made by mixing 65 parts of molten copper with 35 parts of molten zinc.
   e. Alcohol is added to gasoline to make gasohol.
   f. The antifreeze used in a car radiator is a mixture of liquid ethylene glycol and water. The ethylene glycol is present in the greater quantity.
   g. On a windy day, snow and ice disappear into the air without melting.
   h. Sugar, lemon juice, and water make up lemonade.
   i. Rubbing alcohol is a good de-icer for locks because it can dissolve water drops frozen in the keyhole.

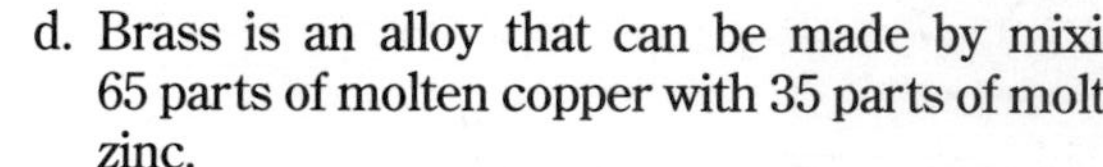
Brass is a solution composed of two metals.

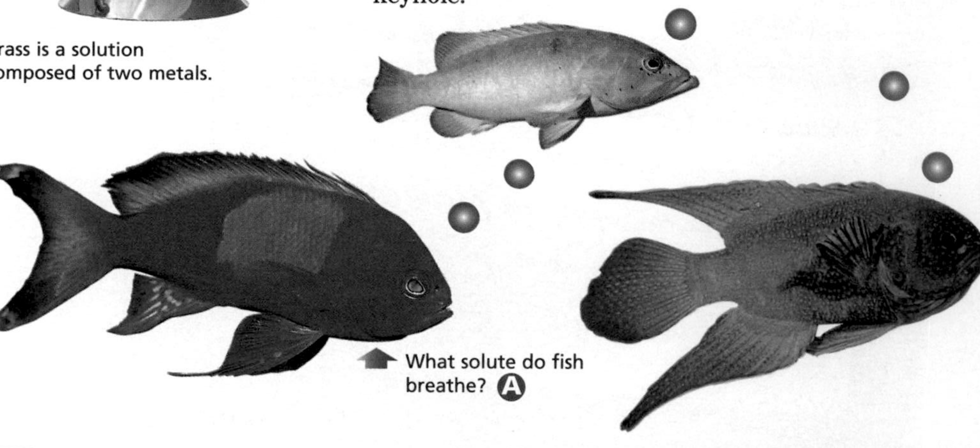
What solute do fish breathe? (A)

Solutions From A to Z

Before students begin this activity, ask them to read through the list of solutions on pages 156–157. They should add any unfamiliar words to their ScienceLog and then look up the definitions of those words. *(Examples include alloy, molten, pulp, DDT, and sparkling water.)*

Divide the class into pairs, and allow them to work through the questions. When students have finished, discuss their answers as a class, and resolve any differences of opinion.

Answers to *Solutions From A to Z*

1.

Type of solutions	Situations (by letter)
solid and liquid	c, h, i, n, p, r, t, u, v, w, y
gas and liquid	a, j, m, o, u
liquid and liquid	b, d, e, f, h, k, v, z
solid and solid	d, z
gas and gas	g, l, q, s, x

Answers may vary; students should be prepared to explain their responses. Note that (d), (z), (h), (u), and (v) have been listed twice. In (d) and (z), the solutions form as liquids and then cool to create solid solutions. In (h), both sugar (a solid) and lemon juice (a liquid) are combined with water to make lemonade. In (u), the effervescent tablet forms both a gas (carbon dioxide) and a solid that dissolve in water. In (v), some insecticides are in solid form when dissolved in water, and others are in liquid form. There may be some confusion when a phase change precedes the solution formation. For example, in (g) and (x), the solid first changes into a gas, and then the two gases form a solution.

2.

Situation	Solute	Solvent
a.	oxygen	water
b.	acetic acid	water
c.	baking soda	water
d.	zinc	copper
e.	alcohol	gasoline
f.	water	ethylene glycol
g.	water vapor (from snow and ice)	air
h.	sugar and lemon juice	water
i.	ice	rubbing alcohol
j.	carbon dioxide	water
k.	food coloring	water
l.	discharged gas	air
m.	sulfur dioxide	rainwater
n.	minerals in soil	acid rain
o.	carbon dioxide	rainwater
p.	limestone	weak acid
q.	water vapor	air
r.	sodium chloride and potassium chloride	ocean water
s.	oxygen gas	nitrogen gas
t.	table salt	water
u.	carbon dioxide and a solid	water
v.	insecticides	runoff water
w.	DDT	ocean water
x.	iodine vapor	air
y.	vitamins	water
z.	copper	gold

In some examples, such as (o), (p), and (u), a chemical change occurs. The resulting mixture is still a solution.

j. When you open a bottle of sparkling water, bubbles of gas suddenly appear throughout the water and rise to the top.
k. A few drops of food coloring can change the color of water.
l. Pulp mills discharge a gas that spreads through the air. The gas can be smelled kilometers away.
m. Sulfur dioxide in the air mixes with water droplets in the atmosphere to form acid rain.
n. Minerals in the ground can be dissolved by acid rain.
o. Rain falling on the ground mixes with carbon dioxide from decaying plants and the roots of plants. The result is the formation of a weak acid.
p. In some places, this weak acid dissolves limestone to make limestone caves.
q. Water from lakes and streams is continually evaporating into the air.
r. Clear ocean water contains a considerable amount of sodium chloride and potassium chloride (two types of salt).
s. Air is made up of almost four parts nitrogen gas to one part oxygen gas.
t. A bit of table salt in water helps heal cuts.
u. When an effervescent tablet is added to water, it fizzes and then becomes clear. The solution tastes a little salty.
v. Runoff water from some farms and forests contains insecticides.
w. There are traces of DDT in ocean water, even in the Arctic and Antarctic Oceans.
x. If iodine crystals are dropped into a dry, warm container, a purple vapor suddenly appears and spreads evenly through the air in the container.
y. When foods are boiled, some of their valuable vitamins are lost in the process.
z. Gold jewelry usually contains some copper to make it stronger and less expensive.

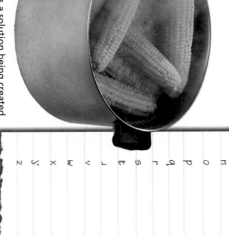

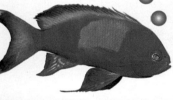

Sparkling water is one solution you can drink.

Situation	Solute	Solvent
a		
b		
c	baking soda	water
d		
e		
f		
g		
h		
i		
j		
k		
l		
m		
n		
o		
p		
q		
r		
s		
t		
u		
v		
w		
x		
y		
z		

How is a solution being created here? B

Theme Connection

Structures

Focus question: How do the sizes of particles relate to the formation of solutions? *(Students should understand that solutions form when the particles of a solute fill in the empty spaces between the particles of a solvent. Because of the small size of the particles and because these particles are uniformly mixed, the individual particles of the components of a solution are not visible. Also, the particles of the solute must be small enough to form a solution.)*

Meeting Individual Needs

Gifted Learners

Explain to students that sometimes when a solute is dissolved in a solvent, the temperature of the resulting solution will be either higher or lower than the temperatures of the starting solute and solvent. Suggest that students research how hot and cold packs make use of this property of solutions.

Homework

The Resource Worksheet on page 9 of the Unit 3 Teaching Resources booklet makes an excellent homework activity to accompany Solutions From A to Z. If you choose to use this worksheet in class, Transparencies 26 and 27 are also available for your use.

Answers to Captions, pages 156–157

A Fish breathe oxygen dissolved in water.

B As the corn is cooked, vitamins are gradually dissolved in the water.

Did You Know...

Many lightweight objects can float on water because of surface tension. Surface tension can be reduced by mixing soap with the water. You can demonstrate this by carefully floating a needle on the surface of a cup of water. Without disturbing the water, slowly add soap or liquid detergent. The needle will sink. (For a detailed explanation, see the margin of page 190.)

Meeting Individual Needs

Learners Having Difficulty

Have students explain what happens when you open a can of soda or other container containing carbonated water. *(The high pressure in the unopened can keeps carbon dioxide dissolved in the water. When the can is opened and the pressure is released, the carbon dioxide accumulates into bubbles that rise to the surface and escape into the air. When the bubbles form, the solution becomes a non-solution.)*

Water: Waste Not, Want Not

Call on a volunteer to read aloud the first four paragraphs of page 158. Involve the class in a discussion of water and its many uses. Help students to realize how precious a resource water is. Ask them to think of substances that pollute the water supply in their area. Then ask them to identify which of these substances dissolve in water. Students can use a table similar to the one on the right side of page 158 to record their findings. *(Sample answers include fertilizer and petroleum products.)*

INDEPENDENT PRACTICE As an extension to this discussion, ask students to do some research to discover where waste water goes in their community. Then they can find out whether any of the wastes from the water are recycled.

Other Solvents Besides Water

Suggest that students copy the table at the bottom of page 158 into their ScienceLog and then add to the list of solvents. Ask them to follow up by doing library research to discover which of the solvents in their table are dangerous.

GUIDED PRACTICE

One interesting demonstration involves placing two watch glasses on an overhead projector. Put some iodine crystals on each watch glass. Add a few drops of water to one watch glass and a few drops of alcohol to the other. Ask students to describe what they observe. *(The iodine will be very soluble in alcohol and not very soluble in water.)* For disposal, add 1/10 molar sodium thiosulfate solution until the iodine solution becomes colorless, then dilute with water to ten times the original volume and pour down the drain.

Answers to *In-Text Questions*

(A) Some sewage-treatment plants can process waste into a sludge that is suitable for agricultural and industrial uses, but this is not widely done at the present time.

(B) The following items could be added to the table of solvents and their uses:

- Dry-cleaning fluid dissolves grease and other dirt on clothing.
- Mineral spirits dissolve oil on axles and bicycle chains.
- Silver dissolves mercury in order to make fillings for teeth.
- Molten cryolite dissolves bauxite during the process of extracting aluminum.
- Gasoline dissolves lead compounds in leaded gasoline.
- Wax dissolves colored dyes in colored candles.

Water: Waste Not, Want Not

Did you notice that water was involved in many of the solutions on pages 156 and 157? In the Middle Ages, people interested in science searched for a universal solvent—a liquid that would dissolve anything. Water comes closest to being this universal solvent.

Think of the many ways water is used each day, such as for cooking, washing, drinking, cleaning, swimming, and manufacturing. After it is used, water often contains many wastes. Some waste materials are easily seen when we inspect waste water, but many other wastes dissolve in water and therefore are invisible.

Make a list of 10 or 12 substances that can pollute the water supply. Look around your school, home, and neighborhood. Decide which of the materials on your list can be dissolved in water.

Substance	Does it dissolve in water?
grease	no
sugar	yes

Much of the waste water ends up at a sewage plant, where it is partially cleaned before it is released into the environment. Where does waste water go in your community? If there is a sewage-treatment facility nearby, visit it and find out how the waste is removed from the sewage. Can any of these wastes be *recycled*? (A)

Other Solvents Besides Water

Water is a great solvent, but as you have just seen, it does not dissolve everything. For example, it does not remove grass stains from clothes very well, nor does it clean greasy hands very well.

There are other solvents besides water. Many of them are useful for doing things that water cannot. The table below lists only a few of these solvents. Perhaps you know of others to add to the table. You can also ask a chemist, a pharmacist, or a hardware store clerk for further suggestions. Add at least six items to the table—more if you can! (B)

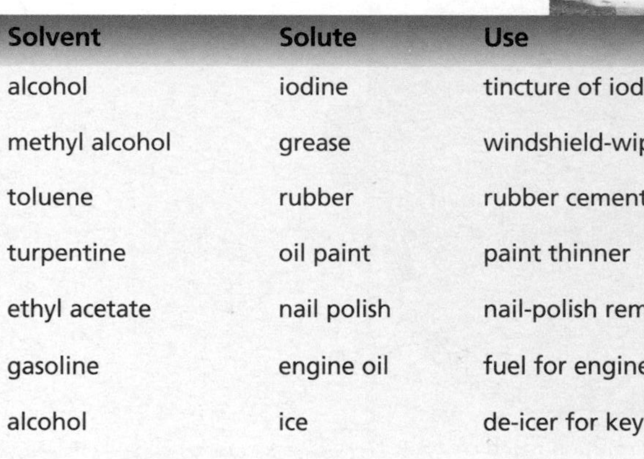

Solvent	Solute	Use
alcohol	iodine	tincture of iodine
methyl alcohol	grease	windshield-wiper fluid
toluene	rubber	rubber cement
turpentine	oil paint	paint thinner
ethyl acetate	nail polish	nail-polish remover
gasoline	engine oil	fuel for engines
alcohol	ice	de-icer for keyholes

Water is a safe solvent. It doesn't burn and isn't poisonous. However, many other solvents are hazardous. Use your library to find out which of the solvents listed in your table can be dangerous.

158

Homework

You may wish to assign A Project: Oil Spills on page 159 as homework.

Waterproof Birds

Water does roll off a duck's back! When a duck or other water bird preens itself, it is actually replacing natural oils that cover its feathers. Since oil does not dissolve in water, the natural oils act as a raincoat that makes the bird waterproof. Without this coating of oil, its wings would become waterlogged, and the bird would not be able to fly or even to float. The bird would be in danger of drowning.

The bird in the photograph below is covered with too much oil. It was caught in an oil spill—a danger to wildlife that is becoming more common every year. How could the bird be cleaned? What will happen to the bird's natural oils in the cleaning process? Ⓒ

A Project: Oil Spills

Our coastlines and wildlife are sometimes harmed by oil spills. The oil is released into the water when oil tankers clean their bilges or when these ships are involved in accidents, such as the one that occurred in Prince William Sound, Alaska, in 1989.

Find out how environmental agencies clean up oil spills on water surfaces and on beaches. How are the affected birds and other animals treated?

159

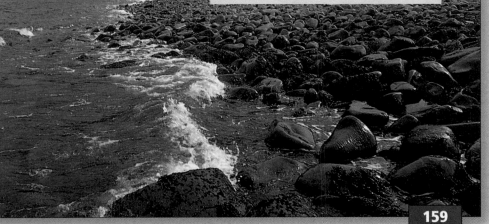

Answers to *In-Text Questions*

Ⓒ Answers will vary. Sample answers include washing it in soapy water or carefully wiping off its feathers with a solvent such as mineral spirits. The bird's natural oils may also be removed during the cleaning process.

ENVIRONMENTAL FOCUS

Explain to students that birds and mammals produce natural oils and waxes that coat their feathers or fur. The coating of natural oils helps keep the animal dry and warm. When the natural oils are removed, the feathers (or fur) lose their insulating properties, and the animal can easily develop hypothermia and die. Birds that have been cleaned by humans cannot be released back into the environment until they have regained their coating of natural oils.

Waterproof Birds

Call on a volunteer to read this section aloud. Then direct students' attention to the photograph. Involve the class in a discussion of the cleanup methods used on the bird.

The petroleum from an oil spill can directly cause hypothermia by soaking and matting fur or feathers. The insulating quality of these body coverings comes from their ability to trap air. For example, the fluffy, tightly packed breast feathers of water birds hold a layer of warm air next to the bird's skin. If these feathers become matted with petroleum, they will no longer be able to trap air and to protect the bird.

Animals can also ingest petroleum as they try to clean or preen themselves. Toxic compounds in the petroleum can damage the liver, kidneys, and other internal organs. Volatile compounds can cause respiratory problems. Animals that have ingested petroleum can also develop a form of anemia.

A Project: Oil Spills

This activity provides students with an opportunity to explore a serious, real-life problem that relates to their study of solutions.

Since this is a research activity, suggest that students do this as an at-home assignment. Encourage them to use their school or local library to research the necessary information. Students' research should include information on the methods of cleaning up oil spills and the treatment of affected wildlife. Students could also contact local and state environmental agencies to discover how these agencies are involved in cleanup activities. If possible, have a representative from such an organization visit your classroom to answer questions.

PORTFOLIO

Students may choose to include the results of their research from the oil-spill project in their Portfolio.

EXPLORATION 2

You may wish to use a cheese grater to make the soap flakes ahead of time.

Divide the class into small groups and distribute the materials.

The following are some procedural hints:

- To be sure the soft-water samples are free of solutes, use clear rainwater or distilled water.
- Use pure soap, not a detergent. Detergents contain foaming agents that will interfere with the results.
- Mix the soap solution thoroughly without excessive shaking. Make sure the solution is very soapy but not sudsy.
- Test the soap solution by adding 10 drops to 10 mL of soft water. This mixture should make soap suds when shaken. If it does not, modify the procedure by having students use 20 drops (instead of 10) in 10 mL of soft water.

The amount of soap suds formed in the different samples of water is used as a rough indicator of how hard the various water samples are. Results may vary depending on the local tap water.

Answers to *Exploration 2*

Answers could include testing equal amounts of soft water and tap water, adding equal amounts of soapy water to both samples, shaking both test tubes the same number of times and with the same force, and using water samples that have the same temperature.

Students may theorize that water from the hot-water tap should have more solutes than water from the cold-water tap because the hot water is heated in a holding tank containing a high concentration of dissolved minerals.

However, if you choose to do this experiment with your students, you may find that the hot water actually contains a lower concentration of solutes than the cold water. This is because heating the water causes a chemical reaction in which many of the solutes precipitate out of the solution.

Two Exploration Worksheets are available to accompany Explorations 2 and 3 on pages 160–161 (Teaching Resources, pages 11 and 13, respectively).

Hard Water

Hard water is *not* frozen water! It is water that contains certain solutes. **Soft water,** such as rainwater or melted snow, is mostly free of dissolved matter. Hard water will not form as many suds when mixed with soap as soft water will. Does your drinking water contain solutes? Here is an easy test to determine whether your drinking water is hard or soft.

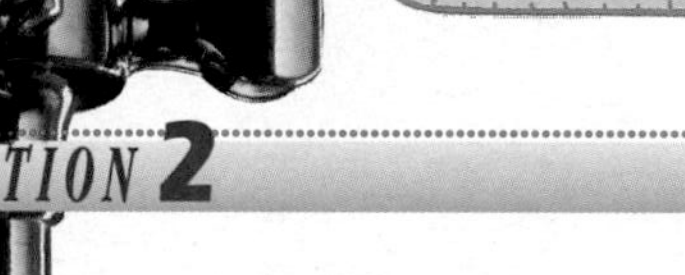

EXPLORATION 2

Tap Water—Hard or Soft?

You Will Need

- a 25 mL (or larger) graduated cylinder
- 10 mL of soft water (distilled water works best)
- 10 mL of tap water (from the cold-water tap)
- 25 mL of soapy water (Add flakes of basic, unscented hand soap to 25 mL of soft water. Mix well, but do not make suds.)
- cup or beaker to hold soapy water
- 2 test tubes with stoppers
- a test-tube rack
- an eyedropper

What to Do 

In one test tube, place 10 mL of tap water. Into the second test tube, pour 10 mL of soft water. Add 20 drops of soapy water to each, and insert the stoppers. Shake the test tubes an equal number of times. Compare the amount of suds formed in each test tube. Is the tap water harder than the soft water? Do you think this is a reliable test?

Make a list of at least three things that you did in this experiment to ensure that your results would be reliable.

Do you think the results would have been the same if you had used water from the hot-water tap?

Collect samples of water from different sources in your city, and test each one for hardness.

CROSS-DISCIPLINARY FOCUS

Mathematics

Pose the following problem to students: It takes 12 mL of dishwashing soap to wash one load of dishes in soft water. Hard water requires 1.5 times more soap to clean the same load of dishes. How much soap would you need to wash three loads of dishes in hard water?
(12 mL × 1.5 × 3 = 54 mL)

Homework

Using a plastic soda container, have students bring in a sample of water from home to test for hardness. They should wear safety goggles and use the same procedure as that described in Exploration 2.

What Not to Do!

Will this procedure give reliable results?

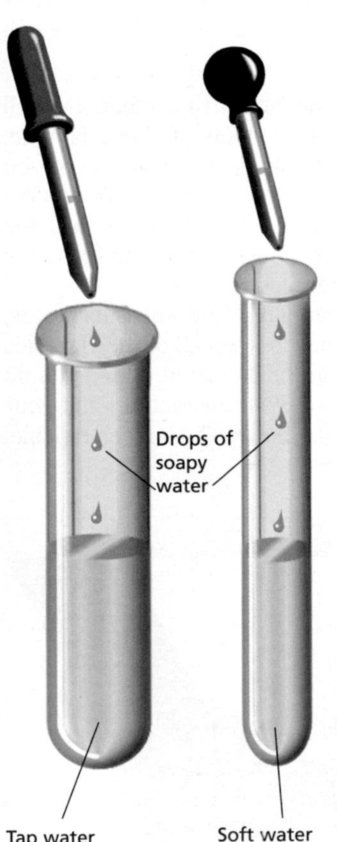

Exploration 3

Solutes in Hard Water

You Will Need

- latex gloves
- flat toothpicks
- sugar
- sodium chloride
- magnesium chloride
- calcium chloride
- iron chloride
- calcium nitrate
- magnesium nitrate
- a 10 mL (or larger) graduated cylinder
- distilled water
- 7 test tubes
- test-tube racks
- soapy water, made according to instructions in Exploration 2
- cup or beaker to hold soapy water
- an eyedropper
- a stirring rod

What to Do

Using the wide end of a flat toothpick, place a few crystals of sugar into a test tube filled with 10 mL of distilled water. Stir until the sugar dissolves. Repeat this procedure for the other six solutes. Test the solutions for hardness, as you did in Exploration 2. Make a table to record which of the solutes reduce sudsing action.

What are some of the solutes that may be responsible for making water hard?

Something to Solve

Does the amount of calcium chloride added to 10 mL of distilled water affect the number of drops of soapy water required to make suds? What do you predict? Try adding different amounts of calcium chloride to test your prediction.

Answer to *What Not to Do!*

The procedure will not give reliable results because different volumes of soap and water are being compared.

Exploration 3

Divide the class into small groups and distribute the materials. You may wish to have students make soapy water in advance using hand-soap shavings. Have students use a separate toothpick for each solute.

Answers to *Exploration 3*

Students will find that magnesium chloride, calcium chloride, calcium nitrate, magnesium nitrate, and iron chloride produce hard water. Since sodium chloride did not affect the hardness of water, students should infer that it is not the chloride that makes water hard, but rather the metallic component of each chloride compound.

Something to Solve

Suggest that students design a mini-experiment to solve this question. Ask them to record step by step what they do, what they predict, and what happens. Point out that they should keep all variables constant throughout their experiment except for the amount of calcium chloride. *(Students should discover that the amount of calcium chloride does have an effect on the amount of soapy water required to produce suds.)*

FOLLOW-UP

Reteaching

Present students with the following scenario: At home, your family complains about how difficult it is to make lots of suds when taking a shower. How could you apply what you have learned from this lesson to explain why this is so? *(The home's water supply probably contains a large quantity of solutes, which keeps the soap from making suds.)*

Assessment

Have students design a demonstration that can be used to teach a group of second-graders the difference between hard water and soft water.

Extension

Bring to class several foods, drinks, or health-care products that are solutions. (Possibilities include soft drinks, mouth wash, after-shave lotion, imitation vanilla flavoring, and cough syrup.) Have the students read the labels of the containers and try to identify the solvents and solutes in each solution.

Closure

Arrange for a visit by a representative from a local water-treatment facility. Ask him or her to describe how impurities are removed from water.

LESSON 3

Soluble, Not Very Soluble, or Insoluble?

FOCUS

Getting Started

Set two watch glasses filled with water on an overhead projector. Into one glass, sprinkle 1–2 mL of powdered sugar. Into the other glass, sprinkle 1–2 mL of cornstarch. Ask the students to observe and describe the changes that occur. *(The powdered sugar dissolves, but the cornstarch does not.)* Explain that the powdered sugar is soluble, but the cornstarch is not. Tell students that they will learn about the factors that affect how things dissolve in water.

Main Ideas

1. Substances vary in their capacity to dissolve in water.
2. The rate of dissolving may be increased by grinding the solute, stirring the mixture, or increasing the temperature of the solvent.

TEACHING STRATEGIES

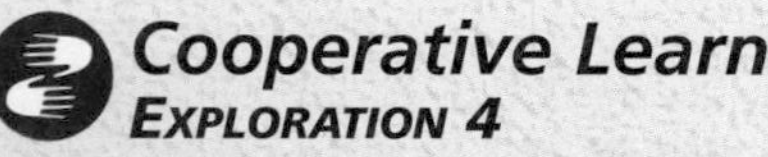

EXPLORATION 4

Remind students that each substance should be allowed to dissolve for 5 minutes.

Students should not ingest any of the materials to be used in Exploration 4.

Cooperative Learning
EXPLORATION 4

Group size: 3 to 4 students
Group goal: to test various substances for solubility in water
Positive interdependence: Assign students roles such as chief investigator, materials coordinator, data recorder, or timer.
Individual accountability: Have each group member record his or her results in a table of solubilities.

Exploration 4 continued ►

LESSON 3

Soluble, Not Very Soluble, or Insoluble?

As you have seen, water comes closest to being a universal solvent because it dissolves a great many substances. However, you know that there are some substances, such as grease and oil, that water cannot dissolve. Are there substances that water can dissolve a little but not completely?

Here are some terms that describe how well a solute dissolves in a solvent (such as water).

- A substance is **soluble** if it disappears in the solvent. This means that it has dissolved completely.
- A substance is *not very soluble* if some of it is still visible in the solvent. This shows that only some of it has dissolved.
- A substance is **insoluble** if none of it disappears in the solvent. This shows that none of it has dissolved.

There are certain factors that affect how well a substance dissolves. One of them is time. When testing how soluble a substance is, you should allow enough time to make sure you get reliable results. Another factor, as you will see later, is temperature. Factors such as these are known as **variables**.

When you compare different substances, you must remember to keep all of the variables except one *constant*, or the same. When you do this, you are *controlling* the variables, and your test results are more likely to be fair and reliable.

EXPLORATION 4

How Soluble Is It?

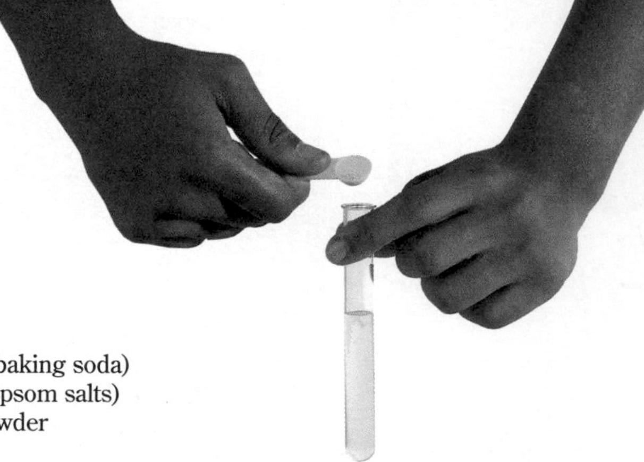

You Will Need

- a 10 mL graduated cylinder
- 20 test tubes or beakers
- test-tube racks
- a stirring rod
- a small spoon or flat toothpicks
- a variety of solid substances to test:

flour	sodium bicarbonate (baking soda)
candle wax	magnesium sulfate (Epsom salts)
cornstarch	flavored drink-mix powder
chalk dust	bouillon cubes
pepper	instant-coffee crystals
salt	instant-tea powder
gelatin	chili powder
sugar	aspirin
coffee	effervescent tablet
tea	solid watercolor paint

What to Do

1. Add a very small sample of one of the solutes listed at left to 10 mL of water, and stir. Does the substance dissolve? If it does, add a second small sample and stir again. If you observe no change, leave it for 5 minutes, stirring occasionally. Why is it necessary to leave it for 5 minutes?

LESSON 3 ORGANIZER

Time Required 2 class periods

Process Skills
observing, classifying, comparing

New Terms
Insoluble—unable to dissolve
Soluble—able to dissolve
Variables—factors that affect the results of a procedure

Materials (per student group)
Exploration 4: 10 mL graduated cylinder; 20 test tubes or 50 mL beakers; test-tube racks to hold 20 test tubes; stirring rod; small spoon or flat toothpicks; 200 mL of water; a pinch or two each of flour, candle wax, cornstarch, chalk dust, pepper, salt, gelatin, sugar, ground coffee, tea, baking soda, Epsom salts, flavored drink-mix powder, instant-coffee crystals, instant-tea powder, chili powder, and solid watercolor paint; 2 bouillon cubes; 2 aspirin tablets; effervescent tablet; safety goggles
Exploration 5: 2 sugar cubes; 250 mL graduated cylinder; 300 mL of water; two 500 mL beakers; stopwatch; optional materials: heat source, stirring rod, safety goggles

continued ►

2. Decide whether the substance is soluble, not very soluble, or insoluble.
3. Repeat the test on the other materials you have gathered. Write down any variable(s) in your experiment. Can you think of anything you could do to make this a more controlled test?
4. In your ScienceLog, record your findings in a table like the one below.

Substance	Soluble	Not very soluble	Insoluble	Was time a factor?	Any other variable(s)?	Special observations

An Experiment Out of Control

John's results did not compare well with those of his classmates. Here's how he performed each test:

TEST 1

He filled a test tube half full with water from the cold-water tap, dumped in some solute, and shook the test tube until his arm got tired.

TEST 2

He poured 5 mL of water into a beaker, added two crystals of another solute, and stirred the mixture once or twice with a stirring rod.

What would you suggest to John to help him improve his experimentation technique?

Detective Work

Soluble or not soluble—that is the question!

Dyani and Mario were having a friendly argument over the experiment they had just finished. Mario thought that baking soda should be entered in their data table under the "Not very soluble" heading. Dyani insisted that baking soda was insoluble. Suddenly, Mario snapped his fingers and said, "I know how we can figure out who's right!" What do you think Mario's idea was?

ORGANIZER, *continued*

Teaching Resources
Exploration Worksheets, pp. 14 and 17
Transparency 28
SourceBook, p. S54

EXPLORATION 4, *continued*

GUIDED PRACTICE Ask: What are the variables in this Exploration? *(Variables include the type of substance, amount of substance used, quantity of water, amount and force of stirring, particle size, and water temperature.)* Ask: Which of these variables should be held constant? *(All variables should be kept the same except for the type of substance being tested.)*

★ **An Exploration Worksheet (Teaching Resources, page 14) and Transparency 28 are available to accompany Exploration 4.**

Answers to *What to Do, pages 162–163*

1. You must wait 5 minutes so that it can dissolve completely.
2. Soluble: magnesium sulfate (Epsom salts), flavored drink-mix powder, salt, sugar, instant-coffee crystals, instant-tea powder, and aspirin
Not very soluble: gelatin, bouillon cubes, coffee, tea, sodium bicarbonate (baking soda), effervescent tablet, and solid watercolor paint
Insoluble: flour, candle wax, chalk dust, pepper, cornstarch, and chili powder
3. To make this a more controlled test, a precise mass of the solutes could be added instead of a very small sample of the substances.
4. Answers will vary. Accept all reasonable responses.

Answers to *An Experiment Out of Control*

In order to compare these mixtures, John should compare equal amounts of water and solute. Also, he should keep the amount of stirring or shaking constant in both tests.

Answers to *Detective Work*

One possibility is to stir the baking soda in distilled water, filter it, and let the remaining liquid evaporate. Any residue that is left is evidence of the partial solubility of the baking soda.

EXPLORATION 5

Have students predict how long an unstirred sugar cube will take to dissolve in water. Divide the class into teams and distribute the materials. Have students begin by recording in their ScienceLog their ideas for speeding up the dissolving time. Remind them that they can do only one thing to speed up the rate.

Check student procedures for safety considerations before allowing them to proceed.

In the Exploration, the amounts of solute and solvent are held constant. The variables that students can change include the amount of agitation, the water temperature, and the size of the sugar particles. Whichever method is chosen, have them record their data, such as number of shakes and temperature, in their ScienceLog.

An Exploration Worksheet is available to accompany Exploration 5 (Teaching Resources, page 17).

Answers to *Exploration 5*

Answers may vary, but students will probably find that temperature is the most important variable affecting the rate at which a substance dissolves.

Answers to *Detective Work*

Possible explanations include the following: The water may have been stirred or heated in one of the beakers, one of the candies may have been crushed or stolen, or some combination of these events may have occurred.

EXPLORATION 5

Let's Have a Sugar-Cube Race!

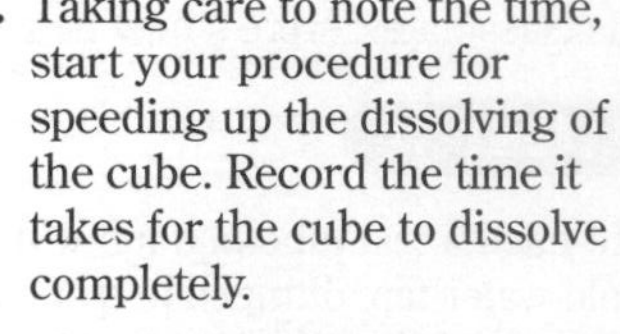

If you drop a sugar cube into a glass of water, how long will it take to dissolve? Will it take 5 minutes, 10 minutes, or even longer? What could you do to speed up the rate at which it dissolves? Before reading further, make a list of things you could try. Record the list in your ScienceLog.

Object of the Race

Dissolve one sugar cube in 150 mL of water as rapidly as possible.

You Will Need

- 2 sugar cubes
- a 250 mL graduated cylinder
- 2 beakers
- a stopwatch

Rules

1. Each team will be given a sugar cube to add to 150 mL of water. Watch it dissolve without disturbing it. Make a note of the time it takes.

Look on page S55 of the SourceBook for some hints about how to speed up the rate of dissolving.

2. Each team must then decide what single thing they will do to the sugar cube or the water to speed up the rate at which the cube dissolves.
3. Each team will get another sugar cube and another 150 mL of water.
4. Taking care to note the time, start your procedure for speeding up the dissolving of the cube. Record the time it takes for the cube to dissolve completely.

What was the single variable you changed to speed up the rate of dissolving? What were the variables of the other teams in the class? Which of these variables is the most important?

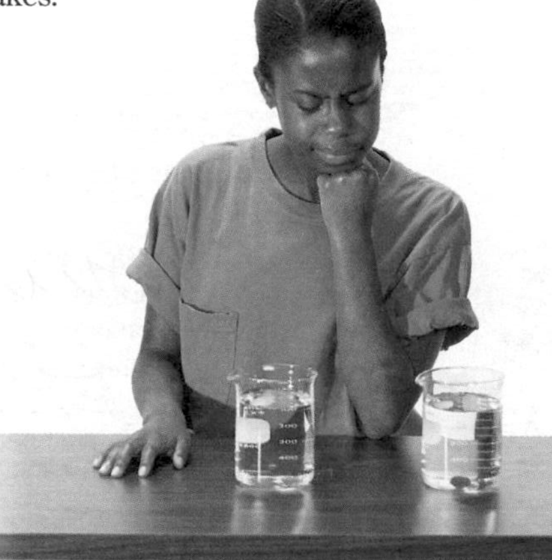

Detective Work

The Case of the Missing Candy

Two identical candies are placed in beakers of water. Fifteen minutes later the experimenter returns to the room and sees that one piece of candy has disappeared.

What do you think could have happened while the experimenter was out of the room? Can you explain why one piece of candy disappeared and one did not?

164

FOLLOW-UP

Assessment

Present students with the following scenario: Cassie wants to find out how long it takes to dissolve a sugar cube at different temperatures. Design an experiment to help her. What variables should she keep the same? *(Volume of water, time allowed to dissolve, and mass of sugar cube)* What do you think Cassie will discover when she completes her experiment? *(She will discover that the sugar dissolves faster at higher temperatures.)*

Reteaching

Divide the class into groups and have them repeat Exploration 4 using vegetable oil and glycerin as the solvents.

Extension

Have students investigate how stains are removed from clothing, furniture, and other items. Have them make a poster illustrating stain-removal techniques and relating the process to the principles of solutions. *(Generally, stain-removal techniques take advantage of the solubility properties of the staining substance. In most cases, the staining substance is not soluble in water but very soluble in some other liquid.)*

Closure

Ask the class to list ways that plants and animals depend on the formation of solutions. Make a table on the board for their answers. Remind students that solutions involve solids, liquids, and gases.

CHALLENGE YOUR THINKING

1. Water: A Sorry Solvent?

How would the world be different if water were not such a good solvent? List at least three ways. Compare your answers with those of your classmates.

2. Van's Variables

Van is trying to determine if water containing detergent will dissolve salt faster than water alone does. When you ask how he plans to perform this experiment, he tells you, "I'm going to add a small sample of salt to 25 mL of water and time how long it takes to dissolve while I stir it. Then I'll add salt to some water containing liquid dish-washing detergent, stir, and check the time again."

What variables must Van control to make sure his test results will be reliable?

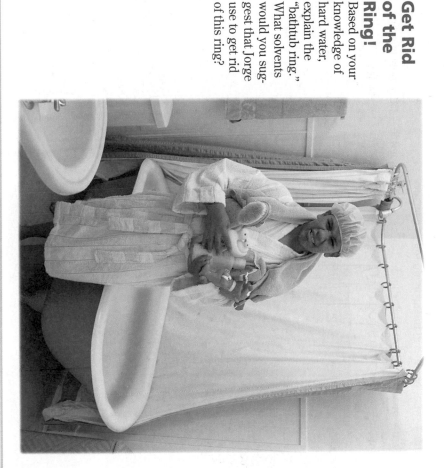

3. Get Rid of the Ring!

Based on your knowledge of hard water, explain the "bathtub ring." What solvents would you suggest that Jorge use to get rid of this ring?

★ **You may wish to provide students with the Chapter 7 Review Worksheet on page 19 of the Unit 3 Teaching Resources booklet.**

Homework

Manufacturers sometimes suggest that coffee makers should be cleaned periodically with vinegar. Have students explain this recommendation. *(Vinegar is used to dissolve the solutes and other particles left behind by the hard water used in making coffee.)*

ENVIRONMENTAL FOCUS

Oil is a liquid that does not dissolve well in water. Because it is less dense than water, it floats on the water's surface. Ask: In the case of an oil spill at sea, what advantages and disadvantages might arise from these characteristics of oil? *(Sample answer: Because all of the oil is at the surface, only the water's surface must be skimmed to collect the spilled oil; otherwise, the water might need to be cleaned at all depths. On the other hand, most ocean life is found near the surface, making it more likely to be affected by the spilled oil.)*

Answers to *Challenge Your Thinking*

1. Answers will vary but should demonstrate an understanding that water-based solutions are everywhere. Sample answers: The ocean's ecosystems would be drastically different, soft drinks might not exist, and inks would not be water-based.
2. Variables that Van must control include the amount of salt added, amount of water, source of the water, and force with which the mixture is stirred.
3. Because tap water often contains many dissolved solutes, some of these solutes may come out of solution and combine with dirt and oil from our bodies and with the soap, forming a residue around the tub. Solvents that Jorge could use to get rid of the ring should dissolve grease in addition to any dissolved matter that is in his water supply. For example, if the water contains dissolved calcium carbonate, Jorge could dissolve it with an acid such as vinegar.

CROSS-DISCIPLINARY FOCUS

Foreign Languages

The words *solution, solvent,* and *solute* are derived from the same Latin word. However, each of these words has had different intermediate forms. Have students consult a dictionary to find out the derivations of these three words. *(They all come from the Latin word* solvere, *meaning "to loosen."* Solution *is derived from Old French. Both* solvent *and* solute *come directly from Latin, although* solvent *comes from* solvens, *and* solute *comes from* solutus. *These are different grammatical forms of the infinitive* solvere.)

Answers to *Challenge Your Thinking,* *continued*

4. Provided that the ice cube consists of pure water, the ice cube melted in water yields pure water, which is not a solution. Only the cup with the sugar and water contains a solution.

5. Sample definitions:
- solute—a substance that dissolves in a solvent to make a solution
- solvent—a substance that dissolves a solute to make a solution
- solution—a homogeneous mixture
- insoluble—the inability of a substance to dissolve or pass into a solution
- Tyndall—The Tyndall effect is a method of identifying a solution; no true solution shows the Tyndall effect.

The following are sample revised answers:

1. The bottle on the left contains a solution (carbon dioxide gas dissolved in water). The open bottle does not contain a true solution because as the bubbles form, the mixture becomes heterogeneous. These pockets of rising gas could scatter light, producing the Tyndall effect.

2. You can't see the sugar because it has dissolved into the iced tea. The particles that make up the sugar separate and spread evenly throughout the liquid, occupying the spaces between the liquid particles.

3. The salt will dissolve, but the pepper will not. Therefore, the salt can be described as *soluble* and the pepper as *insoluble.*

4. What's Your Solution?

The illustration at right shows two cups of water. One cup contains a sugar cube, and the other contains an ice cube. After 15 minutes, both cubes will have disappeared. Will both cups then contain a solution? Justify your answer.

5. Define Your Terms

Here are some entries from *Dr. Punster's Doubtful Dictionary.* Dr. Punster has a great sense of humor, but her definitions don't make good scientific sense. Try rewriting them so that someone who has not read this chapter will know what the words really mean.

- solute—what privates in the army do to the captain
- solvent—having enough money in the bank
- solution—answer to a crossword puzzle
- insoluble—extremely puzzling
- Tyndall—a metal toy

Review your responses to the ScienceLog questions on page 146. Then revise your original ideas so that they reflect what you've learned.

Integrating the Sciences

Earth and Physical Sciences

Limestone is a common rock that is easily dissolved in ground water. As water drips from the ceilings of caves, tiny amounts of limestone are deposited, eventually forming massive columns of limestone that rest on the floor or hang from the ceiling. Have students make their own *stalactites* and *stalagmites* by filling two glasses with a solution of Epsom salts and water. Place the glasses about 10 cm apart and drape a string between the two glasses with both ends of the string in the solution. After two to three days, results should be visible.

CHAPTER 8 Separating Solute and Solvent

1 How can a solute be separated from a solvent?
2 Name some situations in which it is desirable to separate a solute from a solvent.
3 How do you think nature purifies the Earth's water supply?

ScienceLog

Think about these questions for a moment, and answer them in your ScienceLog. When you've finished this chapter, you'll have the opportunity to revise your answers based on what you've learned.

167

CHAPTER 8 Separating Solute and Solvent

Connecting to Other Chapters

Chapter 7 *introduces students to solutions and non-solutions, components of a solution, and the concept of solubility.*

Chapter 8 *explores methods used to separate solutions, focusing on the separation of sugar and salt solutions.*

Chapter 9 *introduces students to the concept of saturation and explores solutions in the environment and the human body.*

Prior Knowledge and Misconceptions

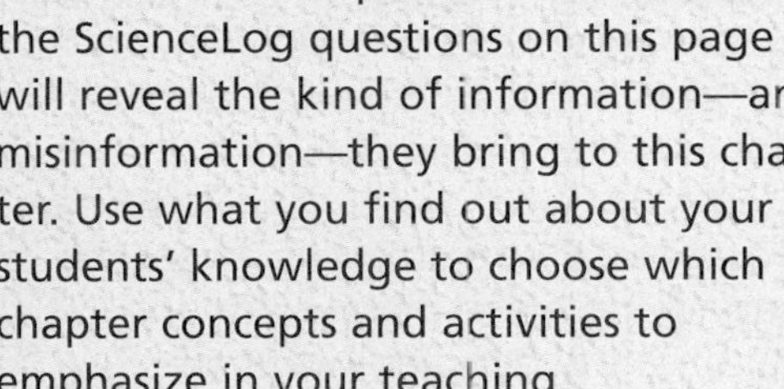

Your students' responses to the ScienceLog questions on this page will reveal the kind of information—and misinformation—they bring to this chapter. Use what you find out about your students' knowledge to choose which chapter concepts and activities to emphasize in your teaching.

In addition to having students answer the questions on this page, you may wish to perform the following demonstration: Before class begins, make or obtain two aqueous solutions, one of them an acid and the other a base. (A discussion of acids and bases can be found in Unit 3 of the SourceBook.) For this example, the base is NaOH dissolved in water; the acid is $FeCl_3$ dissolved in water. Point out to the class that you are holding two different solutions in front of you and that by mixing them, you will cause them to separate. Pour the two solutions together into one beaker. The class will see a dark orange solid immediately congeal in the beaker and sink to the bottom. The products of the reaction are solid $Fe(OH)_3$, commonly known as rust, and a solution of water and table salt. Ask the class for comments about the demonstration and make a list of these comments on the board to find out what students know and what misconceptions they may have about separating the components of a solution, as well as what interests them about this topic.

For safety reasons, make sure that the NaOH solution has a concentration of less than one molar.

The Salt Industry

FOCUS

Getting Started

Put out a bowl of water and instruct students to wet one of their fingers. Then tell students to blow on the wet finger and a dry finger at the same time. Ask: Which finger is losing energy? How can you tell? *(The wet finger is losing energy because it feels cooler than the dry one.)* Explain to students that some water on their wet finger has evaporated. Blowing on the water has given some water molecules enough energy to break away from the rest of the molecules and become a gas. Energy is removed by these molecules, and the liquid that is left behind is cooled. Tell students that in this lesson, they will learn more about evaporation and how it is used to produce salt.

Main Ideas

1. Evaporation is the process by which a liquid changes gradually to a gas.
2. Boiling is the process by which a liquid changes rapidly into a gas. Each liquid has its own boiling point.

TEACHING STRATEGIES

Ask students to cite examples of processes in their everyday lives that involve evaporation and boiling. Then instruct them to silently read the essays about salt preparation on pages 169 and 170 in order to discover the similarities and differences between the two processes.

Answer to *In-Text Question*

Ⓐ The boiling point of water is affected by two factors, temperature and pressure. At lower pressures (such as those at high elevations), water boils at lower temperatures.

LESSON 1

The Salt Industry

Around the world, salt is obtained in several different ways. Many methods use an *evaporation* or *boiling* process to separate the solute (salt) from the solvent (water).

Evaporation is a gradual process in which a liquid changes into a gas or vapor. For example, wet clothes will dry over time by losing water to the surrounding air. The drying, or evaporation, process will go faster if you add heat, as in a clothes dryer.

Boiling is the process in which a liquid rapidly changes into a gas. This occurs when the temperature of the liquid is raised to its boiling point. Water boils at 100°C at sea level. At higher elevations, the boiling point of water is a bit lower. Ⓐ Why do you think this is so? Other liquids have their own characteristic boiling points.

The two reports that follow were created by students from different parts of the world. They describe how salt is produced in their countries. The first report is by two students in Zambia, a country in southern Africa. The second report is by a student in Nova Scotia, a province of Canada.

As you read the reports, count the number of times that the students refer to the evaporation and boiling processes. Think about the ways in which the two processes are similar.

MAKING

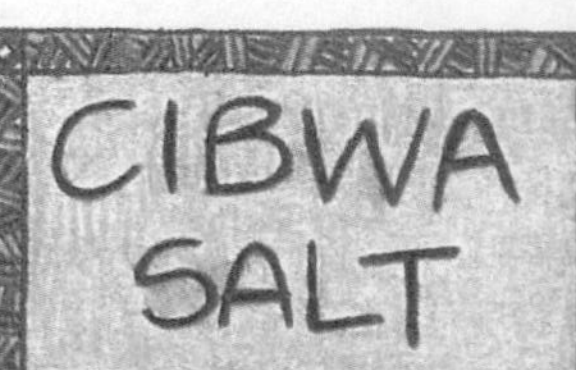

By:
Michael Nosenge
and
John Bavalya

LESSON 1 ORGANIZER

Time Required
1 to 2 class periods

Process Skills
observing, classifying, comparing, analyzing

Theme Connection
Energy

New Terms
Boiling—the process by which liquid rapidly changes into a gas at its boiling point
Evaporation—the process by which a liquid gradually changes into a gas or vapor

Materials (per student group)
Exploration 1: about 20 g of mixture containing equal parts of sand and salt; heat source; beaker of any size; funnel with a piece of filter paper; about 200 mL of water; 1 or 2 stirring rods; several paper towels; tweezers; magnifying glass; metal ring; ring stand; safety goggles; oven mitts

Teaching Resources
Exploration Worksheet, p. 24
SourceBook, p. S67

Cibwa salt is found mainly in the country of Zambia. This kind of salt is found in a sort of grass.

In August, the grass is collected using hoes and then spread on the ground to dry. This grass may remain there for about two months. When it is dry, it turns brown in color.

In September and October, the Cibwa salt is extracted from the grass. People do this by burning the dried grass, which turns it into ash. The salt is in the ash.

This ash is put on a filter made of another kind of grass, and water is poured over the ash into a clay pot. The water dissolves the salt and leaves behind the insoluble impurities of the ash on the filter.

The salt solution in the clay pot is placed over a fire. During the boiling, the water evaporates away. When the water has all evaporated, the salt is in the bottom of the pot in solid form.

The people in our part of Zambia still make salt the same way.

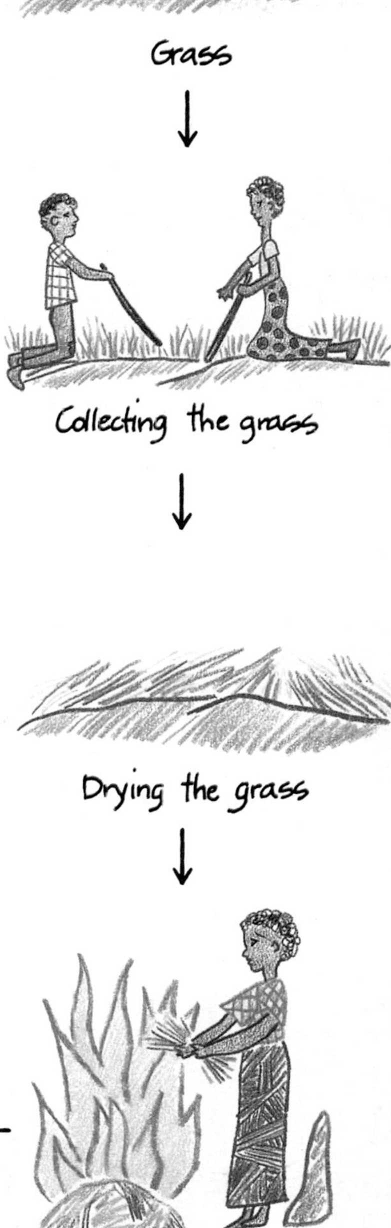

Grass

↓

Collecting the grass

↓

Drying the grass

↓

Burning the dried grass

←

Pouring water over the ash

←

Boiling away the water, leaving the salt behind

Making Cibwa Salt,

pages 168–169

GUIDED PRACTICE Explain to students that the preparation of Cibwa salt from Zambian grass takes advantage of the fact that plants absorb salt from the soil. Plants such as Zambian grass absorb sodium and other ions from the soil. Many important plant nutrients, such as nitrogen, potassium, phosphorus, calcium, and magnesium, occur in the form of ions in the soil. These and other ions are released from weathered rocks and become part of the soil solution. The soil solution consists of the water contained in the soil and its dissolved solids, liquids, and gases. Materials that become a part of the soil solution are available for absorption by plant roots.

The carbon compounds in the grass are converted into carbon dioxide and water when the grass is burned. The remaining residue contains a relatively high concentration of salt, which is then extracted using water. The extract is filtered, and the insoluble material is left behind. Evaporation of the liquid recovers the dissolved salt.

Theme Connection

Energy

Focus question: What role does the exchange of energy play in the different steps of the production of Cibwa salt? *(When the grass is burned, energy is released as new products are formed. As the filtrate is allowed to evaporate, energy is absorbed and the liquid solvent becomes a gas.)*

Homework

Have students find out some other uses for salt besides cooking. *(For instance, salt is used for preserving meat, making soap, and tanning leather.)*

The Salt Plant at Nappan, Nova Scotia

The following is background information that you can share with your students. The Nova Scotian salt deposits were formed by the evaporation of water in a partially enclosed bay over a long period of time. The ancient sea that formed these salt beds existed over 300 million years ago. At that time, hot and dry climatic conditions caused extensive salt beds to be deposited. Over millions of years, the salt beds were buried beneath sediments. Later, tectonic forces in Nova Scotia created a fold in the rock. This raised the ancient salt beds close to the Earth's surface—making it possible for the salt beds to be mined today. The salt-making process described on page 170 involves making a saturated salt solution, which is then pumped out of the ground and processed.

The Nova Scotian salt beds represent only a small part of the world's salt resources. Extensive beds of rock salt also exist in Pakistan, Iran, Germany, eastern Europe, the former Soviet Union, the southern United States, China, and western Africa. Natural brines are found in the Dead Sea region, Austria, France, Germany, India, the United States, and Great Britain. Ocean water is a salt resource available all over the world. Twenty liters (5.2 gal.) of ocean water contains approximately 2.2 kg (1 lb.) of sodium chloride.

Answers to *Comparing Methods, page 171*

The Zambian and Nova Scotian methods of gathering salt are quite different. In Zambia, the salt is taken out of the soil by grass plants and does not need to be mined. Only extraction, filtration, and evaporation are necessary. Both methods are, however, similar in that they require water for the extraction and refining of the salt. In Zambia, water is used to separate the salt from the ash by dissolving the salt. In Nova Scotia, the salt is dissolved in water, and the salt solution is then pumped to the surface.

In Zambia, desalination is achieved by boiling the salt water in clay pots. The water evaporates, leaving the salt behind. In Nova Scotia, the salt is removed without boiling. In both cases, the water escapes into the atmosphere.

The Salt Plant at Nappan, Nova Scotia

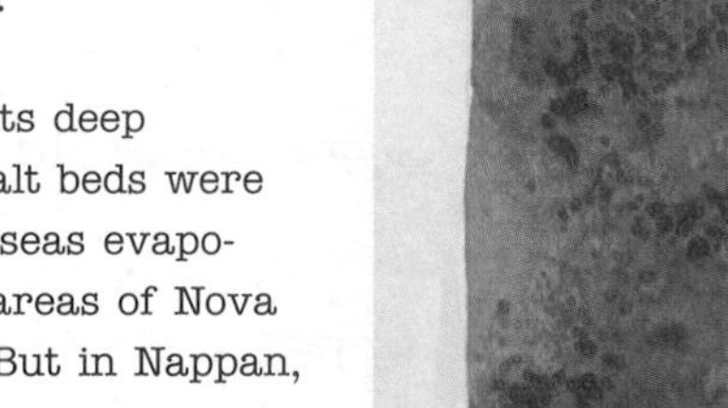

Water evaporates, leaving salt behind

By Johanna Kimball

Much of our area has large salt deposits deep underground. It is believed that these salt beds were formed millions of years ago when salt seas evaporated, leaving the salt behind. In some areas of Nova Scotia, the salt is mined using tunnels. But in Nappan, where I live, a different method is used.

First a hole is drilled down to the salt bed. Then water is pumped from a storage tank down through a pipe. This water dissolves some of the salt, making an underground cavern full of salt water. This salt water is then forced back to the surface, where the water is evaporated, and the salt is left behind.

170

PORTFOLIO

Encourage students to include their responses to Comparing Methods in their Portfolio. Suggest that they highlight key terms and concepts.

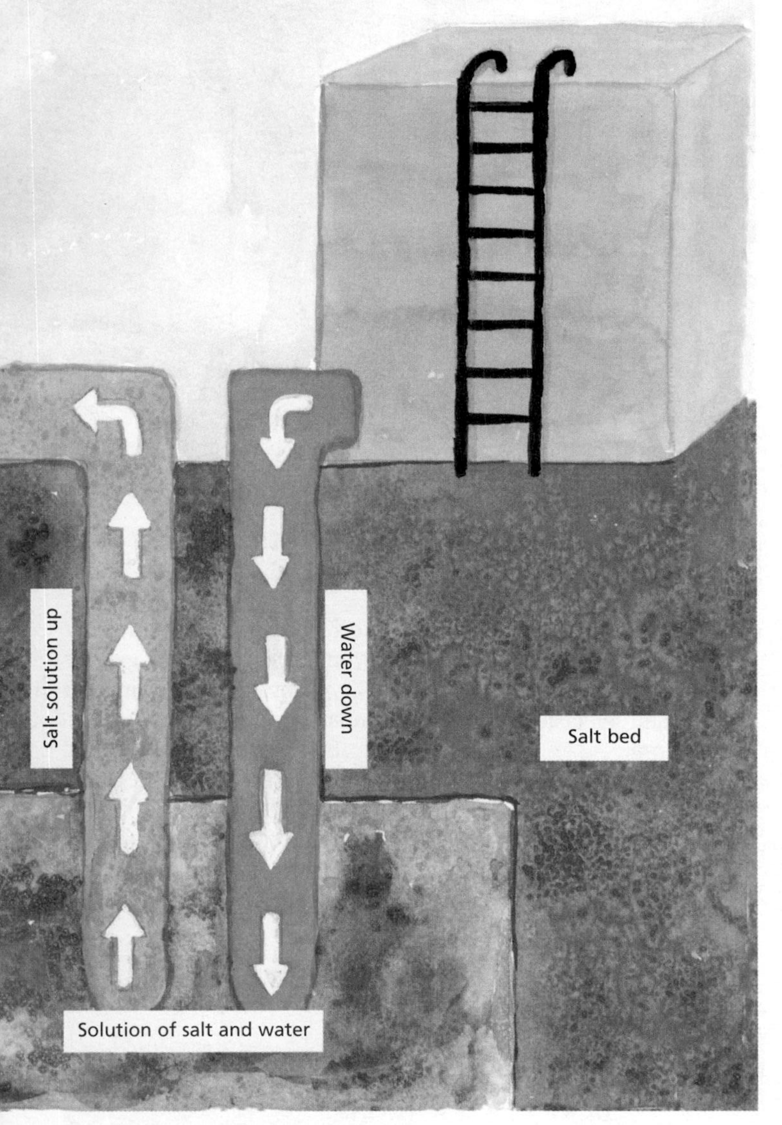

Comparing Methods

Compare the method of salt extraction used in Zambia to the method used in Nova Scotia. What similarities are there in the methods of gathering the salt? How was water used? How was the salt separated from the water? Where did the water go?

Exploration 1

Your Turn to Design

You Will Need

- sand-salt mixture
- a heat source
- oven mitts
- beakers
- a funnel and filter paper
- stirring rods
- paper towels
- tweezers
- a magnifying glass
- a stand
- water

Ask your teacher for any other apparatus that you need.

What to Do

Here's the problem! You have just found a deposit of salt, but it is quite impure. In fact, it's 50 percent sand. Work as a team with several other people to devise a plan to extract the pure salt. Check your plan with your teacher to make sure it's safe, and then carry out your experiment. Report on the success of your plan.

Exploration 1

Divide the class into groups and distribute the necessary materials. Each team should formulate a step-by-step plan to carry out in their investigation.

SAFETY ALERT Check student plans in advance for safety considerations.

One method of extracting the pure salt from the mixture of sand and salt is as follows:

1. Add about 25 mL of water to the mixture. Stir it.
2. Pour the mixture through a filter and collect the clear solution, or filtrate, that passes through the filter paper.
3. Heat the filtrate gently to evaporate the water.
4. Closely examine the substance that remains to determine what it is. *(It is salt.)*

An alternative approach is to have students pour 15 to 20 mL of the filtrate into a jar and let it evaporate naturally.

Point out to the class that normal filtration does not remove dissolved substances; it is not possible to separate solute from solvent by means of filtration because the particles of solute are too small. (Processes such as reverse osmosis, however, use specialized filters that can separate a solute from a solvent.)

★ **An Exploration Worksheet is available to accompany Exploration 1 (Teaching Resources, page 24).**

FOLLOW-UP

Reteaching

Make two columns on the chalkboard. Put the heading "Cibwa salt" in one column and "Nova Scotian salt" in the other column. Have students compare the two processes by listing key terms under each heading. For example, *boiling* fits under the heading "Cibwa salt," while *evaporation* belongs under "Nova Scotian salt."

Assessment

Propose the following situation to students: Imagine that you live far from urban civilization. You have accidentally mixed your salt supply for the next 6 months with flour. How might you recover the salt? *(Answers will vary. The mixture could be added to water. The flour will not dissolve in water, but the salt will. After filtering out the flour, the water in the salt solution can then be boiled off or allowed to evaporate, leaving the salt behind.)*

Extension

As a research activity, have students investigate the source of salt commonly sold in their local area. *(In the northeastern United States, salt is made from natural brine pumped from the ground at Syracuse, New York. In the southeastern and midwestern United States, salt is mined from underground salt domes. In the western United States, salt is evaporated from ocean water.)*

Closure

Explain to students that some medicines have an active ingredient dissolved in water or alcohol. Tincture of iodine and mouthwash are examples. Ask students to propose a way to separate the solute(s) from the solvent in these products.

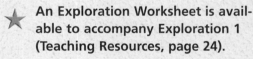

LESSON 2

Desalination and Distillation

FOCUS

Getting Started

Ask students to imagine that the Earth's surface has suddenly been leveled and that sea water now covers the entire globe. Now ask students to imagine that all of the water has evaporated. Explain to students that if all the water were to evaporate, a layer of salt 74 m thick would be left behind, covering the entire planet. (74 m is a little less than the length of a football field.)

Main Ideas

1. Distillation is a process that combines boiling (which creates a gas) with condensation by cooling. During the process, the solute and the solvent are separated from each other and both are retained.
2. In nature, evaporation, condensation, and precipitation (the water cycle) can be compared to the process of distillation.

TEACHING STRATEGIES

GUIDED PRACTICE Call on a volunteer to read the introduction on page 172 aloud. Involve the class in a discussion of the new terms *desalination* and *distillation.* Point out that distillation differs from simple boiling and evaporation in that it includes the additional step of cooling, which changes the vapor back into liquid form.

The Future of Desalination

Call on a volunteer to read aloud the material on desalination. Have students form groups to discuss the questions at the bottom of the page. Reassemble the class and discuss their answers.

LESSON 2

Desalination and Distillation

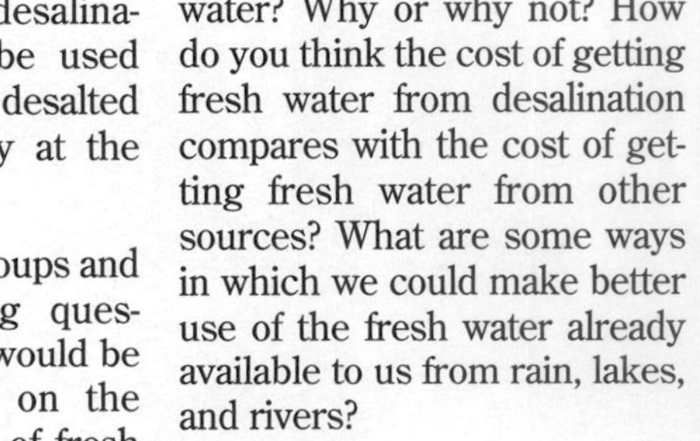

In some parts of the world, fresh drinking water is in short supply. One method of meeting the increasing demand for fresh water is to remove the salt from ocean water. This process is called **desalination**. In this case, it is the solvent (water), not the solute (salt), that is saved.

In some situations, it may be useful to save both the solvent and the solute. To save only the solvent or to save both the solute and the solvent, a process called **distillation** is used. Let's look more closely at both desalination and distillation.

The Future of Desalination

Should we use desalination of ocean water as a way to obtain fresh water? Here are some facts about desalination:

- Desalination can relieve water shortages in dry regions along seacoasts.
- One large desalination plant can produce about 280 million liters of fresh water daily.
- The desalination process requires large amounts of energy, regardless of how that energy is produced.
- Providing desalted water to dry regions far from seacoasts requires a transportation system.
- The desalination plants in use today produce only a small fraction of the world's daily demand for fresh water.
- Nuclear-powered desalination plants might be used someday to produce desalted water and electricity at the same time.

Form into small groups and discuss the following questions: Do you think it would be wise for us to rely on the oceans for our supply of fresh water? Why or why not? How do you think the cost of getting fresh water from desalination compares with the cost of getting fresh water from other sources? What are some ways in which we could make better use of the fresh water already available to us from rain, lakes, and rivers?

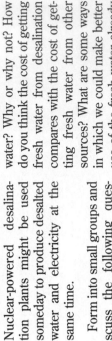

A desalination plant on Curaçao, an island in the Caribbean Sea

172

LESSON 2 ORGANIZER

Time Required 1 to 2 class periods

Process Skills observing, analyzing, comparing, communicating

Theme Connection
Energy

New Terms
Desalination—removing the salts from a substance, usually ocean water, in order to make it drinkable
Distillation—a process in which a solution is heated until it evaporates, leaving behind dissolved materials. The second step involves cooling the vapor back into liquid form and collecting it.
Water cycle—the continuous movement of water from the atmosphere to the Earth and back again. This distillation process purifies the Earth's water supply.

Materials (per student group)
Distillation: various materials for distilling water, such as 2 ring stands, heat source, 250 mL flask with stopper, glass tubing, 250 mL beaker, safety goggles, and oven mitts (See Advance Preparation, page 143C.)

Teaching Resources
Graphing Practice Worksheet, p. 25

Distillation

Distillation involves two steps. In the first step the solution is heated to change one part of the solution into a gas or vapor. The second step involves changing the vapor back into liquid form and collecting it.

Here is a design that Chris used for collecting both the solvent and the solute by distillation. Is there a weakness in his design? Ⓐ

With other students, discuss how Chris's design might be improved using similar equipment. Check your suggestion with your teacher (for safety), and then test your own design. Compare your results with those of other groups.

A Life-Saving Solution

Here's a chance to apply what you have learned about evaporation, boiling, and distillation.

During an expedition to a remote mountainous area, you learn that a person living there urgently needs distilled water to help treat a rare medical condition. You have a teakettle, a camp stove, and a few bottles. Draw a diagram to show how you could use this equipment to obtain distilled water. Explain how your method works. See how many different ways your class can devise to distill water with just the equipment listed.

173

Homework

Many types of vinegar that are sold for use in cooking are distilled before they are bottled. Have students explain why they think this is done. *(The distillation process removes most of the impurities from the vinegar.)*

Answers to *The Future of Desalination, page 172*

It may not be wise for us to depend on desalinated ocean water because of transportation problems and the expense of desalination. In the United States, it costs about $1.00 to produce 3800 L of fresh water from sea water. The same amount of water taken from freshwater sources costs about 45 cents. Students may suggest water conservation as a better way to use the fresh water that is already available.

Distillation

Have students study the distillation apparatus in the illustration on page 173. They can work individually or in groups to locate possible flaws in the design and suggest improvements. After students have completed the exercise, reassemble the class and involve students in a discussion of the pros and cons of the different groups' designs. You may wish to choose one design to test as a demonstration. (The test criterion should be the volume of solvent recovered.)

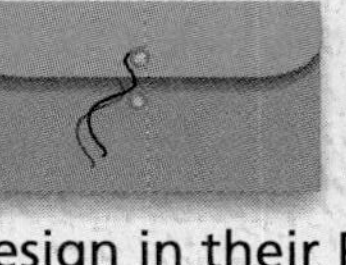

PORTFOLIO

Suggest that students include their distillation design in their Portfolio.

Answer to *In-Text Question*

Ⓐ The main flaw of Chris's design is that it permits distilled solvent to escape from the beaker in the form of steam or vapor. A possible improvement would be to lengthen the condensing tube so that the liquid would have more time to cool before reaching the receiving beaker.

A Life-Saving Solution

Have students draw their distillation setups, and discuss alternatives to the laboratory-type equipment used in the classroom. *(Students could boil the water, condense the vapor, and collect the condensate in the bottles.)*

What Is the World's Largest Distillery?

Call on a volunteer to read aloud pages 174 and 175. Direct students' attention to the illustration of the water cycle.

Using the illustration, ask students to trace the path of a water molecule from the ocean, into the atmosphere, onto the land, and back into the ocean. *(Water evaporates from the oceans, lakes, streams, and soil. Water also enters the air by* transpiration, *a process in which plants give off water vapor into the atmosphere. When water vapor rises in the atmosphere, it expands and cools. As the vapor is cooled, some of it condenses, or changes into liquid water, and forms clouds. Water then falls from the clouds as rain, snow, sleet, or hail.)*

This might be a good time to discuss why the ocean is salty. Students should recognize that water can pick up and dissolve mineral salts as it flows across surface rocks on its way back to the ocean. The salts are then concentrated in the ocean basin by continued distillation of water.

As preparation for the writing and illustrating exercise on page 175, you may want to discuss detours that fresh water can make on its way back to the ocean. For example, a water molecule could be frozen for thousands of years in a glacier before returning to the ocean as meltwater or as part of an iceberg. A water molecule could also seep into the ground and become part of a reservoir of ground water that becomes sealed off for many centuries. Such deposits are sometimes known as *fossil water.* A water molecule could also spend varying amounts of time in the body of a plant or animal.

What Is the World's Largest Distillery?

Surprisingly enough, it is the Earth itself! The Earth's water is involved in a purification process similar to that of a giant distillery. This continuous process is called the **water cycle**.

Each day, vast quantities of water evaporate from the oceans, lakes, rivers, and land. Eventually, this moisture condenses and falls as rain or snow. This is nature's method of purifying the world's water supply. How is nature's method of purifying water similar to the method you developed? Ⓐ

174

Answers to *In-Text Questions*, pages 174–175

Ⓐ Both are two-step processes involving a warming and a cooling stage. Also, when water evaporates, it leaves behind dissolved solutes.

Ⓑ Accept all reasonable responses. Comic strips should include concrete aspects of the water cycle such as evaporation or precipitation.

Theme Connection

Energy

Focus question: What source of energy drives the Earth's water cycle? Explain your answer. *(All of this energy initially comes from the sun. This energy warms the air, land, and water, causing the evaporation of water. Uneven heating results in air and water currents. Water molecules change state as they are carried in these currents.)*

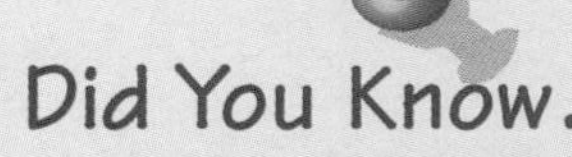

Did You Know. . .

Not only does evaporation increase the salt and mineral content in a body of water, but it also increases the concentration of pesticides and dangerous chemicals. This can have a detrimental effect on aquatic organisms and plants.

Because of the water cycle, all of the water present on the Earth today was present when the Earth was formed. The next time you have a drink of water, stop and ask yourself where the water may have been before. Perhaps that water, or part of it, had been in a royal bath in ancient Egypt. Or perhaps George Washington Carver once drank some of it. Look at the cartoon character that appears on this page. Develop a comic strip showing this raindrop's adventures through time. How many different and unusual places can you imagine? Let your imagination roam, but there is one rule: at some point, you must include the water cycle in your comic strip. (B)

FOLLOW-UP

Reteaching

Have students help you set up several small terrariums. Include moist soil, plants, and a few land snails or earthworms. Cover each terrarium with a glass lid and place it in a warm location. Once finished, students should wash their hands with soap and water. When water droplets appear on the inside of the lid, ask students to draw a terrarium and label the two steps of the distillation process that are occurring. Students should also write a brief description of how the water cycle in the terrarium is similar to that which occurs on Earth.

Assessment

Have students draw a distillation apparatus and label the function of each part in relation to the distillation process.

Extension

Have students research the importance of water in regulating the Earth's temperature. *(Sample answer: Water helps to hold heat near the surface of the Earth.)*

Closure

Have students find out what role the process of dissolving plays in the transportation of vitamins and minerals in the body.

Homework

Assign the Graphing Practice Worksheet on page 25 of the Unit 3 Teaching Resources booklet as homework.

Integrating the Sciences

Earth and Physical Sciences

Share the following information with students: The Great Salt Lake in northwestern Utah is fed by freshwater streams. Once in the lake, water evaporates from the lake's surface rather than draining away. This leaves salt behind. The salt in the Great Salt Lake comes from minerals dissolved in the creeks and rivers that flow into the lake. The salt content varies depending on the amount of water in the lake. At its saltiest, the Great Salt Lake is 27 percent salt. That is almost eight times saltier than the ocean!

ENVIRONMENTAL FOCUS

Ask: How does water interact with the Earth and its organisms to produce solutions? *(As water strikes the Earth in the form of precipitation, it aids in the process of erosion and dissolves substances in the landscape, forming a solution. Dissolved plant and animal matter also becomes part of the solution, helping to cycle nutrients through the environment.)*

Multicultural Extension

George Washington Carver

George Washington Carver helped to revolutionize farming in the southern United States. By introducing such crops as the peanut and the soybean, he helped Southern farmers break their dependence on cotton, which extracted dissolved nutrients from the water in the soil. Because peanuts and soybeans belong to the legume family, they enrich the soil by replenishing it with nitrogen and essential proteins. Have students find out more about George Washington Carver's life and accomplishments.

Another Solution

FOCUS

Getting Started

Fold two or three pieces of white paper towel into a rectangle. While the class observes, hold the edge of the folded towel in a dilute solution of red food coloring. Ask the students to describe what they see. *(The coloring moves along the paper towel and is separated out of the solution.)* Tell the class that they will see how this simple process can be used to investigate some complicated problems.

Main Ideas

1. An inference is a conclusion that attempts to make sense of an observation.
2. A hypothesis is a possible explanation of a question or problem, based on observations. It is a testable statement.
3. A prediction states the expected outcome of a future event.
4. Paper chromatography is a separation method in which the components of a solution are absorbed at different locations along a piece of filter paper.

TEACHING STRATEGIES

Detective Hanamoto's Forgery Case, *pages 176–177*

Call on a volunteer to read aloud the introduction on page 176, pausing before Detective Hanamoto's summary. Discuss inferences and hypotheses. Give examples of both. Have another volunteer read the rest of the section on pages 176–177.

Answer to *In-Text Question*

Ⓐ The similarities include making observations, making inferences, making predictions, formulating hypotheses, making educated guesses, and performing experiments.

LESSON 3

Another Solution

Detective Hanamoto's Forgery Case

Scientists and detectives do many of the same things, such as using scientific equipment and applying investigative methods. Detectives must be good observers, and so must scientists. Detectives search for clues, just as scientists do. Detectives make **inferences** and draw conclusions from the clues; so do scientists. Detectives prepare an explanation, or a **hypothesis,** to explain why certain things happen the way they do; scientists do the same. Both detectives and scientists make educated guesses about something that has happened or that might happen. Detectives working in a forensic lab often conduct experiments and make measurements; so do scientists. In this lesson, you will be asked to do many of the things detectives and scientists do: observe, make inferences, form hypotheses, make predictions, and conduct experiments.

Detective Yori Hanamoto made the following summary of a forgery case he investigated. As you read his report, try to pick out the similarities between detective work and scientific investigation. Ⓐ

> The accused was charged after a complaint was made about an unauthorized withdrawal of money from another person's bank account. The accused claimed that he was given a check for the amount. The police investigation suggested that the accused had altered the check: instead of $7000, the check was now made out for $70,000!

A scientist's job often involves detective work.

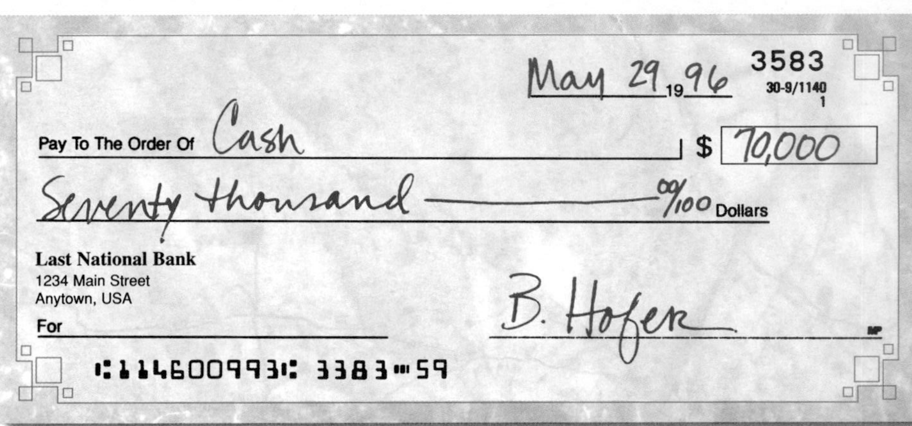

3583
30-9/1140
1
May 29, 19 96
Pay To The Order Of Cash $ 70,000
Seventy thousand 00/100 Dollars
Last National Bank
1234 Main Street
Anytown, USA
For
B. Hofer
⑆114600993⑆ 3383⑈59

LESSON 3 ORGANIZER

Time Required
2 class periods

Process Skills
observing, inferring, predicting, hypothesizing

New Terms
Hypothesis—a testable explanation of why certain things happen
Inferences—conclusions, based on observations or facts, that attempt to make sense of the observations

Materials (per student group)
Exploration 2: metric ruler; strip of filter paper or blank newspaper, about 10 cm × 6 cm; 1 L beaker or other wide-mouth container; about 150 mL of distilled water; equipment to hold paper upright in container, such as pencil and 20 cm of thread; several samples of water-soluble ink or felt-tip pens; pencil; clock or watch; safety goggles

Teaching Resources
Exploration Worksheet, p. 27
Transparencies 29 and 30

After close examination with a magnifying glass, the check still looked genuine. The only way to settle the issue was to perform a test on the ink. If the check was a forgery, there would be a difference in the inks; that is, there would be a difference between the "Seven" and the "ty." There would also be a difference between the comma in "70,000" and the numbers themselves. After I performed the test, there was no doubt: the check was a forgery.

What kind of test could Detective Hanamoto have used? Play detective by performing Exploration 2.

Exploration 2

Analysis of Inks

You Will Need

- a metric ruler
- a strip of filter paper about 10 cm long and 6 cm wide (You can also use ordinary newspaper, but use only the edge so that there is no ink on it to confuse your results!)
- a beaker or other wide-mouth container
- distilled water
- equipment to hold the paper upright in the container
- several samples of washable (water-soluble) inks (Certain felt-tipped pens can also be used.)

What to Do

1. Pour water into the container to a depth of 2 cm.
2. Using a pencil, draw a line at a distance of 3 cm from the bottom of the filter paper or newspaper.
3. Mark small crosses at least 1 cm apart on the line you have drawn.
4. Place a dot of one of the inks on one of the pencil crosses. Put one dot of each of the other inks on each of the other crosses.
5. Let the inks dry, and then suspend the strip by some means so that the bottom 2 cm of the strip is immersed in the water inside the container.
6. After 15 minutes, remove the strip and let it dry. What do you observe?

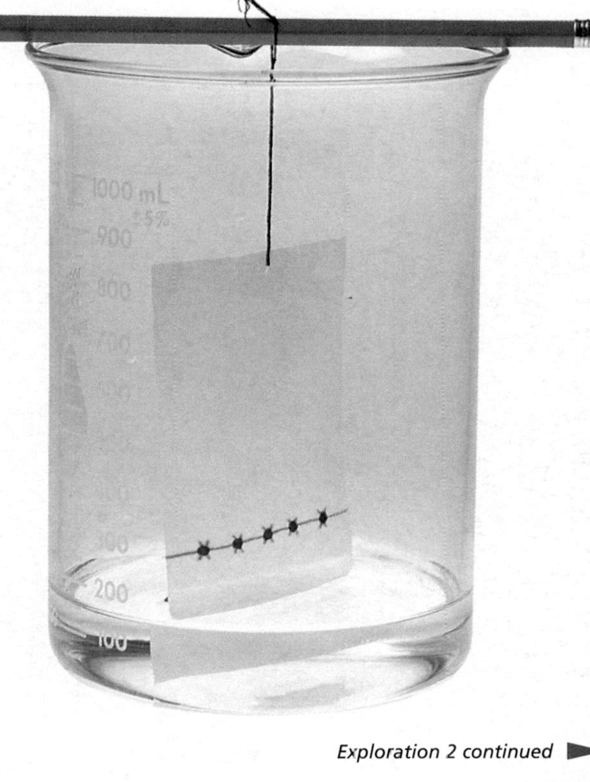

Exploration 2 continued ►

Exploration 2

Make certain that students test only water-soluble inks. Ink from water-soluble transparency markers works well. Capillary tubes or elongated eyedroppers are excellent for transferring drops of ink to the filter paper. Toothpicks also work well. Alternatives to filter paper are paper towels or coffee filters.

Instruct the students to touch the end of a capillary tube or eyedropper to the ink and then to a cross on the filter paper. This way, just the right amount of liquid will be transferred. If using a pen, make a small dot on the filter paper.

INDEPENDENT PRACTICE Have students write their observations in their ScienceLog. Ask students to make an inference about why the inks traveled different distances. After students have completed their investigations, you can share the following analysis of the activity.

The solvent (water) is pulled up the paper by capillary action. The solutes in the solution are carried along at different rates, depending on the solubility of each component and on how much each component adheres to the paper. A component that is very soluble tends to move farther up the paper. Another component that is less soluble, or more strongly attracted by the paper, moves a smaller distance. In this manner, separation of the components occurs. Two identical colors that travel up the paper the same distance at the same rate are probably the same substance. However, if one red spot travels 1 cm up the paper and the other red spot travels 3 cm up the paper, then they are different substances.

Exploration 2 continued ►

An Exploration Worksheet (Teaching Resources, page 27) and Transparencies 29 and 30 are available to accompany Exploration 2.

Multicultural Extension

Plant Dyes and Medicines

Indigenous people from all over the world have developed medical treatments and dying techniques based on plant extracts. Until the invention of synthetic dyes in the late nineteenth century, plants such as indigo were widely cultivated and had great commercial importance. Have students research cultivated or wild local plants that have been used for medicines or dyes. *(Sage and mint may be examples.)*

A simple extract can be made by immersing these plant materials in boiling water. The water-soluble components will dissolve in the hot water—forming a *tea,* or *infusion.* Students can then use the techniques of paper chromatography described in Exploration 2 on this page to investigate the components of the plant extract. Caution students to follow the same safety precautions as they did in Exploration 2.

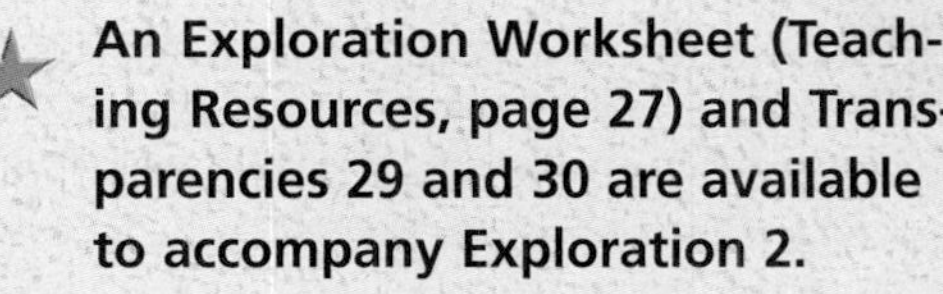

Homework

You may wish to have students answer the in-text question posed on page 176 as homework. (It is the last sentence of the second paragraph.)

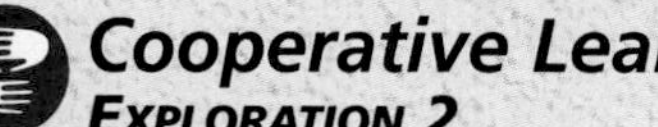

Cooperative Learning
EXPLORATION 2

Group size: 2 to 3 students
Group goal: to investigate the identity of several inks using a method known as paper chromatography
Positive interdependence: Assign each student a role such as group leader, materials coordinator, or timer.
Individual accountability: Have students describe what happened in the investigation and write a brief explanation of the events that they observed.

Answers to
Questions, pages 178–179

1. Students should apply their observations from Exploration 2 to answer this question. Detective Hanamoto must have dissolved some of the ink from the check in a very small amount of water and then used a chromatography technique. If the two inks were different, Detective Hanamoto would have observed that the inks traveled different distances up the paper and at different rates.
2. Most inks are true solutions. Students should suggest testing the solution for the Tyndall effect. They will find that a very dilute solution of ink and water is clear and does not show the Tyndall effect.
3. Sample answer: An *observation* is information gathered using one's senses. Once the observation has been made, an *inference* is a statement that attempts to explain an observation. A *hypothesis* is an explanation of an expected outcome of an experiment, and a *prediction* is a statement of an expected outcome of a future event. Each of these is a distinct step in the process of examining scientific data.
4. Detective Hanamoto acted like a scientist by doing the following:
   - He examined (observed) the check carefully. He also made careful observations of the results of his experiments.
   - He hypothesized that the accused had used a different ink to make the changes on the check. Therefore, testing the ink would reveal differences.
   - From his results, he inferred that the accused was guilty.
   - He predicted that a test would show whether or not the inks were different.
   - He performed an experiment to test the inks.
5. Possible answers include the following:
   - observing the Tyndall effect
   - measuring liquid volumes in investigations
   - classifying solutes and solvents
   - predicting the results of evaporating a solvent
   - experimenting with the composition of inks

EXPLORATION 2, continued

Questions

1. Now explain why Detective Hanamoto *inferred* that the check was written with two different inks.
2. Do you think inks are true solutions? Devise an experiment to find out.
3. Read the following cartoon, which shows Nathan thinking like a detective. Then explain the difference between observations, inferences, hypotheses, and predictions.

4. Did you notice the various ways in which Detective Hanamoto was doing the types of things that scientists do? Complete the table below in your ScienceLog, using specific examples from Detective Hanamoto's summary.

What scientists do	What Detective Hanamoto did
observe	?
hypothesize	?
infer	?
predict	?
experiment	?

5. "Being scientific" means doing the things scientists do. By learning about solutions, you are being scientific. The list below suggests a number of ways that scientists learn about our world. For each one, suggest in your ScienceLog an example of a situation in this unit in which *you* were scientific in the same way.

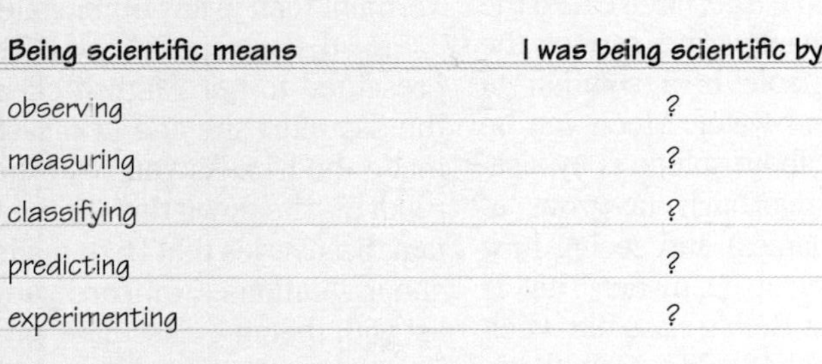

Being scientific means	I was being scientific by
observing	?
measuring	?
classifying	?
predicting	?
experimenting	?

Detective Work at Home 

The method you used to separate the colors of ink is called paper chromatography (*chroma* means "color"). You can use the same method to determine whether food coloring is made of only one color or is a solution of different colors. Try it, and report back to your class.

Detective Work at Home

This activity can be done at school or at home. Reinforce the need for parental permission before performing this activity at home. Also, warn students that nontoxic food coloring can stain skin and clothing. If they do this activity at school, make solutions of mixed colors and have students discover the colors that make up each solution.

Homework

You may wish to have different students test different colors of food coloring for the activity Detective Work at Home.

FOLLOW-UP

Reteaching

Either as a class demonstration or in student groups, have students repeat Exploration 2 using glycerin in place of water. Ask students to explain any differences they observe between the two experiments in terms of different solubilities of the inks in the two solvents.

Assessment

Have students write their own detective story in which they explain how chromatography helped them to solve a crime. (The process of chromatography should be described.)

Extension

Ask students to find the word *forensics* in the dictionary. Then have them research the kind of work a forensic scientist does. Ask them to explain how a knowledge of solutions is important when doing forensic activities.

Closure

Ask students to imagine that they are reporters for a consumer magazine. Their assignment is to write an article about food dyes. Have the reporters tell how their research laboratories used chromatography to find out which foods had few dyes added.

LESSON 4 Sugar Solutions

FOCUS

Getting Started

Pass out samples of maple sugar and display a jar of honey. Tell students that they will see how another property of solutions, the boiling point, is used to make maple syrup and maple candy. They will also learn how bees use evaporation to make honey.

Main Ideas

1. Solutes raise the boiling point of solvents.
2. The greater the amount of solute in a given amount of solvent, the higher the solution's boiling point.
3. In the production of honey, water is evaporated from the nectar; in the making of maple products, water in the sap is boiled away.

TEACHING STRATEGIES

Call on a volunteer to read aloud the introduction and Beth's report on the sugarbush on page 180. One of the purposes of this story is to illustrate that solutions boil at a higher temperature than do pure solvents. Point out that as the solution becomes more concentrated, its boiling point rises.

INDEPENDENT PRACTICE Ask students to make a list in their ScienceLog of the important points listed in Beth's report.

Answers to *In-Text Question*

Ⓐ Beth made the following discoveries: Sap is a solution made up of 98 percent water and 2 percent sugar. The temperature of a solution can be used to indicate its concentration. Increasing the amount of solute in a solution raises the solution's boiling point.

LESSON 4 Sugar Solutions

Maple syrup and maple sugar are made from the sap of the rock maple tree, often called the sugar maple. The sap of the sugar maple is a solution of sugar and water. Much can be learned about solutions by visiting a sugarbush (a grove of sugar maples) and seeing how maple syrup is made. That is just what Beth's class did when they visited Mr. Lowther's sugarbush in Vermont. More maple syrup is produced in Vermont than in any other state.

Read the report that Beth presented to her English class the day after she and her class made the trip. As you read the report, write down three important discoveries that Beth made about solutions. Compare your list with that of a classmate, giving reasons for your choices. Ⓐ

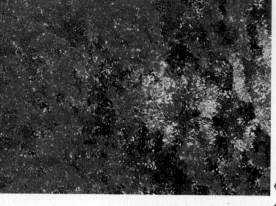

Mr. Lowther's sugarbush in autumn

Mr. Lowther, the owner of the sugarbush I visited, starts tapping his maple trees about the second week of March. If it is a warm spring, he begins sooner. If it's a large tree, Mr. Lowther can bore several holes into it. A small spout, called a spile, is pounded into each hole. In this sugarbush, all of the spiles are connected to a network of plastic tubing that carries the sap to the sugarhouse. There the sap is boiled down to make maple syrup. Mr. Lowther said that a single tree can supply between 40 L and 100 L of sap in a season.

The sugarhouse is an exciting place, full of steam and delicious odors from the boiling sap. I found out that sap consists of only 2 percent sugar and 98 percent water! The sap is boiled in the evaporator until the water content is decreased to 34 percent and the sugar content is increased to 66 percent. Mr. Lowther said that it takes 60 L of sap to make 1 L of maple syrup. Only when the syrup has the proper water and sugar content is it drawn off, strained, and bottled. I asked Mr. Lowther how he knew when the syrup contained 66 percent sugar. He said he uses a thermometer! That's right—when the temperature reaches 104°C, the solution is right for bottling.

Mr. Lowther makes other maple products as well. He prepared a whole batch of maple candy for the class by boiling some syrup in a big pot until the temperature reached 110°C. Then he poured the syrup onto a patch of clean snow, where it hardened. It was great!

A trip to the sugarbush is an interesting way of investigating solutions. We had a great time!

A Trip

Here is a chart of the other maple products Mr. Lowther makes, along with their boiling points.

Maple butter	*113°C*
Maple cream	*114°C–115°C*
Maple wax	*119°C*
Hard maple sugar	*123°C*

LESSON 4 ORGANIZER

Time Required
1 to 2 class periods

Process Skills
observing, analyzing, inferring, predicting

Theme Connection
Cycles

New Terms
none

Materials (per student group)
Exploration 3: 250 mL beaker; 65 g of white sugar; metric balance; 50 mL graduated cylinder; 35 mL of water; 3 drops of maple flavoring; eyedropper; hot plate; stirring rod; candy thermometer, or other thermometer that reads over 100°C; safety goggles; oven mitts

Teaching Resources
Exploration Worksheet, p. 30
Transparency 31
SourceBook, p. S56

Questions to Ponder

Discuss the answers to the following questions with some of your classmates:

1. What method of removing a solute from a solvent is used in making maple products?
2. What effect does sugar have on the boiling point of water? (Pure water boils at 100°C.)
3. What is the relationship between the number of degrees of change in the boiling point of the solution and the percentage of sugar in the solution?
4. How did Beth know that maple wax contains less water than maple cream?
5. At most, how many liters of maple syrup could one tree produce in a season?

Questions to Ponder

Have students work in pairs to discuss and formulate answers to the questions. Have them begin by identifying the steps in the process of making maple syrup. These steps are shown in the illustration on pages 180–181.

Answers to *Questions to Ponder*

1. The method of separating a solute and a solvent in making maple products involves boiling off the solvent (water) and leaving the solute (sugar) behind.
2. The sugar (solute) has the effect of raising the boiling point of water (solvent).
3. The higher the percentage of sugar in the solution, the higher the boiling point of the solution. Adding 66 g of sugar to 100 g of pure water raises the boiling point of the water 4°C.
4. Beth knew that maple wax contains less water than maple cream because the boiling point of maple wax is higher than that of maple cream.
5. According to Beth's report, a single tree can produce 40–100 L of sap in a season. Since about 60 L of sap is required to make 1 L of maple syrup, a tree that produces 100 L of sap will yield about 1.7 L of maple syrup.

Multicultural Extension

Maple Sugar and American Indians

Share the following information with students: Long before European explorers came to North America, American Indians who lived near the Great Lakes and the St. Lawrence River produced maple sugar and maple syrup. Early French and English explorers wrote of the "sweet water" that the American Indians obtained from maple trees. The American Indians heated this water to make maple products.

EXPLORATION 3

You may wish to perform this Exploration as a class demonstration. If artificial maple flavoring is not available, test the effect of sugar on the boiling point of water.

Results may not be exactly 104°C. Possible reasons for this discrepancy are the following:

- Most thermometers are not completely accurate.
- The location of the thermometer in the beaker might influence the results.
- Boiling point is affected by elevation.

Stress that there is a real difference between the boiling point of pure water and that of a sugar solution. The specific temperatures are not as important as the difference.

Students should not ingest any materials during this Exploration.

Answers to *Exploration 3*

A solution similar in concentration to Mr. Lowther's maple syrup should boil at 104°C. To ensure a fair test, the boiling point of pure water should be measured first.

Answers to *Mr. Lowther's Business*

1. Mr. Lowther collects sap from many maple trees.
2. All of the sap from Mr. Lowther's trees is carried by plastic tubing to the sugarhouse.
3. Mr. Lowther speeds up the evaporation of water by boiling the sap.
4. Mr. Lowther stores the product in bottles.
5. Syrup and other maple products are used by humans as food.

EXPLORATION 3

Making Artificial Maple Syrup

You Will Need

- a beaker
- a metric balance
- 65 g of white sugar
- a 50 mL (or larger) graduated cylinder
- 35 mL of water
- 3 drops of maple flavoring
- a hot plate
- oven mitts
- a stirring rod
- a candy thermometer (or one that reads over 100°C)

What to Do

Simply mix the sugar and water in a beaker. While stirring, heat the mixture gently until all of the sugar dissolves. Then add 3 drops of maple flavoring, and stir.

At what temperature do you predict this solution will boil if it has the same percentage of sugar as Mr. Lowther's maple syrup?

If you have a thermometer that reads over 100°C, test your prediction. To make a fair test of your prediction, what else should you do?

The Honey Factory

The honey made by bees comes from *nectar*, a solution that is mainly sugar and water. The bees gather nectar from flowers and then store it in open cells in the beehive. The bees fan their wings in front of the cells to help evaporate the excess water in the nectar. When the solution has reached the right thickness, they close off the open cells with wax. The result is honey. Bees can do what Mr. Lowther does, but without boiling.

Actually, there are many similarities between making maple syrup from sap and making honey from nectar. One similarity is given in the table below. In your ScienceLog, fill in the rest of the steps from Mr. Lowther's operation that correspond to the descriptions of the steps performed by the bees.

Mr. Lowther's business	The honeybees' business
1. Mr. Lowther collects sap from many maple trees.	1. The bees collect nectar from many flowers.
2. ?	2. The bees store the nectar in a central place in the hive.
3. ?	3. The bees speed up the evaporation of water from the thin nectar solution by fanning their wings.
4. ?	4. The bees store the product in cells and seal off the cells with wax.
5. ?	5. Honey is used as a food by bees as well as by humans.

182

FOLLOW-UP

Reteaching

Ask students to suggest ways of making syrup from maple sap without heating. *(Students may suggest using fans or some other means to move the air and speed up the evaporation process.)*

Assessment

Have students design an experiment that determines which of three powdered drink mixes has the highest concentration of sugar. *(Students could determine the boiling point for an equal volume of each liquid—the mixture with the highest boiling point has the highest sugar content.)*

Extension

Divide the class into three groups. Have each group research one of the three plants that are used to produce sugar: sugar beets, sugar cane, and sorghum.

Closure

Most of the maple syrup we buy at the grocery store is not pure maple syrup but a maple-flavored substitute. Have students research the production of maple syrup and report on why artificial substitutes have replaced pure maple syrup in most households.

1. Not So Sparkling Clean

Sometimes there are spots on your dishes after they have been washed. How can you explain this?

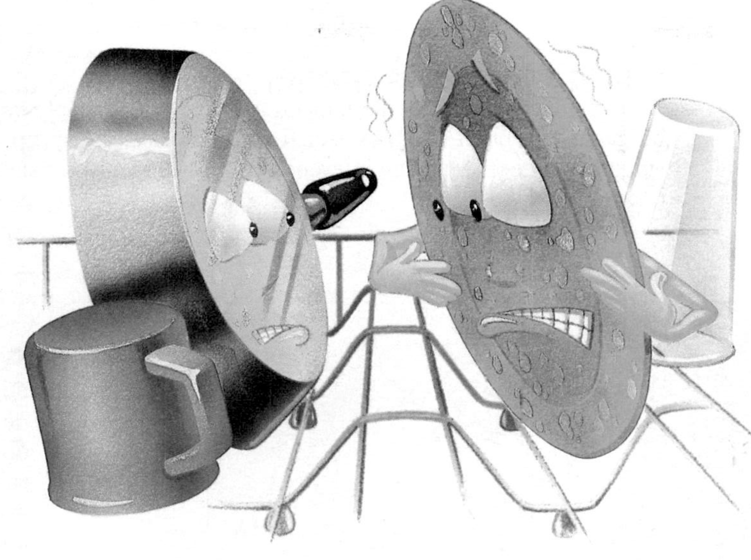

2. In Hot Water

Jane is boiling a beaker of sea water to find out how much salt remains.

a. Describe all of the things Jane might observe.

b. What will happen to the temperature of the boiling liquid as time passes? Why does this happen?

★ **An Exploration Worksheet is available to accompany Exploration 3 on page 182 (Teaching Resources, page 30). In addition, Transparency 31 accompanies The Honey Factory on page 182.**

Homework

You may wish to have students complete the table on page 182 as homework.

Theme Connection

Cycles

In deep water there is a water cycle much like the water cycle in the atmosphere. **Focus question:** What steps may be involved in an oceanic water cycle? *(Sample answer: Heat from within the Earth warms water at the bottom of the ocean. This water rises, bringing nutrients from the ocean's bottom up to the surface. As this water rises, cooler water sinks, creating a cyclic flow of water, nutrients, and other dissolved substances.)*

Answers to *Challenge Your Thinking*

1. Spots on your dishes are an indication that there are dissolved solutes in your water. As the water on the dishes evaporates, the solutes remain behind as a residue.

2. **a.** Jane might observe that the water undergoes a phase change and becomes a gas. When all of the water has evaporated, the dissolved salts will be left behind in solid form.

   b. As time passes, more of the water will evaporate, resulting in a higher concentration of salt in the remaining liquid. The higher the percentage of dissolved solutes, the higher the boiling point.

Answers to Challenge Your Thinking continued ►

★ **You may wish to provide students with the Chapter 8 Review Worksheet on page 31 of the Unit 3 Teaching Resources booklet.**

Cross-Disciplinary Focus

Language Arts

The term *hydroponics* refers to a technique for cultivating plants without using soil. Instead of soil, the plants are grown in a solution containing all of the essential nutrients for plants. Either artificial or natural light sources are used. Discuss hydroponics and hydroponic solutions with the class, highlighting the possible advantages and disadvantages of this technique. Suggest that students write a futuristic short story in which hydroponics plays an important role. Possible story settings could include space colonies, floating cities on the ocean, and so on. Unit 8 provides more information about hydroponics.

Answers to *Challenge Your Thinking,* continued

3. You could use paper chromatography. Different components of the solution will travel different distances up the filter paper and at different rates. This method would not work if the components of the solution were colorless or were both liquids. You could also try boiling off all of the solvent to see whether any dissolved substances remained behind.
4. Accept all reasonable responses. Just as solutes raise the boiling point, they also lower the freezing point. Therefore, the temperature in the freezer wasn't low enough to freeze the maple syrup. (Students will learn more about this in the next chapter.)
5. Accept all reasonable student diagrams. Sample answer: Place a large beaker in a pail, and fill the pail with salt water to a height of 2–3 cm. Cover the top of the pail with clear plastic wrap. Make a depression in the center of the plastic wrap by placing a small rock on top of the plastic wrap. Tape the edges of the plastic wrap to the sides of the pail to make the pail airtight. Place the pail in the sun. As the sun warms the salt water, water vapor will condense on the underside of the plastic wrap. The condensation will run down the depression formed by the rock and will drop into the empty beaker.

The following are sample revised answers:

1. A solute can be separated from a solvent through evaporation, boiling, desalination, and distillation. Desalination is the process of removing salt from water.
2. Two examples include separating salt from sea water to produce drinking water and separating water from maple sap to produce maple products.
3. Each day, vast quantities of water evaporate from the oceans, lakes, rivers, and the land, leaving behind dissolved solutes. Eventually, the pure water vapor condenses and falls as rain or snow.

3. More Detective Work

You are given a clear liquid. It does not show the Tyndall effect when a light is shined through it. Without tasting the liquid, how would you tell whether it is just one substance or a solution? Is there any case in which your method might not work?

4. Salim's Syrup Surprise

Salim wondered what would happen to the artificial maple syrup he made if he put it in the freezer. To his surprise, the syrup would not freeze. Make an inference about what caused Salim's results.

5. A Sunny Solution

On page 172, desalination is described as a process of obtaining fresh water from salt water. The most economical designs make use of solar power (heat from the sun). Come up with a design that does this. Share your design with others by drawing a diagram of it.

Review your responses to the ScienceLog questions on page 167. Then revise your original ideas so that they reflect what you've learned.

184

Meeting Individual Needs

Gifted Learners

Explain to students that the amount of oxygen dissolved in water varies greatly depending on the temperature. Warm water cannot hold as much dissolved oxygen as can cold water. Have students study a map or globe and make predictions about the water temperature and the oxygen concentration in different ocean areas. Have them do library research to check their predictions. Suggest that they research the importance of dissolved oxygen as a factor in the abundance and variety of marine life found in given locations.

CHAPTER 9 Concentration of Solutions

1 **If you place 100 mL of salt water and 100 mL of fresh water in the freezer, which will freeze first? Explain your reasoning.**

2 **What happens to a solution when you add more solute to it? when you add more solvent to it?**

3 **Do you think there's a limit to the amount of solute that can be dissolved in a solvent? If so, how would you describe a solution that has passed that limit?**

4 **What solutions are part of your environment? What solutions are inside you?**

ScienceLog

Think about these questions for a moment, and answer them in your ScienceLog. When you've finished this chapter, you'll have the opportunity to revise your answers based on what you've learned.

185

CHAPTER 9 Concentration of Solutions

Connecting to Other Chapters

Chapter 7
introduces students to solutions and non-solutions, components of a solution, and the concept of solubility.

Chapter 8
explores methods used to separate solutions, focusing on the separation of sugar and salt solutions.

Chapter 9
introduces students to the concept of saturation and explores solutions in the environment and the human body.

Prior Knowledge and Misconceptions

Your students' responses to the ScienceLog questions on this page will reveal the kind of information—and misinformation—they bring to this chapter. Use what you find out about your students' knowledge to choose which chapter concepts and activities to emphasize in your teaching.

In addition to having students answer the questions on this page, you may wish to have the students explain what they think the following words or phrases might mean and on what products they might be found. Accept all reasonable responses. Students may not know the correct answers, but they may be able to make better guesses at the answers after reading this chapter.

1. pH balanced *(This is found on many deodorants and shampoo bottles. pH is a measure of how concentrated an acid or base solution is. Pure water has a pH of 7. A discussion of pH can be found in Unit 3 of the SourceBook.)*
2. from concentrate *(This is found on many juice products. In these cases, the concentrate is juice with most of the water removed.)*
3. tincture *(This is found on many medicinal products, such as tincture of iodine. It denotes an alcohol or water-alcohol solution with a 10 to 20 percent concentration.)*
4. perfume, cologne, and toilet water *(Technically, perfume is a 10 to 20 percent solution, cologne is a 3 to 5 percent solution, and toilet water is the weakest, containing only about 2 percent odorants. The solvent in these solutions is mostly alcohol or water and alcohol.)* You may wish to have students read Perfumes: Fragrant Solutions on page 212.

Getting Colder

FOCUS

Getting Started

If you have access to an ice-cream machine, bring it to class and make ice cream so that students can see how it is made. Ask: Why is salt added to the ice? *(Salt lowers the melting point of the ice so that the liquid's temperature can fall below 0°C.)* Then ask students why a temperature lower than 0°C is needed to make ice cream freeze. *(Because sugar added to the cream lowers the cream's freezing point)*

Main Ideas

1. Solutes lower the freezing point of solvents.
2. The greater amount of solute in a given amount of solvent, the lower the solvent's freezing point.
3. The melting point of ice and the freezing point of water are both 0°C.
4. Adding salt to ice lowers the melting point, often to below the temperature of the surroundings, and therefore causes the ice to melt.

TEACHING STRATEGIES

EXPLORATION 1

Caution students not to use the thermometers to stir with.

Have students work in pairs. When the temperature remains constant (about 0°C), any remaining ice should be removed from the cup.

Two Exploration Worksheets are available to accompany Explorations 1 and 3 on pages 186 and 187 (Teaching Resources, pages 37 and 39, respectively).

Homework

Assign Exploration 2 on page 187 as homework, using the Exploration Worksheet on page 38 of the Unit 3 Teaching Resources booklet as an accompaniment.

Getting Colder

In Chapter 8, Beth discovered that there is a relationship between the boiling point of a solution and the amount of solute (sugar) dissolved in the solvent (water). Would a solute affect the freezing point of a solution as well? The following illustrations suggest an answer to this question.

The salt truck has finally arrived!

Yeah! Let's hope the salt melts the ice enough that we can make it to the game tonight.

What's the message of these illustrations? Here are three experiments to help you find out: one for the classroom (Exploration 1), one for home (Exploration 2), and one already completed for you to analyze (Exploration 3).

EXPLORATION 1

What Is the Temperature of Water When It Freezes?

You Will Need

- a 250 mL graduated cylinder
- a plastic-foam cup
- ice cubes
- a stirring rod
- a thermometer

What to Do

1. Add 150 mL of water to a plastic-foam cup.
2. Now add an ice cube, and stir with the stirring rod. Put the thermometer in the cup. Check the temperature. (Do not use the thermometer to stir the ice water.)
3. Keep adding ice cubes until the temperature will not go any lower. What is the lowest temperature that the water reaches? (A)
4. As soon as the temperature is constant, remove any ice that is left. How much ice did you use to cool the water to its lowest temperature? There is a way you can find out without finding the mass of the ice. (B)

LESSON 1 ORGANIZER

Time Required
2 class periods

Process Skills
observing, hypothesizing, analyzing

New Terms
none

Materials (per student group)
Exploration 1: 250 mL graduated cylinder; large plastic-foam cup; about 10 ice cubes; alcohol thermometer; 150 mL of water; stirring rod

Exploration 2: optional materials: 15 mL measuring spoon (1 Tbsp.); 3 juice glasses, plastic cups, or 250 mL beakers; 45 mL of sugar; 15 mL of salt; stirring rod; 100 mL graduated cylinder; about 250 mL of water (additional optional teacher materials: freezer)

Teaching Resources
Exploration Worksheets, pp. 37, 38, and 39
SourceBook, p. S56

EXPLORATION 2

Effect of a Solute on the Freezing of Water

You Will Need

- a 15 mL measuring spoon (1 tablespoon)
- 3 juice glasses or plastic cups
- 45 mL of sugar
- 15 mL of salt
- water
- a stirring rod

What to Do

Pour 75 mL (5 tablespoons) of water into a glass or cup. Put 45 mL of sugar in the second glass and 15 mL of salt in the third glass. Add enough water to the second and third glasses to make the solutions even with the water level in the first glass. Stir. Put the glasses in the freezer. Examine them every half-hour. In what order do they freeze? Why?

EXPLORATION 3

Mr. Kim's Home Investigation

What to Do

Analyze the results of Mr. Kim's experiment, and draw some conclusions from his findings.

What He Did

Mr. Kim surprised his family by taking over the kitchen one evening to do some experimenting. He collected several thermometers and took the salt, measuring cups, and glasses out of the cupboard. If you analyze the table of results below, you will learn what he did and what information he found. What conclusions can you make about the effect of the solute on the solvent?

Trial	Solutions mixed	Temperature when a crust forms on top	Time to freeze completely
1	5 mL of salt in 100 mL of water	−3.6°C	1 hour, 5 minutes
2	10 mL of salt in 100 mL of water	−7.0°C	1 hour, 30 minutes
3	15 mL of salt in 100 mL of water	−10°C	1 hour, 45 minutes
4	20 mL of salt in 100 mL of water	−14°C	2 hours
5	100 mL of water (no salt)	0°C	48 minutes

Look again at the illustrations on page 186. For each one, write a two-line caption that explains in scientific terms what is illustrated. C

Answers to *In-Text Questions, pages 186–187*

A Students should observe that the temperature of the ice and water will go no lower than 0°C.

B The new water volume will equal the original volume plus the water from the melted ice. Since the mass of 1 mL of water is 1 g, the mass of the ice that was used can be determined.

C The following are sample captions:
- Making ice cream: Adding more salt lowers the melting point of the ice so that the temperature of the liquid water can go below 0°C.
- Antifreeze in the car: Adding antifreeze to the water in the radiator keeps it from freezing by lowering its freezing point.
- The salt truck: Putting salt on ice makes the ice melt at a lower temperature.

EXPLORATION 2

Remind students not to eat or drink any substances during this Exploration.

Answers to *Exploration 2*

The plastic cup filled with water will freeze first, the sugar solution second, and the salt solution third. (Students may be confused by this because there are more milliliters of sugar than of salt. However, the sugar particles are so large that there are fewer particles in 45 mL of sugar than in 15 mL of salt. Therefore, the sugar freezes first because there are fewer particles dissolved in the water.)

Answers to *Exploration 3*

Possible conclusions include the following:
- Adding salt to water lowers its freezing point.
- Adding a solute to a solvent lowers the freezing point of the solvent.
- The more solute in a given amount of solvent, the lower the freezing point of the solvent.

FOLLOW-UP

Assessment

Ask students to create a fact sheet summarizing the important concepts in this lesson.

Reteaching

Ask the class to make a graph of Mr. Kim's experiments. Suppose that Mr. Kim wants to make 100 mL of two salt solutions, one of which freezes at −5°C and the other at −8°C. Have students use their graphs to determine how much salt he should use.

Extension

Have students determine the freezing points of some different types of antifreeze and record their results in a table. Caution students that antifreeze is poisonous.

Closure

Have students design an experiment to determine the best ratio of antifreeze to water for a car driven in temperatures from −10°C to −5°C.

How Much Solute?

FOCUS

Getting Started

Before class, fill two beakers with equal volumes of water. To the first beaker, add one drop of food coloring. To the second, add five drops of food coloring. Hold the two solutions up for the class to see. Ask: These two solutions both contain water and food coloring—why is one darker than the other? *(The darker one contains more coloring, or more solute.)*

Main Ideas

1. A solution becomes more concentrated when more solute is added to it and becomes more dilute when more solvent is added to it.
2. Ways to compare the concentration of two solutions include measuring the masses of a given volume, testing the buoyancy of objects, layering the solutions, and comparing freezing points.

TEACHING STRATEGIES

Call on a volunteer to read page 188 aloud. Involve students in a discussion of the terms *concentration, concentrated,* and *dilute.*

GUIDED PRACTICE At this point, you might want to introduce the concept of density. Explain that density is mass per unit volume. This is directly related to the concentration of solutions. When comparing two solutions, the more concentrated solution will have more particles of solute per unit of volume and will therefore have greater density.

How Much Solute?

Sap is a *dilute* sugar solution. Maple syrup is a more *concentrated* sugar solution. What do these terms mean?

Both beakers in the illustration at right contain the same type of solution. But how could you determine which one contains more solute?

The beakers contain two different **concentrations** of the solute and solvent. The one with *more* solute in the solution is the **concentrated** solution. The one with *less* solute in the solution is the **dilute** solution.

What would happen to the color of solution *A* if you added a lot of water to it? How would adding water to solution *A* change its concentration? Would solution *A* become more concentrated or more dilute? Now suggest two ways of changing the concentration of solution *B*. Would each change make solution *B* more concentrated or more dilute? Ⓐ

By now you may be noticing that the concentration of a solution is a *relative* thing. It applies when you compare solutions made of the same solute and solvent. You can compare the concentration of one solution with that of another having the same solute and solvent by using the terms *more concentrated* or *more dilute.*

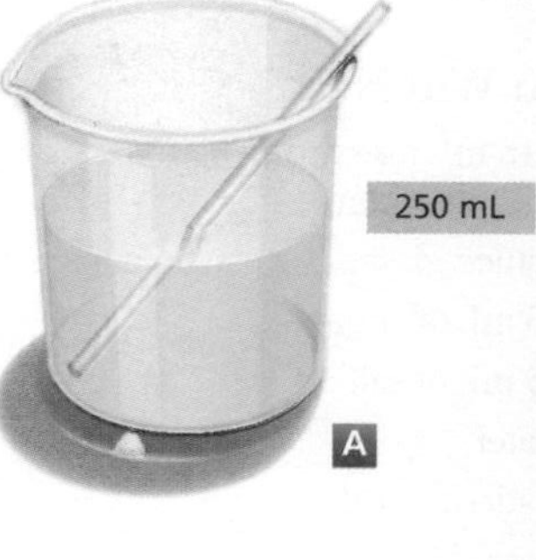

Concentrate on These!

1. Beakers *C* and *D* below contain the same solute and solvent. The amount of solute is the same in both beakers.
   a. What is the difference between *C* and *D*?
   b. After stirring, which solution will be more dilute? more concentrated?
2. Beakers *E* and *F* below contain the same solute and solvent. The amount of solvent is the same in both beakers.
   a. What is the difference between *E* and *F*?
   b. After stirring, which solution will be more dilute? more concentrated?

LESSON 2 ORGANIZER

Time Required
2 class periods

Process Skills
observing, analyzing, comparing, hypothesizing

New Terms
Concentrated solution—a solution with a relatively large amount of solute
Concentration—the amount of solute dissolved in a given amount of solvent
Dilute solution—a solution with a relatively small amount of solute

Materials (per student group)
Exploration 4: about 500 mL each of solutions *A* and *B*; two 500 mL beakers; metric balance; 10 mL graduated cylinder; 2 hard-boiled eggs; 2 test tubes; test-tube rack; several drops of food coloring; clear plastic drinking straw or glass tube; large plastic-foam cup; about 10 ice cubes; metric ruler; alcohol thermometer; stirring rod (additional teacher materials to prepare solutions *A* and *B*: 375 g of salt and 2 L of water for every two student groups)

continued ►

3. Stir to dissolve completely. Then add more solvent to solution *G*. The result is a more ___?___ solution.

4. Stir to dissolve completely. Then boil off some of the solvent in solution *H* to make a more ___?___ solution.

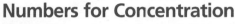

Numbers for Concentration

If you and your classmates were told to prepare a dilute solution of salt and water, would all of the solutions be identical? Probably not. The concentration of each solution would probably differ. In order for everyone to make identical solutions, you would have to be told the amounts of both the solute and the solvent. Another way of stating this is to say that you would need to know the concentration of the solution.

The concentration of a solution is often expressed in "*x* grams of solute dissolved in 100 g of solvent." Here are the concentrations of a few familiar solutions.

Solution	Solute	Solvent
maple sap	2 g sugar	100 g water
maple syrup	194 g sugar	100 g water
vinegar	5 g acetic acid	100 g water
sea water	3.5 g salt and other solutes	100 g water
apple juice	not less than 0.035 g vitamin C	100 g water

Mystery Solutions

It is not always possible to identify which solution is more concentrated just by looking. However, there are other properties or characteristics that can be used to distinguish a concentrated solution from a dilute one. For example, what property or characteristic of maple sap is used to determine the right concentration for maple syrup? Exploration 4 suggests some other ways to discover the relative concentrations of solutions. **B**

ORGANIZER, *continued*

Exploration 5: 25 mL each of 3 salt solutions; 3 large test tubes; test-tube rack; 3 used pencils about 8 cm long and 6 thumbtacks, or 2 plastic drinking straws cut in half, 6 nails, and small ball of modeling clay; metric balance; 25 mL graduated cylinder; metric ruler; scissors (additional teacher materials: 375 g of salt and 3 L of water per class)

Teaching Resources
Exploration Worksheets, pp. 40 and 43
SourceBook, p. S53

Answers to *In-Text Questions, pages 188–189*

A If water were added to solution *A*, it would become less intensely colored, less concentrated, and more dilute. Adding more solvent would make solution *B* more dilute. Adding more solute would make it more concentrated.

B The property of maple sap that is used to determine the right concentration of maple syrup is the boiling point of the solution.

Answers to *Concentrate on These!* *page 188–189*

1. **a.** Solution *D* contains more solvent than solution *C* does.
   b. Solution *D* will be more dilute; solution *C* will be more concentrated.
2. **a.** Solution *F* contains more solute than solution *E* does.
   b. Solution *E* will be more dilute; solution *F* will be more concentrated.
3. Dilute
4. Concentrated

Numbers for Concentration

Have students read the material and study the data table. They should realize that the hypothetical salt solutions discussed in the text will vary.

Meeting Individual Needs

Learners Having Difficulty

Evaluate students' understanding of the data table by asking questions similar to the following:

1. Which solution contains the least solute? *(The apple juice contains only 0.035 g of vitamin C in 100 g of water.)*
2. Which has the highest concentration? *(The maple syrup contains 1[illegible] g of sugar in 100 g of water.)*
3. Which is more dilute, maple syrup or maple sap? *(Maple sap)*

EXPLORATION 4

Set up four stations around the classroom, one for each test. If supplies are plentiful, have a number of setups at each station.

To make the dilute salt solution, dissolve 75 g of salt in 1 L of water. To make the concentrated salt solution, dissolve 300 g of salt in 1 L of water. Stir each solution, allowing time for the salt to dissolve completely. Label the containers *A* and *B*, respectively.

Remind students not to stir with the thermometers in Test 4.

Cooperative Learning
EXPLORATION 4

Group size: 4 students
Group goal: to determine the concentration of two salt solutions
Positive interdependence: Have each student perform one of the tests described in this Exploration.
Individual accountability: Randomly call on individuals from each group to explain what each test shows.

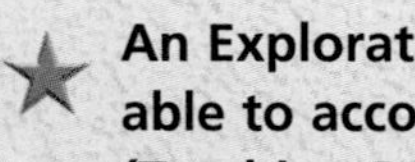

An Exploration Worksheet is available to accompany Exploration 4 (Teaching Resources, page 40).

Answers to *Test 1*

The mass of 10 mL of solution *A* should be 10.75 g. The mass of the same volume of solution *B* (13 g) is greater than that of solution *A*. Solution *B* is more concentrated because its mass is greater than that of an equal volume of solution *A*.

Answers to *Test 2*

The egg will float better in the concentrated salt solution. Concentrated solutions, because of their greater density, generally demonstrate greater buoyancy than do dilute solutions or pure solvents. Buoyancy is a force that opposes the force of gravity.

Explain to students that tests based on buoyancy are useful only when distinguishing between solutions of different concentrations made with the same solute and solvent.

If you used the Did You Know activity on page 157 to demonstrate surface tension, students may wish to know the difference between surface tension and buoyancy. Unlike buoyancy, surface tension is not related to density and is instead the result of the increased attraction among molecules at a liquid's surface. Because water's surface molecules have no additional water molecules above them, they form especially strong attachments with the water molecules that are adjacent to them. Soap forms a suspension at the water's surface and weakens the attractive forces among the water molecules, reducing the surface tension and causing the needle to sink. Even though the soapy water has increased in buoyancy, its density is still less than that of the needle.

EXPLORATION 4

Concentrated or Dilute?

You Will Need

- 2 salt solutions of different concentrations—solution *A* and solution *B*
- 2 beakers
- a metric balance
- a 10 mL graduated cylinder
- 2 hard-boiled eggs
- 2 test tubes and a test-tube rack
- food coloring
- a clear plastic straw
- a plastic-foam cup and ice
- a metric ruler
- a thermometer
- a stirring rod

What to Do

Form into small groups for this Exploration. Your teacher will give each group two salt solutions—solution *A* and solution *B*. One is concentrated; the other is dilute. Perform the following tests on each solution. Explain how each test enables you to distinguish the dilute solution from the concentrated one.

TEST 1

Find the mass of a beaker. Add 10 mL of solution *A* to the beaker and find the new mass. What is the mass of 10 mL of solution *A*? Repeat with 10 mL of solution *B*. How does the mass of solution *A* compare with the mass of solution *B*? Which is more concentrated, solution *A* or solution *B*? Why? Return each solution to its original container.

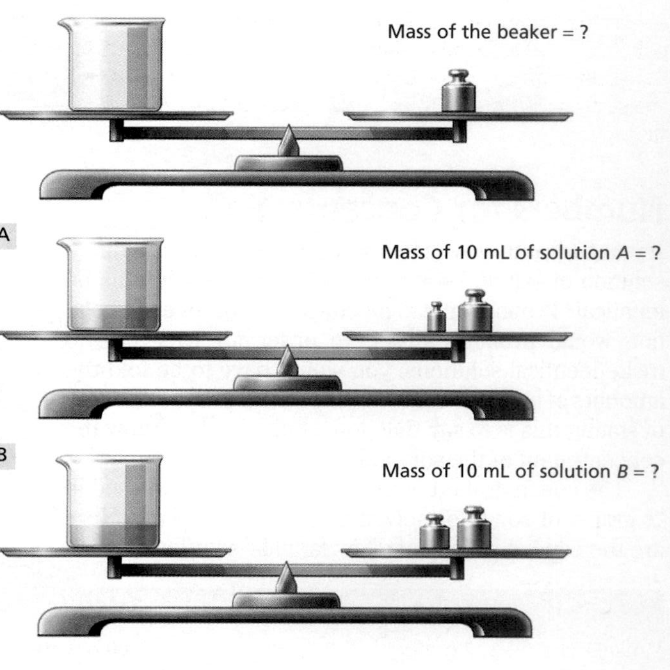

TEST 2

Carefully place an egg into each solution. What happens to the egg in solution *A*? in solution *B*? Which is more concentrated, solution *A* or solution *B*? Why?

190

TEST 3 

Fill two test tubes halfway with each solution. Add several drops of food coloring to test tube *B*. Using a clear plastic straw or glass tube, form two separate layers of the solution inside the straw by following the steps below.

Step 1

Lower the straw about 1 cm into solution *A*. Let solution *A* enter the straw. Now cover the top of the straw securely with your finger.

Step 2

Transfer the straw with the solution from test tube *A* to test tube *B*, and lower the straw 3 cm or more into solution *B*.

Step 3

Lift your finger off the top of the straw to let solution *B* enter. Cover the top of the straw again. Remove the straw from test tube *B*. Which solution is on top? Or did the two mix together?

Reverse the whole procedure, this time putting the straw into test tube *B* first. Can you form layers with the colored solution on the bottom? on the top? Which appears more concentrated, solution *A* or *B*? How did you decide?

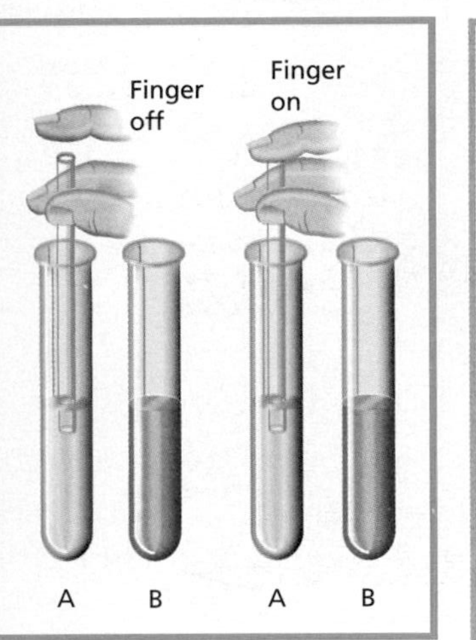

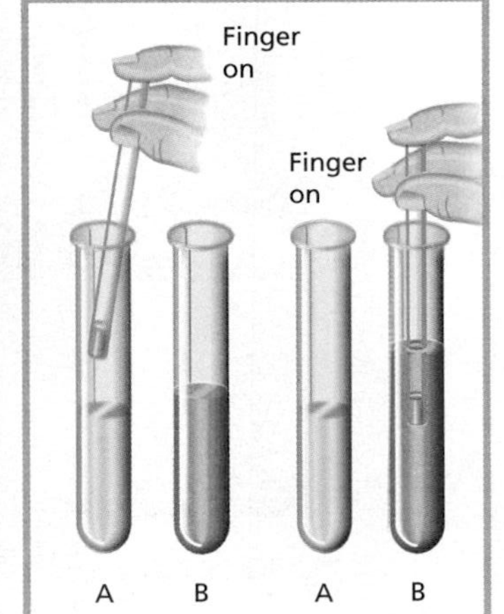

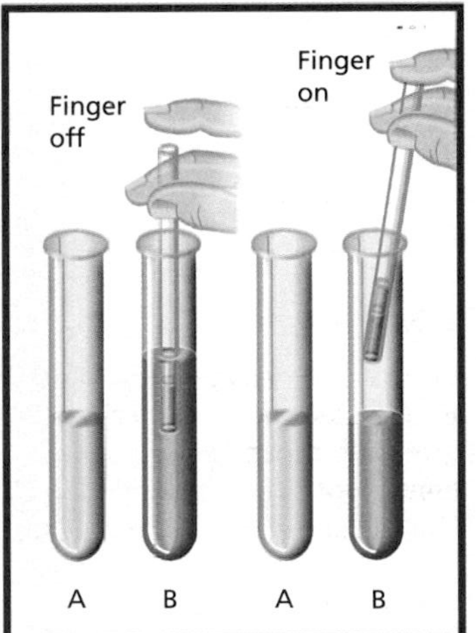

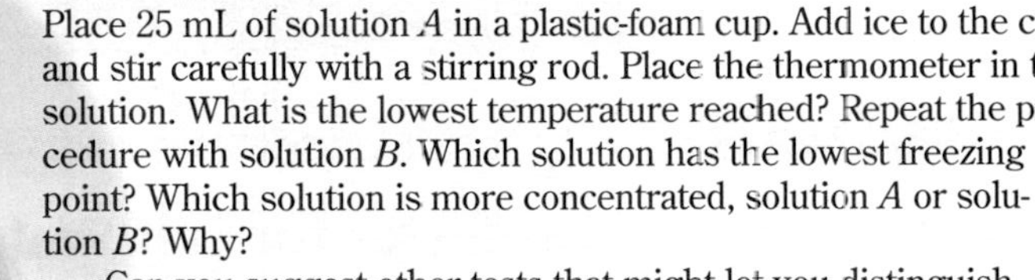

TEST 4

Place 25 mL of solution *A* in a plastic-foam cup. Add ice to the cup and stir carefully with a stirring rod. Place the thermometer in the solution. What is the lowest temperature reached? Repeat the procedure with solution *B*. Which solution has the lowest freezing point? Which solution is more concentrated, solution *A* or solution *B*? Why?

Can you suggest other tests that might let you distinguish between solutions of different concentrations?

Answers to *Test 3*

If students are careful, they will be able to layer (without mixing) the two salt solutions inside the straw or the tube. The dilute salt solution will be the top layer because it contains less solute and is therefore less dense. Students will therefore most likely be able to infer that the heavier, more concentrated solution (solution *B*) will be on the bottom.

Answers to *Test 4*

Students will find that they can achieve a lower temperature reading with the concentrated salt solution. This is because the freezing point of a solution decreases proportionally with the amount of solute it contains.

Other tests to distinguish solutions of different concentrations include comparing boiling points of the solutions and evaporating equal volumes of the solutions and comparing the masses of any solids that remain.

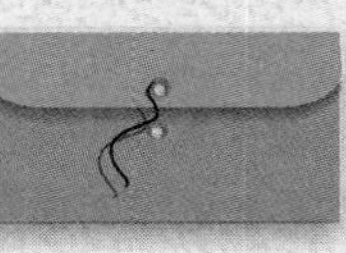

PORTFOLIO

Students may wish to include their results from Exploration 4 in their Portfolio. Encourage them to organize their data from each of the tests so that their results are clear and easy to follow.

Integrating the Sciences

Life and Physical Sciences

Honeybees flap their wings to increase the evaporation of water from their honey, making it a more concentrated solution. Inform students that many structures in organisms serve to regulate the concentration of solutions. For instance, kidneys regulate the concentration of minerals and other dissolved components in blood. Guard cells on the surface of a leaf control the exchange of gases with the surrounding air. Also, many membranes in organisms use osmosis to regulate the concentration of glucose, salt, and other solutes in water.

Solution Fact

Direct students' attention to the photograph on page 192. Some students may have visited the Dead Sea or other bodies of water with high salt concentrations, such as Utah's Great Salt Lake. Ask students to share their experiences concerning floating in salty bodies of water.

Point out that the shape of the image in the advertisement represents the shape of the Dead Sea.

Answer to *Solution Fact*

Higher salt concentrations cause greater buoyancy. The salt solution in the Dead Sea is very concentrated. It exerts a greater buoyant force on the human body than does fresh water and therefore makes it easier for humans to float in the water.

Exploration 5

This investigation explores the relationship between buoyancy and concentration. Use the two salt solutions prepared for Exploration 4 and a sample of pure water as the three solutions to be tested.

Once students have completed this Exploration, you might ask them to layer the three liquids in a clear plastic straw or tube as they did for Test 3 in Exploration 4.

GUIDED PRACTICE Ask students: Is it important that all three pencils used are exactly the same in composition and size? Why or why not? *(Yes, it is important because changing the pencil would alter more than one variable, and you would not know what variables influenced your results.)*

An Exploration Worksheet is available to accompany Exploration 5 (Teaching Resources, page 43).

Solution Fact

Here is an advertisement from the window of a travel agency. Read it, and then discuss the following question with a few of your classmates:

What property of the Dead Sea makes it easy to float in its waters?

Enjoy Life at the Dead Sea on the border between Israel and Jordan!
The Dead Sea
Spend your next vacation on the shores of the Dead Sea, where you can enjoy the sunshine and the water.
We guarantee that you will float.
A great vacation spot for all you sinkers!

Exploration 5

Make a Concentration Tester for Salt Water

You Will Need

- 3 different concentrations of salt solutions
- 3 test tubes
- a test-tube rack
- 3 used pencils about 8 cm long each (or 2 straws, each cut in half)
- 6 thumbtacks (or 6 nails and modeling clay)
- a metric balance
- a 25 mL graduated cylinder
- a metric ruler
- scissors

192

Homework

Have students design a creative travel poster for the Mediterranean Sea, which also has a high concentration of dissolved salt.

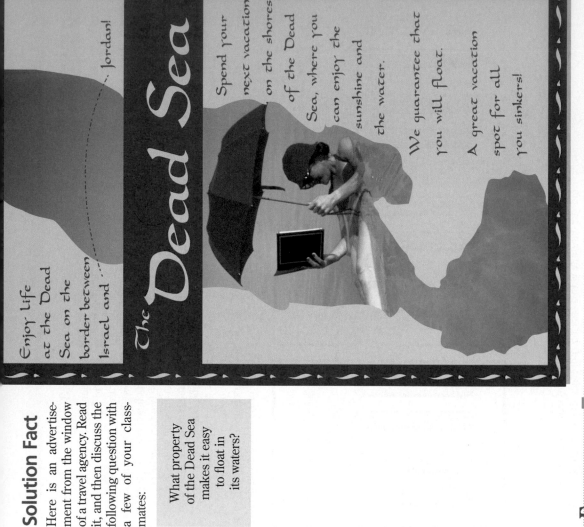

What to Do

1. Make three of one of the types of testers illustrated.
2. Find the mass of 25 mL of each solution. Then pour 25 mL of each solution into its own test tube.
3. Place one tester in each test tube as shown.

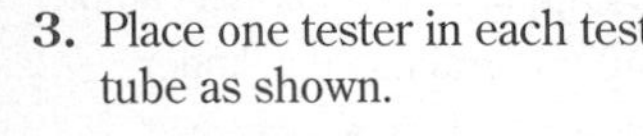

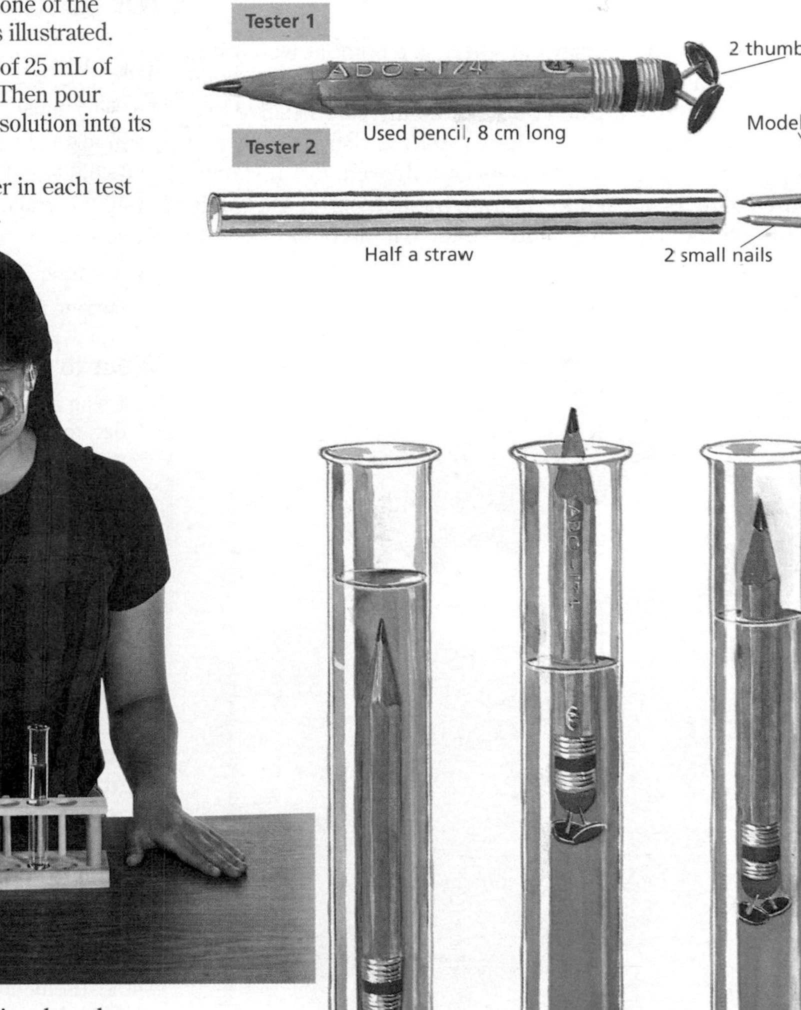

In which solution does the pencil or the straw float the highest? the lowest? in between?

Compare the masses of the solutions. Can you see any relationship between the mass of a solution and its concentration? between the concentration of a solution and how well an object floats in it?

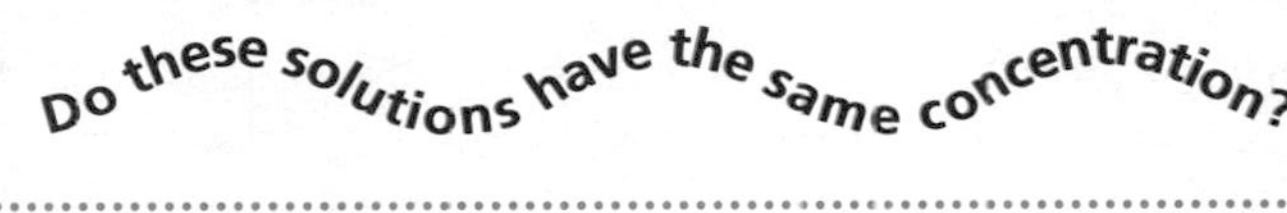

Do these solutions have the same concentration? Ⓐ

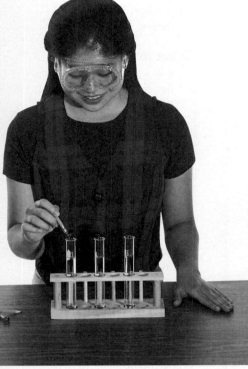

Answers to *Exploration 5*

The pencil (or straw) tester will float the highest in the concentrated salt solution (300 g of salt in 1 L of water) and lower in the dilute salt solution (75 g of salt in 1 L of water). The tester will float the lowest in the pure water sample. The solution with the greater mass has the greater amount of solute and is thus more concentrated. Students should realize that objects have greater buoyancy in more concentrated solutions.

When comparing the masses of the solutions, students will find that the most concentrated solution has the greatest mass because it contains the greatest amount of solute. The mass of 25 mL of the concentrated salt solution is 32.5 g. The mass of 25 mL of the dilute salt solution is 26.9 g. The mass of 25 mL of pure water is 25 g.

Answer to *Caption*

Ⓐ None of the solutions pictured have the same concentration because the testers all float at different heights.

FOLLOW-UP

Reteaching

Have students explore the following tests for concentration:

a. Measure and record the mass of two equal volumes of solutions. Next, place the solutions in separate dishes and let them evaporate. Then find the mass of the solute remaining in each dish.

b. Float a small ruler vertically in each of two solutions. Record the depth to which it sinks in each. (This works best in a tall, narrow container such as a graduated cylinder.)

Assessment

Have students write a paragraph to describe how and why the concentration tester they built in Exploration 5 works. In their description, ask them to include such words as *dilute, concentrated,* and *concentration.*

Extension

Many household products, such as cans of soup or household cleaners, must be diluted before they can be used. Challenge students to list as many concentrated household solutions as they can find, noting the main ingredients and the reasons why each product is stored in concentrated form. Caution students to handle these products with adult supervision.

Closure

Ask students to write and illustrate a brief description of how a solution's characteristics of density, buoyancy, and boiling point are related to each other.

Saturation

FOCUS

Getting Started

Pour cold soda into a glass. Warm a plastic stirring rod in warm water, and place it in the glass of soda. Ask students to observe what is happening. *(More bubbles are coming out of solution near the glass rod.)* Explain that the heat from the rod flowed into the cold solution of carbon dioxide and water and caused the carbon dioxide to become less soluble. Tell students that in this lesson, they will learn more about how temperature affects solubility.

Main Ideas

1. A solution is saturated when no more solute can be dissolved in it.
2. Solubility is expressed as the number of grams of solute that can be dissolved in 100 g of solvent.
3. In most cases, the solubility of solids increases as temperature increases.

TEACHING STRATEGIES

Exploration 6

Divide the class into small groups and distribute the materials. Once students have their solubility values, review their calculations with them. Involve students in a discussion of why their results might vary from the value given in the table on page 195. *(The main source of error will probably occur in determining when the solution was saturated.)*

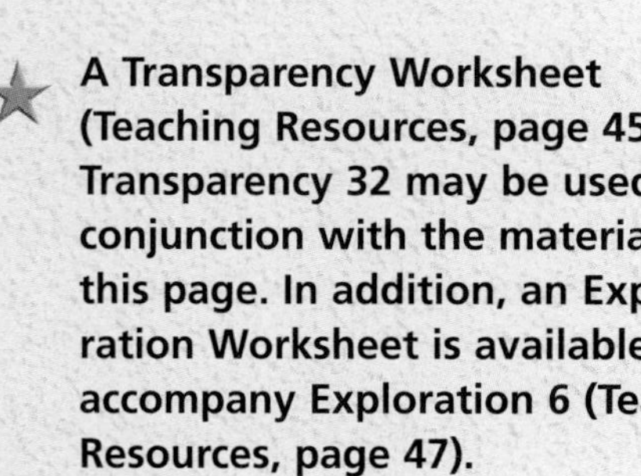

A Transparency Worksheet (Teaching Resources, page 45) and Transparency 32 may be used in conjunction with the material on this page. In addition, an Exploration Worksheet is available to accompany Exploration 6 (Teaching Resources, page 47).

Answers to Comparing Solubility Figures are on the next page. ►

Saturation

A sponge soaks up water easily, but there is a point at which it is unable to soak up any more water. At that point a sponge is said to be saturated. Similarly, when no more solute can dissolve into the solvent, a solution is said to be **saturated**. Earlier, you learned that water does not dissolve all things equally well. For instance, much more solute is needed to saturate the solution if the solute is sugar than if the solute is baking soda. The amount of solute needed to saturate a solution can be measured.

Exploration 6

How Much Solute for Saturation?

You Will Need

- a salt sample
- a graduated cylinder large enough to hold your salt sample
- a 10 mL graduated cylinder
- water
- a test tube
- a stirring rod

What to Do

1. Using a dry graduated cylinder, measure the volume of salt in your sample.
2. Add salt gradually to a test tube containing 10 mL of water until no more salt dissolves. Be sure to stir well after each addition of salt.
3. Measure what is left of your sample of salt. How many milliliters of salt did you use?

Comparing Solubility Figures

One milliliter of salt has a mass of about 1.2 g. What mass of salt did you add to 10 mL of water? What mass of salt would dissolve in 100 mL of water? What mass of salt is needed to saturate 100 g of water? (Remember, 100 g of water is about the same as 100 mL of water.) How does this figure compare with the figure given for salt in the table of solubilities on page 195?

LESSON 3 ORGANIZER

Time Required
2 class periods

Process Skills
analyzing, measuring, communicating, inferring

New Terms
Saturated—the point at which no more solute can dissolve into the solvent in a solution
Solubility—the amount of solute needed to saturate a solution; the number of grams of solute that can dissolve in 100 g of solvent

Materials (per student group)
Exploration 6: about 10 mL of salt; 10 mL of water; 10 mL graduated cylinder; test tube; stirring rod

Teaching Resources
Transparency Worksheet, p. 45
Exploration Worksheet, p. 47
Activity Worksheet, p. 48
Transparencies 32–35
SourceBook, p. S54

Saturation and Temperature

The amount of solute needed to saturate a solution is a measure of the solute's **solubility.** Solubility is often expressed as the number of grams of solute that can dissolve in 100 g of solvent. Do you think the temperature of a solvent affects solubility? The following solubility table may provide the answer. It shows the maximum amount of different solutes (in grams) that can be dissolved in 100 g of water at different temperatures.

Temperature (°C)	Salt (sodium chloride) (g)	Sugar (sucrose) (g)	Baking soda (sodium bicarbonate) (g)	Alum (potassium aluminum sulfate) (g)
0	35.7	179	6.9	4.83
10	35.8	191	8.2	6.90
20	36.0	204	9.6	9.63
30	36.2	220	11.1	14.30
40	36.5	238	12.7	22.20
50	36.8	260	14.5	30.30
60	37.3	287	16.4	46.90
70	37.6	320	—	74.50
80	38.1	362	—	138.80
90	38.6	414	—	—
100	39.2	487	—	—

Making Sense of the Data

1. How does temperature affect the amount of solute that can be dissolved?
2. What substance listed in the table is most soluble at 0°C? least soluble at 0°C?
3. List the given solutes from most soluble to least soluble at 0°C and then at 60°C.
4. What is the solubility of sugar at 0°C? At what temperature is the solubility of sugar approximately double its solubility at 0°C?
5. Look at the solubility of other solutes at 0°C. At what temperature would the solubilities of each be approximately double their solubility at 0°C?
6. What substance shows the smallest increase in solubility as temperature increases?
7. How much sodium chloride must you add to 100 g of water at 50°C in order to saturate the solution?
8. If you added 25 g of alum to 100 g of water and heated the solution to 80°C, would all of the alum dissolve? What would happen if this solution of alum were cooled to 20°C?

Answers to *Comparing Solubility Figures, page 194*

Approximately 3.6 g of salt will saturate 10 mL of water at room temperature.

Approximately 36 g of salt will dissolve in 100 mL of water at room temperature.

Approximately 36 g of salt is needed to saturate 100 g of water.

From the table, the maximum amount of salt that will dissolve in 100 g of water at room temperature (20°C) is 36 g. Student results may vary but should be fairly close to this number.

Saturation and Temperature

GUIDED PRACTICE Direct students' attention to the table of solubilities. Ask: As the temperature in column 1 increases, what do you notice about the quantity of each solute? *(The number of grams of each solute increases as the temperature increases.)* Point out that this is not true for all solutes.

Answers to *Making Sense of the Data*

1. The amount of solute that can be dissolved increases as the temperature increases.
2. The most soluble substance at 0°C is sugar; the least soluble substance at 0°C is alum.
3. The most to least soluble at 0°C are sugar, salt, baking soda, and alum. The most to least soluble at 60°C are sugar, alum, salt, and baking soda. (Note: There is a dramatic increase in the solubility of alum, greater than that of baking soda and salt. Thus, it can be inferred that temperature has a greater effect on the solubilities of some substances than on the solubilities of others.)
4. The solubility of sugar at 0°C is 179 g per 100 g of water. The solubility of sugar is double this amount at a little less than 80°C.
5. The solubility of salt does not double. The solubility of baking soda at between 40°C and 50°C is double that at 0°C. The solubility of alum at about 20°C is double that at 0°C.
6. Salt shows the smallest increase in solubility as the temperature rises.
7. 36.8 g
8. Yes, all of the alum would dissolve. If cooled to 20°C, the solution would be able to hold only 9.63 g. Therefore, 15.37 g (25 g – 9.63 g = 15.37 g) of alum would come out of solution.

Answers to
Picturing the Data

1. The starting points show the solubilities of the solutes at 0°C. Therefore, you know that all of the solutes have different solubilities at 0°C.
2. The line for salt is almost horizontal. This shows that temperature has little effect on its solubility.
3. The lines for sugar and alum are very curved. This shows that temperature has a great effect on their solubilities.
4. The steep curve tells you that the solubility of that substance increases dramatically with temperature.

Answer to
Caption

Ⓐ The graph on page 196 shows the effect of temperature on the solubility of sugar in water.

Homework

You may wish to assign the Activity Worksheet on page 48 of the Unit 3 Teaching Resources booklet as homework. If you choose to use this worksheet in class, Transparency 35 is available to accompany it.

Picturing the Data

Another way of representing the data in the table is to show it in a graph such as the one below. The vertical axis of the graph is a scale of the solute's solubility. The horizontal axis shows the temperature. Use the information in the table of solubilities on page 195 to graph each solute. Use different symbols for each solute (for example, X for salt, * for sugar, O for baking soda, and + for alum).

1. All of the lines start at a different place on the vertical axis. What does this tell you?
2. Which line is almost horizontal? What does this tell you?
3. Some lines are very curved. What does this tell you?
4. One line curves very steeply. What does this tell you?

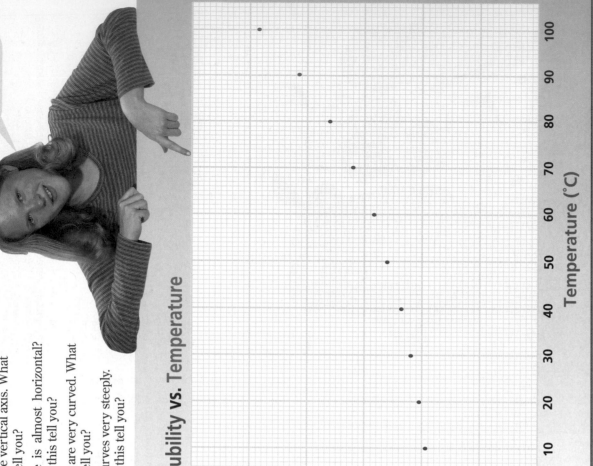

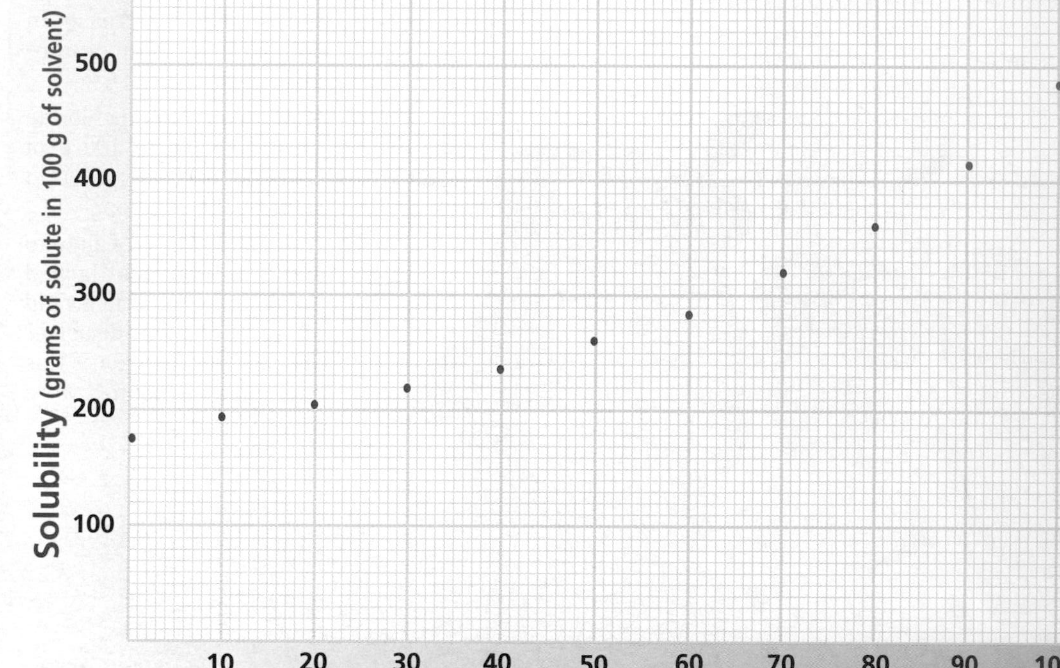

196

FOLLOW-UP

Reteaching

Have students write a paragraph comparing the effects of temperature on the solubilities of salt, sugar, baking soda, and alum in water. (They may refer to the table in their text.)

Assessment

Have students imagine that they are "solutionologists" and that the president of a food company has come to them for advice. The company president says that her company, Mr. Sweetie's, sells artificial maple syrup. The company's workers mixed an unknown amount of sugar with water, heated the mixture until all of the sugar dissolved, and poured the solution into bottles. After the bottles cooled, however, customers complained that sugar crystals had formed in the bottles. Students should write a brief report to the company president telling her what went wrong with her procedure. *(The amount of sugar that is soluble in water at a higher temperature is greater than the amount that is soluble at a lower temperature. When the solution cooled, the excess sugar formed crystals.)*

Extension

To illustrate that gases can be solutions too, have students research what "relative humidity" is. *(Relative humidity is the amount of water vapor in the air compared with the maximum amount of water vapor that the air could contain at the same temperature; it is expressed as a percentage.)*

Closure

Repeat the demonstration described in Getting Started on page 194 and have students explain the relationship between temperature and the solubility of the carbon dioxide. *(As the temperature rises, the carbon dioxide becomes less soluble.)*

LESSON 4 Solutions in the Environment

While reading this lesson about two important solutions, make your own list of words that refer to solutions. The list at right will help you get started. Your list will be useful later on when you work on a crossword puzzle.

- concentrated
- concentration
- dilute
- dissolve
- distillation
- evaporate
- mixture
- saturated
- solute
- solution
- solvent

The Atmosphere

The air around you and around the Earth is called the **atmosphere**. It is a solution. How do you know that it's a solution? Here's how:

- The atmosphere is a mixture;
- like liquid solutions, the atmosphere is clear;
- the atmosphere can be separated into its parts; and
- the concentrations of the parts do not always have to be the same.

The atmosphere is like a huge blanket around the Earth. It is several thousand meters thick. Nevertheless, human actions can change the concentrations of some of its parts. When we drive a car, the gasoline burns. This produces a gas called *carbon dioxide* that becomes one of the solutes in the atmosphere. Burning wood, oil, and other fuels also produces carbon dioxide. Some people think that putting so much carbon dioxide into the air will warm up the atmosphere and therefore slowly warm up our climate. What effect would this have on us? Ⓑ

Carbon dioxide isn't the only gas produced when we burn fuel. Factories that burn coal send many metric tons of *sulfur dioxide* into the air every day. Sulfur dioxide is a very poisonous gas. Unit 1 explains that sulfur dioxide is soluble in water and that when the two are mixed, the solution created is acidic. Rain that falls through air containing sulfur dioxide can produce a solution called acid rain. Many lakes and rivers are becoming more and more acidic because of the acid rain falling into them. Some are so acidic that fish cannot live in them anymore.

LESSON 4 ORGANIZER

Time Required
2 class periods

Process Skills
analyzing, hypothesizing, classifying

Theme Connection
Changes Over Time

New Term
Atmosphere—the solution of air surrounding the Earth

Materials (per student group)
none

Teaching Resources
Theme Worksheet, p. 49
Resource Worksheet, p. 51
Transparency 36

LESSON 4 Solutions in the Environment

FOCUS

Getting Started

Put several leaves into a bowl of vinegar. Ask students to observe the leaves for 5 or 10 minutes. Ask: What caused the change in the leaves? *(Accept all reasonable responses.)* Explain that vinegar is acidic, like acid rain, and that acid rain can have similar effects on plants in the environment.

Main Ideas

1. The atmosphere is a solution.
2. Sulfur dioxide is a pollutant in the atmosphere that can cause acid rain.
3. The ocean is the largest liquid solution on Earth.

TEACHING STRATEGIES

The Atmosphere

Have students copy the list of solution words on page 197 into their ScienceLog. Discuss any words that the students do not understand. Then involve students in a discussion of the new term *atmosphere.* Review why the atmosphere is considered a solution.

When discussing the material on pages 199–200, it may be helpful to point out that both the atmosphere and the ocean are composed of solvents that have solutes dissolved in them. However, because both the atmosphere and the ocean may have particles of other substances floating in them, they may also be considered suspensions.

Answer to *In-Text Question*

Ⓑ It could affect us in several ways. It could cause our oceans to absorb more heat energy or cause a change in the patterns of ocean currents, both of which would profoundly affect the world's weather. Such disruptions in weather patterns would inevitably affect agriculture and our ability to feed ourselves. Sea levels might also rise because the icecaps would begin to melt.

The Ocean

Call on a volunteer to read page 198 aloud. Involve students in a discussion of the solutes in sea water and the importance of the two dissolved gases for water animals and plants.

Sea water is a complex solution that contains many dissolved substances. You might want to explain to students that a measure of the overall amount of all salts present in a sample of sea water is known as *salinity*. The salinity of a sample of sea water can vary depending on where the sample is collected.

Variations in salinity result from such factors as incoming fresh water from river systems and the rate of evaporation.

Meeting Individual Needs

Learners Having Difficulty
To illustrate the principles discussed on page 198, have students evaporate some sea water. If sea water is not available in your location, use the steps given below to make a solution of artificial sea water of standard salinity as defined by the International Association of Physical Sciences of the Ocean.

1. To 500 mL of distilled water, add 12.03 g of anhydrous $MgSO_4$ and stir until completely dissolved.
2. In another 500 mL of distilled water, dissolve 2.19 g of anhydrous $CaCl_2$.
3. In a large container, mix the solutions prepared in steps 1 and 2 above.
4. To the mixed solution prepared in step 3, add 35.1 g of NaCl and stir until dissolved.
5. Add 1.5 g of KCl and stir thoroughly.

Meeting Individual Needs

Gifted Learners
Adaptations of marine organisms to life in salt water is a possible topic for student research or class discussion. Suggest that students research the adaptations to a salty environment that are found in marine fish, marine mammals, and marine birds.

The Ocean

The atmosphere is not the only large and important solution in the world. The ocean is another. Indeed, the ocean is the largest liquid solution on Earth. Parts of it are deep enough to swallow up Mount Everest—the world's highest mountain!

The solvent of this huge solution is water. The ocean contains many different solutes. What are they? This is the sort of question that can be answered by a chemist. Here is what a chemist would find by separating the water from the solutes in some sea water.

Mount Everest
8848 m
Marianas Trench
(Pacific Ocean)
11,036 m

Parts of the ocean are deep enough to swallow up Mount Everest.

If 100 g of sea water is evaporated completely, 3.5 g of dissolved solids will be left behind. Most of this, about 2.4 g, is sodium chloride—table salt. The remainder consists mainly of compounds of magnesium, sulfur, calcium, potassium, bromine, and iodine. Someday the ocean may be a valuable source of these materials, but so far only sodium chloride, magnesium, and bromine are extracted on a commercial basis.

Solids are not the only solutes found in sea water. Two important gases that are dissolved in the ocean are oxygen and carbon dioxide. Without the dissolved oxygen, most of the animals in the ocean could not exist; without the dissolved carbon dioxide, plant life in the ocean would die.

Theme Connection

Changes Over Time
Oceans, rivers, and lakes always have a variety of solutes dissolved in them, including minerals, salts, fertilizers, and pesticides. **Focus question:** What natural processes cause the concentration of these solutes to become greater over time? *(When some of the water in a solution evaporates, the solutes are left behind. As a result, the remaining solution is more concentrated. Therefore, oceans, rivers, and lakes become more concentrated as water evaporates from them over time.)* An accompanying Theme Worksheet is available to be assigned as homework (Teaching Resources, page 49).

Word Hunt

The answers to the clues for this crossword puzzle are probably in the list on page 197. Copy the puzzle into your ScienceLog and complete it there.

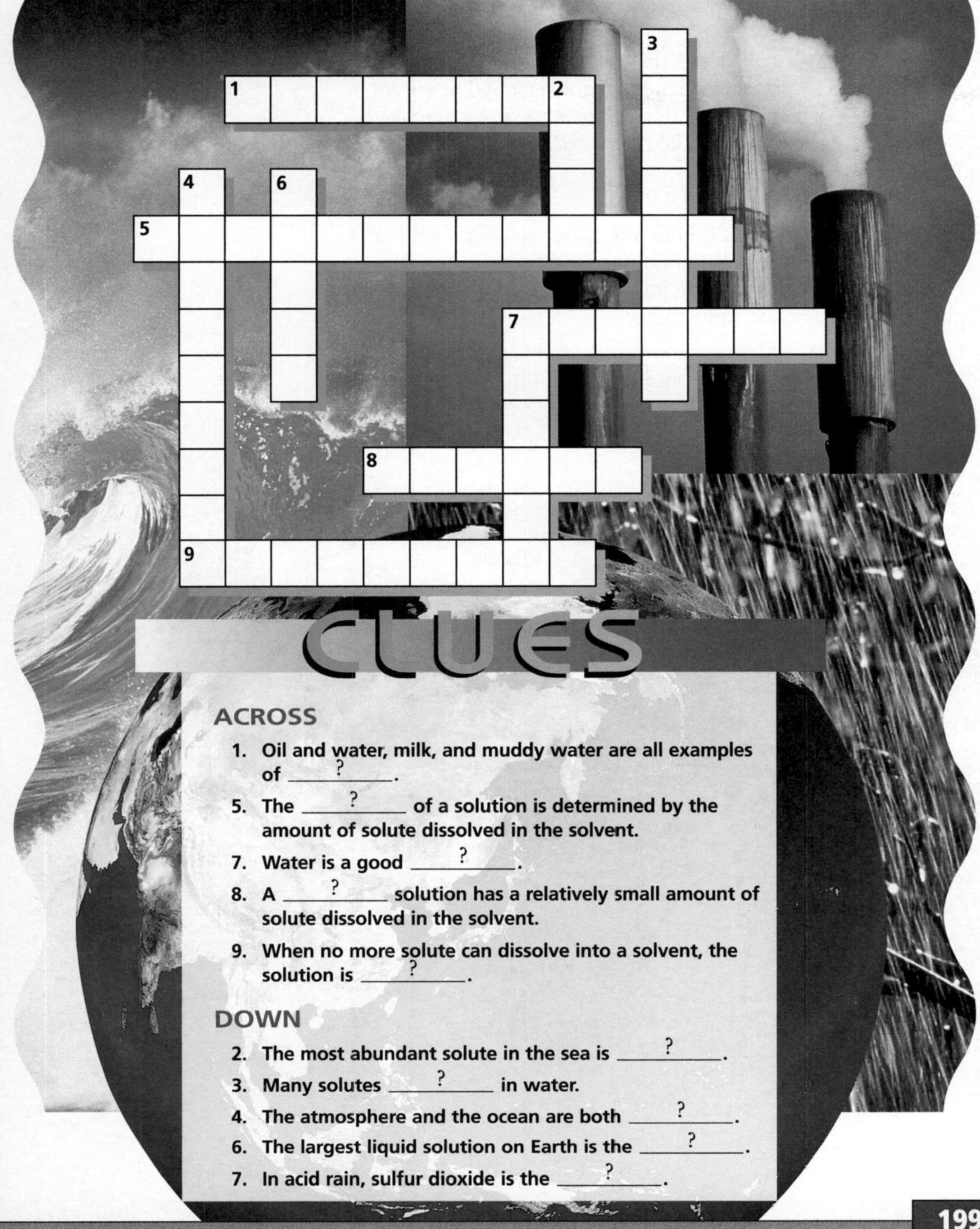

CLUES

ACROSS

1. Oil and water, milk, and muddy water are all examples of ___?___.
5. The ___?___ of a solution is determined by the amount of solute dissolved in the solvent.
7. Water is a good ___?___.
8. A ___?___ solution has a relatively small amount of solute dissolved in the solvent.
9. When no more solute can dissolve into a solvent, the solution is ___?___.

DOWN

2. The most abundant solute in the sea is ___?___.
3. Many solutes ___?___ in water.
4. The atmosphere and the ocean are both ___?___.
6. The largest liquid solution on Earth is the ___?___.
7. In acid rain, sulfur dioxide is the ___?___.

Integrating the Sciences

Earth and Physical Sciences

Water has a great ability to store energy, and it often acts as a temperature regulator for the environment. Have students make a graph showing the average monthly temperatures for several regions. Compare the graphs for a desert, an inland region, and a coastal region. *(Students should see a greater fluctuation in monthly temperatures for regions that are farther from bodies of water.)*

Multicultural Extension

Ocean Travels

Show students a map of the world's ocean currents. Tell them to imagine that they are a small particle traveling in one of the currents. Have them write a story about traveling from region to region. Their story should include descriptions of how the various cultures that they visit interact with the ocean and should indicate the similarities and differences among these cultures.

Answers to *Word Hunt*

Across
1. Mixtures
5. Concentration
7. Solvent
8. Dilute
9. Saturated

Down
2. Salt
3. Dissolve
4. Solutions
6. Ocean
7. Solute

Homework

You may wish to provide students with the Resource Worksheet on page 51 of the Unit 3 Teaching Resources booklet and assign Word Hunt as homework. If you choose to use this worksheet in class, Transparency 36 is available to accompany it.

Meeting Individual Needs

Learners Having Difficulty

Use scrambled words, anagrams, or other word games to focus students' attention on the unit's vocabulary words that are not included in the Word Hunt puzzle. Computer software programs are available that will construct crossword and other puzzles using words of your choice.

Meeting Individual Needs

Second-Language Learners

Have students review the list of solution words shown on page 197. If necessary, suggest that they look back through the unit for the definitions of any words they are unsure of. As a ScienceLog exercise, encourage students to define these terms both in their native language and in English.

Write Away!

Tell students that they can change the endings of the solution words when they are constructing their sentences. You may wish to give students the choice of using a story approach or a purely factual approach in writing their paragraphs.

Gold From the Sea

Involve students in a discussion of the difficulty of communicating ideas without using the proper vocabulary. You may wish to have a thesaurus in the classroom to help students with their search for synonyms.

Answers to *Gold From the Sea*

1. The following is a sample revision of the first sentence: Water is sometimes called the universal *material in which other materials are mixed in such a way that they become invisible,* because so many *dissolved materials* can be dissolved in it.
2. Have students check their local newspaper for the current price of gold, which fluctuates daily. To calculate the worth of the gold in 1 km^3 of sea water, students may need to know that 1 troy ounce equals 31.1 g. For example, if 1 troy ounce of gold costs \$350, then 1 g costs \$350 ÷ 31.1 = \$11.25. Each cubic kilometer of sea water contains 0.004 metric tons, or 4000 g of gold. Therefore, the value of the gold in 1 km^3 of sea water would be 4000 g × \$11.25 = \$45,000.

Write Away!

Take another look at the terms in the list on page 197. Write a short paragraph that contains all of these terms.

Gold From the Sea

Water is sometimes called the universal *solvent* because so many *solutes* can be dissolved in it. Even gold is found *dissolved* in sea water. If just 1 km^3 of this *solution* were *evaporated,* 0.004 metric tons of gold could be recovered. That's a big gold mine!

1. To find out how difficult it is to talk about solutions without using any of the words listed on page 197, rewrite the paragraph above by replacing the italicized words with words or phrases that mean the same thing.
2. Listen to the national news or look in a newspaper for today's gold prices. Can you figure out how much the gold found in 1 km^3 of sea water is worth?

200

FOLLOW-UP

Reteaching

Locate an article that discusses the atmosphere or the oceans. Have students read the article and briefly summarize it using vocabulary words from the unit.

Assessment

Ask students to determine the percentages of oxygen, nitrogen, carbon dioxide, and other gases that make up the atmosphere. Have them present their findings in the form of a bar graph or pie chart. Then have students report on how plants, animals, and decomposers interact to maintain the balance of gases in the atmosphere.

Extension

Have students do some research to find out how much coal a typical coal-fired power plant burns per day and then determine how much sulfur dioxide it produces. *(Burning 1000 tons of coal produces about 90 tons of sulfur dioxide.)* If there are power plants in your area, ask the students to investigate what steps are being taken to reduce sulfur emissions.

Closure

Have students collect magazine and newspaper articles regarding air and water pollution. Then suggest that they use their articles to make a display.

LESSON 5 Solutions in You

You breathe a solution every day. You drink solutions and sometimes even swim in them. Did you know that there are solutions inside you as well? Your body produces a variety of solutions that keep you functioning normally. Read on to find out about some of these internal solutions and the work they do inside your body.

Water Solutions

What substance in your body contributes the most to your weight? Are you surprised that the answer is water? Sixty percent of your body is water. It's little wonder, then, that water solutions are vital to many processes that occur in your body. These solutions aid in digestion, fight disease-causing organisms, maintain proper body temperatures, and much more.

It Starts in the Mouth

Imagine that it's almost dinnertime in your home. As the smell of your dinner comes wafting out of the oven toward you, you realize just how long it's been since lunch. As the aroma fills your nostrils, something begins to happen in your mouth. What is happening?

The flow of fluid you notice results when the salivary glands under your tongue release **saliva,** a solution that is 99 percent water and 1 percent solutes. You produce 1 to 1.5 L of saliva each day. That's more than enough to fill four medium-sized juice glasses! Can you change the quantity of saliva in your mouth by just thinking? Did reading about the dinnertime scenario above change the quantity? Ⓐ

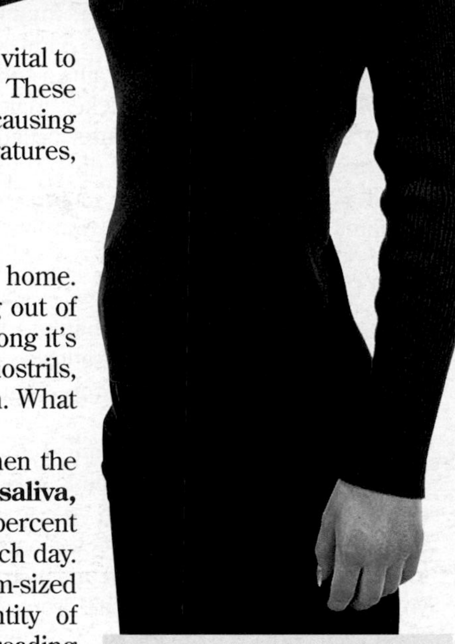

Your body requires about 1.5 L of water a day to maintain its many solutions. Do you drink this much water every day? Do you take in water in any other way besides drinking it? Ⓑ

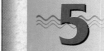

LESSON 5 ORGANIZER

Time Required
1 to 2 class periods

Process Skills
observing, analyzing, comparing, contrasting

New Terms
Perspiration—sweat; a salty solution used by the body for evaporative cooling
Saliva—a digestive solution released by glands under the tongue
Suspension—a substance whose particles are dispersed throughout a liquid but not dissolved in it

Materials (per student group)
none

Teaching Resources
Resource Worksheet, p. 52
SourceBook, p. S57

LESSON 5 Solutions in You

FOCUS

Getting Started

Fill a beaker with 1.5 L of water. Tell students that this represents how much saliva they produce on a daily basis. Also, show them a diagram of the digestive system that includes the salivary glands. Then have them read the material on this page.

Main Ideas

1. Water solutions are vital to many processes that occur in the human body.
2. Key processes that require water solutions are digestion, fighting diseases, and maintaining proper body temperature.
3. The blood in the human body is a suspension.

TEACHING STRATEGIES

It Starts in the Mouth

Show the students how this works by bringing some freshly popped popcorn into class. Allow them to smell the popcorn, and ask them if they can feel their salivary glands working. Explain to them that salivation starts the digestion process.

Answers to *In-Text Questions*

Ⓐ Answers will vary. Let students know that when an enticing aroma is smelled, the mouth will usually start producing saliva.

Ⓑ The human body takes in water not only by drinking it, but also by drinking juices and eating foods. Many foods are composed mostly of water, and the body is able to absorb this water from them.

The Cleansing Solution

To demonstrate how important blinking is to the well-being of the eye, have students try to refrain from blinking. (Tell them to stop the exercise if their eyes start hurting or burning.) Discuss the results with students.

Answers to *In-Text Questions*

Ⓐ Lysozyme combats bacteria that get into the eyes by dissolving the cell walls of the bacteria.

Ⓑ The corneas of a snake's eyes, for instance, are protected by a transparent area of the *general integument.* You can find the answers to this question by consulting books on fish and snake anatomy.

What does saliva do for you? Imagine eating a cracker with a dry mouth. A difficult task, wouldn't you say? Saliva contains a solute called ***mucin*** that lubricates the cracker, making it easier to swallow. Another solute, called salivary ***amylase***, actually begins digestion of the cracker while it's still in the mouth. Saliva also contains a solute called ***lysozyme*** that combats bacteria. Lysozyme provides a natural defense against tooth decay.

Animals have unique solutes in their saliva that help them in their daily lives, too. Find out how the saliva of snakes differs from human saliva. Or do some research to discover the properties of leech saliva that make leeches useful to the medical profession.

The Cleansing Solution

How many times have you blinked while reading this lesson? This is probably a difficult question to answer. Chances are, you weren't thinking about blinking—that is, you weren't doing it consciously.

Your body uses this mechanism to protect your eyes. Each time you blink, your eyelids spread a thin layer of tears over your eyeballs. You see, tears are not just a response to a sad movie, a peeled onion, or a skinned knee. Tears are also an important solution produced by your body to keep your eyeballs from drying out and to wash away foreign particles such as dust. The solutes in tears include salt, mucus, and lysozyme. Think back to the role that lysozyme plays in saliva. What role might this solute play in tears? Ⓐ

Some animals, such as snakes and fish, don't have eyelids and therefore don't blink. Do their eyes have some type of natural protection from drying and from foreign particles? Where could you find an answer to this question? Ⓑ

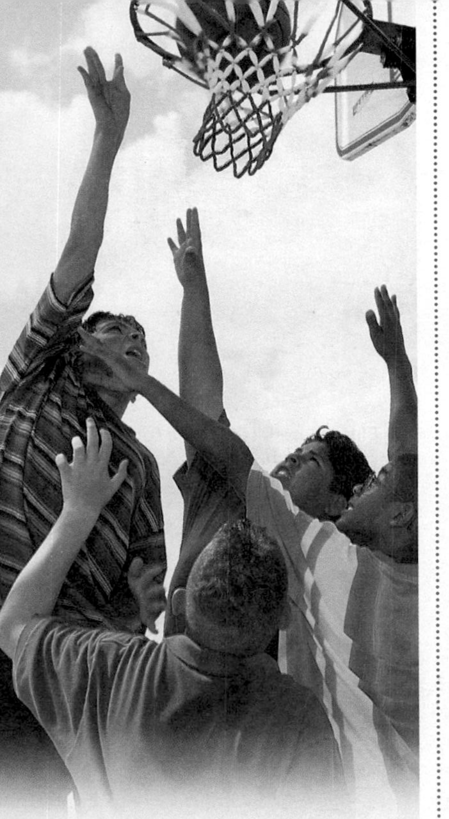

The Cooling Solution

Everybody produces this solution, but some people produce more than others. It helps maintain your body temperature and rid your body of waste products. Can you guess the name of this helpful solution? It's sweat, or more properly, **perspiration.** You can taste one of the main solutes in this water solution—salt—when you lick your lips on a hot day.

When you step out of the shower or bathtub, do you feel cool? This cooling is caused by the water on your body changing from a liquid to a gas—the process of evaporation. How does this help explain how sweating helps to maintain your body temperature? Ⓒ

Blood: A Solution or Not?

For a research project, you and your classmates are going to find out about other body fluids and determine whether they are solutions. You choose blood as your topic, and you're eager to see how the classroom computer can help you.

You have logged onto the computer and have found information about your topic. As you read, you can do further research by "clicking" on the words highlighted in blue. However, some information sites may still be "under construction"—that is, there may be information missing on a particular topic. You might choose to do some further research to help complete this topic.

To begin your search, click on the word blood and read the text that appears on the computer screen.

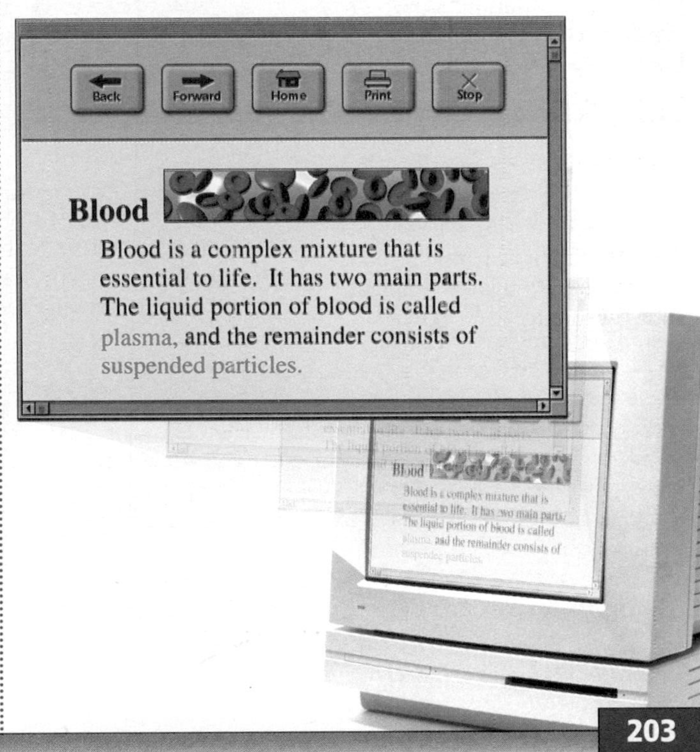

Back
Forward
Home
Print
Stop
Blood
Blood is a complex mixture that is essential to life. It has two main parts. The liquid portion of blood is called plasma, and the remainder consists of suspended particles.

The Cooling Solution

Have students read the paragraph silently. Tell them that perspiration is a mixture of water, salts (mainly sodium chloride), and waste products such as urea and uric acid. Tell them that there is a difference between sweat and body odor. Sweating occurs when the body's temperature rises; body odor occurs when sweat, or perspiration, comes in contact with bacteria present on the skin's surface.

Blood: A Solution or Not?

Once students have read pages 203–205, discuss the terms *plasma, serum, solution,* and *suspension.* Direct students' attention to the activity listed under Solution vs. Suspension on page 205. Explain to them that plasma carries platelets throughout the bloodstream. Discuss situations in which this is important. *(Accept all reasonable responses. Sample answer: When you cut yourself, plasma carries platelets to the site of the cut. Platelets aid in clotting.)* As part of the discussion, have the class reiterate their definitions of *solution* and *suspension.*

Answer to *In-Text Question*

Ⓒ The same process of evaporation is what cools the body. The body produces sweat when it becomes hot. When the sweat evaporates, it absorbs energy from the body, and the body cools.

Multicultural Extension

Blood Banks

One of the most important chemists of the 1940s was the African American scientist Dr. Charles Drew (1904–1950). Dr. Drew developed revolutionary methods for preserving blood. During World War II, his British Blood Bank offered wounded Allied soldiers the blood plasma they needed to survive. After the war, Dr. Drew led an effort to collect and store blood from 100,000 donors for the American Red Cross. Encourage students to research local blood banks and to present their findings to the class.

Did You Know. . .

The main ingredient in antiperspirants is aluminum chloride, which is a type of porous aluminum designed to soak up perspiration. This sometimes has side effects, including itching and irritation.

Answer to *Concept Mapping Corner*

Sample concept map:

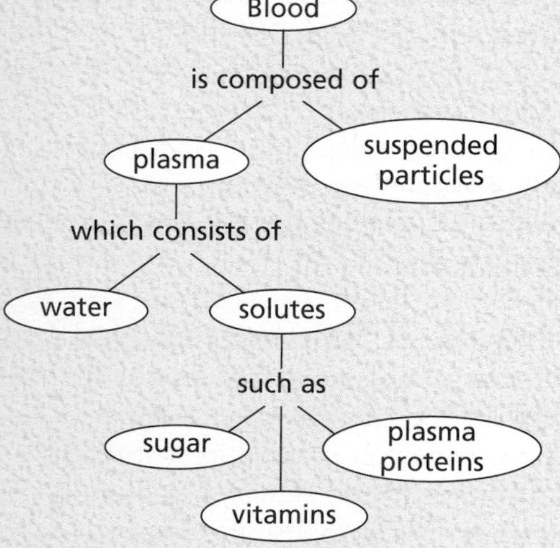

Answer to *In-Text Question*

Ⓐ It is not a good idea to break a blister because by breaking the blister, you may expose the area beneath the skin to germs. This could cause an infection.

Homework

As a homework assignment, have students complete the Resource Worksheet that is available to accompany Concept Mapping Corner (Teaching Resources, page 52).

Cross-Disciplinary Focus

Health

Have students choose a vitamin to research. They should describe the source of the vitamin, whether it is water-soluble, and how the body uses it. When students have completed their research, you may wish to have students work together to determine similarities and differences among the vitamins. *(For instance, all of the B-complex vitamins are water-soluble and have similar functions.)*

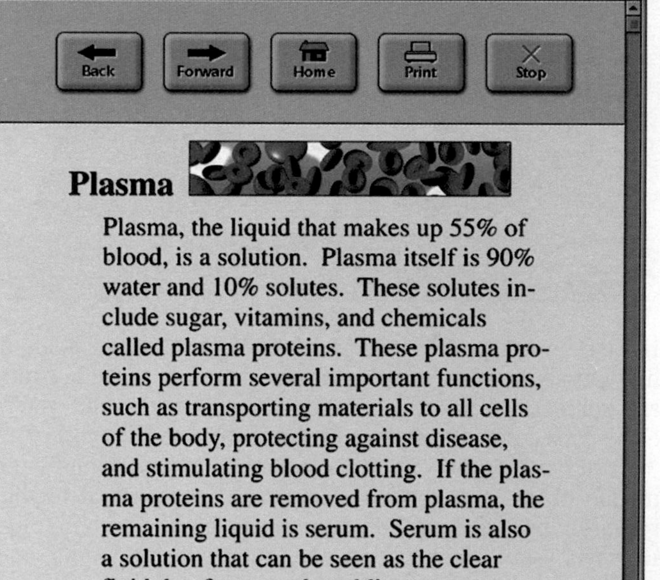

Plasma

Plasma, the liquid that makes up 55% of blood, is a solution. Plasma itself is 90% water and 10% solutes. These solutes include sugar, vitamins, and chemicals called plasma proteins. These plasma proteins perform several important functions, such as transporting materials to all cells of the body, protecting against disease, and stimulating blood clotting. If the plasma proteins are removed from plasma, the remaining liquid is serum. Serum is also a solution that can be seen as the clear fluid that forms under a blister.

This site is under construction. Additional information on the following topics is welcome:

- serum
- hemophilia
- globulins

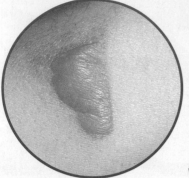

Blisters are filled with a solution called serum.

The plasma protein C1q, magnified 440,000 times

Do you think it is a good or bad idea to break a blister? Explain your reasoning. Ⓐ

Concept-Mapping Corner

Based on what you've learned so far about solutions in the body, create a concept map using the terms shown below. You will have to provide the connecting terms.

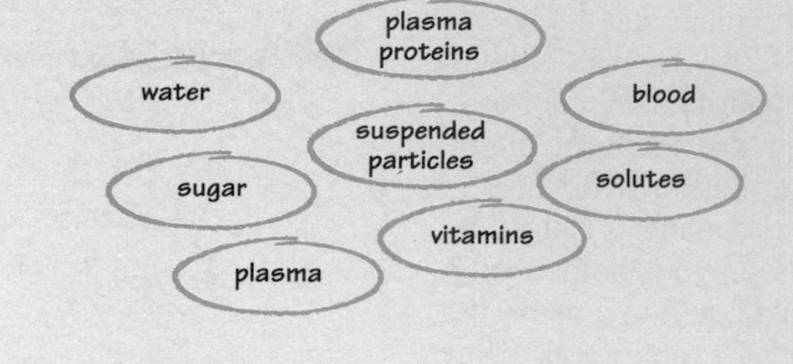

Suspended Particles

Suspended particles make up 45% of blood. These particles (mainly red blood cells, white blood cells, and platelets) can be separated from the plasma by spinning a sample of blood at extremely high speeds in an instrument called a centrifuge. Because these particles are suspended rather than dissolved, blood is known as a suspension.

This site is under construction. Additional information on the following topics is welcome:

- red blood cells
- white blood cells
- platelets
- anemia
- hemoglobin

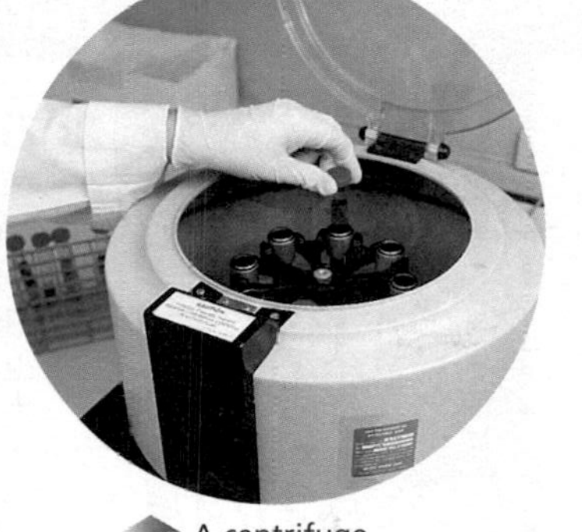

A centrifuge

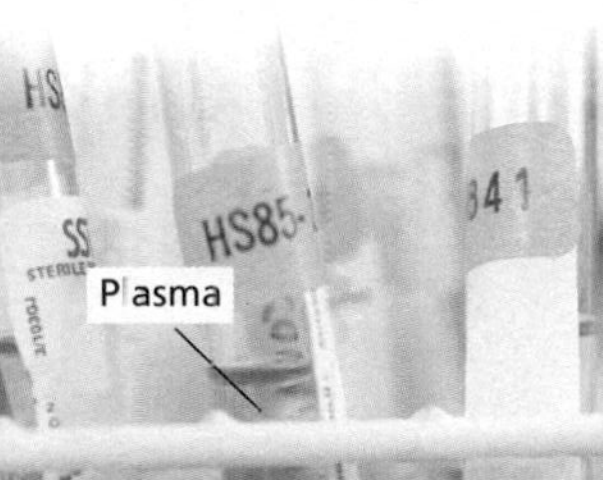

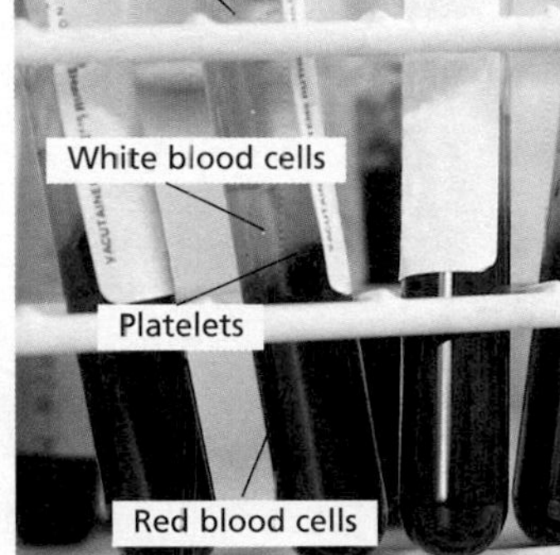

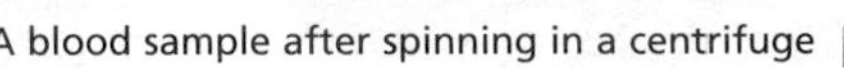

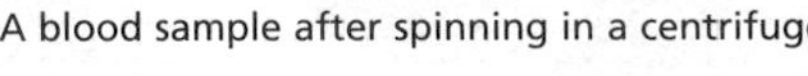

A blood sample after spinning in a centrifuge

Solution vs. Suspension

To better understand the difference between a solution and a **suspension,** try this activity. Dissolve 1 g of sugar in 10 mL of water. Did the sugar dissolve? This represents plasma, a solution. Now add 1 g of cornstarch to the solution. Did the starch dissolve? How do you know? This mixture represents blood, a suspension. In this mixture, the suspended particles are starch. In blood, the suspended particles are mainly blood cells. What is your definition of a solution? of a suspension?

How can this knickknack help us understand what a suspension is?
Let's check out page S57 in the SourceBook to find out!

205

FOLLOW-UP

Reteaching

To further their knowledge, have students form pairs and copy onto index cards the new definitions that they learn from their simulated on-line research.

Assessment

Have students write a short essay that combines everything they have learned about solutions and suspensions. Have them detail various processes in the body that involve solutions or suspensions.

Extension

Have students do more on-line research on the additional topics mentioned at the bottom of the Suspended Particles screen (e.g., red and white blood cells, platelets, anemia, hemoglobin, etc.). Have them share their findings with the class.

Closure

Ask students to name five solutions in the body. Also, have them give examples of suspensions in the body.

Answers to *Challenge Your Thinking*

1. The most dilute solution is red; the most concentrated solution is clear. The order from least to most concentrated is red, green, blue, yellow, and clear.
2. One answer would be to boil the two solutions. The solution with the most dissolved solutes will boil at the highest temperature. Another method would be to compare the freezing points of the two solutions. The solution with more dissolved solutes will have a lower freezing point.
3. The temperatures are frequently much lower in Alaska than in Washington. Therefore, the melting point of ice is not lowered enough to cause melting.
4. **a.** According to the graph, gases become less soluble at higher temperatures. Therefore, a lake would hold more dissolved oxygen in the winter.
   b. The Caribbean can hold more dissolved salt than the Atlantic can because solids are more soluble at higher temperatures.
5. Ice alone will melt when it comes in contact with the chamber holding the cream. Unfortunately, the temperature of melting ice is not cold enough to freeze the cream. Salt, however, lowers the freezing point of water. By adding salt to the ice, you can keep the melting ice at a temperature below 0°C, which is low enough to freeze the cream.
6. A saturated sponge cannot absorb any more water. Excess water runs out of the sponge. Likewise, in a saturated solution, the solvent cannot absorb any more solute. This produces an excess of solute.

Homework

As a homework assignment, have students complete the Activity Worksheet on page 53 of the Unit 3 Teaching Resources booklet.

1. A Rainbow of Results

Maya was given five solutions of sugar in water, each with a different concentration. Four of the solutions had been dyed the following colors by adding food coloring: red, yellow, green, and blue. The fifth solution was clear. She used the layering test to determine their concentrations. The illustration below shows her results.

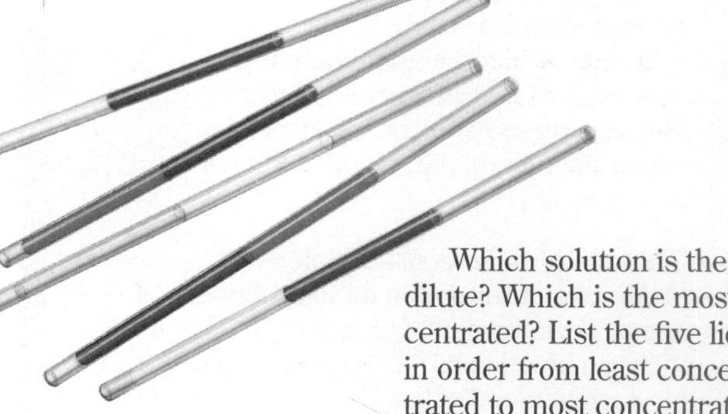

Which solution is the most dilute? Which is the most concentrated? List the five liquids in order from least concentrated to most concentrated.

2. Claim Check

Josh wants to check a claim by Brand X that its canned fruit is packed in very light syrup (a dilute sugar solution). Suggest a way that he can compare the "lightness" of the syrup used by Brand X with that of the syrup used by Brand Y.

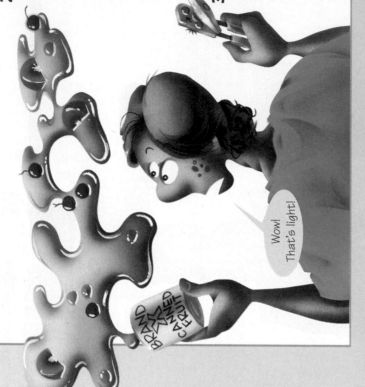

3. Salt vs. Snow

Salt is mixed with sand and placed on roads after snowstorms in many states. This practice does not work as well in Alaska as it does in Washington. Why not?

You may wish to provide students with the Chapter 9 Review Worksheet that is available to accompany this Challenge Your Thinking (Teaching Resources, page 54).

4. Holding Pattern

Using the graph, which shows how temperature affects the solubility of gases and most solids in water, answer the following questions:

a. When can a lake hold more dissolved oxygen—in the winter or in the summer? Why?

b. Liter for liter, which can hold more salt—the Caribbean Sea or the Atlantic Ocean? Why?

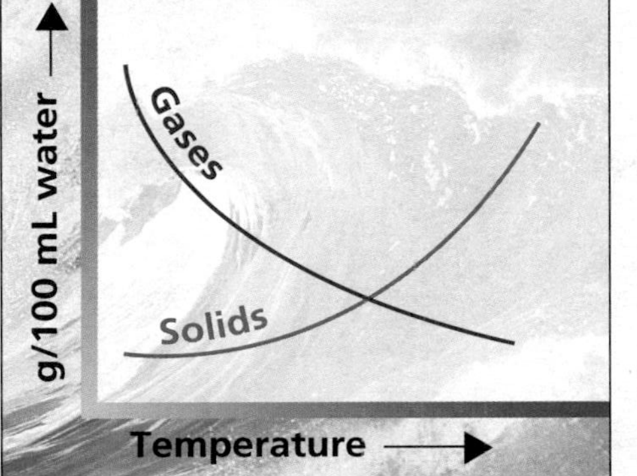

5. The Cold, Hard Facts

To make homemade ice cream, a mixture of salt and ice is placed around the chamber holding the cream. Explain the scientific reason for this.

6. All Soaked Up

On page 194, a cartoon about sponges illustrates the idea of saturation. How is this a way to explain the meaning of the term *saturated solution*?

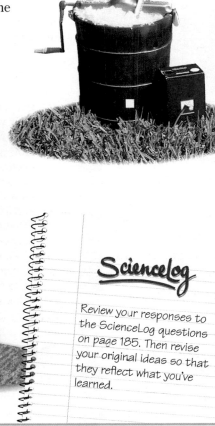

ScienceLog

Review your responses to the ScienceLog questions on page 185. Then revise your original ideas so that they reflect what you've learned.

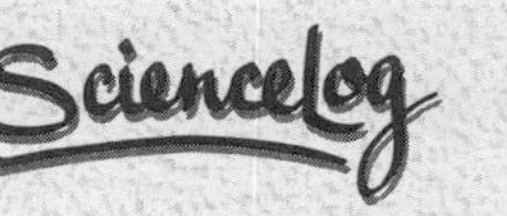

ScienceLog

The following are sample revised answers:

1. The fresh water will freeze first because solutes lower the freezing point of solvents.
2. A solution becomes more concentrated if more solute is added and more dilute if more solvent is added.
3. Yes. When there is more solute present than can be dissolved in a solvent, the solution can be described as saturated.
4. The atmosphere and the ocean are two solutions that are part of the environment. Solutions inside the body include tears, saliva, sweat, and blood plasma.

Making Connections

The Big Ideas

The following is a sample unit summary:

A solution is a homogeneous mixture; a non-solution is not homogeneous because its component parts are still visible. (1) The parts of a solution are the solute and the solvent. The solid is no longer visible because it is in small particles that are uniformly mixed throughout the solution. (2)

Solubility refers to the maximum quantity of solute that will dissolve in a given amount of solvent. To determine whether a substance is soluble, you would attempt to dissolve it in a solvent. A substance is soluble if it disappears into the solvent. This shows that it has dissolved completely. A substance is not very soluble if some of it is still visible in the solvent. A substance is insoluble if none of it disappears into the solvent, showing that none of it has dissolved. (3)

A solute can be separated from a solvent by boiling off the solvent or by allowing the solvent to evaporate. (4)

A solid solute might dissolve faster at higher temperatures and with agitation of the particles. (5)

Evaporation is the gradual change of a liquid into a gas. In the water cycle, water evaporates from the oceans, rivers, lakes, and land. Honeybees also use evaporation to remove the excess water from nectar to make honey. Boiling occurs when a liquid rapidly changes into a gas. To make maple syrup, the excess water is removed by boiling. (6)

The presence of a solute in a solvent can lower the solvent's freezing point and raise its boiling point. This property of solutions can be useful when it is necessary to lower the freezing point of a substance, such as ice on a road, or raise the boiling point of a substance, such as the boiling point of water in a radiator. (7)

A concentrated solution, because it contains more particles of dissolved solute, will have a higher boiling point and a lower freezing point. A solution of greater concentration will also exert greater buoyancy on an object floating in it. (8)

When the maximum point of solubility for a solution is reached, the solution is said to be saturated. Temperature has a direct effect on the solubility of substances. In general, gases become less soluble at higher temperatures, while solids become more soluble. (9)

Solutions in the body include saliva, tears, and sweat. (10)

Blood is considered a suspension because blood cells are suspended, not dissolved, in the plasma. (11)

Making Connections

Unit 3

SourceBook

By reading pages S51–S70 in the SourceBook, you will learn more about different solutions and about the importance of water in making many of them. You will also learn about acids, bases, and salts.

Here's what you'll find in the SourceBook:

The Big Ideas

In your ScienceLog, write a summary of this unit, using the following questions as a guide:

1. How is a solution different from a non-solution?
2. What are the parts of a solution? Why are they not visible?
3. How would you determine whether a substance is soluble, not very soluble, or insoluble?
4. How would you separate a solute from a solvent?
5. How would you make a solute dissolve faster?
6. How is evaporation different from boiling? Which process occurs in the water cycle? in making maple syrup? in making honey?
7. How does the presence of a solute affect the boiling point and freezing point of water? Of what practical value is this?
8. What test could you perform to determine the difference between a concentrated solution and a dilute solution of a given solute in a given solvent?
9. How are the terms *saturated, solubility*, and *temperature* related?
10. What are some solutions in your body?
11. Why is blood a suspension and not a solution?

Checking Your Understanding

1. Suppose that 6 g of salt is mixed with 50 g of water.
   a. What is the mass of the solvent?
   b. What is the mass of the solute?
   c. What is the mass of the solution?

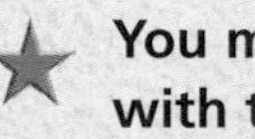

You may wish to provide students with the Unit 3 Review Worksheet on page 62 of the Unit 3 Teaching Resources booklet.

2. Susanne discovered that when she put a little washing soda (sodium carbonate) in water, the solution became warm. She wondered if the amount of washing soda mattered. She found five bottles, filled each one with 100 mL of water, and recorded the temperature of the water in each. Susanne then added different amounts of washing soda to each bottle, stirred each solution, and measured the temperatures again. Her results are shown below.

Bottle	Number of teaspoons* of soda added	Amount of water (mL)	Temperature before (°C)	Temperature after (°C)
A	1	100	22	26
B	2	100	22	30
C	3	100	22	34
D	4	100	22	37
E	5	100	22	39

*1 teaspoon of washing soda has a mass of about 7.1 g.

a. What variables were held constant?
b. What variable changed?
c. What changed in response?
d. What did Susanne find out?
e. Which solution was the most dilute?
f. If washing soda made a blue solution in water, which solution would be the deepest blue?

3. concept map Copy this concept map into your ScienceLog. Then complete the map using the following words: solid, gas, solution, solvent, Tyndall effect, and non-solution.

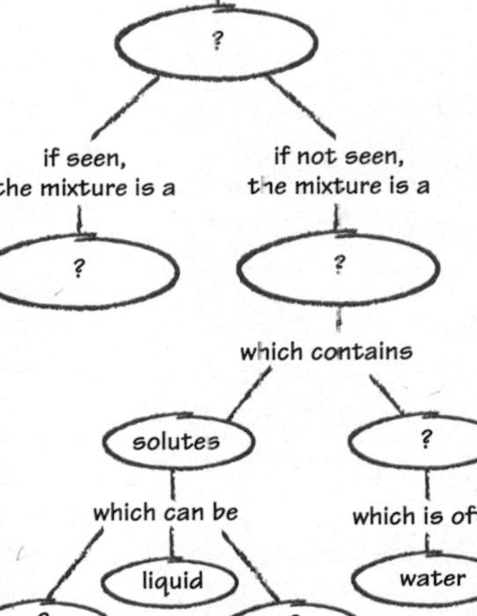

A mixture
can be tested for the
?
if seen, the mixture is a
?
if not seen, the mixture is a
?
which contains
solutes
?
which can be
?
liquid
?
which is often
water

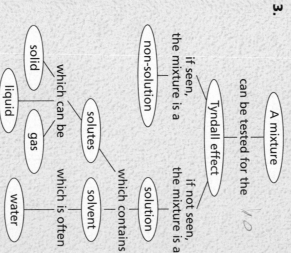

Answers to *Checking Your Understanding, pages 208–209*

1. **a.** The mass of the solvent is 50 g.
   b. The mass of the solute is 6 g.
   c. The mass of the solution is 56 g.

2. **a.** Variables held constant are the starting temperature and the amount of water added.
   b. The variable that changed was the amount of washing soda added.
   c. The temperature after adding the washing soda was the variable that changed in response.
   d. Susanne found out that the addition of washing soda raised the temperature of the solution.
   e. The solution with only 1 teaspoon added was the most dilute.
   f. The solution with 5 teaspoons is the most concentrated and would therefore be the deepest blue.

3.
A mixture
can be tested for the
Tyndall effect
if seen, the mixture is a
non-solution
if not seen, the mixture is a
solution
which contains
solutes
solvent
which can be
solid
gas
liquid
which is often
water

Homework

The Unit 3 Activity Worksheet on page 60 of the Teaching Resources booklet makes an excellent homework assignment. If you choose to use this worksheet in class, Transparency 37 is available to accompany it.

Background

We are unable to drink salt water because the salinity of our bodies is less than the salinity of salt water. When salt water is ingested, water in the stomach lining is moved by osmotic pressure out of the body's cells and into the stomach's contents. Thus, salt water dehydrates the body instead of hydrating it.

Millions of people around the world depend on desalination plants for most of their fresh water. In 1988, there were an estimated 3500 desalination plants in 105 countries, and the number was growing. These plants produce over 11 billion liters of water daily. The older, more common method of desalination is distillation, but more and more countries are turning to the process of reverse osmosis because it requires no heat. Reverse osmosis typically produces 1 L of fresh water for every 2 or 3 L of salt water, and the cost is three to five times more expensive than producing fresh water from natural sources.

Extension

Have students mix salt solutions that have the same concentration as sea water (3.5 g of salt for every 100 g of water). Then have students use the simple distillation method discussed on page 173 to measure the amount of fresh water produced for every 100 g of salt water. Students should use a hot plate instead of a portable stove.

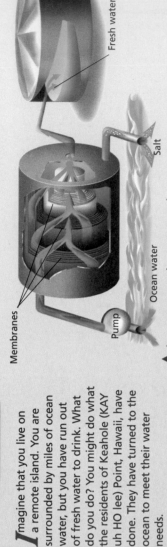

TECHNOLOGY in SOCIETY

Care for a Glass of Ocean Water?

Imagine that you live on a remote island. You are surrounded by miles of ocean water, but you have run out of fresh water to drink. What do you do? You might do what the residents of Keahole (KAY uh HO lee) Point, Hawaii, have done. They have turned to the ocean to meet their water needs.

▲ In reverse osmosis, concentrated salt water passes through the outer membranes while fresh water flows out a central tube.

A 99 Percent Solution

The residents of Keahole Point are not alone. Millions of people live in areas without enough fresh water. In fact, only 1 percent of the Earth's water is free of salt and suitable for drinking.

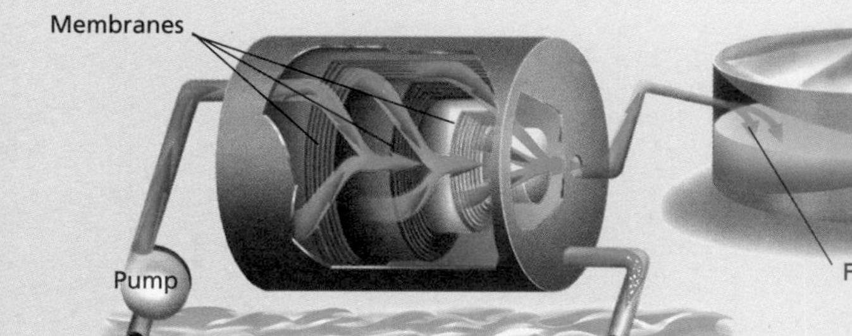

▼ Desalination removes salt from salt water. Plants like this one use desalination to create fresh water from ocean water.

If the salt could be removed from ocean water, potential supplies of fresh water would increase by 99 percent!

Taking Out the Salt

Actually, the process of removing salt from salt water is quite simple and can be accomplished by two basic methods. The first method is called *distillation.* In distillation, the salt water is heated so that the water evaporates and leaves the salt behind. The water vapor is captured and cooled so that it condenses to form fresh water. The other method is called *reverse osmosis.* In this process, water is pumped through a series of thin sheets called membranes. Water and some very tiny particles can pass through the membranes, but salt and other minerals are filtered out. By the time the water has passed through all of the membranes, it is free of salt.

A Very Expensive Glass of Water

Unfortunately, removing the salt from ocean water costs a lot of money. Desalinated ocean water may cost 10 times more than other types of fresh water! Much of the cost comes from the enormous amount of energy needed for distillation or reverse osmosis. Still, in a few parts of the world, it is the only way to create the necessary water supplies. In these places water is a precious commodity.

Water Crisis

Imagine that low water supplies forced your community to create a desalination plant. As a result, your water bill went from $20 per month to $200 per month for the same amount of water. By what percentage did your water bill increase? What lifestyle changes would you make to reduce your water usage?

210

Answers to *Water Crisis*

The new water bill is 1000 percent of the previous bill.

Accept all reasonable responses concerning lifestyle changes. Some common answers might include turning off the shower between rinses or catching rainwater for use on indoor plants. Water-filled plastic bottles can be placed in toilet tanks to reduce the amount of water used when flushing the toilet. Watering the lawn only during cool times of the day will lessen water loss from evaporation. Also, running a washing machine only with a full load of clothes will reduce water use as well.

EYE ON THE ENVIRONMENT

What's in the Water?

Every year, insects and other pests destroy about 10 percent of the crops grown in the United States. To protect their crops, many farmers use pesticides. People also use pesticides on their lawns and gardens. Unfortunately, these poisons, which are designed to kill pests, can also harm people, and they are seeping into our nation's water supplies.

▲ This crop-dusting plane is spreading pesticides over a field of sweet corn. Pesticides kill more than just pests—they often kill beneficial insects as well.

How do pesticides get into our water supplies? The water that doesn't soak into the ground when it rains or when lawns and crops are watered is called *runoff.* This runoff carries the pesticides in the soil to nearby rivers and lakes. Water that soaks into the ground also carries dissolved pesticides with it, polluting water beneath the Earth's surface. Contaminated water may enter human water supplies from any of these places.

Keeping It Clean

Pesticides can usually be removed from water at water-treatment plants. Also, government regulations prevent the use and sale of some particularly dangerous pesticides. But small-scale use of pesticides is hard to regulate, and not all drinking water is treated in water-treatment plants. As more people become concerned about the dangers of pesticides, many have switched to less toxic pest control methods such as biodegradable pesticides. Many farmers and gardeners are controlling pests by releasing the pests' natural predators. For example, ladybugs are released to control aphids. Still others grow pest-resistant varieties of plants. With some insect species, farmers can turn the insects' reproductive cycles against them. The farmers release sterile males to mate with fertile females. After they lay their eggs, the females die, and the eggs do not develop. These alternative methods are often very effective and less expensive than using pesticides.

▲ Using natural predators to control pests is often more effective than using pesticides. The praying mantis is an excellent all-purpose predator.

Do It Yourself!

Go to a garden supply store and study the labels of a few pesticides. Which pests are they designed to control? Note the ingredients and safety precautions listed on the label. What does this tell you about potential dangers to human health?

211

Do It Yourself!

SAFETY ALERT

Students should observe the packages of the pesticides only. The packages should not be opened to perform this activity. Also tell students that they must have adult supervision during this activity. When the research is completed, you might want to take a class poll to see what pests were the most targeted (probably roaches, ants, or mice) and what ingredients were the most common. If the labels do not list the possible harmful effects of the pesticide, you can have students suggest different treatment methods, which should also be listed on the label of each product.

EYE ON THE ENVIRONMENT

Background

Additional alternatives to pesticides that farmers can use include crop rotation to discourage pest infestations and luring insects into traps equipped with attractants. Other methods that gardeners can use include disposing of any plant debris in order to prevent pests from nesting, and turning over the soil in the fall so that pests will freeze during the winter. Gardeners can also cover their plants with cheesecloth or a similar material. This allows rainwater to reach the plants but prevents pests from doing so.

Some pesticides are not only toxic but also *persistent,* which means that they do not break down very quickly. One example is DDT, which was used in the 1940s to control the spread of malaria by mosquitoes. The pesticide is now found everywhere in the world, including Antarctica. One effect of DDT is the weakening of the eggshells of many birds, causing the eggs to crack prematurely, which can kill an unhatched bird. Some bird species have even become endangered or extinct as a result.

ENVIRONMENTAL FOCUS

Discuss some natural ways to eliminate pests around the home. *(You can build a roach trap by putting honey at the bottom of an open jar. Roaches are also repelled by bay leaves.)*

Extension

Have students model the filtering method used in many water-treatment plants. Ask them to place a paper coffee filter in a funnel and to put layers of fine gravel or sand in the filter. Then have them place the funnel over a jar and pour muddy water through the filter system. Students should observe how much cleaner the water is. Ask: What might still be in the water? *(Various solutes)*

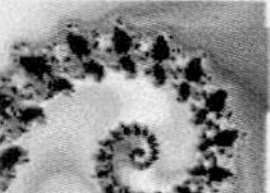

SCIENCE and the *Arts*

Background

When most people hear the word *perfume,* they think of a fragrant liquid they put on their body. However, a perfume can be any substance used as a pleasant fragrance. Many cosmetics contain perfumes. Low-priced perfumes called *oderants* are added to many products, including paper, plastics, and rubber products, to hide unpleasant odors or to make them more attractive to consumers. Many fragrant substances are also burned as incense.

Colognes and *toilet waters* are considered perfumes, but they are weaker solutions than *true perfumes.* Cologne is typically 3 to 5 percent fragrant extract, and toilet water is 2 percent, while a true perfume is 10 to 20 percent. The solvent in all of these is mainly alcohol because most of the extracts are not very soluble in water and because alcohol helps the skin absorb the extract.

Extension

Perfume scents are often grouped according to their most dominant odor. Several major groups are floral, spicy, woody, and herbal. Assign each student a scent to bring to class, and have them cover up the label. Have a class smelling test in which the students smell each scent and vote on which category the scent belongs to. Then tell them what the scent is. Common scents you can use include the following:

- floral: jasmine, lily of the valley, rose, gardenia
- spicy: clove, cinnamon, nutmeg
- woody: sandalwood, cedar
- herbal: clover
- other: vanilla, balsam, patchouli, rosemary

SCIENCE and the *Arts*

Perfume: Fragrant Solutions

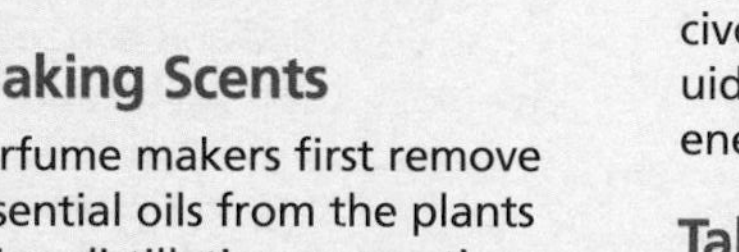

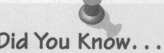

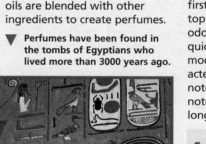

Making perfumes is an ancient art. It was practiced, for example, by the ancient Egyptians, who rubbed their bodies with a substance made by soaking fragrant woods and resins in water and oil. From certain references and formulas in the Bible, we know that the ancient Israelites also practiced the art of perfume making. Other sources indicate that the art was also known to the early Chinese, Arabs, Greeks, and Romans.

Only the E-scent-ials

Perfumes have been made by many cultures for many centuries. Over time, perfume making has developed into a complicated art. A fine perfume may contain over 100 different ingredients. The most familiar ingredients come from fragrant plants or flowers such as sandalwood or roses. These plants get their pleasant odor from their essential oils, which are stored in tiny baglike parts called sacs.

The parts of plants that are used for perfumes include the flowers, roots, and leaves. Other perfume ingredients come from animals and from human-made chemicals.

Making Scents

Perfume makers first remove essential oils from the plants using distillation or reactions with solvents. Then the essential oils are blended with other ingredients to create perfumes.

▼ Perfumes have been found in the tombs of Egyptians who lived more than 3000 years ago.

▼ Not all perfume ingredients smell good. The foul-smelling oil from the African civet is used as a fixative in some perfumes.

Fixatives, which usually come from animals, make the other odors in the perfume last longer. Oddly enough, most natural fixatives smell awful! For example, civet musk is a foul-smelling liquid that the civet sprays on its enemies.

Taking Notes

When you smell a perfume, the first odor you detect is called the top note. It is a very fragrant odor that evaporates rather quickly. The middle note, or modifier, adds a different character to the odor of the top note. The base note, or end note, is the odor that lasts the longest.

Smell for Yourself

Test a number of different perfumes and colognes to see if you can identify three different notes in each. Which scents generally last the longest? Which last only a short time? Write a report on your findings.

212

Smell for Yourself

You may wish to use the same scents mentioned above for this activity. Alternatively, you might bring in some popular colognes to test. These scents will most likely be more complex because they may include as many as 100 different ingredients.

Did You Know. . .

- Other fixatives include castor from the beaver, musk from the male musk deer, and ambergris from the sperm whale.
- The word *perfume* comes from the Latin phrase *per fumum,* which means "through smoke."

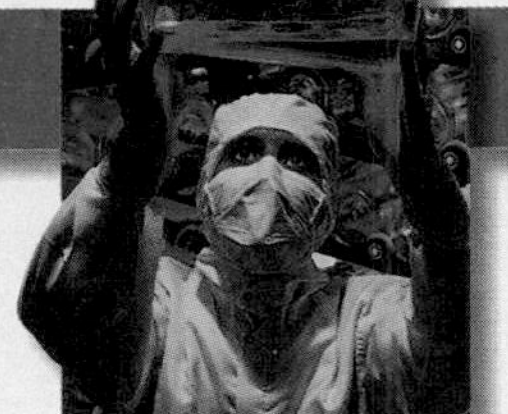

SCIENCE in ACTION

The Dirt on Chemistry

Dr. Elvia Niebla has a Ph.D. in soil chemistry. You might think this is a dirty profession, but Niebla finds it quite fascinating. As the national coordinator of Global Change Research, an environmental research program of the National Forest Service, Niebla really goes places. Her work has taken her to parks, forests, and rivers all across the United States. "I've also done quite a bit of international work as well. I've spent time in Brazil, Germany, and other countries, all on the job."

Cleaning Water With Soil

Much of Niebla's work requires an in-depth knowledge of chemical reactions. "I first started getting into chemistry as a graduate student," says Niebla. "My work focused on how soil can filter water." Now Niebla uses this interest in chemistry in her research of environmental problems. She has done considerable research on water pollution and the role of soil in water purification. Niebla has even researched how soil can be used to purify the outflow from sewage-treatment plants. In one of her research projects, she discovered something unexpected. She discovered that sewage outflow into a stream in Arizona was actually improving the water quality by counteracting toxic ores from mines upstream. Niebla is still surprised by this finding.

A Chemist in Demand

Niebla's research of soil as a natural filter eventually led to her position with Global Change Research. Now she advises national leaders on environmental policy, attends environmental conferences, and even goes to certain congressional hearings. Who would have thought an interest in soil would lead her into politics!

In fact, many of Niebla's interests were developed as she progressed in her educational career. Niebla got her undergraduate degree in zoology, with a minor in chemistry. She then went on to get a master's degree in education. It was not until she began to work on her Ph.D. that she concentrated on soil chemistry. Although she always had an interest in science, she did not always have a clear idea of where it would lead her. She attributes her success to a combination of hard work and study. "At home, my family told me I could be anything I wanted to be—all I had to do was work and study hard."

Niebla has not stopped learning even though she has a Ph.D. Because she wanted to know more about running a laboratory, she went on to earn a degree in business administration. As Niebla puts it, "Now, I not only can do laboratory research, but also can run the laboratory efficiently."

▲ Dr. Elvia Niebla taking a soil sample

What's Your Prediction?

Do you think certain soils would make better filters than others? Do some research to investigate the types of soil and their properties, and make predictions about what would make a good filter. (Hint: Unit 8 includes a section on the types of soils.) How can you find out if your predictions are correct?

Answers to *What's Your Prediction?*

Accept all reasonable responses. Students should recognize that one important feature in determining the filtering ability of a soil is the size of the particles that make up the soil. Soils composed of smaller particles will filter the water more thoroughly. However, if the particles are too small, such as in clay, the water will not drain easily. Students could find out if their predictions are correct by timing how fast a certain amount of water passes through equal amounts of different soils.

SCIENCE in ACTION

Background

Elvia Niebla uses the information gathered by many scientists, including other soil chemists, to determine what environmental problems ought to be studied, and she advises the government on which programs should receive funding. Through her efforts in coordinating research, Niebla provides a direct link between scientific discovery and national environmental policy.

One of the most interesting tools used in soil analysis is the spectrophotometer. To analyze soil, the scientist filters water or another liquid through the soil. The solution that drains through the soil contains chemicals from the soil. The solution is then analyzed using the spectrophotometer. The spectrophotometer can help the scientist identify the chemical components of the soil.

When water containing sewage or other wastes filters through natural soil, microorganisms may break down these substances. For that reason, soils from areas such as temperate forests make excellent natural filters.

Discussion Questions

1. Why does the study of the soil play an important role in controlling pollution? *(Accept all reasonable responses. Students should note that many pollutants seep into the soil and form a solution with water. Acid rain is one example.)*
2. What information might Elvia Niebla provide to Congress? *(Examples include details about pollution, such as where it is occurring and what methods of prevention are recommended.)*

Meeting Individual Needs

Gifted Learners

Have students test the acidity of water from your community water supply or from a local body of water. Students may use pH strips or a commercial testing kit that tests water purity, such as those used to test aquarium water. You may wish to also have students find out what requirements must be met before a water supply is declared safe to drink.

Unit 4 Force and Motion

Unit Overview

In this unit, the concepts of force and motion are developed in ways that allow students to draw from personal experiences, observations, and previous knowledge. In Chapter 10, students identify the forces described in a story about a sailboat race. Next, magnetic, electrical, gravitational, and elastic forces are introduced. In Chapter 11, students learn how force can be measured with force meters. In Chapter 12, students consider the causes and the effects of friction and learn about ways to reduce it. In Chapter 13, students do thought experiments to develop an understanding of Newton's first law of motion. Then students consider the concepts of inertia and forces in pairs.

Using the Themes

The unifying themes emphasized in this unit are **Structures, Systems,** and **Energy.** The following information will help you incorporate these themes into your teaching plan. Focus questions that correspond to these themes appear in the margins of this Annotated Teacher's Edition on pages 221, 228, 259, and 268.

The theme of **Structures** can be a useful teaching tool throughout this unit because the form and function of all structures must take into account the forces that act on the structures. This applies to human-made structures as well as biological ones. For instance, a focus question in Chapter 10 invites students to consider the forces at work in the movement of human arm muscles.

Another useful theme in this unit is **Systems.** Because every force must involve both an action and a reaction, a network of forces can constitute a complex and dynamic interacting system. One example is our solar system, which depends on gravitational forces to maintain its integrity.

Energy is also an important theme when discussing forces. Focus questions in Chapters 12 and 13 challenge students to consider how friction affects energy transfers in a variety of situations.

Using the SourceBook

Unit 4 extends the study of motion by emphasizing observation and description of different types of motion. Newton's laws of motion and gravitation are introduced, along with a discussion of momentum.

Bibliography for Teachers

Adair, Robert K. "The Physics of Baseball." *Physics Today* 48 (5): May 1995, pp. 26–31.

Cash, Terry. *101 Physics Tricks: Fun Experiments With Everyday Materials.* New York City, NY: Sterling Publishing Co., Inc., 1995.

VanCleave, Janice P. *Janice VanCleave's Physics for Every Kid: 101 Easy Experiments in Motion, Heat, Light, Machines, and Sound.* New York City, NY: John Wiley & Sons, Inc., 1991.

Bibliography for Students

Darling, David. *Could You Ever Travel to the Stars?* Minneapolis, MN: Dillon Press, 1990.

Wood, Robert W. *Physics for Kids: 49 Easy Experiments With Mechanics.* Blue Ridge Summit, PA: TAB Books, 1989.

Films, Videotapes, Software, and Other Media

Buoyancy
Film and videotape
Coronet/MTI
108 Wilmot Rd.
Deerfield, IL 60015

Gravity: How It Affects Us
Videotape
Britannica
310 S. Michigan Ave.
Chicago, IL 60604-9839

MMV Force and Motion
Software (Macintosh)
Queue, Inc.
338 Commerce Dr.
Fairfield, CT 06430

Motion
Videotape
Agency for Instructional Technology
Box A
Bloomington, IN 47402-0120

Physical Sciences (Newton's Apple series)
Videodisc
National Geographic Society
1145 17th St. NW
Washington, DC 20036-4688

Unit Organizer

Unit/Chapter	Lesson	Time*	Objectives	Teaching Resources
Unit Opener, p. 214				Science Sleuths: A Day at the Races English/Spanish Audiocassettes Home Connection, p. 1 Math Practice Worksheet, p. 3
Chapter 10, p. 216	**Lesson 1, Forces at Work, p. 217**	2	1. Identify forces as pushes or pulls exerted by one object (the agent) on another object (the receiver). 2. Infer what the effect of a force will be on the receiver, such as changing its shape or motion.	Image and Activity Bank 10-1 Exploration Worksheet, p. 4
	Lesson 2, Different Kinds of Forces, p. 222	2	1. Identify and classify different types of forces. 2. Distinguish between contact and noncontact forces. 3. Analyze the forces at work in several new situations.	Image and Activity Bank 10-2 Exploration Worksheet, p. 5
	Lesson 3, What Makes Things Fall? p. 226	2	1. Understand that all objects exert an attractive force on one another. 2. Recognize how the masses of two objects and their distance from each other affect the gravitational force between them. 3. Distinguish between mass and weight.	Image and Activity Bank 10-3 Discrepant Event Worksheet, p. 11 Activity Worksheet, p. 12 ▼
End of Chapter, p. 231				Review Worksheet, p. 13 Chapter 10 Assessment Worksheet, p. 16
Chapter 11, p. 233	**Lesson 1, The Tools to Use, p. 234**	3	1. Suggest several ways to measure gravitational force. 2. Construct and calibrate a force meter.	Image and Activity Bank 11-1 Exploration Worksheet, p. 18 Exploration Worksheet, p. 20 ▼
	Lesson 2, Estimating the Size of a Force, p. 240	2	1. Estimate, in newtons, the force needed to lift an object. 2. Measure the size of a force in newtons.	Image and Activity Bank 11-2 Exploration Worksheet, p. 23 Exploration Worksheet, p. 24 ▼
	Lesson 3, Measuring More Forces, p. 244	2 to 3	1. Estimate the strength of different muscles in newtons. 2. Measure the strength of different muscles in newtons. 3. Distinguish between measuring mass and weight.	Image and Activity Bank 11-3
End of Chapter, p. 248				Review Worksheet, p. 26 Chapter 11 Assessment Worksheet, p. 29
Chapter 12, p. 250	**Lesson 1, Frictional Force, p. 251**	2 to 3	1. Observe the direction and effects of friction in a variety of situations. 2. Identify factors that influence the magnitude of frictional force.	Image and Activity Bank 12-1 Transparency Worksheet, p. 31 ▼
	Lesson 2, Causes and Consequences of Friction, p. 257	2	1. Describe the cause of friction. 2. Explain ways in which friction may be harmful or helpful, and suggest ways to reduce or increase it.	Image and Activity Bank 12-2 Exploration Worksheet, p. 33 Theme Worksheet, p. 38 Transparency 42 Activity Worksheet, p. 40 ▼
End of Chapter, p. 262				Review Worksheet, p. 42 Chapter 12 Assessment Worksheet, p. 45
Chapter 13, p. 264	**Lesson 1, Thought Experiments, p. 265**	2	1. Predict what happens to an object at rest or an object in motion when unbalanced forces act on it. 2. Identify objects with balanced or unbalanced forces acting on them by observing their motion.	Image and Activity Bank 13-1 Transparency 44 Transparency 45 Transparency 46 Transparency 47
	Lesson 2, Puffball: A Game of Force and Motion, p. 271	1 to 2	1. Predict what happens to an object at rest when balanced and unbalanced forces act on it. 2. Predict what happens to an object in motion when balanced and unbalanced forces act on it.	none
	Lesson 3, What Is Inertia? p. 275	2	1. Analyze the effects of inertia in everyday experiences. 2. Relate the inertia of an object to its mass.	Image and Activity Bank 13-3 Exploration Worksheet, p. 48 Exploration Worksheet, p. 51
	Lesson 4, Forces in Pairs, p. 280	2	1. Identify the reaction force for any action force. 2. Describe how action and reaction forces affect the motion of the two objects involved.	Image and Activity Bank 13-4 Exploration Worksheet, p. 53
End of Chapter, p. 284				Review Worksheet, p. 56 Chapter 13 Assessment Worksheet, p. 59
End of Unit, p. 286				Unit 4 Activity Worksheet, p. 63 ▼ Unit 4 Review Worksheet, p. 64 Unit 4 End-of-Unit Assessment, p. 68 Unit 4 Activity Assessment, p. 72 Unit 4 Self-Evaluation of Achievement, p. 75

* Estimated time is given in number of 50-minute class periods. Actual time may vary depending on period length and individual class characteristics.

▼ Transparencies are available to accompany these worksheets. Please refer to the Teaching Transparencies Cross-Reference chart in the Unit 4 Teaching Resources booklet.

Materials Organizer

Chapter	Page	Activity and Materials per Student Group
10	222	***Exploration 2:** jar or bottle; pail filled with water; drinking glass or bottle; 10–12 paper clips; strong magnet; plastic bread bag, cut into strips; wool sock; 3 sheets of paper; several rubber bands; 3–5 hooked masses; meter stick; large plastic bottle; several sheets of newspaper; cloth towel; large, rectangular board or piece of cardboard; 3–4 large books; about 1 m of string; 2 books; 10–15 drinking straws
	227	**In-Text Activity:** round balloon; several tiny pieces of paper
11	236	**Exploration 2:** wooden dowel, about 30 cm long and 2 cm in diameter; cup hook; cardboard bathroom-tissue tube; 2 paper clips; 2–3 rubber bands; 20 cm of masking tape; three 100 g masses or force meter
	238	**Exploration 3:** coil spring or segment of spring toy; at least 100 g of washers; 100 g, 500 g, and 1000 g hooked masses; 2 meter sticks; force meter
	241	**Exploration 4:** meter stick; 1 kg object or mass
	242	***Exploration 5:** brick; board, about 1 m in length; a few books; small sandbag; 2 pulleys; 3 m rope or cord; force meter
	244	***1. Arm Strength:** 2 luggage cords; two 30 cm dowels; force meter; meter stick; 2 small pieces of masking tape; safety goggles; **2. Forearm Pull:** luggage cord; two 30 cm dowels; force meter; safety goggles; **3. Hand Grip:** bathroom scale; **4. Leg Thrust:** chair; bathroom scale
12	255	**Exploration 1:** 1 or 2 bricks; force meter; board, about 1 m in length; 10 cm × 50 cm sheet of glass; 4–6 plastic drinking straws; 1 m of string; scissors; safety goggles
	257	**The Cause of Friction:** 2 pieces of coarse sandpaper; small piece of metal; a few sheets of different paper; book cover; floor tile; small piece of ice
	258	**Reducing Friction:** half of a stick of margarine or a few spoonfuls of grease; 2 pieces of medium sandpaper; small piece of glass; drop of oil
	258	**The Harm of Friction:** 2 pieces of sandpaper
	259	***Exploration 2, Measuring the Strength of a Paper Towel Using Friction:** 2–5 paper towels; small piece of masking tape; force meter; jar or bottle; hole punch; 2–5 paper reinforcements; **Comparing the Friction of Different Types of Paper:** paper towel; wax paper; 2 small pieces of masking tape; 2 jars of equal size; water to fill jars; wooden dowel; two 30 cm pieces of string; various kinds of paper; different grades of sandpaper; cardboard; hole punch; 2–5 paper reinforcements; **Nail Friction:** claw hammer; 2 to 3 nails; piece of wood; **Brick Transportation:** brick; metric ruler; newspaper or cloth; margarine, grease, or corn or canola oil; roller skate; large plastic tub; 2 drinking straws or dowels; lab aprons; **Measuring Air Friction:** ring stand; ring support; 2 paper clips; 2 pieces of thread; index card; hole punch; masking tape; cardboard strip; clothespin; protractor; **Reducing Friction:** book; jar lid; 15–20 glass marbles; 2 empty paint cans; vegetable, corn, or canola oil; roller bearings; ball bearings; sleeve bearings
13	271	**In-Text Activity:** 1 m × 1.5 m playing area with barrier about 4 cm high and 2 goal areas; plastic-foam or table-tennis ball; 4–6 plastic drinking straws (additional teacher materials: roll of masking tape)
	277	***Exploration 1, Station 1:** metric ruler; 30 cm × 10 cm piece of cardboard; 6 dominoes; **Station 2:** coin; 20 cm × 4 cm piece of thick paper; **Station 3:** raw egg; hard-boiled egg; **Station 4:** thick section of newspaper; meter stick; 60 cm × 5 cm piece of thin wood; safety goggles; **Station 5:** small toy cart with removable passenger; 20 cm of thread; board, about 1.5 m in length; 5–10 books
	283	***Exploration 3, Station 1:** coffee can with slanted holes punched in it and taped edge; string; 250 mL of water; plastic cup; pail (additional teacher materials: hammer; nail); **Station 2:** 1 L or 16 oz. plastic bottle with rubber stopper; 2 effervescent tablets; 3–6 paper towels; 2–6 straws or wooden dowels; 350 mL of water; safety goggles; **Station 3:** roller skates; helmet; elbow, wrist, and knee pads; 2 bags of sand; **Station 4:** long balloon; drinking straw; 2 small pieces of masking tape; 3 m of string; 1 m of string or thread

* You may wish to set up these activities in stations at different locations around the classroom.

Advance Preparation

In-Text Activity, page 271: Build the puffball playing field, using masking tape to attach the edges.

Exploration 3, Station 1, page 283: Use a hammer and nail to punch slanted holes in each can.

Unit Compression

In order to maintain a high level of motivation, interest, and understanding among your students, it is recommended that you cover as much of this unit as possible. However, in the interest of time, you may find it necessary to make certain omissions.

Because Chapter 10 introduces students to the concepts covered in the rest of the unit, all of this chapter should be considered core. However, parts of Chapter 11 may be more easily omitted. For instance, most of Lesson 3, which involves estimating and measuring the forces exerted by various muscles in the body, may be skipped if necessary. The last section (Measuring Mass and Weight, on page 247) should not be skipped, however, because it explores the difference between mass and weight.

In Chapter 12, time could be saved by omitting Exploration 2 on pages 259–260. This is a group of extension activities. Similarly, in Chapter 13, the final lesson discusses action-reaction pairs and could be skipped if necessary. It constitutes a unique approach to Newton's third law, but the unit can stand on its own without it.

Homework Options

Chapter 10
See Teacher's Edition margin, pp. 220, 221, 227, 229, 230, and 232
Activity Worksheet, p. 12
SourceBook, pp. S84

Chapter 11
See Teacher's Edition margin, pp. 236, 239, 246, and 248
Exploration Worksheet, p. 18
SourceBook, p. S86

Chapter 12
See Teacher's Edition margin, pp. 252 and 258
Theme Worksheet, p. 38
Activity Worksheet, p. 40
SourceBook, p. S75

Chapter 13
See Teacher's Edition margin, pp. 267, 273, 278, and 281
Exploration Worksheet, p. 51
SourceBook, p. S79

Unit 4
Math Practice Worksheet, p. 3
Activity Worksheet, p. 63
SourceBook Activity Worksheet, p. 76

Assessment Planning Guide

Lesson, Chapter, and Unit Assessment	SourceBook Assessment	Ongoing and Activity Assessment	Portfolio and Student-Centered Assessment
Lesson Assessment Follow-Up: see Teacher's Edition margin, pp. 221, 225, 230, 239, 243, 247, 256, 261, 270, 274, 279, and 283 **Chapter Assessment** Chapter 10 Review Worksheet, p. 13 Chapter 10 Assessment Worksheet, p. 16* Chapter 11 Review Worksheet, p. 26 Chapter 11 Assessment Worksheet, p. 29* Chapter 12 Review Worksheet, p. 42 Chapter 12 Assessment Worksheet, p. 45* Chapter 13 Review Worksheet, p. 56 Chapter 13 Assessment Worksheet, p. 59* **Unit Assessment** Unit 4 Review Worksheet, p. 64 End-of-Unit Assessment Worksheet, p. 68*	SourceBook Review Worksheet, p. 78 SourceBook Assessment Worksheet, p. 82*	Activity Assessment Worksheet, p. 72* **SnackDisc** Ongoing Assessment Checklists ♦ Teacher Evaluation Checklists ♦ Progress Reports ♦	Portfolio: see Teacher's Edition margin, pp. 219, 223, 229, 238, 242, 254, and 269 **SnackDisc** Self-Evaluation Checklists ♦ Peer Evaluation Checklists ♦ Group Evaluation Checklists ♦ Portfolio Evaluation Checklists ♦

* Also available on the Test Generator software
♦ Also available in the Assessment Checklists and Rubrics booklet

Science Discovery is a versatile videodisc program that provides a vast array of photos, graphics, motion sequences, and activities for you to introduce into your *SciencePlus* classroom. *Science Discovery* consists of two videodiscs: Science Sleuths and the Image and Activity Bank.

Using the *Science Discovery* Videodiscs

Science Sleuths: A Day at the Races
Side A

The Zahlers, a father-and-daughter team competing in the Pine Block Derby, are suspicious when another team's entry scores an unusually high number of points. Mr. Zahler suspects them of cheating. The Science Sleuths must consider all of the evidence to determine if and how someone cheated during the race.

Interviews

1. Setting the scene: Father of Pine Block Derby loser **48540 (play ×2)**

2. Derby loser **49304 (play)**

3. Father of derby winner **49768 (play)**

4. Derby winner **50425 (play)**

5. Referee **50836 (play)**

6. Aerodynamics engineer **51709 (play)**

7. Home video of derby **52160 (play)**

Documents

8. Derby-results chart **53492 (step)**

9. Flyer for Pine Block Derby **53495 (step ×2)**

Literature Search

10. Search on the words: DERBY, PINE, PINE BLOCK, RACES, RACE CARS, SPEED **53499 (step)**

11. Article #1 ("Logging Truck Loses Brakes on Highway 23") **53502**

12. Article #2 ("Technology or Driving Skill? The Argument Continues") **53504 (step)**

13. Article #3 ("Solo to Retire From Track") **53507 (step)**

14. Article #4 ("Tomkins Cold Storage Ends Block Ice") **53510**

Sleuth Information Service

15. Referee's measurements of cars **53512 (step)**

16. Lubrication **53515 (step ×2)**

Sleuth Lab Tests

17. Lubrication effect on speed **53519 (play)**

18. Graph showing weight vs. finish times **53759**

Still Photographs

19. Close-ups of the cars **53761 (step ×11)**

20. X rays of the cars **53774 (step)**

Image and Activity Bank
Side A or B

A selection of still images, short videos, and activities is available for you to use as you teach this unit. For a larger selection and detailed instructions, see the Videodisc Resources booklet included with the Teaching Resources materials.

10-1 Forces at Work, page 217

Sailboat 1451
Differences in air pressure in the atmosphere cause the wind to blow. Sailboats use the wind as a source of energy.

Sail board 1452
Sail boards use the power of the wind.

Kinetic energy; hammering a nail 24720–24877 (play ×2) (Side A only)
The weight of a hammer alone will not drive a nail. When the hammer is swung, kinetic energy is conveyed from the hammer head to the nail.

Pulling a nail 24621–24719 (play ×2) (Side A only)
Use leverage to pull a nail from wood.

Ax 1820
The energy of a falling ax can be used to chop wood. The wedge shape of the ax helps split the wood open.

◀‖ Step Reverse   Play ▶   Pause ‖   Step Forward ‖▶

Parachute 1455
Parachutes provide resistance to slow a free-falling sky diver.

10-2 Different Kinds of Forces, page 222
Magnet lifts spoon 32632–32744 (play ×2) (Side A only)
A magnet lifts a spoon.

Magnetic field 34144–34290 (play ×2) (Side A only)
The compass needle responds to the magnetic field surrounding a bar magnet. Opposite ends of the needle are attracted to each end of the magnet.

Electrostatic energy; Van de Graaf generator 33207–33812 (play ×2) (Side A only)
The blue discharge is ionized air.

Acceleration of a falling body 1709
The effect of acceleration due to gravity is shown by two falling objects of the same mass. The ball shot from the apparatus will hit the ground at the same time as the ball dropped from the same level.

Air track; friction restored 28913–29064 (play ×2) (Side A only)
The glider moves until the air track is turned off. Pressurized air bleeds out through tiny holes on the top of the track. This produces a cushion of air that eliminates friction between the track and the glider.

10-3 What Makes Things Fall? page 226
Mass vs. weight 1736
An object has the same mass on the Earth and on the moon. An object's weight will be six times greater on the Earth than on the moon.

Weightlessness in space 19867–20047 (play ×2) (Side A only)
This person floats as if in space. In space, there is no gravity to give weight to a mass.

11-1 The Tools to Use, page 234
Newton; unit of force 1739–1741 (step ×2)
The SI unit for measuring force (step) Measuring the newton

11-2 Estimating the Size of a Force, page 240
Force 1719–1725 (step ×5)
Objects exert a certain amount of force on the surface they touch or on the ground on which they sit. *Note that the ground or surface exerts an equal and opposite force on each object.* (step) A chair exerts a 50 newton (N) force on the ground. (step) A brick exerts a 20 N force on the surface on which it sits. (step) An empty glass exerts a 1 N force on the surface on which it sits. A 2 L carton of milk exerts a force of 20 N on the surface on which it sits. (step) A liter-sized glass bottle filled with liquid exerts a force of 15 N on the surface where it sits. (step) A young boy exerts a 500 N force on the ground.

11-3 Measuring More Forces, page 244
Strength measurement 1726–1728 (step ×2)
Measure the force exerted by your legs on a bathroom scale by using a backrest and then by pressing the scale against a wall with your feet. (If the scale is in kilograms, multiply your result by 10 to get the force in newtons.) (step) Measure your hand strength by gripping a bathroom scale with both hands. (step) Measure the leaning force exerted by your body by bracing a scale between your shoulder and a wall.

12-1 Frictional Force, page 251
Skier 1713
A skier encounters friction when moving skis across the surface of the snow. Snow allows for less friction than concrete or grass.

Ice skater 1714
Skates minimize frictional force on ice. Less interaction between the skates and the ice means less resistance to the skater's movement. Therefore, small blades are used on ice skates.

Car, aerodynamic 1712
A car is made with a sleek style to reduce the negative effects of air on its performance. Air resistance tends to slow cars traveling at high speeds.

Sliding block 24099–24620 (play ×5)
(play) The angle of repose is the maximum angle at which an object can be tilted before it begins to slide. (play) This wooden block is placed with its widest side down. The board is tilted until the block begins to slide. Note the angle. (play) This wooden block is placed with its narrow side down. The board is tilted until the block begins to slide. Note the angle. Does area of contact influence friction? (play) Note the angle at which this single block of wood begins to slide. (play) Note the angle at which two blocks of wood begin to slide. Does weight influence friction?

12-2 Causes and Consequences of Friction, page 257
Ball bearings 1730
The frictional force between moving parts is reduced by using ball bearings. Wheels spin with less resistance when ball bearings are used at the hub, or center.

Bicycle friction 25088–21819 (play ×3)
The friction between the brake pads and the tire causes resistance against the motion of the tire. Friction between the chain and the sprocket allows the chain to turn the wheels.

Friction creates heat 21820–22323 (play ×2) (Side A only)
The friction between the dowel and the block of wood creates enough heat to make the block burn.

13-1 Thought Experiments, page 265
Path of the moon 1738
The path of the moon would be a straight line if there were no other forces acting on it. The Earth's gravity keeps the moon in orbit around the Earth.

13-3 What Is Inertia? page 275
Air-pressure demonstration 13127–13359 (play ×2) (Side A only)
Air pressure on the paper provides enough weight to resist the applied force and to break the ruler.

Crash testing 25510–25893 (play ×2) (Side A only)
As a car crashes up against a brick wall, it is forced to stop. However, the objects in the car (people) remain in motion and continue to move forward.

13-4 Forces in Pairs, page 280
Equilibrium; tug of war 1747
Equilibrium is maintained on the rope if the forces acting on it are of equal magnitudes.

◀| **Step Reverse**  **Play** ▶ **Pause** ‖ **Step Forward** |▶

UNIT FOCUS

Before beginning the unit, involve students in a discussion of their ideas about force and motion. Ask: Why don't we float away from the Earth? Why is it harder to ride your bike over ice than over asphalt? Why is it difficult to move heavy things? Why is it difficult to stop heavy things that are already in motion? *(Accept all reasonable responses.)* Students may have some knowledge of forces, but many will be surprised by how difficult it is to answer these questions.

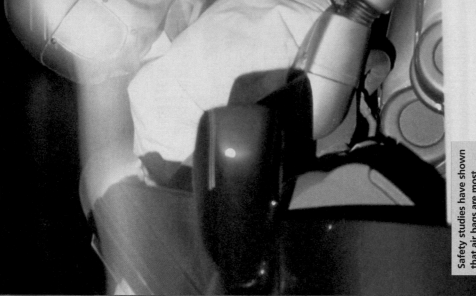

Unit 4 FORCE AND MOTION

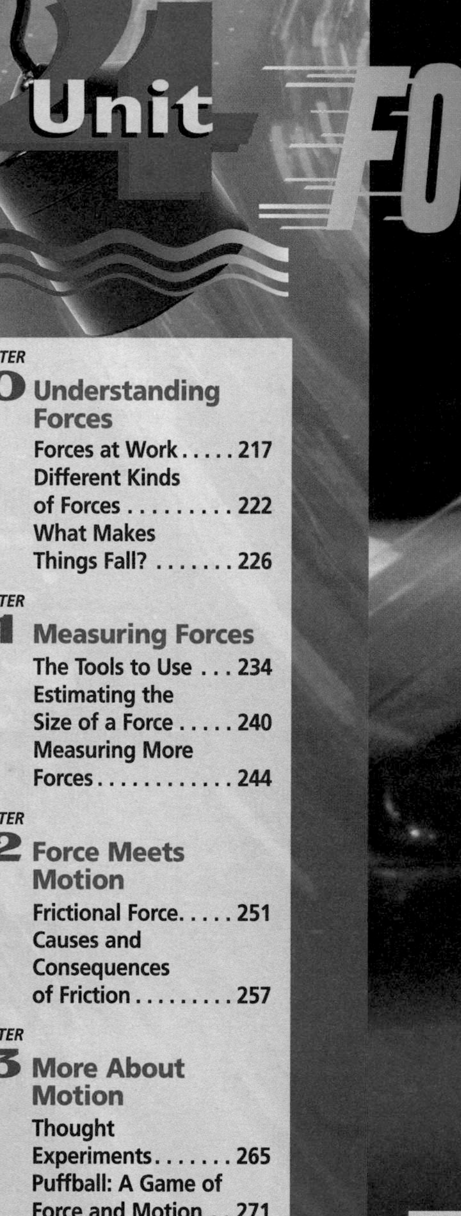

Safety studies have shown that air bags are most effective in preventing injuries during an automobile collision when seat belts are worn by the driver and passengers.

214

Connecting to Other Units

This table will help you integrate topics covered in this unit with topics covered in other units.

Unit 1 Interactions	Most abiotic interactions in a natural community, such as water eroding rock, are the results of forces.
Unit 2 Diversity of Living Things	Adaptations such as wings and slick fur help animals overcome forces such as gravity and air resistance.
Unit 5 Structures and Design	In designing any structure, it is critical to consider all of the various forces that will be acting on it.
Unit 6 The Restless Earth	Forces, gravity, and friction are all involved in the various movements of the Earth's crustal plates.
Unit 7 Toward the Stars	The motion of the stars and planets must be studied in terms of their gravitational pull on each other.

If a car traveling at 30 mph (48 km/h) hits a wall, it will come to a complete stop in less than a tenth of a second. The driver and any passengers, however, will continue traveling forward if not restrained, slamming into the steering wheel, dashboard, or windshield. Even in a low-speed crash, the forces involved can cause life-threatening injuries. The use of seat belts keeps the driver and passengers connected to the car and opposes their continued forward motion, thus helping to prevent injuries.

Air bags offer additional protection in such a collision. Upon impact, an air bag inflates and provides a cushion against the driver's forward motion by allowing the air inside to compress between the driver and the steering wheel. A passenger-side air bag offers a similar cushion between the passenger and the dashboard. What benefits do air bags provide that seat belts don't? Why do you think safety experts advise that air bags be used in addition to—not instead of—seat belts? Ⓐ

Turn the page to further explore the world of forces around you and to learn more about how these forces are related to motion.

215

Using the Photograph

Explain to students that many injuries occur in car crashes because occupants are thrown forward violently. Seat belts prevent many of these injuries by restraining the occupants and keeping them in their seat. It is crucial that the belts "give" a little—like a very stiff rubber band—or the occupant can be injured by the seat belt itself. Other injuries can occur in high-impact collisions because, even if the seat belt gives a little, it does not restrain the entire body. The movement of a person's head could lead to whiplash or other neck injuries. Lap belts alone do not restrain the upper body, which can lead to spinal injury.

Answers to *In-Text Questions*

Ⓐ The gas in an air bag acts as a cushion and stops the forward motion of a passenger. An advantage that air bags have over seat belts is that air bags restrain the entire upper body in a crash, including the head. Nonetheless, seat belts should be worn even in air bag–equipped vehicles because an air bag is insufficient to hold a person securely in the seat during a crash, particularly in a side-impact collision.

Homework

The Math Practice Worksheet on page 3 of the Unit 4 Teaching Resources booklet makes an excellent homework activity.

Understanding Forces

Connecting to Other Chapters

Chapter 10 *offers students hands-on exposure to a wide range of forces.*

Chapter 11 *allows students to measure the strength of forces with commercial and homemade force meters.*

Chapter 12 *explores the causes and consequences of friction in various situations.*

Chapter 13 *discusses the concept of inertia and introduces students to force pairs.*

Prior Knowledge and Misconceptions

Your students' responses to the ScienceLog questions on this page will reveal the kind of information—and misinformation—they bring to this chapter. Use what you find out about your students' knowledge to choose which chapter concepts and activities to emphasize in your teaching.

In addition to having students answer the questions on this page, you may wish to have the students play the following game. Give one student in the class a sentence such as, "The cat rolled the ball." Make sure that the sentence contains a direct object. Then have another student tell the class whether any work is done in performing the action and, if so, what is doing the work. (In this example, the cat performs work by rolling the ball.) Have the second student continue the story by giving another sentence, using the previous object as his or her subject, such as, "The ball hit the wall."

Continue in this way until every student has had a chance to make a sentence. Listen carefully to the students' discussion of the sentences, and use what you find out to assess what your students already know about forces at work, what misconceptions they have, and what about this topic is interesting to them.

CHAPTER 10 Understanding Forces

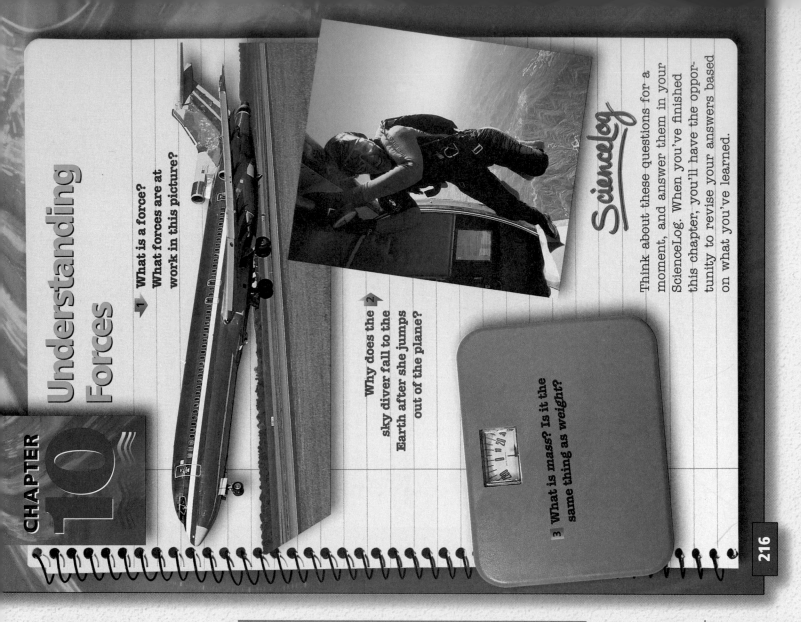

1 **What is a force? What forces are at work in this picture?**

2 **Why does the sky diver fall to the Earth after she jumps out of the plane?**

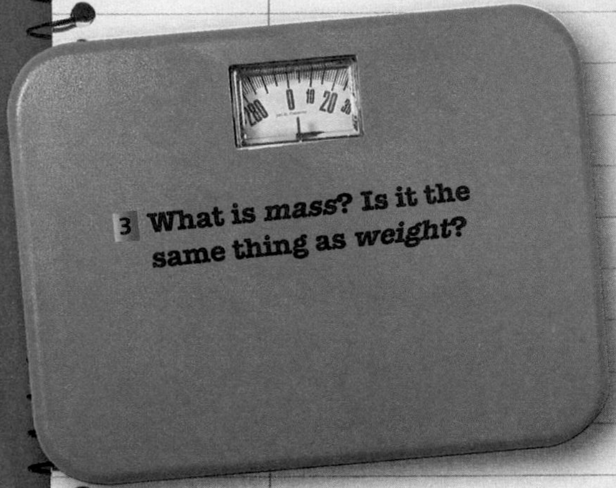

3 **What is *mass*? Is it the same thing as *weight*?**

ScienceLog

Think about these questions for a moment, and answer them in your ScienceLog. When you've finished this chapter, you'll have the opportunity to revise your answers based on what you've learned.

Forces at Work

The Race

The race was about to start, and Kathleen and Kimiko were excited! With difficulty, they dragged their sailboat across the sand and into the water.

A favorable west wind was blowing. The craft was ready to go. Their hearts beating hard, they pulled on the halyards with nimble fingers to raise the sails.

With one hand, Kimiko took control of the tiller. With the other, she tugged on the sheet for the mainsail. Kathleen, sitting amidships, secured the sheet of the jib.

As a gentle breeze pushed on the sails, the craft moved slowly and steadily out beyond the jetty. The full force of the wind stretched the sails, and the craft suddenly sped up. They were off!

The boat shuddered as the choppy waves slapped and sprayed the hull. With the experience of a summer's sailing, the girls sailed toward the starting line. Just as they reached the line, the starter's flag was lowered. The race was on!

Kimiko pushed on the tiller. The rudder twisted against the water. Their course was set. The sails strained and the mast creaked as the boat skimmed and broke the water's surface. Kathleen pushed downward on the centerboard to offset the leeway. Passing the second red buoy on the starboard, Kimiko decided to make a quick turn. She drew the tiller toward herself, hauling in the sheet at the same time, and yelled "Coming about!" The boat swung around until the wind was at the stern. The boom of the mainsail swung suddenly from one side of the craft to the other. In an instant, the boat changed its course.

Kathleen, who was watching another boat coming up rapidly on the leeward side, responded too late to the warning. The boom smacked her arm. She fell over the side and hit the water with a splash. Quickly, almost automatically, Kimiko brought the boat up into the wind. The sails flapped harmlessly, and the boat slowed down . . .

217

LESSON 1 ORGANIZER

Time Required
2 class periods

Process Skills
observing, inferring, analyzing

Theme Connection
Structures

New Term
Force—a push or pull exerted by one object on another

Materials (per student group)
none

Teaching Resources
Exploration Worksheet, p. 4

Forces at Work

FOCUS

Getting Started

Arrange some dominoes in a line so that they will fall in succession. Ask: What will happen if a force is applied to the first domino? *(All of the dominoes will fall over.)* Apply a force to the first domino so that it moves sideways and the other dominoes remain standing. Explain that in this lesson students will learn that forces always have a direction. Replace the first domino, and ask students which direction to push it. When they specify the appropriate direction, push the domino again, allowing the chain of dominoes to fall.

Main Ideas

1. A force is a push or pull exerted by one object on another.
2. A force can alter the shape or the motion of an object.
3. A force may be represented by an arrow because an arrow has both a direction and a size.

TEACHING STRATEGIES

The Race

Call on a volunteer to read aloud the sailing adventure on page 217.

If there are students who have sailed, ask them to describe their experiences and to explain how the sails must be manipulated to catch the wind.

Recognizing Forces
Action words (verbs) may refer to forces (pushes or pulls) or may describe motion, which is caused by a force. Ask students to distinguish between force words and motion words as they write their list of words.

Answers to *In-Text Questions*

Ⓐ amidships—toward the middle of the boat
boom—the horizontal support that stretches the sail; it is attached at one end to the mast
buoy—a marker in the water that indicates the port side of a channel
centerboard—a straight, adjustable board that extends beneath the boat into the water; it increases stability when a boat tips or slips sideways
halyards—ropes used to haul up the sails
hull—the frame or body of a boat
jetty—a wall or pier jutting out into the water
jib—forward sail of a boat
leeward—toward the side of the boat turned away from the wind
leeway—the drift of a boat away from the wind
mainsail—the principal sail on the mast of the boat
rudder—a board in the water to which the tiller is attached; it enables a boat to be steered
sheet—a rope for holding and controlling a sail's position
starboard—the right side of a boat
stern—the back end of the boat
tiller—lever at the back of a craft used for steering

Ⓑ Action words include *moved, sped up, skimmed, broke, fell,* and *swung around*. Forces include *dragged, blowing, beating, pulled, hit, tugged, pushed, stretched,* and *twisted*.

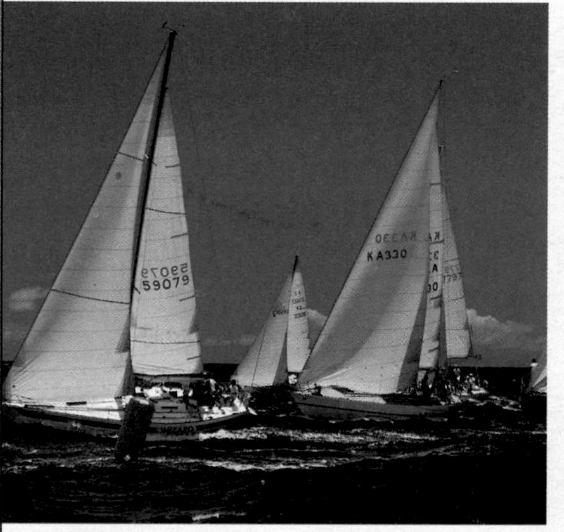

The sailing terms used in this story about Kathleen and Kimiko may be new to you. Look up the definitions for the terms below, or ask someone who sails. Then read the story again. Ⓐ

• amidships	• halyards	• leeward	• sheet
• boom	• hull	• leeway	• starboard
• buoy	• jetty	• mainsail	• stern
• centerboard	• jib	• rudder	• tiller

Recognizing Forces
Sailing is an exciting, fast-paced activity. Many forces are at work as a sailboat skims across the water. Sailors put out tremendous effort and action to keep their boats sailing. Notice how many action words appear in the story: *pushed, hauled, smacked,* and *slapped*, just to name a few. Write a list of all the action words in the story and all the forces that you recognize. Ⓑ

Questions
1. Can there be a force without some object (someone or something) exerting the force and some other object receiving the force?
2. Name some effects that forces have on objects.
3. Is it possible for one object to exert a force on another without ever touching it? Where are there examples of this in the story?
4. Examine the story for pairs of opposing forces, which are forces on the same object that work against each other. How many did you find?
5. How would you define or describe a force? Make sure your definition includes all the things you have discovered about forces.
6. Write an exciting conclusion to the story. Involve forces in it, and then identify them. You might illustrate the events of your conclusion and mark the forces in your drawing.

Answers to *Questions*

1. No. A force is a push or a pull that one body exerts on another. (Objects exerting or receiving forces may need to be inferred. In the phrase, "she fell over the side," students should infer that gravity pulled her down.)
2. Forces are able to change the shape of an object (such as the sail), give motion to an object at rest (such as the sailboat), and alter the speed or direction of an object in motion (such as the craft suddenly speeding up).
3. Yes. One object can exert a force on another without direct contact. In the story, Kathleen lost her balance and fell over the side. The Earth exerted a gravitational force on Kathleen. This is an example of a force that operates at a distance.
4. Several pairs of opposing forces are implied in the story. For example, the force of the wind on the boat is opposed by the frictional force of the boat against the water. When the two forces are balanced, the boat moves at a constant velocity. If the wind's force is less than the water resistance, the boat will slow down.
5. A force is a push or pull that one object exerts on another.
6. Be sure students identify the forces that they include in their stories.

Describing Forces

The passages below are taken from the story of the boat race. Pay careful attention to the underlined words and the words in italics. Then answer the questions, and see what you can discover about forces.

1. What is similar about the forces described in these sentences?
   a. "With one hand, Kimiko took control of the tiller. With the other, she *tugged* on the sheet . . ."
   b. "As a gentle breeze *pushed* on the sails, . . ."
   c. "The boom *smacked* her arm."

   Do you recognize an agent, receiver, and action word for each force? What are they?

2. What do these sentences tell you about the effects of forces?
   a. "The full force of the wind *stretched* the sails, . . ."
   b. ". . . the craft suddenly *sped up*."
   c. "In an instant, the boat *changed* its course."
   d. "She *fell* over the side . . ."
   e. ". . . the boat *slowed down* . . ."

   Did you discover five different effects? What are they?

3. Think of other forces that could cause an object to do the following:
   a. change its shape
   b. begin to move
   c. speed up or slow down
   d. change direction

4. What aspect of force does this quotation describe? "Kathleen pushed *downward* on the centerboard . . ."

In the story of the race, the directions of most of the forces are not described. However, every force has a direction. When scientists draw a force, they use an arrow to indicate the force and its direction. The length of the arrow gives you an idea of the size of the force. The tail of the arrow indicates the point at which the force is exerted.

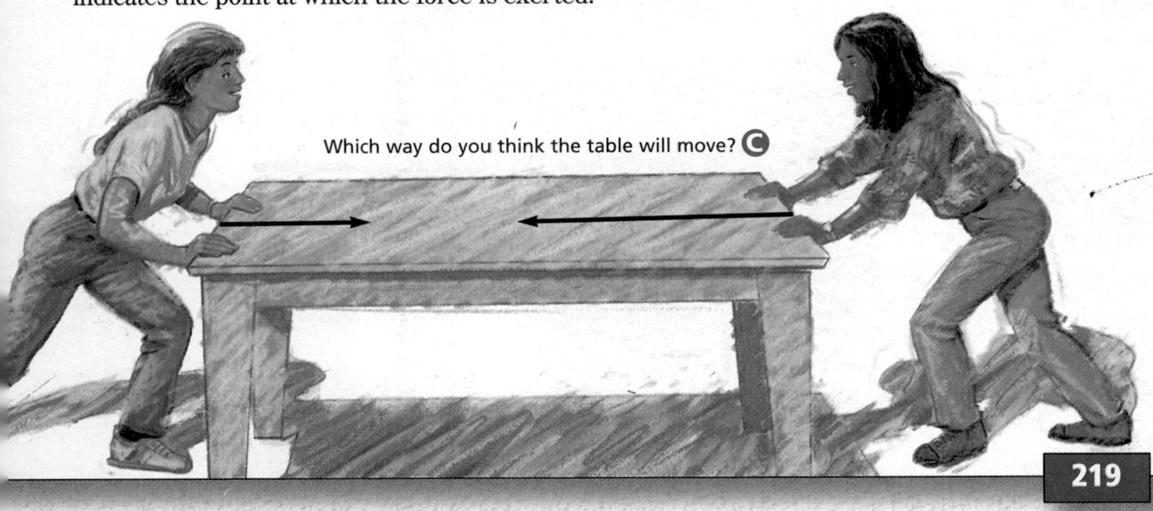

Which way do you think the table will move? Ⓒ

CROSS-DISCIPLINARY FOCUS

Language Arts

Have students consider the following sentence: The *force* of the attack was so great that it *forced* the general to call in more *forces* for support. Ask: Which use of the word *force* is closest to the sense in which it is used in this unit? *(The first one)* You may wish to point out that in science, even everyday words have very strict definitions in order to avoid such confusion. Ask students to consult a dictionary and to make up their own sentences with words such as *motion* or *mass*.

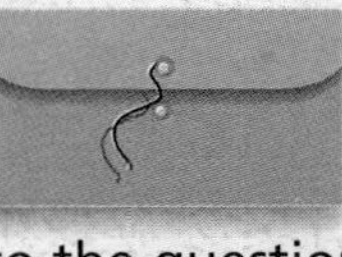

PORTFOLIO

Students may wish to include their answers to the questions on page 218 in their Portfolio. They may later choose to amend their answers to reflect what they have learned in the unit.

Answer to *Caption*

Ⓒ The table in the illustration would move to the left because the force arrow pointing to the left is longer.

Describing Forces

Have students demonstrate opposing forces on a desk or book. Ask them to make a sketch of the demonstration and to use arrows to illustrate the forces that were exerted.

Answers to *Describing Forces*

1. In these sentences, an agent (Kimiko, a breeze, the boom) exerted a push or pull on a receiver (the sheet, the sails, her arm).
2. The effects of forces can do many different things.
   a. Changed the shape of an object
   b. Increased the speed of an object
   c. Changed the direction of an object's motion
   d. Caused an object at rest to fall
   e. Decreased the speed of an object
3. Sample answers:
   a. An air pump inflating a football or wind blowing against a tree
   b. A person opening a door or a baseball pitcher hurling a ball
   c. A person pushing another on a swing or a bicycle's brakes squeezing the rim of the bicycle's wheel
   d. Wind blowing a paper airplane from its course or a hockey puck striking a skate blade during a goal shot
4. The quotation shows that forces have a direction.

Exploration 1

Point out that the forces of gravity and friction are at work in all of the photos. For example, the combined forces of the tacklers and gravity cause the ball carrier to fall. Students should recognize that most of the receivers have several different forces acting on them at the same time.

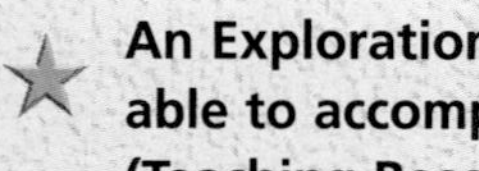

An Exploration Worksheet is available to accompany Exploration 1 (Teaching Resources, page 4).

Answers to *In-Text Questions*

Ⓐ All of the pictures include cases where two or more forces are acting on the same object. For instance, both the cue stick and air resistance are acting on the cue ball, and both gravity and the force of the woman's hand are acting on the car jack. In each case, the object's motion is determined by the combined strengths and directions of the forces acting on it.

Answers to *A Picture Puzzle*

Answers may vary. The following are sample answers:

- **a.** Agent: the cue stick; receiver: the cue ball; effect: the cue ball moves and hits the other balls.
- **b.** Agent: the archer; receiver: the bow; effect: the bowstring stretches, and the bow resists by bending.
- **c.** Agent: the girl's arm; receiver: the balloons; effect: the upward motion of the balloons is halted by the downward pull of the girl's arm.
- **d.** Agent: tacklers; receiver: the player with the ball; effect: the forward motion of the ball carrier is stopped.
- **e.** Agent: the jack; receiver: the car; effect: the car is lifted upward by the jack.
- **f.** Agent: the ax; receiver: the wood; effect: the ax is driven downward into the wood, splitting it.
- **g.** Agent: the parachute; receiver: the race car; effect: the forward motion of the car is slowed.

Exploration 1

A Picture Puzzle

Several forces are illustrated in these pictures. For each picture, determine the object (the agent) that is having a noticeable effect on another object (the receiver). In your ScienceLog, sketch each situation, and draw arrows to indicate the direction of each force and the point where each force is being applied. Make a table showing the agent, the receiver, and the effect of the force for each picture.

Agent	Receiver	Effect
cue stick	cue ball	cue ball moves and hits the other balls

Are there cases where two forces are acting on the same object? If so, what is their combined effect on the object? Ⓐ

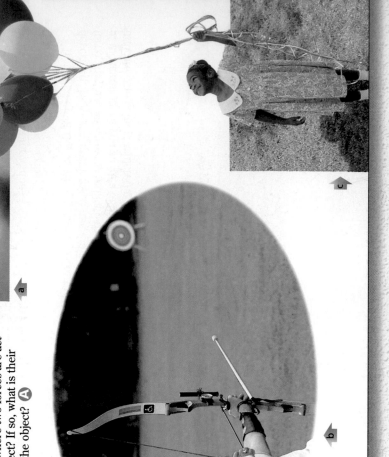

a

b

c

220

Meeting Individual Needs

Second-Language Learners

Have students generate synonyms for the words *force* and *motion.* Slang terms, words in foreign languages, phrases from songs, and other popular sources are appropriate responses. For each word given, the student should use the word in a complete sentence. Compile a list of these words, and use them when introducing concepts in this unit to help students relate the new concepts to their own vocabulary.

Homework

Have students list 10 situations in which forces are acting on them. They should describe the forces in detail, indicating the agents and effects of the forces.

FOLLOW-UP

Reteaching

Have students sit on the floor in a circle. Pick up a soft, foam ball and announce, "I am the agent of force." Then roll the ball to a student. Have students continue rolling the ball to each other. Each time someone new gets the ball, the class should identify the new agent of force. *(The person who rolls the ball)* Ask: Is the receiver of force the same? *(Yes. The ball is the receiver of force.)* Is the direction of force the same? *(No. The direction of force changes each time the ball is pushed to someone new.)* Then ask: What is the effect of each agent of force? *(The ball rolls.)*

Assessment

Perform a series of demonstrations of forces at work. For example, blow on a piece of tissue paper, pull a wagon, squeeze a piece of clay, pick up a book, and push a door. For each demonstration, ask students to identify the agent, the receiver, and the effect of the force.

Extension

Have students select magazine and newspaper articles that describe the causes or effects of forces. Students should carefully analyze the articles and identify the agents, receivers, and effects of the forces. Subject matter might include collisions, weather forecasts, and transportation.

Closure

Take students on a walk around the school. While they are on the walk, ask them to observe and list as many agents, receivers, and effects of forces as possible.

Theme Connection

Structures

Display an illustration of the muscles in a human arm. Indicate one of the larger muscles in the arm and ask the following **focus question:** What forces are at work when this muscle contracts? *(The muscle exerts a force at each of its ends that pulls the bone or connecting tissue toward the middle of the muscle.)*

Meeting Individual Needs

Learners Having Difficulty

Suggest that students write a story using several of the new vocabulary words presented in this lesson. Students may wish to present their stories in cartoon form.

Homework

The Extension activity described on this page makes an excellent homework assignment.

Different Kinds of Forces

FOCUS

Getting Started

Tape paper clips to the bottom of a small toy boat. Place the boat in a shallow pan of water. Ask students if they can think of a way to get the boat to move. Possible suggestions could be pushing or pulling it, blowing on it, and so on. Move a magnet around underneath the pan. This will cause the boat to move. Explain that magnets exert a noncontact force. Tell students that in this lesson they will be learning about many different kinds of contact and noncontact forces.

Main Ideas

1. There are many types of forces, including buoyant, magnetic, electrical, gravitational, elastic, and frictional.
2. Buoyant, elastic, and frictional forces are contact forces. Magnetic, electrical, and gravitational forces are noncontact forces.

TEACHING STRATEGIES

EXPLORATION 2

Set up one experiment at each station around the classroom. For large classes, two setups at each station may be necessary. Instruct groups to move from one station to another at 6-minute intervals. Answers to the questions on this page are provided for each station on the following pages.

 **An Exploration Worksheet is available to accompany Exploration 2 (Teaching Resources, page 5).**

Different Kinds of Forces

You're probably familiar with the many different game booths at a carnival. For this Exploration, you will move among seven science booths or stations and will meet and experience many different kinds of forces.

EXPLORATION 2

A Carnival of Forces

The seven mini-experiments in this Exploration will give you a chance to experience specific forces at work. Each one is set up at its own station. At each station you will find materials arranged as shown on pages 223 to 225. Your purpose at each station is to discover which forces are at work and how they work. As you visit each station, think about the following questions:

- What are the agent, receiver, and effect of each force?
- What name would you give to each force?
- Are there any *noncontact forces*—forces exerted by an agent that does not touch the receiver?

Record your answers, observations, conclusions, and any questions in your ScienceLog.

You Will Need

- an empty jar
- a pail
- water
- a glass
- paper clips
- a strong magnet
- scissors
- a plastic bread bag
- a wool sock
- paper
- hooked masses
- rubber bands
- a plastic bottle
- newspapers and a cloth towel
- a board or long piece of cardboard
- books
- string
- a metric ruler
- drinking straws

LESSON 2 ORGANIZER

Time Required
2 class periods

Process Skills
observing, analyzing, inferring, classifying

New Term
Noncontact forces—forces exerted by an agent that does not touch the receiver

Materials (per student group)
Exploration 2: jar or bottle; pail filled with water; drinking glass or bottle; 10–12 paper clips; strong magnet; plastic bread bag, cut into strips; wool sock; 3 sheets of paper; several rubber bands; 3–5 hooked masses; meter stick; large plastic bottle; several sheets of newspaper; cloth towel; large, rectangular board or piece of cardboard; 3–4 large books; about 1 m of string; 2 books; 10–15 drinking straws

Teaching Resources
Exploration Worksheet, p. 5

STATION 1

What to Do

1. Push the empty jar slowly into a pail of water. Keep its open end facing up.
2. Submerge it; let the water fill the jar.
3. Now take the jar out of the water.

What forces do you feel at each step?

STATION 2

What to Do

1. Bring a strong magnet near the glass.
2. Try to lift the paper clips to the top of the glass. Can you lift the container off the table?

Think about the forces involved.

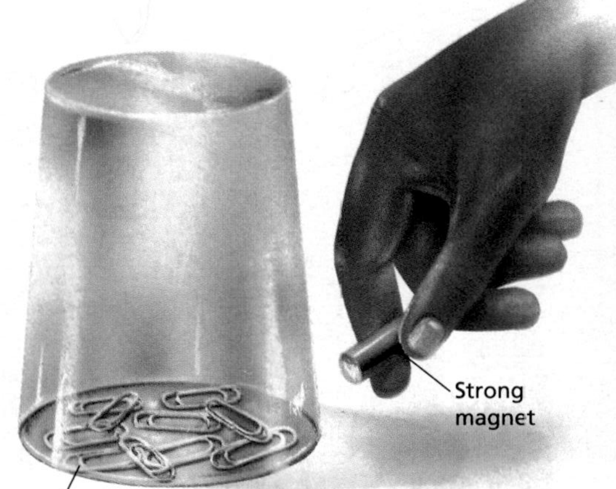

STATION 3 

What to Do

1. With one hand, hold up two long strips cut from a plastic bread bag. Observe them as they hang freely.
2. Now rub both sides of the strips with a wool sock. Allow the strips to hang freely again. Observe any forces at work.
3. Bring one of the strips near a table or a wall. What happens?

Identify the forces acting on the plastic strips.

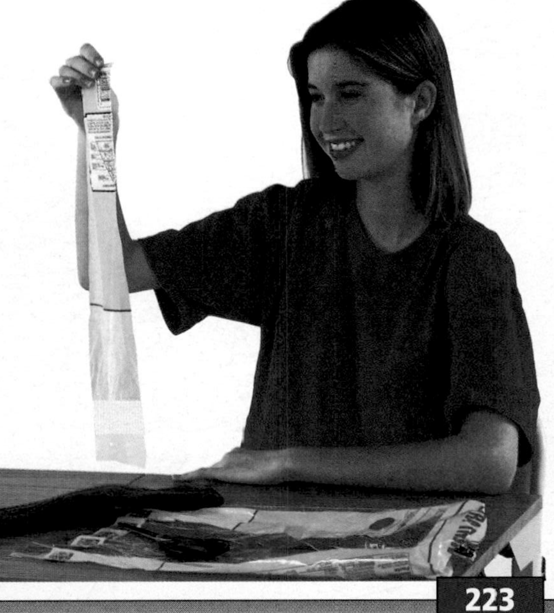

Answers to *Station 1*

As the empty jar is pushed into the water, the water exerts an opposing, upward, buoyant force on the jar. If the jar is released, it pops upward. As the filled jar is lifted upward, the opposing downward force (the weight of the jar and water) is apparent.

The agents are the person's hand and the water, the receiver is the jar, and the effect is the jar being pushed and pulled in different directions at the same time.

The buoyant force is a result of gravity, but it pushes up on objects instead of down. The noncontact force is gravitational.

GUIDED PRACTICE In preparation for upcoming lessons, have students begin thinking about the directions of the forces as well as the agents, receivers, and effects. Have students identify the directions of the forces operating at the different stations.

Answers to *Station 2*

The agent is the magnet, the receiver is the pile of paper clips, and the effect is the magnet attracting the paper clips. If the magnet is strong enough, it may be able to lift the glass off the table just by attracting the paper clips.

The magnetic force is a noncontact force and acts only on specific types of objects.

Answers to *Station 3*

When the plastic strips are rubbed with a wool sock, they move apart from each other. Also, the strips tend to be attracted to whatever else is nearby.

Both strips exert a force on each other, so both strips are agents and receivers. The effect of this force is that the strips move away from each other and toward other objects.

The electrical force is a noncontact force and can be exerted by many different types of objects.

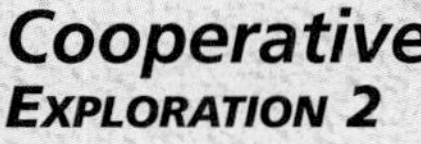

Cooperative Learning EXPLORATION 2

Group size: 3 to 4 students
Group goal: to experience and describe a variety of forces through experimentation
Positive interdependence: Assign each student a role such as primary investigator, timekeeper, recorder, and group leader.
Individual accountability: Students should provide written answers to the questions posed in Which Forces Did You Meet at the Carnival? on page 225.

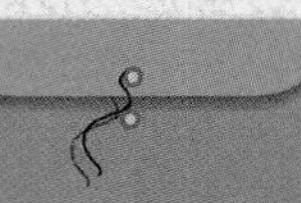

PORTFOLIO

Students may wish to include their observations from one of the stations in their Portfolio. They should evaluate what they learned by performing the activity at that station.

Meeting Individual Needs

Learners Having Difficulty
Review the difference between contact and noncontact forces. Students should understand that a force, such as the force from a magnet, does not have to touch the objects it affects. That is, a magnet exerts a noncontact force.

Station 4

What to Do

1. How fast does paper fall? Why does it fall? Compare the rate of falling for the following:
   a. a single sheet of paper held horizontally
   b. a piece folded into quarters, also held horizontally
   c. a piece crumpled into a ball
2. Hold two pieces of paper together horizontally, and drop them at the same time. Compare the falling rate of two pieces with the falling rate of one—when flat, when folded in quarters, and when crumpled. Why do some fall faster than others?

What are the forces involved at this station?

Station 5

What to Do

1. Suspend a rubber band from a horizontal support.
2. Hang a hooked mass on the rubber band. What happens?
3. Add more masses. How much mass is needed to make the rubber band stretch 10 cm? 20 cm? 30 cm? How much mass do you think would be required to make the rubber band break?

Identify all forces.

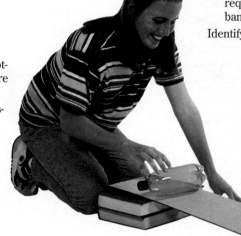

Station 6

What to Do

1. Find out how far a plastic bottle will roll on the floor before stopping.
2. Place several sheets of newspaper on the floor and try again.
3. Repeat the experiment by rolling the bottle over other surfaces, such as a cloth towel spread on the floor.

What is slowing down the bottle?

Answers to
Station 4

The agents are the Earth (exerting a gravitational force) and air resistance, the receiver is the paper, and the effect is that the pieces of paper fall. Some fall more slowly than others because there is an upward force (air resistance) pushing on the pieces of paper. The unfolded sheet of paper falls more slowly because of air resistance. When two sheets that are unfolded or folded into quarters are released together, they fall more quickly than one because they have a greater weight acting against the same air resistance. (A more convincing effect can be achieved by dropping about twenty sheets stapled together instead of two.) The effect is negligible when the papers are crumpled, however, because the air resistance is negligible.

The noncontact force that is evident here is gravitational force.

Answers to
Station 5

The agents are the masses, the receiver is the rubber band, and the effect is that the rubber band is stretched by the downward force of the masses. Also, the rubber band (the agent) exerts an upward, elastic force on the masses (the receivers).

An elastic force occurs when an object is stretched or bent out of shape. The noncontact force is gravity.

Answers to
Station 6

The agents are the Earth exerting gravity and the surface on which the bottle rolls, the receiver is the bottle, and the effect is that the bottle stops rolling. This is due to the force of friction opposing the motion of the bottle. The amount of frictional force depends on the type of material over which the bottle rolls; carpet, for instance, will exert more frictional force than newspaper.

The frictional force occurs only in opposition to a moving object. The noncontact force is gravity.

Answers to
Station 7, page 225

The agents are the hand pulling the rubber band and the rubber band in turn pulling the book; the receiver is the book; and the effect is movement of the book. Also, the table (the agent) exerts a frictional force on the book (the receiver). It takes more force to start two books moving than it does to start one book. The friction between the books and the table opposes the forward motion of the books. The opposing force of friction is lessened by the use of the drinking straws.

The frictional and elastic forces were encountered in earlier stations. Other than gravity, there are no noncontact forces.

Meeting Individual Needs

Second-Language Learners

Give students the opportunity to experiment further with noncontact forces. Provide the following materials: magnets; metal washers; various objects that magnets will and will not attract; and bits of paper, wool, rubber balloons, and string. Tell students to find as many different ways as they can to make the objects move without touching them. Have students summarize their findings in posters.

STATION 7

What to Do

1. Cut a rubber band and tie it securely to a paper clip. Attach the clip to a string tied around a book. How long must the band stretch before the book starts moving?

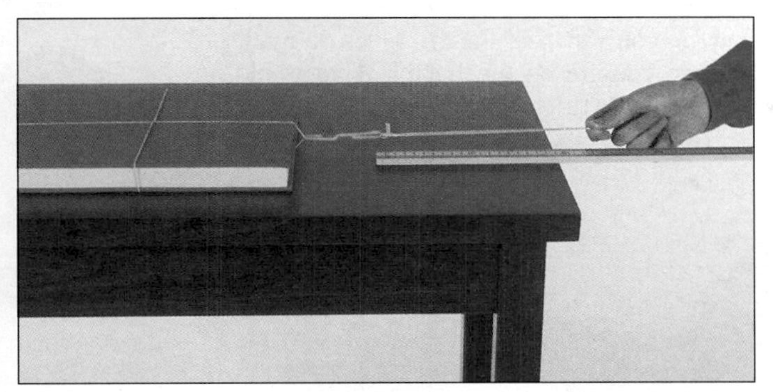

2. Try it again with another book placed on top of the first one. What do you notice?
3. Place the books on drinking straws and repeat the experiment. How has the opposing force changed?

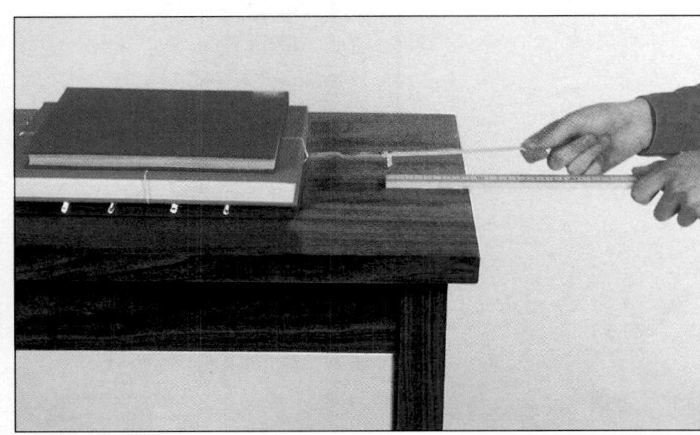

What tends to prevent the books from moving?

Which Forces Did You Meet at the Carnival?

1. Water exerted an upward **buoyant force** on the empty jar at Station 1. Where did you see this force before?
2. A **magnetic force** was present at Station 2. Did the magnet need to touch the paper clips to exert an attractive force on them? Explain.
3. An **electrical force** acted on the charged plastic strips at Station 3. Each strip exerted a force on the other. How is an electrical force like a magnetic force?
4. A **gravitational force** acted on the sheets of paper at Station 4. Is this a noncontact force? What agent is exerting the force? Why did the uncrumpled sheets fall more slowly?
5. The rubber bands at Stations 5 and 7 exerted an **elastic force**. Where else could you find elastic forces?
6. **Frictional forces** were exerted on the bottle and book at Stations 6 and 7. As an object moves in one direction, frictional forces act on the object in the opposite direction. In what direction were the frictional forces at Stations 6 and 7 acting?

Answers to *Which Forces Did You Meet at the Carnival?*

1. Floating is an example of a buoyant force. The sailboat in the story was kept afloat because of the buoyant force from the water.
2. No; the magnet was separated from the paper clips by the glass jar.
3. Each strip exerts a noncontact force on the other strip, but the electrical force is a repellent force, whereas the magnetic force is an attractive force.
4. The Earth exerts a noncontact, gravitational force that pulls the paper to the ground. The uncrumpled sheets of paper fall more slowly because their larger surface areas allow them to be exposed to more frictional force (air resistance).
5. Other examples of elastic forces are the ropes and sail being stretched by the wind in the story and the elastic force of a bow acting on an arrow.
6. The frictional force acted up the incline and in the direction opposite to that of the movement of the dragged books.

Integrating the Sciences

Earth, Life, and Physical Sciences
Explain that the buoyant force exerted on swimming or floating animals depends on the salinity of the water. Allow students to experiment with the buoyancy of several objects in different concentrations of salt water.

FOLLOW-UP

Reteaching

Ask students to name examples of buoyant, gravitational, electrical, magnetic, frictional, and elastic forces. In each case, ask them to identify the agent and the receiver of the force and to state the force's effect.

Assessment

Ask: What are the most obvious types of forces in each of the situations below?

a. a jack-in-the-box *(elastic)*
b. a gymnast on a trampoline *(elastic and gravitational)*
c. deep-sea diving *(buoyant and gravitational)*

Extension

Have students cut a strip of thin cardboard about 2 cm wide by 10 cm long. Fold it in half, and balance it, at the fold, on a pencil point. The pencil point should indent but not puncture the cardboard so that the cardboard can turn easily. Have students put a charge on a comb by running it through their hair. They should hold the comb near one end of the cardboard strip. *(The cardboard turns on the pencil point and moves toward the comb.)*

Closure

Show students a picture of a hot-air balloon. Help students recognize that flying or gliding in the air requires a buoyant force, just as floating in water requires a buoyant force.

What Makes Things Fall?

FOCUS

Getting Started

Use a small spring scale to measure the weight of a handful of paper clips. Then, while the paper clips are still on the scale, hold a strong magnet underneath them so that the scale is pulled down a bit farther. Ask: Do the paper clips really weigh more? *(No)* If not, why does the scale show that they do? *(Because a downward magnetic force has been added to the gravitational force)* Explain that students will learn a little more about weight and gravity in this lesson.

Main Ideas

1. Every object in the universe exerts a gravitational force on every other object.
2. The size of a gravitational force depends on the masses of the objects involved and the distance between them.
3. The weight of an object changes with its location in the universe, but the mass of an object remains constant.

TEACHING STRATEGIES

Call on a volunteer to read page 226 aloud. Involve the class in a discussion of the questions. Sample answers follow on the next page.

What Makes Things Fall?

You are holding a ball in your hand. You let it go. It falls. What is the direction of this motion? Is there a force acting on the ball? If so, what agent exerts this force? What is the direction of the force in relation to you? to the Earth? Does it matter where you are on the Earth? What would happen to the ball if you dropped it into a tunnel that passed through the center of the Earth? Ⓐ

Which way is down? Ⓑ

226

LESSON 3 ORGANIZER

Time Required
2 class periods

Process Skills
observing, analyzing

Theme Connection
Systems

New Terms
Gravitational force—a force that is dependent on mass, and causes objects to move toward each other
Mass—the amount of matter present in an object
Weight—the gravitational force exerted on an object

Materials (per student group)
In-Text Activity: round balloon; several tiny pieces of paper

Teaching Resources
Discrepant Event Worksheet, p. 11
Activity Worksheet, p. 12
Transparency 38
SourceBook, p. S84

Blow up a round balloon to use as a model of the Earth. Rub the *entire* surface of the balloon vigorously with a sweater or cloth, or rub it against your hair. Now hold tiny pieces of paper near its surface at different places. Release the paper.

Does the paper fall toward the balloon? What kind of force is acting on the paper? What is the agent of this force? Ⓒ

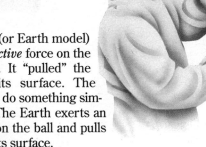

The balloon (or Earth model) exerted an *attractive* force on the pieces of paper. It "pulled" the paper toward its surface. The Earth appears to do something similar to the ball. The Earth exerts an attractive force on the ball and pulls the ball toward its surface.

The forces exerted by the model and the Earth itself are similar in two ways: both are attractive, and both are noncontact forces. However, there is an important difference between the forces that the Earth and the balloon apply.

The attractive force exerted by the balloon is *electrical.* You gave the balloon an electrical charge by rubbing it against your hair or the cloth. The force exerted by the Earth, however, is not electrical. ***It is an attractive force that all objects, because they are made of matter, exert on one another.*** Because the Earth is made of matter, it is able to attract other things made of matter.

The force exerted by the Earth is called gravitational force. This force causes objects to move toward the surface of the Earth in the direction of the Earth's center. The gravitational force that the Earth places on an object is called the **weight** of the object. To find out where the theory of gravitational force came from, read the story on the following page.

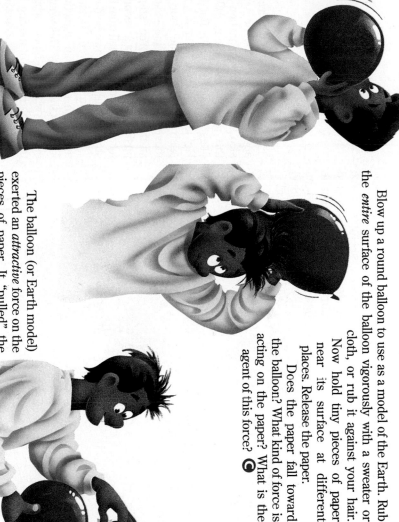

Electrical force

Gravitational force

227

CROSS-DISCIPLINARY FOCUS

Mathematics

If a tabletop weighs 500 N, how much weight would one of the table's four legs have to withstand? *(125 N)*

Homework

Have students find out the root of the word *gravity. (It comes from the Latin word* gravitas, *which means "weight" or "heaviness.")*

In-Text Activity

INDEPENDENT PRACTICE Divide the class into small groups. Distribute balloons and pieces of paper to each group. Ask them to read the top two paragraphs on page 227 and to try the balloon-model of the Earth.

Brainstorm with students about what the word *weight* means to them. Then call on a volunteer to read the remainder of page 227 aloud. Be sure all students are comfortable with the scientific meanings of the terms *gravitational force* and *weight* before proceeding to page 228.

Answers to *In-Text Questions and Caption, pages 226–227*

Ⓐ The ball moves downward when it is released because the Earth exerts a gravitational force on the ball. The direction of the force is downward in relation to a person standing on the Earth and toward the center of the Earth in relation to the Earth. No matter where you are on the Earth, the direction of the force of gravity is always toward the center of the Earth. If a ball were dropped into a tunnel that passed through the center of the Earth, it would fall toward the center, and its momentum would carry it through to the other side. It would oscillate back and forth from one side of the Earth's surface to the other. Air friction would oppose its motion, shortening its path at each oscillation, and it would eventually come to rest at the center of the Earth.

Ⓑ *Down* in this picture means toward the center of the Earth.

Ⓒ The paper falls toward the balloon. This is due to an electrical force exerted by the balloon (the agent).

A Striking Discovery

Explain to students the following thought experiment with which Newton described how satellites orbit the Earth: Begin by drawing successive pictures of stones being thrown from the top of a mountain with greater and greater force. Tell students that each stone is thrown with more force than the one before it. The first stone curves in an arc for 1 km before falling to the ground, the second stone curves 10 km, the third curves 100 km, and so on. Explain that if a stone were thrown with enough force, it would fall in an arc that follows the curvature of the Earth; that is, it would begin to orbit the Earth.

Answers to *In-Text Questions*

Ⓐ The Earth is the agent of the gravitational force, all matter is the receiver of the force, and the effect of the force is that objects move or fall toward the center of the Earth.

Integrating the Sciences

Earth and Physical Sciences

Ask students: Have you ever been at the seashore and observed the rising and falling of the tide? What do you think might cause the ocean to behave in this way? *(Accept all reasonable answers.)* Explain to students that while the Earth is exerting a gravitational pull on the moon, the moon is exerting a pull on the Earth. The water in the oceans is pulled toward the moon, resulting in a high tide on the side of the Earth that is facing the moon. Ask students: What might cause a high tide to be even higher than normal? *(The tides are greatest when the sun and the moon are aligned and both are exerting gravitational forces on the water in the same direction.)*

A Striking Discovery

Why do objects fall? If an object starts to move, isn't there a force causing this movement? The scientist Isaac Newton pondered these questions in the seventeenth century. Legend has it that he was sitting under a tree when an apple bounced off his head and set him thinking!

As Newton was wondering about falling objects, he was also seeking an explanation of the moon's movement. Why does it keep circling the Earth? He already knew that if you push a ball along the floor, it moves in a straight line—even after the initial force stops pushing it. It takes a second force to change the ball from its straight path, as shown below.

Newton wondered, "What if the Earth exerts an attractive force on the moon? This would explain why the moon orbits the Earth instead of moving in a straight line." This constant force would change the straight line to a curving line. Then Newton thought, "If the Earth exerts an attractive force on an object far away, it must also exert an attractive force on objects near its surface. It is this force that causes objects to fall toward the Earth."

Changing the direction of a moving ball

Thus, Newton had an explanation for both the apple falling and the moon following a curved, rather than straight, path. The Earth exerts a gravitational force on every object. What are the agent, receiver, direction, and effect of this gravitational force? Ⓐ

228

Theme Connection

Systems

Focus question: How do forces cause the planets in our solar system to maintain their orbits around the sun? *(The planets orbiting the sun are in constant motion. Without the gravitational force of the sun, they would continue to move in a straight line and fly off into space. However, the sun's gravitational force changes the path of their motion by pulling the planets slightly toward the sun. This causes the path of the planets to curve, which gives the planets their orbits. The amount of the sun's pull is just enough to keep the planets from flying off into space, but not so great that they are pulled into the sun.)*

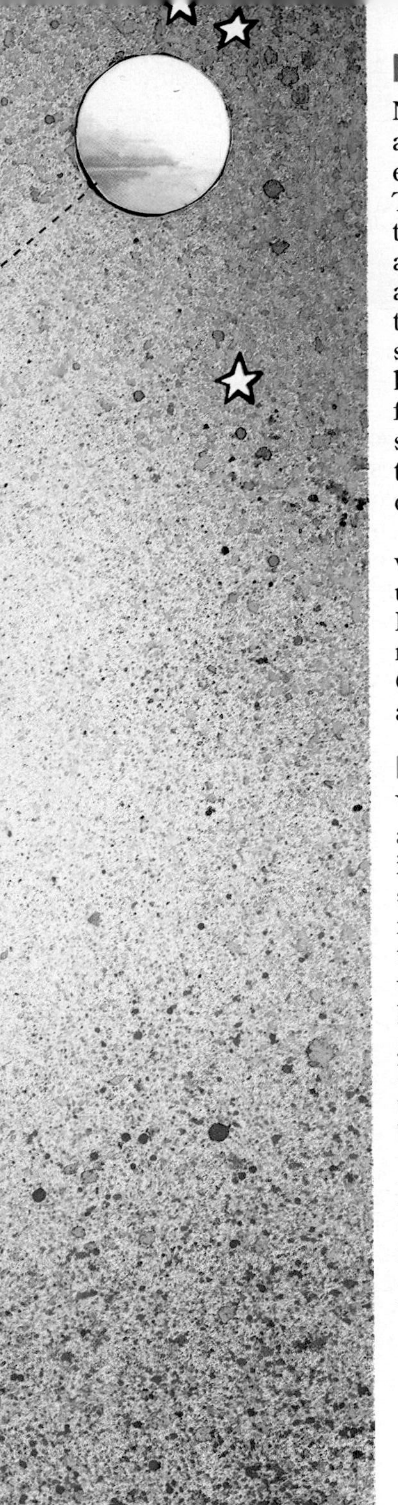

Newton's Law

Newton's conclusion about gravitational force was an important advance in science. He concluded that *every* object in the universe exerts an attractive force on *every other* object in the universe. This law applies to huge bodies of matter like stars and planets: the sun attracts the Earth, and the Earth attracts the sun. It also applies to bodies of very different size: the Earth attracts the ball, and the ball attracts the Earth. It is true of small bodies as well: the table attracts the chair, and the chair attracts the table. With small objects, the gravitational force is extremely small. With large objects, the force is much greater. The size of the attractive force seems to depend on the size of the object. Actually, the strength of the gravitational force depends on how much matter there is in each of the objects attracting each other. The amount of matter in an object is what scientists call its **mass**.

Did Newton ever test his theory? No. That came much later, when Lord Henry Cavendish devised an experiment that measured a small gravitational force between objects and thus proved Newton's theory. He showed that if you double the quantity of matter (mass) in one body, the attractive force also doubles. Cavendish also demonstrated that the farther apart the bodies are, the smaller the force of attraction between them.

Floatfoot's Follies

Write a science-fiction story about gravity (you may want to illustrate it, too). Here is the scenario: Lenny Floatfoot, the mischievous laboratory assistant to Dr. Ethel Earthpull, a well-known gravity researcher, has secretly concocted an anti-gravity spray. The harmless-looking liquid can cancel out the gravitational force exerted by the Earth.

Now Lenny wants to see how the spray works. He chooses your school as a testing ground and sprays it all over the cafeteria and gym. Your principal rushes out of the office just as . . .

Newton's Law

Call on a volunteer to read aloud page 229. Students may have difficulty accepting Newton's law of gravitation. It is difficult to imagine that as the Earth exerts a pulling force on a rubber ball, the ball also exerts a pulling force on the Earth. Ask: If this is so, why does the ball move toward the Earth, but not vice versa? *(The Earth has a much greater mass; thus, the pulling force that the ball exerts on the Earth is not great enough to make it move a measurable amount.)*

Explain that in Cavendish's experiment, large lead balls were brought close to small lead balls. The small balls were attached at either end of a horizontal rod that was hung from above with a wire. The gravitational force between the small and large balls caused the vertical wire to twist by a measurable amount.

A Discrepant Event Worksheet is available as a teacher demonstration to accompany the material on this page (Teaching Resources, page 11).

Answers to *Floatfoot's Follies*

Encourage students to be creative. These stories will help you gauge how well the students understand the idea of *gravitational force*. Students should understand that gravity would no longer hold objects in place. The motion of the revolving Earth would exert a force on the objects, causing them to move tangentially away from the Earth. Suggest that students illustrate their stories with sketches.

If students do not have time to complete this activity in class, it can be done as a homework assignment. The results can be displayed on a bulletin board.

Homework

Inform students that some medical researchers have proposed that patients recovering from surgery can heal faster in a weightless environment. Have students describe an experiment that could test this hypothesis.

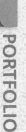

PORTFOLIO

Encourage students to include their story and sketches about Lenny Floatfoot in their Portfolio. Students may wish to highlight references to the agents, receivers, directions, and effects of forces in their stories.

A Massive (or Weighty) Problem

Be sure that students understand the difference between mass and weight. *(The mass of an object remains constant throughout the universe. An object's weight depends on the gravitational forces acting on it. An object could become weightless if the Earth's gravitational force became negligible, but it could not become massless.)*

Answers to *In-Text Questions*

Ⓐ Cavendish's findings showed that the gravitational force between objects decreases as the distance between them increases.

Ⓑ On the moon you would weigh one-sixth of your Earth weight. Since your muscles are used to the Earth's gravity, they would be able to propel you farther on the moon.

Homework

You may wish to assign the Activity Worksheet that is available to supplement the material on this page as homework (Teaching Resources, page 12). If you choose to use this worksheet in class, Transparency 38 is available to accompany it.

A Massive (or Weighty) Problem

You know that the Earth exerts a gravitational force on objects. This force is commonly called the weight of the object. If the Earth were to disappear, an object would have no reason to fall. The object would become "weightless." This does not mean that it would no longer contain any matter. The object contains the same amount of matter with or without the Earth's pull—that is, the object has the same mass. Obviously, mass and weight are really quite different.

How much do you weigh? Suppose you were to take off in a spaceship and travel 6400 km above the Earth (the distance from Washington, D.C., to Paris, France). You would weigh only one-fourth of your weight on Earth. How would Cavendish's findings help to explain this? Ⓐ

Now suppose you go to the moon, where the attractive force is only one-sixth of that on Earth. How much would you weigh there? Why would you be able to jump farther? Ⓑ

Was Galileo just goofing off, or was he proving an important point about gravitational force? Read pages S84–S85 in the SourceBook to find out!

230

FOLLOW-UP

Reteaching

To demonstrate how the moon orbits the Earth, tie a washer to a string. Use a thumbtack to fasten the free end of the string to the eraser at the end of a pencil. Hold the pencil so that the eraser points upward. Push the washer with a strong force so that it circles the eraser. Ask: Why does the washer move in a circle instead of moving in the direction it was pushed? *(The string, similar to gravity, exerts a force that pulls the washer toward the pencil. This changes the motion of the washer.)*

Assessment

Ask the students to explain why the following statement is true: You weigh slightly less at the top of a mountain than you do at its foot. *(You weigh less at the top of a mountain because you are farther from the center of the Earth. Weight depends on your mass and your distance from the center of the Earth.)*

Extension

Have students find out how much they would weigh on the other planets in our solar system. They will first need to find out the gravitational force on those planets.

Closure

Do the following demonstrations: Hang one, then two, and then three washers from a rubber band, pointing out that the stretching of the rubber band allows you to measure the weight of the washers. Then use an equal-arm balance to measure the mass of the three washers. Ask: Which method of measuring the washers would give the same results anywhere in the universe? *(The equal-arm balance would give the same results because it is measuring mass, not weight. On the equal-arm balance, the mass of the washers is balanced against the mass of standard masses. The stretch of the rubber band measures the gravitational pull on an object, which is its weight.)*

1. Mars Mission

You are among the first astronauts to make a space flight to Mars. Your spaceship, *Astro 5*, is well on its way. In outer space far away from the Earth and moon, the gravitational force is almost zero. How do you think zero gravitational force will affect your ability to eat and drink?

Finally you land on the surface of Mars. What does it feel like? Remember that Mars exerts a gravitational force that is about one-third of that of Earth. Compare your weight and mass on Mars with your weight and mass on Earth.

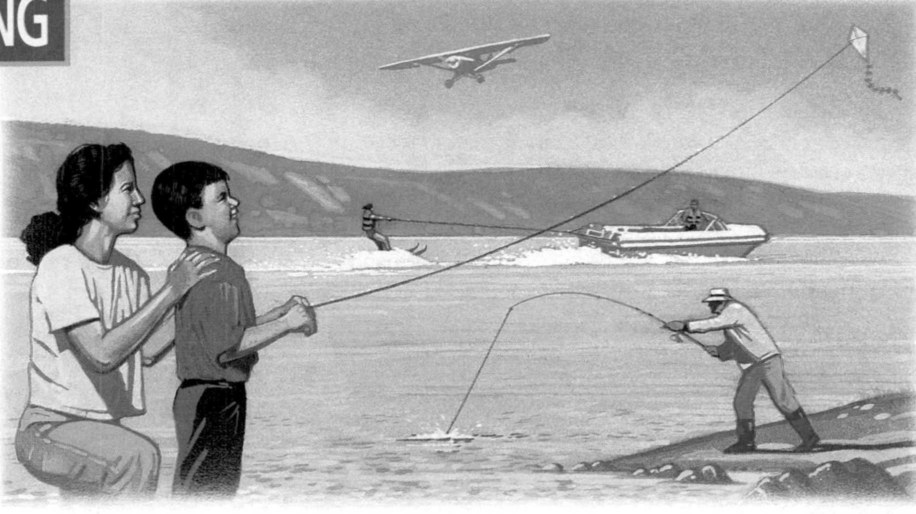

2. A Force Hunt

In the illustration above, many forces are shown. How many can you find? You really *know* a force when you find its agent, receiver, effect, direction, and where it acts. How many forces discovered in the picture do you really *know*? Write down these characteristics for each.

3. That's Some Story!

Write an adventure story in which you "hide" as many special kinds of forces as you can. The hero of your story may be a deep-sea diver or a mountain climber, for example. After you've finished your story, ask a classmate to read it and discover the forces.

★ **You may wish to provide students with the Chapter 10 Review Worksheet on page 13 of the Unit 4 Teaching Resources booklet.**

Meeting Individual Needs

Learners Having Difficulty

Ask: When you are floating in a swimming pool, what are the two main forces that are acting on you? *(Gravitational and buoyant)* Which one is stronger? *(They are equal.)*

Answers to *Challenge Your Thinking*

1. Zero gravitational force might make it difficult to get food and drink into your mouth without it floating around the spaceship. (Students may be surprised to learn, however, that because of special body mechanisms, the actual act of swallowing is not hindered by zero gravitational force.)

   Student answers will vary regarding what it feels like on the surface of Mars. On Mars, a person's mass would be the same, but his or her weight would be one-third of that on Earth.
2. Students should list the agent, receiver, effect, and direction for each force identified. For example, the boat (the agent) pulls on the skier (the receiver). The direction of the force is forward. The force acts at the point where the skier's hands meet the rope handle.
3. Encourage students to be creative. Students should be able to identify the forces in their own stories as well as those in the stories of others.

Meeting Individual Needs

Gifted Students

Explain the difference between the scientific definitions of the terms *speed* and *velocity.* While *speed* is the distance an object travels in a certain amount of time, *velocity* also includes the direction in which it is traveling. For instance, a moving bicycle may have a *speed* of 15 km/hr. To know its *velocity,* you must also know which way it is headed. Ask: How does this distinction vary from the common usage of these terms? *(These words usually mean the same thing in common usage.)*

Answers to *Challenge Your Thinking,* continued

4. Because the rubber band stretches uniformly, a student can make a scale by comparing the stretch caused by several objects with known weights. The rubber band would stretch about one-sixth as far on the moon and one-third as far on Mars as it does on Earth. Aboard *Astro 5* the rubber band would hardly stretch at all, indicating the objects' virtual weightlessness.

5. **a.** The ball would continue traveling in its path until another force stopped it. There would be no downward pull due to the force of gravity.
   b. The left fielder would continue upward until another force acted on him or her.
   c. Without gravity, the batter would not be able to run because the first step would push him or her off the ground. The batter would then continue moving upward until he or she was acted on by some other force.
   d. The coach would continue upward until he or she was acted on by another force.

6. While gravitational force pulls the moon toward the Earth, the moon, like any moving object, has a tendency to travel in a straight line. The combined effect is the moon's constant orbit around the Earth.

Homework

Have students create a set of clear, step-by-step intructions for building a rubber-band scale. Their instructions may include illustrations or photographs.

4. Rubber-Band Scales

Imagine that someone has told you that a rubber band can be used to find the weight of an object. Can you figure out how? How would the rubber band show the differences in the weight of an object when weighed on Earth, the moon, Mars, or aboard *Astro 5*?

5. Ball Game Called Off!

The ball game had started. Suddenly and mysteriously, all gravitational forces disappeared. What would happen if

a. a ball were hit by a batter toward left field?
b. the left fielder made a jump to catch the ball?
c. the batter started to run for first base?
d. the surprised coach jumped out of the dugout?

6. The Falling Moon

Because the Earth exerts an attractive force on the moon just as it does on the falling apple, Newton could imagine that the moon was also falling to the Earth. But, as Newton could see, the moon never falls on anybody's head. What keeps the moon in its orbit? (Hint: Look back at page 228 and review what Newton learned about a ball traveling in a straight path.)

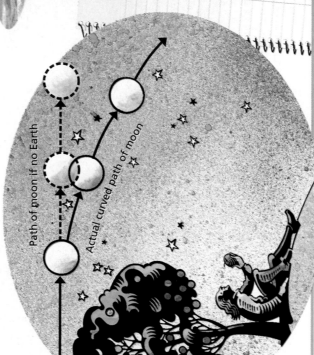

Path of moon if no Earth

Actual curved path of moon

Review your responses to the ScienceLog questions on page 216. Then revise your original ideas so that they reflect what you've learned.

232

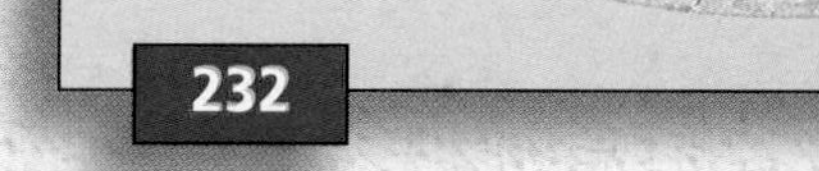

The following are sample revised answers:

1. A force is a push or pull exerted by one object on another object. Gravitational force pulls the airplane toward the Earth. The surface of the runway exerts a frictional force on the airplane's tires. The engine exerts a backward force that pushes the airplane forward.
2. The Earth exerts an attractive force (gravitational force) on the sky diver, pulling her toward the Earth's center. This causes her to fall to the Earth after she jumps from the plane.
3. Mass and weight are not the same. Mass is the amount of matter present in an object, while weight is the gravitational force exerted on that object by the Earth.

CHAPTER 11

Measuring Forces

1 How can you determine the size of a force?

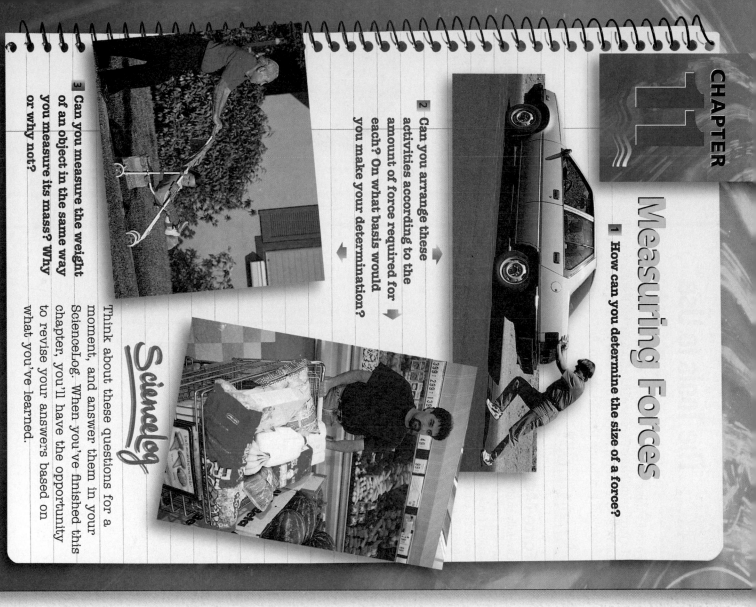

2 Can you arrange these activities according to the amount of force required for each? On what basis would you make your determination?

3 Can you measure the weight of an object in the same way you measure its mass? Why or why not?

ScienceLog

Think about these questions for a moment, and answer them in your ScienceLog. When you've finished this chapter, you'll have the opportunity to revise your answers based on what you've learned.

233

CHAPTER 11 Measuring Forces

Connecting to Other Chapters

Chapter 10 *offers students hands-on exposure to a wide range of forces.*

Chapter 11 *allows students to measure the strength of forces with commercial and homemade force meters.*

Chapter 12 *explores the causes and consequences of friction in various situations.*

Chapter 13 *discusses the concept of inertia and introduces students to force pairs.*

Prior Knowledge and Misconceptions

In addition to having students answer the questions on this page, propose the following scenario to them: An astronaut named Shanequa bet her friend Walter that she could lift a box weighing 900 N. Walter thought that Shanequa could lift only 500 N, so he agreed to the bet. When Shanequa lifted the box successfully, Walter demanded that the box be weighed. So Walter took the box and a bathroom scale, got in a rocket, flew to the moon, and put the box on the scale. It only weighed 150 N! Shanequa immediately exclaimed, "That's cheating!" and called you, the Gravity Gumshoe, to prove it.

Have students finish the story by imagining that they are the Gravity Gumshoe. They should spend no more than 10 minutes writing down an ending to the story. Emphasize that there are no right or wrong answers. Collect and read the papers to find out what students know about measuring forces, what misconceptions they have, and what about measuring forces is interesting to them.

The Tools to Use

FOCUS

Getting Started

Prior to the class, blow up two balloons. Attach one balloon to each end of a drinking straw. Then print *500,000 kg* on each balloon. Begin class by telling students that you will amaze them by exerting enough force to lift 1,000,000 kg single-handedly. Then pick up the apparatus. When students challenge the legitimacy of your demonstration, ask: How do you know the amount of force it takes to lift an object? *(Accept all reasonable responses.)* Explain that in this lesson they will build force meters to accurately measure forces.

Main Ideas

1. Elastic materials that stretch uniformly when masses are added to them can be used to measure force.
2. The unit of force, the newton, is approximately equal to the weight of 100 g on the Earth's surface.

TEACHING STRATEGIES

Exploration 1

Remind students that they will need to record the following information for each of the examples: the effect of the applied force, the effect of increasing the applied force, and how to use the effect of the applied force to measure force. As a class, you may wish to review the completed example to ensure that students understand what they must do.

LESSON 1

The Tools to Use

Forces cannot be seen. But there is a way to discover the size of a force—measure the size of its effect. In this lesson, you will follow a series of Explorations that show you how to make *force meters* and how to use them to measure forces.

Exploration 1

Force Meters

Carefully study the examples on these two pages. In each case, determine what effect is caused by the applied force. If the force were increased, how would the effect change? How could you use the effect of the force as a way to measure the force? These questions have been answered for example (a). For the remaining examples, draw a simple diagram in your ScienceLog showing what would happen if there were an increase in force. Your diagram should also show how you would use the effect to measure the size of the force. If you prefer, you may describe in words the effect of increasing the force and how to use this effect to measure the size of the force.

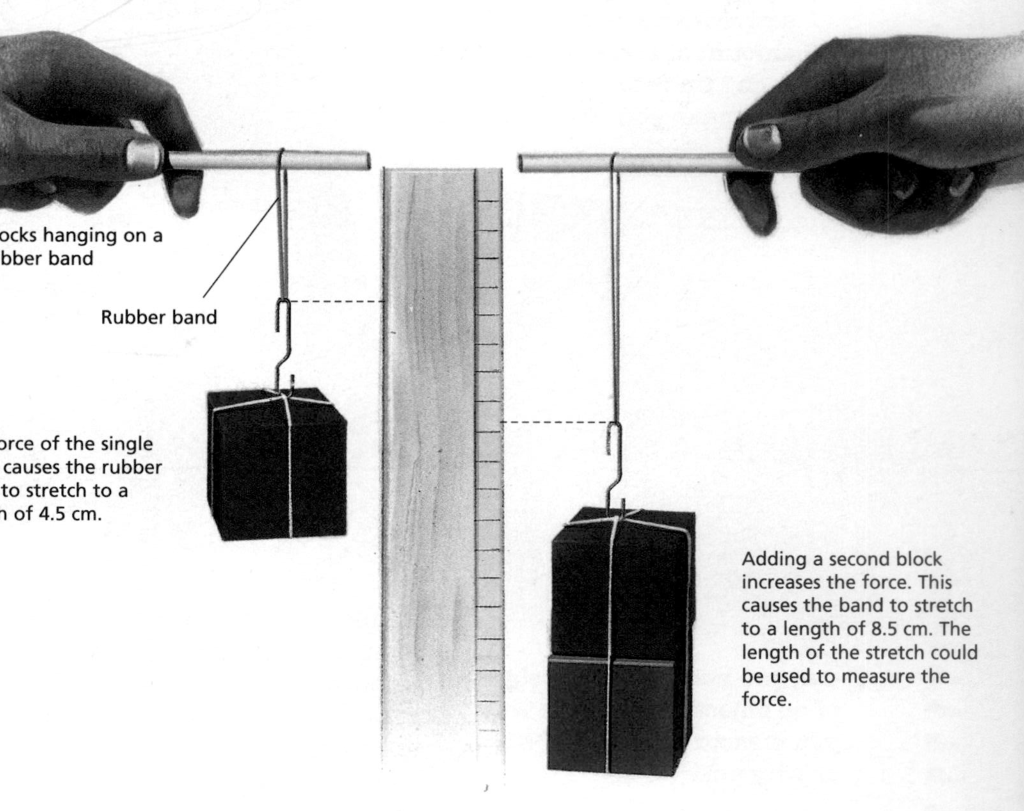

The force of the single block causes the rubber band to stretch to a length of 4.5 cm.

Adding a second block increases the force. This causes the band to stretch to a length of 8.5 cm. The length of the stretch could be used to measure the force.

LESSON 1 ORGANIZER

Time Required
3 class periods

Process Skills
measuring, comparing, hypothesizing, analyzing, predicting

New Term
Newton (N)—a unit for measuring force; one newton is about equal to the gravitational force that the Earth exerts on a 100 g mass

Materials (per student group)
Exploration 2: wooden dowel, about 30 cm long and 2 cm in diameter; cup hook; cardboard bathroom-tissue tube; 2 paper clips; 2–3 rubber bands; 20 cm of masking tape; three 100 g masses or force meter
Exploration 3: coil spring or segment of spring toy; at least 100 g of washers; 100 g, 500 g, and 1000 g hooked masses; 2 meter sticks; force meter

Teaching Resources
Exploration Worksheets, pp. 18 and 20
Transparency 39
SourceBook, p. S86

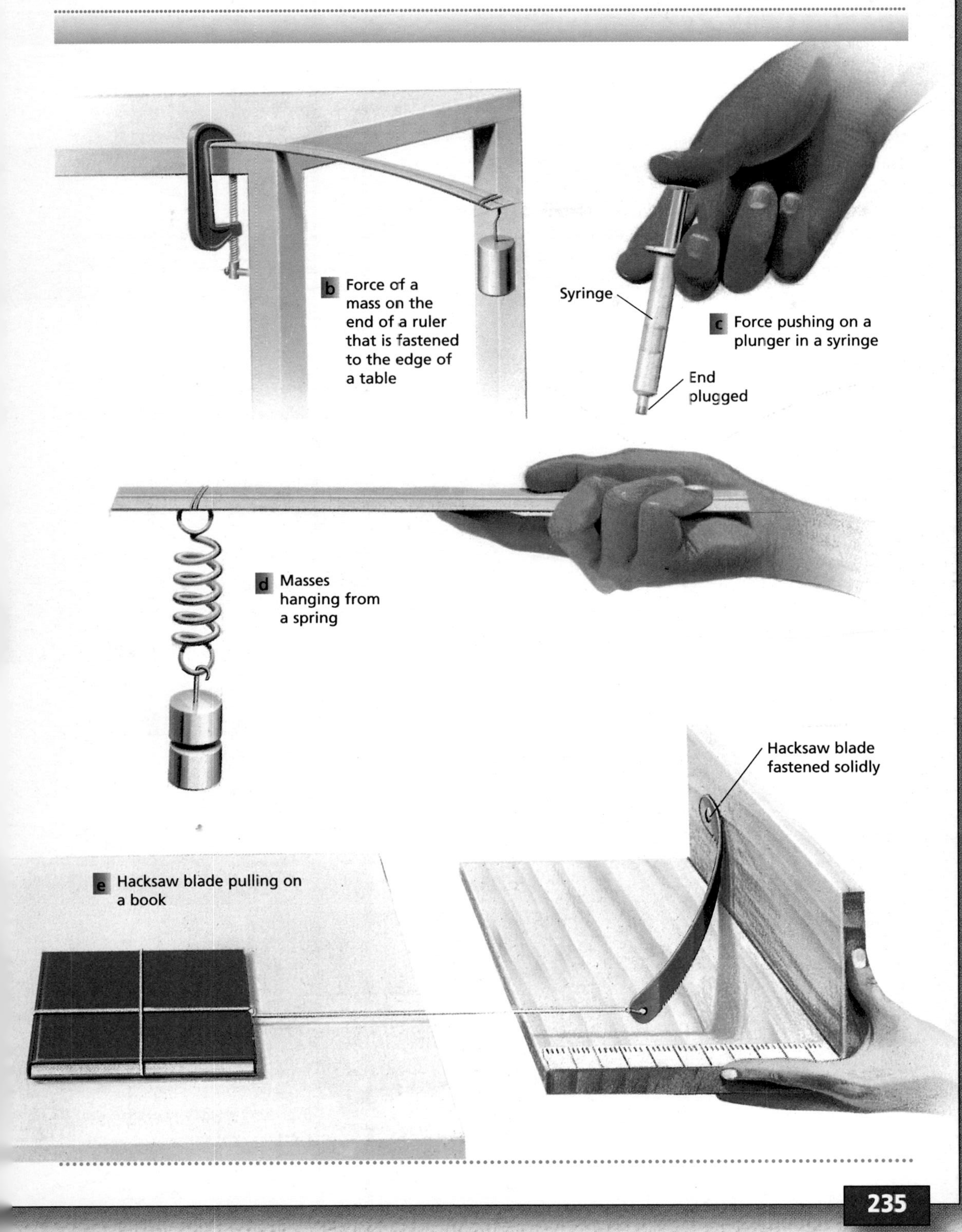

b Force of a mass on the end of a ruler that is fastened to the edge of a table

Syringe

c Force pushing on a plunger in a syringe

End plugged

d Masses hanging from a spring

Hacksaw blade fastened solidly

e Hacksaw blade pulling on a book

Answers to *Force Meters*

b. A mass hung from the end of the ruler bends the ruler. Hanging another mass on the string would increase the force and bend the stick farther. The size of the arc that the stick bends could be used to measure the force.

c. A force pushes down the plunger of the syringe. The greater the applied force, the farther the plunger moves down. The distance it moves down could be used to measure the force.

d. The masses stretch the length of the spring. An increase in force could come from adding more masses, which would increase the length of the spring. This increase could be used to measure the size of the force.

e. The hacksaw blade bends in response to the force. An increase in force could come from placing another book on top of the first book, which would cause the hacksaw blade to bend more as it pulls the books. The amount the blade bends could be used to measure the force.

CROSS-DISCIPLINARY FOCUS

Industrial Arts

Under the supervision of the school's industrial-arts teacher, have students construct the force meter shown in example (b).

Did You Know. . .

In the past, rates of energy usage or production were commonly measured in horsepower, because horses were used to pull carriages. Today the unit of horsepower is given a precise definition. It is about the same amount of energy needed to power ten 75 W light bulbs for one second.

Exploration 2

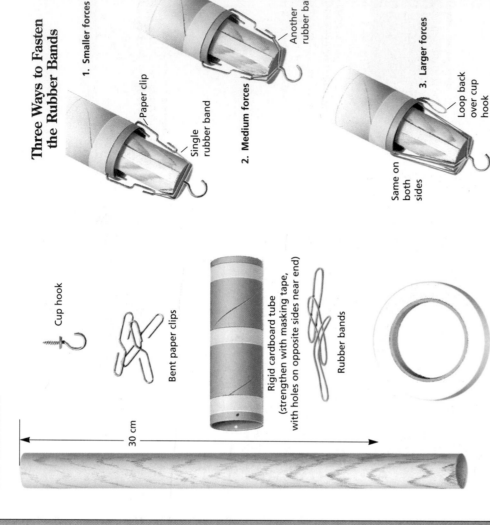

SAFETY TIP Drill a hole in each dowel in advance so that students can screw in the cup hooks more easily.

Each student should make a force meter. The procedures to follow are illustrated in the text. Invite different students to make force meters that measure the three different degrees of forces. Have students refer to the diagrams on pages 236–237.

Other materials students may want to use to design their own meter include wooden blocks, hammer and nails, hacksaw blades, springs, cardboard, and protractors.

Cooperative Learning
Exploration 2

Group size: 2 to 3 students
Group goal: to construct and calibrate a force meter
Positive interdependence: Assign students roles such as materials coordinator, primary assembler, and designer.
Individual accountability: Each student should assemble his or her own force meter based on the design that his or her group created. Students will use their personal force meters in several Explorations later in the unit.

SAFETY TIP Review safety procedures for the laboratory. Make sure that all students are aware of what to do and what not to do in an emergency.

Did You Know. . .

With modern technology, the force of gravity at the surface of the Earth can be measured with an error margin of only 0.0000003 percent. This is as precise as being able to measure the distance around the Earth to within 13 cm!

Exploration 2

Making Your Own Force Meter

Here are plans for constructing a force meter. This method uses rubber bands to measure the size of forces.

Construct the meter, or design one of your own. Instead of using rubber bands, you could use a spring or some other elastic material. After completing your design, find the materials you will need and build the meter.

A Push-Pull Force Meter

You Will Need

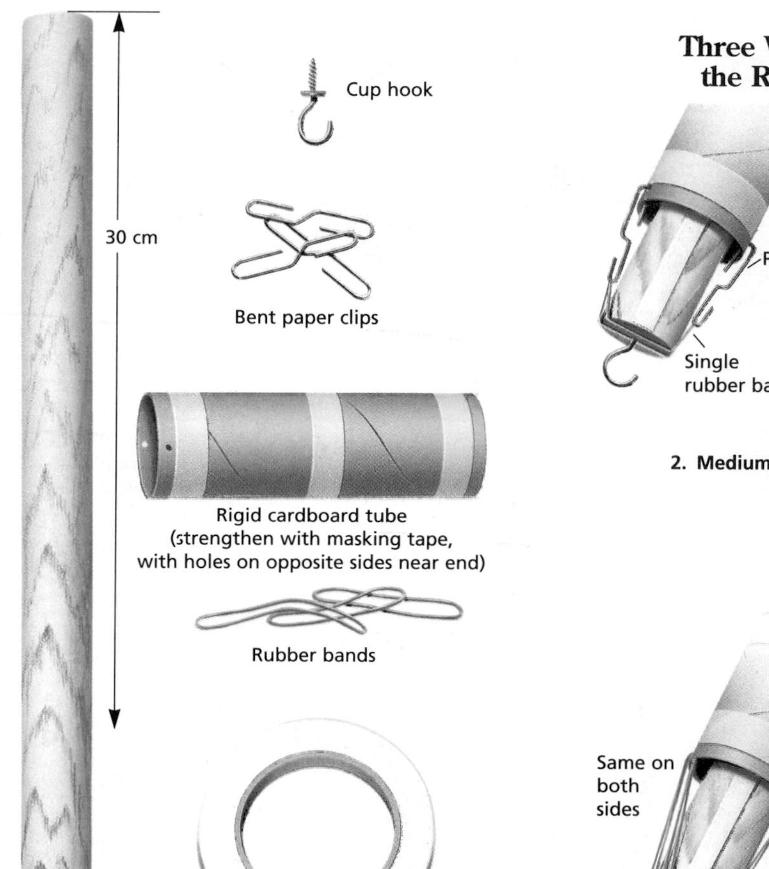

236

Integrating the Sciences

Earth and Physical Sciences

Many of the electrical forces that we encounter on a daily basis are billions of times stronger than the force of gravity. Why, then, is gravity such an important force among planets? *(Gravity acts on all matter. Electrical forces act only on charged matter. Overall, planets are electrically neutral.)*

Homework

You may wish to assign the Exploration Worksheet that accompanies Exploration 2 as homework (Teaching Resources, page 18).

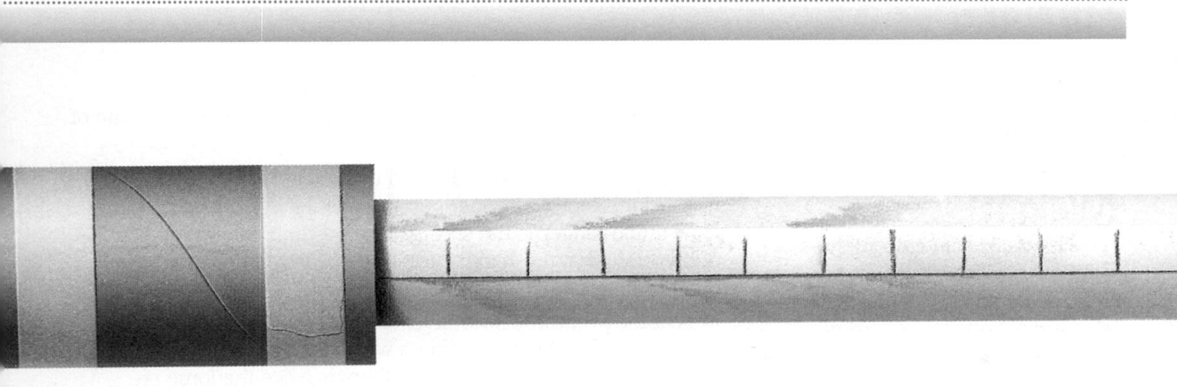

Two Ways to Mark the Newton Scale on Your Force Meter

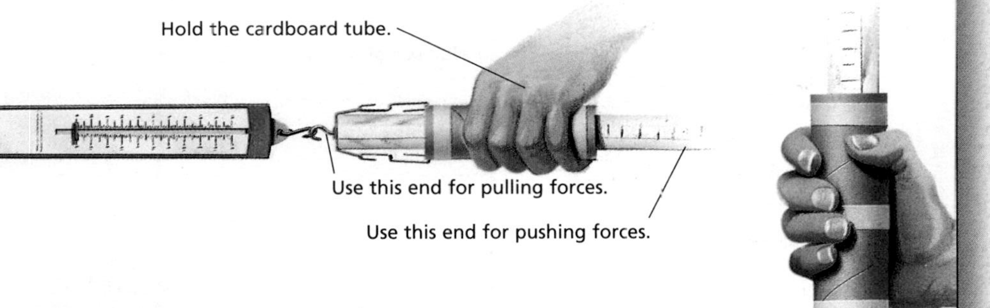

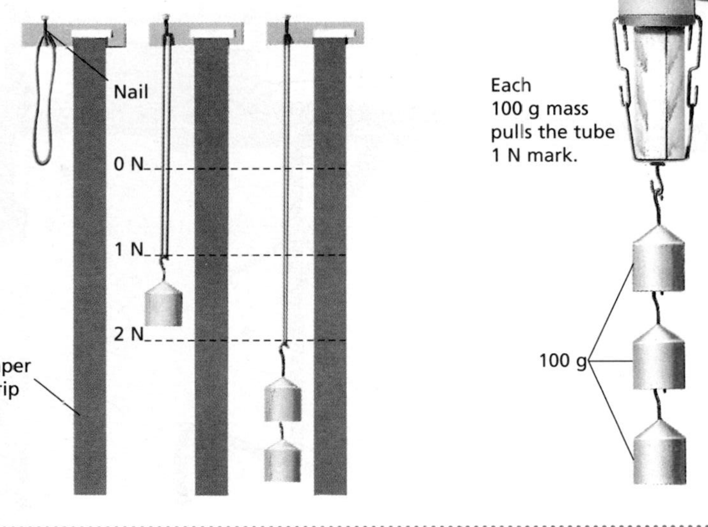

Why Set a Scale?

Every meter needs a scale that will indicate the size of a force according to some standard. The **newton** (N) is the international metric unit (SI) used to measure force. The newton is approximately equal to the weight of a 100 g mass on Earth. In the figure, you can see how the stretch of a rubber band can be used to indicate the size of a force in newtons.

237

GUIDED PRACTICE After students have completed their force meters, reassemble the class and call on a volunteer to read aloud Why Set a Scale? on page 237. Involve students in a discussion of the meaning of the new term *newton.* Direct their attention to the diagram on page 237. The diagram should help students review the idea of calibration. To calibrate their force meters, students will need several 100 g masses or a commercial force meter that measures force in newtons. To illustrate how students can calibrate their force meters, refer them to the diagrams on page 237.

You may wish to have students complete Exploration 3 immediately after constructing their force meters so that they can test how well their meters work.

EXPLORATION 3

Call on a volunteer to read aloud the first paragraph of the Exploration. In the Exploration, students are encouraged to make a formal laboratory report of their experiment. Involve students in a discussion of what is expected of them when writing their reports. Ask students to enter in their ScienceLog their title, purpose, and predictions, as well as the table shown on page 238. Instruct students to perform the Exploration and enter their results, answers to questions, graphs, and conclusions in their ScienceLog.

★ **An Exploration Worksheet (Teaching Resources, page 20) and Transparency 39 are available to accompany Exploration 3.**

Answers to *In-Text Questions*

Ⓐ Students should see the spring increase in length. Actual answers will vary depending on the spring and the size of the washer used.

Ⓑ The force would be about three, four, and five times the force on one washer.

Ⓒ Students should discover that the amount the spring stretches increases along with the number of washers added to it. Their graphs should approximate a straight line at first and then begin to curve and become horizontal as the spring approaches its maximum extension. Springs stretch in a predictable fashion, which makes measuring the force easier.

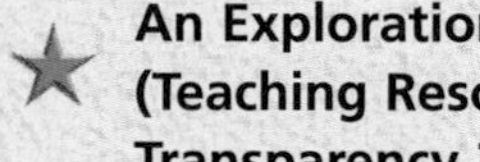

PORTFOLIO

Suggest that students include their procedure and results from Exploration 3 in their Portfolio. Then they may perform a self-assessment to evaluate their use of scientific skills during Exploration 3. You may wish to provide students with one of the Self-Evaluation Checklists available on the SnackDisc.

EXPLORATION 3

Using a Spring to Measure Gravitational Force (Weight)

A spring makes an extremely good force meter. The following experiment will help you discover why. Keep a precise record in your ScienceLog of what you do and discover in the experiment. Then use this experiment to practice writing a formal report. In your report, include a title, purpose, predictions, procedures, results, discoveries, and answers to the questions in this Exploration.

You Will Need

- a spring (a coil spring or a segment of a spring toy, as shown)
- washers
- hanging masses
- 2 meter sticks
- a force meter

What to Do

1. Set up the apparatus as shown, with the spring hanging on one of the meter sticks.
2. Using the other meter stick, read the position of the bottom of the spring. Repeat after adding one washer. What is the result? How much longer is the spring? The position indicates the size of the force that the Earth exerts on one washer—the weight of one washer. Ⓐ
3. Repeat for two, three, four, five, or more washers. Predict the effect each time before adding a washer. (Remember that the gravitational force on two washers is twice the force on one washer. How large would the size of the force be on three, four, or five washers?) Ⓑ
4. Enter your results in a table like the one below. This helps you see all your data clearly. What does your table tell you about the relationship between the amount of stretch and the number of washers? Why does a spring make a good force meter? Ⓒ

Number of washers	Position of bottom of spring	Stretch of spring
0		
1		
2		
etc.		

ENVIRONMENTAL FOCUS

Students may be interested to learn the following information: The rotation of the Earth causes the planet to bulge at the equator. Therefore, you would be farther away from the Earth's center at the equator than you would be at one of the poles, and your weight would be slightly less at the equator than at a pole.

5. Another useful way to study your data is to make a graph like the one to the right. In your ScienceLog, plot the amount of stretch against the number of washers.

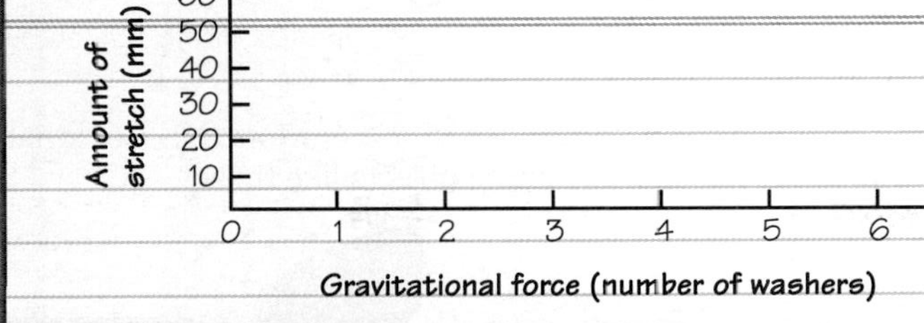

Questions

1. Suppose that an object hung on the spring causes it to stretch 22 mm. What would be the size of the gravitational force on the object (its weight) as measured in washers? Use your graph to find the answer.
2. Choose an object that is not heavy enough to permanently distort the spring. Predict how many washers it will weigh. Now hang it on the spring and find out if your prediction was correct.
3. Could you use the spring to measure the force needed to pull a book across a table? If so, explain how.

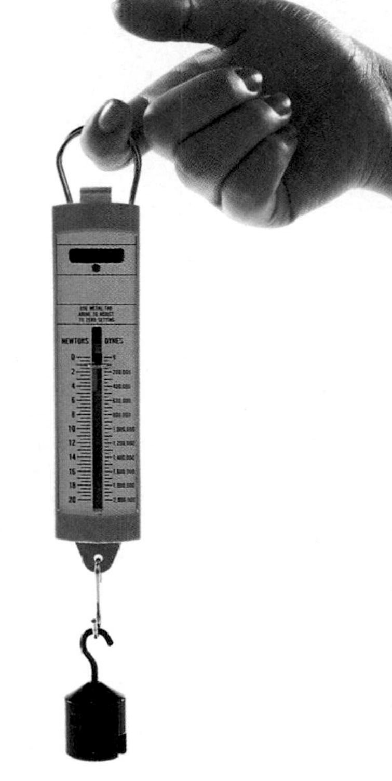

Converting Washers Into Newtons (N)

1. Hold a 100 g mass in your hand. How much force are you exerting to hold it up? Find out by placing the mass on a commercial force meter marked in newtons.
2. Repeat for a 500 g mass and a 1000 g (or 1 kg) mass.
3. How many "washers of force" are equal to 1 N of force? Use the commercial force meter to find out.
4. For the original spring you used, how many millimeters would it have to stretch to indicate a 1 N force?

Answers to *Questions*

1. Students should look at their graphs and interpolate the weight of the object in washers.
2. Again, students can look at their graphs to interpolate the weight of the object in washers based on the amount the spring stretches.
3. The spring can be used to measure the force needed to move the book across the table as long as the book is not so heavy that it causes distortion in the spring.

Answers to *Converting Washers Into Newtons (N)*

1. About 1 N
2. It takes 5 N to hold a 500 g mass and 10 N to hold a 1000 g mass.
3. To find out how many "washers of force" are equivalent to 1 N of force, students should hang washers on the force meter until the meter registers 1 N.
4. Students can look at their graphs to find out how far the spring stretches when the equivalent of 1 N is hung on the spring. They could also hang approximately 100 g of washers on their springs to get the estimate.

Homework

The Extension activity described on this page makes an excellent homework activity.

FOLLOW-UP

Reteaching

Have students use a metric measuring cup to calibrate a pitcher in milliliters, making marks for every 100 mL. Then have them compare this to the calibration of their force meter from Exploration 2. *(Just as known increments of volume are marked on the pitcher, known increments of gravitational force are marked on the force meter.)*

Assessment

Provide students with a spring, three 100 g masses, and a metric ruler. Ask them to use these materials to measure the gravitational force on several objects. *(The 100 g masses could be placed on the spring one at a time to see how far they make the spring stretch. A graph of weight in newtons versus centimeters of stretch could be drawn. Then the objects could be tested to measure how far they make the spring stretch. Their weight in newtons could then be estimated.)*

Extension

Have students design a force meter that can handle much larger loads, such as 20 kg or 50 kg. Students should have their design approved for safety before they build their force meter.

Closure

Have each student compare his or her force meter with a commercial force meter. Discuss how the homemade force meters might be improved.

LESSON 2

Estimating the Size of a Force

FOCUS

Getting Started

Set up a table with several objects similar to those shown in the photograph. Encourage students to actually lift the objects so that they can feel the forces that need to be applied to overcome each object's weight. From this experience, they should be able to better estimate the weights of the objects used in the lesson.

Main Ideas

1. The size of the force needed to lift something can be estimated.
2. The size of a force can be measured in newtons with a force meter.

TEACHING STRATEGIES

GUIDED PRACTICE Copy the list of items from the bottom of page 240 onto the board. Record a few of the students' estimates for each item. Then have the class agree on one answer for each item. Look in the answer box to see how close the estimates were.

LESSON 2

Estimating the Size of a Force

You are now familiar with the use of the newton (N). But what does a newton feel like? Try lifting objects like these:

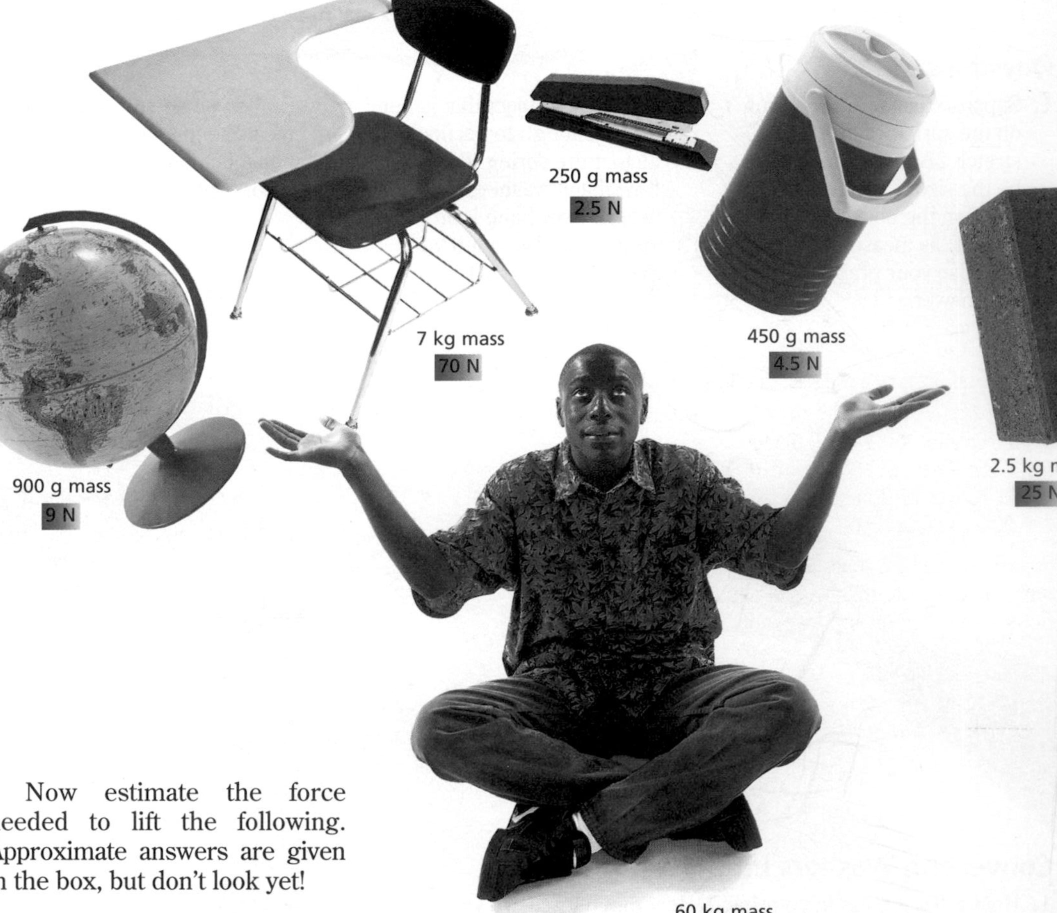

Now estimate the force needed to lift the following. Approximate answers are given in the box, but don't look yet!

a. your science textbook
b. your teacher
c. a pail of water
d. a metal wastepaper basket
e. a dozen eggs
f. a basketball
g. a hammer
h. an iron shovel
i. a concrete block
j. $5 in quarters
k. a newborn baby
l. a compact car

Answers:
a. 9 N | e. 5 N | i. 110 N
b. 700 N | f. 6 N | j. 1 N
c. 60 N | g. 5 N | k. 30 N
d. 12 N | h. 20 N | l. 8000 N

LESSON 2 ORGANIZER

Time Required
2 class periods

Process Skills
measuring, predicting, analyzing

New Terms
none

Materials (per student group)
Exploration 4: meter stick; 1 kg object or mass
Exploration 5: brick; board, about 1 m in length; a few books; small sandbag; 2 pulleys; 3 m rope or cord; force meter

Teaching Resources
Exploration Worksheets, pp. 23 and 24
Transparency 40

EXPLORATION 4

A Finger Exercise

Here's how to feel different sizes of forces with your finger.

You Will Need

- a meter stick
- a 1 kg mass or equivalent

What to Do

1. Balance a meter stick over the edge of a table (at the 50 cm mark).
2. Place a 1 kg mass on the meter stick so that it is centered at the 45 cm mark.
3. Press down at 100 cm to raise the mass. You are using a 1 N force.
4. Press down at 60 cm for a 5 N force.
5. Now shift the 1 kg mass to 0 cm. Press at 100 cm for 10 N, at 75 cm for 20 N, at 60 cm for 50 N, and at 55 cm for 100 N.

With the 1 kg mass at 0 cm, you can calculate the force in newtons with the following formula:

$$\frac{500}{(P - 50)}$$

where P = the pressing point

Think About It!
Can you exert a 100 N force with one finger? with two fingers? What is the maximum size of force each finger can exert? Are all your fingers equal in strength? How do the fingers of your left hand compare in strength with those of your right hand?

EXPLORATION 4

Divide the class into small groups and distribute the materials. Encourage all group members to try the activity so that they have the firsthand experience of feeling different sizes of forces.

The meter stick forms a lever, with the edge of the table serving as the fulcrum. The distance from the force to the fulcrum multiplied by the amount of force should be equal on each side of the balance. For example, 10 N placed 5 cm from one side of the fulcrum would balance 1 N placed at 50 cm on the other side of the fulcrum (10 N × 5 cm = 1 N × 50 cm).

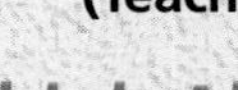

An Exploration Worksheet is available to accompany Exploration 4 (Teaching Resources, page 23).

Think About It!

INDEPENDENT PRACTICE Have students experiment on their own to determine the answers to Think About It! Students should record their findings in their ScienceLog.

Answers to *Think About It!*

Students will probably discover the following: they cannot exert 100 N of force with only one finger; they may be able to exert 100 N of force with two fingers; the maximum force that each finger can exert will vary with the fingers used; their thumbs and index fingers are stronger than their other fingers; and their results will depend on whether students are left-handed or right-handed.

EXPLORATION 5

Set up this activity so that students move from one location to another to measure the 12 situations. Label the locations (a)–(m) as shown on the chart on page 243. Suggest that students work in pairs to measure the forces. Instruct students to copy the chart on page 243 into their ScienceLog and then to predict the size of the force needed in each situation. Suggest that students use masking tape, string, or cord to attach their force meters to the various objects shown in the pictures. As they measure the actual force that must be exerted in each situation, they should record their findings.

Cooperative Learning
EXPLORATION 5

Group size: 2 to 3 students
Group goal: to estimate and measure a number of forces
Positive interdependence: Assign students roles such as materials coordinator, recorder, and primary investigator.
Individual accountability: Randomly choose one student to describe how the force required to lift a telephone receiver could be measured.

An Exploration Worksheet (Teaching Resources, page 24) and Transparency 40 are available to accompany Exploration 5.

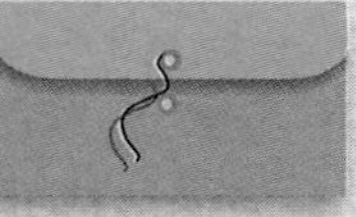

PORTFOLIO
Students may wish to include their measurements from Exploration 5 in their Portfolio. Encourage students to make generalizations about forces based on their findings. Suggest that they identify gravitational and frictional forces that occurred in the Exploration.

EXPLORATION 5

Estimating and Measuring the Size of Forces

Now it's time to test your ability to estimate different sizes of forces.

You Will Need

- a brick
- a board
- 3 books
- a sandbag
- 2 pulleys
- a rope or a cord
- a commercial force meter or the one you made

What to Do

1. Construct a table in your ScienceLog like the one on page 243. As you perform this Exploration, fill in the information to complete your table.
2. Predict the size of the force exerted in each of the situations shown. Record your estimates.
3. Select a force meter that enables you to make an accurate measurement of the force required in each case. Record these measurements also.

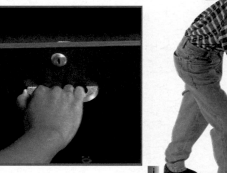

a

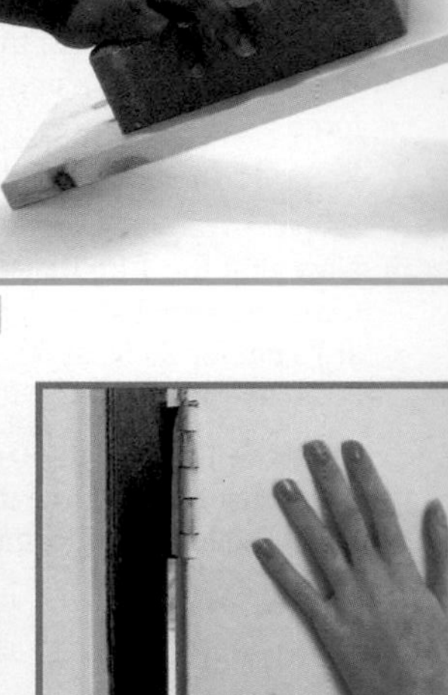
b

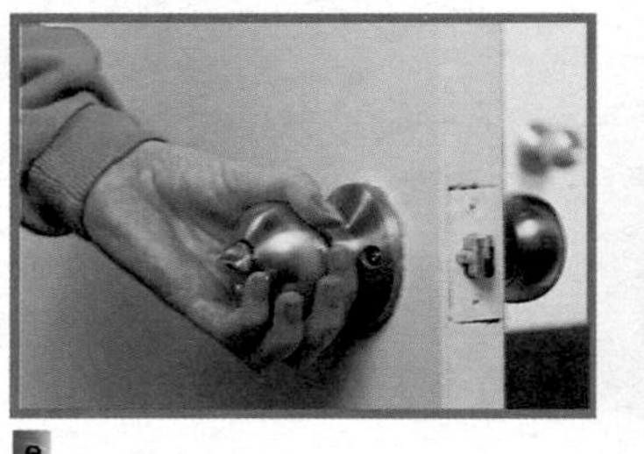
c

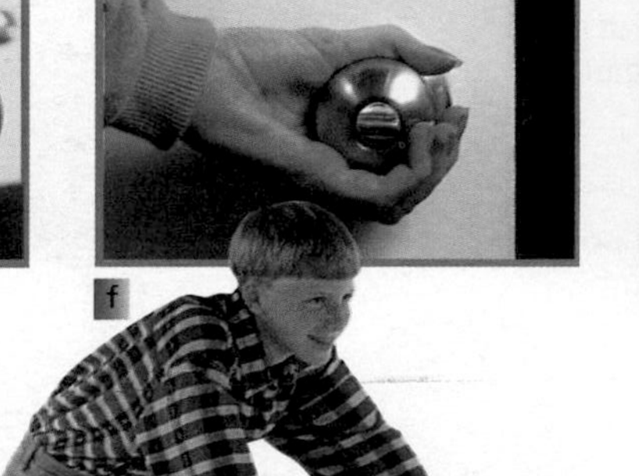
d

e

f

g

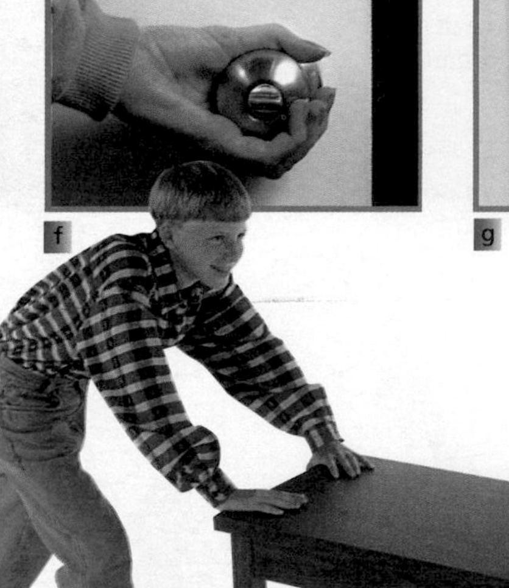
h

i

Integrating the Sciences

Life and Physical Sciences
Show the class a diagram of major human muscle groups. Ask: What do the size and position of these muscles tell us about how they are used? *(Typically, the larger the muscle, the greater the force it exerts and the more often it is used. The position and attachments of a muscle can tell us what its usual function is.)*

Situation	Estimated force (N)	Measured force (N)
a. Lifting a brick		
b. Causing a brick to start moving on a horizontal board		
c. Keeping a brick moving on a horizontal board		
d. Pushing a brick up an inclined board		
e. Pulling a doorknob		
f. Closing a door using the knob		
g. Closing a door by pushing near the hinge		
h. Opening a drawer		
i. Pushing a table or a desk		
j. Lifting a sandbag using a pulley		
k. Lifting a sandbag using two pulleys		
l. Pulling down a screen or blind		
m. Your choice		

Any New Ideas?

You not only practiced measuring forces in this Exploration, but also found out some principles of ***kinetics***, the science of forces. For example, does it take the same force to lift an object as to drag or push it along a surface? What other principles did you discover?

Answers to *Any New Ideas?*

As they work through the measuring activities, students should make the following discoveries: it takes more force to start an object moving than to keep it moving; it takes more force to lift an object vertically than to drag it horizontally; it takes less force to move a brick up an inclined plane than it does to lift it vertically the same distance; and it takes less force to lift an object with two pulleys than it does to lift an object with one pulley.

FOLLOW-UP

Reteaching

Prior to class, fill three shoe boxes with the following: 300 g of sand, three 100 g masses, and 300 g of various objects. Tie a string around each shoe box. Have students examine the contents of each box and estimate the force needed to lift each box. Then have students measure this force with their force meters.

Assessment

Provide students with a force meter and ask them to measure the force needed to open a door, close a door, lift a book, and pull a book across a table.

Extension

Have students research the units that are used to measure force and mass in the British system of measurement. *(Pounds and slugs)* Then ask them to convert these units to SI units of measurement.

Closure

Show students various objects not used in the lesson. Ask them to estimate the force in newtons needed to lift the objects. Then demonstrate the actual force using a force meter.

Measuring More Forces

FOCUS

Getting Started

Show students a picture of an elephant, one of the strongest land animals. Ask the class to guess how much force the elephant's trunk can exert. Record student guesses on the board. Reveal that an elephant can exert close to 2700 N of force using only its trunk. Then tell students that in this lesson they can find out their own strength and compare it with the strength of the elephant's trunk.

Main Ideas

1. Muscle strength can be measured in newtons.
2. A force meter can be used to measure muscle strength.
3. Mass and weight are measured with different kinds of scales and in different units.

TEACHING STRATEGIES

Cooperative Learning
Pages 244–246

Group size: 2 to 3 students
Group goal: to estimate and measure each student's strength
Positive interdependence: Assign students roles such as primary investigator, data recorder, and safety manager.
Individual accountability: Each student should turn in a table showing the estimated and measured strength of the muscles tested at each station. They should also include a brief written report describing the role they played in the activity and how well their group worked together.

1. Arm Strength

Caution students not to do this activity if others are standing nearby, in case they let go of the cord or the cord breaks. Make sure that the cord is very stretchy.

LESSON 3

Measuring More Forces

Here's a chance to measure your own strength. First, guess how many newtons you'll be able to exert. Then see how close your guesses were to your true strength.

1. Arm Strength

How far can you stretch a luggage cord? Ask a classmate to measure the length, in centimeters, that you stretch the cord.

Be Careful: Wear goggles when stretching a luggage cord in case it slips out of your hands as you are stretching it.

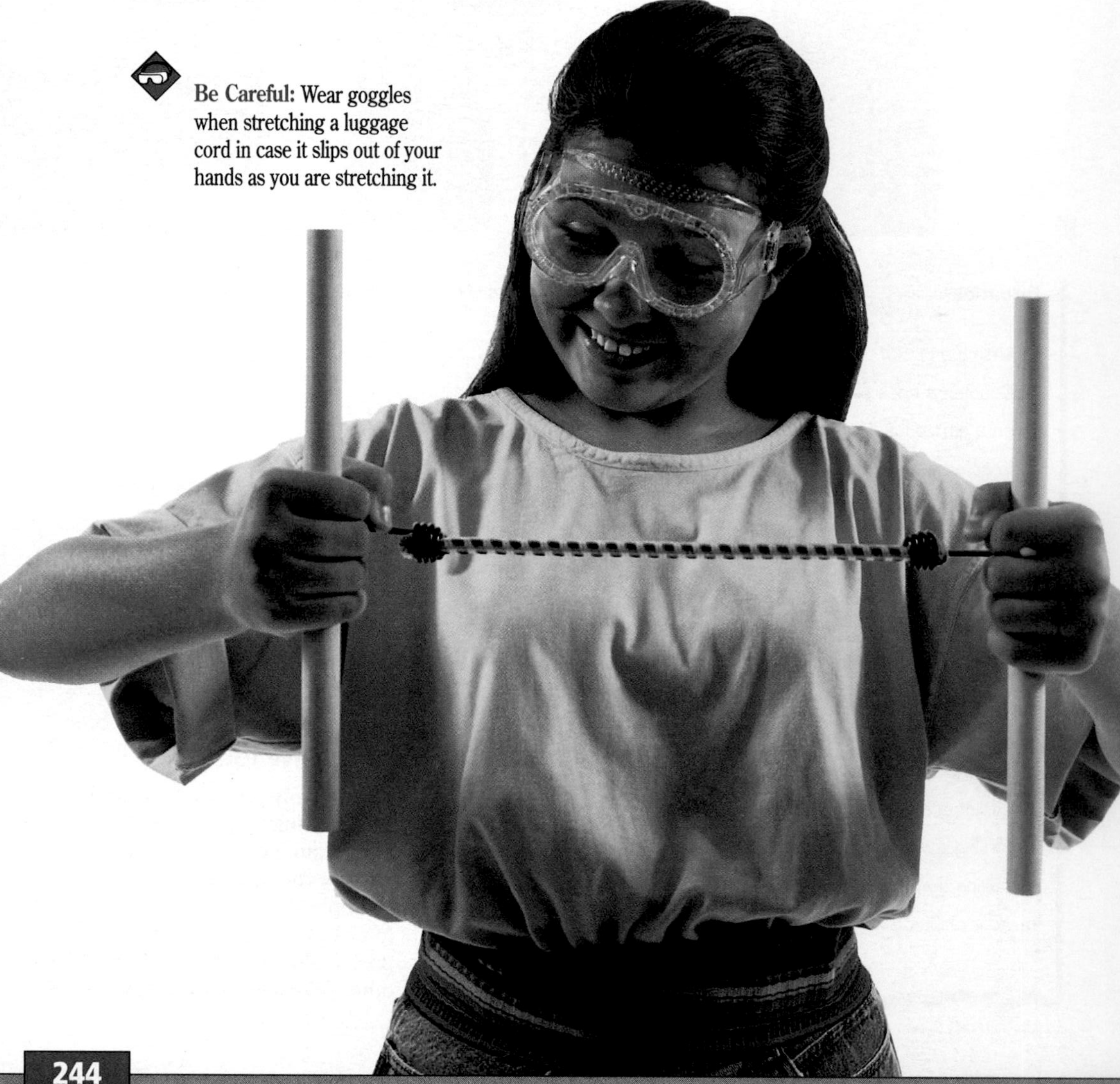

244

LESSON 3 ORGANIZER

Time Required
2 to 3 class periods

New Terms
none

Process Skills
measuring, predicting, analyzing

Materials (per student group)
1. Arm Strength: 2 luggage cords; two 30 cm dowels; force meter; meter stick; 2 small pieces of masking tape; safety goggles; **2. Forearm Pull:** luggage cord; two 30 cm dowels; force meter; safety goggles; **3. Hand Grip:** bathroom scale; **4. Leg Thrust:** chair; bathroom scale

Teaching Resources
none

Find the force you used! Place the cord beside a meter stick as shown below. Pull the cord with a force meter until it matches the length that you were able to stretch the cord. Record the number of newtons shown on the force meter in a table like the one below.

My Strength Record in Newtons

1. Arm strength: ?
2. Forearm pull: ?
3. Hand grip: ?
4. Leg thrust: ?

Repeat this activity using two luggage cords, and record your results.

2. Forearm Pull

Stretch the luggage cord using only your forearm muscles. Use the force meter again to determine the amount of force you exerted. Record your results.

Be Careful: Use caution so that the cord doesn't slip out of your hands.

Test the strength of other muscles; you decide which ones to try!

2. Forearm Pull

SAFETY ALERT Students need to position themselves as shown in the picture. They should wear safety goggles in case the cord slips during the activity.

It is inevitable that students will turn these activities into a competition. However, caution them to take the competition lightly in order to minimize the risk of muscle strain and to prevent the equipment from being broken.

Set up four stations around the room with one activity at each station. If possible, have numerous sets of materials available at each station so that a number of students can work at the same time.

Before students begin, have them copy the table on page 245 into their ScienceLog so that they can keep an accurate record of their results. Suggest that they carefully study the pictures on pages 244–246 so that they will know how to set up and perform each activity.

3. Hand Grip

Remind students to multiply the reading on a metric scale by 10 to find the force in newtons.

4. Leg Thrust

Suggest that one student sit in the chair and another student push down on the chair from the back so that it doesn't move. Again, remind students to multiply the reading on the scale by 10 to find the force in newtons.

INDEPENDENT PRACTICE Encourage students to measure the strength of other muscles. The strength of a little finger, for instance, can be measured by pulling directly on a force meter while holding the rest of the arm still. Students should suggest an additional activity for measuring strength and then perform the activity.

Homework

You may wish to assign the Hand Grip activity for homework.

Integrating the Sciences

Life and Physical Sciences
What are some of the forces that make life on Earth possible? *(Frictional forces, for instance, allow organisms to move around. Gravity holds organisms and the atmosphere near the surface of the Earth.)*

3. Hand Grip

Grip a bathroom scale with both hands. If you use a scale that shows kilograms, multiply the result by 10 to find the force in newtons. If the scale shows pounds, multiply the result by 4.5 to find the force in newtons.

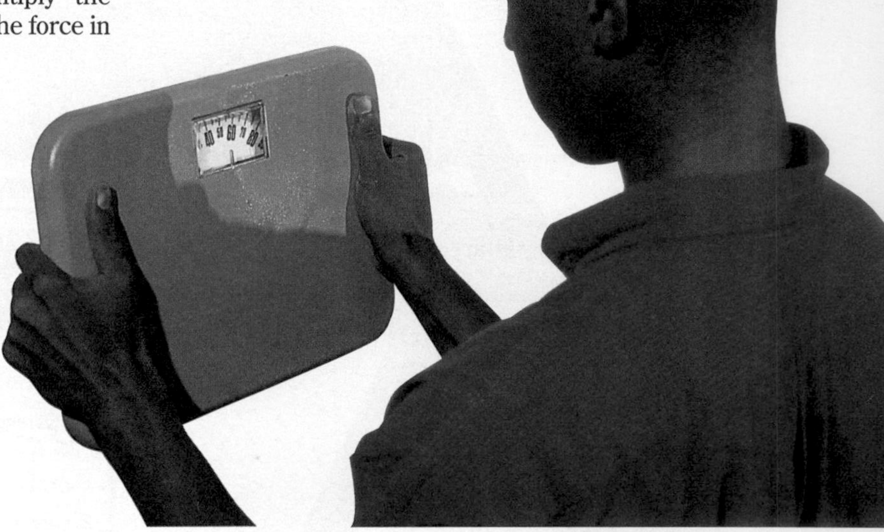

4. Leg Thrust

Using a chair as a backrest, and with a classmate to brace you, push your legs against a bathroom scale that is braced against a wall. Calculate the force you exerted.

246

Measuring Mass and Weight

As you have already learned, the quantity of matter in an object is called its mass. Whether you move an object to the top of a mountain or to the moon, its mass remains the same. The object still has the same quantity of matter in it. Mass is usually measured in grams or kilograms with an equal-arm balance or a pan balance.

Weight, on the other hand, is a measure of the gravitational force that the Earth (or another heavenly body) exerts on an object. The newton is the unit used to measure weight and all other forces. A spring scale is a device that is used to measure weight (see Exploration 5).

The terms *mass* and *weight* are often interchanged and used incorrectly. This can cause some confusion—even for students of science. Carefully study the table and illustration on this page. How can you account for the difference in the weights (N) measured by the spring scale? **A**

Mass is . . .	Weight is . . .
the quantity of matter in an object.	a measure of the gravitational force exerted by the Earth on an object.
constant for an object anywhere in the universe.	varied as you move away from the Earth's surface.
measured with an equal-arm balance or a pan balance.	measured with a spring scale.
measured in grams (g) or kilograms (kg).	measured in newtons (N).

How do measuring devices indicate a constant mass and a changing weight? **B**

Measuring Mass and Weight

GUIDED PRACTICE Before students begin reading, ask them to explain the difference between the terms *mass* and *weight.* Write their responses on the board.

Call on a volunteer to read this page aloud. Pause after the three paragraphs have been read to be sure students are comfortable with the differences between measuring mass and weight. Point out that in everyday life, these terms are often used interchangeably.

Then direct students' attention to the chart. It summarizes the differences between the concepts of mass and weight.

Answers to *In-Text Question and Caption*

A On Earth, where the gravitational force is stronger, the mass on the spring scale has more weight and is pulled down farther.

B The equal-arm balance measures the same mass for a given object at any gravitational force. An object's weight, measured on a spring balance, changes as gravitational force changes.

FOLLOW-UP

Reteaching

Write the following sentence on the board: A 12 N brick has a weight of 120 kg on the Earth and 20 kg on the moon. Ask: What is wrong with this sentence? *(The sentence should read: A 12 kg brick has a weight of 120 N on the Earth and 20 N on the moon.)*

Assessment

Photocopy a diagram of the major muscle groups in the human body. Draw lines to several muscle groups, such as those in the forearms, fingers, and thighs. Ask students how they could find the force in newtons that these muscles can exert.

Extension

With the assistance of a physical-education teacher, have interested students participate in a bodybuilding program. Suggest that they keep track of their improvement by recording how much force their muscles can exert.

Closure

Have students find some world records for weight lifting. If the data are given in kilograms, have them convert the measurements to newtons. If the school has weight-lifting equipment, they can work under the supervision of a physical-education teacher to compare their own strength with the strength of the world-record holders.

CHALLENGE YOUR THINKING

1. Massive Confusion!

Rewrite the sentences, if incorrect, to clear up confusion between *mass* and *weight*.

a. "Oh, no!" Fred shouted. "I've gained! Last week my weight was 46 kg. Now my weight is 48 kg!"

b. If I go far enough away from the Earth's surface, I will get to a place where my mass will be almost zero.

c. An object of 200 N on the surface of the Earth should still measure 200 N in the orbiting space station.

d. In the space station, I would use a spring scale to find the mass of an object.

e. Rachel placed a football on one pan of an equal-arm balance and got a perfect balance with a 250 g mass on the other pan. Naturally, she could expect the same result on the moon.

f. Food is sold in the supermarket by weight—for example, $2.50 for a kilogram of sugar.

g. On the surface of the moon, the weight of a 100 g mass is between 0.1 N and 0.2 N.

2. A Moving Conversation

Consider this conversation:

Moira: How much force does it take to pull the load across the table?

Pedro: A rubber band stretch of 12 cm.

Moira: Would the force be the same if you used a longer rubber band? or a thicker one?

Aaron: How much is a "12 cm force," anyway?

How should Pedro answer Moira's second question? What should he do to answer Aaron's question?

Answers to *Challenge Your Thinking*

1. Sample answers:
   a. "Oh, no!" Fred shouted. "I've gained! Last week my *mass* was 46 kg. Now my *mass* is 48 kg!"
   b. If I go far enough away from the Earth's surface, I will get to a place where my *weight* will be almost zero.
   c. An object of 200 *kg* on the surface of the Earth should still measure 200 *kg* in the orbiting space station.
   d. In the space station, I would *not be able to measure the mass of an object because the spring scale measures weight, and an equal-arm balance would not work.*
   e. This sentence is correct.
   f. Food is sold in the supermarket by *mass*—for example, $2.50 for a kilogram of sugar.
   g. This sentence is correct—the weight would be ⅙ of 1 N.

2. Pedro should tell Moira that the force would be the same because the weight of the object doesn't change. To answer Aaron, Pedro could stretch the original rubber band to 12 cm while it is attached to a force meter. The meter would indicate how many newtons are equal to a "12 cm force."

3. Jane was using a spring scale, which measures weight. As students have learned, the gravitational force on the moon is much less than it is on Earth. Jane would have to place more meat on the spring scale on the moon to equal a certain weight on Earth. Therefore, customers on the moon would get a lot more meat for their money than they would on Earth. This would keep Jane from making a profit. Jane should use an equal-arm balance to measure the mass of the meat, which would be the same on the moon as it is on the Earth.

4. Allow students to present creative or unusual answers. Possible answers include the following:
   a. Calculate your own mass in kilograms and multiply by 10 to determine the force in newtons.
   b. Attach a force meter between your feet and the pedals, and read the scales just as the bike starts to move.
   c. Students should realize that pushing and pulling an object require the same force. One could push the force meter up against the car and read it when the car starts moving, or one could measure the force needed to pull the car by attaching a force meter to the front bumper.
   d. The boulder could be suspended from an enormous force meter, or a lever could be used to measure the force needed.

5. Weights could be estimated if the weight of at least one student were known. Using the seesaw as a balance, students could estimate weights by comparison. For accurate readings, students should be the same distance from the center.

Homework

Have students write three sentences like those in question 1. Their sentences should relate to the topics discussed in this unit and contain at least one error. The next day, have students exchange papers and rewrite the sentences that they receive.

3. Jane's Problem

Everyone liked Jane the butcher. Her meats and prices were good. If you wanted 3 kg of meat, Jane would put it on her spring scale and say, "Here you are, 3 kg of honest weight. At $4.50 per kilogram, that will be $13.50. Make it $13 even, okay?" Out you go, a satisfied customer.

Since space travel had become so popular, Jane decided to open a branch store on the moon. Spring scales were sent from Earth. Weekly meat supplies were rocketed in. Jane set the same prices. Astronauts and tourists flocked to buy Jane's meat. But the store didn't make money. In fact, it lost money from the start. Customers were satisfied—they got plenty for their money. Jane was puzzled.

What was wrong? Help Jane solve her problem.

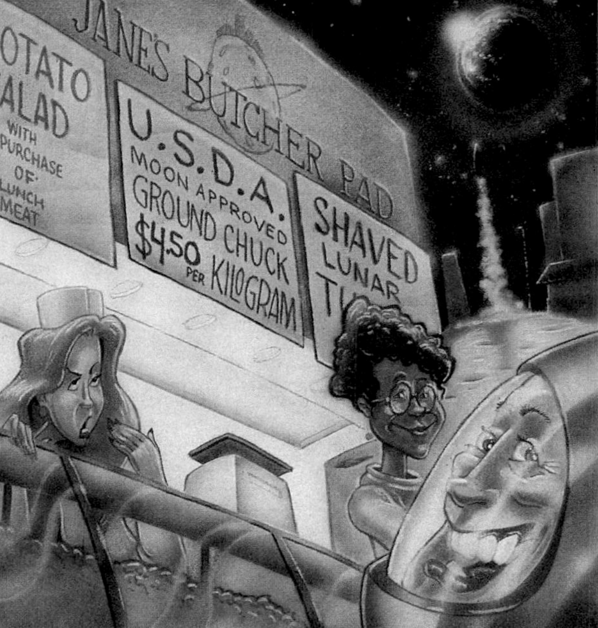

4. Force-Finding Designs

Design an innovative way to measure the force needed to do each of the following tasks. Estimate the force you might measure.

a. perform a chin-up
b. pedal a bike to start it moving
c. push a sports car
d. lift an enormous boulder (approximately 60 cm wide)

5. A New Way to Weigh?

Your task is to find the weight of each member of your class. But you have no scales. There is a seesaw in the park just across from your school. How could you use the seesaw to estimate the weights? (Hints: Does it matter where you sit to balance someone? Would it be helpful to know in advance the weight of some of your fellow students?) Why not test your idea?

Sciencelog

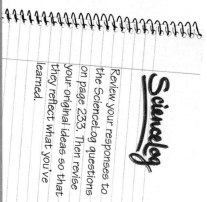

Review your responses to the ScienceLog questions on page 233. Then revise your original ideas so that they reflect what you've learned.

249

★ **You may wish to provide students with the Chapter 11 Review Worksheet on page 26 of the Unit 4 Teaching Resources booklet.**

Sciencelog

The following are sample revised answers:

1. The size of a force can be determined by measuring its effect—how much it displaces the needle on a force meter, for instance.
2. The activities shown could be arranged according to the amount of force required to start each object moving. The stroller with the baby weighs the least and would therefore require the least amount of force; the shopping cart weighs more and would require more force to push than the stroller; the car weighs the most and would therefore require the most force to start moving.
3. Weight and mass cannot be measured in the same way because they are two different things. Weight, the gravitational force exerted on an object, is measured in newtons. Weight can be measured with a spring scale. Mass, the quantity of matter in an object, is measured in grams or kilograms. Mass can be measured with an equal-arm or pan balance.

Chapter 12 Force Meets Motion

Connecting to Other Chapters

Chapter 10 offers students hands-on exposure to a wide range of forces.

Chapter 11 allows students to measure the strength of forces with commercial and homemade force meters.

Chapter 12 explores the causes and consequences of friction in various situations.

Chapter 13 discusses the concept of inertia and introduces students to force pairs.

Prior Knowledge and Misconceptions

In addition to having students answer the questions on this page, you may wish to propose the following scenario to them: A designer of specialty bicycles has asked for your help in designing a new type of bicycle that can be ridden across frozen lakes and other icy surfaces. Describe your idea for creating a successful "icycle." Your description should include a discussion of why normal bicycles cannot be ridden on ice, what modifications you will make, and how those modifications solve the problems that normal bicycles have. Be sure to use the language of forces in your description.

Students may work alone or in small groups to complete this activity. Emphasize that there are no right or wrong answers. Collect the papers, but do not grade them. Instead, read them to find out what your students know about friction and frictional forces, what misconceptions they have, and what about this topic is interesting to them.

CHAPTER 12

Force Meets Motion

1. What causes friction? Identify friction at work in these photographs.

2. How can friction be reduced? How can it be increased?

3. Is friction harmful or helpful? Explain your reasoning.

ScienceLog

Think about these questions for a moment, and answer them in your ScienceLog. When you've finished this chapter, you'll have the opportunity to revise your answers based on what you've learned.

Frictional Force

You saw some examples of friction in Chapter 10 (pages 224–225). Take another look at those examples. What did they tell Ⓐ you about friction? Now read the following Case Studies and answer the questions.

Case Study A

As Jennifer reaches the level field, she gradually slows down.

1. What slows down the skis?
2. Is there a force involved?
3. What is the direction of the force?
4. What name do we give to this kind of force?

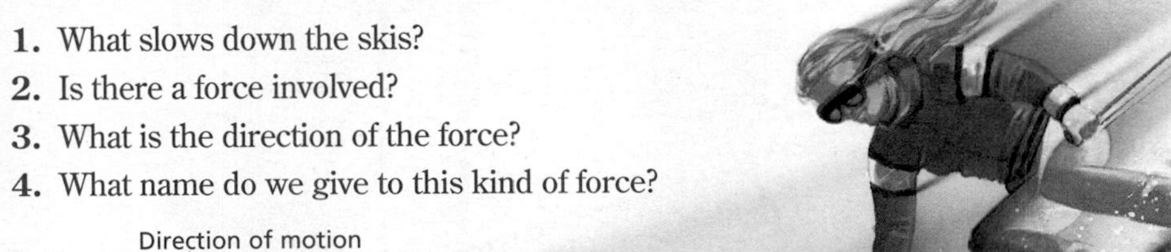

Case Study B

Driving east on a level stretch of highway at 100 km/h, Yoshi shifts into neutral. The car coasts for 300 m before coming to a stop.

1. Identify the frictional force and its direction.

LESSON 1 ORGANIZER

Time Required
2 to 3 class periods

Process Skills
hypothesizing, predicting, analyzing

New Terms
none

Materials (per student group)
Exploration 1: 1 or 2 bricks; force meter; board, about 1 m in length; 10 cm × 50 cm sheet of glass; 4–6 plastic drinking straws; 1 m of string; scissors; safety goggles

Teaching Resources
Transparency Worksheet, p. 31
Transparency 41
SourceBook, p. S75

Frictional Force

FOCUS

Getting Started

Have students imagine a game of tug of war in which one team will be supplied with cleats and the opposing team will wear roller skates. Ask: Will this be a fair contest? Why or why not? *(The first team will have an advantage because there will be much more friction between their feet and the ground, enabling them to win.)*

Main Ideas

1. Friction opposes a starting or continuing motion.
2. Friction is greater when motion is starting than when it is continuing.
3. Friction varies with the kinds of surfaces involved.

TEACHING STRATEGIES

Answer to *In-Text Question*

Ⓐ The examples in Chapter 10 showed that friction depends on the surfaces involved and can be reduced by using rollers. Also, frictional force acts in the direction opposite that of an object's motion.

Answers to *Case Study A*

1. Frictional force between the skis and the snow
2. Yes
3. Opposite to the direction of motion
4. Frictional force

Answer to *Case Study B*

1. The frictional force is exerted by the road against the tires. It acts opposite to the car's motion (i.e., westward).

★ **A Transparency Worksheet (Teaching Resources, page 31) and Transparency 41 are available to accompany this page.**

Answers to *Case Study C*

1. The frictional force is the force exerted by the floor against the bottom of the refrigerator.
2. The frictional force here resists the initial movement of an object instead of acting on an object that is already in motion.
3. The force of Don and Susan together is slightly larger than the frictional force exerted by the floor, so they are able to move the refrigerator together.

Answers to *Case Study Conclusions*

1. The skis, the tires, and the refrigerator all move against the surface below them, resulting in a frictional force.
2. This is particularly noticeable when friction with the ground slows down the skis and the tires.
3. The frictional force below the refrigerator kept it from moving until a larger force—that of Don and Susan—overcame the frictional force.

Answers to *Questions*

Divide the class into small groups to discuss the questions. Reassemble the class and involve the students in a discussion of their responses.

1. Possible responses include frictional forces that slow down skateboard wheels, blades on ice skates, or a book sliding across a table.
2. Examples include the sand against the bottom of the boat that keeps the boat from moving easily as it is dragged, water against the bottom of the boat that slows down its motion, and air friction against the boat that slows down its motion.
3. If there were no frictional force opposing the motion of the skis, they would keep going forever. However, if air friction is considered, then the skier would eventually slow down due to friction between the skier and the air.
4. If Yoshi had been driving on gravel, the car would have stopped sooner due to more friction. If he had been driving on ice, the car would have moved farther before stopping due to less friction.
5. Sample answer: Don could move the refrigerator himself if it were on rollers or on a smoother surface. Both solutions reduce the friction between the floor and the refrigerator.

Case Study C

Don dropped his book behind a large refrigerator. He pulled on one edge of the refrigerator to move it away from the wall, but it did not budge. His friend Susan joined him. Together they just managed to move it.

1. What is the frictional force?
2. How is the frictional force here different from that in the other Case Studies?
3. How do the combined forces that Don and Susan exert on the refrigerator compare with the size of the frictional force?

Case Study Conclusions

The following general conclusions about frictional forces can be drawn from the Case Studies. For each conclusion, find evidence in the Case Studies to support it.

1. When one surface moves against another, a frictional force results. The frictional force works against the direction of motion.
2. If one object is moving over a stationary object, the stationary object tends to slow down the moving object.
3. If a force tries to move an object at rest on a surface, the surface resists this motion with an opposing force of friction. The size of this frictional force matches the force trying to move the object, up to a limiting size. A force greater than this limiting size will move the object.

Questions

Discuss these questions with a group of classmates. Try to reach an agreement about each one.

1. What are some other examples in which frictional forces slow down motion?
2. In the sailboat story at the beginning of this unit, were there any examples of friction? If so, identify them.
3. In Case Study A, what if there had been no frictional forces opposing the motion of the skis?
4. For Case Study B, what would have happened if Yoshi had been driving on gravel? on ice?
5. In Case Study C, how could you help Don move the refrigerator by himself? In your solution, have you reduced, increased, or maintained the limiting size of the force of friction?

Homework

Have students describe three situations in which frictional forces slow down motion. Their descriptions should not be the same as their answers to question 1 above.

Measuring Friction

How large or small are frictional forces? How can they be increased or decreased? You measured friction in Exploration 5 in Chapter 11. Review these measurements to see whether they can help you discover some new ideas about friction. (A)

Janell had some ideas (hypotheses) about friction. She decided to check her new ideas by conducting a series of experiments.

The results of her experiments are summarized on page 254. Look at the results, and then try to solve the Friction Hypothesis Puzzle below. In this puzzle, you will choose the hypothesis that is confirmed by each pair of experiments listed in the left column. For example, which hypothesis is supported by the results of Experiments 2 and 3? Do the experimental results confirm all of Janell's hypotheses? (B)

Friction Hypothesis Puzzle

Experiments designed to test the hypotheses:	Janell's hypotheses:
a. Experiments 1 and 4	1. The heavier the object, the greater the force of friction.
b. Experiments 2 and 3	2. The force of "rolling" friction is greater than the force of "sliding" friction.
c. Experiments 3 and 4	3. Frictional force is constant for an object on a surface, no matter what the area of surface contact is.
d. Experiments 3 and 5	4. The size of the frictional force depends on the kinds of surfaces that are rubbing together.
e. Experiments 3 and 6	5. The force of "starting" friction is greater than the force of "sliding" friction.
f. Experiments 4 and 7	6. The force needed to lift an object is greater than the force needed to drag an object along a level surface.

253

Measuring Friction

Call on a volunteer to read the section called Measuring Friction aloud. Then direct students' attention to page 254. Ask students to read the description of each experiment and to look at the diagrams. Pause after each one to be sure students understand what the experiment was about and how the result was determined.

Answers to *In-Text Questions*

(A) Students may suggest that frictional forces depend on the types of the objects involved.

(B) All of the hypotheses were confirmed except number 2. The force of sliding friction is greater than the force of rolling friction.

Answers to *Friction Hypothesis Puzzle*

a. 6
b. 4
c. 5
d. 1
e. 3
f. 2 (not supported)

CROSS-DISCIPLINARY FOCUS

Mathematics

A basket contains 95 g of marbles, a picture frame weighing 490 g, 2.3 kg of comic books, a pair of shoes weighing 635 g, and a bundle of cassette tapes weighing 0.88 kg. If the force of friction between the basket and the floor is one-half of the basket's weight, how much force should be applied to move the basket? *(At least 22 N)*

PORTFOLIO

Suggest that students include their answers to the Friction Hypothesis Puzzle in their Portfolio. They can include notes to explain their reasoning. Finally, students may wish to create a few friction hypotheses of their own and then design tests to confirm or disprove them.

Multicultural Extension

Trains That Travel on Air

Have interested students find out about the magnetically levitated trains (maglev trains) in Germany, Great Britain, and Japan. This type of train uses magnetic forces to travel at high speeds. The train actually floats above the track. These trains have made test runs at 480 km/h (300 mph), but only low-speed maglev trains are currently in use. The only friction that slows down a maglev train is air friction. Have students find out how these transportation systems suit the needs of the countries in which they are used.

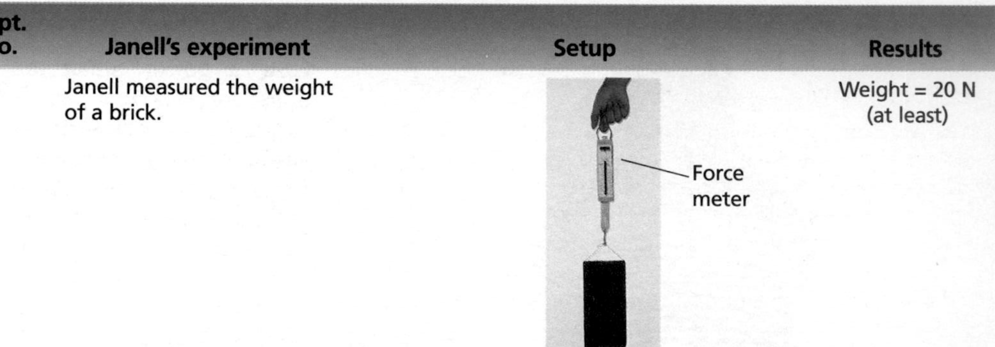

Expt. No.	Janell's experiment	Setup	Results
1	Janell measured the weight of a brick.	Force meter	Weight = 20 N (at least)
2	Janell measured the force needed to start the brick moving on a glass surface.	Force meter; Glass	Force = 6 N
3	Janell measured the force needed to start the brick moving on a wooden surface.	Wood	Force = 10 N
4	Janell measured the force needed to keep the brick moving on wood once it started to move.		Force = 8 N
5	Janell measured the force needed to start two bricks moving on a wooden surface.		Force = 18 N
6	Janell turned the brick on its narrow side and measured the force needed to start it moving on wood.		Force = 10 N
7	Janell placed straws between the brick and the wooden surface and measured the force needed to keep the brick moving.		Force = 1 N

254

CROSS-DISCIPLINARY FOCUS

Foreign Languages

The word *force* is derived from the Latin word *fortis,* meaning "strong." Other words that come from *fortis* include *fortitude, fort,* and *fortified.* Ask: What do these words have in common? *(They all convey the idea of strength in some way. Fort, for instance, often refers to a structure that has been equipped with military defenses.)* Ask: Are there similar words in other languages? *(The Spanish word* fuerte *and the French word* fort, *meaning "strong," for instance)* Encourage discussion.

EXPLORATION 1

Your Turn to Test Friction

Do you still have doubts about Janell's hypotheses? After all, she compared the results of only two or three experiments for each of the hypotheses. Perhaps more proof is needed. It might be valuable to conduct more experiments.

Here is your opportunity to confirm or correct the results of Janell's experiments and to find out whether her hypotheses hold true.

What to Do 

1. Choose one of Janell's hypotheses to test.
2. Design your own experiment.
3. Decide what equipment you will need.
4. Collect the equipment.
5. Perform the experiment.
6. Record your data.
7. Check your results. Did you perform the test enough times? Were the variables controlled?
8. Decide whether the hypothesis you tested is right or wrong.

More About Friction

By now you realize that friction is present all around us. From the cars on the highway to the sneakers on a basketball court, friction is a familiar force. Summarized here are some important scientific points about friction.

Friction . . .

- opposes a starting or continuing motion.
- is greater when motion is started than when motion is continued.
- is greater for heavier weights.

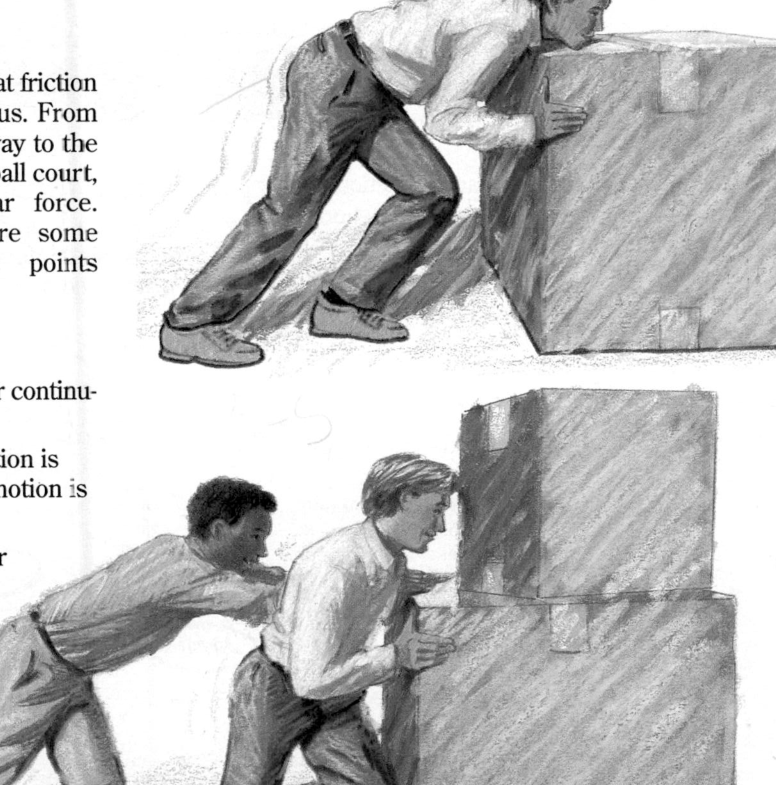

EXPLORATION 1

Divide the class into groups of three or four students. Have the groups design and perform experiments to test Janell's hypotheses. Make sure that each of the hypotheses is chosen by at least one group. Students' results should support Janell's results. Make sure you approve the student designs before they proceed.

This Exploration can be used to emphasize the fact that our understanding of science advances by formulating and testing hypotheses. Experiments may prove or refute hypotheses, or they may cause the hypotheses to be modified. The fact that some hypotheses will be tested several times could be used to emphasize the need for replication of results—if the same conditions are used in a series of experiments, the same results are expected. Explain the idea of experimental error, which can result from imperfect procedures and uncertainties in measurement.

Involve students in a discussion of experimental design, paying particular attention to the control of variables. For example, ask: Would doing experiments 2 and 4 be a fair test of the fifth hypothesis? *(No, because the type of surface should be the same in both experiments)*

More About Friction

Call on a volunteer to read aloud pages 255–256. Then divide the class into small groups to discuss the five questions on page 256. These questions are designed to prepare students for the next lesson. Expect variety in students' answers.

Friction . . .

- varies in size with the kinds of surfaces opposed to one another.
- is reduced when the opposing surfaces are separated by something round.

There are even more things about friction to think about. Consider the following questions and write your ideas in your ScienceLog:

1. What causes friction?
2. What are some ways to reduce friction?
3. What harm can friction cause? Why?
4. Where is friction useful? Why?
5. What substances besides solids can cause friction?

To find out more about what happens when force meets motion, read about my second law on page S81 of the SourceBook.
Isaac Newton

Try a Concept Map!

Copy this concept map into your ScienceLog. Provide the missing information, based on what you have learned so far about friction.

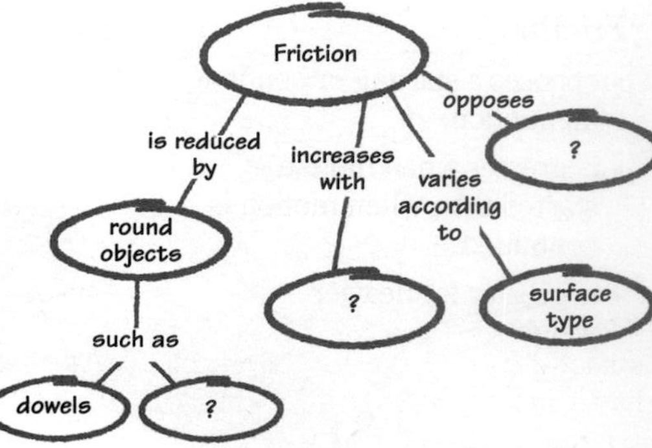

Answers to *In-Text Questions*

1. Friction is caused by the contact between two surfaces.
2. Some ways to reduce friction are to use oils, lubricants, or bearings.
3. Friction can wear down and damage surfaces. It can also create excessive heat.
4. Sample answer: Friction is useful when it provides traction between the wheels of a car or a bike and the ground. It is also necessary for braking.
5. Liquids and gases can cause friction as well.

Answers to *Try a Concept Map!*

Students may respond that round objects include straws, rollers, or wheels; friction increases with an object's mass, surface roughness, or surface area; and friction opposes motion or movement.

FOLLOW-UP

Reteaching

Place a brick or heavy object on your desk. Ask: Is the force of friction greater when the brick starts moving or when the brick is already moving? *(When the brick starts moving)* How can I reduce the friction between the brick and the table? *(Use rollers or oil, or put a smooth surface between the brick and table.)* How can I increase the friction? *(Add more weight, or put a rough surface between the brick and the table.)*

Assessment

Present students with the following hypothesis: Friction is greater between snow and the sole of a shoe than between smooth ice and the sole of a shoe. Ask them to design an experiment to test this hypothesis.

Extension

Have students use the following method to test the frictional force between different surfaces. Use a thumbtack to secure a 2 m length of string to the end of a small block of wood. Tie a bent paper clip to the other end of the string. Dangle this end of the string over the edge of a table. When masses are hooked onto the paper clip, the force of gravity on the masses will pull the block across the table. Tape different kinds of materials to the block and to the table, and record the force needed to move the block for each trial. Try wax paper, aluminum foil, newspaper, cardboard, sandpaper, and strips of cloth. Find out which combinations produce the least and the most friction.

Closure

As a class, explore a children's playground. Students should identify parts of the playground that are designed to reduce or increase friction. *(Reduce friction: slides, places where swings attach to the crossbar; increase friction: steps of the ladder on the slide, seat of the swing)*

Causes and Consequences of Friction

The Cause of Friction

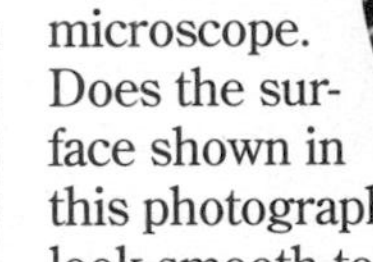

Julie was helping her mother refinish some old furniture. Using sandpaper to remove paint from an old chair made her think about what she had learned about friction. She wondered what would happen when she rubbed the rough sides of two pieces of sandpaper together.

Wow—it was really rough going! She then turned over the two pieces of sandpaper and moved the smooth surfaces over each other. How do you think the results compared? Try Julie's experiment with two pieces of sandpaper. Ⓐ

The rougher the surfaces, the harder it is to move one over the other. The bumps and irregularities on the rough side of the sandpaper interlock and disrupt motion. This causes friction.

But there is friction even when the smooth sides of the sandpaper are rubbed together. If you examine the smoother side with a magnifying glass, you might see that it is not perfectly smooth. Test the smoothness of the surfaces listed below with your finger. Which surfaces are roughest? smoothest? Ⓑ

- piece of metal
- various kinds of paper
- book cover
- window glass
- table top
- floor tile
- wall
- sandpaper
- ice

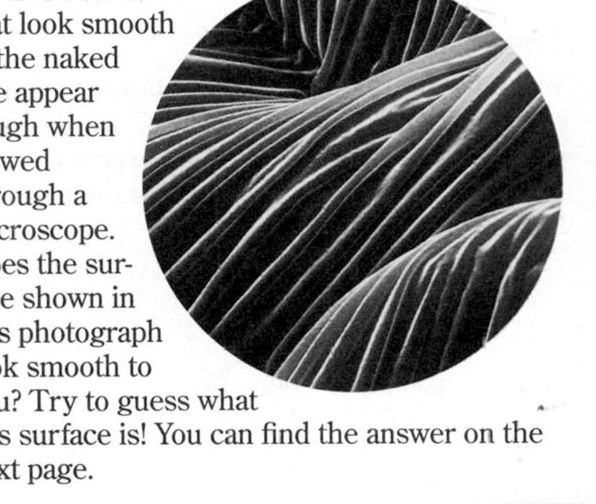

Even surfaces that look smooth to the naked eye appear rough when viewed through a microscope. Does the surface shown in this photograph look smooth to you? Try to guess what this surface is! You can find the answer on the next page.

LESSON 2 ORGANIZER

Time Required
2 class periods

Process Skills
observing, predicting, analyzing

Theme Connection
Energy

New Terms
none

Materials (per student group)
The Cause of Friction: 2 pieces of coarse sandpaper; small piece of metal; a few sheets of different paper; book cover; floor tile; small piece of ice
Reducing Friction: half of a stick of margarine or a few spoonfuls of grease; 2 pieces of sandpaper; small piece of glass; drop of oil
The Harm of Friction: 2 pieces of medium sandpaper
Exploration 2, Measuring the Strength of a Paper Towel Using Friction: 2–5 paper towels; small piece of masking tape; force meter; jar or bottle; hole punch; 2–5 paper reinforcements; **Comparing the Friction of Different Types of Paper:** paper

continued ►

Causes and Consequences of Friction

FOCUS

Getting Started

Have students observe the surfaces of different materials under a compound microscope. Ask: Which surfaces appear the smoothest? Which appear the roughest? Between which surfaces is there the most friction? the least friction?

Main Ideas

1. Friction results when surfaces that are moving past each other interlock.
2. Friction may be reduced by smoothing opposing surfaces, lubricating them, or separating them with ball bearings.

TEACHING STRATEGIES

Answers to *In-Text Questions*

Ⓐ Julie probably found that the smooth surfaces had less friction.

Ⓑ A likely order, from roughest to smoothest, is sandpaper, wall, book cover, various kinds of paper, floor tile, tabletop, piece of metal, window glass, and ice.

ENVIRONMENTAL FOCUS

Ask: Can you think of any environmental reasons why reducing air friction on moving cars might be important? *(Reducing air friction leads to better fuel efficiency. This means that less gasoline is consumed and fewer pollutants are emitted by cars.)*

Reducing Friction

Ask for a student volunteer to read page 258 aloud. As each paragraph is read, ask student volunteers to demonstrate the activity in the paragraph.

Answers to *In-Text Questions*

Ⓐ The friction is reduced when margarine is added to the sandpaper. The friction cannot be eliminated completely.

Ⓑ Students should find that when they put oil on the glass, friction is decreased.

Ⓒ Examples include greasing bicycle parts and car engine parts. The metal parts would rub together, wearing each other down and creating a lot of heat.

Answers to *The Harm of Friction*

As sheets of sandpaper are rubbed together, a considerable amount of heat is generated. The heat generated by friction could cause machine parts to overheat and warp or crack, for instance.

Science "Friction"!

In this creative-writing exercise, students should identify the sources of friction and describe the effects of the lack of friction in everyday situations. Some examples include the following: you cannot walk; you cannot turn a doorknob; everything you touch begins to slide; if you push open a door, it will crash into the wall; the force of the slamming door against the frame will cause the nails, which are no longer held by friction, to pop out and will cause everything on the walls to fall on the floor; and if you try opening a drawer, it comes out in your hands, and the contents fall out and move smoothly along the floor, bouncing off the walls without slowing down.

Homework

The Science "Friction" creative writing activity on this page makes an excellent homework assignment.

Reducing Friction

If we make rough surfaces smoother, we reduce friction. Cover the rough side of two pieces of sandpaper with margarine. Now rub the surfaces together. What happens to the friction? Do you think the friction could be completely eliminated? Ⓐ

Try rubbing your finger on glass. Then place a drop of oil between your finger and the glass. Glass without oil seems quite smooth. Is there any difference with oil? Ⓑ

It is very important to oil or grease the moving metal parts of machines. Can you suggest some examples where this is done? What do you think would happen if the metal parts were not oiled or greased? Ⓒ

Often, metal surfaces that must move or turn against each other are separated by metal balls or rollers. This reduces the friction between the moving parts. Recall that rolling friction is less than sliding friction. The wheels of roller skates, bicycles, motorcycles, and cars turn on metal axles that are separated from the wheels by ball or roller bearings. Adding oil to the bearings further reduces the friction.

The Harm of Friction

What harm does friction do? If you rub two pieces of sandpaper together for 30 seconds, what do you observe? Feel how warm the surfaces are. You may also notice that the surfaces have been slightly worn down.

Friction . . .

- opposes motion.
- produces heat.
- causes surfaces to wear away.

Science "Friction"!

In the following pages you will read about many examples of friction—some harmful, some helpful. Before you look at these pages, try writing or drawing a little science "friction."

Here's the scenario: Your home has been invaded by a UA (unfriendly alien) armed with an AF (antifriction) spray gun. You open the front door to enter your house and . . .

Describe or illustrate what happens in the next few minutes.

The photograph on page 257 shows the surface of transparent tape, magnified hundreds of times.

ORGANIZER, *continued*

towel; wax paper; 2 small pieces of masking tape; 2 jars of equal size; water to fill jars; wooden dowel; two 30 cm pieces of string; various kinds of paper; different grades of sandpaper; cardboard; hole punch; 2–5 paper reinforcements; **Nail Friction:** claw hammer; 2 to 3 nails; piece of wood; **Brick Transportation:** brick; metric ruler; newspaper or cloth; margarine, grease, or corn or canola oil; roller skate; large plastic tub; 2 drinking straws or dowels; lab aprons; **Measuring Air Friction:** ring stand; ring support; 2 paper clips; 2 pieces of thread; index card; hole punch; masking tape; cardboard strip; clothespin; protractor; **Reducing Friction:** book; jar lid; 15–20 glass marbles; 2 empty paint cans; vegetable, corn, or canola oil; roller bearings; ball bearings; sleeve bearings

Teaching Resources

Exploration Worksheet, p. 33
Theme Worksheet, p. 38
Activity Worksheet, p. 40
Transparencies 42 and 43

EXPLORATION 2

Friction Projects

Measurin the Strength of a Paper Towel Using Friction

What to Do

1. Set up your equipment as shown at right. Gently pull the force meter toward you.
2. Now add water to the jar to increase the mass until the paper tears.
3. What is the maximum amount of force the paper towel was able to withstand?
4. Do all paper towels have the same strength? Explain.

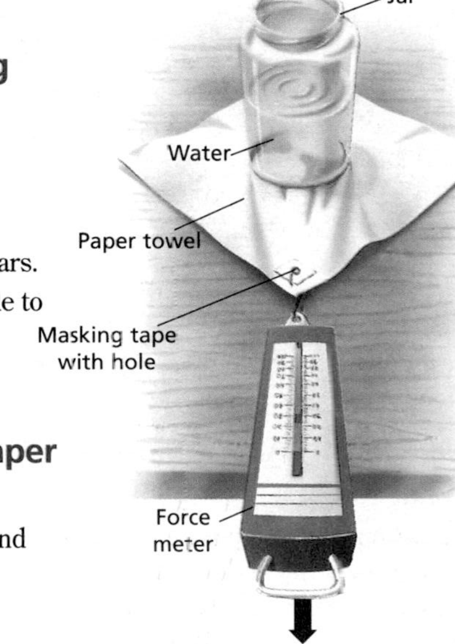

Comparing the Friction of Different Types of Paper

What to Do

1. Set up your equipment as shown below. The pieces of paper and the jars must be of equal size.

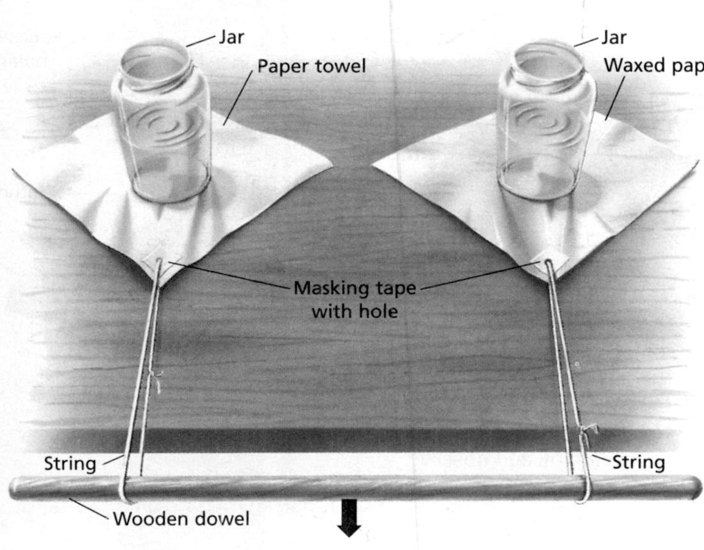

2. Gently pull the dowel toward you. Do the two jars move along evenly? Or does one lag behind, causing the dowel to rotate? Why?
3. Add water to one of the jars until the two move along evenly. What does adding water do? How do the frictional forces compare now?
4. How can you use your data to compare the friction between waxed paper and the table, and between the paper towel and the table?
5. Compare the friction between other surfaces. Use various kinds of paper and cardboard, and use sandpaper of different grades.

Nail Friction

What to Do

1. Use a claw hammer to pull a nail out of a piece of wood.
2. Devise a way to measure the force you used to pull out the nail.
3. Is the force the same for all boards and for all nails? Does the length of the nail matter?

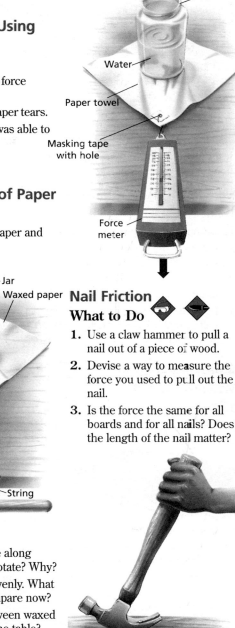

Exploration 2 continued ►

Theme Connection

Energy

Focus question: How does a smooth, sleek, and aerodynamic design improve the energy efficiency of a moving object? *(A sleeker design will reduce the amount of air or water resistance that a moving object encounters.)* A Theme Worksheet is available that describes an excellent activity to accompany this Theme Connection (Teaching Resources, page 38). This Theme Worksheet may also be assigned as homework.

Answers to *Measuring the Strength of a Paper Towel Using Friction*

Students should find that the strengths of paper towels and the forces needed to tear them vary considerably.

Answers to *Nail Friction*

Students may attach a force meter to the hammer handle to measure the force. Both the type of board and the length and shape of the nail will affect the force required to extract the nail.

EXPLORATION 2

For Measuring the Strength of a Paper Towel Using Friction, make sure that students understand the procedure. Point out that increasing the amount of water in the jar increases the jar's mass, in turn increasing the amount of friction between the paper towel and the table. They should measure the amount of frictional force just as the paper tears, using this measurement as an indicator of the paper towel's strength. If students find that the paper towel tears too soon, have students try doubling it up under the jar. If the paper towel doesn't tear, have students try adding 5 or 10 drops of water to its center. (Remember that this should be done to every paper towel tested.) Have 2–5 brands of paper towels available for students to test.

If the paper towel in Comparing the Friction of Different Types of Paper tears, have students try doubling it up as well. Students should pull on the center of the dowel so that it can rotate easily. You may wish to have some marbles or rocks handy in case additional water is not sufficient to equalize the frictional forces.

Answers to *Comparing the Friction of Different Types of Paper*

The jar on the paper towel lags behind because the friction between the paper towel and the table is greater than the friction between the waxed paper and the table. Adding water increases the weight of the jar, and thus the friction between the waxed paper and the table increases as well.

★ **An Exploration Worksheet is available to accompany Exploration 2 (Teaching Resources, page 33).**

Teaching Strategies for Exploration 2 continued ►

EXPLORATION 2, *continued*

In Brick Transportation, have plastic drinking straws or dowels available as rollers and margarine or oil available for lubrication. You may wish to have students use force meters to compare the forces in this activity.

For Measuring Air Friction, have students experiment with different ways of hanging the folded card from the ring stand.

For Reducing Friction, use empty paint cans. As an extension, you may suggest that students fill the empty paint cans with water and note any differences. Small paint cans work well for this activity. Use household oil for lubricating the marbles in this activity. Instead of having students visit local businesses, you may wish to provide samples of these types of bearings for students to inspect in class.

Answers to *Brick Transportation*

The force needed by the various methods will depend on the materials used. Wheels, rollers, and lubrication all act to reduce friction. On land, wheels are used to reduce the friction between vehicles and the surfaces on which they travel. Making the surface of vehicles smoother and lubricating their interior parts are other ways that friction can be reduced. The outside surfaces of land and water vehicles are made as smooth as possible in order to reduce friction from air and water, which can slow these vehicles down.

Answers to *Reducing Friction*

The marbles, which serve as bearings, make it much easier to turn the book and rotate the paint cans. Adding oil reduces the friction further and thus makes it even easier to rotate the paint cans. Sample answers: Roller bearings are long, solid cylinders that roll, and they are found in light and heavy equipment and in wheels. Ball bearings are found in wheels, roller skates, ballpoint pens, and light equipment. A sleeve bearing looks like a thin tube surrounding a solid inner cylinder and is found in light equipment and in automatic transmission bushings.

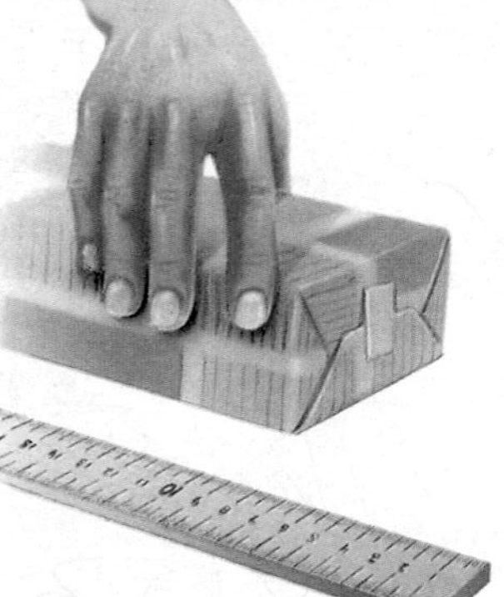

Brick Transportation

What to Do

1. Move a brick a distance of 10 cm in the following ways: dragging only the brick, wrapping it in newspaper or cloth and dragging it, dragging it on rollers, dragging it on a lubricated surface, dragging it on a roller skate, and placing the brick in a large plastic tub and dragging it across the surface of water. Compare the force needed in each of the various methods.
2. Draw conclusions about how friction is decreased in vehicles used for transportation.
3. Is air friction a problem in transportation? If so, give examples.

Measuring Air Friction

What to Do

1. Set up your equipment as shown at right.
2. Pull back the file card so that it is just under position *A*. Let it swing. What happens to the size of the swing? Why?
3. How many swings of the card take place before it swings only as far as *B*?
4. Does the card's shape matter? Try folding the card in half so that it has a V-shape of 90°. Also try angles of 30° and 45°. What do you observe?

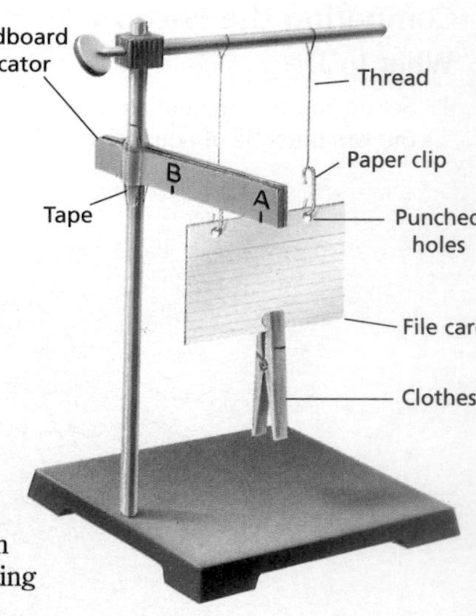

Reducing Friction

What to Do

1. Try rotating a book on a table.

2. Place a jar lid over enough marbles to fill the lid. Place the book on top, and rotate the book.

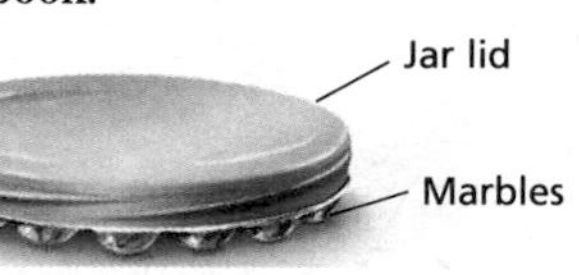

3. Try rotating one paint can over another. Stack one can on top of the other, separating them with glass marbles in the outer ridge of the lid of the bottom can. What happens? Add oil and observe what happens.
4. Go to a garage, auto salvage shop, bicycle shop, or ball bearing company. Get samples of roller bearings, ball bearings, and sleeve bearings. Find out how each type works. Make a list of devices that use the different types of bearings.

260

Answers to *Measuring Air Friction*

The change in the size of the swing will depend on the size of the card; it may take only one or two swings for the card to reach *B*. Students will find that, depending on how the card is hung, a fold will have a pronounced effect on the size of the swing.

Integrating the Sciences

Earth and Physical Sciences

Ask: How is friction involved in the formation of surface waves? *(Waves may form as a result of friction between moving air and moving water.)* Have students experiment by using moving air to make waves in pans filled with different liquids, such as water, milk, and honey.

Friction: Friend or Foe?

Another Picture Puzzle

In the pictures on this page, identify the surfaces that are rubbing together to cause friction.

1. In which situations is the force of friction small? large?
2. In which situations is friction harmful? helpful?
3. How could friction be reduced in the "harmful" examples?
4. How could friction be increased in the "helpful" examples?
5. Think of three other examples of helpful friction.
6. Think of three other examples of harmful friction.

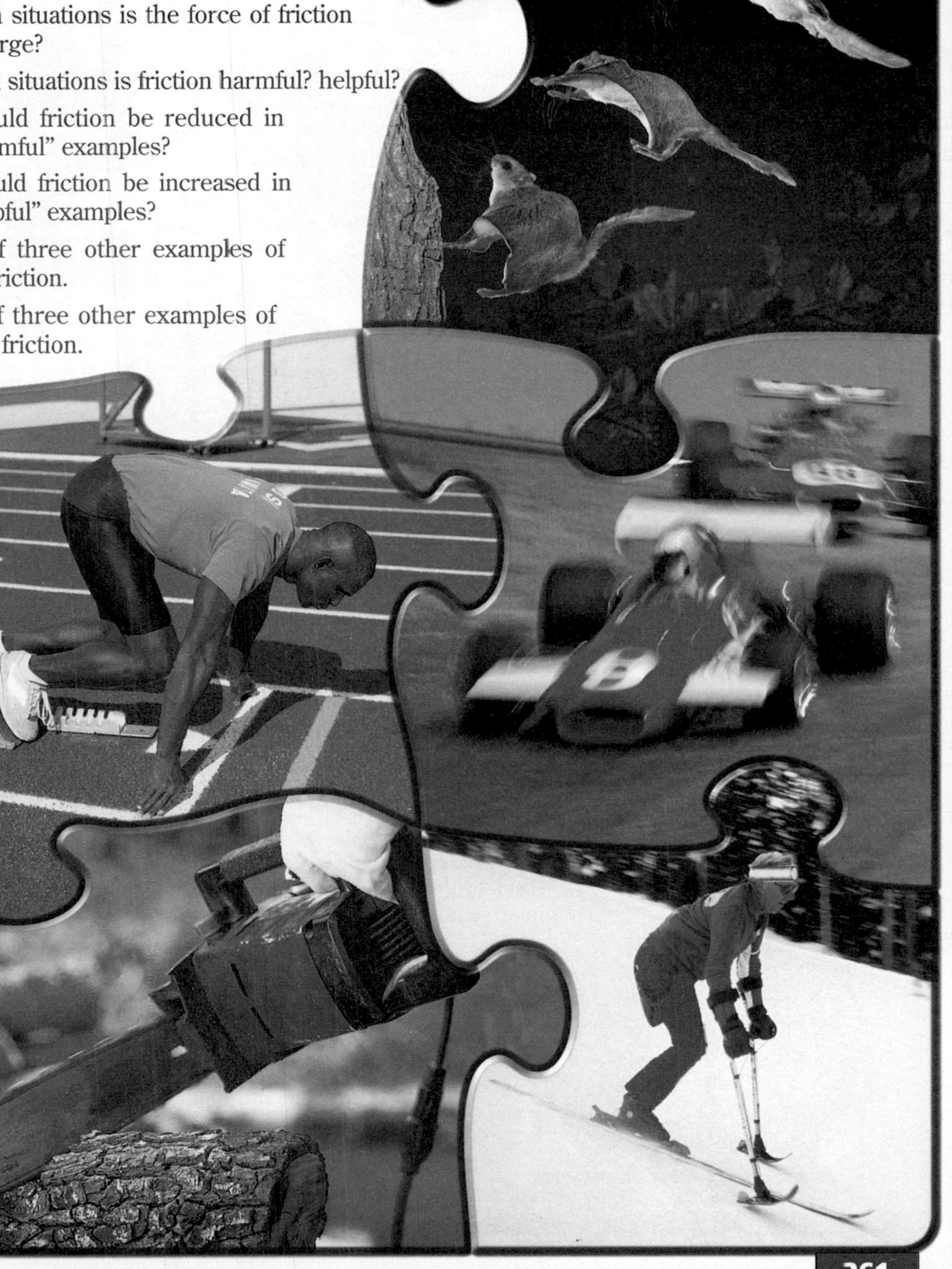

Answers to *Friction: Friend or Foe?*

The surfaces that are rubbing together and causing friction are the bottom of the shoes and the surface of the starting block and track, the squirrel's "wings" and the air, the tires and the pavement, the moving parts of the engine rubbing against each other, the air and the car's body, the snow and the ski and ski poles, and the saw and the log.

1. The force of friction is fairly small between the squirrel and the air and between the ski and the snow; it is fairly large between the shoes and the starting block and between the saw and the log.
2. Friction is undesirable between the ski and the snow and within the engine of the saw; it is helpful between the squirrel and the air and between the shoes and the track.
3. Oil could be added to the saw's engine, and wax could be added to the bottom of the ski.
4. The squirrel's "wings" could be bigger, and the shoes could have cleats.
5. Sample answers: a match and a matchbook, the grip on a tennis racket, and a nail file
6. Sample answers: a rug burn, skates on ice, on a water slide

★ **An Activity Worksheet (Teaching Resources, page 40) and Transparencies 42 and 43 are available to accompany the material on this page.**

FOLLOW-UP

Reteaching

Set up a small ramp. Give students marbles, dice, and modeling clay. Have them roll the marbles and dice down the ramp. Students will see that marbles roll down the ramp more easily than the dice do. Then challenge students to form different shapes with the modeling clay until they find the shape that rolls down the ramp with the least resistance. Students will discover that round balls roll the best.

Assessment

Show students a picture of someone riding a bicycle. Have them identify as many locations as they can where friction is harmful. Then have them identify where friction is helpful.

Extension

Have students investigate the parts of a car that depend on friction for safe operation and the parts of a car in which friction is undesirable. *(Tires, belts, and brakes depend on friction for safe operation. The clutch, treads, pistons, transmission, and bearings are worn down by undesirable friction.)*

Closure

As a class, discuss the following: During the course of a day, when is friction useful? When is it harmful? *(Useful: when walking, running, riding in a car, writing on the chalkboard; harmful: wears out shoes, tires, wheels, machine parts)*

Answers to *Challenge Your Thinking*

1. Student answers will vary. Possible answers include the following:
   Jeremy: . . . by sanding a board.
   Janice: . . . you look at them under a microscope.
   Joanne: . . . putting ball bearings between them.
   Jason: . . . create friction between the wheels of a car or bike and the ground.
   Jeremy: . . . friction can wear down moving objects such as tires.
2. Student answers will vary. Students should be able to support their arguments with logical reasons and clear examples.
3. **a.** Friction between the water and the boat slows down the boat; without this friction, the hovercraft can travel much faster than the boat.
   b. The graphite in the pencil lead acts as a lubricant.
   c. The wheel reduces friction because rolling objects experience much less friction than sliding objects. Were the wheels not somewhat roughened at the point of contact with the surface, however, they would spin in one spot instead of moving across the ground.
   d. Thermal panels prevent space shuttles from burning up in the intense heat caused by friction between the Earth's atmosphere and the space shuttle.
   e. Sliding down a rope can cause excessive heat due to friction between your hands and the rope.
   f. When two pieces of wood are rubbed together, the friction generates so much heat that the wood ignites.
   g. Fluid acts as a lubricant between the ball and the socket. Also, cartilage keeps the bones from rubbing against each other. Another ball-and-socket joint is the shoulder.

★ **You may wish to provide students with the Chapter 12 Review Worksheet on page 42 of the Unit 4 Teaching Resources booklet.**

1. Are You Interrupting?

After discussing the cause of friction, the following conversation was heard:

Jeremy: If you want less friction, make the surfaces smooth, such as . . .

Janice: Even ultrasmooth surfaces have friction. You can see this for yourself if . . .

Joanne: You can get less friction if you keep the surfaces apart by . . .

Jason: Yeah, but roughness actually helps us to . . .

Jeremy: But . . .

Too bad they kept interrupting each other! How would you complete what each person was about to say?

2. The Great Friction Debate

Choose a pro or con side for the following statement: Friction does more harm than good. Write your argument for the side you choose, and include convincing examples. Present your side in a class debate with someone who chose the other side.

Did You Know . . .

An object falling through the air will eventually reach a speed at which the force of gravity is balanced by air friction. This speed, called *terminal speed*, depends on the size and shape of the object. For instance, a sky diver with a closed parachute reaches a terminal speed of about 200 km/h, while a sky diver with an open parachute reaches only about 18 km/h.

Multicultural Extension

Early Uses of the Wheel

One of the most significant inventions in human history was intended to reduce friction: the wheel. Have interested students research the earliest known uses of the wheel. Which cultures developed the wheel? What did they use it for? How did other cultures adapt the wheel for their own special needs? Students should present their findings to the class.

3. Friction Facts

Do you know the following friction facts? Read through them, and answer the questions about them. Afterward, find two friction facts of your own.

a. A hovercraft is a type of vehicle that travels across land or water without touching the surface. Engines and propellers create a cushion of air that the vehicle "floats" on. A hovercraft can travel across water faster than a boat can. Why is this so?

A hovercraft

b. Rubbing the sliding surfaces of a sticky drawer with a pencil lead helps the drawer slide easier. Why?

c. Archeological discoveries suggest that the wheel was first used in Assyria around 3000 B.C. How does the wheel reduce problems with friction?

d. Insulation is a crucial part of a space shuttle. The 33,000 thermal panels on its outer surface protect it during re-entry into the Earth's atmosphere. Why are the panels so important?

e. Sliding quickly down a rope may burn your hands. Why?

f. An ancient method of lighting fires was to rub two pieces of wood together. How does this start a fire?

g. Some parts of the body are ball-and-socket joints, where the round part of one bone fits into the socket or cavity of an adjoining bone. How is friction reduced at these joints? (See the illustration for an idea.) Can you identify other ball-and-socket joints in your body?

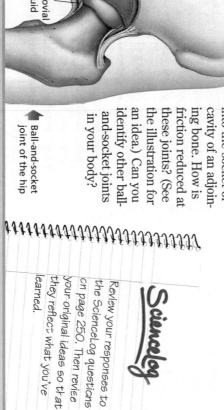

Soft cartilage

Synovial fluid

Ball-and-socket joint of the hip

ScienceLog

Review your responses to the ScienceLog questions on page 250. Then revise your original ideas so that they reflect what you've learned.

263

Did You Know . . .

In 1914, a 16-year-old pilot named Katherine Stinson became the first skywriter. She was also the first woman to fly an airplane in a loop and to have her own flying school.

The following are sample revised answers:

1. Friction occurs when surfaces rub against each other. Friction occurs between the following objects: the chain and the gears, the hands and the handlebars, the tires and the ground, and the feet and the pedals.
2. Friction can be reduced by smoothing opposing surfaces, lubricating the surfaces, or separating them with ball bearings. Friction can be increased by making a smooth surface rougher or by removing lubrication.
3. Friction can be helpful or harmful. For example, automobile brakes use friction to slow a car, but friction can produce unwanted heat, as when the space shuttle reenters the atmosphere.

More About Motion

Connecting to Other Chapters

Chapter 10
offers students hands-on exposure to a wide range of forces.

Chapter 11
allows students to measure the strength of forces with commercial and homemade force meters.

Chapter 12
explores the causes and consequences of friction in various situations.

Chapter 13
discusses the concept of inertia and introduces students to force pairs.

Prior Knowledge and Misconceptions

In addition to having students answer the questions on this page, you may wish to have the students answer the following additional questions. Emphasize that there are no right or wrong answers. (The answers are provided here in case you would like to discuss them with students after they have completed this exercise.) Collect the papers, but do not grade them. Instead, read them to find out what your students know about inertia, what misconceptions they have, and what is interesting to them about this topic.

1. When you are in a car that makes a sudden stop, your body continues to move forward against the seat belt. Why do you think this happens? *(Inertia keeps your body moving at a constant speed. Since the car stops, your body continues to move in relation to the car.)*
2. Some Amazonian hunters use especially long and heavy arrows to hunt in the dense jungle. Why are the arrows so heavy? *(Heavy arrows have greater inertia, so they are deflected less by branches and leaves.)*
3. If everyone on one side of the Earth jumped up in the air at the same time, what do you think would happen to the Earth? *(The Earth would experience an equal and opposite force. But because the Earth is so massive, its inertia would keep it from moving any noticeable amount.)*

CHAPTER 13

More About Motion

1 **Why does the bowling ball continue to roll down the lane after the bowler releases it?**

2 **If the caboose became uncoupled from this moving train, would it immediately stop or continue to move? Explain your reasoning.**

3 **If you push on a wall, does it push you back? Explain your reasoning.**

ScienceLog

Think about these questions for a moment, and answer them in your ScienceLog. When you've finished this chapter, you'll have the opportunity to revise your answers based on what you've learned.

264

Thought Experiments

Everyone is familiar with the motion of objects, but not everyone has the same explanation for how objects move through space or how forces are related to their motions. See what you can discover by conducting a survey.

Look at the questions in the survey below, and answer them on your own. Then ask at least two other people the same questions. How do your views differ from theirs? How are they the same? Ⓐ

A Survey

1. Is a force necessary to start a ball rolling on a level surface?
2. Once the ball is moving, must a force be exerted on the ball to keep it moving?
3. Is a force required to slow the ball down?
4. In the 1500s, Galileo wondered what kept the moon circling around the Earth. What explanation can you give for the moon's continuous orbit?
5. Once the Apollo spacecraft got away from the Earth on its path toward the moon, the astronauts shut off the rocket engines. Is a force necessary to keep the spacecraft going at over 3000 km/h?
6. The path of the Apollo spacecraft was straight; the path of the moon is oval. How would you explain this difference?

To find out whether the answers to your survey are correct, think about the following five thought experiments. Remember, scientists regularly perform thought experiments as they sort out their ideas and develop their hypotheses.

LESSON 1 ORGANIZER

Time Required
2 class periods

Process Skills
observing, inferring, analyzing

Theme Connection
Energy

New Terms
none

Materials (per student group)
none

Teaching Resources
Transparencies 44, 45, 46, and 47

Thought Experiments

FOCUS

Getting Started

Perform the following: Throw a tennis ball horizontally against the wall so that it bounces back toward you. Ask: What would happen if there were no gravity? *(The ball would bounce from wall to wall until air resistance and the loss of energy from hitting the walls finally stopped it.)* Explain that students have just carried out a thought experiment and that they will carry out several more during the lesson.

Main Ideas

1. If balanced forces are acting on an object, its motion will not change.
2. An unbalanced force causes an object to speed up, slow down, or change direction.

TEACHING STRATEGIES

Explain that it is not possible for scientists to perform certain experiments in the real world. For example, on Earth it is difficult to carry out experiments that show how objects move in the absence of gravity. Instead, scientists perform thought experiments.

Call on a volunteer to read aloud the first two paragraphs on page 265. Let students debate the answers to these questions in groups.

Answers to *In-Text Questions*

Ⓐ Students should identify how their views compare with those of the people they interviewed. For instance, students may agree with the other people on the first five questions, but disagree with them on the sixth.

Answers to A Survey are on the next page. ►

Answers to
A Survey, page 265

1. Yes, a force is necessary to start a ball rolling on a level surface.
2. No. However, if friction or gravity is opposing the ball's motion, a steady force that is equal to the opposing force must be exerted on the ball to keep it moving at a constant speed.
3. Yes. On Earth, the force of friction slows down a moving ball. The force of gravity slows down a ball that is moving upward. In outer space, another force would be required to slow down the ball.
4. The moon's motion tends to carry it away from the Earth tangentially, but the Earth's gravity pulls it back just enough to keep it circling the Earth. (You may wish to explain tangential motion to your students.)
5. Once the spacecraft got far enough away from the Earth, there was no friction and only minimal gravitational force opposing its motion. The force of the original thrust of the engines kept it going.
6. The path of the Apollo spacecraft was straight because it was moving directly away from the Earth—opposite to the pull of the Earth's gravity. The path of the moon is not straight because its motion tends to move it away from the Earth tangentially, but that motion is balanced by the Earth's gravitational pull, keeping the moon in orbit around the Earth at all times.

Thought Experiment 1

GUIDED PRACTICE Call on a volunteer to read aloud the first paragraph. Then direct students' attention to each of the lettered diagrams. Draw the diagrams on the board, and ask volunteers to indicate where they think the ball will stop rolling in each case.

Thought Experiment 1 demonstrates that, in addition to an object having a natural state of rest, a moving object has a natural state of steady motion when moving in a straight line.

Transparency 44 is available to accompany Thought Experiment 1.

Thought Experiment 1

More than 350 years ago, Galileo performed an experiment similar to the one below. For the purpose of his thought experiment, Galileo regarded the surfaces involved in his experiment as being so smooth that there was no friction between them. Of course, there are no surfaces like that in the real world.

Diagram a.

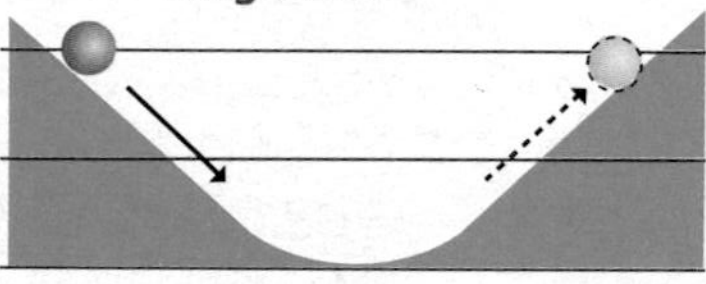

When a ball rolls down the slope on the left in diagram (a), we know that it rolls up the slope on the right to just about the same height—2 m.

Diagram b.

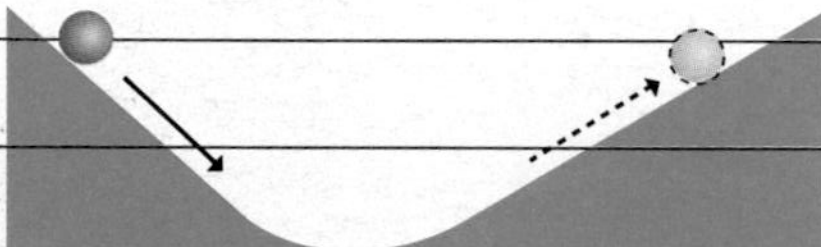

The ball does the same thing in diagram (b), but note how much farther it goes.

Diagram c.

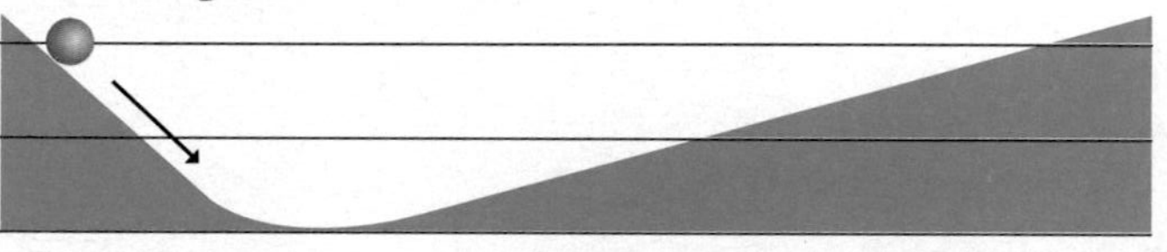

How far would you predict the ball would go in diagram (c)? diagram (d)? diagram (e)? Is any force necessary to keep the ball moving along the horizontal plane in diagram (e)? **A**

Diagram d.

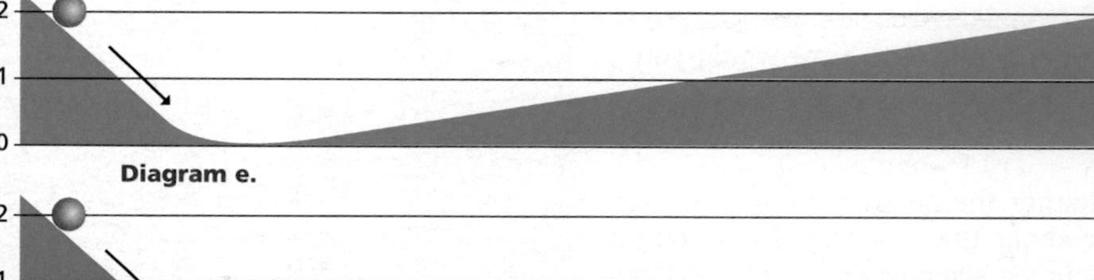

Diagram e.

Galileo said that once an object is moving, it tends to keep moving in a straight line without slowing down or speeding up. But if a force were applied to this moving object, the force could slow it down, speed it up, or change its direction.

266

Answer to
In-Text Question

A In diagrams (a)–(d), in the absence of friction, the ball will roll up to the same vertical height from which it started. In diagrams (b)–(d), as the slope becomes less steep, the ball will roll farther before it is pulled back by gravitation force. In diagram (e), the ball will roll forever on a level surface because there is no frictional force to oppose its motion. No force is necessary to keep the ball in motion.

Thought Experiment 2

Suppose Galileo lived in the twentieth century. He might think about the game of bowling. When a bowler places an initial force on the ball, the ball leaves his or her hand and moves away quickly. It meets an opposing force of friction. Diagrams (a) through (e) show what is likely to happen. Of course, how far the ball goes depends on the surface on which it is moving.

1. In your ScienceLog, match each diagram with the appropriate surface: grass, a concrete road, a frictionless surface, a bowling alley of indefinite length, and living room carpet.
2. If there were no opposing force on the moving ball, what would happen to its motion?

Diagram a. Surface ___?___ (20 m)

Diagram b. Surface ___?___ (50 m)

Diagram c. Surface ___?___

Diagram d. Surface ___?___

Diagram e. Surface ___?___

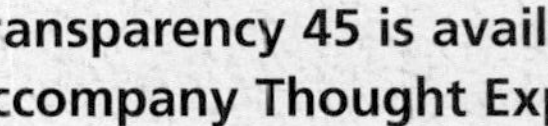

Multicultural Extension

Mae C. Jemison

Share the following information with students: Mae C. Jemison was the first African American female astronaut at NASA. Jemison graduated from Stanford University with a double major in chemical engineering and African/Afro-American studies. She earned a medical degree from Cornell University and qualified as a space-shuttle mission specialist in 1988. In 1992, Dr. Jemison spent a week aboard the space shuttle *Endeavor.*

Homework

As a homework activity, you may wish to assign questions 1 and 2 on this page. The next day, have students work in groups of 2 or 3 to evaluate and improve their answers.

Thought Experiment 2

Call on a volunteer to read aloud the first paragraph on page 267. Suggest that students work individually to answer questions 1 and 2 in their ScienceLog. After a specified amount of time, involve the class in a discussion of their diagrams and what can be inferred from them.

INDEPENDENT PRACTICE Challenge students to add diagrams similar to the ones on this page in their ScienceLog, using surfaces other than those listed in question 1. Make sure you approve the surfaces students select before they proceed. Students may need to experiment with different surfaces to discover the relative amount of friction that each surface tends to exert.

★ **Transparency 45 is available to accompany Thought Experiment 2.**

Answers to *Thought Experiment 2*

1. Surfaces represented by the diagrams include the following:
   a. grass
   b. living room carpet
   c. concrete road
   d. bowling alley
   e. frictionless surface
2. If there were no opposing force on the moving ball, it would keep rolling in a straight line at the same speed forever.

Meeting Individual Needs

Learners Having Difficulty

In order to illustrate the effect of forces on motion, students can have a car race with a partner using small toy cars, two wooden ramps, and materials such as oil, foil, cloth, wax paper, and sandpaper. Ramps should be propped up with books so that they are at the same angle. Have students find out how slow they can make the cars move—the slowest car wins. Students should discuss strategies with their partners. Then they should list the things that made the cars go slower. In a second race, students should attempt to make their cars move as quickly as possible. Students should make a list of what causes the car to move quickly.

Answers to *Thought Experiment 3*

1. In diagram (a), the Earth's gravity stops the ball from going higher.
2. In diagram (b), the ball can be thrown higher because the gravitational force on the moon is only one-sixth of that on Earth.
3. A ball thrown off a spacecraft might be too far from Earth to be affected by its gravity and therefore would drift until another force affected it.

Answers to *Questions*

1. The forces that start the objects moving are gravitational force in Thought Experiment 1, a push by the bowler's hand in Thought Experiment 2, and a push by Tara's hand in Thought Experiment 3.
2. The forces that started the objects in motion were removed when the ball reached a level surface in Thought Experiment 1, when the ball left the bowler's hand in Thought Experiment 2, and when the ball left Tara's hand in Thought Experiment 3.
3. Forces changed the speed or direction of an object in motion in all but the last diagram in each thought experiment.
4. The opposing forces are gravitational force in Thought Experiments 1 and 3 and frictional force in Thought Experiment 2.

Theme Connection

Energy

Focus question: In a soccer game, will the goalie receive as much energy from a moving soccer ball as the player used when kicking the ball? Why or why not? *(No. Energy is lost due to air friction and ground friction as the ball moves toward the goalie. Also, some of the player's energy was turned into heat when the player's foot kicked the ball.)*

Thought Experiment 3

Recall what you learned earlier about gravitational force.

a Earth's surface

b Moon's surface

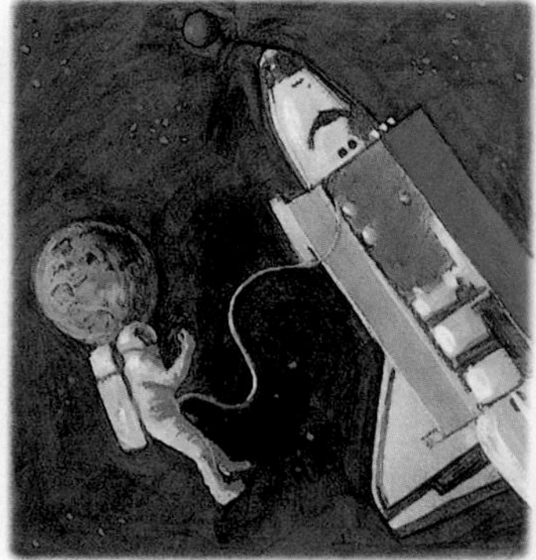

c Space

1. In diagram (a), what stops Tara from throwing the ball higher?
2. Why does the ball go much higher in diagram (b)?
3. What will happen to the ball in diagram (c)?

Conclusions

If you have been working through these thought experiments carefully, some conclusions should already be clear. Check the conclusions shown here. Then record your answers to the accompanying questions in your ScienceLog.

Conclusions	Questions
A force is needed to start an object moving.	1. What force starts the object moving in each thought experiment?
An object may keep moving when this force is no longer exerted on the object.	2. In each thought experiment, when was the force that started the object in motion removed from the object?
If there are no forces on the moving object, it will keep moving in a straight line with steady motion.	3. Which drawings in each thought experiment show that a force is needed to change the direction or the speed of motion?
Objects in motion may be slowed down by opposing forces.	4. What are the opposing forces in each of the thought experiments?

(100 m)

268

Thought Experiment 4

1. You are holding a weight, as in diagram (a). Why doesn't the weight fall?
2. In diagram (b), the added weight, called weight B, takes the place of your hand. Weight A is still not moving. Why?
3. What happens to weight A in diagram (c)? Why?
4. What happens to weight A in diagram (d)? Why?

Diagram a.

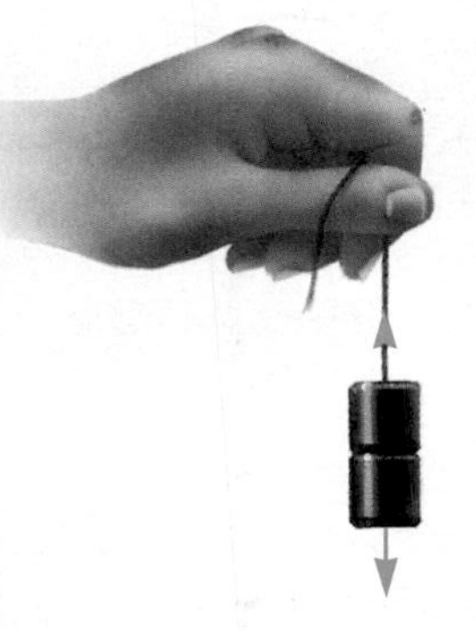

Diagram b.

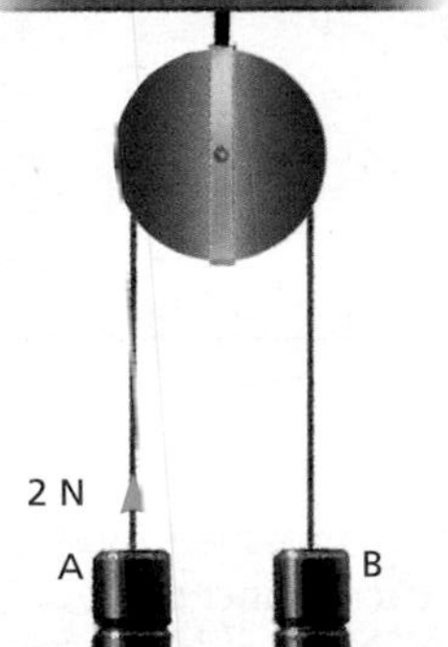

Diagram c.

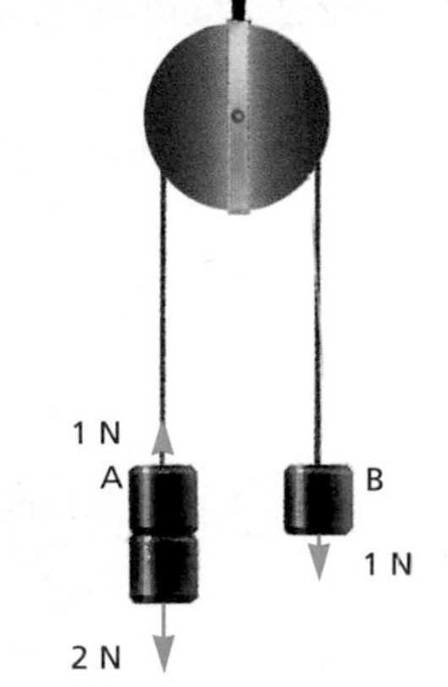

Diagram d.

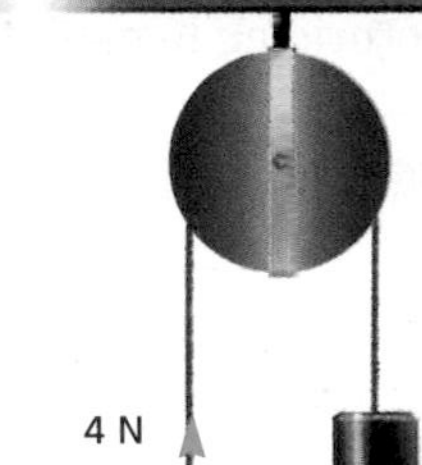

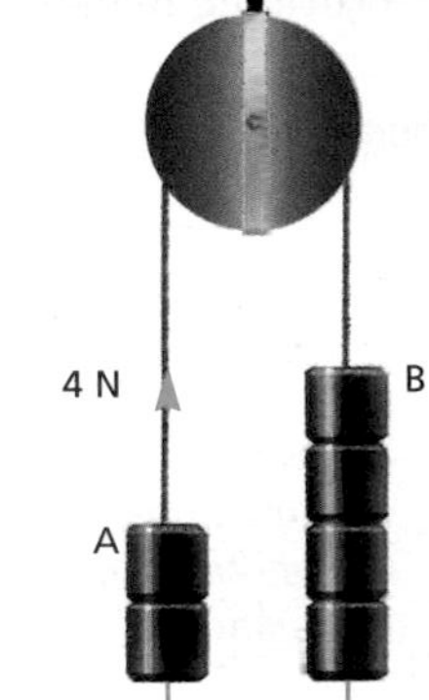

More Conclusions

Have you drawn more conclusions about the thought experiments that are similar to the ones shown here? Answer the accompanying questions in your ScienceLog.

Conclusions	Questions
If forces on an object at rest are balanced, the object stays at rest. It is just as if there were no forces on the object.	1. Do you remember Don's attempt to move the refrigerator? Why did he have so much trouble moving it?
An unbalanced force causes an object at rest to start moving.	2. In which parts of Thought Experiment 4 is there an unbalanced force?

269

Thought Experiment 4

Ask students to read this page silently. Suggest that they look carefully at the diagrams to answer the questions. After a specified amount of time, involve the class in a discussion of their answers. Be sure they understand the meaning of the term *unbalanced*.

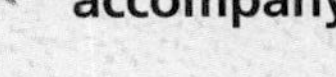

Transparency 46 is available to accompany Thought Experiment 4.

Answers to *Thought Experiment 4*

1. The weight does not fall because the upward force exerted by the hand balances the downward gravitational force.
2. In diagram (b), the upward force exerted on *A* by *B* (*B*'s weight) balances the downward gravitational force on *A* (*A*'s weight). *A* and *B* are equal in weight; therefore, *A* does not move.
3. In diagram (c), *A* moves downward because the gravitational force acting on it (*A*'s weight) is greater than the upward force acting on it (*B*'s weight).
4. In diagram (d), *A* moves upward because the force pulling up on *A* (*B*'s weight) is greater than the downward gravitational force acting on *A* (*A*'s weight).

Answers to *Questions*

1. Don was unable to move the refrigerator alone because the force he exerted on it was less than or equal to the force of friction. Since his efforts were balanced by friction, the refrigerator remained at rest.
2. In Thought Experiment 4, there is an unbalanced force in diagrams (c) and (d). In each case, *A* moves due to this unbalanced force.

Integrating the Sciences

Life and Physical Sciences

You feel colder when the wind is blowing than when the air is still. This is because, although your body heats up the air molecules around you, air that is moving past you carries these heated molecules away. This process results in a *wind chill*.

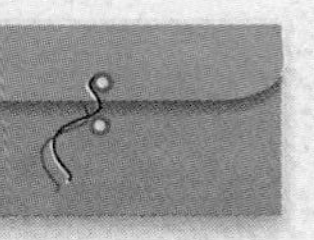

PORTFOLIO

Suggest that students include their answers to the thought-experiment questions in their Portfolio. The tables showing conclusions and questions are useful self-evaluative devices that students may incorporate into their Portfolio.

Answers to *Thought Experiment 5*

1. The frictional force of the air and of the road surface on the tires would slow down the car.
2. The size of the force provided by the car's motor is the same as the size of the opposing frictional force on the car. Therefore, all forces on the car are balanced, and the car moves at constant speed in a straight line.
3. If the speed of the car is steadily increasing, the force provided by the motor is greater than the opposing force of friction. The forces on the car are unbalanced.

Transparency 47 is available to accompany Thought Experiment 5.

Answers to *Questions*

1. In Thought Experiment 5, diagram (b) shows balanced forces.
2. Diagrams (a) and (c) show that an unbalanced force causes an object to change speed. In (a), the force of friction slows down the car. In (c), there is a force provided by the car's motor in the forward direction. This forward force is greater than the force of friction. Therefore, the car speeds up. Unbalanced forces can also cause an object to change direction.

Thought Experiment 5

(The solid arrows refer to forces. The dotted arrows refer to the speed of the car.)

1. A car is moving at 70 km/h. If the driver lets the car coast freely in a straight line until it stops, what would slow it down?

Diagram a.

2. Now, instead of allowing the car to coast, the driver presses the gas pedal just enough to maintain the car's speed at 70 km/h. What is the size of the force that the car's motor is providing to move the wheels along the road?

Diagram b.

3. Suppose that the speed of the car is steadily increasing. How do you think the force of friction and the force provided by the motor compare? Are the forces on the car balanced or unbalanced?

Diagram c.

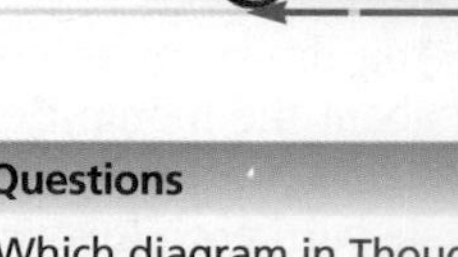

Final Conclusions

Here are some final conclusions that can be drawn from the thought experiments. Answer the accompanying questions in your ScienceLog.

Conclusions	Questions
If forces on an object in motion are balanced, the object will stay at rest or in steady motion in a straight line.	1. Which diagram in Thought Experiment 5 shows balanced forces?
An unbalanced force causes an object to speed up or slow down.	2. Which diagrams in Thought Experiment 5 show this? What else can an unbalanced force do to an object in motion?

270

FOLLOW-UP

Reteaching

Use an equal-arm balance to help students distinguish between balanced and unbalanced forces. Begin by placing a heavy mass on the balance so that one arm drops. Explain that this resembles an unbalanced force because the arm moved. Ask: What caused the arm to start moving? *(Gravity acting on the mass)* Then place exactly enough mass on the other arm to balance the two sides. Explain that this represents balanced forces because the arms stay still. Use other combinations of masses to help clear up any confusion that students may have.

Assessment

Ask students to explain how Galileo concluded that a ball will roll forever if there is no force to cause it to speed up or slow down. *(If a ball rolls down an incline on the left, it tends to roll up an equivalent height on the right. The ball rolls a greater distance as the angle of the incline on the right is reduced. If the plane on the right is horizontal, it will roll forever.)*

Extension

Challenge students to create some force and motion thought experiments of their own. As a class, discuss one or two of these in detail and draw conclusions about class discoveries.

Closure

Have students imagine that they are standing on a bathroom scale inside an elevator. Ask: What happens to your weight as the elevator

- starts up? *(Increases)*
- moves at a constant speed? *(Stays constant)*
- moves up and stops? *(Decreases)*
- starts down? *(Decreases)*
- moves down and stops? *(Increases)*

Puffball: A Game of Force and Motion

Here is a game you can play in class or on your own. As in any game, there are forces involved. The object of this game is to blow a light ball into your opponent's goal. The straws cannot touch the ball, and your hands cannot touch the table.

You Will Need

- a playing area of approximately 1 m × 1.5 m
- a barrier about 4 cm high around the edges of the playing area
- 2 goal areas
- a plastic-foam or table-tennis ball
- drinking straws
- 2 players (or teams of 2 or 3 players)

Play for 15 minutes, and then write down all that you have observed about the effects of two or more forces on the motion of a light ball. Ⓐ

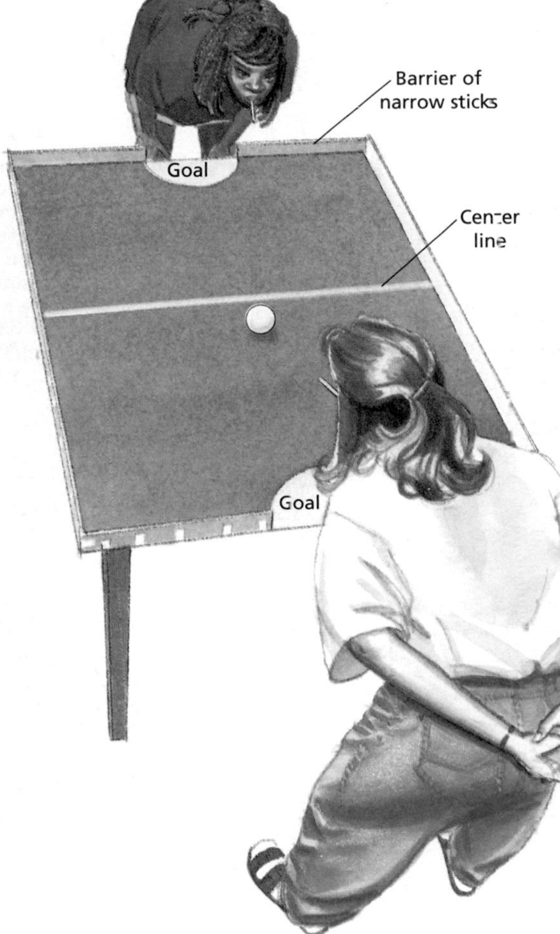

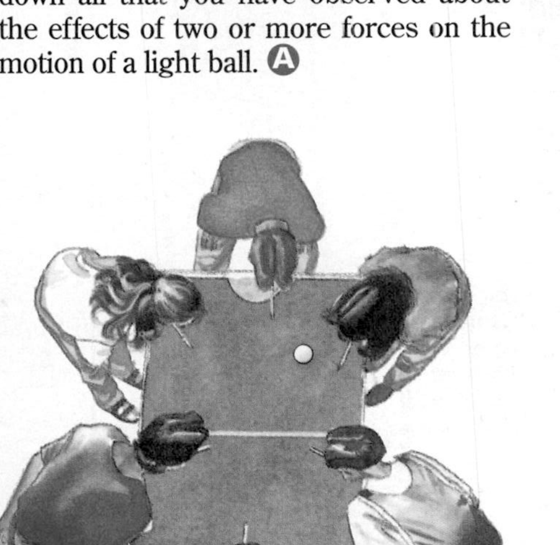

LESSON 2 ORGANIZER

Time Required
1 to 2 class periods

Process Skills
predicting, analyzing

New Terms
none

Materials (per student group)
In-Text Activity: 1 m × 1.5 m playing area with barrier about 4 cm high and 2 goal areas; plastic-foam or table-tennis ball; 4–6 plastic drinking straws (additional teacher materials: roll of masking tape)

Teaching Resources
none

Puffball: A Game of Force and Motion

FOCUS

Getting Started

Take students outside to play tug of war. Caution them against rough play so that no one gets hurt. Students should identify how opposing forces affect the motion of the rope. At first, have two students play against each other. Then, place five students on one end of the rope and one student on the other end of the rope and play again. Try different combinations. Explain to students that in this lesson they will evaluate how different forces applied at the same time can affect the motion of an object.

Main Ideas

1. When balanced forces act on an object at rest, it will remain at rest.
2. An object in motion with balanced forces acting on it will continue in a steady, straight motion.
3. An unbalanced force will cause an object at rest to start moving and an object in motion to speed up, slow down, or change direction.

TEACHING STRATEGIES

GUIDED PRACTICE Before play begins, have the class establish some rules, such as the following: straws cannot touch the ball; keep your hands behind your back; and do not touch the table with your body. Tell students to analyze the forces that are changing the motion of the ball as they play.

Answer to *In-Text Question*

Ⓐ Students will likely observe that when the forces are equal and in opposite directions, there is no effect on the ball. Otherwise, the ball moves or changes direction.

INDEPENDENT PRACTICE After students play the game for a while, ask them to record everything that they have observed about the effects of the forces on the ball. Then involve students in a discussion of what they have discovered about the ball's motion when one or more forces act on it. The game and your discussion set the stage for the second part of the exercise: the analysis of the forces being exerted on the ball and of its resulting motion in different game situations.

Review the two sample game situations on page 272. Make sure students understand that they should draw diagrams in their ScienceLog to represent the situations shown on pages 273 and 274. Instruct them to draw arrows to show the size and direction of the forces acting on the ball in each situation. Remind students that they need to predict who will score. When more than one possible outcome exists, students may choose the winner. Point out to students that the dotted arrows in the diagrams indicate that the ball is already moving when the forces described are exerted.

SAFETY ALERT Be alert for the possibility that students may hyperventilate. The symptoms to watch for include headache, dizziness, and fainting. If students show these symptoms, have them sit quietly with their head between their knees and breathe normally until the symptoms subside. If a student faints, allow the victim to remain where he or she is, and call the school nurse. Contact the school nurse in advance for more information and so that he or she knows that this activity will be taking place.

Meeting Individual Needs

Second-Language Learners

Suggest that students act as television sportscasters and report on the game of puffball on the day after the activity. In a class presentation, students should recap the action in their native language and supplement their commentary with diagrams. The diagrams should have bilingual labels. Place minimal emphasis on mastery of English terms and maximum emphasis on the students' understanding of the diagrams.

Next, work through the situations on pages 273 and 274. A dotted arrow beside the ball indicates the direction in which the ball is moving before Melanie or Pauline begins to blow. No dotted arrow means that the ball is at rest. The direction of the straw in Melanie's or Pauline's mouth indicates the direction of the force.

Your first task is to show the size and direction of the forces. Make diagrams in your ScienceLog, and use solid arrows to show the forces that Melanie and Pauline exert on the ball. Then predict whether a point is going to be scored and, if so, who will score it. In some cases, *you* have to set the conditions and decide the outcome. Two examples have been given to help get you started.

Example 1
The ball is at rest. Pauline is blowing the ball toward Melanie's goal.

It will go in. Pauline gets a point.

Example 2
The ball is at rest. Melanie and Pauline are both exerting a force. You have to decide who is exerting the greater force. Where does the ball go? Who scores the point?

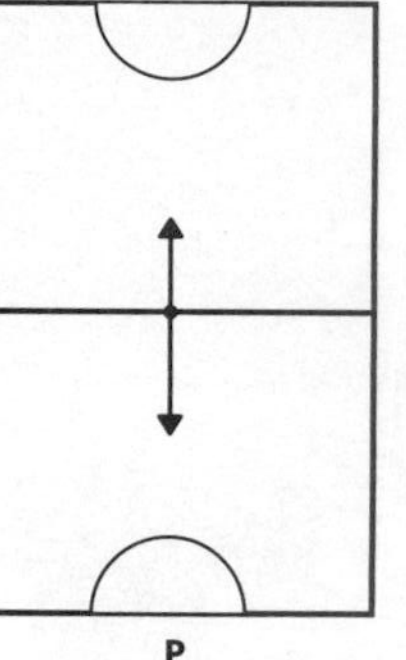

If you decide that Melanie exerts a greater force, the ball will go in Pauline's goal. Melanie scores.

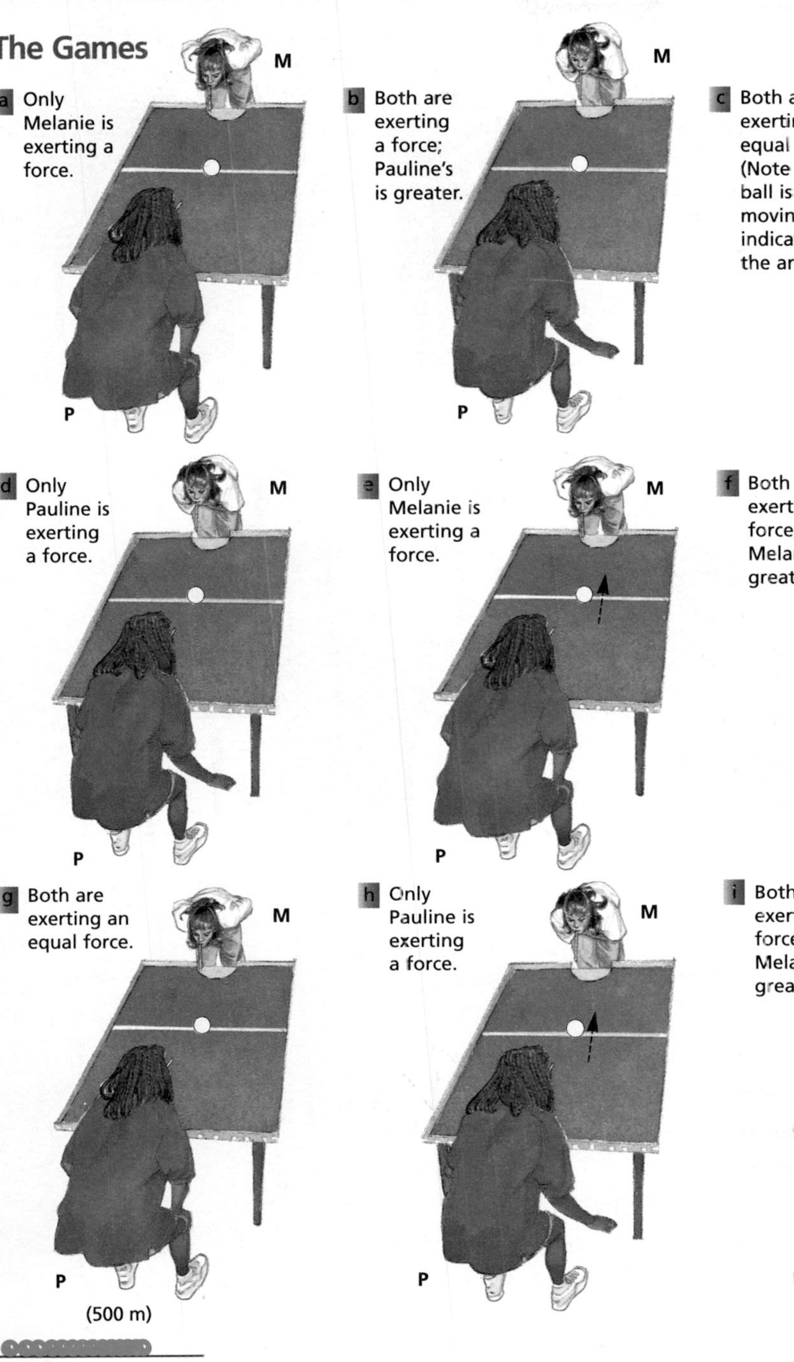
The Games
a Only Melanie is exerting a force.
b Both are exerting a force; Pauline's is greater.
c Both are exerting an equal force. (Note that the ball is initially moving as indicated by the arrow.)
d Only Pauline is exerting a force.
e Only Melanie is exerting a force.
f Both are exerting a force; Melanie's is greater.
g Both are exerting an equal force.
h Only Pauline is exerting a force.
i Both are exerting a force; Melanie's is greater.
M
P
(500 m)

Answers to *The Games*

a. With a single force in the direction of Pauline, the ball moves into Pauline's goal. Melanie scores.

b. Pauline's greater force causes the ball to move in the direction of Melanie, so the ball moves into Melanie's goal. Pauline scores.

c. The girls are exerting equal and opposite forces, but the ball is already moving toward Melanie, so it continues to move in that direction at the same speed. Pauline scores.

d. With a single force in the direction of Melanie, the ball moves into Melanie's goal. Pauline scores.

e. With a single force in the direction of Pauline, the ball, which is moving toward Melanie, slows down. The ball may go into Melanie's goal anyway, it may slow down, or, if Melanie blows hard enough, the ball will stop and then move in the opposite direction into Pauline's goal.

f. Both girls are exerting a force, but Melanie's is greater. However, the ball is already moving toward Melanie's goal. Melanie's force may not be great enough to reverse the ball's direction, and Pauline may score. If Melanie's force is great enough to stop the ball and turn it around, it will go into Pauline's goal.

g. Both girls are exerting an equal force, so the ball remains at rest.

h. With a single force in the direction of Melanie, the ball, which is already moving toward Melanie, speeds up and goes into Melanie's goal. Pauline scores.

i. Both girls are exerting a force, but Melanie's is greater. The ball is already moving toward Pauline. Therefore, the ball speeds up as it moves toward Pauline and goes into Pauline's goal. Melanie scores.

Answers to The Games continued ►

Homework

Have students describe in detail two different puffball scenarios in which more than two players are exerting forces on the ball. Students should explain whether the forces are balanced or unbalanced and give the result of each scenario.

ENVIRONMENTAL FOCUS

Asbestos has been used as a material in many items such as brake pads because it is an excellent absorber of heat. Research has shown, however, that asbestos fibers can cause lung cancer if they are inhaled. Unfortunately, there is no satisfactory substitute for asbestos as a friction material. The work area where people repair and maintain asbestos-containing friction materials is controlled to limit the exposure of workers to asbestos fibers.

FOLLOW-UP

Reteaching

Provide students with a croquet ball and a rubber mallet. Ask them to predict in which direction a moving ball will move after it is hit with the mallet. *(Unless it is hit in the same direction as its motion, it does not travel in the direction it is hit, but at an angle between its original direction of motion and the angle at which it was hit.)*

Assessment

Have two volunteers play a game of puffball. As they do, call on various students to give a running commentary on the forces that are acting on the ball. (See the Safety Alert on page 272 to help you identify the symptoms of hyperventilation.)

Extension

Have interested students do research on different types of jet engines or rocket engines. Students may present their findings in a report to the class.

Closure

Use a hair dryer pointed directly upward to suspend a table-tennis ball in the air. Have students explain the forces acting on the ball. *(The downward force of gravity is balanced by the upward force of the moving air.)*

Answers to *The Games, continued*

j. Even though the girls are exerting equal forces, they are not opposite each other at the table. They are at right angles. The ball moves diagonally across the table. No one scores a point.

k. Since the ball is moving steadily toward Pauline and no additional forces are being exerted, the ball goes into Pauline's goal. Melanie scores a point.

l. The forces being exerted by the girls are equal and opposite, so the ball continues to move to the left at a steady speed. No one scores a point.

m. The ball is moving across the table when Melanie blows on it, so it would move diagonally toward Pauline's corner. No one scores a point.

n. The ball remains at rest because there are no forces acting on it. No one scores a point.

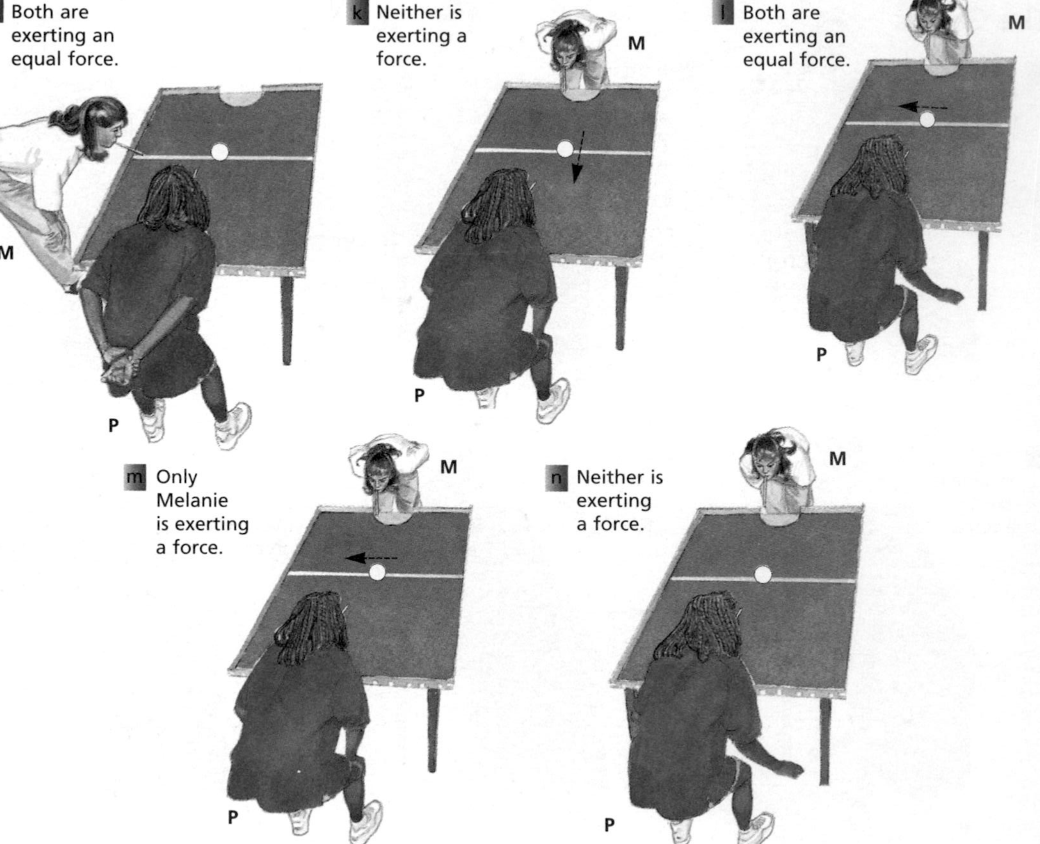
j Both are exerting an equal force.
k Neither is exerting a force.
l Both are exerting an equal force.
m Only Melanie is exerting a force.
n Neither is exerting a force.
M
P

Questions

Check how well you have understood the forces and motion in the puffball game by answering the following questions.

In which of situations (a) through (n) . . .

1. are there essentially no forces on the puffball?
2. are there single forces?
3. are there balanced forces?
4. are there unbalanced forces?
5. is the ball at rest set into motion? Why?
6. does the ball in motion remain in constant motion? Why?
7. does the ball, already in motion, slow down? Why?
8. does the ball, already in motion, speed up? Why?
9. does the ball, already in motion, change direction? Why?
10. does the ball at rest stay at rest? Why?

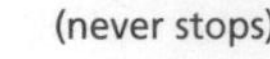

(never stops) →

274

Answers to *Questions*

1. k, n
2. a, d, e, h, m
3. c, g, l
4. b, f, i, j
5. a, b, d, j (A single or unbalanced force acts on the ball at rest.)
6. c, k, l (There are no forces on the ball, or the forces on the ball are balanced.)
7. e, f (There is a single force, or unbalanced forces oppose the motion.)
8. h, i (There is a single force or an unbalanced force in the direction of motion.)
9. e, f, m (There is a single force not in the direction of the motion.)
10. g, n (No forces or balanced forces are acting on the ball at rest.)

What Is Inertia?

You have seen that objects at rest tend to stay at rest. Objects in motion tend to continue in motion, maintaining a constant speed in a straight line. How are these ideas demonstrated in this comic strip? (A)

HORSE AND MOTION — N.R. SHAW

Giddyup, we've got to get to the creek!

HEY!

We're not far from the creek now.

Whoa! We're here!

WHOAAAA!

This tendency is called **inertia**. Inertia is *not* a force. It is a quality or characteristic of an object that causes it to resist any change in its state of rest or motion. Inertia is due to the object's mass. The greater the mass, the greater the inertia—that is, the greater the tendency of an object to remain as it is and the greater the force needed to alter the object's state.

LESSON 3 ORGANIZER

Time Required
2 class periods

Process Skills
predicting, observing, analyzing

New Term
Inertia—the tendency of an object to remain as it is; objects at rest tend to remain at rest, while objects in motion tend to remain in motion at a constant speed and in a straight line

Materials (per student group)
Exploration 1, Station 1: metric ruler; 30 cm × 10 cm piece of card board; 6 dominoes; **Station 2:** coin; 20 cm × 4 cm piece of thick paper; **Station 3:** raw egg; hard-boiled egg; **Station 4:** thick section of newspaper; meter stick; 60 cm × 5 cm piece of thin wood; safety goggles; **Station 5:** small toy cart with removable passenger; 20 cm of thread; board, about 1.5 m in length; 5–10 books

Teaching Resources
Exploration Worksheets, pp. 48 and 51
SourceBook, p. S79

What Is Inertia?

FOCUS

Getting Started

Ask a few volunteers to spin around for a minute or two and then to describe how they feel. *(Dizzy, or as though they are still moving)* Explain that they feel like they are still moving because of inertia. Humans sense motion by the movement of fluid in their inner ears. When the students stop spinning, their inner ear continues to send information to their brain as if they were still spinning, giving the impression of continued motion.

Main Ideas

1. Inertia is the tendency of moving objects to continue in the same direction at the same speed and of resting objects to remain at rest.
2. The amount of an object's inertia depends on its mass.

TEACHING STRATEGIES

Ask: Do you think the horse's unwillingness to move depends on its mass? *(No)* Explain to students that the word *inertia* can be used to describe a mental state—in this case, to describe the stubbornness of the horse. The rider flying over the head of the horse shows the scientific sense of the word. In this case, inertia does depend on the rider's mass. (Point out that the name of the cartoon's author—N. R. Shaw—is a play on the word *inertia*.)

Answer to *In-Text Question*

(A) The horse resists moving at first. When it does move, the rider is nearly thrown off because he tends to stay where he is. At the end, the horse stops suddenly, and the rider tends to stay in motion, flying forward over the head of the horse.

Have You Met Inertia Before?

Call on a volunteer to read page 276. Direct students' attention to the diagrams of the bus on this page. Involve students in a discussion of where they have encountered inertia before. Some possibilities include the following:

- When hurtling around in a roller coaster, your body is thrown from side to side when the roller coaster makes sudden turns, and you are pushed up when the roller coaster suddenly plunges downward.
- When a washing machine spins the water out of clothes, the water flies off in a straight line rather than taking the circular path that the clothes are forced to take.
- When you trip, your foot stops because it hits something, but your body continues moving forward.

Ask students to share some of their experiences of standing up in buses, trains, or subways. Then involve the class in a discussion of the questions that are asked on this page.

Answers to *In-Text Questions*

A Holding onto something would help because it would help put your body in motion along with the bus.

B Your body's tendency to keep moving forward makes you keep moving relative to the bus when the bus stops.

C If a car makes a sharp right-hand turn, you will tend to keep moving in a straight line in relation to the road and to the left in relation to the car. For other examples of inertia, see above.

D Yes. Your body reacts differently to different speeds and turns.

Have You Met Inertia Before?

You are standing unsupported in a parked bus. The bus starts moving forward. What happens to you? What is inertia doing to you? Would holding on to something help? **A**

You were at rest, so you tended to stay at rest. Therefore, as the bus moved forward, you tended to stay where you were. The result was that you fell backward.

Now the bus is moving along. You are standing in the aisle unsupported and are keeping your balance quite easily. Suddenly, the bus driver applies the brakes, slowing down and stopping the bus. What effect does inertia have on you? **B**

You were in motion. When the bus slowed down, you tended to continue forward. As a result, you fell forward.

Suppose you are sitting in the back seat of a car being driven at 80 km/h. Without warning and with tires squealing, the driver swerves around a sharp corner. What happens to you? In what direction do you tend to go in relation to the car? in relation to the road? Can you think of other situations in which you have experienced the effects of inertia? **C**

What does inertia feel like? Try devising some experiments to help you "feel" inertia. You may use devices such as a wagon, a chair with wheels, a bicycle, a skateboard, or a seesaw. For example, try sitting blindfolded on a wagon. Have a couple of friends pull the wagon. Can your inertia tell you when they are starting, speeding up, slowing down, or stopping? **D**

What is momentum? How is it related to inertia? Turn to pages S82–S83 in the SourceBook to find out!

276

Meeting Individual Needs

Learners Having Difficulty

Have students work in groups to design some of their own demonstrations of inertia. If you have a cart, skateboard, or small wagon available, one group could try the experiment suggested in the text. When all of the groups have finished their experiments, have them perform a demonstration for the class.

Warn students to be careful not to hurt anyone with abrupt changes in motion.

EXPLORATION 1

Experiencing Inertia

The five mini-experiments in this Exploration provide other examples of inertia. At each station you will find materials arranged as shown here.

Spend as much time as you need at each station. Then explain how inertia is working in each case. Record your observations and explanations of inertia in your ScienceLog.

STATION 1

Use a ruler to hit the bottom domino sharply toward the barrier. What happens if you hit it gently? Why? Can you reduce the pile completely, one domino at a time? Explain your findings.

STATION 2

Place a coin on a piece of thick paper. Then place them both on the edge of a desk. Remove the paper without disturbing the coin. How can you explain this trick?

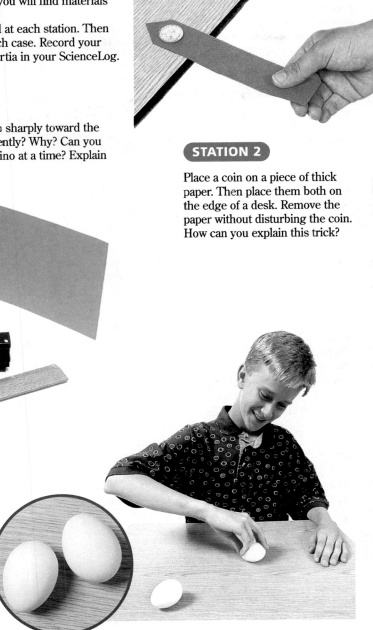

STATION 3

Which egg is hard-boiled? Spin each egg on a smooth surface. (Be careful not to break either egg!) How does inertia help you decide which egg is hard-boiled and which is not?

Exploration 1 continued ►

EXPLORATION 1

Set up five stations around the room. If possible, have two sets of equipment at each station. (You may wish to have a student bring in a small toy cart with a removable passenger from home.) Divide the class into groups of three or four. Ask each of the groups to rotate from one station to the next. When all of the groups have finished making their observations, reassemble the students to discuss their observations.

Cooperative Learning
EXPLORATION 1

Group size: 3 to 4 students
Group goal: to explore the phenomenon of inertia in five mini-experiments
Positive interdependence: Assign students roles such as primary investigator, recorder, timekeeper, and focuser.
Individual accountability: Students should provide written responses to the in-text questions, based on the activities at each station.

★ **An Exploration Worksheet is available to accompany Exploration 1 (Teaching Resources, page 48).**

Answers to *Station 1*

When a strong force is applied, the frictional bonds between the bottom domino and the others are broken. The bottom domino moves, while the others remain where they are because of inertia. When a smaller force is applied, the frictional forces are not overcome, and more than one domino moves. Inertia is resistance to force. The greater the mass, the greater the inertia, or resistance to force. Therefore, when there are fewer dominoes, their inertia is more easily overcome. It should be possible to reduce the pile completely, one domino at a time, because it is possible to move the bottom domino independently of any others above it.

Answers to *Station 2*

If the paper is pulled quickly from under the coin, the coin should stay in place. Its resistance to motion is its inertia.

Answers to *Station 3*

Even though the masses of the two eggs are roughly the same, the hard-boiled egg spins more easily than the raw egg. When spun, the shell and solid contents of the hard-boiled egg respond as one inertial mass. The raw egg, however, has two inertial masses: the shell and its soft contents. The resulting friction between the two slows down the rotation of the raw egg. (If the raw egg is stopped suddenly and quickly released, it will begin spinning again because the soft contents will remain in motion, causing the shell to start spinning again.)

Answers to *Station 4*

Stick *B* will break if it is hit with enough force. The inertia of the newspaper and the air above it holds the stick in place. Therefore, the stick will break rather than move when it is hit sharply.

Answers to *Station 5*

The toy passenger continues to move forward when the cart hits the books because the passenger's inertia tends to keep it in motion. The higher the ramp, the faster the cart will be moving when it hits the books, and the farther the passenger will be thrown. A seat belt made from a wider material, such as a piece of ribbon, would be better than using a piece of thread. The restraining force of the ribbon on the passenger would be spread over a wider area, so the passenger would be less likely to be thrown out. (Use this experiment to reinforce the importance of wearing seat belts in cars.)

Meeting Individual Needs

Gifted Learners

Have gifted students use what they have learned about inertia to speculate about the Earth's inertia. Ask students: Does the Earth have inertia? Does inertia affect the Earth's rotation? What about its revolution? Finally, challenge students to build a model of the Earth that demonstrates this effect of inertia.

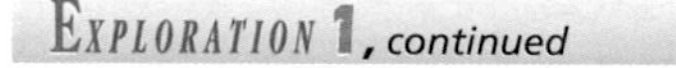

EXPLORATION 1, *continued*

STATION 4

Raise the newspaper by pushing stick *A* gently against stick *B*. Repeat, this time striking stick *B* sharply with stick *A*. Does stick *B* break? Why? (You can use a meter stick for stick *A*, but not for stick *B*!)

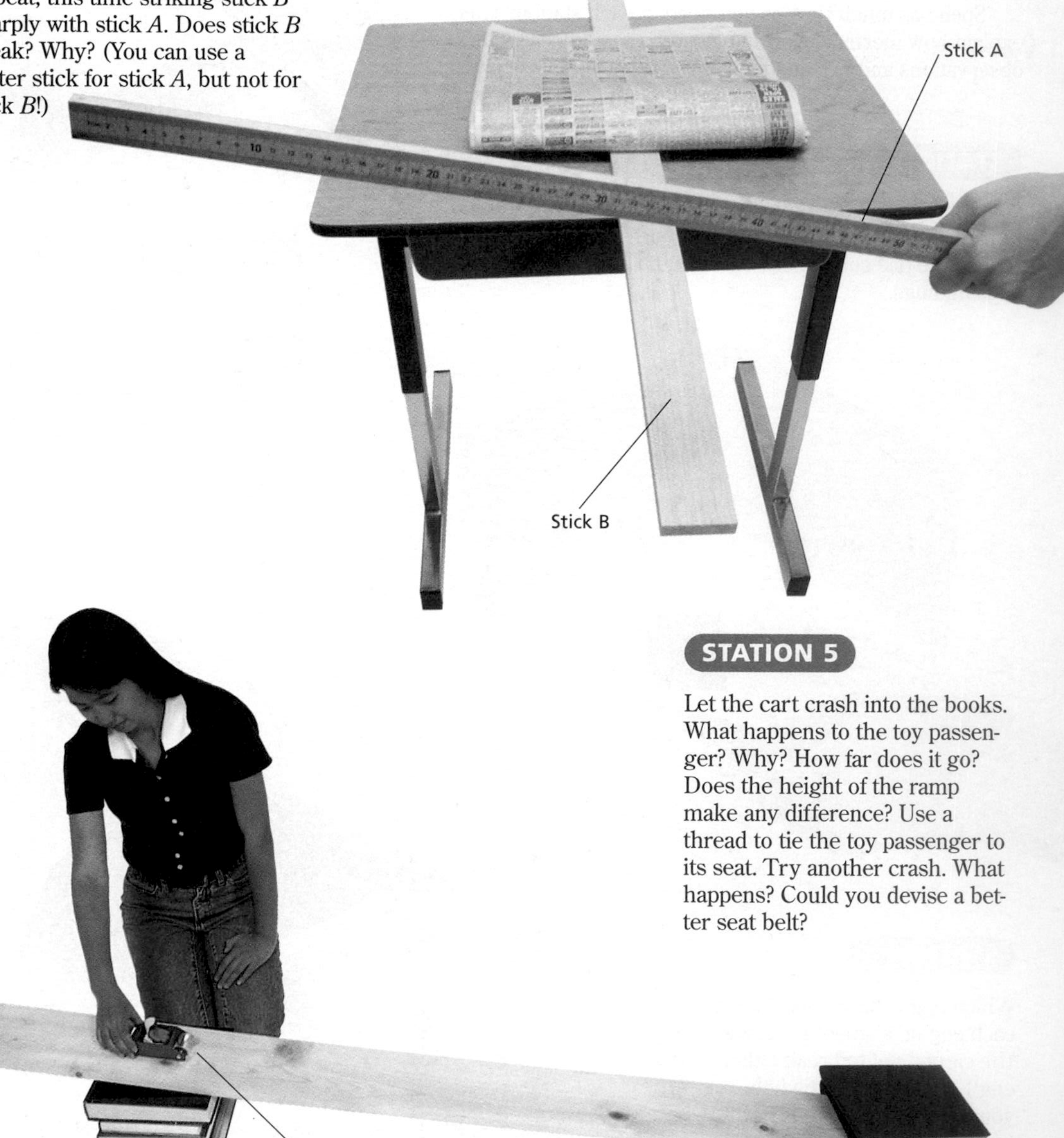

STATION 5

Let the cart crash into the books. What happens to the toy passenger? Why? How far does it go? Does the height of the ramp make any difference? Use a thread to tie the toy passenger to its seat. Try another crash. What happens? Could you devise a better seat belt?

Homework

Have students describe three different demonstrations that could help a third-grader understand the concept of inertia.

Exploration 2

Something to Try at Home

Prove this hypothesis: The greater the inertia of an object, the less effect a force has on the object's motion.

You Will Need 

- a clothespin
- 2 pencils of different sizes
- a small piece of dry spaghetti
- a spoon

What to Do

1. You should do this activity on a flat, uncarpeted floor with an adult present. Be sure no young children are playing nearby.
2. Set up the clothespin, spaghetti, and pencils on the floor as shown in the drawing.
3. Hold down the clothespin by pressing the metal clasp. Using the edge of a spoon handle, break the piece of spaghetti.
4. Substitute other objects of differing masses for the pencils.

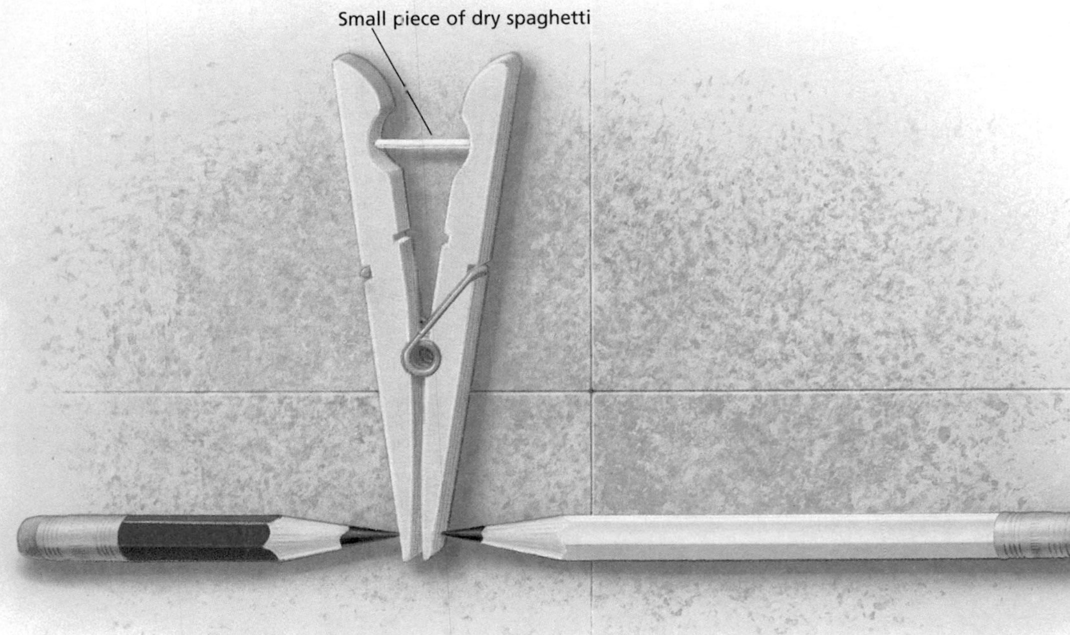

Be Careful:
Do this Exploration only on the floor.

Questions

1. What do you observe?
2. Is the force the same on each pencil? Explain.
3. Can you think of some practical examples of this principle?

279

Exploration 2

GUIDED PRACTICE Remind students that mass and inertia are related. The greater the mass, the greater the inertia. As an introduction to the Exploration, ask the question below. Students should answer prior to the Exploration and then again after completing it.

If the same size force is placed on two bodies with different levels of inertia and thus different masses, would the motion of the two bodies be altered differently? How? *(Yes, the force would have a greater effect on the body with the smaller mass.)*

★ **An Exploration Worksheet is available to accompany Exploration 2 (Teaching Resources, page 51).**

Answers to *Questions*

1. The smaller (less massive) pencil will move much farther than the heavier (more massive) one.
2. The forces applied by the two prongs of the clothespin are the same because both sides of the clothespin spring outward equally. Since the forces on both pencils are the same and the heavier pencil travels a shorter distance, a given force has less effect on a body of greater mass.
3. Some practical examples of this principle are the following: a bicycle is much easier to stop than a train; a puff of air can move a puffball but not a baseball; and it is easier to push an empty box than a full box.

FOLLOW-UP

Reteaching

To reinforce the concept of inertia, take students to the training room of the school gym. Have students observe what occurs when a punching bag is still, when it is hit (unbalanced force), or when two people hit it at exactly the same time (balanced forces). Ask: Is there any frictional force acting on the bag? *(Yes—air friction)* What happens when an unbalanced force overcomes the bag's inertia? *(The bag begins to swing.)*

Assessment

Suspend two cans with ropes from the ceiling or a doorway. One should be empty and the other full of sand. Both cans should have lids. Ask students to push each can in turn. They should also try to stop the moving cans. Ask: Which can has more mass? How can you tell, based on what you know about inertia? *(Students should understand that the force required to get the heavy can to move is greater because of its greater inertia.)*

Extension

Provide students with a roll of paper towels. Ask them to tear off a sheet with a quick jerk and then with a slow pull. Ask them to explain why a quick jerk is more effective.

Closure

The concept of inertia could easily be called the "concept of lazy objects." Name some examples of "lazy objects." *("Lazy objects" would be those that have great mass and therefore great inertia.)*

LESSON 4 Forces in Pairs

FOCUS

Getting Started

Have two students sit down on separate skateboards. Challenge one student to push the other student gently to make the other student roll without moving himself or herself. Because every action has an equal but opposite reaction, the person pushing will always move opposite the direction of the push. Explain that this demonstrates Newton's third law of motion, which students will have a chance to explore in this lesson.

Main Ideas

1. If object *A* exerts a force on object *B* (an action), then object *B* exerts a force on object *A* that is equal in size but opposite in direction (a reaction).
2. Both action and reaction forces can affect the motion of the objects on which the forces are applied.

TEACHING STRATEGIES

GUIDED PRACTICE Using two force meters and two bathroom scales, have two students demonstrate the action and reaction forces illustrated in the first two contests. Contests A and B verify that when one object places a force on another object, the second object puts an equal but opposite force on the first object. In the illustration for Contest A, the force of Ruben's pull is equal to the force of Terry's pull.

Refer students to the photograph at the bottom of the page and ask: Which dog is winning? (*If neither dog is moving, the force of the two dogs is equal even though the sizes of the two dogs are different. Thus, neither wins.*)

LESSON 4 Forces in Pairs

Until now, you have been looking at forces acting on the same object in balanced or unbalanced states. But it is also interesting to consider two forces acting on two different objects at once. Study Contests A, B, and C, and then work through the equations.

Contest A

Ruben and Terry are having a force meter tug of war. Compare the size and direction of their forces. Do you notice that

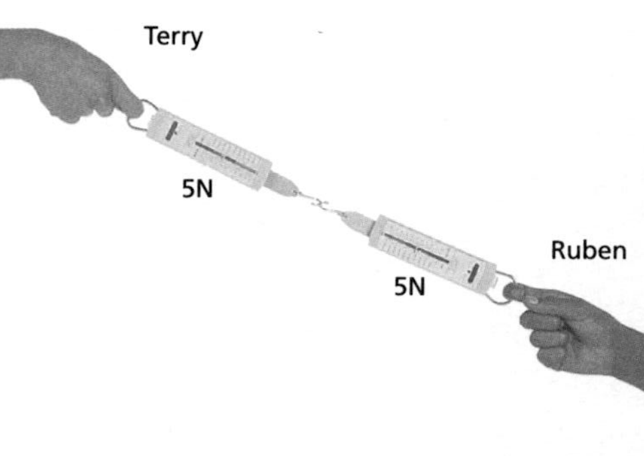

The force of Terry on Ruben	=	The force of Ruben on Terry
(to the left)		(to the right)

Contest B

Compare the size and direction of Tuwa's and Jill's forces. Do you observe that

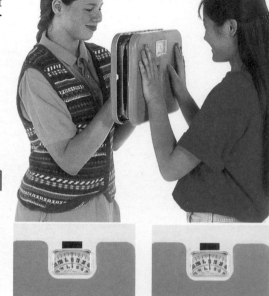

The force of Tuwa on Jill	=	The force of Jill on Tuwa
(to the ?)		(to the ?)

Contest C

Champ is a bigger dog than Buster, but neither one budges in this canine tug of war. Do you conclude that

The force of Champ on Buster	=	The force of Buster on Champ
(to the ?)		(to the ?)

LESSON 4 ORGANIZER

Time Required
2 class periods

Process Skills
observing, inferring, predicting, analyzing

New Terms
Action—a force that acts in a certain direction
Reaction—a force that acts in the direction opposite to that of an action

Materials (per student group)
Exploration 3, Station 1: coffee can with slanted holes punched in it and taped edge; string; 250 mL of water; plastic cup; pail (additional teacher materials: hammer; nail); **Station 2:** 1 L or 16 oz. plastic bottle with rubber stopper; 2 effervescent tablets; 3–6 paper towels; 2–6 straws or wooden dowels; 350 mL of water; safety goggles; **Station 3:** roller skates; helmet; elbow, wrist, and knee pads; 2 bags of sand; **Station 4:** long balloon; drinking straw; 2 small pieces of masking tape; 3 m of string; 1 m of string or thread

Teaching Resources
Exploration Worksheet, p. 53

Do All Forces Act in Pairs?

You may recall that in Exploration 2 of Chapter 10, two strips of plastic were rubbed with a wool sock. They pushed each other apart. *P* and *Q* (at right) are the same distance from the dotted vertical line. What does this suggest about the size of the two forces? **A**

Now use what you learned from Contests A, B, and C to write the "force-pair" equation for *P* and *Q*.

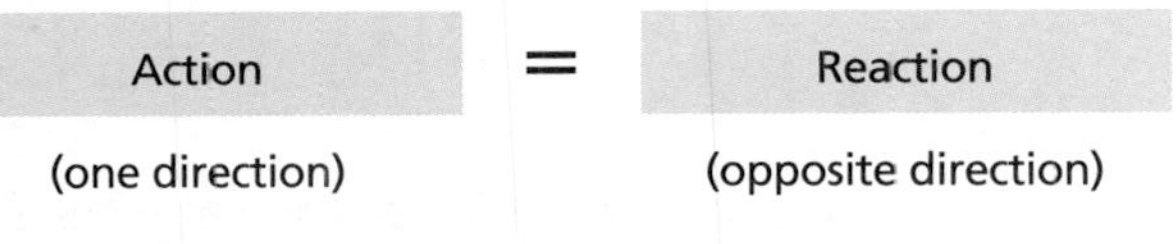
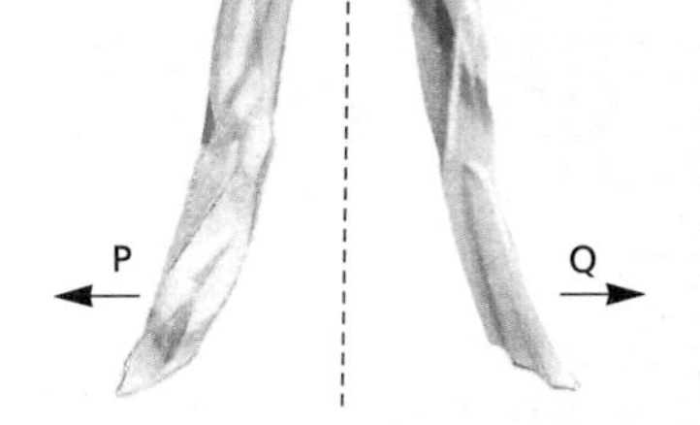

Isaac Newton's Reaction

In addition to his other important discoveries, Newton noticed that forces always occur in pairs. The forces are always equal to each other in size but opposite in direction. He called the forces **action** and **reaction**. Newton's formula states: For every action, there is an equal but opposite reaction.

Action	=	Reaction
(one direction)		(opposite direction)

There is a clearer way to write this:

If *A* puts a force on *B*, then

The force of *A* on *B*	=	The force of *B* on *A*
(in one direction)		(in the opposite direction)

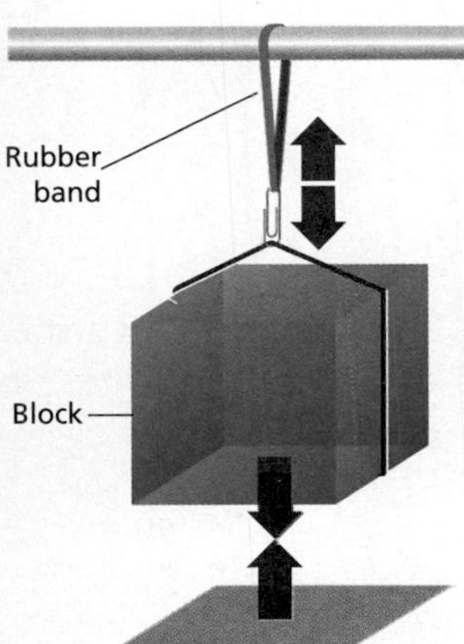

More Action-Reaction Pairs

Example 1

The block puts a force on the rubber band, stretching it downward. The rubber band puts a force on the block, holding it upward. Write the force-pair equation.

Another pair of (noncontact) forces is also at work in this case:

The gravitational force of the Earth on the block	=	The gravitational force of the block on the Earth
(the Earth pulling the block downward)		(the block pulling the Earth upward)

281

Answer to *In-Text Question*

A The fact that *P* and *Q* are equidistant from the dotted vertical line suggests that the two forces are equal. The force-pair equation for this is as follows: the force of *P* on *Q* (to the right) = the force of *Q* on *P* (to the left).

Answers to *Example 1*

Force-pair equation: the force of the block on the rubber band (downward) = the force of the rubber band on the block (upward)

Point out to students that although the forces on the block and the Earth are equal, the resultant motions are not equal. A small force on the block causes it to move, but the Earth has so much inertia that it moves an unmeasurable amount.

Homework

You may wish to have students answer the questions in More Action-Reaction Pairs on pages 281–282 as homework.

Answers to
Example 2

Force-pair equation for the first pair: the force of the block on the table (downward) = the force of the table on the block (upward)

Force-pair equation for the second pair: the force of the Earth on the block (upward) = the force of the block on the Earth (downward)

Answers to
Example 3

If the wall were not pushing on Pat, she would fall forward through the wall. Force-pair equation: the force of Pat on the wall = the force of the wall on Pat

CROSS-DISCIPLINARY FOCUS

Language Arts

Have interested students read a science-fiction story about an unusual application of the principles of force and motion. Then students can write their own science-fiction story. The following concepts could be included in the students' stories: the mass of an object always stays the same, even though its weight changes; friction and gravity depend on your location in space; and even in space, every action force has a reaction force.

EXPLORATION 3, page 283

Set up four stations, with one experiment per station. If possible, have two sets of the necessary materials at each station. Have groups rotate from station to station. Then reassemble the class to discuss their observations. Answers may vary; accept all reasonable responses.

For Station 3, have interested students bring the roller skates and safety equipment to perform this activity. It may be more practical to have one student perform this as a demonstration. Also, have another person nearby to help the student on skates maintain his or her balance.

 **An Exploration Worksheet accompanies Exploration 3 (Teaching Resources, page 53).**

Example 2

What is the force-pair equation for each pair shown at right?
First pair: block and table
Second pair: block and Earth

First pair

Second pair

Example 3

Pat wonders about action-reaction pairs as she stretches for a run. Is the wall she's pushing against *really* pushing back with an equal force? Pat asked some friends to help her find out. (Hint: What would happen to Pat if the wall weren't pushing back with an equal force?)

Here is how Pat and her friends determined that the wall is pushing back with an equal force. What is the force-pair equation?

282

Answers to
Station 1, page 283

Action: the pull of the can and water on the string; reaction: the pull of the string on the can and water; observation: the forces are balanced—when the weight of the can and water decreases due to the holes in the can, both forces of the pair decrease equally in size; force-pair equation: force of water and can pulling down on string = force of string pulling up on water and can

Answers to
Station 2, page 283

Action: the force of the carbon dioxide gas against the cork; reaction: the force of the cork against the carbon dioxide gas; observation: the result is that both the bottle and the cork move, but the cork moves farther because it has less mass; force-pair equation: force of cork on carbon dioxide gas = force of carbon dioxide gas on cork

Making Use of Reaction Force

The four mini-experiments in this Exploration show various uses of reaction force. At each station you will find materials set up as shown here.

Watch what happens in each situation. Identify the action-reaction pair of forces. What is each force doing? Make a sketch in your ScienceLog, drawing in the appropriate forces. Write the force-pair equation for each situation.

STATION 1

Caution: Do this mini-experiment over a pail!

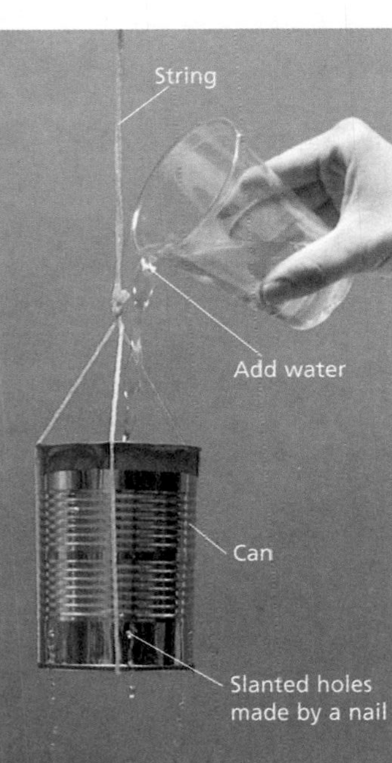

STATION 2

Fill a plastic bottle one-third full with water. Break up an effervescent (fizzy) tablet and put it in the bottle. Firmly insert a rubber stopper. What happens next could be messy, so have paper towels handy.

Be Careful: Do this activity only on the floor. Make sure no one is standing directly in front of the rubber stopper after it is placed in the bottle.

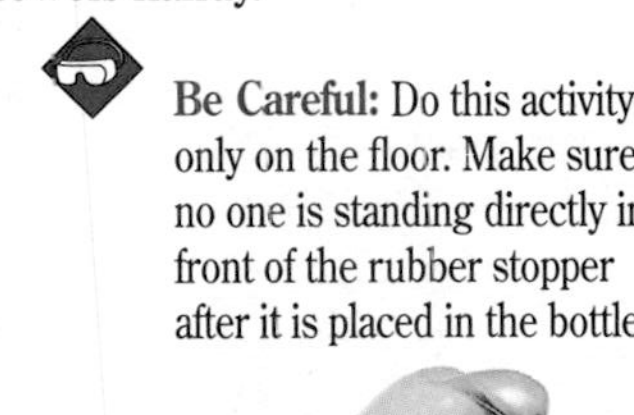

STATION 3

Stand on skates, holding bags of sand. Throw the bags behind you.

Be Careful: Use caution when standing on skates.

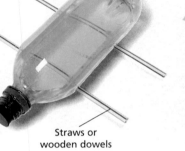

STATION 4

Inflate a long balloon. Tie the end into a knot. Set up the equipment as shown, and have two classmates hold the string by either end. Untie the knot and allow the balloon to move. Will it move if the string is tilted upward?

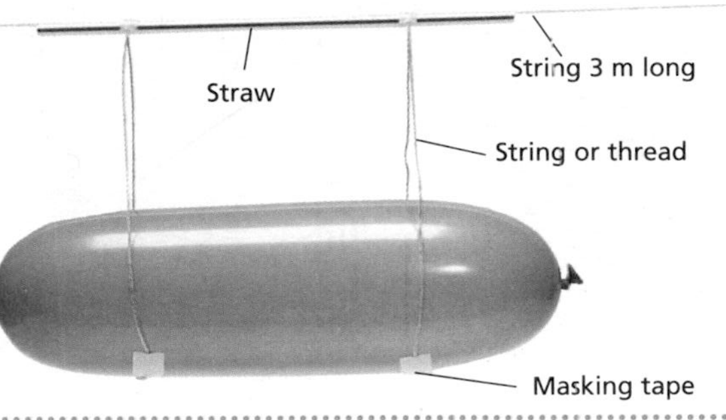

283

Answers to *Station 3*

Action: the push of the skater against the bags of sand; reaction: the push of the bags of sand against the skater; observation: both the skater and the bags of sand move, but the bags of sand move farther because they have less mass; force-pair equation: force of skater on bags of sand = force of bags of sand on skater

Answers to *Station 4*

Action: the force of the balloon against the escaping air; reaction: the force of the escaping air against the balloon; observation: the balloon and straw move in a direction opposite to the escaping air; force-pair equation: force of the balloon on the escaping air = force of the escaping air on the balloon

The balloon should still move if the string is tilted upward.

FOLLOW-UP

Reteaching

Hold a yo-yo by its string and dangle it in front of the class. When the yo-yo stops swaying, ask the class to describe the pair of forces acting on the dangling yo-yo. *(An action force of the yo-yo on the string is pulling the string downward. A reaction force of the string on the yo-yo is pulling the yo-yo upward.)*

Assessment

Have students develop force-pair equations for the following situations. They should explain the observed effect of the force as well.

- a bat striking a ball *(Force of bat on ball = force of ball on bat; the ball moves farther because it has less mass.)*
- a piano falling off a roof *(The force of the Earth on the piano = the force of the piano on the Earth; the piano has less mass than the Earth, so it moves more than the Earth does.)*

Extension

Have students design a toy that is propelled by a reaction force. For example, they could attach a blown-up balloon to the back of a toy boat or car to propel it forward. Make sure that everyone in the room is wearing safety goggles during demonstrations of toys with balloons or rubber bands.

Closure

Take students outside for a final game of tug of war. Allow half of the class to play while the other half identifies force-pairs during the game. Then switch groups so that every student has a chance to play tug of war and to make observations about force and motion during the game.

1. A Slippery Situation

You are stranded in the middle of a large, frozen pond. The ice is so slippery that you can't walk. In fact, you can't even crawl. Devise a way to get off the ice. Compare your ideas with those of your classmates to see who came up with the most creative method.

2. How Changeable Are You?

Scientists are continually revising their theories when new facts are discovered. Take a look again at the survey you carried out on page 265. Have any of your views changed since then? Write down any changes in your ideas.

3. Balancing Act

Which of the following are balanced-force situations? How can you tell? Identify each force of a balanced pair.

a. The elevator and its passengers descended from the 42nd floor at a constant speed.

b. The sweating tug-of-war teams didn't budge in the final 2 minutes of the contest.

c. At the bottom of the hill, the sled traveled 50 m before it stopped.

d. The water skier banked around the curve at a steady speed of 50 km/h.

e. The stunt plane stalled and fell in a "free fall" for 1000 m.

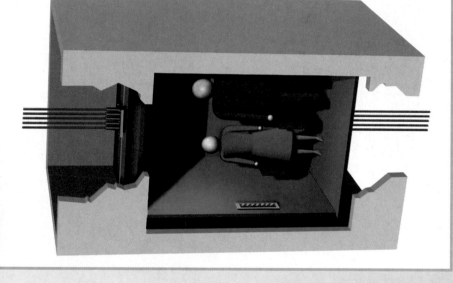

★ **You may wish to provide students with the Chapter 13 Review Worksheet on page 56 of the Unit 4 Teaching Resources booklet.**

Answers to *Challenge Your Thinking*

1. Sample answer: If you threw an object away from you, it would cause your body to move in the opposite direction. (Other possibilities might involve creating friction on the ice in some way.)
2. Students' changes should reflect a better understanding of balanced and unbalanced forces, force pairs, and inertia.
3. Situations (a), (b), and (d) are balanced-force situations, and situations (c) and (e) are not. In (a), the force of the cables holding the elevator (and thus resisting gravity) is equal to the force of gravity pulling the elevator down. In (b), the force of one team pulling is equal to the force of the other team pulling. In (d), the force of the skier's motion to the side is balanced by the force of the boat pulling the skier back to the other side. In (c) and (e), since the speeds of the sled and the plane are not constant, the forces are unbalanced.
4. In the first photo, inertia causes the soil to fly through the air even though the motion of the spade has stopped. In the second photo, inertia causes the acrobat to continue moving through the air after her partner has released her. In the third photo, inertia keeps the table setting from moving to the side even though the table has been removed from under it.
5. If you jerk the bottom ring quickly, inertia will keep the block from falling; it will also keep the force from being transferred to thread *A*. If you pull slowly on the bottom ring, however, it will strain the threads both above and below the block; since both the downward pull *and* the weight of the block put strain on thread *A*, the strain will be more than that on thread *B* (which must resist only the downward pull) and thread *A* will break first.

4. Inertia Identification
Look at the three photographs shown here, and identify the role and effects of inertia in each.

5. Staying Power!
When you pull down slowly on the bottom ring, thread *A* breaks and the block falls. But if you suddenly jerk this ring downward, thread *B* breaks and the block stays hanging. Why? (Try it if you doubt it!)

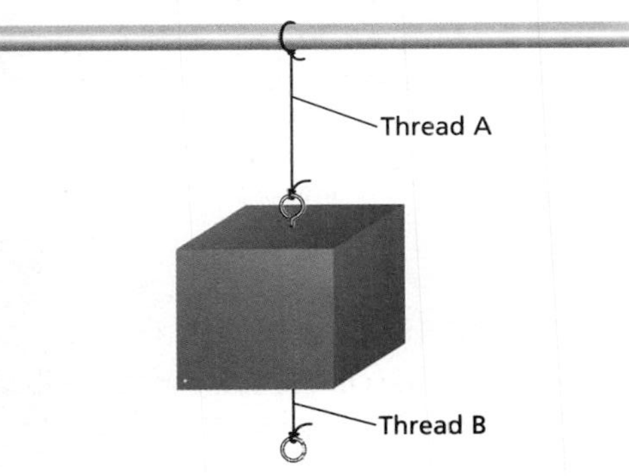

Sciencelog

Review your responses to the ScienceLog questions on page 264. Then revise your original ideas so that they reflect what you've learned.

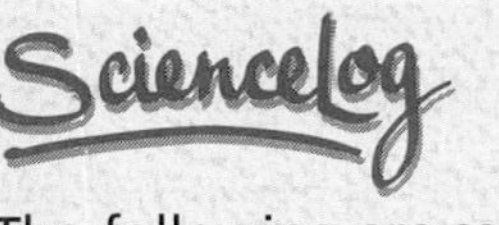

Sciencelog

The following are sample revised answers:

1. Once an object (the bowling ball) is moving, it tends to keep moving in a straight line without slowing down or speeding up until another force acts on it (in this case, the friction of the lane's surface and the impact with the bowling pins).
2. Although there is no longer a force pulling the caboose, it will continue to move until friction or gravity slows it down. This continued motion occurs because of inertia, the tendency of objects in motion to stay in motion.
3. Yes. Forces always occur in pairs, and the forces are always equal in size to each other but opposite in direction. If the wall did not push you back, you would be able to push the wall over.

Making Connections

The Big Ideas

The following is a sample unit summary:

All forces are exerted by an agent that places the force on a receiver. A force always has an effect. All forces occur in pairs. (1)

Gravitational force is the attractive force that exists between all objects. Gravitational force is an attractive force between two bodies of matter. (2)

Mass refers to the quantity of matter in an object. Mass is constant. Weight refers to the gravitational force that the Earth places on an object. It changes with the distance of the object from Earth. (3)

The stretch of a rubber band or spring on a force meter is one way to measure force. (4)

Mass is measured in grams or kilograms. Force and weight are measured in newtons. (5)

Frictional force opposes the movement of an object or slows down an object in motion. Friction increases with heavier masses and rougher surfaces. Roller bearings, oil, and smooth surfaces decrease friction. (6)

An object at rest with balanced forces acting on it will not move, while a moving object with balanced forces acting on it will continue to move at the same speed and in the same direction. With unbalanced forces, an object begins moving, stops moving, or changes direction. (7)

Inertia is a characteristic of all objects. It reflects the fact that bodies at rest tend to stay at rest and bodies in motion tend to stay in motion. Unbalanced forces are required to change a state of motion. (8)

An action-reaction equation represents the concept that every action has an equal but opposite reaction. (9)

Making Connections

Unit 4

Force and Motion

SourceBook

To tackle more knowledge of force and motion, read pages S71–S88 in the SourceBook. You will learn more about observing and describing motion. You will also learn more about the laws of motion. Finally, you will discover how force, gravity, and acceleration are related.

Here's what you'll find in the SourceBook:

The Big Ideas

In your ScienceLog, write a summary of this unit, using the following questions as a guide:

1. What characteristics do all forces share?
2. What is a gravitational force, and what does it do?
3. What is the difference between the mass of an object and its weight?
4. What devices can be used to measure forces?
5. What units are used to measure mass? force? weight?
6. What does frictional force do? What causes it? How can you decrease it?
7. What happens to an object (at rest or in motion) if balanced forces are at work on it? unbalanced forces?
8. What is inertia? How does it relate to forces?
9. What is the meaning of Newton's law of action and reaction?

Checking Your Understanding

1. You began to learn about forces in the story of the sailboat race on page 217. Read the story again, and identify the following:

   a. frictional forces (Name the agent, receiver, and effect.)

Answers to *Checking Your Understanding*

1. Answers to questions (a) through (e) can be found in the tables on page S203. Examples of inertia (f) include the boat opposing the force exerted by Kathleen and Kimiko to drag it through the sand and the inertia of Kathleen being overcome by the force of the boom hitting her. For item (g), when there is an action force, there is an equal and opposite reaction force. Two examples are the weight of the boat on the water and the buoyant force of the water pushing up on the boat.

Answers to Checking Your Understanding continued ►

b. gravitational forces (Name the agent, receiver, and effect.)
c. buoyant forces (Name the agent, receiver, and effect.)
d. balanced forces
e. unbalanced forces
f. examples of inertia
g. pairs of action-reaction forces

2. One of the largest tug-of-war pulls in the world was performed by 1015 high school students in 1976.
   a. If each student literally pulled his or her own weight, what force did the rope have to endure? (Estimate the average weight of a student.)
   b. Can you identify an action-reaction force pair at work? What is its equation?

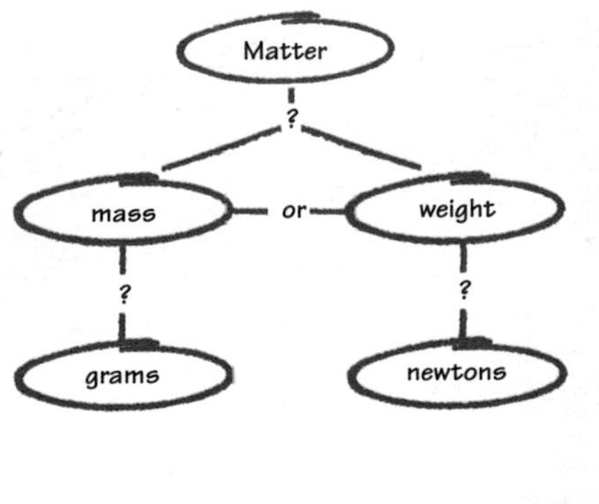

3. Lucy, after experiencing free fall, opens her parachute and drifts steadily downward until her feet hit the ground. She is carried along for several meters before coming to a complete stop.
   a. In each stage of Lucy's fall, describe what forces are acting on her. Give the directions of these forces.
   b. At what points are there unbalanced forces on Lucy? balanced forces?

4. concept map Copy the concept map shown here into your ScienceLog, and replace the missing linking terms.

Matter
?
mass — or — weight
? ?
grams newtons

You may wish to provide students with the Unit 4 Review Worksheet on page 64 of the Unit 4 Teaching Resources booklet.

Answers to
Checking Your Understanding, *continued*

2. a. Because the force of the students pulling in one direction is balanced by the force of the students pulling in the other direction, the force on the rope is measured using one of the forces in this action-reaction pair. (This is similar to Contest C at the bottom of page 280.) Therefore, if each student is assumed to have a mass of 55 kg and thus a weight of 550 N, and each throws his or her total weight into the effort, the force exerted on the rope would be (1015 students ÷ 2) × 550 N/student = 279,125 N.
   b. One action-reaction pair is the force of the students on one side pulling on the rope and the force of the rope pulling on those students. The force-pair equation is as follows: the force of students on one side pulling on the rope = the force of the rope pulling on those students.

3. a. During free fall, the downward force of gravity on Lucy is much greater than the upward force of air resistance on her body. When her parachute opens, the air resistance increases and her velocity decreases. When her feet touch the ground, they exert a force on the ground, and the ground exerts a force on her feet. She is pulled forward by the lateral force of air blowing on her chute before she stops.
   b. When she opens her parachute, she will feel the force of wind resistance against her parachute acting against the force of gravity (unbalanced force). Once she has reached her maximum velocity with the parachute open, she will continue at a steady speed until she reaches the ground (balanced force). At that point, the force of the ground acting upward overcomes the downward force of gravity, stopping her fall (unbalanced force). Air blowing on her chute propels her several meters before friction between her feet and the ground causes her to stop (unbalanced force).

4. The first missing term is *has.* The missing term for both of the remaining two places is *which is measured in.*

WEIRD Science

Background

Fish can reverse their direction of movement without slowing down at all. They can also turn with a turning radius as small as 10 to 30 percent of the length of their bodies. Yellowfin tuna have been reported to swim as fast as 73 km/h. Compare all of this with a typical human-made ship. A ship cruises at about 37 km/h, it must reduce its speed by about 50 percent to turn around, and it requires a circle with a radius of 10 times its hull length!

The mobility of a fish and a ship differ in other ways as well. A ship's motion is limited by the rigidity of its hull, while a fish's body is very smooth and flexible. Robotics technology is still too simple to accurately reproduce the flapping of a fish's tail. For this reason, fish-inspired mechanisms had performed rather poorly until the creation of RoboTuna.

Discussion

Lead a classroom discussion about swimming styles. Ask the students to comment on their own swimming technique. Do they swim more like a fish or a rowboat? How do they turn around? What kind of swimming techniques have other animals adopted? You might also discuss the swimming methods of dogs, jellyfish, frogs, and whales.

Extension

You might want to help students to draw an analogy between motion through water and motion through air by having them build a simple paper airplane. Tell them to observe how far and how fast the plane flies through the air. Then have them alter the plane by adding flaps to the trailing edges of the wings and bending the flaps up or down. A flap increases the drag on a wing, causing the wing to slow down and turn the plane. Encourage students to explore the resistance of the air on different parts of the plane by altering the size and placement of the flaps and the amount that each flap is bent.

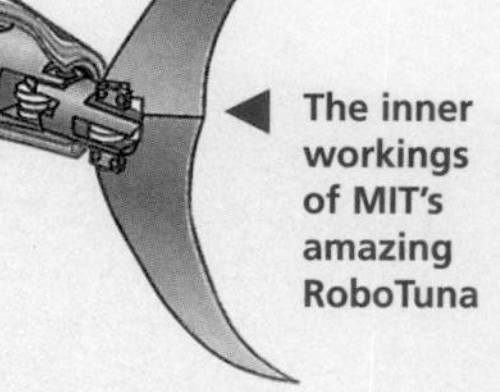

WEIRD Science

Robot Fish

When is a fish tail not a fish tail? When it's the tail of RoboTuna, a truly weird robotic fish designed by scientists at the Massachusetts Institute of Technology.

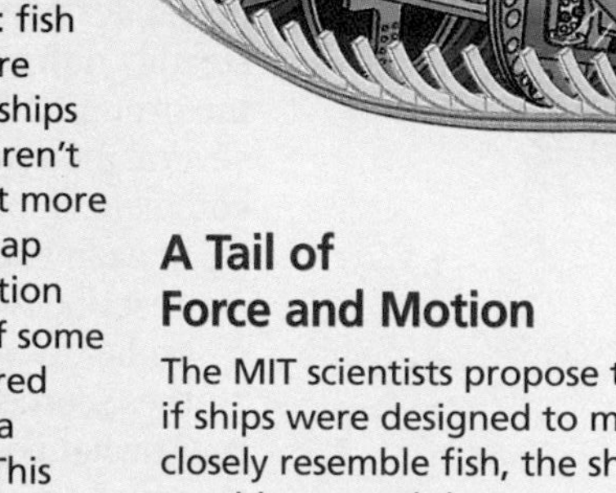

A strut supports the robot, encloses the tendons, and conveys control and sensor information.
Links are connected by aluminum hinges, to which are affixed beams that support ribs spaced 1 in. apart. The ribs and flexible beams hold the skin in place while allowing the body to flex continuously.
A skin of foam and Lycra is smooth enough to eliminate wrinkles or bulges and the stray turbulence they cause.

▶ The inner workings of MIT's amazing RoboTuna

Something Fishy Going On

There's no doubt about it: fish are quicker and much more maneuverable than most ships and submarines. So why aren't ships and submarines built more like fish—with tails that flap back and forth? This question caught the imagination of some scientists at MIT and inspired them to build RoboTuna, a model of a bluefin tuna. This robot fish is 124 cm long and is composed of six motors, a skin of foam and Lycra, and a skeleton of aluminum ribs and hinges connected by pulleys and strings.

▼ Vortices trailing behind a rotating propeller

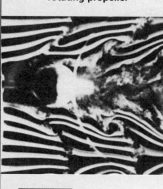

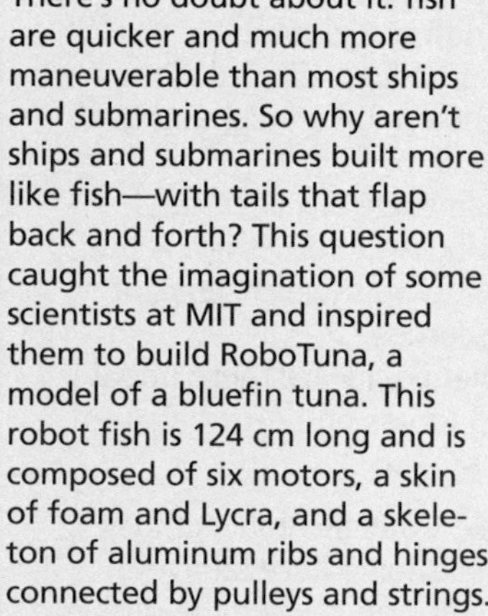

A Tail of Force and Motion

The MIT scientists propose that if ships were designed to more closely resemble fish, the ships would use much less energy and thus save money. A ship moving through water leaves a trail of little whirlpools called *vortices* behind it. These vortices increase the friction between the ship and the water. A fish, however, senses the vortices and responds by flapping its tail, creating vortices of its own. The fish's vortices counteract the effects of the original ones, and the fish is propelled forward with much less effort.

RoboTuna has special sensors that measure changes in water pressure, much as the living tuna senses vortices. Then the robot fish flaps its vortex-producing tail, allowing it to swim like a living fish. As strange as it may seem, RoboTuna may represent the beginning of a new era in nautical design.

Viewing Vortices

Fill a roasting pan three-quarters full of water. Wait long enough for the water to stop moving. Then tie a 6 cm piece of yarn or ribbon to the end of a pencil. Drag the pencil through the water with the yarn or ribbon trailing behind it. How does the yarn or ribbon respond? Where are the vortices?

288

Answers to *Viewing Vortices*

The end of the ribbon or yarn should wiggle back and forth. This is caused in part by the creation of vortices, which trail behind the ribbon or yarn. You might want to put some sand on the bottom of the pan and have students observe how the sand grains move as the pencil passes over them.

Batter Up

If you're a baseball player and you have a batting average of .300, you have a chance of being inducted into the Baseball Hall of Fame. However, a batting average of .300 means that you get a hit only 3 out of every 10 times at bat. This is a very low percentage. Why are baseballs so hard to hit?

Making the Batter's Head Spin

To confuse the batter, pitchers have learned to apply different kinds of spins on the ball. When a pitcher throws a spinning ball, the side of the ball that rotates in the same direction as the ball's motion encounters more *drag* than does the side that rotates opposite to the ball's motion. In this case, drag is the friction between the ball and the air around it. Also, a baseball's stitching changes the course of the ball. The stitching redirects air around the spinning ball, altering the drag on different parts of the ball even more.

Curves and Knuckles

One type of spin used by pitchers is the topspin, in which the ball rotates in the air just like a ball rolls on the ground. Used on a curveball, topspin can cause the ball's path to bend by as much as half a meter as the ball travels from the pitcher to the catcher. Thrown overhand, the curveball bends downward. Thrown sidearm, the curveball bends to the side.

Another trick up a pitcher's sleeve is the knuckleball. This pitch, unlike the curveball, is thrown with only a very small amount of spin. This makes the ball's path extremely erratic. As the ball slowly spins, the position of the seam changes. This causes the drag on the ball to vary slowly, changing the ball's course in midflight.

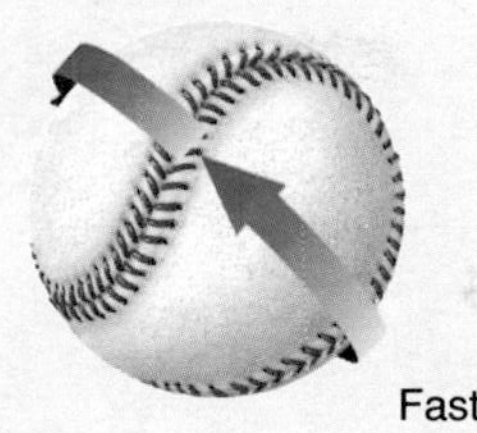

Fastball

Slider

Curveball

Screwball

◀ Some ways to spin a baseball

Keep Your Eye on the Ball

You can study the effects of spin on a ball by applying different spins to a rolling marble on a smooth floor. How much can you alter the path of the marble just by changing its spin? How does this compare with the spin on a baseball?

▼ The crazy dance of the knuckleball

289

Discussion Questions

1. What is the primary difference between a curveball and a knuckleball? *(A curveball is given a lot of spin. A knuckleball is thrown with as little spin as possible.)*
2. What two effects does the spin have on the motion of the ball and why? *(Spin reduces drag so that the ball travels faster. Also, spin causes different parts of the ball to experience different amounts of drag, causing the ball to curve.)*

Answers to *Keep Your Eye on the Ball*

The path of the marble can be altered a significant amount depending on the smoothness of the floor and the amount of spin. A backspin can cause the marble to stop and come back to the thrower. Spins to the left or right can cause the marble's path to bend.

The marble acts differently than a baseball in the air because air friction acts on the entire surface of a baseball, whereas the marble has only one point of contact with the floor at a time.

Background

Before 1919, a baseball pitcher had many other ways of fooling the batter besides spinning the ball. The pitcher could scratch or scuff the ball or even put grease or spit on the ball. These techniques affected the ball's flight in much the same way that the seams do: air traveling past the ball is affected by the irregular surface, causing the ball to curve or wobble in the air. Today these techniques are illegal because they give the pitcher too great of an advantage.

The fastball is another common pitch in baseball. Much of a fastball's effectiveness is due to its incredible speed—the batter has little time to react—but speed is not all there is to a fastball. There are at least four types of fastballs, all of which cause the ball to curve. Two types are thrown with a lot of spin, and two are thrown with only a very small amount of spin.

A spinning fastball can be thrown so that either two or four seams spin around with the ball. This choice is determined by how a pitcher positions the ball in his or her hand before throwing the ball. The four-seam pitch is faster than, but does not curve as much as, the two-seam pitch. The two other types of fastballs are the sinking fastball and the split-finger fastball. These fastballs have much less spin, so the drag is greater. They are also not quite as fast as the other fastballs. However, because their time in the air is longer, there is more time for gravity to influence the path of the ball, causing it to curve downward much more.

Extension

You might have the class observe the motion of a table-tennis ball after altering its surface. Use the following alterations: could include putting tape on it, wrapping rubber bands around it, or painting seams on it.

Science Snapshot

Background

Walter Massey's research interest in physics involves the theory of quantum liquids and solids. In a quantum liquid or solid, physicists model vibrations in the material as little packets of energy called *phonons.* These phonons can interact with one another in complex ways. Massey explained why sound waves are so weakened when passing through certain forms of helium at low temperatures by showing how three of these phonons combine.

In addition to Massey's work with the NSF, he has also founded Inner City Teachers of Science (ICTOS), a program to help prepare students for teaching in under-resourced areas. He has also headed the American Association for the Advancement of Science (AAAS). In 1993, Massey left the NSF to accept an appointment with the University of California system as the senior vice-president for academic affairs and provost, the system's second highest post.

Meeting Individual Needs

Learners Having Difficulty

Have students write a short skit in which Walter Massey is interviewed by a science-education reporter. Massey should be asked questions about his role at the NSF and about his personal background and education.

Multicultural Extension

Walter Massey

Have interested students find out more about Massey's work in science education, administration, and research.

Science Snapshot

Walter Massey (1938–)

When Walter Massey entered college at age 16, he didn't plan on becoming a scientist. He had never taken chemistry or physics or any advanced mathematics. Unprepared in many subjects, he barely made it through the first semester. But with hard work and guidance from his teachers, Massey graduated with a degree in physics. Since then he has become one of the country's most influential figures in science education.

Early Years

Massey was born in Hattiesburg, Mississippi, in 1938. His parents taught him that education was an essential element of success. Massey faced many difficulties, including his own lack of preparation. But Massey persevered and went on to eventually earn a Ph.D. in theoretical physics. He was given a research position at the University of Illinois and then at Brown University, where he eventually became dean of the undergraduate college. Having earned a reputation as an excellent administrator, Massey was appointed director of the Argonne National Laboratory. In 1990, he received a presidential appointment to the directorship of the National Science Foundation (NSF), which is the top position in national science education.

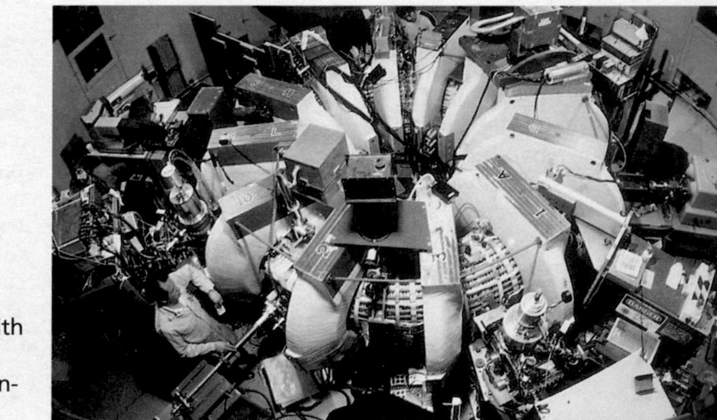

▲ **Fusion research made possible through funding by the NSF**

Reaching Out—Nationwide

The NSF is a government agency that allocates money to researchers in all fields of science and encourages science education across the nation, from elementary schools to doctoral programs. As director of the NSF, Massey worked with politicians to establish the NSF's budget. He then helped determine how this money was to be distributed among the many research proposals submitted to the NSF each year.

In his 2 years as director, Massey helped make the NSF more effective by improving the research grant system and implementing numerous programs to make science research and education more accessible. Massey has also reached out in other ways. He has actively encouraged minority students to pursue careers in science, and he has worked with impoverished students and students with academic problems.

In 1993, Massey resigned from his position with the NSF in order to return to education. He accepted a position as senior vice-president of academic affairs and provost for the University of California system.

Ask the NSF

One part of a scientist's job is to write proposals for experiments and to request funds if necessary. Using one of the Explorations in the text, write a research proposal that specifies the purpose of the experiment, the materials you will need, the cost of the experiment, the time it will take to complete the experiment, and the potential benefits of your research.

Ask the NSF

This activity is designed to help students relate the topics covered here to topics covered in the rest of the unit. You may want to facilitate the proposal-writing process by providing students with a form to fill out. The form could stimulate critical thinking by asking questions such as the following: Will this research address an environmental problem or a health problem? Who will benefit from this research?

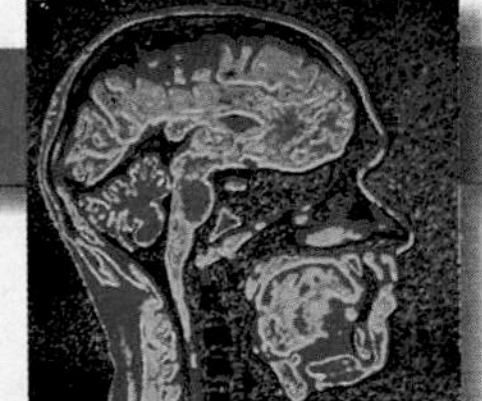

HEALTH WATCH

Muscle Mania

Make a frown. You just used more than 40 muscles. Every day you probably use all 37 muscles in each of your hands and all 33 muscles in your feet. All told, your body contains more than 650 muscles that are responsible for all of your body movements and body functions.

Brawn and Brain

To see how your muscles work, first bend and flex one of your arms as if you were about to arm-wrestle. The flexed muscle is your *biceps.* This muscle, like most other muscles, is striped. The stripes are many small *muscle fibers.* To flex your biceps, each muscle fiber receives a signal from your brain. The signal causes the muscle fibers to contract. When all of the muscle fibers contract at the same time, they pull your muscle into a tighter, more rounded shape. This causes the biceps to pull up on the lower arm. As a result, your lower arm is raised.

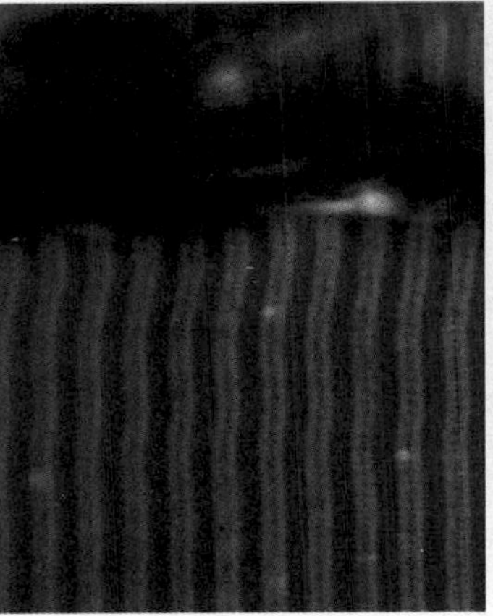

◀ Close-up view of a skeletal muscle showing the striped pattern of muscle fibers

Let's Get Physical

Exercise can increase the size of your muscle fibers but not how many muscle fibers you have. A person becomes stronger as the muscle fibers grow larger. The larger the muscle fibers, the greater the force exerted when the muscle contracts.

Exercises that increase strength are called *resistance exercises.* Resistance exercises, such as weight lifting, can force muscles to work at their physical limit. The strain on the muscles causes them to grow larger. *Aerobic exercises,* on the other hand, are better for building stamina. Exercises such as walking and running increase the heart's ability to pump blood to the muscles. This kind of exercise provides the muscles with more oxygen, which helps them work longer.

Biceps

Triceps

▲ Bending your arm requires muscles to work together. When you bend your arm downward, your triceps flexes while your biceps relaxes. When you bend your arm upward, your biceps flexes while your triceps relaxes.

How to Carry a Book

Put a heavy book in one of your hands. Starting with your arm at your side, slowly bend your elbow until your lower arm is parallel to the floor. Did the book become easier or harder to hold up? Using what you know about gravity and how muscles exert forces, why do you think this is so?

291

Meeting Individual Needs

Learners Having Difficulty

Make a set of flashcards with a common muscle name on one side of the card and the muscle's function on the other side. Then have students draw a card, point to the muscle indicated on the card, and demonstrate its use. Some muscles to include are the biceps, triceps, pectorals, heart, and tongue.

Answers to *How to Carry a Book*

As the lower arm is raised, the book should begin to feel a little heavier. This is because when your arm is at your side, the biceps exerts very little force. When the arm is parallel to the floor, however, the force of gravity is perpendicular to the lower arm, and the biceps must support the entire weight of the object. You might want to draw a simple illustration of this on the board.

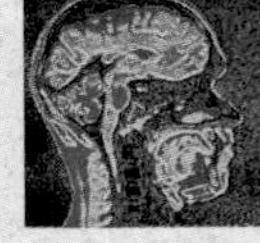

HEALTH WATCH

Background

The skeletal muscle fibers mentioned on this page actually consist of two types: red and white. Red fibers are used for activities that require endurance and continuous exertion. These fibers are developed during aerobic exercises. White fibers, on the other hand, are used when quick bursts of power are needed and so are developed during resistance exercises.

The proportion of red fibers to white fibers in a muscle is determined genetically. However, exercise can alter their relative amounts. The muscles of marathon runners and swimmers are about 80 percent red fibers, while those of sprinters are about 75 percent white fibers. To develop both types of muscle fibers, one should adopt an exercise regimen that includes both resistance and aerobic exercises.

Cross-Disciplinary Focus

Health

Invite a nutritionist or exercise instructor to talk to the class. A nutritionist could explain what materials our muscles need to function properly, why the muscles need these materials, and in what foods these materials can be found. An exercise instructor could lead the class in some simple exercises that help to maintain fitness. He or she could explain what muscles are being used in each exercise and what the functions of those muscles are.

Unit Overview

In this unit, the science of structure is examined from the point of view of technology. In Chapter 14, students explore the integrity of structures and the forces acting on these structures. Students are introduced to some basic design elements used in engineering, including different kinds of beams, trusses, arches, and domes. Students then examine how these design elements influence the durability of various structures. In Chapter 15, students explore the historical development of structure and design. Students then focus their attention on modern urban environments and, by the end of the unit, integrate what they have learned to create their own urban designs.

Using the Themes

The unifying themes emphasized in this unit are **Structures, Changes Over Time,** and **Systems.** The following information will help you incorporate these themes into your teaching plan. Focus questions that correspond to these themes appear in the margins of this Annotated Teacher's Edition on pages 301, 310, 317, 320, 321, 329, and 340.

The various lessons in this unit will allow you to focus on many aspects of the theme of **Structures.** For instance, a discussion of forces and their effects will allow you to discuss not only architectural structures, but biological and geological structures as well. By the end of this unit, students should have a strong enough foundation in this subject to solve problems involving forces that act on everything from a human skeleton to an eroding natural bridge. Specifically, focus questions will allow students to consider what makes a structure rigid or flexible, why arches are able to span such a large amount of space, and how structures are built to withstand weather and other forces of nature.

Changes Over Time is also an important theme in this unit. The first lesson of Chapter 15 takes an extended look at the changes that human structures have undergone throughout history. In addition, students are asked to compare the action of weathering on the pyramids with that on natural rock exposures.

As environmental aspects of design are considered, the theme of **Systems** emerges. This theme is useful in assisting students to consider both a neighborhood's physical characteristics and its residents as parts of an interacting system—as one changes, so does the other.

Using the SourceBook

Unit 5 takes a broad look at design by presenting ways in which other cultures throughout history have solved their own design problems. In addition, the molecular basis of material strength is examined. Other influences on design, including the environment and religion, are also discussed.

Bibliography for Teachers

Hawkes, Nigel. *Structures: The Way Things Are Built.* New York City, NY: Macmillan Publishing Company, 1990.

Kaminetzky, Dov. *Design and Construction Failures: Lessons From Forensic Investigations.* New York City, NY: McGraw-Hill, Inc., 1991.

Mark, Robert. *Light, Wind, and Structure: The Mystery of the Master Builders.* Cambridge, MA: Massachusetts Institute of Technology Press, 1990.

Bibliography for Students

Bates, Robert L. *Stone, Clay, Glass: How Building Materials Are Found and Used.* Hillside, NJ: Enslow Press, 1987.

Darling, David. *Spiderwebs to Skycrapers: The Science of Structures.* New York City, NY: Dillon Press, 1991.

Wilkinson, Philip, and Michael Pollard. *Mysterious Places: The Master Builders.* New York City, NY: Chelsea House Publishers, 1994.

Film, Videotapes, Software, and Other Media

Caracol: The Lost Maya City
Videotape
Coronet/MTI
108 Wilmot Rd.
Deerfield, IL 60015

Chartres Cathedral
Videotape
Britannica
310 S. Michigan Ave.
Chicago, IL 60604-9839

The Puzzle of the Tacoma Narrows Bridge Collapse
Videodisc
John Wiley & Sons
1 Wiley Dr.
Somerset, NJ 08875

Unit Organizer

Unit/Chapter	Lesson	Time*	Objectives	Teaching Resources
Unit Opener, p. 292				Science Sleuths: The Collapsing Bleachers English/Spanish Audiocassettes Home Connection, p. 1
Chapter 14, p. 294	**Lesson 1, Planning the Structure, p. 295**	3	1. Explain the advantages of using a truss or an I-beam instead of a rectangular beam to support a load. 2. Describe how a cantilever is constructed to support a load. 3. Determine that arches and domes provide better support for vertical loads than do beams. 4. Classify bridges by their structural elements.	Image and Activity Bank 14-1 Resource Worksheet, p. 3 ▼ Theme Worksheet, p. 4 Exploration Worksheet, p. 6
	Lesson 2, Response of Structures to Force, p. 304	4	1. Describe behavior of structures and materials by using the following terms: load, response, tension, tensile force, compression, compressive force, shear, and shear force. 2. Explain what is meant by an elastic response in a structure or material. 3. Demonstrate how to strengthen a material by shaping and joining it in various ways.	Image and Activity Bank 14-2 Graphing Practice Worksheet, p. 8 Exploration Worksheet, p. 10 ▼ Exploration Worksheet, p. 14 Activity Worksheet, p. 18 Discrepant Event Worksheet, p. 19 Exploration Worksheet, p. 20 ▼
	Lesson 3, Engineering Structures, p. 314	3	1. Analyze information in order to formulate a solution to a problem. 2. Explain the importance of design in natural and artificial structures. 3. Explain why it is important to analyze the design of structures that have failed as well as the design of structures that have not failed.	Image and Activity Bank 14-3 Exploration Worksheet, p. 26 Transparency 52 Discrepant Event Worksheet, p. 28 Exploration Worksheet, p. 29 Transparency 53 Resource Worksheet, p. 34 ▼
End of Chapter, p. 324				Chapter 14 Review Worksheet, p. 35 ▼ Chapter 14 Assessment Worksheet, p. 38
Chapter 15, p. 326	**Lesson 1, Beauty in Structures, p. 327**	2	1. Describe the architectural characteristics of a particular historical period. 2. Explain how the architecture of the past has influenced the architecture of the present. 3. Demonstrate how culture and technology influence the design of architecture. 4. Describe and compare the influence of Walter Gropius and Frank Lloyd Wright on modern architectural design.	Image and Activity Bank 15-1 Exploration Worksheet, p. 41 Transparency 56 Transparency 57 Exploration Worksheet, p. 43
	Lesson 2, A Question of the Environment, p. 339	2	1. Describe the relationship between urban design and the environment. 2. Recognize the importance of creating an urban environment that conforms to the needs of the people living there. 3. Evaluate how well an urban setting meets social as well as environmental concerns.	Image and Activity Bank 15-2 Exploration Worksheet, p. 44
End of Chapter, p. 344				Chapter 15 Review Worksheet, p. 46 Chapter 15 Assessment Worksheet, p. 48
End of Unit, p. 346				Unit 5 Activity Worksheet, p. 51 Unit 5 Review Worksheet, p. 52 Unit 5 End-of-Unit Assessment, p. 54 Unit 5 Activity Assessment, p. 59 Unit 5 Self-Evaluation of Achievement, p. 62

* Estimated time is given in number of 50-minute class periods. Actual time may vary depending on period length and individual class characteristics.

▼ Transparencies are available to accompany these worksheets. Please refer to the Teaching Transparencies Cross-Reference chart in the Unit 5 Teaching Resources booklet.

Materials Organizer

Chapter	Page	Activity and Materials per Student Group
14	303	**Exploration 1:** 4 index cards (each approximately 7.6 cm × 12.5 cm); 20 cm of narrow masking tape; scissors; metric ruler; set of standard masses
	305	**Exploration 2:** dull scissors; sheet of heavy paper; stick of modeling clay
	307	**Exploration 3:** 2 meter sticks; clamp; force meter from Unit 4 or masses with hooks; several rubber bands
	311	**Exploration 4:** force meter from Unit 4 or masses with hooks; metric ruler; 33 index cards; several small pieces of masking tape; small container of glue; 15 cm of string; 5 sheets of heavy paper (20 cm × 15 cm); small pail filled with sand; piece of cardboard or wood (20 cm × 20 cm); metric scale
	315	**Exploration 5, Activity 1:** 12 cm × 30 cm wooden board with a hole drilled at each end; 10–15 cm wooden post; standard masses; 2 supports (blocks or books); 50 cm of string; 2 eye screws or thumbtacks; **Activity 2:** 6 strips of plywood or corrugated cardboard; nails and a hammer (if wooden boards are used); tacks or glue; standard masses; 2 supports (blocks or books); saftey goggles
	318	**Exploration 6:** 2 large index cards; standard masses or 3–5 paper cups and about 500 cm^3 of sand; 1–2 sticks of modeling clay; protractor; metric ruler; half of an aluminum can; 2 supports (blocks or books)
15	343	**Exploration 3:** materials to construct a model town, such as uncooked spaghetti, string, balsa wood, index cards, cardboard, cloth, and plastic wrap; glue; metric ruler; roll of transparent tape

Advance Preparation

Exploration 4, Investigations 2 and 3, pages 312–313: You may wish to have students glue the cards together ahead of time.
Exploration 6, Activity 3, page 319: Cut tin cans in half vertically and remove their ends if you choose to use tin cans instead of notecards.

Unit Compression

In Chapter 14, the first section of Lesson 1, Meeting of the Council, may be omitted if necessary. Further compression of this chapter is not recommended, but note that several sections may be assigned as reading ahead of time in order to speed in-class discussion.

You may find it necessary to omit Exploration 1, A Journey Through History, from Chapter 15 in order to save time. If you are under severe time constraints, it is possible to skip this chapter altogether. Keep in mind, however, that this deprives students of exposure to the cultural, historical, artistic, and environmental aspects of building design.

Homework Options

Chapter 14	See Teacher's Edition margin, pp. 296, 299, 301, 302, 304, 305, 307, 310, 315, 317, 320, and 325 Theme Worksheet, p. 4 Graphing Practice Worksheet, p. 8 Activity Worksheet, p. 18 Discrepant Event Worksheet, p. 28 SourceBook, pp. S90 and S93
Chapter 15	See Teacher's Edition margin, pp. 329, 333, 336, 339, and 345 SourceBook, p. S95
Unit 5	Activity Assessment Worksheet, p. 59 SourceBook Activity Worksheet, p. 63

Assessment Planning Guide

Lesson, Chapter, and Unit Assessment	SourceBook Assessment	Ongoing and Activity Assessment	Portfolio and Student-Centered Assessment
Lesson Assessment Follow-Up: see Teacher's Edition margin, pp. 303, 313, 323, 338, and 343 **Chapter Assessment** Chapter 14 Review Worksheet, p. 35 Chapter 14 Assessment Worksheet, p. 38* Chapter 15 Review Worksheet, p. 46 Chapter 15 Assessment Worksheet, p. 48* **Unit Assessment** Unit 5 Review Worksheet, p. 52 End-of-Unit Assessment Worksheet, p. 54*	SourceBook Unit Review Worksheet, p. 64 SourceBook Assessment Worksheet, p. 69*	Activity Assessment Worksheet, p. 59* **SnackDisc** Ongoing Assessment Checklists ♦ Teacher Evaluation Checklists ♦ Progress Reports ♦	Portfolio: see Teacher's Edition margin, pp. 298, 309, 320, 323, and 334 **SnackDisc** Self-Evaluation Checklists ♦ Peer Evaluation Checklists ♦ Group Evaluation Checklists ♦ Portfolio Evaluation Checklists ♦

* Also available on the Test Generator software
♦ Also available in the Assessment Checklists and Rubrics booklet

Science Discovery is a versatile videodisc program that provides a vast array of photos, graphics, motion sequences, and activities for you to introduce into your *SciencePlus* classroom. *Science Discovery* consists of two videodiscs: Science Sleuths and the Image and Activity Bank.

Using the *Science Discovery* Videodiscs

Science Sleuths: The Collapsing Bleachers Side B

The South Youth Organization built bleachers for Water Derby Day. One of the bleachers collapsed, and now the city is threatening to take away the organization's funding. The head of the organization thinks that the bleachers were sabotaged by someone who hates the group. The Science Sleuths must analyze the evidence for themselves to determine the true cause of the structure's failure.

Interviews

1. Setting the scene: Youth organizer **222 (play ×2)**

2. Front-left witness **860 (play)**

3. Back-right witness **1672 (play)**

4. Front-right witness **2370 (play)**

5. Engineer **3064 (play)**

6. Police officer **3622 (play)**

Documents

7. Flyer for Water Derby Day **4160 (step ×2)**

8. Police report #1 **4164 (step)**

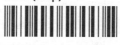

9. Police report #2 **4167 (step)**

10. Police report #3 **4170 (step ×3)**

11. Minutes from Community Center meeting **4175 (step)**

12. Anonymous threatening letter **4178**

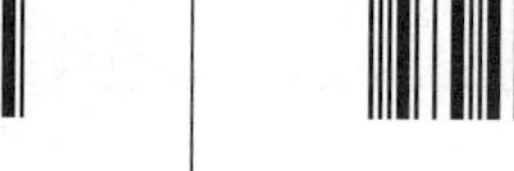

Literature Search

13. Search on the words: BLEACHERS, BUILDING, CONSTRUCTION, DERBY DAY, FOUNDATION, SAND **4180**

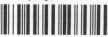

14. Article #1 ("Mexico City's Floating Theater") **4182 (step)**

15. Article #2 ("Wood Tips Column") **4185 (step)**

16. Article #3 ("Bleachers Collapse at Derby Day") **4188 (step)**

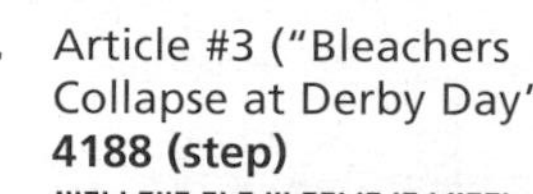

Sleuth Information Service

17. Building plans for bleachers **4191**

18. Materials from collapsed bleachers **4193 (step)**

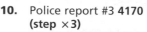

19. Map of Derby site **4196**

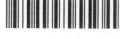

20. Material strength **4198 (step)**

Sleuth Lab Tests

21. Computer simulations of forces on braced and unbraced bleachers **4204 (play)**

22. Simulation to show the effect of vibrating forces on a wet-beach base **4663 (play)**

Still Photographs

23. Derby site **5398**

24. Site of bleachers collapse **5400 (step)**

25. Pieces of broken bleachers **5403 (step ×5)**

Image and Activity Bank Side A or B

A selection of still images, short videos, and activities is available for you to use as you teach this unit. For a larger selection and detailed instructions, see the Videodisc Resources booklet included with the Teaching Resources materials.

14-1 Planning the Structure, page 295

Tree trunk 1981
Trees are some of the oldest and largest standing structures. The large trunk of a tree helps support the incredible weight of the tree.

Skeletal system; human 4154
The skeleton carries the weight of the body and provides solid anchors for muscles.

Building, damaged 1936
This building has suffered earthquake damage. The environment can have a negative effect on structures.

X ray; broken bone 1937
A broken bone is a failed structure. The high-energy waves of X rays can be used to view bone

◀‖ Step Reverse   Play ▶   Pause ‖   Step Forward ‖▶

fractures beneath the skin's surface.

Earthquake demonstration; building without braces 16563–16878 (play ×2) (Side B only)
These buildings have no diagonal bracing. Note their movement during an earthquake.

Earthquake demonstration; building with braces 16879–17105 (play ×2) (Side B only)
Note the effect of an earthquake on a building with cross-bracing.

Earthquake demonstration; sand or rock foundation 17106–17373 (play ×2) (Side B only)
The building in the background is built on rock. The one in front is built on sand.

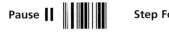

14-2 Response of Structures to Force, page 304

Force, tensile 2002
Tensile force is exerted on the rope by a force on each end. The rope is stretching.

Strength, tensile; measurement 1731–1734 (step ×3)
The tensile strength of a material can be determined by measuring the force necessary to pull a sample apart. (step) Special grips hold the sample at each end. The separation distance between the grips is then increased, and the sample is pulled until a fracture occurs. (step) The specimen under the tensile load begins to tear apart as the maximum strength is approached. The fibers in the rope begin to break near the strength limit. (step) Instruments are connected to the strength-testing machine to measure the load being placed on the sample. The measured load can then be used to find the strength of the material being tested.

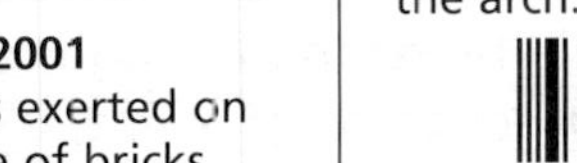

Force, compressive 2001
Compressive force is exerted on the column by a pile of bricks.

Deflection due to force 2003
Deflection is the amount a plank bends when a force is exerted on it.

Force resistance; walls 2004
Thin walls only have to resist vertical forces. Thick walls are required to withstand vertical and horizontal forces.

14-3 Engineering Structures, page 314

Beam 1939
High-strength metal beams support large structures.

Cantilever; traffic signal 1940
A long metal rod acts as a cantilever. The self-supporting limb extends horizontally to support a stoplight above the street.

Cantilever; balcony 1942
Cantilevers can be seen in many balconies. This balcony extends from the house with cantilevers to support its weight.

Arch, stone; Algeria 1953
The stones of an arch support one another. The bottom of each stone in the arch receives pressure from its two neighboring stones. This pressure holds up the arch.

Arch; mosque 1956
Self-supporting arches can be seen all around the world. The bottom of the arch remains in compression to support the mass of the arch.

Arch; geological 1958
Arches can be formed by natural forces. The interaction between neighboring materials within the solid rock supports the weight of the arch.

Aqueduct, Roman 1965
The aqueduct was used to carry water to farms and settlements.

Domes 1959–1960 (step)
Domes are similar to arches. Both semicircular structures support their own mass under the force of gravity. (step)

Bridges 1966–1967 (step)
This bridge is in Scotland. (step) This is a Yugoslavian bridge.

Bridges 1970–1980 (step ×9)
This is a bridge in London. (step) This bridge is in Japan. (step ×2) These bridges are in Venice. (step) This covered bridge is in Germany. (step) This bridge has a simple framework. (step) Large vertical structures support heavy loads that travel across bridges such as this one in Thailand. (step) The two cantilevers on each side of this drawbridge lift to allow large boats to pass through the bridged area. (step) The length of this suspension bridge is held up by the tensile strength of long vertical cables. (step) The Golden Gate Bridge in San Francisco is one of the largest suspension bridges in the world. (step) Here two cantilevers support a third block of wood that acts as a bridge.

15-1 Beauty in Structures, page 327

Structures, large and ancient 1985–1988 (step ×3)
Extremely large structures have been built by humans. These ancient structures have survived years of environmental effects. This is the Porch of Maidens in Athens. (step) Acropolis (step) Dodona Theater (step) This is the Royal Palace in Thailand.

Dome; igloo 1962
Eskimos make good use of on-site materials to build their igloos. Blocks of snow are cut with a saw and stacked to form a dome. Though the walls are ice, the inside air is much warmer than the outside air.

Structures, different 1991–1997 (step ×6)
This is a traditional structure in Sumatra. (step) This school building is a common structure. (step) Skyscrapers offer a great deal of building area for a small land space. (step) Skyscrapers are constructed with a rigid skeleton of steel. They must be carefully designed to resist the forces of gravity, wind, and earthquakes. (step ×2) The Washington Monument, located in Washington, D.C., serves as a memorial to the first U.S. president, George Washington. (step) Built for the world's fair, the Space Needle in Seattle is an engineering marvel.

15-2 A Question of the Environment, page 339

Dome; McMurdo Station 1963
McMurdo Station is a science lab on Ross Island, near Antarctica.

Structures, ancient 1983–1984 (step)
This is Montezuma Castle National Monument, located in southwest Colorado. Ancient structures such as this have survived years of environmental effects. (step) Machu Picchu was built high in the Andes by ancient Incas.

House, elevated; Belize 1989
This traditional house in Belize is elevated to protect it from flooding.

Hut, grass; Guatemala 1990
A traditional structure in Guatemala

Home, earth-sheltered 2045–2046 (step)
Homes are built into the land in order to maintain a constant temperature within the home. The ground acts as a cooling system in the summer and as an insulating system in the winter. (step)

◀| Step Reverse   Play ▶   Pause || Step Forward |▶

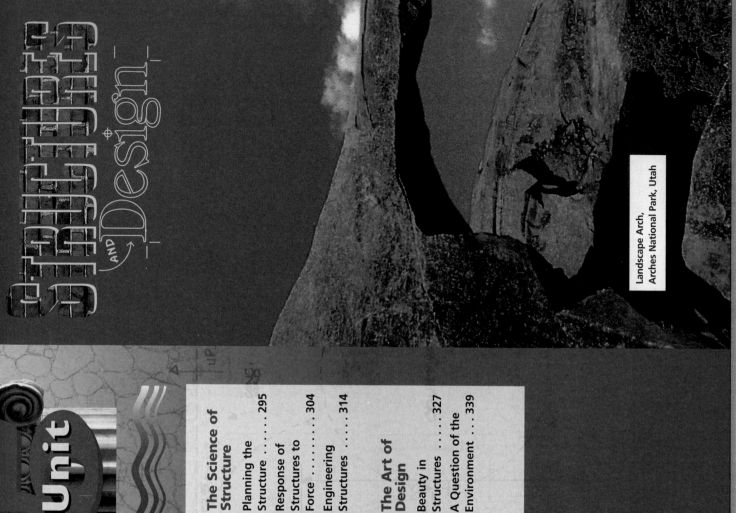

Unit 5 Structures and Design

UNIT FOCUS

Display a photograph, magazine picture, or poster of a large urban area. Call on students to identify the different kinds of structures they see in the illustration. Record their responses on the chalkboard. *(Student responses might include such structures as roads, bridges, houses, stores, churches, piers, and so on.)* Then involve students in a brief discussion of what keeps a building from falling down. *(Most students will recognize that the materials used and the underlying supports, such as beams and columns, keep a building from falling.)*

A good motivating activity is to let students listen to the English/Spanish Audiocassettes as an introduction to the unit. Also, begin the unit by giving Spanish-speaking students a copy of the Spanish Glossary from the Unit 5 Teaching Resources booklet.

Landscape Arch, Arches National Park, Utah

Connecting to Other Units

This table will help you integrate topics covered in this unit with topics covered in other units.

Unit 1 Interactions	Humanity's interactions with the environment depend largely on how and where humans build structures.
Unit 2 Diversity of Living Things	The principles of classification can be applied to structures, such as different bridge designs.
Unit 4 Force and Motion	When a structure fails, it is due to the natural and human-made forces acting on the structure.
Unit 6 The Restless Earth	All structures should be strong enough to withstand earthquakes and other tectonic events.
Unit 7 Toward the Stars	Consideration of how structures withstand forces is critical to the construction of successful spacecraft.

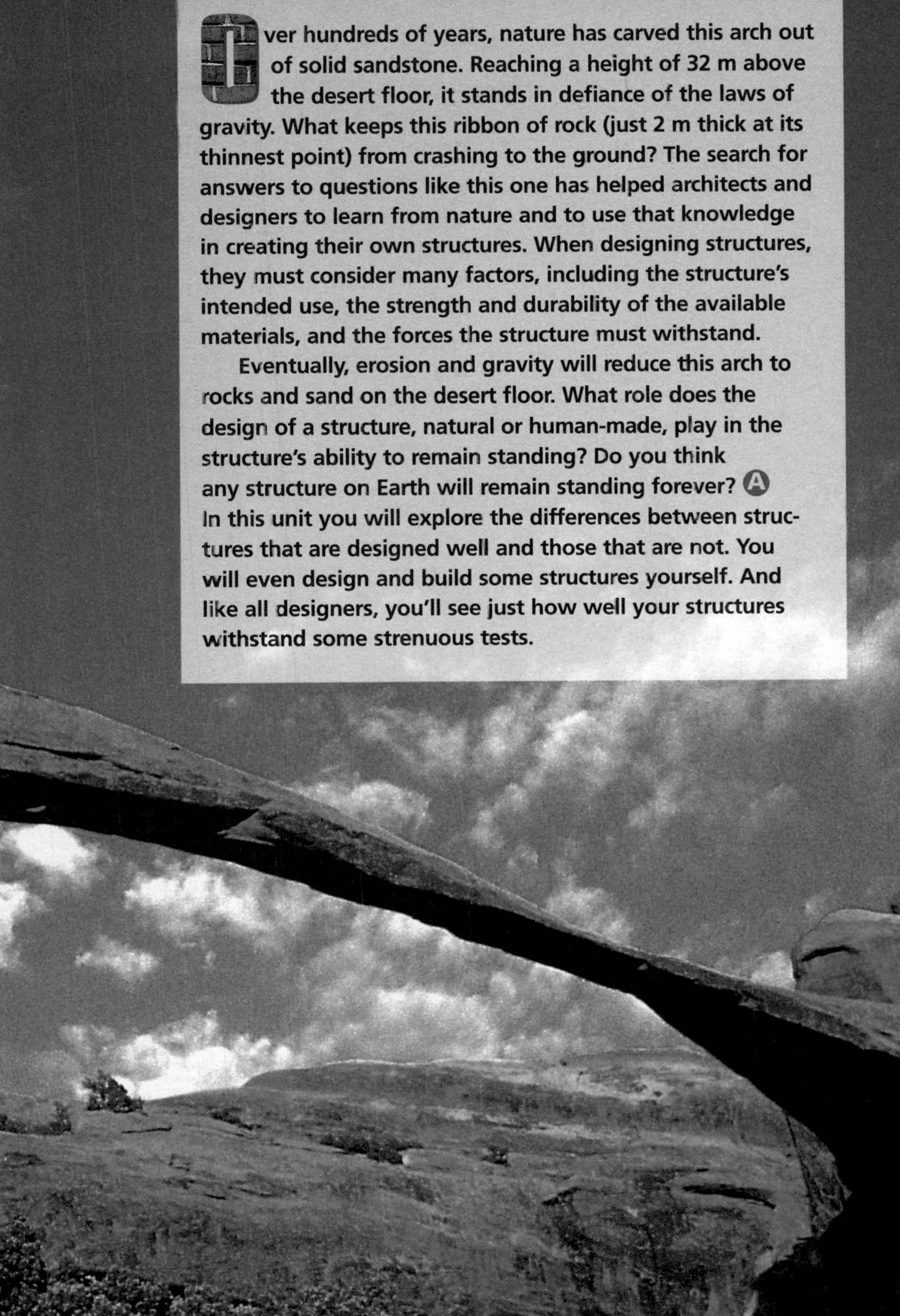

Over hundreds of years, nature has carved this arch out of solid sandstone. Reaching a height of 32 m above the desert floor, it stands in defiance of the laws of gravity. What keeps this ribbon of rock (just 2 m thick at its thinnest point) from crashing to the ground? The search for answers to questions like this one has helped architects and designers to learn from nature and to use that knowledge in creating their own structures. When designing structures, they must consider many factors, including the structure's intended use, the strength and durability of the available materials, and the forces the structure must withstand.

Eventually, erosion and gravity will reduce this arch to rocks and sand on the desert floor. What role does the design of a structure, natural or human-made, play in the structure's ability to remain standing? Do you think any structure on Earth will remain standing forever? Ⓐ In this unit you will explore the differences between structures that are designed well and those that are not. You will even design and build some structures yourself. And like all designers, you'll see just how well your structures withstand some strenuous tests.

293

Using the Photograph

Call on a student to read the text on this page aloud. Have students describe the events that might have led to the creation of this arch. *(Students may describe wind or water flowing under the bridge and creating the space underneath it little by little.)* Ask: Why might a break at the thinnest point cause the whole bridge to collapse? *(Accept all reasonable responses; students will be better able to answer this question after studying arches.)*

Answers to *In-Text Questions*

Ⓐ Students should understand that a structure's design can have a substantial effect on its ability to remain standing. It is unlikely that any structure will remain standing forever.

Chapter 14 The Science of Structure

Connecting to Other Chapters

***Chapter 14* considers the integrity of structures from the standpoint of forces, materials, and design.**

***Chapter 15* offers historical, cultural, artistic, and environmental perspectives on the science of structure.**

Prior Knowledge and Misconceptions

Your students' responses to the ScienceLog questions on this page will reveal the kind of information—and misinformation—they bring to this chapter. Use what you find out about your students' knowledge to choose which concepts and activities to emphasize in your teaching.

In addition to having students answer the ScienceLog questions on this page, you may wish to pose questions such as the following:

1. What are some examples of structures?
2. What is a force?
3. What are some examples of forces?
4. What does the word *design* mean?

Emphasize that you are not looking for right or wrong answers to these questions. If you wish, use these questions to prompt a class discussion and to encourage students to brainstorm for answers to these questions. It may be helpful to record a list of student responses on the board. The responses may help you gauge what students already know about structures and forces, what misconceptions students may have, and what is interesting to them about this topic.

CHAPTER 14 The Science of Structure

1 **What can be learned by studying natural and human-made structures that have not failed?**

2 **What forces might act on this structure? What effects could these forces have?**

3 **What features of this bridge help keep it standing?**

ScienceLog

Think about these questions for a moment, and answer them in your ScienceLog. When you've finished this chapter, you'll have the opportunity to revise your answers based on what you've learned.

LESSON 1 Planning the Structure

Meeting of the Council

The mayor of Zenith could not be heard over the noise in the chamber of the city council.

"Come to order! Come to order!" Mayor Jones pounded the gavel at his desk. The citizens in the gallery and the council members seated in the chamber continued to argue.

Seldom had there been an issue that stirred up so much debate among the citizens of Zenith. ABC Developers was seeking the approval of the Zenith City Council to redevelop the old neighborhood of Riverwood, located across the river from the city's business center.

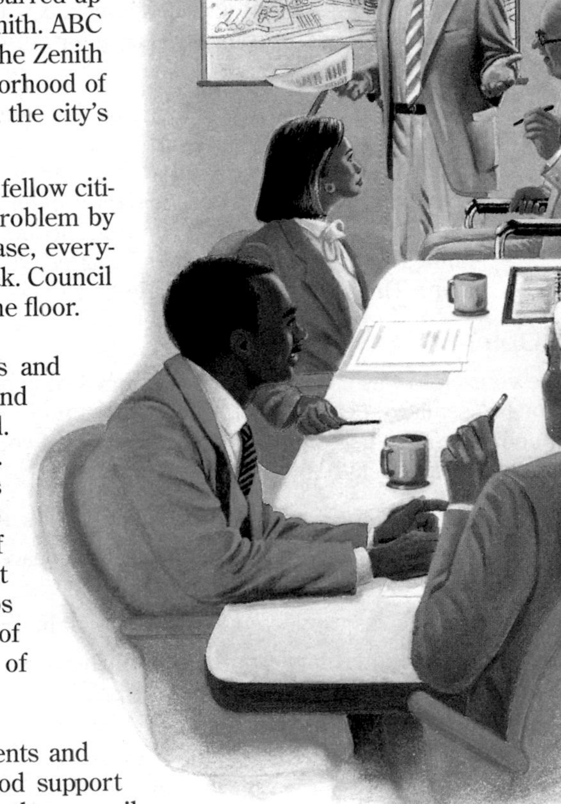

Mayor Jones: Fellow council members and fellow citizens, we will not solve this problem by shouting at one another. Please, everyone will have a chance to speak. Council Member Takashi, you have the floor.

Council Member Takashi: I have received many letters and phone calls from residents and business owners in Riverwood. They support the proposal. With the money ABC has offered them, they will be able to relocate to other parts of the city. The redevelopment will generate many new jobs involving the construction of new buildings. The city of Zenith will benefit greatly.

Council Member Wilson: Mr. Mayor, not all the residents and business owners in Riverwood support ABC's proposal. I present to the council a signed petition from more than 2000 residents and business owners opposing this proposal. I also present a letter from the Zenith Heritage Foundation that opposes the proposal as it now stands. They write that "the development would mean the destruction of many historic homes and buildings. The new buildings would obstruct the view of the beautiful mountains that surround our city."

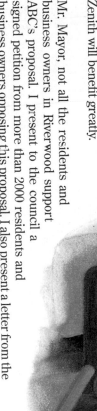

LESSON 1 ORGANIZER

Time Required
3 class periods

Process Skills
analyzing, hypothesizing, predicting

Theme Connection
Structures

New Terms
none

Materials (per student group)
Exploration 1: 4 index cards (each approximately 7.6 cm × 12.5 cm); 20 cm of narrow masking tape; scissors; metric ruler; set of standard masses

Teaching Resources
Resource Worksheet, p. 3
Theme Worksheet, p. 4
Exploration Worksheet, p. 6
Transparency 49
SourceBook, p. S90

LESSON 1 Planning the Structure

FOCUS

Getting Started

Explain to the class that only a few cities were designed before they were built. Tell students that one example is Washington, D.C. George Washington chose the city's site in 1791. He then hired a French engineer, Pierre Charles L'Enfant, to plan the layout of the city. Ask: What are the advantages of planning a city before it is built? *(Sample answer: City planners could set aside places for parks and public buildings, and housing could be planned to reduce the feeling of overcrowding.)*

Main Ideas

1. Opinions may differ about the best solution to a complex problem.
2. All structures must be designed to withstand destructive forces.
3. Problems of design and structure may have a variety of solutions.
4. Valuable information about designing structures can be learned from studying structures that have failed and those that have survived.

TEACHING STRATEGIES

Ask students what is meant by the term *structure* and to identify some of the different kinds of structures they see every day. List their suggestions on the chalkboard. Then ask students to identify ways in which these structures are designed to fulfill their functions. For example, an airplane has wings that enable it to fly, and a house has beams that support its roof. Point out that in this lesson, students will learn about the relationship that exists between design and structure.

Teaching Strategies for Meeting of the Council are on the next page. ►

Meeting of the Council,

pages 295–296

Have students read page 295 silently, and provide them with time to examine the pictures of present-day Riverwood and the proposed redevelopment plan.

GUIDED PRACTICE Help students to identify and summarize the two points of view presented by council members Takashi and Wilson. Have students discuss the merits of both points of view before continuing on to page 296. Using paper ballots, ask them to vote on which point of view they agree with based on the information they have so far. Record the results on the board.

Have students continue reading the debate of the future of Riverwood. When they have finished, invite them to discuss whether or not their previous opinions have changed with this additional information.

As an alternative, have a group of students role-play the city council scenario for the class. When they have finished, involve the class in a discussion of the issues.

Multicultural Extension

Benjamin Banneker

Benjamin Banneker was an astronomer, farmer, mathematician, and surveyor. In 1791, Banneker was an assistant to Major Andrew Ellicott, the surveyor appointed by President Washington to lay out the boundaries of the District of Columbia. Banneker was recommended for this post by Secretary of State Thomas Jefferson. Benjamin Banneker's many accomplishments made him one of the best-known African Americans in early United States history. Suggest that interested students do some research to find out more about him.

Council Member Tufts: Mr. Mayor, Riverwood is an eyesore. The homes and businesses have fallen into disrepair. The woods and marsh are littered with garbage and debris. And the woods are becoming a hazard, especially at night. Something must be done with this area.

Mayor Jones: Many civic organizations have asked to speak to the council about ABC's proposal. I call upon the president of the Environmentalists' Association, Mónica Rodriguez.

Ms. Rodriguez: Mr. Mayor and council members, our organization is concerned about the proposed destruction of the existing marsh and woods. This area is a home for birds and other wildlife and a place where children from all the surrounding neighborhoods come to play. We should preserve this area for our children and for the future.

Mayor Jones: The future of this city belongs to the youth. Let's find out what they would like to do. I suggest we ask all the seventh-, eighth-, and ninth-grade science classes in the city to send their proposals to the council. Then we will invite the people whose homes and businesses might be affected to comment. The council will make the final decision.

Your Opinion

Suppose your class was asked to find a solution to this problem. With your classmates, decide what you think should be done to Riverwood. Have one person record the decisions of your class. Later you will refer back to these decisions.

1. Would you have all the existing structures torn down?
2. Would you alter the environment to make more space for the new buildings? Explain why or why not.
3. What new structures would you like to see built in this area?
4. Will the town benefit more from a new business center or a historic neighborhood? Why do you think so?

An At-Home Assignment

As you learn about structures and design, complete your own plan for Riverwood. This plan should be in the form of a detailed sketch like the one from ABC Developers. Later your class will decide on a plan, and together with your classmates, you can construct a model based on this plan.

296

Did You Know. . .

Spanish settlers in the southwestern United States brought with them not only Spanish architecture but also techniques of irrigation. The Franciscan missionaries directed the construction of canals, called *acequias*, that were used to supply water to missions in many areas. Stretches of one rock-lined *acequia*, Acequia Madre in San Antonio, Texas, are still in use after more than two centuries.

Homework

Propose the following scenario to students: You are an adult who grew up in Riverwood before the redevelopment was implemented. Now, 20 years later, you move back and notice many changes to your old neighborhood. List five good changes and five bad changes.

The present Riverwood

ABC Developers' proposed redevelopment of Riverwood

An At-Home Assignment, *page 296*

As an alternative, you may wish to have the groups that discussed the issues earlier work together to complete this assignment.

INDEPENDENT PRACTICE Have students write three short scripts that could be possible endings to the Meeting of the Council. Each script should be no more than one page long. Students can present their possible outcomes to the class, or the scripts could be posted around the room for all to read.

Meeting Individual Needs

Learners Having Difficulty

Ask students to analyze the following situation: A wooden bridge that was built on a children's playground failed. City officials need to know what caused the bridge to fail. Which of the following probably did *not* contribute to the failure?

a. the materials the bridge was made of
b. the size of the posts holding it up
c. the number of workers who built the bridge *(Correct)*
d. the design of the bridge

Your Opinion, *page 296*

You may wish to divide the class into groups of four or five students to discuss and debate the future of Riverwood. Suggest that each group select a leader to guide the discussion, and have one member take notes. Explain that each group should reach an agreement on a solution to the problem. When the groups have finished their discussions, have the leader of each group present a summary of the group's conclusions to the class.

Answers to *Your Opinion, page 296*

1. Students should discuss both the positive aspects of tearing down the structures (such as removing an eyesore) and the negative aspects (such as destroying historical buildings).
2. Sample answer: I would not alter the environment because saving the existing habitat is more important than having the new buildings be conveniently located.
3. Students may suggest shopping malls, sports complexes, or museums, for instance.
4. Students should evaluate and compare the beneficial aspects of both options.

Structures That Failed

Direct students' attention to the photographs on this page, and call on a volunteer to describe what they show. Involve students in a discussion of why scientists might be interested in studying structures that have failed. Then have students read about the Tacoma Narrows Bridge, pausing before the questions. Ask: Does anyone know where Tacoma is located? If no one responds, explain that Tacoma is a large city in the state of Washington. You may wish to display a map of the United States and call on a volunteer to locate and identify Tacoma.

Students may be interested to know that the Tacoma Narrows Bridge spanned 853 m across Puget Sound. It collapsed after only 4 months of use. The bridge had been designed to withstand winds of 195 km/h, but was destroyed by a wind of only 65.6 km/h. The wind caused the deck of the bridge to move up and down in a wavelike, twisting motion. The distance between the crest and trough of these waves eventually reached 9 m. After several hours, the deck began to twist back and forth until some of the suspending cables snapped, plunging part of the bridge into the water. Because the bridge was closed after it started to sway, no lives were lost.

Answers to *In-Text Questions*

Have students read and discuss the questions. Be sure they understand that all responses are acceptable. Remind them that they will have a chance to review their answers as they work through the unit. Use the following sample responses to guide students' reasoning as they complete the questions:

1. The bridge failed because the design did not provide a frame that was stiff enough to resist moderate winds.
2. The bridge was well constructed, but the design was at fault.
3. The materials were not inferior, but they were too light. (Inform students that, at the time, the engineers believed that the structure of a suspension bridge should be light so that the size of the cables could be minimized.)
4. Oscillations of the bridge's deck and wind were the main factors in the collapse of the bridge.
5. Bridges should be designed with stiffer and heavier decks that are more resistant to twisting and lateral deflection.

PORTFOLIO

Suggest that students record their answers to the questions in Structures That Failed in their Portfolio. Their answers should emphasize analytical reasoning. They may wish to make notes or add information as they increase their knowledge.

Structures That Failed

Before you design new structures for Riverwood, you must ensure that your structures will not fall down. To build sturdy structures that are not likely to fall, engineers carefully study structures that have failed. When the Tacoma Narrows Bridge in Tacoma, Washington, collapsed in 1940, engineers around the world took notice.

The engineers who investigated the collapse asked the following questions:

1. Why did the bridge fail?
2. Was it poorly designed or constructed?
3. Were inferior materials used?
4. What were the forces that caused the bridge to collapse?
5. What changes should be made in bridge design and construction to prevent catastrophes of this kind in the future?

Answer these questions. Get together with two or three of your classmates and record your best answers in your ScienceLog. Be sure to consider all of the forces that affected the bridge.

The Tacoma Narrows Bridge rocked and swayed for several hours before finally crashing into the water.

Picture Puzzles

What can you learn from structures that have failed? Study each of the pictures on this page carefully. For each situation, try to answer the following questions:

1. What was the failure?
2. What might have caused the failure? Could it have been the wind, the weight of snow, the weight of the structure itself, an earthquake, or something else?
3. Was the structure being pulled apart (stretched) or pushed together (compressed) before the failure?
4. How could the failure have been prevented?

A fractured humerus (long bone of the upper arm), as seen on an X ray Ⓐ

Cracked beams are a common sight among the ruins of ancient Greece. Ⓑ

What might have caused this destruction? Ⓒ

Leah drew a concept map to analyze the possible causes of the Tacoma Narrows Bridge collapse. Perhaps you could use a similar diagram to answer the questions above. Ⓓ

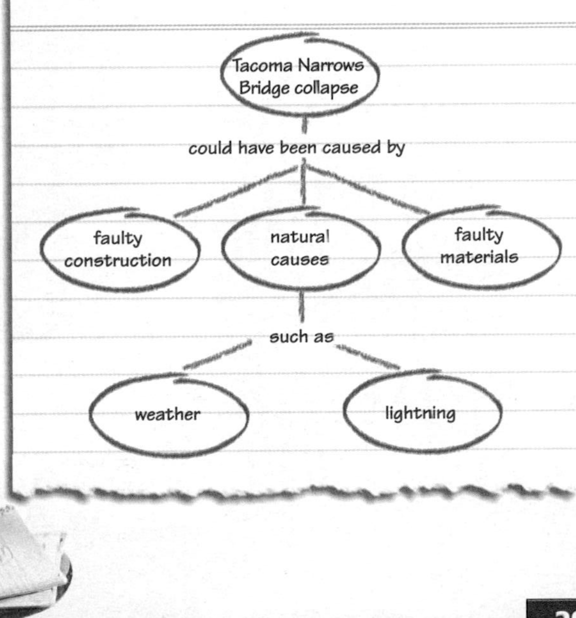

Homework

As a homework assignment, have students think of another example of a structure that failed. After describing the structure, they should write out answers to the four questions on this page.

Picture Puzzles

Have students read the page silently, or call on a volunteer to read it aloud. Point out that the details in a concept map may vary depending on how different individuals interpret the factors contributing to a structure's failure.

Involve students in a discussion of each of the pictures, focusing on the questions presented at the top of the page. Accept all reasonable responses that students support with logical arguments. Answers to the in-text questions may be used to get students started on their concept map.

Answers to *Picture Puzzles*

Ⓐ **1.** The bone broke.
2. The broken bone could have resulted from a fall or from a sporting or automobile accident, for instance.
3. The bone may have been bent, pulled apart, or pushed together.
4. There was probably no way to prevent the bone from breaking except by avoiding the dangerous situation.

Ⓑ **1.** The beams shown in the picture cracked.
2. Factors that could have contributed to the failure may have included the settling of the structure itself, the shifting of the ground under the structure, the weight of the beams, or earthquakes and tremors.
3. Cracks are usually the result of pulling a structure apart.
4. The cracks might have been prevented if the structure had been designed better or built with stronger materials.

Ⓒ **1.** The house was destroyed.
2. The destruction of the house was probably caused by an earthquake. It could also have been the result of a severe storm, such as a hurricane, typhoon, tornado, or flood.
3. The house was torn apart by pushing and pulling.
4. It is possible that the house could have been saved if it had been built of stronger materials and designed to withstand movements of the soil, an earthquake, or a severe storm.

Ⓓ Sample concept map:

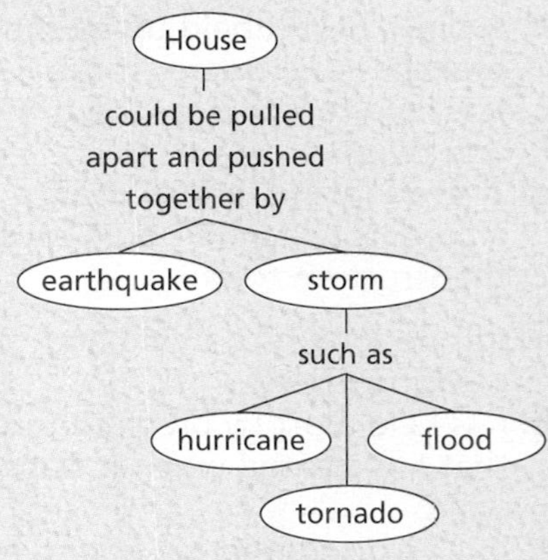

Structures That Haven't Failed Yet

Have students read the page silently. Then direct their attention to the table and discuss what it shows. You may wish to have students construct a table in their ScienceLog and then discuss their ideas. Students should not be expected to recognize all of the major design and structural features of the objects shown in the photographs. The following are some ideas they may consider.

Answers to *Structures That Haven't Failed Yet*

Ⓐ **a.** It was designed as a bell tower for the Pisa Cathedral.
b. The tower is held up by a system of arches and columns.
c. It is held together by mortar, bricks, and stone.
d. A strong earthquake or damage to the tower's foundation could cause the structure to fail. (You may wish to point out to students that the tower leans because its foundation lies on a layer of unstable soil.)

Ⓑ **a.** It is designed to hold back and control huge amounts of water.
b. It is held up by reinforcing rods inside its concrete walls and by the curved design of its structure.
c. It is held together by concrete and steel reinforcing rods and beams.
d. Earthquakes and weathering could cause the structure to fail.

Ⓒ **a.** It is designed to hold and protect eggs and to provide shelter for newly hatched birds.
b. A bird's nest is held up by the branches of a tree or shrub.
c. It is held together by the grasses and twigs.
d. A strong wind or disturbance by a predatory animal could cause the structure to fail.

Ⓓ **a.** It is designed for flying and catching insects.
b. It is held aloft by pressure forces that are created when it flaps its wings.
c. It is held together by bones, muscles, skin, and connective tissues.
d. Crashing into solid objects, severe weather, or illness could cause the structure to fail.

Structures That Haven't Failed Yet

Much can be learned about how to build safe, strong, and beautiful structures by examining those that have not failed yet. Study the following eight pictures of natural and human-made structures that have been successful over the years.

In small groups, discuss the following for each structure:

a. What is it designed to do?
b. What holds it up?
c. What holds it together?
d. What could cause it to fail?

Keep a record of your discussion in a table such as the one shown below.

Structure	Purpose of design	Held up by	Held together by	Possible causes of failure
Leaning Tower of Pisa	bell tower for the Pisa Cathedral	a system of arches and columns	mortar, bricks, and stone	a strong earthquake ?
Hoover Dam				

Hoover Dam, Colorado River, Arizona-Nevada border Ⓑ

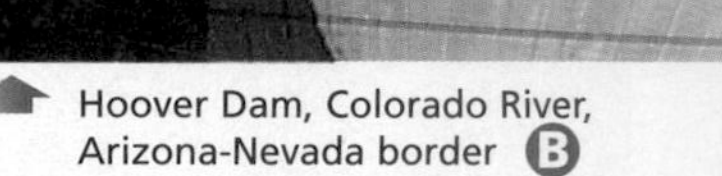

Bird's nest Ⓒ

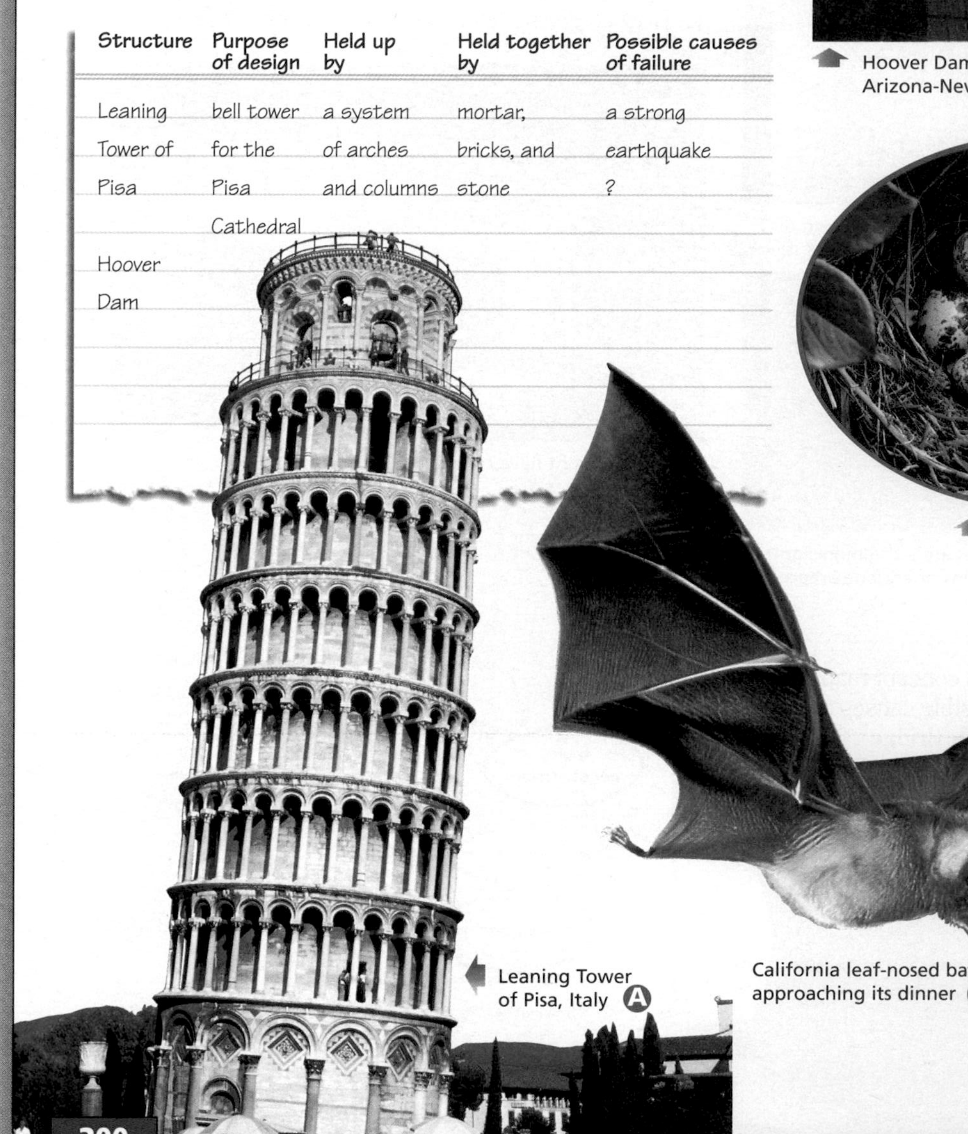

Leaning Tower of Pisa, Italy Ⓐ

California leaf-nosed bat approaching its dinner Ⓓ

Answers to Structures That Haven't Failed Yet continued ▶

★ **A Resource Worksheet (Teaching Resources, page 3) and Transparency 48 are available to accompany Structures That Haven't Failed Yet.**

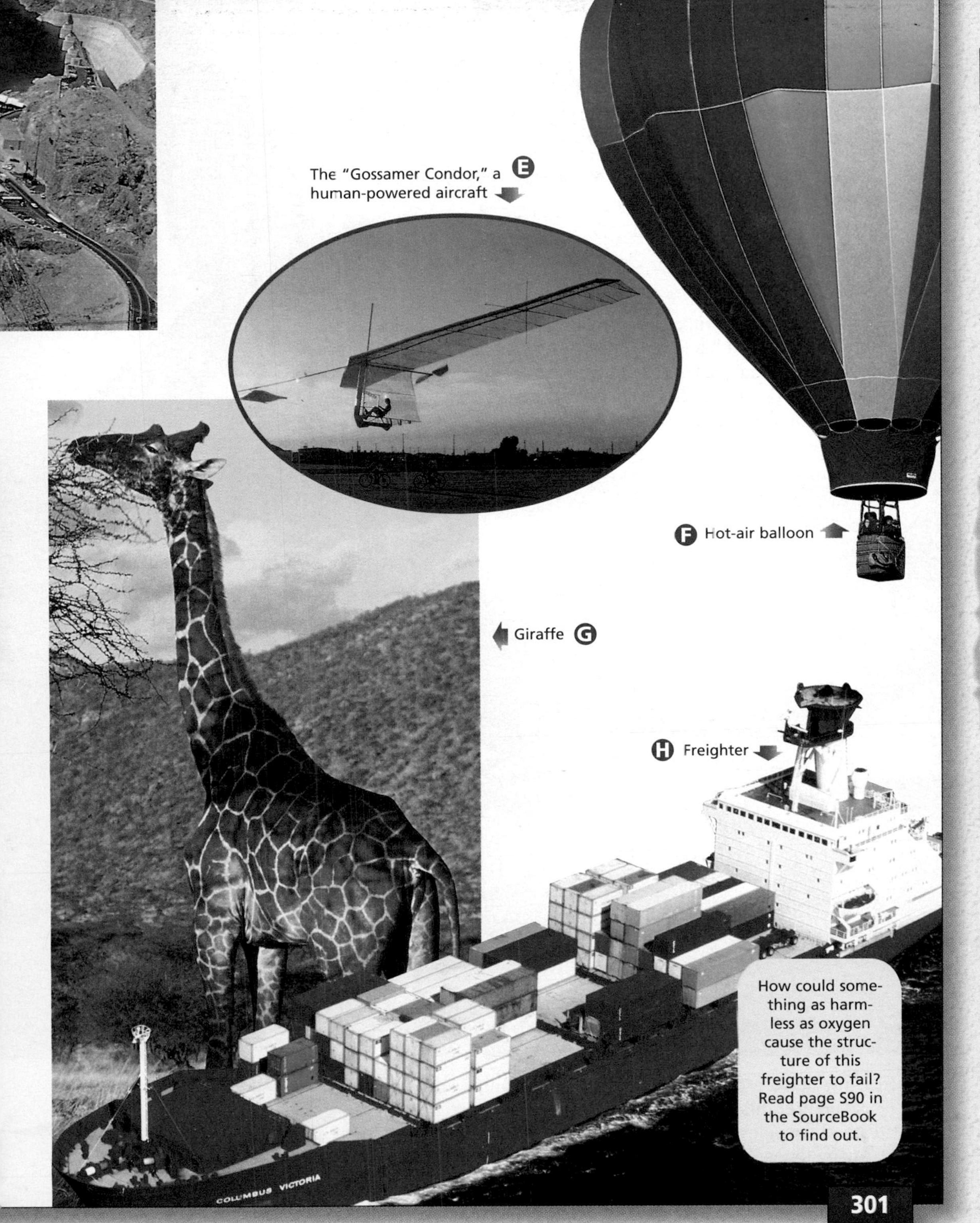

The "Gossamer Condor," a human-powered aircraft (E)

(F) Hot-air balloon

Giraffe (G)

(H) Freighter

How could something as harmless as oxygen cause the structure of this freighter to fail? Read page S90 in the SourceBook to find out.

301

Integrating the Sciences

Life and Physical Sciences
Among the many species of birds, there is an almost endless variety of designs used by birds in creating their nests. Ask students to do research to find out the many ways that birds solve their design problems. Some birds of interest include tailorbirds, who sew leaves together, and weavers, who use elaborate knots and stitches. Ask students to point out any parallels that they can see with human design.

Homework

For homework, you may wish to provide students with the Theme Worksheet on page 4 of the Unit 5 Teaching Resources booklet. It supplements the Integrating the Sciences strategy described above.

Theme Connection

Structures
Focus question: In designing a bridge, what factors might a designer take into account when choosing a system of structural support? *(Designers should take into account all of the forces in the environment that could interact with the bridge structure to cause its failure, such as the amount and type of traffic the bridge is expected to support, the distance spanned, wind, and temperature changes.)*

Answers to *Structures That Haven't Failed Yet, continued*

E **a.** It is designed to fly through the air by human power.
b. It is held up by the force of lift created by the wind passing over and under its wings.
c. It is held together by a framework of supporting structures under a thin fabric skin.
d. Strong winds, severe weather, or a crash could cause the structure to fail.

F **a.** It is designed to float in the air while carrying passengers.
b. It is held up by the buoyancy, or upward force, exerted by the cooler air outside of the balloon.
c. It is held together by the strong cord and thread used to sew together the large panels of nylon or polyester.
d. Strong winds or tears in the fabric could cause the structure to fail.

G **a.** Its structure allows it to reach the leaves in the tops of trees and shrubs.
b. It is held up by a skeleton and strong muscles.
c. It is held together by muscles, bones, skin, and connective tissues.
d. A serious fall or injuries from a struggle could cause the structure to fail.

H **a.** It is designed to float on water in order to transport large amounts of cargo across the ocean.
b. It is held up by the buoyancy, or upward force, exerted by the water on which it rests.
c. It is held together by nuts, bolts, rivets, steel beams, and metal sheets.
d. Strong winds and rough seas or a collision with some other object or ship could cause it to fail.

Learning From Nature

After students have read this page, encourage class discussion of designs in nature. Elicit student responses for additional examples.

Answers to *Learning From Nature*

1. Buildings respond to sunlight by trapping the heat in the material from which they are made. Plants respond to sunlight by capturing the energy from the sun to make the food that enables them to grow.
2. **a.** Trees are made of roots that provide anchorage, branches that support leaves, and leaves that capture sunlight to make food. The roots of a tree are comparable to the foundation of a building. Fibers in the wood support the trunk like beams in a building.
   b. Grass is slender, grows upright, and responds to wind by moving back and forth. Skyscrapers are also designed to sway with the wind to prevent collapse.
   c. A sunflower uses its roots for support and has a thick stem that supports the weight of its heavy flower. A sunflower and street lamp are similar in that they both have a long, straight "stalk" that supports a structure at the top.
   d. A bird in flight uses its different types of feathers for lift and drag. The bones of the bird are hollow to make them lightweight. Airplanes and birds both fly through the air. The wing structures of an airplane and of a bird are similar.
3. Accept all reasonable answers.

Homework

You may wish to assign question 3 as a homework activity. Students could also sketch their natural structure and compare it with a sketch of a human-made structure.

Learning From Nature

As you can see from some of the pictures on pages 300–301, not all designs are created by humans. Designs occur in nature, too! The following activities will help you see how structures in nature compare with human-made structures.

1. Compare the way buildings in the city respond to sunlight with the way plants respond.
2. Describe the structures of

   a. a tree c. a sunflower plant

   b. grass d. a bird in flight

   Compare them with human-made structures. For example, compare the way a blade of grass and a skyscraper respond to a gentle wind. What part of a building is comparable to the roots of a tree? How does a sunflower plant compare with a street lamp? How does a bird compare with an airplane?

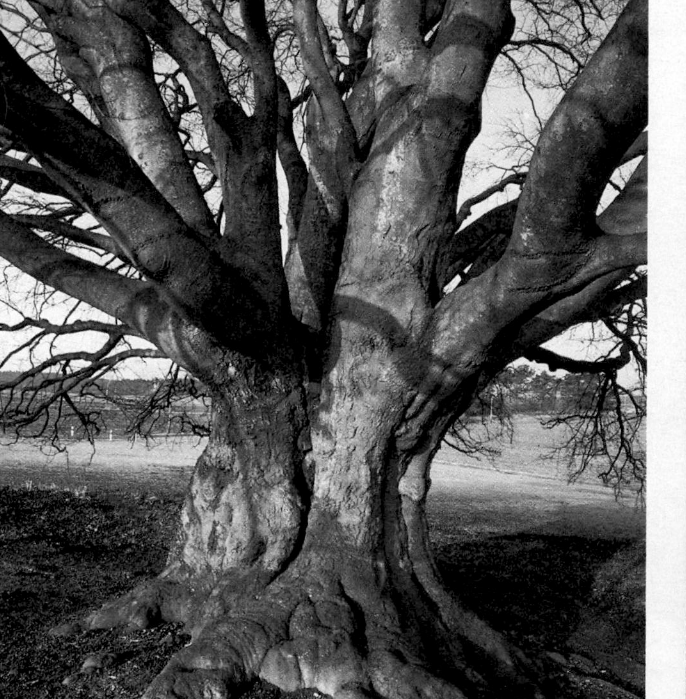

3. Now look around you for another example of a natural structure; this might be something you can see through a nearby window. Or imagine a natural structure you have seen many times. Write a brief description of this structure.

302

Integrating the Sciences

Life and Physical Sciences

Have students investigate examples of coevolution to study the ways that the structure of an organism can serve a purpose for that organism as well as for organisms of another species. An example might be the structure of a flower and the corresponding structure of that flower's pollinator. Have students present their findings in a poster or other classroom display.

Constructing on Your Own

You have examined some structures that failed and others that did not. Can you design a strong, safe structure and build a model of it?

Suppose you decide to include a bridge in your plan for Riverwood. Perhaps you want the bridge to carry automobile traffic across the river between Riverwood and the business center of the city. Or maybe you want a footbridge to cross the river. In either case, your bridge must be strong enough and safe enough for those who will use it.

EXPLORATION 1

Constructing a Model

You cannot design and build the actual bridge, but you can build a model. Try to build the strongest model possible.

You Will Need

- 4 standard index cards, each approximately 7.6 cm × 12.5 cm
- a 20 cm length of narrow masking tape
- scissors
- a metric ruler
- standard masses

What to Do 

1. As shown in the simple diagram below, your model must have the following dimensions. It must cross a gap of 11 cm. To do so, it will have to be at least 12 cm long. The width of the cards should be 6 cm or more.
2. If supporting structures are used, they must leave an unobstructed gap to allow for boat traffic.
3. Your model may rest on books, but it cannot be taped to them.
4. You are allowed to bend, fold, and cut the index cards.

Now comes the fun part! Can you place a mass of 50 g on your bridge without making it collapse? If so, congratulations!

Now continue to add masses to find out which design holds the greatest load. Who had the strongest design? What made it stronger than the rest?

Writing About Your Model

A toy company is requesting suggestions for the construction of cheap, easy-to-make model bridges. You decide to submit your design. Write a report to Trixie's Toy Company that includes the following:

a. a design that works
b. the steps in making the toy bridge
c. the qualities that make it a good design
d. the pitfalls to avoid in building such toy bridges

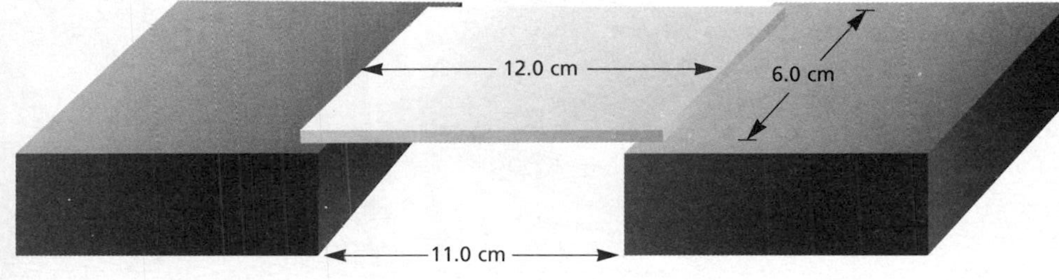

FOLLOW-UP

Reteaching

Students may need additional help in understanding how structures can fail. Have them analyze the teacher's desk. Ask: What could break the top of the desk? What could break the legs? How could you make the desk stronger? Help students realize that different forces may cause structures to fail in different ways.

Assessment

Have students cut pictures from magazines and newspapers that illustrate different kinds of structures. Then have them arrange the pictures on construction paper and write captions to describe the purpose of each structure.

Extension

Tell students to imagine that they have been asked to design a new school. They should work in small groups to think of several proposals. Proposals should include how the project will fit into and make use of the landscape, what materials will be used, and how the design contributes to the function.

EXPLORATION 1

Encourage students to compare and discuss their results as they work to complete the activity. If a hint is necessary, point out that folding the index cards will provide additional strength.

 **An Exploration Worksheet is available to accompany Exploration 1 (Teaching Resources, page 6).**

Writing About Your Model

Call on several volunteers to present their reports to the class. Involve the class in a discussion of the ideas that are presented, and help them to reach a general consensus on the best way of constructing a model bridge.

Meeting Individual Needs

Gifted Learners

As an extension of Exploration 1, students might enjoy constructing a more elaborate bridge. They will need four ring stands (for the supports), cardboard (for the deck), and string (for the cables). Encourage students to try different arrangements of string, which can be threaded through the cardboard. When students have settled on a design that they like, they should then test their designs against forces in the environment such as weight and wind. (A fan can be used to supply the wind.)

To evaluate the bridges, conduct a Science and Technology Olympics. Develop categories such as Strongest Design, Most Efficient Use of Materials, Most Creative Design, and so on. Pick a bronze-, silver-, and gold-medal winner for each category.

Closure

Students may enjoy collecting photographs of structures for a bulletin-board display. After photographs have been collected, ask students to identify the features of the structure that have allowed it to last. *(Answers should include a discussion of materials, design, and appearance.)*

Response of Structures to Force

FOCUS

Getting Started

Stand a cardboard tube from the inside of a roll of bathroom tissue on its end. Balance a book on the tube. Ask the class to guess how many books can be stacked on the tube before it is crushed. Add more books until the tube collapses or the stack tips over. Tell students that builders of structures such as skyscrapers and bridges have to know before they build something just what kinds of forces their structure can withstand.

Main Ideas

1. Forces applied to structures may be classified as either tensile, compressive, or shear.
2. Structures are compressed by their own weight.
3. Materials that respond to the removal of stress by returning to their original shape are said to behave elastically.
4. In an elastic response, tensile force causes an extension that is proportional to the applied force.
5. One measure of strength is the ability of a given material or structure to withstand force without breaking.

TEACHING STRATEGIES

Ask students to consider why it is important to use and understand the correct vocabulary when explaining how to do something or how something works. *(It avoids confusion and misconceptions.)* Then have students read the lesson introduction silently. Ask them to identify any of the words on the list that they used in the description of their model bridge for Trixie's Toy Company.

Homework

Have students complete the Graphing Practice Worksheet on page 8 of the Unit 5 Teaching Resources booklet.

Response of Structures to Force

The designers of Riverwood will need to know whether the structures they design will stand up. Will the structures support their own weight? Can they support a load? How will the structures behave in the wind or during an earthquake? Also, the designers will have to convince the city government that the structures will be safe. To do this, they need to use the right words to communicate their ideas.

Consider your model bridge. What technical words did you use in describing it to Trixie's Toy Company?

Here is a list of words that you can use to describe the response of structures to force. Keep a list of the terms in your ScienceLog. As you learn the meaning of each of these words, write down the definition. Remember to use what you discover in the Explorations to find the best definitions for the terms.

- bending
- compressing
- compression
- compressive force
- contracting
- contraction
- deflection
- extension
- load
- relaxing
- response (respond)
- shear
- shear force
- stretching
- tensile force
- tension

LESSON 2 ORGANIZER

Time Required
4 class periods

Process Skills
hypothesizing, organizing, analyzing

Theme Connection
Structures

New Terms
Compressive force—force that presses a material together
Deflection—the amount that a material bends when a force is applied to it
Elastic response—a material returning to its original shape or size after it has been stretched
Shear force—force that causes bending or twisting in a material
Tensile force—force that causes a material to stretch

Materials (per student group)
Exploration 2: dull scissors; sheet of heavy paper; stick of modeling clay
Exploration 3: 2 meter sticks; clamp; force meter from Unit 4 or masses with hooks; several rubber bands

continued ►

EXPLORATION 2

Three Kinds of Forces: Tensile, Compressive, and Shear

You Will Need

- dull scissors
- heavy paper
- modeling clay

PART 1

1. Study the three structures in Diagram A: a book, a rope, and a suitcase. Do you observe a pair of forces on each? How does each structure respond to the forces? How does each structure change in shape?

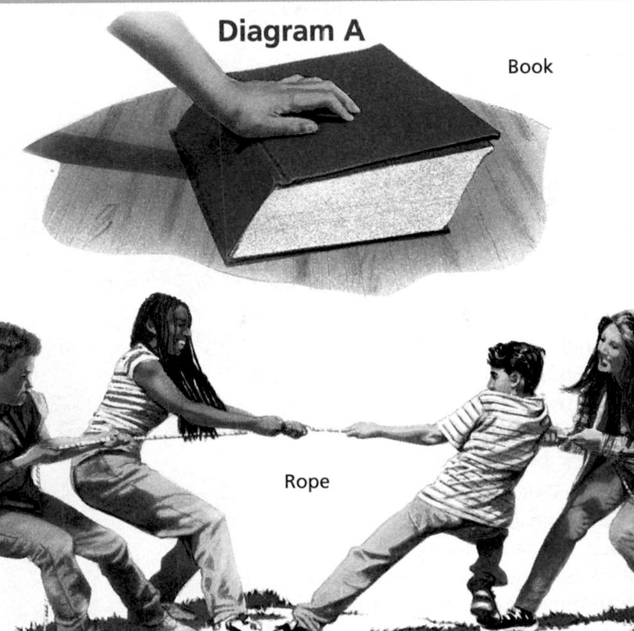

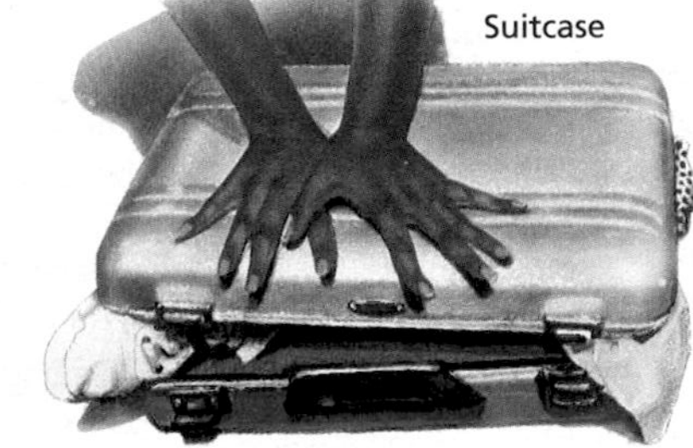

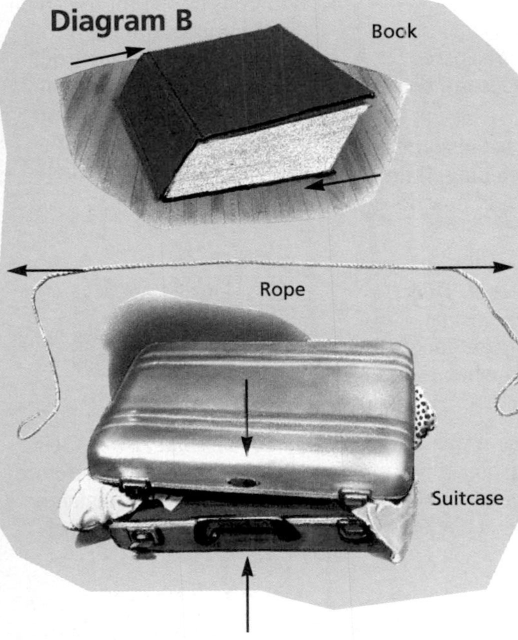

2. In Diagram B, the three structures have been redrawn. The forces have been replaced by arrows. The arrows indicate the directions of the forces. Draw Diagram B in your ScienceLog and correctly label each pair of arrows with one of the following terms:

   tensile force—force that causes stretching

   compressive force—force that causes compression

   shear force—force that causes bending or twisting

Exploration 2 continued ►

EXPLORATION 2

PART 1

GUIDED PRACTICE Involve students in a discussion of the illustrations. With a large textbook or dictionary, demonstrate the application of shear force to the book. You may wish to have a volunteer assist you in this demonstration.

★ **An Exploration Worksheet is available to accompany Exploration 2 (Teaching Resources, page 10).**

Answers to *Part 1*

1. A pair of forces is observed on each object. The force from the hand causes the spine of the book to bend. Pulling on the rope causes it to stretch. The force from the hands causes the suitcase to compress.
2. *Shear force* acts on the book; *tensile force* acts on the rope; and *compressive force* acts on the suitcase.

Teaching Strategies for Exploration 2 continued ►

Homework

Have students make a list of five examples each of tensile, compressive, and shear forces that occur in their daily lives. *(Answers will vary but should reflect an understanding of the differences between these types of forces. Accept all reasonable responses.)*

ORGANIZER, continued

Exploration 4: force meter from Unit 4 or masses with hooks; metric ruler; 33 index cards; several small pieces of masking tape; small container of glue; 15 cm of string; 5 sheets of heavy paper (20 cm × 15 cm); small pail filled with sand; piece of cardboard or wood (20 cm × 20 cm); metric scale

Teaching Resources
Graphing Practice Worksheet, p. 8
Exploration Worksheets, pp. 10, 14, and 20
Activity Worksheet, p. 18
Discrepant Event Worksheet, p. 19
Transparencies 50 and 51
SourceBook, p. S93

EXPLORATION 2, continued

PART 2

You may wish to call on a volunteer to demonstrate each of the activities for the class. Then direct students' attention to the diagrams. Help them to recognize that in each case, forces are acting in opposite directions. Students should conclude that shearing results from opposing forces.

Although the scissor blades are dull, students should be careful when cutting because the pointed ends of the blades may be sharp.

PART 3

Have students turn to pages 298 and 299. The twisting and bending of the bridge deck is observable evidence that shear forces were largely responsible for its collapse. The house on page 299 probably collapsed from shear and tensile forces. There is evidence that parts of the house were pulled or stretched apart. There is also evidence that parts of the house were twisted out of shape. The bone was broken by shear forces. The stone beams were cracked by tensile forces or by being pulled apart. The columns were cracked by compressive forces.

Answers to *Part 4*

1. When the load is removed, the book may return to its original shape.
2. When the force is removed, the rope will contract to its original length.
3. When the load is removed, the contents will expand and the suitcase will return to its original shape.
4. Yes, the column experiences a load because it compresses under its own weight. If the top half of the column is removed, the bottom half will increase slightly in height.

PART 2

1. To get a better idea of shear forces, take a pair of dull scissors and slowly begin to cut a piece of heavy paper. Notice how the paper bends before breaking. When forces bend or twist an object, you know that the object is experiencing shear. In the illustration at right, the arrows represent the shear forces of the scissors' blades.
2. Take modeling clay and form it into a small rectangular block. Grasp the top of the block between two fingers. With the other hand, use two fingers to grab the bottom of the block. Now push the top half and the bottom half in opposite directions. You are exerting shear forces on the modeling clay.

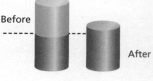

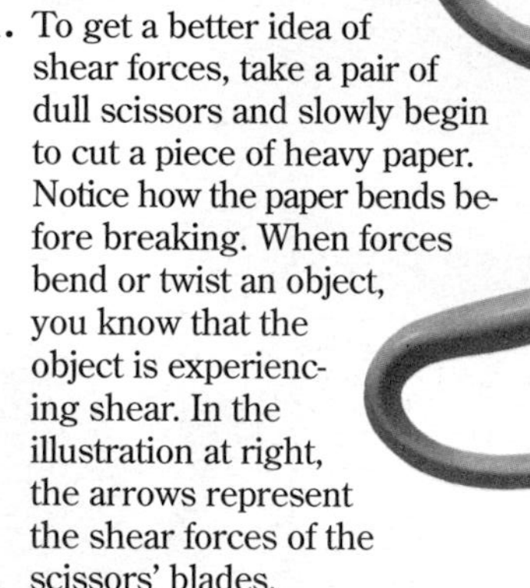

PART 3

The causes of failure in structures are forces that cause tension, compression, and shear. Look again at pages 298 and 299, and identify which forces caused the failure of each structure. For example, the Tacoma Narrows Bridge collapsed from shear caused by wind. Do you see where the bridge twisted and where shear affected the roadway?

PART 4

Look at the structures in Diagram A on page 305. Read the following statements, and answer the questions.

1. The book is under shear. It responds by bending. If the forces are removed, how does the book respond?
2. The rope is under tension. It responds by stretching. If the forces are removed, how does the rope respond?
3. A load is placed on the suitcase and its contents. They respond by compressing. How will the suitcase and its contents respond when the load is removed?

You may find that it was easy to answer these questions. However, the answer to the next question may not be so obvious.

4. If there is no weight placed on the column shown below, does it experience any load? Imagine slicing the column in two. What happens if you remove the top half of the column? How does the bottom half respond?

Before

After

Structures react not only to external forces but also to their own weight. They must therefore be strong enough to withstand external forces and to support their own weight.

Elastic Responses

In Unit 4, a force meter is made using a rubber band. As force increases, so does the length of the rubber band. The stretching is the response of the rubber band to the tensile forces on it. This stretching is said to be an *elastic response*. Why? Because when the force—measured in newtons—is removed, the rubber band relaxes and returns to its original length. Materials used in construction respond in the same way.

Activity 1 of this Exploration is a case study, and Activity 2 is an experiment you can perform yourself.

ACTIVITY 1

Stretching Steel—A Case Study

Does steel really stretch when a strong force is applied to it? Emiko knew that she could increase the tension on the steel strings of her guitar by turning the pegs. But did the strings actually get longer? Emiko went to a mechanical-engineering lab where they had an apparatus to measure the change in length of steel wire. The table below shows Emiko's results for stretching a steel wire 1 m long and 1.0 mm thick.

1. How much would the steel wire stretch when the force pulling it is
   a. 100 N? c. 300 N?
   b. 500 N? d. 900 N?

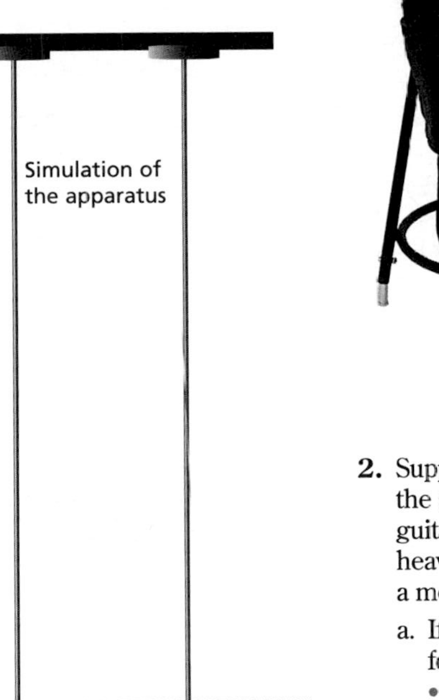
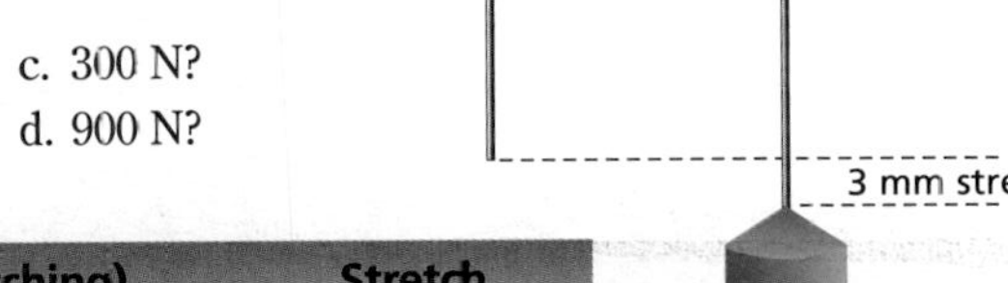

Tensile (stretching) force in newtons (N)	Stretch (mm)
0	0
200	1
400	2
600	3
800	4
1000	5
1200	With this force, the steel string broke.

2. Suppose a 10 m steel wire of the same thickness as the guitar string is used to tow a heavily loaded canoe behind a motorboat.
   a. If the boat puts a steady force of 200 N on the wire,
      - how much will 1 m of the wire stretch? (See the results of Emiko's experiment. She used a wire that was 1 m long.)
      - how much would you expect the 10 m wire to stretch?
   b. If the motorboat suddenly yanked the wire with a force of 1500 N, what would happen?

Exploration 3 continued ►

Homework

Have students complete the Activity Worksheet on page 18 of the Unit 5 Teaching Resources booklet.

Homework

Have students graph the data presented in the table of tensile force on page 307. When grading, check the numbers and labels on the axes, the title, and the accuracy of the line.

EXPLORATION 3

Have students read the introduction to Exploration 3 silently. To refresh their memories, you may wish to demonstrate the use of one of the force meters made in Unit 4. Then refer students to Part 4 of Exploration 2. Point out that the answers to the questions are examples of the elastic response of materials. Then have students speculate why it is desirable for construction materials to be elastic instead of rigid. *(Elastic materials are less likely to fail because they can adapt to changes in the forces exerted on them.)*

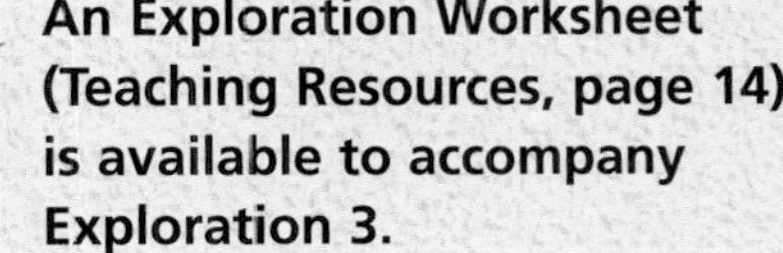

An Exploration Worksheet (Teaching Resources, page 14) is available to accompany Exploration 3.

ACTIVITY 1

Call on a volunteer to read the introduction to the case study. Then direct students' attention to the table. Remind students that force is measured in newtons (N). To evaluate their understanding of what the table shows, ask questions similar to the following:

- Does a steel wire really stretch when a strong force is applied to it? *(Yes)*
- In what increments was force added to the apparatus? *(In 200 N increments)*
- How much did the steel wire stretch for each increment of 200 N force? *(1 mm)*
- How much force was needed to break the wire? *(1200 N)*

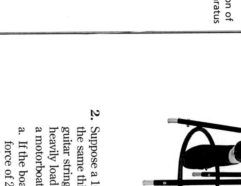

Answers to *Activity 1*

1. **a.** 0.5 mm
   b. 2.5 mm
   c. 1.5 mm
   d. 4.5 mm
2. **a.** 1 mm; 10 × 1 mm = 10 mm
   b. The wire would break.

Teaching Strategies for Exploration 3 continued ►

EXPLORATION 3, *continued*

ACTIVITY 2

You may wish to have students work in pairs or small groups to complete the activity. If space or materials are limited, consider asking two volunteers to conduct the activity as a class demonstration.

Be sure that students understand how to set up the apparatus and what to do. Point out to students that it is important that their measurements be as accurate as possible and that they record their results carefully. Encourage them to make a data table similar to Doug's. Point out that Doug converted centimeters to millimeters in the Deflection column. Explain that this makes it easier to graph the results for step 4 on page 309.

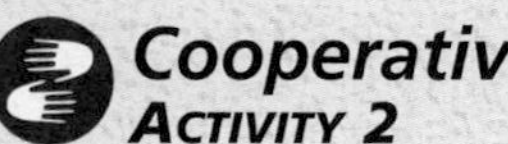

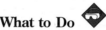

Cooperative Learning ACTIVITY 2

Group size: 2 to 3 students
Group goal: to compare the amount of bending, or deflection, caused by different forces exerted on the end of a model diving board
Positive interdependence: Assign students roles such as primary investigator, recorder, and data analyst.
Individual accountability: Each student should write up the experiment referred to in step 6 on page 309 using a different length of wood extending over the table edge. They should use a format similar to the sample and include a graph of their results.

Call on a volunteer to read step 4 aloud. Then discuss and review the graph on page 309. As students graph their own data, monitor their results by circulating among them. Be prepared to offer help if needed. After students have finished, have them share and discuss their graphs. Before students perform step 6, have them predict what will happen. Then have them complete the activity to test their prediction.

INDEPENDENT PRACTICE Have students repeat Activity 2, but alter the experiment by turning the meter stick on its side. Then have students make bar graphs or charts comparing the strength of a flat meter stick with the strength of a meter stick turned on its side.

ACTIVITY 2

Bending Wood

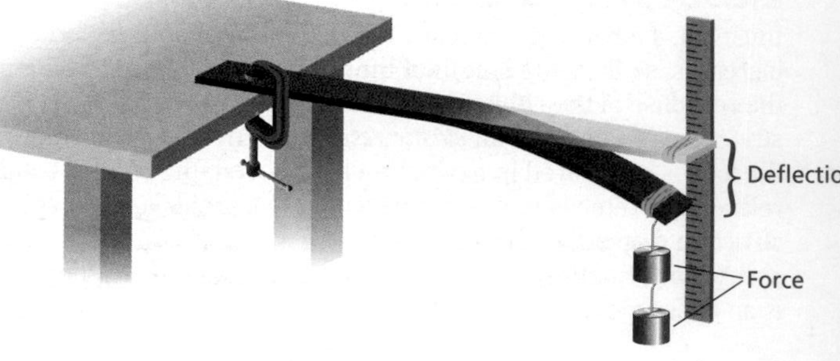

A common example of an elastic response is the motion of a diving board. The diver uses the elastic response of the board to spring into the air.

Your purpose in this Activity is to compare the amount of bending, or **deflection**, with the different forces exerted on the end of a model diving board.

You Will Need

- a piece of wood (like a meter stick)
- a meter stick or metric ruler for measuring
- a strong clamp or other means to fasten the wood to the end of a bench or desk
- standard masses (or use your force meter for applying known forces)
- rubber bands for attaching masses to the meter stick

What to Do 

Caution: Be careful when doing this experiment. Too much force can cause the wood to break and hit you in the face.

1. Set up the apparatus as shown.
2. Hang a mass on the free end of the piece of wood. Measure the amount that the end deflects. Repeat for several different masses.
3. Record your results. See how Doug, a science student, recorded his results.

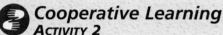

I used a meter stick made of spruce wood. It was 2.5 cm wide and 5 mm thick. I fastened the meter stick to the edge of a table, leaving 90 cm free to bend.

Before putting masses on the end of the meter stick, I first measured the height of the meter stick above the floor. I found the height to be 67.5 cm. Then I measured the height with a mass fastened onto the end of the meter stick. The difference in the heights showed me how much the meter stick deflected. I recorded this deflection in millimeters so it would be easier to graph.

I used 100 g masses to make the meter stick deflect. Because a mass of 100 g has a weight of 1 N, I was able to determine the weight that caused each deflection.

Force (N)	Height at free end from floor before adding masses (cm)	Height after adding masses (cm)	Deflection (mm)
0	67.5	67.5	0
1	67.5	66.1	14
2	67.5	64.7	28
3	67.5	63.2	43
4	67.5	61.8	57
5	67.5	60.5	70

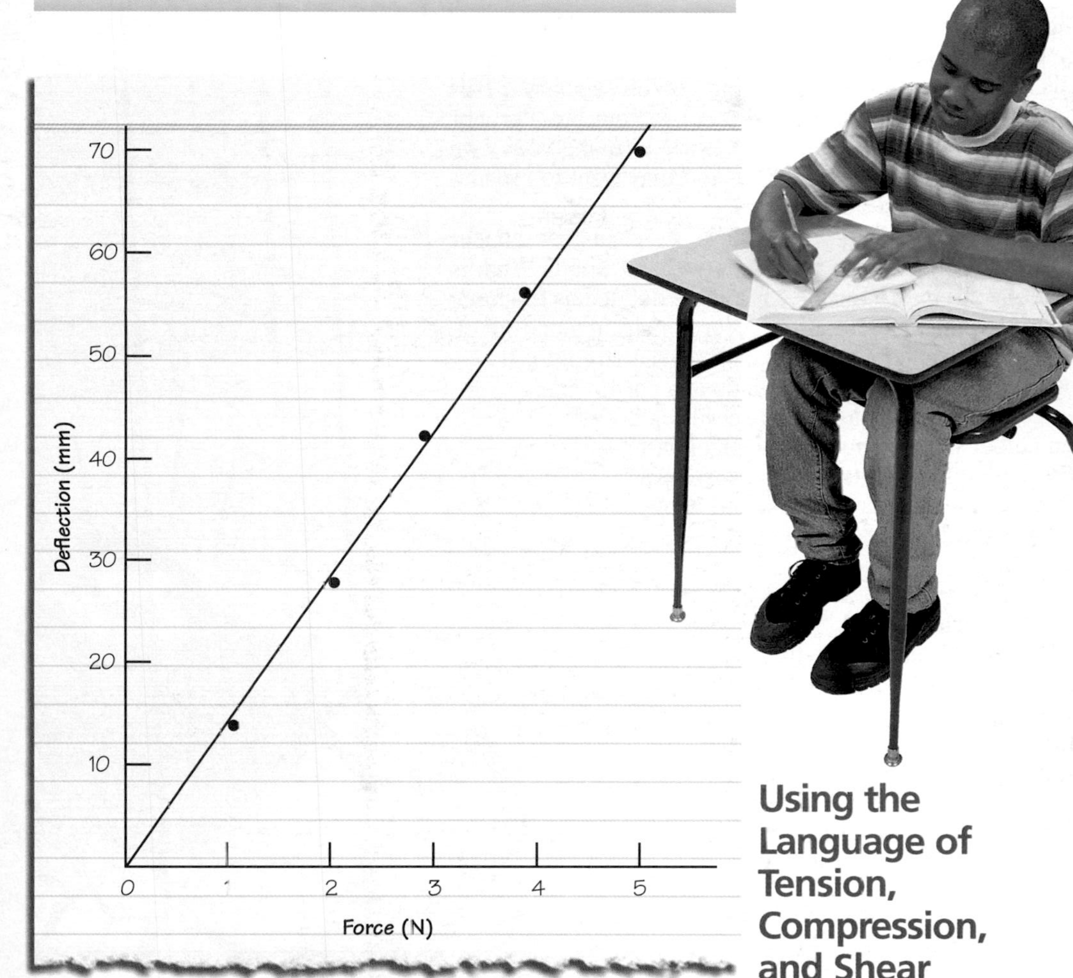

4. Now graph your results. Doug let the vertical axis represent deflection and the horizontal axis represent force. After plotting his points, he drew a line through them. Plot your data and draw your line. Is the line straight or curved?
5. What do you conclude about the relationship between force and deflection?
6. If the length of the piece of wood over the edge of the table were shorter, what difference in results would you expect? Test your prediction by repeating the experiment, but this time clamp the piece of wood at a different position. Plot the new data on the same graph. What do you conclude?

Using the Language of Tension, Compression, and Shear

Did you find definitions for most of the words at the beginning of this lesson? Rewrite your report to Trixie's Toy Company using as many of those words as you can.

Is it easier to describe your model bridge? Using the language of engineers helps you describe structure and force precisely. By mastering scientific language, you can begin to write and think like an engineer!

PORTFOLIO

Suggest that students include their revised reports to Trixie's Toy Company in their Portfolio. They may perform a self-evaluation to assess how well they have used scientific thinking skills during Exploration 3.

Integrating the Sciences

Earth and Physical Sciences

Have students describe the occurrence of shear, compressive, and tensile forces within the Earth's crust. *(At faults, layers of rock are subjected to shear forces when plates slide past each other. Similarly, when plates collide, the edges are subjected to compressive forces. Tensile forces occur at faults where plates are moving away from each other and rock formations are being pulled apart.)*

Answers to *Activity 2*

4. Students should discover that the line on the graph is straight.
5. Students' graphs should show that deflection is proportional to the applied force. In other words, as equal increments of force are applied, the amount of additional deflection is the same each time.
6. Students should discover that deflection is less when the length of the stick projecting over the edge of the table is shorter.

Using the Language of Tension, Compression, and Shear

At this point in the unit, students have encountered all of the words listed on page 304. Encourage them to write original sentences using each of the words. Call on volunteers to share some of their sentences with the class. Be prepared to correct any misconceptions before students continue with the lesson. Then have students rewrite their reports to Trixie's Toy Company. When they have finished, have them share their reports and discuss the merits of understanding and using correct scientific language.

Meeting Individual Needs

Learners Having Difficulty

To reinforce concepts about the effect of forces on structures, suggest that students have a "house of cards" contest. The object is to see who can make the tallest structure using playing cards. Explain that the structures will have to withstand the wind put out by the low setting of a fan. To extend the project, have students test the strength of their card houses by placing increasing masses on the roof of each structure. Besides strength and height, other categories in the contest might include the structure that uses the most cards and the structure with the most attractive design.

Strong Materials

Before students begin reading, ask if they have ever had to decide exactly which kind of material to use in building or making something. Call on volunteers to explain what qualities they looked for in the materials they chose. Now have students read the first two paragraphs of this page silently. Ask them to speculate as to which of the materials—copper wire, nylon fishing line, spider's web, or wood—is the strongest. Next, have them predict the order of strength by listing the items from the strongest to the weakest. Then have students finish reading the page to discover if their predictions were correct.

Most students will be surprised to learn that spider's web, or spider's silk, is second in strength only to nylon. You may wish to explain that spider's silk is the strongest natural fiber known. Even a steel thread of an equivalent size is not as strong.

Answers to *In-Text Questions*

Ⓐ After reading the final paragraph, students should recognize that (c) is the correct order. Therefore, nylon would be the most difficult to pull apart.

Theme Connection

Structures

Focus question: List five natural or human-made structures and identify whether they are rigid or flexible. Explain why their rigidity or flexibility is important to their design. *(Sample answers: Steel beams must be rigid to maintain the structure of buildings; wood must be flexible so that trees can sway in heavy winds without breaking; and a diving board is flexible to allow divers to jump higher.)*

Homework

You may wish to have students read Science and Technology: The Wonder Web on page 349 as homework.

Strong Materials

The designers of Riverwood need to know which materials are best for building the structures they design. The materials they select must be strong but not too expensive. This is a problem you will face many times in your life. Perhaps you will need to select the right fabric to make your own clothes or to upholster furniture. You may want to repair a skateboard or build bookshelves.

Which material do you think is the strongest: copper wire, nylon fishing line, a spider's web, or wood? What is meant by strength? Tensile strength is defined as the force needed to pull materials apart. A cord or string made out of any of these materials has its own tensile strength. If you pull hard enough on the ends of a cord made from one of these materials, the cord will break. Remember our example of the steel wire? It took a 1200 N force to pull the wire apart. Which of the materials mentioned above will break with the least amount of force? the most?

Suppose you have long, threadlike pieces of each material. Suppose also that each "thread" has the same thickness of 1 mm. Does this change your answer? Which is the easiest to pull apart? Which is the most difficult? What is your prediction? Choose the list below that you think places the materials in order from strongest to weakest. Ⓐ

a. copper, wood, nylon, spider's web
b. wood, nylon, copper, spider's web
c. nylon, spider's web, copper, wood
d. copper, nylon, spider's web, wood

Here are the facts! After many tests, it has been found that it takes 820 N to break the nylon line, 190 N to break the spider's web, 110 N to break the copper wire, and 81 N to break the wood. Are these results surprising? How do they compare with your prediction? If you thought differently, remember that the thickness of these materials plays a role in judging their strength. For example, most spider webs that we see have threads less than 1 mm thick.

Integrating the Sciences

Earth and Physical Sciences

Earthquakes can topple buildings, shatter windows, and destroy bridges. Earthquakes occur when rocks under stress shift along a fault (the boundary between two sections of rock that move in different directions). Buildings in areas where earthquakes occur frequently must be able to resist the forces of the shifting, vibrating rocks.

Have students research to discover more about the damage earthquakes can do to buildings, along with some of the precautions that builders can take. Some students may be especially interested in the construction of the large, towering skyscrapers in San Francisco that are built to be resistant to the frequent earthquakes experienced in the Bay Area.

Strong Shapes

Some materials are stronger than others. But the strength of any material can be increased by changing its shape. Recall the bridges you built using index cards. What shapes had the most resistance to bending? Did you discover that a thin, flat card has relatively little strength to support loads that cause it to bend?

The following Exploration should give you many ideas for building models of structures for Riverwood.

EXPLORATION 4

Testing for Strong Shapes

In this Exploration you will test the strength of five different shapes made of different materials. Work with a small group, and then share your discoveries with the rest of your class.

You Will Need

- a force meter or a set of standard masses (10 g or 20 g intervals)
- a metric ruler
- index cards for Investigations 1 through 4
- masking tape for Investigations 1, 4, and 5
- books for Investigation 2
- paste or glue for Investigations 2 and 3
- string for Investigation 4
- heavy paper, a small pail filled with sand, a hard wooden board or piece of cardboard (approximately 20 cm × 20 cm), and a metric scale for Investigation 5

INVESTIGATION 1

Shapes Unlimited

1. Use only a single index card to make each of the five shapes shown below. You may use a small amount of tape to fasten the edges of the index card for shapes (d) and (e).
2. Predict which of the five shapes you think will have the most strength to resist bending.

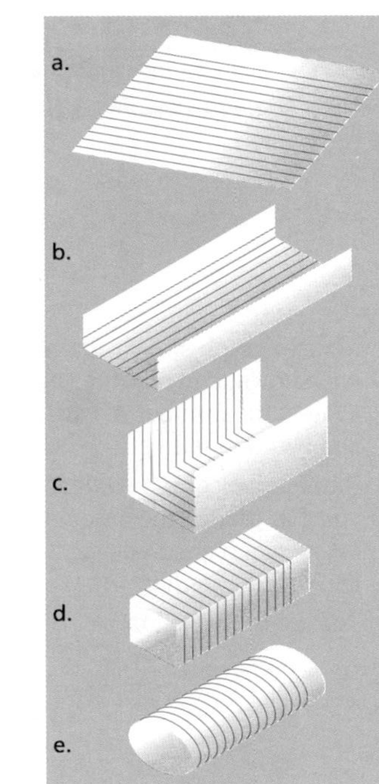

3. With a force meter or masses, apply force to the middle of each shape until it no longer returns to its original position when the force is removed. This happens when a permanent crease or fold is caused by the force.
4. Make and test at least one shape of your own design.
5. Record your results in a table in your ScienceLog. In the left column of your table, sketch each of the shapes. In the right column, record the bending force that was required to cause the shape to buckle.
6. Which of the five shapes was the strongest? Was your prediction correct? How did the shape you designed compare with these five? (B)

Exploration 4 continued ►

Answers to *In-Text Questions*

(B) Students will probably find that the shapes increase in strength in the order that they are pictured, from (a) to (e); the flat shape is the weakest, and the circular tube is the strongest. Have students compare their own shapes to determine who designed the strongest one.

★ **An Exploration Worksheet (Teaching Resources, page 20) and Transparency 51 are available to accompany Exploration 4.**

Strong Shapes

Before students begin Exploration 4, involve them in a discussion of the shapes they used to build their model bridges in Lesson 1. Encourage them to speculate about why it is important for architects and engineers to determine which shapes are the strongest before they begin building structures. *(It is important in order to avoid accidents and wasting materials.)*

EXPLORATION 4

A Discrepant Event Worksheet is available as a teacher demonstration to be used as an introduction to Exploration 4 (Teaching Resources, page 19).

You may wish to have students glue the cards together for Investigations 2 and 3 ahead of time. Paste also works well.

INVESTIGATION 1

You may wish to have students work in pairs or small groups to complete the investigation. Before students begin, have them write their predictions about the strength of the five shapes in their ScienceLog. Then have them test each one. Students will need to determine how to position the card shapes so that weights can be placed on them. One solution is to use the shapes to bridge the space between two books. Suggest that students gently hold the card shapes in place as masses are placed on them. When students finish the investigation, have them discuss and compare their results.

Cooperative Learning
INVESTIGATION 1

Group size: 3 to 4 students
Group goal: to evaluate the strength of different shapes and structures and to create graphs that illustrate the relative strength of different shapes
Positive interdependence: Assign students such roles as materials coordinator, data recorder, assembler, and investigator.
Individual accountability: Each student should provide a summary of his or her findings. This should include a written paragraph that ranks the different shapes, including the shape that was designed individually, in order of strength. Students should also provide a graphic representation that illustrates the relative strength of the different shapes tested.

Teaching Strategies for Exploration 4 continued ►

EXPLORATION 4, continued

INVESTIGATION 2

Point out to students that the term *laminate* means to bond layers of material together. It is a common way to increase the strength of the materials that are used in construction.

You may wish to have students work in pairs or small groups to complete the investigation. Have groups read the instructions and discuss what they are to do. If space or time is limited, have volunteers perform the investigation as a class demonstration. Students should make their own graph from the results.

When students have finished, have them compare and discuss their graphs to determine if all of the results were the same. Help to resolve any differences that may arise.

Meeting Individual Needs

Learners Having Difficulty

Students may be interested in extending their investigation of lamination and corrugation. Provide a piece of plywood and a piece of corrugated cardboard for students to examine. (The piece of plywood should be no larger than 3 mm × 3 cm × 30 cm to avoid the risk of dangerous jagged edges resulting from breakage.) Ask: Which is laminated? *(Plywood)* Which is corrugated? *(Cardboard)* Suggest that students test the strength of the materials. Students should use force meters to see how much force the materials can withstand. Allow them to try to break the plywood and to separate the layers of corrugation on the cardboard. Then have students create posters that depict the benefits of using plywood and corrugated cardboard in building projects. Or have students actually build a bridge or small building out of the materials—they may be surprised at the strength of a corrugated-cardboard or plywood bridge.

INVESTIGATION 2

Lamination

1. Arrange the books and index card as shown below.

2. Place a standard mass on the card so that it sags about 2–3 mm. Measure the sag.
3. Record your results in a table like the one shown below.

Number of cards laminated together	Amount of sag (mm)

4. Laminate (glue together) two cards so that they double in thickness. Using the same standard mass, measure the sag.
5. Repeat the laminations for triple thickness and quadruple thickness. Measure the sag.

6. Make a graph similar to the one shown, and plot the sag against the number of cards laminated together.
7. Extend your graph so that you can predict the amount of sag when six cards are laminated together.
8. Test your prediction.

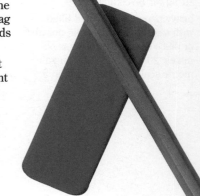

INVESTIGATION 3

Corrugation

1. Laminate three index cards together.

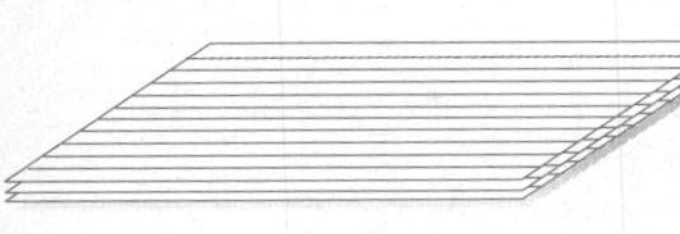

2. Find the force the laminated structure will sustain before sagging 1 cm.
3. Take another three cards and fold the middle card as shown below. Glue the cards together.

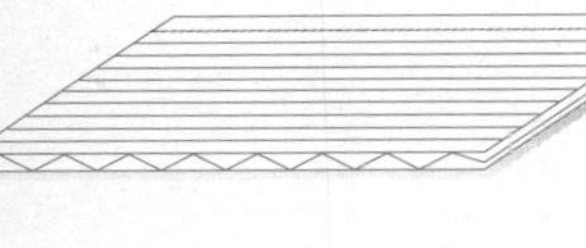

4. Find the force the corrugated structure will sustain before sagging 1 cm.
5. What do you conclude? (A)

INVESTIGATION 4

Tubular Girders

1. Roll an index card into a tubular girder. Hold it together with small pieces of masking tape. Tie a piece of string around the tube.

2. Measure the diameter of the tube.
3. Support the girder as shown (bottom left). Using standard masses or a force meter, find the force required to cause the girder to buckle.
4. Using the same amount of material and keeping the tube the same length, vary the diameter. Repeat for five different diameters. Record your results in a table like the following:

Diameter	Maximum force

5. Make a graph of your findings (maximum force versus diameter).
6. What do you conclude? (B)

INVESTIGATION 5

Tubular Columns

1. Using heavy paper and masking tape, make a tube to be used as a column. The tube should be approximately 20 cm long.
2. Measure the diameter.
3. Construct a platform as shown (top right). Place the pail over the center of the column.

4. Find the force required to buckle the column. You can do this by filling the pail with sand until the column buckles. Weigh the platform, pail, and sand together. This gives you the force required to buckle the column.
5. Repeat two more times, using the same amount of material for the column and keeping it the same height, but varying its diameter. Record your results in a table like the one you made in Investigation 4.
6. What do you conclude? (C)
7. Instead of placing the pail over the center of the column, try placing the pail at different locations on the column to find out the effect of off-center loading. (D)

INVESTIGATION 3

Have students complete the investigation and discuss their results. You may wish to have students glue the cards together and then complete Investigations 4 and 5. When they return to Investigation 3, the glue should be dry enough to proceed with the trials.

INVESTIGATION 4

Have students complete the investigation and compare and discuss their results before making their graphs. Some students might correctly point out that by using the same amount of material, the edges of the card will overlap, thus contributing to the strength of the girder. To account for this, the cards could be trimmed as the diameter decreases so that there is no overlap. Then have students complete their graphs.

You may wish to point out to students that a girder is a supporting structure that is used horizontally. In the next investigation they will examine the strength of a tubular column. A column is used vertically. Encourage students to speculate about whether or not they will achieve the same results with a tubular column as they did with a tubular girder.

INVESTIGATION 5

Point out to students that the results obtained in this investigation are the opposite of what happened with the tubular girders in Investigation 4.

Answers to *In-Text Questions*

(A) The corrugated structure is stronger than the laminated structure.

(B) Assuming the thickness of the girder remains constant, the strength of a tubular girder decreases as its diameter increases.

(C) As the diameter of the tubular column increases, more force is needed to make it buckle.

(D) Off-center loading decreases the amount of force that the column can support and can cause the column to fall over before it has a chance to buckle.

FOLLOW-UP

Reteaching

Have students use the words listed on page 304 to make a crossword puzzle, a word search, or a matching activity between the words and definitions. Have them exchange activities and complete the one that they receive.

Assessment

Provide the class with a diagram of the frame of a house. Have students use labeled arrows to point out places where shear, compressive, and tensile forces are acting on the materials of the frame.

Extension

Have students research one of the following topics: reinforced concrete, adobe, or heavy timber. They should find out how the material is made, what makes it strong, and what it is used for.

Closure

Suggest to students that they make a bulletin board or table display to show how laminated, corrugated, and tubular structures are used in everyday objects.

Lesson 3 Engineering Structures

FOCUS

Getting Started

Show the class a picture or model of flying buttresses on a Gothic cathedral. As you point to the buttresses, ask the class: What do you think these are for? *(Explain to students that they are a type of brace that was first used by builders over 800 years ago to strengthen the walls of European churches. Church builders have also used beams, arches, domes, and other structures in the construction of these large buildings.)* In this lesson, students will look at several kinds of structures still used in buildings today.

Main Ideas

1. A truss framework adds strength and support to a structure.
2. For the same amount of material, an I-beam has greater rigidity and strength than does an ordinary beam.
3. A cantilever is a beam that is fixed at one end.
4. An arch provides more strength against a vertical load than does a simple beam.
5. An arch transforms vertical loads into lateral ones.
6. Bridges are classified according to their structures: arch, cantilever, suspension, or girder beam.

TEACHING STRATEGIES

Beams and Trusses

Most students will have seen wooden or steel beams in houses, malls, or gymnasiums. Direct their attention to the photographs. Point out that the triangular structures are trusses. Ask: What are the trusses supporting? *(The roof)*

Lesson 3 Engineering Structures

Many human-made structures are quite complex. However, they are made up of simpler components that give them their strength and shape. You can include these components in the structures you plan for Riverwood.

Beams and Trusses

Where have you seen *beams* before? What are they made of? Beams can be made of a variety of natural and artificial materials. Have you ever seen a *truss*? A truss is a framework of connected planks or steel bars that add strength and support to a structure. Trusses are frequently found in the roofs of houses and other buildings.

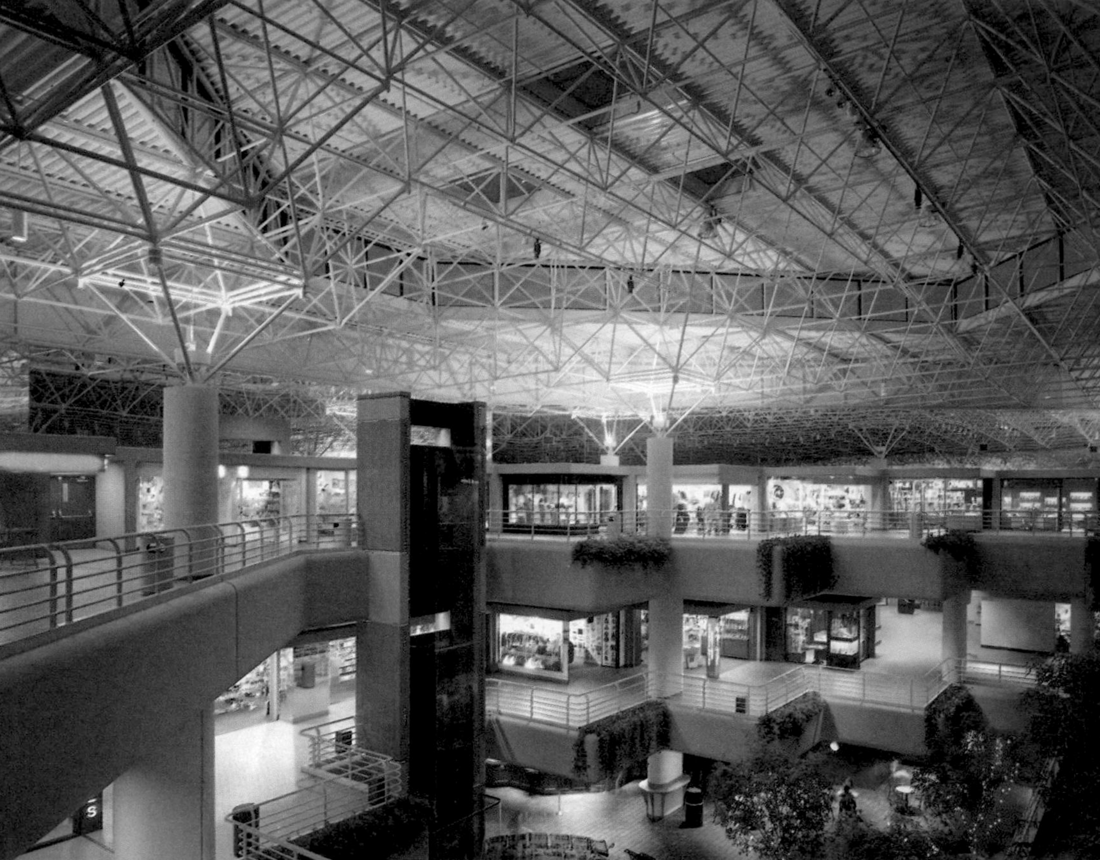

314

LESSON 3 ORGANIZER

Time Required 3 class periods

Process Skills
analyzing, measuring, observing

Theme Connection Structures

New Terms none

Materials (per student group)
Exploration 5, Activity 1: 12 cm × 30 cm wooden board with a hole drilled at each end; 10–15 cm wooden post; standard masses; 2 supports (blocks or books); 50 cm of string; 2 eye screws or thumbtacks; **Activity 2:** 6 strips of plywood or corrugated cardboard; nails and a hammer (if wooden boards are used); tacks or glue; standard masses; 2 supports (blocks or books); safety goggles
Exploration 6: 2 large index cards; standard masses or 3–5 paper cups and about 500 cm^3 of sand; 1–2 sticks of modeling clay; protractor; metric ruler; half of an aluminum can; 2 supports (blocks or books)

Teaching Resources
Exploration Worksheets, pp. 26 and 29
Discrepant Event Worksheet, p. 28
Resource Worksheet, p. 34
Transparencies 52, 53, and 54

EXPLORATION 5

ACTIVITY 1

Comparing a Simple Beam With a Truss

You Will Need

- a wooden board (a thin plywood strip about 10–12 cm wide and 30 cm long) with a hole drilled into each end, or a strip of corrugated cardboard of similar size
- a short wooden post or piece of cardboard (10–15 cm long)
- standard masses
- supports (blocks or books)
- string
- 2 eye screws or thumbtacks

What to Do

1. Set up a beam as shown below.

Put a load of standard masses on the middle of the strip so that it sags noticeably.

2. Convert the beam into a truss, as shown at the top of the next column. If a wooden beam is used, fit the screws through the holes in the beam, and fasten the string to the screws. If cardboard is used, punch a thumbtack into each end, and fasten the string to the thumbtacks. Make sure the string is tight enough to hold the post against the underside of the board.

Place the same load on it as you did before. What do you conclude? (A)

ACTIVITY 2

The I-Beam

You Will Need

- 6 boards, all the same size (boards may be strips of plywood or corrugated cardboard)
- nails and a hammer (if wooden boards are used)
- thumbtacks or glue
- standard masses
- supports (blocks or books)

What to Do  

1. Nail or fasten three boards together as shown below in diagram (a).
2. Nail or fasten three boards together in the shape of an I-beam as shown in diagram (b).
3. Put a load on the middle of the beam shown in diagram (a) so that the beam sags measurably.
4. Now place the same load on the I-beam.
5. What do you observe? Which is stronger, the I-beam or a simple beam made from the same amount of material? (B)

Why do you think one beam is stronger than the other? (C)

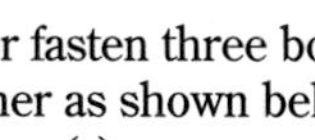

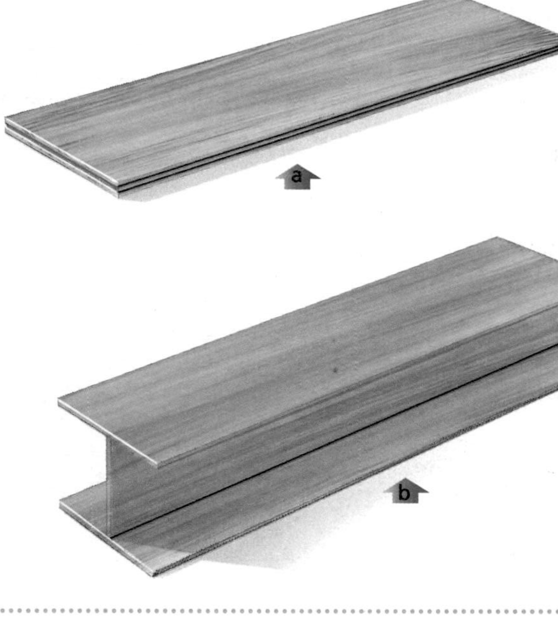

EXPLORATION 5

★ **An Exploration Worksheet is available to accompany Exploration 5 (Teaching Resources, page 26).**

ACTIVITY 1

If a wooden strip is used as a beam, it will need to be very thin in order to sag under a manageable weight. Either a 3 mm piece of plywood or a wooden slat from a window shade works well.

ACTIVITY 2

Have students test the strength of the beams as they did in Activity 1. They should observe that the I-beam sags less or not at all under the load.

SAFETY TIP Make sure that emergency telephone numbers are posted in plain view. All students should know how to dial for help, and they should be aware of any special operating instructions needed to get an outside line (such as dialing 9 first).

Answers to *In-Text Questions*

(A) The beam sags less or not at all when it is converted into a truss. A truss can support more weight than a beam can.

(B) An I-beam can support more weight than can a simple beam made from the same material.

(C) The I-beam is stronger because of its design. You may wish to explain that the force of the weight on the beam is distributed farther from the center of the beam.

Homework

Before students complete Exploration 5, have them make predictions about which beams will sag more in each Activity. They should also explain the reasoning behind their predictions.

Cantilevers

Involve students in a brief discussion of each of the pictures. Students should recognize that each of the structures have two things in common. They are each attached only at one end, and their weight is counterbalanced by the structures to which they are attached.

After students have discussed the illustrations, provide them with time to make their own drawings. When they have finished, call on several volunteers to share their work with the class.

Cantilever Bridges

INDEPENDENT PRACTICE Have students make a sketch of diagram (b), labeling the two cantilevers, the supports for the cantilevers, and the load supported by the cantilevers. Or have students reconstruct the model and add labels.

Answers to *In-Text Questions*

Ⓐ The tree branch is a cantilever because only one end of it is attached to the tree. The roots anchor the tree to the ground, offsetting the weight of the branch. The sign post is a cantilever because only one end of it is attached to the building. The weight of the post and the sign are counterbalanced by the weight of the building. The balcony is actually made up of many cantilevers, with one end of each cantilever attached to the building. The weight of the building offsets the weight of the balcony, keeping the balcony from falling.

Ⓑ Students' drawings should show that a simple beam is supported at both ends and that a cantilever is supported at only one end.

Ⓒ (1) The two cantilevers are the two boards held down by the cans. (2) The cantilevers are supported by the two tables and the weight of the cans. (3) The load is represented by the car and the board that spans the space between the cantilevers.

Ⓓ Cantilevers are able to support heavy loads without a second support because they are attached at one end to a heavier, more solid structure that counterbalances the load. The cantilever bridge in diagram (a) supports its load over a series of cantilevers, each with a greater weight on one end than on the other.

Cantilevers

A *cantilever* is a simple beam that is fixed at one end. Here are some examples of cantilevers. Explain why each fits the definition of a cantilever. Ⓐ

Show the difference between a cantilever and a simple beam by drawing each one supporting a person. Ⓑ

Cantilever Bridges

After single-beam bridges, cantilever bridges were probably the next type of bridge to be built. It is thought that the first cantilever bridge was built in China. Diagram (a) illustrates how they were built. Diagram (b) shows the basic principle of how they work.

Identify the following in diagram (b): Ⓒ

1. the two cantilevers
2. how the cantilevers are supported
3. the load that the cantilevers are supporting

Explain in your own words how cantilevers are able to support heavy loads without collapsing. How does the cantilever bridge in diagram (a) support its load? Ⓓ

a

b

316

★ **Transparency 52 is available to accompany the material on pages 316–317.**

Arches and Domes

By now you know there are many ways to support a load. You can support loads by using columns and beams. You can shape your beams in various ways, forming them into I-beams, for example. You can also transform a beam into a truss. And you can make a beam into a cantilever by anchoring it at one end for support.

Another way to support a load is by making an *arch*. Widespread use of arches was one of the most important advances made in the earliest civilizations. By the time of the Roman Empire, arch bridges and arch aqueducts (for carrying water) had become the major means of spanning distances such as valleys and rivers.

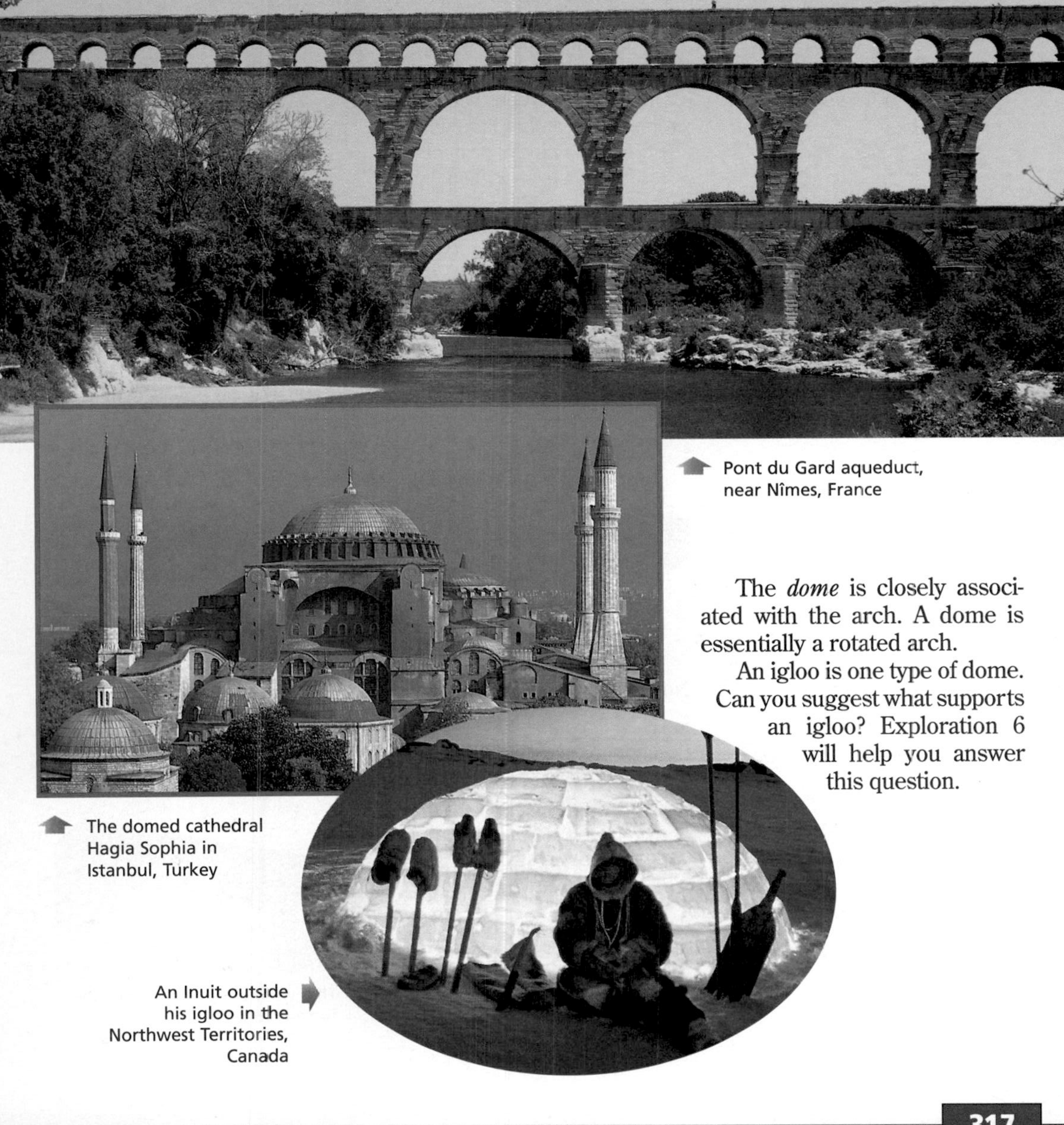

Pont du Gard aqueduct, near Nîmes, France

The domed cathedral Hagia Sophia in Istanbul, Turkey

An Inuit outside his igloo in the Northwest Territories, Canada

The *dome* is closely associated with the arch. A dome is essentially a rotated arch.

An igloo is one type of dome. Can you suggest what supports an igloo? Exploration 6 will help you answer this question.

317

Homework

As a homework assignment, provide students with the Discrepant Event Worksheet on page 28 of the Unit 5 Teaching Resources booklet.

Meeting Individual Needs

Second-Language Learners
Provide students with a small plate, tape, and several strips of heavy paper. Ask students to construct a dome over the plate by using the paper strips and tape. Have students compare their dome to the arches they build in Exploration 6, page 318. Students should realize that a dome is merely a series of arches arranged around a central point. Evaluate their work based on scientific content, not language proficiency.

Arches and Domes

Have students read this section silently. Then direct their attention to the photographs. Ask: What do arches and domes have in common? *(Both have curved surfaces.)* Direct students' attention to the picture of the aqueduct on this page. Ask: What do the arches in the bridge support? *(They support the top of the bridge. The lower arches support the upper arches.)*

Ask students to examine the photograph of Hagia Sophia and ask: What part of the cathedral is a dome? *(The ceiling)* Have a volunteer identify at least one arch in the picture. Then have students speculate about what supports the structure of an igloo. Accept all reasonable responses without comment. Students will have the chance to investigate these kinds of structures further in Exploration 6.

Theme Connection

Structures
Focus question: Why might a designer choose a dome to create a large open space in a structure? *(Students should understand that designers take advantage of the way a dome reacts to the forces acting on it. The downward force of gravity is displaced to the sides so that interior supports are not necessary. Therefore, the dome shape allows for a very high ceiling without interior support.)*

Cross-Disciplinary Focus

Industrial Arts
Point out to students that a geodesic dome is one kind of structure that supports itself. It is strong and easy to set up and take apart. Suggest that they do some research to find out what geodesic domes look like, how they are made, what they are used for, and who invented them. Have students report what they learn to the class.

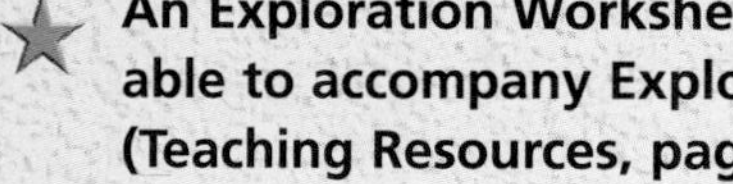

EXPLORATION 6

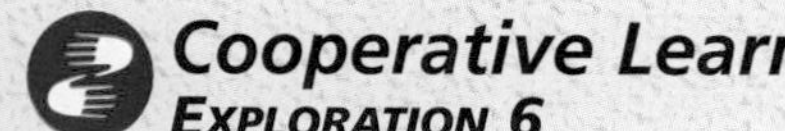

An Exploration Worksheet is available to accompany Exploration 6 (Teaching Resources, page 29).

Cooperative Learning
EXPLORATION 6

Group size: 2 to 3 students
Group goal: to determine the advantages of using an arch for spanning distances
Positive interdependence: Assign students roles such as investigator, engineer, or recorder.
Individual accountability: At the conclusion of each Activity, each student should prepare a summary paragraph or a fact sheet that outlines what has been discovered. These papers can be shared with group members, turned in for evaluation, and then included in student Portfolios.

ACTIVITY 1

Divide the class into small groups, and distribute the materials. The simple beam and the arch should each be constructed from a large index card or heavy paper. Students should use the same-sized card for each test. After students have completed the Activity, reassemble the class and involve them in a discussion of their responses to the questions.

Answers to *Activity 1*

a. The strength of an arch is greater than that of a simple beam.

b. The narrower the arch, the stronger it is. The flatter an arch, the weaker it is.

c. The buttresses hold the arch in place, and they receive a downward as well as an outward force from the arch.

d. The blocks support the arched card by pressing against its ends. They not only receive the compressive force of the load on the arch, but also resist shear forces to hold the arch in place. The blocks support the simple beam by providing a surface on which the ends of the beam can rest.

EXPLORATION 6

Comparing Beams and Arches

In this Exploration you will learn about the advantages of using an arch for spanning distances.

You Will Need

- index cards or heavy paper
- standard masses (or paper cups to be filled with sand)
- modeling clay
- a pencil and paper
- a protractor
- a metric ruler
- a can, with top and bottom removed, cut in half vertically
- supports (blocks or books)

What to Do

ACTIVITY 1

Examine the models below, and then construct a beam and an arch. Find the maximum load that a simple beam can support before it sags 1 cm in the middle. Compare this with the maximum load that an arch can support. Compare high arches with flat arches. You can hold the ends of the arch with blocks or books. These supports are called *buttresses*.

Now think about these questions:

a. How does the strength of an arch compare with that of a simple beam?
b. How does the shape of the arch itself affect its strength?
c. What role do the buttresses play in an arched bridge?
d. In what way do the blocks support the arched card differently than they do the beam?

ACTIVITY 2

Examine the photographs on page 317. Consider the following:

a. the materials from which each arch is made
b. the weight of the arch itself
c. what supports each arch

Arches often sit on the tops of walls. Each wall not only must support the weight of the arch and any load on the arch but also must act as a buttress. In what two directions must a wall exert forces on the arch? (A)

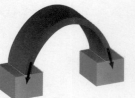

Study the two diagrams of structures below. One shows walls supporting an arch. The other shows walls supporting a beam. Why do the walls in the arched structure need to be much thicker than those in the beam structure? (B)

Thick walls are required to withstand force.

Thin walls are possible since the force is straight down.

ACTIVITY 2

Involve students in a brief discussion of the arches depicted in the photographs on page 317. Students should recognize that the arches are made of bricks, mortar, and stone. As a result, they are extremely heavy. The walls on which the arches rest must support the weight of the arches themselves and any additional weight that the arches support.

Arches can be made thicker in order to give them strength. You might want to point out to students that in the use of domes, the weight is transferred outward as well as downward. Because of this, architects can create a large interior space that is seemingly without support. The photograph of the interior of Hagia Sophia on page S95 will give students a sense of this kind of interior space.

ACTIVITY 3

This part of the Exploration is designed to help you understand how an arch is able to support a load. The basis for creating arches is using a wedge for support. An arch can be thought of as a series of blocks, each wedged into place.

The bottom dimensions of each wedge are smaller than the top dimensions. Imagine trying to push one of the wedges in the arch below downward. What would happen? You can see that such an arch should have great resistance against collapsing from a downward push. **C**

To withstand the downward push, the entire arch must be under load. Without this loading, a downward push on any of the blocks would cause the arch to collapse.

Now make a bridge using wedges formed from modeling clay, as shown. You will need half of a can or a bent card to act as a mold for shaping the arch. With paper, pencil, and protractor, you can make a template for shaping the wedges. Make the template the same diameter as the can plus the desired thickness of the wedges.

Find the load that the clay bridge can support without collapsing. What factors would cause it to collapse? **D**

Reshape the clay into a beam bridge. Use all of the clay. Compare the strength of the beam bridge with the strength of the arched bridge. For a fair comparison, what factors (variables) should be the same for both bridges? **E**

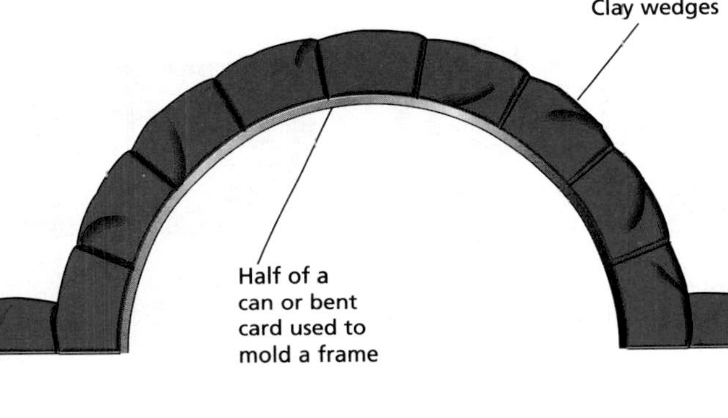

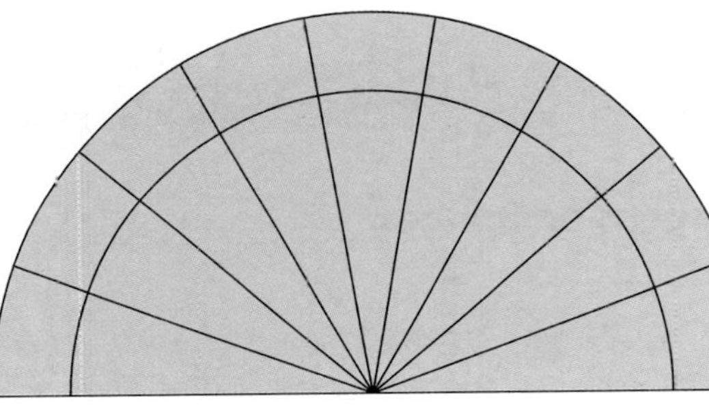

Card or paper template

Answers to *In-Text Questions, pages 318–319*

A A wall holding up an arch must exert both lateral and vertical forces on the arch because the wall must act as both a buttress and a support for the weight of the arch.

B Because a wall supporting an arch must exert forces in two directions, the wall must be much stronger than a wall holding up beams and trusses, in which case the force is acting in only one direction—vertically.

C The downward motion of the wedge would be blocked by the other wedges.

D Factors that could cause the bridge to collapse include weak buttresses, badly shaped wedges, misaligned wedges, or loads that are too heavy and crush or distort the wedges themselves.

E The beam bridge is much weaker than the arched bridge. For a fair comparison, both bridges should be made from the same kind and amount of material and should span the same distance.

ACTIVITY 3

After students have read the explanation of how an arch is constructed, direct their attention to the diagram. Students should recognize that because of the shape of a wedge and the load on the bridge, it cannot be pushed downward unless the entire arch gives way. Help students to recognize that the best way to protect the integrity of an arch is to make sure that the buttresses are strong enough to support it.

If there is a sufficient amount of material, divide the class into small groups and ask each group to make an arch. Instruct the group to make a paper model to determine the shape of the wedges before constructing the clay arch. Remind students to remove the arch from the card or can before testing its strength.

To extend the Exploration, construct a demonstration arch using wooden wedges. Fasten sandpaper to the sides of the wedges to reduce sliding. Depending on the size of the demonstration arch and the loads placed on it, the buttresses may have to provide substantial support. Buttresses made from wooden blocks nailed to a platform are one solution. (Be sure to wear safety goggles when using a hammer and nails.) However, rather than nailing the buttresses, you might have students provide the necessary force.

Meeting Individual Needs

Gifted Learners

Have students construct arches with misaligned or misshapen wedges to determine the effect on the strength of the arch. Students should try building an arch with wedges that are narrower at the top than at the bottom to see how this affects the shape of the arch. Encourage students to experiment with building arches to discover what other factors may affect their strength and shape. Finally, challenge students to build an arch or dome that spans 2 m and stands at least 1 m high using corrugated cardboard, string, and glue. They should work in groups to accomplish this.

Answers to
Bridge Classification

Provide students with a few minutes to examine the four types of bridges in the illustrations on pages 320 and 321. Then involve students in a discussion to share their ideas. Ask: How are downward forces distributed by the various types of bridge structures? *(See the answers below.)*

1. Arch bridges: An arched bridge can be identified by the shape of its main structural feature—an arch. In this type of bridge the forces and stresses caused by loads are carried along the sides of the arch to the buttresses and, eventually, to the ground.
2. Cantilever bridge: A cantilever bridge is characterized by attachment at only one end. The illustration shows double cantilevers that are connected to a beam in the middle. The forces acting downward on one side are counterbalanced by forces acting downward on the other side. Each of the cantilevers is fixed to a supporting structure that carries the downward forces to the ground.
3. Suspension bridge: A suspension bridge is characterized by a cable or series of cables that carry the downward forces to the supports and then to the ground.
4. Girder-beam bridges (page 321): A girder-beam bridge may be essentially a trussed beam. The beam rests on supports that carry the downward forces to the ground. The crossed members of the girder provide strength against shear forces.

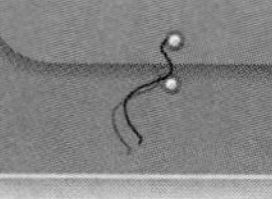

PORTFOLIO

Students may wish to include in their Portfolio sketches of the bridges from the Bridge Classification activity. These may include representations of bridges not pictured in the book. Students should label their drawings and identify the main characteristics of each type of bridge. This can then be used as a guide for future bridge identification.

Transparency 53 is available to accompany the material on this page.

Bridge Classification

Think of the bridges you have seen. Most bridges can be categorized as one of four types: arch, cantilever, suspension, or girder-beam. Examples of these four types are illustrated below. What is the main characteristic of each type?

1. Arch Bridges

a. Simple arch

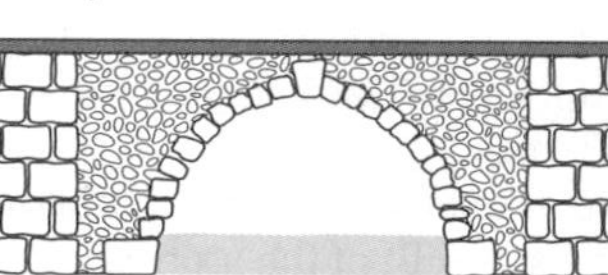

b. Elongated arch

c. Arch with suspended roadway

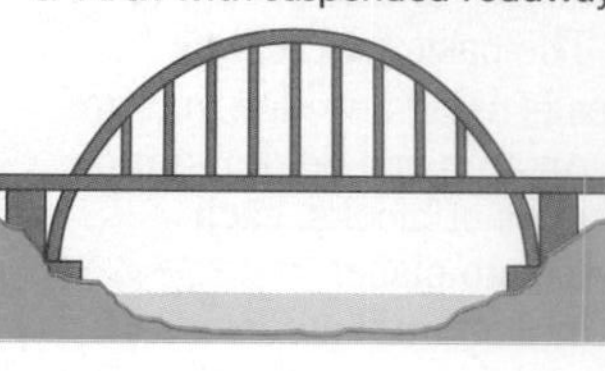

d. Bowstring girder-beam, or tied arch

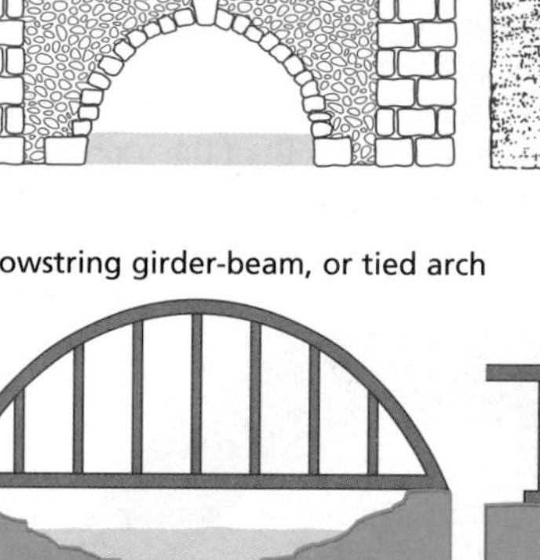

e. Arch with supported roadway

2. Cantilever Bridge

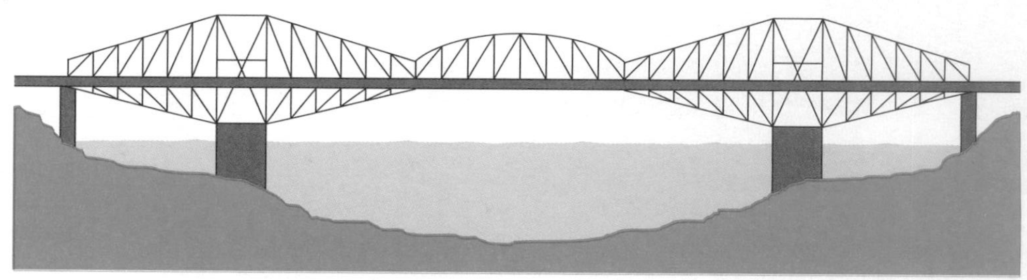

3. Suspension Bridge

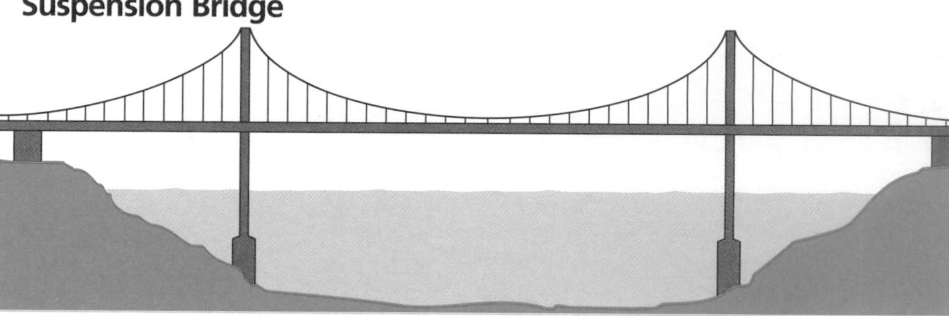

320

Theme Connection

Structures

Focus question: How can bridges be designed to overcome the forces of gravity and of weather? *(Structural strength and a design that redistributes downward forces can help a bridge overcome gravitational forces. Forces due to weather can be overcome by allowing air to flow through the bridge, providing drainage for rainwater, and building joints to accomodate thermal expansion and contraction, for instance.)*

Homework

Have students select a picture of a bridge from an old magazine or newspaper. They should identify what type of bridge it is and the features of the bridge that led them to this conclusion.

4. Girder-Beam Bridges

a. Simple girder

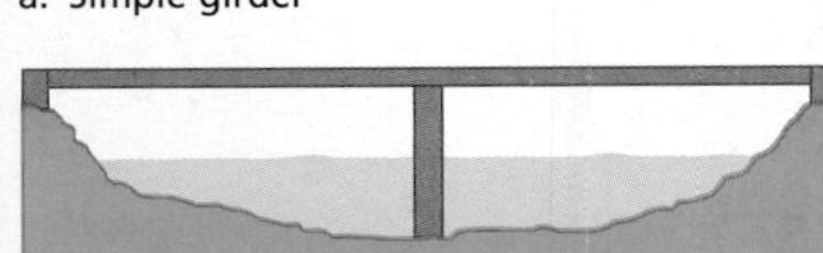

b. Stayed girder

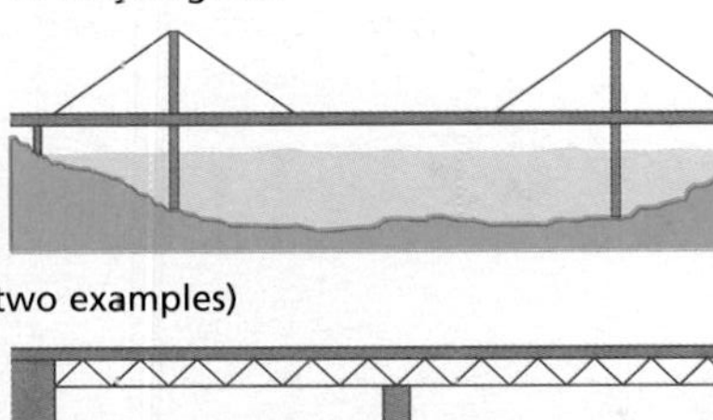

c. Trussed girder (two examples)

Lattice truss

Warren truss

In the next section you will try your hand at classifying some bridges from around the world according to these four basic types.

Famous Bridges

On page 322 are photographs of well-known bridges from different parts of the world. Some are old but are still standing. Others are quite modern. Examine each bridge, and answer the following questions:

a. Which common type does each bridge most closely resemble? (See the section titled Bridge Classification for help.)

b. How is each bridge designed to resist the different forces that might act on it?

A helpful way to record this analysis is to use a table similar to this one. Copy the table into your ScienceLog and fill in the information.

Bridge Identification

Name	Bridge type	Bridge design to resist forces
Bridge of Sighs	Simple arch	Arch supplies strength to support the passageway and people.

Famous Bridges

Direct students' attention to the pictures on page 322. Call on a volunteer to read the captions aloud. Then display a map of the world and ask students to locate and identify the area where each of the bridges is located.

A Resource Worksheet (Teaching Resources, page 34) and Transparency 54 are available to accompany Famous Bridges.

Theme Connection

Structures

Have students study the photograph of the Hooghly River Bridge on page 322. **Focus question:** How is the distribution of downward forces in this bridge different from in a suspension bridge? *(Here, the cables transfer forces directly to the vertical supports. In a suspension bridge, the cables transfer forces to another cable first and then to the vertical supports.)* You may wish to have students diagram the two types of bridges and the forces that act on them.

Answers to *Famous Bridges*

Bridge Identification

Name	Bridge type	Bridge design to resist forces
Bridge of Sighs	Simple or elongated arch	The arch supplies strength to support the passageway and people.
Astoria Bridge	Girder-beam	Although the shape resembles a suspension bridge, the bridge is built like a trussed beam; the forces are transferred to the supports and carried to the ground.
Forth Rail Bridge	Cantilever	The cantilever forces are balanced; the forces acting downward on one side are counterbalanced by forces acting downward on the other side.
Seto Ohashi Bridge	Suspension with truss	The cables and trusses carry the downward forces to the supports and then to the ground; the truss also adds stability to the bridge.
Hooghly River Bridge	Stayed girder	The bridge is built like a trussed beam; the forces are transferred to the supports and carried to the ground.
Tacoma Narrows Bridge	Suspension with truss	The cables and trusses carry the downward forces to the supports and then to the ground; the truss also adds stability to the bridge.

Answers to
Further Analysis, page 323

1. Bridge of Sighs
   a. Every part of the bridge below the walkway is under compression from the weight of the bridge itself and the people walking on it. The two buildings connected to the bridge act as buttresses against the lateral and downward forces of the weight.
   b. None of the bridge is under much tension.
   c. It is unlikely that this bridge would be affected by shear forces.

   Forth Rail Bridge
   a. The supports for the cantilevers and the beams below the roadway are under compression.
   b. The beams that support the roadway from above are under tension.
   c. The roadway may be subject to shear forces from winds. Also, the weight on the cantilevered roadway causes a shear force (bending).

   Hooghly River Bridge
   a. The towers are under compression from the forces directed to them by the cables from which the roadway is suspended.
   b. The cables are under tension from the weight of the roadway.
   c. The roadway could experience shear forces caused by wind or traffic. The towers could also experience shear forces caused by moving water or shifting soil where the towers are anchored.

   Astoria Bridge
   a. The trussed piers are under compression from the roadway they support.
   b. The beams above the roadway are under tension from the weight of the roadway.
   c. The bridge may be subject to shear forces caused by traffic, winds, or earthquakes. The piers anchored to the river bottom may experience shear forces from the moving water or the shifting of the river bottom.

   Seto Ohashi Bridge
   a. The beams under the roadway and the towers are under compression from the weight of the roadway, cables, and traffic.
   b. The cables are under tension from the weight of the roadway suspended from them.
   c. The bridge could experience shear forces caused by moving water or shifting soil.

2. Students' sketches should reflect an understanding of the forces acting on the bridge and of the distribution of these forces within the structure of the bridge.

Astoria Bridge, Oregon

Bridge of Sighs, Italy

Forth Rail Bridge, Scotland

Hooghly River Bridge, India

Seto Ohashi Bridge, Japan

322

The Tacoma Narrows Bridge was rebuilt in 1950. What type of bridge is it? Add this to your bridge identification table. (B)

Compare this photo of the reconstructed Tacoma Narrows Bridge with the photos on page 298. The old bridge pitched and rolled in the wind until it collapsed. What changes were made in the structure of the bridge? How do these changes prevent the bridge from failing again? (A)

Further Analysis

1. Choose two of the bridges on page 322 and consider the following questions for each:
   a. Which parts of the bridge are under compression? What are the compressive forces?
   b. Which parts are under tension? What are the forces causing this tension?
   c. Which parts might experience bending or twisting? What is a possible cause of these shear forces?
2. Visit a bridge and sketch it, showing all the parts that give it strength and support. Label your sketch to show where you think tension, compression, and shear may occur. What type of bridge is it?

323

Did You Know. . .

Soon after the old Tacoma Narrows Bridge was built, people driving on it noticed that it moved in the wind. In fact, it was nicknamed "Galloping Gertie" because of its wavelike motion, and people would come to drive across it just for fun!

Answers to *In-Text Questions and Caption*

(A) Even though both bridges are suspension bridges, the new Tacoma Narrows Bridge was built with a trussed roadway to help keep it stable.

(B) The Tacoma Narrows Bridge is a suspension bridge.

GUIDED PRACTICE Direct students' attention to the picture on page 323. Call on a volunteer to read the information about it aloud. Ask students to turn back to page 298 to compare the two bridges.

FOLLOW-UP

Reteaching

Suggest that students work in pairs or small groups to make a model of one of the bridges illustrated in the text. Have a spokesperson from each group explain how the forces are distributed in their bridge. Display the model bridges around the classroom.

Assessment

Display pictures of local bridges and have students identify them using the classification system in their texts. Then have students identify some of the structures used in the construction of the bridges, such as arches, I-beams, and trusses.

Extension

Point out to students that foundations are the chief means of support in most structures. Ask: How would a foundation built in sand have to be different from a foundation built on solid rock? *(A foundation built in sand would have to account for the sand's instability by starting farther underground or by covering a larger area, for instance.)* How might they be similar? *(They might use the same materials.)*

Closure

Suggest that students make poster diagrams of beams, trusses, and arches. The diagrams should show how the forces are distributed in each of the structures.

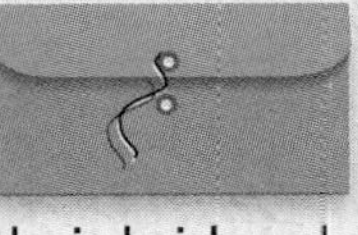

PORTFOLIO Suggest that students include in their Portfolio their bridge sketch from Further Analysis.

1. A Structural Search

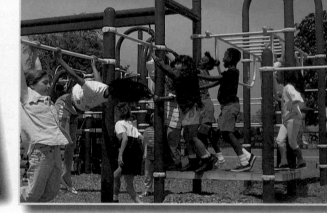

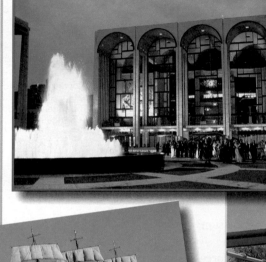

From the photographs below, identify examples of the following structures:

a. simple beams
b. cantilevers
c. trusses
d. arches
e. columns

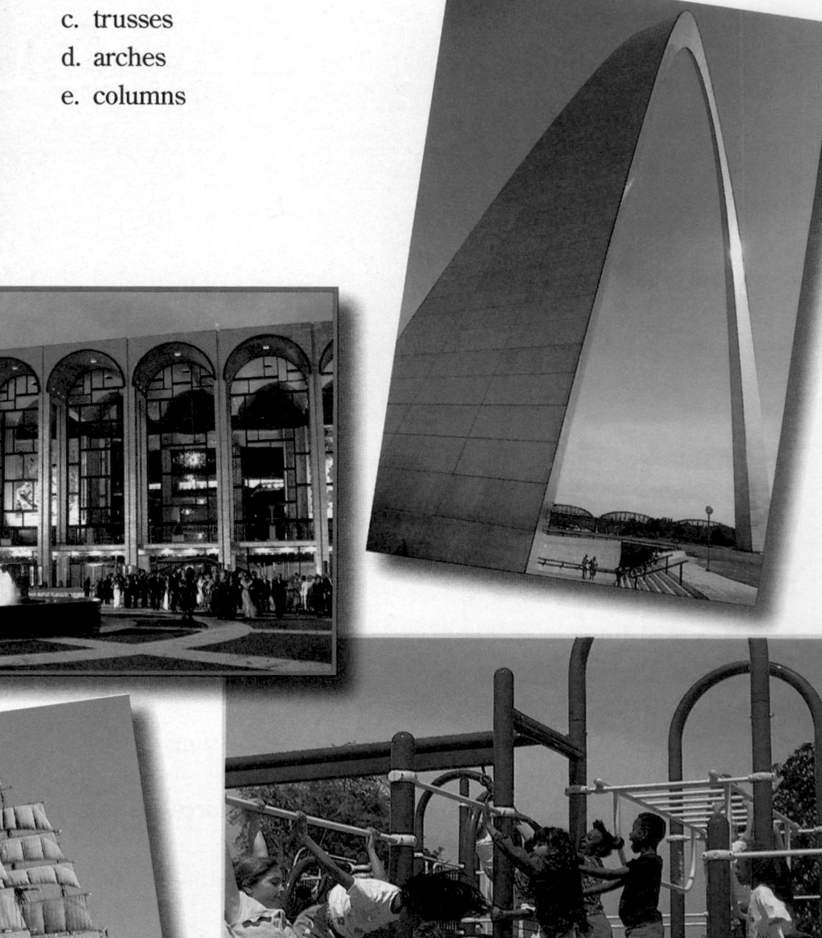

Answers to *Challenge Your Thinking*

1. **a.** Simple beams are shown most clearly in the horizontal bars of the playground equipment.
   b. The sails are mounted on cantilevered supports. Students may notice the cantilever bridge in the background behind the Gateway Arch.
   c. The cables on the sails form a network of trusses. Also, a trussed cantilever bridge is visible in the background behind the Gateway Arch in St. Louis.
   d. The Gateway Arch in St. Louis is a large, simple arch. The building and the playground equipment also include arches.
   e. The vertical bars of the monkey bars are columns. The mast of the sailing ship is also a column, as are the sides of the window frames in the building.
2. Answers will vary according to student experience. Sample answers:
   a. The Empire State Building in New York City probably gets its strength from I-beams and trusses.
   b. The Gateway Arch in St. Louis is attractive because of its grand size and simple design. The structure was built as part of a monument.
   c. The Hoover Dam, located on the Nevada-Arizona border, is very strong. The enormous amount of water that it holds back is evidence of its strength.
3. Answers should demonstrate an understanding of forces and the way structures respond to them. For example, if the student chooses to draw an overpass, he or she should discuss the compressive force of the automobiles on the overpass and the use of columns to direct these forces to the ground.
4. Students may construct tubular columns or girders, corrugated bridges, or arches, for instance. Students may demonstrate the strength of their structures with masses or other objects.

★ **You may wish to provide students with the Chapter 14 Review Worksheet that accompanies this Challenge Your Thinking (Teaching Resources, page 35). Transparency 55 is also available.**

Meeting Individual Needs

Second-Language Learners

Suggest that students make a "picture dictionary" using terms from their native language to identify different kinds of structures. For example, the dictionary might include words for *home, store, office building, theater, bridge, grocery store, restaurant, bakery, hardware store, apartment building, expressway,* and so on. A picture should be drawn to illustrate what each term means. Students should include some of the technical words given on page 304.

2. How About You?

Answer the following questions about structures, and then compare your answers with those given by your classmates:

a. What is the highest structure you have ever been in? Where is it? What gives strength to the structure?

b. What is the most attractive structure you have ever seen? Where is it? What makes it so attractive? What is the structure used for?

c. What is the strongest structure you have been in or on? What evidence do you have of its strength?

3. A Forceful Task

Sketch a human-built structure near your home or school, and then answer the following questions:

a. What forces act on this structure? Where are these forces exerted?

b. How does this structure respond to these forces?

c. What features of this structure give it its ability to withstand these forces?

d. Where does this structure experience compression?

4. Creative Construction

Using three pieces of construction paper or poster board, a pair of scissors, and a small amount of tape or glue, turn each piece of paper into a different kind of strong structure. Each must be stronger than a flat piece of paper or poster board would be. Demonstrate the strength of each of these structures against attempts to compress or bend them.

ScienceLog

Review your responses to the ScienceLog questions on page 294. Then revise your original ideas so that they reflect what you've learned.

325

ScienceLog

The following are sample revised answers:

1. By studying and comparing structures that have not failed, we can learn which designs are the most durable and sturdy. We can use this knowledge to build new structures that will resist failure.
2. Student answers may include the following: The arches and buttresses of the Eiffel Tower experience compression from the weight of the tower and the people on the observations decks. The horizontal beams experience tension from the vertical supports. Shear force may be caused by winds. Compression and tension could result in collapse of the structure; shear forces could cause the tower to sway or even collapse.
3. Student answers may include the following: Cables support the deck of the bridge and direct the forces down the supports and to the ground. Trusses add strength to the horizontal beam of the roadway. The cables are somewhat flexible, preventing them from snapping under the shear forces of wind and earthquakes.

Meeting Individual Needs

Gifted Learners

Hold a competition among students that involves constructing the strongest structure possible with a given set of materials in a certain length of time. For instance, you might give each student 30 craft sticks and 50 twist ties. In 20 minutes, have them create a structure that will protect an egg from bricks placed on top of it. Depending on what materials you choose to use, you should check students' procedures for safety considerations before allowing them to proceed.

ENVIRONMENTAL FOCUS

Ask: Which is better, something that doesn't last very long or something that lasts for many years? *(Many may say that something that lasts for many years is better.)* Point out that many household items, for instance, are now made of substances that will last for many years. In fact, this has became a problem precisely because these items last too long. When these items are thrown away, they may take up space in a landfill indefinitely. Encourage discussion.

Homework

Have students give examples of shear, compressive, and tensile forces that occur as a car is driven down a street. *(Sample answer: A gearshift is under shear forces as it is shifted from one gear to another, the tires are under compressive forces from the weight of the car, and the seat belts are under tensile forces when the brakes are applied due to the weight of the passengers pulling on them.)*

The Art of Design

Connecting to Other Chapters

Chapter 14 *considers the integrity of structures from the standpoint of forces, materials, and design.*

Chapter 15 *offers historical, cultural, artistic, and environmental perspectives on the science of structure.*

Prior Knowledge and Misconceptions

Your students' responses to the ScienceLog questions on this page will reveal the kind of information—and misinformation—they bring to this chapter. Use what you find out about your students' knowledge to choose which concepts and activities to emphasize in your teaching.

In addition to having students answer the questions on this page, you may wish to have them complete the following exercise: In groups of two or three, have students create a tourist brochure for an imaginary travel agency that offers architectural tours—not only around the world, but also through time. Students should describe at least five structures of interest along the tour and point out the architectural significance of these structures as well. Collect the papers, but do not grade them. Instead, read them to find out what students know about various aspects of design, what misconceptions students may have, and what is interesting to them about this topic.

CHAPTER 15

The Art of Design

1 What can a building's design tell you about when it was built?

2 What can architects and planners do to make an urban environment pleasant to view and visit?

ScienceLog

Think about these questions for a moment, and answer them in your ScienceLog. When you've finished this chapter, you'll have the opportunity to revise your answers based on what you've learned.

Beauty in Structures

Our Human-Made Environment—Yesterday and Today

So far you have studied some of the scientific and technical features of structures. But that is not all you must consider when you plan structures for Riverwood. What will your structures look like? Will they blend in with their environment? How will they relate to each other? Engineers, architects, designers, and urban planners often look to the past to help them make these decisions. Look at the sketch below. Do you think the architect who drew this sketch was influenced by the architecture of the past? (A) Why or why not? In Exploration 1 you will take a journey through history. This will help you make decisions about the building style you would like to see in Riverwood.

EXPLORATION 1

A Journey Through History

On the following pages are photographs and illustrations of some important structures that were built at different times and places. You are going to write and illustrate a story about the people and buildings from one of these times in history.

What to Do

1. Team up with a few of your classmates, and look through the next seven pages. Choose one period in history that interests you.
2. Research the period that you have chosen so that you can answer the following questions:
   a. What was the culture of the society like? What kind of government did the society have? Who owned the land? Who owned the buildings? Who built the buildings?
   b. What purposes did the buildings serve?
   c. What structural features are evident in the buildings of this period (columns, beams, trusses, arches, domes, cantilevers)?
   d. What materials were used?
   e. What construction methods were used?

Exploration 1 continued ►

LESSON 1 ORGANIZER

Time Required
2 class periods

Process Skills
observing, analyzing, communicating

Theme Connection
Changes Over Time

New Terms
none

Materials
none

Teaching Resources
Exploration Worksheets, pp. 41 and 43
Transparencies 56 and 57
SourceBook, p. S95

Beauty in Structures

FOCUS

Getting Started

Ask students to name as many different types of architecture as they can. Obtain photos of local structures such as museums, churches, concert halls, and schools. Ask students to group the pictures based on similarities of design. Then ask students which group they like best and why. Then ask them what periods they would choose for designing a new city and why.

Main Ideas

1. Each major period of history is characterized by its own unique architecture.
2. Modern architecture is often based on models from the past.
3. Architectural design is often based on available technology.

TEACHING STRATEGIES

Our Human-Made Environment—Yesterday and Today

Ask a student to read this page aloud. Conduct a class discussion about the picture on this page.

Answers to *In-Text Questions*

(A) The architect who drew this sketch was probably influenced by the architecture of the past, as shown by the pyramid (lower right), the building with columns (upper right), and the cathedral (center), for instance.

Teaching Strategies for Exploration 1 begin on the next page. ►

EXPLORATION 1

Have students read silently pages 327 and 328 to find out what they are to do. Suggest that students work in small groups so that the responsibilities for the project can be divided among group members.

Encourage students to be creative. Point out that including different kinds of illustrations, such as pictures, diagrams, tables, graphs, and charts, can make the story more interesting. Mention that quotations from primary or secondary sources can also make the story seem more real.

Note: The background information on each architectural style that follows is provided as a discussion aid and as a tool to help you evaluate student research.

An Exploration Worksheet is available to accompany Exploration 1 (Teaching Resources, page 41).

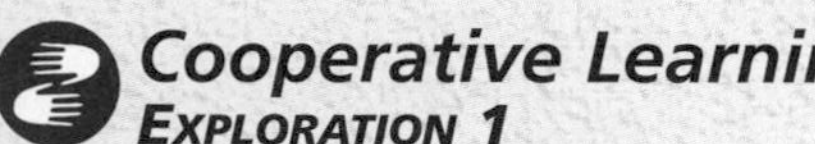

Cooperative Learning
EXPLORATION 1

Group size: 2 to 3 students
Group goal: to research and analyze one period of architecture and to complete an illustrated story about that period
Positive interdependence: Assign students such roles as periodical researcher, book researcher, illustrator, and information compiler.
Individual accountability: Each student should turn in his or her notes from the research they completed. In addition, each student should hand in an annotated bibliography of the sources he or she considered. This should be easy to read, and each entry should include a brief explanation of the content of the work cited.

Nomadic Peoples

Provide students with time to read the material on nomadic peoples and to look at the illustrations.

Nomadic peoples are not a relic of the past. Many nomadic groups exist today. The African Pygmies and Australian Aborigines are examples of nomads who move from place to place in search of food. The Bedouin herders in the deserts of Arabia and northern Africa move from place to place in order to feed and water their herds of camels, sheep, and goats. Gypsies and other nomadic traders exist all over the world; they are merchants and craftspeople who travel to sell their wares.

Since nomads move frequently, their structures must be portable. As a result, they are usually made from easily obtainable and lightweight materials that can be put together and taken apart quickly. Such materials often include animal skins, fabrics, and frames made from wooden poles.

EXPLORATION 1, *continued*

3. Using your knowledge of the period, write your story. Include at least one illustration. You may choose one of the structures shown on these pages or choose one from another source. The length of your story should not be more than two to three pages.

Sources of Information

- Encyclopedias—Look up information on architecture.
- Libraries—Ask the librarian to help you find books on architecture and on the specific period of history that you are studying.
- Your social studies teacher—Ask your social studies teacher to help you find textbooks that describe the period of history that you have selected.

Nomadic Peoples

Nomadic people who formed the earliest human communities did not build permanent dwellings. Their lives involved constant movement in search of food and other materials.

In the past, nomadic Plains Indians depended on the great buffalo herds that once roamed the American Great Plains. Their portable homes were called tepees. What is the shape of the framework that gives support to the tepee? Where else can you see a structure shaped like this? (A)

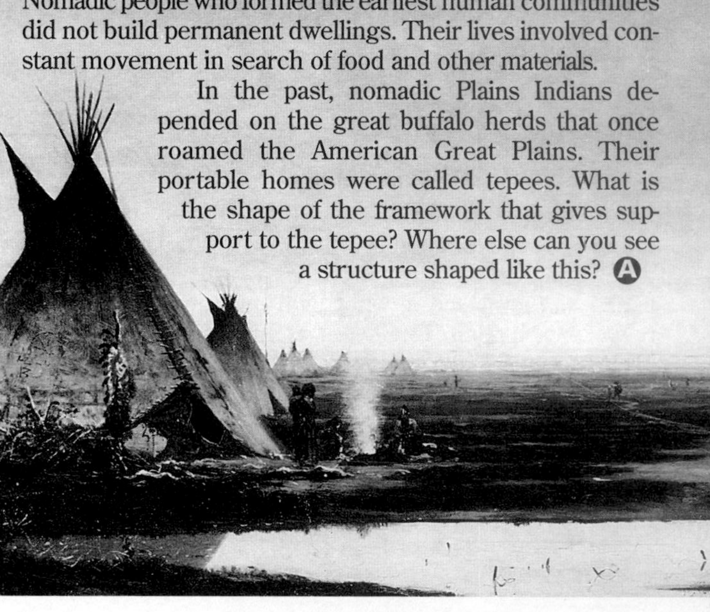

Plains Indians at their buffalo-hide tepees in the late 1800s

Mongolian nomads constructing a *ger*—a felt-covered, collapsible tent

Nomadic cultures still exist. For example, Mongolian nomads make their livelihood by herding sheep and other animals across parts of central Asia. Do you know of any other modern nomadic cultures? (B)

328

Answers to *In-Text Questions*

(A) The tepee is built around a cone-shaped framework. A similar shape is seen in the spires on churches, cathedrals, castles, and government buildings of other cultures.

(B) African Pygmies, Australian Aborigines, Bedouin herders, and gypsy traders

The Earliest Permanent Structures

When people learned to raise crops, they began to settle in one location. Staying in the same location gave them the chance to build permanent structures. The remains of these structures are lasting reminders of the earliest known civilizations.

The remains of Stonehenge

Scientists believe this to be the original configuration of Stonehenge, built in England between 2800 and 1500 B.C. The structure is thought to have been used to determine astronomical occurrences.

The Great Pyramid of Giza and the Great Sphinx were built on the west bank of the Nile River around 2500 B.C. The structures were memorials for an Egyptian king.

329

The Earliest Permanent Structures

Ancient Egypt was one of the world's earliest non-nomadic civilizations, originating along the Nile River about 5000 years ago. Even though the Egyptians built great cities, they are best known for the pyramids they constructed as tombs for their rulers.

Throughout most of its history, ancient Egypt was ruled by kings who were believed to embody the god Horus in human form. Forty-two local provinces called *nomes* were governed by *nomarchs* who were appointed by the king.

Ancient Egyptian society was made up of three main social classes. The upper class consisted of the royal family, wealthy landowners, doctors, high-ranking government officials, priests, and army officers. The middle class consisted mainly of merchants, craftworkers, and manufacturers. The lower, and largest, class consisted mainly of unskilled laborers.

The Egyptians built their houses with bricks of dried mud. The trunks of palm trees were used as columns to support the flat roofs. Many houses had three or more floors.

The oldest and largest stone structures in the world are Egypt's pyramids. They were built from huge limestone blocks, each weighing more than 1.8 metric tons. The ancient Egyptians also built most of their temples from limestone. Many of these structures are noted for the huge decorative columns that were constructed to hold up massive, beamed ceilings.

Theme Connection

Changes Over Time

Focus question: Compare and contrast the effects of weathering on the pyramids to the effects of weathering on natural rock formations. *(The effects are essentially the same in both cases. In general, however, people are more concerned with weathering on the pyramids because their value to us depends on the pyramids' ability to stay intact over long periods of time.)*

Cross-Disciplinary Focus

Social Studies

After discussing the early civilizations, ask students the following question: The earliest civilizations were able to settle in one location and construct permanent buildings after

a. they learned to grow crops. *(Correct)*
b. they constructed a system of roadways.
c. they learned to build pyramids.
d. they developed tools and weapons.

Homework

Have students find out about an early permanent structure and write a short description of it, including the materials from which it was made, its age, and its intended purpose.

Ancient Greece (800–338 B.C.)

The ancient Greeks never organized themselves into a nation. Instead, ancient Greece was made up of many city-states, each consisting of a city or town and the surrounding area. Probably the best known of these city-states was Athens.

Greece's mild climate enabled people to live in simple one- or two-room houses with walls made of sun-dried bricks and floors of hard-packed earth. The wealthy lived in larger, more elaborate, and more comfortable homes built around courtyards.

The architecture for which the ancient Greeks are best known is reflected in their temples, the designs of which centered around the column. The Greeks developed three basic kinds of columns—Doric, Ionic, and Corinthian. Each had its own distinctive decoration and style. The design of a temple consisted of arrangements of columns surrounding a long chamber, which usually housed the sculpted figure of the god or goddess for whom the temple was built. The best-known temples were constructed on the Acropolis in Athens around 450 B.C.

An acropolis was the religious and military center of a city-state, usually built on the top of a hill. The original Acropolis in Athens was partially demolished by a Persian invasion in 480 B.C., but was later rebuilt in even greater splendor. Among the new buildings was the Parthenon, dedicated to the patron goddess Athena. The Erechtheum was built to honor the legendary founders of the city. Another temple honored Athena as the goddess of victory. Two theaters and several minor temples were built on the slopes of the hilltop. The entrance to the Acropolis was a large, roofed gateway called the Propylaea.

Answer to *In-Text Question*

Ⓐ The influence of Greek architecture can often be seen in public buildings. For example, the White House and the Lincoln Memorial in Washington, D.C., were patterned after the magnificent temples of ancient Greece.

Ancient Greece (800–338 B.C.)
The architecture of ancient Greece has had an enormous influence on designers for many centuries. The illustration below shows the Acropolis at Athens, Greece, as it probably looked about 450 B.C. Name some buildings you have seen that appear to be patterned after the buildings of ancient Greece. Ⓐ
Parthenon
Temple
Erechtheum
Propylaea
Ruins of the Erechtheum
330

CROSS-DISCIPLINARY FOCUS

Art

Aesthetics is the study of artistic appreciation. People who study aesthetics try to understand art forms, such as music, poetry, architecture, and the visual arts, in terms of how they make people feel and how they reflect the culture that created them. Have students assess a structure in terms of its aesthetics. They should evaluate the structure by answering questions such as the following:

- How does the structure make you feel?
- Do the building and the environment interact well?
- How will the structure serve the people who use it?
- How does the structure communicate the culture of the people who built it and use it?

Transparency 56 is available to accompany the material on this page.

Ancient Rome (753 B.C.–A.D. 476)

Ancient Rome is characterized by its classical structures. In the model reconstruction below, you can see the famous Colosseum, where thousands of gladiator combats and other events occurred. At the bottom left of the model is the aqueduct that carried water to the citizens of ancient Rome. Remnants of many of the structures are still standing in modern Rome.

Ruins of the Colosseum

★ **Transparency 57 is available to accompany the material on this page.**

Ancient Rome (753 B.C.–A.D. 476)

Ancient Rome was the capital and largest city of the Roman Empire. At its height in the second century A.D., the Roman Empire included all of the land around the Mediterranean Sea, the land that stretched north to the British Isles, and all the land east to the Persian Gulf. It encompassed from 50 to 70 million people and had to accommodate many different customs and languages. The ability of the Romans to hold this diverse empire together is considered an achievement in itself.

At its height, about 1 million people lived in ancient Rome. No other city had ever been as large. Most of the people in Rome lived in crowded apartment buildings that were from three to five stories high. A wealthy family would have lived in a house built around an atrium or courtyard. The atrium was spacious and had an open roof that let in light and air to the surrounding, windowless rooms. Larger houses had a second courtyard called a peristyle, which served as a garden. The poor people in the surrounding farming areas lived in huts made of sun-dried bricks.

Much of ancient Roman architecture was adapted from ancient Greece. This includes the large temples surrounded by rows of columns. The Romans, however, also created many new kinds of theaters that were designed to hold large crowds of people.

The construction of large buildings was made possible by the use of the arch and the invention of concrete. Arches were used to support such structures as bridges and aqueducts. Domes and arched roofs called vaults allowed for the construction of large, open, interior spaces. Arches, vaults, and domes eliminated the need for columns to hold up roofs. Instead, the weight of the roof rested on the outer walls of the building. The invention of concrete provided the strong building material that was needed for the outer walls.

Medieval Europe (500–1500)

Also called the Middle Ages, this period in the history of western Europe spanned ancient and modern times. Life in medieval Europe revolved around the control of land and was based on the system of feudalism. Large areas of land were owned by powerful lords. Small armies of faithful knights defended the land, while peasants farmed it. These feudal estates were bound together by the influence of the Christian church.

Two major types of architecture were developed during the medieval period—Romanesque and Gothic. Romanesque architecture began in the late 700s and reached its peak during the 1000s and 1100s. The most impressive and important Romanesque buildings were churches. A typical Romanesque church had thick walls, heavy curved arches, and rows of columns built close together. There were few windows or openings to let in light. A pointed tower rose where the transept crossed the nave.

Gothic architecture flourished from the mid-1100s to the 1400s. Gothic architects developed a system of construction that enabled them to design churches with thinner walls than had been possible in Romanesque architecture. The ribbed vault or dome was one of the most distinctive characteristics of Gothic churches. Many churches had flying buttresses, which were brick or stone arched supports built against the outside walls. The buttresses strengthened the walls and permitted the use of large stained-glass windows that allowed light to enter the building.

Answer to *In-Text Question*

Ⓐ Students should note any local churches or cathedrals that show similarities to medieval cathedrals.

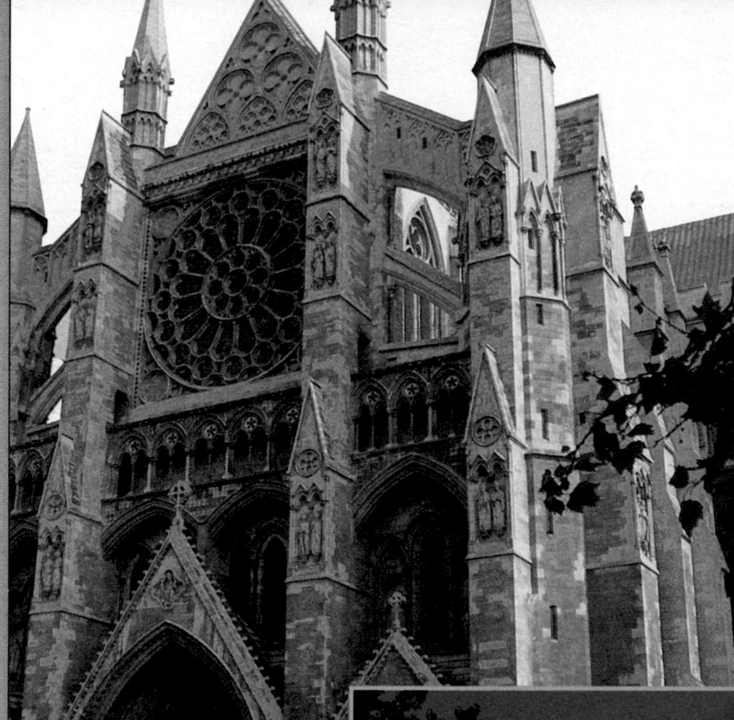

Westminster Abbey, London

Medieval Europe (500–1500)

Medieval Europe is characterized by the famous cathedrals that were built during this period. Is there a church or other building in your area that appears to be patterned after the cathedrals of medieval Europe? In your ScienceLog, make a list of examples you can think of. Ⓐ

Notre Dame Cathedral, Paris

332

Multicultural Extension

Architecture of the United States
Point out to students that many different cultures have influenced the architecture of the United States. Suggest that they do some research to discover what some of these influences have been. Encourage students to share what they discover by preparing an oral presentation for the class. If students need some suggestions, you might point out that Spanish and American Indian culture has influenced the architecture of the West and Southwest. Furniture design has been influenced by the countries of Scandinavia. Architecture of Asian communities in some major cities includes designs that are similar to the architecture of their native countries.

The Renaissance (1300–1600)

Near the end of the Middle Ages, widespread cultural change occurred in Europe. The Renaissance, as it is called, affected science, the arts, and education. Architects and artists, like Donato Bramante and Michelangelo, worked together to create beautiful buildings. Do you see any similarities between Roman and Renaissance architecture? (B)

Tempietto, San Pietro in Montorio, Rome

Palazzo Chiericati, Vicenza, Italy

St. Peter's Basilica, Rome

333

The Renaissance (1300–1600)

The word *renaissance* comes from a French word meaning "rebirth." To scholars and artists of the Renaissance, it meant a rebirth of interest in the classical culture and art of ancient Greece and Rome. The influence of the Renaissance on future generations proved to be immense. Most historians agree that the modern era began with this period of history.

Renaissance architecture began in Italy in the 1400s and gradually spread throughout Europe, eventually reaching the New World. A group of Italian scholars of classical culture is credited with beginning the movement. They greatly admired classical principles of art and architecture. Soon architects began studying the ruins of ancient Rome and Greece and began modeling their designs on classical buildings. They incorporated the use of columns as well as Roman vaults and domes.

The dome of the Cathedral of Florence, designed by Filippo Brunelleschi around 1420, is an early example of Italian Renaissance architecture. The cathedral itself was begun in 1296 in the Gothic style. (See page 351.)

One of the greatest building projects of the late Renaissance was the construction of St. Peter's Basilica in Rome. It was started in 1506 and completed in the 1600s. Ten architects worked on the church during its construction, including Donato Bramante and Michelangelo.

In the early 1500s, the Renaissance spread from Italy to France. The finest examples of French Renaissance buildings are châteaus, or castles, such as the Château de Fontainebleau built in the early 1500s by King Francis I.

Answer to *In-Text Question*

(B) Arches, domes, and columns appear in both Roman and Renaissance architecture.

Homework

Have interested students find out about the work of a Renaissance artist such as Michelangelo or Donato Bramante and share the results of their research with other students.

Modern Civilization

The modern period of architecture is generally considered to have begun in the late 1800s. New materials allowed architects to develop and design the first completely new structural styles in centuries. American architects greatly influenced international architecture for the first time. One example is the skyscraper, which was first developed in the United States.

During the late 1800s and early 1900s, Chicago became the center for modern architecture in the United States. A fire in 1871 destroyed much of the city. The rebuilding of Chicago provided architects with the opportunity to experiment with new designs utilizing the new materials that were then available. The world's first metal-frame skyscraper, the 10-story Home Insurance Building, was constructed in downtown Chicago during this period.

In recent decades, architects have started many new movements with names such as new-brutalism, postmodernism, and deconstructivism. Often these movements have been short-lived.

Postmodernism, which began in the United States in the 1960s, revived the use in modern buildings of historic elements such as classical columns and arches. These were often used in unconventional and playful ways.

PORTFOLIO

Students may want to include their stories from Exploration 1 in their Portfolio. They may wish to evaluate their contribution to the group as well as the final outcome of their research. Encourage them to include sketches, photos, and other illustrations from the architectural period that they studied.

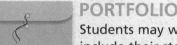

Modern Civilization

Advances in technology have changed how we build structures. Structural steel and elevators have made skyscrapers possible. Prestressed concrete is used to make beautifully curved roofs and walls for stadiums, theaters, and other buildings. Even huge, enclosed stadiums can now be built by applying our modern knowledge of materials and architecture.

Sydney Opera House, Australia

Superdome, New Orleans, Louisiana

New York City skyline with Brooklyn Bridge in foreground

CROSS-DISCIPLINARY FOCUS

Mathematics

Suggest to students that they make graphs to compare different kinds of structures. For example, bar graphs could be used to show the comparative sizes of the tallest buildings, the longest bridges, the largest airports, the biggest city parks, and so on. Encourage students to be creative when they make their graphs. For example, a graph comparing the tallest buildings could be drawn over an image of the Sears Tower.

Conversations With Architectural Pioneers

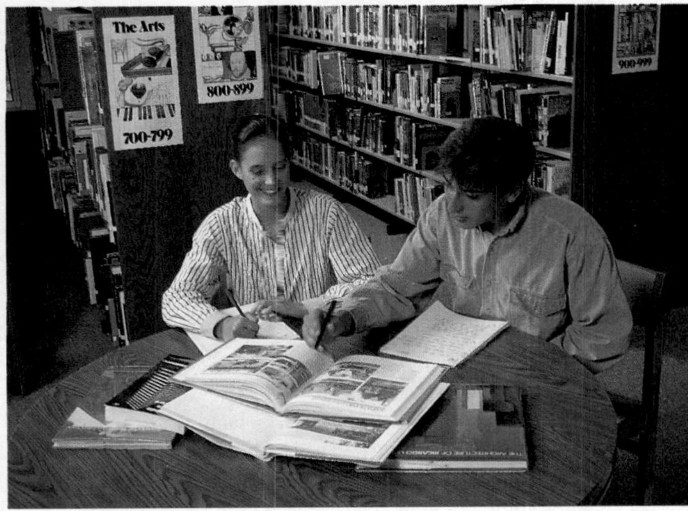

Daniel and Karen are two students who have become very interested in modern architecture. They did some research and created imaginary interviews with two pioneers of modern design. As you read their interviews, make a list in your ScienceLog of the most important ideas. Later, you can apply these ideas to your plans for Riverwood.

Daniel and Karen met Walter Gropius (1883–1969) first. Imagine it is 1951. Gropius is chairman of the Department of Architecture at Harvard University in Cambridge, Massachusetts. Listen in as Karen and Daniel conduct their interview.

Karen: We are pleased that you could take time to meet with us. We have heard so much about you. Many architects in the United States give you credit for developing the modern ideas of architecture and design.

Walter: Thank you. The International Style of architecture and design is the result of ideas from many artists, designers, and engineers from around the world. Much of the pioneering work was done at the Bauhaus School of Design in Germany when I was its director.

Daniel: What was it like to work there?

Walter: It was very exciting. Many young artists and designers worked there, and several of them became famous for their work. We stressed the fundamental elements of design, but we also came up with many new design techniques and ideas. We considered buildings, furniture, books, and much more. Unfortunately, most of us had to leave the country when the Nazis came to power. I first went to England in 1934 and then came to Harvard University in 1937. Here, I have had the good fortune to work with many talented architects.

Karen: What design ideas did you and the other artists and architects work on at the Bauhaus?

Exploration 2 continued ►

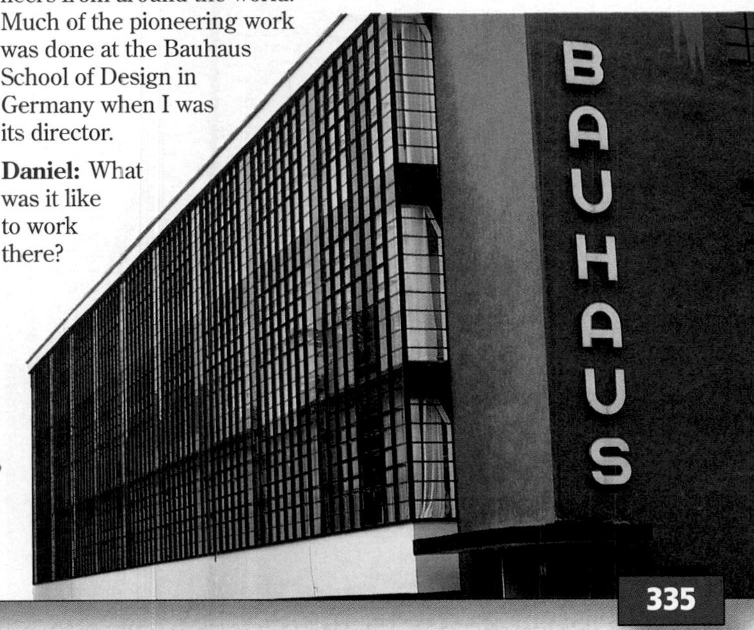

The Bauhaus School of Design, Germany. This building was very unusual when it was built in 1926. How does it compare with buildings in your community that were constructed around that time? Ⓐ

★ **An Exploration Worksheet is available to accompany Exploration 2 (Teaching Resources, page 43).**

EXPLORATION 2

Have students silently read the imaginary interview with Walter Gropius. Or you may wish to have volunteers role-play the interview for the class. When students have finished reading, involve them in a discussion of the interview. You may wish to use questions similar to the following:

- Did Walter Gropius work alone on his designs? *(No, he worked with other architects, designers, and artists.)*
- With what does Gropius link art and design? *(He links art and design with modern technology.)*
- How did Gropius think of structures? *(He thought of structures as forms of sculpture.)*
- To what did Gropius believe designers should adapt their designs? *(He believed they should adapt their designs to mass production.)*

Students may find it interesting to know that the term *International Style* came from the title of a book by the same name. It was written by two American architects in 1932. They had reviewed the architecture of the previous 10 years and concluded that a new international style had developed in several countries. This style was characterized by building designs that had geometric shapes, white walls, and flat roofs. There was little or no exterior ornamentation. Most were constructed from reinforced concrete and had large windows to create a light, airy feeling.

Teaching Strategies for Exploration 2 continued ►

Answer to *Caption*

Ⓐ Students should note whether local buildings from this era show aspects of the Bauhaus style, such as simplicity and the influence of technology.

EXPLORATION 2, *continued*

Encourage students to discuss the photograph at the bottom of this page. Encourage them to suggest words to describe how the building looks and makes them feel—shiny, open, glassy, light, airy, metallic, depersonalized, and so on. Some students may recognize that compared to earlier buildings, these structures use much more glass and metal to create an open and airy feeling.

Then have students read the imaginary interview with Frank Lloyd Wright. When students have finished reading, involve them in a discussion of the interview. You may wish to use questions similar to the following:

- What was Frank Lloyd Wright's lifelong interest? *(He wanted to tackle the problem of the house as a shelter.)*
- How did Wright's views about architecture differ from those of Gropius? *(Gropius believed that architecture should reflect modern technology and be designed for mass production. Wright believed that buildings should blend in with their natural environment and that their construction should include materials from that environment.)*
- What did Wright call his kind of architecture? Why? *(He called it organic architecture because it linked structures with nature.)*
- Which kind of architecture, that of Wright or Gropius, do you think you like better? Why? *(Accept all reasonable answers. Responses should reflect an understanding of both kinds of architecture.)*

Homework

Have students state Wright's two main concerns in building a house and describe how Fallingwater addresses those concerns. *(Answers will vary. For instance, students may note that the use of stone and cantilevers helps the house blend in with its environment and that Wright avoided large windows and other exposed areas, giving Fallingwater the impression of a secure, sheltering building.)*

EXPLORATION 2, *continued*

Walter: We sought to connect art and design with modern technology. We stressed the use of modern materials and machine production. The sculptors at the Bauhaus were particularly important in advancing the idea that human-made structures are a form of sculpture.

Daniel: What have been your goals as a teacher of design and architecture?

Taliesin West, Frank Lloyd Wright's home in Phoenix, Arizona

Walter: I want people to recognize that everything built by humans has a design. Some people use a lot of decoration in their designs. However, I prefer simplicity. I believe designers should be able to adapt their designs to mass production using modern materials.

Daniel and Karen's next stop was in Phoenix, Arizona. There they met Frank Lloyd Wright (1869–1959), the most famous American architect. The year is 1955.

Daniel: When did you begin your career in design, and with whom did you work?

Frank: In 1887 I lived in Chicago and began to work with Louis Henri Sullivan. He was a pioneer in the design of large commercial buildings. In those days, we were learning to build all those skyscrapers. We tried to make them look friendly by including many of the decorative elements of earlier architecture.

After 6 years with Sullivan, I set up my own studio where I could devote myself to designing homes. I wanted to explore the idea of the house as a shelter. That has been a lifelong interest for me. Of course, I continued to design many public buildings, including an art gallery and a university campus.

The Fagus Factory, Germany, designed by Walter Gropius in 1911

336

Karen: It is known that you have opposed the International Style of architecture. How do your views differ from those of Walter Gropius and his followers?

Frank: Modern architecture has been influenced by Gropius's views and my own. I believe homes and buildings should blend in with their natural environment. Their construction should include materials from that environment. I call this "organic architecture" because it links structures with nature.

It seems to me that many architects and designers forget about nature when they design. So there is a definite contrast between my views and those of Gropius. I think a building's relationship to nature is one of its most important characteristics.

Frank Lloyd Wright's Fallingwater house, Bear Run, Pennsylvania (built in 1936). How did Wright use cantilevers to create a home that blends in with its environment? (A)

The land before Fallingwater was built

GUIDED PRACTICE Provide students with a few moments to study and compare the pictures of Fallingwater and the Bauhaus School of Design. Then involve them in a discussion of how the two differ and how they reflect the ideas of their creators. *(The Bauhaus School of Design reflects Gropius's idea of combining mass production and modern technology with design. Wright's Fallingwater reflects his view that architecture should be combined with aspects of nature.)* Have students think about the question posed in the caption.

Answer to *Caption*

(A) Wright used cantilevers to create the parts of the building that look like layers of rock above the waterfall.

Meeting Individual Needs

Gifted Learners
Divide students into small groups. Assign each group the task of redesigning their school building in the fashion of one of the historical periods that they studied in Lesson 1. The groups may decide between drawing and building models of their designs. Each group should also prepare a description of their design that explains how their structures combine beauty and function.

ENVIRONMENTAL FOCUS

Provide each student with a photocopied picture of a natural landscape. Tell them that they have been hired to design a home that incorporates the landscape into the home's design. They should draw their design on a blank sheet of paper and use the copy of the landscape to trace the home's surroundings. Afterward, have them discuss what design features worked well with the landscape.

The Postmodernist Movement

After students have analyzed the pictures on this page, discuss the similarities and differences between the postmodernist architecture and that of Gropius and Wright.

Answers to *In-Text Questions, pages 338–339*

Ⓐ Answers will vary but could include observations that the postmodernist structures combine the old with the new by incorporating attractive features into technologically modern buildings. The building designs may not serve any practical functions. Wright's view of architecture combined nature with design, and Gropius's view combined mass production and modern technology with design.

Ⓑ Most students will probably identify such things as open spaces, plants, decorations, and certain building design features.

The Postmodernist Movement

After creating their imaginary interview with two pioneers of modern architecture, Daniel and Karen visited a major architectural firm in a large city. They were surprised to find out how many important changes there have been in architectural thinking since the pioneering work of Gropius, Wright, and others. Many architects now include a variety of decorative features in their designs for buildings.

Loyola Law School, Los Angeles, California

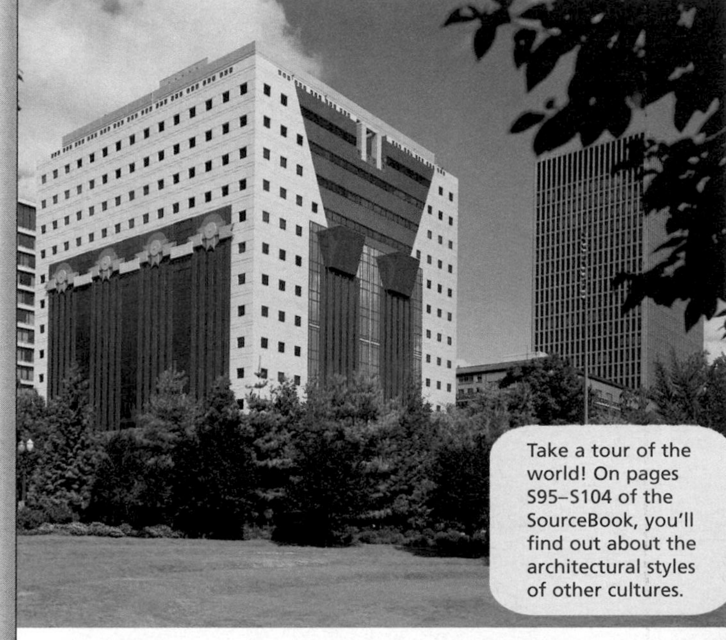

Portland Public Service Building, Portland, Oregon

These features might serve no practical function, but they make the building more interesting to look at. This style is referred to as *postmodernist*.

Examine the photos on this page. What postmodernist features do you see? Describe how you feel about these buildings. In what ways do they differ from those designed by Gropius and Wright? Ⓐ

Find examples of postmodernist architecture in your own environment, and bring some sketches or photographs to class so that you can discuss them with your classmates. Identify the decorative features that have been included, and describe your feelings about these structures.

Take a tour of the world! On pages S95–S104 of the SourceBook, you'll find out about the architectural styles of other cultures.

338

FOLLOW-UP

Reteaching

Ask the class to examine the photograph of Notre Dame Cathedral on page 332. Ask students to name the things they can see that are beautiful and that also serve a function. *(Such structures could include the flying buttresses that provide support for the walls, stained-glass windows that allow light into the cathedral, and the arched doorways and windows that support construction above and that provide openings for people, air, and light below.)*

Assessment

Have students use the information on the historical development of structure and design on pages 328–334 to make a time line. Their time lines should highlight at least one major feature of the architecture from each of the periods.

Extension

Suggest that students write short biographies of some of the most famous modern architects, such as Frank Lloyd Wright, Henry Hobson Richardson, William Le Baron Jenney, Walter Gropius, Le Corbusier, Ludwig Mies van der Rohe, Paul Rudolph, and Louis Kahn. Have students organize their biographies into a booklet, make a cover, and create a title. The booklets could be displayed in the classroom for others to enjoy and read at their leisure.

Closure

Suggest that students work in pairs or small groups to make models of famous buildings. Suggest that they model one of the structures pictured in the lesson or choose a building of their own to model. When students are finished, call on several volunteers to describe their models to the class, pointing out the most important architectural features. Then display the models around the classroom. Each model should be accompanied with a brief description identifying the name of the building and where it is, or was, located.

LESSON 2 A Question of the Environment

The photos on this page illustrate what architects and planners can do to make the urban environment more pleasant. In your ScienceLog, describe the features of these environments that make them pleasant to view and to visit. B

Rivercenter Mall, San Antonio, Texas

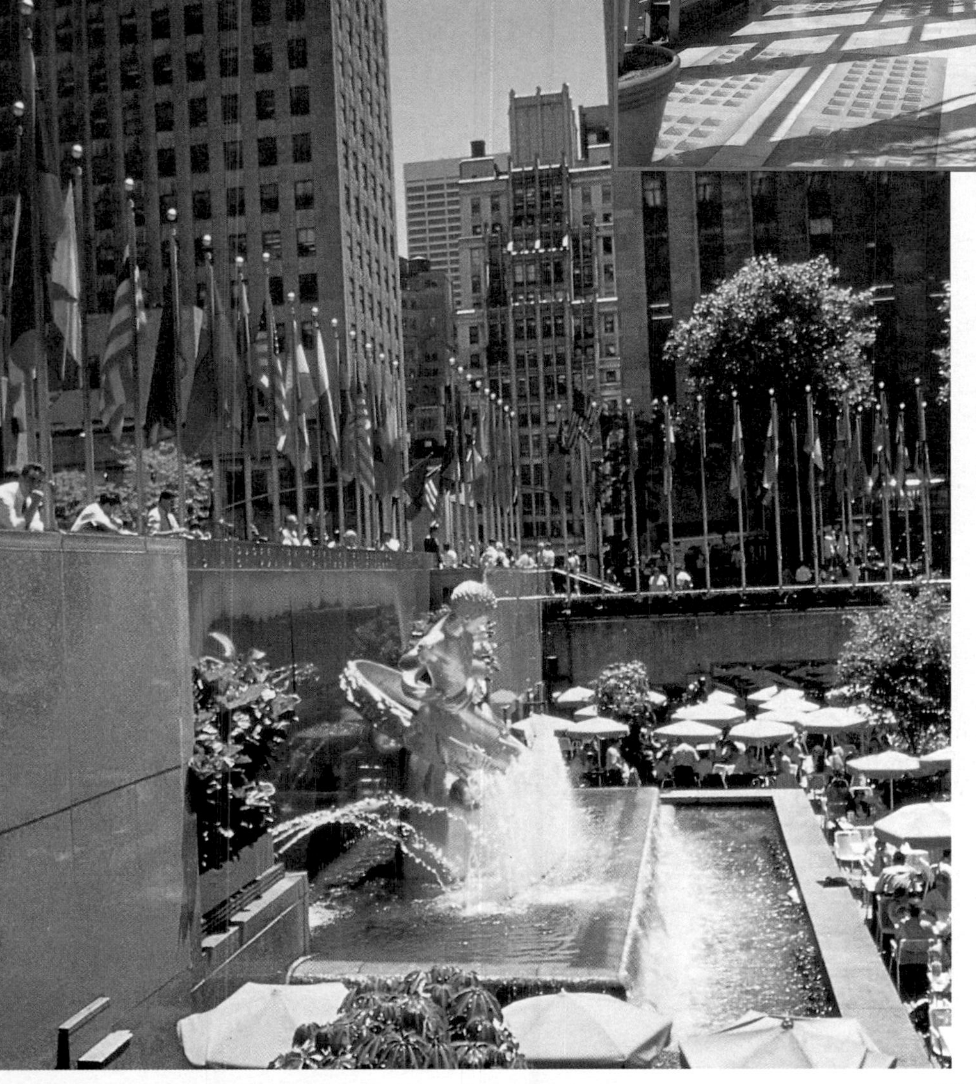

Rockefeller Center, New York City

LESSON 2 ORGANIZER

Time Required
2 class periods

Process Skills
hypothesizing, analyzing

Theme Connection
Systems

New Terms
none

Materials (per student group)
Exploration 3: materials to construct a model town, such as uncooked spaghetti, string, balsa wood, index cards, cardboard, cloth, and plastic wrap; glue; metric ruler; roll of transparent tape

Teaching Resources
Exploration Worksheet, p. 44

LESSON 2 A Question of the Environment

FOCUS

Getting Started

Have students examine the photograph on page 337 of Frank Lloyd Wright's Fallingwater house. Ask: What features of this house make it look like it would be a nice place to live? *(Student responses might include the trees and shrubs, the sound of the waterfall, or the interesting shapes and textures used in the house's construction.)* Tell students that a structure's environment is, in many ways, as important as the structure itself, especially in determining how the people who live and work in the structure feel. Conclude by saying that this lesson is about how buildings and the environment can be combined to produce a safe and pleasant community.

Main Ideas

1. The location and design of structures in an urban setting are social as well as environmental concerns.
2. Most cities have laws that regulate the construction and design of buildings to ensure a pleasant and safe environment.
3. How buildings and open spaces are designed can have an important impact on the social environment of a community.

TEACHING STRATEGIES

Involve students in a discussion of their community. Encourage them to identify what they like and dislike about it. Invite them to speculate about ways in which their community could be improved.

Homework

Have students cut out a picture of a pleasant structure from an old magazine or newspaper. They should describe what features of the structure make it pleasant to view or to visit.

Answers to
In-Text Questions, pages 340–341

Ⓐ The changes to the pictures on this page might involve store signs, outside stairways, trash, telephone poles, plants, and street lights that need to be added or redesigned.

Ⓑ Answers may reflect a deterioration in downtown areas due to overpopulation. In some cities, however, urban renewal projects have led to cleaner, more attractive downtown areas.

Multicultural Extension

Low-Rise Housing
As the world's population has grown and land has become scarce and expensive, different solutions have been proposed to make the most of available space. In recent years, the high-rise apartment has been proposed as one possible answer. However, high-rise living often creates many new problems, including lack of adequate play areas and safety concerns. One possible solution is called urban low-rise housing, in which individual units are linked together. This type of urban planning is not new, however. It was used in North America before the time of European settlers by the Anasazi Indians of the Southwest and in Africa by the Dogon people of Mali. Ask students to research such dwellings used by other cultures and to evaluate whether these could be used successfully today.

Theme Connection

Systems
Focus question: How could including attractive parks and other natural areas help influence residents to take better care of their community? *(Accept all reasonable responses. Many factors interact to determine the involvement of residents in maintaining their community. If residents take pride in their community and find features such as parks useful and attractive, they will be more likely to invest their time and resources in the community's upkeep.)*

What changes would you make in these environments to make them more pleasant? Ⓐ

Did You Know. . .

Some cities incorporate the natural environment in urban design by developing parks, greenbelts, and other preserved areas to help protect wildlife as well as provide residents with beautiful areas to enjoy. In some cases, city planners put parks or recreation areas in flood plains where development could not safely take place.

ENVIRONMENTAL FOCUS

Give students square pieces of paper, and have them use the entire sheet to draw the front view of an apartment unit. Encourage them to be creative and to incorporate features of any architectural style they have studied. When students have finished, tape all the sheets together to form the front view of an entire apartment building. Ask: Is this building attractive? Would you want to live here? Would it fit in with the other buildings in your neighborhood?

Conserving and Restoring the Urban Environment

The downtown area of your favorite city probably looks very different than it did 30 years ago. Ask an adult about these changes. Are there more tall buildings? Have some of the older neighborhoods been torn down and replaced by businesses? Which changes do you think are good changes? Which changes made the downtown area less interesting or less attractive? B

In many cities, citizens have taken action to keep favorite neighborhoods or buildings from being torn down and replaced. Or they have worked to restore run-down neighborhoods or buildings. Think about and discuss the following comments that were made by Phyllis Lambert, an architect and crusader for cities that are more pleasant to live in.

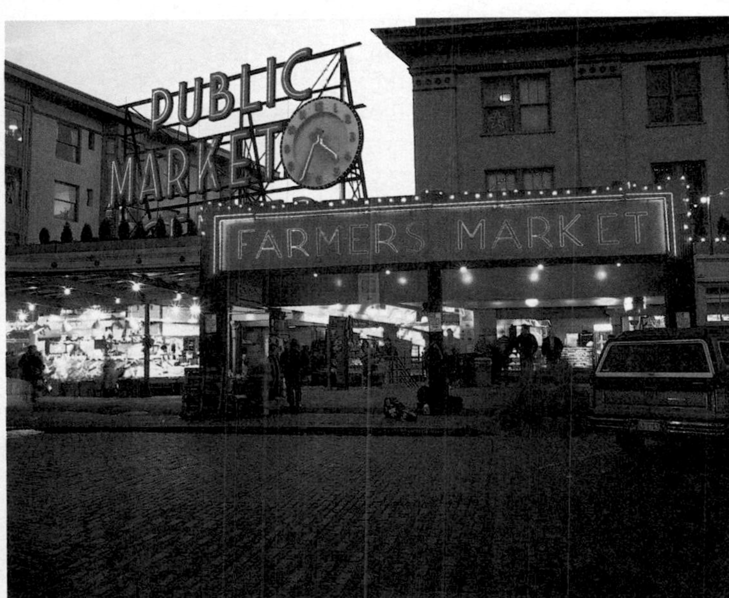

Pike Place Market in Seattle, Washington, was saved and restored as the result of action by a citizens' group.

Architect Phyllis Lambert

It was not the architects who first publicly complained of urban disaster . . . Those who first cried out and pointed to the real problem and created the basic structures of reform have been citizens' groups. These groups are made up of people who are concerned about not having their neighborhoods disrupted . . . Such citizen groups not only make better urban developers than people in government and business, but also become more competent themselves in directing their own lives . . . I consider that my major task . . . is seeing to it that there is an effective means for action and participation by citizens' groups in the decisions that affect our daily lives.

CROSS-DISCIPLINARY FOCUS

Language Arts

Have students complete the following: Phyllis Lambert probably wrote the passage on page 341 to

a. convince people to construct more buildings in the International Style.

b. criticize urban disaster and make suggestions about who should fix it.

c. explain the importance of citizens' groups in planning urban development. *(Correct)*

d. describe why she became a conservationist.

Conserving and Restoring the Urban Environment

Have students read the page silently or call on a volunteer to read it aloud, pausing before the Lambert quotation. Direct students' attention to the photograph of the Pike Place Market and have them suggest how it might have looked before it was restored. Encourage students to express whether or not they like this kind of architecture, and have them provide reasons for their points of view. Discuss with students the positive aspects of restoring run-down neighborhoods. Then provide students with some time to consider the comments by Phyllis Lambert.

Call on several volunteers to summarize Lambert's comments. Then involve students in a discussion of Lambert's ideas and opinions. You may wish to use questions similar to the following:

- According to Phyllis Lambert, who were the first to cry out about the real problems of the cities? *(The people living in the cities were the first to recognize urban problems and to suggest solutions.)*
- How does Phyllis Lambert describe citizens' groups? *(She describes them as groups of people who want to make sure that their neighborhoods are not disrupted.)*
- What does Phyllis Lambert consider to be her major task? *(Making sure that there is an effective means for action and participation by citizens' groups in the decisions that affect their daily lives)*
- Do you agree or disagree with Phyllis Lambert's opinions? Why or why not? *(Answers may vary, but most students will probably agree with Phyllis Lambert's opinions because she takes into account the needs and wishes of the people that are affected by what urban architects do.)*

Challenging Research Activities

Here are some activities that may interest you and your classmates.

1. From your city council representative, find out the laws your city has about the kinds of buildings that can be built in your neighborhood. These are called zoning laws. Look for answers to these questions:
   a. What kinds of buildings are allowed in your neighborhood (commercial buildings, apartment buildings, single-family homes, factories, office buildings)?
   b. Is there a limit on the height of a building?
   c. How close together can builders construct homes and other buildings?
2. Find out about the work of historical preservation groups in your city. What buildings and neighborhoods have been preserved because of their work?
3. Consider the design problems of urban centers. What is meant by the term *inner city*? Find out if your city council has plans to combat the problems of the inner city by changing the environment. If you do not live in a large city, research the history of designs and structures found in Harlem in Manhattan, New York.

Sources of Information

- The library
- Architectural firms
- The planning department of the city government
- Parents and other adults

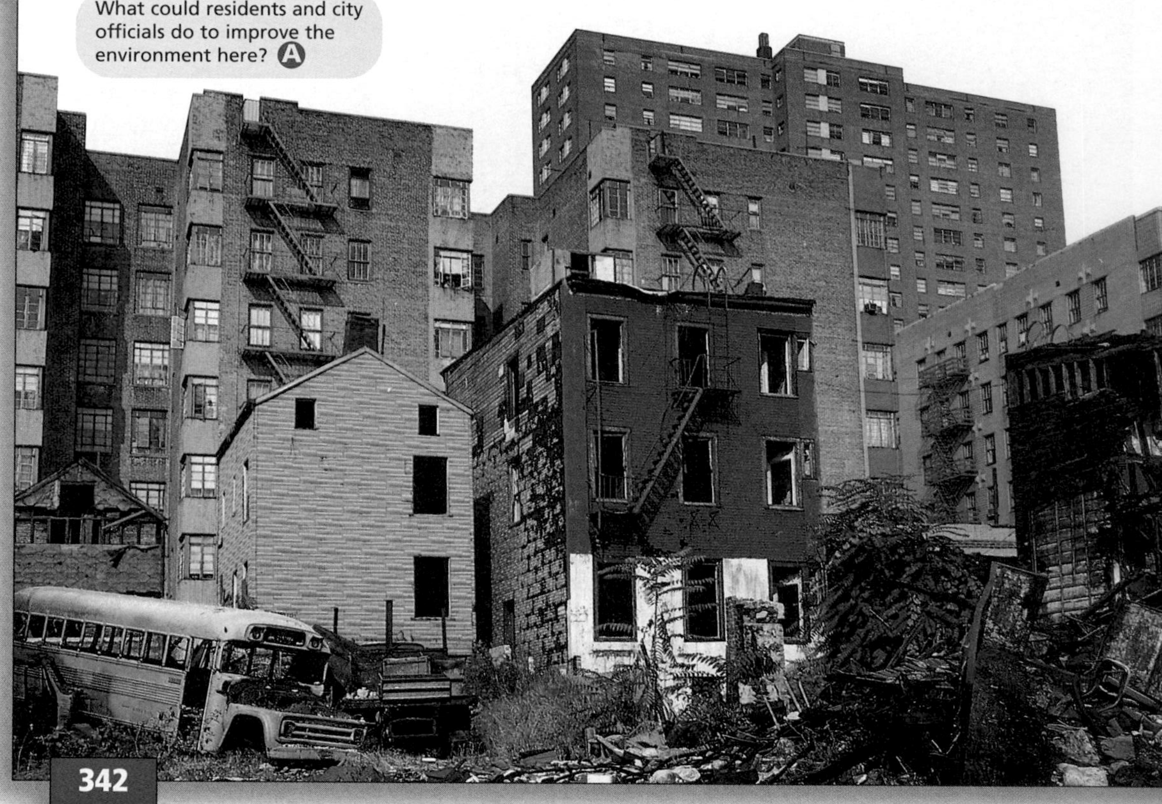

What could residents and city officials do to improve the environment here? (A)

Challenging Research Activities

Before students begin reading and working on the activities, direct their attention to the photograph. Call on a volunteer to read the caption aloud, and invite students to respond to the question it raises.

To be sure that each activity is completed, you may wish to make assignments. Allow students to work in pairs or small groups. When students have finished their assignments, provide time for them to share what they have learned with the class. Students may also enjoy using their information to make displays for the classroom. For example, old photographs showing what their community looked like in the past could be used to make an interesting bulletin-board display. Discuss with students whether or not they think the changes in their community have been improvements. Ask students to give reasons for their opinions.

To extend this material, you may wish to invite a city official or local historian to speak to the class about past, present, or future buildings in the community.

Answer to Caption

(A) Student responses may vary but should indicate an understanding of some of the problems of the inner city as reflected in the photograph—poverty, poor housing, poor sanitary conditions, and lack of recreational areas, for instance.

A Project for Everyone

As an alternative to having students work individually on the suggested projects, you may wish to have them form planning groups made up of three or four members to think of designs for specific areas of the school or community. Have a spokesperson from each group present a proposal to the class. Involve the class in a discussion of the proposal's merits.

To extend the project, suggest to students that they have a contest to see who can think of the best design for a school library, gymnasium, or cafeteria.

A Design for Riverwood

Have students read this section silently. Ask: What are some advantages and disadvantages to being a professional designer or architect? *(Sample answer: You can be very creative in your designs, but you have to meet the needs of many different individuals.)* As an extension, you may wish to have a professional designer from your community visit your classroom.

A Project for Everyone

In a sketch, redesign a familiar environment (your room, the school gym, the classroom, or the school). Decide how the changes you propose would affect other people who share that environment.

A Design for Riverwood

You have now considered many problems in designing and building structures. You know about some of the scientific and engineering ideas involved.

If you were making decisions about a home or other structure, you might be influenced by your knowledge of the history of architecture and design. In making your decisions, you would be more likely to consider the effects on other people. You would want any structure you built to be safe and to meet the standards of zoning laws. In these respects, you now think like a professional designer. More important, you can now play an active role in future decisions about your environment.

It is time to apply some of your new knowledge to the design of Riverwood. As soon as you have a plan for the community, proceed to Exploration 3.

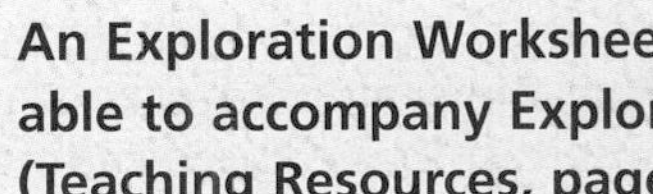

Constructing a Model of Riverwood

1. Submit your plan for Riverwood to your class. Explain how it meets the following criteria:
   a. usefulness
   b. beauty
   c. safety
   d. the interests of people who live in Riverwood
   e. the concerns of those who live in the surrounding neighborhoods
2. As a class, decide which plan best meets the needs of the Riverwood community.
3. Build models of the structures that will appear in Riverwood. As a class, decide
   a. what materials are to be used in building the models (spaghetti, string, balsa wood, index cards, cardboard, or other materials);
   b. what materials are to be used for structures that need enclosure, such as buildings and homes (paper, cloth, plastic wrap, or other materials);
   c. the size of the models;
   d. how the models will be tested for strength; and
   e. the architectural style or styles that will be used.
4. Do the following on your own or with a partner:
   a. Build a model framework for one of the new structures of Riverwood. Test it for strength. Your model must be safe under a load that is two times its own weight.
   b. If your model is not safe, repeat (a) using more material or a different design.
   c. Add walls, windows, and roofs as needed.
   d. Add decorations to your structures in harmony with the architectural style or styles agreed on by your class.

Cooperative Learning EXPLORATION 3

Group size: 2 to 3 students
Group goal: to design a portion of the model city of Riverwood
Positive interdependence: Assign each student a role such as materials coordinator, draftsperson, or team leader.
Individual accountability: Have each student submit a written proposal of the group's design. Students may also include an illustration of their design. Once a plan has been decided upon, the entire class should be involved in making the model of Riverwood.

Establish an area of the classroom where the model can be constructed. Then suggest that different students, or groups of students, take responsibility for modeling a particular part or aspect of the community. For example, a few students might work together on a particular street or area. Or one group of students could be responsible for constructing buildings, another for constructing streets and sidewalks, and another for constructing parks and recreational areas.

★ **An Exploration Worksheet is available to accompany Exploration 3 (Teaching Resources, page 44).**

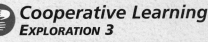

FOLLOW-UP

Reteaching

Have students use pictures from magazines to make posters showing urban buildings and settings designed to make the urban environment more pleasant. Each picture should be accompanied by a caption that describes it.

Assessment

Have students record their ideas about what features are important in the design of an urban environment. *(Answers could include safety, appearance, materials used, interaction with the environment, and practicality.)*

Extension

Students may enjoy designing their ideal home. Remind them to apply the ideas and concepts that they have learned from the unit. Invite volunteers to share their finished designs with the class.

Closure

Suggest that students take a survey to find out what different kinds of architecture are represented by the buildings in their neighborhood or community. They should evaluate the effectiveness of the various building designs in both meeting the needs of the community and appearing attractive. They should summarize their findings in a poster or a short, written report.

Answers to *Challenge Your Thinking*

1. The Seagram Building design, in keeping with the International Style, reflects modern technology. The building's simple angles and uniform glass windows reflect Gropius's interest in simplicity and mass production.
2. • The Parthenon (built in Greece between 447 and 432 B.C.): Students should recognize that the structure resembles those built in ancient Greece. They may even be able to identify it from the model of the Acropolis. The marble columns are typical elements of the period.
   • Renaissance cathedral: Students should recognize from the domes that this building was constructed during the Renaissance. Students may base their answer on the resemblance to the buildings shown on page 333.
   • Postmodern high-rises: Students should identify the high-rise as a modern structure that might be found in any industrialized country. Students may base their answer on modern components such as steel, concrete, and elevators. Students may note that the curves are a postmodern feature because they serve no practical purpose but make the building look more interesting.
3. Answers will vary depending on students' opinions. Sample answer: The setting is clean, new, and stylish, but the drabness and lack of natural elements make it cold and uninviting. Students' sketches may show better landscaping to add style and beauty to this area. They may choose to add a community area in the open space between buildings. This area might have features such as trees, sculptures, benches, or fountains.
4. Citizens can learn about the zoning laws in their city, attend public hearings on city development, write letters to their mayor, and try to establish a cooperative relationship with the city developers.

1. Designer Clues
When it was constructed, the Seagram Building in New York City helped to establish the International Style as a major trend in North American architecture. What clues in the photograph at right tell you that this building might have been designed by Ludwig Mies van der Rohe, who was a colleague of Walter Gropius?

2. Somewhere in Time
Where and when do you think the structures shown below were built? Explain your reasoning.

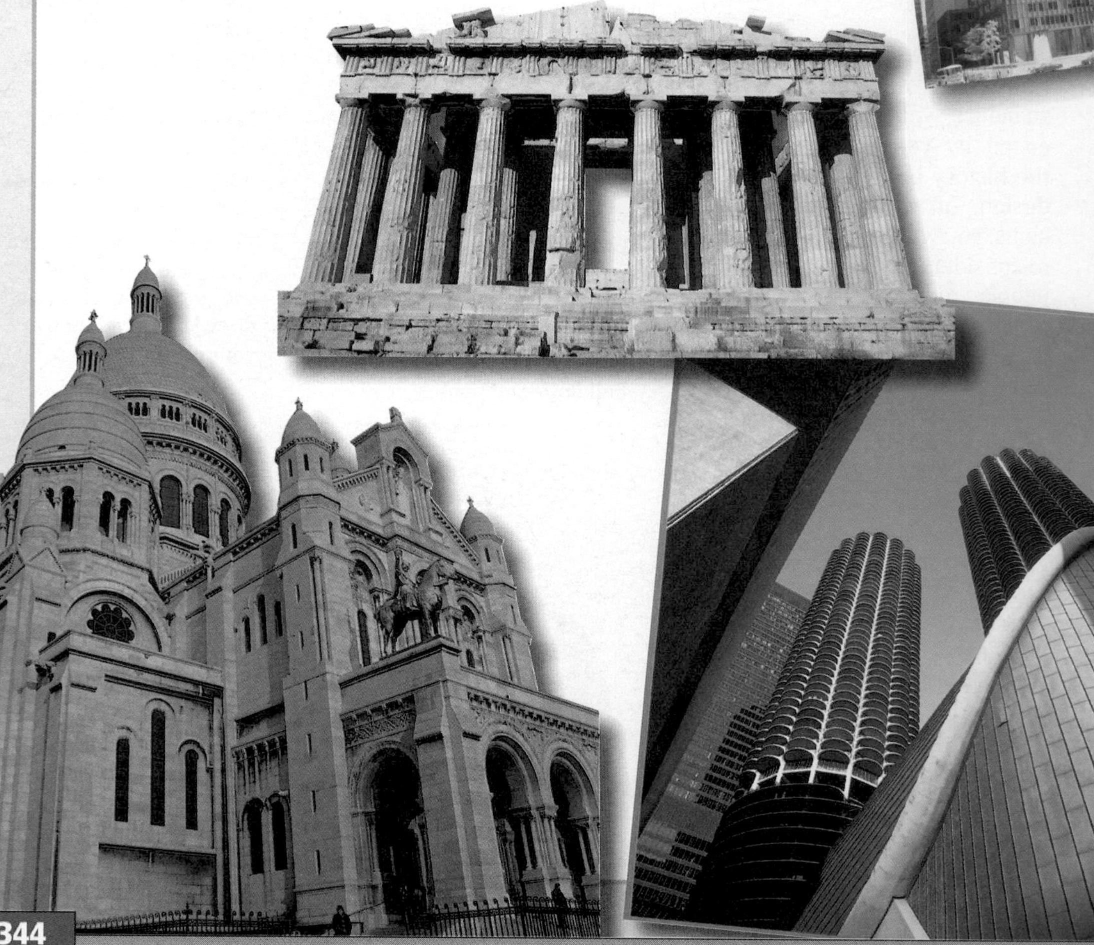

5. The design of the White House was influenced by both ancient Greece and Rome. Students will most likely notice the columns of the curved portico, which extends out into the yard. This design accentuates the landscaping and the fountain.

★ **You may wish to provide students with the Chapter 15 Review Worksheet that is available to accompany this Challenge Your Thinking (Teaching Resources, page 46).**

Integrating the Sciences

Earth, Life, and Physical Sciences
Remind students that the materials used in a structure are often determined by biological and geological factors of the structure's immediate environment. These factors include climate, indigenous animal and plant populations, and the topography of the area. The tepee, for instance, is made of animal hides and wood. The igloo is constructed with blocks of ice and snow. Other examples include log cabins, pueblos, wigwams, and thatched huts.

3. You Are the Urban Planner

Do you think the urban setting shown at right is a pleasant one? List some changes that you think would make this a more pleasant environment. Then make a sketch of how this setting would look if the changes were made.

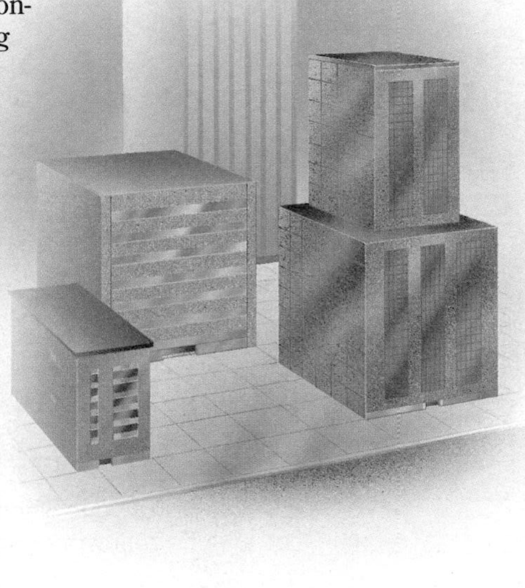

4. Taking Action!

What are some things that citizens' groups can do to help shape the future of their urban environments? Describe at least four actions.

5. What About the White House?

Look at the photo of the White House. Analyze the structure and determine which period of architecture it represents or imitates. Is it attractive to look at? Why or why not? How does it fit into its natural environment?

ScienceLog

Review your responses to the ScienceLog questions on page 326. Then revise your original ideas so that they reflect what you've learned.

Meeting Individual Needs

Second-Language Learners

Explain to the class that structures not only should be safe and functional but also should be aesthetically pleasing and should fit in with their surroundings. Have students use these criteria in designing a common household object, such as a mailbox, door knocker, lampshade, bird feeder, or place mat. Encourage them to be creative and to keep in mind the environment in which the object will be located.

ScienceLog

The following are sample revised answers:

1. Building designs often have specific features that can be identified in terms of historical trends. For example, the design of this cathedral typifies medieval architecture, placing the date of construction between the sixth and sixteenth centuries.
2. Architects can make urban environments more pleasant by using beautiful or unusual building designs and by incorporating natural elements such as plants and sunlight. Furthermore, they can consider how the building's design will blend with the surrounding environment.

Homework

Have students describe and explain the differences between housing built in urban areas and housing built in suburban or rural areas. *(Accept all reasonable responses. Many students will identify income level and the availability of space as factors determining the differences in architectural styles.)*

Making Connections

The Big Ideas

The following is a sample unit summary:

Factors that contribute to the success or failure of a structure include strength and thickness of materials chosen, arrangement and type of design features, and amount and type of environmental forces that interact with the structure. (1) Environmental forces that can contribute to a structure's failure include wind, floods, changes in temperature, earthquakes, and rusting. (2)

Forces affecting a structure can be described as tensile, compressive, or shear. Tensile force causes stretching, as in a cable supporting a bridge; compressive force causes compaction, as in a column supporting a roof; and shear force causes bending or twisting, as in scissors cutting paper. (3, 4) Materials can be strengthened to resist forces by altering their shape (materials formed into tubes or rectangles are stronger), laminating (layered materials are stronger), and adding corrugation or tubular girders. (5)

Some simple elements of design include beams and cantilevers. A beam is a simple structure used to span the distance between two points. A cantilever is a beam that is fixed at one end. Cantilevers can be used in decks, signs, and bridges. (6)

Bridges can be given strength and support through the use of beams, trusses, cables, girders, and columns. (7) Arches and domes are stronger than simple beams because the force of gravity is displaced to the side supports, which are then under compression. The arch or dome is better able to resist the force of gravity. (8)

Architecture of different eras reflects the technology, geography, climate, and cultural traditions of society through the ages. For example, tepees developed by nomadic cultures represent the need for mobility and indicate the types of natural materials that were available. The pyramids of Egypt reflect the importance of the afterlife to the Egyptian society. The carefully planned cities of the ancient Romans show the importance of society, as well as the influence of earlier Greek architecture. Architecture has always been an extension of design. Today, new materials and technology have provided endless possibilities. (9)

Viewing architecture through history shows that a wide range of structures can be both beautiful and enduring. In creating a structure, designers must consider not only the function of the building itself, but also the needs of the community and environment. (10) Through urban planning and group cooperation, citizens can ensure that their urban environments are clean, attractive, and safe. (11)

Making Connections

Unit 5

SOURCEBOOK

Turn to the SourceBook to learn more about materials used in building, what properties make them strong, and what changes can cause them to fail. You'll also explore the variety of architecture found in other cultures.

Here's what you'll find in the SourceBook:

The Big Ideas

In your ScienceLog, write a summary of this unit, using the following questions as a guide:

1. What makes a structure strong?
2. What causes a structure to collapse?
3. How do you distinguish between tensile, compressive, and shear forces?
4. How do the three different forces affect a structure?
5. How can shape, lamination, corrugation, tubular girders, and tubular columns increase the strength of a structure?
6. What is the difference between a beam and a cantilever? Where would you find each?
7. How are bridges given strength and support?
8. Why are arches and domes stronger than simple beams?
9. What are some characteristics of the different architectural periods?
10. What must engineers and builders consider before constructing a building?
11. How can citizens control their urban environments?

346

Checking Your Understanding

1. Study the horse skeleton shown at right. Then answer the questions.
   a. Which parts of this structure function like columns?
   b. Which parts function like beams?
   c. Which parts function like a cantilever?
   d. If the horse had a rider, which parts of the horse would be under compression?
   e. Which parts would be under tension?

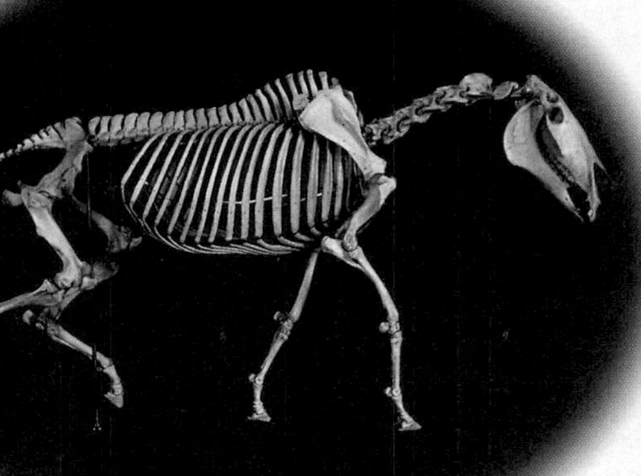

2. Design a solution for this structural problem: Support a glass of water by using three stainless steel knives and three soft-drink bottles. Each knife can touch only one bottle. Using a paper cup, try out your design. Draw the support arrangement you created.

3. concept map Draw a concept map using the following words: nomadic, stone, skyscraper, tepees, modern, architecture, Acropolis, animal skin, ancient Greek, and reinforced concrete.

Answers to *Checking Your Understanding*

1. **a.** The parts of the horse that function as columns are the bones associated with the legs and the cartilage between these bones.
   b. The backbone functions as a beam.
   c. The neck and tail function as cantilevers. Some students might also point out that the ribs function as cantilevers.
   d. The legs and cartilage between the leg bones would be compressed.
   e. The back would be under tension.
2. Solutions to this problem may vary. One solution would be to have the knives act as cantilevers, forming a triangle in which each knife overlaps and supports the knife next to it.
3.

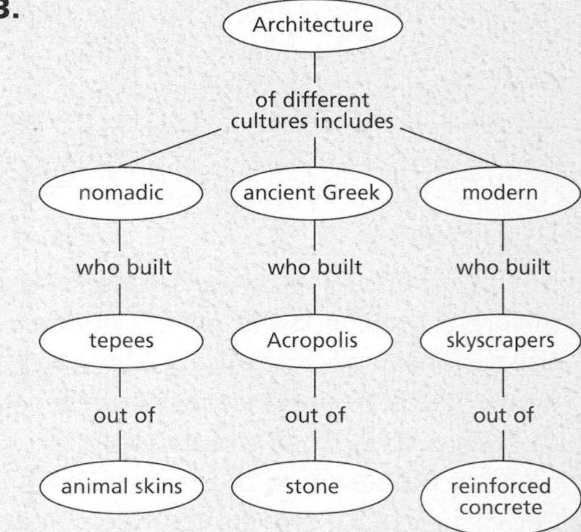

You may wish to provide students with the Unit 5 Review Worksheet that is available to accompany this Making Connections (Teaching Resources, page 52).

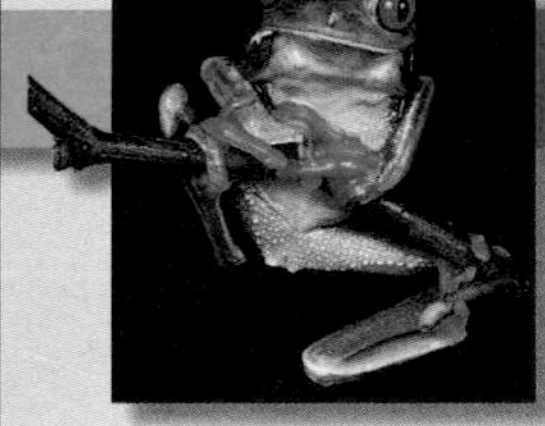

WEIRD Science

Background

Both the materials used to build bridges and the design of bridges have changed over time. The earliest bridges were probably built to cross rivers or similar obstacles. The bridges might have been made of logs or vines. Later, builders began using stone in their bridges. Iron became the preferred material for bridges in the 1700s, and then steel and concrete were introduced in the 1800s.

Okanagan University College has held their competition for the last 12 years. In 1995, they introduced a new category: Team Bridge Building. These entrants have to build a bridge as a team, and they have to build it fast. They are supplied materials at the contest and must plan and build their bridge in $2\frac{1}{2}$ hours. The winner is the team whose bridge has the highest load-to-weight ratio. In 1995, the winning bridge carried 30 times its own weight.

Extension

Have students explore the strengths of different thicknesses of pasta strands. Use spaghetti and vermicelli. Have students bunch 10 pasta strands together to form a tube. They should wrap a piece of tape around the tube at both ends. Then place two tables 10 cm apart, and lay one pasta tube across the gap between the tables. Hang a spring scale from the pasta bridge, and add masses until the bridge breaks. Repeat the process for the remaining pasta tubes. Which kind of pasta works best? You might want to point out to students that real bridge cables are made of smaller cables wrapped together, just like the pasta tubes.

Your Own Secret Recipe

Make sure students are aware that the best bridge will be the one that distributes the weight of its load as evenly as possible. You might want to limit the weight of the bridge and then hold a class contest to see which design holds the most weight.

WEIRD Science

Noodling Around With Bridge Design

A hush falls over the crowd. As the judges look on, the contestant carefully hangs another weight from her model bridge. A faint creaking sound is heard. Suddenly the bridge shatters with a loud snap, sending shards of broken pasta flying in all directions.

▲ Spaghetti bridges on display before being tested

▲ Spaghetti bridge being inspected by a judge

An Unusual Competition

Shards of broken *pasta*? Yes, people have gathered here from all over Canada for a very remarkable event. It is Okanagan University College's annual Spaghetti Bridge Contest.

The rules are simple: using only pasta and glue, build a bridge that spans a distance of 1 m. The bridges are judged on both strength and design, and may compete in one of two areas. In the student competition for Most Functional Bridge, spaghetti bridges must support a weight of 1 kg for 10 minutes. Of those that pass this test, the bridge that weighs the least is the winner.

In the World Open Heavyweight Competition, bridges may be built using any type of pasta, but each bridge's total weight may not exceed 750 g. The competing bridges are loaded with heavier and heavier weights until all but one collapse.

Cannelloni Cantilevers and Tortellini Trusses

The key to success in both competitions is achieving maximum strength with a minimum amount of pasta. Design is crucial. Some people construct elegant suspension bridges, while others develop intricate trusses, relying on networks of pasta strands to give their bridges strength.

In the heavyweight competition, where different types of pasta are allowed, pasta shape becomes important too. Some use flat or tube-shaped pasta (such as fettuccine or cannelloni), while others combine different shapes to invent new, stronger building materials.

Prize-Winning Pasta

One winning bridge in the heavyweight competition was able to support an amazing 131.7 kg. The designer attached pasta cables from the center of the bridge to a high pasta arch above. Thus, the weight was divided among many pasta strands and transferred to the strong, stable arch.

Your Own Secret Recipe

Construct a pasta bridge that spans a distance of 20 cm. Consider different bridge designs, and use one you think will work best. Build your bridge with a flat top so that you can test its strength by putting heavy objects on top of it.

CROSS-DISCIPLINARY FOCUS

Geography

Encourage students to enhance their map-reading skills. Have students gather around a map or road atlas of your region. Then have them identify the location and type of each bridge. *(Point out highway bridges, drawbridges, train bridges, and pedestrian bridges.)* You might point to a bridge on the map and have students discuss what would happen if the bridge were removed. *(Usually, if a bridge were removed, the detour would create a much longer trip.)*

SCIENCE & TECHNOLOGY

The Wonder Web

Would you try to stop a speeding bullet with a spider web? It sounds crazy, but that's exactly what some scientists are trying to do.

Stronger Than a Speeding Bullet

Consider the spider web. How does such a simple, delicate structure made of silk survive the impact of large insects and harsh winds? Some scientists have actually fired tiny plastic-foam bullets into spider webs in search of an answer. Some have used computers to simulate crash landings by insects. Others have studied the silk itself. These scientists have discovered that it's not just a strong fiber that makes a strong web.

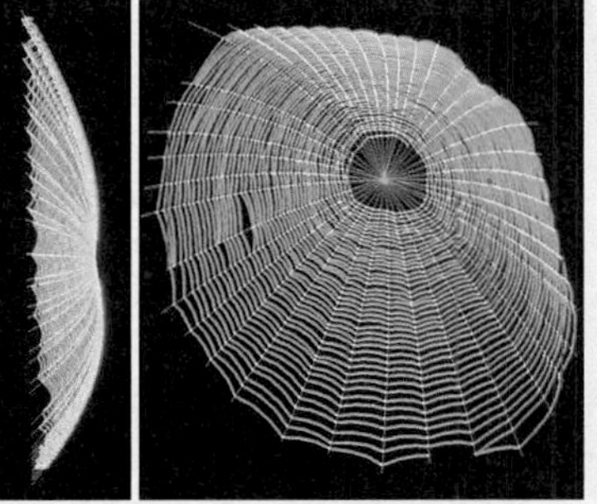

▲ Notice how the web responds to the impact of the insect.

Superthreads

The web's design is the key to its resistance. Webs like the one shown here have two types of silk fibers. Dry, stiff *radial* threads branch out from the web's center like the spokes of a wheel. The radial threads are connected by a spiral of sticky, elastic silk.

When an insect hits a web, the radial fibers stop the insect and keep the web intact. At the same time, the spiral of elastic fiber stretches. As it stretches, it billows like a little parachute against the surrounding air. This acts like a cushion to soften the blow. Together the radial and spiral fibers allow the web to catch heavy insects moving at high speeds.

As with any structure, the success of the web's design depends on its materials. Spider silk is made of both stiff and elastic elements, so it can be both flexible and strong. A spider can make a strand more elastic or stiff depending on its needs.

Learning From Nature

Tennis rackets, parachutes, and bulletproof vests must be made with materials that are both lightweight and impact resistant. The production of these materials often requires extreme heat and dangerous chemicals. In contrast, spider silk is produced naturally at moderate temperatures. The silk is also ideal for medical products like artificial tendons and skin wraps for burn victims.

Unfortunately, spider silk is too fine to be mass-produced. But scientists are not giving up. Through genetic engineering, they hope to develop a silk-producing bacteria or yeast that would produce large populations and that would require little care. Scientists are hoping that further study will soon bring us amazing innovations from the spider's web.

▲ Spinnerets are the silk-producing organs of the spider. Here the silk is being harvested.

Going Further

Many architectural structures reflect natural structures. Look at various types of bridges shown on pages 320–321. Which do you think most resembles a spider's web and why? How do the parts of a bridge work together like threads in a spider's web?

Integrating the Sciences

Life and Physical Sciences

Have students observe some spider webs in your area. They should take photographs or draw sketches of each web. Then have them identify the strongest part of the web and the different types of silk within the web. Remind students to be very careful not to destroy the webs!

Students might find a number of different web designs. The web described in the feature is an example of an *orb web*. Cobwebs commonly found in the corners of rooms are *tangled webs* because their design shows no regular pattern. Other webs that students might find include funnel-shaped webs or flat sheets spun between blades of grass.

SCIENCE & TECHNOLOGY

Background

Spider silk can stretch twice as far as nylon. It is stronger than steel, chemically inert, waterproof, and largely nonallergenic. Spider silk is made of proteins that form into liquid crystals inside a spider's body. The spider excretes the silk through tiny organs called *spinnerets,* which it uses to form the silk into the thin, strong threads of a spider's web.

Spider silk is used by spiders for more than just building a web. Dragline silk is used to drop slowly down through the air. Capture silk is used to snare prey. Spiders even use a special insulating silk for cocoons. Many researchers speculate that spider silk first evolved 400 million years ago when the spider's ancestors began adapting to life in trees and bushes.

Answers to *Going Further*

Students will probably say that a bridge most resembles a spider's web when the bridge's supports—either cables or girders—form a crisscross pattern, such as in a cantilever or suspension bridge. Examples in the text include the Hooghly River Bridge in India (a stayed girder bridge) and the Forth Rail Bridge in Scotland (a cantilever bridge). Photographs of both bridges are on page 322. In each case, the design principle is the same: the structure distributes the force from vehicles to different parts of the structure. In this way, no single cable bears all of the stress by itself.

Extension

The egg is another natural structure whose shape makes it incredibly strong. If it is cushioned, an egg can support many times its own weight. Have students put soft modeling clay on a table and mold it to fit the shape of an egg. Place an egg upright on the mold, and cover the top of the egg with modeling clay as well. Then carefully place a book wrapped in plastic on the egg, holding the book to keep it balanced. If the egg does not crack, add another book. Continue stacking books on the egg until it cracks. Ask: Why do you think the egg is so strong? How might this shape be used in the design of buildings?

Science Snapshot

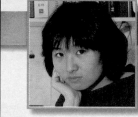

Background

Maya Lin received a bachelor's degree from Yale in 1981 and a master's degree in architecture from Harvard in 1986. In addition to designing the memorial, she codesigned the Museum for African Art in New York City, has exhibited in numerous galleries, and in 1994 was one of 26 people named to serve on the Commission on the Future of the Smithsonian Institution.

The Vietnam War is the longest military operation the United States has ever been involved in. It began in 1957 when a South Vietnamese group known as the Viet Cong first rebelled and ended in 1975 when South Vietnam surrendered. The United States had been sending aid to South Vietnam before 1957. At the time, much of United States foreign policy was based on the "domino theory." It was believed that Communist expansion should be quickly and stubbornly resisted anywhere it appeared, or, like falling dominoes, revolutions would quickly overthrow the world's democracies one by one. While the United States sought to stem Communist expansion, North Vietnam and the Viet Cong sought to reunite North Vietnam with South Vietnam and to remove Western colonial influence.

Multicultural Extension

Identifying Role Models

Chinese Americans like Maya Lin have contributed a great deal to world culture. C. N. Yang and T. D. Lee shared the Nobel Prize for physics in 1957, and I. M. Pei is a well-known architect. Have students think about what culture they would like to know more about, such as German American or African American. Either as a class discussion or as a research project, have students explore the contributions of famous members of their chosen culture.

Discussion Questions

1. What specific features should a memorial possess? *(Accept any reasonable answer. Some responses might include peacefulness, grandeur, or beauty. Students might even describe its construction, such as columns and a dome made of white marble.)*
2. Can you think of any other public sculptures that seem to incorporate unusual or innovative design features? Do you think they are effective monuments? Why or why not? *(Examples of unusual or innovative works include the Gateway Arch in St. Louis, Mount Rushmore in South Dakota, and the Washington Monument in Washington, D.C. Accept all reasonable responses.)*

Now You Try It

You may wish to have students present their designs to the class. Presentations could include written descriptions and sketches of the memorial, as well as background material on their national hero.

Science Snapshot

Maya Lin (1959–)

Sometimes a work of art is so monumental that the artist earns a place in history. Such is the case with Maya Lin, who will always be remembered for designing the Vietnam Veterans Memorial.

Classroom Competition

Maya Lin entered the Vietnam Veterans Memorial design competition as a class project for an architecture course. On the advice of her professor, she visited the future site of the memorial in Washington, D.C. On viewing the site for the first time, she was struck by inspiration: "You couldn't desecrate that land. You had to . . . open the Earth, let it embrace you." She envisioned a monument that would draw in visitors, "make them feel safe" and "protect them from the sounds of the city."

Lin's simple yet elegant design, chosen from 1420 entries, consists of two long, black granite walls built into the ground. The walls come together to form a V, which is about 3 m high at its point. Engraved on the polished walls are the names of the nation's nearly 58,000 dead and missing veterans of the Vietnam War.

Moments of Controversy

Lin's bold and unconventional design created a controversy. Outraged veterans groups called it the "wall of shame" and "an ugly black gash" and fought to replace Lin's design with a more conventional one. Pointing out Lin's Chinese ancestry, some people even suggested that a person of Asian descent shouldn't be allowed to design a memorial for the United States.

Lin responded that critics did not understand the purpose of the design. The memorial was supposed to be a peaceful retreat. The black walls symbolized the boundary between our sunlit world and the quiet, dark world beyond.

▲ The Vietnam Veterans Memorial

A Monument to Remember

The power of Maya Lin's design is evident in the reactions of those who view it. Even casual visitors to the monument describe the experience as intensely emotional. People frequently say that they are overwhelmed at seeing their reflection in the black walls, mingling with the names of the dead.

Lin continues to aim for a sense of "tactility" in her works, meaning that the works should encourage people to touch them. She believes that a work can be more fully appreciated when it is touched.

Now You Try It

Design a memorial to commemorate your current national hero. Think carefully about how the memorial should look and what materials should be used to build it. What feelings do you want the people who visit your memorial to have?

SCIENCE and the Arts

An Architectural Marvel

In the early 1400s the people of Florence, Italy, had an enormous problem on their hands. Their plan for enlarging and rebuilding their cathedral called for a magnificent dome 42 m in diameter. The project had begun more than a hundred years earlier, and now they didn't know how to finish it.

The Enormous Dilemma

The new cathedral was to be the largest in the world. It would establish the power and prestige of Florence over all the other cities in central Italy. It was a matter of great civic pride. But if they couldn't figure out how to build the dome, they would be the laughingstock of the neighboring cities.

The usual method of building a dome, according to techniques known at the time, was to build a supporting structure of wood, lay brick over it, and then remove the wooden structure. But for such a huge dome, there was no way to get timber long enough or strong enough to build the supporting structure.

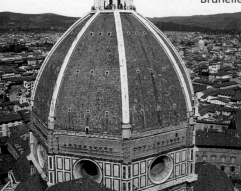

A Creative Solution

In 1418 the cathedral officials offered a cash prize of 200 florins to anyone who could come up with a way to erect the dome. The prize went to Filippo Brunelleschi (broon uh LESS kee), a famous artist and goldsmith. Using models, Brunelleschi demonstrated how the dome could be built. He even included inventions for lifting construction materials into place.

Brunelleschi's design for the dome was a double shell, with ribs formed by Gothic arches. The double-shell construction made the dome relatively light and eliminated the need for a supporting structure.

In 1420 Brunelleschi's proposal was accepted and Brunelleschi was made chief architect for the project. The dome was completed in 1436, ensuring Florence its prestige and Brunelleschi his fame.

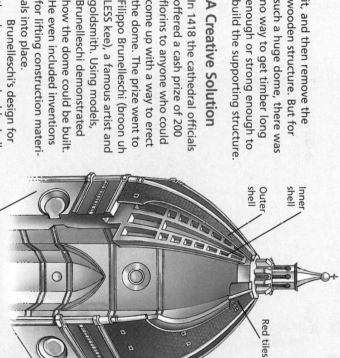

◀ **The dome of the Florence cathedral presented a major engineering problem.**

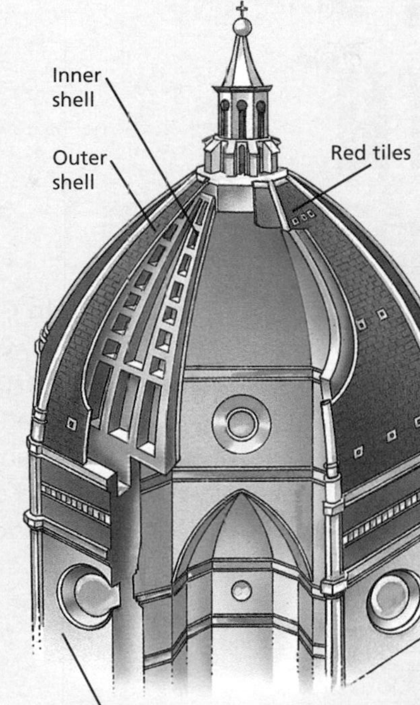

▲ **The dome was built using a double-shell design.**

Share Your Perspective

One of Brunelleschi's contributions was the rediscovery of linear perspective, a method used by Greeks and Romans to show depth on a flat surface. In drawings that use linear perspective, lines are drawn closer together for objects that are farther away. Find examples of linear perspective in art. How could linear perspective help an architect design a building?

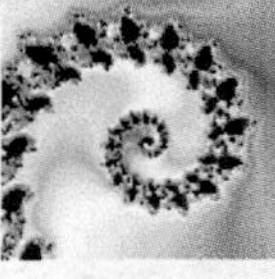

SCIENCE and the Arts

Background

In 59 B.C. Julius Caesar founded a colony on the Arno River called *Florentia*, a Latin word meaning "blossoming." Florence has lived up to its name—it is considered by historians to be the birthplace of the Renaissance. During the period from about 1300 to 1600, it was home to some of the greatest painters, sculptors, architects, and writers in the world. Among them were Leonardo da Vinci, Fra Angelico, Michelangelo, Giovanni Boccaccio, Dante, and Filippo Brunelleschi.

In the 1420s, Brunelleschi and his circle of friends began to articulate a set of ideas that would define the character of Renaissance art and architecture. For example, they believed that the classical models were ideal, especially in sculpture and architecture. Brunelleschi himself is considered by many to have rediscovered the secrets of Roman architecture.

Discussion Questions

Ask students to study the photographs in order to answer the following questions:

1. What does the size of the people on the lantern suggest to you about the size of the structure? *(The dome is very large, about 42 m in diameter.)*
2. What features are added for ornamental effect? Do these structures have any other function? *(The buttresses and arches are attractive, and they also help to support the structure.)*
3. What effect does the use of domes have on the appearance of an interior space? *(Domes allow for a large interior space to exist without supports that would obstruct the view. The large space gives the ceiling an appearance of weightlessness.)*

CROSS-DISCIPLINARY FOCUS

Social Studies

The dome on the Cathedral of Florence was both a religious symbol for the city and a symbol of the city's economic strength. What are some of the uses of domes in other cultures? Can you think of any examples? *(One use is as a place to worship and honor a religion's deity or deities. Examples include the domes of the Pantheon and St. Peter's Basilica in Rome and that of the Hagia Sophia in Istanbul. Domed structures also serve political uses. The Capitol in Washington, D.C., houses the legislative branch of the United States government. In some cases domes are used as tombs, such as the famous Taj Mahal in Agra, India. Domed structures are often used today for sports or recreational structures, such as the Astrodome in Houston, Texas. Finally, domes can be used to make artistic statements. One example is the geodesic domes of Buckminster Fuller. These domes were designed to express the dominance of technology in modern life.)*

Answers to *Share Your Perspective*

Common examples of linear perspective in art include buildings, railroad tracks, or roads that recede toward the horizon. The farther away they are, the smaller they appear to be. Linear perspective is useful because it helps people visualize a three-dimensional scene or object even though it is drawn on a flat surface.

THE RESTLESS EARTH

Unit Overview

In this unit, students are introduced to processes of geological change. In Chapter 16, students are encouraged to explore the mechanisms of plate tectonics as well as mountain-building, earthquakes, and volcanoes. In Chapter 17, students study rock formations and characteristics as well as the processes of forming igneous, sedimentary, and metamorphic rocks. The nature of fossils, hypotheses about mass extinction, and the interpretation of fossil evidence are discussed in Chapter 18. Students also develop their own relative and absolute time scales using familiar events in order to understand the development of the geologic time scale.

Using the Themes

The unifying themes that appear in this unit are **Energy, Systems, Structures, Changes Over Time,** and **Cycles.** The following information will help you incorporate these themes into your teaching plan. Focus questions that correspond to these themes appear in the margins of this Annotated Teacher's Edition on pages 358, 367, 373, 384, 399, 401, and 404.

Energy is important to the understanding of plate tectonics (Chapter 16) because it is the release of energy that causes earthquakes. The exposure of rocks to heat energy causes the formation of metamorphic rocks (Chapter 17), and geologic processes allow us to recapture and use energy from the sun that has been stored in fossil fuels (Chapter 18).

Systems can be discussed in terms of the effects that volcanic activity can have on both atmospheric and biological systems.

Understanding the **Structures** of the Earth's crust is critical to the study of plate tectonics (Chapter 16). The theme of structures appears again in Chapter 17, in which observation of the structure of rocks allows students to classify them and make inferences about their origins.

Changes Over Time can be discussed in conjunction with Chapters 16, 17, and 18. In Chapter 16, students are introduced to the many changes in the Earth's crust, such as mountain-building, movement along faults (earthquakes), and volcanism. Chapter 17 describes patterns of change leading to lithification, intrusions, and the rock cycle. Changes throughout the Earth's history can be discussed in Chapters 16 and 18, which explore how the continents have evolved and continue to evolve through the process of plate tectonics. A study of the fossil record also illustrates how conditions on Earth have changed over time.

Finally, **Cycles** appear in Chapter 17 in the context of the rock cycle. There, students are asked to diagram a rock cycle. Cycles are also important in studying the life cycles of mountains in Chapter 16.

Using the SourceBook

Unit 6 in the SourceBook focuses on rocks and the minerals that compose them. The three types of rocks—igneous, sedimentary, and metamorphic—and their place in the rock cycle are examined. Students learn about the composition and properties of rock-forming minerals. Finally, the unit discusses the Earth's history as recorded in rocks.

Bibliography for Teachers

Barrow, Lloyd H. *Adventures With Rocks and Minerals: Geology Experiments for Young People.* Hillsdale, NJ: Enslow Publications, 1991.

Fortey, Richard. *Fossils: The Key to the Past.* Cambridge, MA: Harvard University Press, 1991.

Roadside Geology Series. Missoula, MT: Mountain Press Publishing Company. (This series includes a number of books, each of which is specific to a state.)

Thro, Ellen. *Volcanoes of the United States.* New York City, NY: Franklin Watts, 1992.

Bibliography for Students

Curtis, Neil, and Michael Allaby. *Planet Earth.* Visual Factfinders Series. New York City, NY: Kingfisher Books, 1993.

Pope, Joyce. *Fossil Detective.* Mahwah, NJ: Troll Associates, 1993.

VanCleave, Janice. *Janice VanCleave's Earth Science for Every Kid: 101 Experiments That Really Work.* New York City, NY: John Wiley & Sons, Inc., 1991.

Films, Videotapes, Software, and Other Media

Continental Drift: The Theory of Plate Tectonics
Videotape
Britannica
310 S. Michigan Ave.
Chicago, IL 60604-9839

The Earthquake
Simulation software (Apple II family)
Focus Media, Inc.
839 Stewart Ave.
Garden City, NY 11530

Fossils: Exploring the Past
Videotape
Britannica
310 S. Michigan Ave.
Chicago, IL 60604-9839

The Rock Cycle
Videotape
Britannica
310 S. Michigan Ave.
Chicago, IL 60604-9839

Unit Organizer

Unit/Chapter	Lesson	Time*	Objectives	Teaching Resources
Unit 6 Opener, p. 352				Science Sleuths: The Misplaced Fossil English/Spanish Audiocassettes Home Connection, p. 1
Chapter 16, p. 354	**Lesson 1, Making Mountains, p. 355**	2 to 3	1. Identify the forces that counteract weathering and erosion. 2. Describe the most recent theory of mountain-building. 3. Construct models to illustrate how folds and faults are formed.	Image and Activity Bank 16-1 Transparency 58 Resource Worksheet, p. 3 Exploration Worksheet, p. 5 Transparency Worksheet, p. 7 ▼
	Lesson 2, The Earth Breaks Apart, p. 362	3	1. Explain current scientific ideas about why earthquakes occur. 2. Describe how seismographs measure earthquake strength and location.	Image and Activity Bank 16-2 Exploration Worksheet, p. 9 Transparency Worksheet, p. 10 ▼ Exploration Worksheet, p. 12
	Lesson 3, Volcanoes—Holes in the Earth, p. 370	2	1. Describe different types of volcanic eruptions. 2. Identify the locations of past and present volcanoes. 3. Infer that volcanoes occur at the edges of plates. 4. Describe the origin of volcanoes using the plate tectonics theory.	Image and Activity Bank 16-3 Resource Worksheet, p. 14 Activity Worksheet, p. 15 Exploration Worksheet, p. 16
End of Chapter, p. 379				Math Practice Worksheet, p. 18 Chapter 16 Review Worksheet, p. 19 Chapter 16 Assessment Worksheet, p. 22
Chapter 17, p. 381	**Lesson 1, Features and Formations, p. 382**	2 to 3	1. Name characteristics that can be used to distinguish one type of rock from another. 2. Devise a classification scheme for rocks based on physical properties. 3. Distinguish between igneous, sedimentary, and metamorphic rocks by using a key.	Image and Activity Bank 17-1 Exploration Worksheet, p. 24 Theme Worksheet, p. 27 Exploration Worksheet, p. 29 ▼
	Lesson 2, Hot Rocks! p. 387	3	1. Explain how differences in the cooling of igneous rocks affect their structure. 2. Recognize the differences between extrusive and intrusive rocks by examining the size of the grains or crystals. 3. Infer how and where igneous rocks were formed by examining the size of the grains or crystals.	Image and Activity Bank 17-2 Exploration Worksheet, p. 31 Resource Worksheet, p. 34 ▼
	Lesson 3, Rocks From Sediments, p. 392	2 to 3	1. Identify possible locations of the deposition of sediments. 2. Explain the role of evaporation in the formation of sedimentary rock. 3. Explain how the number, size, and shape of sediments can be used as clues to the origins of sedimentary rocks.	Image and Activity Bank 17-3 Transparency 63 Exploration Worksheet, p. 35 Activity Worksheet, p. 37 ▼
	Lesson 4, Changed Rocks, p. 396	2	1. Identify heat and pressure as the agents of metamorphism. 2. Explain that metamorphic rocks are usually different in structure and texture from their original form. 3. Infer that any one rock type can be changed into another.	Image and Activity Bank 17-4 Exploration Worksheet, p. 38 ▼ Transparency Worksheet, p. 39 ▼ Activity Worksheet, p. 41
End of Chapter, p. 400				Chapter 17 Review Worksheet, p. 42 ▼ Chapter 17 Assessment Worksheet, p. 45
Chapter 18, p. 402	**Lesson 1, Rocks Reveal a Story, p. 403**	2	1. Describe how fossils are formed. 2. Identify different fossil types. 3. Hypothesize why dinosaurs became extinct. 4. Observe and interpret fossilized tracks.	Image and Activity Bank 18-1 Transparency Worksheet, p. 47 ▼ Resource Worksheet, p. 49 Activity Worksheet, p. 50 ▼
	Lesson 2, Telling Time With Rocks, p. 408	2	1. Explain the concept of relative time. 2. Name the specific divisions of the geologic time scale. 3. Identify when specific events occurred, based on the geologic time scale. 4. Describe how dates in the geologic time scale have been established.	Image and Activity Bank 18-2 Transparency 69 Activity Worksheet, p. 51 ▼
End of Chapter, p. 412				Chapter 18 Review Worksheet, p. 52 Chapter 18 Assessment Worksheet, p. 55
End of Unit, p. 414				Unit 6 Activity Worksheet, p. 58 Unit 6 Review Worksheet, p. 60 Unit 6 End-of-Unit Assessment, p. 63 Unit 6 Activity Assessment, p. 69 Unit 6 Self-Evaluation of Achievement, p. 72

* Estimated time is given in number of 50-minute class periods. Actual time may vary depending on period length and individual class characteristics.

▼ Transparencies are available to accompany these worksheets. Please refer to the Teaching Transparencies Cross-Reference chart in the Unit 6 Teaching Resources booklet.

Materials Organizer

Chapter	Page	Activity and Materials per Student Group
16	361	**Exploration 1:** 2 colors of modeling clay; knife; metric ruler; optional materials: small piece of wax paper
	363	**Exploration 2:** clay block from Exploration 1; knife; metric ruler; optional materials: small piece of wax paper
	377	**Exploration 4:** atlas; world map
17	384	**Exploration 1, Part 1:** magnifying glass; 1 pair of dissimilar rocks per student; **Part 2:** rocks from Part 1; about 30 index cards
	386	**Exploration 2:** magnifying glass; rock samples
	387	**Exploration 3:** metal ring; ring stand; wire gauze; 100 cm^3 of sand; three 250 mL beakers; test-tube tongs; small paper cup; paper towel; flat piece of aluminum foil; 10 g of stearic acid; portable burner; a few ice cubes; 500 mL of water; knife; scoop; two 25 mL test tubes; safety goggles; a few matches; aluminum pie pan; lab aprons; latex gloves
	393	**A Relationship to Water:** optional items: geology field guide or textbook, a variety of sedimentary rocks
	394	**Exploration 4:** 6 mL of soil (with a variety of particle sizes); 10 mL of water; test tube with stopper or small glass jar with tight lid; metric measuring spoon (additional teacher materials: clock or watch, 100 mL graduated cylinder)
	398	**Exploration 5:** 4 pairs of rocks (each pair consisting of an igneous or sedimentary rock and its metamorphosed form); magnifying glass

Advance Preparation

Gather the above materials for the Explorations and other activities.
Exploration 2, page 386: You may wish to screen the rocks ahead of time to be sure that the described traits are visible and that each group has at least one igneous rock, one sedimentary rock, and one metamorphic rock.

Unit Compression

This unit is composed of three distinctive chapters that are strongly interrelated, yet could be taught independently. The relevance of their content to the geology of your region can help guide your selection of material from each chapter and lesson.

The first chapter, Shake, Rattle, and Flow, should be considered essential because it develops an understanding of the theory of plate tectonics and its application to mountain-building, earthquakes, and volcanoes.

If time constraints make it necessary to shorten Chapter 17, Exploration 2 can be omitted, and Exploration 4 can be assigned for homework. A less-preferred option would be to omit Lessons 2, 3, and 4, with the exception of the sections Can Rocks Be Recycled? on page 398 and Those Plates Again!! on page 399.

Undoubtedly, the final chapter on fossils will be of great interest to students. Students might be given an opportunity to choose among the various activities in Lesson 1. Each activity contributes to a deeper understanding of fossils and of the scientists who study them, so selections should be made carefully. Because Lesson 2, Telling Time With Rocks, presents some difficult concepts, you may choose to omit this lesson. If this is necessary, you should at least briefly examine the geologic time scale on page 411.

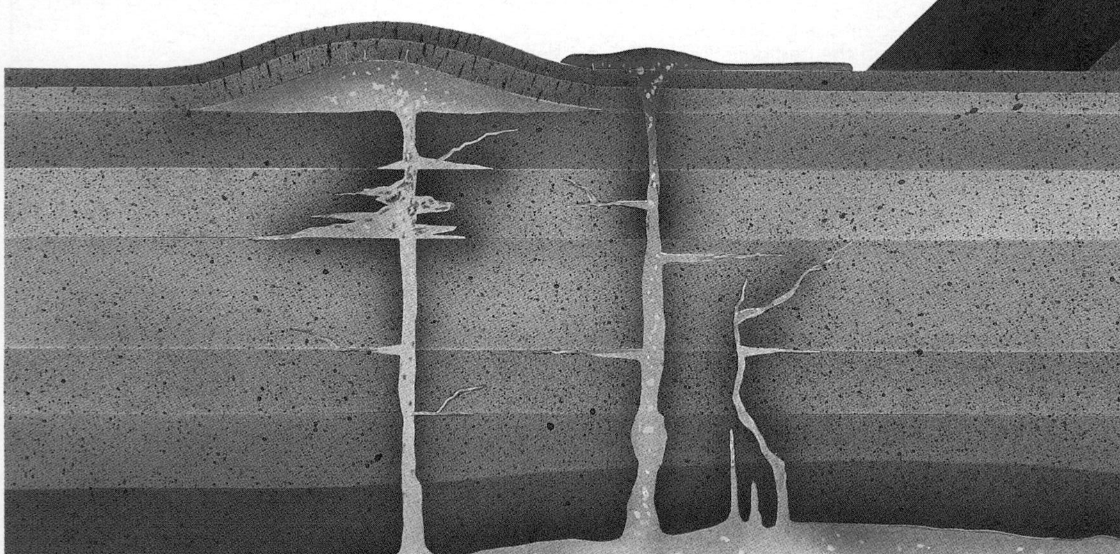

Homework Options

Chapter 16	See Teacher's Edition margin, pp. 358, 361, 364, 366, 371, 373, 375, and 379 Resource Worksheet, p. 14 Activity Worksheet, p. 15 Math Practice Worksheet, p. 18 SourceBook, pp. S108, S110, and S112
Chapter 17	See Teacher's Edition margin, pp. 383, 384, 385, 386, 390, 391, 395, 397, and 399 Theme Worksheet, p. 27 Activity Worksheet, p. 37 Activity Worksheet, p. 41 SourceBook, pp. S115, S117, S120, and S122
Chapter 18	See Teacher's Edition margin, pp. 405, 407, 408, and 411 Activity Worksheet, p. 50 Activity Worksheet, p. 51 SourceBook, pp. S130 and S132
Unit 6	Activity Worksheet: Geology Word Search, p. 58 Unit 6 SourceBook Activity Worksheet, p. 73

Assessment Planning Guide

Lesson, Chapter, and Unit Assessment	SourceBook Assessment	Ongoing and Activity Assessment	Portfolio and Student-Centered Assessment
Lesson Assessment Follow-Up: see Teacher's Edition margin, pp. 361, 369, 378, 386, 391, 395, 399, 407, and 411 **Chapter Assessment** Chapter 16 Review Worksheet, p. 19 Chapter 16 Assessment Worksheet, p. 22* Chapter 17 Review Worksheet, p. 42 Chapter 17 Assessment Worksheet, p. 45* Chapter 18 Review Worksheet, p. 52 Chapter 18 Assessment Worksheet, p. 55* **Unit Assessment** Unit 6 Review Worksheet, p. 60 End-of-Unit Assessment Worksheet, p. 63*	SourceBook Review Worksheet, p. 75 SourceBook Assessment Worksheet, p. 78*	Activity Assessment Worksheet, p. 69* **SnackDisc** Ongoing Assessment Checklists ♦ Teacher Evaluation Checklists ♦ Progress Reports ♦	Portfolio: see Teacher's Edition margin, pp. 365, 377, 386, 390, 398, and 407 **SnackDisc** Self-Evaluation Checklists ♦ Peer Evaluation Checklists ♦ Group Evaluation Checklists ♦ Portfolio Evaluation Checklists ♦

* Also available on the Test Generator software
♦ Also available in the Assessment Checklists and Rubrics booklet

Science Discovery is a versatile videodisc program that provides a vast array of photos, graphics, motion sequences, and activities for you to introduce into your *SciencePlus* classroom. *Science Discovery* consists of two videodiscs: Science Sleuths and the Image and Activity Bank.

Using the *Science Discovery* Videodiscs

Science Sleuths: The Misplaced Fossil Side B

An amateur paleontologist has found a dinosaur bone from the Cretaceous period (65–140 million years ago) in a Tertiary stratum dating to only 10 million years ago. The Science Sleuths must analyze the information in order to verify or disprove the paleontologist's finding.

Interviews

1. Setting the scene: Editor of *Paleontology Today* **5417 (play ×2)**

2. Paleontologist **6335 (play)**

3. Amateur paleontologist **7416 (play)**

4. Hardware-store owner **7950 (play)**

5. Rock-shop owner **8771 (play)**

Documents

6. Abstract of findings with photos 9762 **(step ×2)**

Literature Search

7. Search on the words: DINOSAURS, EVOLUTION, FOSSILS, FOSSIL HOAXES, GOBI DESERT, ONE TON JUNCTION **9766 (step)**

8. Article #1 ("Hoaxes and Hoaxes") **9769 (step ×2)**

9. Article #2 ("Debate on Evolution Continues") **9773 (step)**

10. Article #3 ("Explosions Rock One Ton Junction") **9776**

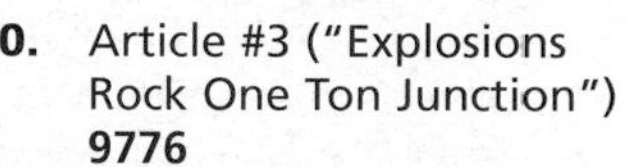

11. Article #4 ("Laborers' Camp Disappears," 1879 article) **9778**

12. Article #5 ("Healing With Bones and Needles") **9780 (step)**

13. Article #6 ("Jefferson University Scientists Go to Gobi Desert") **9783 (step)**

Sleuth Information Service

14. Map of One Ton **9786**

15. Geosurvey of map of fossil site **9788**

16. Historical map of One Ton **9790**

Sleuth Lab Tests

17. Radiometric dating of bone and wood fragments **9792**

18. Identification of metal found in debris **9794**

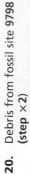

Still Photographs

19. Close-up of bone **9796**

20. Debris from fossil site **9798 (step ×2)**

21. Microscopic photo of wood fragment **9802 (step ×2)**

Image and Activity Bank Side A or B

A selection of still images, short videos, and activities is available for you to use as you teach this unit. For a larger selection and detailed instructions, see the Videodisc Resources booklet included with the Teaching Resources materials.

16-1 Making Mountains, page 355

Plate tectonics; demonstration with globe 15517–16288 (play ×2) (Side B only)

Fold; syncline 884

This shows a very symmetrical (evenly formed) syncline. Most folds are much larger than this and are partially eroded.

Fold; dipping rocks 890

The term *dip* refers to the angle of tilt of the rock formation.

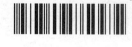

16-2 The Earth Breaks Apart, page 362

Fault 865

Movement along faults offset these plowed rows.

Fault; offset canyon 871

A fault line runs from left to right in the center of the photo. The lower part of the main canyon (center of photo) has been offset by about 90 m to the left along the fault line.

San Francisco earthquake 17970–18795 (play ×2) (Side B only)

◀| Step Reverse Play ▶ Pause || Step Forward |▶

Earthquake detection; seismometer 900–902 (step ×2)
This instrument detects earthquakes and sends electrical currents to seismographs in an earthquake observatory. (step) Seismometer currents become stylus movements that record on seismograph paper the frequency and magnitude of shock waves from an earthquake. (step) This type of seismograph is used in the field by geologists.

Earthquakes; northern California statistics 18798–19205 (play ×2) (Side B only)

16-3 Volcanoes—Holes in the Earth, page 370

Mount St. Helens eruption 1012–1021 (step ×9)
This series of clips shows how Mount St. Helens looked before, during, and after its eruption on May 18, 1980.

Mount St. Helens 19681–20398 (play ×2) (Side B only)
A plume from Mount St. Helens, Washington

Volcano; Paricutín, Mexico 951

Ash, volcanic; Galunggung, Indonesia 953
Ash clouds sometimes cause lightning; there may be discharge within the cloud.

Lava flow; Hawaii 4578–6801 (play ×2) (Side B only)

Pillow lava; forming underwater 19447–19680 (play ×2) (Side B only)
Pillow lava forms when lava pours into water. The surface of the lava forms a thin shell when cooled by the water. The shell fractures, and more lava pours out.

Volcano, cinder cone; Crater Lake 967
Crater Lake is all that remains of what was once an immense volcano. The volcano collapsed, and the caldera filled with water to a depth of 540 m. The cinder cone in the foreground erupted at a later date.

Volcano, shield; Mauna Loa, Hawaii 963
Mauna Loa fills most of this picture. The crater of the volcano Kilauea is on Mauna Loa's slope in the lower left of the image.

Volcano, composite; Mount Shasta, CA 959

17-1 Features and Formations, page 382

Rock classification 540
Sedimentary rocks are formed from particles of other rocks that have been worn down by erosion; igneous rocks are formed from the cooling of magma or lava; metamorphic rocks are formed from conditions of high heat or pressure.

Rock cycle 539
Igneous rocks are formed when magma or lava cools and hardens. Exposure of igneous rocks to excess heat and pressure will cause metamorphism. Erosion can break igneous rocks into sediments that can become sedimentary rock, which can subsequently be metamorphosed.

17-2 Hot Rocks! page 387

Lava bomb 980
Lava bombs are ejected from a volcano in a molten state, but they solidify before hitting the ground.

Ejecta, volcano; Pele's hair 982
These long, strawlike strands of ejected volcanic glass are named after *Pele,* the Hawaiian goddess of Mount Kilauea in Hawaii. Small droplets called Pele's tears are also found.

Basalt 513
The holes in this basalt are called vesicles. They are the result of air bubbles in the lava that formed the basalt.

Basalt plateau 986
The Columbia Plateau in Washington, Oregon, and Idaho is the result of a volcanic eruption. Fissures in the Earth spilled layers of basaltic lava hundreds of meters thick and covering thousands of square kilometers.

Columnar jointing 1003–1005 (step ×2)
This is called Devils Postpile, located in California. (step) As an igneous body cools, shrinkage occurs around uniformly distributed centers. Cracking occurs around each center, causing columnar jointing. (step) These columnar joints have been polished by a glacier.

Neck, volcanic; Shiprock, NM 1002
This is the erosion-resistant core of a volcano that has long since worn away. The linear structure in the foreground is a mass of igneous rock called a dike.

Stock; Devils Tower, Wyoming 1000–1001 (step)
A stock is a small igneous intrusion. The surrounding terrain was eroded away, leaving this erosion-resistant rock structure exposed. (step)

17-3 Rocks From Sediments, page 392

Turbidite demonstration 9761–10487 (play ×2) (Side B only)

17-4 Changed Rocks, page 396

Dike 1008
The rock on the right is a dike; that on the left is sandstone.

18-1 Rocks Reveal a Story, page 403

Fossil, dinosaur 1087
Fossil-bearing beds containing dinosaur remains are found throughout the world. These are in Dinosaur National Monument in Utah.

Fossils 1090–1103 (step ×13)
This is a fossil of a sheeplike animal. (step) Three-toed horse (step) Jawbone and teeth (step) Turtle (step) Fossil ammonite from the Jurassic period (step) Brachiopod (step) Trilobite (step ×3) Stingray (step) Worm burrows (step) Footprint of a three-toed dinosaur (step) If trapped in a sticky substance such as tree resin, organisms can be preserved as fossils. (step) When footprints are covered with sediments, they can be preserved as fossil imprints, like this raccoon footprint.

18-2 Telling Time With Rocks, page 408

Life time line 2805
This time line shows the first appearance of cells, multicellular organisms, animals, dinosaurs, and early humans.

Geologic time 1144–1393 (play, or step through sequence for detail)
Until 500 million years ago, only primitive life-forms existed. The Paleozoic, Mesozoic, and Cenozoic eras (together constituting only one-eighth of the total history of the Earth) saw increasing numbers and complexity of life-forms. Humans appeared very recently.

◀| Step Reverse   Play ▶   Pause || Step Forward |▶

THE RESTLESS EARTH

UNIT FOCUS

Describe the following scenario to students: Two islands exist in the middle of the ocean, hundreds of miles apart. The plants and animals that live on the islands are similar but not identical. Both islands contain fossils, however. The fossils show that millions of years ago the plants and animals that lived on the islands were identical. How could two distant islands have once supported identical plants and animals? *(Students may suggest several possible explanations.)* After students have offered their explanations, tell the students that the islands were once joined together, thus sharing both plants and animals. Tell students that they will have the opportunity in this unit to learn how islands can break apart and move and how fossils like those described can form, as well as other interesting aspects of Earth science.

A good motivating activity is to let students listen to the English/Spanish Audiocassettes as an introduction to the unit. Also, begin the unit by giving Spanish-speaking students a copy of the Spanish Glossary from the Unit 6 Teaching Resources booklet.

THE RESTLESS EARTH

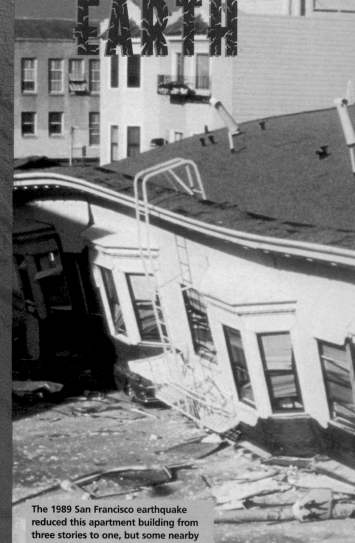

The 1989 San Francisco earthquake reduced this apartment building from three stories to one, but some nearby buildings suffered little or no damage.

352

Connecting to Other Units

This table will help you integrate topics covered in this unit with topics covered in other units.

Unit	Connection
Unit 2 Diversity of Living Things	Fossils reflect changes in biological diversity through time.
Unit 3 Solutions	Sedimentary rocks such as limestone form from the evaporation of solutions.
Unit 4 Force and Motion	Forces such as gravity and friction are important to understanding the processes of plate tectonics.
Unit 5 Structures and Design	Buildings and other structures must often be designed to withstand the forces generated by earthquakes.
Unit 8 Growing Plants	The structure and diversity of ancient plant populations are recorded in fossils.

San Francisco, California. October 17, 1989. At 5:04 P.M. a violent earthquake, measuring 7.1 on the Richter scale, rocks the city. Buildings shudder, sway, and crumble. Frightened citizens fear for their lives, wondering whether the buildings around them will withstand the fierce shaking.

After 15 seconds, the ground becomes still again. As people venture out to inspect the damage, some find puzzling scenes. All around the San Francisco Bay area, demolished houses and office buildings lie shattered next to structures that stand seemingly untouched. Likewise, collapsed sections of highways and bridges adjoin sections that shook but stood firm. Why do you think some structures were able to withstand the earthquake but others were not? Ⓐ

Engineers and architects learn how to make structures more earthquake resistant by studying how the Earth's crust moves and what happens during earthquakes. In this unit you will learn about these things, too. And you will explore volcanoes, rocks, and the ancient history of our planet. These are all part of the study of the dynamic and restless Earth.

Using the Photograph

Call on a student to read aloud the text on this page. Discuss answers to the question posed in the second paragraph (see answer below). Tell students to imagine trying to build a house made of blocks on a kickboard that is floating in water. Ask: How could you make the house stable enough to withstand the movement of the kickboard? *(Answers may include building a shorter structure or using extra materials, such as glue or nails.)* Explain to students that the plates of the Earth's surface are in some ways like kickboards, slowly shifting around and bumping into each other.

Answer to *In-Text Question*

Ⓐ The characteristics of the ground on which structures stand will sometimes determine how strongly earthquakes will affect those structures. More important, however, the design of a building must allow for significant movement in order for the building to withstand an earthquake. Thus, both structural strength and flexibility are critical. Differences in these factors caused some buildings in this photo to collapse while others remained stable.

CHAPTER 16

Shake, Rattle, and Flow

Connecting to Other Chapters

Chapter 16 *examines plate tectonics, with a particular focus on volcanoes, mountains, and earthquakes.*

Chapter 17 *explores the essential characteristics of igneous, sedimentary, and metamorphic rocks.*

Chapter 18 *introduces the geologic time scale as well as the types and formation of fossils.*

Prior Knowledge and Misconceptions

Your students' responses to the ScienceLog questions on this page will reveal the kind of information—and misinformation—they bring to this chapter. Use what you find out about your students' knowledge to choose which chapter concepts and activities to emphasize in your teaching.

In addition to having students answer the questions on this page, you may wish to have students complete the following activity: Have them write a short skit featuring a middle-school student and a third-grader. In the skit, the middle-school student explains to the third-grader what earthquakes are and how they occur. You may wish to have volunteers act out a few of the skits. Emphasize that there are no right or wrong answers in this exercise. Collect the papers, but do not grade them. Instead, read them to find out what students know about earthquakes and plate tectonics, what misconceptions they may have, and what is interesting to them about plate tectonics.

CHAPTER 16
Shake, Rattle, and Flow
How is the Earth's crust similar to a jigsaw puzzle? 1
How do you think the mountain shown on the next page was formed? 2
3 What does a volcanic eruption have in common with an earthquake?
ScienceLog
Think about these questions for a moment, and answer them in your ScienceLog. When you've finished this chapter, you'll have the opportunity to revise your answers based on what you've learned.
354

LESSON 1

Making Mountains

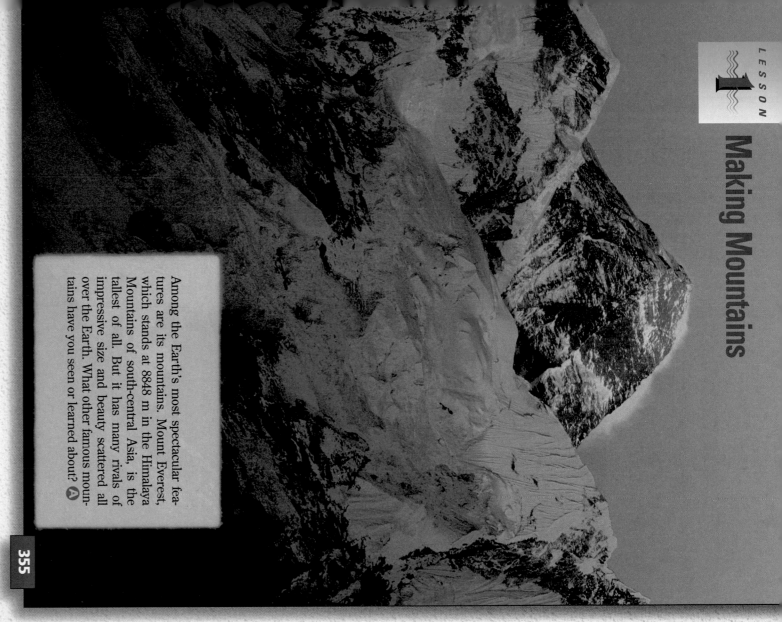

Among the Earth's most spectacular features are its mountains. Mount Everest, which stands at 8848 m in the Himalaya Mountains of south-central Asia, is the tallest of all. But it has many rivals of impressive size and beauty scattered all over the Earth. What other famous mountains have you seen or learned about? Ⓐ

LESSON 1 ORGANIZER

Time Required
2 to 3 class periods

Process Skills
hypothesizing, inferring, observing, organizing

Theme Connection
Changes Over Time

New Terms
Fault—the boundary between two rock sections that have been displaced relative to each other

Plate tectonics—the theory that the Earth's crust is made up of rigid plates that move; the motion of these plates causes continental drift

Materials (per student group)
Exploration 1: 2 colors of modeling clay; knife; metric ruler; optional materials: small piece of wax paper

Teaching Resources
Resource Worksheet, p. 3
Exploration Worksheet, p. 5
Transparency Worksheet, p. 7
Transparencies 58 and 59
SourceBook, p. S108

LESSON 1 Making Mountains

FOCUS

Getting Started

Ask a student to volunteer to be a mountain explorer. Then cover the student's eyes with a bandanna. Have the explorer feel the surface of a relief map or globe and locate mountain ranges. Ask: Are mountains all the same size? *(No)* Are they bunched together or spread out evenly? *(Bunched together)* Remove the blindfold. Conclude by saying that in this lesson students will learn about how mountains are formed and why they are bunched together in certain areas of the Earth's surface.

Main Ideas

1. The external forces of weathering and erosion are offset by internal forces that cause the land to be uplifted.
2. The plate tectonic theory states that mountains are formed when plates collide.
3. Rocks under stress either bend to form folds or break to create faults.
4. Mountain-building is an ongoing process.

TEACHING STRATEGIES

GUIDED PRACTICE As a review, ask volunteers to give definitions of *weathering* and *erosion* and to write their responses on the chalkboard. *(Weathering is the process by which rock is worn down by water, wind, or ice. Erosion occurs when weathered fragments of soil, rock, and other materials are carried away.)*

Answer to *In-Text Question*

Ⓐ Encourage students to mention famous mountains from around the world or from your area.

The Rise and Fall of Mountains

GUIDED PRACTICE Use the following questions to ascertain students' prior knowledge of the mountain-building process:

- What are mountains made of? *(Rock, sometimes covered by soil, snow, or ice)*
- How are mountains formed? *(Rocks are uplifted by tectonic forces or volcanism.)*
- Do mountains last forever? *(No, they wear down by weathering and erosion.)*

Have students read the paragraphs that follow on this page and the next. Then divide them into pairs or small groups to discuss the questions. Reassemble the class, and have the groups report their ideas.

Answers to *In-Text Questions*

A Mountains probably were not formed this way; however, to people in ancient times, this may have seemed like a plausible explanation. The myth explains earthquakes and volcanoes as the actions of giants trapped beneath the mountains.

B The hills and mountains probably formed as a result of plate movements. The valleys resulted from erosion.

Answers to *Agricola's Proposal*

Call on volunteers to read aloud the text and monologue on the next page. Then involve students in a discussion of Agricola's ideas, which are summarized in statements 1 through 5.

1. Students should identify this proposal as one that makes sense in some cases. However, the most obvious weakness in Agricola's idea is that it does not provide an explanation for how new mountains are formed. Specifically, Agricola did not explain uplift.
2. Students should recognize that this proposal does not make sense. The piling up of sand could create dunes or small hills, but the formation of mountains would be unlikely.
3. Students should recognize that this proposal does not make sense. There is no evidence that underground winds exist. If they did exist, it seems very unlikely that they would have the force to push rocks into mountains.
4. Students should recognize that this proposal does not make sense. Earthquakes, although powerful, cannot force tons of rock up as high as mountain ranges.
5. Students should identify this proposal as one that makes sense in some cases. Mountains can be formed by volcanic eruptions, although such an explanation would account for only one kind of mountain. For example, Mount Shasta in northern California and Mount Vesuvius in Pompeii, Italy, were formed by volcanoes.

The Rise and Fall of Mountains

Throughout the ages, people have marveled at and wondered about mountains. Some of the ideas and questions that people have had about mountain formation are presented on this page and the next. Read each of the ideas, and discuss them with your classmates.

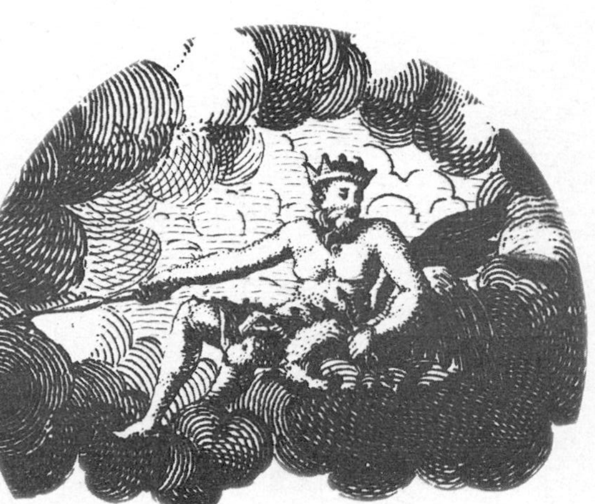

Back to the Greeks

The people of ancient Greece and Rome believed that mountains were formed by giants who wanted to reach the heavens. The giants piled one peak on top of another to form a stairway to the sky. Zeus, the king of the Greek gods, struck down the peaks with his thunderbolts and scattered the remains into rugged mountain chains. Then Zeus imprisoned the giants beneath the mountains. During their struggle to escape, the giants broke the Earth's crust and hurled liquid rock at Zeus. Can you imagine the mountains being formed in this way? What explanation does the myth give for earthquakes? for volcanoes? **A**

Agricola's Proposal

In 1546, the German scholar and scientist Georgius Agricola suggested that mountains came into being in five different ways:

1. through the eroding action of water
2. through the piling up of sands along seacoasts and in desert lands
3. through underground winds
4. through earthquakes
5. through volcanic fires

Agricola believed, however, that most mountains were formed by the eroding action of water. Read how he described it. Do you think any of Agricola's ideas make sense? Are there ideas you disagree with? Why?

Do you think the hills and mountains at Death Valley, California, were formed by the eroding action of water? **B**

It is evident that many and indeed most mountains owe their origin to the action of water. The little brooks first wash away the surface soil and then cut into the solid rock. By carrying it away grain by grain, the water can finally cut even a mountain range in two, removing great blocks of rock in the process.

In a few years, a small stream can dig a deep depression or river bed across a level or gently sloping plain. Over the course of years these stream beds reach an astonishing depth, while their banks rise majestically on either side. From those banks small fragments of rock are continually detached by rain or frost. Large rocks that have cracks and fissures fall because of their grand weight and land in the stream below. In this way the steep cliffs gradually recede and convert into gentle slopes, and so the original plain converts into a series of elevations and depressions. The elevations are called mountains and the depressions valleys.

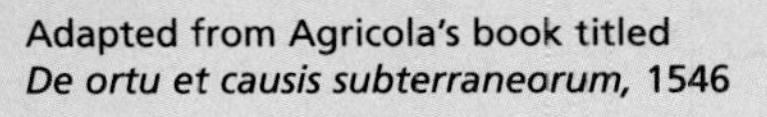

Adapted from Agricola's book titled
De ortu et causis subterraneorum, 1546

Tani's Idea

Someday there will be no mountains to climb. When we studied erosion and weathering in science class, the teacher told us that mountains are worn down about 30 cm every 1000 years. Since Mount Everest is the tallest mountain in the world, at 8848 m, by the time it is worn down to sea level, all the other mountains will be gone, too.

Do you agree with Tani's reasoning? If not, how would you convince her that she is wrong? According to her figures, how long would it take for Mount Everest to wear down? C

357

Answers to *In-Text Questions*

C Tani is incorrect because processes of uplift are also continually occurring. As old mountains are worn down, new ones are formed. It would take between 31 million and 32 million years for Mount Everest to wear down if no uplifting processes occurred.

Meeting Individual Needs

Second-Language Learners

Provide groups of students with materials to build small stream tables. You may include aluminum roasting pans, sand, pebbles, dirt, watering cans, and any other suitable materials. Students should use the materials to discover how water can form deep canyons and cut through rock. Encourage students to experiment with the materials, such as varying the shape of the sand or soil, the amount of water, and the speed at which the water flows. Students should present their conclusions in a short presentation.

Cross-Disciplinary Focus

Mathematics

Present students with the following problem: Mount Shasta, a volcanic mountain in the Cascade Range of northern California, is 4.317×10^3 m high. What is its height in centimeters?

a. 43,170 cm
b. 431,700 cm *(Correct)*
c. 4,317,000 cm
d. 43,170,000 cm

A Modern View of Mountain-Building

Direct students' attention to the map on this page, and have students identify the continents and oceans. Be certain that students notice the arrows on the map. Spend some time discussing how movement occurs at the different kinds of boundaries. You may wish to introduce the terms *convergent* and *divergent* to describe boundaries where plates are moving toward and away from each other, respectively.

Some students may have difficulty understanding that crust is found under the oceans as well as the continents and that both continents and oceans ride on plates. You may wish to demonstrate this concept using tangible objects, such as pieces of cardboard.

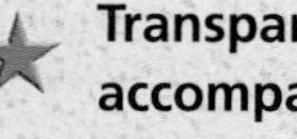

Transparency 58 is available to accompany the material on this page.

Homework

Have students answer the following question in one or two paragraphs. They may consult the SourceBook or other reference materials for information. What is Pangaea? *(Pangaea is Wegener's name for the hypothetical supercontinent that existed on Earth 200 million years ago.)*

Meeting Individual Needs

Learners Having Difficulty

Have students create a relief map that illustrates the tectonic plates. Each plate should be made out of a different color of modeling clay. Allow students to develop their own key for indicating what happens at each of the plate boundaries. Have students use a toothpick to draw rough outlines of the continents on their map. Suggest that they use another toothpick to indicate where their hometown is located.

A Modern View of Mountain-Building

Plate Tectonics

In 1911, a German scientist named Alfred Wegener proposed a theory that led to our current understanding of the way mountains are formed. Wegener suggested that the continents of the world were once joined together but over time have split and drifted apart. This theory of *continental drift* evolved into our current understanding of the movement of the continents, the theory of **plate tectonics**.

According to this theory, first proposed by Canadian geologist J. Tuzo Wilson, the outer crust of the Earth (the *lithosphere*) is broken into seven large, rigid plates and several smaller ones. The continents and oceans ride on top of these plates. It is the motion of the plates that causes continental drift. The map below shows the boundaries of the plates. The arrows indicate the direction of movement of the plates.

Alfred Wegener

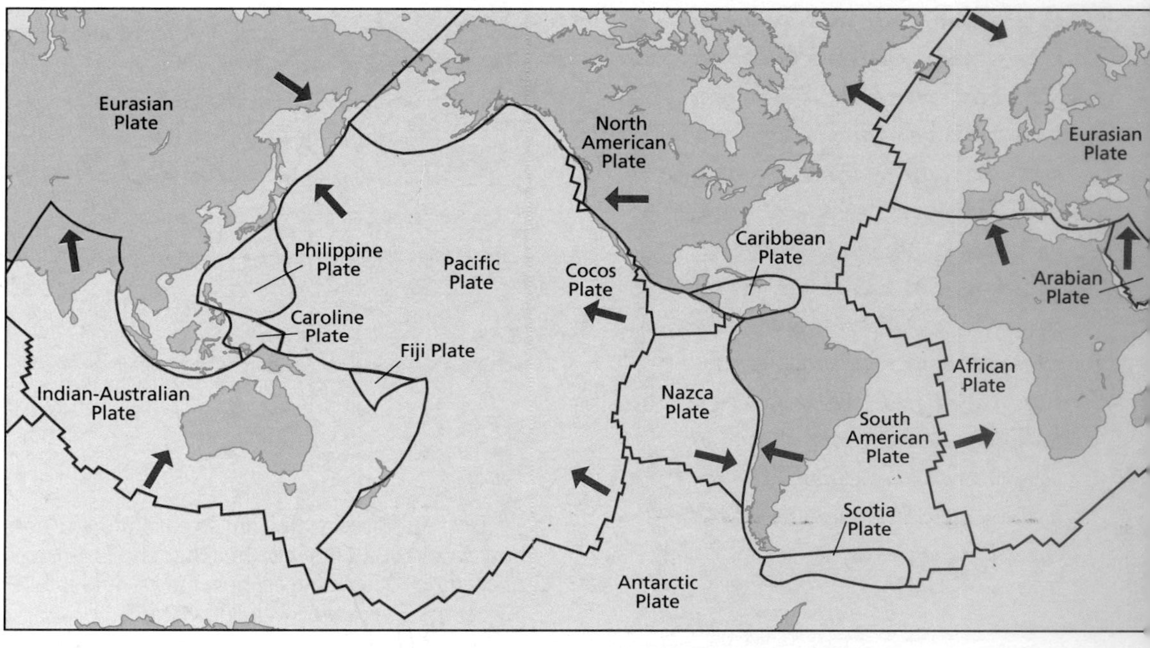

When the plates move, what happens at the boundaries of the plates? Do the thought experiment on page 359 to find an answer to this question.

Theme Connection

Changes Over Time

Focus question: Using the map on this page, how do you predict the map of the Earth will look 300 million years from now? (*North America will be next to Asia, the Atlantic Ocean will be much larger, Africa will have moved away from the Middle East, and the Mediterranean Sea will have disappeared and will have been replaced by a mountain range.)*

An Earth-Moving Thought Experiment

Think of the Earth as a jigsaw puzzle with seven huge pieces and several smaller ones.

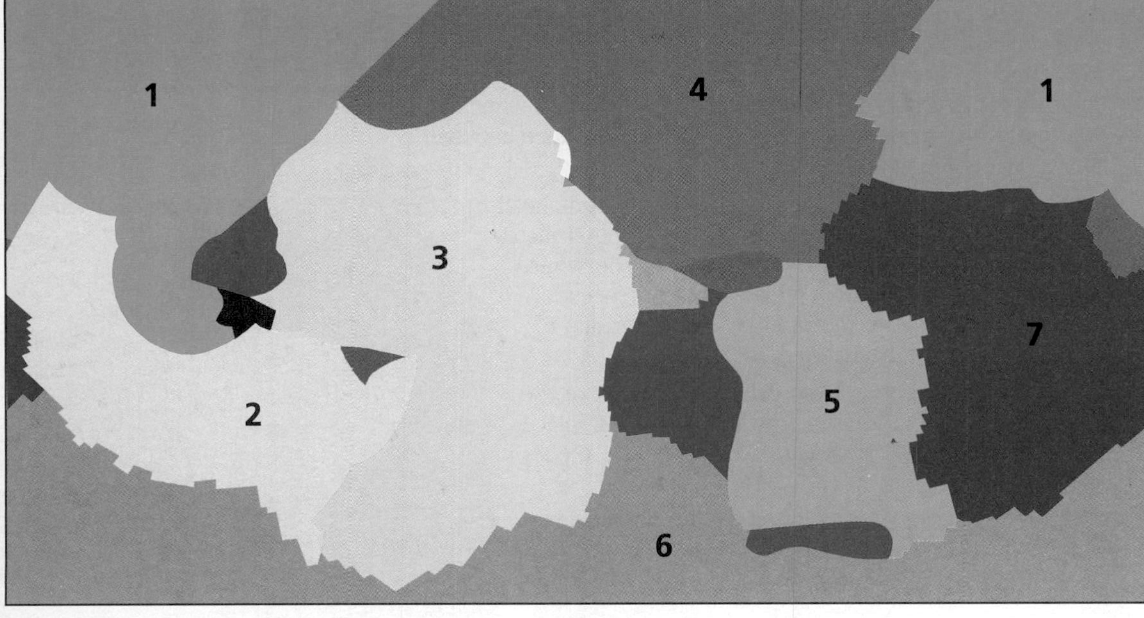

- Compare the jigsaw pieces shown above to the map of the Earth on page 358. What does each puzzle piece represent? Why are there two plates with the number 1?
- Find the common border between piece 4 and pieces 1 and 7. Is this border on land or in the ocean?
- Scientists know that molten rock that comes from deep in the Earth forms new crust along this border. In other words, the crust is spreading outward from this border. What effect will this movement have on pieces 4, 1, and 7?
- As new crust is formed, what happens to other pieces in the jigsaw puzzle? What is happening at the boundaries of the jigsaw pieces? Work with two or three of your classmates to predict the changes that may occur in the plates during the next million years. Draw these changes on a map in your ScienceLog. Where will each plate be? How will the puzzle change?

According to Wegener, the continents once fit together to form a supercontinent called Pangaea. Read more about it on page S108 of the SourceBook.

Meeting Individual Needs

Gifted Learners

Pose one of the following "what if" questions to students:

- What if the substances forming the continental crust were much harder and more resistant to erosion?
- What would happen if the average temperature inside the Earth's mantle were much higher?
- What would the Earth's surface be like if there were 20 smaller crustal plates instead of the existing major ones?
- What would the Earth be like if it were a solid ball without a molten mantle and core?

Have students write paragraphs or outlines to answer these questions. Encourage them to consider the effects these changes would have on living organisms. *(Accept all reasonable responses.)*

An Earth-Moving Thought Experiment

You may wish to take this opportunity to discuss the problem of representing the spherical Earth on a flat piece of paper. An atlas may provide examples of different ways to solve this problem. Be sure to point out that the Mercator projection used on page 358 shows Europe and North America larger than their actual size relative to the other continents. Use a globe to help illustrate this concept.

★ **A Resource Worksheet (Teaching Resources, page 3) is available to accompany An Earth-Moving Thought Experiment.**

Answers to *An Earth-Moving Thought Experiment*

- Each puzzle piece represents a large, rigid plate. (To help students understand how two pieces can have the same number, trace the puzzle pieces, cut out the rectangle, and form a cylinder by matching the pieces at the side edges.)
- The common border between piece 4 and pieces 1 and 7 is in the Atlantic Ocean.
- Pieces 4, 1, and 7 will be forced apart by the new crust forming at their common boundary. The Atlantic Ocean will get wider.
- As new crust is added at the common boundary of plates 4, 1, and 7, the other plates will collide or slide over, under, or past each other. At the boundaries, plates are moving toward, away from, or past each other. Sample predictions: Plates 7 and 2 will move up (north), plate 1 will move to the right (east), plate 4 will move to the left (west), and plate 3 will decrease in size.

Making Use of the Theory
Have students read the material on this page. Involve them in a discussion of the theory of plate tectonics and the accompanying questions.

Answers to *Making Use of the Theory*

- If Africa had drifted away from South America at a rate of 10 cm per year, it would have taken 60 million years for the continents to get to their present positions. (6000 km × 100,000 cm/km × 1 year/10 cm = 60,000,000 years)
- When an oceanic plate collides with a continental plate, it slopes under the continental crust. Puzzle pieces involved are 3 and 1, 3 and 2, and 3 and 4—near Alaska. The various disruptions that may occur are mountain-building, earthquakes, volcanoes, folds, and faults. Invite students to speculate about what happens. You need not go into detail about earthquakes or volcanoes at this time.
- This may be happening at the boundaries of pieces 1 and 7 and pieces 1 and 2.

Rocks Under Stress
Evidence of folds and faults in rocks suggests that tremendous forces must exist beneath the ground. The models that students will make in the following Exploration should help explain how folds and faults occur.

Making Use of the Theory
Review the following observations and explanations made by geologists. Then use what you've learned about plate tectonics to answer the questions.

- As the new crust made of younger rock is formed, the ocean floor spreads. This causes the plates to move apart. They move from 5 cm to 20 cm per year. Estimate how long it has taken the continent of Africa to drift away from South America. Check an atlas or a world map to determine the current distance between these two continents.
- When an oceanic plate drifts up against a continental plate, the oceanic plate is forced under the continental plate. Look at the map on the previous page to see where this may be happening. Which puzzle pieces are involved? This type of motion may cause various disruptions. What might they be?
- When two continental plates collide, the crust of the Earth buckles and moves upward. Locate a place where this may be happening. Which puzzle pieces are involved?

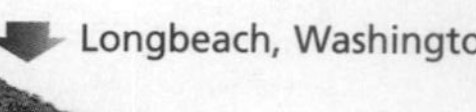

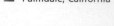

Palmdale, California

Longbeach, Washington

Bear Island, in the Arctic Ocean

Black Hills, South Dakota

Rocks Under Stress
It is hard to believe that there are forces strong enough to push the Earth's crust into mountains. Yet when geologists study existing mountains, they find rocks that have been bent, folded, crumpled, and broken, as in the pictures shown on this page. Try the following Exploration to find out how folds and **faults** (boundaries between rock sections that move relative to each other) occur in the Earth's crust.

EXPLORATION 1

Folds and Faults

Models can help explain how folds and faults are formed in the crust.

PART 1

You Will Need

- 2 colors of modeling clay
- a knife
- a metric ruler

What to Do

1. Cut or flatten the clay into strips approximately 1 cm × 3 cm × 10 cm. Make two strips of each color.
2. Stack the strips, alternating the colors.
3. Place one narrow end of the block against a wall. With your hand, apply pressure to the other narrow end.

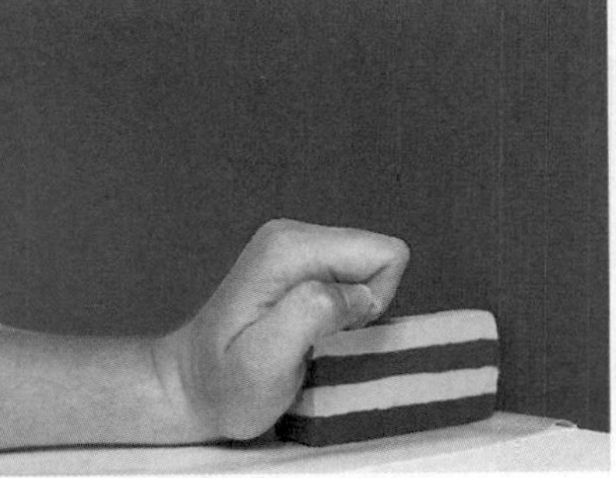

4. Compare your model to the pictures on the previous page. What process does this model illustrate? What do the layers of clay represent? Where do the forces that deform rocks originate? Ⓐ

PART 2

What to Do

1. Look at the photograph above. Why do you think the fault in the rock formed as it did? Use another clay model to demonstrate your answer.
2. What forces do you think caused the fault to form?
3. If 1 cm in the photograph corresponds to 75 cm in the actual rock, find out how far the rocks have moved along the fault.

Drawing Conclusions

Rocks deep within the Earth are subjected to extreme pressures and temperatures. As a result, they sometimes behave like clay, bending when forces are applied. On or near the surface, however, rocks tend to break rather than bend. This is because the pressures and temperatures are not as great on the surface of the Earth. In this Exploration, you have found out that as rocks bend to form folds and break to form faults, mountains are sometimes created. Mountains can be formed in other ways as well. For example, Mount Shasta in northern California was formed by the action of volcanoes.

Mountain-building appears to be an ongoing process. Refer back to Tani's idea about mountains on page 357. Could you use what you have learned in this Exploration to change Tani's mind? Explain your reasoning. Ⓑ

FOLLOW-UP

Reteaching

Have students use two index cards to simulate the following situations:

a. two plates colliding, causing buckling and folding

b. two plates colliding, with one plate being pushed under another

c. two plates sliding by each other

Assessment

Have students describe in words or with pictures how the formation of folds and faults can result in mountains. *(At a fold, pressure can push rock upward. At a fault, one plate pushes above the other.)*

Extension

Have students find out about mountains located under the oceans. Ask: Do the forces that build mountains on land also build them under water? *(Yes)*

Closure

Have students locate relief maps, photographs, or other information on nearby mountain areas. Ask them to determine approximately when these mountains were formed.

EXPLORATION 1

Divide the class into small groups and distribute the materials. The modeling clay can be rolled flat using a rolling pin or wooden dowel. In addition, you may wish to provide wax paper so that students can protect their work areas. If possible, have each group make both a fold model and a fault model.

★ **An Exploration Worksheet, a Transparency Worksheet (Teaching Resources, pages 5 and 7, respectively), and Transparency 59 are available to accompany Exploration 1.**

Answers to *In-Text Questions*

Ⓐ This model illustrates folding. The layers of clay represent layers of rock. The forces might originate from two colliding plates or from one plate slipping under another.

Ⓑ This Exploration demonstrated how new mountains can be formed by folds or faults in the Earth's crust. Tani's idea doesn't account for the formation of new mountains.

Answers to *Part 2*

1. The fault formed when the rock broke and one side moved away from the other.
2. The fault in the picture resulted when compressive and shear forces caused the rocks in the crust to break and slide past each other.
3. Students should measure the distance along the fault between the two dark and thick lines. They should measure from the top edge of the band on the left to the top edge of the band on the right, a distance of approximately 0.3 cm. Therefore, the distance of offset is 0.3×75 cm, or 22.5 cm.

Homework

The Closure activity described on this page makes an excellent homework activity.

LESSON 2

The Earth Breaks Apart

FOCUS

Getting Started

Take students to a recreational area on the school grounds. Select two small teams of students for a game of tug of war. Allow students to begin the game. Ask: What force is exerted on the rope? *(Tension)* What would happen if the rope snapped in half during the game? *(The students holding the rope would fall downward and backward.)* What might make the students fall harder or faster? *(More tension on the rope)* What would be released when the rope broke? *(Energy)* Explain to students that they will learn in this lesson how the tug-of-war game resembles the interaction of tectonic plates, which can cause earthquakes.

Main Ideas

1. Earthquakes occur when plates slide past each other, collide, or move apart.
2. The elastic rebound theory explains how rocks spring back into their original shape after they have been deformed by seismic forces.
3. Seismographs provide a method for determining an earthquake's strength.

TEACHING STRATEGIES

A Picture Study

The photograph was taken after the San Francisco earthquake in 1906. The scene is along the San Andreas fault, where a slip of several meters occurred.

Answer to *In-Text Question*

Ⓐ Answers will vary. For example, students may pose questions about the event itself, the observer's feelings during the event, or the results of the event.

LESSON 2

The Earth Breaks Apart

A Picture Study

Imagine that you find an old photo album while visiting a grandparent or an older neighbor. The picture shown here was included among the others. In your ScienceLog, make a list of the questions you would ask your grandparent or neighbor about the photograph. What answers do you think he or she might give? Ⓐ

LESSON 2 ORGANIZER

Time Required 3 class periods

Process Skills
observing, inferring, hypothesizing, analyzing

Theme Connection
Energy

New Terms
Elastic rebound theory—explains how rocks spring back to their original shape after they have been deformed by tectonic forces
Epicenter—the position on the surface of the Earth directly above the focus of an earthquake
Focus, or hypocenter—the point at which stress breaks the friction lock between two plates of the Earth's crust
Magnitude—the strength of an earthquake
Richter scale of magnitude—scale used to measure the strength of an earthquake based on the amplitude of seismic waves
Seismographs—instruments that record the vibrations of the Earth during an earthquake

continued ►

EXPLORATION 2

Another Fault Model

In Exploration 1, you made models that helped explain how mountains form when folds and faults are created. In this activity you will make another fault model that will help explain how earthquakes occur.

You Will Need

- the clay block from Exploration 1
- a knife
- a metric ruler

What to Do 

1. Separate the layers of clay carefully, and reshape each layer into rectangles measuring 1 cm × 3 cm × 10 cm.
2. Restack the layers as before.
3. Cut the block lengthwise into two equal pieces.
4. Holding the left portion of the block in your left hand, push the other block forward with your right hand until the blocks slide apart.

Questions

1. What does each block of clay represent?
2. What does each layer represent?
3. What is simulated by the boundary between the two clay blocks?
4. What would happen to a fence that crossed the fault?
5. Compare your model with the fault in the photograph on page 361.

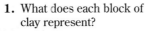

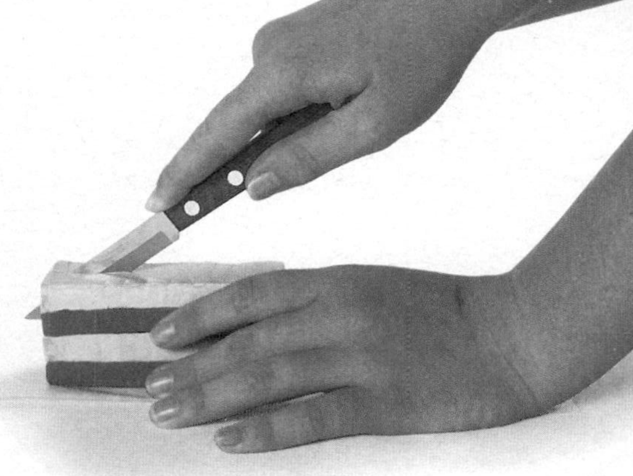

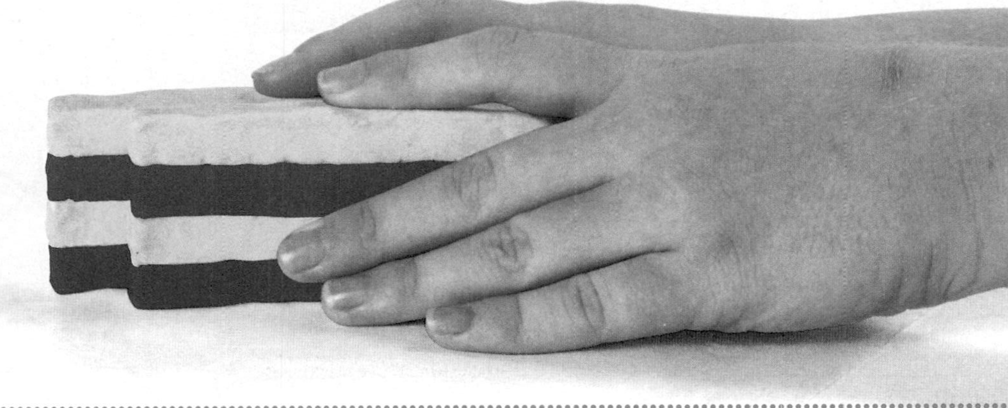

EXPLORATION 2

The procedure for this Exploration is similar to that used in Exploration 1. However, students should understand that this model represents a type of fault that contributes to earthquakes.

You may wish to have students cut the clay with the edge of a ruler instead of a knife.

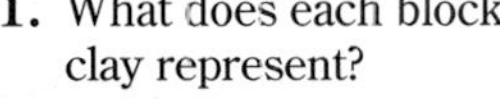

An Exploration Worksheet (Teaching Resources, page 9) is available to accompany Exploration 2.

Answers to *Questions*

1. The blocks of clay represent sections of the Earth's crust.
2. Each layer of clay represents a layer of rock in the Earth's crust.
3. The boundary between the clay blocks simulates the fault where the plates are moving against each other in opposite directions.
4. If a fence crossed the fault, it would be broken or torn apart. (Refer students to the photograph on the previous page.)
5. The model shows that the two portions of clay have moved in opposite directions from each other. This is similar to the photograph on the previous page.

ORGANIZER, continued

Seismologists—people who study earthquakes

Shock, or seismic, waves—stored energy that is released in the form of intense vibrations during an earthquake

Materials (per student group)

Exploration 2: clay block from Exploration 1; knife; metric ruler; optional materials: small piece of wax paper

Teaching Resources

Exploration Worksheets, pp. 9 and 12

Transparency Worksheet, p. 10

Transparency 60

SourceBook, p. S110

Another Picture Study

Have students study the photograph of the stream crossing the San Andreas fault in central California. The stream bed provides evidence that the ground moved along the fault in a direction perpendicular to the stream bed. The area at the top of the photograph is located on the Pacific plate, which is slowly moving northward (to the right in the photograph) relative to the North American plate (in the foreground). This type of long-term movement along a fault produces recurring earthquakes.

What Causes Earthquakes?—An Elastic Theory

Have students read the material and study the diagrams on page 365.

GUIDED PRACTICE Students may have some difficulty relating the sketches on the following page to the factual information. You may wish to use the clay model from Exploration 2 and a rubber band to perform a demonstration of the changes involved. One way to do this is to anchor the rubber band to each clay portion with toothpicks. When the portions are moved in opposite directions to one another, tension will be produced in the rubber band. This is analogous to the energy stored in rocks along a fault, which are held in place by friction. After sufficient tension builds up, the rubber band will break and snap back, much like the rocks after the friction lock is broken. For the purposes of the demonstration, you may wish to release one end of the rubber band from its anchor in the soft modeling clay, rather than stretching it to its breaking point.

Homework

Tell students to imagine that they have just been in an earthquake and to describe their experience in their ScienceLog. Emphasize that they should include observations using as many senses as possible. *(Accept all reasonable responses.)*

Another Picture Study

Starting at the bottom left of the photo, trace the course of the stream as it crosses the fault. Does this picture provide you with any clues about what might have happened to the fence shown on page 362? Explain. Ⓐ

What Causes Earthquakes?—An Elastic Theory

What do earthquakes and rubber bands have in common? Think about it for a moment. When you stretch a rubber band, energy is stored. What happens to the energy when you let go? Can rocks be stretched? Do rocks store energy? H. F. Reid, an investigator of the 1906 San Francisco earthquake, answered "yes" to both questions. He developed the **elastic rebound theory** to explain the cause of earthquakes.

Look at each diagram on the next page and read the related facts to examine this theory further!

★ **A Transparency Worksheet (Teaching Resources, page 10) and Transparency 60 are available to accompany What Causes Earthquakes?—An Elastic Theory.**

The diagrams represent aerial views of the boundary between two plates, P and Q.

Diagram A

- P and Q are blocks of crust slipping in opposite directions along a fault.
- A street crosses the fault.
- The rocks forming blocks P and Q are locked by friction that is created by the pressure of surrounding rock layers. The friction keeps the blocks from moving.
- The rocks store the energy of the friction just like a stretched rubber band stores energy.

Diagram B

- As stress builds up, the rocks begin to strain.
- The street bends with the rock.
- The rocks continue to build up energy from the pressure.

Diagram C

- Finally, the stress breaks the friction lock at the **focus,** or **hypocenter**.
- On each side of the focus, the rocks spring back into their original shape. This is similar to the way a rubber band returns to its original shape after you have stretched it.
- The stored energy is released in the form of intense vibrations known as **shock waves** or **seismic waves**. This release is what we call an earthquake.

Diagram D

- The street is displaced at the fault.
- The blocks have moved in opposite directions.

Can You Help Michael?

To determine whether her students understood the elastic rebound theory, Ms. Morris asked them to use the theory to explain the photograph of the stream on page 364. Michael was absent from school when Ms. Morris explained the theory, and now he needs some help.

Imagine that you are a member of Ms. Morris's class and that Michael has come to you for help with the assignment. Write down what you might say or do to explain the theory to him. B

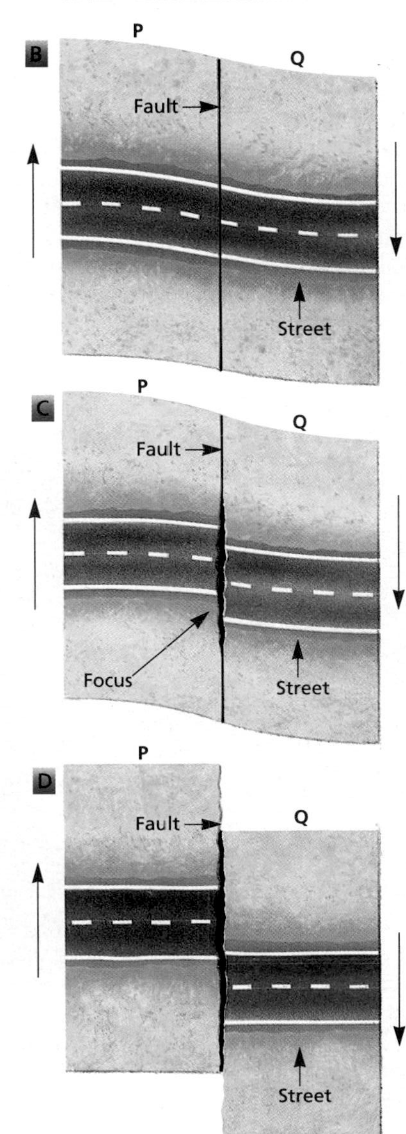

365

Did You Know. . .

In the event of an earthquake, there are several safety precautions that students should take. If they are indoors, students should stand in a doorway or crouch under a table until the shaking stops. If they are outdoors, they should move away from tall trees, buildings, and power lines.

Diagrams A–D

Students may have difficulty interpreting the diagrams. To aid their understanding, ask students to use their clay models of faults from Exploration 2.

Place the two portions of clay in their original positions. Place a row of toothpicks directly opposite each other on the inner edge of each piece of clay. Have students gently slide the clay portions along each other, as in Exploration 2. As the stress builds up, the model illustrates diagrams B and C. When the clay portions slip, the model represents diagram D. The toothpicks that were opposite each other change position relative to one another.

The meaning of the term *focus,* or *hypocenter,* should be clarified because students may be familiar with the term *epicenter* from media reports of earthquakes. The *epicenter* is the point on the Earth's surface directly above the focus. The focus can be far below the surface of the Earth.

You may wish to inform your students that much of what we know about the Earth's interior comes from monitoring the two types of seismic waves that travel out from the focus of an earthquake. By studying these waves, scientists have discovered that the Earth has a core of very dense material and that the inner part of this core is solid.

Answers to *In-Text Questions,* pages 364–365

A Students should conclude that this photograph supports their inference about the fence on page 362 because the shifting of the stream's course shows the same pattern as the shifting of the fence line.

B Answers will vary but should demonstrate an understanding of the various aspects of the elastic rebound theory.

PORTFOLIO

Students may wish to include their essay from Can You Help Michael? in their Portfolio. Allow them to make notes in the margins if they feel they need to correct or amend their explanation.

Measuring the Strength of Earthquakes

Call on a volunteer to read this page aloud. Review the terms *magnitude, seismologists, Richter scale of magnitude,* and *seismographs.* Many students will already be familiar with these terms. Involve them in a discussion of what the Richter scale is and what the numbers mean.

To explain how seismographs record vibrations, refer to the schematic diagrams on page S204. There is a network of seismic recording stations throughout the United States. They are operated by major universities and by the United States Geological Survey.

Homework

Have students design a working seismograph that would use household objects. Their design should be accompanied by an explanation of how the seismograph works.

Measuring the Strength of Earthquakes

Have you ever felt the Earth vibrate? Perhaps you have felt the ground shake as a train passed by. Earthquakes also cause the ground to vibrate. As you can imagine, huge amounts of energy are released when the Earth moves like this.

As you probably know, every earthquake is given a number. What does this number mean? The number describes the **magnitude,** or strength, of the earthquake. But what does this number really mean? Measuring the energy released in an earthquake is a long, complicated process. To enable rapid calculation of the strength of an earthquake, **seismologists** (people who study earthquakes) have adopted the **Richter scale of magnitude,** shown below.

The number expressing the magnitude of an earthquake is calculated by means of a formula. The information contained in the formula is provided by instruments known as **seismographs.** These instruments record the vibrations of the Earth during a quake. An increase of just one number in the measurements corresponds to a tenfold increase in earthquake strength. For instance, an earthquake that is 7.4 on the Richter scale is about 10 times stronger than an earthquake with a magnitude of 6.4.

A seismologist in the field

The Modified Richter Scale of Magnitude

Magnitude	2.0 to 3.4	3.5 to 4.2	4.3 to 4.8	4.9 to 5.4	5.5 to 6.1
Estimated damage	Not felt, but recorded	Felt by a few	Felt by most	Felt by everyone in the affected area	Moderate to slight damage
Estimated number of earthquakes recorded each year	More than 150,000	30,000	4800	1400	500

Some Earth-Shaking Questions

1. Refer to the table below. What is the magnitude of the strongest earthquake ever recorded by seismographs?
2. Which earthquake magnitude occurs most often?
3. An earthquake that measures ___?___ on the Richter scale will bend bridges and railroad tracks.
4. What does the prefix *seismo,* as in *seismograph,* mean? Find other words that use this prefix, and explain them.
5. Look at the diagram to the right. What do you think the **epicenter** of an earthquake is?

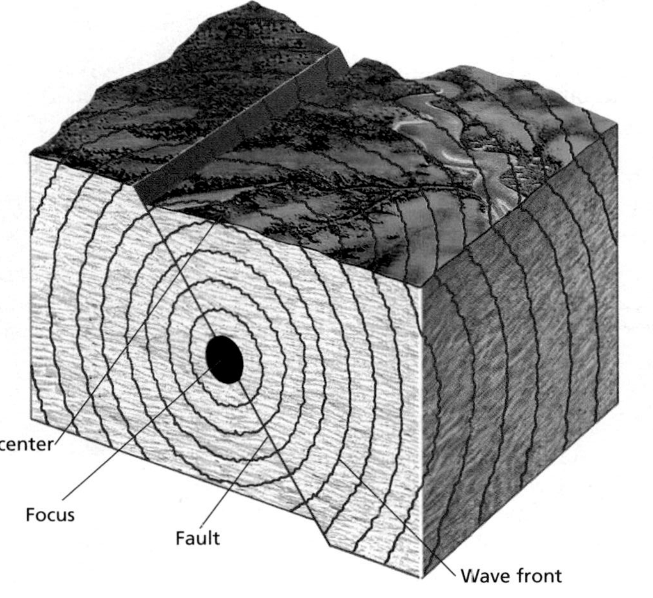

6.2 to 6.9	7.0 to 7.3	7.4 to 7.9	8.0 to 8.6	9.0	10.0
Widespread damage to most structures	Serious damage; railway tracks and bridges bent	Great damage	Very great damage	Would be felt in most parts of the globe	Would be felt all over the Earth
100	15	4	Occur infrequently	Possible but never recorded	Possible but never recorded

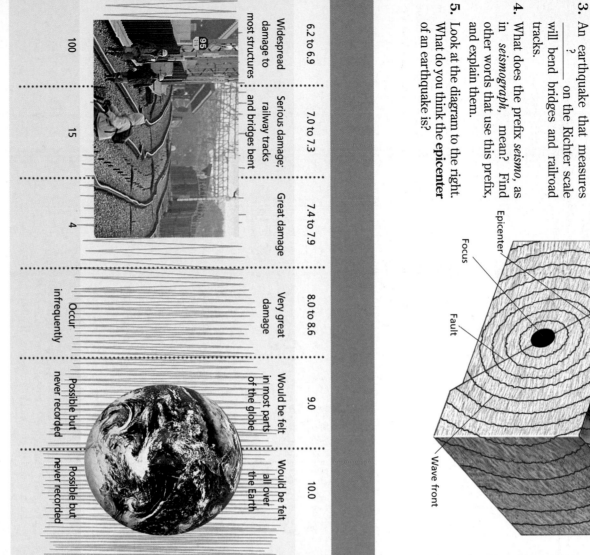

Answers to *Some Earth-Shaking Questions*

1. 8.0 to 8.6
2. 2.0 to 3.4
3. 7.0 to 7.3
4. *Seismos* is a Greek word meaning "earthquake"; *seismo* is the combining form of this word. For example, a seismogram is a record of an earthquake recorded by a seismograph. Other words with this prefix include seismology (the study of earthquakes), seismography (the branch of seismology dealing with the mapping and description of earthquakes), and seismic (about or from earthquakes).
5. The epicenter is the position on the surface of the Earth directly above the focus of an earthquake.

Theme Connection

Energy

Pluck a guitar string. **Focus question:** How is hearing a plucked guitar string similar to feeling an earthquake? *(In both examples, energy travels in waves to sensory receptors on our bodies. The plucked guitar string creates sound waves in the air that we hear, and an earthquake creates waves in the Earth's crust that we feel.)*

Answers to
In-Text Questions

Ⓐ Answers will vary; students should recognize that areas closer to plate boundaries are more likely to experience earthquakes.

Ⓑ Model buildings could be tested by subjecting them to shaking or other vibrations. Models without the earthquake-resistant features should also be tested for comparison.

EXPLORATION 3

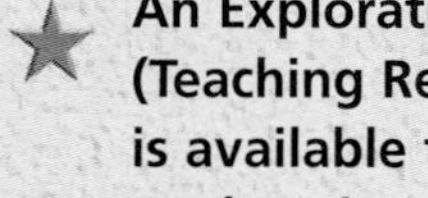

An Exploration Worksheet (Teaching Resources, page 12) is available to accompany Exploration 3.

Answers to
Part 1

1. Answers will vary. For instance, students may state that having flexible pipes is the most effective feature because it preserves the connections to outside sources of water and gas.
2. Although most of these features could be used in any type of building, some buildings might be too oddly shaped or too small to incorporate these features.
3. Features such as the steel beams would have to be incorporated into a building's initial design; it would be impractical to add such features later.
4. Answers will vary; students should recognize that areas closer to plate boundaries are more likely to experience earthquakes and thus are in greater need of such features.

Could Your Town Be Next?

In areas where earthquakes are common, buildings have been designed to withstand intense shock waves. But earthquakes also occur in areas where they are unexpected. Major earthquakes rocked New Madrid, Missouri, in 1811 and 1812. In 1886, Charleston, South Carolina, was flattened by an earthquake. In fact, some seismologists say that more than half of the United States is at risk. Do you suppose that an earthquake could happen in your state? Explain your reasoning. Ⓐ

Several cities and states are becoming concerned about the damage that would occur if an earthquake struck. This has led to building codes that require varying degrees of earthquake protection. How do engineers try to keep buildings from falling during earthquakes? In the next Exploration, you'll find out. You'll also get the chance to come up with some earthquake-resistant designs of your own!

Aftermath of the 1886 earthquake in Charleston, South Carolina

EXPLORATION 3

Earthquake-Resistant Designs

PART 1

The diagram on the next page shows some of the design features that engineers and architects use to make buildings that better withstand earthquakes. After you have examined all of the features, answer the following questions:

1. Which feature do you think is most effective in limiting damage caused by earthquakes?
2. Could all of these features be used for all types of buildings? Why or why not?
3. Could all of these features be added to an existing building to make it stronger? Why or why not?
4. Do you think there should be a building code in your area that requires any of these features to be included in new building designs? Explain your answer.

PART 2

With two or three of your classmates, think of other ways to make a building safer during an earthquake. Make sketches of a building that uses your earthquake-resistant features. Share your designs with the rest of the class. You might even construct a model of an earthquake-resistant building that uses your new features along with the ones shown in the diagram. How could you test such a model? Ⓑ

Mass Damper
The mass damper is a 6-ton weight built into the top of the building. Motion sensors detect swaying and send a message to a computer system. The computer system then directs hydraulic actuators (shown in red) to shift the weight of the mass damper in order to counteract the building's movement during an earthquake or high winds.
Active Tendon System
In this system, as with the mass damper, sensors detect the motion of the building during an earthquake. A computer directs the tendons to lengthen or shorten in order to push or pull the building into its correct alignment.
Base Isolators
Base isolators act as shock absorbers against the force of an earthquake. Each base isolator (approximately 60 cm tall and 60 cm wide) consists of layers of rubber and steel wrapped around a lead core. This arrangement of materials absorbs the energy of seismic waves that would otherwise travel up through the building.
Cross-braces
Cross-braces lend strength by counteracting the push-and-pull pressures that occur at the sides of a building during an earthquake. These cross-braces are made of steel, which is strong but flexible enough to stretch considerably before breaking.
Flexible Pipes
The swaying and rocking of a building during an earthquake can cause pipes to break. When gas lines are broken in this manner, devastating fires can result. Pipes with flexible joints are better able to twist and bend without breaking during an earthquake.

369

FOLLOW-UP

Reteaching

Have students look at the map of the world on page 358. Ask: Where would you suggest that someone avoid living if he or she did not want to experience an earthquake? Explain. *(Responses should identify areas near plate boundaries.)*

Assessment

Ask students to use clay and toothpicks to make a model of railroad tracks crossing a fault line. Have them show what might happen during an earthquake. *(The students should move the clay portions in opposite directions from each other, bending or breaking the tracks.)*

Extension

Have students research local earthquakes (or earthquakes of interest to them) by contacting their library, a science museum, a nearby university, or some other community agency. Have them collect information on the frequency of the earthquakes, their dates of occurrence, their magnitude, the location of their epicenters, and the effects they had on people and property.

Closure

As a class, assess the likelihood of an earthquake in your area. Then discuss possible safety precautions that people should take. Your class may wish to extend the discussion of earthquake safety precautions by watching a videotape or by inviting a guest speaker on this topic.

Volcanoes—Holes in the Earth

FOCUS

Getting Started

The video *Volcanoes of the Kenya Rift* highlights several volcanoes in eastern Africa, exploring not only their present-day characteristics but also how they formed in the past and what the area may look like in the future. The video is available from Coronet/MTI Film & Video, 108 Wilmot Road, Deerfield, IL 60015.

Main Ideas

1. Volcanoes generally occur along plate boundaries and are created when plates collide or move apart.
2. Volcanic eruptions occur in different ways and produce different results.

TEACHING STRATEGIES

A Survivor's Story—Mount St. Helens, May 18, 1980

Create an atmosphere in the classroom in which students can relate to the words of survivor David Crockett. For example, ask students to imagine what it would be like to be in the middle of a volcanic eruption. Elicit descriptive expressions such as hot, dark, suffocating, and so on. Then call on a volunteer to read the survivor's words.

INDEPENDENT PRACTICE You may wish to have students find other magazine or newspaper reports of the eruption. If students have access to a college or large public library, they may be able to find articles from newspapers in cities located near the eruption site. These will contain more details of the effect of the eruption on the nearby communities. Have students present their research through oral reports, bulletin-board displays, role-playing, a simulated newscast, or other means.

LESSON 3 Volcanoes—Holes in the Earth

A Survivor's Story—Mount St. Helens, May 18, 1980

At the moment of the eruption, David Crockett was there.

He shouted into his recorder, "The road exploded in front of me . . . it's hard to breathe . . . my eyes are full of ash."

Grabbing his equipment, he left the car and raced up the near ridge which served as a wall between him and the blast. Since it had been clear-cut of timber, he was not threatened with falling timber but a blizzard of ash turned day into night. "Right now I think I'm dead," he recorded, coughing and fighting for his breath. It was then he realized that the deadly blast had gone over him, leaving him uninjured. He dropped to his knees and thanked God for his miraculous escape.

(From *Mount St. Helens: A Sleeping Volcano Awakes,* Marian T. Place, 1981)

LESSON 3 ORGANIZER

Time Required
2 class periods

Process Skills
observing, analyzing, communicating

Theme Connection
Systems

New Terms
Lava—magma that emerges on the Earth's surface
Magma—molten rock that triggers earthquakes and creates volcanoes as it rises within the Earth

Materials (per student group)
Exploration 4: atlas; world map

Teaching Resources
Resource Worksheet, p. 14
Activity Worksheet, p. 15
Exploration Worksheet, p. 16
SourceBook, p. S112

Have students make a list of 10 observations from the photograph of Mount St. Helens. Encourage them to focus on details. Discuss these observations as a class, concentrating not only on the eruption itself but also on making inferences about the eruption's effects on the surrounding area.

Homework

You may wish to assign as homework the Resource Worksheet and the Activity Worksheet on pages 14 and 15, respectively, of the Unit 6 Teaching Resources booklet.

Primary Source

Description of change: excerpted from *Mount St. Helens: A Sleeping Volcano Awakes,* by Marian T. Place, 1981
Rationale: to illustrate the experience of an eruption survivor

A Sleeping Giant Awakens

Call on a volunteer to read this page aloud. Direct students' attention to the photographs. Use analogies to make the figures describing the effects of the eruption of Mount St. Helens more meaningful. For example, to help students visualize the size of the debris column, locate a familiar place about 15 km away from your school. To help students appreciate the size of the crater, identify nearby locations that are approximately 1 km and 2 km away.

Answers to *In-Text Questions*

A *Dormant* means inactive.

B Accept all reasonable responses. You may wish to explain to students that the scientific predictions about a possible eruption are based on a variety of evidence. Scientists monitor volcanoes such as Mount St. Helens first by studying regional geological activity and then by geochemical investigations and dating that reveal the volcano's history. Once the seismic data are known, scientists monitor the shape of the volcano for contractions and expansions by using tilt-meters and surveying techniques. Changes in horizontal and vertical distances are measured by laser-beam technology. Swelling of the ground and earthquakes usually precede, accompany, and follow volcanic eruptions. These changes can be recorded using seismic instruments.

A Sleeping Giant Awakens

For several weeks in the spring of 1980, tremors and escaping gases warned of an impending eruption of beautiful Mount St. Helens in western Washington State. But the volcano had been dormant for more than 120 years. **A** (What does the word *dormant* mean?) Few people who lived nearby thought it would really erupt.

However, many scientists were confident that Mount St. Helens would indeed erupt soon. As early as 1975, in fact, some geologists had predicted an eruption before the end of the century. How do you think they could make such a prediction? **B**

When Mount St. Helens did erupt, the explosion was tremendous. It released 500 times the energy of the atomic bomb dropped on Hiroshima in 1945. Much of the mountain was blown away. One cubic kilometer of debris—rock, ash, and steam—shot up in a column 15 km to 18 km high. The eruption left a crater 1 km wide, 2 km long, and 0.3 km deep. The forest in the immediate area of the volcano was completely destroyed, leaving a barren wasteland. Fifty-eight people were killed in the eruption, along with thousands of forest animals.

Mount St. Helens before the eruption

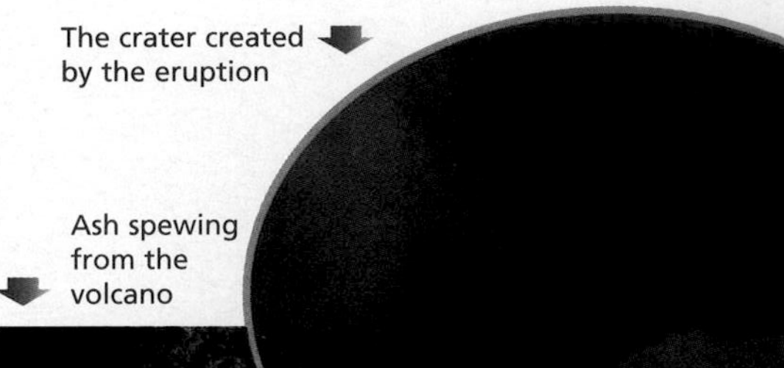

The crater created by the eruption

Ash spewing from the volcano

Forest area destroyed by the blast

372

Meeting Individual Needs

Second-Language Learners

Have students tell the story of a volcanic eruption using a comic-strip format. Students can create or gather a series of drawings or photographs to illustrate the sequence of events. Then encourage them to write a simple dialogue or commentary to accompany the pictures.

Cross-Disciplinary Focus

Social Studies

Involve students in a discussion of how people perceive risk in terms of places to live. Accept all reasonable answers to the following questions:

- Why did some of the residents ignore the predictions?
- How much time will it take before they will again feel safe living close to a volcano?
- Why might people choose to live in volcanic areas?

New Life on Mount St. Helens

Now, more than 15 years later, the land around the volcano is repairing itself. Grass and native plants have pushed through the layers of ash. Many animals, including the largest elk herd in Washington, are again making their homes on the mountainside. Scientists predict that within 50 years, the area will once more be a healthy—although young—forest.

A Volcanic Word Search

No doubt you have studied or read about volcanoes before. You may also have seen films of eruptions or visited a volcano. Make a list of all the words you can remember that are associated with volcanoes. As you read the following pages, add new words to your list. Then make a word-search puzzle in your ScienceLog like the one started here. The words can run in straight lines in any direction—up, down, across, and diagonally. Fill in all the unused boxes with other letters, and then give the puzzle to a friend to solve.

V						
	O					
		L		C		
			C	R		
				A		
				T	N	
				E		O
				R		

Life returns to Mount St. Helens

373

ENVIRONMENTAL FOCUS

The video *Return to Mt. St. Helens* chronicles the changes that have occurred in that area since the time of the eruption. It discusses the damage caused by the eruption, the re-emergence of biological communities, and the ways that scientists have used the area as a living laboratory. It is available from Coronet/MTI Film & Video, 108 Wilmot Road, Deerfield, IL 60015.

Did You Know. . .

Pumice, a volcanic rock, is sold commercially for use in smoothing, scouring, and polishing various household items.

New Life on Mount St. Helens

Encourage a class discussion of the following questions: Do you think that humans should try to speed up the recovery of the local ecosystems by importing plants and animals? Why or why not? *(Students should consider both positive and negative aspects of human intervention. Although the land may be available for use sooner, importing non-native plants and animals might create an unstable biological community.)*

You may wish to refer students to Unit 1 for another look at the recovery of Mount St. Helens's surroundings in the years following the eruption.

A Volcanic Word Search

Suggest to students that they write all the new terms about volcanoes in their ScienceLog. The descriptions of famous volcanoes on the following three pages provide some of the vocabulary associated with volcanoes. These pages also provide background information on the different types of eruptions that give rise to cones. Some insight is provided into the death and destruction that can result from a volcanic eruption. Any terms describing the effect of volcanic eruptions on world climate could also be included. Then, at the end of the lesson, students can create their own word-search puzzles and exchange them with their classmates.

★ **A Resource Worksheet and an Activity Worksheet (Teaching Resources, pages 14 and 15, respectively) are available to accompany the material on this page.**

Homework

You may wish to assign A Volcanic Word Search as homework.

Theme Connection

Systems

Focus question: How can a large volcanic eruption change global weather patterns in the years that follow the eruption? *(Sample answer: Ash and dust blown into the atmosphere can reflect sunlight and, as a result, cool the planet.)*

Other Famous Volcanoes

Encourage students to research the names of famous volcanoes throughout the world. Invite them to compete individually or in a team to compose the longest list of names.

Students should use the material on the next few pages to compile a list of "Facts About Volcanoes" (e.g., volcanoes can be of various shapes). Memorization of dates, names, and characteristics of individual volcanoes is not encouraged.

Volcanoes are commonly classified according to their eruption status, in addition to descriptions of their shape and composition. The following terms are used:

- *Active*—a volcano that has erupted recently, is currently erupting, or is expected to erupt in the near future. Mauna Loa and Mount St. Helens are active volcanoes.
- *Dormant*—a volcano that has had no recent eruption but that could erupt in the future. Mount Rainier, in Washington, is a dormant volcano.
- *Extinct*—a volcano that has not erupted in recorded history and is not expected to erupt in the future. Mount Mazama, under Crater Lake in Oregon, is extinct.

Vesuvius, Italy

You may wish to share the following information with students. Composite volcanoes such as Vesuvius, located in southern Italy in the city of Pompeii, form large mountains that rise 1800 to 2400 m above their bases. They are also called *stratovolcanoes.*

The eruption of Vesuvius was recorded by an eyewitness, the Roman writer Pliny, who was only 17 years old at the time of the eruption. His account was the first of many writings on the subject. There are numerous magazine articles available with color photographs of the remains of the Roman cities of Pompeii and Herculaneum, which were covered by ash and mud in this historic eruption. You may want to locate some of these articles for classroom use or have students locate them as a research project.

Other Famous Volcanoes

Vesuvius, Italy

The first well-documented eruption of a volcano was that of Vesuvius in A.D. 79. When the long-dormant volcano erupted, a cloud of hot ash rolled down the mountain. The nearby cities of Pompeii and Herculaneum were buried under almost 7.5 m of ash and volcanic rock. Sixteen thousand people were killed.

Vesuvius, like Mount St. Helens, is a *composite volcano.* Such a volcano is formed when alternate layers of lava and rock fragments build up around the volcano's vent and form a cone-shaped tower. Vesuvius is still a highly explosive volcano.

Artist's impression of the eruption of Vesuvius

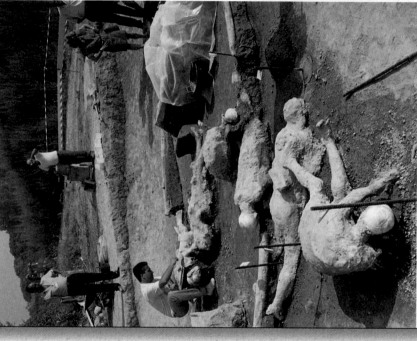

Tourists look at ancient casts of bodies that were buried under ash and volcanic rock from the eruption of Vesuvius in A.D. 79.

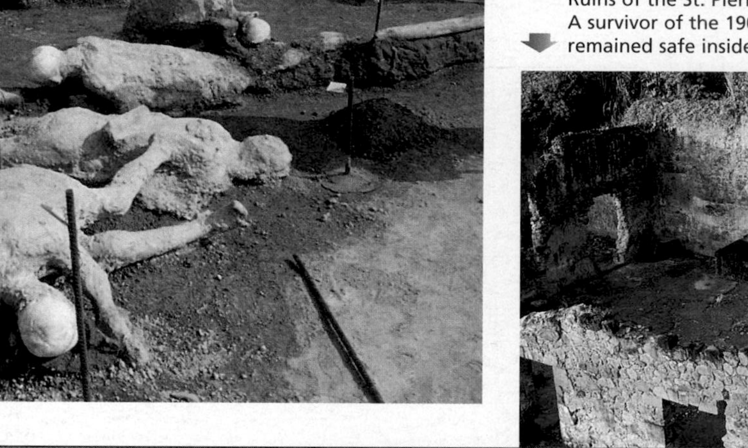

Mount Pelée, Martinique

Just before 8:00 A.M. on May 8, 1902, this volcano on the island of Martinique in the French West Indies exploded. A huge cloud of hot gases, believed to have had a temperature of at least 1000°C, rose and hung in the air for a moment and then roared down the slope to engulf the town of St. Pierre. Within minutes, the city was in ruins and all but two of the inhabitants were dead.

Ruins of the St. Pierre, Martinique, town jail. A survivor of the 1902 Mount Pelée eruption remained safe inside one of the cells.

374

Mount Pelée, Martinique

Share the following information with students: Mount Pelée is an example of a lava dome volcano. Thick, pasty lava forms domes in the shape of steep-sided, craggy knobs or spines over a volcanic vent. Other lava domes can be short, steep-sided lava flows called *coulees.*

Multicultural Extension

Martinique

Have interested students research the cultural traditions of Martinique in terms of its history and the ways that historical patterns affect people's lives today. Students should present their findings to the class in the form of a poster or an oral report.

Paricutín, Mexico

On February 20, 1943, two farmers in western Mexico were preparing to plant corn when they noticed white fumes rising from a hole in the field. As night fell, rocks and cinders began to spew from the hole. By morning, these cinders had formed a cone. Within 6 days, the cone had grown to more than 150 m tall! After a few months a crack opened in the cone, spilling lava that buried a nearby town. This volcano continued to be active until 1952, when the cone reached a height of 410 m. Paricutín is a type of volcano known as a *cinder cone,* which is usually smaller than other types and is active for only a few years.

Eruption of Paricutín

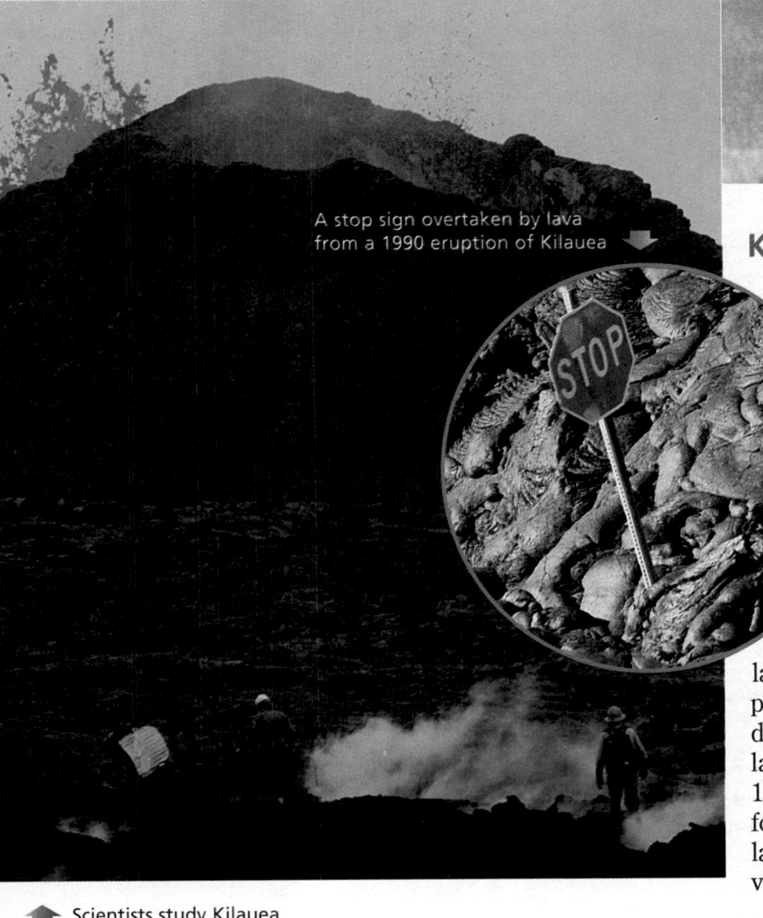

A stop sign overtaken by lava from a 1990 eruption of Kilauea

Scientists study Kilauea

Kilauea, Hawaii

With the Hawaiian Volcano Observatory located on its western rim, Kilauea is perhaps the most studied volcano in the world. The low, broad mountain was formed by thousands of layers of lava from past eruptions. Kilauea, a *shield volcano,* erupts often but usually gently because the lava is runny and does not plug the vent. However, destruction can result when lava overflows the rim. In 1955, lava flowed from Kilauea for 88 days, overrunning a village and more than 3.5 km^2 of valuable cropland.

CROSS-DISCIPLINARY FOCUS

Language Arts

Have students write headlines and brief newspaper articles about one of the volcanoes described. Encourage extra research and illustrations for the news articles.

Homework

Have interested students make a scientific illustration or model of a volcano. Their volcano should have interesting aspects labeled and explained (e.g., ash clouds, lava).

Paricutín, Mexico

Call on a volunteer to read the paragraph. Ask: Why do you think rocks and cinders spewed from the hole? *(Students may suggest that they were propelled by escaping gases.)* Inform students that lava pours out of the top of a cinder cone after excess gas in the magma below the Earth's surface has escaped.

Kilauea, Hawaii

Students may be interested in the following information: The Hawaiian Islands are a cluster of shield volcanoes and are some of the largest volcanoes in the world. Mauna Loa, Hawaii, is 8500 m from the ocean bottom to its peak. Short and broad, shield volcanoes are built almost entirely of lava flows. From a main vent, lava pours out on top of previous flows. The typical result is a broad, gently sloping cone about 4500 to 6000 m high and 5 to 6 km wide.

Surtsey, Iceland

You may wish to inform students that new crust is continually formed at the mid-Atlantic ridge. Volcanoes occur in and around Iceland because Iceland is located on the mid-Atlantic ridge. The formation of new crust by volcanic activity in this area works to push the adjacent plates in opposite directions.

Mount Pinatubo, Philippines

Call on a volunteer to read the material about the recent volcanic eruption in the Philippines. Involve students in a discussion of the details of the eruption. Then have them compare and contrast this eruption with the one that occurred at Mount St. Helens.

Answers to *In-Text Questions*

Ⓐ Sample answer: Scientists can learn about the geological and biological changes that occur after a volcano stops erupting.

Ⓑ The scientists were probably recording temperature changes, seismic activity, and other geological aspects of the Earth's crust.

Surtsey, Iceland

Surtsey, an island off the coast of Iceland, was born when an underwater volcano erupted in November 1963. Spectacular explosions and huge clouds of steam were produced when sea water mixed with hot volcanic material. During the next three and a half years, erupting volcanic material formed the island. After the final eruption of lava in 1967, the island covered more than 2.6 km^2. Surtsey is now the site of a long-term biological research program. What do you think scientists can learn by studying Surtsey? Ⓐ

Surtsey, 1964

A satellite image showing the worldwide spread of particles from Mount Pinatubo

A village covered by ash from Mount Pinatubo

Mount Pinatubo, Philippines

After being dormant for 600 years, Mount Pinatubo erupted in June 1991. Ash and gas blasted into the atmosphere and spread around the world in days. The floating gas caused spectacular sunsets worldwide but also blocked sunlight. The reduction in sunlight may have caused lower global temperatures for years after the eruption.

Fields and villages were buried under the heavy ash. Heavy rains turned ash on Pinatubo's slopes into a mudslide that swept away roads, bridges, and villages. Almost 900 people died, and 42,000 homes were destroyed.

But the loss could have been much worse. Scientists had been monitoring the volcano with computers and were able to warn citizens of an impending eruption. Almost 200,000 people were evacuated from dangerous areas. What do you think the scientists were monitoring with their computers? Ⓑ

Are Volcanoes All Bad?

Volcanoes are obviously destructive, but we wouldn't be here without them! Volcanic activity was responsible for the creation of the Earth's planetary crust. And gases from the Earth's first volcanoes formed the planet's early atmosphere.

Do some research to find out some of the other benefits provided by volcanoes. Share your findings with your classmates. Are you surprised by what you found? Do the benefits provided by volcanoes have any impact on your life? (C)

EXPLORATION 4

Where Are the Volcanoes?

Where are the active volcanoes in the United States? Does it surprise you that Mount St. Helens lies on the west coast? Explain. (D)

Why do volcanoes occur only in certain places? Scientists have an answer, and this Exploration will help you determine what that answer is.

On a map of the world, locate as many famous volcanoes as you can. Add at least five names to the list given at right. Then sketch a map of the world in your ScienceLog, and mark the locations. (E)

- Fujiyama (Japan)
- Mayon (Philippines)
- Hekla (Iceland)
- Etna (Italy)
- Katmai (Alaska)

The more volcanoes—active, dormant, or extinct—you can locate, the more easily you can determine the scientists' answer.

When you have completed your map, describe, in general terms, where volcanoes occur. Compare your map to the map on page 358 showing the Earth's plates. Describe any similarities between the maps. Why do you think volcanoes occur only in certain places? Can you determine the part of the world that scientists call the "ring of fire"? (F)

EXPLORATION 4

Students may work individually or in small groups to complete this Exploration. Have world maps, almanacs, and atlases available. When the activity has been completed, reassemble the class to discuss their findings and the final two questions of the Exploration.

An Exploration Worksheet (Teaching Resources, page 16) is available to accompany Exploration 4.

Answers to *In-Text Questions*

(C) Answers will vary. Students may identify benefits such as increased soil fertility due to volcanic ash deposits, which might affect their lives by improving the soil in nearby areas.

(D) Active volcanoes occur in Hawaii and in the continental United States along the West Coast. After their studies of earthquakes and plate boundaries, students should not be surprised by the fact that Mount St. Helens is located on the West Coast.

(E) Other famous volcanoes:
- Aconcagua (located in the Andes in western Argentina)
- Colima (southwestern Mexico)
- Haleakala (island of Maui)
- Lassen Peak (northern California)
- Mauna Kea (island of Hawaii)
- Popocatepetl (central Mexico)
- Stromboli (Italy)

(F) Students should conclude that volcanoes and plate boundaries are associated because the separating of plates causes volcanoes. Volcanoes and earthquakes tend to occur in the same locations. The "ring of fire" is a broad circle of volcanic activity surrounding the Pacific plate.

PORTFOLIO

Encourage students to place the map they create in Exploration 4 in their Portfolio. As they gather and organize information about volcanoes, earthquakes, and shifting plates, they can add to or alter their map.

The Causes of Volcanoes—Those Plates Again!

INDEPENDENT PRACTICE As a synthesis exercise, ask students to rewrite this material in their own words as if they were explaining the causes of volcanoes to a friend or family member.

Answers to *In-Text Questions*

Ⓐ The gases cannot escape gradually. Thus, pressure builds up until the upper layers of the crust violently burst open.

Ⓑ Surtsey was formed in this way.

Ⓒ Accept all reasonable responses. Students should explain the differences between active, dormant, and extinct volcanoes. They should also propose a plausible sequence of events leading up to the eruption of a volcano.

FOLLOW-UP

Reteaching

Ask students to explain why earthquakes and volcanoes tend to occur in predictable locations around the world. *(Movement at plate boundaries causes both earthquakes and volcanoes.)*

Assessment

Have students decide whether these statements are accurate. Students should correct any of the incorrect statements.

a. Volcanoes can occur only in certain places. *(True)*
b. Volcanoes occur when plates move apart or come together. *(True)*
c. Once volcanoes erupt, they become dormant. *(False. Volcanoes can erupt several times.)*
d. Scientists cannot predict when volcanic eruptions will occur. *(False. The presence of earthquakes, ash, and gas, as well as the use of detection equipment, all help predict volcanic eruptions.)*
e. Volcanoes can be either quiet or explosive. *(True)*
f. Volcanic eruptions occur only on land. *(False. Volcanic eruptions also occur in the ocean.)*
g. The violence of an eruption is partly determined by the gases in the magma. *(True)*
h. Volcanic eruptions can affect the world's climate. *(True)*

Extension

Have students research various volcanic rocks and their past and present uses. Examples include pumice, basalt, and obsidian.

Closure

Present students with the following scenario: You are a journalist assigned to the site of a volcanic eruption. After days of reporting on death and destruction, you have been told to write an article with this opening statement: While volcanoes often destroy both lives and property, they also benefit the Earth. Complete the article.

The Cause of Volcanoes—Those Plates Again!

When the Earth's crust shifts restlessly or a volcano erupts somewhere, the explanation usually lies in the movement of the Earth's plates.

Most volcanoes on land are formed when two plates collide. The collision usually drives one of the plates under the other. Upon reaching the hotter temperatures of the Earth's interior, the underlying plate begins to melt. This melted rock, which is known as **magma,** is lighter than the rock around it. Thus, it slowly rises, triggering earthquakes and creating volcanoes. The magma that emerges on the surface is called **lava**. The severity of the eruption is determined by the gases in the magma. If the gases can escape gradually, as they do in shield volcanoes, there is usually no explosion. Why do you think composite volcanoes (such as Mount St. Helens) usually erupt with violent explosions? Ⓐ

Magma that emerges on the Earth's surface is called lava.

Some volcanoes are formed when two plates move apart. This usually occurs beneath the surface of the ocean. When the plates separate, magma moves up from below to fill the gap. The lava that pours out builds up the ocean floor, sometimes creating underwater mountain ranges. What land formation have you studied that could have been formed this way? Ⓑ

Still other volcanoes, called hot-spot volcanoes, do not occur at plate boundaries. Instead, they are found at points in the Earth's crust where large amounts of magma are being formed. The magma rises and pushes *through* the plate, forming a volcano. The volcanoes on the Hawaiian Islands are hot-spot volcanoes.

A Field Trip

Imagine that you are taking a field trip to a volcano with your class. You stand on the slopes, looking up toward the peak and around at the surrounding countryside. How might you tell if the volcano is active, dormant, or extinct?

Suddenly you hear rumblings, and the ground begins to move. In your ScienceLog, record the sequence of events inside the volcano as you think they would occur. Ⓒ

CHALLENGE YOUR THINKING

1. Sharpen Your Logic Skills

Use the basic facts presented to deduce the answers to the questions that follow. Refer back to the map on page 358 and consult an atlas or world map as necessary.

Facts:	Questions:
• The Atlantic Ocean is getting wider. New crust is being formed underneath the ocean.	• What is happening to the plates at the mid-Atlantic boundary? What evidence of crustal movement would you expect to find in Iceland?
• Plates collide.	• Find three examples of colliding plate boundaries. What kinds of activities will occur here?
• When plates collide, the continental crust thickens and buckles as the plates are compressed.	• Locate the Himalayas and the Rocky Mountains. Could they have been formed this way?
• Plates slide by each other.	• Find two places where plates are sliding by each other. What do you predict will happen there?

2. In My State?

Seismologists say that more than half of the United States may be at risk for an earthquake. Look at this map of the United States and the map on page 358, and answer the following questions:

- Which states do you think are at risk? Which are safe? Explain your reasoning.
- Do you think your state is among those at risk? Why or why not?
- Do you think there are parts of the United States that may never experience an earthquake? Explain your answer.

CROSS-DISCIPLINARY FOCUS

Language Arts

A number of popular works of fiction have been written on the subject of traveling to the interior of the Earth. Perhaps the most famous example is Jules Verne's *Journey to the Center of the Earth,* published in 1864. You may wish to have students read excerpts from this book as a basis for a class discussion. Then ask students to write their own stories about a journey to the center of the Earth.

Homework

The Math Practice Worksheet on page 18 of the Unit 6 Teaching Resources booklet makes an excellent homework activity to accompany the material on this page.

★ **You may wish to provide students with the Chapter 16 Review Worksheet on page 19 of the Unit 6 Teaching Resources booklet.**

Answers to *Challenge Your Thinking*

1. • The plates are slowly separating. You would expect to find volcanoes in Iceland.
   - Possible student answers include the following: colliding plates in Alaska, the Caribbean, the Mediterranean Sea, the Philippines, and along the west coast of South America. Earthquakes, mountain-building, and volcanoes will probably occur here.
   - Yes. The Himalayas began to form when the plate carrying India ran into the Eurasian plate. The Rocky Mountains were formed when colliding crustal plates caused oceanic crust to slide under continental crust, causing a folding of the continental crust.
   - Sliding boundaries occur along the west coast of North America and across the Indian Ocean between the African and Eurasian plates. Earthquakes can be expected in these areas.
2. Student answers may vary.
   - Most students will conclude from the map that the western states are more at risk than central and eastern states because the western states are closer to the continental fault.
   - Students should assess the likelihood of earthquakes in their state and support their argument with a reasonable explanation.
   - Many students will conclude from looking at the map that central and eastern states may never experience earthquakes. Others may be aware that some central and eastern states have experienced very powerful earthquakes. Charleston, South Carolina, for example, experienced a large earthquake in 1886.

Answers to *Challenge Your Thinking,* *continued*

3. The volume of land under strain should be calculated by multiplying length (1000 km) by depth (100 km) by width (2 × 50 km).
   1000 km × 100 km × 100 km = 10^7 km^3
   Then convert the answer into cubic meters.
   10^7 km^3 × 10^9 m^3/km^3 = 10^{16} m^3
   Energy released:
   100 J/m^3 × 10^{16} m^3 = 10^{18} J, or 10^{15} kJ
4. The student's story or illustration should demonstrate an understanding of how his or her chosen type of volcano is formed. For example, a drawing of Paricutín should show the development of a cinder cone.
5. Scripts will vary but should cover a range of facts about the eruption.

The following are sample revised answers:

1. The outer crust of the Earth's surface is broken up into pieces that fit together like a puzzle. These pieces are called tectonic plates.
2. This mountain might have been formed by the buckling of the Earth's crust that takes place at the boundaries of colliding tectonic plates. It also might have been formed by the action of volcanoes.
3. Both volcanic eruptions and earthquakes are caused by shifting tectonic plates.

Multicultural Extension

Geoffrey Jean Baptiste Lislet
Lislet (1755–1836) was one of the first widely known European geologists of African heritage. Versed in botany, zoology, philosophy, and astronomy, Lislet served as correspondent to the French Academy of Science for over a decade and was skilled at mapping geological resources. He also established one of the first scientific societies. You may wish to have interested students learn more about Lislet's life and the problems he faced when founding a new scientific society.

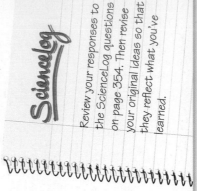

3. A Lot of Energy

About 100 J of energy is released from each cubic meter of rock at the time of an earthquake—the equivalent of one firecracker per cubic meter.

This may not seem like very much energy, but consider the following situation: Suppose the fault is 1000 km long, extends 100 km downward, and disturbs streets as far as 50 km on either side of the fault. How many cubic meters of rock would be under strain? How much energy would be released when the rocks break the friction lock?

4. A Hot Task

Write a short story or draw a colorful picture that illustrates the formation of one of the following:

- the island of Surtsey
- Paricutín in Mexico
- a hot-spot volcano

5. This News Just In

You are a member of a television news team covering the Mount St. Helens eruption, or you are someone being interviewed by the news team. Pick a role: play an anchor, news reporter, science reporter, local citizen, geologist, safety expert, or engineer. Write a script for a news broadcast, and then act it out with your classmates.

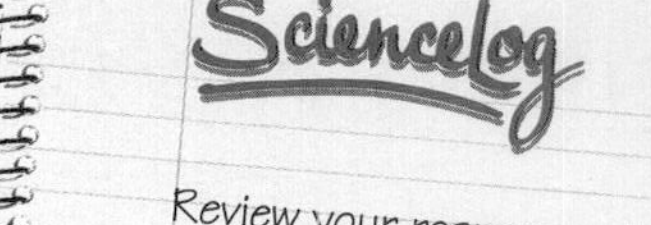

Review your responses to the ScienceLog questions on page 354. Then revise your original ideas so that they reflect what you've learned.

Integrating the Sciences

Earth and Life Sciences
Explain to students that water, a material necessary for all life, can be distributed unevenly due to the presence of mountain ranges. Have students investigate and describe in an oral or written report the effect of mountains on the average precipitation of nearby regions and on the types of life that these regions can therefore support. *(Deserts often exist downwind of mountain ranges as a result of the rain-shadow effect, which describes how air loses moisture as it passes over mountains.)*

CHAPTER 17

The Role of Rocks

1 How would you classify these rocks?

2 How do you think each of these rocks was formed?

3 Do you think these rocks could be changed into different rocks? Why or why not?

ScienceLog

Think about these questions for a moment, and answer them in your ScienceLog. When you've finished this chapter, you'll have the opportunity to revise your answers based on what you've learned.

The Role of Rocks

Connecting to Other Chapters

Chapter 16 *examines plate tectonics, with a particular focus on volcanoes, mountains, and earthquakes.*

Chapter 17 *explores the essential characteristics of igneous, sedimentary, and metamorphic rocks.*

Chapter 18 *introduces the geologic time scale as well as the types and formation of fossils.*

Prior Knowledge and Misconceptions

Your students' responses to the ScienceLog questions on this page will reveal the kind of information—and misinformation—they bring to this chapter. Use what you find out about your students' knowledge to choose which chapter concepts and activities to emphasize in your teaching.

In addition to having students answer the questions on this page, you may wish to assign a "free write" to assess prior knowledge. To do this, instruct students to write for 3 to 5 minutes on the subject of rocks. Tell them to keep their pens moving at all times, writing in a stream-of-consciousness fashion. Emphasize that there are no right or wrong answers in this exercise. It may be best to ask students not to put their names on their papers. Collect the papers, but do not grade them. Instead, read them to find out what students know about rocks, what misconceptions they may have, and what about rocks is interesting to them.

Features and Formations

FOCUS

Getting Started

Have students examine the photographs and write down as many observations as they can about the rocks shown. Discuss their observations as a class. *(Students may note the color, texture, size, shape, and stacking arrangement of the rocks, for instance.)* Explain to students that this lesson will allow them to learn more about the way rocks are formed and classified.

Main Ideas

1. Rocks are extremely diverse in most physical characteristics.
2. Rocks exhibit a variety of textural and compositional characteristics by which they can be classified.
3. Earth scientists classify rocks into three groups (igneous, sedimentary, and metamorphic) based on the way the rocks were formed.

TEACHING STRATEGIES

Diversity in Rocks

Ask students to share with the class any descriptions of unusual rock formations that they have seen, including the location, size, color, and other characteristics of the formations.

Features and Formations

Diversity in Rocks

In Dartmoor, England, a bizarre granite formation called a tor rises awkwardly from an open field. Not far away, in Northern Ireland, is another unusual rock formation. Giant's Causeway, as it is known, formed when lava cooled into huge hexagonal columns of a type of rock called basalt. Obviously, rock formations come in different materials, sizes, and shapes. But you haven't seen anything yet.

Tor in eastern Dartmoor, England

A section of Giant's Causeway, Northern Ireland

Sunrise at the Pink Cliffs in Bryce Canyon, Utah

Consider the odd figures shaped from sandstone and other types of rock that line the beautiful Pink Cliffs of Bryce Canyon National Park in Utah. And look at the strange formations of limestone deep within Carlsbad Caverns in New Mexico. The hanging formations, called stalactites, have been growing in length for thousands of years. Stalagmites are the formations that grow up from the floor of the cave.

These examples are just a few of the interesting and unusual rock formations that can be found on Earth. Have you ever seen any interesting rock formations firsthand? Share your experiences with the class. Ⓐ

LESSON 1 ORGANIZER

Time Required
2 to 3 class periods

Process Skills
observing, inferring, classifying

Theme Connection
Structures

New Terms
Igneous rock—a rock formed by the cooling and solidification of magma or lava
Metamorphic rock—a rock formed when another rock is changed by extreme heat or pressure
Sedimentary rock—a rock formed by precipitation or from the weathered pieces of other rocks

Materials (per student group)
Exploration 1, Part 1: magnifying glass; 1 pair of dissimilar rocks per student; **Part 2:** rocks from Part 1; about 30 index cards
Exploration 2: magnifying glass; rock samples

Teaching Resources
Exploration Worksheets, pp. 24 and 29
Theme Worksheet, p. 27
Transparency 61
SourceBook, p. S115

Stalactites and stalagmites in Carlsbad Caverns, New Mexico

What do you think these fantastic formations have in common with the individual rocks shown below? Have you ever seen rocks like these before? B

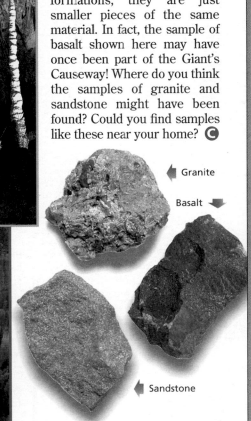

Although these rocks are tiny compared with the grand formations, they are just smaller pieces of the same material. In fact, the sample of basalt shown here may have once been part of the Giant's Causeway! Where do you think the samples of granite and sandstone might have been found? Could you find samples like these near your home? C

Granite

Basalt

Sandstone

If you know what any rock, large or small, is made of and how it was formed, you can determine what type of rock it is. In Exploration 1, you will begin your study of rock classification by examining rock samples.

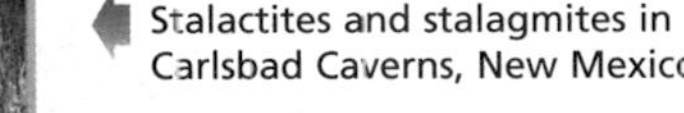

Integrating the Sciences

Earth and Physical Sciences

Have interested students find out how solutions and the process of evaporation relate to the formation of stalactites and stalagmites. *(The sediments that make up these formations were deposited when the water in which they were originally dissolved evaporated and left them behind.)*

Meeting Individual Needs

Second-Language Learners

Use language-independent visual aids to emphasize the main ideas of this lesson. For instance, you may wish to bring a large rock to class. Using a heavy towel to cover the rock, carefully break off pieces with a hammer to demonstrate the process by which a rock can be "born." (The towel keeps small pieces of flying rock from posing a danger to you or your students. You and any students nearby should wear safety goggles during this demonstration.)

Answers to *In-Text Questions, pages 382–383*

A Encourage students to describe in detail any interesting rock formations they have seen.

B The individual rocks were probably once part of larger formations like those shown on pages 382–383.

C The granite may have broken off a tor like that pictured on page 382, and the sandstone may have originated in the Pink Cliffs. In some regions, these rocks may be available locally.

Homework

In preparation for Exploration 1 on page 384, have students bring in two small rocks from near their home. They should make a list of five ways that the rocks are different and three ways that the rocks are similar.

Exploration 1

Invite students to bring in rocks that differ significantly in appearance, or provide such samples yourself. Divide the class into small groups and distribute two dissimilar rocks to each student for examination. Answers throughout Exploration 1 will vary depending on the rock samples available.

PART 1

Collect observations from each group, and write them on the chalkboard. Draw students' attention to the number of materials that make up the rock and the way the materials fit together, in addition to the size, shape, orientation, and color of these materials.

Question 5 can be the subject of a class discussion. Properties may include color, composition, texture, luster, and shape.

PART 2

The rock classification scheme should reflect those properties that the students found most helpful in identifying their rocks. No one scheme is necessarily better than another. Ask: If you were devising a rock classification scheme that could be used anywhere in the world, what might you need to do differently? *(A much wider variety of rocks would have to be included in the analysis.)*

Provide index cards for each group so that they can make appropriate labels for display purposes.

GUIDED PRACTICE Spend some time reviewing classification systems. Use a collection of common objects such as shoes, cars in the parking lot, or colored pencils to demonstrate how to devise a hierarchical classification scheme.

An Exploration Worksheet (Teaching Resources, page 24) is available to accompany Exploration 1.

Homework

You may wish to assign the Theme Worksheet on page 27 of the Unit 6 Teaching Resources booklet as homework.

Exploration 1

Getting Acquainted With Rocks

Imagine picking up a rock just outside your home. What does it look like? Where did it come from? How did it form? Studying the rock can give you clues about its age, its formation, and the changes in the Earth it has experienced.

Examining rocks can give you information about many of the things you wish to know about the Earth.

PART 1

Examining Rocks

You Will Need

- a magnifying glass
- a collection of different rocks

What to Do

1. Divide into groups of four. Each group should have at least eight rocks to study.
2. Select two rocks from those given to your group. Study each rock and record your observations on a sheet of paper. Your descriptions should be clear enough so that your classmates can identify the two rocks.
3. When everyone in your group has finished writing the descriptions, mix up the rocks on a table. Try to pick out the two that you wrote about. Can you find them?
4. After finding your own rocks, put all of your group's rocks back on the table, and mix them up again. This time, exchange descriptions with another member of your group. Find the rocks described on your classmate's paper. Were you successful? If not, perhaps the descriptions need to be improved. If you were successful, could you suggest anything to add to or delete from the descriptions? Repeat this exchange with each member of your group.
5. As a group, compare the descriptions written by each member. Which properties of the rocks did each describe? Which observations were the most helpful? the least helpful?
6. Using all the observation sheets of your group, make up a new sheet listing those properties of the rocks that were most helpful in identifying them. Below is an example.

Property (characteristic)	Observations related to property
Color	Ranged from plain gray to black, white, and red.
Composition	The gray rock was smooth. The mixed rock was made up of small rocks stuck together. The red rock had shiny crystals and was rough.

PART 2

Classifying Rocks

You Will Need

- rocks from Part 1
- index cards

What to Do

1. Join with another group so that you now have 16 rocks to work with. Choose one property that half of the rocks share, and divide the rocks into two groups according to this property. Using an index card, make a label naming the property for each group.
2. Determine how each of the two large groups can be divided into smaller and smaller groups that have more similar characteristics. Make labels each time you subdivide the groups.
3. Continue until you have separated all the rocks into either groups having only one rock or groups whose properties are all the same.
4. Describe your classification system in your ScienceLog. Then display your rocks and labels in the classroom.

Theme Connection

Structures

Focus question: How does the structure of a rock help determine its usefulness to humans? *(A rock's structure determines many of its physical characteristics, such as hardness and breakage pattern. A rock's usefulness to humans is determined by these characteristics. For instance, a diamond has a structure that makes it extremely hard. Thus, diamonds are widely used by industries to cut, grind, or shape other materials.)*

The History of Rock

Each rock you examined has a history to tell. The crucial points of the history include how the rock was formed, where it was formed, and how it ended up in your hands.

The observations you made about the rocks you gathered are clues to discovering their history. In this section, you will learn to interpret the data and reconstruct the events that led to the rocks' formation. Consider the descriptions below. Each reveals the history of a different kind of rock.

Deep within the Earth, molten rock, called magma, cooled and hardened. Crystals formed that were easily visible; an **igneous** rock was born.

Particles of sand and soil were carried by wind and water and were eventually deposited on an ocean bottom. Increasing pressure from the accumulated weight of the particles, combined with chemicals from the sea, cemented the particles together to form a **sedimentary** rock.

Heat and extreme pressure reorganized the crystals of an existing rock to change it into a new type of rock, a **metamorphic** rock.

Rock Groups

You probably would not have much trouble classifying rock bands into groups according to the type of music they play. But can you classify rocks into groups? There are so many different kinds of rocks that scientists needed ways to organize them. Read the three descriptions above again. On what basis have geologists classified rocks? Perhaps a clue will help: *ignis* is the Latin word for "fire." (A)

385

Homework

Have students write a short story about a brother and sister, Rafael and Rhonda. They are geologists who spent their childhood in a mountainous area. The stories should tell about some of the things that these two did as children and about why they decided to become geologists.

The History of Rock

Call on a volunteer to read the text at the top of this page. Have students describe in their own words the meanings of the terms *igneous, sedimentary,* and *metamorphic.*

Rock Groups

Igneous rocks form when molten rock cools. Sedimentary rocks usually form when older igneous, sedimentary, and metamorphic rocks break apart due to weathering and erosion. Under appropriate conditions of heat and pressure, igneous and sedimentary rocks can be changed into the third rock group, metamorphic rocks.

Some students will be familiar with the three groups of rocks. The next few lessons deal with the rock types and the differences among them.

INDEPENDENT PRACTICE Have students create posters illustrating the three possible histories of rocks. Posters should include the type of rock, how it formed, and where it is likely to be found.

Multicultural Extension

World Ceramics

Ceramics have been made by people all over the world for thousands of years. The materials used in ceramics come from the Earth, so the characteristics of a ceramic tradition are often closely linked to the environment as well as to the local culture. Have students investigate the clay deposits and ceramic traditions of either an individual culture or a particular geographic region.

Answer to *In-Text Question*

(A) Geologists have classified rocks on the basis of the rocks' origins.

EXPLORATION 2

The basic characteristics used for rock identification are mineral composition, texture, grain or crystal size, shape, size, and orientation of material in the rock. The identification key works well for igneous and sedimentary rocks, but not as well for metamorphic rocks. You may need to screen a variety of rocks ahead of time for characteristics described in the key. Each group should have three to five rocks, with each of the three rock types represented at least once.

Students may need extra assistance for this Exploration. To clarify different parts of the key, display rocks that illustrate the various characteristics. Students can then compare their samples with the rocks on display.

You may wish to assign a number to each of the rocks used in your classroom. Keep a record of each rock's type and the group that examined each rock.

★ **An Exploration Worksheet (Teaching Resources, page 29) and Transparency 61 are available to accompany Exploration 2.**

Homework

Ask students to devise a "rock tour" of their area. They should observe and describe buildings and other features of their neighborhood environments that are made from rock. Whenever possible, students should also identify the type of rock observed and give explanations for their rock identifications.

EXPLORATION 2

Rock Group Identification

It can be difficult to decide whether a rock is sedimentary, igneous, or metamorphic, but the identification key shown will help you to classify the rocks.

You Will Need

- a magnifying glass
- rock samples

What to Do

In your ScienceLog, make a table like the one shown. As you study each rock, use the table to record your observations.

Sample number	Observations	Type of rock
1	flat sheets	?

Choose one rock to begin your study. Read the two sentences under number 1 in the identification key, and pick the sentence that best describes your rock. The direction at the end of the sentence refers you to the next set of sentences. Read these sentences, and again select the one that best describes your rock. Continue this process until you reach either sedimentary (S), metamorphic (M), or igneous (I). Repeat this procedure for eight different rocks.

You may find that some rocks do not seem to fit readily into any category, and they may not look like the examples shown. Expert help may be required. Remember, identifying rock types is not easy!

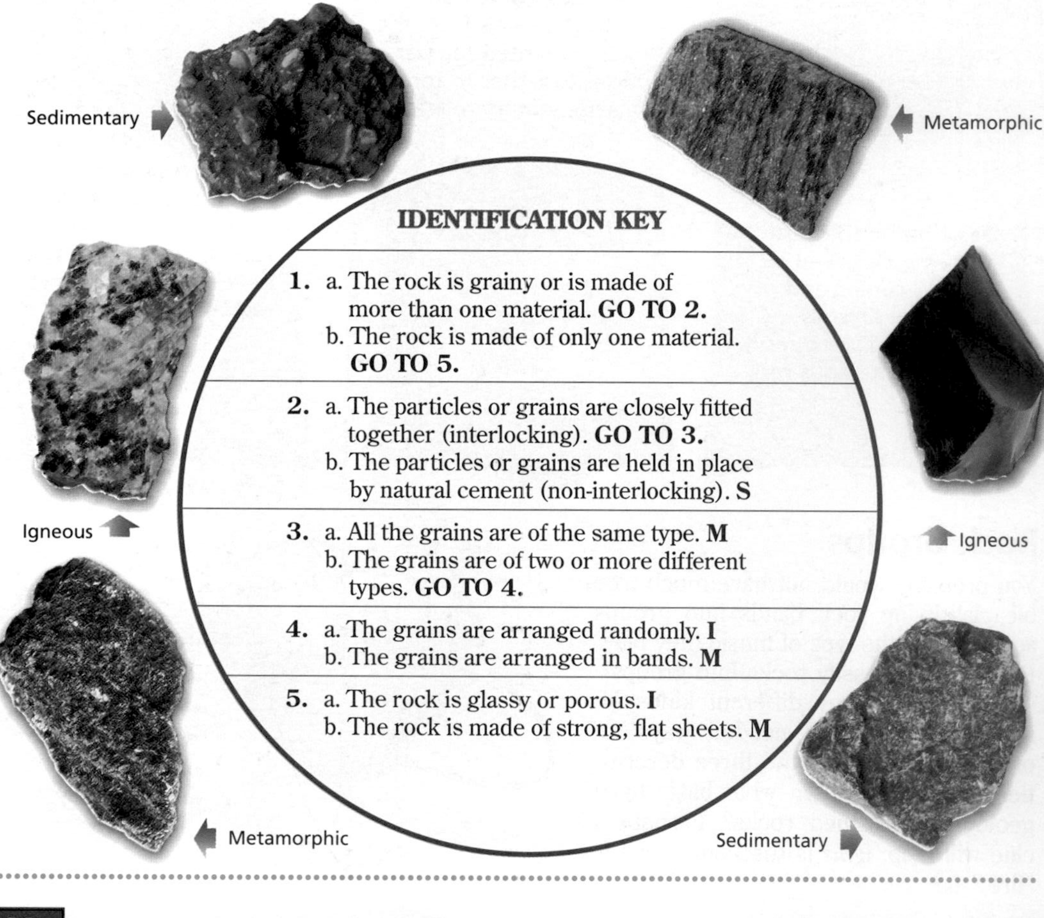

IDENTIFICATION KEY

1. a. The rock is grainy or is made of more than one material. **GO TO 2.**
   b. The rock is made of only one material. **GO TO 5.**
2. a. The particles or grains are closely fitted together (interlocking). **GO TO 3.**
   b. The particles or grains are held in place by natural cement (non-interlocking). **S**
3. a. All the grains are of the same type. **M**
   b. The grains are of two or more different types. **GO TO 4.**
4. a. The grains are arranged randomly. **I**
   b. The grains are arranged in bands. **M**
5. a. The rock is glassy or porous. **I**
   b. The rock is made of strong, flat sheets. **M**

386

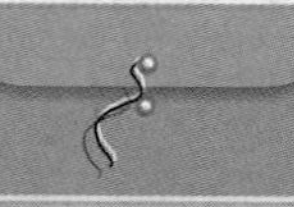

PORTFOLIO

Students may want to include their procedures and results from Exploration 2 in their Portfolio. Suggest that students record any questions they had as well.

FOLLOW-UP

Reteaching

Ask students to make a list of everything they have eaten in the last two or three days. Then have them organize their list into groups of similar food items (such as drinks, desserts, salty foods, etc.). Ask them to describe their method of organization in detail. Help them understand that just as food items can be placed into groups of similar items, rocks can be classified into groups with common characteristics.

Assessment

Ask students to compare the classification scheme they developed in Exploration 2 with the method that scientists use to classify rocks.

Extension

Have students research rock resources in their state. On a map, have them locate the sites of quarries, gravel mines, and other similar operations.

Closure

Present students with the following scenario: Imagine that you are in the business of selling rocks. Make a list of the rocks you sell and all of the ways your rocks could be used. Now write an advertisement to promote the sale of your rocks.

Hot Rocks!

Exploring Igneous Rocks

Before doing Exploration 3, look back at the rocks on page 386. Two of the rocks shown there are igneous. At one time both of these rocks were molten. One seems to be made of different types of particles, while the other is smooth and seems to be made of only one material. Why do two rocks of the same type look so different? Performing the Exploration that follows will help you answer this question.

EXPLORATION 3

A Simulation: Forming Igneous Rocks

This activity will help you understand how igneous rocks are formed.

Be Careful! Since you will be using hot materials, you must use caution. Safety goggles must be worn! Use care when handling stearic acid.

You Will Need

- latex gloves
- a metal ring
- a ring stand
- wire gauze
- sand
- 3 beakers
- 2 test tubes
- test-tube tongs
- a small paper cup
- a piece of paper towel
- a flat piece of aluminum foil
- an aluminum pie pan
- stearic acid
- a portable burner
- ice
- water
- a knife
- a scoop

What to Do

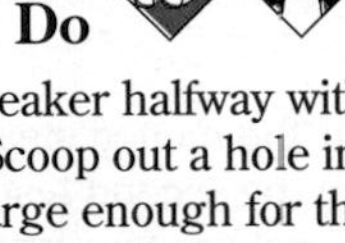

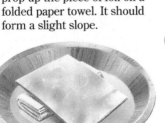

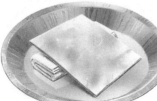

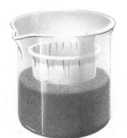

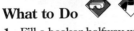

1. Fill a beaker halfway with sand. Scoop out a hole in the sand large enough for the paper cup, and place the cup in the hole.

2. In the center of the pie pan, prop up the piece of foil on a folded paper towel. It should form a slight slope.

3. Fill a second beaker halfway with ice and water.
4. Set up the ring stand. Place the wire gauze on top of the metal ring and the burner below it. Fill the third beaker halfway with water and place it on the wire gauze.

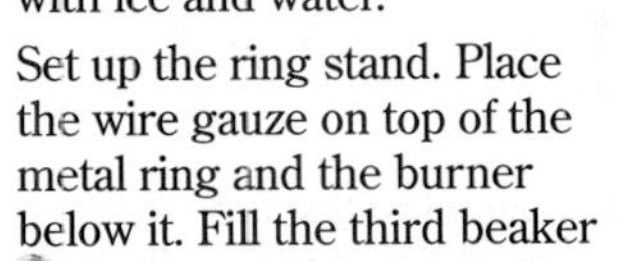

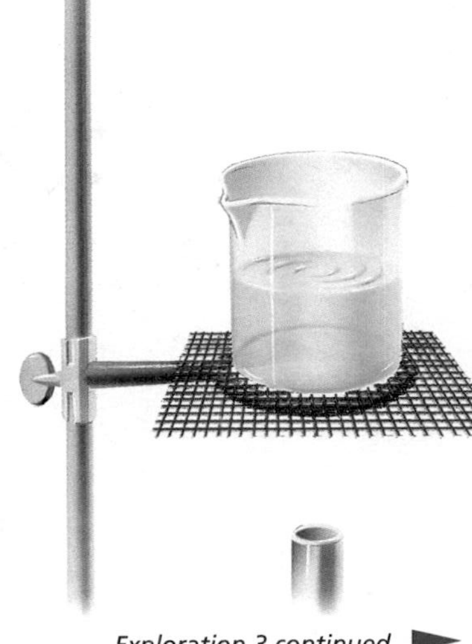

Exploration 3 continued ►

LESSON 2 ORGANIZER

Time Required 3 class periods

Process Skills
observing, inferring, analyzing

New Terms
Extrusive, or volcanic, rocks—igneous rocks formed from the molten lava that flows onto the surface of the Earth
Intrusive, or plutonic, rocks—igneous rocks formed from slowly cooling magma deep within the Earth's crust

Materials (per student group)
Exploration 3: metal ring; ring stand; wire gauze; 100 cm³ of sand; three 250 mL beakers; test-tube tongs; small paper cup; paper towel; flat piece of aluminum foil; 10 g of stearic acid; portable burner; a few ice cubes; 500 mL of water; knife; scoop; two 25 mL test tubes; safety goggles; a few matches; aluminum pie pan; lab aprons; latex gloves

Teaching Resources
Exploration Worksheet, p. 31
Resource Worksheet, p. 34
Transparency 62
SourceBook, p. S117

Hot Rocks!

FOCUS

Getting Started

Perform the following demonstration for students: Melt several candles in a double boiler. Pour the melted wax into a mold, and allow the wax to cool. Have students make notes about their observations. Explain to students that this demonstration simulates the formation of one of the types of rocks mentioned in Lesson 1. Ask: What type of rock formation does this simulate? *(Igneous rock)*

Main Ideas

1. Igneous rocks are formed when molten rock cools.
2. The size of crystals in igneous rocks is determined by the rate of cooling.
3. Igneous rocks that form on the surface of the Earth are called extrusive, or volcanic, rocks.
4. When cooling of molten rock occurs deep within the crust, the igneous rock formed is called intrusive, or plutonic, rock.

TEACHING STRATEGIES

Exploring Igneous Rocks

Have students read the opening paragraph and then look at the photographs on page 386. Students may hypothesize that the rocks on page 386 have different features because they formed under different conditions. If possible, provide samples of obsidian, granite, rhyolite, and pumice for students to observe and handle. Name each sample, and explain that they are all igneous rocks. Ask: How were these rocks formed? *(From the cooling of magma or lava)*

Teaching Strategies for Exploration 3 are on the next page. ►

EXPLORATION 3

Before beginning the Exploration, discuss safety precautions with students. Safety goggles should be worn, and students must take care to avoid burns.

WASTE DISPOSAL ALERT Dispose of stearic acid in an approved incinerator. Discharge, treatment, or disposal may be subject to federal, state, or local laws.

Remind students to follow the directions in the sequence they are given so that the paper cup in the sand and the sloping surface are ready for the melted stearic acid.

Once the stearic acid is melted (its melting point is 28°C), students must quickly pour it down the slope and into the paper cup. This Exploration works well as a cooperative learning activity.

Cooperative Learning
EXPLORATION 3

Group size: 3 to 4 students
Group goal: to simulate the formation of igneous rocks
Positive interdependence: Assign each student a role such as group leader, materials coordinator, recorder, or safety officer.
Individual accountability: Each student should be able to answer any of the questions from Interpreting Your Observations.

★ **An Exploration Worksheet (Teaching Resources, page 31) is available to accompany Exploration 3.**

5. Using the scoop, fill each test tube halfway with stearic acid. Stand the test tubes in the beaker of water. Heat the water until all of the stearic acid melts.

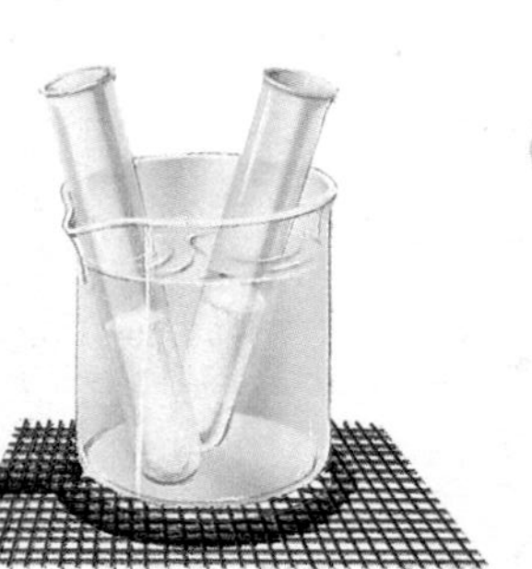

6. Using the test-tube tongs, carefully pour a small portion of melted stearic acid from one test tube onto the sloping foil, allowing it to run down the slope.

7. Quickly pour the rest of the stearic acid from the first test tube into the paper cup sitting in the sand. Observe both samples of stearic acid as they cool. Record your observations in your ScienceLog.

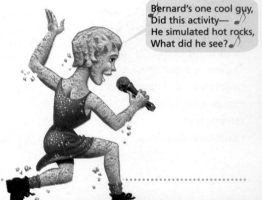

8. Pour the melted stearic acid from the second test tube into the beaker of ice water. Record your observations.

9. When the stearic acid in the ice water is completely cool, cut it in half with the knife.

Interpreting Your Observations

1. Compare the stearic acid in the paper cup with the stearic acid poured down the slope and the stearic acid poured into the ice water. In your ScienceLog, record the differences you observe. What do you think caused these differences?
2. Take a close look at the stearic acid that you cut in half. Is there a difference between the appearance of the material in the center and the material on the edge? If so, what do you think caused the difference?
3. This experiment is meant to simulate the formation of igneous rocks. What is simulated by each of the following?
   a. the melted stearic acid in the cup
   b. the sand
   c. the melted stearic acid that was poured down the sloping foil
   d. the melted stearic acid that was poured into the ice water

Bernard's one cool guy,
Did this activity—
He simulated hot rocks,
What did he see?

Answers to
Interpreting Your Observations

1. The stearic acid in the cup has the largest crystals. The stearic acid on the slope cools rapidly and may not form crystals at all. The stearic acid in the ice water forms a lump with a glossy surface due to rapid cooling. Different rates of cooling account for the differences in appearance.
2. The stearic acid on the edge has smaller crystals; it cooled more quickly because it was closer to the ice water. The inner material cooled more slowly because it was protected by the outer material.
3. **a.** The melted stearic acid in the cup simulates molten rock cooling deep within the Earth's crust.
   b. The sand simulates the surrounding rock that prevents heat from escaping quickly.
   c. Pouring the melted stearic acid on the slope is similar to lava pouring out onto land surfaces.
   d. Pouring the melted stearic acid into the ice water is similar to molten rock flowing into the ocean.

Bernard's Observations

Here is what Bernard wrote in his ScienceLog. What would you add to make his entry more complete? **A**

This was one of the most interesting experiments we've done. When the melted stearic acid cooled quickly by being poured into cold water, it became a rubbery blob. Our teacher said that when lava from a volcano flows over the Earth's surface, it cools down fast, and the rock formed has really small crystals. When the stearic acid cools more slowly, bigger crystals are formed. I think this must be what happens when magma cools slowly under the Earth's crust.

Basalt

Granite

Pumice

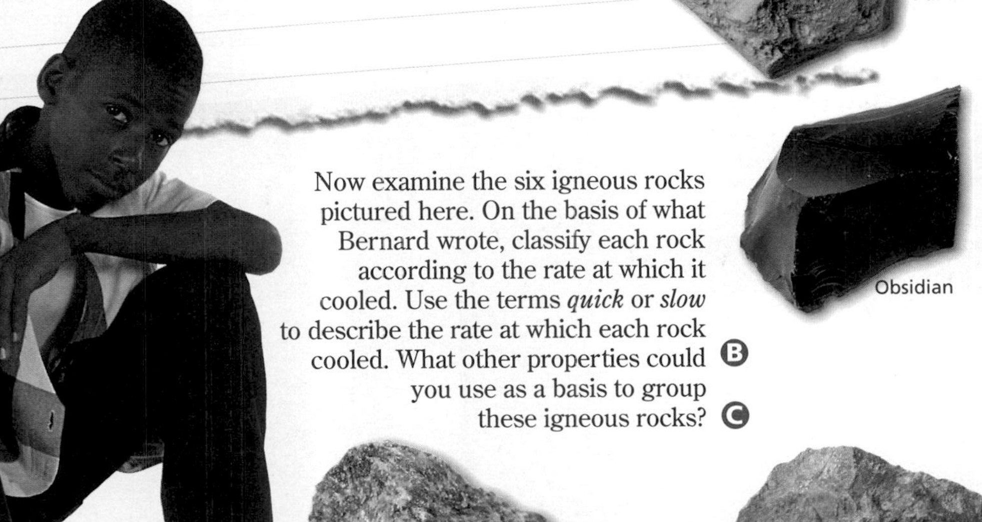

Now examine the six igneous rocks pictured here. On the basis of what Bernard wrote, classify each rock according to the rate at which it cooled. Use the terms *quick* or *slow* to describe the rate at which each rock cooled. What other properties could you use as a basis to group these igneous rocks? **B** **C**

Obsidian

Gabbro

Rhyolite

389

Did You Know. . .

Igneous rock makes up more than two-thirds of the Earth's crust.

Bernard's Observations

Call on a volunteer to read aloud the text at the top of this page. Involve students in a discussion of what to add to make Bernard's report more complete.

Direct students' attention to the photographs of the rocks on this page. Invite them to make inferences about the rate at which each rock cooled, based on Bernard's report and their own results from Exploration 3.

Answers to *In-Text Questions*

A Students should suggest that he include what took place when the stearic acid hardened both in the paper cup and on the slope.

B **Basalt** has small crystals, so it must have cooled quickly. (Basalt is formed when highly fluid, low-silicate magma escapes to the surface as lava and cools quickly. Oceanic crust, formed from lava flows at midocean ridges, is mostly basalt. Basalt is dark because it contains the minerals iron and magnesium.)
Granite has large crystals, so it must have cooled slowly. (Granite is formed from viscous, high-silicate magma that hardens slowly beneath the Earth's surface.)
Pumice has very small crystals or no crystals, so it must have cooled very quickly. (Pumice forms from lava containing hot gases. The network of holes formed by escaping gases makes the rock so lightweight that pieces of pumice will usually float on water.)
Obsidian has no visible crystals, so it must have cooled very quickly. (Obsidian forms when lava cools so quickly that no visible crystals can form, in effect creating natural glass. Although chunks of obsidian appear dark, thin slices are transparent and nearly colorless.)
Rhyolite has small crystals, so it must have cooled quickly. (Rhyolite is formed when a viscous, high-silicate lava cools quickly at or near the Earth's surface.)
Gabbro has large crystals, so it must have cooled slowly. (Gabbro is formed by highly fluid, low-silicate magma that hardens slowly beneath the Earth's surface. Gabbro is similar to basalt in composition.)

C Students may suggest that these rocks could be grouped by color or texture.

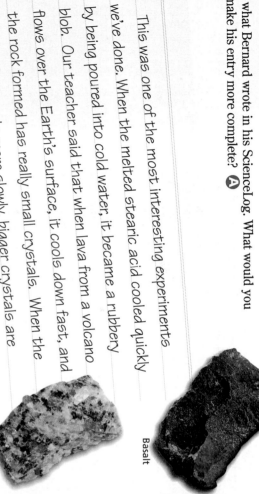

Fiery Language

Encourage students to develop definitions by using their own words rather than a textbook's glossary or a dictionary. After the unit is completed, students can brainstorm a number of ways to state the definitions and then develop a final definition for each term.

The Formation of Igneous Rocks

Call on a volunteer to read the text aloud. Then direct students' attention to the photographs on this page. Involve students in a discussion of the photographs, using the photos to illustrate points discussed in the text.

- **Devils Tower** (top right) was formed in the interior of an ancient volcanic cone. The cone eroded, leaving the plug as a free-standing tower. Point out that igneous rock formed within a volcano is considered extrusive because it has left the Earth's crust, even though it has not yet emerged onto the surface.
- **Ropy lava** (bottom right) is also called *pahoehoe,* a Hawaiian word meaning "ropy."
- **Pillow lava** (bottom left), like ropy lava, cools quickly at the surface as the molten rock comes into contact with water. The interior of these rocks, however, will sometimes continue to flow after the exterior has hardened.

Have students copy the illustration on the next page into their ScienceLog. Assign the letters *a–f* to the labels, and have students place the letters in the correct positions. Point out that the labels may be used more than once. Note that *a, c,* and *e* refer to extrusive igneous rocks and *b, d,* and *f* refer to intrusive igneous rocks.

Fiery Language

Many terms are used in connection with igneous rocks. You have seen some already: lava, magma, and molten. Recall that magma is molten rock inside the Earth and lava is molten rock on the surface of the Earth. Beginning with these words, compile a dictionary of rock terms in your ScienceLog. Add to your dictionary as you come across new terms.

The Formation of Igneous Rocks

One way igneous rocks form is when molten rock flows onto the surface of the Earth. Rocks formed in this way are called **extrusive** or **volcanic** rocks. Basalt is the most abundant rock of this type on Earth. Other volcanic rocks include rhyolite, obsidian, and pumice. These extrusive rocks are made of tiny crystals.

When magma erupts from underwater volcanoes, it can form rounded shapes about 1 m across. These formations are called pillow lava. Referring back to your observations in Exploration 3, do you think the cooled pillow lava would have small or large crystals? Why do you think so? Ⓐ

These photographs show some of the different forms of igneous rocks. There are many, many more.

Devils Tower in Wyoming is a basalt formation. Do its hexagonal columns remind you of any rock formations you've seen previously in this chapter? Ⓑ

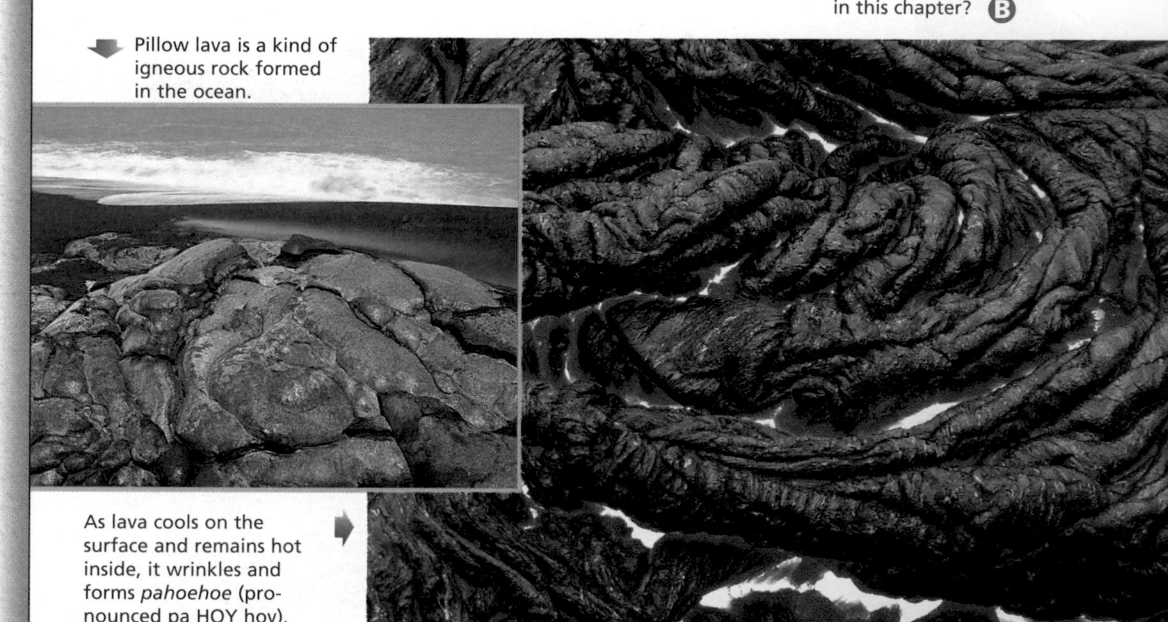

Pillow lava is a kind of igneous rock formed in the ocean.

As lava cools on the surface and remains hot inside, it wrinkles and forms *pahoehoe* (pronounced pa HOY hoy), Hawaiian for "ropy lava."

Answers to *In-Text Questions and Caption,* *pages 390–391*

Ⓐ The surface of pillow lava would have small crystals because it cools quickly. The interior, however, may be composed of larger crystals because the interior cools more slowly than the surface does.

Ⓑ It resembles Giant's Causeway, pictured at the beginning of Lesson 1.

Ⓒ The crystals in rocks formed from magma would be relatively large.

Ⓓ Students should identify those rocks that resemble the photos of granite and gabbro on page 389.

Ⓔ Refer to page S204 for a correctly labeled illustration.

Homework

You may wish to assign the activity described in Fiery Language as homework.

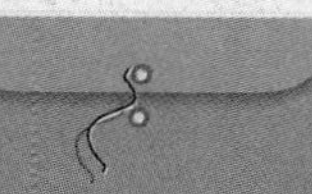

PORTFOLIO

Encourage students to include their drawings from The Formation of Igneous Rocks in their Portfolio.

Volcanic rocks that form from lava tend to cool quickly. Magma, however, tends to cool slowly deep within the Earth's crust.

Recall the stearic acid experiment from Exploration 3. Based on your experimentation, what size do you think the crystals of rocks formed from magma would be? (C) Rocks formed in this way are known as **intrusive** or **plutonic** rocks. Granite is the most abundant example. Gabbro is another common plutonic rock. Look again at the photos of granite and gabbro on page 389. Do any of your class samples look like these? (D)

The illustration below shows the formation of igneous rock by volcanic activity. Copy the drawing into your ScienceLog, and place the following labels in the correct positions: (E)

This giant granite boulder seems to balance precariously in the Erongo Mountains of Namibia, Africa.

- extrusive (volcanic) igneous rocks
- rocks formed from the slow cooling of magma
- rocks formed from the fast cooling of lava
- rocks with large crystals
- rocks with small crystals
- intrusive (plutonic) igneous rocks

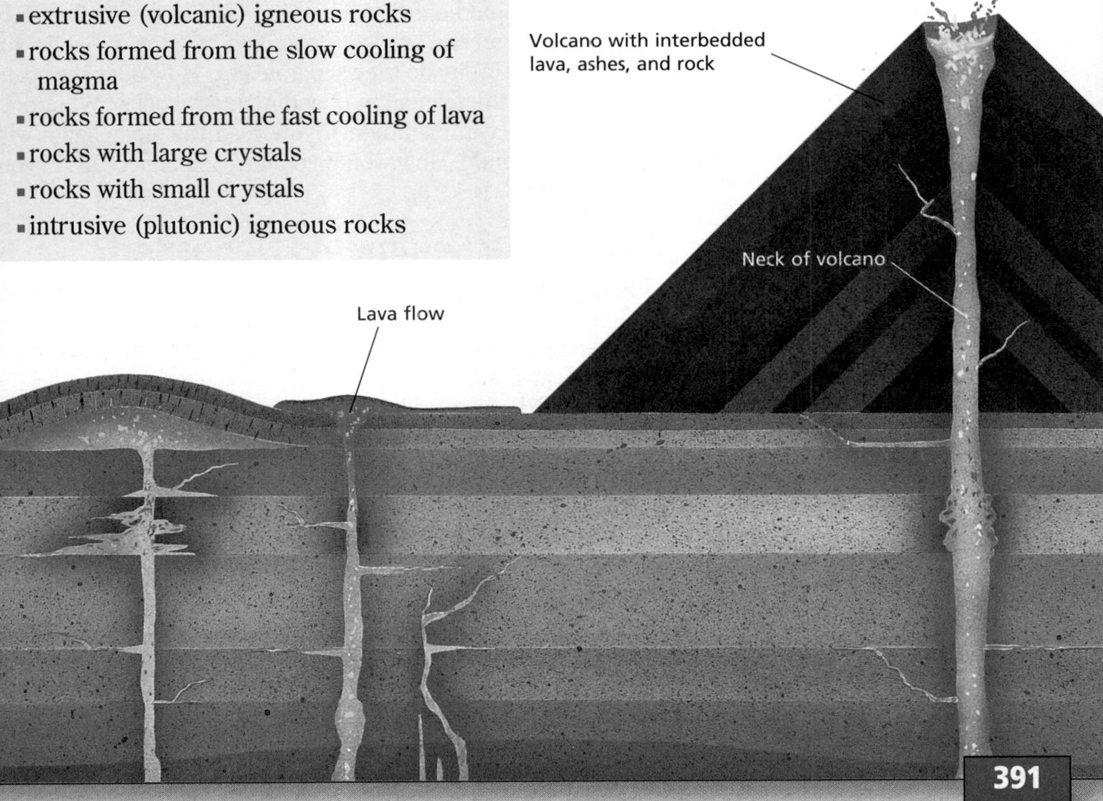

Did You Know. . .

Devils Tower, in northeastern Wyoming, was established as the nation's first national monument in 1906 by President Theodore Roosevelt. It stands 264 m tall.

Integrating the Sciences

Earth and Physical Sciences

Explain to students that water is recycled and transported in our atmosphere by convection. Convection is also a major cause of the movement of magma beneath the Earth's crust. Ask students to research how convection works and to make a poster illustrating convection both in the atmosphere and in the Earth's interior. *(Air and water molecules rise when heated, just as molten material rises when heated.)*

FOLLOW-UP

Reteaching

Provide small groups of students with modeling clay, cardboard, and other materials. Assign each group a term from the lesson (such as *extrusive rock, magma, granite,* or *obsidian*), and have students create models to demonstrate that term.

Assessment

Have students observe samples of extrusive and intrusive igneous rocks, study their characteristics, and then write a description of one of the rocks, including inferences about its formation and origin.

Extension

In some regions, hot igneous rocks produce a source of energy called geothermal energy. Have interested students do research on geothermal energy and prepare a report for the class. Students should focus their research on where and how geothermal energy is harnessed for human use.

Closure

Ask students to bring in samples of what they think may be igneous rocks. Then give them the following key to use in identifying the rocks:

1. Coarse-grained: particles clearly visible to the unaided eye
   - **a.** Mostly light-colored minerals: granite
   - **b.** Mostly dark-colored minerals: gabbro
2. Fine-grained: size of sugar grains or smaller
   - **a.** Frothy, porous: pumice
   - **b.** Solid, may have rounded bubbles filled with crystals
     - **i.** Light-colored, gray, pinkish, purplish: felsite
     - **ii.** Dark-colored, black, greenish, brownish: basalt

★ **A Resource Worksheet (Teaching Resources, page 34) and Transparency 62 are available to accompany the activity on this page.**

Homework

The Extension activity described on this page may be assigned as homework.

Rocks From Sediments

FOCUS

Getting Started

Obtain samples of different colors of sand. Have students layer sand of different colors in a large beaker. Ask: What would result if the pieces of sand were pressed or cemented together? *(A rock with layers of different colors would be formed.)* Explain that the sand is a sediment and that students will learn in this lesson how layers of sediment form sedimentary rock.

Main Ideas

1. Sedimentary rocks are formed from sediments transported by wind, water, or ice to the deposition site.
2. Sediments suspended in water settle out at different rates, forming layers.
3. Sedimentary rocks can be formed by evaporation, cementation, compaction, and crystallization.

TEACHING STRATEGIES

Hanging Out With Granite Man

Have volunteers read the parts of Granite Man and Sam Sandstone to the rest of the class. Use the in-text questions that follow the dialogue to spur class discussion.

Where Are Sediments Deposited?

Explain to students that the shelf is the gently sloping, submerged edge of a continent that extends to a depth of about 200 m. The slope is the region of steep incline between the shelf and the ocean bottom.

Have students draw the diagram in their ScienceLog and add the appropriate labels. Point out that the movement and deposition of sediment generally follow a downhill trend in response to gravity.

Transparency 63 is available to accompany the material on this page.

Rocks From Sediments

Hanging Out With Granite Man

"Yeah, yeah, yeah . . . so I was on my way to the concert when I ran into my old friend Sam Sandstone. 'Where ya been?' I asked him. 'Up the river,' he says. So I say, 'No way, man! Tell me about it.' So he starts groovin' on these rocks that fall into the river. Now, the river is one swift dude. It carries these rocks, like FAST, all the way to the ocean. On the way, the heavy rock dudes drop to the bottom first, the pebbles settle out just a little later, and the sand goes swirling and twirling like crazy on top. Then the river dumps these cruisers into the ocean, where they pile up and get pressed together. 'Solid?' I ask him. 'Solid, man,' Sam says. 'These dudes have their own group, and they are tight. They are their own kind of rock—ya know what I'm sayin'?' 'Yeah, Sam,' I say, 'that's too cool.' 'My sediments exactly,' says Sam."

Although Granite Man should check out current fashion trends, his story needs no updating. Think about the process described by Sam. What do you think sediments really are? Why do they separate out? How does the speed of a river contribute to the formation of sedimentary rocks? What does the ocean have to do with sediment? Now read the story again with these questions in mind. How would you have explained this in your own words? Ⓐ

Where Are Sediments Deposited?

Examine the illustration to the left. Think about why sediments get deposited in layers. Sketch the drawing in your ScienceLog, and label the locations where sediments might be deposited.

Where would you expect to find larger materials, such as stones, pebbles, and gravel? Where might sand be deposited? Where would you expect to find deposits of fine materials such as silt and clay? Ⓑ

LESSON 3 ORGANIZER

Time Required 2 to 3 class periods

Process Skills
observing, inferring, predicting

New Terms
Cementation—the process by which the spaces between particles of loose sediments are filled with a cementing agent
Compaction—the process by which pressure of overlying sediments or pressure from the Earth compresses the particles of sediments
Lithification—the process by which soft sediments become hard rock

Materials (per student group)
A Relationship to Water: optional items: geology field guide or textbook, a variety of sedimentary rocks
Exploration 4: 6 mL of soil (with a variety of particle sizes); 10 mL of water; test tube with stopper or small glass jar with tight lid; metric measuring spoon (additional teacher materials: clock or watch, 100 mL graduated cylinder)

Teaching Resources
Exploration Worksheet, p. 35
Activity Worksheet, p. 37
Transparencies 63 and 64
SourceBook, p. S120

A Relationship to Water

The formation of sedimentary rocks is closely associated with water. One type forms when water carries soil, pebbles, and other particles to the ocean floor. These deposits of sediments become rock. The second method involves chemicals dissolved in water. Certain chemicals, such as calcium carbonate, can be dissolved in water. When the water evaporates or the chemical precipitates, sedimentary rocks can form.

Several sedimentary rocks are pictured here. However, try to work with actual rocks if samples are available. Then consider the following questions:

1. Which of the rocks are obviously made up of sediments?

2. What kinds of sediments (shells of sea animals, clay, sand, pebbles, etc.) make up each of these rocks?

3. Which rocks do not appear to be composed of sediments? (These are the kind formed by changes in the water's solutes.)

Limestone
Conglomerate
Gypsum
Coquina
Sandstone
Shale
For more help with rock classification, look on pages S118–S123 in the SourceBook, or check the library for a geologist's field guide.

Answers to *A Relationship to Water*

1. Coquina and conglomerate are obviously made up of sediments. Shale and sandstone are also made up of sediments.
2. Shale is made up of silt and clay. Sandstone is composed of sand. Coquina contains fragments of fossils and shells cemented together. Conglomerates contain pebbles and rock fragments cemented together.
3. Gypsum and limestone do not appear to be composed of sediments. Gypsum forms from the precipitation of sulfates. Limestone forms from the precipitation of carbonates.

Answers to *In-Text Questions*, page 392

(A) Sediments include solid material that is physically transported by water, ice, or wind, as well as dissolved substances that are chemically precipitated from water. Types of sediments include rocks, pebbles, sand, and clay. Due to their greater mass, the heavier particles settle to the bottom first. A swift river can transport larger particles of sediment than a slower river can. The ocean is the final resting place for the finest sediments.

(B) Stones, pebbles, and gravel deposits could be found upstream or at river bends. Sand could be deposited at river bends, in lakes and river deltas, at beaches and barrier islands, and on the ocean shelf and slope. Deposits of fine material such as silt and clay could be found in lakes and river deltas and on the ocean shelf and slope.

EXPLORATION 4

This Exploration is intended to help students understand how sediments form layers. Part 1 focuses on clastic sediments, and Part 2 focuses on chemical sediments. Explain to students that clastic sediments are made up of rock particles that are broken and weathered from pre-existing rock. Rocks such as conglomerate, breccia, sandstone, and shale are formed by the deposition of clastic sediments. Chemical sediments are precipitated from a solution, primarily the ocean. They include limestone, dolomite, halite, and gypsum.

An Exploration Worksheet (Teaching Resources, page 35) is available to accompany Exploration 4.

PART 1

Students should discover that the largest particles settle out first. If their predictions were not correct, students should draw a corrected sketch. You may wish to have students let their test tube's contents settle overnight. This activity works well as a cooperative learning exercise.

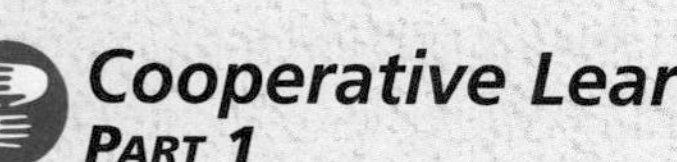

Cooperative Learning *PART 1*

Group size: 2 to 3 students
Group goal: to investigate the process of sediment deposition
Positive interdependence: Assign each student a role such as group leader, materials coordinator, and experimenter.
Individual accountability: Have each student submit a sketch predicting the results of the experiment.

PART 2

Your students may have studied evaporation in Unit 3. If so, review this concept, as well as distillation and desalination.

Answer to *In-Text Question*

Ⓐ Answers will vary. You may wish to inform students that many large land areas were once covered by salty, inland seas. The water in these seas evaporated, leaving salt deposits behind.

EXPLORATION 4

Forming Sedimentary Rocks

Earth scientists tell us that certain sedimentary rocks are formed from layers of sediment. In other cases, evaporation plays an important role in the formation of sedimentary rocks. The two parts of this Exploration will help you find out more about both processes.

PART 1

You Will Need

- about 6 mL of soil that contains a variety of particle sizes
- 10 mL of water
- a test tube with a stopper or a small glass jar with a tight lid

What to Do

1. Place the water into the test tube or jar, and pour the soil into the water.

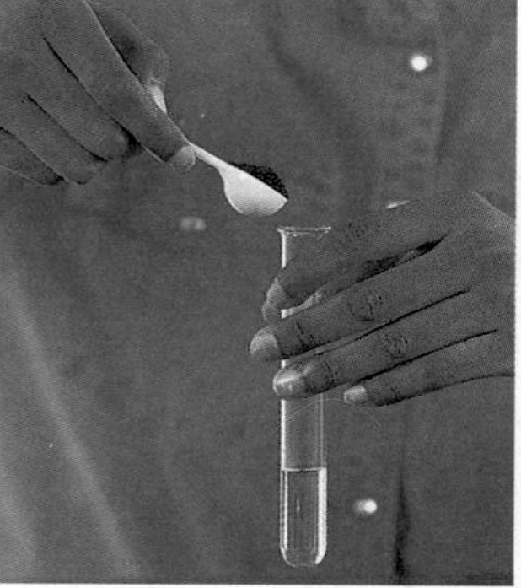

2. Before shaking the mixture, predict what will happen. Which particles will settle out first—the largest or the smallest? Sketch what you think the results will be.
3. Now shake the mixture vigorously for 1 minute. Observe the results. Does it resemble your sketch? Was your prediction correct? If not, draw a corrected sketch.

PART 2

Carmen read that if 100 g of sea water evaporates completely, 3.5 g of dissolved solids remain behind. She was curious to know whether the solid material left behind was just like table salt. Carmen collected some sea water and left it in the sunlight for several days until all the water had disappeared. She discovered that the material that remained was not as white as table salt. She decided to find out why.

Carmen told her teacher, Mr. Delaney, about her experiment the next day. He told her that what she had collected was mostly ordinary table salt. However, other compounds that had been dissolved in the sea water also remained behind after evaporation.

Carmen asked Mr. Delaney how her results might explain the formation of salt beds that have been found in certain parts of the world. How would you answer Carmen's question? Ⓐ

1
I think I'll try this.
2
3
4
5
It looks like table salt!
6
Besides table salt, other compounds are dissolved in the sea water.
How could my results explain the way salt beds are formed?

ENVIRONMENTAL FOCUS

What is ground water? By definition, it is water that soaks into the ground and fills the spaces in sand, gravel, and rock. Ground water is the source of more than half of the drinking water in the United States. When ground water is withdrawn, the land that was above it can settle. So much ground water has been withdrawn in some areas that local land elevations have been lowered by more than 3 m. Ground water is being used in most areas much faster than nature can replenish it. Encourage class discussion.

Did You Know. . .

When salt water evaporates, the salt left behind may eventually become embedded in limestone and shale. When the deposits become deep enough, the salt may flow like magma into surrounding rock layers. As the salt flows, enormous caverns called salt domes are created.

A Cliffhanger

Patrick and Nancy discussed the last Exploration they did in class.

Nancy said, "I'm confused. I can see how the layers formed in the test tube, but understanding how pieces of dirt end up as rock on the side of a cliff is beyond me."

"Maybe it would help if we tried to make some rocks," Patrick suggested. "Let's ask Mr. Delaney to help." Their teacher then suggested that they read the information below for some hints about what to do.

Sediment Becomes Rock

When soft sediments become hard rock, **lithification** has occurred. One form of lithification is called *cementation.* In this process, the spaces between particles of loose sediments are filled with a cementing agent that is carried in solution by the water. As the cementing agent fills the gaps between the particles, the loose materials are cemented together to form sedimentary rocks.

During *compaction,* the pressure of overlying sediments compresses the space between the particles. As the particles are pressed closer together, they stick to each other and form sedimentary rocks.

There are still other lithification methods. *Drying* and *crystallization* are two of them. How do you think these processes of lithification work? (B)

Rocky Solutions

Patrick and Nancy asked their classmates to help them come up with some ideas to simulate lithification. Gary remarked that popcorn can be made to stick together by mixing it with corn syrup. Juana had watched a construction crew mix concrete the previous weekend, and she thought this was a good simulation. Randy suggested pressing small marbles of modeling clay together.

What type of lithification was each student thinking about in proposing these ideas? (C) Come up with your own lithification model and give it a try!

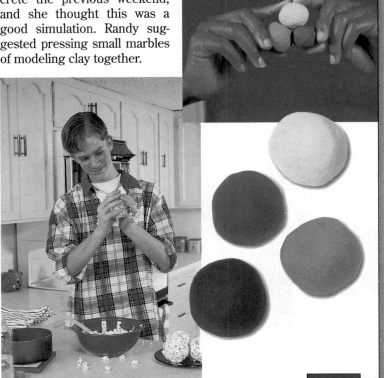

Sediment Becomes Rock

Ask a student to read this page aloud. Involve students in a discussion of the new terms *lithification, cementation,* and *compaction.*

Rocky Solutions

GUIDED PRACTICE Ask students to compare and contrast these various processes of lithification. Then have the class discuss the ideas suggested by Gary, Juana, and Randy. Encourage students to think of their own ideas and to relate these ideas to the processes of lithification.

Answers to *In-Text Questions*

(B) Drying is a process in which liquids evaporate, leaving behind solid sediments. Crystallization is a process in which solids form in a saturated solution.

(C) Gary's method simulates cementation, Juana's simulates drying, and Randy's simulates compaction.

Homework

You may wish to assign the Activity Worksheet on page 37 of the Unit 6 Teaching Resources booklet as homework. (If you choose to use this worksheet in class, Transparency 64 is available to accompany it.)

FOLLOW-UP

Reteaching

Bring a sample of each type of sedimentary rock to class so that students can compare rocks formed by cementation, compaction, crystallization, and drying. Then lead students in a game of "This Is Your Life" for each rock. Trace the history of each rock on the chalkboard so that students can visualize how the processes of sedimentary rock formation are similar and different.

Assessment

Bring in samples of sedimentary rocks. Have students study the samples and explain how each may have become lithified. *(Sandstone—cementation of sand; shale—compaction of mud and silt; gypsum—drying or evaporation of water containing sulfates; limestone—crystallization of a solute; conglomerate—cementation of pebbles and rock fragments; coquina—cementation of fossils and shells of marine organisms)*

Extension

Have students research and write a report on the different forms of limestone.

Closure

Have students experiment with making "rocks."

1. Mix 500 g of coarse sand, a handful of pebbles, and 150 g of cement in a milk carton. Slowly add water and stir until a porridgelike mixture is formed. Store the container until the mixture is hard (1 to 3 days).
2. Follow the same procedure as above, but omit the pebbles to make a rock resembling sandstone.

LESSON 4

Changed Rocks

FOCUS

Getting Started

Show the class a bag of unpopped popcorn and a bowl or two of popped popcorn. Share the popped popcorn with the class. Tell the class that the unpopped popcorn has been changed by heat into something different, the popped popcorn. Explain that in this lesson students will learn about the changes that take place during the formation of certain kinds of rocks. This change, or metamorphosis, can be compared with the change that occurs when unpopped popcorn is exposed to heat and is popped.

Main Ideas

1. Rocks subjected to heat and pressure beneath the Earth's surface will change their original texture and composition to become metamorphic rocks.
2. Rocks can be changed into another form when conditions in their environment change.

TEACHING STRATEGIES

The Meaning of Metamorphic

Involve students in a discussion of metamorphosis by directing their attention to the pictures of the caterpillar and butterfly.

Encourage students to think of other ways in which rocks might change. *(Erosion, breakage)* Ask: How are these kinds of changes different from the changes shown in the photos on this page? *(Erosion and breakage do not cause changes in color and composition, for instance.)*

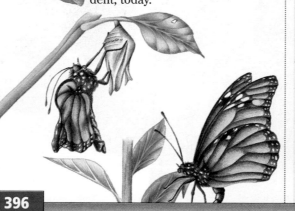

LESSON 4

Changed Rocks

The Meaning of Metamorphic

Metamorphosis means "change." You might have used the word to describe the life cycle of frogs or insects. For example, when a caterpillar spins a cocoon and comes out as a butterfly, metamorphosis has occurred. How do you think metamorphosis applies to a rock? Ⓐ

Metamorphism, the term that designates change in rocks, occurs when the environment changes, such as when there are movements in the Earth's crust. How could igneous and sedimentary rocks change into new kinds of rocks? Read Ⓑ what Charles Lyell, a nineteenth-century geologist, might tell Rocky, a curious student, today.

396

An Interview With ...

CHARLES LYELL
1797–1875

ROCKY: Mr. Lyell, how did you become interested in studying rocks?

CHARLES: Well, I became a geologist in a roundabout way. I was born and raised in Great Britain. I planned on studying law—in fact, I even went to Oxford University for that! But I had a great fascination for geology and spent most of my free time studying rocks and geological theories. My hobby soon became a career as I became more and more interested in the way in which rocks are formed. A scientist named James Hutton had proposed several theories on the subject just before I was born. I found these theories absolutely fascinating.

ROCKY: What were these theories?

CHARLES: The one that interests me the most is about **intrusions.** You see, Mr. Hutton suggested that magma sometimes pushes up into the cracks in rocks. When the magma cools, it forms igneous rocks, such as granite.

ROCKY: That's why we call granite an intrusive rock—because the magma intrudes, or is forced, into the rock, right?

LESSON 4 ORGANIZER

Time Required
2 class periods

Process Skills
observing, inferring, analyzing, organizing

Theme Connection
Cycles

New Terms
Dikes—intrusions that cut vertically across existing rocks
Intrusions—magma that has pushed into cracks in existing rocks
Sills—intrusions that form in horizontal cracks in existing rocks

Materials (per student group)
Exploration 5: 4 pairs of rocks (each pair consisting of an igneous or sedimentary rock and its metamorphosed form); magnifying glass

Teaching Resources
Exploration Worksheet, p. 38
Transparency Worksheet, p. 39
Activity Worksheet, p. 41
Transparency 65
SourceBook, p. S122

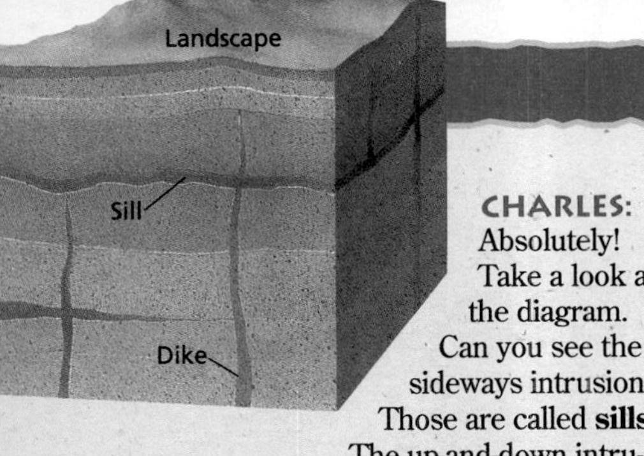

CHARLES: Absolutely! Take a look at the diagram. Can you see the sideways intrusions? Those are called **sills.** The up-and-down intrusions are called **dikes.** These are exactly what Mr. Hutton was talking about.

ROCKY: Very interesting. But how have you used Mr. Hutton's theory in your own study of geology?

CHARLES: I'm so glad you asked. You see, Hutton found that the rocks around the intrusions had changed.

ROCKY: Changed? Could you explain that a bit?

CHARLES: Well, magma is very hot. And when it intrudes into rock, I think the heat affects the sedimentary or igneous rock surrounding the intrusion. This rock doesn't become hot enough to melt. However, certain chemical changes take place. For instance, when magma cools, it gives off gases and fluids that can move into the cracks or tiny pores of the surrounding rock.

ROCKY: Then what happens?

CHARLES: Those rocks change into something entirely new! Because of the changes these rocks go through, I call them metamorphic rocks. As I'm sure you can see, I am utterly convinced that Hutton's theory is correct.

Can you locate the dike in this photograph? (C)

Hutton's theory has never been disproved, and the special name that Lyell gave to changed rocks is still used today. Here are some pictures showing original rocks and their metamorphosed forms. What differences do you see between the metamorphic rocks and the original rocks? (D)

Homework

Have students choose an object from home that is made from rock, and observe it carefully. Then have them do the following:

- Describe the rock and the object made from it.
- Classify the rock as igneous, sedimentary, or metamorphic.
- State your reasons for classifying the rock as you did.
- Write a paragraph that describes how the rock was formed.

(Observations should include important identifying properties, the reasons for the classification should be logical, and the description of how the rock formed should agree with the text.)

An Interview With . . .

CHARLES LYELL

Have students develop their own definitions for the terms *intrusion, sill,* and *dike.* Students should write their definitions in their ScienceLog.

Answers to *In-Text Questions and Caption, pages 396–397*

(A) Students should understand that a metamorphic rock has been changed from its original form.

(B) Accept all reasonable responses; encourage students to read the interview for a more complete answer.

(C) The dike is the vertical stripe of red rock.

(D) These rocks exhibit differences in color, composition, texture, grain size, and breakage patterns, for instance.

Integrating the Sciences

Earth and Physical Sciences

Review with students the difference between a physical change and a chemical change. Ask: In general, is the metamorphosis of limestone to marble a physical change or a chemical change? *(It involves significant changes in color and composition. These are signs that new chemicals have formed, so the metamorphosis can best be considered a chemical change.)*

Multicultural Extension

Haiku Poetry

Haiku is one of the oldest forms of poetry in Japanese culture, and it often reflects the Earth's beauty and the natural environment. Haiku poetry consists of three lines composed of 5, 7, and 5 syllables, respectively. Have students write haiku poetry to describe the color, texture, composition, and aesthetic qualities of metamorphic rocks. *(Sample answer:*

Rocks are beautiful.
My rock is picture perfect,
Glowing in the dust.)

EXPLORATION 5

To prepare students for the Exploration, involve them in a discussion of the features they will look for in the rock pairs. *(Possible observations: color, general appearance, composition, crystal size, the way the crystals are locked together, presence of banding or layering, and distortion of layers)* Encourage them to include all the characteristics noted in the rock identification key on page 386.

Metamorphism is characterized by new minerals, new textures, new structures, or a combination of these. The metamorphic rock is often so changed that it is difficult to identify the original rock.

An Exploration Worksheet (Teaching Resources, page 38) is available to accompany Exploration 5.

Can Rocks Be Recycled?

Ask students to read page 398 and answer the questions with a partner. After a specified amount of time, reassemble the class to discuss their responses. Emphasize that all rock types can be changed into another form.

A Transparency Worksheet (Teaching Resources, page 39) and Transparency 65 are available to accompany Can Rocks Be Recycled?

Answers to *Can Rocks Be Recycled?*

1. Igneous rocks are formed when molten rock cools and hardens.
2. Exposure of igneous rock to excess heat and pressure will cause metamorphism. Erosion can break igneous rocks into sediments that can become sedimentary rock, which can subsequently be metamorphosed.
3. Sedimentary rocks are formed by compaction or cementation of sediments, by crystallization of solutes, or by drying of a solution.
4. Sediments are eroded from exposed rocks.
5. Exposure of sedimentary rock to extreme heat and pressure can turn it into magma. When cooled and solidified, the magma becomes igneous rock.

EXPLORATION 5

Rocks: Before and After

You know that metamorphic rocks are changed igneous or sedimentary rocks. In this activity, you will compare original rocks with their metamorphosed forms.

You Will Need

- 4 pairs of rocks, each pair consisting of an igneous or sedimentary rock and its metamorphosed form
- a magnifying glass

What to Do

1. For each pair, compare the properties of the two rocks. Record your observations for each pair in your ScienceLog.
2. When you have finished your examination, check the distinguishing property or properties that you identified against the rock identification key on page 386. Do your observations include all these characteristics?
3. How closely do the metamorphic rocks resemble the original rocks that they were formed from?

Can Rocks Be Recycled?

Do you suppose that metamorphic rocks can be changed back into their original forms? Can igneous rocks be changed into sedimentary rocks? In other words, can rocks be recycled? Discuss the following questions with a classmate, and decide whether you think rocks can be recycled:

1. How are igneous rocks formed?
2. What conditions change igneous rocks into metamorphic rocks?
3. How are sedimentary rocks formed?
4. What is the source of sediments?
5. What conditions change sedimentary rocks into igneous rocks?

What pattern, or cycle, have you discovered? Using the words and phrases listed at right, write a description in your ScienceLog of how the recycling of rocks occurs. Ⓐ

- plate tectonics
- molten rock (magma or lava)
- metamorphic rock
- heat
- cooling
- uplift
- igneous rock
- erosion
- pressure
- sedimentary rock
- melting
- cementing
- sediments
- intrusions

398

Answer to *In-Text Question*

Ⓐ Sample description: *Cooling* of *molten rock* forms *igneous rocks.* Weathering and *erosion* change igneous rocks into *sediments. Cementing* of these sediments produces *sedimentary rocks.* When these rocks are buried and then exposed to extreme *heat* and *pressure* due to *plate tectonics,* they can form *metamorphic rock,* or they can *melt.* Melted rock can rise up into cracks and cool, causing an *intrusion.* If the rocks are not completely melted, they can be *uplifted* by huge crystal movements. Once on the surface, the rocks can be weathered, repeating the cycle.

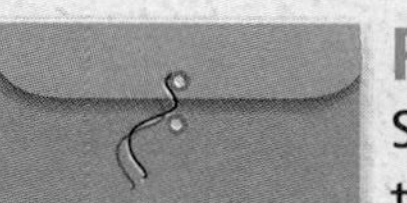

PORTFOLIO

Some students may wish to include their answers to Can Rocks Be Recycled? in their Portfolio. Encourage peer review between pairs of students, in which the reviewer's notes are included in each student's Portfolio along with the original answers.

Those Plates Again!!

The theory of plate tectonics supports the idea that rocks can be recycled. The collision of plates creates conditions that cause the formation of both igneous and metamorphic rocks. Look at the diagram below. It shows the collision of a continental plate with an oceanic plate.

With a partner, find the locations of each of the following occurrences:

1. The oceanic plate is moving underneath the continental plate.
2. Parts of the continental plate are under great pressure because of the collision. This leads to the formation of metamorphic rock.
3. The rocks of the oceanic plate are being heated to a high temperature and are melting.
4. Magma is rising in cracks in the continental plate, forming new igneous rock.
5. Igneous and metamorphic rocks erode to become sediments that then are carried to the bottom of the ocean. This is the first step in forming sedimentary rock.

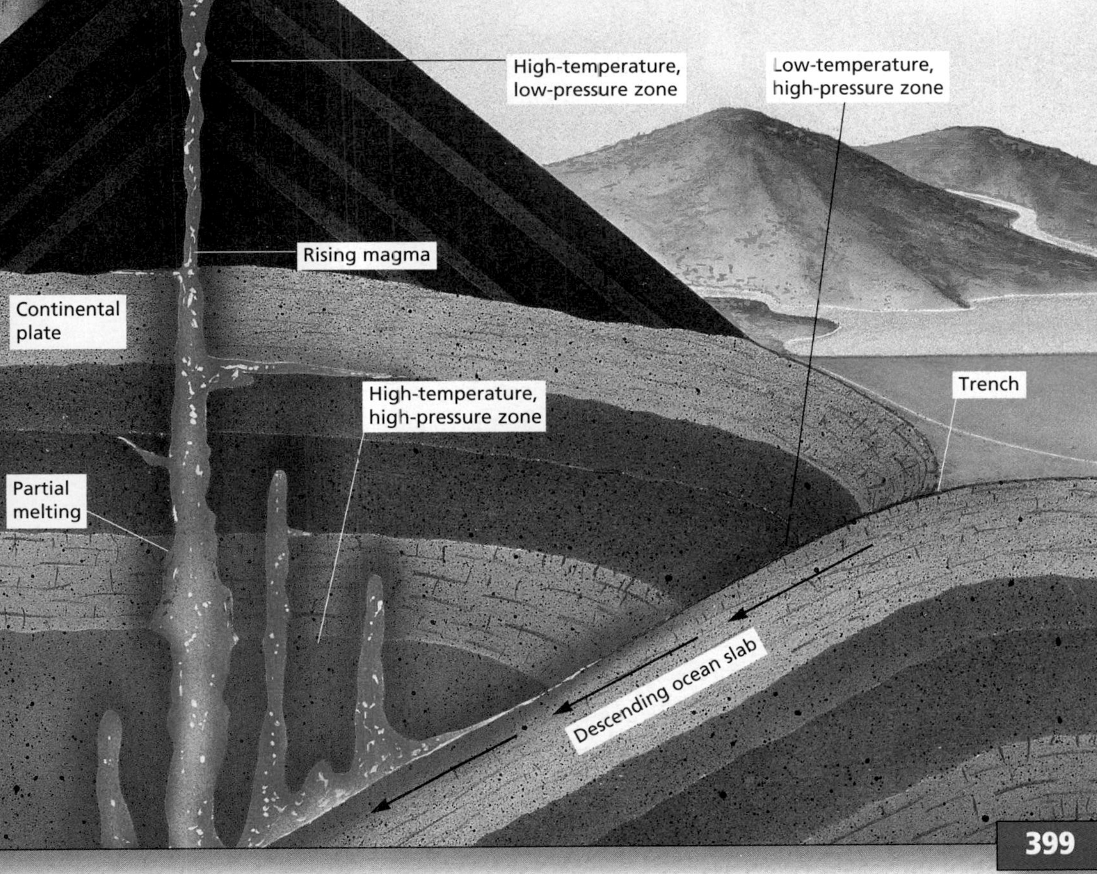

399

Those Plates Again!!

Draw an outline on the chalkboard similar to the illustration on this page. Call on students to come forward and place the numbers in their correct locations. (A correctly labeled diagram appears on page S205.)

Theme Connection

Cycles

Focus question: How does the formation of igneous, sedimentary, and metamorphic rocks form a cyclic pattern of change? *(Students should understand that the processes forming each of these rock types occur continuously, are interrelated, and form a pattern called the rock cycle. Have students make a diagram of the rock cycle in their ScienceLog.)*

ENVIRONMENTAL FOCUS

Ask: How can humanity's use of the environment, such as clear-cutting rain forests for farmland, affect the rock cycle? *(As the trees are cut down, the soil erodes away much more easily, exposing more rock. This exposed rock is then subject to weathering and erosion, which create more sediment.)*

Homework

You may wish to assign the Activity Worksheet on page 41 of the Unit 6 Teaching Resources booklet as homework.

FOLLOW-UP

Reteaching

Ask students to describe the characteristics they would look for in order to identify a metamorphic rock. *(Possible answers include obvious banding patterns and alternating layers of dark and light minerals.)*

Assessment

Ask students to imagine a conversation between James Hutton and Charles Lyell. Then have students write a dialogue in which Lyell tells Hutton exactly why he agrees with the theory of intrusions and why he decided to name a third group of rocks *metamorphic.*

Extension

Certain metamorphic rocks, such as schist and gneiss, can be produced from many different types of rock. Have students research the variety of original rocks that can be changed to schist and gneiss and the conditions that produce the change in each case. *(Shales, impure sandstones, and basalt can metamorphose into schists; conglomerates, shales, and granites can metamorphose into gneiss.)*

Closure

Have students write a short description that compares and contrasts the baking of a cake with the formation of a metamorphic rock. *(Both involve changes in texture and color, for instance, but the formation of a metamorphic rock involves more extensive chemical changes than does the baking of a cake.)*

Answers to *Challenge Your Thinking*

1. Sample answer: I can barely remember my first days. As young limestone, I grew slowly from the precipitation of carbonates. Eventually, I became a full-grown rock. Just when things were calming down, however, a rumbling occurred, and an intrusion of molten rock formed very close by. The heat and pressure were so strong that I eventually changed into a completely different rock. I became marble. Many years later, I was dug up and loaded into a truck with a bunch of my marble friends. Since I had a crack in my side, though, I couldn't be used for marble tiles, and the people at the quarry gave me to a teacher. The teacher brought me to this classroom to be looked at by students, and here I am!

2. Rock 1: sedimentary
Rock 2: metamorphic
Rock 3: metamorphic
Rock 4: igneous

3. 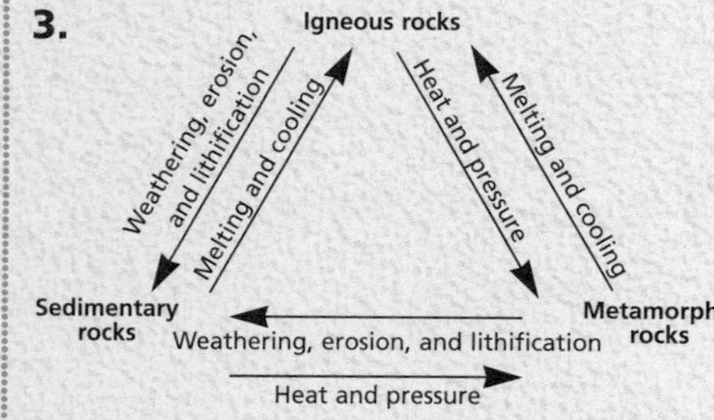

4. Students should note that some sedimentary rocks (such as shale, sandstone, coquina, and conglomerate) are composed of visible particles of sediment. These are formed by compaction and cementation. Sedimentary rocks (such as gypsum and limestone) that are not made of visible particles are created by drying or crystallization.

5. Students' comic strips or diagrams should reflect an understanding of how plate tectonics results in both the movement of molten rock and the buildup of heat and pressure. These two results can cause the formation of igneous and metamorphic rocks, respectively.

6. Sample questions: What kinds of rocks do you deal with most often? What methods do you use for observing the rocks in your area? What are the most important things for a junior-high-school student to know about your field?

CHALLENGE YOUR THINKING

1. And the Rock Said . . .

Choose a rock from the class collection. Examine the rock carefully, noting its physical characteristics. Now imagine that this rock can talk. What would it tell you? Write a poem or short story in which the rock tells its history, including how it was formed and how it got to your classroom. Also include in your poem or short story the rock's predictions for its future.

2. What's in a Name?

Nigel is having trouble classifying some rocks. Using the following information, help Nigel determine whether the rocks are igneous, sedimentary, or metamorphic. (Go back to the identification key on page 386 for more help.)

Rock 1: This rock is grainy and has a lot of different particles that seem to fit together loosely. I think I see half of a pebble in there!

Rock 2: This rock is grainy too, but the grains fit together tightly. There are at least four different grains here—two of them are gray, one is white, and one is black. You can see that each color has its own band.

Rock 3: This rock isn't grainy at all but is made of several flat sheets.

Rock 4: I like this rock because it's smooth and glossy. It's only made of one material, and it isn't grainy either.

3. The Recycling Cycle

Copy this diagram into your ScienceLog. Do not write your answers in the book. For each arrow, use the groups of words that follow to fill in the processes that occur as each type of rock is recycled. One answer has been completed for you.

- weathering, erosion, and lithification
- heat and pressure
- melting and cooling

You will have to use each group of words more than once.

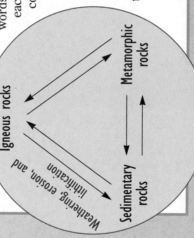

Igneous rocks
Weathering, erosion, and lithification
Sedimentary rocks
Metamorphic rocks

You may wish to provide students with the Chapter 17 Review Worksheet on page 42 of the Unit 6 Teaching Resources booklet. Transparency 66 is available to accompany this worksheet.

CROSS-DISCIPLINARY FOCUS

Art

Materials from the Earth have been used by artists for a long time. Prehistoric rock carvings and paintings made with crushed mineral pigments have been found throughout the world. Marble, a metamorphic rock, is used by sculptors, as are other rocks, such as granite and soapstone. Ask students to design a work of art using materials from the Earth. In conjunction with the design, have them write an explanation of how they took advantage of the specific properties of the material.

4. True Clues

Take another look at the rocks you and your classmates examined in Exploration 1, page 384. Can you tell which rocks are sedimentary? Choose one of the sedimentary rocks and study it carefully. Try to determine its composition, as well as the process by which it became lithified. What clues can you use to make your determination?

5. Comic Relief

Create a colorful comic strip or a series of diagrams that shows how the movement of continental plates can cause igneous and metamorphic rock to form.

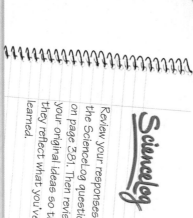

6. Good Evening, Ladies and Gentlemen

Imagine that you are a talk-show host. Choose one of the following people to be a guest on your program:

- a geologist who lives in the Rocky Mountains
- a rock collector who is organizing a field trip for junior-high-school students
- an expert on plate tectonics

Make a list of questions you would ask your guest about rocks.

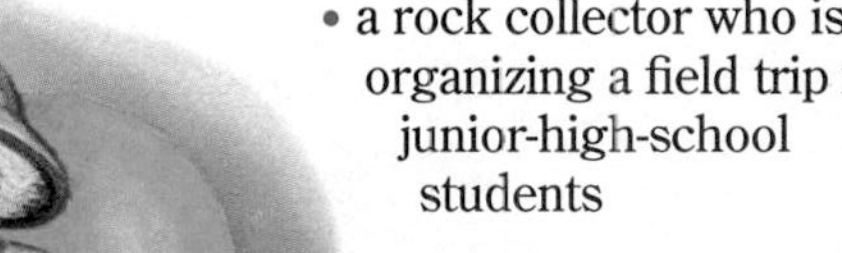

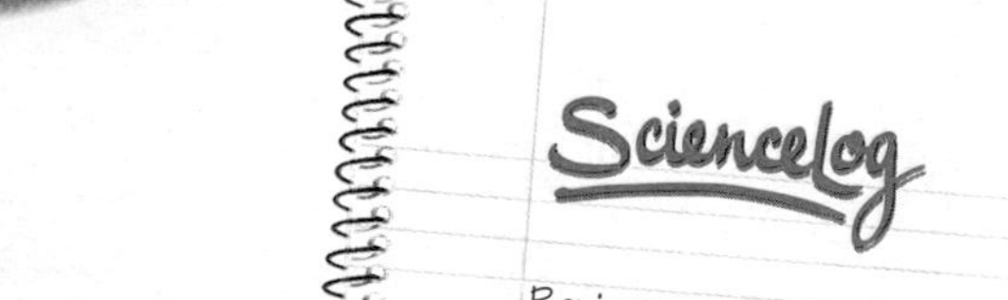

Sciencelog

Review your responses to the ScienceLog questions on page 381. Then revise your original ideas so that they reflect what you've learned.

401

Theme Connection

Changes Over Time

The video *How Solid Is Rock?* describes the constant changes that occur in and below the Earth's surface. It also draws a parallel between these changes and the changes that are continually occurring in the Earth's atmosphere. The video is available from Britannica, 310 South Michigan Avenue, Chicago, IL 60604-9839.

Sciencelog

The following are sample revised answers:

1. The rocks should be classified according to their appearance and characteristics. The conglomerate rock is sedimentary, the red slate is metamorphic, and the obsidian is igneous.
2. Sedimentary rock is formed through the lithification of sediments. Specifically, this conglomerate rock was formed by the cementation of pebbles and rock fragments. During cementation, a cementing agent that was dissolved in water seeped between the sediments and bonded them together. Metamorphic rock forms when extreme heat and pressure reorganize the crystals of an existing rock to change it into a new type of rock. This gray slate was reorganized from shale. Igneous rock is formed when magma cools and hardens. The magma cooled very quickly to form this obsidian.
3. Yes; the collision of tectonic plates creates heat and pressure that can change sedimentary and igneous rocks into metamorphic rock. In addition, processes such as erosion and lithification can form sedimentary rock from igneous and metamorphic rocks.

Meeting Individual Needs

Learners Having Difficulty  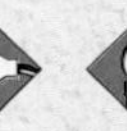 

Using a hot plate, heat a few spoonfuls of sugar in a nonstick frying pan, stirring frequently. Once the sugar has caramelized, pour it onto wax paper to cool. Provide students with a cube of sugar and a piece of the caramelized sugar for comparison. Students should describe the metamorphosis that has taken place, compare the appearances of the two substances, and relate the change to the formation of metamorphic rock.

Fossils—Records of the Past

Connecting to Other Chapters

Chapter 16
examines plate tectonics, with a particular focus on volcanoes, mountains, and earthquakes.

Chapter 17
explores the essential characteristics of igneous, sedimentary, and metamorphic rocks.

Chapter 18
introduces the geologic time scale as well as the types and formation of fossils.

Prior Knowledge and Misconceptions

Your students' responses to the ScienceLog questions on this page will reveal the kind of information—and misinformation—they bring to this chapter. Use what you find out about your students' knowledge to choose which chapter concepts and activities to emphasize in your teaching.

In addition to having students answer the questions on this page, you may wish to have students complete the following activity: Divide the class into four or five groups. Have each group discuss and develop an answer to one of the following questions:

1. What is a fossil?
2. How does a fossil form?
3. What killed the dinosaurs?
4. How can we determine the age of a fossil?
5. Why do we study fossils?

Have each group present its answer to the class. Encourage class discussion by allowing students to ask each other questions. During the discussion, take note of what students know about fossils, what misconceptions they may have, and what about fossils is interesting to them.

CHAPTER 18

Fossils—Records of the Past

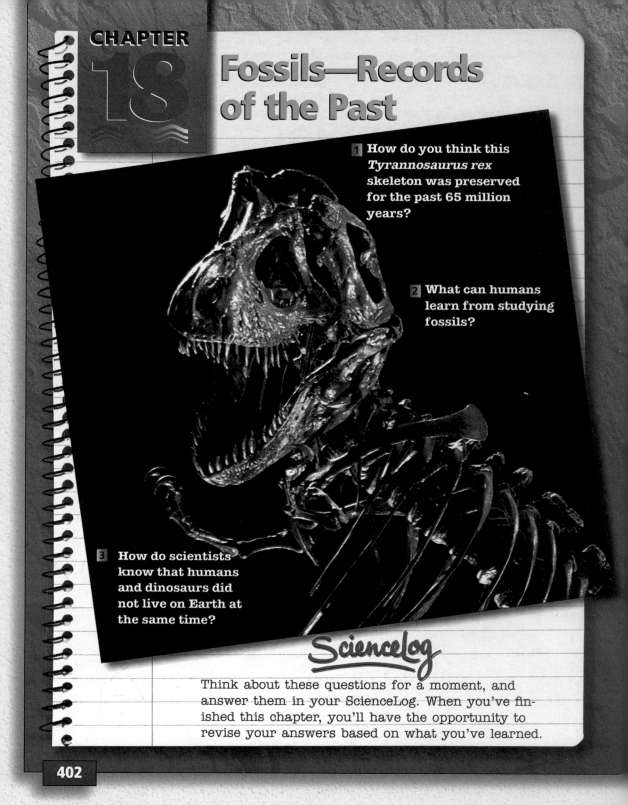

1 How do you think this *Tyrannosaurus rex* skeleton was preserved for the past 65 million years?

2 What can humans learn from studying fossils?

3 How do scientists know that humans and dinosaurs did not live on Earth at the same time?

ScienceLog

Think about these questions for a moment, and answer them in your ScienceLog. When you've finished this chapter, you'll have the opportunity to revise your answers based on what you've learned.

402

LESSON 1 Rocks Reveal a Story

Imagine the following scene: Dense clumps of trees and huge ferns crowd the many swamps and river banks. Amphibians large and small crawl on the land and swim in the warm waters. The air is alive with the sounds of giant insects, but no birds can be seen. Arthropods with many legs scurry across the mudflat. Can you imagine that such a scene ever took place in your state? How do we know that such events actually happened? (A)

Three hundred million years later, the scene has changed dramatically. In this unit, you have been learning about the forces that caused some of these tremendous changes. Now you are going to learn how scientists use fossils to reconstruct the events of the past. All kinds of fossils are found across the United States. Have you ever collected fossils or seen any in museums? (B)

What does fossil evidence suggest about your state's history? Can you discover more about it? Consult an encyclopedia or your librarian for more information.

LESSON 1 ORGANIZER

Time Required
2 class periods

Process Skills
observing, analyzing, inferring, hypothesizing

Theme Connection
Energy

New Terms
Fossil—the remains of prehistoric animals and plants preserved in rock
Paleontology—the study of fossils, ancient life-forms, and their evolution

Materials (per student group)
none

Teaching Resources
Transparency Worksheet, p. 47
Resource Worksheet, p. 49
Activity Worksheet, p. 50
Transparencies 67 and 68
SourceBook, p. S130

LESSON 1 Rocks Reveal a Story

FOCUS

Getting Started

With your students, put together a terrarium using horsetails, ferns, and club mosses for plants and insects and a frog or salamander for animals. Tell the students that the terrarium is a good model of what North America looked like millions of years ago. There were ferns and horsetails as big as the trees of today and amphibians as large as modern crocodiles. Explain that scientists study this ancient time by investigating the fossils of ancient life-forms.

Main Ideas

1. Fossils can provide information about ancient plants and animals.
2. Fossils can be classified into four different groups.
3. The environment of an organism determines the type of fossilization.
4. Many hypotheses have been proposed to explain the extinction of dinosaurs.

TEACHING STRATEGIES

Call on a volunteer to read the text on this page aloud. Encourage students to discuss the in-text questions. Direct their attention to the illustration. Elicit comments from students about fossils and what they can tell about the history of the Earth.

INDEPENDENT PRACTICE Encourage students to do some library research on the geological history of their state and to identify any characteristic fossils of plants or animals.

Answers to *In-Text Questions*

(A) Students will learn in this lesson that fossils provide evidence of past life-forms.

(B) Encourage students to share stories of collecting fossils and visiting museums.

How Are Fossils Formed?

Involve students in a discussion in order to determine their prior knowledge about fossils. Identify and correct any misconceptions before proceeding.

Types of Fossils

Students may have difficulty understanding how a mold becomes a cast. Help them to understand that, in nature, the mold is filled with mineral matter that has been deposited by ground water in a very slow process. The identification exercise at the bottom of the page can be used to reinforce student familiarity with the different fossil types.

GUIDED PRACTICE If possible, provide fossil samples that illustrate the four different types of fossils. Call on a volunteer to read aloud the material about fossil types. Ask other students to describe how each fossil was formed, referring to what was just read. Imprints, molds, and casts can be demonstrated with plaster, clay, or putty.

★ **A Transparency Worksheet (Teaching Resources, page 47) and Transparency 67 are available to accompany the material on this page.**

Answers to *In-Text Questions*

Ⓐ Accept all reasonable responses. Students will learn more about this topic from the material that follows.

Ⓑ The *Chasmosaurus,* the reptile, and parts of the *Archaeopteryx* are examples of the first category of fossils. The fossils of the *Archaeopteryx* and the trilobite are impressions, casts, or molds. The fossil of the cycad frond was formed by black layers of carbon. The footprints are examples of the fourth category of fossils.

How Are Fossils Formed?

What are fossils? Can you describe how they are formed? Write down some of your ideas, and when you finish, share them with your classmates. Did you change your mind about any of your ideas after your discussion? Ⓐ

Types of Fossils

A **fossil** is the mark or the remains of a prehistoric plant or animal. Most fossils are preserved in rock. Rocks that contain fossils lie undisturbed beneath the surface of the Earth for millions of years. Finally, through weathering and erosion, the fossil-bearing rocks are exposed on the surface.

Fossils that are preserved in rock fall into four groups:

1. bones, shells, or other physical remains
2. impressions, molds, or casts (casts are formed when a mold is filled with sediment that hardens into stone)
3. black layer of carbon in the shape of an organism
4. nests, tracks, trails, or other evidence left by an animal

a *Chasmosaurus*

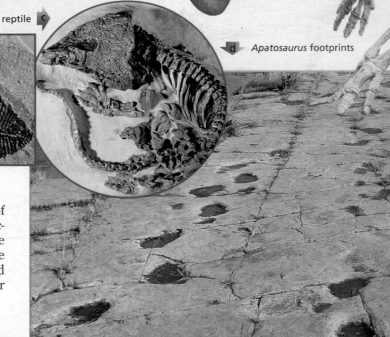

b Cycad frond

c Primitive reptile

d *Apatosaurus* footprints

At least one example of each of these groups is pictured on these pages. Examine each picture and identify the group it belongs to. Record your conclusions in your ScienceLog. Ⓑ

Theme Connection

Energy

Remind students that the energy we get from fossil fuels originally came from the sun. **Focus question:** How did this energy transfer take place? *(Solar energy was trapped in plants during photosynthesis. The remains of these plants became fossilized in sedimentary rock. The petroleum products derived from these rocks can then be burned to release that energy.)* You may wish to have interested students diagram this series of steps.

e Archaeopteryx

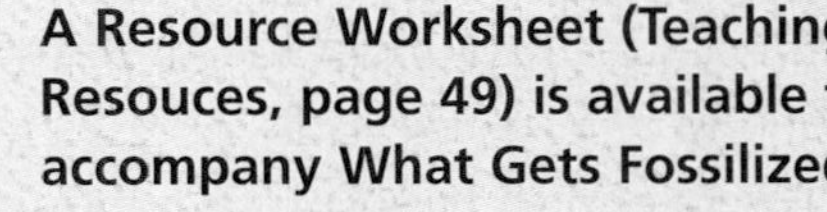

f Trilobite

What Gets Fossilized?

Examples of life are all around us, yet the chance of a single organism being preserved in the geologic record is very unlikely. Why is this so?

The remains of organisms can be destroyed in many ways. If they are left lying on the ground, they can be consumed by scavengers or bacteria. Bones or shells can be crushed by overlying sediments, or they can be eroded along a seashore or in a streambed. Why, then, do some survive in fossil form?

If an organism has hard body parts, or if it is covered by some protective material like soft mud or sand shortly after its death, the remains have a better chance of being preserved.

In your ScienceLog, sort the list below into these three categories: good chance for fossilization, fair chance for fossilization, and poor chance for fossilization. Explain your reasoning for each item. C

a. a tree growing in a swamp
b. deer tracks in the sand
c. a seashell on a beach
d. a horse drowned while crossing a river's flood plain
e. a jellyfish in the ocean
f. a freshwater clam in a stream
g. dinosaur tracks in mud
h. a small animal caught in a tar pit
i. a reindeer killed by disease on an island
j. an evergreen growing on a rocky ledge

Something to Think About

In which rock group would you expect to find the greatest number of fossils? the fewest fossils? Give an explanation for your answers. D

Future Fossils

Imagine that you have been transported several million years into the future. Discuss with your classmates what you might find in the rocks that exist in your area. What are some human-made objects that might survive? Which ones might not? What could a scientist of the future learn about our civilization from these fossils? E

A Classroom Collection

Do you live in an area where fossils are found? Perhaps you have seen fossils embedded in rock. The fossils could be as small as a leaf or as large as a dinosaur. Look carefully at the rocks around your neighborhood. Also, look at the stones used in constructing large buildings—sometimes fossils can be seen there, too. If possible, make a display of fossils brought in by your class.

What Gets Fossilized?

Call on a volunteer to read this section aloud. You may wish to divide the class into small discussion groups to do the classification exercise. Students need to determine the conditions that must exist for organisms to become fossils. Students should conclude that an organism's environment is the key to whether it will be fossilized.

Show the class several objects, such as shells, twigs, bones, and leaves. Ask students to consider good and poor environments for fossil formation. Suggest that they think about the effects of natural agents that tend to destroy soft and hard body parts.

After students have grasped the importance of the environment to fossilization, have them do the classification exercise.

★ **A Resource Worksheet (Teaching Resouces, page 49) is available to accompany What Gets Fossilized?**

A Classroom Collection

Encourage students to look for fossils embedded in rocks or stones in their neighborhood. Good places to look for fossils are outcrops of sedimentary rock. Local museums or universities may be able to suggest nearby fossil-hunting sites.

Homework

The activity described in A Classroom Collection makes an excellent homework assignment.

Answers to *In-Text Questions*

C **a.** Poor—The bacteria present in the swamp will decompose the wood before it can be fossilized.
b. Poor—The tracks in the sand are easily washed or blown away.
c. Fair—Shells can be buried and preserved.
d. Fair—The rapid deposition of sediments may bury the horse's remains.
e. Poor—The jellyfish's lack of hard body parts makes it unlikely to be fossilized.
f. Fair—If the stream is slow-moving, the clam may be buried in sediments.
g. Fair—Fine sediments may preserve the tracks, depending on the location of the mud.
h. Good—Tar acts as a preservative.
i. Poor—Bacteria are likely to decompose the reindeer.
j. Poor—There is little opportunity for a protective covering to preserve the remains of the tree.

D The greatest number of fossils would be found in sedimentary rocks. The fewest would be in igneous and metamorphic rocks. The extreme heat and pressure that cause these latter rocks to form would destroy most organic material before it could be preserved.

E Remind students that the geologic processes occurring today will affect the preservation of any human-made objects. Direct groups to think of objects that might be preserved (objects made of metal or plastic, for example) and others that might not (such as food and other soft or perishable items). Students should be prepared to defend their choices.

What Killed the Dinosaurs?

Before students read the hypotheses about dinosaur extinction, ask them to offer their own extinction hypotheses. Record each suggestion on the chalkboard without evaluation. When students have exhausted their list of ideas, refer them to the hypotheses in the text.

Ask if all of the hypotheses listed in the text are equally plausible. Emphasize that good hypotheses can be tested and often lead to further questions. Although students may not have the background to decide conclusively whether the hypotheses are testable, review the process of hypothesis formation and testing.

Answers to *In-Text Questions*

Ⓐ Sample answer: The dinosaurs lost their food resources due to new, competing species.

Ⓑ The only way to test these extinction hypotheses is to examine the fossil record and other geological data.

- Radiation-induced changes in rocks of that age would support this hypothesis.
- This hypothesis is hard to test because plant materials (including any poisons) from that time have decomposed.
- Evidence of climatic changes could be sought in the fossil record.
- Evidence of a massive plant extinction could be sought in the fossil record. Remains of the dust cloud could be found in sedimentary rock layers. Evidence for this theory exists in the form of an iridium layer in the rock record at the Cretaceous-Tertiary boundary. However, critics argue that dinosaurs were dying off before the asteroid struck and that evidence from volcanic eruptions suggests that the dust clouds created by the impact would not have had long-term effects.
- This hypothesis is difficult to test. Although dinosaur remains could be examined for signs of death by disease, many diseases would not be detectable in the fossil remains.

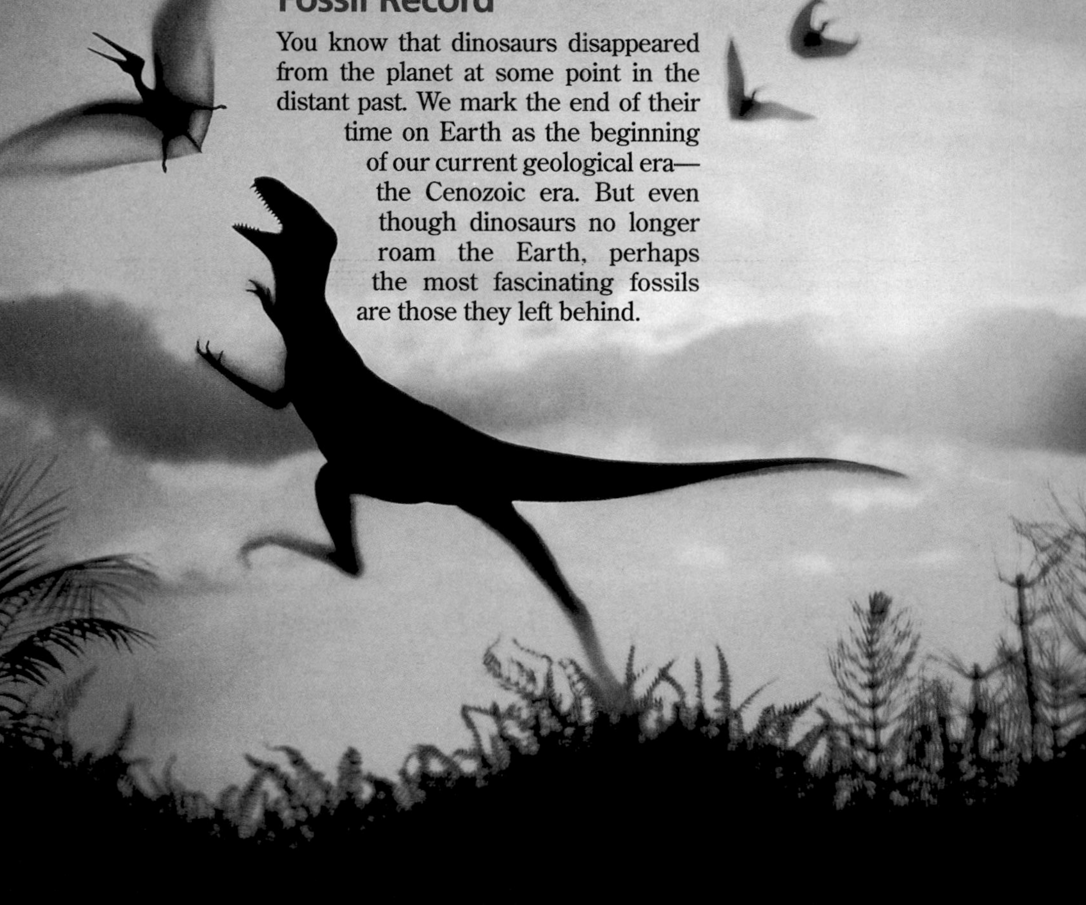

Interpreting the Fossil Record

You know that dinosaurs disappeared from the planet at some point in the distant past. We mark the end of their time on Earth as the beginning of our current geological era—the Cenozoic era. But even though dinosaurs no longer roam the Earth, perhaps the most fascinating fossils are those they left behind.

What Killed the Dinosaurs?

Many hypotheses have been proposed to explain the sudden disappearance of dinosaurs 65 million years ago. Can you think of any yourself? Ⓐ Science relies on hypotheses that can be tested. A hypothesis will begin to gain acceptance only if it seems correct even after a great deal of data has been collected and examined. Many hypotheses have been suggested to explain the extinction of dinosaurs. Some of them are listed here.

- Radiation from an exploding star burned them.
- Flowering plants developed lethal poisons that the dinosaurs ate.
- The climate changed in ways that dinosaurs could not cope with.
- A huge asteroid collided with the Earth, creating a vast dust cloud that blocked the sun. This lessened photosynthesis and drastically lowered the temperature.
- An epidemic of some sort killed them.

How many of these hypotheses could be tested? How likely do you think they are? Ⓑ

406

Meeting Individual Needs

Gifted Learners

Ask students to describe the "typical" dinosaur. Students may use terms such as *coldblooded* and *fierce.* However, many paleontologists believe differently. The giant size of some dinosaurs probably meant that they were too large for their body temperatures to vary like those of modern reptiles. Instead, dinosaurs are believed to have been warmblooded animals, just as mammals are. In addition, scientists have discovered nesting grounds where certain dinosaurs laid and cared for their eggs, and other evidence suggests that some species traveled in herds. Have interested students prepare a report based on articles from recent periodicals. Students should focus on the behavior and ecology of dinosaurs as well as on their anatomy.

You Be the Scientist

Paleontology is the study of fossils, ancient life-forms, and their evolution. One way that paleontologists learn about ancient animals is to examine their fossilized footprints. Answer the questions below to learn more.

What Tracks Can Tell Us

1. Have you ever made footprints in mud, wet sand, or snow? How do they change when you walk, run, or hop? Sketch your ideas and compare them with a classmate's sketch.
2. Would your footprints in mud, sand, or snow look exactly like those of your classmates? Why or why not?
3. Suppose you are a paleontologist whose task is to interpret the fossilized footprints shown on this page. How would you arrange these footprints to create a logical sequence? What story do the footprints tell? The following questions can help you discover some things about the animals who made these tracks:
   - How many animals were involved?
   - In which direction did each animal move?
   - What can you tell about the size of each animal?
   - How many legs did each animal have?

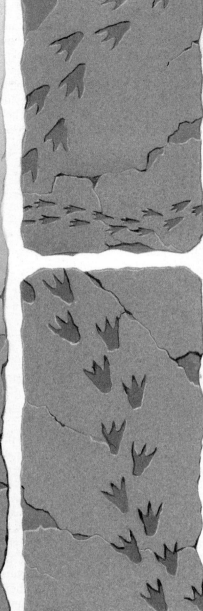

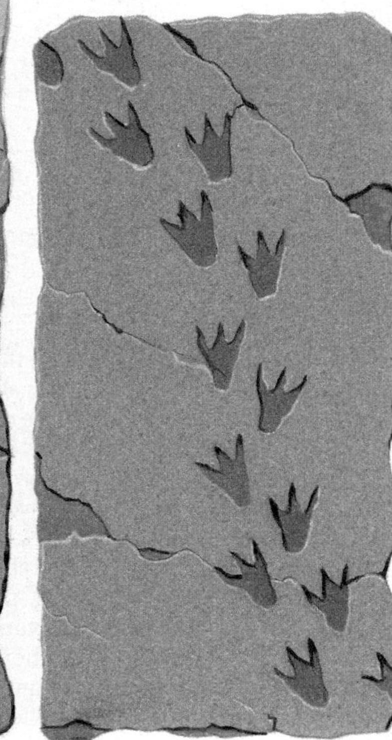

Answers to *What Tracks Can Tell Us*

1. Footprints show differences in spacing and weight distribution when different types of movement are involved. Students' sketches should reflect these differences.
2. No; students should recognize that their classmates may have feet of different sizes and may take shorter or longer steps.
3. Sample answer: Two animals were involved, one larger than the other. They were moving in the same direction, and then their paths intersected. We can't be sure, but these animals could have been dinosaurs that walked on two legs. The blocks should be arranged so that the one in the upper-right corner replaces the block below it. The block in the lower-right corner should be moved and turned so that the footprints on it lead away from the cluster of footprints on the large block. One possible story is that the paths of a large animal and a small animal crossed. There was a struggle, and the large animal walked away.

CROSS-DISCIPLINARY FOCUS

Art

Have students redraw the footprints shown on this page so that they tell a different story. You may wish to have students examine each other's work.

Homework

The Activity Worksheet on page 50 of the Unit 6 Teaching Resources booklet makes an excellent homework activity. If used in class, Transparency 68 is available to accompany it.

FOLLOW-UP

Reteaching

Have students make molds, casts, or impressions of everyday objects in clay or plaster. Then have them explain how ancient organisms might have been fossilized in similar ways.

Assessment

Bring in photographs or a collection of fossils. Have students study the items and classify them according to the four types of fossils.

Extension

Have students research the changes in the Earth's surface from the age of the dinosaurs to modern times. They should focus on how the moving continents may have affected living species.

Closure

Take a field trip to a natural history museum, or show a video about the age of dinosaurs. Emphasize how careful research and investigation of the fossil record have aided scientists in understanding the history of the Earth as well as the development of different species.

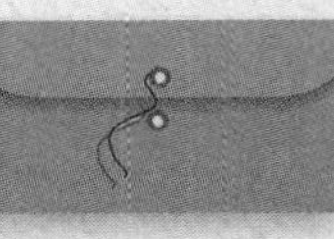

PORTFOLIO

Students may include their answers to What Tracks Can Tell Us in their Portfolio.

Telling Time With Rocks

FOCUS

Getting Started

Obtain a cross-sectional slice of a small tree trunk. Before class, count the rings, and mark several rings that represent historical events that the students would recognize. Then ask students if they know how to tell the age of a tree. *(Many will know to count the rings.)* Point out the rings you have marked, and discuss the historical events with the students. Explain to students that in this lesson, they will learn about how rock layers, much like tree rings, can be used to find out about the age of rocks and the fossils they contain.

Main Ideas

1. A relative time scale shows a sequence of events in order. It does not provide information about the amount of time involved.
2. The events of the Earth's history have been placed in order on the geologic time scale.
3. The subdivisions of the geologic time scale are based on recognizable changes, such as changes in life-forms and episodes of mountain-building.

TEACHING STRATEGIES

A Relative Time Scale

Help students understand that a relative time scale gives only one kind of information: whether one event occurs before or after another. An exact history (an *absolute* time scale) requires specific dates. Working individually or in pairs, have students list events (a) through (i) in chronological order.

Homework

You may wish to assign the activity described in A Relative Time Scale as homework.

LESSON 2

Telling Time With Rocks

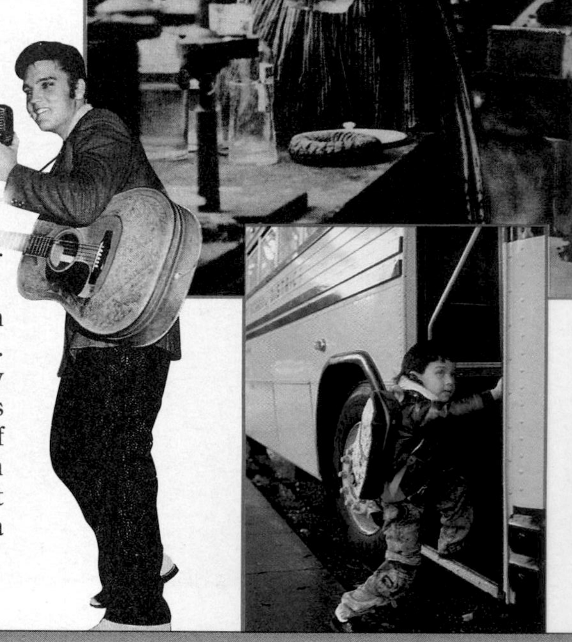

The Rock Record

You know that geologists are interested in events that took place long before humans were around to record them. You also know that the rocks of the Earth's crust contain evidence of these events. In this lesson, you will find out how geologists have organized this evidence into a geological history of the Earth. The first step is to learn what is meant by *relative time.*

A Relative Time Scale

In your ScienceLog, place the following events in order, putting the most recent event at the top of your list: Ⓐ

a. You started kindergarten.
b. Columbus discovered the New World.
c. Elvis hit the top of the charts.
d. The first humans landed on the moon.
e. Marie Curie won the Nobel Prize for her discovery of the elements radium and polonium.
f. The Berlin Wall was torn down.
g. Magellan set sail to explore the world.
h. You were born.
i. Today.

Did you have trouble deciding the order of events? Is the time span between all the events the same? Did most of these events occur recently or before you were born? Ⓑ

When you list events in the order in which they occurred, you make a relative time scale. You do not need the exact dates; you simply need to know what happened first. This gives you a scale based only on the sequence of events. Such a scale does not tell you how much time is involved or exactly when an event occurred. However, it does show whether a given event happened before or after another.

LESSON 2 ORGANIZER

Time Required
2 class periods

Process Skills
organizing, predicting, classifying, analyzing

New Terms
Epoch—a subdivision of a period
Era—a subdivision of the geologic time scale
Geologic time scale—a time scale that covers the Earth's history from its origin to the present
Period—a subdivision of an era

Materials (per student group)
none

Teaching Resources
Activity Worksheet, p. 51
Transparencies 69 and 70
SourceBook, p. S132

The Geologic Time Scale

Just as you ordered the historical events on page 408, geologists and paleontologists have ordered the geological events recorded in sedimentary rock. In doing this, they developed the **geologic time scale.** On this scale, geological and biological events recorded in rocks are placed in the order in which they occurred.

Subdividing the Geologic Time Scale

The geologic time scale started out simply. Then, as more data became available, the scale became more complex. First, rock formations and layers were placed in order. Next, geologists faced the task of subdividing this time scale. To see how they approached this task, look back at the relative time scale you made earlier. Classify the events into various groups, calling each group an "era." For example, you might include Columbus's discovery of the New World and Magellan's voyage in "the era of exploration." Do some eras contain only one event? If so, think of at least two more events to add to each era. Now subdivide the eras into smaller groups, giving each its own name. **C**

On what basis did you make your subdivisions? Geologists usually subdivide their time scales based on the types of life-forms present at certain times. They call each subdivision an **era**. The eras are shown in the box below.

Eras are then subdivided into **periods**. For instance, the Mesozoic era consists of three periods: the Triassic, the Jurassic, and the Cretaceous. Periods can be further divided into **epochs**.

The dividing line between any two time units—for example, between one era and the next or one period and the next—is based on a recognizable change in the fossil record. For instance, the extinction of the dinosaurs separates the Mesozoic era from the Cenozoic era. You should keep in mind that the dividing line is never sharp or sudden. Rather, the division spans the time that it took for a major change to occur.

Eras

Cenozoic ("recent life")	the current era
Mesozoic ("middle life")	the era of the dinosaurs
Paleozoic ("ancient life")	the era of the earliest complex life-forms on Earth

The time before the Paleozoic era is called Precambrian. Only very primitive life-forms existed then.

The movement of continental plates created the Appalachian Mountains at the end of the Paleozoic era.

Answers to *In-Text Questions, pages 408–409*

A The probable sequence, from the present to the past, is as follows: i, a, f, h, d, c, e, g, b.

B The time span between the events is not the same. Most of these events occurred before students were born.

C Possible divisions include the era of exploration, the era of electronics, my personal era, etc. Students should group events by similarities or common factors.

The Geologic Time Scale

Call on a volunteer to read this section aloud. Emphasize the analogy between the sequencing activity on the previous page and the development of the geologic time scale by scientists.

★ **Transparency 69 is available to accompany The Geologic Time Scale.**

Subdividing the Geologic Time Scale

You may wish to perform this section's activity as a class. Have a volunteer read the text to the class. Elicit ideas from the students about how the different events presented on the previous page might be grouped. Have students add more events to each group and then subdivide them again. Dates are still needed before an accurate picture can be presented.

The specific names of eras, periods, and epochs are not meant to be memorized. The information is presented so that students can use the time scale in a later exercise. Emphasize that each dividing line between subdivisions is not a sharp break, but is a transition zone.

INDEPENDENT PRACTICE Have students create a relative time scale that organizes a sequence of events. Encourage students to draw on their personal experiences as well as their knowledge of changes in culture, politics, and other aspects of society. Topics that may interest students include the civil rights movement, the Olympic Games, American history, space exploration, World War II, and music.

Answers to *Use Your New Knowledge!*

1. The geologic time scale is subdivided into eras, periods, and epochs.
2. Large units of time are subdivided in order to provide more specific information. Such units are easier to use when communicating information.
3. The division is based on a recognizable change, such as a change in life-forms.
4. The dividing line is not sharp, but rather is a zone of transition.
5. Actual dates are missing.

Clocks in Rocks

The concept of radioactive dating is complex and has been presented here only to explain how dates were finally assigned on the geologic time scale. You may wish to inform students that radioactive elements decay to become new substances, called *daughter* substances. Radioactive elements decay at a steady pace. Thus, by measuring what proportion of a radioactive element has decayed into the daughter substance, scientists can determine the age of the rock containing the radioactive element.

Integrating the Sciences

Earth and Life Sciences

Have students label the eras and periods on adding-machine tape to create a geologic time scale. Have them use an encyclopedia or the illustration on the next page to include some of the life-forms that existed at various points along the scale. Encourage them to think about how changes on the Earth may have allowed changes in the life-forms. *(For instance, changes in global temperature or humidity may have limited the range of some species and allowed the spread of others.)*

Use Your New Knowledge!

Based on this table and what you read about the geologic time scale, answer these questions in your ScienceLog:

1. How is the geologic time scale subdivided?
2. What is the purpose of subdividing the large units of time?
3. How do geologists decide where one division ends and a new one begins?
4. How sharp is the dividing line between subdivisions?
5. What is still missing from the geologic time scale? (Hint: Check the next paragraph.)

Geologic Time Scale

Era	Period	Epoch
Cenozoic	Quaternary	Holocene (recent)
	Tertiary	Pleistocene Pliocene Miocene Oligocene Eocene Paleocene
Mesozoic	Cretaceous Jurassic Triassic	
Paleozoic	Permian Pennsylvanian Mississippian Devonian Silurian Ordovician Cambrian	

Clocks in Rocks

The geologic time scale has no dates—it merely puts the events of the Earth's history in sequence. How scientists finally found an accurate way to date the Earth is an interesting story. The debate over the age of the Earth had been raging for hundreds of years. Various techniques for dating the Earth had been tried, and estimates ranged from 10 million to 100 million years old. As you will find out, none of these estimates were even close!

The debate was finally resolved near the beginning of the twentieth century. Several scientists contributed to the solution. In 1895, a French physicist named Henri Becquerel discovered that uranium salts were the source of a previously unknown form of energy. Shortly thereafter, two other French physicists, named Marie and Pierre Curie, applied the term *radioactivity* to Becquerel's discovery. In 1905, Ernest Rutherford and Harriet Brooks, two physicists in Canada, suggested that the age of a rock could be determined by measuring the amount of radioactivity emitted by the rock. Rutherford first determined the age of a rock in his laboratory the next year. At last, geologists had a method for putting dates on the geologic time scale. Now scientists estimate that the Earth is at least 4.5 billion years old!

To find out more about radioactivity and rock age, turn to page S133 in the SourceBook.

Harriet Brooks
An early researcher of rock age

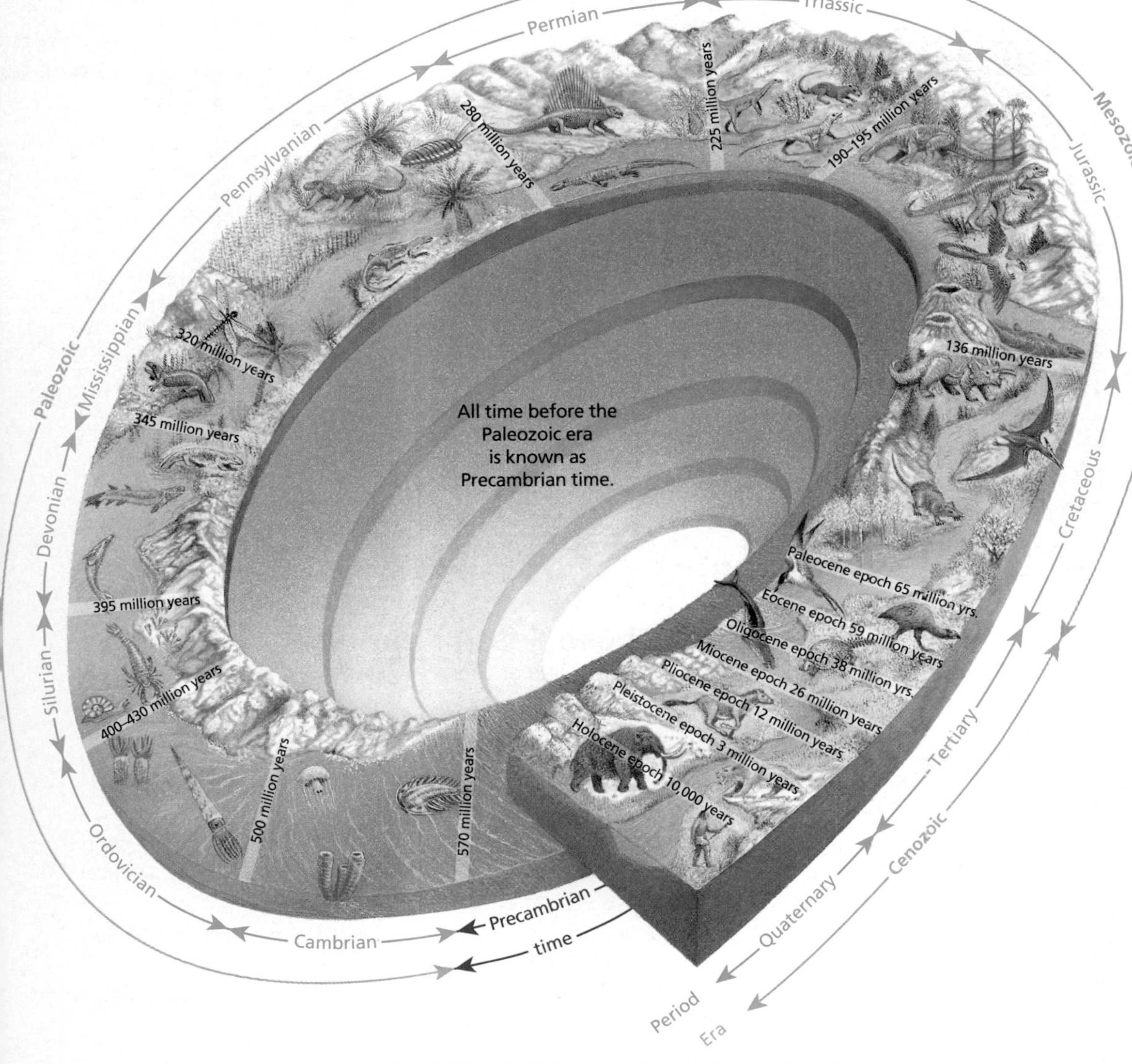

Relative and absolute time and the history of the Earth

Investigating the Geologic Time Scale

Use this diagram, which shows the eras, periods, and epochs and the times when they occurred, to answer the following questions:

1. Which was the longest era? the shortest era?
2. Which division of time provides the least fossil evidence?
3. When did plants first appear on land? When did animals first appear on land?
4. In which period did the first forests grow? How long ago was it?
5. Suggest some changes your grandchildren might see when they study the geologic time scale.

411

Answers to *Investigating the Geologic Time Scale*

1. The longest era is the Paleozoic; the shortest is the Cenozoic.
2. The rocks from Precambrian time provide the least fossil evidence because few life-forms existed during that time.
3. Plants first appeared on land during the Silurian period. Animals first appeared on land during the Devonian period.
4. The first forests began growing in the late Mississippian period, about 320 million years ago.
5. Due to improvements in dating techniques and new fossil finds, the dates that appear here may be revised, and our knowledge of the life-forms that existed at various times may improve.

Homework

You may wish to assign the Activity Worksheet on page 51 of the Unit 6 Teaching Resources booklet as homework. If you choose to use this worksheet in class, Transparency 70 is available to accompany it.

FOLLOW-UP

Reteaching

Have students make a relative time scale of events from their own lives. Then ask them to convert it to an absolute time scale by adding specific dates.

Assessment

Have students create an alternate illustration of the geologic time scale by using adding-machine tape or another continuous source of paper. A convenient scale is 1 cm for 1 million years.

Extension

Divide the class into groups. Assign each group a different geological period to investigate. Have students find out the absolute time span for their period and what kinds of organisms lived during their period. The groups should then present their findings to the class in the form of an illustrated poster. Display the posters in chronological order.

Closure

Have students plan a field trip to a fossil-collecting area. Local natural history museums may have specific information on nearby areas. Check with the museum personnel about any regulations regarding fossil collecting.

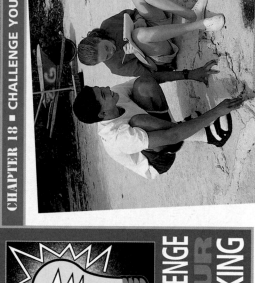

Answers to *Challenge Your Thinking*

1. **a.** It is theorized that at one time all of the Earth's continents were connected, forming one large continent. (Students may know that scientists call this continent Pangaea.)
   b. These sedimentary rocks were formed on the ocean floor. As layers of pebbles, soil, and other particles collected, the remains of marine organisms also became trapped in the layers of sediment.
   c. Few organisms existed in Precambrian time.
   d. Most plants do not have hard body parts. Their cells decompose easily, leaving no exoskeleton or inner skeleton to be preserved.
   e. Magma, the molten rock that becomes igneous rock, is found deep in the Earth where few organisms exist. Also, the extreme heat involved destroys most organic material.
2. Students will have a variety of hypotheses as well as a variety of opinions on the likelihood of each hypothesis. Answers should demonstrate an understanding of how geologists use the fossil record to test their hypotheses. For information about how to evaluate the likelihood of a hypothesis, see the Annotated Teacher's Edition margin on page 406.
3. **a.** These are the fossilized physical remains of an organism.
   b. The organism was an animal—specifically, a dinosaur. (Students may be interested to know that this is a *Coelophysis,* a meat-eating dinosaur that may have been cannibalistic.)
   c. The dinosaur remains were most likely protected from destruction, weathering, and decomposition by a layer of soft mud or sand.
4. One hundred million years ago, the Earth was in the Cretaceous period. Students' descriptions will vary but should reflect the information they have learned about that time period.
5. Sample concept maps for each are on pages S205–S207.

1. Explain These Fossil Findings

Give one or more possible explanations for each of the following statements:

a. Fossils of similar species are found in Africa, Asia, South America, and North America.

b. Some sedimentary rocks contain a large number of marine fossils.

c. Rocks of Precambrian time have few fossils.

d. Fossils of plants are less common than fossils of animals.

e. Fossils are seldom found in igneous rocks.

2. The Case of the Disappearing Dinosaurs

Look back at page 406 and review the hypotheses about how dinosaurs became extinct. With two or three of your classmates, propose some additional hypotheses. Share your hypotheses with your classmates, and write all of them down in your ScienceLog. Then answer the following questions:

a. Which of these hypotheses could be tested?

b. How likely do you think each hypothesis is? Explain your reasoning.

★ **You may wish to provide students with the Chapter 18 Review Worksheet on page 52 of the Unit 6 Teaching Resources booklet.**

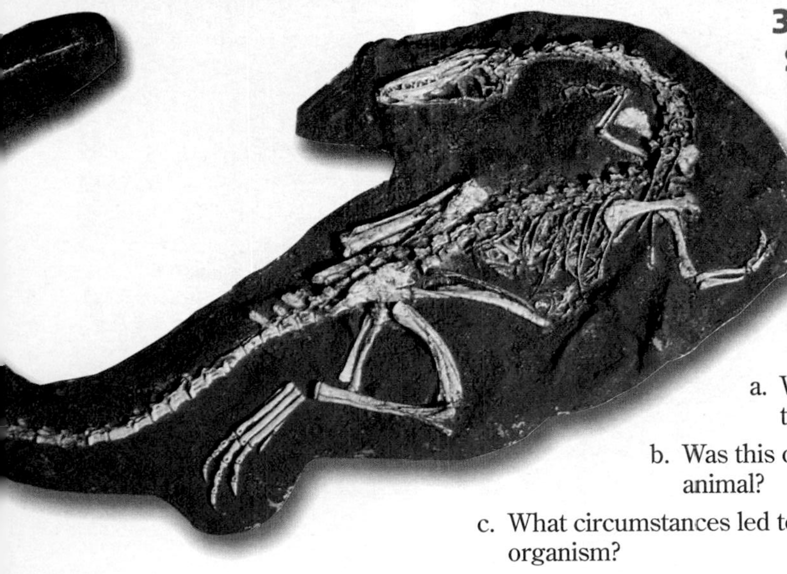

3. Test Your Skill as a Paleontologist

Imagine that you are a paleontologist and just discovered the fossil shown here. Answer the following questions in your ScienceLog:

a. What type of fossil is this?

b. Was this organism a plant or an animal?

c. What circumstances led to fossilization of the organism?

Refer back to the information on pages 404–405 if you need help.

4. Life in an Ancient Ocean

Fossil evidence leads scientists to believe that 100 million years ago the western plains of the United States were covered by a warm, shallow sea. Check the geologic time scale to discover what animals existed then. Imagine that you were an animal living on the coast of that sea. Write a short story or draw a colorful picture that illustrates what life and your surroundings were like.

5. Mapping the Past

Create a concept map to illustrate one of the following:

a. the types of fossils and the examples shown on pages 404–405

b. the chances of fossilization for the organisms listed on page 405

c. the division of the geologic time scale into eras, periods, and epochs

Review your responses to the ScienceLog questions on page 402. Then revise your original ideas so that they reflect what you've learned.

The following are sample revised answers:

1. The dinosaur skeleton was probably covered by a protective layer of material such as soft mud or sand. This kept it from being destroyed.
2. By studying fossils, humans can learn about life-forms that once inhabited the Earth and about how biological communities have changed throughout Earth's history.
3. By studying the layers of sediment in which fossils are found, geologists have been able to develop a relative time scale that puts biological and geological events in order. Geologists also have various methods for dating fossils. Using these techniques, geologists have ruled out the possibility that humans and dinosaurs coexisted.

Multicultural Extension

Cultural and Scientific Creation Stories

Using the library or an on-line database, obtain several myths and folklore stories from African, Greek, Western European, Mexican, and American Indian cultures about how the Earth and its life-forms were created. Share these stories with students, and ask them to compare the stories to each other and to the scientific explanation of the Earth's history.

Making Connections

The Big Ideas

The following is a sample unit summary:

Mountains are formed when continental plates collide and the Earth buckles upward (uplift) or from the action of volcanoes.

Earthquakes are caused by the release of energy from the friction lock between blocks of crust slipping in opposite directions. The energy is released in the form of shock waves that originate at the focus and radiate outward. (1)

The elastic rebound theory explains how plates of rock locked by friction return to their original form after being strained by the pressure of surrounding rock layers. As the crustal plates spring back into their original shape, the stored energy is released as intense vibrations called an earthquake. (2)

Buildings may have features such as flexible pipes, cross-braces, base isolators, mass dampers, and active tendon systems to help them withstand earthquakes. (3)

Movement of the Earth's crust causes volcanoes to occur. When plates separate, volcanic rock fills the gap. When plates collide, some of the rock involved melts into magma that rises to the surface, causing a volcano.

The theory of plate tectonics describes the locations of sliding boundaries between Earth's floating crustal plates. At these boundaries, earthquakes and volcanoes occur. (4)

Igneous rocks are made either of one material and are glassy or porous, or they are made of interlocking grains that are randomly arranged. Igneous rocks are formed from molten rock. Sedimentary rocks are made of several kinds of rock grains that are cemented, or lithified, together. These rocks are formed by chemical precipitation or the erosion and weathering of other rocks. Metamorphic rocks may be made of one material and arranged in sheets or may be made of several different materials with grains that are closely interlocked or arranged in bands. Metamorphic rocks form when heat and pressure from the environment chemically change an existing rock. (5)

Rocks are recycled through erosion and weathering into sedimentary rock, through melting into igneous rocks, and through chemical changes into metamorphic rock. (6)

Fossils are formed by preservation of bones or shells; formation of a cast mold, impression, or organism-shaped carbon layer; or preservation of a trail left by an animal. (7) Fossils are most commonly found in sedimentary rocks because the heat and chemical changes that form other rock types often destroy delicate remains. (8)

A relative time scale is a list of events in the order that the events happened. Actual dates are not included. (9)

After the Precambrian time, the geologic time scale is divided into three major eras: Paleozoic, Mesozoic, and Cenozoic. The eras are further divided into periods, and the periods are divided into epochs. (10)

Unit 6 Making Connections

THE RESTLESS EARTH

SOURCEBOOK

For more information on plate tectonics, rocks, and fossils, look at pages S107–S136 in the SourceBook. You will find interesting information about rocks and the stories they reveal. Read to discover more about our restless Earth.

Here's what you'll find in the SourceBook:

UNIT 6

The Big Ideas

In your ScienceLog, write a summary of this unit, using the following questions as a guide:

1. How does the theory of plate tectonics explain mountain-building, earthquakes, and volcanoes?
2. What is the elastic rebound theory, and how does it relate to earthquakes?
3. How are buildings designed to withstand the shock of earthquakes?
4. What does the plate tectonics theory tell us about the location of earthquakes and volcanoes?
5. How do you distinguish among igneous, sedimentary, and metamorphic rocks? How is each formed?
6. How are rocks recycled?
7. How are fossils formed?
8. Where are fossils the easiest to find? Why?
9. What is a relative time scale?
10. How is the geologic time scale divided?

Checking Your Understanding

1. Fill in the blanks for each of the following sentences:
   a. The formation of ___?___ and ___?___ can result in mountains. Mountains can also be formed by the action of ___?___.

b. When an earthquake occurs, a ___?___ measures its magnitude. The actual place where the earthquake occurs is called the ___?___, and the place where it strikes the Earth's surface is called the ___?___.

c. The change in location of continents can be explained by the theory of ___?___ ___?___. This theory also helps explain why ___?___, ___?___, and ___?___ occur at plate boundaries.

d. In a volcano, ___?___ is released either through small cracks on the crust or through a large vent. When it emerges, it is called ___?___. A ___?___ volcano no longer erupts with this material.

2. How good are you at making connections? How are the words in each pair related to each other?

a. plate tectonics and earthquakes

b. the Cenozoic era and the extinction of dinosaurs

c. volcanoes and the formation of igneous rock

d. shale and slate

3. Write a story that traces the formation of one of the following items:

- a sedimentary rock
- a fossil of a prehistoric turtle
- the Himalaya Mountains
- a volcanic island

4. concept map Copy the concept map into your ScienceLog. Then complete the map using the following words: basalt, pumice, metamorphic, shale, rocks, igneous, slate, limestone, sedimentary, marble.

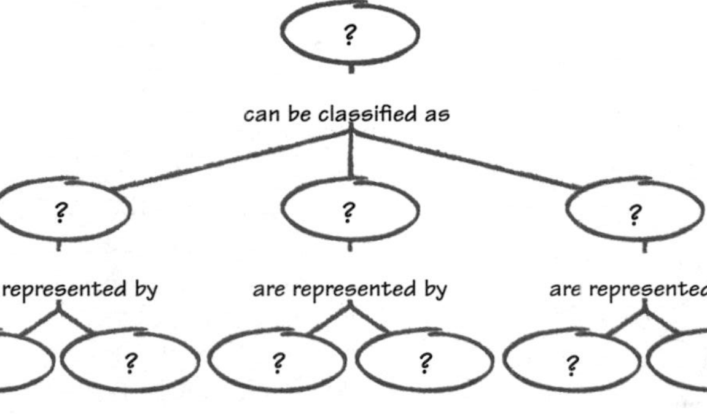

415

Homework

You may wish to assign the Unit 6 Activity Worksheet: Geology Word Search as homework (Teaching Resources, page 58).

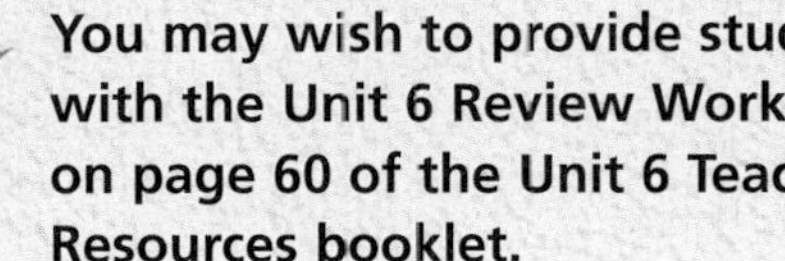

You may wish to provide students with the Unit 6 Review Worksheet on page 60 of the Unit 6 Teaching Resources booklet.

Answers to *Checking Your Understanding*

1. a. Folds; faults; volcanoes

b. Seismograph; focus or hypocenter; epicenter

c. Plate tectonics; for the next three blanks, possible answers include mountains, earthquakes, volcanoes, folds, and faults.

d. Magma; lava; dormant

2. Sample answers:

a. The movement of crustal plates is explained by the theory of plate tectonics. When two plates meet and slide past each other, an earthquake occurs.

b. The extinction of the dinosaurs separates the Mesozoic era from the Cenozoic era, which is the era of the most recent life-forms on Earth.

c. Molten rock cools to form igneous rock. Molten rock reaches the Earth's surface through the eruptions of volcanoes; after erupting, the molten rock cools to form igneous rocks such as basalt or pumice.

d. Shale is a sedimentary rock. Heat and pressure from surrounding rock can cause chemical changes in shale, thus creating slate, a metamorphic rock.

3. Answers will vary. Discussion of the formation of sedimentary rock should include erosion, weathering, transport by water or wind, and lithification. Discussion of fossilization of a prehistoric turtle should include finding the preserved remains of the shell or the formation of a cast. Students should discuss the turtle's place on the geologic time scale. Discussion of the Himalaya Mountains should include the theory of plate tectonics, uplift, and mountain-building. Discussion of a volcanic island should include the theory of plate tectonics, magma, and a quiet or explosive eruption.

4.

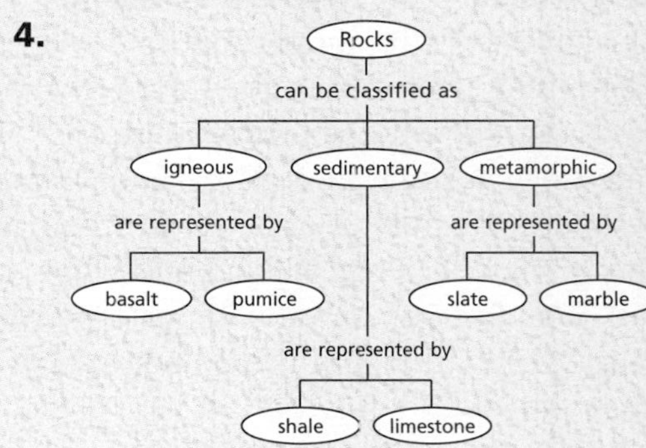

Science Snapshot

Background

Paleontology is the study of extinct organisms, including their structure, behavior, environment, and evolution. It requires an understanding of the geological context in which fossils are found, the physical processes by which fossils are formed, and the biology and ecology of the organisms themselves. Therefore, paleontology integrates Earth, physical, and life sciences.

Jack Horner has received numerous honors for his work as a paleontologist, including an honorary doctorate from the University of Montana and a MacArthur Foundation Fellowship (commonly known as the "Genius Award"). In addition to his work linking dinosaur behavior to that of birds, he has also been instrumental in supporting the view that dinosaurs were warmblooded and that dinosaur babies might have had some kind of downy body covering.

Critical Thinking

Present the following situation to your students: Suppose that a paleontologist unearths a fossil bed that contains no whole skeletons but resembles a jumbled pile of bones. The bones from different animals and from different parts of the animals are randomly mixed together in the fossil bed. How might this fossil bed have formed? *(Students might suggest that the bones were left behind by a predator or carried and mixed together by a stream. Encourage students to think of ways that a paleontologist could test these hypotheses.)*

Science Snapshot

Jack Horner (1946–)

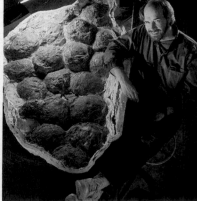

The mother nuzzles her babies in the nest. She feels the warmth of their bodies as their downy skin tickles her nose. A worried yelp from another adult startles her. A baby has crawled out of a nearby nest and scampered away. Its parent, who made the noise, quickly recaptures the baby and brings it back to safety.

A New Look at Old Ideas

Would you believe the story above is about a *dinosaur*? People usually think of dinosaurs as huge, scary, coldblooded, scaly beasts. But people's ideas have begun to change, thanks in part to the work of Jack Horner. "I mainly study dinosaur behavior," Horner says. "If I find fossils of several nests close to each other, that tells me that the dinosaurs that built those nests may have lived together in a group.

▼ Cast of a dinosaur hatchling

"My research shows that dinosaurs were really more like birds than lizards. For example, dinosaurs guarded their eggs and took care of their young, just like birds do. Based on this and many other findings, I think that birds probably evolved from dinosaurs." Horner also suggests that dinosaurs may have been *endothermic*, which means they regulated their body temperature more like birds than like lizards.

"By studying the prehistoric world, we can learn how dinosaurs and many other living organisms interacted with their environment over hundreds of millions of years. That information may help us understand how all living organisms today—including humans—affect and are affected by the environment."

A Rocky Beginning

As a child, Horner collected rocks at his father's quarry. He found his first dinosaur bones when he was seven or eight years old. "I've always liked the detective work that's involved in paleontology. You can't study a living dinosaur, so you have to figure out everything using clues from the past."

In school, Horner had difficulties because of a learning disability called *dyslexia*. Dyslexia made it difficult for him to read the papers and textbooks that most teachers used in their classes. Because of this, he earned poor grades throughout his life. But that didn't keep him from pursuing his interests or from wanting to learn. "It's just a matter of curiosity. When you ask questions about your surroundings—no matter where you live—you will be surprised how much science you can explore."

▲ Jack Horner with lambeosaur eggs near Choteau, Montana

Bird Brains

Observe the behavior of birds in your area. Focus on one or two species of birds. Note their eating habits, the sounds they make, and their interactions with other birds. In accordance with Horner's theory, what do your observations suggest about the possible behavior of dinosaurs?

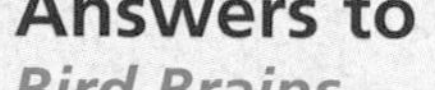

Meeting Individual Needs

Gifted Learners

Have students complete the following activity: Write a research proposal for a paleontology project. Include a specific description of what you plan to investigate (such as the nest-building behavior of dinosaurs), how you plan to proceed (including where you will dig and what you hope to find), and the importance of your investigation, both to paleontology and to the world at large.

Answers to *Bird Brains*

One thing students should notice is that birds are very active creatures, in contrast to our traditional view of dinosaurs as huge and sluggish. They might also want to compare bird behavior with modern reptile behavior, such as that of a turtle or lizard. The differences are primarily due to birds being warmblooded and reptiles being coldblooded. Also, reptiles tend to be solitary creatures, while birds often live in flocks.

SCIENCE and the Arts

The Cullinan Diamond

The largest diamond ever discovered was found in Pretoria, South Africa, by a diamond miner who first thought it was a big chunk of glass. It weighed 3106 carats—about half a kilogram—and measured 5 × 6 × 10 cm. The diamond was named the Cullinan Diamond after the founder of the mine. It was sold for $750,000 by a province in South Africa and was later presented to King Edward VIII of England in 1907 for his 66th birthday.

Cutting the World's Largest Diamond

The Cullinan Diamond was sent to Joseph Asscher, a famous diamond cutter in Amsterdam. The task of cutting it was truly nerve-wracking. The first cut on large diamonds is often made by cleaving the diamond along its grain. The difficulty with this is that the grain is invisible. Asscher studied the diamond for months and finally decided on his line of cleavage. In preparation for the cleaving, he cut a V-shaped groove along that line.

◀ Diamonds are usually found in igneous rock called kimberlite.

The tension was almost unbearable as he lifted his mallet to strike the metal rule he held along the groove. If his line of cleavage was not correct, the world's largest diamond could shatter completely. He brought the hammer down and, strangely, the metal rule broke in two! The diamond was still intact, but Asscher's nerves were shattered. He went to a hospital to recover.

When he finally had the strength and courage to try again, Asscher brought his physician with him. This time, his mallet struck cleanly, and the diamond cleaved perfectly. Asscher himself did not learn of his success until some time afterward, for he had fainted the moment he struck the blow!

The Cullinan Diamond was eventually cut into 9 major gems and 96 smaller ones, plus a number of fragments. The largest gem, the Cullinan I, or Star of Africa, is mounted in the imperial scepter of the British government. It is on display in the Tower of London.

The Diamond Cutter's Art

A diamond is judged by the four C's: color, cut, clarity, and carat weight. Perfect cutting can bring out the full beauty and value of a diamond. The traditional cuts are designed to maximize the brilliance and "fire" of the diamond. One of the most popular cuts is the brilliant cut, used in many rings. In this cut, there are 58 facets, or flat surfaces, including the tiny one at the bottom called the culet.

▼ The British imperial scepter showcases the largest gem from the largest diamond ever discovered, the Cullinan Diamond. This gem is about the size of an extra-large chicken egg.

Find Out for Yourself

Do some research on a favorite gemstone or precious metal. Where is it found? What is its chemical composition? How is it formed? How is it cut or polished?

CROSS-DISCIPLINARY FOCUS

Art

Present students with the following scenario: Due to an accident at the Tower of London, the imperial scepter has been broken. The British government has asked the class to submit designs for new uses for the Star of Africa. In creating their designs, students should keep in mind the size and shape of the diamond. Students should write a short description or draw a picture of the item they design. *(Encourage students to be creative. They might mount the stone in a piece of jewelry or use it for another purpose entirely, such as a drill bit, a gift to another country, or part of a public sculpture. Make sure they consider the shape of the stone and its new setting.)*

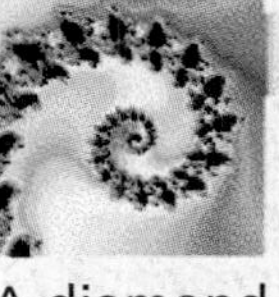

Background

A diamond is the hardest natural substance on Earth. The only material hard enough to cut a diamond is another diamond. Diamonds are a pure form of the element carbon. They have been crystallized in such a way that the carbon atoms form a tetrahedral structure. Their structure gives the diamonds their extreme hardness.

Diamonds are found in kimberlite, an intrusive igneous rock that forms when iron-magnesium-silicate magma is injected into overlying rocks. The injected material hardens in the shape of a cylinder or pipe. Not all kimberlite pipes contain diamonds, however. Several hundred kimberlite pipes are known to exist, but only a few contain diamonds.

Discussion Questions

After students have read the feature, involve them in a discussion of what they have read. You may wish to pose the following questions:

1. What is involved in the process of cutting a diamond? *(Students should mention the need to study the diamond, the task of cleaving the diamond along its grain, and the eventual cutting of the large diamond into smaller ones.)*
2. Where is the Star of Africa mounted and displayed? *(It is mounted in the imperial scepter of the British government. The scepter is on display in the Tower of London.)*

Answers to *Find Out for Yourself*

Sample response: Emerald is a type of beryl (beryllium aluminum silicate) that has very small amounts of chromium in its crystals, giving the gem its color. It is a metamorphic gem that forms at the edge of igneous intrusions in other minerals, such as limestone. It is harder than quartz but softer than topaz, and it is found in many locations around the world. Many of the finest emeralds come from Colombia. Emeralds can be cut in a variety of ways; one popular cut is the step-cut.

EYE ON THE ENVIRONMENT

Background

A mineral consists of a single chemical element or compound. A mineral is a natural, inorganic, and solid material. Some of the most common minerals that are mined include copper, iron, and nickel. (Coal is mined as well, but it is not considered a mineral because it is an organic substance.)

Typically, deposits of minerals form in the Earth's crust as magma cools. In this process, dense minerals sink to the bottom of the liquid magma and solidify, forming what later become layers of ore within the hardened magma. Thus, minerals can be considered igneous rocks. However, some other minerals, such as certain lead, copper, and zinc ores, form in response to heat and pressure and are more appropriately considered metamorphic rocks.

Debate

Divide the class into groups of three to five students each, and assign each group one of the following roles: mining company executive, mine vehicle operator, retailer of pots and pans, environmentalist concerned with forest ecosystems, farmer whose land contains mineral deposits, or water-pollution specialist. Have each group consider the following proposal: Despite mounting costs and greater consumer demand for mined materials, the mining industry should face stricter environmental regulations to protect remaining natural resources. Once students have had time to develop their responses to this proposal, have a representative from each group participate in a panel debate on the proposal.

Answers to *In Your Opinion*

Answers will vary. Students should support their opinions on this issue with a reasonable and thoughtful discussion. Students should recognize the following issues: extent of damage to the environment, economic feasibility of restoring the environment after mining, possible job opportunities for displaced miners, and current demand for mined materials.

EYE ON THE ENVIRONMENT

Mining the Earth's Resources

Imagine a hole 4 km in diameter and almost 1 km deep. That's wider than 35 football fields and more than twice as deep as the Empire State Building is high! The largest human-made hole, located in Bingham, Utah, is just that size. Why would anyone want to dig a hole that large? The answer is copper. Out of this hole comes 227,000 metric tons of copper each year. That much copper makes a lot of pots, pans, pipes, wire, and pennies.

Many Ways to Mine

Mining occurs on the surface and below the surface of the Earth. Subsurface mining requires digging deep shafts into the Earth where the materials are extracted. Surface mining, such as open-pit and strip mining, takes place on the Earth's surface. Huge machines strip away rocks, trees, and grasses before the mining begins. In an open-pit mine, machines dig large holes in the ground. Metal ores, sand, gravel, or building stone are then removed from the pit. Copper and iron ores as well as marble, sandstone, and limestone are often mined this way. In a strip mine, huge bulldozers strip away the top layers of Earth's surface to get at the desired resource. Coal is often mined in this way.

After the deposits are extracted from the Earth, unwanted rock and other materials are dumped. The desired materials are shipped to processing plants where they are refined into usable resources. Some quarries and open-pit mines fill with water over time. Mining companies sometimes restore the land they have mined; however, only about one-third of the mined land in the United States has been restored.

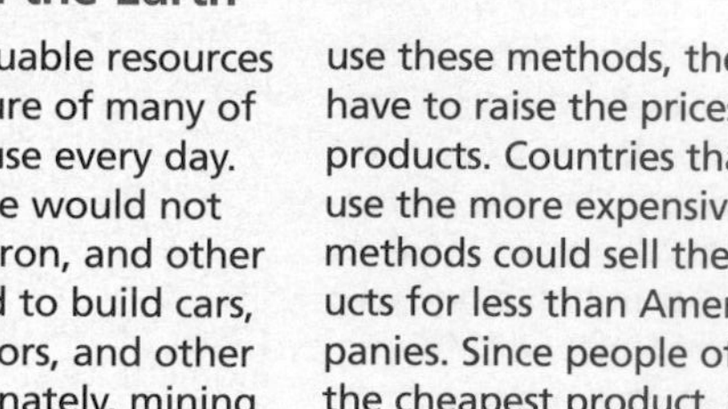

▲ The largest open-pit copper mine produces huge amounts of copper along with some gold and other metals.

The Impact on the Earth

Mines provide valuable resources for the manufacture of many of the products we use every day. Without mines, we would not have the copper, iron, and other materials we need to build cars, stereos, refrigerators, and other products. Unfortunately, mining is one of the most damaging things people can do to the Earth. Mines in the United States produce more waste than all American cities and towns together! Mining dramatically increases erosion, air pollution, and water pollution in the surrounding environment.

We have the technology to decrease the environmental damage caused by mining, but are we willing to pay the price? Cleaner and less destructive methods of mining cost more money. If American companies use these methods, they will have to raise the prices of their products. Countries that don't use the more expensive mining methods could sell their products for less than American companies. Since people often buy the cheapest product, American companies could lose a lot of business.

In Your Opinion

The mining industry not only provides needed resources; it also supports hundreds of thousands of jobs. Is protecting the environment worth people losing jobs and industry losing business? Answer this question for yourself and then express your ideas in a class debate.

418

Meeting Individual Needs

Learners Having Difficulty

Provide each student with a paper napkin, a few toothpicks, and a peanut-butter cup. Students should wash their hands before beginning this activity. Instruct them to remove as much of the mineral deposit (the peanut-butter filling) as they can while causing as little damage as possible to the surface layer of soil (the top covering of chocolate). Once they have completed this task, discuss with students their strategies and the difficulties they encountered. Help them to see the connection between this activity and the problem the mining industry faces in trying to remove underground layers of minerals without disturbing local soils and ecosystems. You may wish to have students brainstorm improvements to their peanut-butter extraction techniques and to give them the opportunity to try out their new ideas. You may also wish to ask students to identify ways in which their model fails to represent a real mining site.

TECHNOLOGY in SOCIETY

Reconstructing Artifacts

A one-of-a-kind fossil tooth is transported thousands of miles and delivered to the lab. A scientist leans over the million-year-old treasure, delicately brushing away a fleck of dirt from its fragile surface. Suddenly the tooth cracks and crumbles into dust, and a critical specimen is lost forever.

Technology Meets Fragile Fossils

Scientists have many methods of preserving fossils, including coating them with hardeners and cushioning them in padded containers. Nonetheless, stories like the one above are not uncommon. Fossils are extremely delicate—the more they are handled and shipped, the more likely they are to fall apart.

One solution to this problem involves state-of-the-art technology. Using a powerful network of computers and three-dimensional laser scanners, fossils and ancient artifacts can be copied and reconstructed. To do this, the fossil is first placed on a rotating table. A laser beam jumps from point to point on the surface of the object and transfers information to a powerful computer. The table turns slightly, and the laser continues its detailed exploration of the fossil's surface. After several hours, the computer has constructed a nearly perfect three-dimensional map of the object.

Reconstructing the Evidence

A second device uses the map of the object to build an exact replica of the fossil or artifact. The replica is made of sturdy materials so that it won't shatter or decompose like the original. People can study such replicas without the fear of damaging an irreplaceable specimen. Also, more than one scientist can study the same object at the same time.

A Solution to a Controversy

This new technology has helped solve another problem as well. In 1990 a federal law went into effect that required many museums to return cultural remains to Native American groups. These include bones, pieces of art, and other artifacts. These remains of ancient civilizations provide great insight into how people lived long ago. Unfortunately, when museums give back these specimens, the opportunity to study these resources is lost. But with this new technology, bones and artifacts can be copied before they are returned to their proper owners, allowing scientists to study them for years to come.

▼ **Computer scanning process of an 8000-year-old human skull. Once scanned, a model will be made to its exact specifications.**

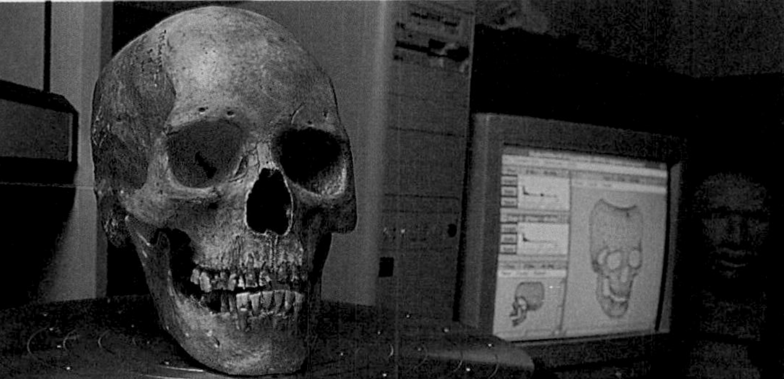

Replication Sensation

How hard is it to make replicas by hand? Make a sturdy model of a bone, eggshell, or other object using modeling clay. The replica should accurately reproduce the size, shape, and features of the original object. How close is your replica to the original? What are the disadvantages of your model?

Answers to *Replication Sensation*

Students should recognize that their models may not represent the full detail of the original. In addition, clay models may be altered or destroyed fairly easily.

TECHNOLOGY in SOCIETY

Background

The technological accomplishments described in this feature were made possible by integrating equipment from several different sources. Eventually, other museums and universities may have equipment that will enable them to create three-dimensional copies of fossils and artifacts. For now, however, the cost is too high. John Kappelman, a physical anthropologist, is in charge of the first such project at the University of Texas. He hopes that one day all researchers will have access to any fossils they need; today most fossils can be studied only by visiting the museums where the fossils are kept.

Kappelman's project has not solved all of the disagreements raised by the Native American Graves Protection and Repatriation Act (NAGPRA) of 1990. First of all, it takes a long time to scan each specimen. One lab cannot possibly keep up with the demands of all the museums wishing to take advantage of this technology. Also, conflicting interpretations of NAGPRA have caused museums and American Indian groups to disagree over many aspects of the repatriation process. In addition, the law applies only to American museums that receive federal funding. Many museums are not covered by this law and have been reluctant to give up their collections.

Extension

Encourage students to think of other ways in which this technology could be used. You may wish to divide the class into small groups and have each group submit a plan for a creative use of this technology. *(Other possible uses include making architectural models or prototypes of machinery.)*

Multicultural Extension

Archaeology

Have students investigate the archaeological traditions of a specific group of American Indians. Students should focus on the cultural artifacts that document the history of the group. What types of artifacts are they, and what purposes did they serve? In what ways are the artifacts functional, and in what ways are they artistic?

Unit 7 TOWARD the STARS

Unit Overview

In this unit, students examine the structure of the universe. In Chapter 19, students seek their own explanations for common astronomical events and explore the theories used to explain these events. In Chapter 20, students investigate how the tilt of the Earth's axis and its orbit around the Sun create the seasons. In Chapter 21, students investigate the solar system by examining the discoveries of three space probes. In Chapter 22, students learn about the composition and life span of stars and about the big-bang theory. They also take an imaginary journey through the universe.

Using the Themes

The unifying themes emphasized in this unit are **Cycles, Systems, Changes Over Time, Energy,** and **Structures.** The following information will help you incorporate these themes into your teaching plan. Focus questions that correspond to these themes appear in the margins of this Annotated Teacher's Edition on pages 429, 434, 447, 478, and 484.

Several opportunities arise to use the themes of **Cycles** and **Changes Over Time** in this unit. These themes can be discussed together in terms of the many abiotic and biotic cycles caused by the revolution of the Earth around the Sun. In Chapter 19, a focus question addresses the changes that occur in the time of sunset during the course of a year, and a focus question in Chapter 20 challenges students to hypothesize what biological variations occur because of the Earth's revolution and axial tilt.

Because celestial bodies are parts of interacting **Systems,** this theme can be used to help students understand the influence of these bodies on one another. In addition, a focus question in Chapter 19 offers students a chance to think about the models that early astronomers developed to explain these interactions.

Energy and **Structures** are also important themes in this unit. Specifically, the method by which stars generate energy is considered in Chapter 22, as is the structural similarity between our solar system and the Milky Way.

Using the SourceBook

Unit 7 focuses on space and space exploration. Students learn about the composition and energy production of the Sun and about other members of the solar system. The unit highlights stars, galaxies, and some of the stranger objects in space. The unit concludes with a brief look at cosmology.

Bibliography for Teachers

Dolan, Terrance. *Probing Deep Space.* New York City, NY: Chelsea House Publishers, 1993.

Lancaster-Brown, Peter. *Skywatch: Eyes-On Activities for Getting to Know the Stars, Planets, and Galaxies.* New York City, NY: Sterling Publishing Co., Inc., 1993.

McSween, Harry Y. *Stardust to Planets: A Geological Tour of the Solar System.* New York City, NY: St. Martin's Press, 1993.

Bibliography for Students

Carlisle, Madelyn W. *Let's Investigate Magical, Mysterious Meteorites.* Hauppauge, NY: Barron's Educational Series, Inc., 1992.

Couper, Heather, and Nigel Henbest. *How the Universe Works.* Pleasantville, NY: The Reader's Digest Association, Inc., 1994.

Jones, Brian. *Space Exploration.* Milwaukee, WI: Gareth Stevens Children's Books, 1990.

Films, Videotapes, Software, and Other Media

Exploring Our Solar System
Videotape
AIMS Media
9710 DeSoto Ave.
Chatsworth, CA 91311-4409

A Field Trip to the Sky
CD-ROM or software (Macintosh or MS-DOS)
Sunburst Communications
101 Castleton St.
P.O. Box 100
Pleasantville, NY 10570

Space
Film or videotape
Coronet/MTI
108 Wilmot Rd.
Deerfield, IL 60015

Star Recognition: A Graphic Guide to the Heavens
Videotape
Britannica
310 S. Michigan Ave.
Chicago, IL 60604-9839

Unit Organizer

Unit/Chapter	Lesson	Time*	Objectives	Teaching Resources
Unit Opener, p. 420				Science Sleuths: The Lost Mining Probe, p. 420 English/Spanish Audiocassettes Home Connection, p. 1
Chapter 19, p. 422	**Lesson 1, What Is Astronomy? p. 423**	1	1. Suggest a definition for astronomy. 2. Identify some of the objects observed and studied by astronomers.	Image and Activity Bank 19-1
	Lesson 2, An Ancient Science, p. 427	4	1. Evaluate models of the universe as put forth by Aristotle, Aristarchus, and Ptolemy. 2. Explain the difference between heliocentric and geocentric models of the solar system.	Image and Activity Bank 19-2 Transparency Worksheet, p. 3 ▼ Exploration Worksheet, p. 5 Activity Worksheet, p. 10 ▼ Transparency Worksheet, p. 11 ▼ Transparency Worksheet, p. 13 ▼ Theme Worksheet, p. 15
End of Chapter, p. 436				Chapter 19 Review Worksheet, p. 17 Chapter 19 Assessment Worksheet, p. 20
Chapter 20, p. 438	**Lesson 1, A Scientific Revolution, p. 439**	2 to 3	1. Explain the motions of celestial bodies by using the Copernican model of the solar system. 2. Explain the apparent retrograde motion of Mars. 3. Explain what is meant by the plane of the ecliptic. 4. Draw a scale model of the distances of the planets from the Sun.	Image and Activity Bank 20-1 Exploration Worksheet, p. 22 ▼
	Lesson 2, Motions and Their Effects, p. 445	4	1. Explain why the amount of daylight changes throughout the year. 2. Describe how the tilt of Earth on its axis creates the seasons. 3. Identify the position of Earth during the summer and winter solstices and the vernal and autumnal equinoxes. 4. Illustrate how a planet orbits the Sun by drawing an ellipse.	Image and Activity Bank 20-2 Exploration Worksheet, p. 27 Transparency Worksheet, p. 29 ▼ Exploration Worksheet, p. 31
End of Chapter, p. 452				Discrepant Event Worksheet, p. 34 Chapter 20 Review Worksheet, p. 35 Chapter 20 Assessment Worksheet, p. 38
Chapter 21, p. 454	**Lesson 1, Visitors From Space, p. 455**	2	1. Identify the differences between a meteor, a meteorite, and a comet. 2. Explain why many meteorite-impact craters are visible on the Moon but not on Earth. 3. Describe the composition of meteors and comets. 4. Explain the origin of comets. 5. Describe the relationship between comets and meteor showers.	Image and Activity Bank 21-1
	Lesson 2, The Space Probes, p. 460	2	1. Describe in general terms the distances that separate the planets. 2. Identify some of the principal features that distinguish one planet from another. 3. Explain how technology has been used to gather data about the planets.	Image and Activity Bank 21-2
	Lesson 3, Colonizing the Solar System, p. 465	1	1. Compare and contrast Mars and Earth. 2. Discuss the possibility of establishing a colony on Mars.	Image and Activity Bank 21-3
End of Chapter, p. 470				Activity Worksheet, p. 41 Chapter 21 Review Worksheet, p. 42 Chapter 21 Assessment Worksheet, p. 45
Chapter 22, p. 472	**Lesson 1, Messages From the Stars, p. 473**	2	1. Describe the size of the stars, using comparisons to Earth and the Sun. 2. Explain the relationship between the color of a star and its temperature. 3. Describe the life span of stars.	Image and Activity Bank 22-1 Transparency Worksheet, p. 47 ▼ Transparency Worksheet, p. 49 ▼ Exploration Worksheet, p. 51 Exploration Worksheet, p. 53
	Lesson 2, Earth's Place in the Universe, p. 480	2	1. Describe Earth's place in the solar system, the Milky Way galaxy, and the universe. 2. Explain what astronomical units and light-years are and how they are used.	Image and Activity Bank 22-2 Math Practice Worksheet, p. 55 Transparency Worksheet, p. 56 ▼ Activity Worksheet, p. 58
	Lesson 3, A Likely Beginning, p. 487	1	1. Describe the big-bang theory as one way of explaining how the universe began. 2. Identify cosmic microwaves and the expansion of the universe as two observable consequences of the big bang.	Exploration Worksheet, p. 59
End of Chapter, p. 490				Chapter 22 Review Worksheet, p. 60 Chapter 22 Assessment Worksheet, p. 63
End of Unit, p. 492				Unit 7 Activity Worksheet, p. 65 ▼ Unit 7 Review Worksheet, p. 66 Unit 7 End-of-Unit Assessment Worksheet, p. 69 Unit 7 Activity Assessment Worksheet, p. 75 Unit 7 Self-Evaluation of Achievement, p. 78

* Estimated time is given in number of 50-minute class periods. Actual time may vary depending on period length and individual class characteristics.

▼ Transparencies are available to accompany these worksheets. Please refer to the Teaching Transparencies Cross-Reference chart in the Unit 7 Teaching Resources booklet.

Materials Organizer

Chapter	Page	Activity and Materials per Student Group
19	429	**Exploration 1, Part 2:** small ball of modeling clay; paper-towel tube; piece of cardboard, about 20 cm $\times$ 20 cm; optional materials: binoculars
20	440	**Exploration 1:** compass; protractor; meter stick; metric ruler; set of star charts; tube of glue or roll of tape; 4 sheets of unlined paper; 9 marbles or small plastic-foam balls, 1–2 cm in diameter; stick of modeling clay
	446	***Exploration 2, Part 2:** plastic-foam ball, about 10 cm in diameter; rubber band; small cup; 2 toothpicks; light source; protractor; model of the ecliptic from Exploration 1; compass
	450	**Exploration 3:** letter-sized sheet of corrugated cardboard or stack of several sheets of cardboard; unlined paper; 2 pushpins; 40 cm of string; metric ruler
21	455	**Meteors and Meteorites:** aluminum pie pan; 5–10 mL of water; about 200 cm^3 of flour
22	475	**Exploration 1:** nail, about 5 cm in length; Bunsen burner or portable burner; incandescent light source with clear bulb; 2 fresh D-cell batteries; 1 weak D-cell battery; flashlight bulb; about 30 cm of small-gauge electric wire; pair of tongs; set of star charts; safety goggles
	479	**Exploration 2:** 2 round, white, medium-sized balloons; 2 plastic-foam balls, about 1 cm in diameter; red, yellow, and black permanent felt-tip markers
	489	**Exploration 4:** large, round balloon; black permanent felt-tip marker

* You may wish to set up these activities in stations at different locations around the classroom.

Advance Preparation

Except for gathering the above materials, there is no advance preparation required for this unit.

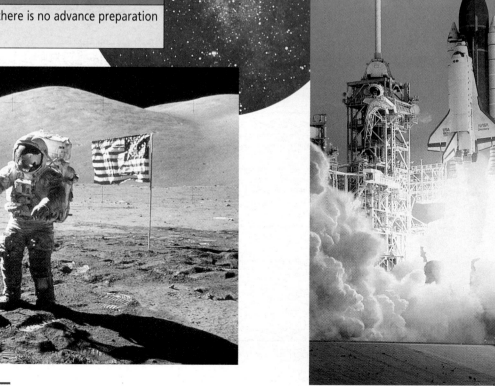

Unit Compression

Ideally, you will have enough time to complete all four of the chapters in this unit. Students will find this approach valuable because the concepts build on each other in succession. However, the chapters are also independent enough that they can be taught individually if necessary. Thus, it is possible to leave out any of the chapters in order to save time.

Because the first two chapters of the unit introduce the study of astronomy and the nature of moving celestial bodies, they should be given priority over the last two. If your students have some prior knowledge of the subject, however, you may wish to modify your teaching plan accordingly.

As you progress through the unit, keep in mind that assigning students to read sections ahead of time will speed in-class discussion, as will having them attempt to answer questions on their own before discussing the questions in groups. Even if time is short, your students will find this to be an engaging and stimulating unit.

Homework Options

Chapter 19	See Teacher's Edition margin, pp. 423, 426, 429, 430, 431, 434, and 435 Exploration Worksheet, p. 5 Activity Worksheet, p. 10 Theme Worksheet, p. 15 SourceBook, p. S141
Chapter 20	See Teacher's Edition margin, pp. 441, 448, and 453 SourceBook, p. S152
Chapter 21	See Teacher's Edition margin, pp. 457, 458, 464, and 470 Activity Worksheet, p. 41 SourceBook, pp. S144 and S152
Chapter 22	See Teacher's Edition margin, pp. 477, 480, 485, 486, and 488 Math Practice Worksheet, p. 55 Activity Worksheet, p. 58 SourceBook, pp. S147 and S157
Unit 7	Unit 7 Activity Worksheet, p. 65 SourceBook Activity Worksheet, p. 79

Assessment Planning Guide

Lesson, Chapter, and Unit Assessment	SourceBook Assessment	Ongoing and Activity Assessment	Portfolio and Student-Centered Assessment
Lesson Assessment Follow-Up: see Teacher's Edition margin, pp. 426, 435, 444, 451, 459, 464, 469, 479, 486, and 489 **Chapter Assessment** Chapter 19 Review Worksheet, p. 17 Chapter 19 Assessment Worksheet, p. 20* Chapter 20 Review Worksheet, p. 35 Chapter 20 Assessment Worksheet, p. 38* Chapter 21 Review Worksheet, p. 42 Chapter 21 Assessment Worksheet, p. 45* Chapter 22 Review Worksheet, p. 60 Chapter 22 Assessment Worksheet, p. 63* **Unit Assessment** Unit 7 Review Worksheet, p. 66 Unit 7 End-of-Unit Assessment Worksheet, p. 69*	SourceBook Review Worksheet, p. 80 SourceBook Assessment Worksheet, p. 84*	Unit 7 Activity Assessment Worksheet, p. 75* **SnackDisc** Ongoing Assessment Checklists ♦ Teacher Evaluation Checklists ♦ Progress Reports ♦	Portfolio: see Teacher's Edition margin, pp. 425, 457, 463, and 478 **SnackDisc** Self-Evaluation Checklists ♦ Peer Evaluation Checklists ♦ Group Evaluation Checklists ♦ Portfolio Evaluation Checklists ♦

* Also available on the Test Generator software
♦ Also available in the Assessment Checklists and Rubrics booklet

Science Discovery is a versatile videodisc program that provides a vast array of photos, graphics, motion sequences, and activities for you to introduce into your *SciencePlus* classroom. *Science Discovery* consists of two videodiscs: Science Sleuths and the Image and Activity Bank.

Using the *Science Discovery* Videodiscs

Science Sleuths: The Lost Mining Probe Side B

It's the future. The World Space Police have gotten a tip that an illegal mining probe may have been launched into a restricted area of space. The Science Sleuths must analyze the evidence for themselves to determine whether the probe was sent and, if so, exactly where it is now.

Interviews

1. Setting the scene: World Space Police officer **9813 (play ×2)**

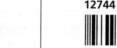

2. Anonymous tipster **10387 (play)**

3. Attorney for accused company **11090 (play)**

4. Planetary seismologist **11740 (play)**

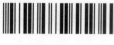

Documents

5. Fax from planetary geologist **12725 (step ×2)**

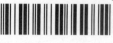

6. Solar system mining regulations **12729 (step)**

7. World Space Police report **12732 (step ×2)**

8. Advertisement for mining probe **12736 (step ×2)**

9. Anonymous data from probe **12740 (step ×2)**

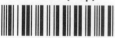

10. Data from the Moon **12744 (step ×2)**

11. Data from Mercury **12748 (step ×2)**

12. Data from Mars **12752 (step ×2)**

13. Data from asteroids **12756 (step)**

Literature Search

14. Search on the words: GEOLOGICAL PROBE, INTERPLANETARY MINING, MOONS, PLANETS, SOLAR SYSTEM **12759**

15. Article #1 ("Miners Itching to Send New Probe") **12761 (step)**

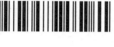

16. Article #2 ("Interplanetary Mining to Expand") **12764 (step)**

17. Article #3 ("Moons, Rings, and Tides") **12767 (step ×2)**

Sleuth Information Service

18. Planet and satellite table **12771 (step ×17)**

19. Temperature-conversion chart **12790 (step)**

20. Comets **12793 (step ×2)**

21. Halley's comet **12797 (step ×2)**

Image and Activity Bank Side A or B

A selection of still images, short videos, and activities is available for you to use as you teach this unit. For a larger selection and detailed instructions, see the Videodisc Resources booklet included with the Teaching Resources materials.

19-1 What Is Astronomy? page 423

Space-shuttle takeoff 20048–20256 (Side A only) (play ×2)
A fisheye lens shows the space shuttle taking off.

Hubble telescope animation 17629–17929 (Side A only) (play ×2)
Simulation of the Hubble telescope in space

Aurora borealis; time-lapse 19012–19389 (Side A only) (play ×2)
This time-lapse sequence was taken with an all-sky camera. The aurora borealis appears as luminous ribbons of light in the night sky. The phenomenon is visible in a zone surrounding the north magnetic pole.

Mercury 1580
Mercury is the planet closest to the Sun in our solar system.

Pluto 1625
Pluto, the farthest known planet from the Sun, appears very bright.

Nebulae 1630–1631 (step)
The Horsehead Nebula in Orion
(step) The Great Nebula in Orion

Step Reverse | Play | Pause | Step Forward

19-2 An Ancient Science, page 427
Stars; time-lapse 17975–18215 (Side A only) (play ×2)
The stars "move" across the night sky above Devils Tower, Wyoming. The rotating Earth makes it appear as if the field of stars rotates about Polaris (the North Star).

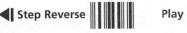

Astronomy, historical 1665
Time line showing six major astronomers and their discoveries

20-1 A Scientific Revolution, page 439
Earth's rotation 16128–16575 (Side A only) (play ×2)
Rotation of the Earth as seen from space

20-2 Motions and Their Effects, page 445
Earth-Sun relationship 1672
This shows the Earth's position during the year as it moves around the Sun. The solstices and equinoxes are depicted.

Seasons; sunlight and Earth's tilt 1673–1675 (step ×2)
The Northern Hemisphere is tilted toward the Sun, receiving more solar radiation during the summer. (step) The Northern Hemisphere tilts away from the Sun and receives less solar radiation during the winter. (step) The amount of solar radiation is equal in both hemispheres during the spring and fall equinoxes.

Moon 1588
Full view of Earth's moon

Moon phases 1657–1664 (step ×7)
New moon (step) Waxing crescent (step) First quarter (step) Waxing gibbous (step) Full moon (step) Waning gibbous (step) Third quarter (step) Waning crescent

Solar eclipse 16793–17093 (Side A only) (play ×2)
This time-lapse sequence of a total eclipse of the Sun by the Moon shows a ring of light, the Sun's corona.

21-1 Visitors From Space, page 455
Meteorites, antarctic; collection 1576–1578 (step ×2)
Searching for meteorites (step) Photographing a meteorite found on the ice (step) Studying a meteorite

Crater, meteor 1654–1656 (step ×2)
Manicouagan crater in Quebec, Canada (step) Meteorites form craters when they collide with the surface of the Earth. (step) Meteor crater on Earth

21-2 The Space Probes, page 460
***Voyager's* view of Neptune and Saturn 17458–17628 (Side A only) (play ×2)**
Voyager passing Neptune and Saturn

Jupiter; Red Spot 17371–17457 (Side A only) (play ×2)
The Great Red Spot is a permanent atmospheric disturbance on Jupiter. It rotates just like a storm on Earth. The atmosphere is made up mostly of ammonia gas. The dense clouds are frozen ammonia crystals.

Jupiter 1597
Full view of the planet Jupiter

Io 1602–1603 (step)
Io, one of Jupiter's moons, is three-fourths of the size of the planet Mercury. (step) Io has an atmosphere made up mainly of sulfur dioxide.

Saturn 1609–1610 (step)
Full view of Saturn (step) The clouds on Saturn are yellow and light orange.

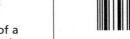

Saturn 1612
The rings of Saturn consist of frozen water and are very thin, no more than 3 km thick.

Dione 1614
The surface of Dione, one of Saturn's moons, is covered with craters.

Enceladus 1615
This is a view of Enceladus, one of Saturn's moons. The grooves are evidence that the crust was deformed by internal heat.

Uranus 1616–1617 (step)
Uranus has a pale, blue-green color, as seen on the left. The right image uses false color to bring out details that are otherwise invisible. Note the rings surrounding the pole. (step) Uranus has nine major rings.

Miranda 1619
Miranda is the smallest and closest of Uranus's five moons.

Neptune 1621–1622 (step)
Full view of Neptune (step) The Great Dark Spot on Neptune is partially encircled by a thin layer of clouds made up of frozen methane crystals. The spot's diameter is as large as that of the Earth.

Venus 1582–1583 (step)
Full view (step) Venus is covered by thick, dense clouds that are driven by fierce winds.

Venus flyover; topography 17094–17370 (Side A only) (play ×2)
Computer reconstruction of the surface of Venus

21-3 Colonizing the Solar System, page 465
Mars 1591–1596 (step ×5)
View of Mars (step) View of the south pole of Mars (step) The surface of Mars as seen by the *Viking 2* spacecraft. (step) The surface of Mars is covered with rocks and sandy soil. (step) Olympus Mons, a volcano on Mars (step) Phobos, one of Mars's two moons

22-1 Messages From the Stars, page 473
Sun model 1678
This cross section shows the Sun's core, radiation zone, convection zone, photosphere, chromosphere, and corona.

Sunspots 1679
Sunspots are areas of cooler, darker material pulled up to the Sun's surface from inner regions by the Sun's strong magnetosphere.

Sun 1579
The Sun consists mainly of hydrogen and helium gases. It is 150 million kilometers from Earth.

Solar flare 18440–18709 (Side A only) (play ×2)
Solar flares are sudden eruptions of hydrogen gas on the surface of the Sun.

22-2 Earth's Place in the Universe, page 480
Solar system 1677
Model of the solar system

Galaxy, Andromeda 1636–1637 (step)
The Andromeda galaxy is the closest spiral galaxy to our own, the Milky Way. (step)

◀| Step Reverse  Play ▶  Pause || 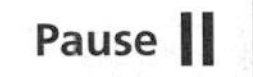 Step Forward |▶  

Unit 7 TOWARD the STARS

UNIT FOCUS

Challenge students to name as many different kinds of celestial bodies as they can (moons, planets, stars, galaxies, comets, asteroids, meteors, and so on). Keep track of their responses on the chalkboard. Then ask students how many of these objects they have actually seen. Put a check next to each item on the list that is identified. Call on volunteers to describe each of the objects for the class.

Ask students to raise their hand if they have ever identified a constellation in the night sky. Call on volunteers to name the constellations they have seen. Keep track of their responses by listing them on the chalkboard. Then call on volunteers to describe what they think about when they look at the night sky. Encourage specific responses.

A good motivating activity is to let students listen to the English/Spanish Audiocassettes as an introduction to the unit. Also, begin the unit by giving Spanish-speaking students a copy of the Spanish Glossary from the Unit 7 Teaching Resources booklet.

Unit 7 TOWARD the STARS

In NASA's 1.3-million-gallon underwater simulator tank, astronauts practice installing a camera on a model of the Hubble Space Telescope. The tank provides conditions that simulate the weightlessness experienced in space.

Connecting to Other Units

This table will help you integrate topics covered in this unit with topics covered in other units.

Unit 2 Diversity of Living Things	Classification schemes are used to arrange objects such as stars and planets into a logical order.
Unit 3 Solutions	Observations of astronomical objects depend on knowing how arriving light is affected by the atmosphere.
Unit 4 Force and Motion	Gravitational forces are responsible for nearly all of the motions of heavenly bodies.
Unit 5 Structures and Design	Studying the structures of stars and galaxies helps us learn how these objects evolve.
Unit 6 The Restless Earth	Volcanism and other geologic features are factors in determining where observatories will be built.

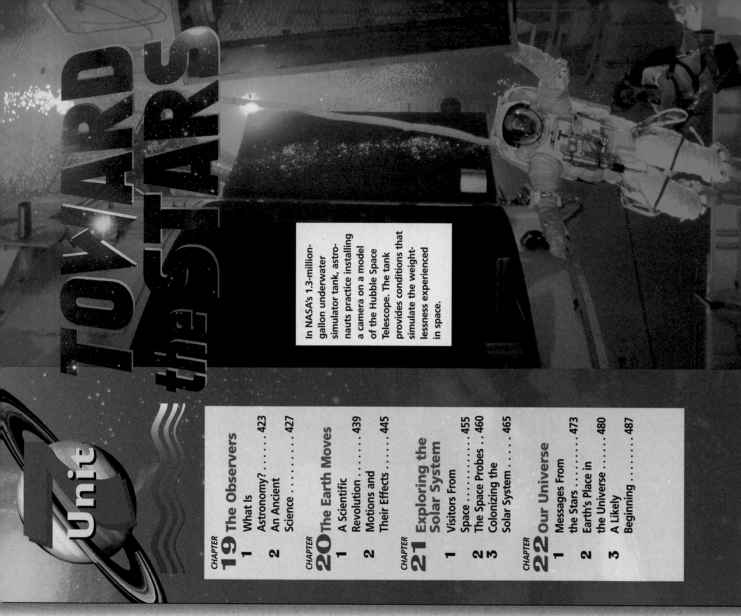

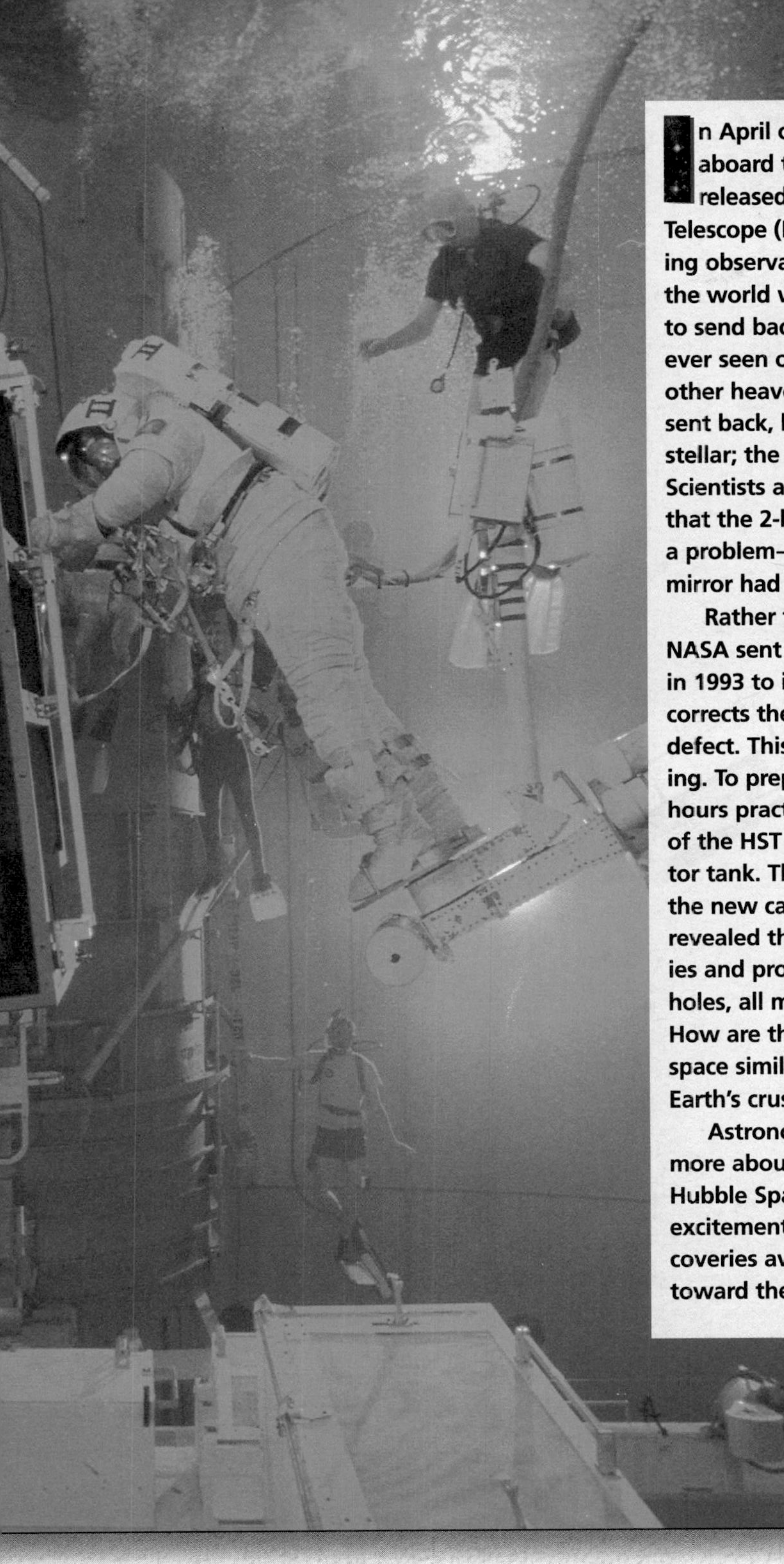

In April of 1990, astronauts aboard the space shuttle *Discovery* released the Hubble Space Telescope (HST) to serve as an orbiting observatory. Astronomers around the world waited eagerly for the HST to send back the sharpest images ever seen of planets, galaxies, and other heavenly bodies. What the HST sent back, however, was less than stellar; the images were blurred. Scientists at NASA quickly discovered that the 2-billion-dollar telescope had a problem—its main light-collecting mirror had been incorrectly made.

Rather than replace the mirror, NASA sent astronauts back into space in 1993 to install a new camera that corrects the problems caused by the defect. This was no easy undertaking. To prepare, astronauts spent hours practicing on a life-sized model of the HST in an underwater simulator tank. The practice paid off. With the new camera, the HST has clearly revealed the shapes of distant galaxies and provided evidence of black holes, all millions of light-years away. How are these images from distant space similar to fossils found in the Earth's crust? (A)

Astronomers are excited to learn more about the universe from the Hubble Space Telescope. Join the excitement by reading on. Many discoveries await you in your journey toward the stars.

Using the Photograph

After students have read the text on this page, you may wish to point out the two meanings of the word *stellar.* Then explain to students why the universe, to astronomers, is like a time machine. Tell them that light travels at a finite speed, so if the light they are seeing took a year to reach Earth, then they are seeing light that was emitted a year ago by that object. Therefore, it is as if astronomers are looking back in time. The farther into space they probe, the further back in time they see. Light from quasars, the farthest objects in the universe, is thought to have been emitted very early in the life of the universe.

Answer to *In-Text Question*

(A) The study of fossils is similar to the study of distant objects in space because, in general, the deeper the fossils are found in the ground, the older they are, just as the deeper into space an object is, the "older" the light that we receive is. The principle of *stratification* explains that sedimentation will cause newer rocks to overlay older rocks. Therefore, tools or animals trapped in the higher layer of rocks are younger than those trapped farther down.

Critical Thinking

Ask students to state some similarities and differences between weightlessness in space and weightlessness in water. Then tell them that water and space are similar in that humans can't breathe in either environment. But in water, the sense of weightlessness arises because the downward force of gravity can be balanced by the upward, buoyant force of the water. In space, gravitational forces are sometimes so weak that they would not measurably affect you, and you would drift freely. In water, you feel resistance from the water whenever you move. It may also be harder to see for great distances because water scatters light. In space, on the other hand, there is virtually no resistance or scattering of light because space is a vacuum.

CHAPTER 19 The Observers

Connecting to Other Chapters

Chapter 19
explores the science of astronomy and discusses the observations of some early astronomers.

Chapter 20
describes the various motions of celestial bodies and some of the effects of those motions.

Chapter 21
introduces meteors, meteorites, and comets and examines what a colony on Mars might be like.

Chapter 22
examines Earth's place in the universe and introduces the big-bang theory.

Prior Knowledge and Misconceptions

Your students' responses to the ScienceLog questions on this page will reveal the kind of information—and misinformation—they bring to this chapter. Use what you find out about your students' knowledge to choose concepts and activities to emphasize in your teaching.

In addition to having students answer the ScienceLog questions on this page, you may wish to have them perform the following exercise: Divide the class into groups of 3–4 students and tell them that each group represents a team of astronomers working together in an observatory. Have each team put together a skit that demonstrates what the duties of each member of the team are, what equipment the team will use to observe the sky, and what the research goal of the team is. You may wish to have volunteers perform their skits for the class. Emphasize that there are no right or wrong answers for this exercise. You may wish to encourage the class to discuss each skit. The performances should help you evaluate what students already know about astronomy and about what astronomers do, what misconceptions students may have, and what is interesting to them about this topic.

CHAPTER 19
The Observers
1 Will the night sky over your home appear the same 6 months from now as it will tonight? Explain your answer.
An ancient view of the planets traveling around the Earth
SEPTEM
PLANETARVM
ORBES
2 Some early astronomers thought that the Sun revolves around the Earth, while others believed that the Earth revolves around the Sun. What evidence could astronomers on each side use to support their belief?
ScienceLog
Think about these questions for a moment, and answer them in your ScienceLog. When you've finished this chapter, you'll have the opportunity to revise your answers based on what you've learned.
422

LESSON 1

What Is Astronomy?

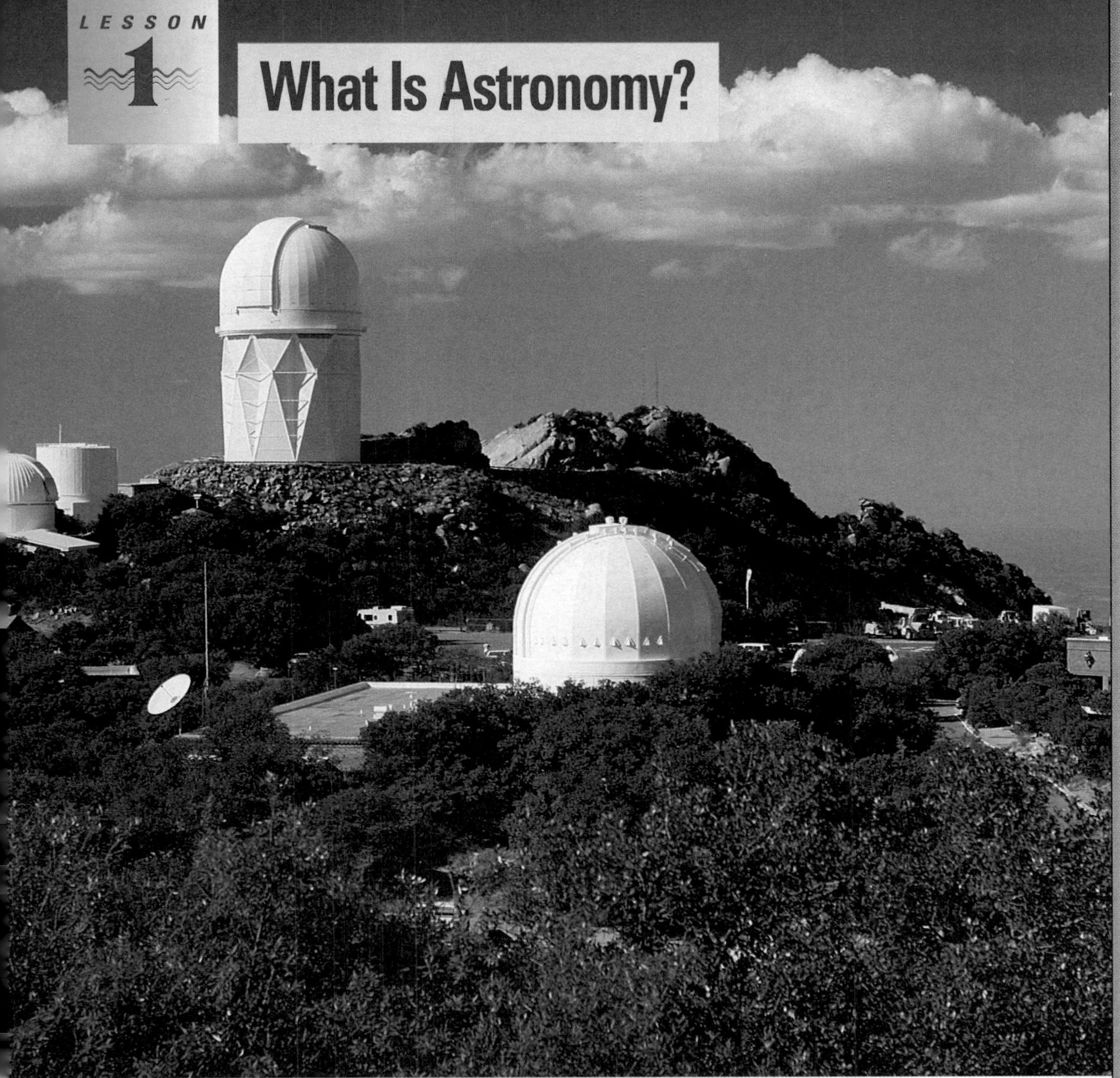

Imagine!

You are an astronomer at Kitt Peak National Observatory (above), 95 km southwest of Tucson, Arizona. You are using a telescope to take photographs of the heavens. The next morning while studying the photographs, you discover a most amazing sight.

Quickly, you call other Kitt Peak astronomers for consultation. Together, you decide that you have photographed two galaxies colliding in outer space. Even more exciting, it looks as though there's a dark, mysterious "black hole" within the galaxies. A breakthrough? You and your fellow astronomers write an announcement about your discovery for the media.

LESSON 1 ORGANIZER

Time Required
1 class period

Process Skills
analyzing, classifying, measuring

New Term
Astronomy—the study of the universe and the objects in it

Materials (per student group)
none

Teaching Resources
none

LESSON 1

What Is Astronomy?

FOCUS

Getting Started

Several days prior to beginning the lesson, tell students to go outside on a clear night and observe the sky. Have students take careful notes about what they see. Before you begin the lesson, have students share what they observed. Ask: What kind of scientist studies the objects you saw? *(An astronomer)* Conclude by telling students that in this lesson they will learn more about what astronomers do and how they make their observations.

Main Ideas

1. Astronomy is the study of the universe and its components.
2. New discoveries in astronomy are being made continuously.
3. Scientists who study the universe and Earth's place in it are called astronomers.

TEACHING STRATEGIES

Imagine!

Before students begin reading, direct their attention to the photograph, and call on a volunteer to describe what it shows. Ask students if they have ever used a telescope. If so, call on a volunteer to explain what he or she saw when using one.

If any of your students have visited an observatory, have them share their experiences with the class. Then have students read pages 423 and 424.

Homework

Have students answer the following question: If you could study anything about the universe, what would you choose and why? *(Accept all reasonable responses.)*

GUIDED PRACTICE When students have finished reading, call on a volunteer to summarize the newspaper article. Then direct students' attention to the questions at the bottom of the page. Point out that each of the new terms mentioned in the questions will be discussed later in the unit.

Meeting Individual Needs

Second-Language Learners

Have students begin a bilingual dictionary of astronomical terms. As they progress through the unit, they should add to their dictionaries. Students should include the English term, the term in their native language, and an illustration and definition of the term. Terms might include *planet, comet, star, meteor, galaxy, solar system, satellite, orbit, ecliptic,* and *asteroid.*

Integrating the Sciences

Life and Physical Sciences

Astronomers have detected a planet orbiting around a star other than the Sun. Ask the class to think about all of the different types of living organisms they know. Then ask: Do you think any of these planets might harbor some form of life? What anatomical structures do you think are necessary for life as we know it? *(Answers might include a nervous system, structures for locomotion, reproductive structures, etc.)* You might have students draw their conception of a life-form having only these structures.

The Tucson Tribune

Since 1952 January 1, 1996 75 cents

Enormous Dark Object Discovered by Astronomers

By Alan Moore
HRW News Service

TUCSON, ARIZONA — Using the powerful telescope at Kitt Peak National Observatory, astronomers have discovered a huge, mysterious object within two colliding galaxies. If the object is a black hole, it is 10 to 100 times as massive as any black hole previously known or even believed possible.

The object was discovered while astronomers were studying a galaxy named NGC 6240, which is some 300 million light-years from Earth. The Kitt Peak astronomers estimate that the object must be 100 billion times as massive as the Sun. This means that its mass would be as great as the mass of some entire galaxies. Despite such a large mass, the object occupies less than one-millionth of the space occupied by the Earth's galaxy, the Milky Way.

NGC 6240 as seen by the telescope at Kitt Peak National Observatory

The discoverers say that if the object isn't a black hole, it might be a dormant quasar of the kind from which galaxies were born more than 10 billion years ago, when the universe was still young. The Kitt Peak astronomers acknowledge that many more observations must be made to reconcile established theories with the new discovery.

How important is your discovery? What is a galaxy? a black hole? a light-year? a quasar? One simple answer is that they are all part of the science called **astronomy**.

424

What do you think astronomy is? In your ScienceLog, write as many phrases as you can to complete the statement, "Astronomy is ___?___." The photos on this page will help you add to your list. Ⓐ

Artist's impression of the Milky Way galaxy. The red circle shows the location of our solar system (which is too small to be seen in this image).

Saturn's rings (color-enhanced)

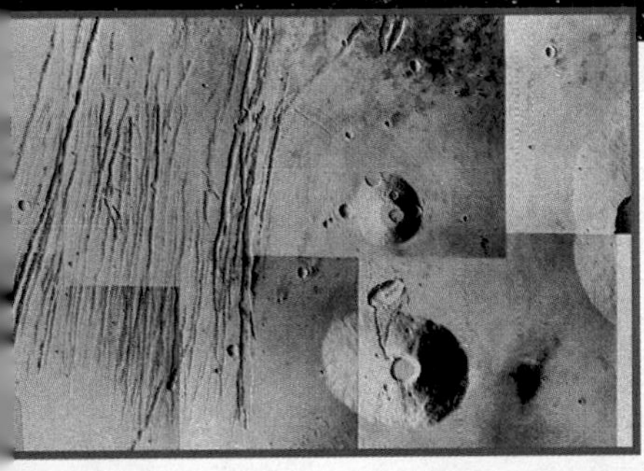

Volcanoes on Mars

The shuttle *Discovery*, with the Hubble Space Telescope aboard

Star cluster

Astronaut on our nearest neighbor—the Moon

425

Did You Know. . .

Helium, the second lightest element in the universe, was first discovered in the Sun by astronomers before it was found on Earth. The word helium comes from the Greek word for "Sun."

Answer to *In-Text Question*

Ⓐ Possible responses include the following: Astronomy is calculating the distance to stars, studying the stars with a telescope, exploring outer space, or studying galaxies.

GUIDED PRACTICE After reading the newspaper article on the discovery of a possible black hole, direct students' attention to the photographs on page 425. Involve them in a discussion of what each one shows. Invite students to describe in a sentence or two what is depicted in each photograph.

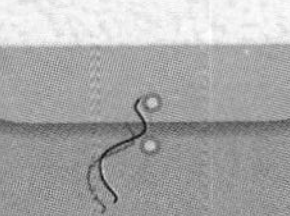

PORTFOLIO

Students may choose to include their responses to the in-text question in their Portfolio. Suggest that students complete the phrase at least five different ways. They may add to or annotate their sentences in their Portfolio as they complete the unit.

CROSS-DISCIPLINARY FOCUS

Language Arts

The names for stars consist of two words. The first word is a letter of the Greek alphabet that corresponds to how bright the star is compared with the others in its constellation. The second word is the possessive form of the Latin name of the constellation. For example, Alpha Centauri, the closest star to the Earth other than the Sun, is the brightest star (since alpha is the first letter in the Greek alphabet) in the constellation Centaurus. Some stars may have more than one name. For example, the second-brightest star in Orion is Beta Orionis, but it is more commonly referred to as Rigel, from the Arabic word meaning "foot." Most of these common names come from Arabic. Have students look up some Arabic names for bright stars. Students should determine what the Arabic names mean and whether the stars have any other names. (It may be helpful to introduce students to the Greek alphabet first.)

Ask an Astronomer

You may wish to have students write their questions on slips of paper and display them on a bulletin board. Group them under categories that pertain to the universe as a whole, to the solar system, or to human technology. As students complete the lessons, have them answer the questions that apply. If some questions remain unanswered after the unit is completed, have volunteers do some research to discover and share the answers with the class.

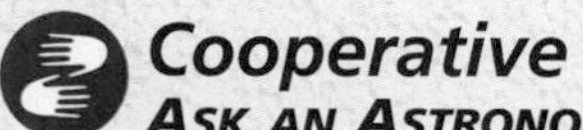

Cooperative Learning
Ask an Astronomer

Group size: 2 to 3 students
Group goal: to generate thought-provoking questions about astronomy
Individual accountability: Assign each student a role such as team leader, recorder, and researcher.
Positive interdependence: Have each student select a different question from the group's list of questions and answer it in writing.

Answers to *Ask an Astronomer*

Sample additional questions: What is inside the Sun? Is there life on other planets? How did telescopes change our picture of the universe?

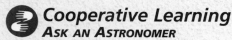

Homework

The Closure activity described on this page makes an excellent homework activity.

Ask an Astronomer

What do astronomers do? If you could ask an astronomer some questions about the universe and Earth's place in it, what would you ask? In small groups, make a list of questions that interest you. The questions here will get you started.

Share your group's questions with the other groups. As you study this unit, you'll discover answers to many of these questions.

What is the universe?
How did the universe begin?
If not, what will happen to it?
Will the Sun last forever?
How old is our planet?

426

FOLLOW-UP

Reteaching

Students may enjoy writing a poem to describe the science of astronomy. Have volunteers share their poems with the class. Then have students work together to organize their poems in a booklet, design a cover, and think of a title.

Assessment

Have students create a concept map about astronomy based on what they have learned in this lesson.

Extension

Students may enjoy doing research to discover more about careers in astronomy. Have them share what they learn by presenting oral reports to the class or by writing fact sheets.

Closure

Suggest to students that they look for articles on astronomy in magazines and newspapers. They could select articles to make a bulletin-board display entitled Astronomy Today. Have students select an article and summarize it in a brief oral report.

LESSON 2 An Ancient Science

Observing the Heavens

Did you know that astronomy is the oldest science? Since ancient times, humans have looked toward the stars. In fact, ancient observers were skilled viewers of the skies. By careful observations, early Babylonians, Egyptians, Aztecs, and Mayas could accurately predict the locations of heavenly bodies. They relied on the skies for answers to questions about when to plant crops and when the seasons would change.

You are like the ancient astronomers when you look up at the sky on a clear night. What you see is what they saw. Make a list of all the heavenly bodies these ancient stargazers might have seen.

Did you include in your list NJYLCRQ, KMML, QRYPQ, QFMMRGLE QRYPQ, and the QSL (or SUN, if you crack the simple code)? Ⓐ Ancient astronomers saw all of these. In the following Exploration, you will think about and investigate some of the motions of these heavenly objects. Part 1 has a list of observations that all sound like they might be true; some may even be observations you have made yourself. The trick is, some are false. Your mission is to discover which statements are not true. In Part 2, you will have the chance to check out some of these statements experimentally.

Ruins of El Caracol in the ancient Mayan city of Chichén Itzá, Mexico. More than a thousand years ago, Mayas studied the planet Venus through openings at the top of the tower.

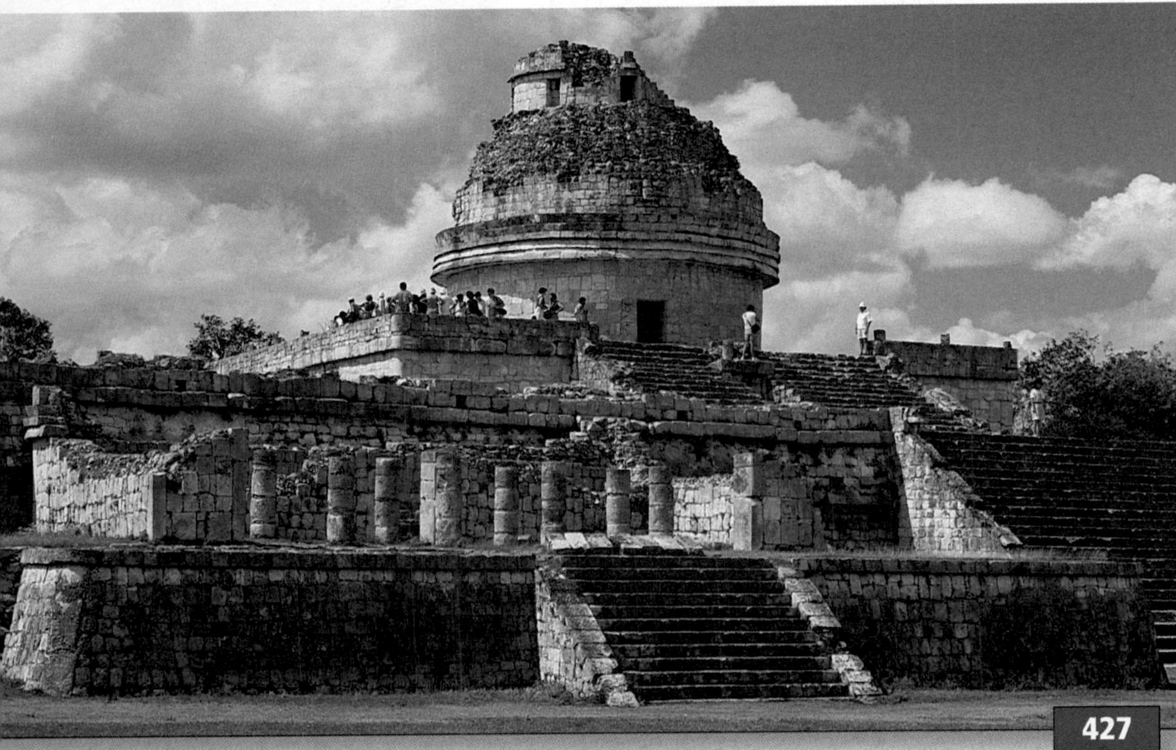

LESSON 2 ORGANIZER

Time Required
4 class periods

Process Skills
observing, measuring, inferring

Theme Connections
Cycles, Systems

New Term
Retrograde motion—motion in which a planet appears to make a loop in the sky, going backward for a while before continuing its forward motion

Materials (per student group)
Exploration 1, Part 2: small ball of modeling clay; paper-towel tube; piece of cardboard, about 20 cm × 20 cm; optional materials: binoculars

Teaching Resources
Transparency Worksheets, pp. 3, 11, and 13
Exploration Worksheet, p. 5
Activity Worksheet, p. 10
Theme Worksheet, p. 15
Transparencies 71–74
SourceBook, p. S141

LESSON 2 An Ancient Science

FOCUS

Getting Started

Build a model telescope. Insert a cardboard tube into a slightly larger one, and make sure that the tubes can slide back and forth. Secure a double concave lens into the exposed end of the smaller tube with modeling clay. Secure a double convex lens into the far end of the larger tube. Allow students to practice focusing on distant objects by sliding the tubes back and forth. (Warn students not to use the telescope to look directly at the Sun.) Explain that this telescope is a model of early tools used by astronomers to study the universe.

Main Ideas

1. Many observable astronomical events can be explained by the movements of the planets, stars, and other heavenly objects.
2. Two models used to explain the motions of heavenly objects are the heliocentric and geocentric models of the solar system.

TEACHING STRATEGIES

Observing the Heavens

Have students work in groups to list the kinds of heavenly bodies that ancient observers might have seen in the night sky. Students' lists will probably include the coded words listed below. Encourage students to describe the motions associated with each one, whether real or apparent.

★ **A Transparency Worksheet (Teaching Resources, page 3) and Transparency 71 are available to accompany Observing the Heavens.**

Answer to *In-Text Question*

Ⓐ The coded words are *planets, Moon, stars, shooting stars,* and *Sun.*

Exploration 1

Divide the class into groups of three students to complete Part 1, A Celestial Quiz.

Answers to *Part 1*

1. False. The Sun rises at different locations on the horizon depending on the time of year.
2. True. The exact number of minutes by which the length of daylight changes depends on the latitude and on the time of year. The rate of change is greater the farther one is from the equator. Also, the rate of change slows down around the time of the solstices.
3. False. Unless you live within 23.5 degrees north or south of the equator, the Sun is never directly overhead.
4. False. Due to Earth's tilt on its axis, the summer's noon shadow is shorter than the winter's noon shadow in the Northern Hemisphere. This is because the angle at which the Sun's rays strike Earth's surface is closer to perpendicular in the summer than in the winter.
5. True. Shadows shorten as the Sun rises and lengthen after noon.
6. True. Stars appear to rise and set at various times of the night because of Earth's rotation.
7. True. The position of the Big Dipper seems to change because of Earth's rotation.
8. True. The Moon appears to be full when Earth is almost exactly between the Sun and the Moon. In this orientation, the Moon rises in the east at about the same time that the Sun sets in the west.
9. True. Each evening the Moon will appear east of its previous position.
10. True. It takes the Moon the same amount of time to turn once on its axis as it takes it to circle the Earth.
11. False. Stars vary in color; they can be bluish, white, yellow, or red.
12. True. Although the size of the Moon does not change, it appears to be larger when it is near the horizon. The apparent difference in size is an optical illusion caused by the atmosphere.

Exploration 1

Heavenly Motions

PART 1

A Celestial Quiz

Work in groups of three. Decide if each statement is true or false. Is there complete agreement on each statement among members of your group?

1. Each day the Sun rises due east and sets due west.
2. Every 7 days, the length of daylight changes by several minutes.
3. In your area, there are times of the year when the Sun is directly overhead.
4. Your winter shadow at noon is shorter than your summer shadow at noon.

5. Shadows shorten as the morning progresses.
6. Like the Sun, many other stars appear to rise and set at various times of the night.
7. If you look at the Big Dipper at two different hours of the same night, its orientation (position) in the sky will have changed.
8. If you see the Moon in the early evening in the east, it must be a full Moon.
9. On two consecutive evenings at the same hour, the Moon will be farther east on the second evening.
10. You always see the same side of the Moon from your position on Earth.
11. All stars that you see appear to be the same color.
12. The Moon appears larger when it is seen on the horizon than when it is seen higher in the sky.

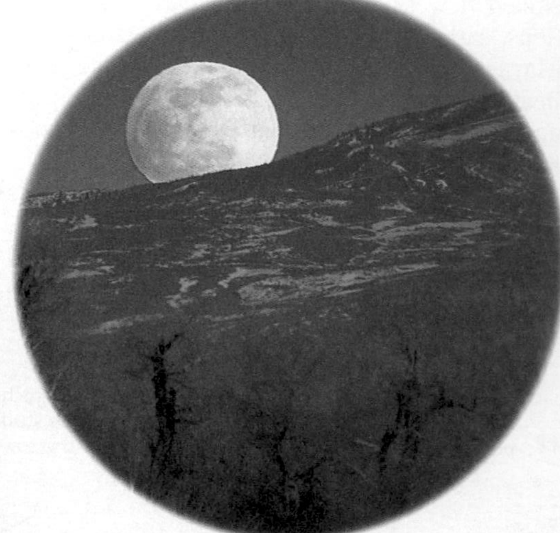

13. The only time you can see the Moon is at night.
14. The Moon is closer to Earth than are the stars.
15. Human behavior is affected by the phase or shape of the Moon.
16. Not all stars are the same brightness.
17. The star patterns, or constellations, change shape from night to night.
18. Planets are indistinguishable from stars when seen with the naked eye.
19. Some constellations are seen only in certain seasons.
20. Planets wander in and out of the constellations.

Did everyone agree on these statements? Which ones created the greatest debate? In the next part of this Exploration, you will check out some of these statements as an observer. From your investigations, you may be able to suggest other observations that could be added to the Celestial Quiz.

13. False. The Moon can often be seen during the day if its path is not too close to the path of the Sun.
14. True. The mean distance from Earth to the Moon is 384,403 km. The mean distance from Earth to the Sun is about 150 million kilometers. The nearest star, excluding the Sun, is Alpha Centauri, which is 4.3 light-years away.
15. Answers will vary. Certain scientific studies purport to show a relationship between the phases of the Moon and human behavior, but there is no agreement on the scientific validity and merit of these studies.
16. True. One way scientists classify stars is by their brightness.
17. False. No change in star patterns can be detected from night to night. However, over thousands of years, changes can be seen as stars move relative to each other.
18. False. Planets can be detected because they move relative to the more distant stars. Also, stars "flicker" but planets do not.

PART 2

Checking It Out

Choose at least two of the following observational activities to help you verify statements from the Celestial Quiz. When reporting your conclusions, indicate which of the statements from the Celestial Quiz you have confirmed or disproved. What other observations have you made?

1. The first sundial that was used to determine the passage of time was a simple upright stick. Using modeling clay, mount a paper-towel tube on a piece of cardboard, as shown below. For several days during the next week, place it outdoors at the same time each day. Be sure to put the cardboard in exactly the same position each time. Draw the shadow's position and length at that time, and record the date for each day. Does the direction of the shadow change? Does the length of the shadow change? What does the shadow indicate about the Sun's position in the sky?

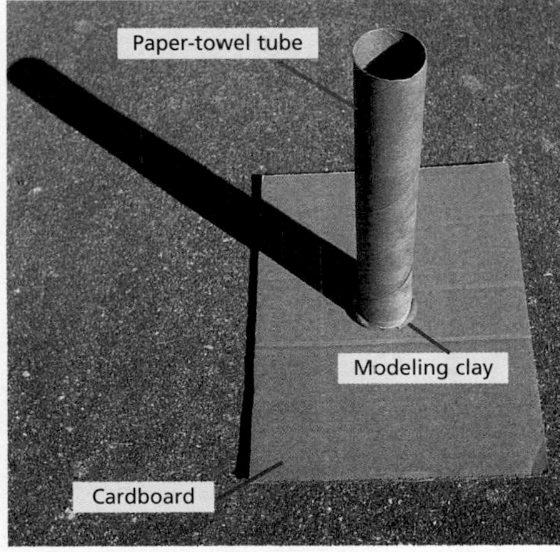

2. Use the shadow stick to find out how shadow length and position change during the day. What direction does the shadow point in early morning? at noon? a little before the Sun sets? Will there ever be a time when the Sun is directly overhead and there is no shadow? How does the length of the shadow vary?

Felipe observes the rising Sun appearing at the top of a building

3. Observe the rising or setting Sun. **Caution:** *Don't look directly at the Sun.* Several times during the week, record the time when the Sun first appears above or disappears below the horizon (or the top of a building). Make sure you view the Sun from the same position each time. Use some fixed object, such as a tree, to determine the exact point at which the Sun rises or sets. Does this point change? If it changes, in what direction does it change? Would this direction be the same in spring as in fall? Does the time of rising or setting change? By how much?

Exploration 1 continued ►

Answers to *Part 2*

1. The shadow's direction and length will change slightly over several days during most of the year. Little change would be expected if the activity is performed near the winter or summer solstice. Observing the shadow demonstrates that the Sun's position in the sky changes continuously because its rays must come from a different direction to cast a different shadow. (This activity does not verify or disprove any of the statements from Part 1.)
2. The shadow should shift from west to east during the course of the day. In the United States, there is never a time when the Sun is directly overhead. This activity verifies statement 5 and disproves statement 1 from Part 1. (Statement 3 cannot be verified unless the observations are continued for a year, or at least through two solstices.)
3. The point at which the Sun rises and sets changes from day to day. It seems to move from south to north from December to June and from north to south from June to December. The times of rising and setting may vary up to a few minutes a day, depending on geographical location and the time of the year. This activity verifies statement 2 and disproves statement 1 from Part 1.

Answers to Part 2 continued ►

19. True. Some constellations can be seen only during certain seasons due to Earth's tilt on its axis and its revolution around the Sun.
20. True. Against the background of stars, planets appear to wander through the constellations as they revolve around the Sun. Because they are so much closer to us than are the stars, planets appear to move more quickly across the sky as they travel around the Sun.

Homework

The Exploration Worksheet on page 5 of the Unit 7 Teaching Resources booklet makes an excellent homework activity to accompany Exploration 1.

Theme Connection

Cycles

Have students find out how the time of sunset changes during the course of a year. **Focus question:** What cyclic pattern is evident? *(The Sun sets earlier in the winter and later in the summer; be aware of changes due to Daylight Saving Time.)* Imagine a country where everyone sets their clocks to six o'clock at sunset. What kinds of problems might this introduce? *(Accept all reasonable responses.)* Students might be interested to know that, until recently, Saudi Arabia did just that.

Exploration 1, continued

Answers to
Part 2, continued

4. The Moon seems to move from west to east when viewed at the same time over several nights. The Moon changes from new moon to full moon and back again in 29.5 days. If performed before the Moon is full, this activity proves statement 9 from Part 1.

5. Stars are not all the same brightness or color. They may appear blue-white, white, yellow, or red. The brightest stars are Sirius, Canopus, Alpha Centauri, Arcturus, Vega, Capella, and Rigel. Canopus and Alpha Centauri are not visible from the mid-northern latitudes. This activity verifies statement 16 and disproves statement 11 from Part 1.

6. A star sighted near the western horizon will seem to set in the west as the Sun does. Most stars seem to rise in the east and set in the west. Stars near Polaris (the North Star) remain above the horizon and appear to circle Polaris in a counter-clockwise direction. This activity verifies statement 6 from Part 1.

7. The Big Dipper will not change shape, but it will change orientation. The Big Dipper and other star groups seem to revolve about the North Star. This activity verifies statement 7.

8. The star's position does change; it rises about 4 minutes earlier each night. The star's position moves to the east each night. This activity does not verify or disprove any of the statements from Part 1.

9. The features on the visible portion of the Moon do not change from night to night. Prominent features on the Moon include dark areas called *maria* and craters, such as the crater Tycho. Notice the numerous white streaks running radially outward from the crater. This activity verifies statement 10 from Part 1.

4. Over the period of a week, chart the position and shape of the Moon. Do this at the same time each night. As the month progresses, in what direction does the Moon move? How does its shape change from night to night?
5. On a clear night, spend some time observing the stars. Are all stars equally bright? Are they all the same color? Using a star chart, try to identify the brightest stars.
6. Sight a star near the western horizon. Make note of its position at various times on the same night. Does it set as the Sun does? Do stars rise in the east?
7. Locate the constellation of the Big Dipper in the night sky. Draw what you see on a piece of paper. An hour or two later, observe the Big Dipper again and draw what you see, using the same piece of paper. Did its shape change? Did its orientation change? Is there a point in the sky about which it and other stars seem to be revolving?

Mika lines up two fixed points with a star to see if the star moves

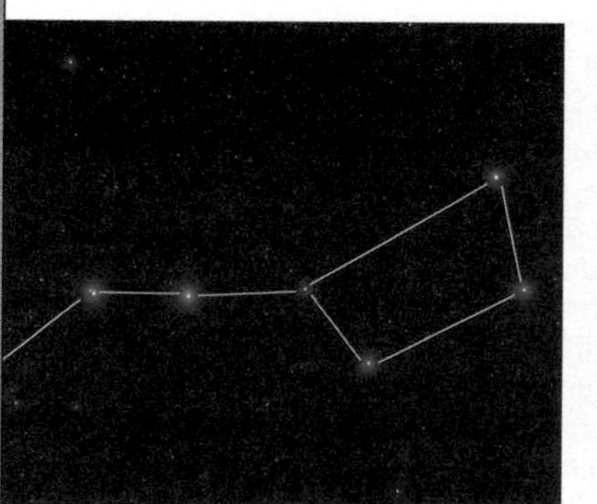

The Big Dipper

8. Line up two fixed points, such as a house and a tree, with a star. Each night, make a sighting at the same time from the same spot. What do you observe? Does the star's position appear to be different from night to night? If so, in what direction does it move?
9. With or without binoculars, carefully observe the features of the Moon. Make a diagram of some of the features you see. Do the features change from night to night, or do they remain the same?

Think About It

Look over the statements in the Celestial Quiz again. Are there some statements not yet verified? What observational activities could you plan in order to check them out? Are there other statements you could add to the Celestial Quiz? How would you verify these new statements? Are there still other observational projects you would like to carry out? What are they?

Answers to
Think About It

The statements in Part 1 that were not verified or disproved by the activities in Part 2 include 3, 4, 8, 12, 13, 14, 15, 17, 18, 19, and 20. (Encourage students to develop and share strategies to either verify or disprove these statements.)

Homework

The Activity Worksheet on page 10 of the Unit 7 Teaching Resources booklet makes an excellent homework activity once students have completed Exploration 1. If you choose to use this worksheet in class, Transparency 72 is available to accompany it.

A Celestial Picture Study

The following illustrations and their accompanying stories will add to the large number of observations you have already made. Team up with another person to study each illustration, read its story, and consider the discovery questions. How many new ideas can you discover about the motion of the heavenly bodies?

Picture Study 1

American Indians in Wyoming's Bighorn Mountains used a calendar wheel (shown at right) to predict the day when the Sun would reach its highest point in the sky. According to legend, "the growing power of the world" is strongest on the day the Sun is highest. On this day, the rising Sun was in line with two piles of rocks, or cairns, of the calendar wheel.

- Is the Sun higher in the sky at one time of the year than another? Explain.
- Does the Sun rise and set at different positions on the horizon from day to day? Explain.

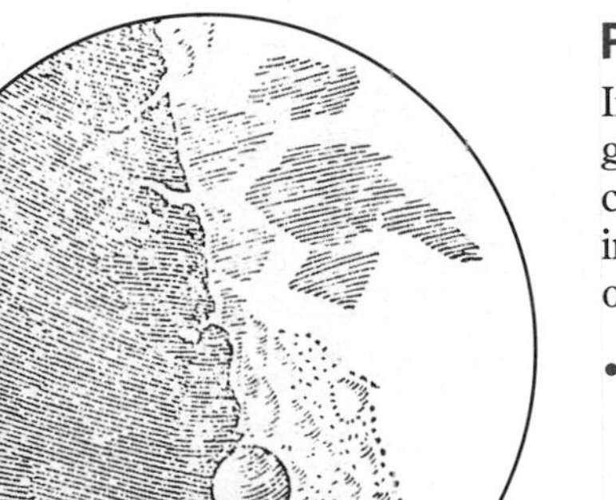

Picture Study 2

In 1610 a famous scientist named Galileo made the diagram of the Moon shown at left by using a new invention called a telescope. (You will read more about Galileo later in the unit.) Below his diagram is a recent photograph of the Moon in a similar phase.

- The face of the Moon you see at night looks like the one that Galileo saw almost 400 years ago. Explain why this is so.

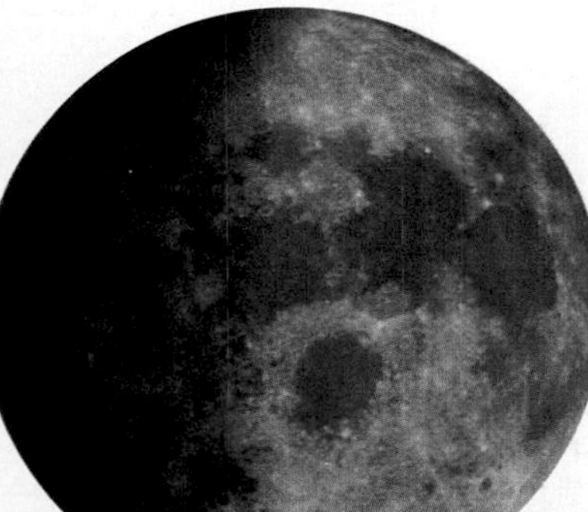

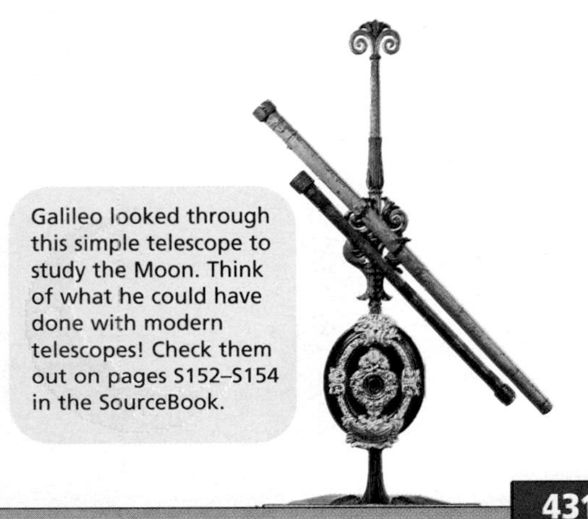

Galileo looked through this simple telescope to study the Moon. Think of what he could have done with modern telescopes! Check them out on pages S152–S154 in the SourceBook.

A Celestial Picture Study

These picture studies introduce concepts and ideas that will be discussed in greater detail later in the unit. Emphasis should be on promoting observational skills instead of on giving explanations or making inferences.

Picture Study 1

Students may be interested to know that calendar wheels such as the one shown here are about 600 years old. Their exact purpose is not certain, but it is believed that they were used as crude calendars to predict the summer solstice. The people who made them were probably the ancestors of such American Indian groups as the Crow, Gros Ventre, and Blackfeet.

Answers to *Picture Study 1*

- Yes, the Sun is higher in the sky at certain times of the year. The Sun is highest in the summer and lowest in the winter.
- The position on the horizon where the Sun rises and sets changes from day to day. (It moves southward from June to December and northward from December to June.) By observing these changes over a period of time, early astronomers were able to predict with some accuracy where the Sun would rise and set.

Picture Study 2

The strongest of Galileo's telescopes magnified only about 33 times, but that was enough for him to make some startling observations. For example, he was the first person to see Saturn's rings, four of Jupiter's moons, and spots on the surface of the Sun.

Answers to *Picture Study 2*

- Galileo's diagram of the Moon is similar to the Moon we see at night because the same side of the Moon is always visible from our position on Earth. This is because the Moon circles the Earth at the same rate that it rotates on its axis. Moreover, the Moon does not experience the same erosive forces that the Earth does (storms, flooding, volcanism, etc.) and so has not changed significantly in the past 400 years.

★ **Two Transparency Worksheets (Teaching Resources, pages 11 and 13) and Transparencies 73 and 74 are available to accompany A Celestial Picture Study.**

Homework

Have students answer the following question: Of all the topics in science that you have studied, which do you think would be most useful to an astronomer and why? *(Accept all reasonable responses.)*

CROSS-DISCIPLINARY FOCUS

Social Studies

Have students research how some past civilizations used their observations of the movements of the Sun, Moon, and stars to create calendars or astronomical monuments. Examples include Stonehenge in southern England, the stone calendar representing the cosmology of the Aztecs, and the Mayan Toltec Caracol, which was probably used as an observatory.

Answers to
Picture Study 3

- A time-lapse photograph of the planets in our solar system shows that some of them, particularly Mars, appear to make strange looping motions that neither the Sun nor the Moon make. This kind of motion is called *retrograde motion.* It is an optical illusion caused by the difference in the orbital speeds of the planets. Retrograde motion will be discussed in greater detail later in the unit.
- The photograph shows that the orbits of the planets follow more or less the same path. This indicates that the planets orbit the Sun in similar planes. That is, if you looked at the solar system on edge, the planets, with the exception of Pluto, would appear to be in the same "slice" of space. This is called the *ecliptic plane.*

Answer to
In-Text Question

Ⓐ At this time, nine planets are known to exist.

Picture Study 4

You may wish to point out to students that the cardinal directions are printed on each star chart. Remind students that to use a star chart they must hold it above their heads with the cardinal directions aligned. The circle enclosing the chart represents the horizon.

Answers to
Picture Study 4

- The location of Polaris (the North Star) is the same in both charts. Both charts indicate the cardinal directions and the horizon. Many of the constellations appear on both charts.
- Students should observe that there are a number of constellations that appear on only one chart. All of the constellations have different positions and orientations on the charts.

Picture Study 3

The early observers noted objects in the skies that they called wanderers. These are the planets. At that time, only five were known. How Ⓐ many do we know about now? The time-lapse photograph at right, taken in a planetarium, shows the motions of several planets over a long period of time.

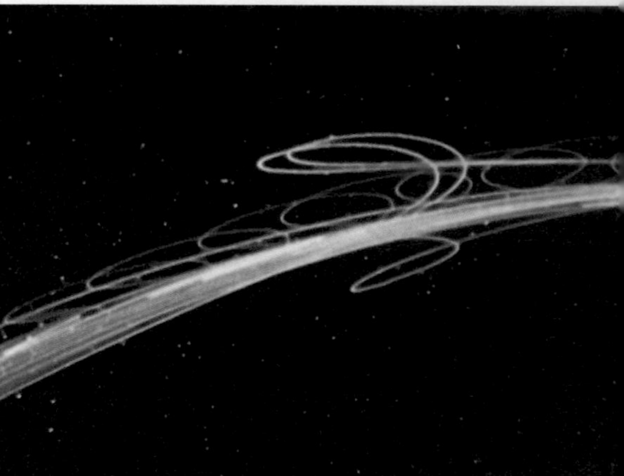

- What strange motions do planets seem to have that the Sun and Moon do not?
- Does each planet seem to follow more or less the same path that the other planets do? What does this indicate?

Picture Study 4

Star charts show the positions of stars and constellations for a particular time of the year. Study the star charts for both a summer and a winter sky.

432

Meeting Individual Needs

Gifted Learners

Have students create a scale model of the calendar wheel shown on page 431. The actual wheel is 21.35 m in diameter and has 28 spokes leading from the center hub. The hub is 3.66 m in diameter.

Picture Study 5

If the lens of a camera is held open for a time while aimed at the night sky, each star will appear in the photograph as a "star trail."

- Do stars appear to revolve? If the arrow points to the beginning of a star trail, in what direction are the stars revolving?
- What other observations regarding stars can you make from this photograph?

Making Sense of the Skies

If you had lived 2000 years ago, when there were no telescopes or spaceships, and the only means of observing celestial objects was with the naked eye, how would you have explained observations like the ones made so far in this unit? With a friend, weave your explanations into a theory that explains as many of your observations as possible. Can your theory explain the reasons for night and day? the star trails? why the calendar wheel works? Are there observations that your theory cannot explain? Does this mean that your theory is not useful?

In Exploration 2, you will be able to compare your theory with those of a number of early thinkers and observers.

EXPLORATION 2

The Early Observers

Living in Greece and Egypt over 2000 years ago were many keen observers and thinkers. How did they explain their observations of the heavens? As you read about three of them, think of the observations on which they may have based their ideas.

1. Aristotle (384–322 B.C.)

Aristotle (pronounced air us TAHT ul) was the most influential thinker of his time. He suggested that everything in the universe is composed of four elements: earth, air, fire, and water. He proposed that each element has a natural resting place, with earth at the center, surrounded by spheres of water, air, and fire.

His model of the universe contained a total of eight spheres revolving around Earth. The first seven spheres carried the Moon, the Sun, and the five known planets. Aristotle proposed that beyond these seven spheres, the stars were tiny lamps fixed to another revolving sphere.

He theorized that the sphere carrying the Sun revolved around Earth each day, causing day and night. The motion was much like following the threads on a corkscrew. With each revolution around Earth, the Sun was carried higher (or lower) in the sky. He proposed that this movement causes the seasons.

- How is this theory based on everyday observations? B
- Try drawing a diagram based on Aristotle's ideas.

ARISTOTLE

Exploration 2 continued ►

Answers to *Picture Study 5*

- In a time-lapse photograph of the night sky, stars seem to revolve counterclockwise around a central point, forming circular paths. (It is Earth's rotation that causes these patterns in the night sky. You may wish to point out that a similar photograph taken in the Southern Hemisphere would show the stars revolving clockwise.)
- Careful examination of the photograph shows that the star trails vary in brightness and color. The varying widths of the star tracks indicate that the stars are of different brightnesses. Stars that make complete circular paths do not set (i.e., they don't go below the horizon). The length of the various trails indicates how long the photograph was exposed.

Making Sense of the Skies

To summarize what students have learned so far, you may wish to involve them in a discussion of their observations and explanations. Then provide time for students to work with a partner to develop a theory that explains many of the observations they have made so far. Call on volunteers to share their theories with the class. *(Accept all reasonable ideas.)*

Answer to *In-Text Question*

B To an observer on Earth, it appears that Earth is motionless and that everything else moves around it. Two observations that might give the impression that celestial objects are attached to spheres revolving around Earth are (1) the Sun moving across the sky, while Earth does not seem to move and (2) seeing the stars near the axis of rotation (near Polaris) move in a circle each night.

Exploration 2 continued 

EXPLORATION 2

Divide the class into pairs of students. Suggest that they answer all of the questions raised and draw each of the models as described.

1. Aristotle (384–322 B.C.)

You may wish to share the following information with students: Aristotle was born in Stagira, a small town in northern Greece. When he was about 18, Aristotle entered the Academy, Plato's school in Athens. He studied there for the next 20 years. Around 334 B.C., Aristotle founded his own school, the Lyceum.

Aristotle's view that everything was made up of combinations of the four elements (air, earth, fire, and water) was accepted for nearly 2000 years. It was not until the seventeenth century that scientists began to insist that direct observation and experimentation, rather than abstract philosophy, should be the basis of scientific thought.

EXPLORATION 2, *continued*

2. Aristarchus of Samos (3rd century B.C.)

You may wish to share the following information with students. Based on clever reasoning and observation, Aristarchus was the first to arrive at the radical conclusion that Earth revolved around the Sun and that the Sun, not Earth, was the center of the solar system. During his lifetime, Aristarchus was reviled for his ideas and was threatened with prosecution for alleged impiety against the prevailing views of the time.

Answer to *In-Text Question*

Ⓐ Good ideas and reasoning are often disregarded when they do not blend neatly with prevailing notions of what is true.

3. Ptolemy (2nd century A.D.)

You may wish to share the following information with students: Ptolemy, also known as Claudius Ptolemaeus, made most of his astronomical observations in Alexandria, Egypt. He compiled his ideas in a 13-volume work that became known as the *Almagest,* a Greek-Arabic term that means "the greatest."

Much of what Ptolemy taught was pure astrology. He cataloged existing beliefs about how the positions of the planets and stars controlled everything on Earth. In spite of his mysticism, Ptolemy was a tireless observer. He cataloged the celestial latitude and longitude of over 1000 stars. He also discovered the irregularity of the Moon's orbit. The *Almagest* survived the collapse of Greece and endured the fall of the Roman Empire to become the backbone of the Roman Catholic Church's view of the universe. The heavens, as defined by Ptolemy, were perfection, with Earth at its center. This view prevailed until the sixteenth century.

2. Aristarchus of Samos (3rd century B.C.)

Aristarchus (pronounced air us TAR kus) combined mathematics with good observation to come up with another theory that is surprisingly close to our current understanding of the solar system. He suggested that the Sun is the center of the universe—not Earth. He proposed that Earth and the other planets revolve around the Sun, causing the cycle that we call the year. Day and night, Aristarchus suggested, are caused by the rotation of Earth on its axis. He was the first to measure the distance to the Sun and Moon. He concluded that the Sun is much farther away and much larger than previously thought. The stars, he reasoned, are unimaginably far away. Although Aristarchus's theory explained observations as well as Aristotle's theory did, it was not accepted by the people of his day and was forgotten for nearly 2000 years.

- What do you think about Aristarchus's theory?
- Why do you think good ideas and reasoning are sometimes disregarded in favor of other ideas? Ⓐ
- Try drawing a diagram that illustrates Aristarchus's ideas.

3. Ptolemy (2nd century A.D.)

PTOLEMY

Ptolemy (pronounced TAHL uh mee) supported the ideas of Aristotle. He reasoned that if Earth really moved, the tremendous winds caused by Earth's movement would blow birds off their perches and leaves off the trees. Ptolemy supported Aristotle's idea that the Sun revolves around a stationary Earth; he noted that every day, we see the Sun actually moving across the sky.

But Ptolemy recognized that Aristotle's model had two serious flaws. First, it could not explain why the wanderers, or planets, varied in brightness from year to year. If they orbited Earth at fixed distances, their brightness should not vary greatly. Second, the paths that the planets traveled were more erratic than the paths suggested by Aristotle's theory. A planet such as Mars, for instance, appears to make a loop in the sky, going backward for a while before continuing its forward motion. This curious backward motion is called **retrograde motion**.

As shown in this planetarium photograph, Mars appears to make a looping motion in the sky (as seen from Earth).

INDEPENDENT PRACTICE Encourage students to research and discover more about the early astronomers discussed in the Exploration. Students should discover as many biographical facts as they can, along with information about the scientific contributions that these astronomers made. Students should share their information with the class.

Theme Connection

Systems

Focus question: How did ancient astronomers develop a model of our solar system based only on observations with the naked eye? *(They kept track of observations through time. This allowed them to develop theories about the Earth and other bodies moving through space.)* The Theme Worksheet on page 15 of the Unit 7 Teaching Resources booklet makes an excellent homework activity to accompany this Theme Connection.

Ptolemy's solution was to add another motion to a planet. He suggested that a planet not only revolves in a circular orbit around Earth but also makes smaller revolutions about points on its main orbit—like a small wheel turning on the circumference of a larger wheel. To get a good picture of Ptolemy's idea, imagine the rocking seat of a Ferris wheel actually making a complete turn as the Ferris wheel itself turns. The motion of the seat is like the motion of a planet in Ptolemy's model.

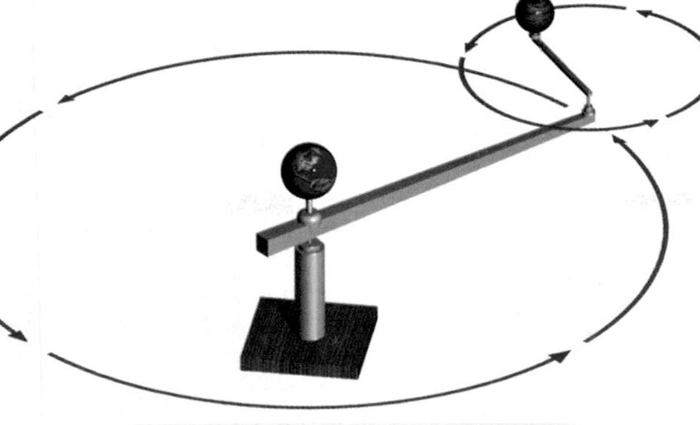

Ptolemy gave Mars two motions. He thought that Mars moves in small circles as it moves around Earth in a larger orbit. In what way could a Ferris wheel be considered a model of these motions?

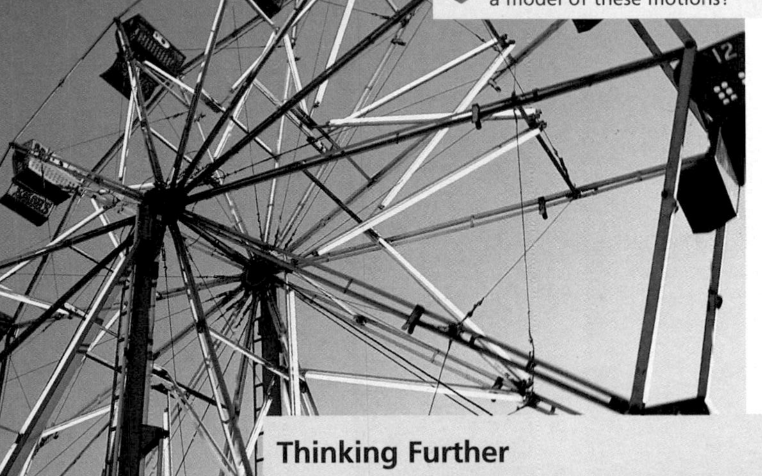

- Study the model at the top of this page, which shows how Ptolemy thought Mars moved. How does Ptolemy's "solution" explain the apparent changes in brightness and the apparent backward (retrograde) motion of a planet such as Mars, as viewed from Earth?
- Can you think of another explanation for retrograde motion?

Thinking Further

1. Is the theory you proposed on page 433 similar to any of the early Greek and Egyptian observers' theories? Which theory seems to best explain your observations of the heavens? Why?
2. Imagine that you live at the time of Aristotle, but your ideas match those of Aristarchus of Samos. What could you say to convince Aristotle that Earth revolves around the Sun? Remember, they had no telescopes or spaceships back then.
3. For 1500 years, people believed that Earth was the center of the universe. Why do you think this was such a popular view?
4. Aristarchus of Samos proposed a *heliocentric* model of the solar system. Aristotle and Ptolemy both proposed *geocentric* models. Based on their theories, what do you think these terms mean?

Answers to *In-Text Questions*

- When Mars was on the far side of a small circle, it would appear faint. Then, as Mars came around its epicycle, it would appear to be closer and brighter and to be moving in the opposite direction. Retrograde motion in this model occurs when the planet's motion on the small circle is opposite to its motion around the Sun.
- Another explanation for retrograde motion could be that Earth is moving faster in its orbit than Mars is. Much like a faster car on the inside lane of a racetrack, as Earth passes Mars, Mars seems to go backward.

Answers to *Thinking Further*

1. Accept all reasonable responses. Students may recognize that the Earth-centered theory proposed by Aristotle and Ptolemy is the best explanation for their observations, even though it is incorrect.
2. Accept all reasonable responses. Students might try to explain the Earth's revolution in terms of seasonal changes.
3. It explained the motions of the Sun and stars in the simplest way.
4. *Heliocentric* means that the Sun is at the center of the solar system. *Geocentric* means that Earth is at the center of the solar system.

Homework

Have students complete the Extension activity described below as homework.

FOLLOW-UP

Reteaching

Suggest that students make an astronomical calendar showing events that can be observed for several months or throughout the year. Suggest that their calendar include information on the phases of the Moon, when and where planets will be visible, when eclipses will occur, and so on.

Assessment

Present students with the following scenario: Imagine that an alien has just arrived from a distant planet. The alien knows a lot about the universe but doesn't know what it looks like from Earth.

Ask students to explain what the alien is likely to see in terms of the motion of heavenly bodies. Suggest that students role-play this scenario after they have written their responses as an essay.

Extension

Have interested students research the contributions of women astronomers. Some possibilities include Helen S. Hogg and Maria Mitchell. Have students present their findings in a report to the class.

Closure

Have students graph the time at which the Sun rises and sets each day. This information can be found in most newspapers.

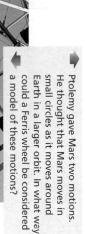

Answers to *Challenge Your Thinking*

1. Reasons for thinking that the Earth is flat include the following:
   - It looks flat from our point of view.
   - If the Earth were spherical, we might fall off when our rotation takes us sideways or upside down.

   Reasons for thinking that the Earth is spherical include the following:
   - We can't see all the way across it, even with a telescope.
   - As ships disappear over the horizon, the top of the mast is the last to disappear.
   - As you approach a point on the horizon, objects behind that point rise into view.
   - Earth's shadow on the Moon during an eclipse is curved.
   - The Sun's angle above the horizon at noon is lower at more northern locations. The Sun does not rise above the horizon at the North and South Poles at certain times of year.
2. By itself, the observation could be used to support either theory. However, the amount of change as you move a given distance north on a spherical Earth is greater than the amount of change that would occur if Earth were flat. Therefore, the observation supports the spherical-Earth theory better. From the diagrams, you can see that Observer 1 will see Polaris just above the horizon, while Observer 2 will see it at about 45 degrees to 50 degrees above the horizon. Observer 4 will see Polaris as being more directly overhead than Observer 3 will.
3. The scene on the left would be seen by Observer 2. The middle scene would be seen by Observer 3. The scene on the right would be seen by Observer 1. The middle scene is closest to what Observer 4 would see. Observer 4 would see almost the same view that Observer 5 would see.
4. Answers will vary. Sample questions: Aristotle, how do you explain the apparent retrograde motion of the planets? Ptolemy, why do the position of the stars change from season to season? Aristarchus, what are the stars, and is there a pattern to their movement?
5. Answers will vary depending on the student's choice. Concept maps should show logical connections and a clear hierarchy of ideas that are relevant to the question.

1. The Flat Earth

Many early astronomers suggested that Earth is flat. For example, some early Greeks pictured Earth as a flat, circular disk floating in a sea of water (at right). Another early observer considered Earth to be a flat object freely suspended in space. What may have been their reasons for thinking Earth is flat? What are some reasons for thinking Earth is spherical?

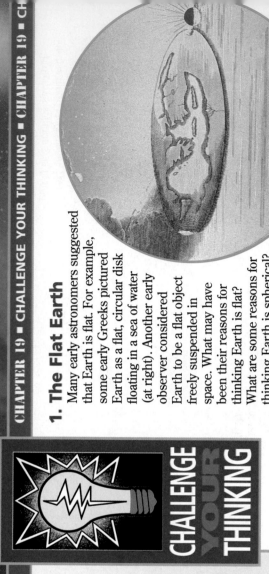

2. Follow That Star

The farther north you travel, the higher Polaris (the North Star) will appear above the horizon. At the North Pole (N.P.), it will be directly overhead. Does this observation support a flat-Earth or a spherical-Earth theory? Study the diagrams to the left, and explain how they help support your answer.

Multicultural Extension

Myths of the Sky

Many of the names of celestial objects come from myths in other cultures, especially those of ancient Greece and Rome. Assign each student a planet or moon of our solar system, and have him or her research the story associated with the body's name.

3. Polaris Positions

The scenes below show Polaris at various positions above the horizon. Copy the scenes into your ScienceLog, and label each one with the observer (from the spherical-Earth diagrams in question 2) who could have seen such a view. Which scene below is closest to what observer 4 would see? Would the view that observer 5 sees be much different from that seen by observer 4?

4. Back From the Past

Through time travel, Aristotle, Aristarchus, and Ptolemy will speak to your class. Make a list of questions that you would ask each one of them based on what you know about their ideas.

5. Mapping the Stars

concept map Create a concept map based on the answers to your favorite question from Exploration 1, page 428. Do some research or check the SourceBook if you need more information to create your map.

Review your responses to the ScienceLog questions on page 422. Then revise your original ideas so that they reflect what you've learned.

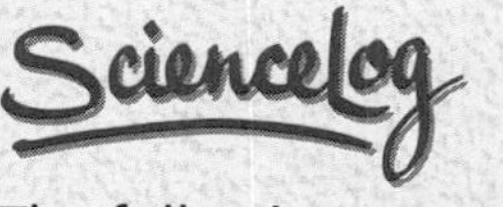

The following are sample revised answers:

1. No. As the Earth revolves around the Sun, the portion of sky seen above a certain point will change throughout the year.
2. Appearances suggest that the Sun moves around the Earth. The Earth appears to be motionless, and the Sun appears to move from one side of the Earth to the other. However, the retrograde motion of Mars and seasonal changes in the night sky indicate that the Earth and other planets are revolving around the Sun.

★ **You may wish to provide students with the Chapter 19 Review Worksheet on page 17 of the Unit 7 Teaching Resources booklet. It accompanies this Challenge Your Thinking.**

The Earth Moves

Connecting to Other Chapters

Chapter 19
explores the science of astronomy and discusses the observations of some early astronomers.

Chapter 20
describes the various motions of celestial bodies and some of the effects of those motions.

Chapter 21
introduces meteors, meteorites, and comets and examines what a colony on Mars might be like.

Chapter 22
examines Earth's place in the universe and introduces the big-bang theory.

Prior Knowledge and Misconceptions

Your students' responses to the ScienceLog questions on this page will reveal the kind of information—and misinformation—they bring to this chapter. Use what you find out about your students' knowledge to choose concepts and activities to emphasize in your teaching.

In addition to having students answer the ScienceLog questions on this page, you may wish to have them listen to the following scenario: Tarla, an intergalactic traveler who needs chocolate to fuel his starship, has come to our planet demanding that we give him Earth's entire crop of cocoa beans. If we fail to do so, Tarla has threatened to stop the Earth from rotating. Ask: How would we know if Tarla stopped the Earth from rotating? What consequences would a nonrotating Earth have for our town and for people on the other side of the Earth? It may be helpful to record a list of student responses on the board and then lead a class discussion. The responses should help you evaluate what students already know about planetary motion, what misconceptions students may have, and what is interesting to them about this topic.

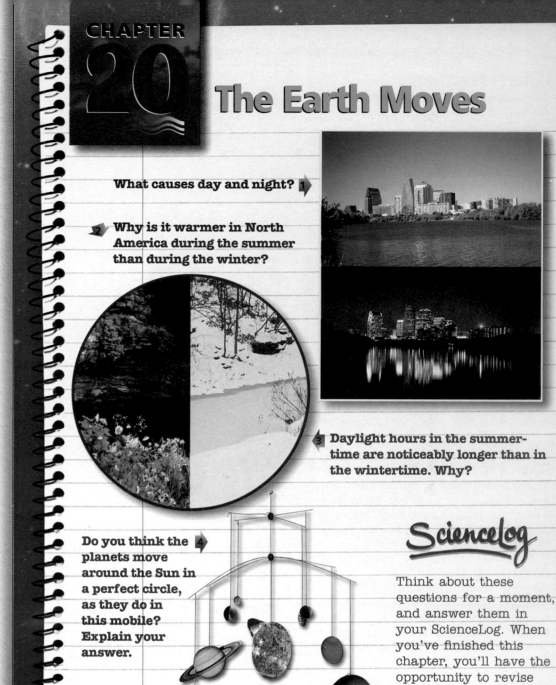

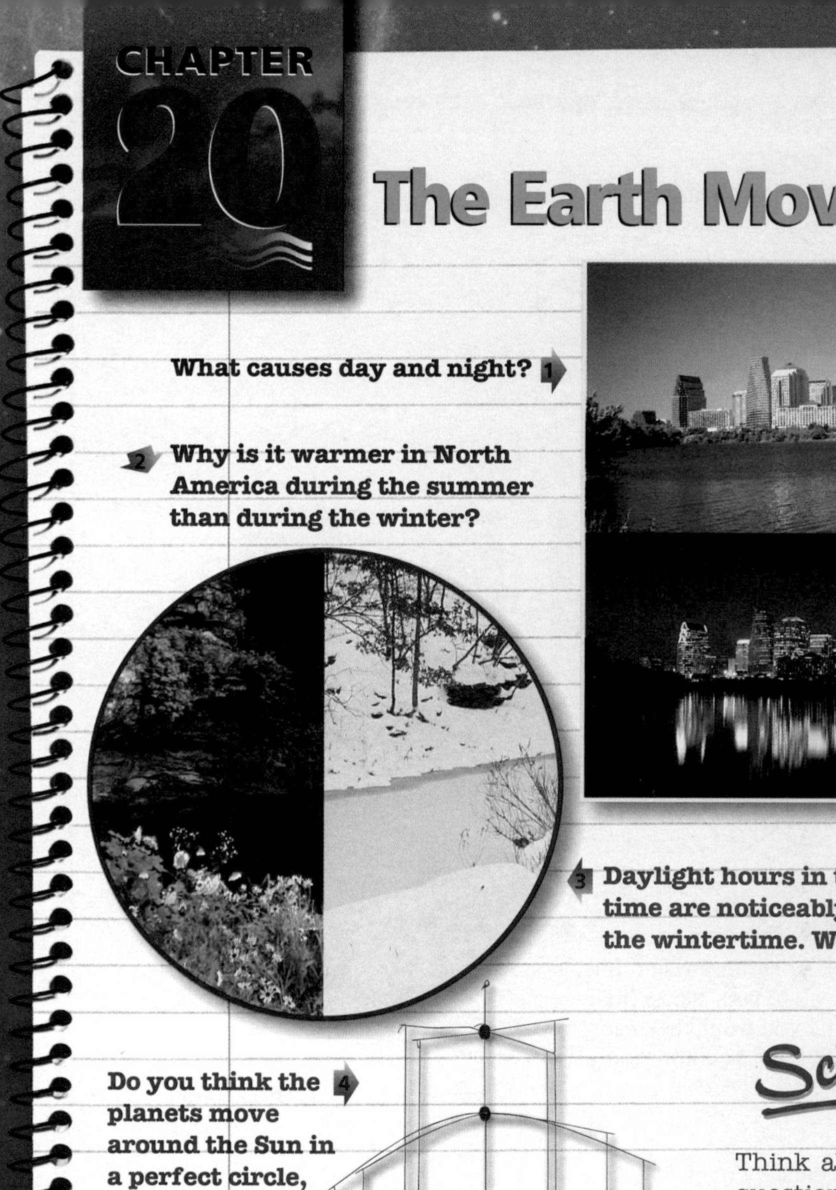

CHAPTER 20
The Earth Moves
What causes day and night? 1
2 Why is it warmer in North America during the summer than during the winter?
3 Daylight hours in the summertime are noticeably longer than in the wintertime. Why?
Do you think the planets move around the Sun in a perfect circle, as they do in this mobile? Explain your answer. 4
ScienceLog
Think about these questions for a moment, and answer them in your ScienceLog. When you've finished this chapter, you'll have the opportunity to revise your answers based on what you've learned.
438

LESSON 1 A Scientific Revolution

The Trial of Galileo

It is June 1632. The atmosphere in the courtroom is electric. The greatest scientist of the time, Galileo Galilei, is on trial for teaching the views of Copernicus—that the Sun, rather than Earth, is the center of the solar system. In this model, Earth is just one of a number of planets that revolve around the Sun. This is the second time Galileo has been on trial for his views. After the first trial, he was ordered by the Roman Catholic Church not "to hold or defend" Copernicus's theory. At the second trial, Galileo must answer to charges that he has willfully disobeyed this order.

The verdict—guilty. Galileo is sentenced to house arrest for the remainder of his life. Legend has it that as Galileo was led sadly from the courtroom, he was heard to mutter, "And still the Earth moves."

Seventeenth-century painting of Galileo on trial for his views about the solar system

439

LESSON 1 ORGANIZER

Time Required 2 to 3 class periods

Process Skills
analyzing, inferring, classifying

New Terms
Constellations—ancient groupings of stars used for navigation and for marking the seasons, among other uses
Plane of the ecliptic—the plane in which Earth orbits the Sun
Revolution—the motion of one object around another object in a circle, ellipse, or similar curve
Rotation—the turning of an object on its own axis
Zodiac—circular backdrop of 12 constellations through which the planets appear to move, centered on the ecliptic

Materials (per student group)
Exploration 1: compass; protractor; meter stick; metric ruler; set of star charts; tube of glue or roll of tape; 4 sheets of unlined paper; 9 marbles or small plastic-foam balls, 1–2 cm in diameter; stick of modeling clay

Teaching Resources
Exploration Worksheet, p. 22
Transparency 75
SourceBook, p. S152

LESSON 1 A Scientific Revolution

FOCUS

Getting Started

Ask students what planet they live on. Then ask them where that planet is located. *(The solar system)* Ask: Why do we call it that? Help students realize that we call it the solar system because the Sun, a star, is at the center of it. The Latin word for Sun is *Sol.* Explain that our acceptance of a Sun-centered system is relatively new; students will see that people in many parts of the world changed their view about the solar system from an Earth-centered to a Sun-centered system fewer than 400 years ago.

Main Ideas

1. Religious beliefs and scientific ideas are sometimes in conflict.
2. The Copernican model of the solar system can be used to explain the motions of the planets and the apparent motions of the Sun and stars.

TEACHING STRATEGIES

GUIDED PRACTICE Remind students that they read about Aristarchus in the previous lesson. Call on a volunteer to briefly explain the model that Aristarchus developed. *(The Sun-centered theory)* Ask students if they recall what happened to Aristarchus's theory. *(It was forgotten for nearly 2000 years.)* What theory was supported instead? *(Ptolemy's Earth-centered theory)*

The Trial of Galileo

Have students read about Galileo's trial to discover what they are to do in his defense. You may wish to have them work in small, cooperative-learning groups to research answers to the questions on page 440.

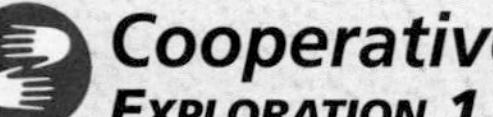

Cooperative Learning
Exploration 1, Part 1

Group size: 2 to 3 students
Group goal: to construct a model that illustrates the motions of the planets in relation to the Sun and the zodiac
Positive interdependence: Assign each student a role such as group leader, materials organizer, and investigator.
Individual accountability: Each student should be able to orally explain the model, using the questions to guide their explanation.

Answers to *In-Text Questions*

A
- Galileo (1564–1642) was an Italian astronomer and physicist. He is considered to be the first to effectively use the telescope to study astronomy. Galileo became convinced that the theories of Copernicus were correct, and he spent much of his life trying to prove them.
- Copernicus (1473–1543) was a Polish astronomer who revived and expanded the ideas of Aristarchus. Like Aristarchus, Copernicus speculated that Earth and all the other planets revolved around the Sun. Galileo considered himself a Copernican because he supported the idea that the planets revolve around the Sun.
- A major part of Roman Catholic philosophy of the time was the notion that Earth and humans were the center of all things. Removing Earth from the center would make Earth and humans seem less significant.

B Sample questions: Did anyone else hold the same views as Galileo? What did the general population believe about the motions of the stars and planets?

C The six known planets at the time of Galileo were Mercury, Venus, Earth, Mars, Jupiter, and Saturn. Today, Uranus, Neptune, and Pluto have been added to the list.

D Students should recognize that the Earth will be drawn 10 cm from the Sun in their model because it is 1.0 AU away from the Sun.

Imagine that you are living at the time of Galileo and have been asked to be part of a team of lawyers preparing to defend him in the trial. The task is great, and much needs to be done. Here are some questions that may need to be answered: **A**

- Who is Galileo? What is his background?
- Who was this man Copernicus? Where did he come from? What were his ideas? Why does Galileo consider himself a Copernican?
- Why would Church officials care whether the Sun went around Earth or vice versa?

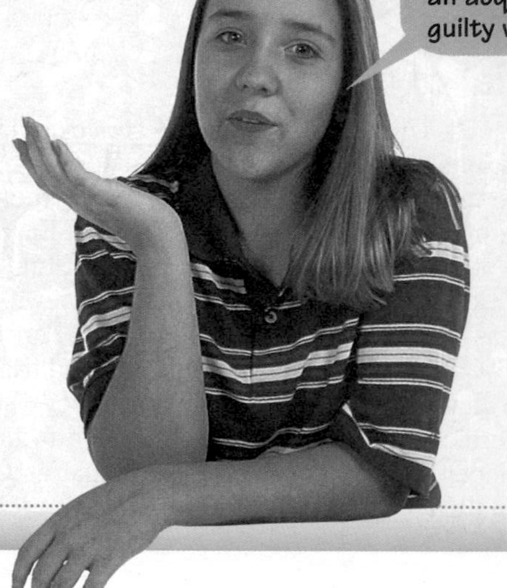

Are there other questions that may be important in working toward an acquittal (not guilty verdict)? B

EXPLORATION 1

Copernicus's Model: The Planets and the Ecliptic

To prepare for the trial, you and your team need to make a model of the solar system based on Copernicus's ideas. You want to see whether the Copernican theory explains celestial motions that the theories of Aristotle and Ptolemy could not explain.

PART 1

The Model and Its Use

At the time of Galileo, only six planets were known. **C** How many are known now? Galileo taught that the six planets orbited the Sun. Your model will show the orbits of the first four planets you would see if you started at the Sun and traveled outward. In the table at right, the distance of each planet from the Sun is expressed in terms of Earth's average distance from the Sun. This distance is referred to as one astronomical unit (AU), and it is equal to 150 million kilometers. (Try writing out this figure.) The time it takes for each planet to revolve once around the Sun is given in Earth days.

Your model will have a scale of 1 AU = 10 cm. Using this scale, Earth would be 10 cm from the Sun on your model. Why? **D**

Observers from the earliest times noticed that when seen from Earth, the planets moved against a background of 12 different groupings of stars, called **constellations**. These constellations make up what is popularly known as the **zodiac**. The early observers knew that the planets did not wander randomly through the heavens. The planets appeared to move counterclockwise through the zodiac, which consists of the constellations Leo, Virgo, Libra, Scorpius, Sagittarius, Capricornus, Aquarius, Pisces, Aries, Taurus, Gemini, and Cancer. Your model would not be complete without the 12 constellations of the zodiac.

Planet	Average distance (AU)	Time it takes to revolve once around the Sun (in days)
Mercury	0.4	88
Venus	0.7	225
Earth	1.0	365
Mars	1.5	687

An Exploration Worksheet (Teaching Resources, page 22) and Transparency 75 are available to accompany Exploration 1.

You Will Need

- a pencil
- a compass
- a protractor
- a meter stick
- star charts
- glue or tape
- four sheets of unlined paper, fastened together to form one large rectangle
- several small plastic-foam balls or marbles to represent the planets
- modeling clay to hold the plastic-foam balls or marbles in place

What to Do

1. Turn your glued or taped paper so that its longer direction is up and down. Using a compass, draw a circle with a radius of 15 cm. The center of the circle should be in the lower half of the paper so that the circle comes within 4 cm of the bottom of the page. Which planet would have the orbit you just drew? Now draw other circles to represent the orbits of the remaining planets from the table on page 440.
2. Where is the Sun in this model?
3. Using a protractor and ruler, divide the circles into 12 equal segments. Each segment will form an angle of 30 degrees at the center of the circles. Extend each line 4 cm beyond the orbit of Mars.
4. In the segments outside the orbit of Mars, place the names of the constellations that form a backdrop to the moving planets. Refer to the diagram at right to place the zodiac constellations in their proper order. Place Taurus at the top of your model, as shown in the diagram. Each constellation will occupy one 30-degree segment, although in reality they vary in size. Of course the stars making up these constellations are actually a tremendous distance beyond the orbit of Mars. Have you come across the names of these constellations before? Where?
5. Examine a star chart for this month. Which constellations of the zodiac are visible in the night sky? Where must Earth be placed on your model in order for those constellations to be visible this month? Use a marble or plastic-foam ball to represent Earth, and place it in its proper location on the model. Use a small piece of modeling clay to hold it in place.

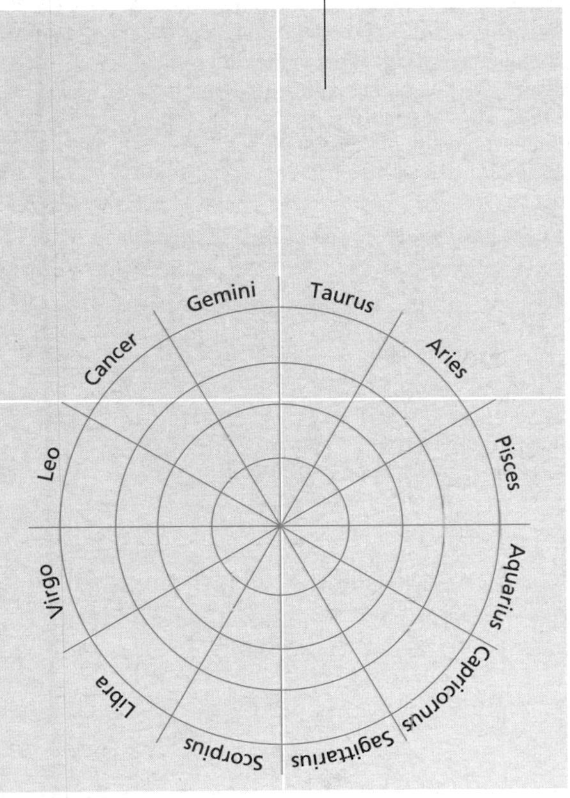

Exploration 1 continued ►

Homework

Have students write a paragraph that answers the following question: What were the major obstacles faced in changing from a geocentric to a heliocentric viewpoint? *(The primary obstacles were religion and tradition, both of which made the acceptance of new viewpoints difficult. The religion of the time held that the Earth must be the center of the universe, and long-held beliefs are difficult to dispel because they are deeply ingrained in society's entire outlook on the world.)*

EXPLORATION 1, *continued*

The following strategies and answers for each step will assist you in guiding students through this Exploration:

1. Students should use the information in the table on page 440 to determine the size of the "orbits" in their model. Suggest that students draw their circles on the lower half of the paper. The top of the paper may be used for Part 2. You may wish to point out that based on the scale they are using, one-tenth of an AU is equal to 1 cm. Therefore, the orbit of Mercury will have a 4 cm radius, the orbit of Venus will have a 7 cm radius, the orbit of Earth will have a 10 cm radius, and the orbit of Mars will have a 15 cm radius. The outermost circle in the model represents the orbit of Mars.
2. Since the diagram is a Copernican, or heliocentric, model of the solar system, the Sun is at its center.
3. Remind students that there are 360 degrees in a circle. Since there are 12 constellations, each one will occupy one-twelfth, or 30 degrees, of each circle. You may wish to demonstrate how to use a protractor to divide the circumference of a circle into 12 equal sections.
4. Students may recognize that some of the names of the constellations are also the names of the birth signs used in astrology. They may have seen these names in newspaper columns, in books on horoscopes, on birthday cards, and so on. Be sure students understand that astrology is based on mysticism and is a pseudoscience. Astronomy is a science based on observation.
5. Astronomy magazines, such as *Sky and Telescope* and *Astronomy*, include monthly star charts that students can use to learn the correct location of Earth in their models.

Exploration 1 continued ►

EXPLORATION 1, continued

Have students work in groups to develop answers to the questions on this page. To save time, you may wish to have each group work on one or two questions. Then reassemble the class to discuss the answers to all of the questions.

Answers to *Interpreting the Model*

1. Galileo taught that the Sun was the center of the solar system and that the planets revolved around the Sun. This idea caused Galileo problems with the Roman Catholic Church because it removed Earth and humans from the center of all things.
2. The stars appear to change their position throughout the year. Different constellations are seen at different times of the year.
3. Since the Martian year is almost twice as long as Earth's, it will have moved halfway around its orbit and be on the opposite side of the Sun from Earth after one Earth year. Mars will not be visible from Earth because the Sun will be in the way.
4. To show night and day on the model, "Earth" must be rotated on its axis in a west-to-east direction. Rotation, or the turning of an object on its own axis, is different from revolution. Revolution occurs when one object circles or moves around another, as when Earth revolves around the Sun. As a result of Earth's west-to-east rotation, the star trails appear to be going in a counterclockwise direction.
5. As Earth rotates on its axis, constellations appear to rise in the east and set in the west, just as the Sun does. As Earth revolves around the Sun, new constellations will appear in the eastern sky, and old ones will disappear from the western sky.
6. In this model, the paper represents the plane of the ecliptic. The Sun and planets, with the exception of Pluto, lie more or less within this plane.
7. Most students will probably agree that this is a good model because it can be used to explain many of the daily and yearly astronomical events observed from Earth.

EXPLORATION 1, continued

Interpreting the Model

1. From this model, infer the ideas that Galileo taught and that eventually caused him problems with Church authorities.
2. On the model, demonstrate the motion of Earth over one year. To make the model consistent with observation, the ball must be moved in a counterclockwise motion around the Sun. What might you observe about the stars' positions because of Earth's motion?
3. Use another plastic-foam ball or marble to represent Mars. Place Earth and Mars in their respective orbits, with the constellation Taurus as their backdrop of stars. Move Earth through one year. In this time, how far has Mars moved? After the planets have moved this distance, will Mars be visible from Earth?
4. Any model must show why the Earth experiences night and day. How would you move the ball to demonstrate this? The motion you just showed is called **rotation.** What is the difference between rotation and **revolution?** Refer to the picture of the star trails on page 433 and the observations you made of the Big Dipper. In what direction must Earth be rotating to cause these effects?
5. As the year progresses, new constellations of the zodiac appear in the eastern sky, while others disappear in the western sky. Use your model to explain this observation.
6. Copernicus (and later Galileo) inferred that the Sun and the planets lie in the same plane. In other words, they inferred that the solar system is essentially flat. This plane is called the **plane of the ecliptic**. Where is the plane of the ecliptic in your model?
7. In your opinion, is this a good model? Could it be used as an exhibit in the retrial of Galileo? Does it explain the retrograde motion of Mars in a satisfactory way? Is it a simpler explanation than the two motions that Ptolemy suggested? Part 2 of this Exploration will help you discover some answers.

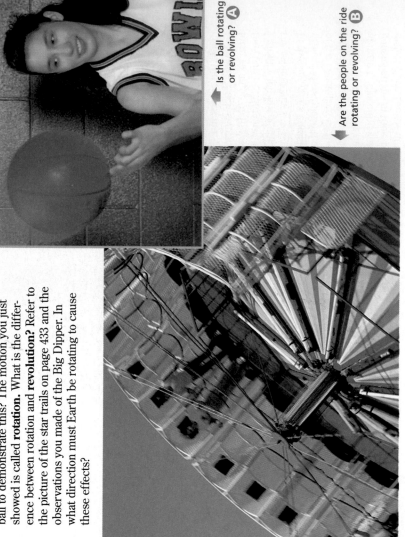

Is the ball rotating or revolving? (A)

Are the people on the ride rotating or revolving? (B)

442

Answers to *Captions*

(A) The ball is rotating.

(B) The people on the carnival ride are revolving.

Meeting Individual Needs

Learners Having Difficulty

Provide students with tops, marbles, and balls. Have them use the materials to demonstrate to the class the difference between rotation and revolution. Encourage students to include a poem, song, or mnemonic device to help them remember how these terms differ.

PART 2

The Retrograde Motion of Mars

What causes Mars to appear to go backward in its path for a while before continuing in its original direction? A Copernican would say that it is simply the result of the faster Earth catching up with and passing Mars. This causes Mars to appear to loop backward in the sky when viewed against the more distant stars.

You can demonstrate this by following the progress of Earth and Mars over a period of 100 days on your model. Every 10 days on the model, note where these planets lie among the more distant stars.

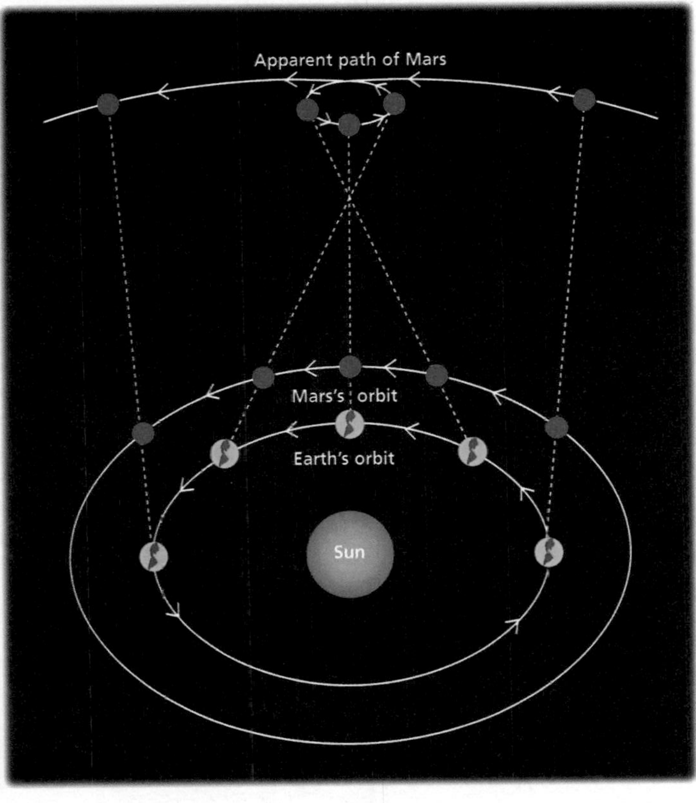

Exploration 1 continued

EXPLORATION 1, continued

PART 2

Before students begin reading, remind them that they learned about the retrograde motion of Mars in Chapter 19. Call on a volunteer to define retrograde motion. *(The apparent backward movement of an object)* Determine if students can recall Ptolemy's explanation for this phenomenon. If they can't, refer them to Chapter 19. *(Ptolemy proposed that Mars not only revolved around Earth, but also revolved in a smaller orbit, as shown on page 435. He theorized that this created retrograde motion.)* Then involve students in a discussion of how Copernicus might explain why Mars appears to reverse direction and go backward in its orbit for a while. *(Accept all reasonable responses.)* Have students read the introduction to Part 2 to discover if their ideas were correct.

Remind students to refer to the diagrams in their texts as they determine the locations of Earth and Mars during a period of 100 days.

INDEPENDENT PRACTICE Have students write an astronomer's log to accompany Part 2. Students should make their entries from Day 1 of the astronomer's observation and proceed at 10-day intervals through 100 days of observation. When the log is complete, it should provide a series of simple statements that describe retrograde motion.

Exploration 1 continued

Multicultural Extension

Astronomy in Past Cultures

Many contributions to our knowledge of the heavens came from past cultures. Have students investigate one of these cultures and present the results of their research to the class. Cultures to consider include the Aztecs, Mayas, Celts, ancient Egyptians, Babylonians, the Moguls in India, the ancient Chinese, and the ancient Greeks.

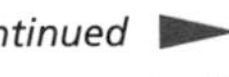

Did You Know. . .

The four brightest moons of Jupiter are called the Galilean moons after Galileo, who first saw them through his telescope. The four moons are Io, Ganymede, Callisto, and Europa. Galileo fought to name them after his patrons, the Medici family, but he was unsuccessful. Instead, the moons were named after Roman mythological figures associated with the god Jupiter.

CROSS-DISCIPLINARY FOCUS

Language Arts

Challenge students to write newspaper articles about important events in the history of astronomy, such as the trial of Galileo; the discovery of Uranus, Neptune, or Pluto; the first spacecraft to land on Mars; the first shuttle flight; or the *Voyager 2* flyby of Saturn or Jupiter. When students have finished, have them organize their work in an astronomy newsletter.

Answers to
In-Text Questions

4. Students should recognize that the distance Earth moves on their model every 10 days can be calculated in the following manner:
   63 cm ÷ 365 days = 0.17 cm/day
   0.17 cm × 10 days = 1.7 cm every 10 days

5. The distance Mars moves on the model can be calculated in the same way. If necessary, remind students that the circumference of a circle is determined by the formula: π × diameter. (Use 3.14 for the value of π.)
   3.14 × 30 cm = 94 cm orbit
   94 cm ÷ 687 days = 0.14 cm/day
   0.14 cm × 10 days = 1.4 cm every 10 days

7. As students connect the positions of Earth and Mars, they should see that as Earth moves past Mars, Mars appears to reverse direction against the backdrop of stars. The model shows that Mars continues in its orbit without actually reversing direction. Students should conclude that the retrograde motion of Mars as observed from Earth is an optical illusion created by the different rates at which the two planets are revolving.

Defending Galileo

Reassemble the class and involve them in a discussion of what they have learned so far. Completing Exploration 1 should have helped students realize that the Copernican model provides a plausible explanation for retrograde motion. A geocentric model cannot adequately explain Mars's retrograde motion.

What to Do

1. The top of your page will represent where the stars appear to be when viewed from Earth. Call this the *star line*. In reality, this is a tremendous distance beyond the orbit of Mars. As Mars and other planets move, they will appear to move among the stars along this line.
2. Using a pencil, draw a dot to represent the point at which Earth's orbit crosses the boundary between Taurus and Aries, as shown in the diagram below. This is the starting position of Earth. Label this position as Earth 1.

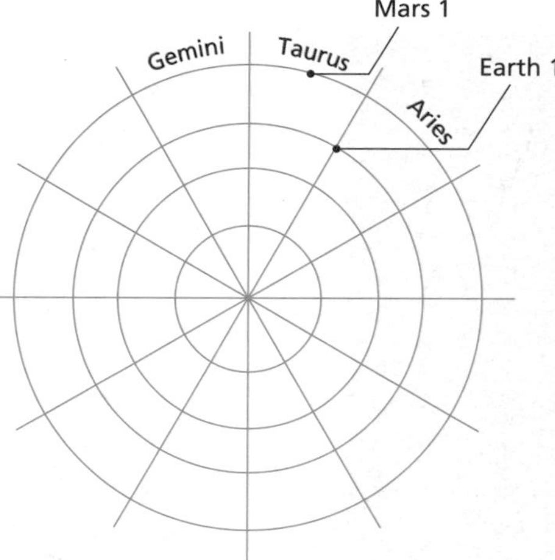

3. Draw Mars on its orbit so that it is halfway between the Taurus-Aries and the Taurus-Gemini boundaries. Label this position as Mars 1, as shown above.
4. In 365 days, Earth will travel once around the Sun—a distance of about 63 cm on your model. How far will it travel in one day? Where will Earth be on the model in 10 days? Can you convince yourself that it has moved a distance of 1.7 cm from its original position? Using a compass, measure off this distance and label this position as Earth 2. Repeat this step to locate Earth's position in 20 days, 30 days, 40 days, and so on up to 100 days. Label these positions as Earth 3, Earth 4, and so on up to Earth 11. Be sure that Earth is revolving counterclockwise around the Sun.
5. Where will Mars be after 10 days, 20 days, and so on up to 100 days? Taking into account the larger orbit of Mars and the fact that Mars takes about twice as long as Earth to make the journey around the Sun, you can discover that Mars will move 1.4 cm along its orbit in 10 days. Mark off 1.4 cm segments from the original position of Mars, moving in a counterclockwise direction. Label the ends of these successive segments as Mars 2, Mars 3, and so on up to Mars 11.
6. Draw a line connecting Earth 1 with Mars 1. Extend this line to the star line at the top of the page. This is where Mars will appear to be for an observer on Earth.
7. Repeat this for each new position of Earth and Mars. Where does Mars appear on the star line for each position of Earth and Mars? Does Mars appear to change direction when viewed against the background of stars? Does it really go backward?

Defending Galileo

Are you ready to develop your defense of Galileo? How can what you just discovered be used to convince a jury that Earth moves around the Sun—not the other way around? Practice your defense. Try explaining to an imaginary jury why retrograde motion is a natural consequence of Earth's motion. In the pages that follow, you will investigate in greater detail the effect of the motions of Earth as suggested by Copernicus and later by Galileo. This information may provide crucial evidence for the retrial of Galileo.

FOLLOW-UP

Reteaching

Have students work in pairs to draw a cartoon or comic strip in which Galileo explains the Copernican model of the solar system. Display the finished cartoons around the classroom for others to enjoy.

Assessment

Have students form small groups, and challenge them to create a human model to demonstrate Mars's apparent retrograde motion. Students will play the roles of the background stars, Earth, and Mars. To assess their understanding, call on each student to explain the different parts of the model.

Extension

Suggest that students do some research to discover the meaning of the terms *celestial equator* and *celestial poles.* Have them make a poster diagram to explain the terms to the class.

Closure

Point out to students that the names of many constellations are associated with heroes, heroines, and beasts in ancient mythology. Suggest that they choose one of the constellations and do some research to discover the story of the character it represents. They should report their findings to the class.

Motions and Their Effects

Many Journeys in a Lifetime

How many times have you traveled around the Sun? According to the Copernican theory, you make this journey once a year. The journey covers a distance of 942 million kilometers. If you were a tour guide for this journey, what sights would you point out? What major events would you experience? Besides the changing constellations of the zodiac, what other signposts would tell you where you are on your journey? For example, think about changes you experience throughout the day and year.

Referring to the model of the plane of the ecliptic that you made in Exploration 1, write down three observations you would point out as tour guide for this trip. Then try Exploration 2, in which you will model such a journey. Ⓐ

LESSON 2 ORGANIZER

Time Required
4 class periods

Process Skills
analyzing, inferring, measuring

Theme Connection
Changes Over Time

New Term
Axis—a line on which an object rotates. Earth's axis runs through its center from the North Pole to the South Pole.

Materials (per student group)
Exploration 2, Part 2: plastic-foam ball, about 10 cm in diameter; rubber band; small cup; 2 toothpicks; light source; protractor; model of the ecliptic from Exploration 1; compass
Exploration 3: letter-sized sheet of corrugated cardboard or stack of several sheets of cardboard; unlined paper; 2 pushpins; 40 cm of string; metric ruler

Teaching Resources
Exploration Worksheets, pp. 27 and 31
Transparency Worksheet, p. 29
Transparency 76

Motions and Their Effects

FOCUS

Getting Started

Place a lamp without a lampshade in front of the class. Tell students that the lamp represents the Sun and that you represent the Earth. To demonstrate the motion of the Earth around the Sun, begin by rotating in place. Point out to students that part of the time the Sun shines on your face, and part of the time it shines on the back of your head. Next, as you continue to rotate, begin to revolve around the lamp. Ask: Which movements represent Earth's motion in a day? *(The rotation)* Which represents Earth's motion in a year? *(The revolution)*

Main Ideas

1. The tilt of Earth's axis causes the seasons to change.
2. In the Northern Hemisphere, the Sun appears at its highest and lowest points during the summer and winter solstices, respectively.
3. During the vernal (spring) and autumnal equinoxes, the Sun is directly above the equator.
4. The planets revolve around the Sun in elliptical orbits.

TEACHING STRATEGIES

Many Journeys in a Lifetime

Have students calculate the millions of kilometers they have traveled, as the Earth revolved around the Sun, since the day they were born. *(A 14-year-old has traveled about 13.2 billion kilometers in his or her lifetime: 942 million km/year × 14 years.)*

Answer to *In-Text Question*

Ⓐ Observations might include sunrises and sunsets, changing phases of the Moon, changing seasons, changing lengths of daylight throughout the year, changing positions of the planets, and the retrograde motion of Mars.

EXPLORATION 2

Answers to the questions posed in each of the four sections of this Exploration are answered on pages 447–448.

GUIDED PRACTICE Display a globe and have volunteers identify and locate the equator, the tropic of Cancer, and the tropic of Capricorn. Ask students to identify the hemisphere in which each one lies. Also tell students that a position in a north-south directon on the globe is given by *latitude*. Latitude is the angle formed by imaginary lines drawn from a point on the globe to the center of the globe and from the center of the globe to the equator. The North Pole has a latitude of 90° north, the equator is at 0°, and the South Pole is at 90° south. Point out that in this Exploration they will learn more about the significance of these imaginary lines.

Have students read pages 446–448 and then review each of the pictures. For the successful completion of the Exploration, it is important that students understand what is illustrated in each picture.

Note that students must position the toothpicks such that, if extended, they would meet in the center of the plastic-foam ball.

You may wish to have groups share one or two light sources set up in the classroom. Another option would be to set up four separate stations for groups to use. Each station should represent one of the positions in the Exploration. If this method is used, caution students not to change the orientation of the globes. Also, be sure that the light source from one station does not affect the results at other stations.

Cooperative Learning
EXPLORATION 2

Group size: 3 to 4 students
Group goal: to create a model to study the Earth's motion around the Sun
Positive interdependence: Assign each student a role such as principal investigator, task leader, materials organizer, and reporter. Students should alternate roles for each of the four positions described in the Exploration.
Individual accountability: Each person in the group should complete drawings of both the winter and summer solstices.

EXPLORATION 2

A Trip Around the Sun

You Will Need

- a large plastic-foam ball
- a rubber band
- a small cup
- 2 toothpicks
- a light source
- your model of the ecliptic from Exploration 1
- a protractor
- a compass

What to Do

1. Form into groups of four or five. Set up the model as shown below. The rubber band around the large plastic-foam ball represents the *equator* of Earth. The toothpicks represent the axis about which Earth rotates. Earth's **axis** is an imaginary line that runs from the North Pole to the South Pole, straight through the center of Earth. The axis is perpendicular to the plane of the equator. The light source represents the Sun. The light source should be placed level with the plastic-foam ball.

What season is represented in this setup? Ⓐ

2. You'll be setting up four situations that represent different positions on your journey at specific times of the year. At each position of Earth, you'll have several missions to accomplish and some thought-provoking questions to consider.

3. When setting up the models, keep two things in mind:
   - The motion of Earth around the Sun is counterclockwise when viewed from over the North Pole of Earth, as you can see from the diagram below.

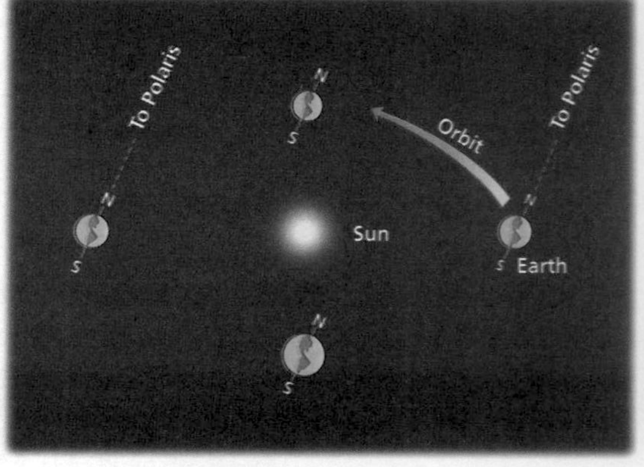

   - Regardless of Earth's position around the Sun, Earth's axis always points in the same direction. At any position along Earth's orbit, Earth's axis is parallel to where its axis was (or will be) at any other position in the orbit. Earth's axis is tilted at an angle of $23\frac{1}{2}$ degrees from the perpendicular, toward the plane of the ecliptic (as shown below). The axis always appears to point toward Polaris, the North Star.

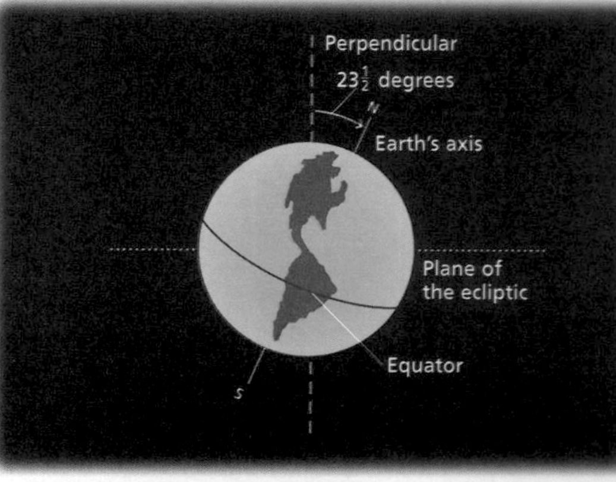

Answer to Caption

Ⓐ The season represented is summer in the Northern Hemisphere. (This is also winter in the Southern Hemisphere.)

★ **An Exploration Worksheet is available to accompany Exploration 2 (Teaching Resources, page 27).**

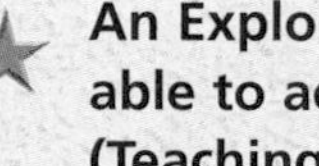

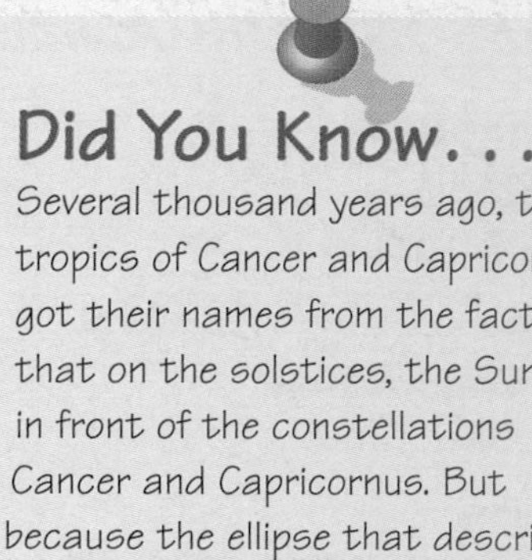

Did You Know. . .

Several thousand years ago, the tropics of Cancer and Capricorn got their names from the fact that on the solstices, the Sun was in front of the constellations Cancer and Capricornus. But because the ellipse that describes the Earth's orbit rotates slowly with respect to the stars (a motion called *precession*), the Sun is now in front of Gemini and Sagittarius at the solstices.

POSITION 1

December 21 or 22—The Winter Solstice

In this position, Earth's axis is tilted at an angle of $23\frac{1}{2}$ degrees from the perpendicular so that the Northern Hemisphere is pointing away from the Sun. Notice which part of Earth is pointed directly toward the Sun. Do this by inserting a toothpick in a position where it will not cast a shadow. The toothpick should be positioned so that if extended, it would pass through the center of the plastic-foam ball. This latitude is $23\frac{1}{2}$ degrees south of the equator—a latitude called the *tropic of Capricorn.* At this time of the year, the constellation Sagittarius cannot be seen from Earth because it is behind the Sun.

1. In your ScienceLog, use a compass to draw a diagram of Earth similar to the one shown here. Label the equator and the tropic of Capricorn on your diagram.
2. Demonstrate a day by rotating Earth one full rotation. Which of Earth's poles is in darkness 24 hours a day? How does the length of the day change as you approach the equator? What two reasons can you give for the winter season in the Northern Hemisphere?
3. If Sagittarius is on the other side of the Sun, away from Earth, what constellations must be in Earth's night skies? Set up your model of the plane of the ecliptic (from Exploration 1) to represent this situation.

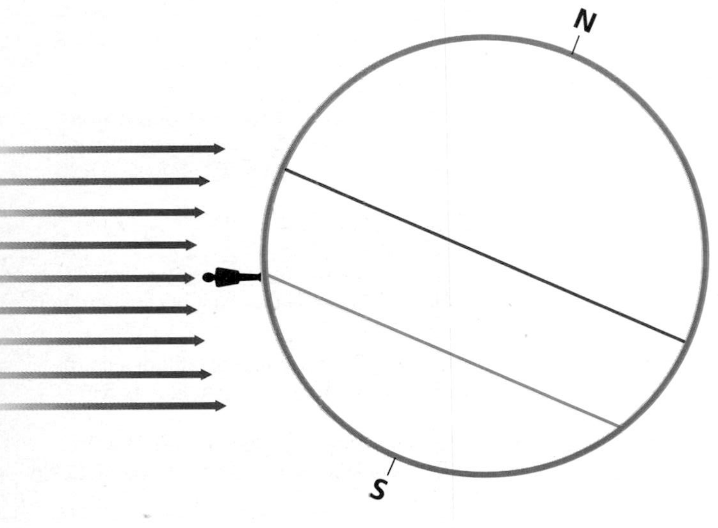

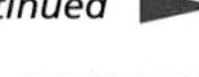

POSITION 2

March 20 or 21—The Vernal Equinox

Move Earth counterclockwise one-quarter of the way around the Sun. Be careful to keep the axis pointing in the same direction at all times. Revolving Earth this far means that one-quarter of a year has passed. In this position, people all over Earth are experiencing equal hours of day and night. Can you see why?

1. At what latitude is the Sun directly overhead at noon? Find out by inserting a toothpick so that it does not make a shadow. Draw a diagram like the one you drew representing Earth at the winter solstice. Label the latitude at which the Sun is directly overhead at noon.
2. Rotate the model Earth in this position so that the opposite side of the model is now experiencing sunlight. Is the Sun still directly overhead at the same latitude?
3. Go to your model from Exploration 1 and set it up to represent the vernal equinox. What constellations are visible in the night sky?

Exploration 2 continued ►

Theme Connection

Changes Over Time

Focus question: As the Earth orbits the Sun, what biological changes occur to the organisms on Earth? *(During a single revolution around the Sun, the tilt of the Earth toward the Sun creates seasonal changes in climate as well as changes in the amount of daylight. Organisms must adapt to these changing conditions. For example, when the hemisphere they inhabit is tilted away from the sun, animals may migrate, grow thicker fur, or hibernate.)*

CROSS-DISCIPLINARY FOCUS

Language Arts

Students may want to look up the literal meaning of the following astronomical terms: equinox *(equal night),* solstice *(sun stand),* umbra *(shade),* and penumbra *(almost shade).*

Answers to *Position 1*

2. The North Pole will remain in darkness for 24 hours because it tilts so far away from the Sun that the Sun's rays do not strike it. In contrast, the South Pole is in continual daylight because it is tilted toward the Sun.
   As you approach the equator from the North Pole, there are more hours of daylight.
   The winter season in the Northern Hemisphere is the result of (1) fewer hours of sunlight during the days and (2) the tilt of the North Pole away from the Sun, causing the rays of sunlight to be less direct than at any other time of the year.
3. Students should infer that Leo, Cancer, and Gemini will be visible; they may include Virgo and Taurus as well. According to star charts, Cancer, Gemini, Taurus, Aries, and Pisces are most likely to be seen in the night sky on the winter solstice.

Answers to *Position 2*

Because the Sun is directly over the equator, every part of the Earth experiences equal hours of day and night.

1. A toothpick inserted into the equator will not cast a shadow because the Sun (the light source) is directly overhead. Students' diagrams should be similar to the illustration on page 447, except that the person should be standing on the equator.
2. Caution students that as they rotate the Earth, they must be careful to keep the axis pointing in the same direction. They should discover that when they rotate Earth in its vernal-equinox position, the Sun is always directly overhead at the equator.
3. Students should infer that Scorpius, Libra, and Virgo will be visible. According to the star charts, Leo, Cancer, and Gemini will also be visible.

Exploration 2 continued ►

EXPLORATION 2, *continued*

POSITION 3

Students may be interested to learn that during the summer solstice, there are 13 hours and 13 minutes of daylight at 20 degrees latitude, 14 hours and 30 minutes of daylight at 40 degrees latitude, and 18 hours and 30 minutes of daylight at 60 degrees latitude.

Answers to *Position 3*

1. When a toothpick is inserted into the tropic of Cancer, it does not cast a shadow, proving that the Sun is directly overhead. Direct rays are the hottest.
   Earth will be on the opposite side of the Sun from where it was during the winter solstice.
2. The South Pole will remain in darkness for 24 hours because it tilts away from the Sun and the Sun's rays do not strike it.
   The North Pole will be in continuous daylight because it is tilted toward the Sun.
   The warmer temperatures in the Northern Hemisphere are caused by (1) the tilt of the North Pole toward the Sun (so that the daylight hours are longer) and (2) sunlight striking Earth more directly there than at any other time of the year.
3. From their model, students may suggest that Cancer, Gemini, or Taurus is behind the Sun. The actual answer is Gemini. According to star charts, Scorpius, Libra, and Virgo are visible.

POSITION 3

June 20 or 21— The Summer Solstice

Earth has moved so that it is now directly opposite to where it was on December 21 or 22. Earth's axis should be parallel to what it was on that date. This is the first day of summer—the longest day of the year in the Northern Hemisphere.

1. The Sun's hottest rays are hitting Earth $23\frac{1}{2}$ degrees north of the equator at a latitude called the *tropic of Cancer*. How could you prove that this is so? Draw a diagram like the one you drew representing Earth at the winter solstice, only this time label the tropic of Cancer. Will Earth be on the same side of the Sun as it was for the winter solstice?
2. Which pole is in darkness 24 hours a day? in sunlight 24 hours a day? What two reasons can you give for the warmer temperatures in the Northern Hemisphere?
3. Go to your model of the plane of the ecliptic and set it up to represent the summer solstice. What constellation is behind the Sun? What constellations are visible in Earth's night sky?

POSITION 4

September 22 or 23— The Autumnal Equinox

In this position on your journey around the Sun, the Sun's hottest rays are again hitting the equator at noon, as they were on March 20 or 21. This means that on this day, the Sun is directly overhead at noon. Everyone on Earth is again experiencing equal hours of daylight and darkness.

1. During the complete day, is the Sun always directly over the equator at noon? How could you find out?

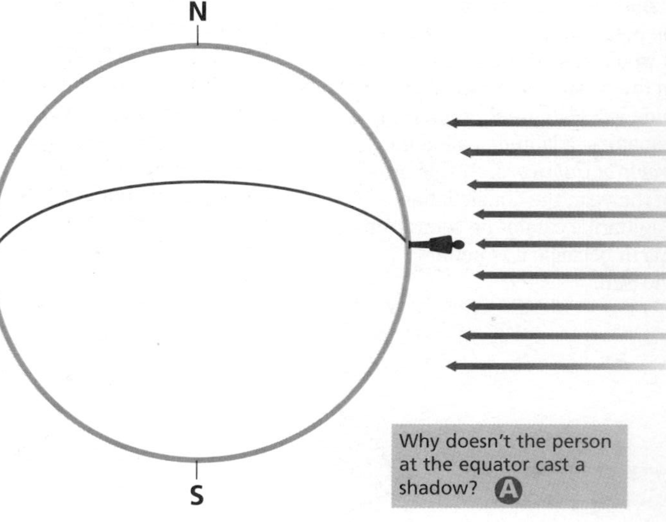

Why doesn't the person at the equator cast a shadow? **A**

2. If you were at the North Pole looking toward the Sun, would you see it rise above the horizon during the day, as it does where you live?
3. During a complete year, through what range of latitudes do the Sun's hottest rays move? Where would you most like to live?
4. Set up your model of the plane of the ecliptic to represent the autumnal equinox. What constellations are visible in the night sky at this point in your journey?

Now that you have made this trip, you know enough to be the tour guide on the next trip. What are the major sights and events you would point out on your next journey around the Sun? Record what you plan to highlight.

Answers to *Position 4*

1. During a complete day, the Sun is always directly overhead at the equator at noon. This can be proved by inserting a toothpick into the equator and noting that it does not cast a shadow.
2. The Sun at the North Pole moves along the horizon, but does not rise above it.
3. During a complete year, the Sun's hottest rays, or perpendicular rays, move from $23\frac{1}{2}$ degrees south latitude (the tropic of Capricorn) to $23\frac{1}{2}$ degrees north latitude (the tropic of Cancer), for a total of 47 degrees.
4. Students should infer that Taurus, Aries, Pisces, and Aquarius will be visible.

★ **A Transparency Worksheet (Teaching Resources, page 29) and Transparency 76 may be used as a follow-up activity to Exploration 2.**

Homework

Have students research what kinds of seasonal or daily variations in weather occur on other planets in the solar system. Have them write a one-page summary of what they learn. *(Mercury, for instance, has days that last 90 Earth days and thus experiences large daily fluctuations in temperature.)*

Kepler's Puzzle

The models you made in Explorations 1 and 2 lend support to Copernicus's theory. But during Galileo's time, those who thought that Copernicus's theory was valid still faced one serious problem: If the orbits of the planets are perfect circles, why does the observed brightness of Mars differ so much each time Earth comes between it and the Sun? The variation is much greater than it would be if the orbits were perfect circles.

In 1605, a German astronomer named Johannes Kepler was trying to discover the true shape of planetary orbits. For three years, Kepler attempted to fit the information known about Mars's orbit to circles of all sizes. At that time, everyone assumed that the planets traveled in perfectly circular paths because the circle was thought to represent perfection. But no matter how hard he tried to make it work, no circle would fit the data. Kepler had fallen into the trap of anticipating what he believed to be the correct answer, thus blinding himself to possible alternatives. Has this ever happened to you?

In frustration Kepler finally resorted to an idea that he had previously disregarded—a shape called an *ellipse,* which looks like a stretched-out circle. When the information known about Mars's orbit was applied to this shape, it fit perfectly. Mars's orbit did not describe a circle, but an ellipse instead!

With this discovery, Kepler was able to describe the paths of all revolving objects bound by gravitational forces. His discovery applies to comets, planets, moons, human-made satellites, and even distant stars that have been caught up in each other's gravitational forces.

In the next Exploration, you will explore the shape for which Kepler at first had such low regard. Then you will be able to answer the question that had long baffled astronomers: Why does the brightness of Mars vary?

Kepler musing over possible shapes for orbital paths

Answer to
Caption, page 448

A The person at the equator doesn't cast a shadow because the Sun is directly overhead.

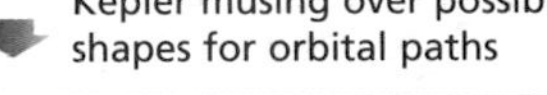

Did You Know...

Astronauts left a mirror on the Moon that is used by Earth-based telescopes to measure the distance from the Earth to the Moon. Telescopes can do this by bouncing a laser beam off the mirror and timing how long it takes to come back. This method allows the distance to the Moon to be measured to within a few centimeters!

Kepler's Puzzle

Have students read about Kepler and his attempts to discover the shape of orbital paths. Remind students that Ptolemy's theory accounted for this phenomenon by suggesting that Mars circled in a small orbit of its own as it revolved around the Earth. Mars would appear brighter if it were near Earth on Mars's own orbit. Students should realize that the Copernican model they have been working with does not support Ptolemy's ideas.

The following background information may be of interest to students: Johannes Kepler (1571–1630) was a German astronomer who constructed a model of the solar system based on a very complex theory involving five different geometric shapes. He sent a copy of his theory to the Danish astronomer Tycho Brahe in Prague. Brahe invited Kepler to come to Prague and work as his assistant.

Within a year, Brahe died, and Kepler was given complete access to Brahe's lifetime accumulation of observations and data. Kepler soon discovered that Brahe's work supported the ellipse as the correct orbital shape of the planets.

Eight years later, Kepler published two laws based on Brahe's observations of the orbit of Mars. The first law stated that every planet moves in an elliptical orbit with the Sun at one of the foci. The second law stated that all of the planets speed up in their orbits as they approach the Sun and slow down as they move away from it.

Mars varies in brightness for two reasons. One, which could be explained by Copernican theory, is that Mars and Earth each orbit the Sun at different speeds. As a result, both planets are on the same side of the Sun at certain times, while at other times they are on opposite sides of the Sun. Mars is brighter when it is closer to Earth.

Mars also varies in brightness when both planets are in opposition, that is, when Earth is between Mars and the Sun. If orbits were circular, the distance at opposition would always be the same, and hence the brightness would be constant. Since orbits are actually elliptical, the distance at opposition varies from 56 million to 100 million kilometers. This causes a marked difference in brightness that could not be explained by the Copernican model, which involved circular orbits.

EXPLORATION 3

If necessary, point out to students that the string must be longer than the distance between the two thumbtacks. As they draw, students should hold the pencil so that the string is always taut.

Encourage students to describe the shape of their ellipses. Help them to recognize that an ellipse is shaped like an oval.

An Exploration Worksheet is available to accompany Exploration 3 (Teaching Resources, page 31).

Answers to *In-Text Questions*

1. Ellipses with different shapes can be drawn by changing the distance between the two thumbtacks.
2. The closer together the thumbtacks are, the more circular the ellipse becomes. The farther apart the thumbtacks are, the longer the ellipse becomes. A circle will result if the thumbtacks are placed on top of each other (in effect, if only one thumbtack is used).
3. **a.**

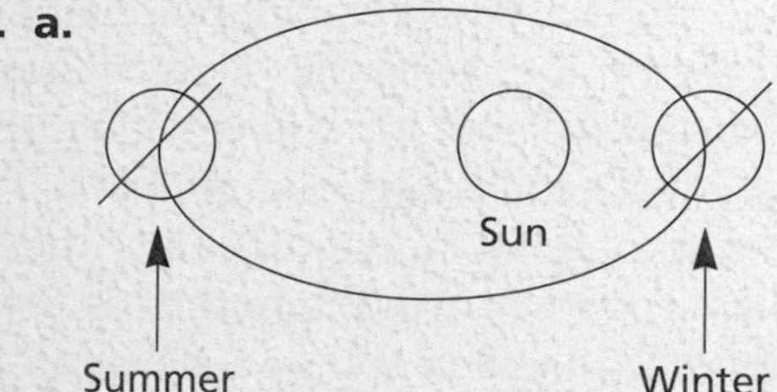

   b. It is winter in the Northern Hemisphere (even though Earth is closest to the Sun at this time) because Earth's axis is tilted away from the Sun. As a result, the Sun's rays do not strike the Northern Hemisphere directly and there are fewer hours of daylight.
4. Kepler's discovery shows that differences in Mars's brightness as seen from Earth are due to Mars's elliptical orbit. Mars appears brightest when Mars and Earth are closest together. At their greatest distance apart, Mars appears faintest.
5. Earth's orbital speed increases during the summer and fall (from the summer to winter solstice) since Earth is getting closer to the Sun at that time. It decreases during the winter and spring (from the winter to summer solstice) as Earth gets farther from the Sun. Because of Earth's elliptical orbit, the time interval between the vernal equinox and the autumnal equinox is longer than that between the autumnal equinox and the next vernal equinox. The Earth is closest to the Sun during winter in the Northern Hemisphere—about 147,100,000 km (as opposed to 152,100,000 km at its greatest distance from the Sun). Therefore, Earth completes the semi-ellipse from the autumnal equinox to the vernal equinox faster than it does the opposite semi-ellipse. If Earth's orbit were circular, Earth's orbital speed would be constant.

EXPLORATION 3

Investigating an Ellipse

You Will Need

- a sheet of corrugated cardboard or a stack of several sheets of cardboard
- unlined paper
- 2 pushpins
- a 40 cm piece of string
- a metric ruler

What to Do 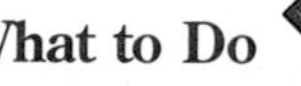

1. Press the pushpins into a sheet of unlined paper that has been placed on top of the cardboard. Loosely tie the string around the pushpins, as shown in the photograph at right. Place a pencil through the string and trace out an ellipse. Your tracing is the shape that planets travel as they orbit the Sun. Using the same piece of string, can you make ellipses with different shapes? How?
2. Measure the long axis of several ellipses. What is the relationship between the positions of the pushpins and the shape of the ellipse? How must the pushpins be positioned to make a circle?
3. At the position of one of the pushpins, draw a diagram representing the Sun. This location is a focus of the ellipse. (The plural of focus is *foci*. An ellipse has two foci.) When Earth is closest to the Sun, it is 147 million kilometers from the Sun. At this distance, it is winter in the Northern Hemisphere. At its farthest distance, it is 152 million kilometers from the Sun. Earth reaches this position during the summer in the Northern Hemisphere.
   a. Place this information on your diagram. Label the summer and winter positions of Earth.
   b. How do you explain that when Earth is closest to the Sun, it is winter in the Northern Hemisphere? (Think of what you found out at Position 1 in Exploration 2.)
4. How does Kepler's discovery explain why Mars changes in brightness as seen from Earth?
5. Through his observations, Kepler made another discovery about planets moving in their elliptical orbits: The planets do not move at constant speeds. A planet's speed increases as it gets closer to the Sun. Compare the speed of Earth in its orbit during spring, summer, fall, and winter. If Earth's orbit were circular, what might you infer about its speed?

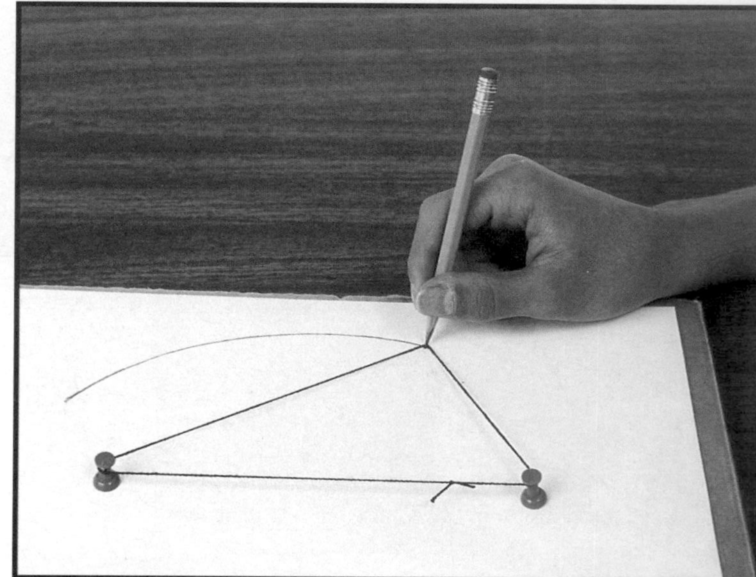

Drawing an ellipse

450

Meeting Individual Needs

Learners Having Difficulty

As an extension to Exploration 3, have interested students model the orbit of another planet, such as Mercury or Venus. Students can draw their ellipses on cardboard and then cut along the line drawn. They should then tape the border of the cutout to another piece of cardboard. A marble can then be used to represent a planet. By tilting the cardboard, students can make the model planet revolve around its orbit, rotating on its axis as it goes.

The Church has been wronged. The ideas of Aristotle must be upheld.
I wrote and taught what I thought reflected the reality I saw.
GALILEO

A Mock Trial

You have all the physical evidence needed for the retrial of Galileo. Which were the more persuasive arguments in support of the Copernican theory? What questions would you ask of the following witnesses who have been asked to testify at the retrial? Are there other witnesses you would call? What other witnesses could the prosecutor call?

Set up a mock retrial of Galileo with different class members playing the different roles. Form at least six small teams. Each team will be responsible for filling out the details of one or more roles and for choosing which team member will play that role. Those not playing a role will be the jury. Besides Galileo, his lawyer, and the prosecutor, include at least the following three roles. You may, of course, add other roles to the list.

- Church official: I believe absolutely in the teaching of Aristotle. Earth is at the center of all things; therefore, humans are at the center of all things.
- Merchant: What do I know of moving planets or suns? I believe what I *see*. I see the Sun moving, so I believe it moves.
- Astronomer/Scholar: Galileo is very persuasive in his arguments. But I would like to examine both theories more carefully.

So now you know your world revolves around the Sun. Read pages S140–S141 in the SourceBook to find out more hot stuff about the Sun!

451

A Mock Trial

In reality, Galileo would not have had a lawyer of his own. Questions would have been asked by an Inquisitor who had been chosen by the Roman Catholic Church. The Inquisitor may have tried to act impartially, but he would certainly have thought that Galileo was wrong.

Since scientists were sharply divided on the issue of Earth's position in the universe, you might encourage students to include at least one other astronomer and a student of Kepler. The astronomer might say something like the following: "There is no doubt that Galileo is a good experimenter, but his conclusions contradict the unsurpassed work of Aristotle and other reputable scholars. Most other scientists support these scholars' theories."

Having students participate in this mock trial should help them to summarize and analyze the main concepts presented in Lessons 1 and 2.

FOLLOW-UP

Reteaching

Have the class work in groups of four or five students. Ask them to use foam spheres and a lamp to show how the phases of Venus depend on the orbit of Venus around the Sun. (The chart of orbits on page 444 will help them.) Have them explain how the phases of Venus would support Galileo's model of the solar system.

Assessment

Prepare several questions about the Earth and its travels around the Sun during one year. Put one question each on separate index cards. Questions could include the following: Why does the North Pole receive 24 hours of daylight during the summer? How do we know that the Earth's orbit is an ellipse? Why are the days shorter in the winter?

Pass out the question cards and have students answer in the form of a written explanation, drawing, or demonstration.

Extension

Challenge students to do some research on the orbits of Mars, Earth, Venus, and Mercury. Then have them use this information to make a poster diagram showing the elliptical orbits of the four planets. Their diagrams should indicate how far each is from the Sun and how close each is to the Sun.

Closure

In 1979, Pope John Paul II declared that the Roman Catholic Church may have been in error when it condemned Galileo. A church commission was instructed to reevaluate Galileo's case. In 1983, the commission reached its conclusion. Have interested students do research using an on-line database to discover what the conclusion was. Have students share their findings with the rest of the class.

1. Prove It!

The observations below are all true. Using plastic-foam balls, toothpicks, and any other materials you think you'll need, show why each observation is true. Work in groups so that you can share ideas as scientists often do when they work on a problem. Then, with the aid of a diagram, explain in writing why each is true.

The Observations:

a. The Sun rises in the east and sets in the west.

b. The Sun is higher in the sky in the Northern Hemisphere during the summer than during the winter.

c. During the summer in the Northern Hemisphere, the North Pole receives 24 hours of daylight.

d. The Moon goes through phases from new Moon (when it is not visible in the night sky) to full Moon (when the full face of the Moon can be seen).

Crescent Moon

Quarter Moon

Gibbous Moon

Full Moon

e. Each night the Moon rises later (by approximately 50 minutes) than it did on the previous night.

f. The same face of the Moon is always visible from all parts of Earth.

2. A Cosmic Crossword Puzzle

Several new terms were introduced in this chapter, including *constellations, zodiac, rotation, revolution, axis, equator, ecliptic, tropic, solstice, equinox,* and *ellipse*. Create a crossword puzzle that contains many of these words. You can add other words that you found in this chapter as well.

452

Answers to *Challenge Your Thinking*

1. **a.** Because Earth's rotation is from west to east, the rays of sunlight cross Earth's surface from east to west, making the Sun appear to rise in the east and set in the west.
   b. The Sun appears higher in the sky when the Northern Hemisphere is tilted toward the Sun (summer) because the Sun is closer to being directly overhead than when the Northern Hemisphere is tilted away (winter).
   c. Since the North Pole is tilted toward the Sun during the summer months, the pole receives 24 hours of sunlight a day during this time.
   d. When the Moon is between Earth and the Sun, we "see" only its darkened half. As the Moon's orbit takes it around Earth, we see a larger and larger portion of the lighted surface, until the Moon is full (when Earth is between the Moon and the Sun), and we see its entire lighted face.
   e. Since the Moon orbits Earth once every $27\frac{1}{3}$ days, it is approximately $\frac{1}{27}$ of its way farther around Earth each time Earth completes one rotation (24 hours). To reach the point where the Moon rises, Earth must rotate a little farther each day. This takes about $\frac{1}{27}$ of a day (about 50 minutes).
   f. The Moon rotates once on its axis in exactly the same time it takes to make one revolution around Earth. Both motions occur in a west-to-east direction.
2. Student puzzles will vary but should demonstrate an understanding of the definitions of these concepts.
3. A total solar eclipse occurs only when the Moon's orbit takes it exactly between the Sun and Earth. A total eclipse does not occur every month because the Moon's orbit is oriented at a slight angle to the plane of the ecliptic.
4. Ptolemy's two misconceptions, which do not appear in the diagram on page 443, are that (1) Mars makes small revolutions about points on its main orbit and (2) Mars orbits around the Earth instead of the Sun.
5. In the Southern Hemisphere, the stars appear to revolve clockwise. Australian students would be wearing summer clothing when it is winter in the Northern Hemisphere.

Did You Know. . .

Because the axis of the Earth is aimed toward a point in the sky very near to Polaris, the North Star, it is the only visible star in the sky that does not move. Historically, many people thought that this unique characteristic allowed the star to impart "virtue" to a compass needle so that the needle would always point north.

3. A Brief Break From the Sun

In July of 1991, a complete eclipse of the Sun was visible in Hawaii and Mexico. With a diagram or your model, show how a total eclipse of the Sun occurs and why one does not occur every month.

4. Misconceptions in Motion

The explanation of retrograde motion that you just studied differs from Ptolemy's explanation. Compare the diagram illustrating his explanation (page 435) with the diagram on page 443. What misconceptions did Ptolemy's model illustrate that do not appear in the newer understanding?

5. G'day Mate!

You've made many observations of what happens in the Northern Hemisphere as a result of the motions of heavenly bodies. What happens in the Southern Hemisphere? Which way will the stars appear to revolve there—clockwise or counterclockwise? On days when you wear a winter coat to school, what would Australian students be most comfortable wearing?

Review your responses to the ScienceLog questions on page 438. Then revise your original ideas so that they reflect what you've learned.

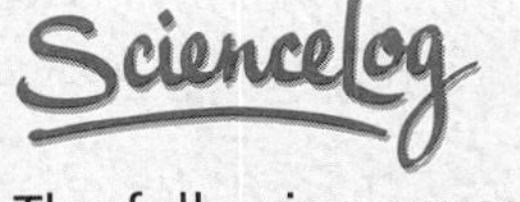

The following are sample revised answers:

1. Day and night are caused by rotation of the Earth on its axis. A location on Earth will experience "day" when it is facing toward the Sun and "night" when it is facing away from the Sun.
2. Because of the Earth's tilt, sunlight strikes the Earth more directly in the Northern Hemisphere during the summer than at any other time of the year. Also, daylight hours are longer during the summer, contributing to higher temperatures.
3. When one hemisphere of the Earth is tilted toward the Sun, as the Earth rotates, that side of the Earth will be exposed to the Sun for a longer period of time.
4. No, the planets do not move around the Sun in a perfect circle. Kepler proved that the planets orbit the Sun in an ellipse (a stretched-out circle). If the planets orbited in perfect circles, there would be less change in the observed brightness of Mars throughout the year.

★ **The Discrepant Event Worksheet on page 34 of the Unit 7 Teaching Resources booklet works well as a teacher demonstration after students have completed Chapter 20. In addition, you may wish to provide students with the Chapter 20 Review Worksheet that is available to accompany this Challenge Your Thinking (Teaching Resources, page 35).**

Homework

Besides the circle and the ellipse, the paths of objects around the Sun can trace out a parabola or hyperbola. Have students research parabolas and hyperbolas and develop a classification scheme for these orbits. *(An object in a parabolic or hyperbolic orbit around the Sun will pass the Sun only once before leaving the solar system. Many comets are thought to have such orbits.)*

Integrating the Sciences

Earth and Physical Sciences

Sir Isaac Newton lived in England during the seventeenth and eighteenth centuries, and he knew about Kepler's ideas of planetary motion. Challenge students to learn about Newton's discoveries. Have students work in small groups to create a presentation about gravity. Each group should have a separate topic related to gravity, such as the history of Newton's theory, using Newton's formulas, demonstrating the law of universal gravitation, and experiencing gravity on Earth and in outer space.

ENVIRONMENTAL FOCUS

Both the rotation of the Earth on its axis and the revolution of the Earth around the Sun affect life on Earth in a number of ways. Discuss some of these effects with the class, such as the cycles of the tides, the seasons, and the diurnal and nocturnal life cycles of organisms. Ask: What would life be like without some of these phenomena? *(Accept all reasonable responses.)*

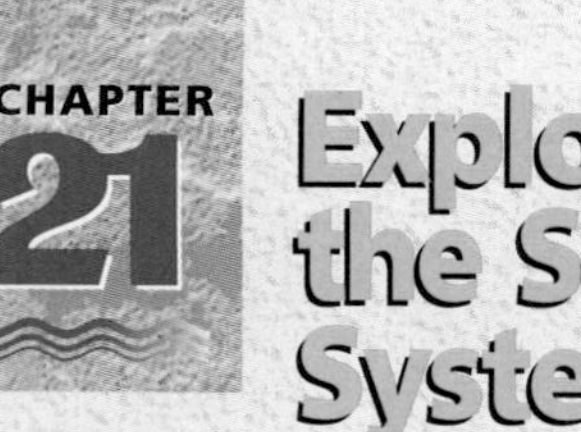

Exploring the Solar System

Connecting to Other Chapters

Chapter 19
explores the science of astronomy and discusses the observations of some early astronomers.

Chapter 20
describes the various motions of celestial bodies and some of the effects of those motions.

Chapter 21
introduces meteors, meteorites, and comets and examines what a colony on Mars might be like.

Chapter 22
examines Earth's place in the universe and introduces the big-bang theory.

Prior Knowledge and Misconceptions

Your students' responses to the ScienceLog questions on this page will reveal the kind of information—and misinformation—they bring to this chapter. Use what you find out about your students' knowledge to choose concepts and activities to emphasize in your teaching.

In addition to having students answer the ScienceLog questions on this page, you may wish to have them perform the following exercise: Divide the class into groups of 3–4 students, and tell them that the year is 2030 and the Earth is so overcrowded that it can no longer support its inhabitants. The United Nations has agreed to lead an effort to colonize another planet. Ask: Which planet would you recommend? To solve this problem, each group has been specially appointed to determine which planet is most suitable for colonization and to design a preliminary plan for the colonization process. You may wish to remind students to consider such factors as the weather on the new planet, the distance people would have to travel, and the solidity of each planet's surface. You may wish to have volunteers share their plans with the class. Emphasize that there are no right or wrong answers for this exercise. Have students answer using their best estimates of any needed information. The responses should help you evaluate what students already know about the solar system, what misconceptions they may have, and what is interesting to them about this topic.

CHAPTER 21

Exploring the Solar System

1 Why were comets once called hairy stars?

2 If you took a trip through the solar system, about how long do you think it might take to get from Earth to Saturn? What do you think you would see along the way?

3 How similar to Earth are other planets in our solar system? Do you think humans will ever live on another planet? Explain your reasoning.

ScienceLog

Think about these questions for a moment, and answer them in your ScienceLog. After you've finished this chapter, you'll have the opportunity to revise your answers based on what you've learned.

LESSON 1 Visitors From Space

Meteors and Meteorites

The photo above shows evidence of visitors from space. Do you know what these visitors are?

Try viewing the Moon through binoculars. What is its most distinguishing characteristic? Is there any evidence of water or of wind and dust storms? What do you suppose caused the craters on the Moon? Much can be learned about our nearest neighbor by simple observation. (A)

The Moon's features are similar to those on Earth. Compare the picture of the crater in Arizona, shown above, with the picture of a Moon crater, shown on the right. How are they similar? Do you see any differences? Could the two craters have been formed in the same way? You can simulate one possible cause of the craters in the following way: From a height of about 1 m, drop several milliliters of water onto a bed of flour in a pie pan. What do you observe? What could have been "dropped" on Earth and on the Moon to create the craters there? (B)

Pieces of rock or iron that sometimes enter Earth's atmosphere are called **meteors**. If they are large enough, they crash to Earth instead of burning up completely in the atmosphere. Meteors that reach the ground are called **meteorites**. The craters are visible evidence of their impact, even though the meteorite itself may not be found. That's because the heat generated on impact can vaporize a meteorite. How does your flour-and-water activity simulate this part of crater formation? (C)

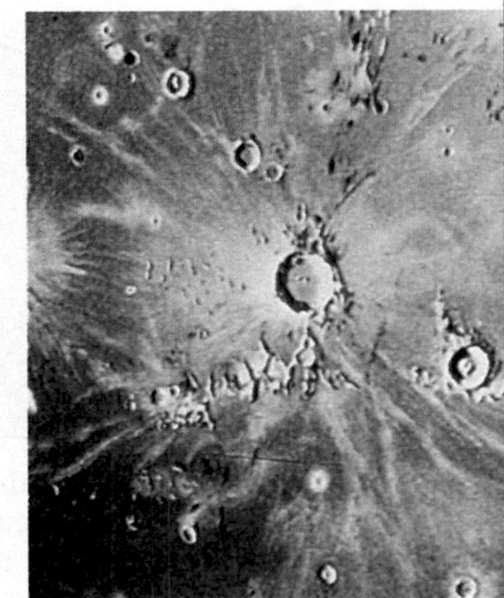

Moon crater

LESSON 1 ORGANIZER

Time Required
2 class periods

Process Skills
observing, analyzing, inferring, measuring

New Terms
Asteroids—rocks and boulders that have been observed in the solar system. They are different from comets in that they have no tails.
Comets—frozen masses, consisting of ice, dust, and other material, that slowly orbit the Sun
Meteorites—meteors that have made impact with Earth's surface
Meteors—lumps of rock or metal that sometimes enter Earth's atmosphere

Materials (per student group)
Meteors and Meteorites: aluminum pie pan; 5–10 mL of water; about 200 cm³ of flour

Teaching Resources
SourceBook, p. S144

LESSON 1 Visitors From Space

FOCUS

Getting Started

Invite class members to discuss what shooting stars are like, where they come from, and what they may be made of. Point out that in this lesson they will discover whether their ideas are correct.

Main Ideas

1. Meteors are lumps of rock or metal that sometimes enter Earth's atmosphere.
2. A meteorite is a meteor that strikes Earth's surface.
3. Meteorites form craters on Earth that resemble those on the Moon.
4. Most of the evidence of meteorite impacts on Earth has been destroyed by weathering and erosion.
5. Comets are frozen masses of water, dust, and other materials.

TEACHING STRATEGIES

Answers to *In-Text Questions*

(A) The Moon has many craters. There is no clear evidence of wind or water on the Moon. Accept all reasonable suggestions for what may have caused the Moon's craters.

(B) The craters are all deep and round. The Moon's craters appear rougher than the one in Arizona, however. Students should observe that the water forms craterlike depressions in the flour. They may infer that objects from space may have caused craters on the Moon and Earth.

(C) The water may not be visible because it soaked into the flour.

Much of the evidence of meteorite impacts on Earth has been destroyed through weathering and erosion by air and water. How could weathering and erosion be simulated in the flour-and-water activity? Why do there seem to be many more craters on the Moon than on Earth? (A)

Have you ever seen a "shooting star"? If you have, what you saw was not really a star at all, but something within Earth's atmosphere. (Recall that stars are located great distances away from Earth.) What observation might support the statement that "shooting stars" are a nearby phenomenon? What you saw were (B) rock particles or "dust" that had entered the atmosphere. The heat created as the particles streaked through the atmosphere caused the surrounding air to glow white-hot, and the meteor burned up before it could reach Earth's surface.

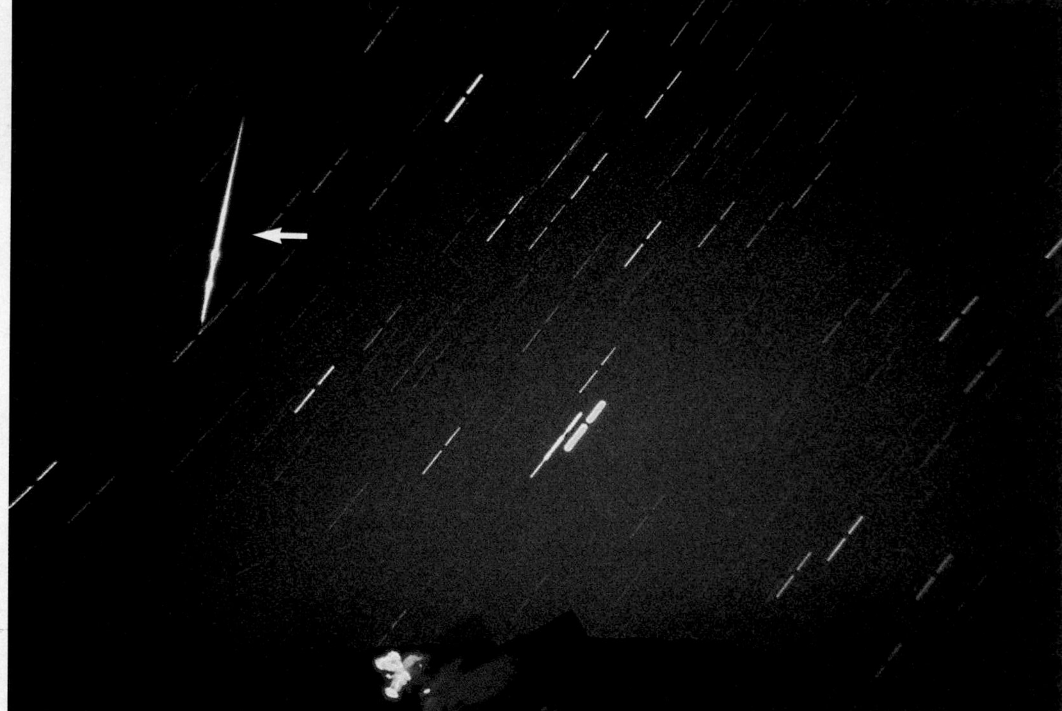

During certain times of the year, meteor showers are quite common. The arrow in the time-lapse photograph above points to a meteor observed during a meteor shower. (Star trails can be seen in the background.) Many "shooting stars" can be observed in a short time period during a meteor shower because Earth is passing through debris left behind by another visitor from the outer reaches of the solar system. What might this visitor be? Read on.

Answers to *In-Text Questions*

(A) Students can simulate weathering and erosion by gently blowing on the flour and spraying it with water. They should conclude that the absence of weathering and erosion on the Moon has left its craters intact, while erosion has destroyed most of the craters on Earth. Students may also point out that the Earth's atmosphere prevents most meteorites from reaching the planet's surface.

(B) Since "shooting stars" move quickly against the background of the actual stars, they must be much closer than the stars.

ENVIRONMENTAL FOCUS

Ask: How does the atmosphere protect life on Earth? *(Most meteors are burned up by frictional heating in the atmosphere. Atmospheric gases absorb and reflect solar radiation, protecting life from damaging X rays and from much of the Sun's ultraviolet light. These gases also act as insulators to keep the surface of the Earth warm enough to sustain life.)*

The Hairy Star

The ancient observers were familiar with "hairy stars"—stars with tails. From the photo at the right, can you see why it might be called a hairy star? Of course, what they saw were not stars at all, but other visitors from the outer reaches of the solar system. If you knew Greek, you might know that the ancient observers were talking about comets. The word *comet* comes from the Greek word for "hair." The material in the tail of a comet gets left in space and can be responsible for meteor showers.

Throughout history, comets were thought to be symbols of bad luck and disaster. To look at one was to invite something terrible to happen. For instance, the comet of 1066 appeared shortly before King Harold II of England was overthrown by William the Conqueror. The comet was seen as a bad omen for the losing side. The arrival of this comet is recorded in a part of the Bayeux Tapestry, shown below. Notice the comet at the top of the tapestry and the reactions of the people.

A comet

ISTI MIRANT
HAROLD

Events of the Norman Conquest were embroidered into the Bayeux Tapestry in the 11th century A.D.

Today, comets are regarded as objects of interest. Some amateur and professional astronomers spend a lot of time trying to discover new ones, which are then named after the discoverer. As a result of this interest, scientists now know the answers to questions such as those below. As you read further, see if you too can discover answers to these questions.

- Do comets revolve around the Sun as do the other members of the solar system?
- Why do comets have "tails"? Rocks and boulders called **asteroids** have been observed in the solar system, but they have no tails.
- Are comets and asteroids made of the same material?

What eventually happens to a comet?

457

Multicultural Extension

International Space Research
Point out to students that the United States conducts joint research with other countries through organizations such as the European Space Agency (ESA). Suggest that students find out what international organizations are involved in space programs and what their plans are for future research projects.

PORTFOLIO

Have students place their answers to the questions from The Hairy Star in their Portfolio. Students can write their answers in paragraph or outline form.

Homework

Have students explain in their own words the differences among meteors, meteorites, minor planets, asteroids, comets, planets, moons, and stars.

The Hairy Star

Have students read page 457 silently, and then direct their attention to the photograph of the Bayeux Tapestry. Help them to locate the comet, and involve them in a discussion of the people's reactions to it. Encourage students to speculate why people were sometimes afraid of comets. *(Possible responses include the following: They did not understand what comets were; they were afraid that a comet might hit Earth; or they believed that comets were bad omens.)*

As students continue reading the lesson, they should discover the answers to some of the questions at the bottom of page 457. At this time, use the responses provided below to guide students toward the correct answers.

Answers to *In-Text Questions*

- Comets travel around the Sun in an elliptical path. Some make the journey in less than 7 years. Others may travel in such a large orbit that it takes thousands of years to complete the orbit.
- The head, or nucleus, of a comet probably consists of ice, dust, and other frozen substances. As a comet approaches the Sun, heat causes the outer layers of the nucleus to vaporize. This vaporization releases the dust and gases that form the tail. A comet's tail may extend for 160 million kilometers across space. Solar wind pushes the comet's tail so that it is always pointing away from the Sun. (See page 458.)
- Asteroids differ from comets in that they are made mostly of rocky materials. As a result, they do not form cometlike tails. Most asteroids are found between the orbits of Mars and Jupiter, forming a band called the asteroid belt.
- Answer to caption: Eventually, a comet will vaporize or burn away, but this may take millions of years.

Comets—What Are They?

You may find the following information helpful: Today astronomers believe that comets exist in two forms. In the more familiar form, they are visible to the naked eye and have long tails. In the second form, the volatile gases have been burned off and the rocky comet is almost indistinguishable from an asteroid. In this form, comets can be seen only with the aid of a telescope.

In 1577, the Danish astronomer Tycho Brahe declared that comets travel in paths far beyond the Moon, and not just within Earth's atmosphere. Toward the end of the eighteenth century, Isaac Newton concluded that comets travel in elongated elliptical orbits. During the same period, Edmund Halley tracked the orbits of 24 comets. In 1950, the Dutch astronomer Jan H. Oort suggested that a cloud of comets exists at the outer reaches of the solar system. There, about 100 billion comets—the vast majority of the known comets—are located, thousands of times farther away than the planet Pluto is from Earth. Held weakly by the Sun's gravitational pull, comets in the Oort cloud follow huge orbits that take thousands of years to complete.

The Comet of a Lifetime

Share the following information with students: By calculating the orbit of a comet he observed in 1682, Halley proved that it was the same one that astronomers had seen in 1531 and 1607. He correctly predicted its return in 1758. The comet, named after Halley, has returned at approximately 76-year intervals ever since. On May 21, 1910, Earth is believed to have passed through the tail of Halley's comet.

Homework

Have students answer the following questions: How old will you be when Halley's comet makes its next appearance? Why do you think this comet is sometimes called "the comet of a lifetime"? *(During an average person's lifetime, they would have the opportunity to see the comet only once.)*

Comets—What Are They?

The tail of a comet is a clue to its makeup. Think for a moment about what substances, when vaporized by the Sun's heat, can be made to glow. Did you name ice? If so, you've discovered one part of a comet's makeup.

In the outer reaches of the solar system, trillions of frozen ice masses are held in a slow orbit around the Sun by gravitational forces. Besides frozen water, these ice masses consist of dust and frozen gases. The orbits of some of the ice masses send them speeding toward the Sun, at which time they may be seen by observers on Earth. As the comet nears the Sun, the rise in temperature turns the ice to steam and causes dust particles to be swept away from the head of the comet. These substances form the comet's tail. The tail becomes visible when the gas glows and the particles reflect and scatter light.

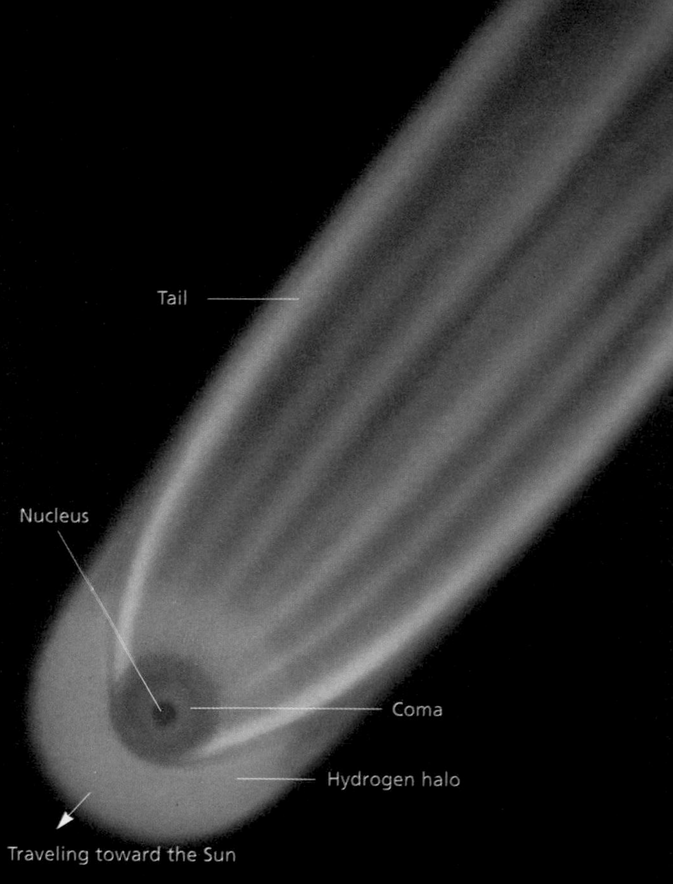

The parts of a comet

Edmund Halley

The Comet of a Lifetime

In 1682 Edmund Halley (pronounced HAL ee) observed the comet that now bears his name. After calculating its orbit around the Sun, Halley predicted that the comet would return in about 76 years. Few believed him, and he did not live to see his prediction come true, but it did. The return of the comet in 1758 proved that comets are members of the solar system and that they revolve around the Sun.

Did You Know. . .

The first reported sightings of Halley's comet were made by Chinese astronomers around 240 B.C.

Halley's comet was last seen passing Earth's orbit in 1986. Examine the diagram below, which shows the orbit of Halley's comet. You'll learn a great deal about comets by using the diagram to answer the questions at right.

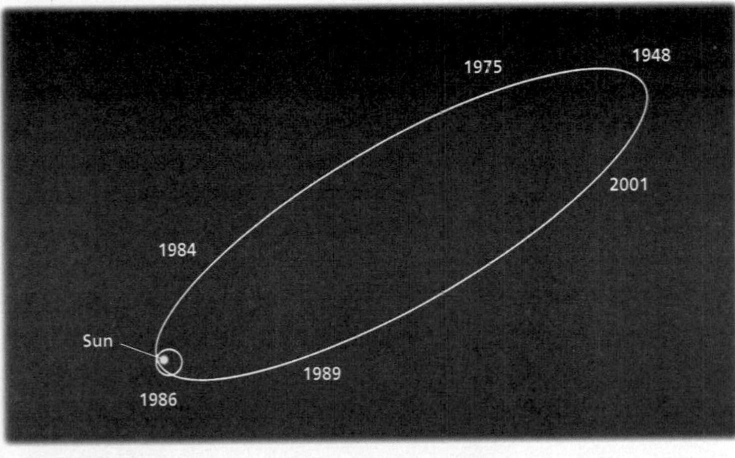

The orbit of Halley's comet. Earth's orbit is the small ellipse at the left. Why do you think Earth's orbit looks circular in this view? (A)

1. Does the comet's orbit appear to be consistent with Kepler's discovery of the paths of revolving bodies?
2. At what point in its orbit does the comet travel at its greatest speed? at its slowest speed?
3. When will Halley's comet be seen again by observers on Earth?
4. Why can't Halley's comet be seen with the naked eye at the present time?

Halley's comet seen in the Massachusetts night sky, 1986

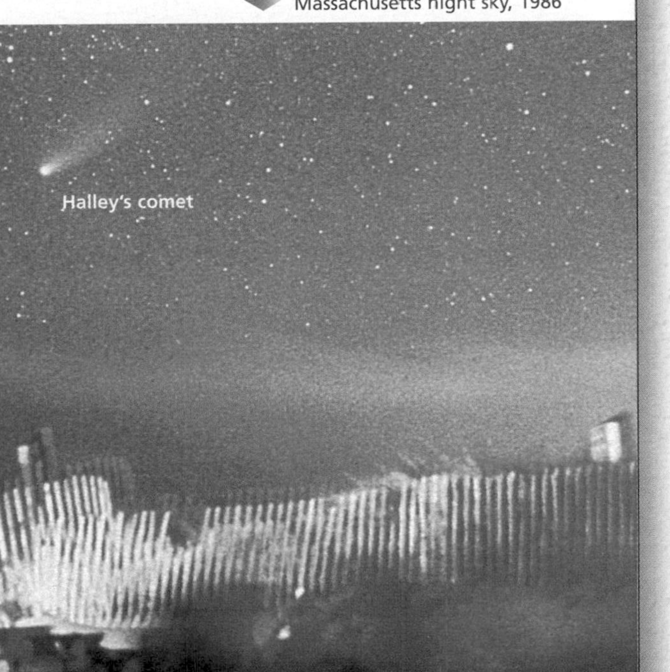

Answers to *In-Text Questions*

1. Yes, because its orbit appears to be elliptical.
2. The comet travels at its greatest speed when it is closest to the Sun and at its lowest speed when it is farthest from the Sun.
3. Students may answer 2062 because 1986 + 76 = 2062. (Explain that the comet has an orbit of *about* 76 years, and scientists predict that Halley's comet will actually return in 2061.)
4. The comet is presently making its way to the outer limits of the solar system, where its material is not being vaporized and illuminated by the Sun.

Answer to *Caption*

(A) Earth's orbit is only slightly elliptical and appears comparatively circular when seen in contrast with the much larger ellipse of Halley's comet's orbit.

FOLLOW-UP

Reteaching

Have each student write one question about the lesson on an index card. Have students use the index cards to quiz each other. Students can keep score by counting the number of questions they answer correctly.

Assessment

Ask students to make a diagram of the orbit of a comet, showing its relationship to the planets, the asteroid belt, the Sun, and the Oort cloud, and showing the position of the comet's tail as it circles the Sun. Call on volunteers to explain their diagrams to the class. (Include the asteroid belt, the Oort cloud, and the position of the comet's tail as it circles the Sun only if you mentioned these points in the discussions. They were presented in the Teaching Strategies but not in the pupil's text.)

Extension

Point out to students that on March 13, 1986, the European Space Agency's *Giotto* spacecraft passed within 600 km of Halley's comet and took 2000 photographs of it. Suggest that they do some research to find out what the space agency discovered about the comet.

Closure

Have students do some research to discover when they can expect to view a meteor shower in their area. If possible, take a nighttime field trip to observe the night sky. Students should keep track of the number of meteors they observe. Have them discuss the experience on the following school day. *(Annual meteor showers include Perseids: July 27–August 17; Orionids: October 15–25; Leonids: November 14–20; Geminids: December 9–13; Quadrantids: January 1–6; Lyrids: April 19–22; and Eta Aquarids: May 1–8.)*

The Space Probes

FOCUS

Getting Started

Read aloud short passages from Orson Welles's classic radio broadcast, "The War of the Worlds." Ask students to describe how they think people felt when they heard the broadcast in 1938. Then tell students that the broadcast terrified many listeners, who thought that Martians had actually landed on Earth and attacked New Jersey. Explain that in this lesson they will learn more about the characteristics of Mars (including its potential for supporting life-forms) as well as the other planets in our solar system.

Main Ideas

1. The planets are separated by vast distances.
2. Each planet has its own unique features and characteristics.
3. Technology has provided up-close views of the planets and their satellites.

TEACHING STRATEGIES

The Voyager Discoveries

Have students read page 460. When they have finished reading, discuss the diagram of the flight paths of *Voyager 1* and *Voyager 2*. Note that the green area at the center of the diagram represents the orbits of Earth, Mercury, and Venus around the Sun.

Answer to *Caption*

Ⓐ All four of the planets pictured are much larger than Earth.

The Voyager Discoveries continued ►

LESSON 2

The Space Probes

The Voyager Discoveries

Early voyagers sailed the seas and discovered new lands. They returned to report their findings. The Voyager spacecraft also discovered new lands, but they reported by sending back images and information that have kept scientists busy for years.

Your flight plans—the paths of *Voyager 1* and *Voyager 2*. The white labels show the positions of the planets on the dates indicated.

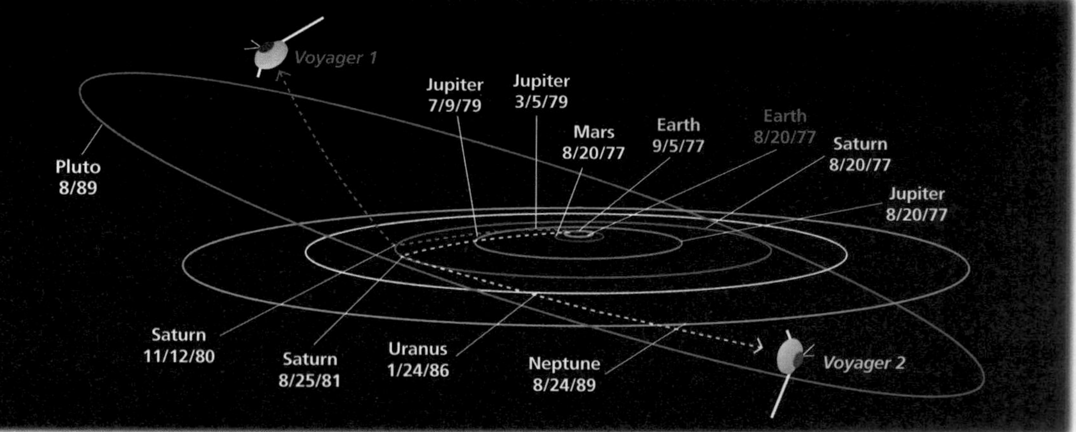

Where you're heading: the giant planets. Compare their sizes with Earth's size. Ⓐ

Imagine going on a journey that takes you from Cape Canaveral in Florida, past Mars, and finally near the giant planets of Jupiter, Saturn, Uranus, and Neptune. You would see sights that human eyes have never seen before.

If either Voyager spacecraft had a crew and you were a member, how would you respond to the sights and experiences as they unfold? Use the picture study on pages 461 and 462 to write a personal account as you and your spacecraft encounter different members of the solar system. Your log could focus on the times and events shown on these pages.

LESSON 2 ORGANIZER

Time Required
2 class periods

Process Skills
observing, analyzing, inferring, classifying

New Terms
none

Materials (per student group)
none

Teaching Resources
SourceBook, p. S152

Day 1

Blastoff

How would you feel to be leaving on such a journey?

Day 215

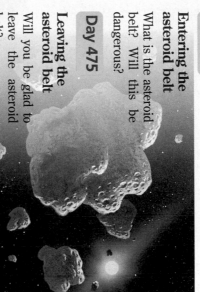

Crossing the orbit of Mars

Examine the flight paths of *Voyager 1* and *Voyager 2* on page 460. Note where Mars was on August 20, 1977—the day *Voyager 1* left Cape Canaveral, Florida. Will you be able to see Mars as you cross its orbit? Will *Voyager 2*?

Day 295

Entering the asteroid belt

What is the asteroid belt? Will this be dangerous?

Day 475

Leaving the asteroid belt

Will you be glad to leave the asteroid belt?

Day 570

Looking at Jupiter from a distance

What do you see? Describe Jupiter to someone who has not seen it before. How many different features can you describe?

Jupiter from a distance of 30 million kilometers

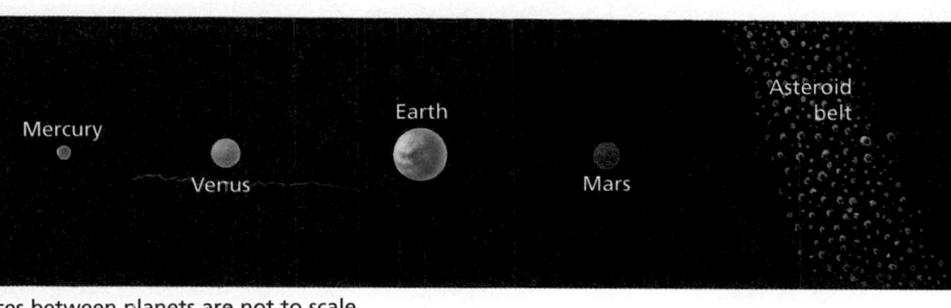

Distances between planets are not to scale.

Answers to *In-Text Questions*

Day 1: Words such as *excited, anxious, nervous, exhilarated,* and *thrilled* would probably be appropriate.

Day 215: Using the data from page 440 about Mars's and Earth's rates of revolution around the Sun, students will see that in 215 days, Mars will have traveled almost one-third of the way around its orbit, positioning it too far from *Voyager 1* or *Voyager 2* to be visible.

Day 295: The asteroid belt lies between the orbits of Mars and Jupiter. It is the region where most of the asteroids in the solar system are found. The possibility of striking an asteroid makes passing through the area dangerous.

Day 475: A sense of relief should be felt when leaving the asteroid belt.

Day 570: From a distance, Jupiter may be described as having an atmosphere of swirling orange and cream-colored bands. Embedded in the bands is a giant red spot that looks like a huge storm.

The Voyager Discoveries, *continued*

GUIDED PRACTICE You may wish to use discussion questions such as the following:

- When was *Voyager 1* launched? (Voyager 1 *was launched on August 20, 1977; point out that* Voyager 2 *was launched on September 5, 1977.)*
- Did the Voyager spacecraft travel in the same direction? *(They left Earth in the same direction but later traveled in opposite directions.)*
- Which planets did each spacecraft fly by? (Voyager 1 *flew by Jupiter and Saturn.* Voyager 2 *flew by Jupiter, Saturn, Uranus, and Neptune.)*
- About how long did it take *Voyager 2* to reach Uranus? *(About 8.5 years)*
- According to the diagram, what happened to the spacecraft? *(They left the solar system and continued on into space.)*

Be sure that students understand how they are to use the picture study on pages 461–462. This activity can be done individually, in pairs, or in small groups. As students create their logs, encourage them to include any additional questions they may have for later research and discussion. You may wish to gather materials for a resource center on space exploration for students to use as they work. Provide class time for students to share and discuss their ideas.

Integrating the Sciences

Life and Physical Sciences

The Search for Extraterrestrial Intelligence (SETI) program was cut from NASA's budget in October of 1993 for financial reasons. The effort used radio telescopes to target nearby stars in the hope of detecting evidence of life around other stars. Hold a class discussion about the value of searching for life on other planets.

Answers to
In-Text Questions

Day 630: The Great Red Spot looks very much like what it is—a giant storm. (It covers an area about 26,000–40,000 km long by 15,000 km wide. Its diameter is two to three times that of the Earth.)

Day 650: A satellite is a heavenly body that revolves around a planet or some other heavenly body. The surfaces of the satellites shown are different in both color and markings. Io is seen with plumes of sulfuric vapor from a volcanic eruption. Europa's surface looks like a cracked eggshell and is covered with a layer of ice. Ganymede is covered with impact craters, rocks, and ice.

Day 662: Accept all reasonable answers, which might include data analysis, preparatory work for the next encounter, and spacecraft maintenance. (Point out that the trip from Jupiter to Saturn will take over 2.5 years.)

Day 1567: Saturn and Jupiter are both very large planets. Each appears to be covered with a banded atmosphere. However, Jupiter's atmosphere appears to be made up of swirling clouds, while that of Saturn appears to be smooth and calm. The most prominent feature of Saturn is its ring system. The most prominent feature of Jupiter is the Great Red Spot.

Meeting Individual Needs

Second-Language Learners

Have students make poster diagrams of the solar system. Their diagrams should include the following: the inner planets, the outer planets, the Sun, and the asteroid belt. Explain that each part of the diagram should be identified with a bilingual label that presents an interesting fact about it. When students finish, review their work for science content, with minimal emphasis on language proficiency.

Day 630

Observing the Great Red Spot

What words could describe your view? What is the Great Red Spot? How large is it?

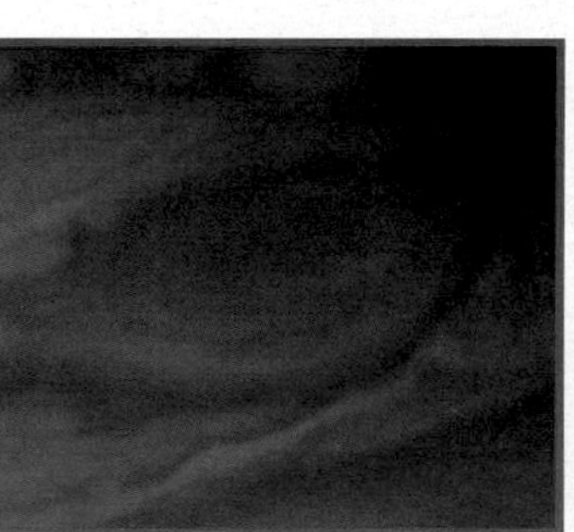

The Great Red Spot

Day 650

Examining some satellites of Jupiter

Jupiter has 16 satellites, three of which are larger than Earth's Moon. What is a satellite? How are the three satellites shown here different? Which one appears to be volcanically active? Which one could be covered with a thick layer of fractured ice?

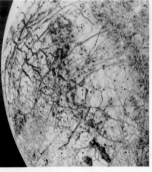

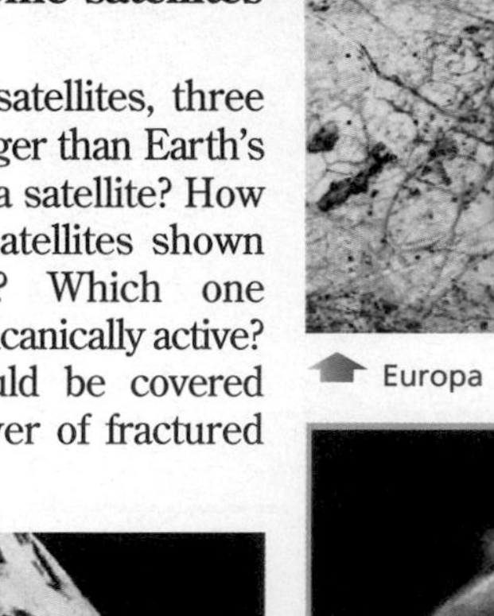

Europa

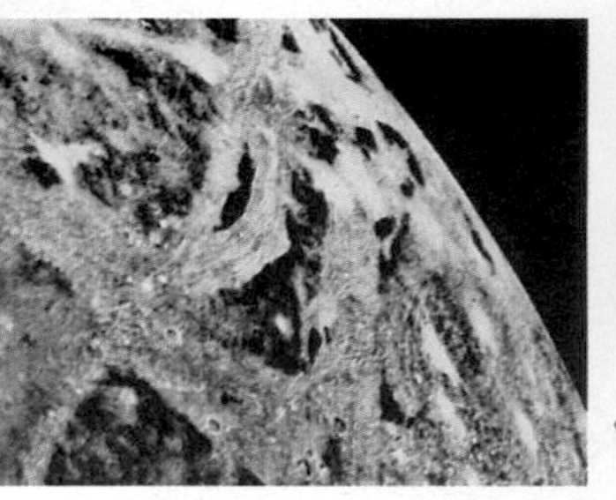

Ganymede

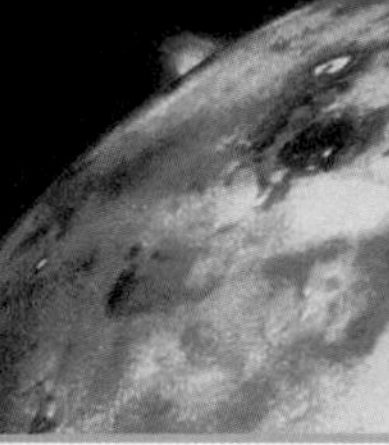

Io

Day 662

On the way to Saturn

What will you do for the next 905 days?

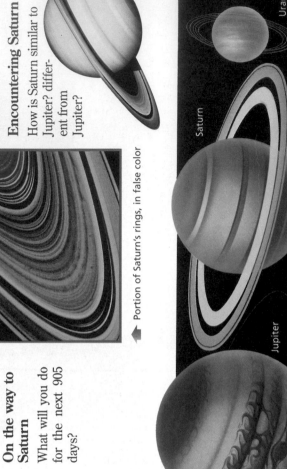

Portion of Saturn's rings, in false color

Day 1567

Encountering Saturn

How is Saturn similar to Jupiter? different from Jupiter?

462

Did You Know. . .

Amino acids are the building blocks of proteins and of life. Some meteorites are known to contain amino acids. Some scientists have speculated that life on Earth began with the arrival of amino acids from outer space!

Research Activities

1. In his book *Cosmos,* astronomer Carl Sagan wrote a similar account of what a Voyager captain might say if the Voyager spacecraft had a crew. His account makes interesting reading.
2. Before leaving the solar system forever, *Voyager 2* made two more encounters: Uranus on January 24, 1986, and Neptune on August 24, 1989. Find out what you can about these distant planets.
3. *Voyager 1* and *Voyager 2* will drift forever among the stars. Their chances of ever encountering another celestial body are almost zero. Perhaps in the future an alien from a distant star system will find one of the spacecraft and wonder about the world that sent it on its epic voyage. That alien will be able to learn something of Earth by listening to a gold-plated record that carries "The Sounds of Earth." The record begins with greetings in over 60 languages. The alien will be able to hear rolling thunder, a roaring volcano, the crashing surf, and gurgling mud. Living sounds include those of whales, chimpanzees, and Chuck Berry singing the rock song "Johnny B. Goode."

   Design a wordless message for another spacecraft that will eventually leave the solar system. What would you place in the message to give aliens some idea of what life is like on Earth?

Uranus seen as a crescent. Where must the Sun be in this picture? (A)

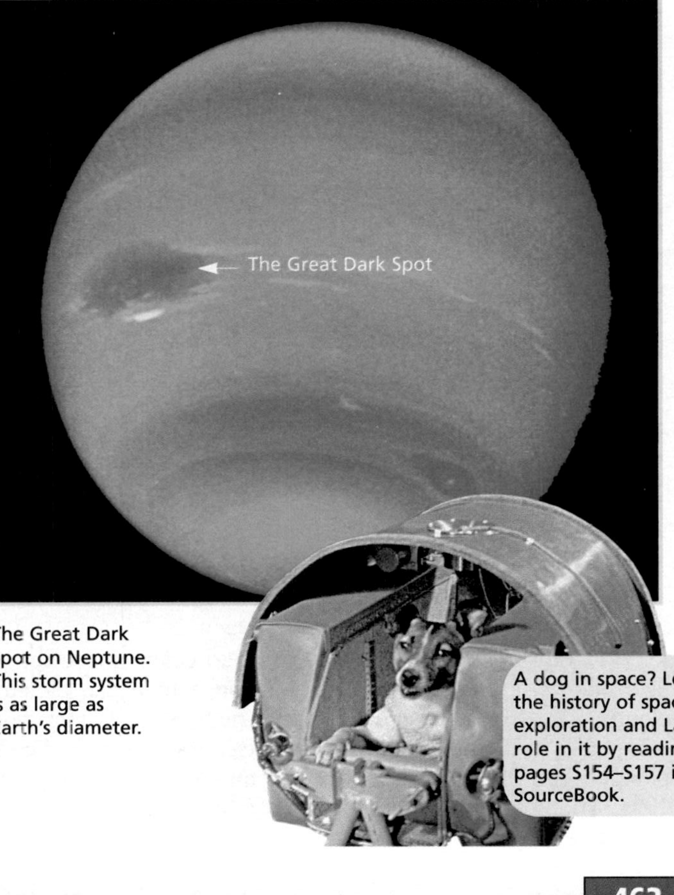

The Great Dark Spot on Neptune. This storm system is as large as Earth's diameter.

A dog in space? Learn the history of space exploration and Laika's role in it by reading pages S154–S157 in the SourceBook.

463

Answers to *Research Activities*

1. The account written by Carl Sagan begins on page 151 of *Cosmos.* This book is an excellent resource about the solar system and outer space. Have students complete a book report on *Cosmos* and present it to the class with visual aids.
2. Uranus is about 2,866,900,000 km from the Sun. It has a diameter of about 50,800 km. Fifteen satellites circle the planet. It also has a ring system. Uranus is the only planet that is tilted on its side. It makes one complete orbit around the Sun every 30,685 Earth days. *Voyager 2* discovered that Uranus has a relatively quiet atmosphere with few clouds and no evidence of storm systems.

   Neptune is about 4.5 billion kilometers from the Sun. It has a diameter of about 49,500 km. Eight satellites circle the planet. Neptune is an extremely cold planet with an upper atmospheric temperature of –220°C. Its atmosphere is extremely turbulent, with storm systems comparable to those on Jupiter.
3. Students may enjoy organizing their "space messages" into a booklet for others to enjoy.

Answer to *Caption*

(A) In the picture of Uranus, the Sun must be on the left because the left side of Uranus is illuminated.

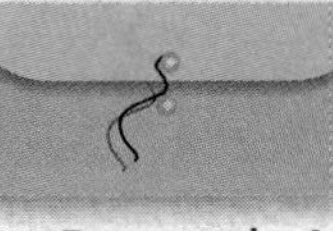

PORTFOLIO

Encourage students to include their responses to Research Activities in their Portfolio.

Meeting Individual Needs

Gifted Learners

Have students find out about some of the problems faced by explorers on the Earth. Then have them write a one-page report of a hypothetical explorer of another planet in the solar system. They could focus on the trip to the planet or on what difficulties the explorer would face after landing.

Magellan—Mapping a Distant World

Ferdinand Magellan, a Portuguese explorer in service to Spain, mapped the coastlines of the Americas in the early 1500s. The spacecraft *Magellan,* launched into space in 1989, orbited and mapped a more distant land—a planet called Venus. The information that *Magellan* sent back to Earth enables us to know more about the surface of Venus than we know about that of Earth. That's because much of Earth is covered with water and ice, which hide its surface features.

The surface of Venus is hidden from observers on Earth by a thick layer of clouds. Some early astronomers suggested that rivers, streams, and perhaps even living things may lie beneath the clouds. Could this be true? Before giving your opinion, examine the information regarding Venus on page 465.

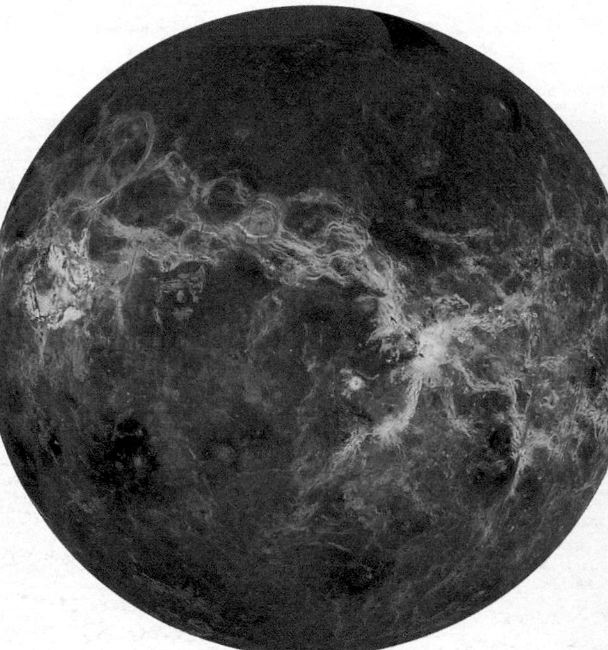

Computers created this image of Venus using information sent back to Earth by *Magellan.*

How did *Magellan* penetrate the cloud cover to see the surface below? It used microwave radiation, which can penetrate clouds. The reflected signals were sent back to Earth, where computers formed them into images. Its work finished, *Magellan* plunged into the thick, hot Venusian atmosphere in October of 1994. The spacecraft was crushed by the pressure of Venus's atmosphere.

EXPLORATION 1

Analyzing the Data

Scientists have interpreted a great deal of information sent back by *Magellan.* Using the images on this page, be a scientist and interpret the data yourself. Perhaps you can answer these questions:

1. What surface features do you recognize on Venus that are similar to those on Earth?
2. How is the surface of Venus similar to or different from that of its nearby neighbor, Mercury? Can you think of explanations for your observations?

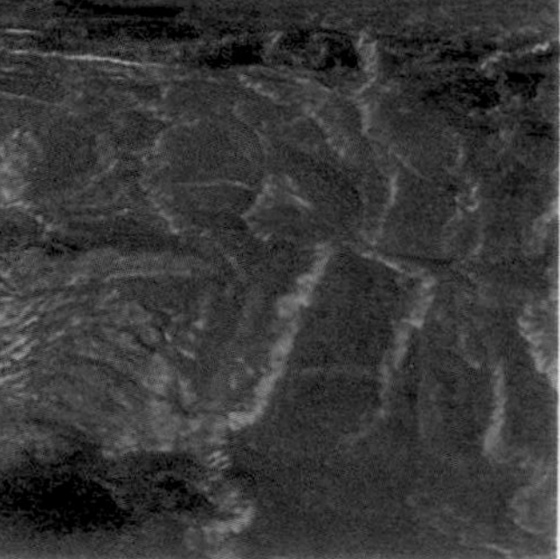

The surface of Venus

The surface of Mercury

464

Magellan—Mapping a Distant World

Have students read about the *Magellan* space probe. If possible, display some pictures of Venus taken by *Magellan,* and call on students to describe the planet's appearance.

You may wish to share the following information with students: The *Magellan* probe was launched in May 1989 and reached Venus 15 months later. During 1991, the probe mapped more than 90 percent of the Venusian surface. Venus is similar to Earth in size, density, and probably composition, but its surface is covered by a dense atmosphere of carbon dioxide. The planet's average temperature is about 482°C. Surface water, if there ever was any, has long since boiled away, leaving behind a landscape that is completely arid and incapable of supporting life. Venus has few craters but seems to be covered with fractures, ranging from elaborate networks of fine cracks to giant canyons thousands of kilometers long.

Answers to *Exploration 1*

1. Students may note that the surface of Venus has canyons, valleys, and plateaus that look similar to those on Earth.
2. The surface of Mercury is covered with craters and is almost free of cracks. The cracked surface of Venus is relatively free of craters. The dense atmosphere of Venus (which exerts a surface pressure 90 times greater than that of Earth's atmosphere) causes most meteors to burn up or break apart before hitting the Venusian surface. The almost total absence of an atmosphere on Mercury allows meteors to constantly bombard its surface. The cracks in the Venusian surface may be due to past and present volcanic activity, which is absent on Mercury.

FOLLOW-UP

Reteaching

Have students make planetary data tables. The tables should include such information as the distance of each planet from the Sun, each planet's diameter, the number of Earth years it takes each planet to orbit the Sun, and the number of satellites revolving around each planet.

Assessment

Have students write a short essay expressing their views on whether the United States should continue to spend money on the space program or whether the money should be spent elsewhere.

Extension

Suggest that students do some research on space probes that were not mentioned in the lesson, such as *Surveyor 5, Pioneer X,* and *Mariners IV, IX,* and *X.*

Closure

Remind students of the Getting Started discussion concerning Orson Welles's radio broadcast of "War of the Worlds." Welles received a lot of criticism for panicking the nation. Have students write a letter to an imaginary newspaper explaining why an invasion from Mars is unlikely. Encourage students to research and incorporate information about Mars attained from the Voyager probes.

Homework

The Extension activity on this page may be assigned as homework.

LESSON 3 Colonizing the Solar System

Project Mars

Of all the planets, Mars offers the greatest hope of sustaining a human colony. Why is this so? Why not a colony on Venus, Jupiter, or Pluto? Examine the following table of information to find some answers.

Mars	Venus	Jupiter, Saturn, Uranus, and Neptune	Pluto
• very cold; average temperature: – 23°C • extremely dry • evidence of flowing water in the past • icecaps • thin atmosphere of carbon dioxide • former atmosphere may have reacted with soil and rocks	• surface temperature: 500°C • very dense atmosphere of carbon dioxide • atmospheric pressure 90 times that on Earth • clouds of sulfuric acid	• gaseous • consist mainly of hydrogen and helium • violent storms	• $\frac{1}{100}$ the volume of Earth • extremely cold: – 230°C • receives little sunlight • surface may be covered with layer of frozen methane

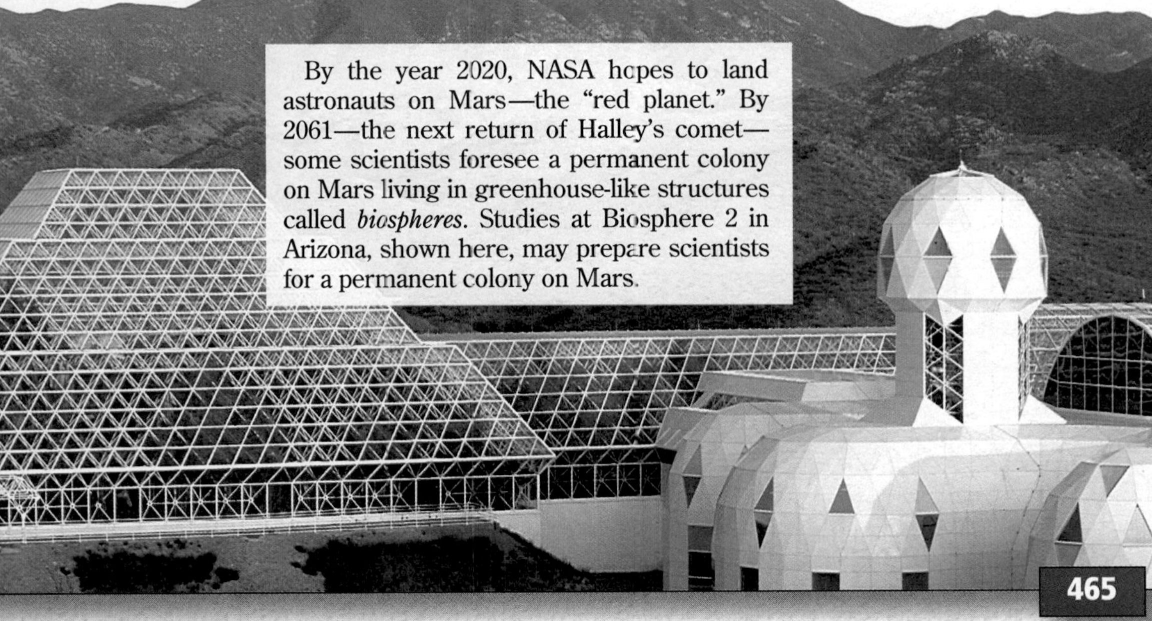

By the year 2020, NASA hopes to land astronauts on Mars—the "red planet." By 2061—the next return of Halley's comet—some scientists foresee a permanent colony on Mars living in greenhouse-like structures called *biospheres*. Studies at Biosphere 2 in Arizona, shown here, may prepare scientists for a permanent colony on Mars.

LESSON 3 ORGANIZER

Time Required
1 class period

Process Skills
observing, analyzing, hypothesizing

New Terms
none

Materials (per student group)
none

Teaching Resources
none

LESSON 3 Colonizing the Solar System

FOCUS

Getting Started

Ask: Why would humans want to establish a colony on another planet? Have students discuss and list the advantages and disadvantages of establishing a colony on Mars. *(Accept all reasonable responses. Although a colony on another planet would provide living space for many people, it would be expensive and time consuming to establish.)*

Main Ideas

1. Permanent space stations and a colony on Mars will probably be a reality in the next century.
2. Mars is more like Earth than any of the other planets.

TEACHING STRATEGIES

Project Mars

Call on a volunteer to read aloud pages 465–466. Provide students with time to study the data table at the top of page 465. Then involve them in a discussion of why Mars is the best candidate for colonization. Help students recognize that, based on the information in the table, the Martian environment is most similar to that of Earth.

Project Mars continued ►

Project Mars, *continued*

Have students identify some of the problems that will need to be overcome if humans are to live on Mars. *(Shelter, food, water, and oxygen will have to be provided.)*

You may wish to remind students that Biosphere 2 was built with private funds in the desert outside Oracle, Arizona. It covers an area about the size of two and a half football fields. The biosphere contains 3800 species of plants and animals in five different ecosystems: a desert, savanna, rain forest, marsh, and ocean. In September 1991, eight people entered Biosphere 2 to live in it for 2 years. They emerged in September 1993. The project has been controversial. Some scientists believe that it was done more for publicity than for scientific discovery.

Encourage students to discuss and evaluate the technological developments that must take place before a colony on Mars could be established.

Answers to *In-Text Questions*

Ⓐ Accept all reasonable responses. Students should recognize that each of the steps depends on the success of the one preceding it.

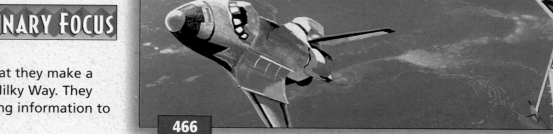

CROSS-DISCIPLINARY FOCUS

Mathematics

Suggest to students that they make a scale diagram of the Milky Way. They should use the following information to prepare their scale:

- The Milky Way galaxy is about 100,000 light-years across.
- It is about 10,000 light-years thick at the center.
- It is about 3000 light-years thick where Earth is located.
- The center of the galaxy is about 30,000 light-years from our solar system.

There is even speculation that it may be possible to "green" Mars by releasing the gases that are locked up in the rocks and soil and by melting some of the water in the permafrost and icecaps. If this is to happen, many more space developments must take place, such as those shown at right. What do you think of these plans? Of what value is each? How would you add to or modify them? Ⓐ

1998: Permanent space station built to orbit Earth, using space shuttles to transport people and materials
2010: Permanent space station on the Moon
1998–2014: Numerous robot landings on Mars
2014–2020: First astronauts land on Mars
2020–2030: First research colony established on Mars

Perhaps the events will proceed as described in this hypothetical news story:

> The first Earthlings will arrive on Mars almost exactly four centuries after the Pilgrims landed near Plymouth Rock. After a three-month flight from Earth, the *Mayflower,* a nuclear-powered rocket not yet built, will deposit them on the Martian surface. They will live in a prefabricated colony designed to shelter 12 to 14 astronauts for a year.

Would you like to be one of these Earthlings? What training would you have to undergo? This may be something that you'd like to investigate further.

Artist's conception of a permanent space station built to orbit Earth. Space shuttles would be used to transport people and materials.

Primary Source

Description of change: excerpted from "The Red Planet May Be the Next Giant Step for Mankind," by Brad Darrach and Steve Petranek, in *Life,* May 1991, pp. 26–34
Rationale: to introduce the idea of establishing a research colony on Mars

EXPLORATION 2

Researching the Solar System

In this Exploration, you will choose from three different kinds of projects. Each requires you to research a part of the solar system. In Activity 1, you can be part of a team that's planning a mission to Mars. In Activity 2, you can play the role of an advertising agency and develop a travel poster for one part of the solar system. Or in Activity 3, you can create an imaginary alien that could survive in conditions that exist somewhere else in the solar system.

ACTIVITY 1

Planning a Mission

Form into groups of four and be part of a planning team for a mission to Mars. Your task is to create some guidelines and procedures that will ensure the success of the mission. At right are some questions you might want to consider.

1. Should vast sums of money be spent on this venture? Explain your reasoning.
2. Who will be part of the first 12-member team to spend a year on Mars?
3. What will be some of the first tasks to be done?
4. Where will their water and oxygen come from?
5. Where will they obtain food?
6. What materials will have to be recycled?
7. What essential materials must be brought from Earth?

Exploration 2 continued ►

Integrating the Sciences

Earth and Life Sciences

As students explore the possibilities of colonizing Mars, have them compile a fact sheet about life on Earth. Students should answer the following questions:

- What environmental factors are needed to sustain life?
- What life processes are necessary for all living things to survive?
- How are living things different from nonliving things?
- How do living things interact with each other and with nonliving things?

Students may review the first two units and check with other reference books to find answers.

EXPLORATION 2

Have students decide which of the three projects they wish to do. You may decide to allow them to work in pairs or small groups. Encourage students to determine a way to share the results of their project with the class.

ACTIVITY 1

Suggest that each member of a group assume responsibility for a part of the project. Tasks might include gathering resource materials for the project, drawing a diagram of the living quarters on Mars, planning the trip, and deciding what to take on the mission.

Answers to *Activity 1*

1. Some students may feel that the money should be spent because of the benefits that could be gained from the mission. Others may feel that the money should be spent to solve problems on Earth before establishing colonies on Mars.
2. Determinations should be based on what an individual has to contribute to the overall mission.
3. The first tasks to be done should include providing oxygen, heat, shelter, food, and water for the people on the mission.
4. Water and oxygen will have to be brought from Earth initially. Eventually, the mission should become self-sustaining, with an established water cycle and a carbon dioxide and oxygen cycle. Plants would be necessary for establishing these cycles.
5. At first, food will have to be brought from Earth. Later, food will be grown while on the mission.
6. Oxygen, water, and all organic material should be recycled. Ideally, everything should be recycled.
7. Oxygen, water, food for people, food for plants, seeds, clothing, fuel or means to produce energy, building materials, and medicine are some of the essential materials that would have to be brought from Earth.

ACTIVITY 2

Developing a Travel Poster

The year is 2065. The colonizing of Mars has been a success, and an increasing number of people are visiting this outpost as tourists. The usual length of stay is 1 year. Travel agencies are advertising this and other excursions throughout the solar system.

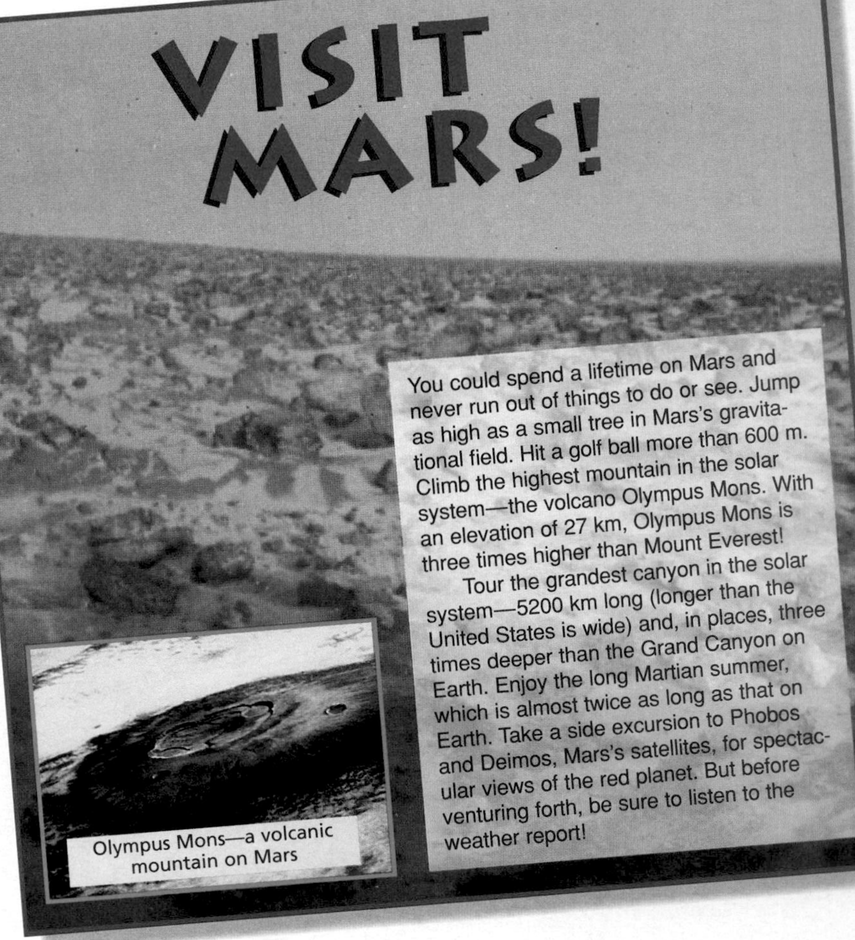

VISIT MARS!
You could spend a lifetime on Mars and never run out of things to do or see. Jump as high as a small tree in Mars's gravitational field. Hit a golf ball more than 600 m. Climb the highest mountain in the solar system—the volcano Olympus Mons. With an elevation of 27 km, Olympus Mons is three times higher than Mount Everest!
Tour the grandest canyon in the solar system—5200 km long (longer than the United States is wide) and, in places, three times deeper than the Grand Canyon on Earth. Enjoy the long Martian summer, which is almost twice as long as that on Earth. Take a side excursion to Phobos and Deimos, Mars's satellites, for spectacular views of the red planet. But before venturing forth, be sure to listen to the weather report!

Olympus Mons—a volcanic mountain on Mars

Research a component of the solar system. You could choose a planet or one or more of the many satellites. Comets, asteroids, and the Sun would make interesting studies as well. Develop a travel poster that would encourage others to visit the site (or perhaps would keep them away).

ACTIVITY 2

Encourage students to use the sample poster as a model for their own posters. Challenge students to be creative. For example, some students may wish to advertise trips through the solar system instead of to just one place. Others may wish to include pictures from magazines or drawings of their own. Some students may want to include poems or make collages.

To extend the Activity, suggest that students make travel brochures. Display several brochures for them to use as models.

Point out to students that they will have to do some research in order to determine what to include in their posters. Explain that the information in their posters must be based on facts but that this requirement doesn't preclude creativity. Their posters should be informative and interesting. Display the completed posters around the classroom.

Cooperative Learning
EXPLORATION 2, ACTIVITY 2

Group size: 3 to 4 students
Group goal: to research a component of the solar system and develop a travel poster for that area
Positive interdependence: Assign each student a role such as group leader, research assistant, editor, and artist.
Individual accountability: Have each student in the group be prepared to give an oral presentation of the group's poster. Call on an individual at random to present the poster.

Meeting Individual Needs

Learners Having Difficulty

Suggest that students create a sales and advertising campaign that promotes Earth as the greatest and most hospitable planet in the galaxy. Students should develop a logo, some sample advertisements, and a commercial (either on videotape or audiotape). Students should compare features of Earth with features of other planets. Allow students creative freedom in developing their sales materials.

Multicultural Extension

The Space Program

Suggest that students write biographies of African Americans who have participated in the space program, such as Dr. Mae Jemison, the first African American woman selected for spaceflight, Guion S. Bluford, Jr., the first African American to fly in space, and Robert McNair, the second African American to fly in space. Students may also write about other people who have been involved in the space program, such as Franklin Chang-Diaz (Costa Rican) and Rodolfo Neri (Mexican).

ACTIVITY 3

Create an Alien—Plant or Animal

Whisper-thin winds hiss along a dry, dusty canyon. Deadly ultraviolet radiation pours from an unshielded Sun. Nighttime cold reaches –80°C. Perfect weather for a fellow like the Martian waterseeker. Its parasol tail can lift three meters in Mars's low gravity, shading it from ultraviolet sunburn. The long snout can probe for pockets of ice under dried-up channels. And the giant ears, needed to hear well in the thin air, also serve as blankets: In Mars's frigid nights the waterseeker stays snug by clamping its ears tightly around its whole body.

The text above describes an imaginary alien animal living on Mars. Research a part of the solar system and create an imaginary alien plant or animal that is adapted to the conditions there. Describe why it is well adapted to its environment. A good way to start is to think of a living plant or animal that is familiar to you. Consider what adaptations would be necessary for it to survive in a more hostile world.

469

ACTIVITY 3

Students must make their alien's characteristics suitable to the environment in which it will live. Therefore, students must research what the environment is like.

When students have finished their work, have them organize their work into a booklet, design a cover, and think of a title. Then have them display the booklet where class members may enjoy it at their leisure.

Primary Source

Description of change: excerpted from *National Geographic Picture Atlas of Our Universe,* by Roy A. Gallant, p. 45

Rationale: to provide an example of a creature with extraterrestrial adaptations

CROSS-DISCIPLINARY FOCUS

Social Studies

Read the following excerpt from a speech given in 1960. Have students write a letter to John F. Kennedy to respond to his ideas.

"I believe this nation should commit itself . . . before this decade is out, of landing a man on the Moon and returning him safely to Earth. No single space project in this period will be more exciting or more impressive . . . or more important for the long-range exploration of space; and none will be so difficult or expensive to accomplish."

FOLLOW-UP

Reteaching

Plan a trip to the school library to instruct students on how to conduct research. Librarians are often available to give guided tours of the library. To test what students have learned from your trip, send them on a scavenger hunt to locate certain pieces of information relevant to this lesson.

Assessment

Have students write stories about a day in the life of a Martian pioneer. Explain that the stories may be entertaining but should be based on facts about the Martian environment and the problems of living on Mars.

Extension

Challenge students to discover what scientists have learned about Mars from space probes that have already been there. The Mariner and Viking missions collected data about the Martian environment.

Closure

Suggest that students create a model of a biosphere that could be used for a colony on Mars. Supply students with a variety of materials such as clay, papier-mâché, containers, pipe cleaners, tape, and construction paper to make their colonies look as realistic as possible.

Answers to *Challenge Your Thinking*

1. **a.** The illustration at the bottom of page 481 of the pupil's text shows the orbits of Neptune and Pluto.
   b. Given Uranus's distance from the Sun, the difference between summer and winter temperatures that would result from heating by the Sun's rays would be very small. However, due to its tilt on its axis, each of Uranus's poles has constant daylight for nearly 42 years and thus has very long summers. (Uranus's year is 84 Earth years.) Each pole also experiences nearly 42 years of constant darkness, so Uranus has very long winters as well. The equatorial regions should experience four definite seasons.
   c. If Earth were much smaller or more massive, it would have a much denser atmosphere due to its greater gravity, and it would have a different mixture of atmospheric gases (far more hydrogen, for example). These characteristics would make life far different from that on Earth today, if not impossible. If Earth were much larger or less massive, it might have no atmosphere due to its lesser gravity. This would leave it dry and lifeless, like Mars. If Earth were much closer to the Sun, it would be much hotter, and water would be vaporized. This would result in a dry, lifeless surface similar to that of Mercury or Venus. If Earth were much farther from the Sun, it would be much colder, and its water would be frozen. This, too, would leave it dry and lifeless.
2. The tape should be longer than 600 cm. Pluto should be placed at 600 cm, Neptune at 456.3 cm, Uranus at 291.3 cm, Saturn at 144.8 cm, Jupiter at 78.9 cm, Mars at 23.1 cm, Earth at 15.2 cm, Venus at 10.9 cm, and Mercury at 5.9 cm.
3. **a.** Student answers will vary. Students might add such advances as the invention of the television, computers, the first space shuttle flight, or the smallpox vaccine.
   b. Accept all reasonable answers.

1. Planet Peculiarities

Here are some unusual facts about the members of our solar system. For each fact, there is a question to consider or an activity to do.

a. Pluto is usually, but not always, the outermost planet. For 20 of the 248 years it takes to orbit the Sun, Pluto's orbit lies inside that of Neptune.

> Illustrate this by drawing a diagram of the possible orbits of Neptune and Pluto around the Sun.

b. All planets except one rotate about an axis that is somewhat close to being perpendicular to the plane of the ecliptic. For example, you know that Earth's axis is tilted only $23\frac{1}{2}$ degrees from perpendicular. But Uranus's axis (at left) is tilted 98 degrees! Perhaps Uranus was hit by a huge object in the past, and the blow changed its tilt.

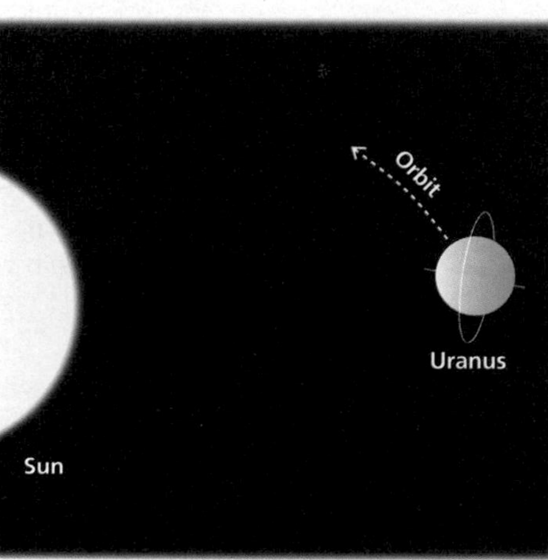

> Speculate about Uranus's rather unusual seasons, which result from this tilt.

c. Earth can be considered a miracle planet. It is just the right size and the right distance from the Sun to support life as we know it.

> What would be the consequences if Earth were much larger? much smaller? closer to the Sun? farther from the Sun?

The "miracle planet"

Homework

The Activity Worksheet on page 41 of the Unit 7 Teaching Resources booklet makes an excellent homework activity after students have completed Chapter 21.

2. Create a Solar System

Model the solar system on a long piece of adding-machine tape. Since Pluto is the most distant planet, at 6,000,000,000 km from the Sun, how long of a piece of tape will you need if you use a scale of 1 cm = 10,000,000 km? Check in the SourceBook for the average distances of the planets from the Sun. Then start at one end of the tape and indicate the position of each planet along the tape. Add other information to the tape, such as the observations from your journey on pages 461 and 462.

3. Repeat Performances

Halley's comet is sighted every 76 years. Examine the list of human achievements that had been accomplished by the time of each return.

Year of return	Technologies/advances
1607	The telescope was invented. Three years later Galileo discovered four of the moons of Jupiter using the new invention.
1682	Halley predicted the return of the comet in 76 years. Popular mode of transportation: walking.
1759	Horses and sailing ships were the main modes of transportation. Halley's prediction of the comet's return was confirmed.
1835	Age of travel by steam-powered trains began.
1910	Three years before, the Wright brothers made their historic flight in their airplane at Kitty Hawk, North Carolina. The flight lasted 10 seconds.
1986	*Voyager 2* encountered and photographed the planet Uranus after a journey lasting 9 years. Four years later, in 1990, the most complex telescope ever built, the Hubble Space Telescope, was placed in orbit.

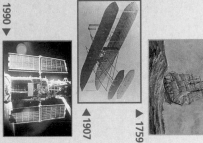

▲ 1759

◀ 1907

1990 ▶

a. What other examples could you add under Technologies/advances for any of the dates? Consider such categories as medicine and communications.

b. When Halley's comet returns in 2061, how far do you think humans will have traveled in the solar system and beyond? By what means? What technologies or advances do you think you might be able to add for that date?

ScienceLog

Review your answers to the ScienceLog questions on page 454. Then revise your original answers so that they reflect what you've learned.

471

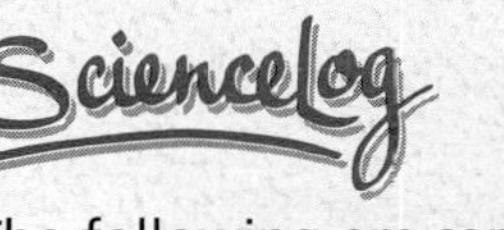

ScienceLog

The following are sample revised answers:

1. Comets were once called "hairy stars" because as the frozen gases at the head of the comet vaporize in the heat of the Sun, the gases and dust released in a stream behind the comet's head give the appearance of flowing hair or a tail.
2. If you took the same route as the Voyager spacecraft, it would take a little over 4 years (1567 days) to reach Saturn. Along the way you would cross the orbit of Mars, pass through an asteroid belt, and pass Jupiter and Jupiter's satellites.
3. The other planets are not similar to our own. The planets vary greatly in temperature, amount and content of atmosphere, and surface features. Student answers to the second question will vary according to their opinion. Students who answer "yes" should recognize that Mars offers the greatest hope of sustaining a human colony.

★ **You may wish to provide students with the Chapter 21 Review Worksheet on page 42 of the Unit 7 Teaching Resources booklet. It accompanies this Challenge Your Thinking.**

Our Universe

Connecting to Other Chapters

Chapter 19
explores the science of astronomy and discusses the observations of some early astronomers.

Chapter 20
describes the various motions of celestial bodies and some of the effects of those motions.

Chapter 21
introduces meteors, meteorites, and comets and examines what a colony on Mars might be like.

Chapter 22
examines Earth's place in the universe and introduces the big-bang theory.

Prior Knowledge and Misconceptions

Your students' responses to the ScienceLog questions on this page will reveal the kind of information—and misinformation—they bring to this chapter. Use what you find out about your students' knowledge to choose concepts and activities to emphasize in your teaching.

In addition to having students answer the ScienceLog questions on this page, you may wish to have them perform the following exercise: Divide the class into groups of 3–4 students and tell each group that they represent an interplanetary travel service with flights to destinations all over the universe. Have students put together a rough schedule of flights for a customer who has business conferences to attend on Mars, near Alpha Centauri (the closest star other than the Sun), in the Andromeda galaxy, and in the Sombrero galaxy. Have students include an estimate of how long each trip will take and what sort of in-flight entertainment is available for passengers. Emphasize that this is not a research project and that there are no right or wrong solutions for this assignment. You may wish to have volunteers present their travel schedules to the class and to use these presentations to stimulate discussion. Afterward, point out to the class that if a ship could travel at the speed of light, it would take several minutes to reach Mars, 4.3 years to reach Alpha Centauri, 2.2 million years to reach the Andromeda galaxy, and 40 million years to reach the Sombrero galaxy. The presentations and discussion should help you evaluate what students already know and what misconceptions they may have about the scale of the universe.

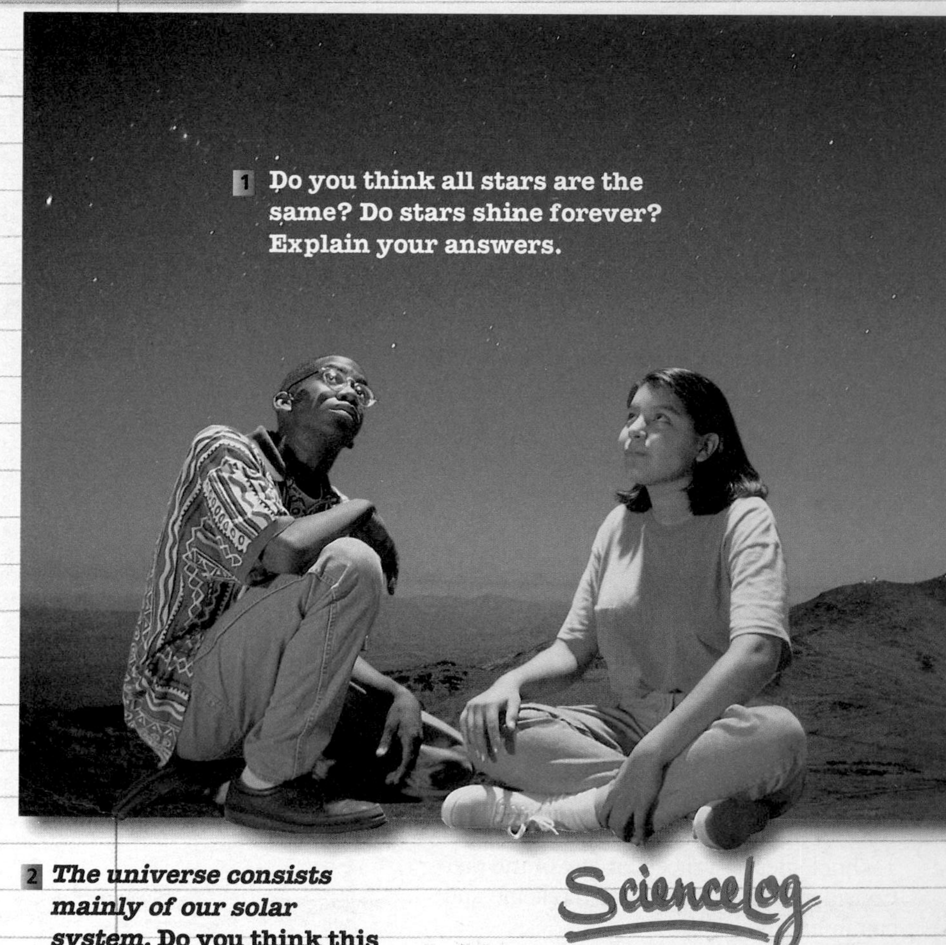

CHAPTER 22
Our Universe
1 Do you think all stars are the same? Do stars shine forever? Explain your answers.
2 The universe consists mainly of our solar system. Do you think this is a true statement? Explain your reasoning.
3 What is meant by "the big bang"?
ScienceLog
Think about these questions for a moment, and answer them in your ScienceLog. When you've finished this chapter, you'll have the opportunity to revise your answers based on what you've learned.
472

Messages From the Stars

Seeing the Light

Stars communicate with us through the light they emit, so you can find out a great deal about stars by studying their light. Look again at the photograph of star trails on page 433 and think about the following questions:

- Are all stars equally bright? Explain your answer.
- Do stars differ in the color of their light? If so, what could be the reason for this?
- Exactly what are stars? What causes them to glow?

You'll discover answers to these questions in the reading and Exploration that follow.

Hot gases rising from the Sun's surface

The Sun under magnification

The Closest Star

You are very familiar with the Sun as a source of light and heat. Did you also know that it's a star? For living things on Earth, the Sun is a special star because it supplies the energy necessary for life. However, compared with the stars that you observe on a clear night, the Sun is just a typical star. In fact, it is smaller and less bright than many stars you see! On a clear night, identify several bright stars. If you were watching the Sun from the same distance that separates you and these stars, you would see that many stars are larger and brighter than the Sun. The largest stars are larger than the orbit of Jupiter.

LESSON 1 ORGANIZER

Time Required
2 class periods

Process Skills
observing, analyzing, measuring, classifying

Theme Connection
Energy

New Terms
Black hole—a small and dense object formed from a massive collapsing star. It has a gravitational pull so strong that even light cannot escape.

Neutron star—a very small star consisting of the remnants of an exploded star. It contains the mass of several Suns, compressed to the size of a small asteroid.

Supergiant—a star even larger than a giant star that is very near the end of its life

Supernova—a high-mass star that explodes, producing a bright light and leaving behind a neutron star or black hole

continued ►

Messages From the Stars

FOCUS

Getting Started

Prior to the lesson, create a "canstellation" to show to the class. First cut both ends out of a large coffee can. Cover any jagged edges with tape. Trace the star pattern of Orion onto black construction paper. Poke holes in the pattern to represent the stars, making sure that the hole sizes vary to show the relative brightness of the stars. Tape the paper to one end of the can. For added effect, paint the inside of the can black. Darken the room and project the star pattern onto the wall by inserting a flashlight into the can. Ask students to explain what a constellation is. *(A collection of stars)* Ask: What constellation is this? *(Orion)* Explain to students that stars can differ in size and color, and that in this lesson students will learn what these differences mean in terms of a star's energy and life span.

Main Ideas

1. Stars vary in size, color, and temperature.
2. The color of a star can indicate its size, temperature, and life span.
3. Stars pass through stages that depend in part on their mass.
4. The life span of a star is billions of years.

TEACHING STRATEGIES

Seeing the Light

GUIDED PRACTICE Involve students in a discussion of what a star is, and help them formulate a definition.

Answers to Seeing the Light and Teaching Strategies for The Closest Star are on the next page. ►

★ Two Transparency Worksheets (Teaching Resources, pages 47 and 49) and Transparencies 77 and 78 are available to accompany The Closest Star.

Answers to
Seeing the Light, page 473

Students may not know the answers to these questions at this point. You may wish to have them revise their answers later. The following are sample revised answers:

- Stars differ in brightness because of their distance from the Earth, their size, and their energy output.
- Stars differ in the color of their light because their surface temperatures are different.
- Stars are huge balls of glowing gas. Nuclear reactions taking place within the core of a star cause it to glow.

The Closest Star, *pages 473–474*

Monitor students' understanding of how a star produces heat and light by involving them in a discussion of the process. You may wish to point out that the nuclear reaction that occurs in a star is called fusion because it is a coming together, or fusing, of two hydrogen atoms to form one helium atom. This is the opposite of nuclear fission, which takes place in nuclear reactors on Earth. Nuclear fission occurs when atoms break apart.

Encourage students to always use caution when studying and observing the Sun. Never look directly at it; looking into the Sun can cause permanent damage to the eyes.

Answer to
Caption

Ⓐ Students may find it easiest to locate Orion's belt first. Then they may be able to find Betelgeuse, which is the bright star between the belt and the top-left corner.

To get an idea of the size of the Sun, compare it with the size of Earth. How much larger than Earth is it? Is it 100 times as large? 1000 times? Actually, it is capable of holding more than a million Earths!

All stars, including the Sun, are made up chiefly of hydrogen gas, along with some helium gas. Although we haven't sampled the core of any star, scientists think that heat and light are produced in stars in the following way: When the gases in a star reach a tremendously high pressure and temperature, *nuclear reactions* occur within the star's core. In these reactions, hydrogen atoms fuse together to form another substance—helium. In doing so, a tremendous amount of energy is released. Some of that energy becomes the heat we feel and the light we see from stars. Humans have duplicated this process in the hydrogen bomb. If each star releases so much energy, imagine the amount of energy released by all of the stars in a constellation!

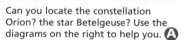

Can you locate the constellation Orion? the star Betelgeuse? Use the diagrams on the right to help you. Ⓐ

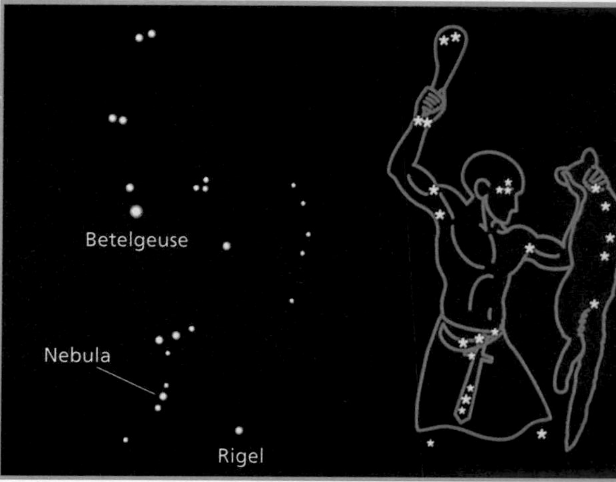

Two diagrams of the constellation Orion

One easily recognizable constellation is Orion, the Hunter, shown above. Orion's most distinguishing characteristic is a group of three stars that makes up the belt of the Hunter. Find Orion on the star chart for winter skies on page 432. Better still, locate it in your night sky. In the Northern Hemisphere, Orion is visible all winter.

ORGANIZER, *continued*

White dwarf—the final stage in the life of low-mass stars like the Sun

Materials (per student group)
Exploration 1: nail, about 5 cm in length; Bunsen burner or portable burner; incandescent light source with clear bulb; 2 fresh D-cell batteries; 1 weak D-cell battery; flashlight bulb; about 30 cm of small-gauge electric wire; pair of tongs; set of star charts; safety goggles

Exploration 2: 2 round, white, medium-sized balloons; 2 plastic-foam balls, about 1 cm in diameter; red, yellow, and black permanent felt-tip markers

Teaching Resources
Transparency Worksheets, pp. 47 and 49
Exploration Worksheets, pp. 51 and 53
Transparencies 77 and 78
SourceBook, p. S147

Can you find several very bright stars in Orion? Each one is brighter and much more massive than the Sun. The reddish star in Orion is called Betelgeuse (BET ul jooz). What does this red color tell you about Betelgeuse? Exploration 1 will help you answer this question.

EXPLORATION 1

The Temperature of a Star

You Will Need

- a large nail
- a Bunsen burner or a portable burner
- a light source with a clear bulb
- star charts
- 3 D-cell batteries (2 fresh, 1 weak)
- a flashlight bulb
- small-gauge electric wire
- a pair of tongs

What to Do

Be Careful: To avoid burns, use extreme care when handling hot materials.

1. Using a pair of tongs, heat a nail over a flame until the nail begins to glow. What color is it when it first begins to glow? Can you make it glow any other color? How?
2. Now turn on the clear light bulb. What color is the filament as it glows? Which has a higher temperature, the nail or the filament in the light bulb? What can you do to make the nail glow as brightly as the filament in the light bulb?

Exploration 1 continued ►

★ **An Exploration Worksheet accompanies Exploration 1 (Teaching Resources, page 51).**

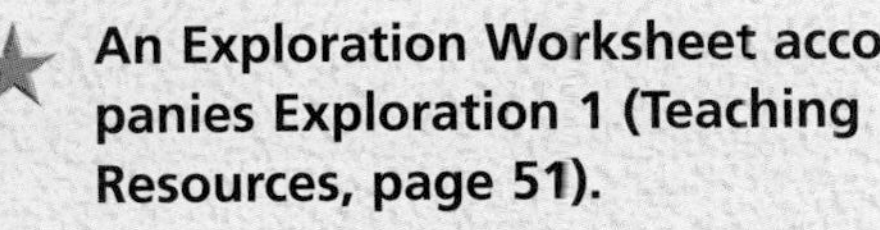

CROSS-DISCIPLINARY FOCUS

Art

Provide students with a detailed star map. Have them draw a new constellation of their choice and tell what organism or object the constellation represents. You may have them add new star names as well.

Did You Know. . .

Sunspots are dark spots on the Sun's surface that are cooler than the areas around them. The amount of sunspot activity on the Sun's surface rises and falls in an 11-year cycle. Sunspots are also associated with solar flares that spew out clouds of charged particles toward the Earth, disrupting radio reception around the world.

EXPLORATION 1

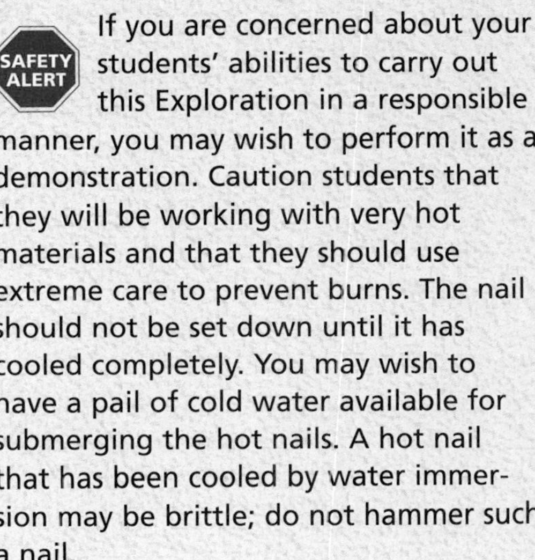

Divide the class into small groups and distribute the materials.

SAFETY ALERT If you are concerned about your students' abilities to carry out this Exploration in a responsible manner, you may wish to perform it as a demonstration. Caution students that they will be working with very hot materials and that they should use extreme care to prevent burns. The nail should not be set down until it has cooled completely. You may wish to have a pail of cold water available for submerging the hot nails. A hot nail that has been cooled by water immersion may be brittle; do not hammer such a nail.

SAFETY TIP This would be a good time to review the safety procedures that students should follow in the event of a fire. If a student's clothes are on fire, that student should stop, drop to the floor, and roll. If a student sees that someone else's clothes or hair is on fire, he or she should grab the nearest fire blanket and use it to extinguish the flames. Have students role-play such a situation so that they may practice the proper procedures.

Answers to *Exploration 1, pages 475–476*

1. The nail should turn from red to orange as it gets hotter. If enough heat is applied, it will eventually turn white.
2. The filament in the clear light bulb is white as it glows. The filament is white and therefore has a higher temperature than the red nail. By adding more heat to the nail, it will eventually glow white.
3. Students should discover that with the weak battery, the filament burns with a dull red color; with one fresh battery, the filament burns with a bright red or orange color; with two fresh batteries, the filament becomes almost white. Cooler objects emit red light. As an object becomes hotter, its color gradually changes from red to orange to white.

Answers to *Questions*

1. The color of a star is an indication of its surface temperature. Stars with relatively high surface temperatures are blue-white or white; stars with relatively low surface temperatures are red or orange.
2. The order of the stars, from highest to lowest surface temperature, is Sirius, Procyon, Capella, Aldebaran, and Betelgeuse. You may wish to point out to students that the color of the hottest stars is blue-white.
3. Most stars appear white or bluish white. (Have students compare and discuss their observations.)
4. Students' responses may vary, but possibilities include Aldebaran, Altair, Arcturus, Betelgeuse, Capella, Castor, Deneb, Polaris, Pollux, Regulus, and Vega.

History of a Star

Have students read pages 476–478 silently. Be sure they understand that the answers to the questions can be found by identifying the corresponding numbers in the margins on pages 477–478.

Answers to *History of a Star*

1. A red giant is a very large and relatively cool star that is nearing its final stages of life. Stars become red giants as they use up their hydrogen and begin burning other fuels.
2. Neutron stars and black holes begin as very massive stars that explode into supernovas after their red-giant stage.
3. A supernova is an exploding star that results from a very massive star using up its nuclear fuel and collapsing in on itself. When it explodes, it remains extremely bright for several weeks. It leaves behind a neutron star or black hole.
4. Stars produce heat and light energy from nuclear reactions that occur in their cores. First, hydrogen atoms are fused to create helium. When hydrogen is used up, atoms of other elements will be fused. During the process, huge amounts of energy are released as heat and light.
5. Eventually the Sun will run out of nuclear fuel and will begin to expand in size until it reaches the red-giant stage. In time, it will cool and contract under its own gravitational forces until it becomes a white dwarf.
6. The greater the mass of a star, the smaller and denser it will become until it finally collapses. The most massive stars will explode into supernovas and then become black holes. Other, very massive stars will become neutron stars after they are supernovas. Stars with much less mass, like the Sun, may collapse and become white dwarfs.

3. What is the relationship between the temperature of the glowing object and its color? If you're not sure, try the following: Attach a weak D-cell battery to a clear flashlight bulb using conducting wires. Observe the color of the filament. Feel the temperature of the bulb. Now attach a fresh D-cell battery to the bulb and repeat your observations. Finally, attach two fresh D-cell batteries and make similar observations. Note the color of the filament. Feel the temperature of the bulb. What conclusion can you draw about the relationship between the color of light and the temperature of the object emitting the light?

Questions

1. Now do you know what information we can get from the color of a star? What color are the stars with relatively high surface temperatures? with relatively low surface temperatures?
2. Sirius, the brightest star in the sky, appears bluish white. Aldebaran is orange. Procyon appears yellow-white. Capella is yellow. Betelgeuse is red. Arrange these stars in order from highest to lowest surface temperature.
3. Spend some time looking at the stars. How many can you see that are bluish white? red? yellow? What color are most stars?
4. Using a star chart, identify the names of some of the more distinctive stars.

History of a Star

If you've spent time around a campfire or warmed yourself in front of a fireplace, you know that when the fire dies down, the embers will glow for a while and the ashes will eventually cool. Is this the fate of stars as well? Do they have a life span in which they produce light and heat and then die? If so, what remains? Read on for answers to these questions and the ones below. Use the numbers in the margins of the section titled The Life of a Star to help you.

1. What is a red giant?
2. How are neutron stars and black holes formed?
3. What is a supernova?
4. How do stars produce the heat and light energy that they give off?
5. What is the eventual fate of the Sun?
6. What is the relationship between the mass of a star and the way it will end?

The arrow indicates a supernova in the Large Magellanic Cloud, one of the two nearest galaxies to our own.

Integrating the Sciences

Earth and Life Sciences

Today we know that stars generate their energy by banging together the nuclei of elements until they fuse. This idea was not proposed until the 1930s. In fact, throughout the nineteenth century, scientists thought that the Sun shone because it was shrinking. Why do you think shrinking might make the Sun shine? *(When a gas is compressed, it heats up. If the Sun shrank, it would heat up and radiate the energy into space.)* Can you think of any other reasons why the Sun might be hot? *(Many chemical reactions give off heat.)*

The Life of a Star

Can stars be seen in the daytime sky? Sometime during the
next 10,000 years, a star may explode in the constellation of
Orion. The exploding star, or **supernova,** will be so bright (3)
that it should be visible during the day. That star is
Betelgeuse.

Betelgeuse was once a bluish and very massive star
with a mass 20 times that of the Sun. It was a **supergiant.** (4)
But now, after burning for 10 million years and exhausting
its hydrogen fuel, it is nearing the end of its life span.
Betelgeuse has expanded to a diameter 700 times that of
the Sun. It is now a **red supergiant.** To observers on Earth, (1)
Betelgeuse appears red—the color emitted by the cooler
surface gases that are far away from the core.

The star Betelgeuse

The Crab nebula is the remnant of a supernova that was observed in1054.

As its nuclear fuel is used up, Betelgeuse will collapse in on itself. This is due
to the extreme gravitational forces of its
(3) great mass. The resulting shock wave will
send much of Betelgeuse into space. This
is the supernova that will be visible from
Earth.

What remains of Betelgeuse will be a
(2) very dense body called a **neutron star.** A neutron star contains the mass of several Suns compressed to the size of a small asteroid (30 km in diameter). (Ceres, the largest known asteroid, is 1000 km across.) A thimbleful of a neutron star would have a mass of 100 million tons! And what happens to the material thrown out into space? It mixes with the stuff from which new stars and planets are formed.

Homework

Have students write a one-paragraph narrative comparing the life of a living organism with the life of a star. Alternatively, you might wish to have students express their thoughts in the form of a chart.

Meeting Individual Needs

Gifted Students

Challenge students to observe the night sky and to locate faint, hazy patches of light called nebulae. Suggest that they document the location of the nebulae in the sky and research the following questions:

- What are nebulae made of?
- Where did they come from?
- How large are they?
- How far away are the nebulae that we see?
- How are nebulae classified?

The Life of a Star

Determine student understanding of the material by involving them in a discussion of what they have read. If necessary, review the major points. Divide the class into groups to discuss the answers to the questions on page 476. Then reassemble the class and call on volunteers to share their ideas.

Students may be interested in learning how stars are formed. Explain that scientists do not fully understand the process but speculate that stars begin as a cloud of gas made up mostly of hydrogen and dust. These materials have been added to and mixed with the remains of stars that have exploded and have been mixed with the gases thrown out from the surfaces of giant stars.

As a result of gravity, a part of the giant interstellar cloud begins to contract into a ball. It continues to contract for millions of years as gravity pulls it together. Eventually the pressure of the gas at the center of the ball is so great that it becomes extremely hot. When the temperature reaches approximately 1,100,000°C, nuclear fusion begins to occur, and huge amounts of energy are produced. As the gases that surround the core of this new star get hotter, they begin to glow.

The mass of the contracting interstellar cloud determines the kind of star that will evolve. A mass about one-tenth of that of the Sun produces a red, faint star. A mass about 50 times that of the Sun produces a blue, bright star.

INDEPENDENT PRACTICE Have students explore the concept of density by manipulating the following items: small paper cups, sand, crushed cornflakes, ball bearings, and a metric scale. Students should discover that different materials have different masses even when they occupy the same volume of space. Encourage students to link that information to the different densities of stars.

The Life of a Star continued ►

The Life of a Star, *continued*

The greater a star's mass, the brighter it is, the higher its temperature is, and the faster it uses up its nuclear fuel. For example, a star with a mass about 10 times that of the Sun will exhaust its fuel supply in a few million years. A star with a mass about one-tenth of that of the Sun will take hundreds of billions of years to run out of fuel.

Once out of fuel, the star may eventually collapse into a black hole. When a star becomes a black hole, its size decreases considerably. For example, the Sun currently has a diameter of 1,392,000 km. To become a black hole, its mass would have to be compressed to a diameter of less than 6 km. Such an increase in density would create a tremendous increase in gravitational pull at the surface of the Sun so that even light could not escape it. However, the Sun will not become a black hole because it does not have enough mass.

Students may wonder how scientists know there are black holes if they cannot see them. Explain that one way to determine where there is a black hole is to observe its gravitational effect on other objects. For example, in the early 1970s astronomers discovered a star that orbited what seemed to be an invisible object in the constellation Cygnus. The best explanation was that the object was a black hole; it was named Cygnus X-1. A black hole may also attract and retain nearby comets, planets, and other heavenly bodies.

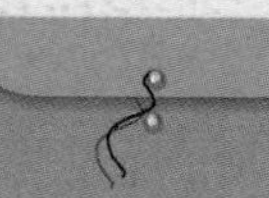

PORTFOLIO

Students may wish to include their autobiography or drawings from Checking It Out in their Portfolio. Encourage students to leave wide margins or space at the end of their stories for additions or comments.

If the mass (and therefore the gravitational force) of the col- (6)
lapsing star is great enough, it may collapse even further to form
a smaller and denser object called a **black hole**. The gravitational (2)
forces of a black hole are so extreme that nothing—not even light—can escape from it. Since no light can escape from this super-dense object, it cannot be seen in the usual ways—hence its name, a black hole.

Our star, the Sun, will not end its life in a giant explosion or (5)
supernova. This is a fate that is reserved for more massive stars
such as Betelgeuse. As the Sun's hydrogen fuel runs out and it (6)
starts to burn other fuels, the Sun will expand in size. Eventually, it will expand past the orbits of the inner planets and perhaps
even past that of Earth. When this happens, the Sun will have (1)
become a red giant.

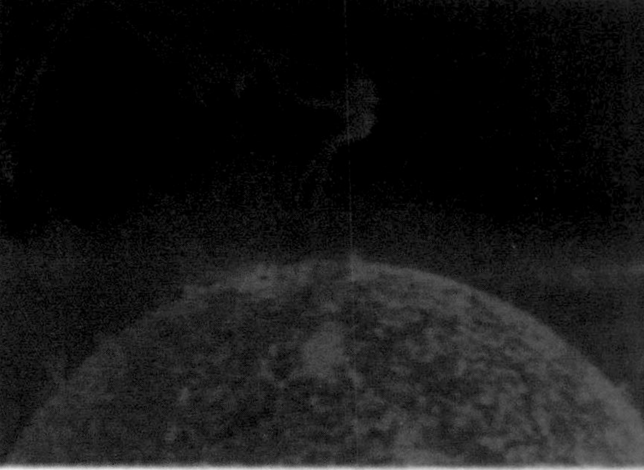

Solar eruption

By the time the Sun has become a red giant, the oceans and atmosphere of Earth will have boiled away. In time,
as the Sun goes through the remain- (4)
der of its nuclear fuel, it will cool and contract under its own gravitational forces until it becomes a **white dwarf**—a dim but extremely dense star. However, this scenario will not happen for 5 billion years. Perhaps by then humans will have migrated to another planet, to the moons of Jupiter, or to a star system far beyond this one—one with a quieter star.

Sirius, the Dog Star

You can be thankful that the Sun is a yellow star and not a bluish white star such as Sirius, the Dog Star. Sirius is the brightest star in our night sky. If the Sun and Sirius were viewed from equal distances, Sirius would be 23 times brighter
than the Sun, even though it has only twice the Sun's mass. Sirius (6)
is using its fuel at such a rapid rate that its life span is estimated to be less than a billion years. The Sun, on the other hand, is already 4.5 billion years old and is expected to be around for another 5 billion years.

Checking It Out

With a partner, discuss the answers you found to the six numbered questions on page 476. In your own words, describe what you learned.

478

Theme Connection

Energy

Focus question: How do stars produce energy? *(Stars produce energy from a series of nuclear reactions in their cores. During this process, huge amounts of energy are released as heat and light. Eventually a star runs out of nuclear fuel and begins to expand until it becomes a red giant. When the giant star begins to collapse under its own gravitational force, it may explode as a supernova and release tremendous amounts of energy into space.)*

★ **An Exploration Worksheet (Teaching Resources, page 53) is available to accompany Exploration 2 on pages 479–480.**

EXPLORATION 2

Simulating the Life of a Star

You Will Need

- 2 white balloons
- 2 small, white, plastic-foam balls
- red, yellow, and black felt-tip markers

What to Do

Form into teams of three or four to demonstrate the history of two stars—one the size of the Sun and the other a much larger star. In your simulation, you will use balloons to represent each star at various stages of its life. Decide how to use the remaining materials listed to the left. Then figure out how to sequence the steps of the simulation. The illustrations below may help you plan your simulation. The following questions also may provide some hints.

Questions

1. How large will each star be early in its life? How large will each be at various stages in its life? How will you represent this in your simulation?

The Life of a Smaller Star (such as the Sun)

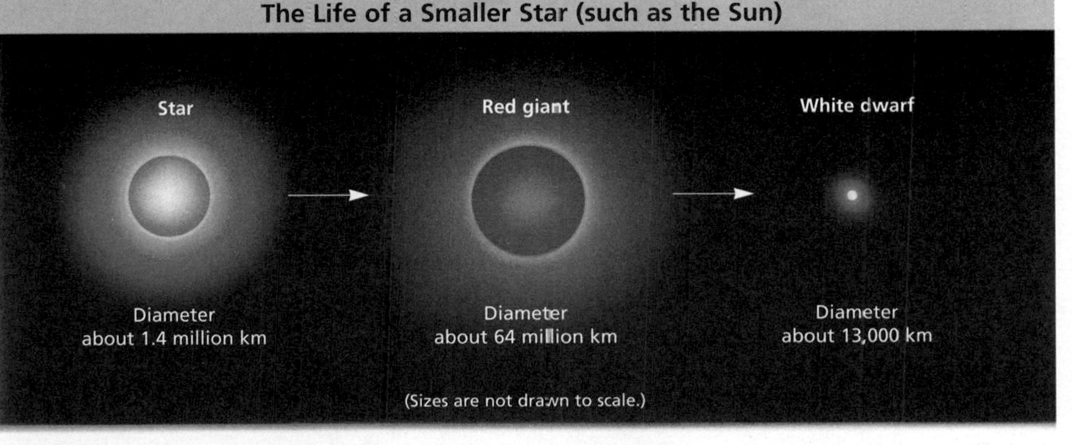

The Life of a Larger Star (such as Betelgeuse)

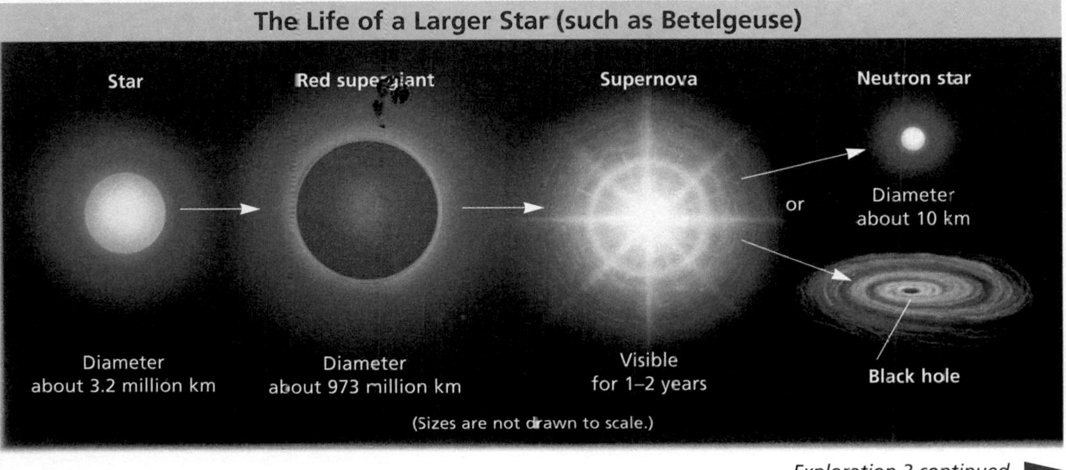

Exploration 2 continued ►

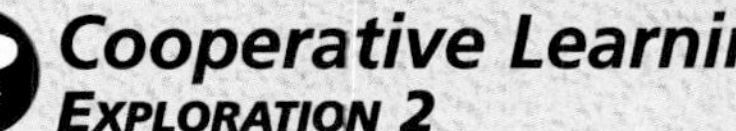

EXPLORATION 2, *pp. 479–480*

Cooperative Learning
EXPLORATION 2

Group size: 3 to 4 students
Group goal: to simulate the life spans of two stars of different sizes
Positive interdependence: Assign each student a role such as team leader, materials coordinator, designer, and presenter.
Individual accountability: Each student should be able to give an oral description of his or her group's simulation.

Once students are satisfied with their procedures and you have approved the procedures for safety, have them conduct simulations within their own group. The questions are intended to guide student's thinking as they plan their simulations. Definitive answers are not called for.

The following briefly describes one way to conduct each of the simulations:

A smaller star (the Sun)

- Color one balloon yellow.
- Place a small, white ball inside the balloon.
- Inflate the balloon slowly so that it starts out small and remains that way for some time.
- Expand the balloon completely to represent a red giant; color the balloon red.
- To represent a white dwarf, release the air and take out the ball inside.

A larger star

- Color one ball black, and place it in the balloon.
- Blow up the balloon to a large size.
- After a short time, blow up the balloon some more. Color it red.
- Use a pin to explode the balloon.
- Throw away the balloon, but keep the black ball to represent a black hole.

FOLLOW-UP

Reteaching

Have students return to the questions on page 473 and write a brief answer for each of them. (See the answers in the margin on page 474.)

Assessment

Have student groups create relative time lines showing a star's life from its origin to the white-dwarf stage. The time lines should be annotated with descriptions of each stage.

Extension

Suggest that students research the composition of the Sun's interior and what is happening on its surface.

Closure

Suggest that students do some research on the Milky Way galaxy: where it is, how big it is, how many stars it includes, and where the Sun and the solar system are located within it. Have students share their findings with the class.

Earth's Place in the Universe

FOCUS

Getting Started

Create a word search on the board using the names of all celestial bodies that students have studied in this unit. The words could include *planets, moons, constellations, stars, meteors, comets, supernovas, black holes,* and *galaxies.* Have students call out the words as they locate them. When the word search is completed, ask students: What contains all of the objects listed? *(The universe)* Conclude by saying that in this lesson, students will learn about the size of the universe and Earth's place in it.

Main Ideas

1. The solar system occupies only a tiny part of the universe.
2. Despite the huge number of objects it contains, the universe is mainly empty space.
3. Distances of space are measured in astronomical units and light-years.
4. The light from stars in our galaxy was emitted thousands of years ago. The light from other galaxies was emitted millions or even billions of years ago.

TEACHING STRATEGIES

At the Speed of Light

Ask students what the term *universe* means to them. List their ideas on the chalkboard. Then ask them what they think their place is in the universe. Ask: Where do the Earth, Sun, and solar system fit in? Help them to conclude that all matter is part of the vast physical universe. Point out that in this lesson they will learn about the size and composition of the universe.

Homework

The Math Practice Worksheet on page 55 of the Unit 7 Teaching Resources booklet makes an excellent homework activity.

EXPLORATION 2, *continued*

2. How and when in your simulation will you use the markers?
3. What does each plastic-foam ball represent in the simulation?
4. Will the star explode in a supernova? How would you represent this in your simulation?
5. How long will each star exist? How will you represent this in your simulation?

When your team has perfected the simulation, demonstrate it to the rest of the class. Ask for feedback on your simulation. What improvements could you make? Also, give feedback on the other teams' simulations.

Earth's Place in the Universe

At the Speed of Light

The stars remind you how much larger the universe is than the solar system. But just how large is the universe? To answer that question, you'll be taking another imaginary trip in the next Exploration. Only on this trip, you'll be traveling at the speed of light—300,000 km/s (kilometers per second)! At this speed, time takes on a new dimension. When traveling at this speed, you will age much less than if you had stayed on Earth.

The ticket has been bought for the trip, and you're ready to go. The travel brochure tells you that each time your distance from Earth increases by a multiple of 100, you will stop. At these stops, you'll have a chance to look back in the direction you came from—toward the solar system. You'll be looking back at Earth's place in an immense universe.

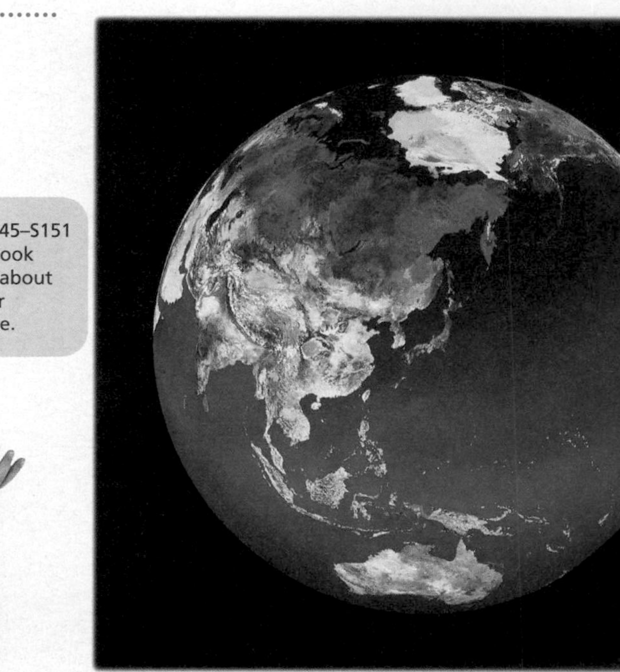

A look back at Earth

Read pages S145–S151 in the SourceBook to learn more about stars and other objects in space.

LESSON 2 ORGANIZER

Time Required
2 class periods

Process Skills
observing, analyzing, hypothesizing, inferring

Theme Connection
Structures

New Terms
Light-year—the distance light travels in one year (about 63,240 AU, or 9.5 trillion kilometers)
Local Group—a family of galaxies, of which the Milky Way is a member
Milky Way galaxy—our home galaxy; the galaxy to which our solar system belongs

Materials (per student group)
none

Teaching Resources
Math Practice Worksheet, p. 55
Transparency Worksheet, p.56
Activity Worksheet, p. 58
Transparency 79

EXPLORATION 3

Heading for the Stars

STOP 1

You have traveled about 8 minutes since leaving Earth

In this time, you have traveled a distance equal to that from Earth to the Sun. This is how long it takes light, traveling at a speed of 300,000 km/s, to reach Earth from the Sun. The distance from Earth to the Sun, you'll remember, is 150 million kilometers—one astronomical unit (AU). Looking back toward Earth, can you see the four inner planets in their orbit around the Sun? What are they?

1. How many planets are 1 AU or less from the Sun?
2. Would all of the planets be on the same side of the Sun as Earth?

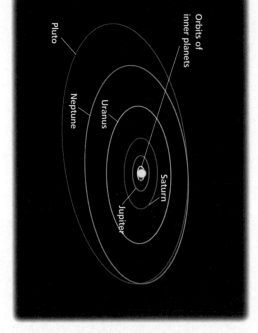

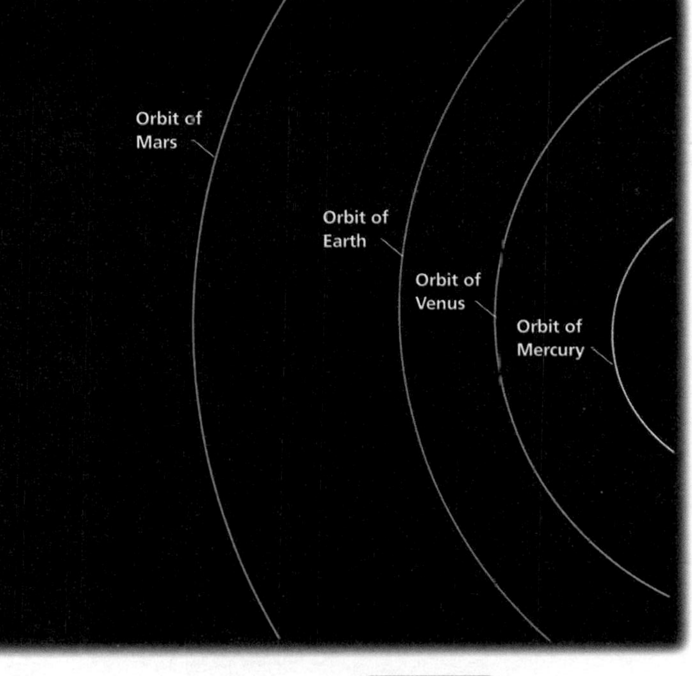

STOP 2

You have traveled 14 hours since leaving Earth

You are 100 AU from Earth. The inner planets are hidden by the brightness of the Sun. All of the outer planets are visible except for the outermost planet, which is too small to be seen. Which planet is that?

1. What does 100 AU mean?
2. If you are looking down over the north poles of the planets, in what direction would you expect them to be moving around the Sun?

Exploration 3 continued ►

Integrating the Sciences

Life and Physical Sciences

Even though the Sun appears yellow in the sky, it emits most of its light in the green part of the visible spectrum. This is precisely the frequency of light to which human eyes are most sensitive. This may be a coincidence, but some scientists have suggested that this is due to natural selection. Ask: Why would natural selection favor good eyesight in the green part of the spectrum? *(It would be an advantage to be able to see best in the range of light that is emitted most.)*

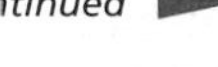

A Transparency Worksheet (Teaching Resources, page 56) and Transparency 79 are available to accompany Exploration 3.

Meeting Individual Needs

Learners Having Difficulty

A concrete example is often useful to help students understand how fast light travels. Point out to students the following fact: If you were traveling at the speed of light, you could get to the Sun and back in about 16 minutes.

EXPLORATION 3

Remind students that the last imaginary trip they took was A Trip Around the Sun in Chapter 20, Lesson 2. This time, they will be traveling much greater distances. Point out that no one can really travel at the speed of light, but in order to travel the great distances in this imaginary trip, it is necessary to pretend that traveling at the speed of light is possible.

STOP 1

Divide the class into small groups to read, analyze the pictures, and respond to the questions. They should recognize that the inner planets are Mars, Earth, Venus, and Mercury.

Answers to *Stop 1*

1. Earth, Venus, and Mercury are 1 AU or less from the Sun. By definition, Earth is 1 AU from the Sun. Venus and Mercury are less than 1 AU from the Sun because they are closer to the Sun than Earth is. (Actually, the distance from the Sun to Mercury is 0.39 AU, to Venus 0.72 AU, and to Mars 1.52 AU.)
2. Rarely do the planets line up on the same side of the Sun; they probably would not do so on this occasion.

STOP 2

Suggest that students look at the diagram and identify the outer planets. You may wish to point out that the five outer planets are much farther from the Sun than are the four inner planets. Except for Pluto, the outer planets are also much larger. The planet that is usually farthest from the Sun, Pluto, cannot be seen at this stop because it is too small. Note that until the year 2000, Pluto is closer to the Sun than is Neptune.

Answers to *Stop 2*

1. 100 AU is 100 times the average distance from Earth to the Sun, or 15 billion kilometers.
2. The planets are revolving counterclockwise around the Sun.

EXPLORATION 3, continued

STOP 3

Explain to students that astronomers usually use astronomical units to measure distances within the solar system and use light-years to measure distances outside the solar system. You may wish to point out that a light-year is equal to about 9.5 trillion kilometers. If students are curious as to how to arrive at this figure, invite one or more volunteers to come to the chalkboard and calculate it. Or have students use their calculators to see who can arrive at the answer first.

365 days × 24 hr./day × 60 min./hr. × 60 sec./min. × 300,000 km/sec. = 9.5 trillion kilometers

Answer to Stop 3

63,240 AU × 0.16 = 10,118.4 AU

10,118.4 AU × 150,000,000 km = 1,517,760,000,000 km from Earth

Answers to Stop 4

1. You are 1,011,840 AU from Earth.
2. It takes about 9 years.

Did You Know. . .

The Magellanic Clouds, two companion galaxies of the Milky Way, were once thought to be 75,000 light-years away. This is so close that if the Milky Way were rotated in space, its outer arms would touch these galaxies. Today better measurements give the distance as almost three times the earlier estimate. (See page 484.)

STOP 3

You have traveled 0.16 years (58 days) since leaving Earth

Looking back at the Sun from a distance of 10,000 AU, you see just one very bright star in a sea of blackness. Just as astronomers cannot see any planets around other stars (because any planets that may be there do not reflect enough light), you can no longer see the planets around the Sun.

The distance that you have now traveled is so great that kilometers and even astronomical units are becoming cumbersome. A new unit is needed—the **light-year.** A light-year is the distance that light travels in 1 year. That distance is equal to 63,240 AU. Since you have traveled 0.16 years at the speed of light, you have been traveling for 0.16 light-years.

At this stop, how far (in kilometers) are you from Earth?

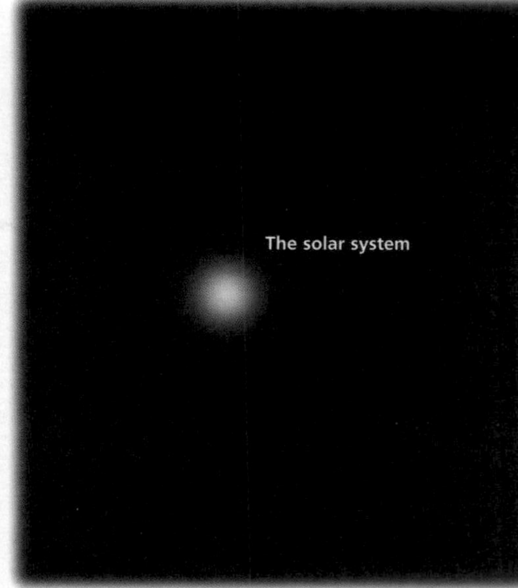

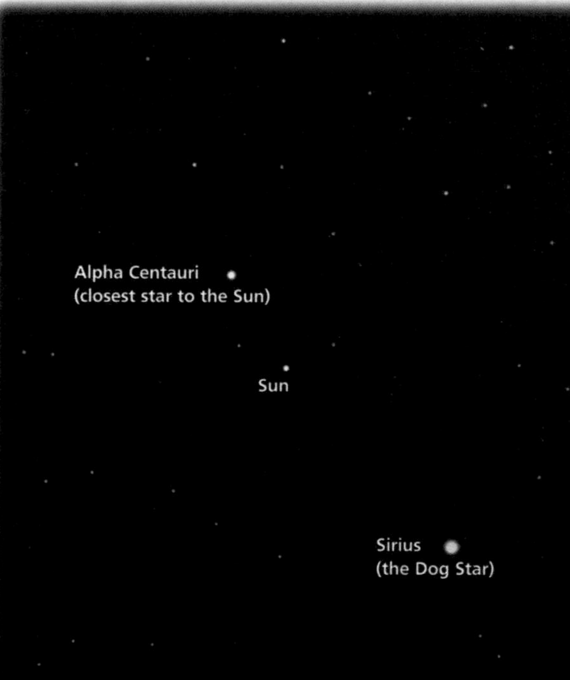

STOP 4

You have traveled 16 years since leaving Earth

You have traveled a distance of 16 light-years. Looking back toward the Sun, you can see other stars. You can see Sirius—the brightest star seen from Earth's Northern Hemisphere (except for the Sun). From this distance, you might notice that what you're seeing is actually two stars: Sirius and its companion white dwarf. The closest star to Earth (except for the Sun), Alpha Centauri, can also be seen. It is only 4.3 light-years from Earth. On Earth, it can be seen anywhere south of Miami, Florida.

1. In AUs, how far are you from Earth?
2. Sirius is about 9 light-years from Earth. How long does it take for light from Sirius to reach Earth?

STOP 5

You have traveled 1600 years since leaving Earth

You are now at a distance of 1600 light-years from Earth. Some of the stars that were so familiar to you on Earth are now behind you. There is the constellation of Orion with the red star Betelgeuse. You can also see the Pleiades, a cluster of stars in the constellation Taurus. Had you lived 150 million years ago, during the time of the dinosaurs, you would not have been able to see this group of stars. They are a cluster of young stars that formed relatively recently from clouds of dust and gas.

Examine the star chart on page 432. Compare the stars shown in the star chart with those that you would see 1600 light-years from Earth.

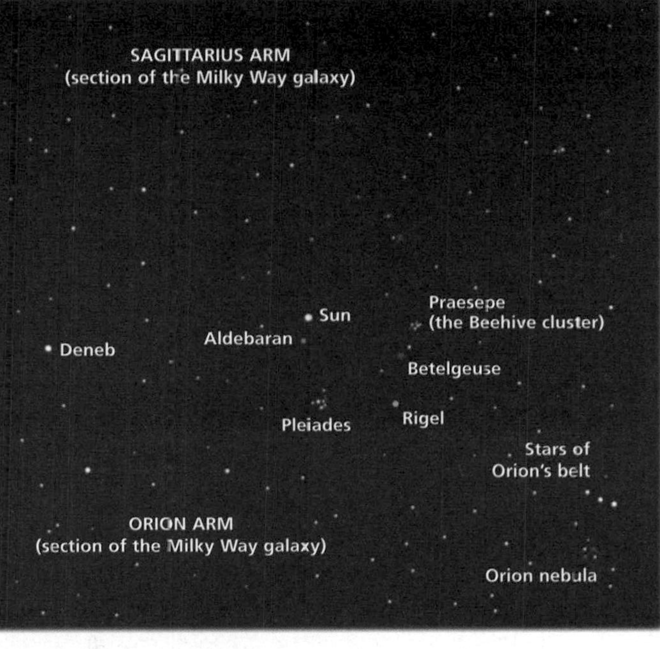

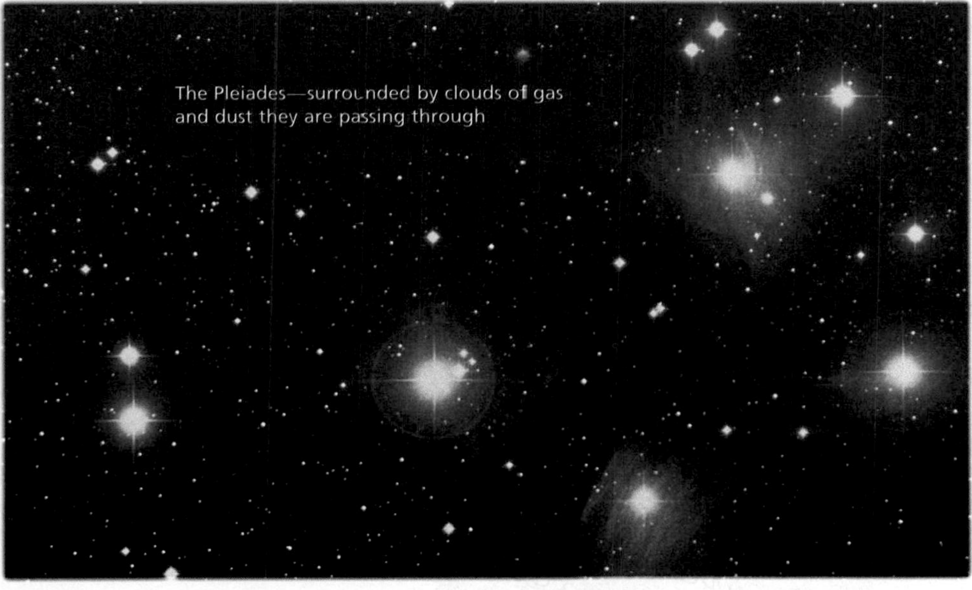
The Pleiades—surrounded by clouds of gas and dust they are passing through

Exploration 3 continued ►

STOP 5

Have students locate the area of space on the winter star chart (on page 432) that is shown in the illustration on page 483. Helpful points of reference include the Orion constellation and the Pleiades.

As students compare the star chart with the illustration, they should recognize that because they are looking back at the stars, right and left are reversed. That is, a star that appeared off to their right on Earth will now appear off to their left.

Students may be interested to know that the Pleiades is one of many star clusters in the Milky Way galaxy. It belongs to a category called open clusters, or galactic clusters, which have from 10 to a few hundred stars and include some of the youngest stars in the galaxy.

Without a telescope, the six brightest stars in the Pleiades can easily be seen as a closely spaced group. With the aid of a telescope, as seen here, they can clearly be seen surrounded by clouds of dust and gas. The earliest recorded reference to this star cluster is in ancient Chinese records dating from 2357 B.C.

The ancient Greeks named these stars for the seven daughters of Atlas featured in an ancient myth. According to the myth, Zeus saved them from the pursuit of the giant Orion by transforming them into a group of celestial doves. But only six sisters are in the Pleides. The seventh sister is said to have been stolen by Canopus. Canopus is the second-brightest star in the sky, found in the southern constellation of Carina.

EXPLORATION 3, *continued*

STOP 6

Point out to students that a portion of the Milky Way galaxy can be seen without a telescope. Ask if any of them have seen the milky-looking band of stars stretching across the night sky. Call on volunteers to describe what it looks like. Point out that in addition to stars, the Milky Way galaxy contains huge dust and gas clouds that block out the light from the stars behind them.

Students may be interested to know that the Milky Way galaxy is about 100,000 light-years across and about 10,000 light-years thick at the center. It becomes much flatter toward the edges. Our solar system is located about 33,000 light-years away from the center of the galaxy. The distance between the stars in our section of the galaxy averages about 5 light-years. Stars at the center of the galaxy are about 100 times closer.

The stars, gas, and dust in the Milky Way orbit the center of the galaxy, very much like the planets orbit the Sun. For example, the Sun orbits the center of the galaxy once every 250 million years. Almost all of the bright stars in the Milky Way orbit in the same direction. For this reason, the entire galactic system appears to rotate on its own axis.

Answers to *Stop 6*

1. Just as an island is cut off from other land areas, a galaxy is cut off from other galaxies. A galaxy is surrounded mostly by empty space in much the same way that an island is surrounded by water.
2. If the diameter of the galaxy is about 100,000 light-years across, its radius is about 50,000 light-years. The Sun is about two-thirds of the way from the center of the galaxy to its edge. Two-thirds of 50,000 is about 33,000 light-years.
3. Most of the stars appear to be concentrated in the center of the galaxy.

STOP 6

You have traveled 160,000 years since leaving Earth

At a distance of 160,000 light-years from Earth, a surprising thing has happened: as you have entered emptier space, all of the stars that you have been observing have merged into a spiral-shaped structure. This is the **Milky Way** galaxy, our home galaxy. A galaxy is a huge system of stars, dust, and gases held together by gravity.

From a distance of 160,000 light-years from Earth, you see the Sun as just one of billions of stars. The Sun is located near a spiral arm called the Orion Arm (named for the constellation Orion), about two-thirds from the center of the Milky Way. You now realize that on a clear night on Earth, all of the brighter stars you could see were relatively close to Earth, near the same spiral arm as the Sun.

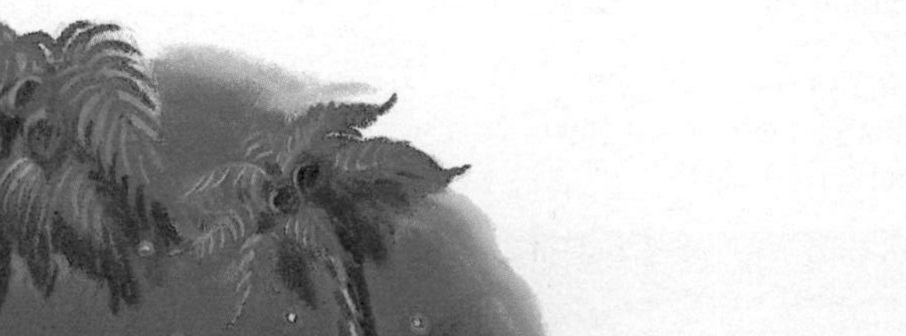

An "island universe"

1. A galaxy has been described as an island universe. What does this mean?
2. The Milky Way galaxy is 100,000 light-years across. Approximately how far, in light-years, from the center is the Sun?
3. Where do most of the stars seem to be concentrated in the Milky Way galaxy?

Theme Connection

Structures

Focus question: How is the structure of the solar system similar to the structure of the Milky Way galaxy? *(The solar system consists of one star with nine planets and several smaller bodies revolving around it, while the Milky Way galaxy is composed of billions of stars that revolve about a central point. The solar system is a very small part of the extremely large Milky Way galaxy.)*

STOP 7

You have traveled 16 million years since leaving Earth

You are now 16 million light-years from Earth. The Milky Way galaxy is now seen as part of a family of galaxies called the **Local Group.** All of the galaxies are held together by mutual gravitational forces, just as gravitational forces in our solar system hold the planets in orbits around the Sun. You may be familiar with one of the galaxies—the Andromeda galaxy, or M31—since it can be seen from Earth with the naked eye. It is the farthest object that can be seen from Earth without the aid of a telescope.

1. Andromeda is 2.2 million light-years from Earth. Astronomers study it by observing the light it emits. How long ago did that light leave Andromeda?
2. Studying distant galaxies has been described as looking back into the past. What does this mean?
3. You decide to pause on your trip to write a letter back home. When will your letter arrive if it travels at the speed of light?
4. How would you address your letter? In other words, what is your cosmic address back home?
5. What are the two most important ideas this trip has conveyed to you?

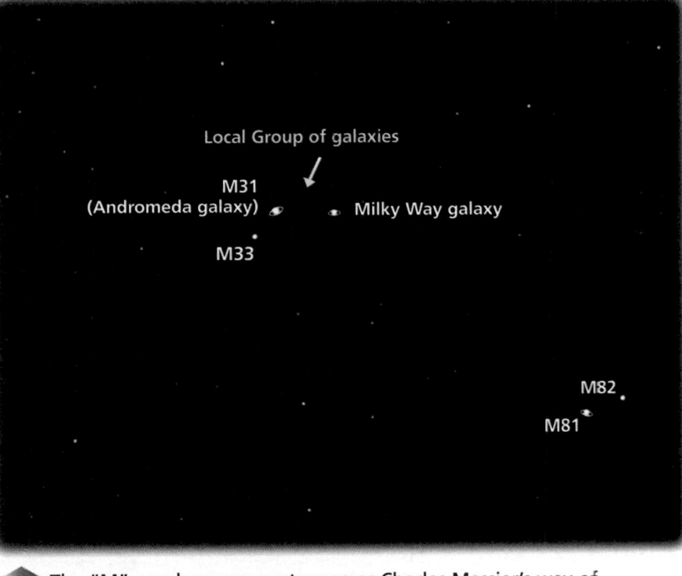

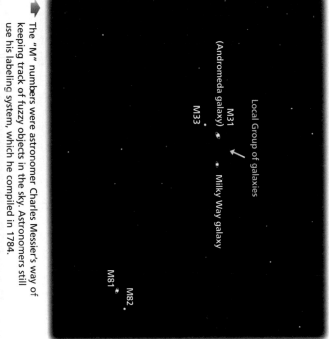

The "M" numbers were astronomer Charles Messier's way of keeping track of fuzzy objects in the sky. Astronomers still use his labeling system, which he compiled in 1784.

The Andromeda galaxy, a typical spiral galaxy

Exploration 3 continued

CROSS-DISCIPLINARY FOCUS

Mathematics

Astronomers have found evidence of a black hole in one of the giant galaxies in the Virgo cluster of galaxies. This black hole is thought to be millions of times more massive than the Sun. Theoretically, the size of a black hole is proportional to its mass. Ask: How much larger than the Sun would such a black hole be? *(Millions of times larger, or about the size of our solar system)*

Homework

The Activity Worksheet on page 58 of the Unit 7 Teaching Resources booklet makes an excellent homework assignment after students have completed Exploration 3.

STOP 7

Students may be interested to learn that the Local Group consists of 3 spiral galaxies, including the Milky Way, the Triangulum Spiral (M33), and the Andromeda Spiral (M31); 4 irregular galaxies, including the Large Magellanic Cloud and the Small Magellanic Cloud; and about 25 elliptical galaxies, most of which are relatively small. The Local Group, in turn, is part of a larger grouping called the Virgo Cluster, which contains thousands of galaxies of all types.

Answers to *Stop 7*

1. Since a light-year is the distance light travels in one year, the light reaching Earth from the Andromeda galaxy left there about 2.2 million years ago.
2. Since galaxies are so far away, astronomers are observing what they looked like when the light that is being seen on Earth left them. For example, when astronomers look at the Andromeda galaxy, they are seeing what it looked like 2.2 million years ago.
3. The letter will reach Earth 16 million years from now.
4. The cosmic address may include the following elements:
   Local Group
   Milky Way Galaxy
   Orion Arm
   Solar System
   Earth
   United States of America
   State, Town, Street
5. Answers will vary. Have students share and discuss their ideas. Their responses should indicate an understanding that objects in the universe are separated by vast distances and that our solar system is one very small part of that universe.

EXPLORATION 3, continued

Answers to *Stop and Think*

Accept all reasonable responses. Students should understand that although the chances of any given planet having life on it are very slim, there are probably billions of planets in the universe. Life-forms in other solar systems may be similar to those on Earth if we assume that similar laws of nature hold true.

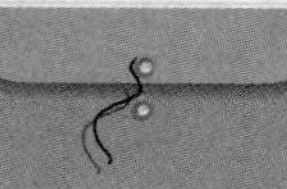

PORTFOLIO

Encourage students to include their responses to Stop and Think in their Portfolio. They may write a brief outline or a series of phrases comparing and contrasting their opinions with those of their partners.

A Conversation With Huck

You may wish to introduce this material by having students describe how the night sky appears to them. Encourage students to be as scientific or as poetic as they wish. When students have finished, call on volunteers to share what they wrote with the class.

Homework

The activity described in A Conversation With Huck makes an excellent homework assignment.

Stop and Think

This is where your journey into the universe stops. As you look around you from the depths of space, your view takes in 100 billion galaxies, consisting of possibly 1 billion trillion stars. In his book *Cosmos,* Carl Sagan described it this way:

> A handful of sand contains about 10,000 grains, more than the number of stars we can see with the naked eye on a clear night. But the number of stars we can see is only the tiniest fraction of the number of stars that are. Meanwhile the Cosmos is rich beyond measure: the total number of stars in the universe is greater than all the grains of sand on all the beaches of the planet Earth.

Around one of the stars, the Sun, you know that life exists on a planet called Earth. Only one other star has been found to have a planet orbiting it.

With a partner, discuss whether you think life exists elsewhere in the universe. What are your reasons? Would you like to know for sure? If we assume that the laws of nature hold true on other stars and planetary systems as they do here, should other life-forms be similar? Why or why not?

There are more stars in the universe than there are grains of sand on all the beaches on Earth.

A Conversation With Huck

> "We had the sky, up there, all speckled with stars, and we used to lay on our backs and look up at them, and discuss about whether they was made, or only just happened."
> —Huck Finn in Mark Twain's *Huckleberry Finn*

When you look up in the sky at night, you also see it speckled with stars. But in studying this unit, you have had a closer look at them than Huck did. What do you suppose Huck was wondering about as he looked up at the sky? Join in a discussion with Huck. Write down what you and he might say.

486

Primary Sources

Description of change: excerpted from *Cosmos,* by Carl Sagan
Rationale: to highlight the vast quantity of stars in the universe

Description of change: excerpted from *Huckleberry Finn*, by Mark Twain
Rationale: to raise the question of how the universe began

FOLLOW-UP

Reteaching

Suggest that students make a travel itinerary or time line for the trip described in the lesson. The itinerary should provide a brief description of each stop and indicate how much time it has taken to get to each point along the journey.

Assessment

Have students make diagrams to illustrate where Earth fits into the universe. Explain that their diagrams should begin with Earth and the Moon and end with the Local Group.

Extension

Explain to students that the most distant objects that have been observed from Earth are called quasars. Suggest that they do some research to discover what quasars are, how they are detected, and how far away they are.

Closure

Invite an astronomer from a local university to come to your class and talk with students. Have students write down questions in advance that they would like to ask the astronomer about his or her work, recent discoveries, or objects in the universe.

LESSON 3

A Likely Beginning

The Big-Bang Theory

Scientists have a theory that answers part of Huckleberry Finn's question, the one you just discussed. But like all theories, it can be modified as predictions are tested by experiments and observations. Read the paragraphs below about the big-bang theory. As you read, pretend that you're a newspaper editor who has to write headings for each paragraph. In your ScienceLog, write down your three- or four-word headings that summarize the message of each paragraph.

___?___

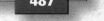

According to this theory, the entire universe was once concentrated into a very small volume that was extremely hot and dense. Then, about 10 billion to 20 billion years ago, a sudden event called the **big bang** caused the universe to begin expanding. Matter was sent outward in all directions.

___?___

A few million years later, the universe cooled, and the hydrogen and helium that were formed in the big bang clumped together as a result of gravitational forces. New stars then condensed within these clumps to create what we now see as galaxies. Using a large telescope, astronomers have detected very bright energy sources that are billions of light-years from Earth. One theory is that these energy sources, called **quasars,** are galaxies in the early stages of formation.

Supernova remnant

___?___

Oxygen, carbon, and iron—all of which are found on Earth—were formed within the stars. Some of the more massive stars eventually exploded in events called supernovas. These explosions sent material far out into space, where it mixed with other matter to form new stars and planets.

LESSON 3 ORGANIZER

Time Required
1 class period

Process Skills
observing, analyzing, inferring, classifying

New Terms
Big bang—a cosmic event that took place 15–20 billion years ago and created the universe
Quasars—bright energy sources billions of light-years from Earth that are possibly galaxies in their early stages of formation

Materials (per student group)
Exploration 4: large, round balloon; black permanent felt-tip marker

Teaching Resources
Exploration Worksheet, p. 59
SourceBook, p. S157

LESSON 3

A Likely Beginning

FOCUS

Getting Started

Have students take out a sheet of paper and write a short paragraph explaining their beliefs of how the universe began. Be sure to honor all explanations. Have volunteers share their theories with the rest of the class.

Main Ideas

1. The universe may have begun in the big bang—a cosmic event that created the universe and all of the matter in it.
2. The big-bang theory provides an explanation for the expansion of the universe.
3. According to this theory, we are all made of elements first created within the stars.

TEACHING STRATEGIES

The Big-Bang Theory

Have students complete this exercise on their own. When they have finished, call on volunteers to share their headings with the class, and involve the class in a discussion of each of the paragraphs. As an alternative, you may wish to have a volunteer read each paragraph aloud, have students think of and write a heading, and then have them discuss their headings before going on to the next paragraph.

Answers to The Big-Bang Theory are on the next page. ►

Answers to
The Big-Bang Theory, pages 487–488

Although headings will certainly vary, responses should reflect the ideas presented in the following sample headings:
- A Cosmic Expansion
- Stars and Galaxies Form
- Planets Form From Exploding Stars
- Formation of the Sun
- An Echo From the Big Bang
- The Expanding Universe

Meeting Individual Needs

Gifted Learners

When we see light emitted from a distant star that is moving away from us, the light is slightly reddened. Astronomers use this effect, called the *Doppler effect*, to measure how fast the star is moving. A similar effect occurs with sound. Discuss with students the sound of an approaching train or car. What happens to the sound? *(It gets louder, and the pitch of the sound rises as well.)*

Homework

One variation of the big-bang theory is the *pulsating-universe theory.* Have students research this theory and present their results to the class. They may wish to consult the SourceBook for help. *(The pulsating-universe theory states that the universe expands and contracts in cycles.)*

____?____

Approximately 5 billion years ago a new star, the Sun, and a system of planets formed from gas and dust particles in space, which mixed with heavy elements that were created inside stars. One of the planets that formed—Earth—is the home of humans. Another planet—Jupiter—never reached high enough temperatures to start nuclear reactions in its core, even though it is massive and made of the same stuff as the Sun.

Jupiter, made of the same stuff as the Sun

____?____

There are a number of predictions suggested by this theory. If the universe began from an expansion of super-hot material, some of that heat should still be detectable. By now, this radiation should have the same wavelengths as microwaves. In the 1960s two American physicists, Arno Penzias and Robert Wilson, were working on a way to get rid of static on their microwave antenna system. The static seemed to be coming from the sky. After trying everything they could think of to eliminate the static (they even tried cleaning bird droppings from the antenna of their radiotelescope), they realized that they may have found the radiation left over from the birth of the universe. They may have been listening to the faint echo of the big bang that occurred 10 to 20 billion years ago! Experiments and observations have so far supported the big-bang theory. As so often happens in science, this discovery illustrates how progress is sometimes made through chance instead of through carefully planned experiments.

An artist's impression of how the big bang may have looked

____?____

Since the galaxies are spreading apart, astronomers reasoned that a big bang may have occurred. In addition, the more distant galaxies are moving apart at a faster speed than the closer ones. In 1925, an American astronomer named Edwin Hubble discovered evidence of an expanding universe.

488

Exploration 4

Model of an Expanding Universe

All models are imperfect because they cannot duplicate the real thing. After you construct the following model, think about its uses. Does it help you visualize a difficult idea? How is this model a good one? What are its limitations?

You Will Need

- a large round balloon
- a marker

1

2

3

What to Do

1. Blow up a balloon only partway. This will represent the universe.
2. With a marker, place dots on the balloon, some close to each other, some far apart. The dots will represent the galaxies that are part of the universe. Measure or estimate the distances between the different galaxies.
3. Blow up the balloon some more and observe what happens to the dots.

Questions

1. Did the dots all separate by the same amount?
2. Which ones separated faster than others? slower than others?
3. Discuss this model in terms of the big-bang theory. What does the model illustrate about the theory? What predictions of the theory are illustrated by the model?
4. How is this model an imperfect picture of the universe?

489

Exploration 4

Divide the class into pairs and distribute the materials. Suggest that one student blow up the balloon while the other observes what happens.

Have students read the Exploration and then answer any questions they may have before starting the activity. Then have students complete the activity and answer the questions. When they have finished, invite students to share and discuss their ideas.

Answers to *Questions*

1. The dots did not all separate by the same amount. The farther apart two dots were from one another, the farther they separated.
2. Dots that were farther from one another separated faster than did dots that were side by side.
3. This model shows two things: that all matter in the universe is moving outward from a point and that every part of the universe must be moving away from every other part. It illustrates the idea that the universe is expanding and that more distant galaxies are moving away from each other at a faster rate than are nearby galaxies.
4. The model suggests that the universe is hollow, with galaxies only at its outermost edge.

★ **An Exploration Worksheet is available to accompany Exploration 4 (Teaching Resources, page 59).**

FOLLOW-UP

Reteaching

Have students work in groups to create murals or other classroom displays to illustrate the big-bang theory and its consequences.

Assessment

Suggest that students write an explanation for the following statement: We are all part of the stars. Have volunteers share their explanations.

Extension

Explain to students that there are scientific theories other than the big-bang theory to explain how the universe began. Suggest that they do some research to discover what some of these other theories are. Have them share what they learn by making oral presentations to the class.

Closure

Suggest to students that the universe is contracting instead of expanding. Have them write a brief explanation of what this might mean. *(All of the matter in the universe may contract to such a density that it suddenly expands again. This process of expanding and contracting may happen over and over again. This could be illustrated by the balloon model.)*

Answers to *Challenge Your Thinking*

1. • Galaxy—a huge grouping of stars that is held together by gravitational forces
   • Light-year—the distance that light travels in one year
   • Black hole—a celestial object with an extremely dense mass that exerts such a large gravitational force that even light cannot escape it
   • Stars—gaseous balls made mostly of hydrogen and helium that produce light and heat by nuclear fusion
   • Milky Way—the galaxy in which the Sun and its solar system are located
   • Quasar—a very distant body that emits energy at an extremely high rate
   • Observation—an event or object that is detected by human senses
   • Theory—explanation that makes sense of a wide range of observations. Theories are useful if they suggest predictions that can be tested through observations and experimentation.

2. Student interviews should reflect the following information:
   • William Herschel (1738–1822) founded stellar astronomy, the study of the region beyond our solar system. He also discovered Uranus and two of its satellites as well as two of Saturn's satellites.
   • Annie Cannon (1863–1941) discovered many new stars, including a double star. She analyzed the brightness and characteristics of hundreds of thousands of stars and was a staff member of the Harvard Observatory.
   • Henrietta Leavitt (1868–1961) studied certain types of stars called *Cepheid variables.* These stars grow brighter and darker in cycles, and the time it takes one of these stars to complete a cycle is related to its brightness. Her work helped later astronomers study the size of the universe. She was also a staff member of the Harvard Observatory.
   • Edwin Hubble (1889–1953) was the first person to show that galaxies other than the Milky Way exist. He demonstrated that these galaxies are in motion and that the universe is expanding. He was a member of the staff of the Mount Wilson Observatory in California, and the Hubble Space Telescope is named after him.
   • Karl Jansky (1905–1950) was the first person to detect radio waves coming from beyond our solar system. He made this discovery while studying static in transatlantic radio transmissions. His work led to the development of *radio astronomy,* which uses radio waves to study celestial bodies that are not visible with optical telescopes.
   • Robert Goddard (1882–1945) worked extensively with both liquid- and solid-propelled rockets and was responsible for developments in both spacecraft and missile propulsion. He was posthumously awarded both the Congressional Gold Medal and the Langley Gold Medal.
   • Antony Hewish (1924–) was awarded the Nobel Prize in physics for his discovery of pulsars. Observations made by Jocelyn Bell (1943–), his graduate assistant, included radio signals that fluctuated in certain patterns. Together, Hewish and Bell determined that these represented a new type of star, called a pulsar. Since then, more than 300 pulsars have been identified. (Students will find additional information about Bell by researching her under her married name, Burnell.)

CHALLENGE YOUR THINKING

1. Read All About It

The following newspaper article is similar to the reading that started the unit. This time, certain terms are underlined. In your own words, describe your understanding of the underlined terms.

> Astronomers reported today the discovery of an almost inconceivably massive object within two colliding galaxies 300 million light-years from Earth. If the object is a black hole, it is by far the largest one ever detected. Its mass is estimated to be equivalent to all of the stars in the Milky Way. If the object isn't a black hole, it might be a dormant quasar. More observations are needed to make sense of this new observation in terms of current theory.

2. All in a Light-Year's Work

Below are the names of some scientists and their astronomical contributions. Learn more about one of them by doing library research. Present your findings as if you were interviewing him or her about the work he or she did.

William Herschel—discovered the planet Uranus
Annie Cannon—classified a quarter of a million stars by the light they emitted
Henrietta Leavitt—devised another way of measuring vast cosmic distances
Edwin Hubble—suggested that the universe was expanding
Karl Jansky—first discovered that stars emit radio waves
Robert Goddard—built and launched the world's first liquid-propelled rocket
Jocelyn Bell and Antony Hewish—discovered the first pulsar in 1967

Annie Cannon

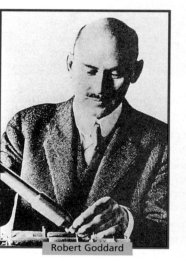

Robert Goddard

Jocelyn Bell and Antony Hewish

490

3. Past, Present, and Future

Some of the stars that we associate with a particular constellation are not part of the same star group at all. Examine the Big Dipper as it was 100,000 years ago and as it is now. Draw a diagram of what you think it will look like in another 100,000 years. Which stars are obviously not part of the same star group? Why do they all look equally far away from Earth?

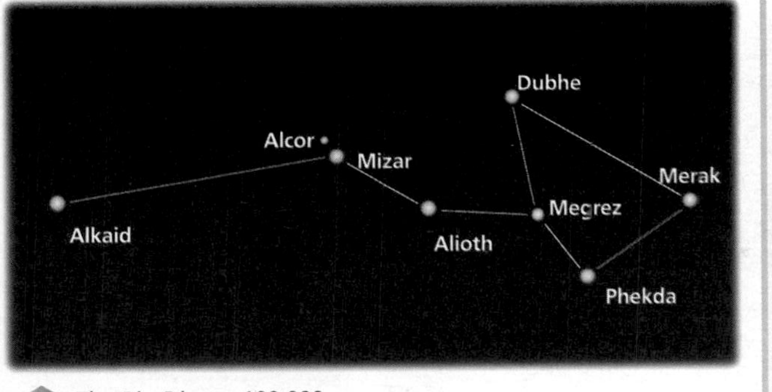

The Big Dipper 100,000 years ago

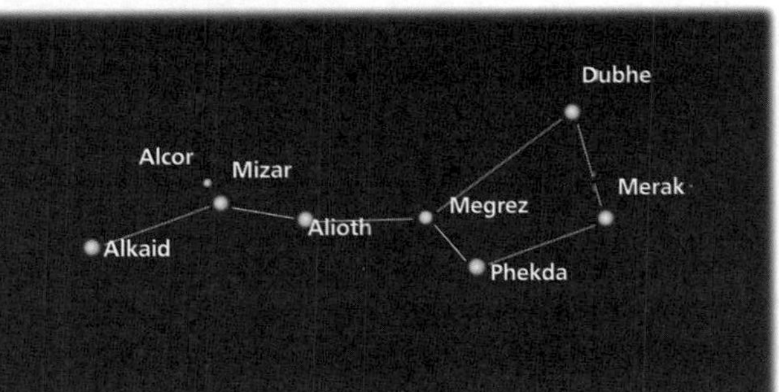

The Big Dipper today

4. Give It Some Sirius Thought

Examine the orbit of the white dwarf around the star Sirius (below).

a. What is a white dwarf?

b. What is the shape of its orbit?

c. Where is it moving the fastest? the slowest?

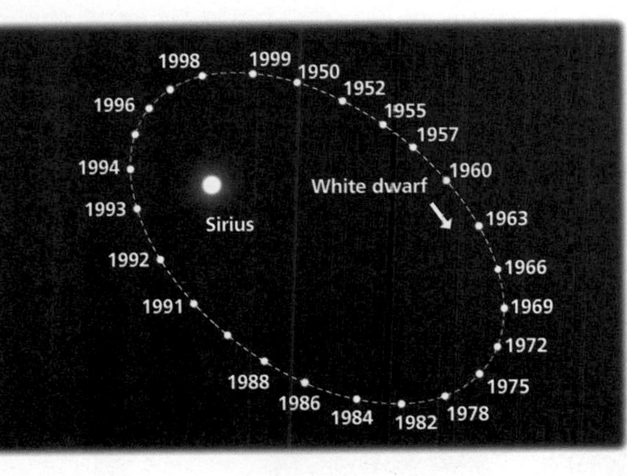

Review your answers to the ScienceLog questions on page 472. Then revise your original ideas so that they reflect what you've learned.

Answers to *Challenge Your Thinking*, *continued*

3. Alkaid and Dubhe are not part of the same star group as the others because their positions have changed greatly in relation to the others. Diagrams of these stars' future positions should show the stars twice as far along their respective trajectories. The pairs Mizar and Alcor and Alioth and Megrez appear to be the most closely related to each other. They all look equally far from Earth because they are so far away that we cannot visually distinguish their distances and because they all appear about the same size and brightness.

4. a. A white dwarf is the small and very massive remnant of a star that was once up to eight times the size of our Sun. Once the star used up its hydrogen fuel for nuclear fusion, it began to expand, appearing first as a red giant. When the gases that expanded to produce the red giant were expelled by the star, a small, massive core called a white dwarf remained.

b. Elliptical

c. The white dwarf moves the fastest as it makes its closest approach to Sirius. It moves the slowest when it is in the portion of its orbit that is the farthest from Sirius.

★ **You may wish to provide students with the Chapter 22 Review Worksheet on page 60 of the Unit 7 Teaching Resources booklet. It accompanies this Challenge Your Thinking.**

Sciencelog

The following are sample revised answers:

1. All stars are not the same. They vary in size, shape, and age. Stars do not shine forever. Stars use up their nuclear fuel and eventually become white dwarfs, neutron stars, or black holes, depending on their size.
2. Our solar system is just a small part of the universe. In fact, our solar system is just one small part of the Milky Way galaxy, which is part of the Local Group of galaxies. Beyond this group are billions more galaxies.
3. The big bang is the name that scientists gave to the event in which the universe may have been formed. According to the big-bang theory, a sudden cosmic event caused the universe to begin expanding. Matter was sent outward in all directions. This matter condensed to form stars and other bodies.

Making Connections

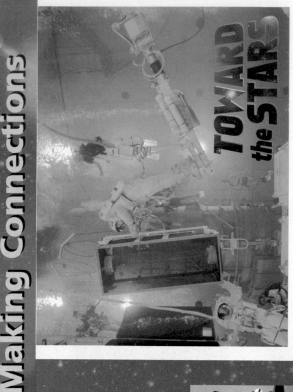

The Big Ideas

The following is a sample unit summary:

Astronomy is the study of the Earth's place in space, of the observable make-up of the universe, and of the theories that make sense of our observations of the universe. (1) We learn about the solar system and the universe by direct observation with the naked eye, telescopes, space probes, and other technology. (2)

In the heliocentric model, the Sun is at the center of our solar system, with Earth and other members of the solar system revolving around it. In the geocentric model, Earth is the stationary center of the solar system, with the Sun and other members of the solar system revolving around it. The heliocentric model is the better model because it better explains observable phenomena. (3)

We see different stars in the sky at different times of the year because Earth revolves around the Sun, which makes the Sun appear to pass through the 12 constellations of the zodiac. As Earth orbits the Sun, stars located on the side of the Sun opposite the Earth are obscured. (4)

The seasons are caused by changes in the angle at which the Sun's rays strike Earth's surface (due to the tilt of Earth's axis) as Earth revolves around the Sun as well as by changes in the amount of daylight. (5)

The solar system is made up of the Sun and the bodies revolving around it, including the nine planets. The universe consists of solar systems, galaxies, comets, asteroids, meteors, and all of the space between these bodies. (6)

Stars are bodies of gases that give off tremendous amounts of energy in the form of light and heat. Stars produce energy because of nuclear reactions in their cores. Eventually, a star will run out of nuclear fuel and begin to expand in size until it becomes a red giant. In time, it will cool and begin to collapse under its own gravitational forces. The greater the mass of the star, the smaller and denser it will become, until it finally collapses. If its mass is large enough, the collapsing star will explode into a supernova, after which it may become a neutron star or a black hole. Stars with less mass, like the Sun, may collapse into white dwarfs. (7)

The universe is so immense that special units of measurement must be used to determine distances between objects in space. (8)

Our cosmic address may include the elements Local Group, Milky Way Galaxy, Orion Arm, Solar System, Earth, United States of America, State, Town, and Street. (9)

The universe may have begun about 10–20 billion years ago in an event called the big bang. Matter was hurled outward in all directions, and as it cooled, it clumped together. Nuclear fusion occurred within the clumps once the clumps became large enough, and they formed stars. Groups of stars formed galaxies. Evidence that this happened includes cosmic microwaves (radiation from the explosion) and the expansion of the universe. (10)

Making Connections

Unit 7

SOURCEBOOK

You have been introduced to astronomy and to human exploration of the universe. On pages S137–S160 in the SourceBook, you can take a closer look at members of the solar system and at how humans are continuing their investigations of space.

Here's what you'll find in the SourceBook:

The Big Ideas

In your ScienceLog, write a summary of this unit, using the following questions as a guide:

1. What is astronomy?
2. How do we learn about the solar system and the universe?
3. What are the essential differences between the heliocentric model and the geocentric model of the solar system? Which is the better model, and why?
4. Why do you see different stars in the sky at different times of the year?
5. What causes the seasons?
6. What objects make up the solar system? the universe?
7. What are stars, and what is their fate?
8. Why are special units used to measure distances between objects in space?
9. What is your cosmic address?
10. How might the universe have begun? What evidence do we have?

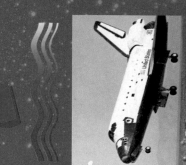

492

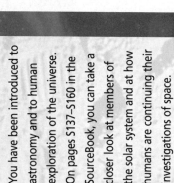

Checking Your Understanding

1. Although the other planets lie near the plane of Earth and Sun (the plane of the ecliptic), they are not exactly on this plane.

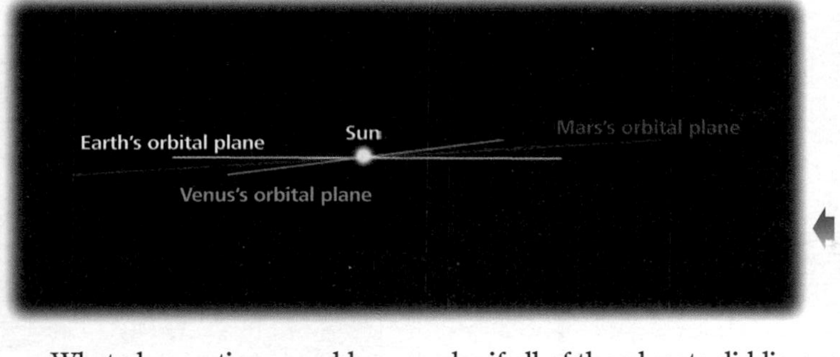

A side view of part of the solar system, showing the orbits of three planets

   a. What observations would you make if all of the planets did lie exactly on the same plane?
   b. Earth, the Moon, and the Sun do not all lie exactly on the plane of the ecliptic. If they did, what observations would you make on a regular basis?

2. Where on Earth would you be if
   a. the star Polaris were directly overhead?
   b. the star Polaris never rose above the horizon?
   c. the Sun were directly overhead once a year?
   d. the Sun were directly overhead twice a year?
   e. the stars at night appeared to revolve clockwise over your head?
   f. the stars at night appeared to revolve counterclockwise over your head?
   g. there were equal hours of daylight and darkness?
   h. there were 24 hours of sunlight?

3. Earth's axis is tilted at an angle of $23\frac{1}{2}$ degrees toward the plane of the ecliptic. What would the consequences be if the tilt were
   a. 0 degrees—no tilt at all? **b.** 45 degrees? **c.** 90 degrees?

4. 

Using the terms shown at right, create a concept map in your ScienceLog. You will have to provide the linking terms.

comets
Halley's comet
star
galaxies
the Moon
the Sun
Milky Way
satellites
solar system
the universe
seven planets

Homework

You may wish to assign the Unit 7 Activity Worksheet as homework (Teaching Resources, page 65).

You may wish to provide students with the Unit 7 Review Worksheet on page 66 of the Unit 7 Teaching Resources booklet. It accompanies this Making Connections.

Answers to *Checking Your Understanding*

1. a. The planets would pass directly in front of one another and pass directly behind the Sun. Our view of one or more planets would be blocked by another planet at times. If we looked at the solar system from the side (in profile), the planets would appear to move back and forth along a single line. The planets would travel along the same arc in the sky.
b. There would be a total eclipse of both the Sun and the Moon every month.

2. a. The North Pole
b. The Southern Hemisphere
c. On the tropic of Cancer or on the tropic of Capricorn
d. Anywhere between the tropics
e. The Southern Hemisphere
f. The Northern Hemisphere
g. On the equator, or anywhere on Earth at the time of equinox
h. Anywhere north of the Arctic Circle or south of the Antarctic Circle during their respective summers

3. a. There would be equal hours of daylight and darkness every day; that is, there would be a perpetual equinox. There would be only one season. The Sun's hottest rays would be hitting the equator all of the time. Glaciation would occur as the poles cooled, and an ice age would begin.
b. The Sun's hottest rays would hit Earth as far north as 45 degrees north in the summer and 45 degrees south in the winter. Summers would be much warmer and winters would be much colder farther to the north and south. More people on Earth would see the Sun directly overhead. Everyone north of 45 degrees north or south of 45 degrees south would experience 24 hours of daylight at some point during the summer and would experience 24 hours of darkness at some point during the winter.
c. Each hemisphere would be in sunlight for about six months of the year and in darkness for about six months of the year. The seasons would be extreme.

4. See the sample concept map on page S208.

TECHNOLOGY *in* SOCIETY

Background

The Russian Salyut program, which began in 1971, set the stage for the launch of the *Mir* space station on February 20, 1986. Improvements over previous stations include a more sophisticated communications system, greater power capability, a more elaborate computer-control system, and six bays where space vehicles can dock.

The cooperative project for constructing the international space station, *Alpha,* began after the United States and Russia realized that the cost was becoming prohibitively expensive. When *Alpha* is complete, scientists will use it to study things such as the effect of gravity on everything from small mammals to protein molecules to fire. Scientists hope that the study of human cells under weightless conditions will provide insight into treatments for cancer and other diseases.

Discussion Questions

1. Can you think of any other uses for a space station? *(Answers could include its use as a rest stop for travel further out into space or for manufacturing in a weightless environment. Some materials, such as crystals, have superior quality when produced in a weightless environment. Scientists could also study plant growth in the absence of gravity.)*
2. One intended use of the space shuttle is to transport materials into orbit to build a space station. What advantages and disadvantages will construction workers face in building a space station in a weightless environment? *(Every object will have to be tethered so that it does not float away. The construction will require special tools because workers will be in cumbersome spacesuits. Also, if a worker pushes something, he or she will be pushed in the opposite direction, so he or she will have to be securely anchored. One advantage is that workers could easily move very massive objects like girders.)*

TECHNOLOGY *in* SOCIETY

Stationed in Space

On a June day in 1995, the space shuttle *Atlantis* docked at the Russian space station, *Mir,* and picked up three passengers. These passengers, one from the United States and two from Russia, had completed a 3-month stay at the space station. Their mission was only a prelude for much bigger things to come.

An International Effort

By 2002 the United States, Russia, Canada, Japan, and the European Space Agency intend to build an international space station called *Alpha*. Seventeen countries are currently making parts for the station, which will be assembled in orbit. The purpose of the *Atlantis-Mir* mission, the first in a series of six missions, is to develop construction techniques for assembling *Alpha* in space.

The space station will be made up of cylindrical rooms called *modules,* and *nodes* will connect the modules. Each of these components will be built on the ground and then assembled 274 km above the Earth. The current plan calls for 38 assembly trips. When finished, the space station will include the United States laboratory, the European Space Agency laboratory, Japan's experiment module, and a habitation module for a six-person crew.

Life Aboard

One of the strange things about living in space is the lack of gravity. Unless fastened down, everything inside the space station, including the astronauts, will float! The designers of *Alpha's* habitation module have come up with some intriguing solutions. For example, each astronaut will sleep in a sack similar to a sleeping bag. The sack will keep the astronauts from floating around while they sleep. Astronauts will shower with a hand-held nozzle that squirts water onto their body. After showering, the water droplets will be vacuumed up so that they won't float around. Other problems being studied include how to prepare and serve food, how to design an effective toilet, and how to dispose of waste.

▼ The *Mir* space station

▲ The proposed *Alpha* space station, showing space shuttle linkup

Ready to Go

Workers are already building the United States laboratory module, which is scheduled for launch in 1998. Other modules will be built by different nations until the space station is completed.

Addressing the Gravity of the Situation

Create blueprints for devices that will help the space station crew cope with the lack of gravity. Pick a problem to solve such as brushing teeth, disposing of wastes, getting exercise, and washing hair.

494

CROSS-DISCIPLINARY FOCUS

Social Studies

Lead a class discussion about what laws might govern life in the space station. Who should determine the rules? What laws unique to a space-station environment might be needed?

Addressing the Gravity of the Situation

In creating blueprints for devices, encourage students to be as creative and as detailed as they like. Suggest that they pay close attention to their daily routine and note what devices they use regularly. Have them consider how these devices would work in space.

Weird Science

Holes Where Stars Once Were

In the deep reaches of outer space, an invisible phantom lurks, ready to swallow up everything that comes near it. Once trapped in its grasp, matter is stretched, torn to pieces, and crushed into oblivion. Does this sound like a scene from a horror story? Guess again! Scientists have a name for just such a phenomenon. They call it a black hole.

Born of a Collapsing Star

Certain stars can become black holes. As a star runs out of fuel, it cools and eventually collapses under the force of its own gravity. If the collapsing star is large enough, it may shrink so much and become so dense that it becomes a black hole. One spoonful of the matter in a black hole is so dense that its mass would be greater than the mass of the Earth itself! The resulting gravitational attraction is enormous—so enormous, in fact, that even light cannot escape its effects.

If you could look inside a black hole, scientists predict that at its center you would see a *singularity,* a tiny point of incredible density, temperature, and pressure. The area around the singularity is called the *event horizon*. The event horizon represents the boundary of the black hole. Anything, including light, that crosses the event horizon will eventually be pulled into the black hole.

Searching for the Invisible

Since we can't see black holes, how do we know they exist? Scientists look for the effect of the black hole on the matter around it. As matter comes near the event horizon, the matter begins to swirl around the black hole. The swirling matter forms a huge disk many light-years across called an *accretion disk.* Matter from the center of the disk is pulled into the black hole in much the same way that water swirls down a drain.

The Story of M87

For years scientists had theorized about black holes but hadn't actually found one. Then in 1994 scientists found something strange about 50 million light-years away, at the core of a galaxy called M87. Scientists detected a disk-shaped cloud of gas with a diameter of 60 light-years, rotating at about 2 million kilometers per hour. When you consider that a mass more than

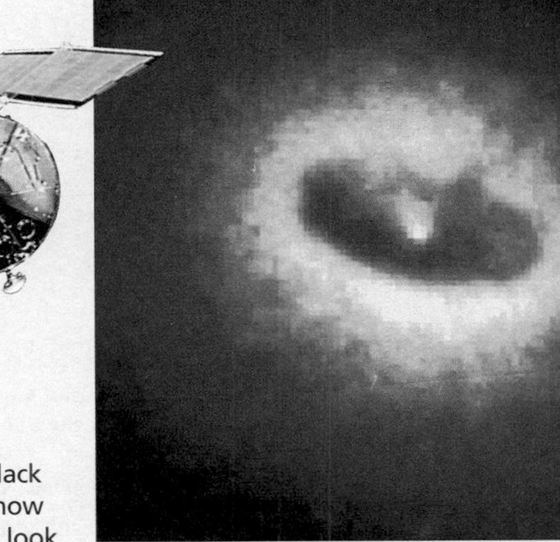

▲ A photograph of M87 taken by the Hubble Space Telescope

2 billion times that of the sun was crammed into a space no bigger than our solar system, it is obvious that something was pulling in the gases at the center of the galaxy.

The black hole in M87 was the first ever discovered. Many astronomers believe that such objects lie at the heart of many galaxies. Some scientists suggest that we have a giant black hole at the center of our own Milky Way. But don't worry. If that's the case, the Earth is too far away to be pulled in.

Modeling a Black Hole

Make a model to show how a black hole pulls in the matter surrounding it. On your model indicate the singularity, event horizon, and accretion disk.

495

Modeling a Black Hole

Encourage students to use a variety of materials and to clearly mark the important parts of the black hole with toothpicks or flags. Students will need to research the terms not included in this feature. *(The event horizon is the boundary of the black hole, and the singularity is the center of the black hole.)*

Weird Science

Background

Scientists believe that two types of stars can turn into black holes. When an extremely large star, about 8 to 25 times as massive as the Sun, runs out of fuel and dies, it usually explodes in a supernova. A star that is more than 25 times as massive as the Sun may collapse without exploding. If the core of either type of star is at least three times more massive than the whole Sun, the core will collapse under its own gravity and become a black hole.

Many scientist also believe that black holes may have formed at the center of many galaxies in the early universe. As a galaxy forms, gases and stars rotate around its center in a swirling mass. Some of this matter may become so closely packed in the center of the forming galaxy that it eventually condenses into a single lump. The gravity of the resulting mass causes it to condense even further until it is so dense that a black hole is formed. These black holes typically are millions or even billions of times as massive as the Sun.

CROSS-DISCIPLINARY FOCUS

Foreign Languages

Each of the following is how the term *black hole* is written in a different language. Have students identify what the languages are: *chyornaya dyra* (CHOR nah yah dih RAH, *Russian*), *trou noir* (troo NWAH, *French*), *Schwarzes Loch* (SHVART ses lock, *German*), and *agujero negro* (ah goo HEH roh NEH gro, *Spanish*). Challenge students to find out the literal English translation for these words. *(They all translate literally into "black hole.")* Then encourage students to share their knowledge of foreign languages by providing translations for other astronomical terms.

SCIENCE & TECHNOLOGY

Background

A radio telescope collects radio waves, just as an optical telescope collects visible light. However, to bring radio waves into sharp focus, a radio telescope must be much larger than an optical telescope because radio waves have much longer wavelengths than do visible light waves.

While some objects that give off relatively few radio waves can be observed better by optical telescopes, objects that produce more radio waves can be observed better by radio telescopes. However, for objects that produce both visible light and radio waves, radio telescopes produce images with a higher resolution than do optical telescopes.

Radio telescopes do have one disadvantage: they can create images from only one wavelength at a time. Using a radio telescope is like looking at an object through a lens that lets you see only green light. If you wanted to see the object in another color, like red or blue, you would have to change the lens. This is exactly what astronomers have to do with radio telescopes.

What Do They See?

Provide students with some popular astronomy books and magazines. You may wish to contact NASA's Space Science Institute, which operates the Hubble Space Telescope. The telescope takes numerous photographs daily that are eventually released to the public.

SCIENCE & TECHNOLOGY

The Ultimate Telescope

▲ The VLA near Datil, New Mexico. Only a few of the 27 radio telescopes of the VLA can be seen in this photograph.

The largest telescope in the world doesn't depend on visible light, lenses, or mirrors. Instead, it collects radio waves from the far reaches of outer space. This radio telescope, called the Very Large Array (VLA), works its marvels in a remote desert in New Mexico.

From Radio Waves to Computer Images

A radio telescope collects the radio waves given off by most objects in space. A bowl-shaped dish called a reflector collects the waves and focuses them onto a small radio antenna hung over the center of the dish. The antenna converts the waves into electric signals. The signals are relayed to a radio receiver, where they are amplified and recorded on tape that can be read by a computer. The computer combines the signals and uses them to create an image of the source of the radio waves.

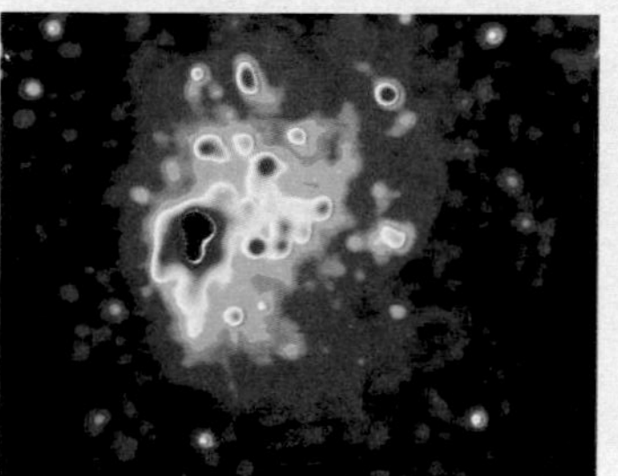

▲ This false-color radio map created using the VLA is of a small, irregular galaxy close to our own.

A Marvel at "Seeing"

Radio telescopes have some distinct advantages over optical telescopes. They can "see" objects that are as far as 13 billion light-years away. They can even detect objects that don't give off any light at all. They can be used in any kind of weather, can receive signals through atmospheric pollution, and can even penetrate the cosmic dust and gas clouds that occupy vast areas of space. However, radio telescopes need to be large in order to be accurate.

Telescope Teamwork

The VLA is an array of 27 separate radio telescopes mounted on railroad tracks and linked by computers. Each of the 27 reflectors is 25 m in diameter. When they operate together, they work like a single telescope with a diameter of 47 km!

Radio waves from all 27 dishes are amplified a million times by receivers, and then they are relayed to a control room at the center of the VLA. Here, computers turn the waves into images that are displayed on a TV monitor. Using the VLA, astronomers have been able to explore distant galaxies, pulsars, quasars, and possible black holes.

A system of telescopes even larger than the VLA has been used. In the Very Long Baseline Array (VLBA), radio telescopes in different parts of the world all work together. The effect is a telescope that is almost as large as the Earth itself!

What Do They See?

Find out about some of the objects "seen" by the VLA, such as pulsars, quasars, and possible black holes. Prepare a report or create a model of one of the objects and make a presentation to your class. Use diagrams and photographs to make your presentation more interesting.

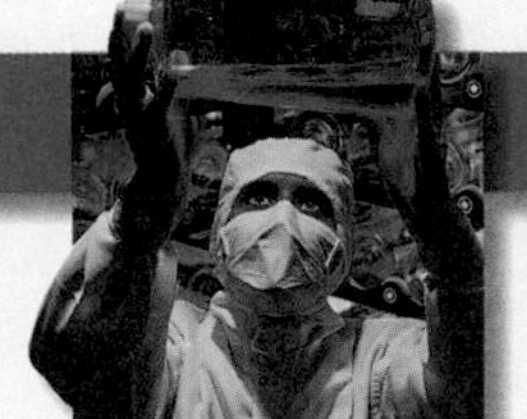

SCIENCE in ACTION

New Frontiers for Geophysicists

Bob Grimm may have one of the most interesting jobs in our solar system. He is one of a few scientists working in the relatively new field of planetary geophysics. His research takes him off the Earth and out to the other planets.

Why Bother?

Geophysicists use the science of physics to study the Earth and its atmosphere. They also try to answer questions about the origin and history of the Earth. Planetary geophysicists study the geophysics of other planets. Bob Grimm explains the importance of his work: "It is only by questioning that we will learn, and I feel that by asking questions about outer space, we will learn more about our own world."

Probing the Planets

Grimm studies information about the planets that has been sent to Earth by planetary probes like *Magellan, Voyager 2*, and *Galileo*. In order to piece together what the surface of a planet looks like, he studies its geologic structures. Using computer models and mathematical equations, he analyzes these surface structures in order to understand how they formed.

Grimm also programs and uses sophisticated computers that enable him to view a planet's surface from many different angles. "I am the end user of robot spacecraft that have gone out on missions and sent back photographs of the planets."

▲ Bob Grimm (right) studies planetary pictures taken by spacecraft launched from Earth.

The Earth Connection

Planetary geophysicists relate what they see happening on other planets to the Earth. "We try to see the evolution of Earth in a bigger context—where it came from and where it's going," Grimm says. "The information we obtain . . . can give us valuable clues to the way our own planet works."

The field of planetary geophysics has been a rewarding one for Bob. "The sense of exploration . . . really appeals to me. Every day, by looking at new pictures of planets, I do what no one has ever done before. It's like a hunt—I try to figure something out to bring some relationships together, and soon I have a story to tell!"

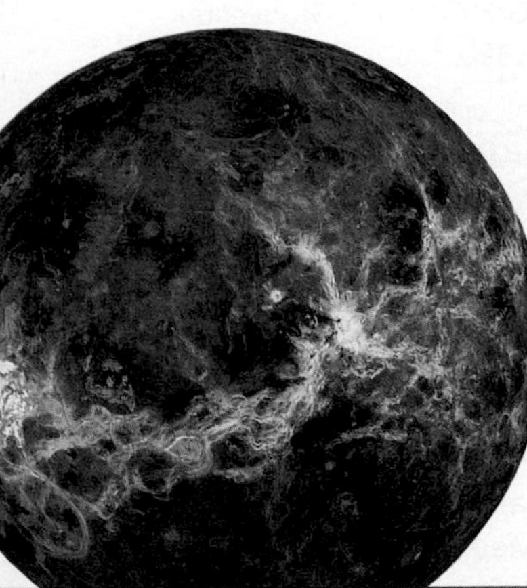

◀ This image of Venus was created using signals transmitted by the *Magellan* spacecraft. The *Magellan* spacecraft orbited Venus for 4 years, until October 12, 1994, when its operations ended.

A Project Idea

Try a planetary-mapping experiment. Observe the full Moon and make a detailed sketch of it. Locate craters and distinguish light areas from dark areas. Then compare your sketch to a map of the Moon. Find out what the light and dark areas are. What does the number of craters tell you about the geologic history of the Moon and planets?

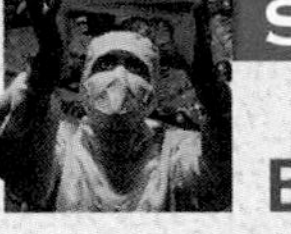

SCIENCE in ACTION

Background

Geophysics includes such fields as seismology, meteorology, and hydrology. Geophysicists study many topics, from the shape of the Earth's surface to the processes at work deep in the Earth's interior. They scrutinize the Earth's structure, temperature variations, climatic patterns, volcanic activity, gravitational field, and magnetic field.

Integrating the Sciences

Earth, Life, and Physical Sciences

Students may enjoy analyzing satellite photographs of Earth, especially those taken by a Landsat satellite. Suggest that they identify rivers, mountains, forests, and other geological features, as well as cities, highways, and other human-made structures.

Discussion Questions

1. Would the constellations as we see them on Earth look different from another planet in our solar system? *(Not really; the stars are too far away for the difference in positions to have any observable effect.)*
2. How can knowledge of other planets help us understand our own planet? *(Planetary geophysicists can study processes occurring on other planets that would be difficult if not impossible to reproduce on Earth. Examples include strong gravitational fields, extremely high or low temperatures, and violent atmospheric patterns.)*

Answers to *A Project Idea*

The dark areas are called *maria* (singular, *mare*), which is Latin for "seas." They look like bodies of water but are really very old basalt flows from early volcanic activity. The lighter areas are called the *highlands* of the Moon. The presence of minerals such as aluminum makes these areas lighter than the maria. The maria are smooth because lava has covered many of the craters. The number of craters indicates that the Moon has been bombarded by meteors many times.

You may wish to relate the following information about planetary histories: Mercury has been bombarded by meteors many times. Earth, Venus, and Mars have atmospheres that protect them from many meteors. Craters cover more of the surface of Callisto, a satellite of Jupiter, than that of Earth's satellite, the Moon.

GROWING PLANTS

Unit Overview

In this unit, students examine the relationship between the environment and plant life. In Chapter 23, students consider proposals for the construction of a biosphere on Mars, design their own procedures for testing the germination rate of seeds, and explore the characteristics of soil. In Chapter 24, students examine the interaction of plants and water, transpiration, and plant nutrition. The chapter continues with an examination of the anatomy and life cycle of flowering plants. In Chapter 25, students evaluate the use of pesticides on plants and consider safe alternatives. Then students complete a project for the growth and use of plants in interior decoration, outdoor landscaping, or vegetable gardening.

Using the Themes

The unifying themes emphasized in this unit are **Structures, Changes Over Time,** and **Cycles.** The following information will help you incorporate these themes into your teaching plan. Focus questions that correspond to these themes appear in the margins of this Annotated Teacher's Edition on pages 508, 522, 532, and 550.

As students become familiar with the anatomy of plants, the theme of **Structures** can be a very useful teaching tool. Specifically, in Chapter 24, a focus question encourages students to consider the types of structures within a plant that are involved in water transport.

Later in the unit, the theme of **Changes Over Time** also emerges. This theme can be discussed both in terms of the changes that take place during the life cycle of a plant and in terms of the changes that many plants can effect on the process of soil formation.

Finally, discussions of a plant's life cycle evoke the theme of **Cycles.** In Chapter 25, a focus question challenges students to consider the role of plants in natural cycles such as water, atmospheric, and nutrient cycles.

Using the SourceBook

Unit 8 emphasizes the internal structure of plants. It also gives students a closer look at plant growth. Finally, the unit explores the various ways that humans use plants—for food, clothing, and medicine.

Bibliography for Teachers

The Big Book of Gardening Skills. Pownal, VT: Storey Communications, Inc., 1993.

Clarke, Graham, and Alan Toogood. *The Complete Book of Plant Propagation.* London, England: Ward Lock Limited, 1992.

Huxley, Anthony. *Green Inheritance: The World Wildlife Fund Book of Plants.* New York City, NY: Four Walls Eight Windows, 1992.

Bibliography for Students

Adair, Gene. *George Washington Carver.* New York City, NY: Chelsea House Publishers, 1989.

Handelsman, Judith F. *Gardens From Garbage: How to Grow Plants From Recycled Kitchen Scraps.* Brookfield, CT: The Millbrook Press, 1993.

Stidworthy, John. *Plants and Seeds.* New York City, NY: Gloucester Press, 1990.

Films, Videotapes, Software, and Other Media

Encyclopedia of Landscape Plants
Videodisc
Videodiscovery
1700 Westlake Ave. N, Suite 600
Seattle, WA 98109-3012

Living Trees/The Living Soil
Videotape
AIMS Media
9710 DeSoto Ave.
Chatsworth, CA 91311-4409

Plant Reproduction
Videotape
Britannica
310 S. Michigan Ave.
Chicago, IL 60604-9839

Adaptations of Plants
Film or videotape
Coronet/MTI
108 Wilmot Rd.
Deerfield, IL 60015

Botanical Gardens
Software (Apple II family or Macintosh)
Sunburst Communications
101 Castleton St.
P.O. Box 100
Pleasantville, NY 10570

Unit Organizer

Unit/Chapter	Lesson	Time*	Objectives	Teaching Resources
Unit Opener, p. 498				Science Sleuths: Green Thumb Plant Rentals #2 English/Spanish Audiocassettes Home Connection, p. 1
Chapter 23, p. 500	**Lesson 1, A Garden in Space, p. 501**	1	1. Explain what a biosphere is. 2. Describe what a biosphere on Mars might be like. 3. Identify the basic needs necessary for the survival of living things.	none
	Lesson 2, The Beginning of the Garden, p. 503	2	1. Explain why growing plants are fundamental to living in a biosphere. 2. Describe how to conduct a test to compare seed germination rates. 3. Describe the growth of a plant from a seed to maturity.	Image and Activity Bank 23-2 Exploration Worksheet, p. 3 Transparency 81 Exploration Worksheet, p. 4
	Lesson 3, Breaking Ground, p. 507	4	1. Identify different types of soil and the characteristics of each. 2. Investigate the water-holding capacity and percolation rate of soil. 3. Identify the factors needed for plant growth. 4. Design and perform an experiment to investigate plant growth.	Image and Activity Bank 23-3 Transparency Worksheet, p. 6 ▼ Theme Worksheet, p. 8 Exploration Worksheet, p. 10 Exploration Worksheet, p. 11 ▼ Exploration Worksheet, p. 13 ▼
End of Chapter, p. 515				Chapter 23 Review Worksheet, p. 14 ▼ Chapter 23 Assessment Worksheet, p. 16
Chapter 24, p. 517	**Lesson 1, Water, Water, Everywhere! p. 518**	3	1. Describe some interactions between water and plants. 2. Identify the relationship between the structure and function of leaves and roots. 3. Explain the process of osmosis. 4. Explain the importance of leaves in plant transpiration.	Image and Activity Bank 24-1 Transparency Worksheet, p. 18 ▼ Transparency Worksheet, p. 20 ▼ Exploration Worksheet, p. 22 Exploration Worksheet, p. 28
	Lesson 2, Giving Plants a Hand, p. 524	2	1. Identify the three main nutrients found in most fertilizers, and explain the purpose of each. 2. Explain what soil pH is and why it is important to plant growth. 3. Explain how plants can be grown in nutrient solutions.	Discrepant Event Worksheet, p. 31 Graphing Practice Worksheet, p. 32 Exploration Worksheet, p. 34 Exploration Worksheet, p. 35
	Lesson 3, Flowers and Pollination, p. 528	2 to 3	1. Identify the parts of a flower. 2. Describe the process of pollination. 3. Describe the life cycle of a flowering plant. 4. Explain the difference between self- and cross-pollination. 5. Explain how cross-pollination is used as a plant-breeding technique.	Image and Activity Bank 24-3 Exploration Worksheet, p. 38 ▼ Exploration Worksheet, p. 40 Activity Worksheet, p. 42
	Lesson 4, Plants From Plant Parts, p. 534	2	1. Identify several methods of vegetative reproduction. 2. Describe and use vegetative-reproduction techniques to grow new plants. 3. Explain how roots increase in length and penetrate the soil.	Image and Activity Bank 24-4 Exploration Worksheet, p. 43 Exploration Worksheet, p. 44
End of Chapter, p. 538				Activity Worksheet, p. 45 Chapter 24 Review Worksheet, p. 46 Chapter 24 Assessment Worksheet, p. 49
Chapter 25, p. 540	**Lesson 1, Medicine or Poison? p. 541**	1	1. Identify several poisonous pesticides and their effects on people and animals. 2. Suggest safe alternatives for some commonly used pesticides. 3. Evaluate the positive and negative aspects of using poisonous pesticides or safe alternatives.	Image and Activity Bank 25-1 Exploration Worksheet, p. 51
	Lesson 2, Make a Green World! p. 544	6	1. Describe the best conditions for growing specific kinds of plants. 2. Describe how to grow and use plants in indoor or outdoor environments. 3. Analyze the environment in order to determine the kinds of plants that should grow best in it.	Exploration Worksheet, p. 52 Exploration Worksheet, p. 54 Exploration Worksheet, p. 56
End of Chapter, p. 552				Activity Worksheet, p. 58 ▼ Chapter 25 Review Worksheet, p. 60 Chapter 25 Assessment Worksheet, p. 62
End of Unit, p. 554				Unit 8 Activity Worksheet, p. 65 Unit 8 Review Worksheet, p. 66 Unit 8 End-of-Unit Assessment Worksheet, p. 69 Unit 8 Activity Assessment, p. 75 Unit 8 Self-Evaluation of Achievement, p. 78

* Estimated time is given in number of 50-minute class periods. Actual time may vary depending on period length and individual class characteristics

▼ Transparencies are available to accompany these worksheets. Please refer to the Teaching Transparencies Cross-Reference chart in the Unit 8 Teaching Resources booklet.

Materials Organizer

Chapter	Page	Activity and Materials per Student Group
23	504	**Exploration 1:** package or collection of seeds from fruits, vegetables, flowers, etc.; various materials to germinate seeds, such as moist paper towel, jar, jarful of potting soil, etc.
	505	**Exploration 2:** germinated seeds from Exploration 1; about 200 mL of potting soil; small container for planting, such as small cup, pot, or milk carton; metric ruler; metric scale; about 50 mL of water; optional materials: fertilizer
	508	**Exploration 3:** 500 mL of soil from vacant lot, garden, or student's backyard; magnifying glass; about 50 mL of water; several sheets of newspaper
	509	**Exploration 4:** jar, at least 5 cm in diameter and 10 cm high, with lid; about 100 mL of water; 200 mL of soil sample from Exploration 3 or from commercial source; magnifying glass; spoon
	510	**Exploration 5:** 2 large, plastic-foam cups; 2 or 3 paper towels; 500 mL glass, beaker, or wide-mouth bottle; about 500 mL of soil sample from Exploration 3; about 500 mL of sand; stopwatch or clock; 500 mL of water; 250 mL beaker; 100 mL graduated cylinder; several 2 cm $\times$ 5 cm pieces of cardboard
	512	**Exploration 6:** a few seedlings from Exploration 1; various materials for transplanting and testing plant growth, such as several large plastic-foam cups, several cupfuls of garden soil, about 1 L of water, spoon or craft stick, etc. (See Advance Preparation below.)
24	518	***Exploration 1, Part 1:** a few germinated seeds; magnifying glass; **Part 2:** two 500 mL beakers or containers; 1 L of water; at least 20 g of salt; several pairs of identical plant pieces, such as 5 mm thick carrot slices, fern fronds, lettuce leaves, and geranium leaves; **Part 3:** none; **Part 4:** 15 cm stalk of celery, cut at both ends; a few drops of blue food coloring; clear glass or 250 mL beaker; about 100 mL of water; **Part 5:** geranium or other plant with large leaves; clear plastic sandwich bag; small piece of masking tape
	522	**Exploration 2, Part 1:** variety of leaves; magnifying glass; **Part 2:** glass slide with coverslip; microscope; eyedropper; about 5 mL of water; large leaf such as geranium or lettuce
	525	**Exploration 3:** about 50 mL distilled water; several 20 mL soil samples from different areas; 1 piece each of blue and red litmus paper for each sample; small paper plate
	526	**Exploration 4:** 200 mL of rainwater or distilled water; solute, such as small packet of hydroponic nutrient powder, at least 2 mL vinegar, few spoonfuls of gardener's lime, etc.; 250 mL beaker or plastic container; baby-food jar or clear plastic cup; about 100 mL vermiculite or clean sand; 10 mustard seeds untreated with fertilizer or fungicide; eyedropper; metric ruler; latex gloves; lab aprons; safety goggles (See Advance Preparation below.)
	530	**Exploration 5:** several different cut flowers; single-edged razor blade in holder or sharp knife
	535	**Exploration 7:** several mature plants, such as impatiens, coleus, geranium, Swedish ivy, snake plant, African violet, or begonia; about 1 L of water; scissors; 1–2 L of potting mix with equal volumes of perlite and vermiculite; large, shallow container such as baking pan; several large sheets of clear plastic wrap; 1–2 L of general-purpose soil mixture; large, deep container such as plastic garden pot; metric ruler
	536	**Exploration 8:** several sprouted mung beans, radish seeds, or mustard seeds untreated with fertilizer or fungicide; fine-tipped waterproof marker; metric ruler; several paper towels; about 100 mL of water
25	541	**Exploration 1:** books, magazines, and other resource materials on gardening; glue or tape
	545	**Exploration 2:** graph paper; pencil; metric ruler
	546	**Exploration 3:** tracing paper; pencil
	548	**Exploration 5:** various items to begin projects, such as graph paper, pencil, tracing paper, and a metric ruler

* You may wish to set up these activities in stations at different locations around the classroom.

Advance Preparation

Several of the exercises in this unit require plants that have been growing for certain lengths of time. It may be helpful to begin growing these plants in advance. For more details, read the following information about specific Explorations:

Exploration 2, page 505: Save the germinated seeds from Exploration 1, page 504, in order to carry out Exploration 2.

Exploration 4, page 509, and Exploration 5, page 510: Set aside soil remaining from Exploration 3, page 508, in order to carry out Explorations 4 and 5. Make sure the soil sample used for Exploration 5 has had a chance to dry.

Exploration 6, page 512: Four to six weeks should be allowed for sufficient plant growth to occur.

Exploration 1, page 518: Germinated seeds are required for this Exploration.

Exploration 4, page 526: Each student group will need to choose one solute to test. It may be helpful to have the groups make their choice in advance so that the necessary materials can be obtained.

Exploration 8, page 536: Sprouted mung beans or mustard seeds are required for this Exploration.

Unit Compression

There are two primary objectives for this unit. First, students should be able to design and carry out original scientific investigations about plant function and growth. Second, students should develop their skills by applying scientific knowledge of plants to a design-intensive activity.

To accomplish these ends, you may find that some activities are more pertinent than others. For example, Exploration 6 on page 512 should be considered essential for achieving the first objective. Similarly, Exploration 5 on page 548 is indispensible for achieving the second objective. Depending on time constraints and individual class characteristics, you may wish to focus student attention only on activities in the unit that you feel are essential for achieving these goals.

Homework Options

Chapter 23
See Teacher's Edition margin, pp. 502, 504, 508, 510, 512, and 513
Theme Worksheet, p. 8
SourceBook, pp. S170 and S171

Chapter 24
See Teacher's Edition margin, pp. 519, 522, 524, 525, 529, 532, 533, 536, and 539
Graphing Practice Worksheet, p. 32
Activity Worksheet, p. 42
Activity Worksheet, p. 45
SourceBook, pp. S167, S168, and S169

Chapter 25
See Teacher's Edition margin, pp. 542, 545, 548, and 552
Exploration Worksheet, p. 52
Activity Worksheet, p. 58
SourceBook, p. S174

Unit 8
Unit 8 Activity Worksheet, p. 65
SourceBook Activity Worksheet, p. 79

Assessment Planning Guide

Lesson, Chapter, and Unit Assessment	SourceBook Assessment	Ongoing and Activity Assessment	Portfolio and Student-Centered Assessment
Lesson Assessment Follow-Up: see Teacher's Edition margin, pp. 502, 506, 514, 523, 527, 533, 537, 543, and 551 **Chapter Assessment** Chapter 23 Review Worksheet, p. 14 Chapter 23 Assessment Worksheet, p. 16* Chapter 24 Review Worksheet, p. 46 Chapter 24 Assessment Worksheet, p. 49* Chapter 25 Review Worksheet, p. 60 Chapter 25 Assessment Worksheet, p. 62* **Unit Assessment** Unit 8 Review Worksheet, p. 66 Unit 8 End-of-Unit Assessment Worksheet, p. 69*	SourceBook Review Worksheet, p. 82 SourceBook Assessment Worksheet, p. 86*	Unit 8 Activity Assessment Worksheet, p. 75* **SnackDisc** Ongoing Assessment Checklists ♦ Teacher Evaluation Checklists ♦ Progress Reports ♦	Portfolio: see Teacher's Edition margin, pp. 504, 506, 512, 522, 530, 531, and 542 **SnackDisc** Self-Evaluation Checklists ♦ Peer Evaluation Checklists ♦ Group Evaluation Checklists ♦ Portfolio Evaluation Checklists ♦

* Also available on the Test Generator software
♦ Also available in the Assessment Checklists and Rubrics booklet

Science Discovery is a versatile videodisc program that provides a vast array of photos, graphics, motion sequences, and activities for you to introduce into your *SciencePlus* classroom. *Science Discovery* consists of two videodiscs: Science Sleuths and the Image and Activity Bank.

Using the *Science Discovery* Videodiscs

Science Sleuths: Green Thumb Plant Rentals #2 Side B

The plants from Green Thumb Incorporated (GTI), a plant-rental company, are dying at Varisystems (VSI). The owner of GTI suspects a rival company of sabotage. The Science Sleuths must evaluate the evidence to determine why the plants are dying.

Interviews

1. Setting the scene: Vice-president of GTI **12807 (play ×2)**

2. Plant and soil technician for GTI **13691 (play)**

3. Office manager for VSI **14718 (play)**

4. Custodial Engineer for VSI **15525 (play)**

5. Botanist **16110 (play)**

6. The Galloping Gardener **16400 (play)**

Documents

7. GTI pamphlet **17316 (step ×3)**

8. Letter from GTI **17321 (step)**

9. Soil-preparation checklist **17324**

10. VSI file **17326 (step ×4)**

11. Soil chart **17332**

Literature Search

12. Search on the words: APHIDS, GTI, PLANT DISEASE, PLANTS, VARISYSTEMS, WATER **17334**

13. Article #1 ("Farmers Protest in Favor of Insecticide") **17336 (step)**

14. Article #2 ("Some Greens to Get the Blues Out") **17339 (step ×2)**

15. Article #3 ("Caring for Houseplants") **17343 (step ×2)**

Sleuth Information Service

16. Transpiration **17347 (step)**

17. Potted-plant soil **17350 (step)**

18. Herbicide **17353 (step)**

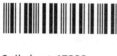

19. Climate chart for building **17356**

Sleuth Lab Tests

20. Overwatering and under-watering tests **17358 (step ×12)**

21. Tap-water analysis **17372**

22. Chemical analysis of soil composition **17374**

23. Water pH **17376**

24. Soil pH **17378 (step)**

Still Photographs

25. Plant at VSI **17380**

26. Plant at GTI greenhouse **17382**

27. VSI plant leaves **17384**

28. GTI plant leaves **17386**

29. VSI plant roots **17388**

30. GTI plant roots **17390**

31. VSI plant roots—microscopic view **17392**

32. GTI plant roots—microscopic view **17394**

33. VSI plant stem **17396**

34. GTI plant stem **17398**

35. Agar plates of VSI and GTI plant soil **17400 (step)**

36. Plants growing at VSI **17404 (step)**

◀| Step Reverse   Play ▶   Pause || Step Forward |▶

Image and Activity Bank
Side A or B

A selection of still images, short videos, and activities is available for you to use as you teach this unit. For a larger selection and detailed instructions, see the Videodisc Resources booklet included with the Teaching Resources materials.

23-2 The Beginning of the Garden, page 503

Seed germination 37899–38604 (Side B only) (play ×2)
Seeds germinate when sufficient moisture and oxygen and an appropriate temperature are present.

Seed germination; lab setup 389–424 (step ×35)
These experiments demonstrate germination under different sets of conditions. Experimental conditions include the following: light vs. no light, water vs. no water, heat vs. no heat, soil vs. no soil, and space vs. no space.

23-3 Breaking Ground, page 507

Soil formation; layering 591
Soil builds up layer by layer.

Plant health 3075–3076 (step)
Conditions that affect growth (step) Environmental causes of poor plant health

Plant health; levels of light 3078–3081 (step ×3)
Low light (step) Medium light (step) High light (step) Very high light

24-1 Water, Water, Everywhere! page 518

Osmosis experiment 2523
When celery is placed in salt water, it loses water, causing it to wilt. At the same time, the salt solution is slightly diluted by water from the celery.

Movement into and out of cells; lab setup 340–343 (step ×3)
An aerosol can diffuses scent and propellant throughout a room. (step) A celery stalk in red food coloring demonstrates water moving into and out of cells. (step) Use these materials for an experiment involving an apple in salt water and fresh water. (step) Place slices of apple in fresh water and in salt water. What happens to each slice?

Plant structure; effects of water 3070–3072 (step ×2)
Loss of rigidity in a dry plant (step) Recovery from watering (step) The movement of water through a plant stem

Water and plants; lab setup 308
This setup demonstrates the capillary action of plants. The four strips (polyester, cotton, linen, and wool) have different rates of water movement.

Plant structure 3064–3065 (step)
Water moves from the roots up the plant by diffusion. (step) Diffusion through root hairs

Water and plants 309
Placing plastic bags on a plant demonstrates transpiration.

Photosynthesis equation 3060
Plants use light as an energy source to convert carbon dioxide and water into sugar and oxygen. Sugars can be used immediately or can be converted into starch or cellulose.

Photosynthesis; oxygen production by plants 38606–38959 (Side B only) (play ×2)
Oxygen gas released by an elodea plant during photosynthesis

Photosynthesis; compared with aerobic cellular respiration 3059
Chemical differences between photosynthesis and aerobic cellular respiration

Plant structure; stoma 3063
Structure of a stoma (breathing pore)

24-3 Flowers and Pollination, page 528

Butterfly, copper 3440
Butterflies feed on nectar that they suck up with their tongue. Butterflies are important pollinators of long, thin, tuberous flowers.

Honeybee 3459
The sterile female worker bees shown here gather nectar and pollen that they carry back to their hives to feed to the queen and her workers. The nectar is stored as honey in wax honeycombs. Many plants rely on bees for pollination.

Honeybee; dance 51738–52188 (Side B only) (play ×2)
Bees follow aerial scent trails to and from flowers. When a bee finds a new source of nectar, it returns to the hive and performs a coded dance that conveys the direction and distance of the nectar source to the other bees.

Flower; diagram of parts 3047–3049 (step ×2)
Cross section of a flower (step) Structures of a flower (unlabeled) (step) Structures of a flower (labeled)

Pollen; microscopic sample from gerbera daisy 3015–3016 (step)
Fertilization occurs when two half-sets of chromosomes are combined, one from a male organism and one from a female organism. In flowers, the male half of the chromosomes is carried by pollen. (step) Pollen is covered by a pitted shell to protect it while traveling through the air (by wind or insect) on its way to another flower. The pits also help it to adhere to insects.

Flowering plant; life cycle 3050–3055 (step ×5)
Seed, seedling, buds, and flowers

Cross-pollination 4234
Cross-pollination fertilizes plants with another's pollen.

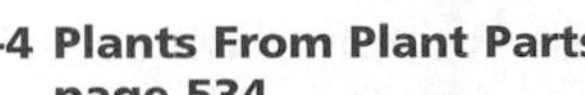

24-4 Plants From Plant Parts, page 534

Plant structure; root cap 3073
Root growth

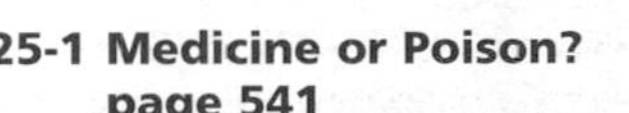

25-1 Medicine or Poison? page 541

Crop-duster 2606
A crop-duster spraying corn with pesticides

Beetle, Japanese 3419
Japanese beetles are serious crop pests, feeding on 200 kinds of plants. They were accidentally introduced in 1916 on iris roots. They have been partially controlled by parasitic wasps and flies that lay eggs on the beetles' larvae.

Beetle, fiery searcher ground; eating hornworm 3415
Fiery searcher ground beetles hunt caterpillars at night, keeping down the population of crop-destroying moths. If the crops are sprayed with insecticide, the beetles die with the moths, and when the moths return, there are no beetles to eat them.

Cockroach 3353
Cockroaches do not transmit human diseases. Some can survive for months by eating dust. They have been poisoned so much that many are becoming immune to pesticides. Cockroaches have not changed much in the last 250 million years.

◀| Step Reverse Play ▶ Pause || Step Forward |▶

UNIT FOCUS

Have students name as many different kinds of plants as they can. Keep track of their suggestions on the chalkboard. Then ask students to suggest some ways in which the plants might be classified. *(Categories might include trees, shrubs, flowering plants, fruits, and vegetables.)* Encourage students to discuss how these different kinds of plants contribute to their lives. *(Responses might include food, shade, beauty, flowers, building materials, and medicines.)* Ask students if they have ever grown plants. Call on volunteers to share their plant-growing experiences with the class.

A good motivating activity is to let students listen to the English/Spanish Audiocassettes as an introduction to the unit. Also, begin the unit by giving Spanish-speaking students a copy of the Spanish Glossary from the Unit 8 Teaching Resources booklet.

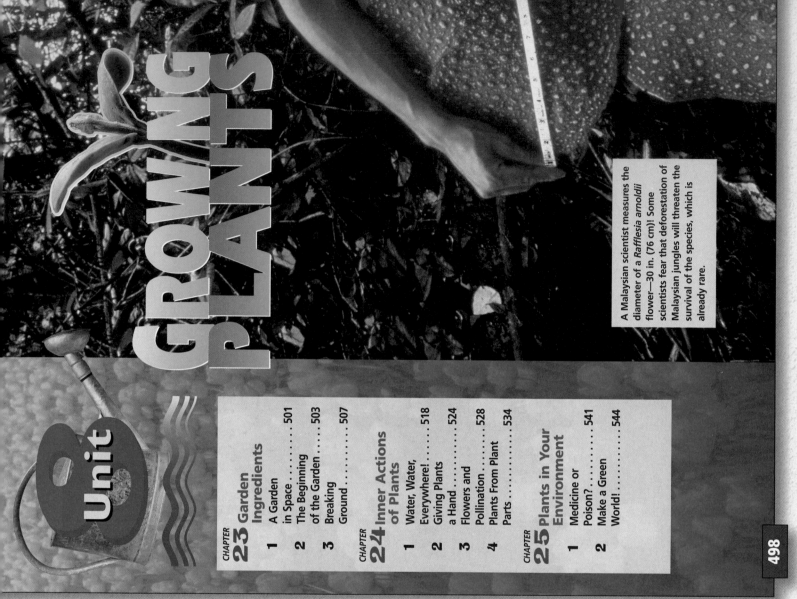

Unit 8 GROWING PLANTS

A Malaysian scientist measures the diameter of a *Rafflesia arnoldii* flower—30 in. (76 cm)! Some scientists fear that deforestation of Malaysian jungles will threaten the survival of the species, which is already rare.

Connecting to Other Units

This table will help you integrate topics covered in this unit with topics covered in other units.

Unit 1 Interactions	Human activities such as landscaping and gardening have profound effects on our environment.
Unit 2 Diversity of Living Things	Just like animal species, plant species evolve through the process of natural selection.
Unit 3 Solutions	Plants are dependent on the dissolved nutrients that they obtain from water in the soil.

In a Malaysian jungle, a gigantic flower blooms, emitting an odor similar to that of rotting flesh. The rare plant *Rafflesia arnoldii* produces this flower, which is the largest in the world—up to 1 m across with a mass of 7 kg.

The rafflesia has no roots or stems and does not perform photosynthesis. Instead, the parasitic plant feeds off of its host, a wild grapevine. Scientists don't know exactly how the rafflesia is planted; one theory is that the seeds are lodged in the vine by animals that have eaten or stepped on the rafflesia fruit. Under the right conditions, a rafflesia bud emerges from the vine after 18 months. It grows there for another 9 months before finally opening into the huge, stinking blossom. The smell attracts carrion flies. What do you think the flies do for the flowers? Ⓐ

After just 4 days, the bloom wilts into a slimy mess. If the flower was female and was pollinated, thousands of seeds will emerge months later from a half-rotten fruit. Some of these seeds may one day produce more of the mysterious blossoms. As unusual as this species is, *Rafflesia arnoldii* shares many characteristics with other plants. You'll learn about these characteristics in this unit. Turn the page and watch your knowledge about plants grow!

499

Using the Photograph

Have students make a diagram of this flower in their ScienceLog. Then have them label the parts of the flower that they can identify. If students do not know the names of certain flower parts, encourage them to think of a name that makes sense to them.

Later, as students progress through the unit, have them refer to their diagram and make changes or additions to it based on their new knowledge. Ask: What does it mean for a flower to be female? *(The flower has only female reproductive parts.)*

Answer to *In-Text Question*

Ⓐ Students should recognize that the flies probably help to pollinate the flower, much like bees, butterflies, and wasps help to pollinate other types of flowers.

Did You Know . . .

Rafflesia arnoldii is named after Sir Thomas Stamford Raffles (a British statesman and explorer and the founder of Singapore) and his companion Dr. Joseph Arnold (a surgeon and naturalist), who documented the flower's existence in 1818.

Connecting to Other Units, *continued*

Unit 5 Structures and Design	The unique structures of roots, stems, and leaves allow them to perform specialized functions.
Unit 6 The Restless Earth	Geologic processes are responsible for many of the different soil types found around the world.
Unit 7 Toward the Stars	An understanding of how plants live and reproduce is essential for the establishment of a space colony.

Garden Ingredients

Connecting to Other Chapters

Chapter 23
introduces students to plants and begins to explore the needs of plants, as well as their role in our lives.

Chapter 24
discusses plants in more detail, including the biotic and abiotic factors necessary for a plant to survive.

Chapter 25
introduces students to the use of plants in both agriculture and landscaping.

Prior Knowledge and Misconceptions

Your students' responses to the ScienceLog questions on this page will reveal the kind of information—and misinformation—they bring to this chapter. Use what you find out about your students' knowledge to choose which chapter concepts and activities to emphasize in your teaching.

In addition to having students answer the questions on this page, you might want to have students generate their own questions about the survival needs of plants, including the need for different types of soil. Their questions might address topics that they are already familiar with or topics that they want to learn more about. Also, you might have them choose one of their questions to pose to the class. Emphasize that there are no right or wrong answers to this exercise. Collect the students' questions and read them to find out what students know and what misconceptions they may have about the needs of plants. After students have studied this chapter, you may wish to have them return to the questions they generated here and provide answers for them.

CHAPTER 23

Garden Ingredients

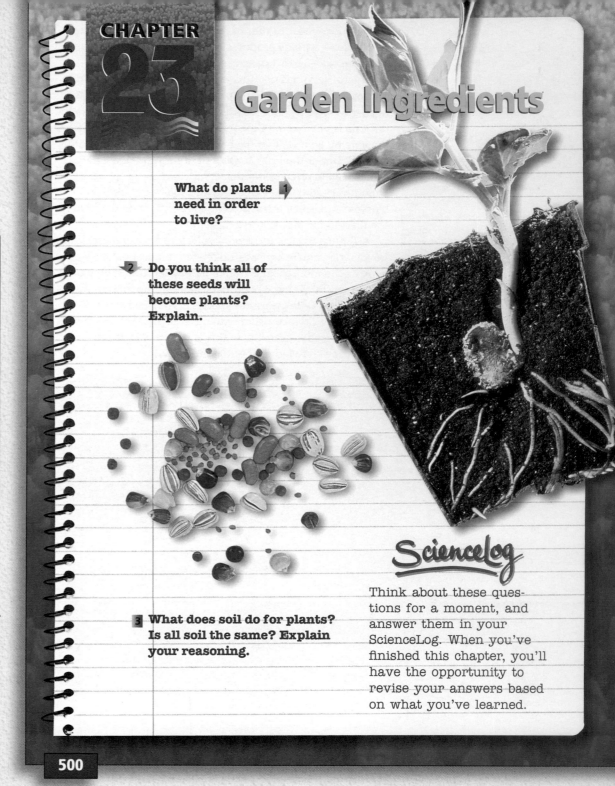

What do plants need in order to live? 1

2 **Do you think all of these seeds will become plants? Explain.**

3 **What does soil do for plants? Is all soil the same? Explain your reasoning.**

ScienceLog

Think about these questions for a moment, and answer them in your ScienceLog. When you've finished this chapter, you'll have the opportunity to revise your answers based on what you've learned.

A Garden in Space

By A.D. 2140, fusion-powered rockets had made it technically possible to colonize space. The expansion of Earth's population to nearly 10 billion people made emigration to a space colony an attractive possibility.

For several decades, experiments were conducted to determine how to sustain life in space. Some people favored the establishment of orbiting space colonies. The colonies would be close enough to Earth to permit regular visits between space-dwellers and their friends and relatives on Earth.

Other people favored the establishment of colonies on Mars. Travel time to Mars had been reduced, at the closest approach of Mars to Earth, to less than a month.

On January 23, 2152, the secretary-general of the United Nations, Elena Kwame Tao, announced her administration's plans to establish a colony on Mars. The announcement read:

Proposals Invited

During the next 15 years, a vast network of fully enclosed biospheres will be built on Mars. Groups interested in establishing a colony are invited to apply to the Working Committee for the Colonization of Mars, General Assembly of the United Nations.

Proposals to the Working Committee must incorporate detailed plans for the establishment of the colony. This includes the design and contents of the biosphere that will house the colony.

Could this story really happen someday, or is it only science fiction? The illustration on page 502 shows what such a man-made **biosphere** might look like. The cutaway view allows you to see the areas inside the structure. A biosphere is an area or region where conditions are suitable for living things to survive. What must all living things, including plants, have for survival, regardless of where they are? In your ScienceLog, make a list of what you think is essential. Ⓐ

LESSON 1 ORGANIZER

Time Required
1 class period

Process Skills
analyzing, organizing, communicating

New Term
Biosphere—an area or region where conditions are suitable for living things to survive

Materials (per student group)
none

Teaching Resources
SourceBook, p. S171

A Garden in Space

FOCUS

Getting Started

Ask: Why can't people live on Mars? *(Responses could include lack of water and air, dust storms, extremes of temperature, and lack of food.)* Why can't plants grow on Mars? *(Students should recognize that plants need many of the same things that people do.)* Why would the first settlers on Mars want to grow plants? *(Accept all reasonable responses, but stress the importance of food and oxygen production by plants.)*

Main Ideas

1. All living things have basic needs that are necessary for their survival.
2. In order for people to colonize other planets, an environment similar to the Earth's must be established.

TEACHING STRATEGIES

GUIDED PRACTICE Direct students' attention to the illustrations on pages 501 and 502. Encourage them to discuss what is shown.

Students may be interested to learn that Biosphere 2, a structure similar to the one pictured on page 502, was built near Oracle, Arizona. From September 1991 through September 1993, eight people lived in the biosphere. Their assignment was to live in it for 2 years to determine if a similar biosphere could be used as a model for a space colony on Mars.

Answers to *In-Text Questions*

Ⓐ Students should recognize that most living things need food, air, light, and water in order to survive.

Think About It!

Divide the class into groups. Have students brainstorm answers to the questions and work together to compose reports to share with the class. You may wish to have them supplement their ideas with further research. Student responses will vary but should reflect an understanding of the basic needs of living things and the necessity of devising a means for meeting those needs in their Mars biosphere.

Homework

The biosphere pictured here does not include a polar ecosystem. For homework, have students briefly discuss the advantages and disadvantages of including a polar ecosystem in a biosphere such as this one. *(Sample answer: While a polar region would include species that might not be present elsewhere, the cold temperatures might be costly and difficult to maintain.)*

FOLLOW-UP

Reteaching

Have students make a poster that illustrates the biotic and abiotic needs of living things.

Assessment

Have students make a sketch or model showing the floor plan of their proposed biosphere. Or have students design a detailed drawing or model of one particular area inside the biosphere.

Extension

Suggest to students that they research NASA's plans to build a space station and space colonies.

Closure

Have students write poems or draw comic strips about what it might be like to live in a colony on Mars.

Pretend that you and a group of your friends have decided to make a biosphere proposal to the United Nations. You should consider the questions at right and discuss them with others:

Think About It!

- How will you control temperatures?
- What will happen to wastes?
- Where will fresh water and fresh air come from?
- How will the plants survive? What foods will you need? (Make a list of 10 nutritious items.)
- What is the natural source of each of the foods on your list?
- What will your biosphere have to include so that you can have a continuous supply of food?
- List everything you would need in order to support the food-producing plants and animals.

Write a report based on your answers. Then present your report and compare it with those of other groups. You might want to change some of your answers after listening to the other groups.

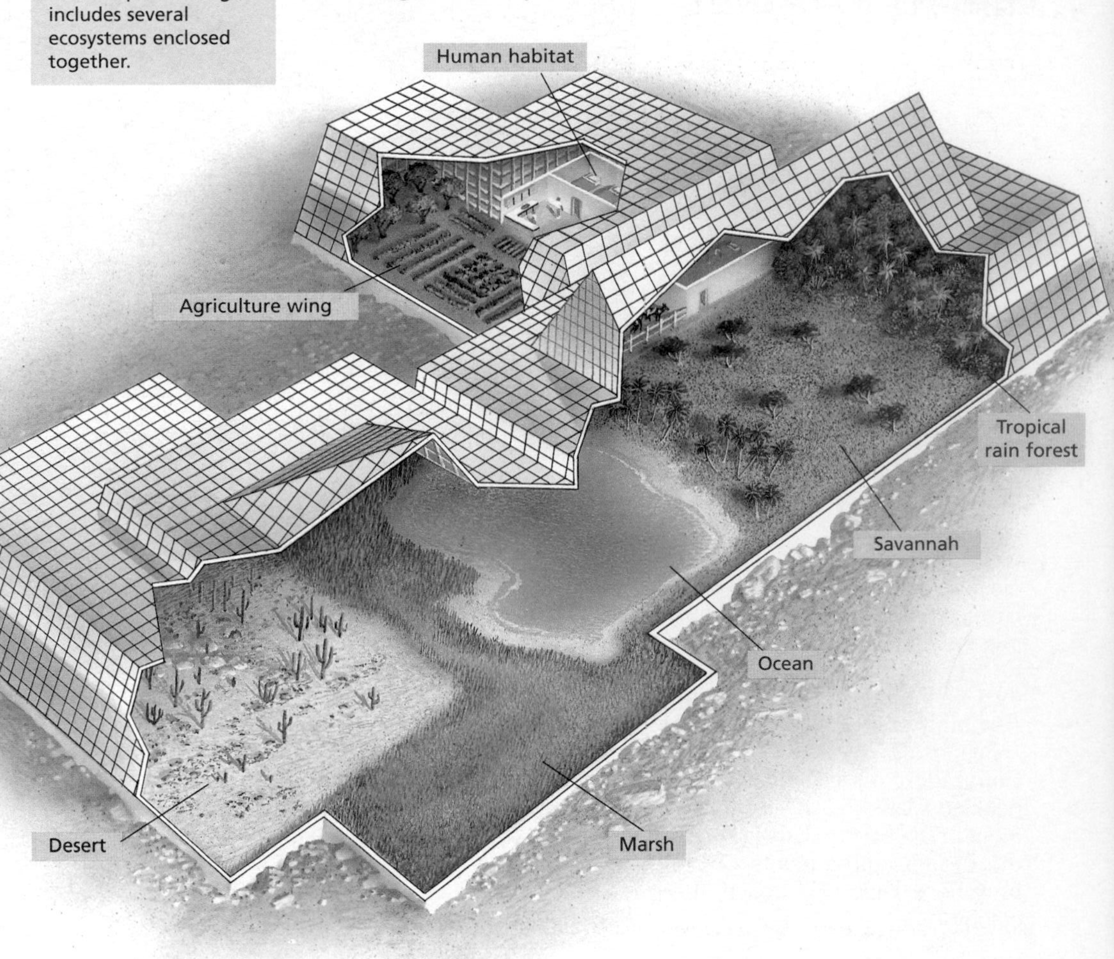

This biosphere design includes several ecosystems enclosed together.

LESSON 2 The Beginning of the Garden

Germination of Seeds

Growing plants is a fundamental part of living in an experimental biosphere. But what are the needs of one living thing—a plant—here on Earth? Will the plant's needs be the same in the Mars biosphere? The best way to start your investigation is to grow a plant of your own.

As shown below, Barbara collected seeds from flowers such as nasturtiums, zinnias, and marigolds. You may do the same, or you can use seeds from a store-bought package. In either case, you and your classmates should pick one kind of seed and grow the same kind of plant.

Barbara is uncertain whether she can rely on her seeds to develop into plants. In fact, some of the seeds you collect or buy may not **germinate**. (The word *germinate* means to sprout, or to begin to develop.) Imagine how important it is for a farmer to know what percentage of seeds will germinate and what size the crop will be. To find out, a farmer may test samples of seeds before planting them. You will test seeds for their germination rates so that you won't be disappointed when you begin to grow your plant.

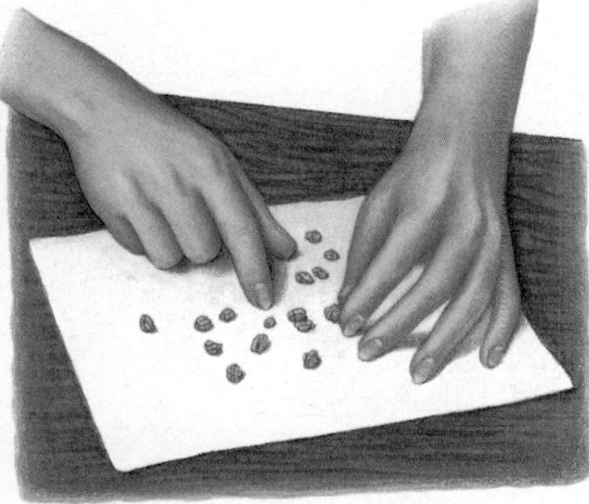

1 Pick the first seeds that form.

2 Dry the seeds.

3 Clean and keep the heaviest, plumpest seeds.

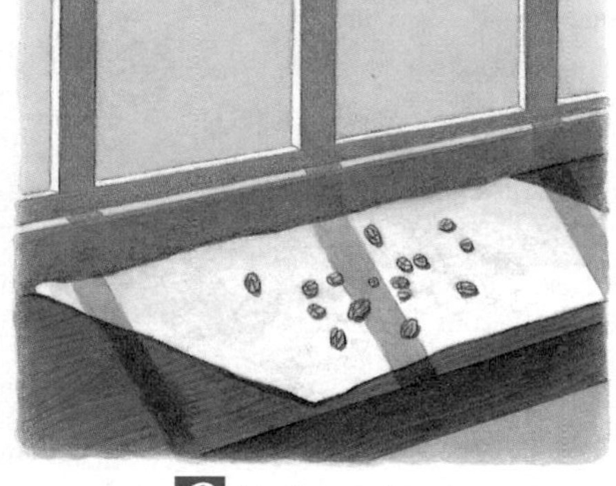

4 Keep the seeds in a cool, dry place.

LESSON 2 ORGANIZER

Time Required
2 class periods

Process Skills
contrasting, comparing, measuring, predicting

New Term
Germinate—to sprout or begin to develop

Materials (per student group)
Exploration 1: package or collection of seeds from fruits, vegetables, flowers, etc.; various materials to germinate seeds, such as moist paper towel, jar, jarful of potting soil, etc.

Exploration 2: germinated seeds from Exploration 1; about 200 mL of potting soil; small container for planting, such as small cup, pot, or milk carton; metric ruler; metric scale; about 50 mL of water; optional materials: fertilizer

Teaching Resources
Exploration Worksheets, pp. 3 and 4
Transparency 81
SourceBook, p. S170

LESSON 2 The Beginning of the Garden

FOCUS

Getting Started

Ask students if they have ever had a garden or planted seeds in containers. Call on a volunteer to explain how he or she prepared the soil and planted the seeds. Then involve students in a brief discussion of what factors are necessary in order for seeds to germinate. *(Warmth and moisture)*

Main Ideas

1. Given the right conditions, seeds will develop into plants.
2. In a given sample, it is unlikely that all of the seeds will germinate.
3. Completing a successful experiment requires careful observations and record keeping.

TEACHING STRATEGIES

Germination of Seeds

Have students read the first paragraph on page 503. Let them discuss whether the needs of plants on Earth are the same as those in a Mars biosphere. Help students to conclude that the needs of plants are the same wherever they grow. Then have them read the rest of the page.

To encourage the development of scientific thinking, point out the last sentence of the second paragraph on this page and ask: Why should the same kind of seed and plant be used? *(By keeping these variables constant, more reliable comparisons can be made.)*

You may wish to point out that many seeds used in agriculture are guaranteed by the commercial supplier to achieve a certain rate of germination. This allows farmers to predict more accurately how many seeds they will need.

EXPLORATION 1

Encourage students to devise a plan to test the germination rate of seeds. Remember to have students reserve enough germinated seeds (untreated with fertilizer or fungicide) to satisfy the requirements for Exploration 2.

Make sure that you have checked all materials and procedures for safety considerations before students carry out their plans.

Cooperative Learning
EXPLORATION 1

Group size: 2 to 4 students
Group goal: to plan and carry out an experiment to test seed germination rates
Individual accountability: Assign each student a role such as a group leader, primary investigator, artist, or recorder.
Positive interdependence: Have each student keep a written record of each step of his or her experiment.

Seed Analysts

Ask students why a seed analyst's job is important. *(Farmers, foresters, and others rely on information about germination rates to prepare their farms, forests, and gardens.)* Have one student explain how to read the table. For example, students should understand that for sweet corn 90 percent of the seeds planted will grow into plants.

Answers to *Questions*

1. Responses will vary with the type of seeds used in Exploration 1.
2. Temperature, light, water, nutrients, space, and the kind of plant may affect the germination rate.
3. Answers may vary but should include the concept of holding all variables constant except for the one being tested.
4. Accept all reasonable responses.

An Exploration Worksheet (Teaching Resources, page 3) is available to accompany Exploration 1. Transparency 81 is available to accompany Seed Analysts.

EXPLORATION 1

Testing for Germination Rate

You Will Need

- a package or collection of seeds
- materials of your choice

What to Do

Plan your own procedure for testing what proportion of the seeds will germinate. Remember, seeds need moisture and warmth to begin to grow. To get you started, here are some factors to consider as you plan your germination test:

- the kind of seeds to use (fruits, vegetables, flowers, etc.)
- the number of seeds to use in the test
- the substance to place them in (in a jar, on a moist towel, in potting soil, etc.)

Before carrying out your test, discuss with your classmates how you plan to do it. Exchange advice and ideas. Keep a record of your test in your ScienceLog.

1. Begin by stating the kind of seeds tested and where you got the seeds. Be sure to note how long they have been stored. Will the number of seeds you use affect your results?
2. Describe the procedure used to test the seeds, and include a sketch of your seed samples.
3. Report the results.

Seed Analysts

Germination rates of seeds that you buy in the store are tested by *seed analysts*. The table below lists the percentage of seeds that germinated in one test. Seed analysts are responsible for finding out information such as that provided in the table.

Seed Germination Levels

Percentage	Kind of seed
90%	sweet corn
80%	beans, broccoli, cabbage, turnip, mustard, cantaloupe, cucumber, pumpkin, popcorn, watermelon, radish
75%	peas, beets, tomato, onion, asparagus, swiss chard
70%	lettuce
65%	spinach, leek, pepper, eggplant, chives
60%	carrot, parsnip
55%	celery, parsley
50%	dill, sage, summer savory, thyme
35%	watercress

Questions

1. How do the percentages shown here compare with the results from your test in Exploration 1?
2. What do you think may cause variations in the germination rate of seeds?
3. Suggest ways you could find out if a specific factor affects germination rate.
4. You must select a seed analyst to join the biosphere. Create a want ad for a newspaper that begins, "A successful applicant . . ."

Food for thought . . . You probably eat plants or plant parts—even seeds such as peanuts—every day. Read pages S171–S174 in the SourceBook to discover other ways we use plants.

CROSS-DISCIPLINARY FOCUS

Social Studies

Suggest that students make a career handbook of professions that involve working with plants. The book might include information on horticulture, botany, farming, agricultural engineering, plant breeding, and so on. Suggest that students include a description of each career and provide information on the kind of education a person must have to enter each field.

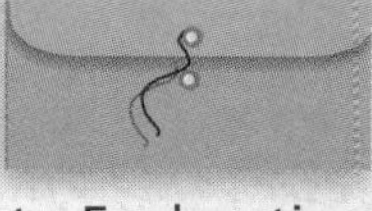

PORTFOLIO

Suggest that students include their responses to Exploration 1 in their Portfolio. Have them perform a self-evaluation of their work using a checklist from the SnackDisc.

Homework

You may wish to have students answer the questions on this page as a homework assignment.

Growing Your Own Plants

Have you ever grown plants before? With three or four classmates, make a list of plants that you have grown or have seen others grow. Group the plants under the following headings: Vegetables, Flowers, Indoor plants, and Others. Were the plants healthy? How and where did you plant them? What did you do to help the plants grow? Ⓐ

In the following Exploration, you will start growing your own plant from one of the germinated seeds. Keep a daily account in your ScienceLog. Carefully record everything you do and everything you observe about the development of the plant. Keep in mind that gardeners learn from the experience of others. Talking to other gardeners is one way to become a better gardener. Another way to learn is to read articles and books written by experienced gardeners.

A professional gardener at a commercial greenhouse can give you some tips on growing healthy plants.

EXPLORATION 2

Your Plant

You Will Need

- germinated seeds from Exploration 1
- potting soil
- pots or cups for planting
- fertilizer (optional)
- a metric ruler
- a metric scale

What to Do 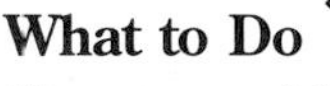

Choose one of the seeds. Look closely at the seedling, and be gentle because the newly formed roots and leaves are easily damaged. Plant the seedling in soil that has been thoroughly broken up. Soil mixtures from a garden store would be useful. Think about the position that the root should be in. Then make a small hole in the soil. Put in the seedling, and gently fill in the space around it with soil. The leaves should be showing as if they had just broken through the soil. Lightly pack the soil so that it supports the seedling in an upright position. Add water to make the soil moist. Some fertilizer may be added as well. Record what you do and observe as your plant grows.

Things to Observe and Record

- Record the planting conditions. Include the medium (garden soil) that the seedling is planted in, the kind and size of the container, and where the container is located.
- Record when and how much you water and fertilize the seedling.

Exploration 2 continued

EXPLORATION 2

Have students use the seeds they germinated in Exploration 1. If possible, have each student plant two or more germinated seeds in case one of them fails to grow. Have students begin germinating their seeds well in advance so that they have fully germinated by the time the class is ready to begin this Exploration.

Supply all students with the same kind of container. An inexpensive container can be made from cardboard milk cartons by cutting off the top third and poking a few holes in the bottom. You may wish to provide potting soil for them to use. All students should use potting soil from the same source.

Have students examine one of the germinated seeds with a magnifying glass and determine which part of the sprout is the leaf and which part is the root. Demonstrate for the class how to plant one of the germinated seeds. Caution students to be very careful not to break the delicate root or leaf.

If students are using commercial potting soil, it may have some nutrients in it already, and adding fertilizer may be toxic to a young plant. Consult the package directions for more specific recommendations.

Encourage students to make periodic progress reports discussing variations in growth rate, size, and color among the plants they grew.

Exploration 2 continued

An Exploration Worksheet is available to accompany Exploration 2 (Teaching Resources, page 4).

Answers to *In-Text Questions*

Ⓐ Students should identify factors such as soil, fertilizer, moisture, and light that affect plant growth.

Meeting Individual Needs

Learners Having Difficulty

Give students the opportunity to participate in a plant-propagation project by making a "grocery-store garden." They should use plants and plant parts from fruits and vegetables they can buy at the grocery store, such as sweet potatoes, the leafy tops of carrots, the tops of pineapples, and so on. When the plants have grown substantially, have students display them for the class.

Growing Your Own Plants

Have students share and discuss gardening stories and their lists of plants. Invite them to prepare a master chart on the chalkboard to classify the plants they have grown as either vegetables, flowers, indoor plants, or other plants. Encourage each student to contribute at least one plant to the chart. To add interest to the project, bring a couple of seed catalogs to class for students to browse through. Then have students form small groups or choose partners to complete Exploration 2.

Exploration 2, *continued*

Depending on the type of plant used, you may wish to have students consider other aspects of the plant, such as the size of the leaves or the rate of growth. Also, when weighing the seedlings in their containers, students should try to be consistent in the amount of moisture present in the soil. For instance, they may wish to weigh each plant just before or after its regular watering.

FOLLOW-UP

Reteaching

Have students become plant experts by learning all that they can about one type of plant that was grown for Exploration 2. The experts can serve as resources for other students who are trying to grow the selected plants.

Assessment

Have students make bulletin-board displays or murals to show how a plant develops from a seed to a mature plant. Suggest that they use pictures or drawings of their own plants as well as pictures from magazines.

Extension

Have interested students start growing some perennial plants. These can be started indoors and then transplanted outdoors. Some of the easiest perennials to grow include black-eyed Susan, Shasta daisy, coral bells, baby's breath, lupine, and poppy. Encourage students to find out which plants grow best in their area.

Closure

Students may enjoy setting up a reading center that focuses on the variety, care, and growth of plants. The materials in the center could include gardening magazines, seed catalogs, library books about plants, newspaper articles about plants and gardening, stories involving plants, and plant poetry.

- Record the date when you plant the germinated seed, and sketch what it looks like.
- Record changes in the plant as it grows, including
  - —height and other dimensions, such as the circumference of the main stem;
  - —the number, color, and shape of the leaves;
  - —the number of buds and flowers;
  - —characteristics of the flowers; and
  - —signs of pests and diseases.

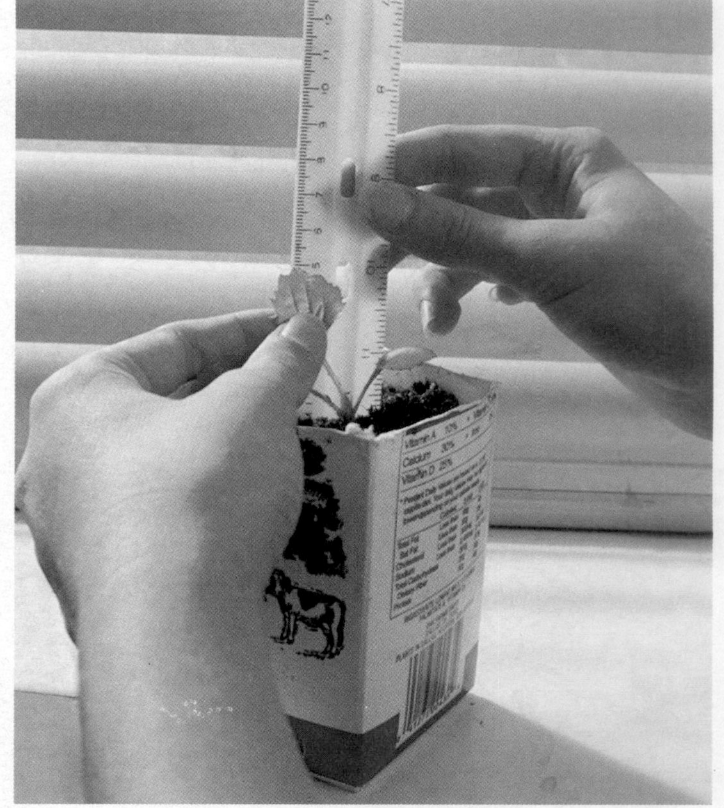

Special Instructions

1. If there are enough germinated seeds available, plant two or three of them. This gives you a better chance of success if something happens to one of them.
2. After the seedlings are planted and the soil is moistened, measure the mass of the container and its contents. Later you can see how growth affects the plant's mass. When you weigh the container a second time, make sure the same amount of moisture has been added to the soil.
3. Record any changes in growing conditions, such as the amount of light or heat at the location where your plant is growing.
4. When your plant is fully developed, compare your results with those of your classmates.

Other Things to Do

1. Use a series of photographs to show your plant's growing stages.
2. Draw a series of sketches of your plant as it grows.

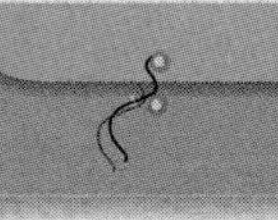

PORTFOLIO

Have students add their ScienceLog entries from Exploration 2 to their Portfolio. Allow peer groups to evaluate each other's work with regard to number of observations made, original ideas introduced, methods used to record data, and general appearance.

ENVIRONMENTAL FOCUS

Ask: What factors of the nonliving environment influence the germination of a seed? *(Moisture, warmth, and space influence the germination of plant seeds. Seeds germinate best when the temperature is warm, the soil or growth medium is moist, and the seeds have plenty of room to grow.)*

Breaking Ground

Soil—An Important Factor

"If you have good soil and enough rain, anything can grow." This statement may be a slight exaggeration. However, it emphasizes the importance of two factors, soil and water, to a plant's growth.

You are familiar with soil; you can see it in a garden, the park, the football field, vacant lots, flower pots, and other places. But have you ever thought about questions such as the following?

- Exactly what is soil?
- Where does it come from?
- Are there different kinds of soil?
- If there are, how do they differ from one another?
- What is the best kind of soil for growing plants?
- Is sand considered a kind of soil?
- Is there any danger that our soil might disappear or be damaged? If so, how?

Discuss some of these questions with your classmates. Then consider whether soil will be important in the Mars biosphere. If so, what will the soil be like, how will it be obtained, and how will it be conserved so that it can continue to be useful? Ⓐ

Soil in a garden

Soil in the park

Soil in a football field

Soil in a vacant lot

LESSON 3 ORGANIZER

Time Required
4 class periods, plus a 4–6 week growing period

Process Skills
organizing, comparing, analyzing

Theme Connection
Changes Over Time

New Terms
Humus—a part of fertile soil that is derived from the decomposition of living things
Loam—a mixture of sand, silt, and clay
Nutrients—chemicals needed for the functioning and growth of living things
Percolation rate—the speed at which a certain volume of water passes through a sample of soil
Texture—the feel of soil

Materials (per student group)
Exploration 3: 500 mL of soil from vacant lot, garden, or student's backyard; magnifying glass; about 50 mL of water; several sheets of newspaper
Exploration 4: jar, at least 5 cm in diameter and 10 cm high, with lid;

continued

Breaking Ground

FOCUS

Getting Started

Before class, obtain half of a liter of soil and place it in a clear plastic bag. To begin the lesson, hold up the bag and ask students to identify what is inside it. *(Many students may reply "dirt.")* Students may be interested to know that it takes between 100 and 1000 years for 1 cm of soil to form. In 1 mL of soil (a pinch), there may be 2 billion microorganisms.

Main Ideas

1. Soil is made up of rock and mineral particles and the decaying remains of living things.
2. Soil provides support for the roots of plants and holds the water, air, and nutrients needed for plant growth.
3. Soils with fine particles have a greater water-holding capacity than soils with coarse particles.
4. Factors affecting plant growth include soil type, temperature, nutrients, water, and light.

TEACHING STRATEGIES

Soil—An Important Factor

Have students read the material and discuss possible answers to the questions. Accept all reasonable responses. Students will have numerous opportunities to refine their ideas about soil as they progress through the lesson.

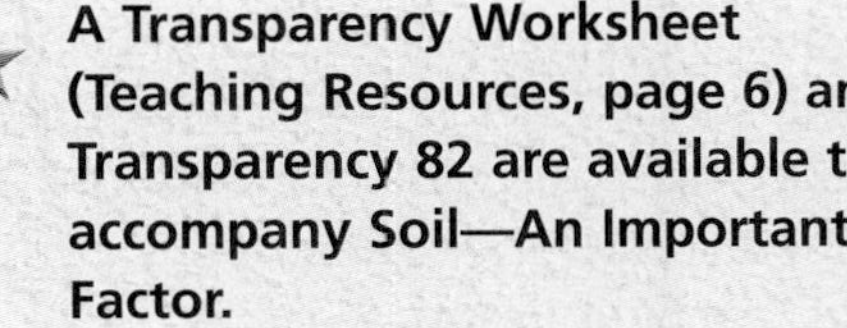

A Transparency Worksheet (Teaching Resources, page 6) and Transparency 82 are available to accompany Soil—An Important Factor.

Answers to *In-Text Questions*

Ⓐ Sample answers: Soil would likely be important in the Mars biosphere. It could be brought from Earth and replenished periodically.

Exploration 3

Divide the class into small groups and distribute the materials. Groups should read and complete the Exploration. Discuss with students the three identifying characteristics of soil—composition, texture, and color. After students have written their soil descriptions, challenge each group to identify a soil sample from someone else's description.

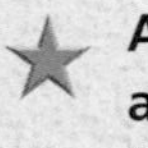

An Exploration Worksheet is available to accompany Exploration 3 (Teaching Resources, page 10).

Answers to *What Is Soil?*

Accept all reasonable responses. Sample answer: Soil is a mixture of tiny rocks and other objects along with the remains of living things.

Soil Particles

This is an introduction to Exploration 4. It presents the idea that particle size determines the texture of soil.

INDEPENDENT PRACTICE Familiarize students with the concept of texture by having them perform a simple activity on their own. Provide students with small samples of rocks, pebbles, and sand. Encourage them to observe and manipulate these materials and to compare their textures.

Theme Connection

Changes Over Time

Focus question: Why are plants an important factor in the development of healthy soil? *(Plants can break down rocks into soil, and plants provide nutrients for the soil when they decompose.)* For homework, provide students with the Theme Worksheet on page 8 of the Unit 8 Teaching Resources booklet.

Exploration 3

A Close Look at Soil

You Will Need

- a half of a liter of soil obtained from your backyard, a vacant lot, or a garden
- a magnifying glass
- water
- newspaper

Keep the soil in a sealed plastic bag or jar. Label it with your name. You will use this soil in many of the activities that follow.

What to Do

1. Along with particles of various minerals (the substances that rocks are made of), soil contains **humus,** a material derived from the decomposition of once-living things. Humus is an important part of fertile soil. Spread a small amount of your soil sample on a piece of newspaper. Use a magnifying glass to examine it closely. Do you see mineral particles—pebbles, sand, or even finer particles? What evidence can you find of things that once were living? Try to identify what these things were.
2. Add a small amount of water to your soil—just enough to make it slightly damp. Now rub some of the soil between your fingers. Does it feel smooth or gritty? What other words can you use to describe how the soil feels? This characteristic of soil is called **texture**.
3. What color is your soil—black, gray, reddish, yellowish, or some other color? The color of soil depends on many factors, including exposure to air, the climate of the environment, and the amount of humus present. Darker soils are usually rich in humus.
4. After making close observations of your soil sample, write a complete description of it. Form a group with three other students. Mix up your soil descriptions and distribute them among the members of your group. Try to match the description you receive with the appropriate sample.

What Is Soil?

You have been chosen to write a definition of soil for a new science dictionary. Drawing on your recent experience with soil, write a definition consisting of one or two sentences. Compare it with definitions that others have written and with those in various dictionaries. Of all these definitions, which do you think is best? Why?

Soil Particles

Even soils that look the same may be made from different sizes of particles. Particle size determines the texture of soil. In the next Exploration, you will investigate particle size.

ORGANIZER, *continued*

about 100 mL of water; 200 mL of soil sample from Exploration 3 or from commercial source; magnifying glass; spoon

Exploration 5: 2 large, plastic-foam cups; 2 or 3 paper towels; 500 mL glass, beaker, or wide-mouth bottle; about 500 mL of soil sample from Exploration 3; about 500 mL of sand; stopwatch or clock; 500 mL of water; 250 mL beaker; 100 mL graduated cylinder; several 2 cm × 5 cm pieces of cardboard

Exploration 6: a few seedlings from Exploration 1; various materials for transplanting and testing plant growth, such as several large plastic-foam cups, several cupfuls of garden soil, about 1 L of water, spoon or craft stick, etc. (See page 512.)

Teaching Resources

Transparency Worksheet, p. 6
Theme Worksheet, p. 8
Exploration Worksheets, pp. 10, 11, and 13
Transparencies 82–84

Exploration 4

Sorting the Soil

You Will Need

- a jar at least 5 cm in diameter and 10 cm high, with a lid
- water
- soil (from your sample or from a commercial source)
- spoon

What to Do

1. Fill the jar halfway with soil. Add water until the jar is nearly full. Put the lid on securely. Shake vigorously for 1 minute.
2. Place the jar on a tabletop or desk, and let the soil settle for 15 minutes. Do separate layers form as the soil settles? How many layers do you see in your jar?

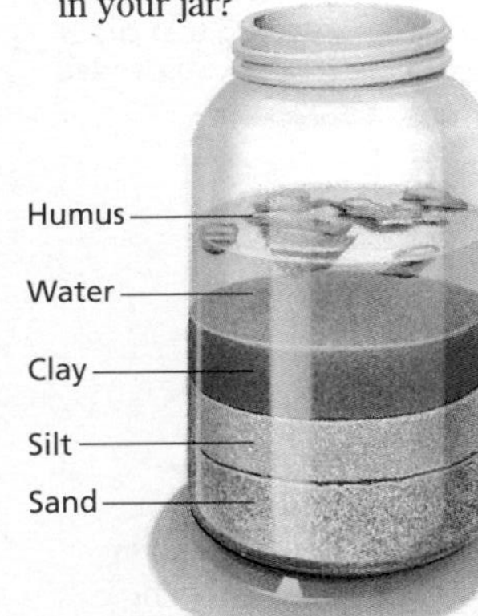

Use the following information to identify the components of your sample:

- humus: floats on top; derived from the decomposition of living things
- clay: very fine particles
- silt: slightly larger particles
- sand: still larger particles
- even larger particles

Compare your jar with those of your classmates. What differences do you observe?

3. Let your water-soil mixture settle for 24 hours, and then observe the layers. In your ScienceLog, make a sketch or a graph showing the proportion of each of the three bottom layers in the jar. It is this proportion that determines the soil type.
4. Look at the soil classification table below. What are the sizes of the particles in your jar? Particles of clay and silt must be magnified to be seen individually.
5. Slowly pour off the water and humus. Spoon off the clay, and touch it to feel its texture. Observe the textures of the silt and sand particles as well. Compare your observations with those given in the table.

Soil Classification Table

Particle	Texture
sand (0.02 mm–2 mm)	Feels gritty when rubbed between fingers. Not sticky when moist.
silt (0.002 mm–0.02 mm)	Feels smooth and powdery when rubbed between fingers. Not sticky when moist.
clay (less than 0.002 mm)	Feels smooth and sticky when moist. Forms hard clots when dry. May remain suspended in water for a long period of time.

Soil and Plants

How does soil interact with plants? Why is this interaction important to plants? In what other ways do plants and soils interact?

Have you ever pulled up a plant by its roots? In this case, what was the soil doing for the plant? The roots of the plant form a complex network. Water and other **nutrients** (chemicals needed for plant functioning and growth) enter the plant from the soil through the smallest roots. Thus, soil performs two important functions for plants:

- It provides support.
- It holds essential nutrients that can be taken in by the network of tiny roots.

Think About It!

A mixture of sand, silt, and clay is called **loam**. A typical mixture is 40 percent sand, 40 percent silt, and 20 percent clay. Mixtures with larger portions of one ingredient are called sandy loam, silty loam, or clay loam. How do you think different soil mixtures hold water and other nutrients?

509

Exploration 4

Explain that in the previous Exploration, students discovered that soil was made up of several different kinds of substances. In this Exploration, they will take a closer look at some of these substances and determine how much of their soil sample is made up of each substance. Divide the class into small groups and distribute the materials. Monitor what groups are doing by circulating among them and making yourself available for assistance.

Students should observe that the soil samples, once they are well shaken, settle into layers according to particle size. The illustration on this page represents an ideal situation—actual soil samples will probably have different (and less dramatic) results. Suggest that students examine and compare the layers formed by each other's soil samples. Encourage them to identify differences in the samples. For example, some samples may settle out into more than three layers, or similar layers of different samples may have different thicknesses and colors. Help students establish an accurate record of the characteristics of their own soil sample by using magnifying glasses and making sketches.

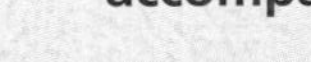

An Exploration Worksheet (Teaching Resources, page 11) and Transparency 83 are available to accompany Exploration 4.

Soil and Plants

Before students read this material, ask them how soil interacts with plants. Accept all reasonable responses. Then have students read the material to see if their ideas are correct. Students may find it interesting to know that the nutrients in soil that are used by plants include nitrogen, potassium, calcium, phosphorus, sulfur, magnesium, iron, manganese, zinc, copper, chlorine, and boron.

Integrating the Sciences

Earth and Life Sciences

Suggest that students research how soil is formed. They should consider the contributions of weathering and erosion, climate, surface features, rivers, streams, and time. Then have them perform some simple demonstrations that model the effect of mechanical and chemical weathering on rocks. For example, rubbing sandstone against a surface leaves grains of sand behind, and crumbling some rocks, such as mica or talc, causes particles to break off.

Think About It!

Have students read the material silently, or call on a volunteer to read it aloud. Involve students in a brief discussion of the question. Accept all reasonable responses, encouraging students to provide reasons for their ideas. Point out to students that they will have an opportunity to test their ideas in the following Exploration.

Soil and Water

Call on a volunteer to read the material in the left-hand column of page 510. Involve students in a discussion that will further expand their definition of soil. Next, discuss the questions about how soil and water interact. Have students record their answers so that they can review, evaluate, and revise them after completing Exploration 5.

EXPLORATION 5

Divide the class into pairs or small groups and distribute the materials. Have groups read the steps of the Exploration before they begin. Be sure that they use dry soil samples. Point out that in step 3 they use the weight of a cup of water to pack down the sand or soil in the cup. They should not pour any water into the soil or sand until step 5. Students should use the same glass of water to pack each of their samples. Instruct students to measure the level of sand and soil in each cup to make sure that the amounts used are equal. The volume of water that flows through each cup should be measured with a graduated cylinder. Remind students that it is important to perform the experiment carefully and accurately if the results are to be meaningful.

After the Exploration, assemble the class and involve them in a discussion of their results. Although their results will vary because different soil samples were used, students should discover that more water was held by the soil than by the sand.

★ **An Exploration Worksheet (Teaching Resources, page 13) and Transparency 84 are available to accompany Exploration 5.**

Soil and Water

When you started your study of soils, you wrote a preliminary definition of soil. Your definition was probably more about what soil is than what it does. Now expand your first definition to include the uses of soil. (A)

Do you agree with the statement, "Providing water to the roots of plants is the most important function of soil"? What are some arguments that might support this statement? (B) Consider the following questions about how soil and water interact. Use what you already know about soil to come up with the best answers. (C)

- How much water can soil hold?
- Can some soils hold more water than others?
- Which holds more water, a soil mixture or pure sand?
- Does soil or sand allow water to pass through at a faster rate?

EXPLORATION 5

The Amount of Water That Soil Holds

You Will Need

- 2 large plastic-foam cups
- paper towels
- 2 glasses, beakers, or wide-mouth bottles
- your soil sample
- sand
- a stopwatch or clock
- a graduated cylinder
- small strips of cardboard

What to Do

Before starting this Exploration, let your soil sample air-dry by spreading it out on a paper towel for 24 hours.

1. Using a pencil, punch three holes in the bottom of each cup.

Remember to record all of your measurements in your ScienceLog. A table like the one below might be helpful.

Quantities	Soil	Sand
volume of water added	? mL	? mL
volume of water that passed through	? mL	? mL
volume of water held	? mL	? mL
time for water to reach the bottom of the cup	? seconds	? seconds

510

Answers to *In-Text Questions*

(A) Sample definition: Soil is made up of small rocks and humus and provides plants with water and nutrients.

(B) Soil holds water around the roots so it is available to the plants. Without soil, many plants would not exist.

(C) Accept all reasonable answers. Many students may predict that sandier soil will hold less water.

Homework

For homework, have students find out what *peat* is and what it is used for. *(Peat is partly decomposed plant matter that can be used in gardening or as fuel. It usually forms in bogs or swamps and absorbs moisture well.)*

2. Place two circles of paper towel in the bottom of each cup to cover the holes.
3. Fill one cup three-quarters full with dry sand and the other cup three-quarters full with dried soil from your sample. Place a glass of water on top of the sand or soil to pack it together. This gets rid of air spaces. Use the same glass of water for each cup.
4. Place each cup in the mouth of a beaker, bottle, or glass so that the cup is supported above the bottom of the container.
5. Add 250 mL of water to the sand. Time how long it takes for the first drop of water to drip into the container.
6. Repeat step 5 for the soil sample. Be sure to pour water into the soil at the same rate at which you poured water into the sand.
7. When water has stopped dripping from both cups, measure the volume of water that passed through each cup by pouring the water into a graduated cylinder. How much water was held by the sand? by the soil?

Making Sense of Your Data

1. Compare the amount of water held by the soil with that held by the sand. Was the prediction you made at the start of this lesson accurate? Which do you predict would hold more water, sandy clay or silty clay? Why is the capacity to hold water an important property of soils?
2. Did your soil hold more or less water than that of your classmates? Compare your results with those recorded by your class. At the same time, compare the soil types. Create and complete a table like this one in your ScienceLog.

Name	Soil type	Water-holding capacity (measured in milliliters)

3. The time it takes for water to pass through a sample of soil indicates its **percolation rate**. Which had a faster percolation rate, sand or soil? Why do you think this is so? Compare the percolation rates of the soil samples of your classmates. Which types of soils have slow percolation rates? fast percolation rates? Why is percolation rate an important characteristic of soil?
4. What could be added to clay soil to improve it for gardening? What could be added to sandy soil, like that in the photograph below? How would the soils be improved for growing plants?
5. What would you use to make a soil mixture suitable for the Mars biosphere? How much clay or sand would you use?

Answers to *Making Sense of Your Data*

1. Sand holds less water than soil does. In general, silty clay will hold more water than will sandy clay. The water-holding capacity of soil is important because plants must be able to obtain water from soil even during periods of little rain.
2. Different types of soil hold different amounts of water. However, all of the soil samples will probably hold more water than pure sand will. Soils with fine particles tend to hold more water than soils with coarser particles.
3. Sand has a greater percolation rate than most soils because it has fewer fine particles. Soils with more coarse particles have a higher percolation rate than soils with more fine particles. Percolation rate is important because plants require air to remain healthy; water retained in the soil can force out all of the air. Students may also point out that if water cannot drain through the soil, flooding may occur. The opposite is also true: there may be a problem if too little water is retained.
4. Clay soil can be improved for gardening by adding sand, silt, and humus. Sandy soil can be improved by adding clay, silt, and humus. In both cases, the resulting mixture provides a soil that retains enough water to maintain the healthy growth of plants, but not so much that all of the air is forced out.
5. Answers may vary but should reflect an understanding that loam—about 40 percent sand, 40 percent silt, and 20 percent clay—would work well. Additional amounts of humus could also be added to help the plant grow.

Did You Know. . .

Like eggs, most seeds are storehouses of nourishment. A young plant inside a seed is dependent on this nourishment until it can grow out of the soil and produce its own food. If a seed is buried too deeply, the plant will exhaust its food supply before it can emerge from the soil and will most likely die.

Exploration 6

To monitor each group's ideas and to encourage students to plan their investigations carefully, ask students to submit summaries of their investigations for your approval before the investigations begin. Explain that each summary should include a statement of the hypothesis that is being tested, a detailed description of the procedure, and a list of the materials that will be used. Point out that the summary should clearly state how the chosen variable is to be tested.

Make sure that you have checked all materials and procedures for safety considerations before students carry out their investigations.

Each group of students should use the same method to evaluate plant growth. The best measurement is likely to be the one that includes more than one characteristic.

Allow sufficient time for adequate plant growth. Four to six weeks may be needed. After all of the results are recorded, allow class time for each group to briefly present their conclusions and to respond to any questions about procedures and results.

Cooperative Learning
Exploration 6

Group size: 2 to 3 students
Group goal: to set up an experiment to investigate the effect of different variables on plant growth
Positive interdependence: Assign each student a role such as group leader, materials coordinator, or recorder.
Individual accountability: Have each student write a paragraph summarizing his or her group's results, with comments on how the results compared with those of other groups.

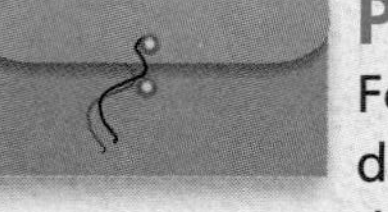

PORTFOLIO

For Exploration 6, students may organize their preparatory work, notes taken during the Exploration, and final reports into one entry for their Portfolio. Encourage students to include photos or sketches of their plants.

Exploration 6

Growing the Biggest and the Best

This is a long-term project. Your objective is to find out what it takes to grow the biggest and the best plant in the classroom. You will want to find out how to get the best results in each of these areas:

- type of plant to use
- type of soil
- amount of water
- amount of sunlight
- temperature
- distance between plants
- size of container
- fertilizer (plant food)

As you start this project, you will need to organize your activities. The class should divide into 15 groups. Each group will choose one investigation from those listed in the table at right. The groups can pool their results later. Doing this will help you find out how to obtain the best results possible. Then you will all be on the way to growing the biggest and the best!

Your class is going to investigate the effects of five variables on the growth of three types of plants. How will you judge the amount of plant growth? Make a decision now. Here are some factors to consider:

- height of plant above soil
- number of leaves
- size of leaves
- mass of plant

The best idea may be to keep a record of as many observations as possible. Make these observations every second or third day, depending on how fast your plant grows, and record them in your ScienceLog. Record your observations in several different ways: a written description, measurements in a table, graphs, life-size or scale drawings of the plant, or some other means.

Experiment	Plant 1	Plant 2	Plant 3
Investigating different kinds of soil	Group 1	Group 2	Group 3
Investigating different amounts of humus in the soil	Group 4	Group 5	Group 6
Investigating different amounts of water	Group 7	Group 8	Group 9
Investigating different amounts of sunlight	Group 10	Group 11	Group 12
Investigating use of fertilizers	Group 13	Group 14	Group 15

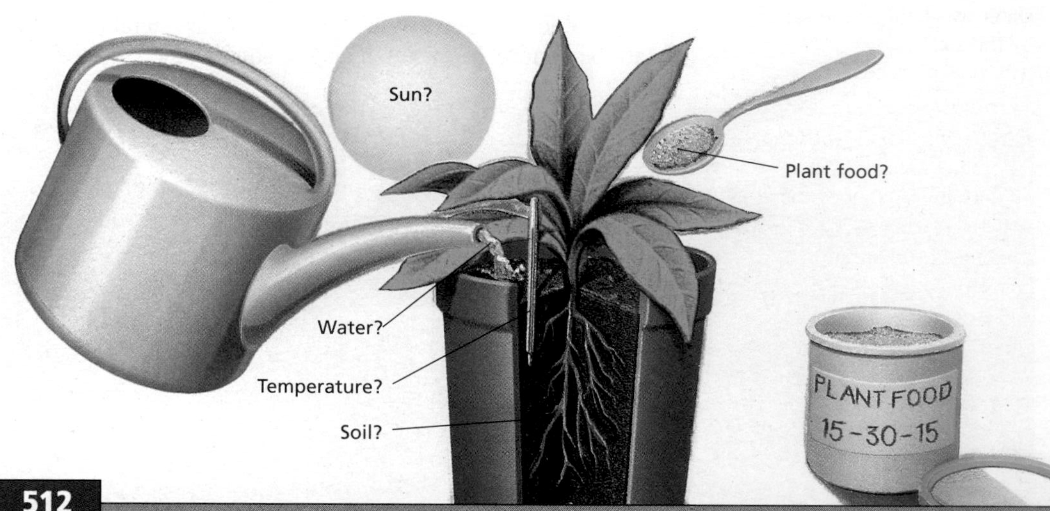

512

Second-Language Learners

Suggest that students take pictures to document the steps and stages of Exploration 6. Students should also photograph the changes in their plant over time. Encourage them to display their photos in chronological order with a brief bilingual caption beneath each photo. During the presentation of the finished reports, students can use their photo display as a visual aid.

Homework

You may wish to assign all or part of Exploration 6 as homework.

Wait a Minute!

Were you ready to start? Just a word of caution—you have to be careful that you don't do what Alex did! He was investigating the effect of the amount of water on plant growth. After looking at the illustration below, would you say his results are reliable? How would you alter his experiment?

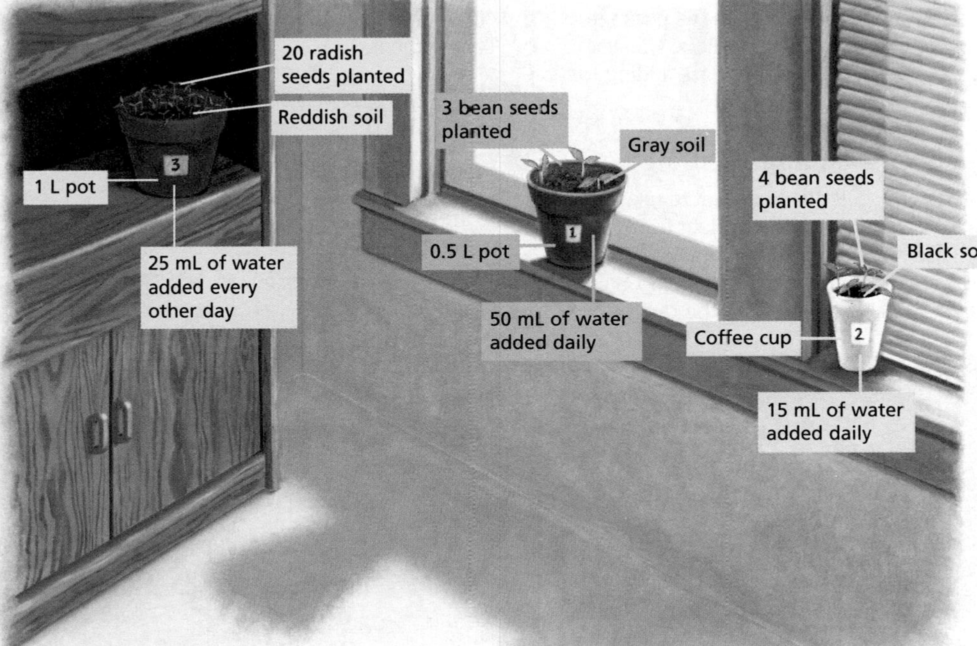

Alex varied the amount of water in the three containers, but he didn't keep the other variables constant. Remember to keep everything the same in each of your trials, except the variable whose effect you wish to measure.

General Hints 

1. All groups should use the same kind and size of container. Large plastic-foam cups are good.
2. All groups should use the same kind of garden soil, except those who are investigating soils. (Soil will be provided for you.)
3. All groups will use seedlings that have grown since the germination tests in Exploration 1. By now they may have several leaves. This is a good time to transplant them so that they will have room to grow.
4. Water the plants when the soil appears to be dry. Keep a record of the amount of water added. Remember to add the same amount of water to each container, unless you are part of the watering experiment. The amount of water required may be different for different species of plants.

Hints for Each Investigation

Investigating different kinds of soil: Use sand, clay, silt, loam, or subsoil (found about 50 cm to 60 cm beneath the topsoil).

Investigating amount of humus: Add different amounts of compost, peat moss, or bagged manure to soil that has only a small amount of natural humus in it.

Investigating amount of water: Remember that both the amount of water and the time between waterings must be considered.

Exploration 6 continued ▶

Homework

To encourage critical thinking, have students create their own scenario based on the one described in Wait a Minute! Students should set up the scenario so that at least one variable is being tested incorrectly. After they have completed this activity, have students trade papers and analyze each other's work.

EXPLORATION 6, *continued*

Wait a Minute!

Provide students with time to read the material and study the illustration. Then involve them in a brief discussion of Alex's experiment. Help students conclude that Alex's results are unreliable because he did not control all of the variables. The variables he should have controlled, but did not, include the location of the plants, the number of germinated seeds in each container, the size of the container, the amount of sunlight, the temperature, and the type of soil. Students should be able to suggest ways in which each of these variables could have been controlled. Students should recognize that the only factor that Alex should have varied is the amount of water because that is the variable he was testing.

General Hints

Have students read the section. Be prepared to answer any questions they may have. You may wish to provide students with the potting soil and containers that they are to use. The tops of milk cartons can be cut off to make containers. Be sure students poke holes in the bottoms of their containers to provide drainage. Either commercial potting soil or good garden soil can be used. Students who are doing the soil investigations will need several different kinds of soils.

Hints for Each Investigation

Be sure students read the material that applies to them. You may wish to monitor the soils used by groups that are investigating different kinds of soils. You may also wish to monitor the amount and kind of humus used by groups that are investigating the amount of humus in soil. If bagged manure is used, be sure that it is pasteurized, composted manure. Composted manure will have little or no odor.

Exploration 6 continued ▶

EXPLORATION 6, continued

Hints for Transplanting Seedlings

Have students read the hints and encourage them to study the photographs. You may wish to demonstrate the procedures for the class. As students actually begin transplanting, monitor how they are progressing and offer assistance as needed. Have them work on newspaper so that spilled soil can be cleaned up easily.

Judging the Results

Involve students in a discussion of each of the questions. Their responses will vary depending on their experiences with the investigations.

Answers to *Judging the Results*

1. Students should identify problems and successes in the different procedures. They should note the results of each experiment.
2. Characteristics that students may choose to monitor could include size, color, weight, and number of leaves.
3. Answers will vary depending on individual experiences.
4. Answers will vary depending on individual experiences.
5. Encourage students to consider the special needs of the different kinds of plants.
6. Students should conclude that all plants require the right amount of sunlight and water in order to grow properly, but that different kinds of plants may have different requirements.
7. Students should agree with this statement.

Investigating amount of sunlight: Look for locations with plenty of sunlight and locations with very little sunlight, or vary the amount of time that the plants are placed in direct sunlight.
Investigating fertilizers: Use liquid fertilizer, and vary either the amount used or the time between applications. Read the directions on the package before using any fertilizer. The package will also tell you what types of nutrients you are adding to the soil.

Hints for Transplanting Seedlings

1. Put some of the soil that you want to use in the bottom of a container. Lift one plant from beneath the roots, being careful not to hurt the roots. (A spoon or craft stick may help.) Some soil will be removed along with the roots.

2. Put the plant on the soil in the container and hold the plant upright. Fill the container with more of the same kind of soil. Make the soil firm enough to support the plant but not so firm that there is no space for air.

3. Add enough water (at room temperature) to moisten the soil. Do not pour water directly on the plant. Keep the plant at a constant temperature in a fairly dark place for a few hours. This helps it to recover from the shock of being transplanted. It can then be given the amount of light that you have decided to use in your experiment.

Judging the Results

1. Carefully listen to or read each group's report. What can you learn from their experiments?
2. How might you determine exactly what is the *best* plant growth?
3. How might you improve the design of your experiment?
4. What factors were difficult or impossible to control in your experiment?
5. What would you recommend to a future bean grower? radish grower? mustard-plant grower?
6. Which recommendations would be useful for growing any plant?
7. The best conditions for growing one kind of plant are not necessarily the best conditions for growing other kinds of plants. Do you agree with this statement? Support your answer with evidence from the experiments.

FOLLOW-UP

Reteaching

Suggest that students make poster diagrams to show the characteristics of the very best soil for gardening. Their diagrams should show the importance of each characteristic and indicate whether the characteristic would be supplied by sand, silt, clay, or humus.

Assessment

Have students create a concept map using the following words: soil, humus, clay, silt, sand, loam, plants, nutrients, and percolation rate. Suggest that students illustrate their concept maps if they wish.

Extension

Explain to students that compost is the most commonly used form of organic fertilizer. Have interested students form groups to research how compost is made. Students may go to the library or consult a local nursery. The students may then make compost. Have these students explain the procedure to the rest of the class.

Closure

Invite a representative from a local nursery or botanical garden to your class to give a talk on plant growth. Encourage students to share the results of Exploration 6 with your guest. Students may then ask the speaker for tips on how to get better results when growing plants.

1. How Does Your Garden Grow?
Prepare an article for a gardening manual. You need to explain the instructions for growing a seedling from a seed so that an eager beginning gardener like Katrina will easily understand what to do.

2. Dirty Words!
Try your hand at the crossword puzzle below. Don't write in your book; copy the puzzle into your ScienceLog.

CLUES

Across
3. A test of a hypothesis
5. A material in soil that increases the rate at which water passes through soil
7. The time it takes for water to pass through soil (two words)
10. Material derived from once-living things; enriches the soil
12. Chemical needed for growth and functioning

Down
1. A system that supports life
2. This is what the plant begins as
4. The chief water entrance of a plant
6. To sprout
8. A mixture of clay, silt, and sand
9. The way soil feels in your fingers
11. Fine soil particles

★ **You may wish to provide students with the Chapter 23 Review Worksheet that is available to accompany this Challenge Your Thinking (Teaching Resources, page 14). Transparency 85 is also available for your use.**

Answers to *Challenge Your Thinking*

1. Answers will vary. Students should include the selection of seedlings and soil, careful transplanting of the seedling, addition of nutrients (fertilizers), and provision of sunlight, water, and room to grow.

2. **Across**
3. experiment
5. sand
7. percolation rate
10. humus
12. nutrient

Down
1. biosphere
2. seed
4. root
6. germinate
8. loam
9. texture
11. silt

Multicultural Extension

Plants From Around the World
Have students work in groups to make murals, bulletin-board displays, or posters that show plants that grow in different parts of the world. For example, divide the class into seven groups, one group for each continent. Each group could focus on the plants that are common on their assigned continent. Students should also describe how the plants are used by people who live in these areas.

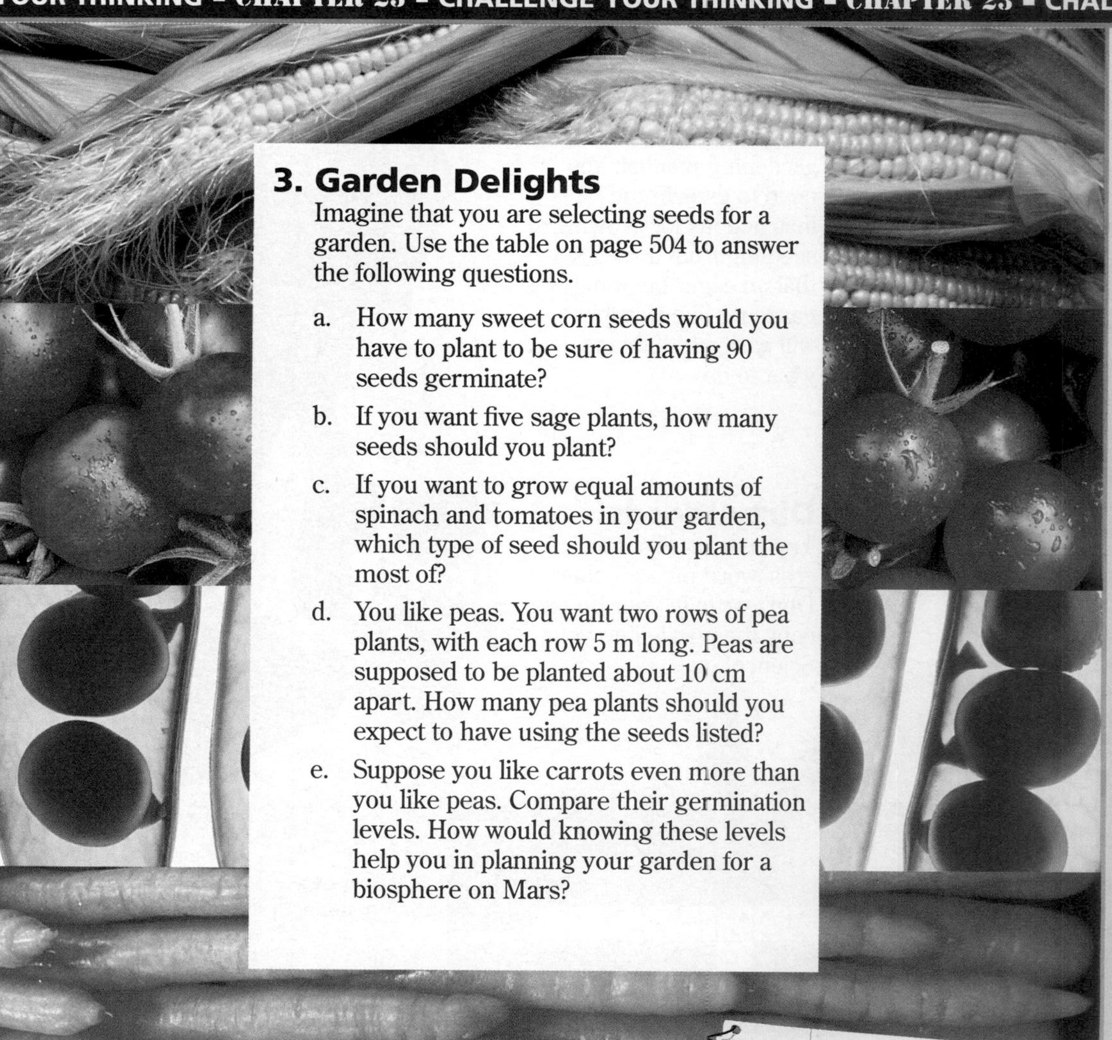

3. Garden Delights

Imagine that you are selecting seeds for a garden. Use the table on page 504 to answer the following questions.

a. How many sweet corn seeds would you have to plant to be sure of having 90 seeds germinate?

b. If you want five sage plants, how many seeds should you plant?

c. If you want to grow equal amounts of spinach and tomatoes in your garden, which type of seed should you plant the most of?

d. You like peas. You want two rows of pea plants, with each row 5 m long. Peas are supposed to be planted about 10 cm apart. How many pea plants should you expect to have using the seeds listed?

e. Suppose you like carrots even more than you like peas. Compare their germination levels. How would knowing these levels help you in planning your garden for a biosphere on Mars?

4. Soil Is . . . ?

After reading this chapter, how would you complete this statement? Use as many words or additional sentences as you need.

Sciencelog

Review your responses to the questions on page 500. Then revise your original ideas so that they reflect what you've learned.

Answers to *Challenge Your Thinking*, continued

3. a. 100 seeds
b. 10 seeds
c. Spinach
d. 75 plants
e. Peas have a germination rate of 75 percent, while carrots have a germination rate of 60 percent. Considering the germination rate will help a person predict how many plants can be grown from a certain number of seeds, whether at home or on Mars. (A person who likes carrots more than peas should take 25 percent more carrot seeds than pea seeds to Mars in order to end up with more carrot plants.)

4. Answers will vary. The following are sample answers:
- Soil is an anchor for plants and a provider of the water, air, and minerals needed for plant growth.
- Soil is good for plants if it contains the right amount of humus, a variety of particle sizes, and water.
- Soil is made up of rock, mineral particles, and the decayed remains of organisms.

Sciencelog

The following are sample revised answers:

1. Like many organisms, plants need food, air, and water. Plants also need light in order to grow. Students may answer that plants need soil. (You may wish to inform students that they may be surprised by what they learn in Chapter 24 about plants growing without soil.)
2. It is unlikely that all of these seeds will become plants. For a given number of seeds, only a certain number will germinate. The percentage of plants that germinate for a certain type of seed is called the *rate of germination*. Different types of seeds have different rates of germination.
3. Soil provides support for the plants and holds the water, air, and nutrients that are needed for plant growth. There are many types of soil. Different types of soil have different levels of clay, silt, sand, and humus.

CHAPTER 24

Inner Actions of Plants

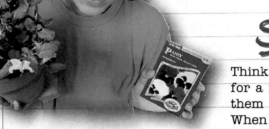

1 How do plants get the water they need?

2 Do you think plants can grow without soil? Why or why not?

3 Which came first, the flower or the seed? Explain your answer.

4 What is the first part of a new plant to develop? Why is this part so important?

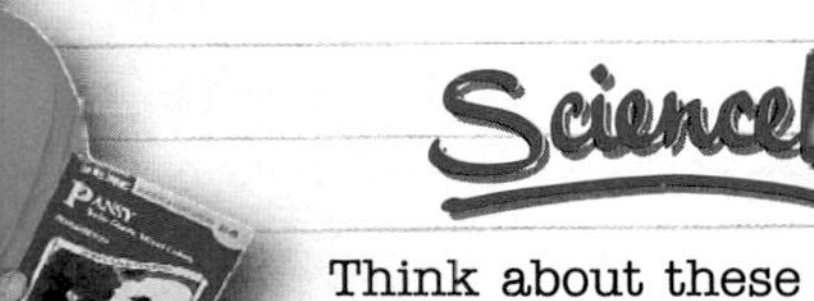

ScienceLog

Think about these questions for a moment, and answer them in your ScienceLog. When you've finished this chapter, you'll have the opportunity to revise your answers based on what you've learned.

517

CHAPTER 24

Inner Actions of Plants

Connecting to Other Chapters

Chapter 23 ***introduces students to plants and begins to explore the needs of plants, as well as their role in our lives.***

Chapter 24 ***discusses plants in more detail, including the biotic and abiotic factors necessary for a plant to survive.***

Chapter 25 ***introduces students to the use of plants in both agriculture and landscaping.***

Prior Knowledge and Misconceptions

Your students' responses to the ScienceLog questions on this page will reveal the kind of information—and misinformation—they bring to this chapter. Use what you find out about your students' knowledge to choose which chapter concepts and activities to emphasize in your teaching.

In addition to having students answer the questions on this page, you may wish to perform the following demonstration for them: A day or two before you are ready to discuss this activity with the class, place a white carnation in a glass of water that contains blue or black food coloring. (The petals of the flower will become colored.) Just before class, place the carnation in a beaker of colorless water and present the carnation to the class along with an uncolored (white) carnation.

Have students propose ways that the dark carnation could have become colored. They should try to explain the process in terms of the parts of the plant. Make a list of students' suggestions on the board. Have students discuss the list. Listen carefully to the discussion to find out what students know about plant processes, what misconceptions they may have, and what about this topic is interesting to them.

LESSON 1 Water, Water, Everywhere!

FOCUS

Getting Started

About 20 minutes before class, place one slice of a peeled potato in a beaker of tap water and another slice in a beaker of concentrated salt solution (150 g of salt in 1 L of water). Explain to students that a potato is actually a swollen underground stem where the food produced by photosynthesis is stored. Have students examine the soaked slices and describe what they see. *(The potato slice in the salt solution should be spongy.)* Explain to students that this is a process known as *osmosis.* Tell students that they will learn more about osmosis in this lesson.

Main Ideas

1. Water enters a plant through its roots and exits through its leaves.
2. Water moves from one plant cell to another by osmosis.
3. A plant's structure permits the distribution of water throughout the plant.

TEACHING STRATEGIES

EXPLORATION 1

INDEPENDENT PRACTICE Have volunteers identify the leaves, stems, and roots of a plant. Encourage students to hypothesize how water and food get from one part of the plant to another.

PART 1

Suggest that students place a drop of water on the roots and then observe them. Students should recognize that the tiny root hairs greatly increase the surface area through which water can be absorbed.

Answers to Give Your Advice are on the next page. ►

LESSON 1 Water, Water, Everywhere!

You know that soil and water are important for the natural growth of plants on Earth. Consider the Mars biosphere. Will plants on Mars need soil and water just as plants on Earth do? While soil may not be absolutely necessary, water must always be present for successful plant growth. How does a plant take in water? How does water travel through the plant? How is the water released again into the Earth's environment or into a closed environment such as a biosphere?

This Exploration will give you an opportunity to collect evidence of the interactions between plants and water.

EXPLORATION 1

Water and Plants Interact

PART 1

Entry

Water enters plants through their roots.

You Will Need

- germinated seeds
- a magnifying glass

What to Do

Look at the newly formed root of a seedling. Notice the root hairs, which are extensions of the root cells. Use a magnifying glass to look at them closely.

Each root hair adds to the total surface area of the root. In what way might this be useful to the plant? Imagine that the water in the soil forms a thin layer around each root hair.

Give Your Advice

Simone wants to plant a garden in an area of hard-packed clay soil. Ray has sandy soil in his garden. Tell these gardeners why their soils need improvement and what they can add to their soils to give the roots of their plants better access to water in the soil.

LESSON 1 ORGANIZER

Time Required
3 class periods

Process Skills
observing, organizing, inferring, analyzing

Theme Connection
Structures

New Terms
Osmosis—movement of water particles across a semipermeable membrane to areas where water particles are less concentrated
Photosynthesis—the process in which green plants use sunlight to convert water and carbon dioxide into food
Stomata (singular, *stoma*)—narrow openings in the leaves of a plant through which gases (oxygen, carbon dioxide, water vapor) pass into and out of the plant
Transpiration—the movement of water out of a plant through the stomata in its leaves

Materials (per student group)
Exploration 1, Part 1: a few germinated seeds; magnifying glass;

continued ►

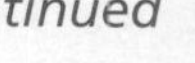

PART 2

Which Way?

The surface of a root hair is a membrane that water can pass through. Water from outside the root hair passes through the membrane. It enters one of the many cells that make up the root. By crossing several membranes inside the root, the water moves to other cells in the plant. But water must be able to move out of cells as well as into them. What makes water move in a particular direction through cell membranes? Try this activity and see.

You Will Need

- 2 containers
- water
- salt
- plants from which pieces may be removed

What to Do

1. Put some water into container *A* and a concentrated salt solution into container *B*.
2. Into each container, place several different plant pieces, like a slice of carrot 5 mm thick, a fern frond, a lettuce leaf, and a geranium leaf. Be sure to place the same kinds of things into each container.

3. After 20 minutes, observe and compare the plant pieces in the containers. Look again after 1 or 2 hours. Do they resemble the plants that they were taken from?
4. Which plant pieces remained crisp or became firmer? Which ones became spongy or limp?
5. In which direction did the water move in container *A*? in container *B*?

You should have noticed that in container *A*, water moved from the container into the plant cells. In container *B*, water moved out of the plant cells and into the salt solution. This movement of water is called **osmosis**. What determines the direction that the water will move? It is the concentration of solutes that determines water's movement.

Water moves through cell membranes from areas where water is more concentrated to areas where water is less concentrated. What does this mean? You know from Unit 3 that the amount of solute in a solution determines how dilute or concentrated the solution is. Now think about a solution in this way: Water (the solvent) and any solute dissolved in it are made of tiny particles.

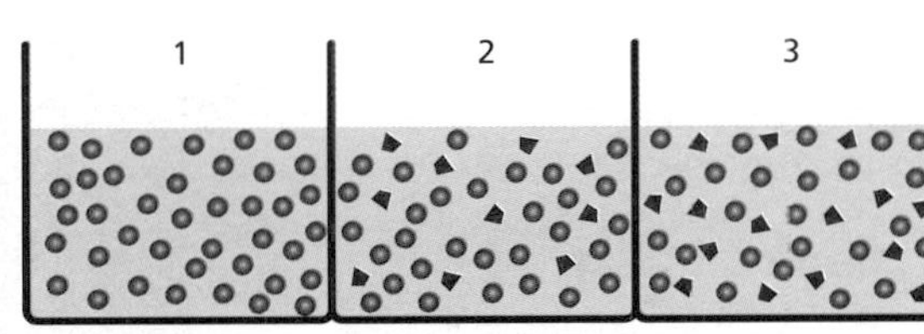

Look at the diagram above. In section *1*, all of the particles are water particles. In section *2*, there are some particles of water and some particles of solute. That means that the concentration of water is lower in section *2*. In section *3*, there are many particles of solute, so the concentration of water is even lower than in section *2*. Predict what would happen if the walls between these sections were cell membranes. Would water move through the cell membranes? In what direction would it move? (A)

Think about the observations that you made when you put the pieces of plants into containers *A* and *B*.

Exploration 1 continued ►

EXPLORATION 1, continued

Answers to *Give Your Advice, page 518*

Simone's clay soil is composed of very fine particles that can become packed into a hard, heavy mass. In that condition, air spaces are few, and water cannot be absorbed into or drained through the soil. Adding some sand can improve drainage.

Ray has sandy soil through which water passes quickly, preventing plants from obtaining sufficient moisture. Adding some clay soil and some humus will help retain the necessary moisture.

Answers to *Part 2*

Students should observe that the plant material in container *A* remained crisp or became firmer. The plant material in container *B* became spongy or limp.

In container *A*, water moved from the container into the plant cells. In container *B*, water moved out of the plant cells and into the salt solution.

Answers to *In-Text Questions*

(A) Students should conclude that in the diagram on this page, water would move from section *1* to section *2* to section *3*.

★ **Two Transparency Worksheets and an Exploration Worksheet (Teaching Resources, pages 18, 20, and 22, respectively), as well as Transparencies 86 and 87, are available to accompany Exploration 1.**

ORGANIZER, *continued*

Part 2: two 500 mL beakers or containers; 1 L of water; at least 20 g of salt; several pairs of identical plant pieces, such as 5 mm thick carrot slices, fern fronds, lettuce leaves, and geranium leaves; **Part 3:** none; **Part 4:** 15 cm stalk of celery, cut at both ends; a few drops of blue food coloring; clear glass or 250 mL beaker; about 100 mL of water; **Part 5:** geranium or other plant with large leaves; clear plastic sandwich bag; small piece of masking tape

Exploration 2, Part 1: variety of leaves; magnifying glass; **Part 2:** glass slide with coverslip; microscope; eyedropper; about 5 mL of water; large leaf such as geranium or lettuce

Teaching Resources
Transparency Worksheets, pp. 18 and 20
Exploration Worksheets, pp. 22 and 28
Transparencies 86 and 87
SourceBook, p. S167

Homework

For homework, have students investigate the existence of water on the surface of Mars. *(Mars has icecaps, which are probably composed partly of water, and channels on the planet's surface, which may be indications of past water movement.)*

Answers to
In-Text Questions

Ⓐ The movement of water out of the plant cells caused the parts to become limp or spongy.

Ⓑ Students may recognize that gardeners can use a knowledge of osmosis to determine when plants need water (when the plants begin to wilt) and to keep plants fresh after they have been harvested.

Answers to
Ben's Blunder

The leaves turned yellow because the concentration of fertilizer in the soil caused water to move out of the plant. Ben should have diluted the fertilizer.

PART 3

Call on a volunteer to read aloud the first paragraph of Part 3. Direct students' attention to the illustrations of the impatiens plant. Ask students to respond with a show of hands if they have ever had a plant that wilted like the one in the illustration. Call on a volunteer to explain what he or she did and what happened as a result. Involve students in a discussion of the questions.

Answers to
Part 3

Students should recognize that once water is supplied to the soil, the concentration of water particles becomes greater outside the root cells. Therefore, the movement of water will be from the soil into the cells.

PART 4

Students should observe that water and food coloring traveled up tiny tubes in the celery stalk. You may wish to inform students that these tube-shaped cells in the celery are called *xylem.* Have students cut the celery stalk in half horizontally and vertically to examine the path of the colored water. Suggest that they use a magnifying glass to examine the celery stalk in detail.

In container *A*, the water inside the plant cells had dissolved substances in it, such as minerals that plants need. The water in container *A* did not have these dissolved substances in it. Water moved from container *A* into the plant cells just as it would move from section *1* to section *2* in the diagram.

In container *B*, there were fewer dissolved substances in the water inside the plant cells than in the water outside that had a lot of solute (salt) dissolved in it. Water moved from the plant cells into the solution, just as it would move from section *2* into section *3* in the diagram. How did the movement of water out of the plant cells affect the plant parts? Ⓐ Water always moves from areas where water particles are more concentrated to areas where they are less concentrated.

Knowing about osmosis can be useful to gardeners, while not knowing about it can cause problems. Think of some situations in which gardeners need to understand osmosis. Ⓑ

Ben's Blunder

Ben put some powdered fertilizer on top of the soil of his potted plant. Then he poured a little water over it. A few days later, the leaves of his plant wilted and turned yellow. Why did this happen? What should Ben have done differently?

PART 3

A Quick Recovery

A plant that has been deprived of water appears limp and wilted. A wilted impatiens usually recovers quickly once water is provided. Explain why water moves into the root cells and then to all of the plant's cells. Where is the concentration of water particles greater? In what direction is the water moving?

The appearance of the recovered impatiens shows that water has moved through the plant. Water has made the shape of the plant parts normal again.

PART 4

Special Routes

You Will Need

- a stalk of celery (cut at both ends)
- blue food coloring
- a clear container
- water

What to Do

Put a few drops of blue food coloring and a piece of celery into a container of water. Observe the celery periodically for several hours. Describe the top end of the celery. What evidence is there that water and food coloring have moved up the celery stalk? Sketch the stalk of celery, showing any features that you think are important.

The celery plant has special cells that serve as tubes for the movement of liquids. Can you see them?

Answer to
Part 4

The appearance of blue food coloring in the celery above the water level is evidence that the water and food coloring have moved up the celery stalk.

Integrating the Sciences

Life and Physical Sciences

Remind students that all matter, including plants, air, soil, and water, is made up of particles. When particles of water move from an area of higher concentration to an area of lower concentration, osmosis occurs. When other substances behave that way, we usually say that diffusion has occurred. Challenge students to use the concept of diffusion in order to explain how perfume can be smelled from across a room.

PART 5

Exit

When water reaches the leaves of a plant, it is used in the food-making process called **photosynthesis**. Water is also released from the leaves into the air. It is easy to collect some of this water.

You Will Need

- a geranium or other plant with large leaves
- a clear plastic bag
- tape

What to Do

Cover one of the leaves of the plant with a plastic bag. Tape it so that no air can enter or leave. Leave the plant in the sunlight or under a bright lamp for a half-hour. Look closely at the inner surface of the bag. What do you see?

The movement of water out of the plant through the leaves is called **transpiration**.

Think About It!

1. A leaf has been compared to a wet towel on a clothesline.
   a. What conditions help a towel to dry?
   b. Would these conditions also increase the transpiration rate of a plant?
   c. How could you help prevent a plant from drying out while it is transpiring?
   d. What beneficial effect does transpiration have for plants on a hot day?
2. In the Mars biosphere, the plants' water supply will have to be maintained. How would you retrieve the water lost by plants through transpiration?
3. Eli and Tanya are farmers. They plant a field of corn with 50,000 seeds. The minimum germination rate for corn seeds is 90 percent. Suppose that two-thirds of the germinated seeds grow into mature corn plants. One corn plant gives off about 200 L of water by transpiration while it is growing.
   a. How much water could all of the corn plants together lose through transpiration while they are growing?
   b. What soil type would you consider to be the best for growing corn? Why?
4. Where in the United States does transpiration cause problems for farmers? What can the farmers do about it?

Answers to *Think About It!*

1. **a.** Dry, warm, windy conditions help the towel to dry by increasing the rate of evaporation.
   b. Yes
   c. Answers may vary, but possible responses include spraying the plant with water, placing the plant in a cooler or shadier environment, keeping the plant in a humid or moist location, and keeping the plant out of the wind and away from drafts.
   d. Because evaporation of water absorbs heat, transpiration helps cool the plant.
2. Student responses may vary but should indicate that the transpired water vapor would have to be condensed into liquid water and then collected in some way. For example, the transpired water vapor might condense on the roof of the biosphere and flow down the sides into pipes that would carry it to collection vessels.
3. **a.** $50,000 \times 0.90 = 45,000$ germinated kernels
   $45,000 \times \frac{2}{3} = 30,000$ mature corn plants
   $30,000 \times 200$ L $= 6,000,000$ L of water that could be transpired by all of the corn plants while they are growing
   b. A loam that holds water well would be good for corn. Such a soil would contain a small amount of sand and larger amounts of clay, silt, and humus. (You may wish to point out that different soil types work well in different environments.)
4. Dry areas in the Southwest and Midwest often suffer from substantial water loss. Irrigation and soil treatment could help prevent this problem.

Integrating the Sciences

Life and Physical Sciences

Students might be interested to learn that only about 1 percent of the water that reaches the leaves of many plants is used by the plants in chemical reactions. The rest evaporates, or transpires, from the leaves as water vapor. Leaves produce heat as a result of the chemical processes that occur there, and the process of transpiration cools the leaves. The evaporation of water from the leaves also helps pull water from the roots up through the stem to the leaves.

PART 5

Students should see that tiny droplets of water form on the inside of the plastic bag. They should conclude that water taken in by the roots of the plant has traveled up through the stem and out through the leaves.

You may wish to briefly review what students know about photosynthesis. During photosynthesis, carbon dioxide, water, and light energy from the sun combine in the presence of chlorophyll and certain enzymes to produce glucose and oxygen.

Water and Plants

Have students read this material silently, or call on a volunteer to read it aloud. Involve students in a brief discussion of how the activities they have completed support the statements and conclusions about the way water is useful to plants. For example, students observed that water caused a wilted plant to regain its shape. They observed that water vapor evaporates from leaves and cools them. They also learned that plants use water during photosynthesis.

Investigate

Encourage students to research photosynthesis and respiration, and provide time for them to share and discuss what they learn. Through their research, students should discover that respiration takes place in the cells of all living things in order to release the energy required for basic life processes. In many ways, the process of respiration is the reverse of photosynthesis. During respiration, glucose combines with oxygen to form carbon dioxide and water.

EXPLORATION 2

PART 1

This activity will be more dramatic and interesting if students are provided with a variety of different leaves to look at. Consider using an African violet, geranium, maranta, begonia, and spider plant. Students could also bring in leaves from home to examine.

★ **An Exploration Worksheet is available to accompany Exploration 2 (Teaching Resources, page 28).**

Homework

Have students write a short answer to the following question: Why can overwatering be bad for a plant? *(Accept all reasonable responses. When roots are exposed to too much water, they may be unable to obtain oxygen from the air in the soil. In addition, water allows the growth of microorganisms that may begin to decompose the roots.)*

Water and Plants

Consider the information that you have gathered about water and plants. You know that water enters a plant's roots, that it moves upward through the plant, and that some of the water escapes through the leaves. Water and plants interact continuously.

In what ways is water useful to a plant?

- Water helps the plant hold its shape.
- Water carries essential dissolved substances throughout the plant.
- Water has a cooling effect as it evaporates during transpiration.
- Water is used by plants for photosynthesis (making food), which takes place in the leaves. For photosynthesis to occur, plants also need carbon dioxide and sunlight. Oxygen is then released by the plant.

Investigate

You might want to investigate the process of photosynthesis by using the resources of a library. If you do, include a look at the process called *respiration* as well. Many people are surprised when they learn that plants carry out respiration. The process of respiration that occurs in plant cells is similar to respiration that occurs in our own cells; it involves taking in oxygen that is then used to release energy from stored food. Carbon dioxide is released during respiration.

Your recent observations have shown that leaves are important to plants. The next Exploration gives you a chance to look at the leaves of a variety of plants and to consider how a leaf's structure suits a plant's needs.

EXPLORATION 2

A Look at Leaves

PART 1

Outer Looks

You Will Need

- a variety of leaves
- a magnifying glass

What to Do

1. Examine the upper and lower surfaces of each kind of leaf.
2. Describe each leaf. You may either draw the features of each leaf surface or write a list of words to describe its features.
3. Exchange your list or drawing with others to see if they recognize the leaves that you described.

PORTFOLIO

Students may wish to include their findings from Investigate in their Portfolio. Encourage students to do research outside their textbooks and to place their notes in their Portfolio.

Theme Connection

Structures

Focus question: What structures does water pass through as it moves from the soil into a plant and back out into the environment? *(Plants absorb water from the soil through root hairs. Water diffuses into the roots and then moves up special tubes until it is distributed throughout the plant. When water reaches the leaves of the plants, it is either used in photosynthesis or released as water vapor.)*

Think About It!

1. In what ways are all of the leaves alike?
2. How do you think the basic shape of leaves equips them for photosynthesis?
3. Can you tell the upper and lower surfaces of the leaves apart? How? Why do you think the surfaces are different?

Now look for clues to answer this question: How do most substances enter and exit a leaf?

PART 2

A Closer Look

You Will Need

- a glass slide
- a coverslip
- a microscope
- a leaf (A large geranium leaf or a piece of romaine lettuce is suitable.)

What to Do 

1. Carefully peel the thin layer from the lower surface of a leaf. (Hint: If the leaf is crisp, bend a piece of it until it snaps, and then separate the layer from the leaf's surface.)
2. Place the sample on a slide, with the lower surface of the leaf facing upward. Put one drop of water on top of the sample. Gently set the coverslip over your sample, and examine it under the microscope.

Compare what you observe on your slide with the picture shown below. Do you see what look like narrow openings? These are **stomata** (singular, *stoma*). What substances are likely to pass through these openings?

Maranta

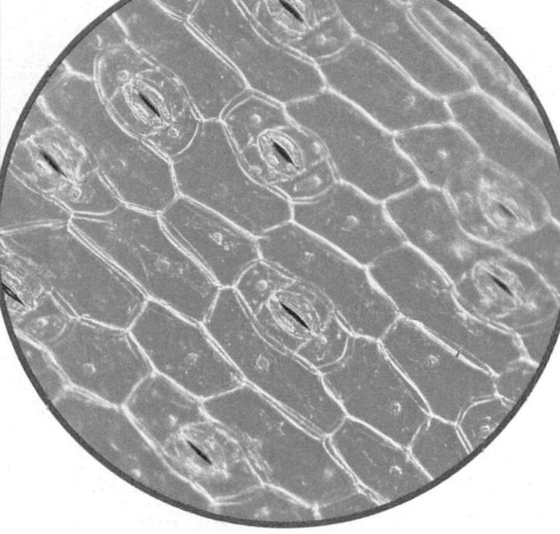

Willow

Some More to Think About

Look at the leaves of the plants you are growing. Are they all in perfect condition? Look at the leaves of other available plants.

1. Make a record of any damage you find on the leaves. This could include holes, color changes, or wilting.
2. Caterpillars and other living things find plant leaves very attractive. Explain why you think this is so.
3. What kinds of problems do you predict that your plants will have if their leaves suffer damage from other living things?

523

FOLLOW-UP

Reteaching

Suggest that students make a model of a plant from clay, papier-mâché, or some other material to show how water flows from roots to leaves. Display the finished models.

Assessment

Have students write and illustrate a paragraph explaining how the potatoes in the Getting Started activity on page 518 demonstrated the process of osmosis.

Extension

Explain to students that not all plants live in an environment where water is abundant. Have interested students find out about plants that grow in very dry climates and about how these plants are adapted to obtain and conserve water.

Closure

Using a microscope, suggest that students examine a cross section of a leaf. If possible, provide them with a prepared slide. Suggest that they use a biology text or some other source to identify the different kinds of cells and tissues they observe.

Answers to *Think About It!*

1. Accept all reasonable responses. Most leaves have veins and are green.
2. The basic shape of leaves provides a large surface area to collect the sun's energy.
3. Answers will vary but may include differences in the color, texture, luster, and characteristics of the veins. The upper surface of a leaf is usually smoother than the lower surface. The veins on the lower surface tend to stand out more. The upper surface is usually a darker green because more chlorophyll is concentrated on the upper surface, where sunlight strikes the leaves.

PART 2

You may wish to demonstrate how to prepare a slide for the class. Staining the tissue with a little iodine starch-test reagent may make it easier for students to distinguish the cells. Encourage students to identify stomata. Students should conclude that gases such as oxygen, carbon dioxide, and water vapor can pass through these openings.

Answers to *Some More to Think About*

1. Findings will vary depending on the plants used. Students should illustrate their observations.
2. Leaves offer a good source of food for many types of immature and adult insects.
3. A plant's ability to produce food will be limited if many of its leaves suffer damage. If enough leaves are lost, the plant will be unable to carry out photosynthesis.

Giving Plants a Hand

FOCUS

Getting Started

You may be able to purchase a hydroponic tomato at a local supermarket. Have students guess what might be special about the tomato. Explain that the tomato is special because it was grown without soil.

Main Ideas

1. Nitrogen, phosphorus, and potassium are the soil nutrients needed most by plants.
2. Plants may develop deficiencies if a soil nutrient is lacking or if the soil's pH is too high or too low.
3. Liquid solutions of nutrients can replace soil for plant growth.

TEACHING STRATEGIES

What's in a Number?

Point out that the order in which the nutrients in fertilizers are listed is always nitrogen, phosphorus, and potassium.

Answers to *In-Text Questions*

Ⓐ 15-30-15 has 15 percent nitrogen, 30 percent phosphorus, and 15 percent potassium.

Ⓑ Phosphorus is the chemical in the fertilizer that is present in the largest amount. Without it, plants would not develop healthy roots and stems, and water and nutrients could not be absorbed properly.

Ⓒ Discolored leaves may indicate a nitrogen deficiency.

Homework

The Graphing Practice Worksheet on page 32 of the Unit 8 Teaching Resources booklet makes an excellent homework activity once students have read What's in a Number? on pages 524–525.

LESSON 2

Giving Plants a Hand

What's in a Number?

I'm going to use 10-52-10 now, and then switch to 20-20-20.
How about 15-30-15 later?

Are the people on the other side of the fence speaking in code? It certainly looks that way! Actually, they are describing the kinds of fertilizers they use. Each set of numbers refers to the percentages of the three chemical substances that are most essential for plants. What does 10-52-10 mean? A 10-52-10 fertilizer contains 10 percent nitrogen (N), 52 percent phosphorus (P), and 10 percent potassium (K). The other ingredients are inactive substances like clay or chemical substances like sulfur that make the fertilizer easy to use. 20-20-20 contains 20 percent N, 20 percent P, and 20 percent K. What does 15-30-15 contain? Ⓐ

These number groupings are called the N-P-K ratings of fertilizers. Why did one person on the other side of the fence recommend using 10-52-10 first? This fertilizer is especially good for seedlings and for plants that have just been transplanted or repotted. Which chemical is present in the largest amount? This chemical helps plants develop strong root and stem structures. What would happen if a plant lacked this chemical? Ⓑ

Each mixture has its special use. For fertilizing leafy trees and shrubs, 28-14-14 is ideal. The nitrogen helps branches and leaves develop. Nitrogen must also be present for plants to have their usual green color. What symptoms do you think would indicate a nitrogen deficiency? Ⓒ

LESSON 2 ORGANIZER

Time Required
2 class periods, plus a 1-week growing period

Process Skills communicating, measuring, organizing, analyzing

New Term
Hydroponics—growing plants in nutrient solutions instead of soil

Materials (per student group)
Exploration 3: about 50 mL distilled water; several 20 mL soil samples from different areas; 1 piece each of blue and red litmus paper for each sample; small paper plate

Exploration 4: 200 mL of rainwater or distilled water; solute, such as small packet of hydroponic nutrient powder, at least 2 mL vinegar, few spoonfuls of gardener's lime, etc.; 250 mL beaker or plastic container; baby-food jar or clear plastic cup; about 100 mL of vermiculite or clean sand; 10 mustard seeds untreated with fertilizer or fungicide; eyedropper; metric ruler; latex gloves; lab aprons; safety goggles

Teaching Resources
Discrepant Event Worksheet, p. 31
Graphing Practice Worksheet, p. 32
Exploration Worksheets, pp. 34 and 35

Potassium strengthens a plant, especially when fruits or vegetables are forming. It also increases a plant's resistance to pests and diseases. Potassium may also play a part in storing or releasing nitrogen, according to the plant's needs.

Nitrogen, phosphorus, and potassium are not the only substances needed by plants. There are many others, but these three nutrients are needed in the greatest quantities.

Increase Your Knowledge

1. Find several different kinds of fertilizers by using N-P-K ratings. Look at containers in gardening centers and read catalogs from gardening stores. Match each fertilizer with its recommended use.
2. Investigate the composition and use of substances called *natural* fertilizers, such as manure and compost.
3. Earlier you chose a soil mixture for the Mars biosphere. Now consider what fertilizers you would take. Would you take
   a. chemical fertilizers only?
   b. natural fertilizers only?
   c. both types of fertilizers?

   Explain the reasons for your choice.

The composition of fertilizers is not the only chemistry that is important to gardeners. Plants have their own special pH needs as well. Some grow best in acidic soil, which has a pH value less than 7. Others require an alkaline soil, which has a pH value greater than 7. For example, cabbage grows best in slightly alkaline soil, while rhododendrons thrive in acidic soil. Many plants grow best in soil with a pH of 6.5. Is this soil acidic, neutral, or alkaline? (D) The pH of the soil affects whether nutrients will dissolve in the water around the plant. That's why gardeners must be aware of the nutrients in a fertilizer and of the pH of their soil.

What do you think is the pH of the soil in this cabbage patch? (E)

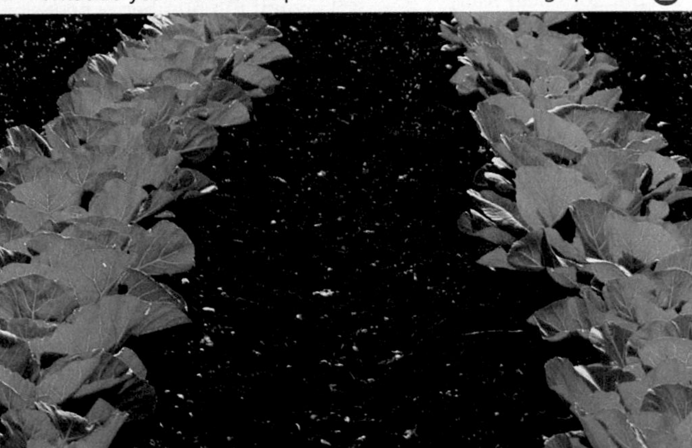

EXPLORATION 3

The Litmus Test

You Will Need

- distilled water
- soil samples from different areas
- blue and red litmus paper

What to Do

Perform a simple test on a sample of soil. Add a little distilled water to a small amount of soil until it makes a thick paste. Place pieces of red and blue litmus paper on the moist soil and record your observations. If the soil is acidic, the blue litmus paper will turn red. If the soil is alkaline, the red litmus paper will turn blue.

So far, you have been learning about how to grow plants in soil. Can plants be grown without soil? Read on and find out.

Homework

Have students create a pamphlet that describes in simple language the variations in and uses of fertilizers. The pamphlet should be addressed to someone interested in gardening. *(Answers should reflect an understanding of the N-P-K ratings discussed on pages 524–525.)*

Meeting Individual Needs

Gifted Learners

Have interested students conduct research on the effect of ground pollution on plant growth. With your supervision, students can conduct experiments of their own design to test the effect of ground pollutants on plants. Suggest that they try to find answers to the following questions: What kinds of ground pollution affect plants? Where are the pollutants absorbed into the plant? Have students share what they discover with the class.

Increase Your Knowledge

GUIDED PRACTICE Have students read the paragraph on soil pH. You may wish to draw a pH scale on the chalkboard (see page 60) to help students visualize what *acidic* and *alkaline* mean.

Answers to *Increase Your Knowledge*

1. Answers will vary. (Students should evaluate at least three different fertilizers.)
2. The N-P-K ratings of manure and compost samples vary. These samples are often high in nitrogen.
3. Accept all reasonable responses. Students may say that they would take chemical fertilizers because these weigh less and would take up less space on the trip and in the biosphere. Some students may point out that there will be waste from humans and other organisms in the biosphere that could be recycled as natural fertilizer.

Answers to *In-Text Question and Caption*

(D) A pH of 6.5 is slightly acidic.

(E) The pH of the soil in a cabbage patch should be around 8, or slightly alkaline.

EXPLORATION 3

Divide the class into small groups and distribute the materials. Soil samples should be from different areas of the community. Have students complete the Exploration and compare their results to discover how the pH of soils from different areas varies.

★ **As an introduction to Exploration 3, you may wish to perform the teacher demonstration described in the Discrepant Event Worksheet on page 33 of the Unit 8 Teaching Resources booklet. An Exploration Worksheet is also available to accompany Exploration 3 (Teaching Resources, page 34).**

Plants Without Soil—The Way of the Future?

Have students read and discuss the material on hydroponics. (They may be aware of this topic if they did the extension activity on hydroponics in Unit 3.) Encourage them to speculate about whether hydroponics would be a good way to grow plants in a biosphere. Then have them complete the Exploration that follows.

EXPLORATION 4

This Exploration helps students understand that plants can get the nutrients and water they require from either soil or solutions of nutrients. The Exploration is easy to set up and results can be seen within a week. Variables such as the temperature and nature of the nutrient solution can be easily controlled.

Divide the class into small groups and distribute the materials. Each group should investigate one solution.

If students have trouble identifying a solution of their choice that they wish to test, you may wish to suggest pure tap water or distilled water with chalk, baking soda, fertilizer, or sugar added to it. Chalk and baking soda can be used as substitutes for gardener's lime. In addition, students may need help figuring out how to create a quarter-strength nutrient solution; suggest that they dilute a full-strength solution by a factor of four.

SAFETY ALERT Make sure that all students submit their proposed solutions to you for a safety review before they begin the Exploration.

INDEPENDENT PRACTICE Have students create concept maps about plant nutrients, fertilizers, and soil pH. Students should include nitrogen, phosphorus, and potassium in their maps along with the effects of these nutrients on plants.

Plants Without Soil—The Way of the Future?

Growing a plant without soil sounds impossible! But if you replace soil with something else that can perform the functions of soil, then the plant should grow. To grow a plant without soil, you must find a way to anchor the plant in an upright position and to provide water, air, and other essential nutrients to it.

Today some plants are being grown without soil. This process is known as **hydroponics**. Instead of soil, a solution supplies the important nutrients. Perhaps hydroponics would be a useful way to grow food in your biosphere. Consider using this method for at least some of the crops when you write your proposal.

Here is an Exploration that investigates the growth of mustard plants in different nutrient solutions.

EXPLORATION 4

Which Nutrient Solutions Are Best for Growing Mustard Plants?

Before starting, form into teams of three or four and choose one of the nutrient solutions listed below to investigate.

You Will Need

- rainwater or distilled water
- rainwater or distilled water with hydroponic nutrient powder dissolved in it (See the package for directions.)
- rainwater or distilled water with quarter-strength hydroponic powder dissolved in it
- acidic water—rainwater or distilled water with vinegar added (at least 10 mL of vinegar per liter of pure water)
- alkaline water—rainwater or distilled water saturated with gardener's lime
- a solution of your own choice
- a beaker or plastic container to hold the nutrient solution
- a baby-food jar or clear plastic cup
- vermiculite or clean sand
- mustard seeds
- a dropper
- latex gloves

What to Do

1. Fill a baby-food jar or a clear plastic cup about halfway with a rooting medium such as vermiculite or well-washed sand.
2. Slowly add your nutrient solution until the top of the rooting medium is damp.
3. Place 10 mustard seeds in the thin film of water on the surface of the rooting medium. Put your jar in moderate sunlight. Each day, add nutrient solution with a dropper to keep the top of the rooting medium damp.

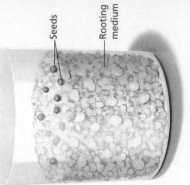

4. At the same time each day, count the number of seeds that have germinated. Record your findings in a table like this one.

Day	0	1	2	3
Number sprouted	0	0	3	8

Cooperative Learning
EXPLORATION 4

Group size: 3 to 4 students
Group goal: to investigate the effect of different nutrient solutions on plant growth
Positive interdependence: Assign each student a role such as chief investigator, materials coordinator, recorder, or data collector.
Individual accountability: Have students write a brief summary of their findings.

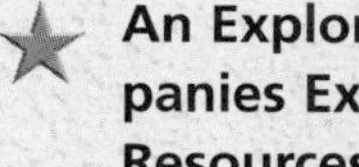

An Exploration Worksheet accompanies Exploration 4 (Teaching Resources, page 35).

Meeting Individual Needs

Gifted Learners

Have students propose and carry out an additional experiment that applies to hydroponics. (This can be an extension of question 6 on page 527.) For example, students may wish to test the effects of different types of rooting mediums. Be sure that students submit their proposals to you for a safety review before they begin their investigation.

5. Once your seeds have germinated, measure the height of the five largest seedlings at the same time each day. Then calculate their average height.

6. Plot a graph of the average height of a seedling against the time in days.

Analyze Your Findings

Share your results from the Exploration with the members of other teams.

1. Did the average time for germination of the seeds vary with the kind of nutrient solution used? If so, list the nutrient solutions in order, from least time taken for seed germination to most time taken.
2. How was seedling growth affected by different types of nutrient solutions? Which solutions produced the most plant growth? Which produced the least plant growth?
3. How did the amount of hydroponic powder dissolved in the water affect plant growth?
4. Suppose your class wished to present a report of your findings to another class that is also doing this Exploration. Design a table and record the most important data found by each of the teams in your class. Will the other groups of students clearly understand the way you have set up and labeled the parts of your table?
5. Draw some conclusions from your class data to present to the other class.
6. Have you thought of other questions that you might investigate about hydroponics? What are some of them? Choose one of these questions and design an experiment that will help you answer it.

A plant scientist studies lettuce that is growing hydroponically on polystyrene boards.

527

Analyze Your Findings

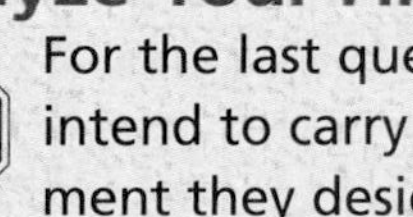

For the last question, if students intend to carry out the experiment they design, they should have you review their plans for safety considerations before proceeding.

Answers to *Analyze Your Findings*

1. Answers will vary. Seed germination rate will probably be about the same for all of the nutrient solutions because germination depends little on the nutrient content of the medium.
2. Seedling growth will probably be the best in the nutrient solution made according to the label directions. It will probably be the worst in the saturated lime solution.
3. Students will probably observe that seedling growth is less when less than the recommended amount of hydroponic solution is used. If more than the recommended amount is used, the plants may grow too fast and become spindly.
4. Student tables should illustrate the most important information from each group.
5. The data will vary depending on the results of student investigations and the types of nutrient solutions used.
6. Sample questions: Do the nutrients supplied by the hydroponic solution need to be different at different stages of the plant's life? Can hydroponic powder be dissolved in something other than water and still be effective?

FOLLOW-UP

Reteaching

Bring in labels from four or five different fertilizers. Have students identify the nutrient content of each fertilizer using only the numbers provided. Then discuss what each nutrient does for the plant.

Assessment

From an agricultural extension agent, fertilizer dealer, or university agriculture department, obtain plants or photographs of plants showing the effects of nitrogen, potassium, and phosphorus deficiencies. Ask the class to be "plant doctors" and to diagnose the plants' nutrient problems by drawing on the information they learned in this lesson.

Extension

Have students create a presentation about the great potential of hydroponics for the future of agriculture. Suggest that the students make their presentation to the class.

Closure

Have students draw a map of their school's neighborhood or of the school grounds. At as many locations as possible, have students identify the pH of the soil. Then have them list some of the plants they found growing in each location.

Flowers and Pollination

FOCUS

Getting Started

In order to help students begin thinking about flowers, encourage discussion about the following question: What is a flower? Have students think about different types of flowers, the function of flowers, and where and when flowers grow. Record student ideas on the board.

Main Ideas

1. Pollination is the beginning of the process by which flowering plants produce fruits and seeds.
2. The flowers of some kinds of plants are self-pollinating; those of other kinds must be cross-pollinated.
3. Carefully controlled cross-pollination can result in new varieties of plants.

TEACHING STRATEGIES

New Plants From Seeds

Have a volunteer read the lesson introduction on this page. Then, using the illustration on page 530, have students discuss the questions.

Answers to *In-Text Questions*

1. The petals
2. The sepals
3. The pistil (stigma, style, and ovary) make up the female parts of a flower. It is here that the seeds and fruit develop.
4. The stamen (anthers and filaments) make up the male parts of a flower. This is where pollen is produced.
5. Seeds form in the ovary.
6. For seeds to form, pollen must enter the ovary, and fertilization must take place.

Flowers and Pollination

New Plants From Seeds

You have been growing plants from seeds. You know that seeds are the parts of plants that are produced specifically to make new plants. You have worked with seeds in germination tests and have seen how plants can grow from seeds. A moment's thought will probably remind you of things you've known for years about flowers and seeds. But where did the seeds come from? the flowers? How are flowers and seeds related to each other?

Look at the illustration of a flower on page 530. See if you can answer these questions:

1. What parts of a flower are often brightly colored and may attract bees, birds, or bats?
2. What parts at the base of the blossom enclose the flower bud before it opens and are often green?
3. Which parts are known as the female parts of the flower? What do they do for the plant?
4. Which parts are known as the male parts of the flower? What do they do for the plant?
5. Where do the seeds form?
6. What has to happen in order for seeds to form?

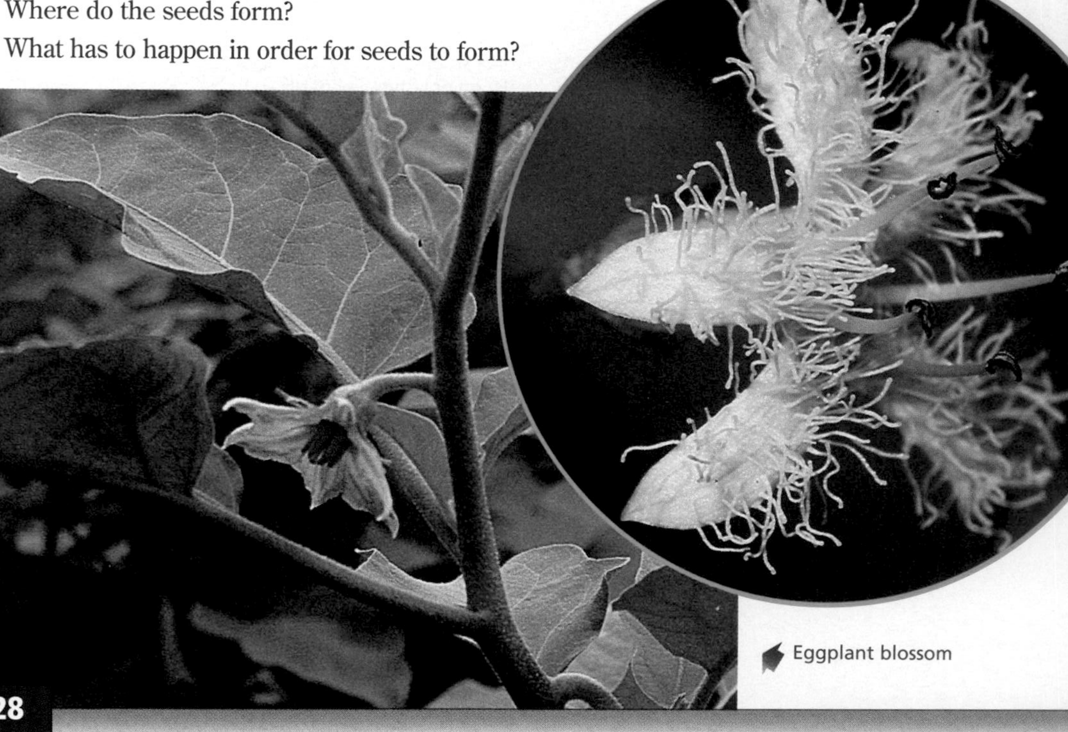

Squash blossoms

Buckbean blossom

Eggplant blossom

528

LESSON 3 ORGANIZER

Time Required
2 to 3 class periods

Process Skills
observing, classifying, analyzing

Theme Connection
Changes Over Time

New Term
Pollination—the transfer of pollen from the tip of the anther to the tip of the stigma of the same or another flower

Materials (per student group)
Exploration 5: several different cut flowers; single-edged razor blade in holder or sharp knife

Teaching Resources
Exploration Worksheets, pp. 38 and 40
Activity Worksheet, p. 42
Transparencies 88 and 89
SourceBook, p. S169

Pollination and Seed Production

Pollination is the transfer of pollen from the tip of the *stamen* of a flower to the tip of the *pistil* of the same or another flower. After this, fertilization (and later fruit and seed production) can occur.

In nature, pollination is often aided by the action of wind and insects. Sometimes people can help make pollination successful. This is especially useful for plants grown indoors. How could you make certain that plants are pollinated in the biosphere? **A**

Most flowers have stamens and pistils in the same blossom. Only a slight movement is required to pollinate plants such as tomatoes, peppers, eggplants, peas, or beans. Some plants, however, have separate male and female blossoms. These include cucumbers and squash. You can recognize female blossoms by the miniature fruit *(ovary)* that appears beneath the flower.

Cucumber blossoms

Sweet pea blossoms

Pepper blossoms

Tomato blossom

529

Did You Know. . .

The belief that a four-leaf clover is good luck originated around 200 B.C. among the Druids, a group of people who lived on the British Isles. The Druids believed that possession of a four-leaf clover, a rarity at the time, allowed the bearer to see demons and to counteract their evil influences with incantations.

Meeting Individual Needs

Second-Language Learners

Provide students with a complete flower, such as a lily or a tulip, and ask them to make a bilingual poster or diagram to show what the different parts of the flower look like. Each part should be identified with a bilingual label. Allow students to refer to a bilingual dictionary, if necessary. When students have finished, review their work for science content, with minimal emphasis on language proficiency.

Pollination and Seed Production

GUIDED PRACTICE Have students read the material silently. When they have finished, involve them in a discussion of pollination. Be sure students understand that pollination and fertilization are not the same thing. In fertilization, a sperm cell in a grain of pollen travels through the pistil of a flower down a pollen tube to unite with an egg cell in the ovary.

If possible, bring a few flowers to class and demonstrate how easily pollen can be shaken from the anthers.

Answer to *In-Text Question*

A In a biosphere, pollen could be transferred by having people shake the plants, by providing artificial wind, or by using a colony of bees.

Multicultural Extension

Herbs and Spices

Spices usually come from plant seeds, while herbs come from leaves or stalks. For instance, cilantro and coriander come from the same plant: cilantro, an herb, is the leaf, and coriander, a spice, is the seed. The herbs and spices used in a country can be indicative of the plants of that region. Have students choose an herb or spice and research how it is produced and where it comes from. Have them create an information sheet that includes the origin of the herb or spice, a recipe for a dish from another country that uses that herb or spice, and an interesting fact about the culture of that country. You may wish to assemble the information sheets into a booklet.

Homework

Have students find out the derivation of the word *pollinate.* (Pollinate *comes from the word* pollen, *which is derived from the Latin word meaning "dust."*)

EXPLORATION 5

You may wish to introduce the concept of perfect and imperfect flowers. Explain that a flower that contains either male or female parts but not both is called an *imperfect* flower. If the flower contains a stamen but no pistil it is called a *staminate* flower. If it contains a pistil but no stamen it is called a *pistillate* flower. A flower that contains both male and female parts is called a *perfect* flower.

If possible, provide students with a wide variety of flowers to examine. Lilies and other flowers with distinct parts work well. Set up stations around the room so that students can travel from one station to the next, investigating a different flower at each station. If there are examples of both perfect and imperfect flowers, be sure that all students have a chance to examine at least one of each type. Monitor the activity by walking among students and confirming their observations. Encourage students to share their results with one another.

When students have completed the activity, involve them in a discussion of how the specific features of the flowers increase their chances of becoming pollinated. For example, pollen is easily knocked from anthers on long filaments. Insects easily collect pollen from short stamens that lie flat against the petals. A tall pistil easily catches pollen from the wind. A short pistil among tall stamens is easily covered with pollen that drops from the anthers.

★ **An Exploration Worksheet (Teaching Resources, pages 38–39) and Transparencies 88 and 89 are available to accompany Exploration 5.**

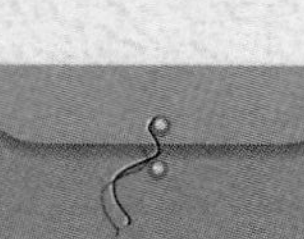

PORTFOLIO

Encourage students to include their observations and sketches from Exploration 5 in their Portfolio.

EXPLORATION 5

Parts of Flowers

You Will Need

- a variety of cut flowers
- a razor blade or sharp knife

What to Do

Be careful: Use sharp instruments carefully. Razor blades should be single-edged and in a holder. Return sharp instruments to a safe place after use.

1. Study each flower.
   a. Determine whether the flower contains male parts, female parts, or both.
   b. Identify the stamen and pistil. If necessary, carefully dissect the flower with a razor blade or sharp knife.
2. Sketch and label the flower.
3. What features of each flower increase its chances for pollination in nature?

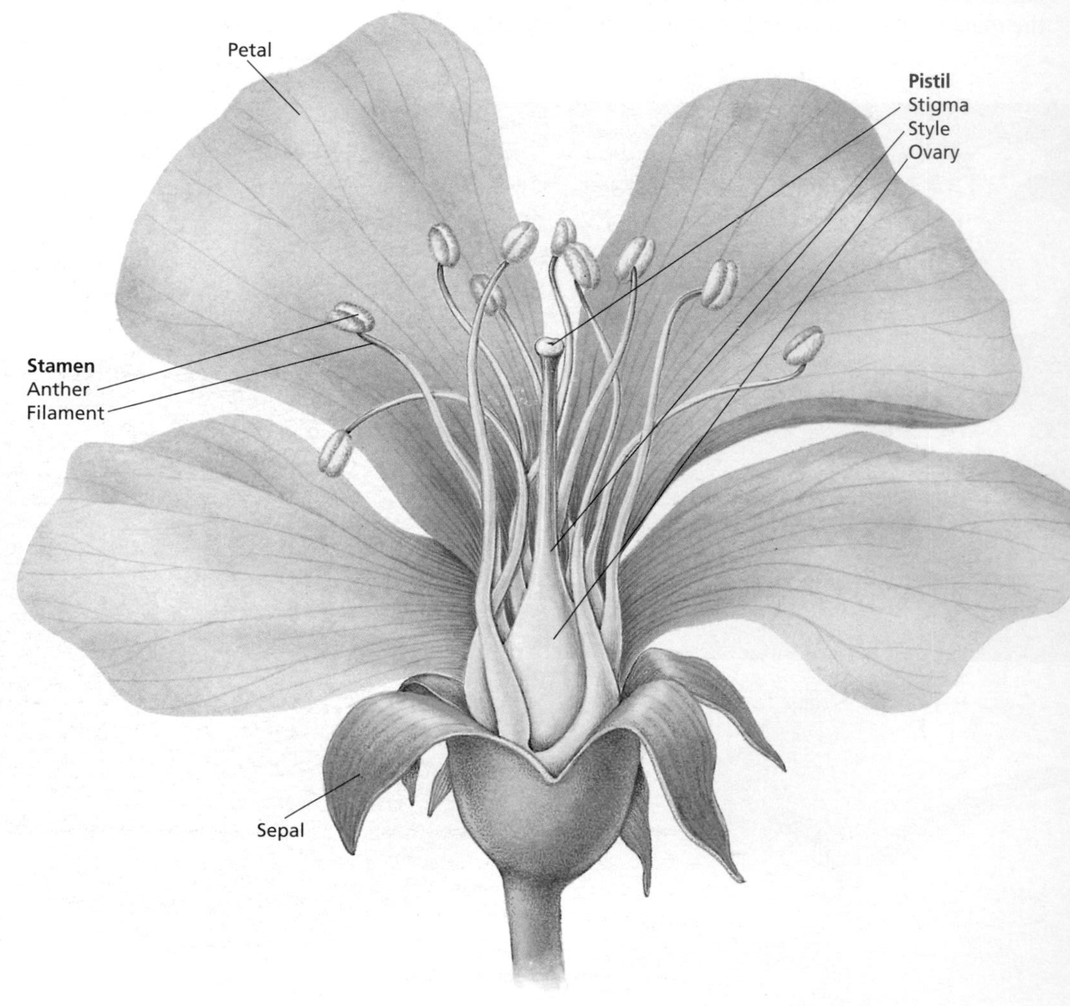

530

ENVIRONMENTAL FOCUS

Ask: How do flowers attract bees? *(By scent or coloration)* Inform students that, unlike humans, bees can see ultraviolet light. Many flowers have special markings that reflect only ultraviolet light. Thus, a flower that seems colorless to us might look completely different and very attractive to a bee!

A Picture Story

In your own words, describe how seeds form and develop into new plants. Use this plant life-cycle diagram as your source of information.

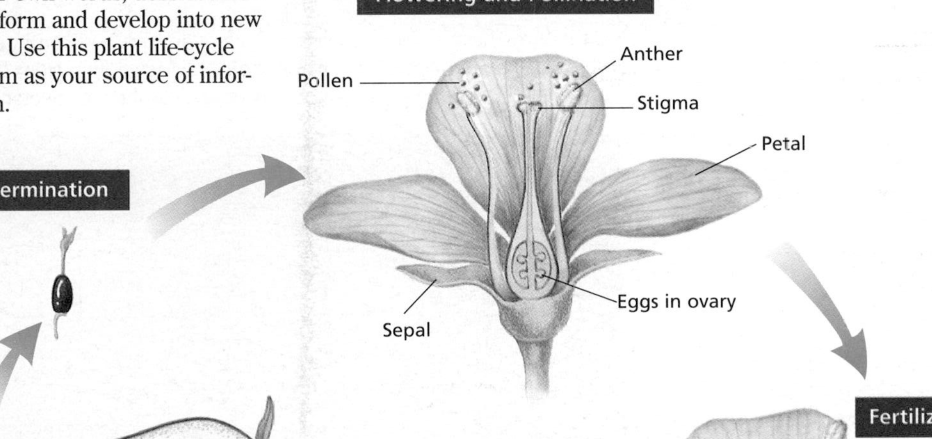

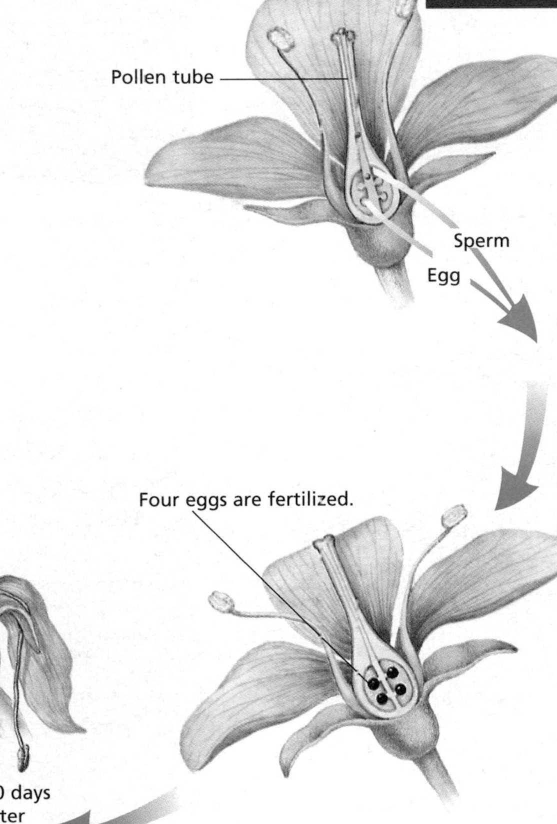

531

EXPLORATION 5, *continued*

A Picture Story

GUIDED PRACTICE Before students begin writing, you may wish to have them review the diagram as a class. Begin the discussion with the germination of a seed, and call on students to continue the explanation until the entire cycle has been discussed. If necessary, elaborate on parts of the process. For example, you may wish to explain to students that the pollen grain produces the pollen tube and the sperm cell. The eggs develop within the ovary. An egg and a sperm unite within the ovary to form a seed. Point out that the fleshy part of a fruit is the enlarged wall of the ovary that was once a part of the flower.

Answers to *A Picture Story*

Students' descriptions of how seeds form and develop into new plants should include the following steps:

- A seed germinates and produces a seedling.
- The seedling grows into an adult plant that produces flowers.
- Pollen grains from the anthers of the flower are transferred to the stigma during pollination.
- A pollen grain grows a pollen tube down the pistil. The sperm passes through the tube to reach the egg in the ovary.
- During fertilization, the sperm unites with the egg.
- The fertilized eggs develop into seeds as the ovary develops into a fruit.
- The fruit continues to grow until it ripens and falls from the plant.
- The fruit eventually releases the seeds, and the process begins again.

CROSS-DISCIPLINARY FOCUS

Health

You may wish to have students compare reproduction in flowering plants with reproduction in mammals.

PORTFOLIO

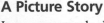

Encourage students to place their responses to A Picture Story in their Portfolio. Allow students to write their stories as outlines, paragraphs, brief phrases, or any other logical organization. Prior to placing their stories in their Portfolio, have students assess each other's stories as a form of peer review.

EXPLORATION 6

For Part 1, be sure students understand that the plastic bag must be removed briefly in order to transfer the pollen. Also, you may wish to emphasize that two different trees (or plants), not just two different flowers, must be involved for cross-pollination to occur.

Answers to *Part 1*

This illustration shows an example of cross-pollination because the pollen came from one tree and the egg came from another.

Answers to *Part 2*

This illustration is an example of self-pollination because pollen from the anthers of a flower is transferred to the stigma of the same flower. Examples of perfect flowers in which self-pollination occurs include buttercup, wild rose, petunia, morning glory, and lily.

Answers to *Part 3*

This is an example of cross-pollination because pollen from the anthers of one plant is transferred to the stigma of another plant. You may wish to point out that this example differs from the example in Part 1 because the holly plant has imperfect flowers; the flowers of one plant are all male and the flowers of the other plant are all female.

Answers to *Part 4, page 533*

This is an example of self-pollination because pollen from the anthers of one flower is transferred to the stigma of another flower on the same plant. Point out that this example differs from the example in Part 2 in that a single cucumber plant has both male and female flowers. Examples of imperfect flowering plants that self-pollinate include pumpkin, squash, willow, and watermelon.

EXPLORATION 6

What Kind of Pollination?

Look at the examples of some of the ways that pollination takes place. Decide which show *self-pollination* (pollen and egg coming from the same plant) and which show *cross-pollination* (pollen and egg coming from different plants).

PART 1

Peach Tree

Jack performed this experiment, which was recommended in his gardening book.

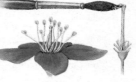

1. Pinch off the petals and anthers.

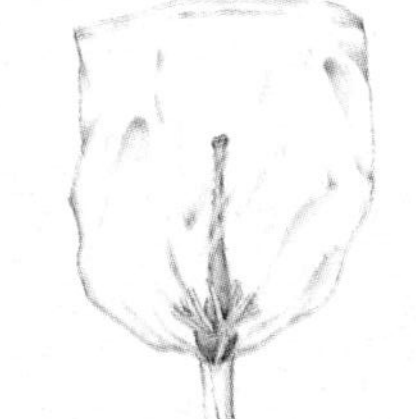

2. Cover the pistil with a bag.
3. Get anthers from another tree.
4. Using a brush, transfer pollen from the anthers to the stigma, and re-cover the pistil with the bag for a short time.

PART 2

Nasturtium Plant

Nasturtiums have perfect flowers. (Perfect flowers have both male and female parts.) A bee or hummingbird carries pollen from the anthers to the stigma of the same flower as it probes for nectar.

PART 3

Holly Plant

Each plant has only male flowers or only female flowers. Pollen must be transferred by the wind or by insects.

Male plant

Female plant

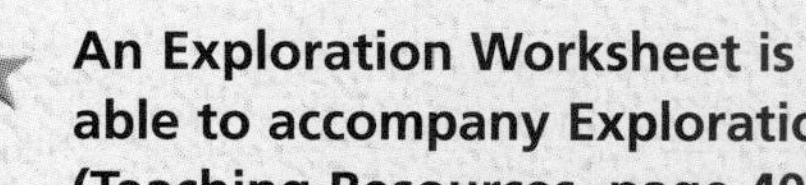

Theme Connection

Changes Over Time

Focus question: What changes occur during the life cycle of a flowering plant? *(A seed germinates and develops into a seedling. The seedling grows into a mature plant, which then produces flowers. The flowers are pollinated and produce seeds. The seeds are scattered, and the cycle begins again.)*

★ **An Exploration Worksheet is available to accompany Exploration 6 (Teaching Resources, page 40).**

Homework

As a follow-up to Lesson 3, have students complete the Activity Worksheet on page 42 of the Unit 8 Teaching Resources booklet.

PART 4

Cucumber Plant

Cucumber plants have separate male and female flowers, called *imperfect* flowers, on the same plant.

Suggestion: Try your hand at making certain that either self-pollination or cross-pollination takes place. Save any seeds that are produced. Plant them, and enjoy the results of your plant-breeding technique.

Cross-Pollination as a Plant-Breeding Technique

Cross-pollination often requires a delicate touch. A scientist transfers pollen from the anthers of a flower on one plant to the stigma of a flower on another plant. Here is what might be done with two varieties of wheat, wheat *S* and wheat *W*.

Wheat *S*	Wheat *W*
▪ resists disease	▪ is subject to certain diseases
▪ survives low temperatures	▪ dies at low temperatures
▪ has a high yield	▪ has a low yield
▪ matures in 110 days	▪ matures in 90 days

Wheat *S* has one quality that a farmer might like to change. What is it? Before you read further, propose a way to get a wheat sample with the desirable qualities of both *S* and *W*.

Here is one method. It is called *cross-breeding.* Ⓐ

1. Grow wheat *S* and wheat *W*.
2. When flowers appear, remove the anthers from the flowers of wheat *S*. Cover the flowers with small bags, as shown in the photograph below.
3. Several days later, transfer pollen from the anthers of the flowers of wheat *W* to the stigmas of the flowers of wheat *S*.
4. At harvest time, collect the seeds from wheat *S*.

Plants grown from these seeds will have various qualities of wheat *S* and wheat *W*. What combinations do you think might be produced? Ⓑ

Why are bags placed on the flowers during cross-pollination? Ⓒ

FOLLOW-UP

Reteaching

Students may enjoy making a model of a flower that includes all of the parts labeled in the diagram on page 530. Use clay, papier-mâché, or a variety of other materials. Encourage students to be creative.

Assessment

Play a game of "pollination pursuit" to assess student understanding of the life cycle of flowering plants. Enlarge and transfer the pictures from the plant cycle diagram on page 531 onto poster board. Label each structure using only letters. Make two stacks of index cards: one stack labeled with the letters and the other stack labeled with the names of each of the structures. Divide the class into two teams. Take turns calling on each team. Roll a die. If the number rolled is even, draw a card from the name stack, and ask one team to state the letter of the feature named. If the number rolled is odd, draw a card from the letter stack and have the team name the feature labeled with that letter. Give each team 30 seconds to come up with a team answer. For extra credit, have the teams name the function of each structure.

Extension

Explain to students that inside every seed is a tiny embryo. Then suggest that they complete the following activity: Soak some dry lima beans until they double in size. Separate the two halves of one of the seeds, and locate the embryonic plant with its tiny leaves and root. Draw and label a diagram of the parts of the seed embryo.

Closure

Show students the video *Plant Reproduction*. (See Films, Videotapes, Software, and Other Media on page 497A.)

Cross-Pollination as a Plant-Breeding Technique

Point out to students that cross-pollination can be used to develop varieties of plants with special characteristics. These plants are called *hybrids.*

Homework

Encourage students to complete the activity suggested at the end of Exploration 6 for homework. They should have you check their plans for safety considerations before proceeding.

Answers to *In-Text Questions and Caption*

Ⓐ Sample answer: Farmers might like to shorten the growing season of wheat *S* from 110 days to 90 days. Accept all reasonable suggestions for ways of getting a new wheat sample.

Ⓑ Wheat that resists disease, survives low temperatures, has a high yield, and matures in 90 days might be produced.

Ⓒ This prevents unwanted pollen from coming into contact with the plant's stigma.

Plants From Plant Parts

FOCUS

Getting Started

About two weeks prior to beginning this lesson, place several cuttings of coleus, ivy, or another houseplant in water until roots begin to appear. Ask the class whether all plants must begin to grow from seeds. *(No)* Many students will realize that it is possible to grow many plants from stem cuttings or from other parts of the plant. Show the rooted cuttings to the class and explain how they were produced. Finally, tell the class that in this lesson they will learn about several ways of growing plants without seeds.

Main Ideas

1. Many plants are capable of reproducing vegetatively.
2. Roots are the first structures to form in most methods of vegetative reproduction.

TEACHING STRATEGIES

Have students read the introduction to the lesson. As a class, discuss the methods of vegetative reproduction described.

Answer to *In-Text Question*

Ⓐ **A.** Grafting
B. Layering
C. Division
D. Cuttings

Answers to *Biosphere Decision*

Depending on the types of plants, new plants can be grown from mature plants without planting seeds. The decision must depend on the types of plants that are chosen.

Plants From Plant Parts

Can you grow a new plant without a seed? You can if you use roots, stems, leaves, buds, or twigs from a mature plant. This way of growing new plants is called *vegetative reproduction*. Read the brief descriptions below of the methods used in each example. Look at the illustrations and identify the method shown. Tell which part of the plant is being used. Remember, for each kind of plant, only some plant parts are suitable for vegetative reproduction. Ⓐ

The methods used to grow new plants from plant parts include the following:

1. Division—The roots of the plant are divided and planted separately.
2. Layering—Plants send out new roots from points on their stems that have been left in contact with the ground. These parts are separated from the mature plant and used to grow new plants.
3. Cuttings—A piece of a plant is cut off and replanted. This is successful only if the stem will readily grow new roots.
4. Grafting—A bud or twig from an existing plant is transferred to another plant that already has a root system.

A

B

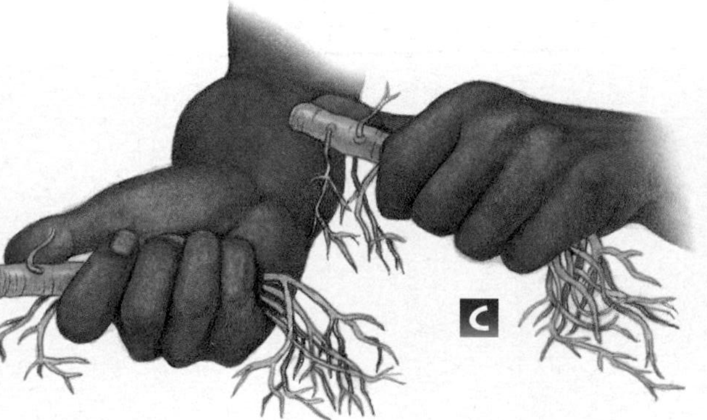

Biosphere Decision

The members of the biosphere proposal committee must decide what kinds of plants to take and whether seeds, seedlings, or mature plants should be chosen. Some members think that they will need plants in all three stages. What do you think?

534

LESSON 4 ORGANIZER

Time Required
2 class periods, plus a 1-week growing period

Process Skills
comparing, observing, analyzing

New Terms
none

Materials (per student group)
Exploration 7: several mature plants, such as impatiens, coleus, geranium, Swedish ivy, snake plant, African violet, or begonia; about 1 L of water; scissors; 1–2 L of potting mix with equal volumes of perlite and vermiculite; large, shallow container such as baking pan; several large sheets of clear plastic wrap; 1–2 L of general-purpose soil mixture; large, deep container such as plastic garden pot; metric ruler
Exploration 8: several sprouted mung beans, radish seeds, or mustard seeds untreated with fertilizer or fungicide; fine-tipped waterproof marker; metric ruler; several paper towels; about 100 mL of water

Teaching Resources
Exploration Worksheets, pp. 43 and 44
SourceBook, p. S168

EXPLORATION 7

Vegetative Reproduction

You Will Need

- mature plants
- scissors
- potting mix (equal volumes of perlite and vermiculite)
- a large, shallow container
- clear plastic wrap
- a large, deep container
- general-purpose soil mixture

What to Do

1. Water the plants the day before taking cuttings from them. This will save the plants from added stress.
2. You will plant several cuttings in one container to save space and material. The potting mix should be moist but not wet, or the cuttings might rot.
3. Fill the shallow container with the mix. Then dig small holes about 2 cm deep with your finger or a pencil. Space the holes about 4 to 5 cm apart.
4. Take a stem or leaf cutting as described below. Always hold the plants by the leaf, not by the stem, to avoid damaging the plant.

   For stem cuttings, cut the stem with scissors just below the third or fourth pair of leaves, as shown at right. Strip off the lower pair of leaves so that two or three sets of healthy leaves are left. The new roots will probably grow from the places where the leaves were removed.

   For leaf cuttings, cut off a leaf with its stem attached. For a snake plant, cut off a 5 cm section from the middle of a leaf.
5. Carefully place the cutting into a hole in the potting mix. Fill in the hole so that the cutting is held securely.
6. Put a piece of clear plastic over the whole container until the new plants have grown roots.
7. When the new plants begin to grow, carefully remove them from the shallow container and transplant them into a deeper container filled with the general-purpose soil mixture. Remember to take the proper precautions when transplanting.

Malik takes a stem cutting from an ivy

Try some of these plants.

Stem Cuttings	Leaf Cuttings
impatiens	snake plant
coleus	African violet
geranium	begonia
Swedish ivy	

EXPLORATION 7

One day before doing the Exploration, ask students to water the plants that will be used. Have students read and discuss what they are to do. A volunteer should demonstrate how to plant a cutting. When students finish planting their own cuttings, have them cover the containers with clear plastic wrap and place the cuttings where they will receive indirect sunlight. Allow the cuttings to remain in the potting mix for at least a week before examining the plants for roots. Suggest that students use a magnifying glass to examine several of the newly rooted plants. Ask them to observe what the roots look like and to determine where growth occurs. Encourage students to transplant some of their cuttings into flowerpots containing potting soil.

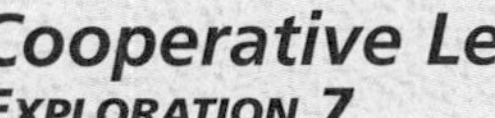

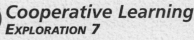

Cooperative Learning EXPLORATION 7

Group size: 2 to 3 students
Group goal: to investigate plant growth from stem and leaf cuttings
Positive interdependence: Assign each student a role such as group leader, plant manager, or materials coordinator.
Individual accountability: Have students keep a day-by-day account in their ScienceLog of their procedures and observations.

★ **An Exploration Worksheet is available to accompany Exploration 7 (Teaching Resources, page 43).**

Root Growth

For every method of vegetative reproduction described, what plant part must develop first (if it is not already there)? The root is the first part of the plant to emerge from the seed as it germinates. When a new plant forms from part of an old one, the roots must develop if the new plant is to live. The next Exploration will help you understand how roots grow.

Germinating corn seed

EXPLORATION 8

The Growth of a Root

You Will Need

- sprouted mung beans, radish seeds, or mustard seeds
- a fine-tipped waterproof marker
- a metric ruler
- paper towels
- water

What to Do

Examine mung beans, radish seeds, or mustard seeds that have already sprouted on moist paper towels. They should have roots. Using a fine-tipped marker, gently put marks 2 mm apart along the entire length of one root. Watch for several days as the root lengthens. Then measure the distance between the marks on the root.

1. In your ScienceLog, draw the root, showing the distances you recorded between marks.
2. Where is growth taking place in the root? What evidence leads you to think so?
3. Look for root hairs on the new root. Water enters the plant through the root hairs. What other purpose could root hairs serve as the growing root moves through the soil?

536

Root Growth

Have students read the introductory paragraph silently. Then direct their attention to the photograph, and have them identify what it shows. *(A germinating seed with a stem and roots)* Point out to students that the primary root is covered with tiny root hairs. Call on a volunteer to remind students about the function of the root hairs for a young plant. *(To increase surface area for water absorption and provide tiny pockets of air around the root)* Have students discuss how the roots in the photograph are similar to or different from the roots that grew on their cuttings in the previous Exploration.

EXPLORATION 8

Of the three choices, mung beans will provide the largest roots and will probably be the easiest to work with. Remind students to keep the seedlings moist. You may wish to have students perform the Exploration with all three seeds and then compare the results to determine which roots grew the fastest. Once the seeds have germinated and the roots have been marked, two or three days should be long enough for noticeable growth to occur.

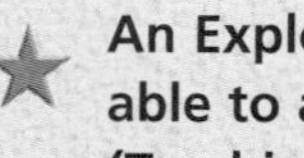

An Exploration Worksheet is available to accompany Exploration 8 (Teaching Resources, page 44).

Answers to *Exploration 8*

Students should discover that the marks closest to the root tip are the farthest apart, indicating that the root grows near the tip.

Students should further conclude that the root hairs are able to enter almost any crevice in the soil and extract water and minerals.

Students may suggest that root hairs help to stabilize the plant.

CROSS-DISCIPLINARY FOCUS

Art

Students may enjoy using plant products and plant parts, such as seeds, stems, dried flowers, leaves, and roots, to make works of art. Encourage students to be creative. Invite volunteers to share their work with their classmates, or set up an area of the classroom as an art gallery to display students' work.

Homework

Have students write a short paragraph that describes the function of roots, including an example. *(Accept all reasonable responses. Roots serve to anchor the plant. They also store food and transport materials within the plant.)*

Growing Through the Soil

How does a root grow in length? Why aren't roots damaged when they push through hard soil? Examine the photograph below for answers.

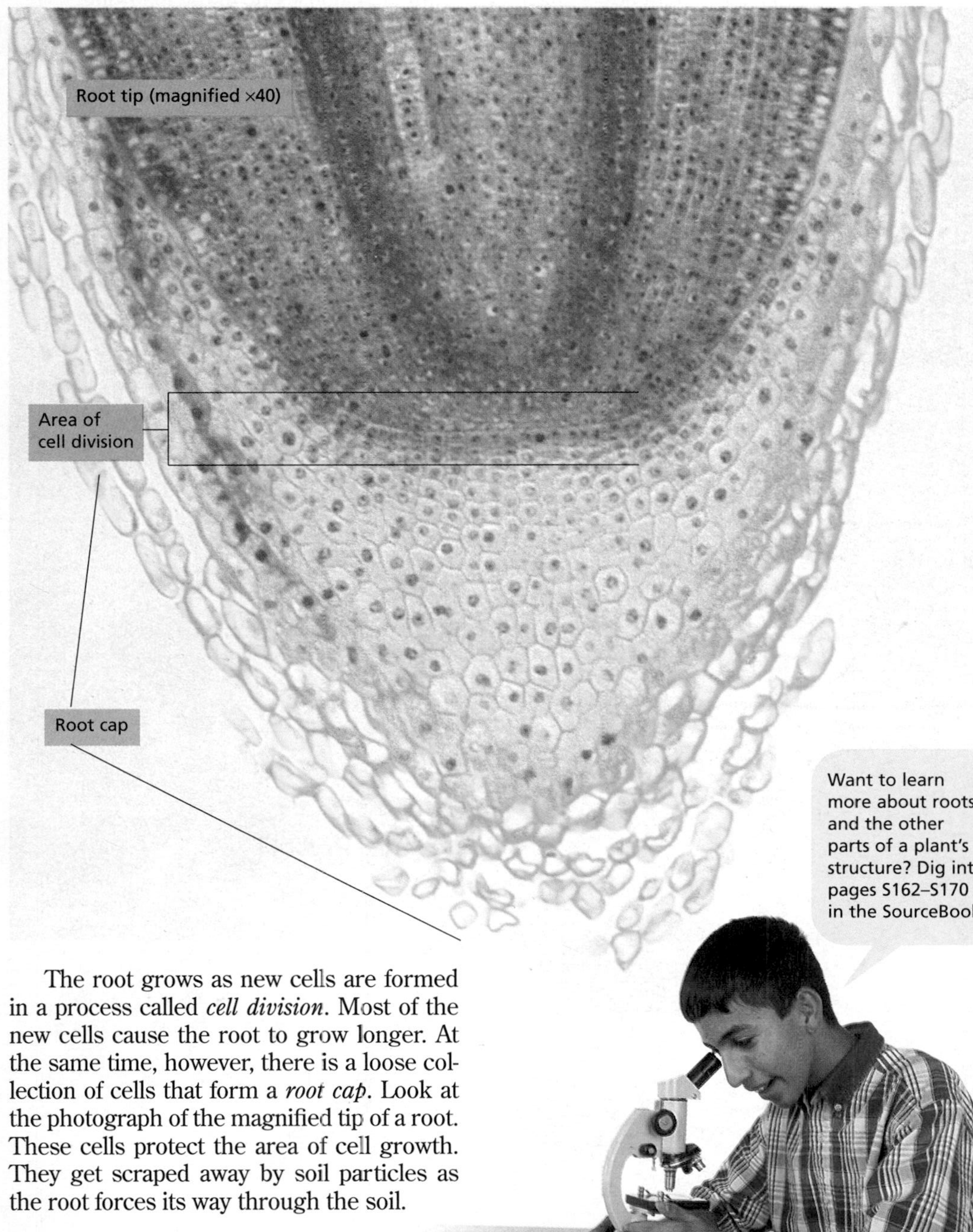
Root tip (magnified ×40)
Area of cell division
Root cap
Want to learn more about roots and the other parts of a plant's structure? Dig into pages S162–S170 in the SourceBook!

The root grows as new cells are formed in a process called *cell division.* Most of the new cells cause the root to grow longer. At the same time, however, there is a loose collection of cells that form a *root cap.* Look at the photograph of the magnified tip of a root. These cells protect the area of cell growth. They get scraped away by soil particles as the root forces its way through the soil.

Growing Through the Soil

Encourage students to respond to the questions before they read the answers at the bottom of the page. Students might find it interesting to use a magnifying glass to look for the root cap at the end of a root on a germinated seed. If possible, have students observe a root cap under a microscope.

Meeting Individual Needs

Gifted Learners

Explain to students that plants often respond to their environment with movements called *tropisms.* Two examples of *tropisms* are geotropism and phototropism. Have students discover the meaning of these terms. *(Geotropism is a response to gravity; phototropism is a response to light.)* Then have them design and perform two activities to demonstrate how geotropism and phototropism function in plants.

FOLLOW-UP

Reteaching

Suggest that students survey adult family members and friends to find out if they have ever grown plants vegetatively. Have students make a list of the plants, the methods used, and the success rate of the methods surveyed.

Assessment

Have students write a description of one of the methods of vegetative reproduction. They should not identify which method they are describing. Then have them exchange papers and identify the method described on the paper they receive.

Extension

Suggest that students make poster diagrams to compare the ways that different roots and stems grow. Point out that they will have to do some research. Resources could include biology books, botany books, and encyclopedias. Display the finished poster diagrams around the classroom.

Closure

Some students may enjoy trying their hand at propagating plants using one of the methods described on page 534. Suggest that they do this at home with an adult's supervision. Have students present an initial outline and then make weekly reports on their progress. Students should display the rooted plants in the classroom for others to examine.

CHALLENGE YOUR THINKING

1. Let There Be Light

Light affects plant development. Choose one of the following ideas and outline an experiment that would support or disprove the idea:

a. Does the shape of a tree have much of an effect on the amount of light it gets? Think about a pine tree and a maple tree.

b. Billy stated, "I think plants will grow faster in red light than in green light, and I am going to try to prove it."

c. Poinsettias bloom naturally in December. Dionne said, "If I could give a poinsettia plant the light conditions it would normally get in November and December, then I could make it bloom in the summer or any other time I want."

2. Out of Order!

a. In your ScienceLog, redraw this illustration, putting things in order. Be careful! Most, but not all, of the parts are out of place.

b. Label all of the stages shown. Two are labeled for you.

c. Why is this called a cycle?

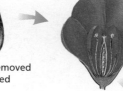

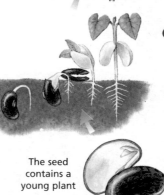

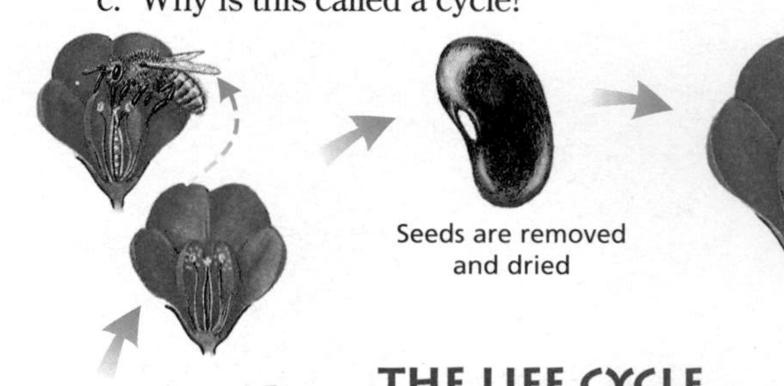
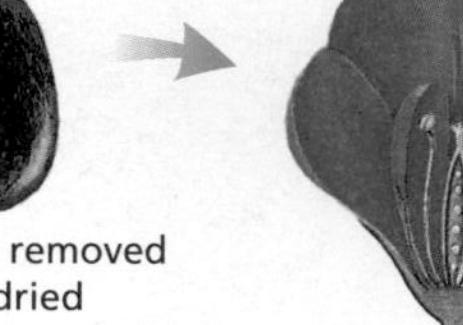

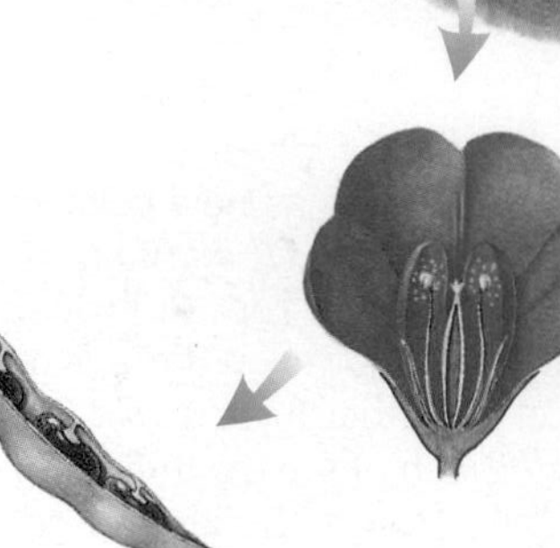
Seeds are removed and dried
THE LIFE CYCLE OF A BEAN PLANT
The seed contains a young plant

★ **You may wish to provide students with the Chapter 24 Review Worksheet that is available to accompany this Challenge Your Thinking (Teaching Resources, page 46).**

CROSS-DISCIPLINARY FOCUS

Mathematics

Have students make graphs comparing the largest crop-producing states in the United States. For example, bar graphs could be used to compare the five largest producers of corn, wheat, or soybeans. To extend the activity, have students compare the largest crop-producing nations in the world.

Answers to *Challenge Your Thinking*

1. Answers will vary. Students should make sure that only the variable being tested is allowed to change. They should also test a large sample or repeat their experiments several times. Suggestions of the important factors to study include (a) shape of the tree or the leaves, (b) color of the light, and (c) amount and intensity of light as the time of the year changes.
2. For a corrected diagram, see page S209. This is called a cycle because it has no beginning and no end. The plant life cycle continues over and over again through the production of seeds.
3. In grasses, new growth occurs at the base of each blade of grass. As evidence, students may note that grass continues to grow after the tops of the blades have been cut off by a lawn mower. To test this answer, a blade of grass could be marked at regular intervals. Then areas of growth would be visible as the marks spread apart over time.
4. Student answers will vary. Growing plants without soil might be a more efficient use of valuable space and would eliminate the need to transport soil to Mars. However, growing plants without soil would probably require the transportation and use of more chemicals and water and would require the presence of an expert in hydroponic gardening.
5. Just as in plant cells, water moves across human cell membranes from areas where water is more concentrated to areas where it is less concentrated. If a person eats a lot of salt, water will flow out of cells and into the area between cells, where the water concentration has been lowered by the presence of salt. This will result in the accumulation of fluid in the body, a condition known as *edema*. A reduction in the amount of dietary salt can lower the amount of salt in the area between the cells, therefore raising the water concentration and causing the water to move back into cells.

3. What's the Point?

Where do you think the area of cell growth (the growth point) is on a blade of grass? What evidence can you give to support your answer? Suggest how you would find out if your answer is correct.

4. Pros and Cons

As a member of the biosphere proposal committee, you have been asked to consider the practicality of growing plants without soil and to list possible reasons for and against using that method in the Mars biosphere. Prepare your lists for discussion.

5. A Swell Question

Sometimes too much fluid can accumulate between the cells in a person's body tissues. A common result of this is swelling, often around the ankles. Why might a person with this condition be told to reduce the amount of salt in his or her diet? (Hint: Think about water movement through plant cells.)

Review your responses to the ScienceLog questions on page 517. Then revise your original ideas so that they reflect what you've learned.

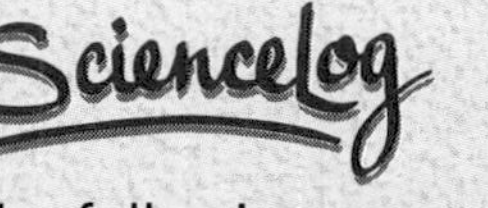

The following are sample revised answers:

1. Plants get water through their roots. Root hairs on the roots increase the total surface area of the root, allowing maximum access to water in the soil or in a hydroponic solution. Water travels through the plant in cells that serve as tubes. Water moves across cell membranes as a result of osmosis.
2. Plants can grow without soil as long as the functions performed by soil are provided in some other way. For example, the plant must be anchored in an upright position and nutrients must be available. The process of growing plants in a nutrient solution is called hydroponics.
3. Accept all reasonable responses. Students might assert that the flower came first because flowers produce seeds. Or students might say that the seed came first because flowers grow from seeds. Actually, because the pattern is cyclical, neither came first; the development from seed to flower to seed is continuous.
4. The first part of the plant to develop is the root. Roots must develop if the new plant is to live because roots provide stability and transfer water and essential nutrients to the plant.

Homework

The Activity Worksheet on page 45 of the Unit 8 Teaching Resources booklet makes an excellent homework activity after students have completed Chapter 24.

Plants in Your Environment

Connecting to Other Chapters

Chapter 23
introduces students to plants and begins to explore the needs of plants, as well as their role in our lives.

Chapter 24
discusses plants in more detail, including the biotic and abiotic factors necessary for a plant to survive.

Chapter 25
introduces students to the use of plants in both agriculture and landscaping.

Prior Knowledge and Misconceptions

Your students' responses to the ScienceLog questions on this page will reveal the kind of information—and misinformation—they bring to this chapter. Use what you find out about your students' knowledge to choose which chapter concepts and activities to emphasize in your teaching.

In addition to having students answer the questions on this page, you may wish to have them complete the following activity: Have students write a short skit in which three travelers are shipwrecked on an otherwise uninhabited island. They need to ensure that they have a regular food source. One traveler is familiar with what plants need in order to grow. Another traveler knows which plants are poisonous and which are edible. The third traveler is skilled in planting techniques. The skit should explore how the three travelers work together to establish a garden that will accommodate their food needs. You may wish to have volunteers act out a few of the skits. Emphasize that there are no right or wrong answers in this exercise. Collect the skits and read them to find out what students know about planting and maintaining a garden, what misconceptions they may have, and what about this topic interests them.

CHAPTER 25

Plants in Your Environment

1 Could this garden be a dangerous place? Explain.

2 How could this ladybug protect the garden above?

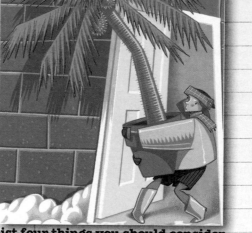

3 List four things you should consider when choosing plants to decorate your environment.

ScienceLog

Think about these questions for a moment, and answer them in your ScienceLog. When you've finished this chapter, you'll have the opportunity to revise your answers based on what you've learned.

Medicine or Poison?

Plant Medicines: Use With Caution!

As Mom came in the back door, three voices delivered some bad news.

"The cat's sick," said Maria.

"Poisoned!" Angela called out from the living room.

"We phoned the vet, and she said it could be from something we used to kill pests or treat diseases of plants in our yard!" This was from Carlos, who was getting dinner ready.

After getting back from the veterinarian's office, Mom said, "The vet says the cat will be all right this time, fortunately. But we must not let this kind of thing happen again. We'll have to be more careful about using poisons, or we'll be hurting ourselves next."

The Herrera family did something about their concerns. They began by learning about the situation. In this Exploration you will help the Herrera family identify some potential problems with pesticides and some alternatives that they might use.

Exploration 1

Danger in the Yard

PART 1

Identifying the Issue

Maria made a chart showing which chemicals had been used in the yard. The area numbers correspond to parts of the yard as shown on Maria's map (page 542).

Area number	Plants	Pest/ disease	Treatment
1	marigolds	slugs	metaldehyde bait
2	roses	aphids	dimethoate
		Japanese beetles	diazinon
3	daisies	spittlebugs	malathion
4	lilacs	leaf miners	diazinon
5	phlox	powdery mildew	benomyl

1. Find out if these pesticides (substances used to kill pests) and treatments can have effects on other living things. You might get help from people at the following locations:
   a. the Environmental Protection Agency
   b. a garden-supply store
   c. an agricultural institute or university
2. Find out how chemicals used in agriculture can affect living things, especially if the chemicals get into drinking water or are breathed in with the air.

Exploration 1 continued ►

LESSON 1 ORGANIZER

Time Required
1 class period

Process Skills
observing, analyzing, measuring

New Terms
none

Materials (per student group)
Exploration 1: books, magazines, and other resource materials on gardening; glue or tape

Teaching Resources
Exploration Worksheet, p. 51
SourceBook, p. S174

Medicine or Poison?

FOCUS

Getting Started

If possible, obtain photographs of insect-infested or diseased plants. To begin the lesson, show the photographs to your students and ask them what they think should be done about the problems. *(Many will suggest spraying the plants with pesticides; other students may disagree and say that pesticides are harmful to people or to the environment.)* Explain to the class that pesticides are very useful when they are used properly, but that they can cause serious health problems if they are used carelessly. Conclude by saying that in this lesson students will learn about several alternatives to pesticide use.

Main Ideas

1. Plant pests and diseases are often treated with poisonous chemicals.
2. Safe and effective alternatives to poisonous pesticides are available.

TEACHING STRATEGIES

Exploration 1

Other useful sources of information may include local cooperative extension services and college or university departments of entomology or plant pathology.

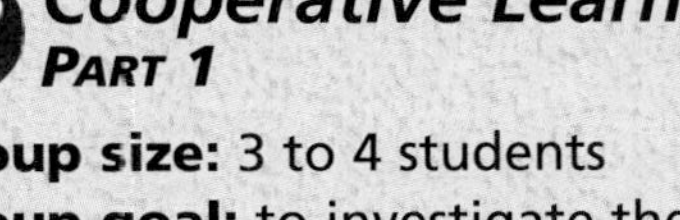

Cooperative Learning PART 1

Group size: 3 to 4 students
Group goal: to investigate the effects of pesticides on the environment
Positive interdependence: Have each student consult a different source for information about pesticides.
Individual accountability: Have each group present the information they have gathered in a booklet. Each student should contribute to his or her group's booklet.

Exploration 1 continued ►

EXPLORATION 1, continued

Students could work in small groups to read the Exploration and then do the library research needed to answer the questions. Involve the class in a discussion to share what they have discovered. This is an excellent opportunity for students to use their problem-solving and decision-making skills.

Students' analyses should include consideration of the fact that one of the fastest growing areas of pesticide use is for residential lawns. Students should also consider the advantages of using pesticides, such as improving the quality and appearance of food, reducing the labor costs in food production, and reducing the cost of the final product.

INDEPENDENT PRACTICE If the class becomes divided on the use of pesticides versus other alternatives, involve students in a debate of the issues.

An Exploration Worksheet is available to accompany Exploration 1 (Teaching Resources, page 51).

Answer to *In-Text Question*

Ⓐ Sample questions: Can the chemical affect the organisms in the fish pool? How long will the chemical remain in the water of the fish pool? Will the chemical have different effects on different plants? Should the picnic table be covered when sprays are being used? Is it safe for children to play on the swings while sprays are being used?

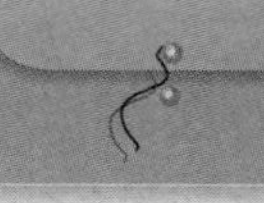

PORTFOLIO

Suggest that students summarize their responses to each part of Exploration 1 and place the summaries in their Portfolio.

EXPLORATION 1, *continued*

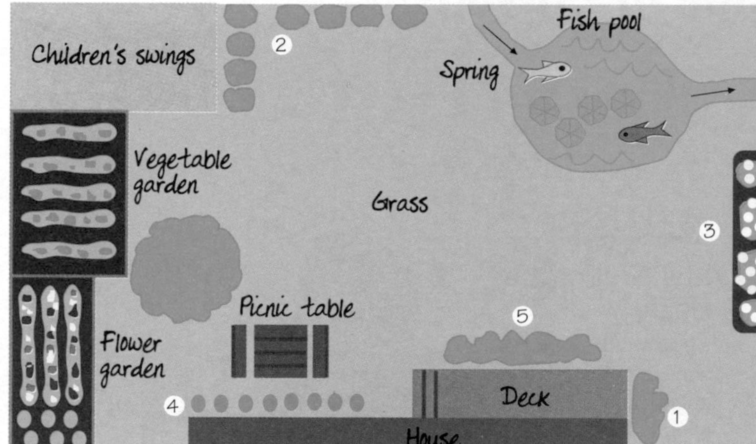

3. Look at Maria's map of the yard. Consider questions like these:
   a. Would a spray be carried by air currents? Where might it land?
   b. How long will the chemical remain on the plants? in the ground?
   c. How should I rinse the sprayer I use? Where should I put the rinse water? Where should I put the empty poison bottle?

Add five more questions that might be asked. Ⓐ

The Herrera family felt that they were ready to attack the problem. Mom said, "We must not spray poisons thoughtlessly, but we do want to get rid of pests and diseases from our plants. I want to find out if there are other ways of dealing with them."

Carlos said, "Janice's father uses ladybugs. He said they eat pests on the roses. He doesn't use chemicals in his yard."

Mom said, "Ask him for any other ideas he has. We could find some safe alternatives."

PART 2

Identifying Alternatives

1. How can the Herrera family eliminate plant pests and diseases without using poisons? Two alternatives are illustrated here.

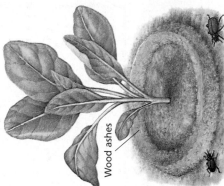

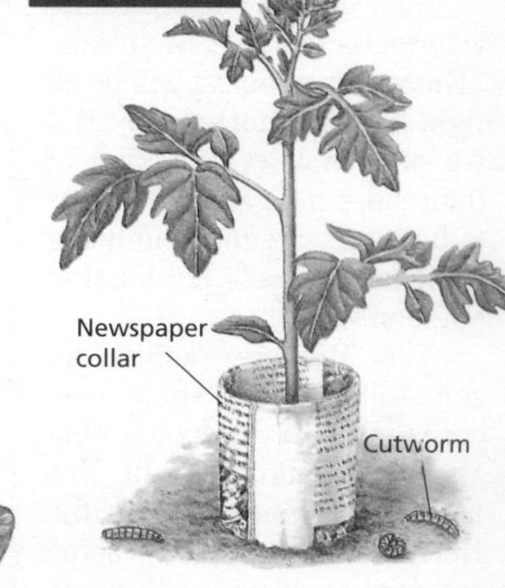

Find at least five other ways to help plants stay healthy without using poisons. Ask gardeners you know for alternative methods. Read a magazine devoted to organic gardening.

2. Make a collage of all the methods you discover.

542

Integrating the Sciences

Life and Physical Sciences

Have students use what they learned in Unit 3, Solutions, to describe how chemicals can end up in drinking water and how water can be purified. *(The chemicals are dissolved in the water. In this process, the individual particles of the chemicals are dispersed among those of the water. The chemicals are then carried with the water. Water can be purified by distillation.)*

Homework

Have students make a map of their own yard or of a nearby park or natural area. They should use Maria's map on this page as a model. Then have them list three questions about the use of chemicals in the area shown on their map.

PART 3

Researching

Read gardening magazines and books. You need to answer questions like these:

- Can the use of all garden poisons be avoided?
- Are there safe ways to use some poisons to kill pests?
- Do alternatives to poisons have some disadvantages?

PART 4

Decision Making

1. Work in groups to discuss alternative treatments for the Herreras' pest and disease problems.
2. Decide what action you would recommend for the future treatment of pests and diseases.
3. If you still find it necessary to use some poisons in your garden, name them and list the precautions that you would take to avoid possible dangers from them.

PART 5

Taking Action

The Herreras decided that they could eliminate some poisons from their yard. They found a collection depot where they could take these poisons, and they gladly got rid of them. Then they discovered how to safely use and store the chemicals they kept. Next season, the Herreras will be prepared. The cat and all of the family will be safer.

PART 6

Evaluating

Make a list of questions to ask the Herreras next year about their new ways of caring for plants.

FOLLOW-UP

Reteaching

Review research methods with students. Go over sources of information such as encyclopedias, magazines, newspapers, and books, and explain to students how to access these sources. Students may also need a review of note-taking methods.

Assessment

Suggest that students visit a local gardening center. Have them make a list of the names of the pesticides sold. Ask them to read the labels to determine which pesticides are poisonous and which are safe to use. Students should prepare fact sheets to show what they have learned.

Extension

Point out to students that one of the major drawbacks of pesticides is that they can contaminate the food chain. Suggest that they research how this happens.

Closure

Students may enjoy talking with a local gardener to discover which pests are most common in their area. Invite a guest speaker to visit the class, and encourage students to ask the expert their questions.

LESSON 2 Make a Green World!

FOCUS

Getting Started

Ask students to name all of the uses of plants that they have learned about so far. Write their responses on the board. *(Uses could include feeding people and other organisms, producing oxygen, preventing soil erosion, and making areas more attractive to people.)* Have students describe plants that they have in or around their homes and the benefits they gain from these plants. Explain that in this lesson students will put together all of their knowledge about plants and will design a project using what they know about growing plants.

Main Ideas

1. A plant's environment provides the necessary substances for it to live and grow.
2. The growth and use of plants to create, improve, or change an environment requires careful planning.

TEACHING STRATEGIES

A Design Project

Have students read the material silently. Explain that after they have completed Explorations 2, 3, and 4, students who are interested in the same project will form teams to work on it.

Small-Scale Designs

Have students describe the environment in an imaginary Mars biosphere. They should recognize that light, temperature, moisture, and nutrients will still be needed for plants to grow, no matter where the plants are.

LESSON 2 Make a Green World!

A Design Project

Now you have a chance to do some designing on a larger scale. You will choose a project to work on with other members of a team. The project requires you to plan for the growth and use of plants in one of the following situations:

a. indoors—at home or school
b. outdoors—around your home or another building
c. in a vegetable garden

Explorations 2, 3, and 4 will introduce you to these projects. Choose the project that suits you best after you have completed the Explorations.

African violet—soft, filtered light

Arrowhead vine plant—medium light

Small-Scale Designs

You have cared for plants and observed them closely. A plant's environment provides its living conditions. Plants have special needs for light, temperature, and moisture. An African violet grows well in soft, filtered light, while a cactus thrives in full sunlight. People who have learned to provide the right environment may grow both of these plants in the same house, but probably not on the same windowsill. There can be a variety of environments in one room.

What kind of environment will there be in the imaginary Mars biosphere? Consider the surroundings that your biosphere plants will need.

Cactus—full light

Geranium—medium light

LESSON 2 ORGANIZER

Time Required
6 class periods

Process Skills
analyzing, communicating, observing

Theme Connection
Cycles

New Terms
none

Materials (per student group)
Exploration 2: graph paper; pencil; metric ruler
Exploration 3: tracing paper; pencil
Exploration 5: various items to begin the projects, such as graph paper, a pencil, tracing paper, and a metric ruler

Teaching Resources
Exploration Worksheets, pp. 52, 54, and 56

EXPLORATION 2

A Greener Room

Han-Ling and Brent worked together on a plan for their classroom. They studied the conditions that the room provided for green plants so that they could select plants that would grow well there.

What They Used

- graph paper and a pencil
- a metric ruler

What They Did

1. They noted where the windows, doors, and furniture were located in the room.
2. They looked outside and found where trees, shrubs, and buildings partially blocked the sunlight.

   Complete their study. You will begin with step 3.

3. In your ScienceLog, draw a detailed floor plan showing what their classroom might look like. A sample sketch is shown below. You may add your own details.

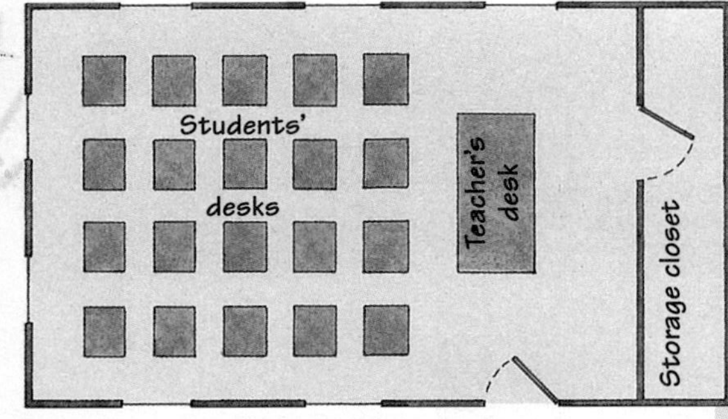

4. Using the following descriptions, determine what areas receive low, medium, high, and very high light. Indicate these on the floor plan.
   - low light—the light level that normally exists in a shaded window
   - medium light—the light level of an area within 3 m of a sunny window that is partially blocked by buildings or trees
   - high light—the light level that exists in front of an unobstructed window
   - very high light—the light level of an area that gets 3 to 5 hours of direct sunlight every day
5. Han-Ling used the chart below to choose these six plants for the room: heart-leaf philodendron, red ivy, aluminum plant, prayer plant, coleus, and snake plant.
   a. Indicate in your drawing where you would advise Han-Ling to put these plants.
   b. The classroom has space for four more plants. Help Brent select the next four plants. Then indicate in your ScienceLog where he should put them.

Genus name	Common name	Light level
Adiantum	maidenhair fern	medium
Aloe	aloe	very high
Asparagus	asparagus fern*	medium
Coleus	coleus*	very high
Dracaena	dragon tree	medium
Hedera	ivy*	high
Hemigraphis	red ivy	low
Maranta	prayer plant	medium
Mimosa	sensitive plant*	high
Philodendron	heart-leaf philodendron*	low
Pilea	aluminum plant*	medium
Sansevieria	snake plant	low
Saxifraga	strawberry-geranium*	very high
Yucca	false agave	high

* can be used in hanging containers

EXPLORATION 2

Have students read and complete the Exploration. Be sure they understand that they are to complete the drawing of Brent and Han-Ling's classroom, not their own classroom. Students may add more classroom furniture as well as outdoor vegetation. Students should also indicate the path of the sun in their illustrations. Their own classroom will be used for one of the team projects later in the lesson. Have students share their finished drawings and their choices of plants.

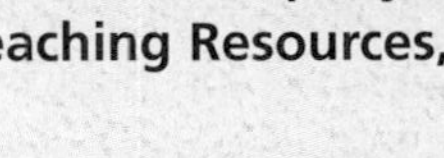

An Exploration Worksheet is available to accompany Exploration 2 (Teaching Resources, page 52).

Homework

You may wish to assign all or part of Exploration 2 as homework.

Multicultural Extension

Gardening in Miniature

Explain to students that bonsai is the ancient Japanese art of creating dwarf trees and shrubs. Have interested students research how trees can be made to grow in miniature and present an oral report to the class. Suggest that they bring some samples or some pictures to show the results of bonsai techniques.

CROSS-DISCIPLINARY FOCUS

Language Arts

Have students create a booklet about strange and unusual plants. If possible, they should include a picture of each plant—either a photocopy or a drawing of their own—along with a description of the plant that includes information about where it grows. Unusual plants might include carnivorous plants, plants that grow in strange shapes or sizes, and plants that have developed unusual adaptations or structures. Display the booklet for other students to read and enjoy.

EXPLORATION 3

GUIDED PRACTICE Provide students with some time to analyze and compare the two landscape drawings. Involve them in a discussion of how the drawings are similar and different. Ask students to consider the significance of each feature, such as the drainage of water on the property, the areas that are sunny and those that are shady, and the areas that are windy and those that are protected from the wind.

An Exploration Worksheet is available to accompany Exploration 3 (Teaching Resources, page 54).

Answers to *Exploration 3*

The following are sample answers:

1. Shrubs appear along the front of the house in both drawings. The driveway and sidewalk are in the same place.
2. Flowering shrubs, evergreen shrubs, a walkway, a fence, and flowering trees or dwarf fruit trees have been added along the north side of the property. A fence and hedge have been added along most of the southern edge of the property. Shade trees have been planted in front of the house. A deck has been added to the back of the house. A pond has been placed in the low, wet area of the backyard. A vegetable garden, rose garden, and play area have been created along the back of the property. A large shade tree has been planted in the southeast corner of the lot. Flower beds have been planted along the deck and the rear of the house.
3. Responses will vary but may be similar to the following:
   a. If you want to sit in the sun, you might choose the rose garden, the middle of the back lawn, or the deck.
   b. The front walkway near the house or the deck near the back of the house might be ideal places to talk to a friend out of the winter wind.
   c. The shade tree at the southeast corner of the lot or the dwarf fruit trees along the north edge of the property might be good places to enjoy a summer breeze.
4. **A.** The designer planned a shady garden along the north edge of the property.
   B. The designer planned a sunny garden on the northeast corner of the property.
   C. The designer planned a rose garden on the east side of the property.
5. The gardens and pond are designed to provide beauty. The trees provide shade for comfort. The play area was provided for entertainment.

EXPLORATION 3

Planning a Landscape

Think about the area around the outside of a house. How could you design a yard that would take advantage of the natural features of the landscape? This Exploration will help you make your plans.

You Will Need

- tracing paper and a pencil

What to Do

Compare the professionally landscaped lot with the original plans.

1. What features are the same?
2. What changes were made?
3. Trace the final plans. Use lowercase letters to show where you could
   a. sit in the sun,
   b. stand and talk to a friend out of the winter wind, and
   c. find a breeze in summer.
4. Use capital letters to show where the designer planned
   A. a shady garden,
   B. a sunny garden, and
   C. a decorative garden.
5. What human factors, such as beauty and comfort, did the designer consider when planning the landscape?

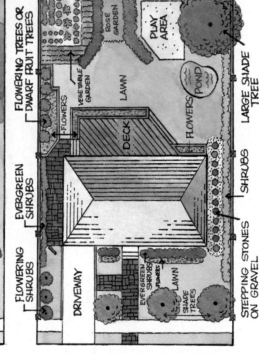

The original plans

A professional landscaper's final plans

546

EXPLORATION 4

Theo's Garden

The landscape shown in Exploration 3 includes a vegetable garden. Look again at that area of the plan. Describe the light and wind conditions that you would expect to find there. Then read how Theo planned his vegetable garden.

Theo and several of the neighbors in his apartment building rented spaces for gardens in a field outside of town. Theo planned to grow the things he likes to eat.

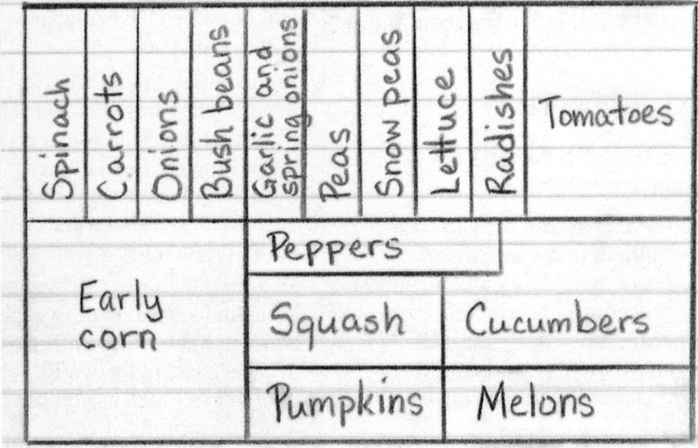

1. He drew an accurate diagram of his garden by letting 1 cm on the paper represent 50 cm of the actual garden plot. The diagram he drew shows the plants he intends to grow. Measure the length and width of the diagram. How large will his garden plot be? A garden this big may be large enough to provide food for four people.

2. Theo showed the diagram to his neighbor, Mrs. Chandra, who always grows her own vegetables. Mrs. Chandra suggested that Theo do the following:
   a. Plant the garden well away from any large trees.
   b. Plant the garden 1 m or more from a road or sidewalk.
   c. Avoid areas at the bottom of the hill.
   d. Buy young tomato, pepper, and melon plants instead of starting with seeds.

   Discuss this advice with several other students in your class. Speculate on the reasons Mrs. Chandra had for each suggestion.

3. Consider the advantages of having a garden that supplies you with fruit and vegetables. But remember, benefits always have costs. What are the costs of growing your own food?

EXPLORATION 4

Divide the class into small groups. Ask them to read the Exploration, discuss the questions, and formulate answers. After a specified amount of time, reassemble the class to share their findings.

An Exploration Worksheet is available to accompany Exploration 4 (Teaching Resources, page 56).

Answers to *Exploration 4*

Students should realize that the garden in the final plan in Exploration 3 on page 546 is in a sunny area because of the absence of any trees or structures around it. The top map on page 546 indicates that there is little wind and that the drainage is good.

1. The garden measures about 3.5 m × 5.5 m.
2. **a.** The garden should be away from trees so that it will not be shaded by them.
   b. The garden should be planted away from roads and sidewalks to avoid damage from people and vehicles.
   c. Avoid low-lying areas so that the garden will not become flooded during rainfalls. Higher elevations provide better drainage.
   d. These plants take a long time to grow from seeds. Therefore, it is better to begin with plants that have been commercially grown. Also, the growing season in many parts of the country is not long enough to grow these plants outside from seeds.
3. Answers may vary, but possible costs include hard work, having to stay home during the summer to take care of the garden, paying for water and other supplies (such as fertilizer and garden tools), and having to harvest the vegetables by hand.

Exploration 5

Have students read the project rules on page 548. To ensure that they understand what they are to do, provide time for them to ask questions or to ask for clarification.

Point out to students that at this time they should be prepared to use all of the experience and knowledge they have gained throughout this unit. Suggest that they will add to their knowledge by doing some of the suggested research and by conducting interviews.

You may wish to ask the school librarian to make resource material available to students, either in the classroom or at special areas in the school library. Organizing the project materials for display will also require your attention.

Allow three full class periods for teams to work together on their projects. Monitor what students are doing by occasionally visiting each group and asking questions about the progress of their project.

CHOICE 1

If students seem to be having trouble deciding how to divide the work among members of the team, you may wish to suggest the following: Two students can measure the classroom or school interior and can make a scale drawing. One or two students can visit local flower shops and nurseries to obtain helpful information. A couple of the students can do the library research on common indoor plants, their needs, and their origins. Everyone can request cuttings from various sources and can begin to grow their own plants. All members should meet to share their information and work together to design a final plan.

Your Team Project

Now it is time to choose one of the three projects described in Explorations 2, 3, and 4.

The following rules apply to each project:

1. Each project will be a team effort. All members of the team will make contributions.
2. When the project is completed, it will be presented in an exhibit. The exhibit can include posters, photographs, illustrations, charts, diagrams, models, and other things that you think will help explain what you have learned.
3. Each project must include interviews with people who have helped you. The interviews should be written up and included with the exhibit.
4. Each member of the team will obtain information for the project from at least two written sources (books, pamphlets, magazine articles, etc.) other than encyclopedias. This information will be included in the project report, along with a list of the sources of information used.

Let's find out more about the choices you have.

CHOICE 1

A Greener Room or a Greener School

1. Develop a plan for your classroom or some other area inside your school. Follow the procedure described in Exploration 2. Be sure to share the work among the members of the group.
2. Find out more about indoor plants by visiting flower shops and nurseries where you can see the plants and ask questions about them. Talk to your parents and neighbors about the plants they have.
3. Read books, brochures, and pamphlets on indoor plants. Public libraries, flower shops, and nurseries have information to lend you.
4. In the lesson titled Plants From Plant Parts, you learned how to grow plants from cuttings. Request free cuttings from parents and neighbors or from flower shops and nurseries, and grow your own indoor plants.
5. Many indoor plants grow outdoors in tropical or semitropical parts of the world. Find out as much as you can about the origins of your favorite indoor plants.

Dumb cane (below), a common houseplant, grows much larger as a wild plant in the rain forests of Hawaii (left).

Did You Know...

Dumb cane gets its name from the fact that someone chewing on the stem can become temporarily speechless. The plant contains chemicals called oxalates that cause swelling of the tongue and throat. (This is a good opportunity to remind students to exercise caution when they encounter unknown plants.)

Homework

You may wish to assign all or part of Exploration 5 as homework.

CHOICE 2

Landscape Gardening

Here is your chance to redesign the school grounds. Each member of your group should participate in creating the overall plan. Then individual gardens and spaces should be assigned to members of the team to plan in more detail.

1. Decide how you would like to make use of the school grounds. What are your goals in redesigning the landscape?
2. Measure the dimensions of the site. On graph paper, sketch any existing features, as one student did below.
3. Lay tracing paper over the sketch and begin to redesign the area. Make a rough outline of your own design. This will be a preliminary plan. It should be redone after you have researched plants and landscape gardening.
4. Your research should include the following:
   a. visits to landscaped lots
   b. interviews with landscape architects and gardeners (Large nurseries usually employ them.)
   c. visits to nurseries
   d. reading about landscape design (Borrow books from a public library, or pick up pamphlets from your local nursery.)
5. Your landscape design should include plans for each of the following:
   a. a variety of trees and shrubs
   b. each of these kinds of flower gardens:
      i. a shady garden
      ii. a sunny garden
      iii. a cool garden (near a fountain, pond, or stream)
      iv. a hot, dry garden

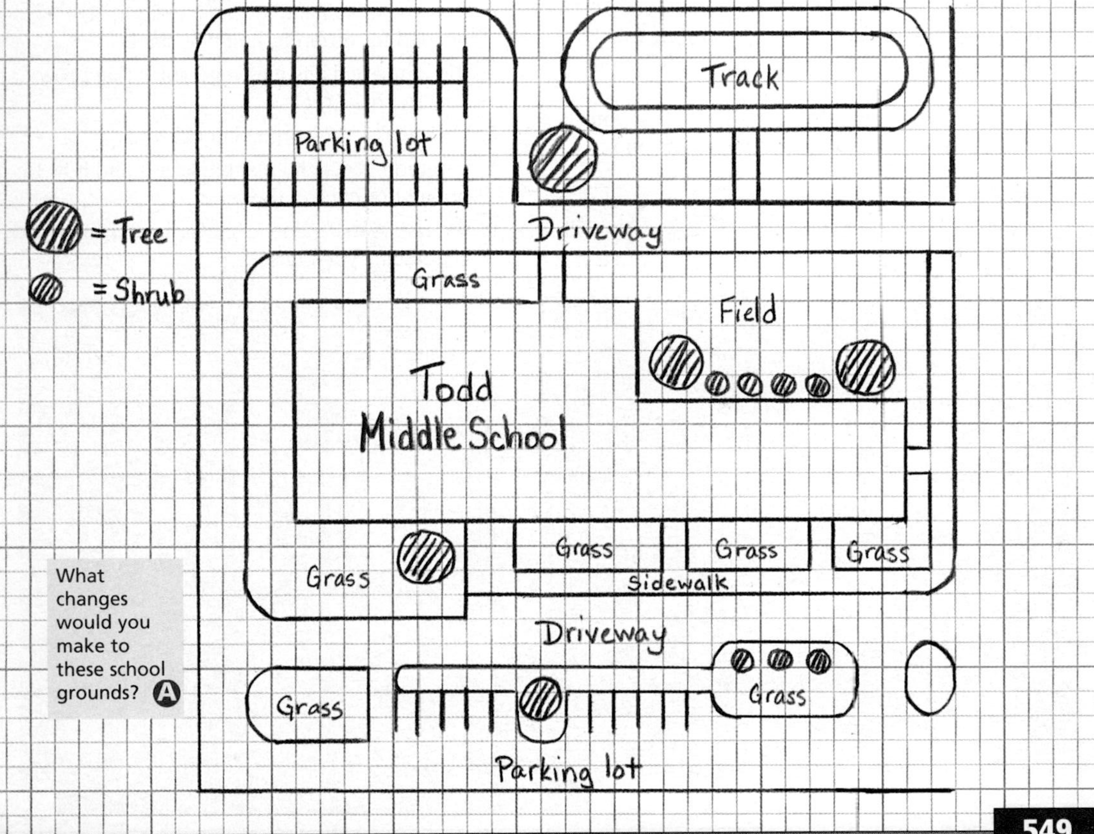

What changes would you make to these school grounds? (A)

ENVIRONMENTAL FOCUS

Have students make a concept map using the following terms: nutrients, plants, dissolve, chemicals, beneficial, animals, harmful, toxic pesticides, water, and soil. *(Answers should demonstrate an understanding that many types of chemicals can be dissolved in water and can affect the health of organisms in both positive and negative ways.)*

CHOICE 2

Suggest to students that they decide what their goals are in redesigning the landscape of the school grounds. They should record these goals so that they can refer to them as they work on the project. Point out that as the project progresses, they may discover that changes in their original goals have to be made. Explain that this is a normal part of any problem-solving process. Remaining flexible and open to new ideas is important if the best results are to be obtained.

Point out that steps 2 and 3 can each be accomplished by a few students. Step 4 provides another good place for members of the team to divide the work. One or two students can research each of the items and report back to the group. Also, mention that they are not confined to using only the items listed in step 5. They will probably want to include additional items in their final plan.

Answer to *Caption*

(A) Students may suggest adding more trees or rearranging parking areas.

CHOICE 3

Growing Your Own Food

If you prefer growing edible plants to decorative ones, this project is for you. You will design a garden. Later, if you wish, you can carry out the plan and grow your own fruits and vegetables.

Each member of the group should learn about three different kinds of fruits and vegetables. The garden you plan should include the three kinds you have selected and any other vegetables and fruits that you would like to grow.

The following procedure is suggested for this project:

1. Select a location for the garden. It should get at least 3 hours of full sun during the growing season.

   You may select one large area to be divided among the members of your group or separate areas for each member. Each member should have an area of at least 10 m^2. If your group decides to divide one large area among the members, leave a 1 m path around each member's area.

   Draw a sketch of the garden site that you have selected. Measure its dimensions and include these on your sketch. Include nearby buildings and trees in your sketch, and note the amount of light that each part of the garden will get during the growing season.

2. Study the soil at your garden site. What kind of soil do you have? How much of your soil is clay? sand? What difference does this make?

   Use a soil-testing kit or ask an experienced gardener to determine if your soil is rich in nutrients.

3. Research for this project should include the following:

   a. reading books on fruit and vegetable gardening
   b. interviewing experienced gardeners
   c. visiting nurseries and gardens

   Have each member research and report on one or more of the following topics:

   a. the light, water, and nutrient needs of different plants
   b. planting instructions for different kinds of plants
   c. the type of soil and level of acidity (pH) needed for different kinds of plants
   d. the kinds of fruits and vegetables that grow best in your area
   e. chemical versus organic fertilizers
   f. common pests and plant diseases and how to prevent or control them
   g. the nutritional value of the various fruits and vegetables you chose

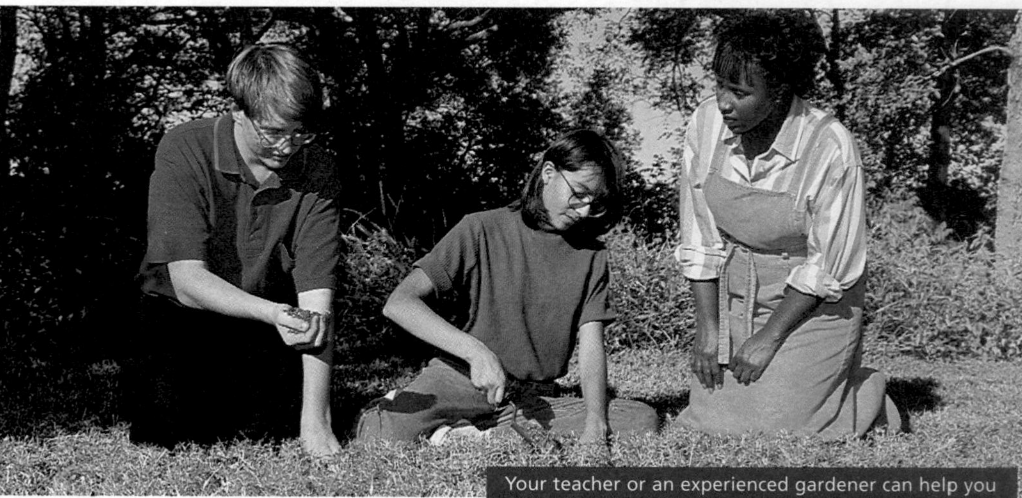

Your teacher or an experienced gardener can help you determine the soil type at your garden site.

CHOICE 3

If students are planning to actually plant a garden, this activity will have to be completed by the time planting season begins in their area. If students are having difficulty deciding where to select a location for their garden, suggest that they consider a part of the schoolyard, a vacant lot, or an area on the property where one of the group members lives. Point out that unless the team actually plans to grow a garden, the location can be anywhere that is accessible.

Again, help students to divide work responsibilities, if necessary.

CROSS-DISCIPLINARY FOCUS

Health

Students interested in health and fitness may enjoy researching the healthful benefits of specific fruits and vegetables. Suggest that they organize their information in a chart to display in the classroom. The information should include the vitamins and minerals found in each fruit and vegetable listed in the chart and the benefits of those vitamins and minerals.

Theme Connection

Cycles

Focus question: What role do plants play in natural cycles such as the water cycle? *(Plants absorb water from the ground and transpire it into the air. Plants also play a role in cycles involving atmospheric gases by absorbing carbon dioxide and releasing oxygen. In addition, plants play a role in nutrient cycles by absorbing nutrients during their lifetime and releasing the nutrients when they decompose.)* You may wish to have students diagram one or more of these cycles in their ScienceLog.

Share your research with your group. Use the information that the group has gathered to learn more about the three plants you have selected.

4. Develop a design for your garden, and sketch it. Use graph paper. In your written report, include your findings about topics (a) through (g) in step 3. At the end of the report, list all the sources of information you used, including books, pamphlets, interviews, and visits.
5. The exhibit should include your plan and report, along with photographs, illustrations, samples, and other materials that will help you communicate what you have learned to other prospective gardeners.

Testing Your Design

You have completed a design. Why not find out how it works? The summer ahead will give you a perfect chance to test your design. In the fall, you can enjoy the results!

Design for Outer Space: Biosphere Proposal

Now it is time to submit your proposal for the Mars biosphere.

1. Review the biosphere report that you wrote earlier. Make any changes necessary so that the report reflects what you now know about growing plants. Share your ideas with the other members of your group.
2. With your group, list the problems that you would expect to face when building a biosphere.
3. Discuss how to prepare for the problems that you have listed.
4. Write your own proposal for growing plants and providing food for the people who will live in the biosphere.

Design for Outer Space: Biosphere Proposal

Remind students that their biospheres will be located on Mars. Students may wish to find out about the Martian environment in order to design a more successful and practical plan.

In order to complete their proposals, students will have to draw on everything that they have learned in the unit. Suggest that it may be helpful if they review particular lessons as they write their proposals. You may wish to display the proposals around the classroom for students to read and enjoy.

FOLLOW-UP

Reteaching

Suggest that students investigate the meaning of the terms *long-day plant, short-day plant, perennial,* and *annual.* They should share what they learn by making fact sheets available to the class.

Assessment

Present students with this scenario: The library in your school needs some new plants. The librarian has asked your class for some suggestions. What factors will you advise the librarian to take into consideration when choosing plants? Refer to the chart on page 545 to make some specific plant recommendations for the library.

Extension

Explain to students that plants have adapted in many ways to different climates around the world. Suggest that they do some research to find out how plants have adapted to conditions in different biomes. Biomes to research include deserts, tropical forests, tundra, deciduous forests, and temperate forests. To extend the activity, have students work together to make vegetation maps of the United States or the world for display in the classroom or school library.

Closure

Have students make a list of three ways that people make use of plants. The uses that students list should be only those that were unfamiliar to them before they began this lesson.

Answers to *Challenge Your Thinking*

1. Answers will vary. Students should demonstrate a strong understanding of alternatives to pesticides.

2. • African violets require moderate light levels and should be placed on the table.
   • The geraniums need high light levels and should be moved to the windowsill.
   • Red ivy requires low light levels and should be placed away from the windows—perhaps on the table in the lower-right corner of the illustration.
   • The cactus requires high light levels and should be placed near a window.
   • The orange tree requires plenty of space and high light levels and should be placed outdoors.
   • The coleus requires high light levels and should be placed near the window.

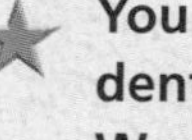

You may wish to provide students with the Chapter 25 Review Worksheet that is available to accompany this Challenge Your Thinking (Teaching Resources, page 60).

Homework

The Activity Worksheet on page 58 of the Unit 8 Teaching Resources booklet makes an excellent homework activity after students have completed Chapter 25. If you choose to use this worksheet in class, Transparency 90 is available to accompany it.

1. Alternative Alert

Design a newsletter for the Environmental Gardening Group that points out specific alternatives to treating plant diseases and pests with poisons. Illustrate your letter with drawings of sick and healthy plants. Be sure to explain the benefits of environmentally safe pest control.

2. What's Wrong With This Picture?

Amy wants all of her plants to grow strong and healthy, but someone put the plants in the wrong places! For each plant shown, describe the environment that it needs. Then explain how Amy should rearrange the plants.

Did You Know. . .

Of the Seven Wonders of the Ancient World, only one was ever alive: the hanging gardens of Babylon. They are thought to have been built by King Nebuchadnezzar II for his wife, who came from a mountainous region, in order to make her feel more at home.

3. Back to Nature

Many people are using more "natural" ways of gardening, including the following:

- using animal manure instead of chemicals for fertilizers
- avoiding the use of manufactured pesticides
- creating landscapes around their homes by growing only plants that are naturally part of the local environment

Think about these methods, and then make a list of the advantages of each. Do you think there are any disadvantages to any of these methods?

The landscaping around this Texas house incorporates plants that are naturally part of the local environment, such as buffalo grass.

4. A High-Rise Garden?

Milo has an apartment on the 26th floor. He wants a garden on the balcony.

a. Prepare a list of questions to ask him in order to find out how he intends to have a garden at such a great height.

b. Suppose that Milo is choosing his apartment now and can move into any one he likes. Tell him some conditions to look for if he wants a successful balcony garden.

Review your responses to the ScienceLog questions on page 540. Then revise your original ideas so that they reflect what you've learned.

Answers to *Challenge Your Thinking, continued*

3. Answers will vary. Sample advantages include the following:
- Using manure could alleviate the need for potentially harmful chemicals.
- Avoiding manufactured pesticides prevents the contamination of water supplies, avoids killing beneficial insects, and avoids the development of pesticide-resistant insects.
- Creating natural landscapes saves water, reduces the necessity of adding nutrients to the soil, and increases the chance of the plants' survival because they are naturally suited to the environment.

Disadvantages include having to learn how to garden organically and having to put in extra time.

4. a. Sample questions: What size plants will you grow? How will you get them up there? Is your garden for decoration, for food, or for both?

b. Milo should look for an apartment with a balcony that is well shielded from the wind and that receives a good amount of sunlight.

The following are sample revised answers:

1. The use of toxic chemicals to rid the garden of pesticides and microorganisms might make this garden a dangerous place for people and animals.
2. Ladybugs are a natural pest controller. They eat other insects that can be harmful to garden plants.
3. Answers will vary. Possible answers include the amount of light available in the area, the amount of space available for the plant, the plant's temperature requirements, and how the plant will fit into the existing landscape.

Making Connections

The Big Ideas

The following is a sample unit summary:

A biosphere must provide light, air, water, space, warmth, and food for living things to survive. It must also allow for waste disposal and for the recycling of resources. (1)

Soil is a mixture of rock and mineral particles, clay, silt, sand, and humus. Soil provides support for the plant, and it supplies nutrients and water to the roots. (2)

Water and nutrients are absorbed across the cell membranes of root hairs. Water concentrated outside the root moves into and through the plant by osmosis.

Leaves provide the main surface where food is made during photosynthesis. Excess water is released through the stomata in leaves in a process called transpiration. Transpiration helps cool the plant by eliminating the heat released during chemical processes. (3)

The N-P-K rating identifies the percentage of nitrogen, phosphorus, and potassium in fertilizer. Nitrogen helps give plants their green color, and it helps branches and leaves develop. Phosphorus helps plants develop strong root and stem structures. Potassium strengthens a plant and increases resistance to pests and disease. (4)

Plants can be grown in a nutrient solution by using a technique called hydroponic gardening. The plant must be anchored in an upright position, and the solution must provide the proper amount of nutrients, water, and air. (5)

Pollination is the transfer of pollen from the stamens of a flower to the pistil of the same or another flower. Pollen is produced in an anther, which is the structure at the tip of a stamen. Bright petals and the production of nectar attract animals and insects to the flower. The structure and placement of the stamens and anthers cause the animals and insects to brush against the anthers in order to reach the nectar. Pollen that adheres to them may be carried to another flower (cross-pollination) and be deposited on the stigma of a pistil. (6)

Vegetative reproduction is a method of growing a new plant from part of an existing plant. Division, layering, cuttings, and grafting are means of vegetative reproduction. Roots are the first parts of the plant to develop during vegetative reproduction; they must form for the new plant to live. (7)

Instead of using pesticides, you can use safe alternatives. A cardboard collar or wood ashes around the stem can prevent damage from pests. You can also prevent damage from pesticides by properly disposing of poisonous wastes and contaminated water. (8)

Light level, moisture, and temperature should be considered when growing indoor plants. The plants should also be arranged attractively. In landscaping, the conditions of the ground, the weather, the position of structures, and beauty should be considered. For a vegetable garden, the richness of the soil, types of food grown, season, and climate of the garden should be considered. (9)

Making Connections

Unit 8

SourceBook

You have learned how plants grow and how to help them grow healthy and strong. Next you will explore the functions of plant parts and the many different uses that humans have for plants. Read pages S161 to S176 in the SourceBook to increase your understanding of plants.

UNIT 8

Here's what you'll find in the SourceBook:

The Big Ideas

In your ScienceLog, write a summary of this unit, using the following questions as a guide:

1. What must a biosphere provide in order for living things to survive?
2. What is soil, and what does it do for plants?
3. How are water and other nutrients taken in and transported within plants? What functions do leaves perform?
4. What does the N-P-K rating of fertilizer tell you?
5. How can you grow plants without soil?
6. What parts of flowers are involved in pollination?
7. What is vegetative reproduction? What role do roots play in this process?
8. How can you prevent damage from pesticides?
9. What factors should you consider when growing indoor plants? outdoor plants? a vegetable garden?

Checking Your Understanding

1. Answer these questions about Matthew's experiment.
   a. Matthew had two plants of the same kind. He placed one on a windowsill and the other in a closed cupboard. What hypothesis was he testing?
   b. After 2 weeks, Matthew noticed that the leaves of the plant in the cupboard were not as green as the leaves of the plant on the windowsill. The plant in the

554

cupboard was also taller and more spindly than the other plant. What would be a good conclusion for this experiment?

2. Megan measured the growth of a leaf over a period of time. Her results are shown in the table at right.

a. Plot this data on a graph.

b. On day 19 the length of the leaf was 52 mm, and on day 21 it was 53 mm. What conclusion would you draw from these facts?

c. What value would you expect for day 23?

3. Describe how the following changes would affect life inside the Mars biosphere:

a. Eliminate the plants that are edible by humans.

b. Eliminate all animal life, except humans.

c. Eliminate the ocean.

d. Use chemical sprays to control pests.

4. concept map Draw the concept map at right in your ScienceLog, and complete it using the following terms: flower, stomata, stamen, leaf, plant, root hair, pistil, and root.

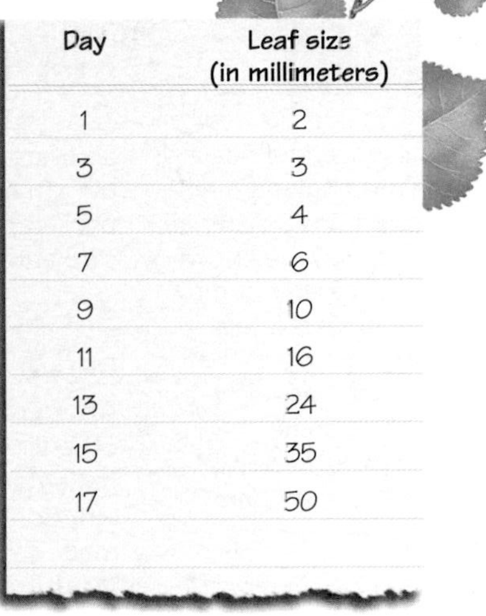

Day	Leaf size (in millimeters)
1	2
3	3
5	4
7	6
9	10
11	16
13	24
15	35
17	50

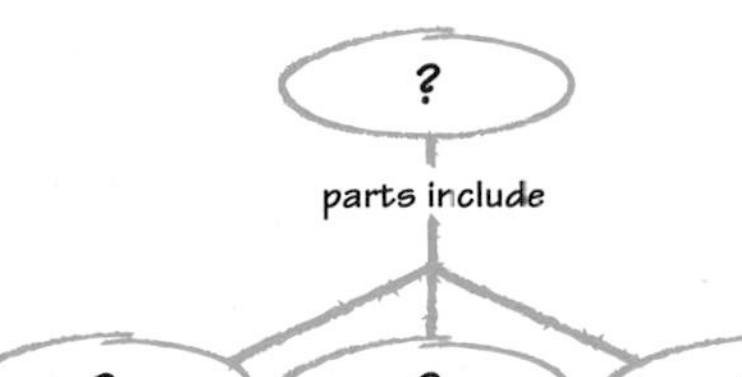

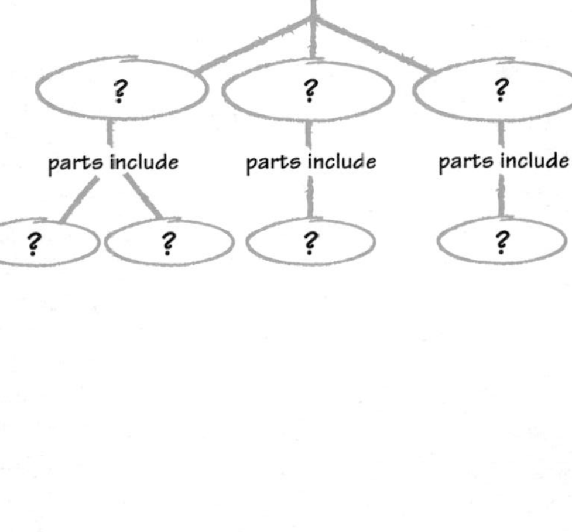

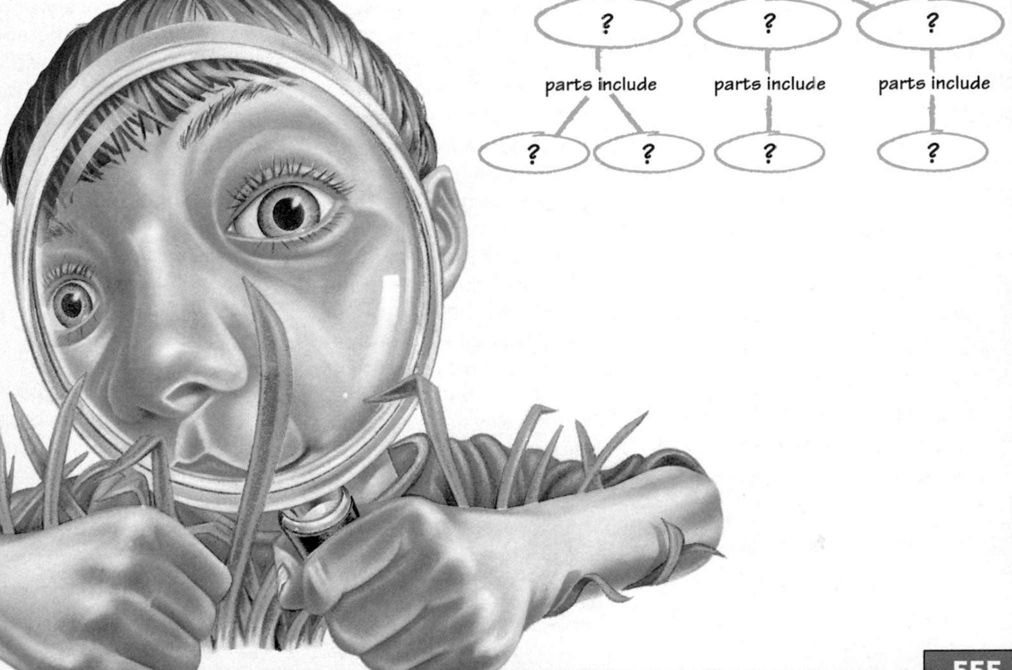

Answers to
Checking Your Understanding

1. a. Matthew was trying to test the hypothesis that plants need light for healthy growth.

b. Lack of light causes leaves to lose some of their color and causes stems to grow longer and thinner.

2. a.

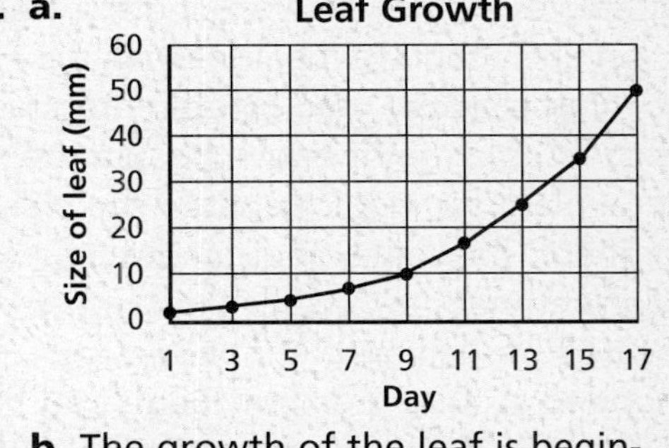

b. The growth of the leaf is beginning to slow down.

c. A little more than 53 mm

3. Answers will vary. Students should recognize that all biotic and abiotic features in a biosphere are dependent on each other. The loss of one biotic or abiotic factor affects every other organism within the biosphere. Students should also note that everything put into the food supply, such as pesticides sprayed on plants, will filter through every system in the biosphere.

4.

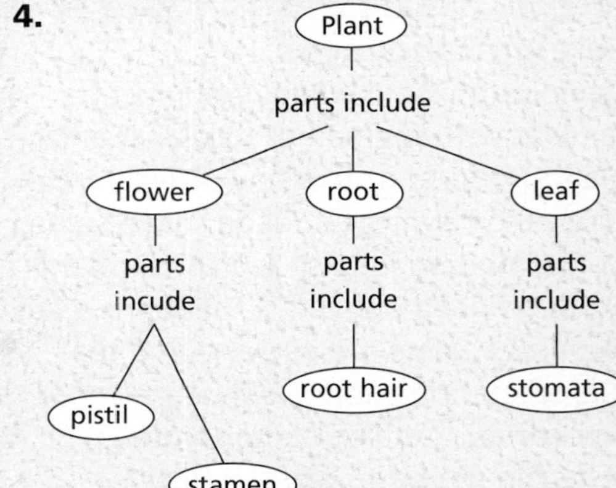

Homework

The Unit 8 Activity Worksheet that accompanies this Making Connections makes an excellent homework assignment (Teaching Resources, page 66).

Meeting Individual Needs

Learners Having Difficulty

Have students use their experiences and do research to create a 30-minute television show about gardening. Students should be creative about what topics they cover on their show, but they should be sure to include information about maintaining a garden and growing specific kinds of plants. Students should either videotape their television show or present it live to the class.

★ **You may wish to provide students with the Unit 8 Review Worksheet that is available to accompany this Making Connections (Teaching Resources, page 66).**

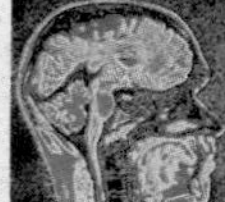

HEALTH WATCH

Background

Some biologists estimate that there are 265,000 species of flowering plants in the world. Of these, less than half of 1 percent have been studied for their potential medicinal qualities. Because there are so many species, efficient strategies are necessary to find the plants most likely to have medicinal value.

One strategy used by ethnobotanists is to assume that if native people use a local plant for medicine, then the plant probably has some medicinal value. Many ethnobotanists seek out native healers or shamans. Ethnobotanists hope to acquire the knowledge that has taken the shamans years to accumulate. With these insights, the researchers can then decide which plants they should collect and study.

Some of the most useful drugs developed from plants used by indigenous peoples include aspirin for reducing pain and inflammation, codeine for easing pain and suppressing coughs, and quinine for combating malaria.

CROSS-DISCIPLINARY FOCUS

Foreign Languages

Ethnobotanists like Paul Cox must be familiar with the names of plants and plant products from many different cultures. From what languages have some of our modern English words for plant products been derived? *(There are many examples. Some are ginseng and tea [Chinese], cinnamon [Hebrew], alcohol and coffee [Arabic], bamboo [Malay], pistachio [Persian], cashew [Tupi], quinine [Quechua], and banana [West African, as in Mande].)*

ENVIRONMENTAL FOCUS

The destruction of the world's rain forests is driving many plant and animal species to extinction, some of which are unknown to modern science. Have interested students do some research to identify rain forests in different parts of the world. Suggest that they make a map to show the locations of these rain forests. How quickly are they disappearing?

HEALTH WATCH

Ancient Wisdom of the Samoan Healers

In Samoan culture, the healer is one of the most valued members of the community. After all, the healer possesses the knowledge to treat disease. In some cases, Samoan healers know about ancient treatments that Western medicine has yet to discover. Recently, some researchers have turned to Samoan healers to ask them about their medical secrets.

A Samoan Remedy

Paul Cox is an *ethnobotanist.* He travels to remote places to look for plants that can help cure diseases. He seeks the advice of shamans and other native healers in his search.

In 1984, Paul Cox made a trip to Samoa to observe healers. While there he met 78-year-old Epenesa, a well-known Samoan healer. She was able to identify over 200 medicinal plants and astounded Cox with her knowledge. Epenesa had an accurate understanding of human anatomy, and she dispensed prescriptions with great care and accuracy.

Blending Polynesian and Western Medicine

After Cox spent months observing Epenesa treat patients, she gave him her prescription for yellow fever—a tea made from the wood of a rain-forest tree called *Homalanthus nutans*. Cox brought the yellow-fever remedy to the United States, and in 1986 the National Cancer Institute (NCI) began studying the plant. They found that the plant contains a virus-fighting chemical called *prostratin.* Further research by NCI indicates that it may have potential as a treatment for AIDS.

Another compound from Samoan healers treats inflammation. The healers apply the bark of a local tree, *Erythrina variegata,* to the inflamed skin. Only one of the two varieties of this tree is helpful, the healers said. When a team of scientists evaluated the bark, they found that the healers were absolutely correct. The active component from the bark, *flavanone,* is now under development as a medicine that Western doctors may prescribe some day.

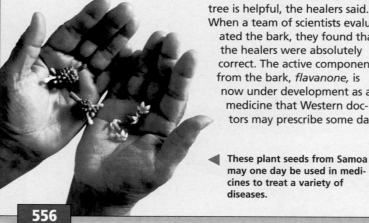

These plant seeds from Samoa may one day be used in medicines to treat a variety of diseases.

Botanist Paul Cox interviews a Samoan healer about remedies made from local plants.

Preserving Their Knowledge

When the two healers who provided Cox with information leading to the discovery of prostratin died in 1993, generations of medical knowledge died with them. The healers' deaths point out the urgency of recording their ancient wisdom before all of the healers are gone. Cox and other ethnobotanists must work hard to gather information from healers as quickly as they can.

The Feel of Natural Healing

Next time you have a mosquito bite or a mild sunburn, consider a treatment that comes from the experience of Native American healers. Aloe vera, another plant product, is found in a variety of lotions and ointments. Find out how well it works for you!

Homework

American Indians have traditionally used many plants to help cure diseases. Suggest that students find out what plants in your area can be used for this purpose. Have students consult a field guide and then discuss what illnesses can be treated with these plants.

The Feel of Natural Healing

Encourge students to read the ingredients list on products that they use frequently. See if they can identify plant products or other naturally occurring ingredients.

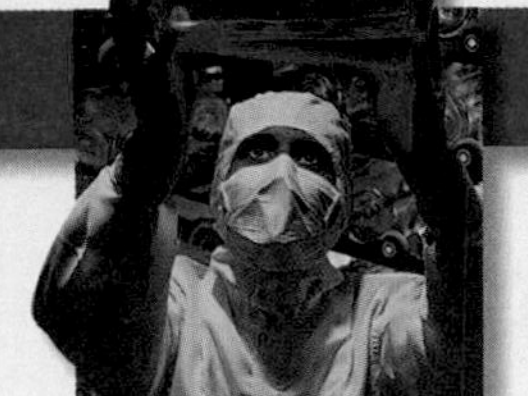

SCIENCE in ACTION

A New Portrait of Human Evolution

Bonnie Jacobs uses special snapshots that let her see back in time. If you look at these snapshots, you might see an ancient grassland, desert, or rain forest. And if you're lucky, Jacobs's snapshot might even show you a glimpse of the place where our human family may have gotten its start.

A Paleo What?

Bonnie Jacobs describes her work as a *paleobotanist:* "A paleobotanist is someone who studies fossil plants. That means you could study fossilized leaves, wood, pollen, flower parts, or anything else that comes from a plant.

"I have always had an interest in ancient things," she says. In school, Jacobs was fascinated by fossils, ancient cultures, and geology. Although she remained interested in these areas, her work took a new direction while she was living in Africa. "In 1981 I became involved in a project in Kenya. I ended up studying plants in the context of human evolution."

Fossil Plants and Human Evolution

Jacobs's current project involves the study of fossilized plants in rocks that range from 14 million years old to about 4 million years old. Scientists theorize that humans first evolved about 6 million years ago.

"In addition to fossils of plants, the rocks also contain fossils of bone—and some are from human ancestors. I think it would be very exciting to figure out what the environment was like when these human ancestors were evolving.

"Each site gives me a picture, a little snapshot, of what the plant community was like when those plants became fossils. Knowing about the plant communities allows me to picture what the landscape was like. When I have several pictures of the landscape at different times, I can reconstruct how the area evolved over millions of years.

"Main ideas about the causes of human evolution have a lot to do with changes in the landscape. For instance, most scientists who study human evolution make an assumption that there was a big change from forested to more open environments just before the origin of the human family. And that assumption needs to be tested. The best way to test that is to go back to the plants themselves."

◀ **"I think what makes science cool is that bit by bit, whatever the science is, you learn about how the world works."**

It's an Adventure!

In doing her research, Jacobs has traveled to many different places. "When you are in a foreign country, you have all kinds of encounters with people," she says. "A colleague we worked with in Kenya, Kiptalam Chepboi, grew up in the area where we do fieldwork. He took me to a sweat-bee hive. Sweat bees don't sting. You can take a honeycomb out from under a rock ledge, pop the whole thing in your mouth, and suck the honey out without worrying about getting stung. That was one of the neatest things I did out there."

The Modern Record

Make your own plant fossil. Press a leaf or other plant part into a piece of clay. Fill the depression with plaster of Paris. Write up a paleobotany report using your plaster fossil as the main subject. What does the fossil tell you about the environment it came from?

557

The Modern Record

You may wish to provide students with clay and plaster of Paris to complete this activity in class. Encourage students to model their paleobotany report after a real science report, as if they had actually set out on an expedition and found that particular fossil. If possible, obtain copies of a short paleobotany article from a local college or university library for students to look over and emulate.

SCIENCE in ACTION

Background

Paleobotanists study not only the fossils of plants but also what these fossils tell us about ancient climates, ecosystems, and the evolution of species. For example, Bonnie Jacobs studies the edges of leaves to see if features of the leaf edge such as smoothness or roughness can be used to predict the amount of rainfall that an area received.

Jacobs is one of a group of experts from all over the world that has studied in the Lake Baringo region of Kenya. Researchers studying the evolution of humans and of a wide variety of plants, mammals, fish, and microorganisms have joined forces with specialists in geology and dating techniques to gather information from the area. Fossils from this area are from 1 to 15 million years old and are among the most informative and important fossils ever found.

Critical Thinking

Have students imagine the following scenario: You are a paleobotanist digging for fossils in a hot, dry area of Africa. You find a fossil that is about 2.3 million years old and that resembles the stalk of a banana tree. What does this tell you about how the climate has changed? What could have caused this? *(The climate was probably more humid than it is today. Students' ideas about the cause of this change may vary considerably. Accept all reasonable responses. For example, students might propose that wind patterns have shifted to blow dry air instead of humid air to the area.)*

Integrating the Sciences

Earth and Life Sciences

Have students describe the processes that are involved in the fossilization of a leaf. *(Students may describe the leaf falling off a tree and landing in mud or sediment. The leaf is then covered up by layers of earth. Gradually, water passing through the decomposing leaf deposits minerals. Over time, the minerals replace the organic material of the leaf, and a fossil is formed.)*

SCIENCE & TECHNOLOGY

Background

Sewage is water that contains waste matter produced by people. Only about one-tenth of this waste is solid matter. Untreated sewage contains harmful chemicals and disease-producing bacteria. Even treated sewage can be harmful because the waste product, called sludge, often contains dangerous concentrations of toxic chemicals. In addition, organic wastes can be converted into potentially harmful inorganic compounds such as sulfates, phosphates, and nitrates. These compounds may cause algae blooms, which can result in the destruction of pond or lake ecosystems.

About 80 percent of the nation's sewage comes from industrial sources. The cost of treating this sewage has increased steadily over the years, partially due to the fact that modern sewage contains contaminants that are more difficult to eliminate, but also because government regulations have become tougher. It has been estimated that solar aquatic water purification should cost about one-half to one-third less than conventional waste-treatment methods.

CROSS-DISCIPLINARY FOCUS

Mathematics

Have students read their monthly home water bills to find out how much water their household uses. Each student should find the average amount of water consumed per person in his or her household. To do this, divide the total water bill by the number of people living in the house. As a class, add together the amount per person from each home and divide by the total number of people in the class. Finally, give students the population of your city or town, and have them estimate your entire city's monthly water consumption. Ask students what sources of water consumption are not included in this estimate. *(Public swimming pools, businesses, city parks, and water fountains, for instance)*

SCIENCE & TECHNOLOGY

Sunshine and Sewage

What happens to the water and waste when you flush the toilet? Easy, it goes down the drain and into the sewer pipes. Then what happens? In most cases, the sewage is piped to a sewage-treatment plant where much of the waste is separated from the water and most of the dangerous microorganisms are killed. This process is expensive and requires the use of chemicals. Then the water is dumped into a nearby stream or the ocean. The water may be used again as drinking water, but only after it has gone through another purification process, using filters and more chemicals.

Solar Aquatics

One scientist has developed an interesting way to turn waste water back into clean water without chemicals. In Rhode Island, Massachusetts, Indiana, and Vermont, an experimental process of sewage treatment has had some promising results. The process, called *solar aquatic water purification*, uses food chains and natural processes such as photosynthesis to break down sewage. The process yields little pollution and leaves water cleaner than chemically treated water. And it's even less expensive!

Putting Plants and Animals to Work

In solar aquatic water purification, raw sewage is fed into a series of clear, fiberglass tanks that allow sunlight to enter. The tanks contain many different kinds of bacteria and algae. The bacteria digest nutrients in the sewage. The algae consume the waste products given off by the bacteria. The stuff left over then flows to the next series of tanks, where snails consume the algae. In subsequent tanks, the process continues with mollusks, fish, and about 120 types of plants.

During the final purification step, water passes through a carefully planned marshland ecosystem created in wooden troughs. After running the course of the troughs, clear, clean water emerges. The water can then be used or discharged into a nearby pond or lagoon. The entire process takes about 5 days and produces about 78,000 L of clean water per day.

▲ John Todd hoists beakers of treated and untreated water at his award-winning solar aquatic water treatment plant in Sugarbush, Vermont.

The Real Plus of Sewage-Treatment Plants

The plants and animals form a food chain that ultimately leaves behind clean water. Apart from their role in the food chain, plants absorb such contaminants as phosphorus, cadmium, and lead. The plants also provide an added bonus. Trees and flowering plants that grow in the tanks can be sold or replanted elsewhere.

Broad Horizons

In time, scientists hope to apply the solar aquatic process on a larger scale. Some water treatment plants process 238 million liters of water daily. Calculate how many solar aquatic water purification systems it would take to process that much water. Why might that be a serious drawback to using the system on a large scale?

558

Integrating the Sciences

Earth and Life Sciences

Traditional sewage treatment and solar aquatic water purification both model the natural processes by which wastes are broken down in nature. Have students construct a large poster, mural, or bulletin-board display that ties together processes such as the food chain, the carbon dioxide–oxygen cycle, the nitrogen cycle, and the water cycle. The emphasis of the display should be on the purification of water by natural processes. Some research may be necessary.

Answer to *Broad Horizons*

A solar aquatic process produces about 78,000 L of water every day, whereas a chemical treatment plant produces over 238 million liters of water every day. By dividing 238 million by 78,000, you will find that it would take over 3000 solar aquatic systems to purify as much water as one chemical treatment plant. However, you could accurately evaluate the comparative efficiency of these plants only by comparing the size and operating cost of each plant.

EYE ON THE ENVIRONMENT

A Rainbow of Cotton

Think about your favorite T-shirt. Chances are, it's made of cotton and brightly colored. The fibers in cotton plants, however, are naturally white. They must be dyed with chemicals—often toxic ones—to create the bright colors seen in T-shirts and other fabrics. To help solve this problem, an ingenious woman named Sally Fox had an idea: What if you could grow the cotton *already colored*?

Learning From the Past

Cotton fibers come from the plant's seed pods, or *bolls*. Bolls are a little bigger than a golf ball and open at maturity to reveal a fuzzy mass of fibers and seeds. Once the seeds are removed, the fibers can be twisted into yarn and used to make many kinds of fabric.

Sally Fox began her career as an *entomologist*, a scientist who studies insects. She first found out about colored cotton while studying pest resistance. Although most of the cotton grown for textiles is white, different shades of cotton have been harvested by Native American farmers for centuries. These types of cotton showed some resistance to pests, but had fibers too short to be used by the textile industry.

In 1982, drawing on the techniques of generations of plant and animal breeders, Sally began a long, slow task. She patiently used cross-breeding to produce strains of cotton that were both colored *and* long-fibered. Her cotton, officially registered under the name FoxFibre®, has earned her high praise.

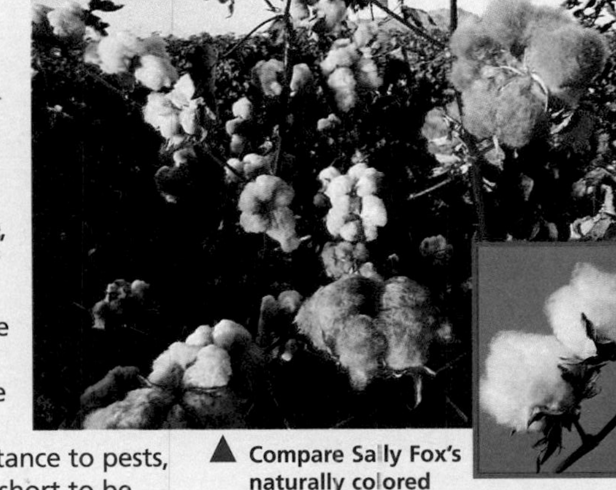

▲ Compare Sally Fox's naturally colored cotton (above) with common cotton (right).

Solutions to Environmental Problems

The textile industry is the source of two major environmental hazards. The first hazard is the dyes used for cotton fabrics. The second is the pesticides that are required for growing cotton. These pesticides, like the dyes, can cause damage to both living things and natural resources such as water and land.

Sally's cotton represents a solution to both of these problems. First, since the cotton is naturally colored, no dyes are necessary. Second, the native strains of cotton from which she bred her plants passed on natural resistance to many common pests. Thus, fewer pesticides are necessary to grow her cotton successfully.

Sally Fox's efforts demonstrate that, with ingenuity and patience, science and agriculture can combine in new ways to offer realistic solutions to pressing environmental problems.

◀ Sally Fox in front of a harvest of colored cotton

Some Detective Work

Like Sally's cotton, many types of plants and breeds of domesticated animals have been created through artificial selection. Do a little research to find out where and when your favorite fruit or breed of dog was first established.

559

EYE ON THE ENVIRONMENT

Background

Cotton has many properties that make it a superior material for fabrics. Cotton fabrics are durable, washable, and comfortable. Cotton clothing keeps you cool in the summer because it breathes well and allows the moisture to evaporate quickly. Some other uses for cotton include fine yarns, carpets, and blends with other fabrics.

Sally Fox grows cotton in Arizona and New Mexico. She sells her cotton to such companies as Fieldcrest, Esprit de Corps, and L. L. Bean. Fox is continuing to refine her crop by trying to breed fire-resistant varieties as well as new colors.

Multicultural Extension

Cash Crops in History

Cotton is an example of a *cash crop*. Cash crops are grown for sale rather than for immediate use on the farm. Such crops often play a vital role in the economy of a society. Suggest that each of your students choose a region or country he or she is interested in and do research on its cash crops. Some interesting areas include the American South and Midwest, the Russian plains, Ireland, and various countries of Central and South America. Students should find out what the crops are, how dependent the people of the country are on the success of the crops, and what role the crops have played in the country's history. Then conclude the activity by making a class map and chart showing the world's cash crops.

Meeting Individual Needs

Learners Having Difficulty

Give students samples of different fabrics to touch and observe. Fabrics could include cotton, linen, polyester, nylon, silk, leather, or suede. Then have students classify the fabrics based on their origin—plant, animal, or synthetic. Ask: What are the advantages of fabrics made from plants? Compared with the other fabrics, what are the disadvantages? *(Students should recognize that plant-based fabrics have a different texture, come from a renewable resource, and are cheaper than animal-based fabrics. They should also recognize that plant-based fabrics may be less durable and more prone to stains than the other fabrics.)*

Answers to *Some Detective Work*

One example is the dachshund. It was initially bred in Germany in the 1700s, where it was used to hunt badgers. Its tubular body and short legs are ideal for following prey into burrows. The word *dachshund* is German for "badger hound."

SourceBook

Unit 1

INTERACTIONS

In This Unit

Now that you understand some of the interactions among organisms and their environment, consider these questions.

1. What is the biosphere?
2. How are substances recycled within an environment?
3. What is a biome, and which one do you inhabit?
4. What can we do to help solve major environmental problems?

In this unit, you will take a closer look at populations and communities, how they change over time, and what we can do to help preserve our air, water, and soil.

THE BIOSPHERE

All life exists within the **biosphere**, a thin layer that surrounds the Earth. In this layer, all living organisms are born, grow, reproduce, and die. The figure below shows the full vertical dimensions of the biosphere. Yet most of Earth's millions of species of organisms live in a much narrower range, only 200 m thick, that extends 100 m above and below the Earth's surface. As you can imagine, the biosphere has many different environments each of which supports many different organisms. The same biosphere that has a warm marine environment for jellyfish also has a cold, icy environment where polar bears live.

To make it easier to understand organisms and their environments, the biosphere is divided into **ecosystems**. An ecosystem consists of all the *biotic* (living) and *abiotic* (nonliving) things in an area. The kinds of living organisms and their interactions, along with the physical environment—like the soil, water, and weather—all contribute to an ecosystem. An ecosystem may be as small as a single decaying log in a forest or as large as the Amazon River!

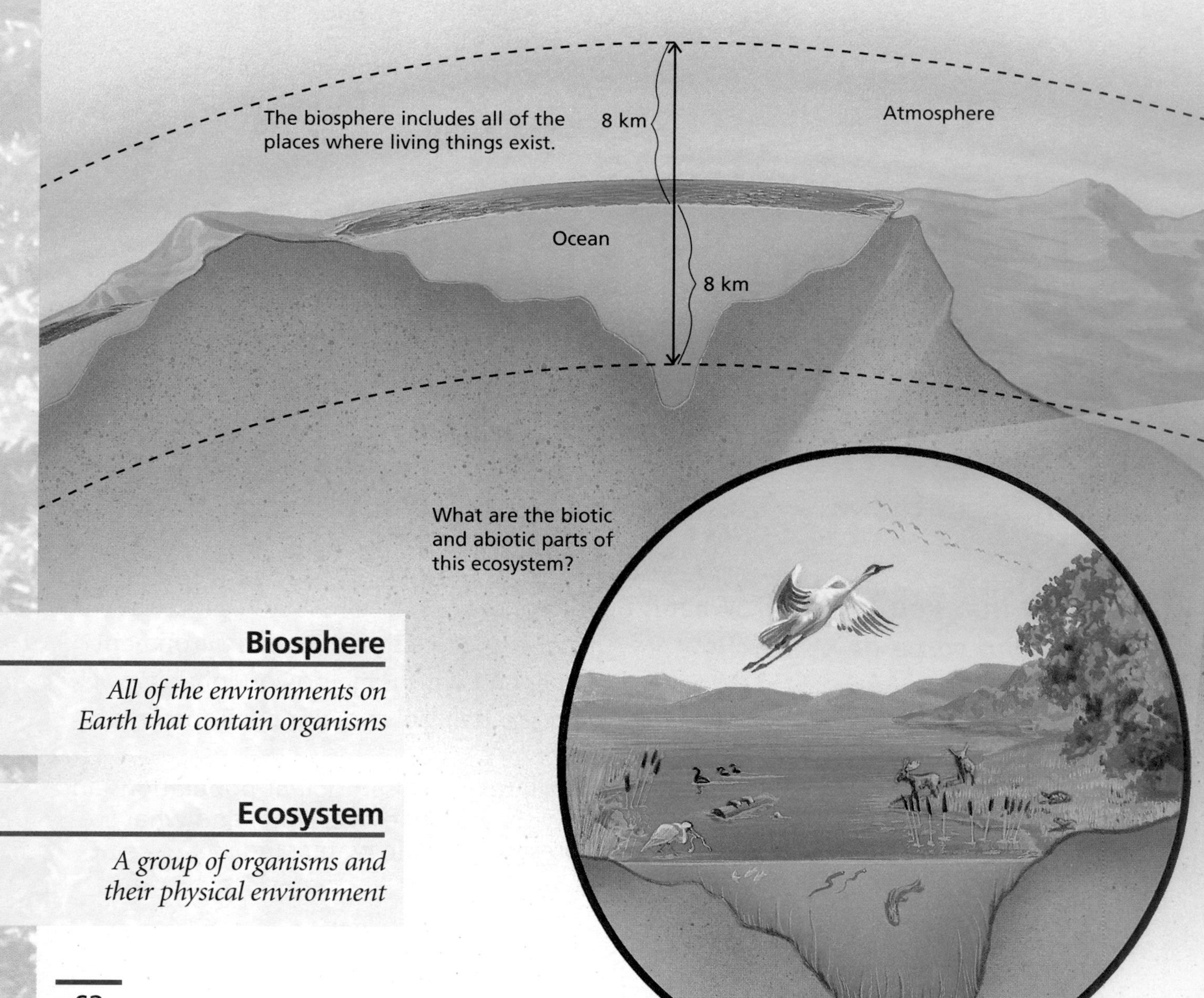

The biosphere includes all of the places where living things exist.

What are the biotic and abiotic parts of this ecosystem?

Biosphere

All of the environments on Earth that contain organisms

Ecosystem

A group of organisms and their physical environment

S2

Think of the different ecosystems that you visit and interact with. Did you include places like the park, the beach, and your own backyard? Each of these places contains biotic and abiotic factors that interact in unique ways. The interactions within an ecosystem involve many cycles and pathways that link organisms to each other and to their abiotic environment. In this section, you will examine some of these interactions.

A family of brown bears in Katmai National Park, Alaska

Populations and Communities

When a coyote preys on a jackrabbit or a caterpillar eats an oak leaf, we say that these organisms are interacting. However, an ecosystem does not contain only one coyote or one oak leaf. Instead, individual organisms exist as members of interacting *groups*. Two types of groups found in ecosystems are populations and communities.

Populations What is the population of your classroom? You would probably answer this question with a number. But how would you arrive at that number? Would you include the teacher? the potted plants? the fish in the aquarium? In an ecosystem, a **population** represents the number of organisms that belong to the same *species*. So if your classroom contains plants and fish, it contains several populations.

In addition to the species of the organism, a population is described by its location and a particular point in time. Was the population of your classroom the same 2 years ago? 10 years ago? Consider the Alaskan brown bears in the photo above. Since this photo was taken a while ago, these bears belong to an earlier population than do the brown bears that are currently found in Alaska.

Together, the individuals in a population form a *reproductive group*. For a species to survive, it must have enough individuals for reproduction to exceed losses due to death or disease. With too small a population, the species may disappear. Organisms in small populations may not be able to find each other, or they may be unable to ensure the successful growth of their young. For example, the wild populations of whooping cranes and California condors became so small that the animals could no longer replace their losses in nature. Fortunately, successful breeding programs for these two species have begun to increase their numbers.

You can see California condors in some zoos only because people have established breeding programs for them.

Population

A group of organisms of the same species

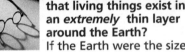

DID YOU KNOW . . .

that living things exist in an *extremely* thin layer around the Earth? If the Earth were the size of a basketball, the zone containing life would be thinner than a page from this book!

Members of a population also *compete* with one another for resources such as food and living space. You have seen competition among members of a population if you have ever watched alley cats fight over territorial rights or observed a flock of pigeons peck aggressively at bread crumbs. The resources that a population needs to survive, like food, soil nutrients, and living space, are called **limiting factors**. These factors determine how large a population can grow.

Limiting factors

Environmental elements that stabilize population size and keep species from producing too many offspring

Characteristics of Populations One important characteristic of a population is that the number of individuals in it constantly changes. A population grows as new members are born and as adults migrate in from different populations. A population may become smaller as individuals die or migrate to other populations.

Suppose a biologist is studying a prairie dog colony that has 10 individuals. During one year, 3 animals are born and 5 die. The difference between 3 births and 5 deaths represents a *net loss* of 2 prairie dogs. However, by counting the prairie dogs, the biologist discovers that the population actually grew by a total of 5 individuals. Where could the other 7 prairie dogs have come from?

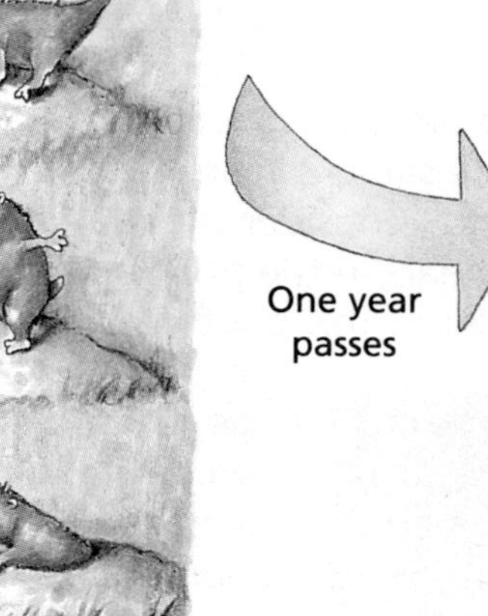

Births and deaths alone do not explain the annual changes in the size of a prairie dog colony. New animals periodically join the colony, while others leave.

Population density

The number of individuals of the same species living in a given area

Another important characteristic to consider is the number of individuals that occupy a given area. This is called the **population density** of a species. For example, the population density of humans on Manhattan Island in New York City is about 26,000 people per square kilometer. Notice that the population density is not just the *number* of organisms in the population; it also includes the *size* of the area in which the individuals live.

Look again at the prairie dog colony. Can you see that each member of the colony had more living space before the population grew? What are the advantages of having more living space? Are there advantages to living in a dense population? If a population becomes too dense, it begins to decline because its environment cannot provide enough of its basic needs. Some members of the population may move elsewhere, while others may die due to lack of food or shelter.

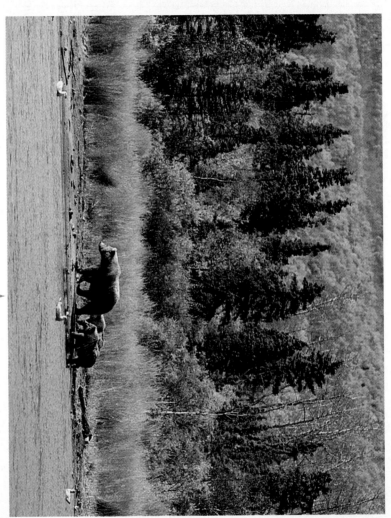

The community shown in this photograph is composed of several different populations, including grizzly bears, spruce trees, aspen trees, several kinds of grasses, ducks, and many unseen organisms.

Communities A **community** consists of all of the different populations living and interacting in the same area. For example, all of the plant and animal populations in a forest belong to the same community. Similarly, the populations of fish and other marine organisms found on a coral reef form their own community. The populations of a community depend on one another for food, other nutrients, and sometimes even a home. Studying communities allows you to learn about the interactions among many different populations that exist together in an ecosystem.

Community

All of the populations that live in an area

Interactions in Communities
Populations of different species in a community often have close, permanent relationships with one another. Many times these relationships result in harm being done to individual members of one of the populations. The predator-prey relationship is one of the most common within a community. A *predator* is an organism that kills and eats other organisms. The organism eaten by a predator is the *prey*. Lions, for example, are predators that eat many different prey, including zebra and antelope.

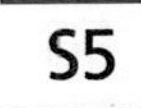

Predator-prey relationships are common in communities.

Human host

Beef containing tapeworms

Adult tapeworm in human intestine

Immature tapeworms burrow into muscle

Cow host

Tapeworm eggs

The life cycle of the beef tapeworm alternates between cows and humans.

Another relationship that occurs in communities is called *parasitism*. A *parasite* is an organism that lives in or on another living organism (the host) and is harmful to that organism. A parasite is different from a predator. The parasite does not need to kill the host organism to benefit from it. In fact, the death of the host usually results in the death of the parasite. The life cycles of many parasites, such as tapeworms, often involve more than one host.

Not all of the interactions that occur in communities result in the death or weakening of organisms. There are some that cause no harm. In a relationship called *commensalism*, one organism benefits while the other is neither benefited nor harmed. In many cases, one organism provides a home for another organism. For example, the clown fish lives among the poisonous tentacles of the sea anemone. The tentacles do not harm the clown fish, yet they protect it from larger predators. When two different organisms interact and *both* benefit, the relationship is known as *mutualism*. The bacteria that live in your large intestine are a good example of this type of relationship. You give the bacteria food and a home. In return, the bacteria produce vitamins that you cannot make for yourself.

DID YOU KNOW . . .

that a tapeworm lacks a digestive tract? Tapeworms have evolved into such efficient parasites that they don't need a digestive tract. A tapeworm simply attaches itself to the inside of a host's intestine and absorbs digested nutrients directly through its skin.

The relationship between a clown fish and sea anemone is an example of commensalism.

Energy Pathways

You are already familiar with two types of energy pathways that exist in ecosystems—food chains and food webs. No matter how complex the food web, a community always needs more producers than consumers. Producers provide the energy necessary for themselves and the consumers. An **energy pyramid**, like the one shown below, is used to represent the energy levels of a food web.

The organisms at each level in an energy pyramid use much of their energy just to stay alive. Only about 10 percent of an organism's energy can be passed on to the next level when it is eaten. Therefore, only a few animals are found at the top of the energy pyramid. These animals—the *top carnivores*, such as lions, wolves, eagles, and bass—exist in smaller populations. They depend on large populations of other animals to survive.

Energy pyramid

A diagram that shows how energy decreases at each feeding level in a food chain

The energy levels of this pond ecosystem are illustrated by an energy pyramid.

Each successive layer of an energy pyramid is smaller than the one below it. The bottom of the pyramid represents the producers. Herbivores make up the next level of the pyramid. The upper levels of the pyramid consist of carnivores, some of which eat herbivores. The carnivores at the very top of the pyramid eat other carnivores. There are more herbivores in the community than carnivores, but fewer herbivores than producers. What do you think would happen if there were more carnivores than producers?

Ecosystem Cycles

Food and energy are not the only resources an organism must have to survive. Substances from the abiotic environment are also essential for the survival of the organisms in an ecosystem. These materials, such as water, carbon, oxygen, nitrogen, and mineral nutrients, are used by living things and then returned to the environment. In this *recycling* process, there is a constant exchange of substances between the living and nonliving parts of an ecosystem.

The Water Cycle The evaporation of water and the condensation of water vapor both play a major role in the water cycle. Organisms also contribute to the water cycle by using and releasing water.

Water vapor eventually cools and forms clouds from which rain, sleet, or snow fall back to the ground.
The sun heats water on the Earth's surface and causes it to evaporate into the atmosphere as water vapor.
Animals drink water or extract water from their food and return it to the environment through waste products.
Plants use water to make food. They also release water back into the atmosphere.
S8

Cutting down too many trees in a forest can disrupt the water cycle and decrease the amount of water returned to the atmosphere. It can also cause an increase in soil erosion. Without enough water and nutrient-rich soil, the entire forest may eventually die. Tragically, this now seems to be occurring in many tropical rain forests on Earth.

The Carbon Dioxide–Oxygen Cycle The gases carbon dioxide and oxygen are used in chemical reactions in organisms. Plants use carbon dioxide and sunlight to make food and oxygen. Both plants and animals need oxygen to convert the energy stored in food. When they do, carbon dioxide is released.

The cutting of too many trees disrupted the water cycle of this forest. The crops that replaced the trees cannot return as much water to the cycle.

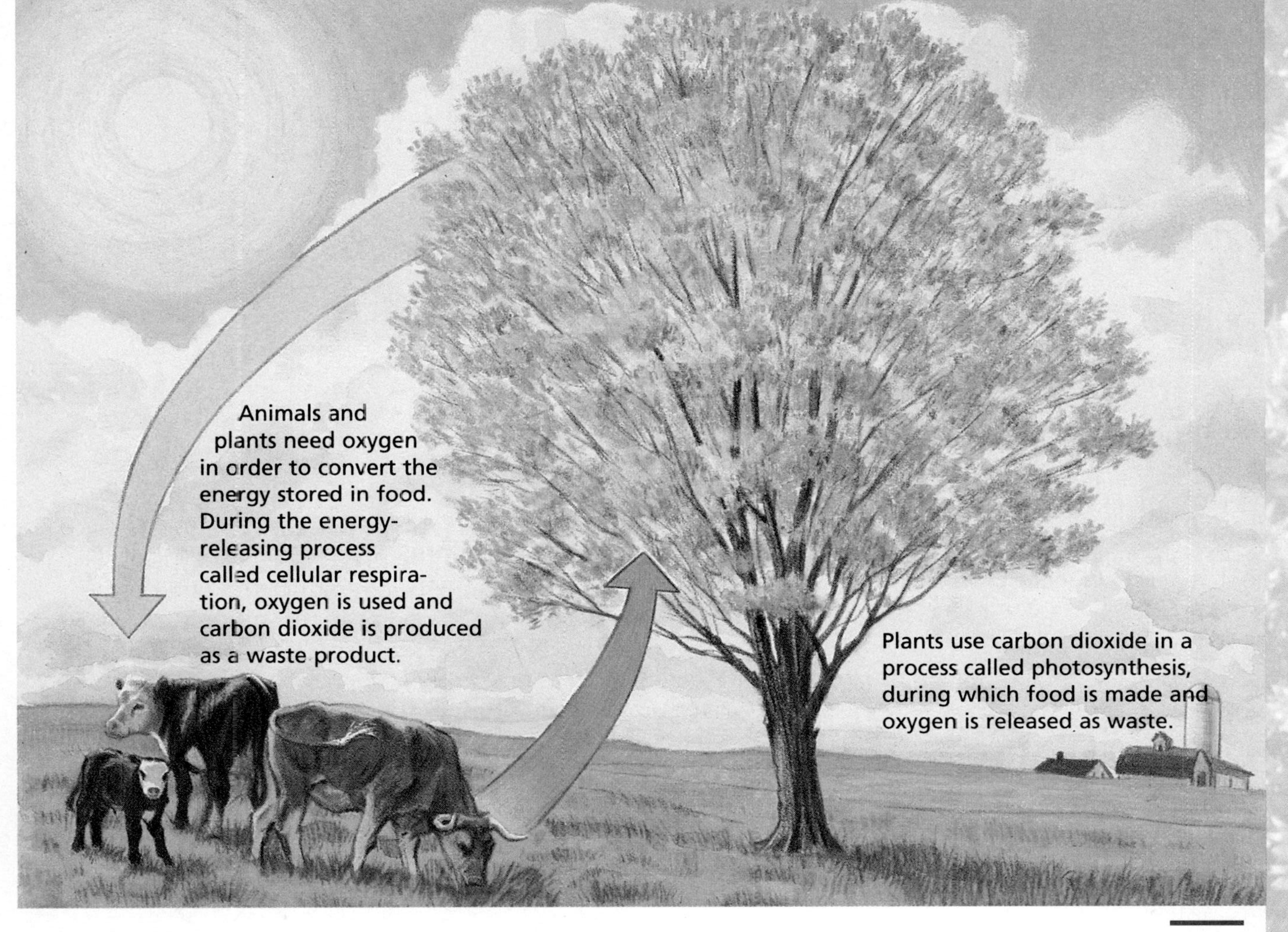

Animals and plants need oxygen in order to convert the energy stored in food. During the energy-releasing process called cellular respiration, oxygen is used and carbon dioxide is produced as a waste product.
Plants use carbon dioxide in a process called photosynthesis, during which food is made and oxygen is released as waste.

The Nitrogen Cycle Nitrogen gas makes up about 80 percent of the Earth's atmosphere. Organisms use nitrogen in the production of proteins. However, most organisms cannot use nitrogen directly from the atmosphere. Fortunately, bacteria that live in the roots of certain plants and in the soil are able to change nitrogen gas into ammonia and other nitrogen compounds, such as amino acids, that can be used by the plants. When these plants die, other bacteria release the nitrogen compounds, which are absorbed by other plants and enter the food web of an ecosystem.

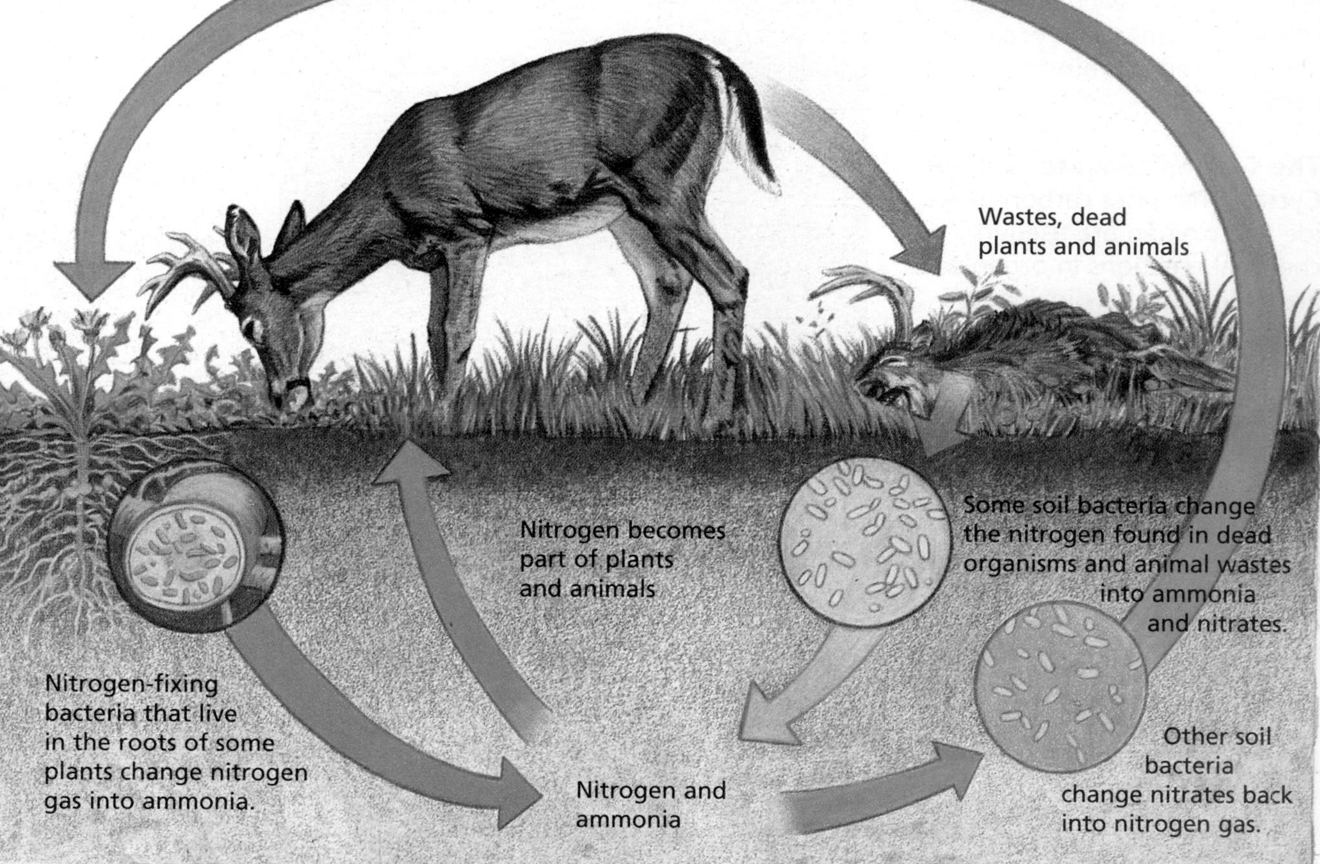
Nitrogen in the atmosphere
Wastes, dead plants and animals
Some soil bacteria change the nitrogen found in dead organisms and animal wastes into ammonia and nitrates.
Nitrogen becomes part of plants and animals
Nitrogen-fixing bacteria that live in the roots of some plants change nitrogen gas into ammonia.
Nitrogen and ammonia
Other soil bacteria change nitrates back into nitrogen gas.

DID YOU KNOW...

that lightning in the atmosphere plays a role in the nitrogen cycle? The energy found in lightning changes some of the nitrogen gas in the atmosphere into ammonia, which then falls to the Earth and enters the soil.

SUMMARY

The biosphere is the thin layer of the Earth that includes all organisms and the places where they live. The biosphere contains the living (biotic) and nonliving (abiotic) factors that make up Earth's ecosystems. Groups of organisms of the same species in ecosystems are called populations. Populations combine and interact to form communities. Within a community, predator-prey interactions occur, as do other relationships, such as commensalism, parasitism, and mutualism. The transfer of energy through an ecosystem can be represented by an energy pyramid. Water and essential elements, such as carbon, oxygen, and nitrogen, circulate in ecosystems through several complex cycles.

Succession and the Biomes

Have you ever wondered how a scene such as the one pictured on the right came to be? Natural geological processes gradually molded the land into its present shape. The plant and animal communities that you see here also developed slowly over time. The makeup of these communities changed in response to a variety of limiting factors, such as climate, soil composition, disease, and natural disasters.

In this section, you will take a close look at the process by which living communities change over time. You will also look at some of the geographic areas that cover major parts of the Earth and that share similar climates.

The Sonoran Desert of Organ Pipe Cactus National Monument in Arizona

Ecological Succession

Compare the two photographs below. Both were taken at Glacier Bay National Park in Alaska. The view on the left shows the land that is closest to the glacier. The glacier has only recently receded, and the land near it is rocky and lifeless. In the photograph on the right, the land has been exposed for hundreds of years. You can see the forest of spruce trees and can imagine the kinds of animals that inhabit this forest.

The glaciers of Alaska's Glacier Bay formed during the last ice age. Since that time, a slow, steady melting has taken place, exposing land that was covered by ice for thousands of years. During the next few centuries, a gradual process of change will occur in Glacier Bay as communities of organisms become established on the bare rock. Over time, one community will gradually replace another in an ecological process called **succession**.

Succession

A series of changes that occurs in the communities of an ecosystem

Within Glacier Bay National Park in Alaska, there are barren, rocky environments (left) and lush forests (right).

S11

Primary Succession Imagine an underwater volcano violently erupting. Huge rivers of lava pour out of the ocean floor until, suddenly, a new island breaks through the surface of the water. This brand-new island is barren of plant and animal life. But not for long! Living organisms will invade this new environment as the process of succession begins. Succession that starts in an area where the land has never had organisms living on it is called *primary succession*. In addition to land formed by lava flows and exposed by the retreat of glaciers, sand dunes are also places where primary succession occurs. The types of communities that come and go during primary succession are different in each of these areas, but they all follow similar patterns of development.

Now let's take a closer look at succession as it occurs around Glacier Bay, Alaska. Traveling toward the glacier is like traveling backward in time. As you move away from the glacier, you can investigate the changes to the land that resulted from succession. Communities representing each stage of succession thrive in different parts of the bay area. Standing at the edge of the glacier, you can see that primary succession begins soon after bare, lifeless rock is uncovered by the melting of glacial ice.

A Right below your feet, you can see succession beginning at the base of the glacier. *Pioneer species* are the first living things to appear in a previously lifeless area. The harsh habitat of these pioneer species is characterized by high winds, extreme cold, bright sun, and heavy rain. There is no soil at this stage. *Lichens*, which consist of a fungus and an alga growing together, are one of the first pioneer species to appear. They often completely cover the surface of rocks. In order to get necessary nutrients, lichens produce acids that break down the hard rock. As dead lichens collect and decay, they mix with small pieces of the rock to form a thin crust—the beginning of soil.

B A short distance away from the lichen-covered rocks, *mosses* (another pioneer species) grow in the new soil formed by lichens. Mosses are simple plants that form thick mats. Their decaying parts add to the mixture of dead lichens and rock particles. Mosses also trap bits of dust and dead matter brought in by the wind.

More organic matter is added by a variety of animal species, such as insects and birds, that are a part of the moss community. This organic matter enriches the developing soil.

C If you walk away from the glacier and cross over the areas of lichen and moss, you will see that the soil becomes richer. The land no longer looks as barren as it did near the end of the glacier. Meadows of grasses and wild flowers have replaced the pioneer species. Wind, birds, and small mammals bring seeds for new plants into this meadow area. The roots of these plants break up rock and trap decaying organic matter that helps form rich layers of soil. Eventually, the soil can support the growth of

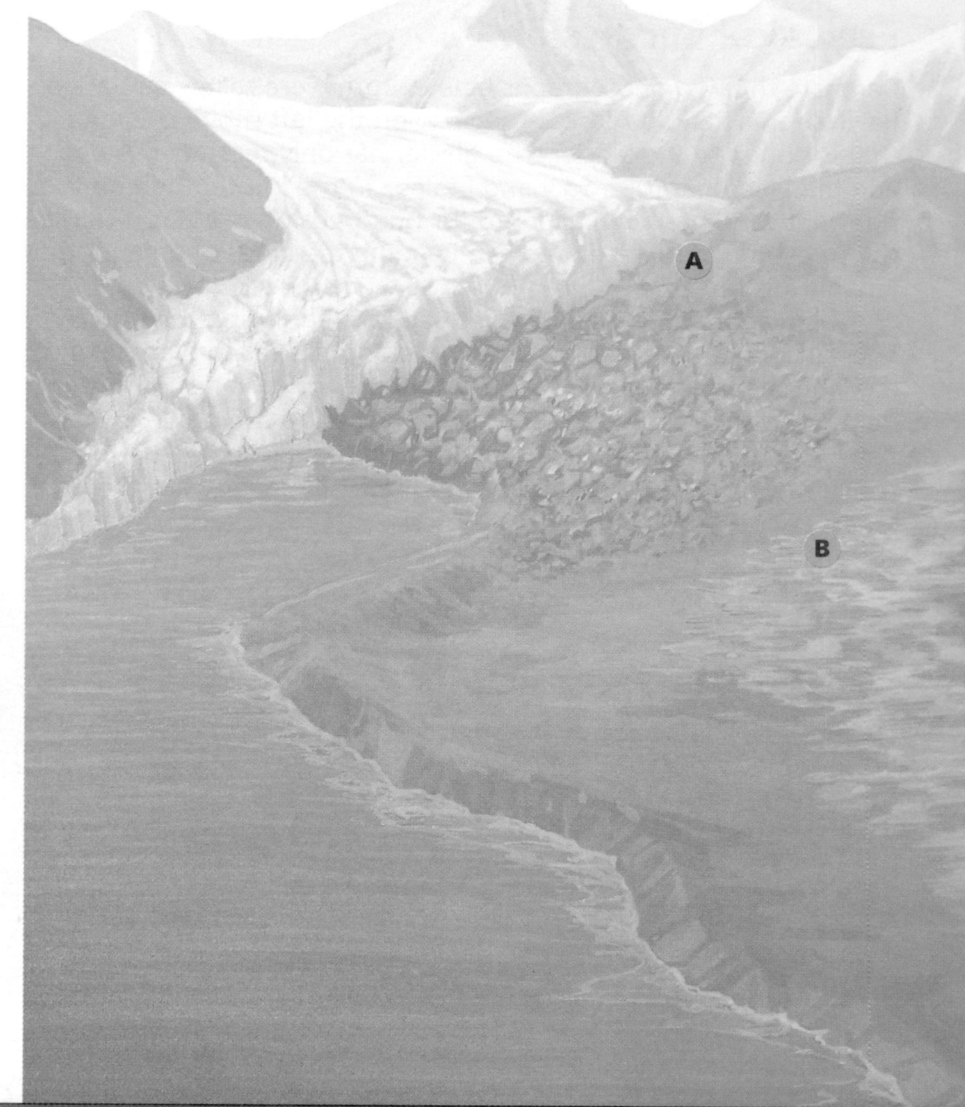

small trees such as willow and alder, which thrive in the bright sunlight of the open meadows.

D If you leave behind the grassy meadows and small trees, you find that these communities are replaced by forests of white spruce. Spruce seedlings grow well in the shade that the smaller trees provide, so the seedlings eventually replace the mature willows and alders. The spruce trees represent the final stage of succession. The community that exists at this stage is called a **climax community**. It includes the population of white spruce, some smaller trees and shrubs, and the animals that find shelter and food in such a forest.

Secondary Succession

Natural disasters, such as fires, floods, diseases, and swarms of insects, can disturb a climax community. Human activities, like agriculture and forestry, can damage these living communities as well. Once the disturbance is over, the climax community will slowly reestablish itself. This type of succession, which occurs after an original community has been disturbed or destroyed, is called *secondary succession*.

Because soil is already present, the process of secondary succession takes place more quickly than does primary succession. The pioneers for secondary succession are usually grasses and other weed-like species of flowering plants, whose seeds are brought in from the surrounding area by wind, mammals, or birds. Grasses and weeds grow quickly, and most have root systems that protect the soil of a disturbed area from erosion. Often, seeds from the plants that originally lived in the area are lying undisturbed beneath the ground. When the environment is able to support the seedlings, these seeds will germinate and grow. As the plant communities are reestablished, the animals that once lived there return as well.

Climax community

The final stage in the succession of communities

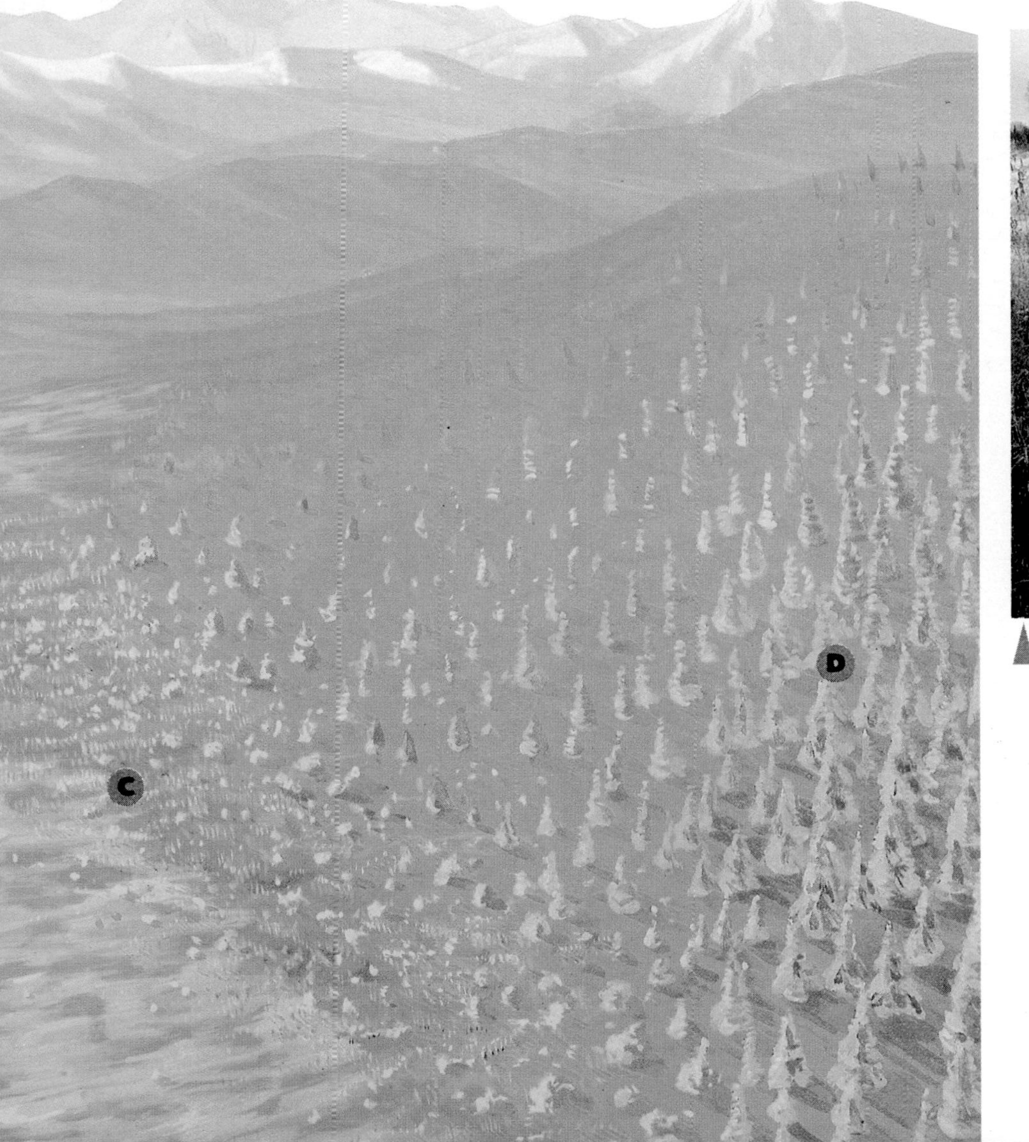

Secondary succession is occurring in this abandoned field. Just a few years earlier, the field was used to grow crops.

S13

Terrestrial Biomes

Large areas of Earth's continents share similar climates and soil types. These geographic regions are called **biomes**. The biomes differ in the types of plants and animals found in them. Scientists usually identify these regions by their climax plant species. The six major terrestrial (land) biomes are indicated on the map shown here.

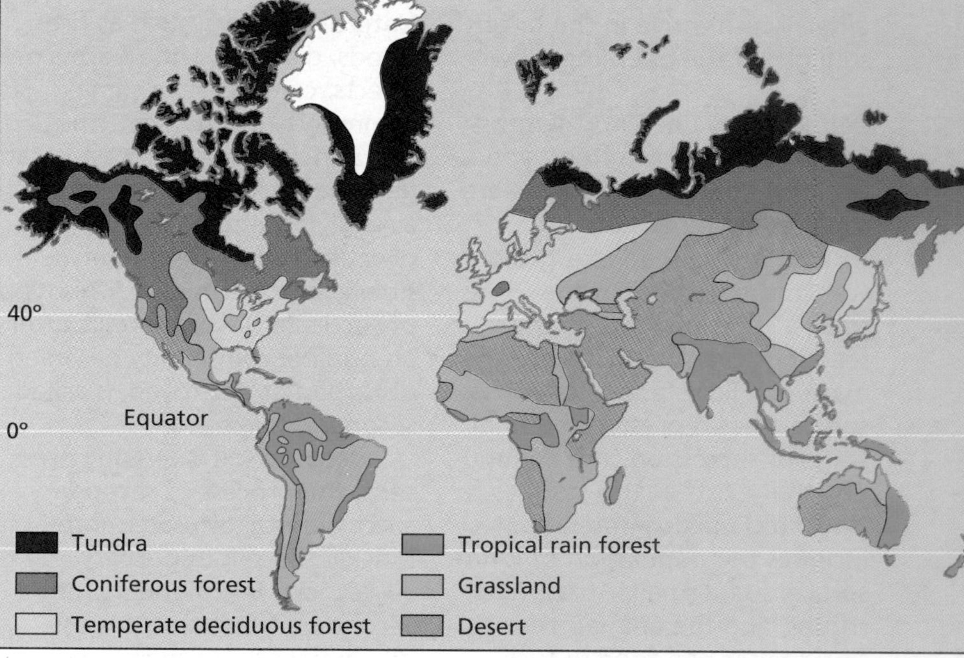

Use the color key to locate the Earth's major terrestrial biomes.

Biomes

Large areas with similar climates and life-forms

The Tundra Just south of the Arctic icecap is the freezing, harsh biome known as the *tundra*. A land of extreme cold, high winds, and very little rain or snow, the tundra experiences winters with temperatures as low as −40°C! The tundra receives the least amount of sunlight of any of the Earth's terrestrial biomes. Life on the tundra is limited to hardy species that are adapted to the cold, harsh climate or are able to migrate to warmer regions when necessary.

Plant life on the tundra includes ground-hugging species of grasses, lichens, mosses, and flowering herbs that can reproduce during the short growing season. Small willow and birch trees grow in some areas. Animal inhabitants include migratory birds, musk oxen, wolves, grizzly and polar bears, lemmings, and arctic hares and foxes.

DID YOU KNOW...

that the tundra is one of the most fragile biomes on the planet?
The food webs of the tundra are simpler than those of other biomes and are therefore easily disrupted.

The Forests Three kinds of forest biomes cover a large portion of the Earth. They differ in climate, amount of rain, temperature range, and growing season. These and other limiting factors, such as soil type and humidity, determine the kinds of communities that inhabit each forest biome.

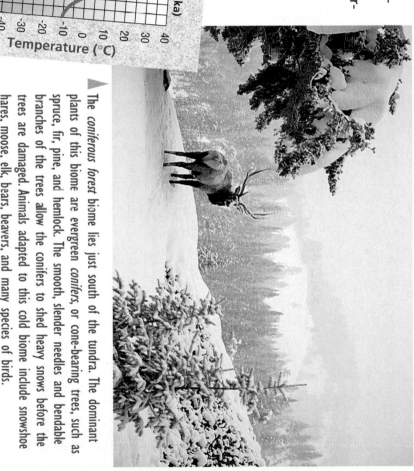

The *coniferous forest* biome lies just south of the tundra. The dominant plants of this biome are evergreen *conifers*, or cone-bearing trees, such as spruce, fir, pine, and hemlock. The smooth, slender needles and bendable branches of the trees allow the conifers to shed heavy snows before the trees are damaged. Animals adapted to this cold biome include snowshoe hares, moose, elk, bears, beavers, and many species of birds.

The *temperate deciduous forest* biome has a humid climate marked by distinct seasonal changes. Deciduous trees, like oak, hickory, beech, maple, and elm, are the dominant plant species of this biome. Animal life includes deer, bears, snakes, rabbits, squirrels, and many birds and insects.

The *tropical rain forest* biome is found along and near the equator. The hot, wet climate of the tropical rain forest supports the greatest variety of life found anywhere on Earth. New species of plants and animals are discovered each year. Some may never be seen by humans because they live up to 60 m above the ground in the canopy of tree branches. The animal species in tropical rain forests include monkeys, squirrels, bats, snakes, deer, many types of rodents, many types of birds, and a variety of different insects.

Tropical Rain Forest
(Manokwari, New Guinea, Indonesia)
Precipitation (cm): 0 5 10 15 20 25 30 35 40 45 50
Temperature (°C): -60 -50 -40 -30 -20 -10 0 10 20 30 40
J F M A M J J A S O N D
Months

Grasslands The grassland biome is usually found at the same latitude as the deciduous forest biome, except that the grasses require less rainfall than trees do. In North America, the grasslands are called *prairies*. Great herds of bison once roamed these prairies. Because grassland soils are rich in nutrients, much of the world's best farmland is found in this biome. Today much of the original vegetation has been replaced by fields of grain such as wheat and corn, and great herds of cattle and sheep graze in place of bison.

Although great herds of bison no longer exist, a variety of small animals, such as jackrabbits, ground squirrels, prairie dogs, and mice, still inhabit North American prairies. They are prey for grassland predators such as coyotes, foxes, snakes, and hawks.

Temperate Grasslands
(Lawrence, Kansas)
Precipitation (cm): 0 5 10 15 20 25 30 35 40 45 50
Temperature (°C): -60 -50 -40 -30 -20 -10 0 10 20 30 40
J F M A M J J A S O N D
Months

Desert communities contain plants such as cacti, small trees, woody shrubs, and many wildflowers. The plants that grow there have many adaptations for getting and conserving water. Many desert animals are nocturnal, meaning that they are active only at night, when it is cooler. Kangaroo rats and other rodents, bats, snakes, toads, birds, and insects populate the desert.

The Desert Deserts occupy geographic areas that receive less than 25 cm of rainfall per year. Rainfall in deserts is rare; some deserts do not receive any rain for several years at a time. Although not all deserts are hot, the hottest places on Earth are found in deserts. For example, Death Valley, California, a part of the Mojave Desert of the southwestern United States, had a record high temperature of 57°C (135°F)—in the shade!

DID YOU KNOW...

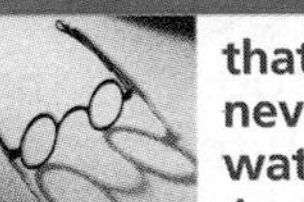

that some desert animals never need to drink water?
American desert ground squirrels can get all of the water they need from the seeds that they eat.

Aquatic Biomes

Oceans and seas contain salt-water ecosystems. Lakes, ponds, rivers, and streams contain freshwater ecosystems. Along coastlines, salt water and fresh water come together to form a special ecosystem called an *estuary*. All of these ecosystems make up the aquatic (water) biomes of Earth. Together, the aquatic biomes cover more than three-quarters of the Earth's surface. Temperature, sunlight, salinity, water flow, and the amount of available nutrients are important abiotic factors that determine how organisms are distributed in the aquatic biomes.

The Marine Biome Covering about 70 percent of the Earth's surface, the marine biome is the largest biome on Earth. Because it is so large, factors such as temperature, light, and depth of water vary from region to region. The *intertidal zone*, which is closest to the shore, is rich in nutrients that have washed into the ocean from shoreline communities. During low tide, this region is exposed to air. Organisms living in the intertidal zone must be adapted for life in the water as well as out of the water. One of the most productive intertidal ecosystems, the estuary, occurs in bays, salt marshes, or swamps where the salinity of the water is between that of sea water and fresh water. Estuaries are breeding grounds for many marine animals and support large and diverse communities of organisms, such as fishes, clams, snails, oysters, and crabs.

Sea stars have adaptations for life in the intertidal zone.

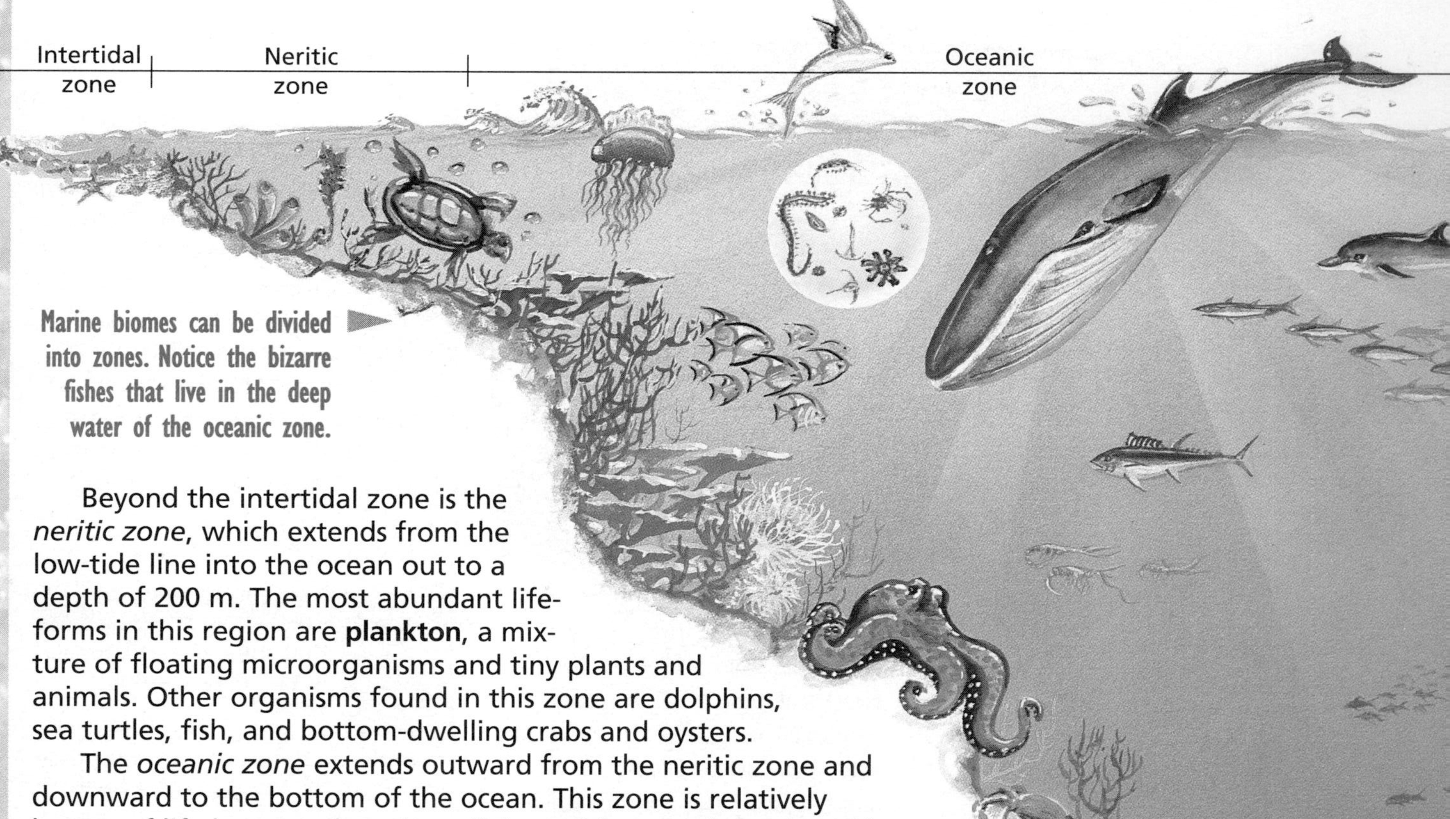

Marine biomes can be divided into zones. Notice the bizarre fishes that live in the deep water of the oceanic zone.

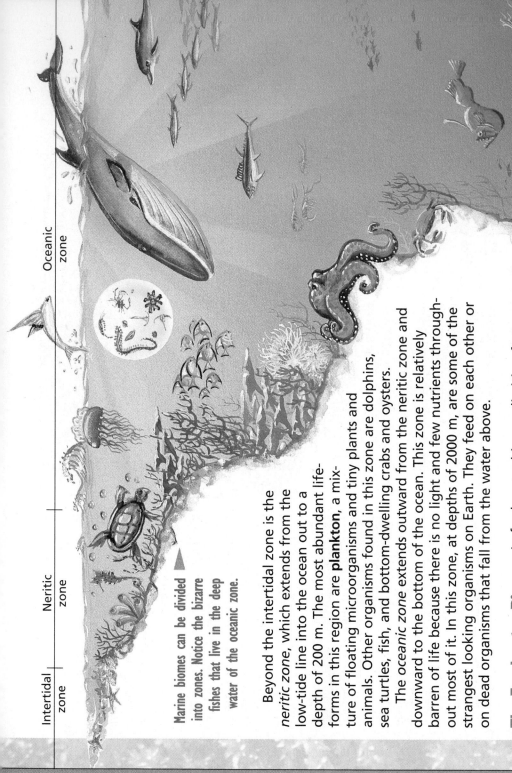

Beyond the intertidal zone is the *neritic zone*, which extends from the low-tide line into the ocean out to a depth of 200 m. The most abundant life-forms in this region are **plankton**, a mixture of floating microorganisms and tiny plants and animals. Other organisms found in this zone are dolphins, sea turtles, fish, and bottom-dwelling crabs and oysters.

The *oceanic zone* extends outward from the neritic zone and downward to the bottom of the ocean. This zone is relatively barren of life because there is no light and few nutrients throughout most of it. In this zone, at depths of 2000 m, are some of the strangest looking organisms on Earth. They feed on each other or on dead organisms that fall from the water above.

The Freshwater Biome In freshwater biomes, limiting factors include rate of water flow, oxygen content, available nutrients, light, and temperature. These factors vary greatly between streams, ponds, and lakes. Lakes and ponds have nutrient-rich water that supports the growth of surface algae and floating plants. Many species of fish, insects, reptiles, birds, and mammals live in or near lakes and ponds.

Life is sparse, however, in swiftly flowing streams. Cold water and the currents that carry away both the nutrients and organisms are the main limiting factors. Plants that attach to rocks live in swift streams, but floating plants and algae merely pass through with the current. The animals of this ecosystem are either excellent swimmers, like trout, or they live on the underside of rocks, like crayfish and some insect species.

Plankton

Small organisms that float near the surface of the ocean

S18

SUMMARY

Succession is the process by which one community is replaced by another over time. Primary succession occurs in areas that have not previously supported life. Secondary succession occurs in areas where an existing community was disturbed or destroyed. The biosphere includes several large geographic areas that are characterized by their climates and the types of plants that live there. The six major terrestrial biomes are tundra, grassland, desert, coniferous forest, deciduous forest, and tropical rain forest. The major aquatic regions are the marine and freshwater biomes.

HUMANS AND THE ENVIRONMENT

▲ These male and female passenger pigeons were painted by John James Audubon in 1824.

Prior to the mid-1800s, an estimated 3 to 5 billion passenger pigeons inhabited the forests of eastern North America. But in 1914, the last known passenger pigeon died at the Cincinnati Zoological Gardens. At that moment, this beautiful species of pigeon disappeared forever. How could so large a population become **extinct** so quickly?

In this case, the cause was the migration of an animal species into the birds' ecosystem. This species consumed the oak and beech forests in which the birds nested. The invading species also preyed on the birds. As the population of passenger pigeons declined, they became unable to reproduce as quickly as they once had. Soon all of the passenger pigeons were gone—their species extinct.

Who were the invaders that caused this extinction? They were human settlers who used the wood of the forests to build their homes and sold the birds to city dwellers to eat as delicacies. Though the extinction of a species is a natural process, human activities in recent years have greatly increased the rate at which species become extinct.

Extinct

Refers to a species that no longer exists

The Use and Misuse of Natural Resources

Natural Resources

All natural substances that humans can remove from the environment for their own use

Pollution

Contamination of the environment

Human societies need a steady supply of food and materials from the Earth and its biosphere. These organisms and substances are called **natural resources**. Some of our natural resources can be replaced, but others cannot. *Renewable resources* can be replaced at about the same speed that we use them. These include trees, crops, livestock, fish, wildlife, water, oxygen, nitrogen, and carbon. *Nonrenewable resources* cannot be replaced at the same speed that we use them. They include soil, minerals, metal ores, and fossil fuels.

Pollution is the process of making the environment unclean with waste products such as poisonous gases, liquid chemicals, radioactive materials, heat, garbage, and even noise. All of these waste materials are referred to as pollutants. You find them in the air you breathe, the water you drink, and the soil where your food grows.

Air Pollution Burning fuels and burning trash are the major sources of *air pollution*. Automobiles, factories, power plants, and other industries are the largest contributors of this pollution. Most coal and petroleum fuels contain sulfur, which combines with oxygen when burned, releasing poisonous sulfur dioxide gas into the air. Breathing air containing sulfur dioxide gas can cause health problems such as lung disease. Sulfur dioxide also appears to be involved in the production of acid rain.

DID YOU KNOW...

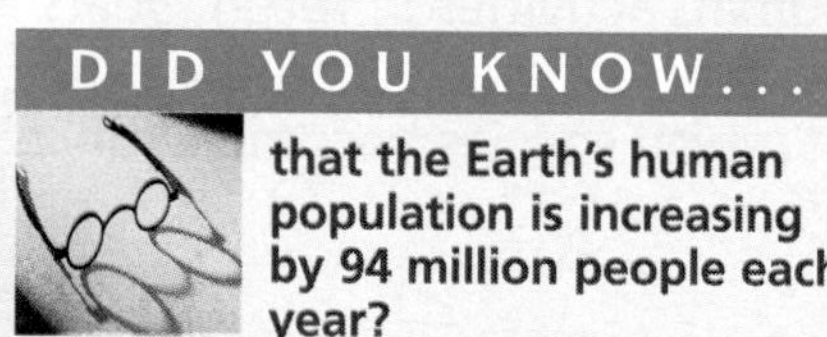

that the Earth's human population is increasing by 94 million people each year?

This is equivalent to 260,000 people each day, 11,000 each hour, or about 3 people each second.

S20

Water Pollution One of the most dangerous kinds of *water pollution* occurs when *industrial wastes* and untreated *sewage* are dumped directly into rivers, lakes, and oceans. Sewage can add harmful bacteria and viruses to water supplies. It also carries household chemicals, such as phosphates from detergents, that allow algae and water plants to grow out of control. Another source of water pollution is the chemicals used to control pests on crops and lawns. These chemicals frequently wash off the land and into water supplies, where they harm both aquatic organisms and the animals that eat these species.

Soil Pollution A third form of pollution is the buildup of harmful substances in the soil. Some *soil pollution* comes from pollutants that fall out of the air. Most, however, is caused by chemicals used to control pests and plant diseases. Another cause of soil pollution is leaking toxic wastes. Toxic chemicals are often buried in the ground in containers that are supposed to remain sealed. However, many of these chemicals, including nuclear wastes, have leaked into the ground, poisoning the soil and entering ground-water supplies.

A disturbing factor in the pollution of the Earth's environments is the amount of waste produced by humans. Americans alone produce about 4 trillion kilograms of wastes each year. In 1992, about 1 billion kilograms of toxic materials were released into the air by American industries—about 4.5 kilograms for every person in the United States.

The Consequences of Pollution

From a distance, the picture on the right looks perfect. The dark green of tall trees surrounds the deep blue of a crystal-clear lake. But everything is not perfect. There are few birds or frogs around, and the fish are almost gone. This lake, like thousands of others in eastern North America, is dying because its waters have become an unfit habitat for many of the organisms that once lived there. What caused this destruction? Acid rain, an apparent consequence of industrial pollutants, has contaminated the lake.

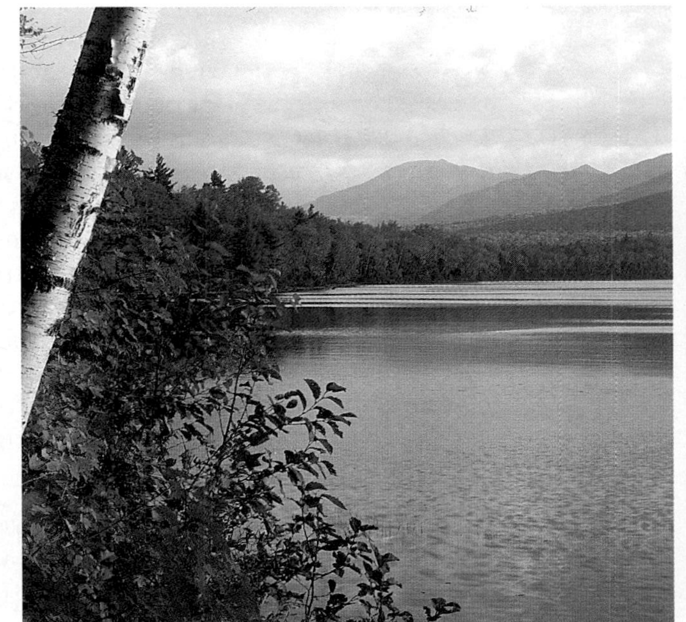

Many lakes in the Adirondack Mountains of New York are suffering from acid-rain contamination.

There will never be another Tasmanian wolf (above). Once it was an important predator in its native Tasmania, an island south of Australia. The giant panda (right) is endangered and may soon follow the Tasmanian wolf into extinction.

Since the 1700s, over 60 animal species have disappeared from the United States alone. In the 1700s, the extinction rate was about one species every fifty years—today it is one species every year! In 1994, there were over 700 plant and animal species in the United States that were **endangered**; the number worldwide is over 1100 species, according to the U.S. Department of Commerce.

Why should we be concerned if organisms become extinct? After all, species become extinct naturally. However, as the rate of extinction increases, the variety of life on Earth decreases. New species will develop, but the rate of that development is relatively slow. Remember that every organism occupies a niche in a food chain. If we eliminate one species, an important food chain could collapse and threaten the food web of an entire ecosystem. Also remember that humans are at the top of many food chains, so when we cause the extinction of a species, we may be endangering ourselves. It has been said that each time we eliminate a species, we lose a little of ourselves. After all, we all occupy the same biosphere.

Endangered
In danger of becoming extinct

Conservation
The protection and wise use of natural resources

What Can We Do?

As humans, we have the knowledge and the ability to change the environment to meet our needs. Along with this ability comes a responsibility to all other life on the planet. It is true that many laws have been passed over the last 30 years to help clean up polluted ponds, lakes, and streams and improve the quality of the air. But legislation alone is not enough. It will also take a commitment from individuals like you to make the future better for all forms of life on Earth. But what can you do? Probably the single most effective thing that we *all* can do is practice **conservation**—the wise use of materials and energy. By conserving materials, recycling solid wastes, and saving water, we can keep the Earth cleaner and can guarantee that we and other organisms will have enough fresh water in the future. By conserving energy, we can reduce air pollution and extend the dwindling stores of fossil fuels in the Earth.

Many states require the exhaust of motor vehicles to be tested for pollution levels. Vehicles that give off too many pollutants must be repaired.

Solid Waste Conservation Solid waste is any material that is thrown away that is not a liquid or a gas. It includes everything from banana peels to plastic bags to old cars. The amount of waste produced in the United States has almost doubled since 1960. Today we live in a "throw-away" society. Many products are designed to be used once and then discarded. As a result of the growing amount of garbage that Americans generate, the United States is running out of landfill space, the waste disposal facilities where wastes are dumped each day. Some scientists claim that by the year 2000 more than 25 states will be completely out of landfill space.

One important thing we can do to reduce the amount of waste that goes to landfills is produce less waste. Cutting back on waste is really just a matter of using good common sense. For example, don't buy products that have unnecessary packaging, use both sides of a piece of paper, and don't take plastic utensils from restaurants if you are not going to use them. Another useful strategy is to **recycle**, or reuse materials that would otherwise be discarded. Many communities recycle materials like bottles, cans, and newspapers to companies that use them to make new items.

Community recycling centers provide one way for citizens to recycle solid wastes such as aluminum cans and newspapers.

Recycle

To reuse materials rather than throwing them away

Water Conservation The goal of water conservation is to preserve our limited supply of clean, fresh water. Humans use half of Earth's freshwater supply to irrigate agricultural plants. We also use water to manufacture products and to generate electricity. In addition, an average person in the United States uses almost 300 L of water daily! What does a person do with so much water each day? Only about 2 L is used for drinking. The rest is used for such things as flushing toilets, bathing, food preparation, watering lawns, washing cars, and doing laundry.

There are many ways that people can conserve water. One way is to waste less water. Leaking faucets and long showers are common ways that people waste water.

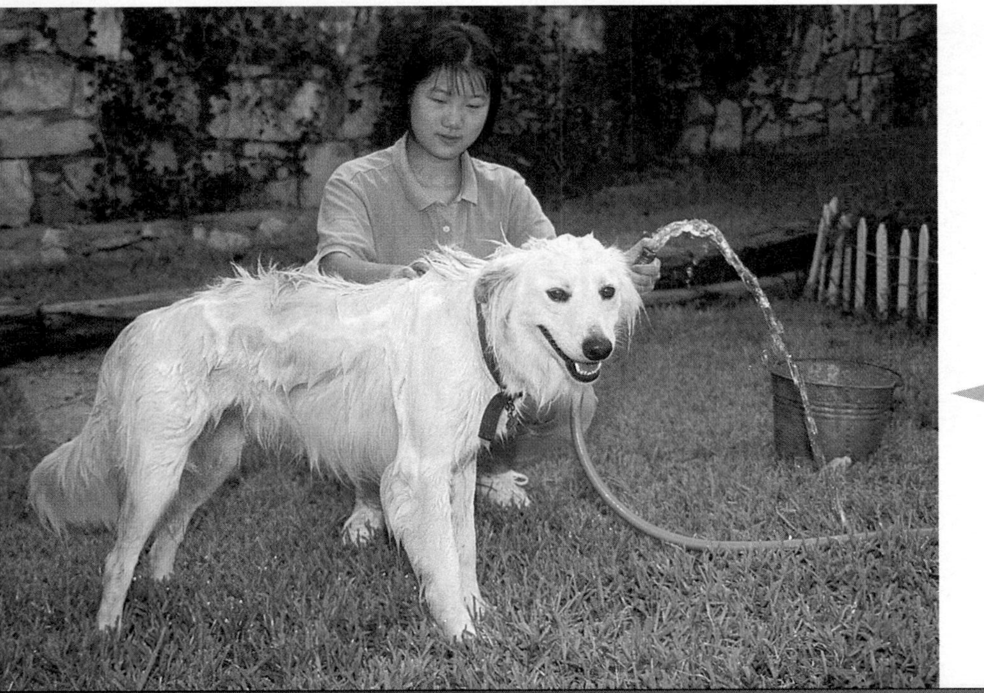

Using water foolishly is a waste of an important natural resource.

DID YOU KNOW...

that by recycling paper you help to keep the carbon dioxide–oxygen cycle going?

Recycling wastepaper saves trees. Trees use carbon dioxide and release oxygen that all organisms use to make energy.

S23

Another way to conserve water is to protect watersheds, areas of land where water either enters the ground or is captured by streams and rivers. Clearing forests and grasslands for farming and for building cities and towns allows water to run off instead of soaking into the ground. This reduces the supply of ground water and causes the flooding of rivers and lakes.

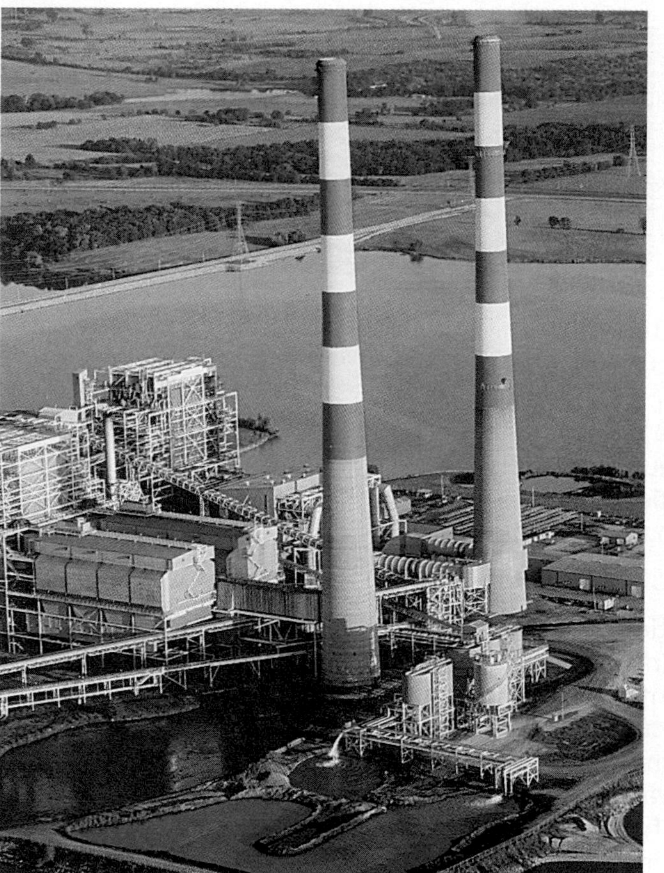

Most power plants use fossil fuels to generate electricity. This facility burns coal.

Energy Conservation Almost 90 percent of the energy that powers today's machines comes from *fossil fuels* such as coal, oil, and natural gas. Most of the air pollution we have today is caused by the burning of fossil fuels, mainly in motor vehicles and industry. Besides being an unclean energy source, fossil fuels are also a nonrenewable resource. We are using fossil fuels at a much faster rate than they can be replaced by nature. Some scientists claim that much of the Earth's oil may be used up in your lifetime.

Because of the decreasing reserves of fossil fuels and the role that these fuels play in polluting the environment, it is important to develop and use other sources of energy, such as solar energy, wind energy, and geothermal energy. Since our reliance on these alternative forms of energy may never totally replace the use of fossil fuels, it is also important that we use existing fossil fuels more wisely. What can you do to conserve fossil fuels? For one thing, do not waste electricity—most of the generators that produce electricity burn fossil fuels to operate. Another way to reduce fossil-fuel consumption is to bike, walk, or ride a bus instead of driving.

SUMMARY

Humans require a variety of materials from the environment for their activities. Many of these activities cause pollution of the air, water, and soil, which harms both wildlife and humans. The result has been acid rain, the loss of habitat for wildlife, and the extinction of many species. The extinction of organisms disrupts food webs, which could lead to the collapse of entire ecosystems. An important solution to environmental pollution is conservation. By reducing packaging, recycling materials, not wasting water, and becoming less dependent on fossil fuels for energy, we can all play a part in producing a cleaner environment.

Unit CheckUp

Concept Mapping

The concept map shown here illustrates major ideas in this unit. Complete the map by supplying the missing terms. Then extend your map by answering the additional question below. Write your answers in your ScienceLog. **Do not write in this textbook.**

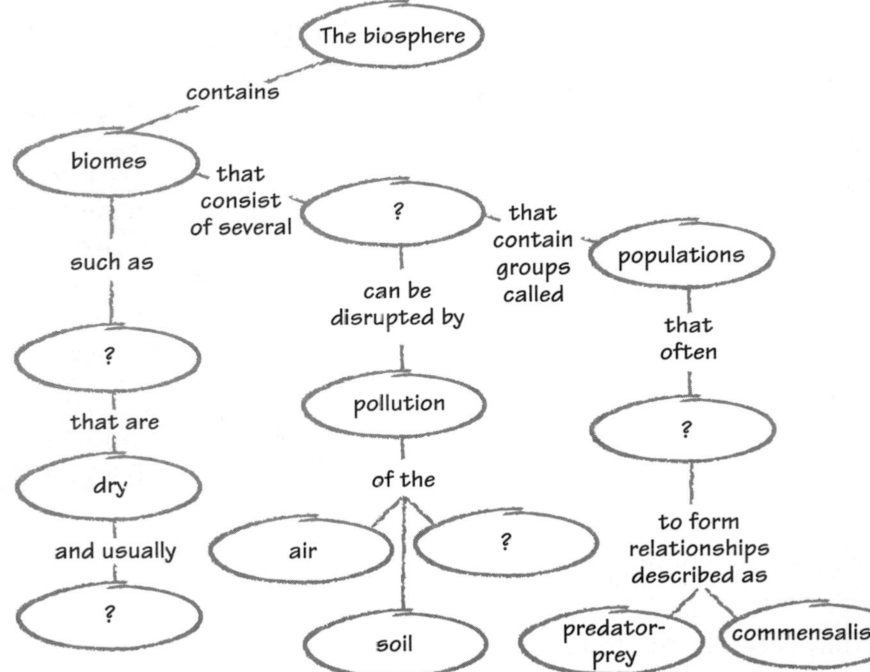

Where and how would you connect the terms *sulfur dioxide*, *tundra*, and *community*?

Checking Your Understanding

Select the choice that most completely and correctly answers each of the following questions.

1. Which is NOT a part of the biosphere?
   a. ground water
   b. bacteria
   c. the sun
   d. clouds
2. All of the different populations living and interacting in the same area form a
   a. biome.
   b. community.
   c. dense population.
   d. limiting factor.
3. Which is not directly involved in the nitrogen cycle?
   a. evaporation
   b. bacteria
   c. ammonia
   d. dead plants
4. Typical pioneer organisms in primary succession are
   a. trees.
   b. glaciers.
   c. grasses.
   d. lichens.
5. The process of making the environment unclean with wastes is called
   a. conservation.
   b. pollution.
   c. extinction.
   d. urban development.

Interpreting Photos

Name the biome the scene in this photograph represents, and describe where ecosystems such as this one are generally found on Earth.

Critical Thinking

Carefully consider the following questions, and write a response in your ScienceLog that indicates your understanding of science.

1. How might space travel and the building of undersea cities affect the population density of the human species?
2. Explain the meaning of the following statement: The carbon dioxide–oxygen cycle enables you to run, talk, and even read this question.
3. A farmer plows a field, covers it with plastic, and then leaves the area for 5 years. When the farmer returns, he finds that the field is just as he left it. Explain how covering the field prevented succession from occurring.
4. Give two reasons why you would expect there to be more day-active animals in a tropical rain forest than in a desert.
5. How might the absence of sanitary landfills during the Middle Ages have contributed to the spread of bubonic plague, a disease carried by rats?

Portfolio Idea

A lake in the middle of a growing rural town has become overgrown with algae, resulting in the death of many of the lake's fish. Imagine that you are a county official assigned to investigate the problem. Think about the steps you might take to determine the cause of the overgrown lake. After determining the cause, what suggestions would you offer to help make the lake "healthy" again? Write down your investigative methods, findings, and solutions in the form of a report that you could send to the town's mayor.

Diversity of Living Things

In This Unit

Now that you have been introduced to some of the many different kinds of organisms and how they are classified, consider these questions.

1. What are the five kingdoms of living things, and what characteristics do the members of each group share?
2. What are the major subgroups of organisms in each kingdom?
3. What is the mechanism by which organisms change over time?
4. How have advancements in agriculture and medicine affected the diversity of life on Earth?

In this unit, you will take a closer look at the diversity of living organisms and ways in which this diversity may have developed.

The Diversity of Modern Life

Did you know that the inside and outside of your body is inhabited by many living things that are so small they can be seen only with powerful microscopes? Would you have imagined that a forest of giants over 30 m tall could consist of algae instead of trees? If you did not know these things, you are not alone. Most living things go unnoticed. Close to 2 million species have already been identified and named, and some scientists estimate there may be as many as 100 million different kinds of organisms living on Earth today! Let's look at some of the variety of Earth's organisms. As you know, scientists group organisms into large categories called *kingdoms*. The organisms in each of the five kingdoms share fundamental similarities.

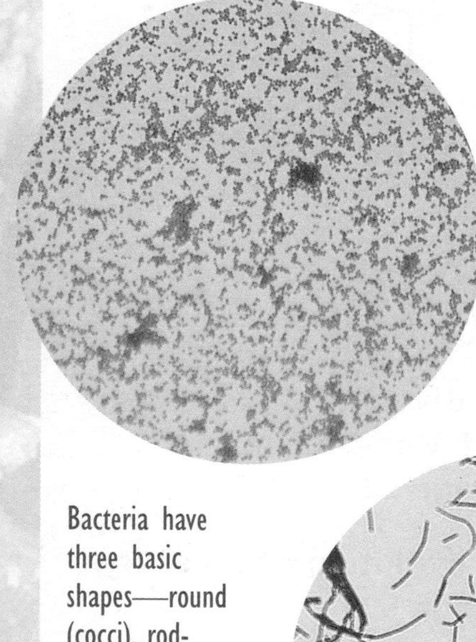

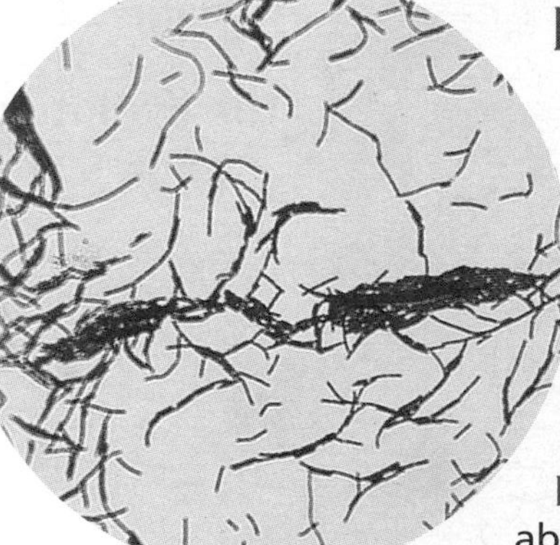

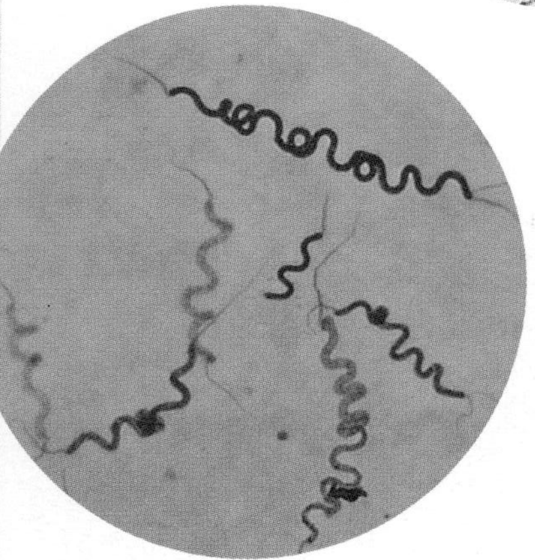

Bacteria have three basic shapes—round (cocci), rod-shaped (bacilli), and spiral-shaped (spirilli). Many live in pairs, chains, or clusters.

Kingdom Monera

The smallest and simplest life-forms are placed in the kingdom Monera. All monerans are single-celled organisms that often live in chains or clusters of cells. They are different from all other living things because the cell of a moneran has no nucleus. The bacteria and the cyanobacteria are the two major groups in this kingdom.

Bacteria Bacteria are single-celled organisms that absorb food molecules from their surroundings, such as the surface of your skin or the inside of your intestines. Bacteria live in almost every habitat on Earth. You may already be familiar with the bacteria that can invade your body and cause disease. However, many bacteria are beneficial. For example, bacteria in the intestines are necessary for you to properly digest your food. Bacteria also play a huge role in the environment by helping to decay dead organisms. This process is essential for putting nitrogen back into the soil.

Cyanobacteria Cyanobacteria often live in fresh water. Some species are responsible for the smelly, green scum that you may see on polluted ponds and streams. Other cyanobacteria are the food-producing organisms in a lichen.

DID YOU KNOW...

that recent discoveries may indicate the need for a sixth kingdom? One group of monerans, known as archaebacteria (AHR kee bak TIRH ee uh), are quite different from other monerans at the cellular level. Similar to ancient living organisms, they are able to live in extremely harsh environments such as hot springs and hydrothermal vents.

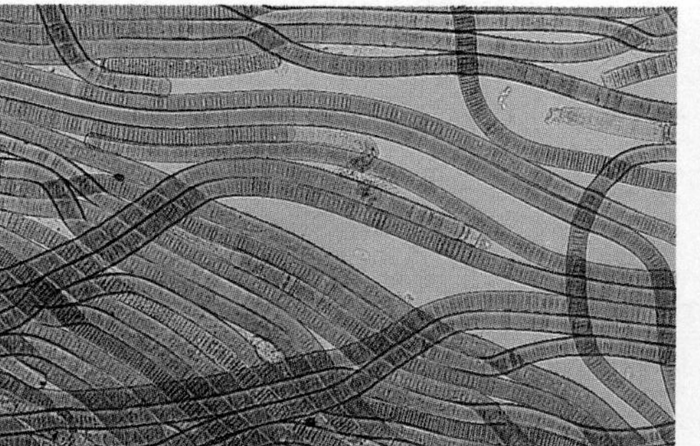

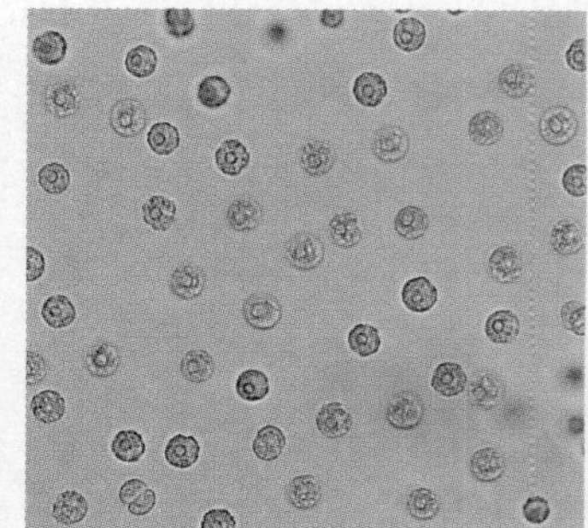

Cyanobacteria contain chlorophyll and produce their own food by photosynthesis. They have different shapes and often attach to one another in long chains or clusters.

S28

Kingdom Protista

The more complex single-celled organisms are placed in the kingdom Protista. Also included in this kingdom are some simple, multicellular organisms. Unlike the monerans, protists have much larger and more complex cells that contain nuclei and other organelles. Although some multicellular protists are quite large, their cells do not form tissues or organs. The protists are divided into several groups that share similar characteristics. Some of these groups contain the animal-like protists called *protozoans*. The others contain the plantlike protists called *algae*.

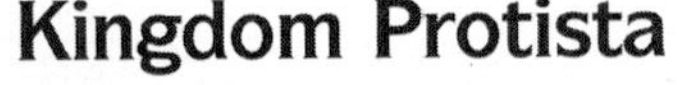

The amoeba is a protozoan that uses a flowing motion to move and pursue its prey.

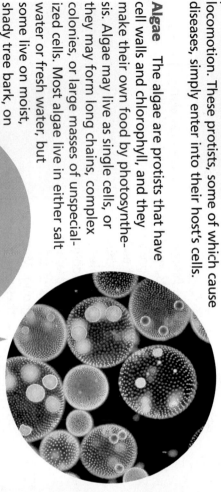

Paramecia are protozoans with hairlike *cilia* that work like tiny oars.

Protozoans These protists are single-celled organisms with no cell walls or chlorophyll, so they are more animal-like than plantlike. Protozoans *ingest* their food (take it in), as you do, and *digest* it internally. Because they must find their food, most protozoans must be able to move about. Some protozoans, however, have no means of self-locomotion. These protists, some of which cause diseases, simply enter into their host's cells.

A brown alga called kelp forms magnificent underwater "forests."

Algae The algae are protists that have cell walls and chlorophyll, and they make their own food by photosynthesis. Algae may live as single cells, or they may form long chains, complex colonies, or large masses of unspecialized cells. Most algae live in either salt water or fresh water, but some live on moist, shady tree bark, on wet concrete and rocks, or with a fungus as part of a lichen. Algae are divided into five major groups based on their color and structure.

Volvox is a colonial green alga.

Euglena is a unique kind of alga because it has characteristics of both protozoans and green algae. Can you name some of these characteristics?

Kingdom Fungi

DID YOU KNOW...

that penicillin is produced by a fungus?
This widely used drug, which has bacteria-killing properties, is produced by the fungus *Penicillium notatum*.

Some fungi (FUN jeye) make us sneeze, others dirty our bathtubs, and many kinds spoil our food. These are some of the ways that fungi affect our lives. But what, exactly, are fungi? Most members of the kingdom Fungi are multicellular organisms with a complex cell structure, cell walls, and no chlorophyll. You can find them growing both inside and on top of their food. Fungi eat by releasing digestive chemicals into their food and absorbing the digested nutrients into their bodies. The fungi are divided into groups based on their body structure and how they produce reproductive structures called *spores*.

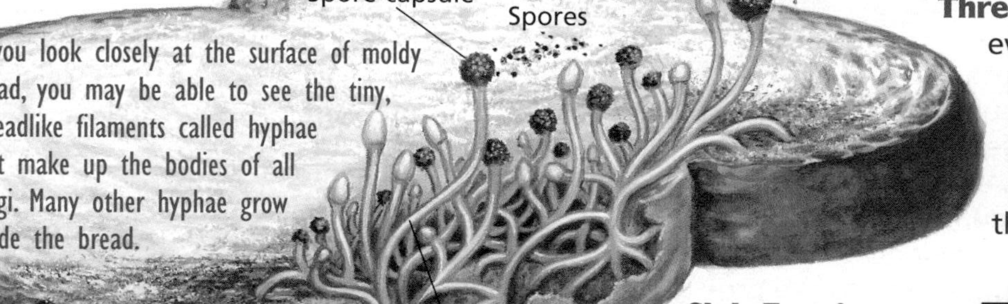

If you look closely at the surface of moldy bread, you may be able to see the tiny, threadlike filaments called hyphae that make up the bodies of all fungi. Many other hyphae grow inside the bread.

Threadlike Fungi Have you ever noticed a black or gray powdery substance growing on bread? These are the spore capsules of bread mold, an example of a threadlike fungus.

Club Fungi
Mushrooms, puffballs, and bracket fungi are members of this group. These familiar structures that grow on the surface of soil or on dead trees are made up of closely packed hyphae. They come in a variety of shapes and colors.

Bracket fungi often grow on rotting logs.

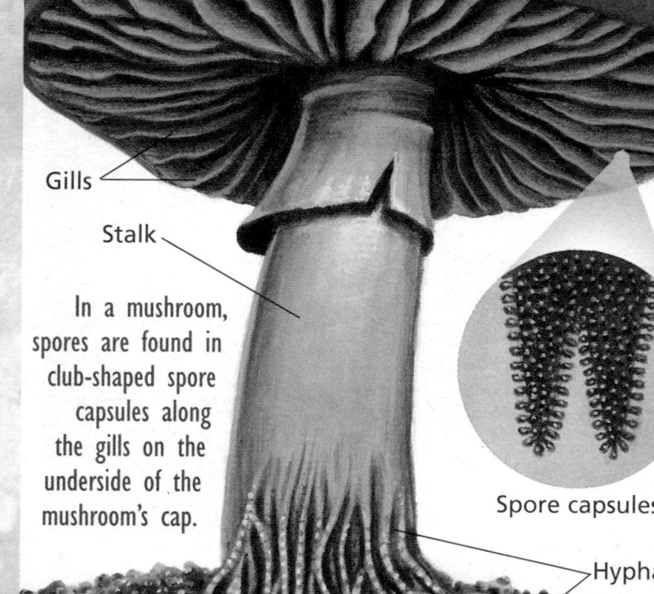

In a mushroom, spores are found in club-shaped spore capsules along the gills on the underside of the mushroom's cap.

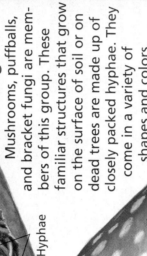

Sac Fungi
Yeasts, mildews, and morels (a type of edible fungus that resembles a wrinkled mushroom) are members of this group. Many of these fungi are parasites that attack and cause diseases in several kinds of trees, such as elm and chestnut. Some yeasts even cause diseases in humans.

The spores of sac fungi are found inside tiny cups that form on the outside of the host.

S30

Kingdom Plantae

Organisms like mosses, ferns, begonias, and maple trees are members of the plant kingdom. All plants are multicellular, have a complex cell structure, have cell walls, and contain chlorophyll. Most plants live on land. Unlike the cells of the multicellular algae and fungi, plant cells specialize to form many different types of tissues and organ structures. Both the largest and longest-living things on Earth are plants. Plants are divided into two main groups based on whether or not they have water-conducting (vascular) tissue.

Nonvascular Plants Have you ever noticed a bright green carpet covering a forest floor or growing on a rotten log? This carpet is made of plants that do not have water-conducting tissue. The *nonvascular plants* must live in moist environments, such as a shaded forest floor. Because they cannot conduct water internally, they cannot grow very large. The *mosses* and the *liverworts* are two types of plants in this group.

Notice that individual liverworts (left) and mosses (below) grow together in tightly packed mats. This helps them to conserve water and also to reproduce. The tiny capsules that appear at the top of moss plants contain spores.

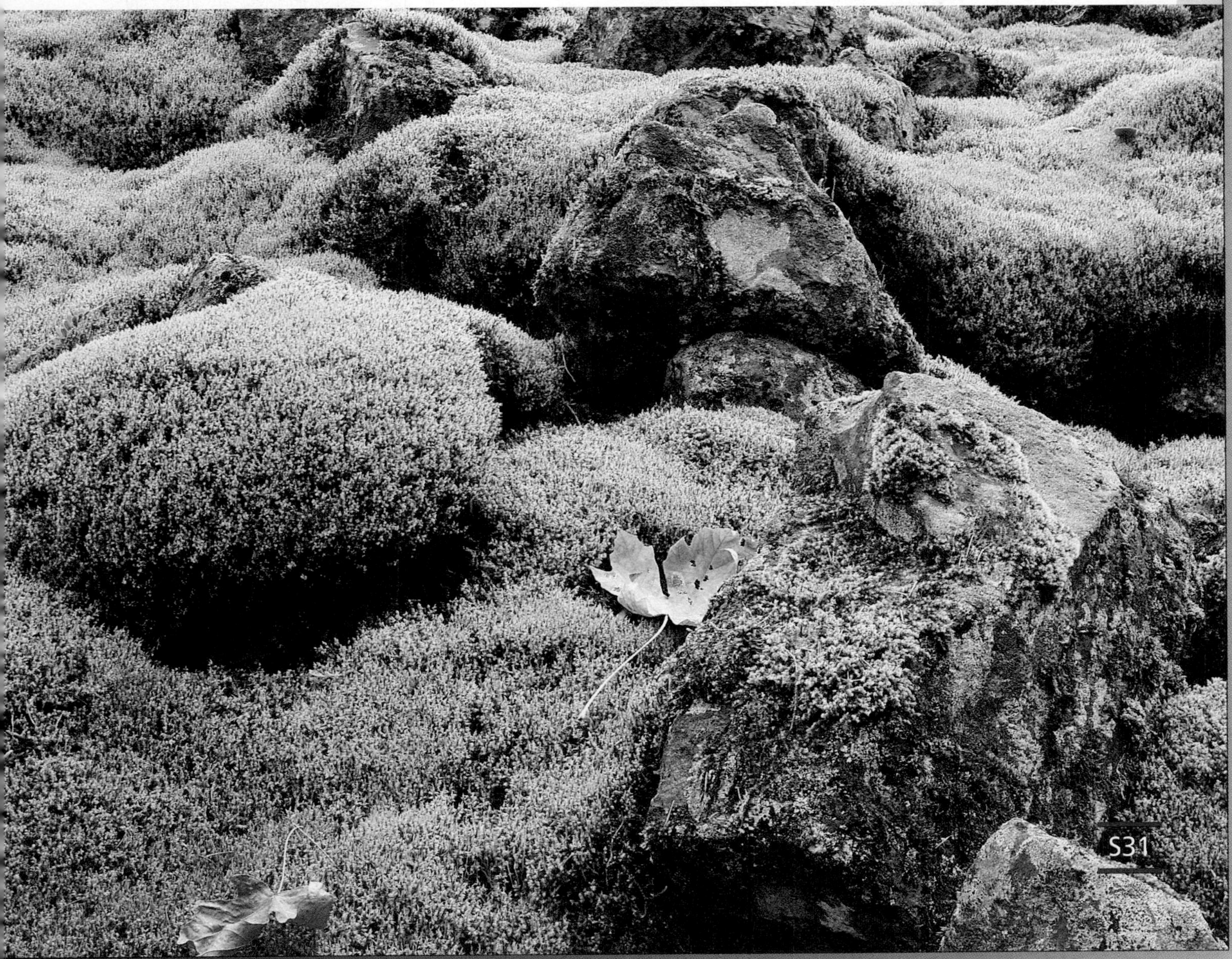

S31

Vascular Plants Most plants have *vascular tissue* that conducts water, minerals, and food molecules throughout their bodies, which are composed of roots, stems, and leaves. Vascular tissue not only allows these plants to live in drier environments but also serves as a support system for holding up their stems and leaves. Without vascular tissue, plants could never grow more than 10 to 20 cm tall. There are five groups of vascular plants.

Club mosses are vascular plants that resemble mosses. However, the stems of club mosses are much sturdier due to their vascular tissue, and they grow larger than the mosses.

Horsetails have hollow stems with many vertical ridges and joints. A whorl of needle-like leaves appears at each joint. Their stems feel gritty due to the presence of silica, the material in sand. Because horsetails are gritty and coarse, early settlers often used them to scour pots and pans.

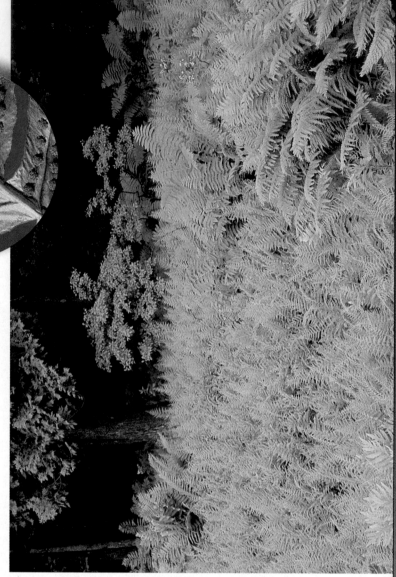

Ferns have distinctive leaves, called *fronds*, that consist of a strong central vein and many pairs of leaflets. Their stems remain underground. The horizontal rows of bumps that appear on the underside of fern fronds are not a disease or bugs—they are structures that contain spores.

Fruit

A ripened ovary that contains the seeds of a flowering plant

Gymnosperms, whose name means "naked seed," are one of the two types of *seed plants.* Seed plants are by far the most common and most successful kind of plants. Most gymnosperms are *conifers* (cone-bearing plants) that produce seeds on the dry scales of their cones. All conifers, such as these spruce trees (right), have small needle-like leaves and most stay green all year. Other conifers are pine, redwood, juniper, fir, larch, hemlock, and cypress trees. Some bristlecone pines in California are believed to be nearly 5000 years old, making them among the oldest living things in the world. Cycads and ginkgo trees (far right) are other kinds of gymnosperms.

Angiosperms, whose name means "covered seed," are also seed plants. Usually called *flowering plants,* angiosperms are the largest, most diverse group of modern plants. Their seeds are produced inside of a special structure, called an *ovary,* which is part of a flower. Although their flowers are not always as obvious or showy, plants such as grasses, palms, and shade trees are also flowering plants.

The ovary of a flower ripens into a **fruit** that surrounds and protects maturing seeds. Fruits also help to disperse seeds (spread them around) once they are mature. You have eaten the fruit of many angiosperms, like the orange tree above.

DID YOU KNOW...

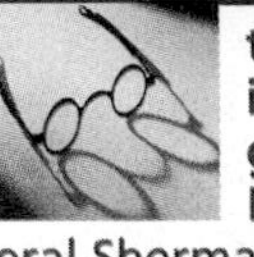

that the largest tree in the world is a gymnosperm? Named in honor of General Sherman of Civil War fame, this giant sequoia in California stands more than 80 m tall and measures 30 m around the bottom of its trunk. The man who discovered the tree had served under General Sherman during the war.

Kingdom Animalia

Into which kingdom are you classified? You are not a moneran, because you are made of many cells and most of your cells have a nucleus. You are not a protist, because you are multicellular and have complex tissues and organ systems. You are not a plant, because you do not have cell walls and you cannot make your own food. You are not a fungus, because you digest your food inside of your body with a complex digestive system. Therefore, by a process of elimination, you must be a member of the only remaining kingdom—the animal kingdom.

The animal kingdom is divided into nine major groups. Most of these groups are composed of invertebrates—organisms such as worms, insects, and oysters. Humans belong to another group of animals that includes fish, amphibians, reptiles, birds, and mammals—the *vertebrates*.

Sponges are the simplest of the animal groups. Most adult sponges live in salt water, attached to the sea bottom or to some other object. A sponge's body is composed of two layers of body cells. These body cells cling to a network of tiny spikes or fibers that surrounds the hollow central cavity of the sponge. The walls of a sponge have many tiny pores through which water and tiny bits of food are drawn into the sponge's central cavity. Water and wastes leave through a large opening on top.

Invertebrates Invertebrates are animals that have no backbones. Invertebrates either have an outside covering for support or are supported by the water in which they live. The members of each invertebrate group share characteristic body plans. As you read about the invertebrates below, notice the differences between each group.

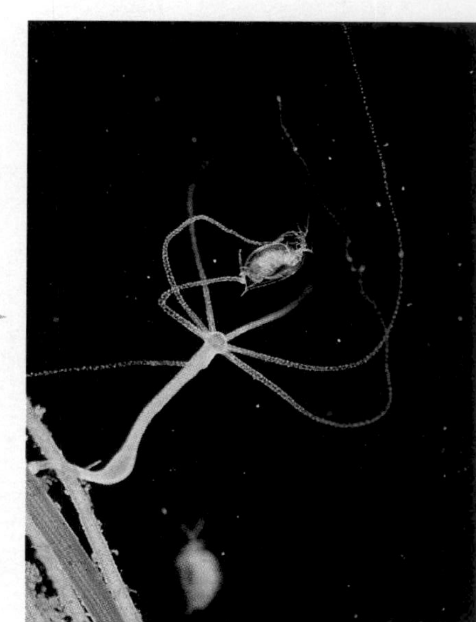

Coelenterates (sih LEHN tuhr ayts), such as this hydra, are simple animals whose bodies are composed of two specialized cell layers (tissues) that are separated by a jelly-like substance. Hydras, jellyfish, sea anemones, and corals are all types of coelenterates. All live in water and have a hollow saclike body that has a single opening through which food enters and wastes are expelled. The opening of this hydra is surrounded by tentacles lined with stinging cells for capturing and paralyzing live food.

Flatworms have a flattened body with one body opening, a digestive system with many lobes, and a simple nervous system. Many flatworms are parasites that live inside of other animals. Blood flukes, liver flukes, and tapeworms are some flatworm parasites of humans. *Turbellarians* are free-living flatworms, the majority of which live in marine environments. Planarians (freshwater turbellarians), like the one shown here, are distinctive among flatworms because they have two eyespots that are used to detect light.

Roundworms are named for their rounded body shape. They also have a straight digestive tube and two body openings—a mouth for taking in food and another for expelling wastes. Most roundworms, such as these nematodes, are free-living organisms that live in pond water or in moist soil. Some roundworms, however, are parasites of animals or plants. For example, ascaris worms and hookworms live in the intestines of humans and other mammals, and root-knot nematodes live in the roots of plants.

Segmented worms have rounded bodies with two body openings, but their bodies are divided into a series of segments. The earthworm is a common example of the segmented worms, which also include leeches and marine tube worms. Earthworms have well-developed body systems that include organs such as a simple "brain" that connects to a nerve cord, and a set of five primitive "hearts."

S35

Mollusks (MAHL uhsks), such as this octopus, are soft-bodied animals with well-developed organ systems. Many mollusks have hard shells made of calcium that protect their bodies. Clams and oysters, for example, are covered by a two-part, external shell, whereas snails secrete a single, spiraling shell. Octopuses and squids have tentacles and no external shell. They swim very fast and shoot out an inky substance that confuses their predators.

Exoskeleton

A hard outer covering for protection that is typical of arthropods

Arthropods are by far the largest group of animals. Arthropods, such as this beetle, are characterized by multiple body segments and jointed *appendages*. They live successfully on dry land and in water, and many have complex life cycles. Among the arthropods are lobsters, crabs, shrimp, insects, spiders, centipedes, and millipedes. Arthropods have a hardened **exoskeleton** that protects their well-developed organ systems.

Regeneration

The ability of some organisms to grow new body parts

Echinoderms (ee KEYE noh duhrms), also known as the "spiny-skinned" animals, include sea stars, sea urchins, sand dollars, and sea cucumbers. Echinoderms live in the ocean and have bodies protected by a spiny skeleton that lies under their skin. Many echinoderms have flexible body branches, or *arms*. Echinoderms, like the sea star shown here, move by using special structures called tube feet, which resemble suction cups. Sea stars are known for their ability to replace missing arms by a process called **regeneration**.

DID YOU KNOW...

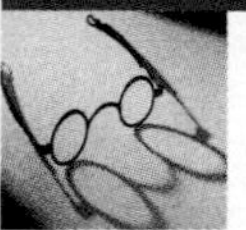

that the insects are the largest group of organisms on Earth? One scientist estimates that there are about 200 million insects alive at any one time for each person on Earth.

Vertebrates Vertebrates are animals that have a backbone. This backbone is part of an internal skeleton, called an **endoskeleton**, which provides support for the body and aids in its movement. All vertebrates have a body that includes a head, and most have appendages. There are seven living groups of vertebrates in kingdom Animalia.

Endoskeleton

An internal skeletal system that functions to support the body of vertebrates

Ectotherm

An organism that has a body temperature that changes according to the temperature of the environment

Jawless fishes, like the sea lamprey wrapped around this trout, have jawless mouthes adapted for sucking body fluids from other fishes; elongated, snakelike bodies with no appendages; gills for obtaining oxygen; and flexible skeletons made of cartilage. Like all fishes, they are **ectotherms** and live only in water.

Cartilaginous fishes are animals such as sharks. Sharks have skeletons made of cartilage, two pairs of fleshy fins, gills, and strong jaws with many rows of teeth for tearing and eating flesh. Stingrays and skates are other members of this group, which generally live in the ocean.

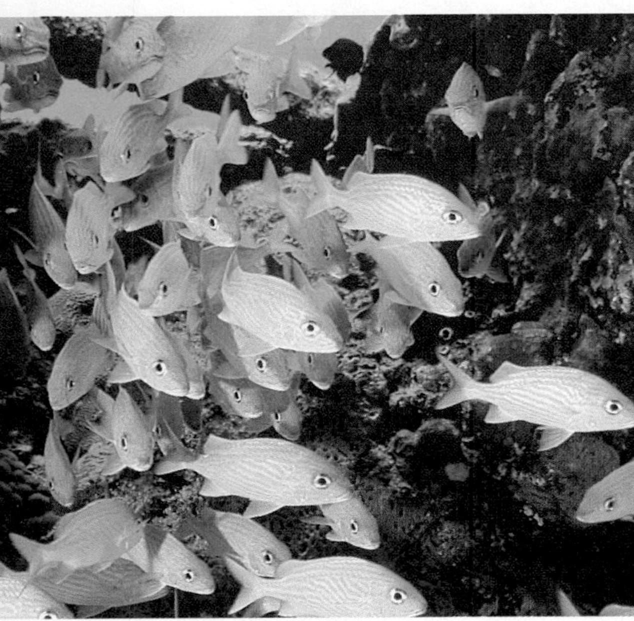

Bony fishes include fishes like catfish, goldfish, flounder, eels, trout and the blue-striped grunts shown here. Bony fishes are the most numerous, varied, and successful kind of fishes. They have gills and skeletons made of bone. Most have a streamlined body that is tapered at both ends and two pairs of fanlike fins. Their fins and body shape allow them to move easily through the water.

Amphibians include frogs, toads, and salamanders. The name *amphibian* means "double life." Most amphibians live part of their life in water and part on land. Most amphibians have two pairs of legs as adults, enabling them to move on land. Like the fishes, amphibians lay their eggs in water and are ectotherms. Amphibians usually have smooth skin that must be kept moist. Most have gills when they are young and lungs when they mature.

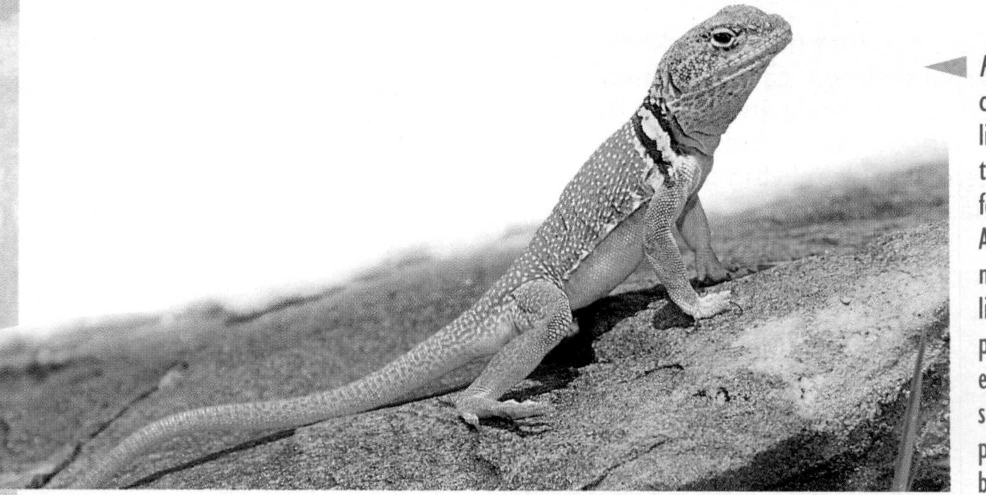

Reptiles include turtles, snakes, lizards, crocodiles, and alligators. Many reptiles, like this yellow-headed collard lizard, have two pairs of strong legs with clawed toes for digging, climbing, and moving on land. At one time, reptiles called *dinosaurs* dominated Earth. Reptiles are truly adapted to life on land. They are covered by hard plates or scales that prevent water loss by evaporation. And they lay eggs that are surrounded by a tough, leathery shell that prevents them from drying out. Reptiles breathe with lungs and are ectotherms.

Birds have light, hollow bones and enlarged lungs as an adaptation for flying. Birds, such as this great blue heron, can be recognized by a body covering of feathers and a pair of wings, which gives most of them the ability to fly. In addition, they have scaly legs and feet with clawed toes, and they lay eggs covered by a hard shell. Unlike the fish, amphibians, and reptiles, birds are **endotherms**. This allows birds to occupy many different habitats on Earth.

Mammals have a very advanced nervous system, which includes a highly developed brain and keen senses. They feed their young with milk from mammary glands, have hair on their bodies, and breathe with lungs. Like birds, mammals are endotherms. They are found in habitats on land, in the air, or in water. One small group of mammals lays eggs, but all others give birth to live young.

Endotherm

An organism that has a body temperature that remains constant despite temperature changes in the environment

SUMMARY

According to the current system of biological classification, living things can be separated into five kingdoms: Monera, Protista, Fungi, Plantae, and Animalia. The members of each kingdom have major characteristics in common. The organisms of each kingdom can be further divided into smaller groups whose members have even more characteristics in common. Some of these characteristics include body design, types of habitat, manner of obtaining food, and methods of locomotion and reproduction.

The Evolution of Diversity

Imagine going on a long trip and arriving in a strange new place where even the animals and plants look odd. You find giant tortoises grazing like cattle, seafaring lizards, and unusual, flightless birds. Yet other life-forms look much like organisms with which you are familiar. This was the experience of Charles Darwin when he visited the Galápagos Islands near South America in the 1830s.

Since Darwin's time, scientists have gathered much information about the organisms that once existed. From this information, they have developed explanations as to how life changes over time. Scientists have even performed experiments that indicate life might have originated naturally from substances that occur on Earth. The concept that all life that exists on the Earth has evolved from other organisms is a fundamental theme in biology.

DID YOU KNOW...

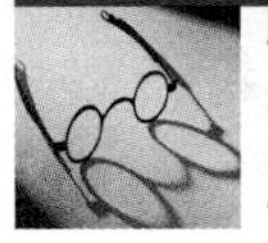

that at one time Charles Darwin, the founder of modern evolutionary theory, considered becoming a minister?
He even completed a degree in theology at Cambridge University when he was in his early twenties.

The marine iguana is one species that Darwin found living on the Galápagos Islands.

A History of Life on Earth

If you could travel back to prehistoric times, what kinds of organisms would you see? Depending on the time of your arrival, you might see horses the size of dogs, flying reptiles, forests of giant ferns, and even giant insects. Biologists estimate that all the species alive today make up only 1 percent of all the species that ever lived. This means that 99 percent are now extinct! Some of the extinct species are preserved in the rocks of the Earth's crust as **fossils**. To get the most out of your trip, you could not stay very long at any one particular point in time, since the Earth's history covers an amazingly long period of time—about 4.5 *billion* years!

Fossil

The preserved remains or imprint of an organism that once lived

Earth's Calendar During our lives, we mark the passing of each year on a calendar. The study of human history, however, is broken into longer units of time, such as decades, centuries, and millennia. To us, a millennium, or 1000 years, is a very long period of time, since very few people live more than a century. The Earth, as you know, is estimated to be about 4.5 billion years old, which is more than 45 million (45,000,000) centuries of time! How much time is this? If we could squeeze the entire history of the Earth into one calendar year, we would be living in the last fraction of a second of December 31!

Scientists divide the history of Earth's living organisms into several time periods called *eras*. Each era is further divided into periods. Eras are named for the kinds of organisms that were abundant during that time. Each one ends with a major change in the predominant group of organisms that existed.

The Earth's history, as recorded in its rocks, reveals that the diversity of life today has evolved from simple to complex forms. This gradual progression is illustrated in the table shown below. As you study the illustration, think back to the five kingdoms that you have recently read about.

An abundance of fossils is found in the rocks that formed during the Cenozoic, Mesozoic, and Paleozoic eras. However, these fossils represent only 13 percent of Earth's history. Eighty-seven percent of Earth's history occurred before these eras, in an interval of time known as *Precambrian* time. During the Precambrian, life-forms first began to inhabit the Earth. Unfortunately, there are not many fossils of these organisms because most were single-celled or had soft bodies. These kinds of organisms do not make good fossils.

Relative time scale

Era	Period	Million Years Ago	Organisms
Cenozoic	Quaternary	0.01 3	Complex human societies develop. Many extinctions occur.
	Tertiary	12	Modern mammals, birds, and sea life appear. Early hominids appear. Grasslands appear.
		26	Bats, monkeys, and whales appear. Plants resemble modern species.
		38	Primitive apes, elephants, horses, and camels develop. Forests of gymnosperms and angiosperms spread.
		59	Mammals spread rapidly. Fruit-bearing trees are common.
		65	Primates appear. Flowering plants thrive.
Mesozoic	Cretaceous	136	Flowering plants and trees appear. Dinosaurs become extinct by the end.
	Jurassic	190-195	First feathered birds and large dinosaurs appear.
	Triassic	225	Dinosaurs and first mammals appear.
Paleozoic	Permian	280	Ferns, conifers, and seed ferns common. Many invertebrates become extinct.
	Pennsylvanian	320	Giant club mosses are common. Flying insects appear.
	Mississippian	345	First reptiles appear. First gymnosperms appear.
	Devonian	395	Amphibians and non-flying insects appear.
	Silurian	400-430	Fish and shell-forming sea animals appear. First plants appear on land.
	Ordovician	500	First vertebrates appear in the sea.
	Cambrian	570	First red and green algae appear. Trilobites abundant. Clams, snails, and segmented worms appear.
Precambrian time		~4,500	Bacteria, cyano bacteria, and algae predominate. Marine invertebrates appear.

Evolution by Natural Selection

Charles Darwin served as ship's naturalist on the HMS *Beagle*, a British survey and mapping ship. His job on the ship was to study and collect the different kinds of plants and animals he found in the many places where the ship stopped, including the Galápagos Islands. In addition, he collected many fossils and made notes of everything he saw. During the long, five-year voyage, Darwin carefully studied these notes and collections and gradually became convinced that life on Earth had changed over time.

Developing a Theory When he returned home, Darwin began to classify his collections and to wonder *how* organisms could change over millions of years. He knew that farmers selected and bred animals and plants with certain traits, but he did not know what could cause such selection in nature. Then Darwin read an essay by the noted economist Thomas Malthus. The essay discussed Malthus's observations that the size of a human population gets larger when there is enough food and other necessities and gets smaller when there is not. Malthus predicted that factors such as starvation, disease, crime, and war would prevent the human population from growing too large.

Darwin realized that *all* organisms must compete for food, water, and other necessities in order to survive. Mathus called this competition a *struggle for existence*. This idea became the key to Darwin's theory of evolution by **natural selection**. He reasoned that the individuals that are better able to compete are the ones most likely to survive and leave more offspring, offspring with the parents' characteristics. Those that are not able to compete successfully are less likely to live long enough to produce offspring.

Natural selection

The process by which only the organisms that are best suited to their environment survive, passing their traits on to their offspring

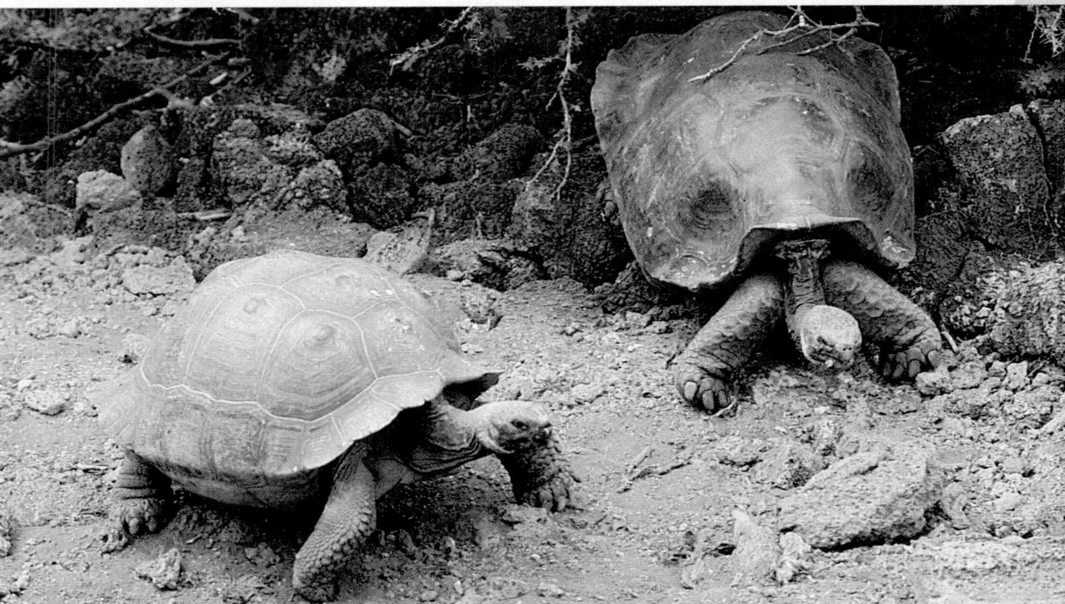

These giant tortoises must compete for a limited supply of food.

Darwin worked on his theory for more than 20 years. But before he finished, he received an essay from a man named Alfred Russel Wallace that stated the main points of the theory that Darwin had worked on for so long. Wallace was also a naturalist and had worked in the jungles of Indonesia. Darwin quickly finished a paper on his theory, and in 1858, both papers were presented jointly at a scientific meeting in London. The next year, Darwin published his famous book *On the Origin of Species by Means of Natural Selection*, in which he clearly presented the vast body of evidence that supports his theory.

Four Parts of the Theory There are four main parts to the theory of evolution by natural selection. The parts presented here are a bit different from the way Darwin stated them. Our knowledge of genes explains how variations can be transmitted from parent to offspring—something Darwin did not know.

Variation: All members of the same species are somewhat different from one another; in other words, individuals are not exactly alike in all of their traits. These differences are called *variations.* Darwin believed that variations were caused by some agent that could be passed from parent to offspring, but he did not know what that agent was. Today we know that variations result from sections of DNA called *genes.* New variations can appear when genes change slightly by a process called **mutation.** Some variations, like the color of these Bengal tigers, may be very noticeable. But many less noticeable variations, such as those in size, weight, and running speed, can also occur.

Mutation

A change in the DNA of a gene

DID YOU KNOW...

that mutations can be produced by environmental agents?
X rays, ultraviolet radiation, and substances such as benzene, asbestos, and other industrial pollutants have been known to cause changes in genes.

Overproduction of Offspring: Each species produces many more offspring than can survive and reproduce. For example, female fishes lay enormous numbers of eggs. If all of these eggs hatched and the young survived, the waters of the Earth would quickly be overrun with fishes. Plants, such as this dandelion, also produce large numbers of seeds, but not all of them grow into adult plants.

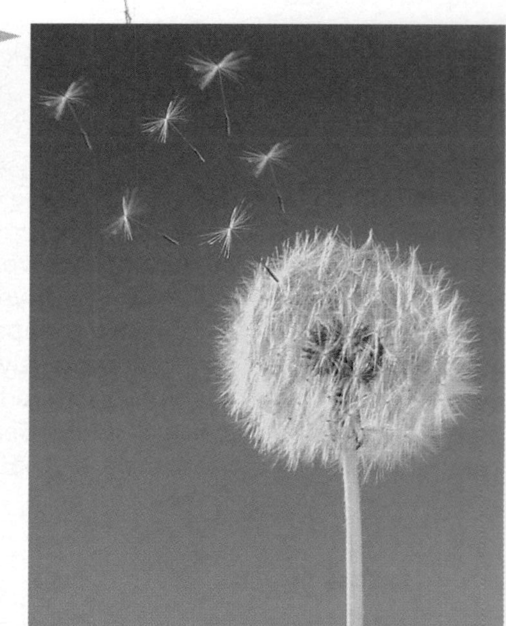

S42

Struggle for Existence: The overproduction of offspring leads to a struggle for survival. All organisms must compete for a limited amount of food, water, and living space. During this struggle, some individuals will get enough of these necessities to survive and reproduce, and some will not.

Natural Selection: Individuals with certain variations compete more successfully than individuals that lack those variations. Which of these peppered moths would more likely be passed over by a bird looking for a meal? In other words, which moth is better camouflaged? Would the same be true if the moths lived on a soot-covered tree? As you can see, the environment determines which variations in an organism are advantageous. Individuals with traits that make them successful in their environment have the best chance to survive and reproduce. Offspring that inherit beneficial variations also have a better chance of surviving, as long as the environment remains the same. Thus, in a population, the percentage of individuals with advantageous variations increases over time.

As many slight, yet beneficial, changes accumulate in organisms over millions of years, significant changes in structure and function may occur. In this way, species that are adapted to certain environments develop.

Adaptation Through variation and natural selection, species may *adapt* (change and become better suited) to new conditions in their environments. Consider what happened to the peppered moths in England during the Industrial Revolution. Where tree trunks were darkened by pollution, populations of mostly light-colored moths were gradually replaced with ones dominated by dark-colored moths. Because the variation for dark color existed within the peppered moth population, this species was able to adapt to and survive a change in its environment.

S43

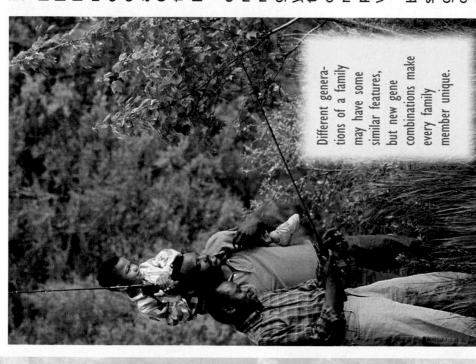

Different generations of a family may have some similar features, but new gene combinations make every family member unique.

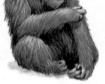

Patterns of Diversity

Have you ever noticed similarities between members of the same family? What did you notice? How would you describe the similarities? You might describe similar facial features, such as eye color and nose shape. You probably would not describe similar features such as backbones, two arms, two legs, and the ability to walk upright. Of course these features are common among family members, but they also indicate how all people on Earth are related.

Being related means sharing an *ancestor*. For example, your parents are your ancestors. You may know or have pictures of some of your other recent ancestors, such as your grandparents, great-grandparents, or great-great-grandparents. Do you notice any similarities between these ancestors and yourself? You also have other ancestors—ones through which you are related to other mammals, vertebrates, invertebrates, and even protists. When organisms share the same ancestor, we call this ancestor a *common ancestor*.

Determining common ancestry is the key to building modern classification systems. It also helps biologists to work out pathways of evolution and to create evolutionary trees like the one shown below.

This evolutionary tree indicates that the theoretical path leading to the evolution of humans has included several common ancestors. Can you locate where they are?

Common ancestry is determined by studying and comparing such things as homologous structures, embryonic development, and biochemical molecules. Let's take a look at what these things mean.

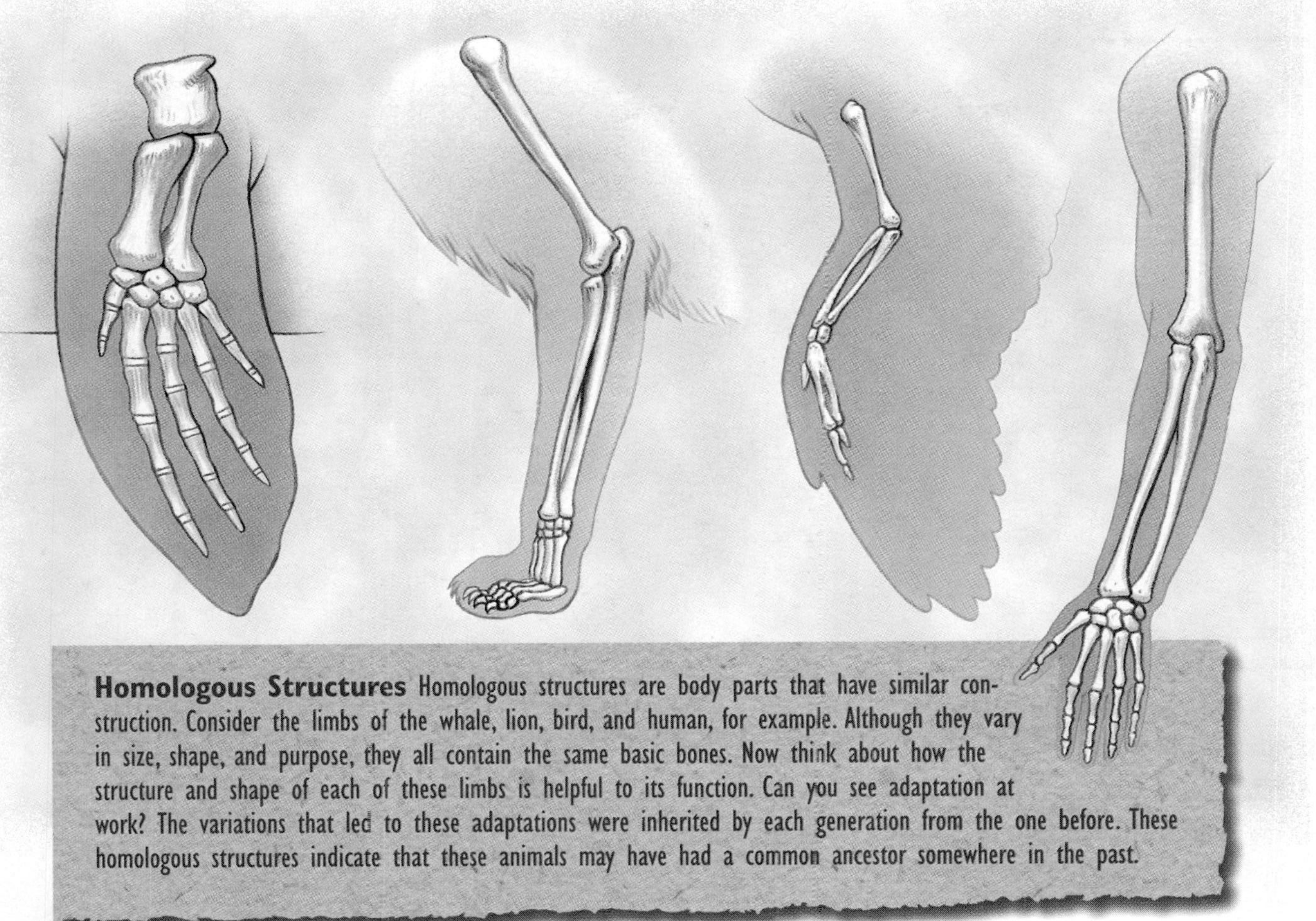

Homologous Structures Homologous structures are body parts that have similar construction. Consider the limbs of the whale, lion, bird, and human, for example. Although they vary in size, shape, and purpose, they all contain the same basic bones. Now think about how the structure and shape of each of these limbs is helpful to its function. Can you see adaptation at work? The variations that led to these adaptations were inherited by each generation from the one before. These homologous structures indicate that these animals may have had a common ancestor somewhere in the past.

Embryonic Development
Relationships among animals can also be seen by comparing their development before birth. Most multicellular organisms, such as humans, begin as a single cell called an egg. When the egg is fertilized, it begins to divide rapidly, forming first a mass of cells and then an embryo. The embryo changes shape gradually as it develops. Eventually, it takes the basic shape of its parents.

Note that the chicken embryo and the human embryo are similar in appearance. This indicates that chickens and humans have a more recent common evolutionary ancestor than do fish and chickens or fish and humans.

The DNA of a chimpanzee and the DNA of a human are very similar. This scientist is teaching a chimpanzee how to use sign language.

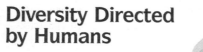

Biochemical Molecules
Homologous structures and embryonic development were used by Darwin and other early biologists to help them trace the evolution of life and classify it. However, these methods were too general to establish the closest relationships between organisms. Today these relationships are determined by studying the biological molecules that make up organisms. The more closely related the organisms are, the more similar their molecules will be. The most important molecule that scientists study is DNA, because this molecule contains the instructions for making all of the others. These instructions are written in a code that scientists have learned to read. By comparing the code sequence in the DNA of different organisms, scientists are able to determine how closely related the organisms are.

The fact that every living thing on Earth has DNA is the most convincing evidence that all life on Earth may have a common ancestor.

Diversity Directed by Humans

If you could put together a better pet, what would it be like? If you could design a better vegetable, how would it taste? Believe it or not, people have been able to do just that by a process called *artificial selection*.

S46

Selective Breeding Humans first began to domesticate plants and animals around 9000 B.C., or about 11,000 years ago. At that time, people began to collect seeds of wild grasses, like wheat, to grow them for food. They also tamed certain wild animals, such as cattle, for doing work. Gradually, they started to alter these wild varieties by selecting and raising the offspring of individuals with the most desired traits.

Improving plants and animals by choosing and reproducing individuals with certain desirable traits is called **selective breeding**. Most domestic animals—from cats and dogs to horses, sheep, and pigs—are the result of thousands of years of selective breeding by humans. So are most cultivated varieties of ornamental and food plants.

By selecting and growing wild wheat plants with the largest seeds, domesticated varieties that always produce large seeds were eventually developed. Compare the grain size in modern bread wheat (top) with one of its wild ancestors (right).

Selective breeding

The process of making hereditary improvements in animals and plants

Cattle have been selectively bred for centuries. Individual bulls and cows are selected for qualities such as strength, size, and meat and milk production, and they are bred in a controlled manner. Calves that have the most desirable combinations of traits are selected to eventually parent the next generation. There are more than 40 different cattle breeds today. Shown is a herd of Brahman cattle, which thrive in the hot, humid climate of the southern United States.

Resistance Humans have tried to control many species of insects and disease-causing bacteria with chemicals. Now many insects and bacteria are *resistant* to these chemicals. When we spray or dust insect pests with chemicals called *pesticides*, most of the insects are killed. However, due to genetic diversity, a few individuals survive the treatment because they have genes that make them resistant to the particular chemicals that were used. These individuals survive and go on to reproduce, passing this resistance on to their offspring. Resistance keeps scientists busy trying to produce new pesticides to control insects.

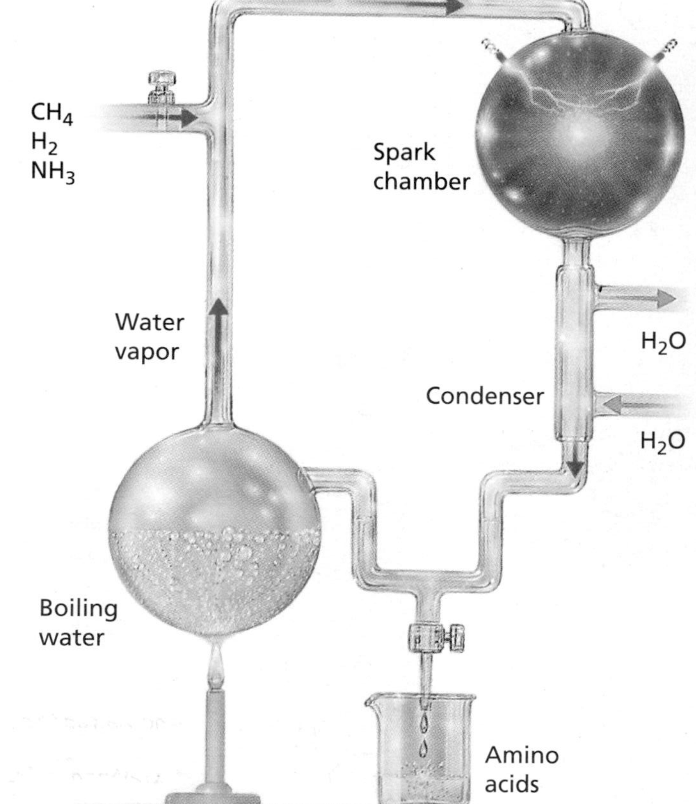

In 1953, two American scientists used an apparatus similar to this one to heat a mixture of nitrogen, methane, hydrogen, ammonia, and water. When ignited with an electrical spark, the mixture reacted to form amino acids, the molecules that build proteins.

The Origin of Life

In studying how the first cells could have appeared, it occurred to some scientists that life itself may have developed naturally on Earth. This probably could not have happened if the Earth had always been the same as it is now. Scientists think that an oxygen-rich atmosphere would have prevented the first complex biological molecules from forming. But about 4 billion years ago, it appears that conditions on Earth were very different. Scientific evidence suggests that the Earth's surface was covered by erupting volcanoes and shallow, hot seas. There was no oxygen in the atmosphere, and the daily weather forecast would have included torrential rain and violent electrical storms.

In 1923 a Russian biochemist named Alexander Oparin announced his theory that life could have begun when carbon-containing molecules formed in an early atmosphere of extremely hot gases such as hydrogen, ammonia, methane, and water vapor. These molecules, he said, collected in hot, shallow seas and gradually came together to form the major biological molecules found in living things. After millions of years, the first cells appeared. No one knows exactly how these molecules might have come together to form cells, but the possibility has been demonstrated by several experiments.

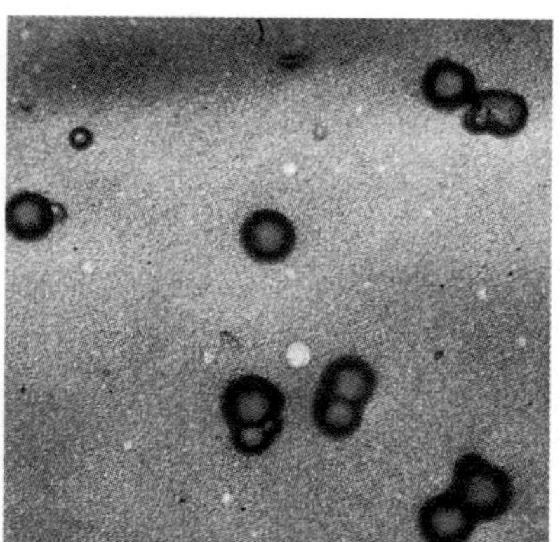

Biochemists have observed that some of the biological molecules they make in the laboratory tend to gather together into tiny spheres that resemble primitive cells. Many scientists think that over millions of years, similar microscopic spheres could have developed the organization necessary to become living cells.

SUMMARY

The history of the Earth's living organisms is divided into time periods called eras. The simplest forms of life appeared before the earliest of these eras. The most complex organisms, including humans, live in the current era, the Cenozoic. Observations made by Charles Darwin led to the development of the theory of evolution by natural selection, which explains how organisms change over time. Homologous structures, similarities in embryonic development, and biochemical similarities are used to classify organisms by their evolutionary relationships. Humans have been able to change living things by processes such as selective breeding. Experiments have shown that life may have developed naturally on Earth about 4 billion years ago.

Unit CheckUp

Concept Mapping

The concept map below illustrates major ideas in this unit. Complete the map by supplying the missing terms. Then extend your map by answering the additional question below. Write your answers in your ScienceLog. **Do not write in this textbook.**

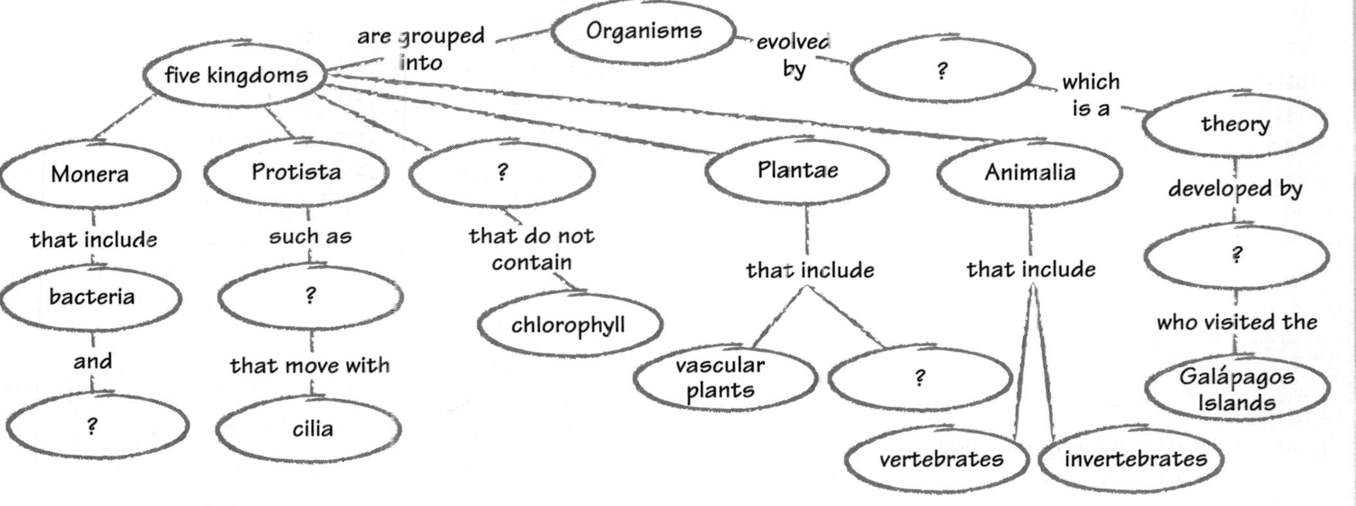

Where and how would you connect the terms *spores, algae,* and *backbones*?

Checking Your Understanding

Select the choice that most completely and correctly answers each of the following questions.

1. Which is NOT a kingdom name?
a. Protista
b. Animalia
c. Protozoa
d. Fungi

2. An animal with a backbone is called
a. a segmented worm.
b. a protist.
c. an echinoderm.
d. a vertebrate.

3. Which is a component of fungi but NOT of plants?
a. chlorophyll
b. hyphae
c. a nucleus
d. a cell wall

4. During what interval did most of the Earth's history occur?
a. Cambrian era
b. Precambrian time
c. Mesozoic era
d. Paleozoic era

5. Similarly constructed limbs on different vertebrates are called
a. homologous structures.
b. biochemical similarities.
c. wings.
d. mutational structures.

Interpreting Graphs

Bacteria are grown in a test tube that contains a limited amount of food. This graph illustrates how the population of bacteria changes over a two-day period. What do you predict will happen to the size of the population after day two? Support your prediction with an explanation.

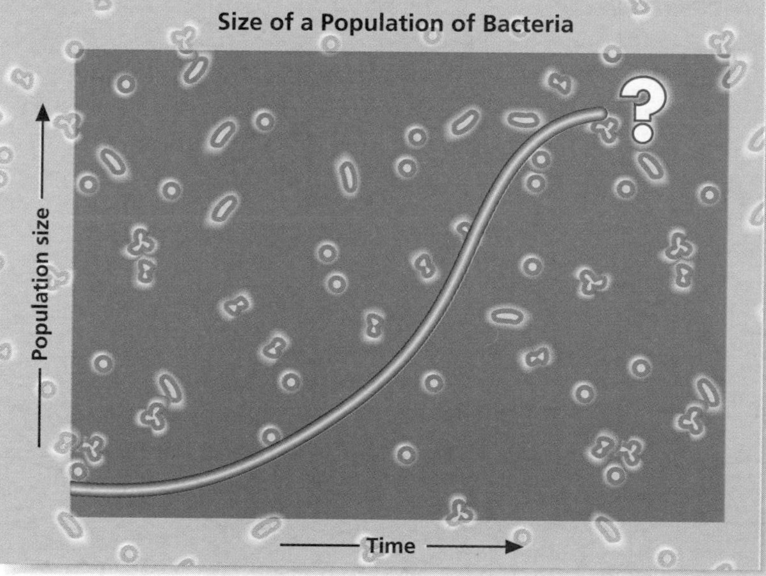

Critical Thinking

Carefully consider the following questions, and write a response in your ScienceLog that indicates your understanding of science.

1. A biologist notices a large number of short plants growing near a swampy area. Although she thinks they are nonvascular plants, she is not sure. She collects one of the plants to bring back to her laboratory, where she will look at a slice from its stem under a microscope. Why does she wish to observe the plant's stem? Why is her guess that the plants are nonvascular a logical assumption?

2. Explain why an earthworm is able to crawl faster than a planarian can move.

3. There are many species of small birds called finches that live on the Galápagos Islands. Scientists think that the many different kinds evolved from a single species of finch that flew to the islands years ago. Is it more likely that this ancestral species arrived from Europe, North America, or South America? Explain the reason for your choice.

4. You discover a new species of protozoan but are unsure where to place it on the protist evolutionary tree. What kind of observations would you make about the protist, and what kind of analyses would you perform on it to help you classify this organism?

5. The wing of a bird and the wing of an insect have the same function, yet they are NOT homologous structures. Explain why they are not.

Portfolio Idea

Create a table that summarizes the major characteristics of the organisms in each of the five kingdoms. Start by placing the kingdom names as headings in your table. Then compare and contrast the organisms in each kingdom to come up with the characteristics you will have in your table. Make your table large enough to include a colored drawing or a magazine clipping of a representative organism for each kingdom.

S50

SOLUTIONS

In This Unit

Now that you have been introduced to solutions, consider these questions.

1. How can you tell solutions from other types of mixtures, and what are the properties of each type of mixture?
2. How are different types of mixtures used to manufacture useful products?
3. What are some properties of acids and bases?
4. What happens when acidic and basic solutions interact?

In this unit, you will take a closer look at the properties and interactions of solutions and non-solutions.

Solutions, Suspensions, and Colloids

What do a pizza and a soft drink have in common? Besides tasting good, both are mixtures of different substances. Recall that a mixture is a combination of two or more substances that keep their own characteristics after they are mixed. For example, even after you bake a pizza, you can separate the parts of the topping—olives, peppers, cheese, and tomato sauce. These ingredients do not combine chemically and thus can be sorted. Having parts that can be separated by a physical means, such as sorting, is a property of mixtures.

Separating the parts of a soft drink, however, is not as easy as separating the parts of a pizza. As you know, a soft drink is a type of mixture called a solution. In a solution, one substance *dissolves,* or "disappears," into the other, making the mixture look like a single substance. To separate the parts of a solution, you must use physical properties other than appearance, such as boiling point. Many familiar substances are mixtures. Some are solutions, and some are not. In this section, you will learn more about solutions and two types of non-solutions—*suspensions* and *colloids.*

A pizza is a mixture of many different substances. Because the major parts of the topping are not dissolved in a solvent, the parts may be separated by physical means. This boy is removing certain parts of the topping and setting them aside.

Solutions

You have just learned that a **solution** is a mixture in which one substance is dissolved in another. Every solution has two parts: a *solvent* and a *solute.* The solvent of a solution is the material in which the solute is dissolved. A solution usually has more solvent than solute.

Like the soft drink in the photo below, many solutions contain more than one solute, or dissolved substance. A soft drink is a mixture of water, flavorings, colorings, and carbon dioxide. In a soft drink, water is the solvent, and the flavorings, colorings, and carbon dioxide are solutes. How can you tell that a soft drink is a solution of more than one substance? Before a soft-drink bottle is opened, the liquid inside looks like a single substance. But when you open the bottle, a gas bubbles out of the liquid, giving the soft drink its fizz. These bubbles contain carbon dioxide.

The carbon dioxide gas in a soft drink bubbles out when the bottle is opened and the pressure is released.

Solution

A mixture that appears to be a single substance but is composed of particles of a solute distributed evenly among the particles of a solvent

Types of Solutions If you were asked to think of an example of a solution, you would probably name something that is a liquid. However, solutions can also be gases or solids. Because the solvent makes up most of a solution, solutions exist in the same state (solid, liquid, or gas) as their solvents. Liquid water, for example, is a very good solvent. For this reason, many familiar solutions are liquids. The table above lists some familiar examples of different types of solutions and their components.

Familiar Examples of Solutions

Example	Type of solution/ solvent	Type of solute
air	gas	gases
soda water (carbonated water)	liquid	gas
rubbing alcohol (alcohol in water)	liquid	liquid
tap water	liquid	gases, solids
sea water	liquid	gases, solids
sugar water	liquid	solid
dental fillings (silver-mercury amalgam)	solid	liquid
brass	solid	solid

Describing Solutions In order to describe a solution, you must know the solvent and solute it contains. A salt solution, for example, is a mixture made by dissolving a salt in water. You must also know the relative amounts of each of the materials in the solution. If a small amount of solute is dissolved in a large amount of solvent, the solution is said to be **dilute.** On the other hand, a large amount of solute dissolved in a small amount of solvent makes a **concentrated** solution. The photos below illustrate some differences between dilute and concentrated solutions.

Dilute

Describes a solution that is made by dissolving a small amount of solute in a large amount of solvent

Concentrated

Describes a solution that is made by dissolving a large amount of solute in a small amount of solvent

When a few grams of the salt copper sulfate dissolve in a liter of water, a dilute copper sulfate solution is formed.

When 10 times the amount of copper sulfate dissolves in a liter of water, a concentrated copper sulfate solution is formed.

DID YOU KNOW...

that 14 karat gold is a solution of solids? It is 58 percent gold and 42 percent silver and copper. The gold is the solvent, and the silver and copper are solutes.

Saturated

Describes a solution that contains all the solute that it can hold under existing conditions

Solubility

The amount of a solute that will form a saturated solution when dissolved in a standard amount of a specific solvent at a specific temperature

Saturation You may have observed that there is a limit to how concentrated a solution can be. For example, consider what happens if you continue adding sugar to iced tea. Eventually, no more sugar will dissolve in the tea. When a solution becomes so concentrated that it can hold no more solute, the solution is said to be **saturated.** A saturated solution contains all the dissolved solute it can hold. Any additional solute will settle to the bottom of the solution's container.

What can you do if your tea is still not sweet enough after you have added all the sugar it will hold? Just heat it up. In most cases, raising the temperature of a solution will increase the amount of solute it can hold. For example, you can dissolve more sugar in hot tea than in iced tea. Recall that the hotter the temperature of a substance is, the faster its molecules move and the more kinetic energy they have. Increasing the kinetic energy of the molecules in tea by heating it enables the tea to hold more sugar molecules. So if you like your tea really sweet, drink it hot instead of cold.

To make a saturated solution, add solute to the solvent until some undissolved solute remains at the bottom of the container.

Heating and stirring a saturated solution will enable the undissolved solute to dissolve.

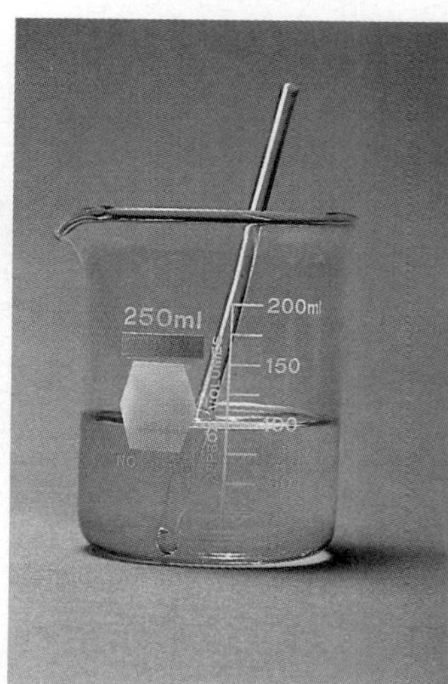

If the solution is then cooled and the extra solute stays dissolved, the solution is said to be *supersaturated.* Such solutions contain more dissolved solute than they normally would at that temperature.

Solubility The maximum amount of sugar that a glass of iced tea will hold is related to the sugar's ability to dissolve in water, which is the solvent in tea. **Solubility,** you may recall, is the amount of solute needed to saturate a solution under certain conditions. As these conditions change, the solubility of a solute changes. One condition that affects solubility is temperature. As the temperature of tea changes, so does sugar's solubility in it. Another condition that affects the solubility of a solute is the type of solvent. For example, sugar is much less soluble in alcohol, another common solvent, than it is in water. You also know that the larger the glass of tea, the more sugar it will hold. Therefore, the amount of solute that will dissolve in a solvent depends on three factors—the temperature, type, and amount of the solvent.

Solubilities are usually given in grams per 100 g of solvent. Look at the figure on the right to compare the solubilities of two different solutes in water. The amounts of sugar and salt shown beside the beaker will make saturated solutions in 100 g of water at 60°C. As you can see, it takes more sugar than salt to make a saturated solution at the same temperature. Therefore, sugar is said to have greater solubility in water than salt.

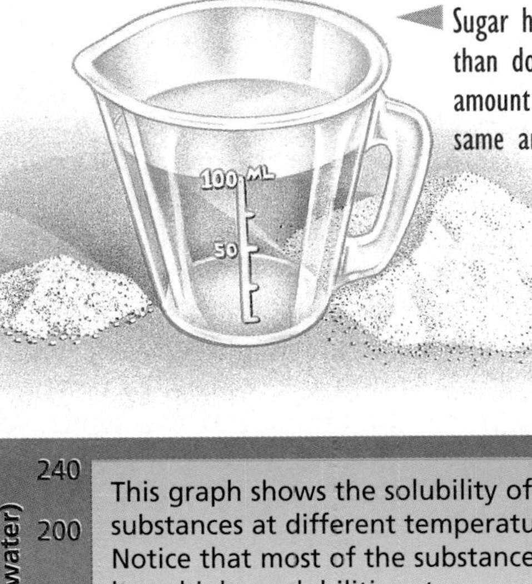

Sugar has greater solubility in water than does salt. Therefore, a greater amount of sugar will dissolve in the same amount of water.

As the graph on this page indicates, the solubility of most solids in water increases as the temperature of the water increases. However, the opposite is true for the solubility of gases in water. Gas molecules have so much energy that they are hard to hold in solution. Slowing them down by cooling them makes them more likely to stay mixed with the solvent. Gases, such as oxygen and carbon dioxide, are therefore more soluble in *cold* water than in *hot* water. This is why more gas escapes when you open a warm soft-drink can than when you open a cold can.

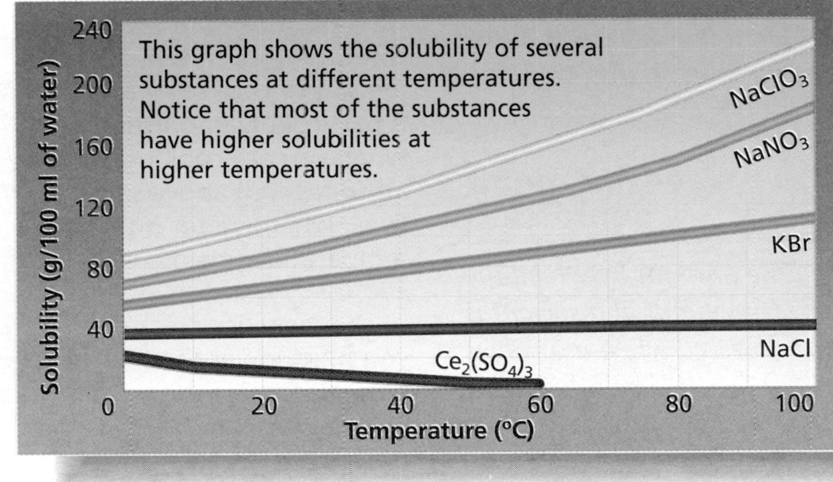

Rate of Dissolving As you know, it takes time for a solute to dissolve in a solvent. If you want to dissolve a solid in a liquid, you can do one or more of the three things listed to the right to speed up the process.

Not all the methods for making a solid dissolve faster in a liquid would work for gases. Since gases become less soluble in water as the water gets warmer, keeping an opened soft-drink bottle in the refrigerator (rather than letting it warm up) enables the liquid inside to keep its fizz longer.

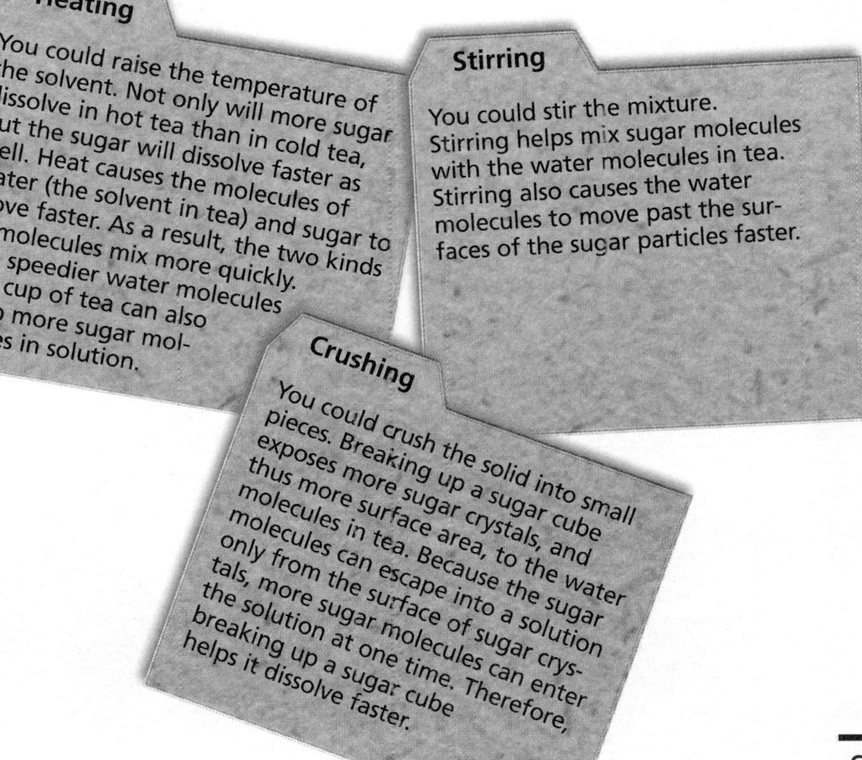

Heating
You could raise the temperature of the solvent. Not only will more sugar dissolve in hot tea than in cold tea, but the sugar will dissolve faster as well. Heat causes the molecules of water (the solvent in tea) and sugar to move faster. As a result, the two kinds of molecules mix more quickly. The speedier water molecules in a cup of tea can also keep more sugar molecules in solution.
Stirring
You could stir the mixture. Stirring helps mix sugar molecules with the water molecules in tea. Stirring also causes the water molecules to move past the surfaces of the sugar particles faster.
Crushing
You could crush the solid into small pieces. Breaking up a sugar cube exposes more sugar crystals, and thus more surface area, to the water molecules in tea. Because the sugar molecules can escape into a solution only from the surface of sugar crystals, more sugar molecules can enter the solution at one time. Therefore, breaking up a sugar cube helps it dissolve faster.

Practical Solutions Have you ever wondered why cooks put salt in the water they cook with or why both ice *and* salt are needed to make homemade ice cream? The reason is that the salt changes certain physical properties of water in a way that makes cooking and making ice cream easier.

Boiling-point elevation

The amount by which the boiling point of a solvent is raised when a solute is dissolved in it

Freezing-point depression

The amount by which the freezing point of a solvent is lowered when a solute is dissolved in it

As you know, water boils at 100°C (212°F) at sea level. When you put salt in water, the water must be heated to a higher temperature before it will begin to boil. In other words, adding salt to water increases the boiling point of the water. The amount by which the boiling point is increased is called the **boiling-point elevation.** The more salt you add, the more the boiling point is elevated. By raising the temperature at which your cooking water boils, you can cook food faster. Adding salt to cooking water is particularly important when you are cooking at altitudes well above sea level, where water boils at temperatures below 100°C.

To make ice cream, you must lower the temperature of the milk, sugar, and other ingredients enough to make them freeze. At sea level, water freezes at 0°C (32°F), but this temperature is not low enough to freeze ice cream. Adding salt to the ice in an ice-cream freezer solves this problem. As the ice melts, the salt dissolves in the water, lowering its freezing point. The remaining ice then cools the water and salt solution to subzero temperatures. The amount by which the water's freezing point is lowered is called the **freezing-point depression.** The more salt you add, the more the freezing point is depressed. The lower temperature of the salted water enables the ice cream to freeze.

DID YOU KNOW...

that using the properties of a solution can increase driving safety?
In the winter, salt is often spread on roadways. As the salt dissolves in the ice water, it lowers the freezing-point of water so that it remains liquid at lower temperatures.

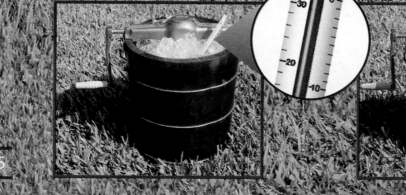

Compare the normal freezing point of water with the freezing point of the salt and water solution. This difference is the freezing-point depression.

1 Before salt is added to the ice surrounding an ice-cream freezer, the temperature of the ice is 0°C.

2 After salt is added, the temperature of the ice drops below 0°C.

S56

Suspensions

Do you ever use Italian dressing on your salad? If so, you know that the herbs settle to the bottom while the bottle sits on the shelf. You must therefore shake the bottle before using the dressing. After you shake the bottle, the herbs sink to the bottom again. Muddy water contains particles of clay that act the same way. After muddy water stands for a while, the clay particles eventually settle. These mixtures are not solutions; they are suspensions. A **suspension** is a mixture that consists of a fluid (liquid or gas) and particles that gradually settle out of the mixture.

Suspensions are common in the pharmaceutical industry. For example, liquid medicines that require shaking before being taken are suspensions. Some paints are suspensions of solid pigments in a liquid. These pigments must be finely ground so that the paint does not separate too readily. You may have noticed that paint stores have special machines that shake cans of paint in order to mix the components.

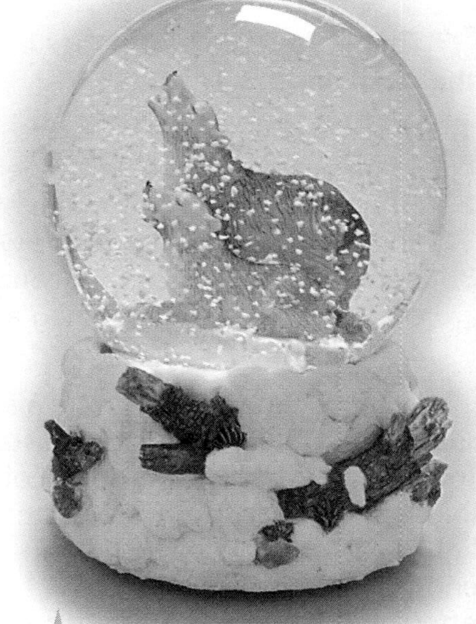

A winter-scene knickknack is a suspension of plastic snow in water. No matter how hard you shake it, the snow will still settle.

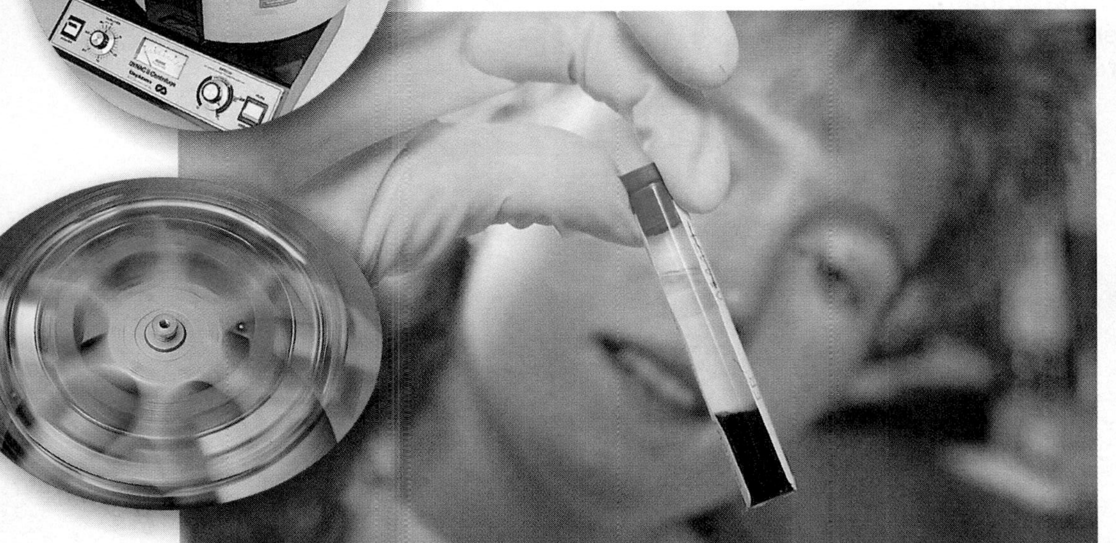

An instrument called a centrifuge can be used to quickly separate the solid and liquid components of blood. A centrifuge spins at a very high speed, simulating intense gravity and causing the tiny cells and other solid components of blood to settle.

Suspension

A mixture containing particles of one substance that are dispersed in a fluid and that slowly settle out of the mixture

Colloids

What do lipstick, milk, mayonnaise, cherry-flavored gelatin, and smoke have in common? All of these substances are colloids. A **colloid** is a mixture in which the particles, while they do not dissolve, are small enough that they do not settle out or sink to the bottom of their containers. The particles in colloids are usually too small to see without a microscope. You do not have to look far to find a colloid. In fact, the cytoplasm in your body's cells is a colloid. Many of the foods you eat are also colloids. Jelly, for example, is a colloid that tastes great on toast!

Colloid

A mixture containing particles that are not dissolved but are too small to settle

There are three basic types of colloids, categorized according to their makeup.

1. *Gels* are colloids that consist of liquid particles spread out in a solid. Some examples of gels are gelatin, jelly, and lipstick.
2. *Emulsions* are colloids made of two liquids. Milk, mayonnaise, and hand cream are oil-and-water emulsions.
3. *Aerosols* are colloids that contain either solid or liquid particles suspended in a gas. Fog and smoke are examples of aerosols.

Because they appear to be a single substance, colloids are often mistaken for solutions. Unlike solutions, colloids contain particles that scatter light. As you know, this phenomenon is known as the *Tyndall effect* and is one important way to distinguish colloids from solutions. Light scattering can be used to determine the size, shape, and concentration of the particles in a colloid or suspension. These qualities affect the amount of light that is scattered as it passes through such mixtures.

There are many examples of colloids in everyday life. The toothpaste shown here is a gel.

The jar at the left contains a mixture of water and sodium chloride. The jar at the right contains a mixture of gelatin in water. Which mixture is a colloid and which is a solution?

SUMMARY

Mixtures contain two or more substances that can be separated by sorting or some other physical means. Solutions are mixtures in which one substance, called the solute, dissolves in another, called the solvent. Solubility is the amount of solute that can dissolve in a standard amount of solvent at a certain temperature. Heating and stirring a mixture and crushing a solid will increase the rate at which a solute dissolves in a liquid. A mixture with particles large enough to settle is called a suspension. Colloids are mixtures in which the particles are too small to settle.

Acids, Bases, and Salts

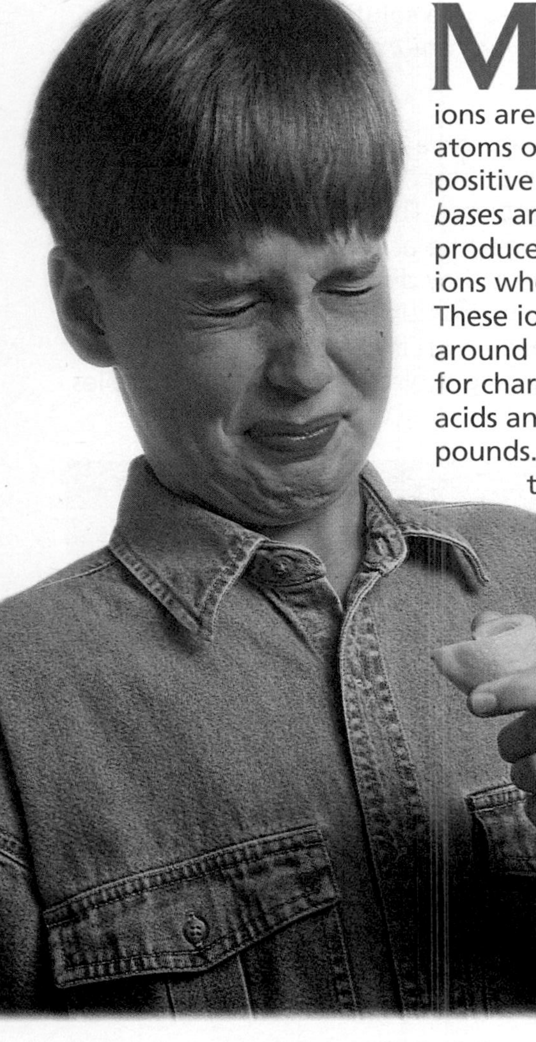

Many chemical compounds are soluble in water. When some compounds dissolve in water, ions are formed. Recall that *ions* are atoms or groups of atoms that have a positive or negative charge. *Acids* and *bases* are chemical compounds that produce certain positive and negative ions when they are dissolved in water. These ions, which are free to move around the solution, are responsible for characteristics that distinguish acids and bases from other compounds. For example, when solutions of acids and bases react, chemical compounds called *salts* are formed.

Acid

A compound that produces hydrogen ions (H^+) when dissolved in water

What characteristics of acids are indicated by the images shown here?

Characteristics of Acids

When you think of an acid, you may think of a substance that "eats" through things. Some acids do react vigorously with metals and other substances, appearing to eat through them. However, not all acids are that strong. Many foods and beverages, such as oranges and carbonated drinks, contain acids that are too weak to eat through solid materials. Several characteristics of acids are illustrated in the photos above.

The Formation of Acids An **acid** is a compound that produces hydrogen ions (H^+) when dissolved in water. Solutions that form when acids dissolve are termed *acidic.* How can you predict if a substance will form an acidic solution when it dissolves in water? The best way is to know its chemical formula. Look at the formulas for some common acids in the table to the right. Notice that each of these acids has hydrogen at the beginning of its formula.

Some Common Acids

Name	Formula
hydrochloric acid	HCl
sulfuric acid	H_2SO_4
nitric acid	HNO_3
phosphoric acid	H_3PO_4

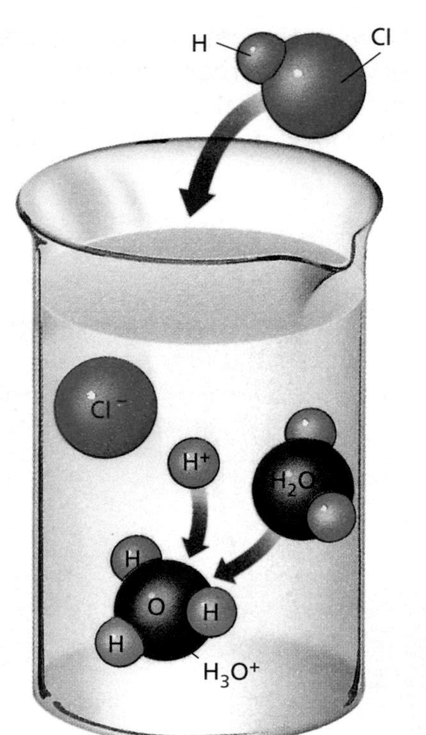

Hydronium ions have one oxygen atom and three hydrogen atoms. The number of hydronium ions in solution determines the strength of an acid.

Never touch or taste a concentrated solution of a strong acid.

When an acid such as hydrochloric acid (HCl) dissolves in water, it separates into hydrogen ions (H^+) and chloride ions (Cl^-). Because a hydrogen atom has one proton and one electron, an H^+ ion is just a proton. However, protons are not normally found uncombined in nature. When H^+ ions are produced in a solution, each combines with a water molecule to form a *hydronium* ion, or H_3O^+. Any solution that contains H_3O^+ is acidic.

The Strength of Acids Why are some acids stronger, or more acidic, than others? The strength of an acid depends on the number of hydronium ions that it produces. The more hydronium ions that the acid produces, the stronger the acid is. *Strong acids* completely break down into ions when they dissolve in water. Therefore, strong acids produce many hydronium ions. Weak acids, however, do not completely break down into ions, so they produce fewer hydronium ions. The following table contains some examples of strong and weak acids.

Strength of Some Common Acids

Strong acids	Weak acids
hydrobromic acid—HBr	boric acid—H_3BO_3
hydrochloric acid—HCl	carbonic acid—H_2CO_3
sulfuric acid—H_2SO_4	hydrofluoric acid—HF
nitric acid—HNO_3	phosphoric acid—H_3PO_4

Many weak acids were originally found in living organisms and are therefore called *organic acids.* For example, acetylsalicylic acid—the active ingredient in aspirin—was first discovered in willow trees, which belong to the genus *Salix.* Most organic acids contain a –COOH group as part of the molecule. This group breaks apart into $–COO^-$ and H^+ ions. The table below lists some organic acids with which you may be familiar. The most obvious characteristic of these acids is their sour taste.

Organic Acids

Name	Chemical formula	Found in or produced by
acetic acid	CH_3COOH	vinegar
citric acid	$C_5H_7O_5COOH$	citrus fruits
formic acid	HCOOH	ants
lactic acid	C_2H_5OCOOH	sour milk
malic acid	$C_3H_5O_3COOH$	apples
acetylsalicylic acid	$C_6H_4(OH)COOH$	aspirin
tartaric acid	$C_3H_5O_4COOH$	soft drinks

Acid has eroded the features of these marble statues.

Concentrated solutions of strong acids are very *corrosive*. These solutions react vigorously with some metals, forming hydrogen gas and dissolving the metal. They can even eat through human skin and can cause severe burns. Therefore, concentrated solutions of strong acids must be handled with great care.

In addition to reacting with metals, acids also react with substances such as limestone and marble, releasing carbon dioxide. This reaction causes the deterioration of marble buildings and statues, like those shown to the left. In nature, the reaction of carbonic acid with the calcium carbonate in limestone leads to the formation of caves and sinkholes.

DID YOU KNOW...

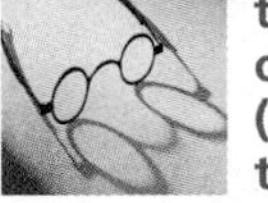

that some limestone caves form when thermal (hot) ground water rises toward the surface? Ground water near hot spots, places where hot, liquid rock rises toward the surface, can contain sulfuric acid, which reacts with and dissolves limestone even better than the carbonic acid in rainwater does.

Formation of a Limestone Cave

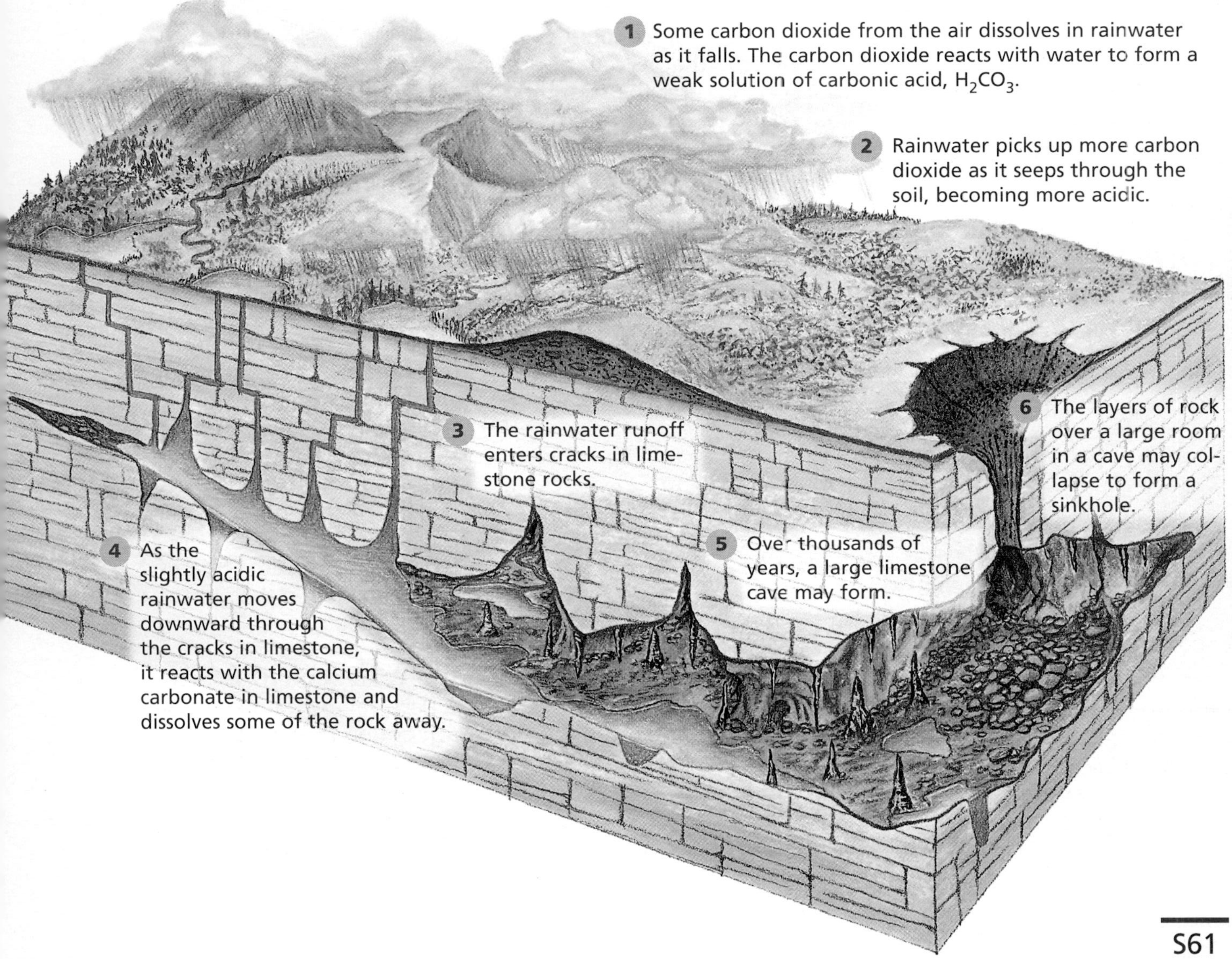

1 Some carbon dioxide from the air dissolves in rainwater as it falls. The carbon dioxide reacts with water to form a weak solution of carbonic acid, H_2CO_3.

2 Rainwater picks up more carbon dioxide as it seeps through the soil, becoming more acidic.

3 The rainwater runoff enters cracks in limestone rocks.

4 As the slightly acidic rainwater moves downward through the cracks in limestone, it reacts with the calcium carbonate in limestone and dissolves some of the rock away.

5 Over thousands of years, a large limestone cave may form.

6 The layers of rock over a large room in a cave may collapse to form a sinkhole.

One important use for sulfuric acid in industry is in the manufacture of plastics.

Uses of Acids

Acids are very important to many industrial processes. Sulfuric acid, for example, is so widely used that its rate of production indicates the strength of our economy. Sulfuric acid is a colorless, oily, and dense liquid that is used in the manufacture of many products, such as fertilizers and automobile batteries. It is used in tanning leather for your shoes, in treating the paper on which the words of this book are printed, and even in making the vanilla flavoring for cookies, cakes, and ice cream. Sulfuric acid is also used in steel production and many other manufacturing processes.

Nitric acid, another strong acid, was first used by the alchemists of the Middle Ages to dissolve metals. They named it *aqua fortis,* which is Latin for "strong water." Today, the uses of nitric acid range from making explosives such as dynamite, which contains nitroglycerin, to the manufacture of ammonium nitrate, a widely used type of solid fertilizer.

Hydrochloric acid, which is often sold as *muriatic acid,* is used to clean concrete and to balance the acid content of swimming pools. Like other strong acids, concentrated hydrochloric acid must be handled with extreme care.

Base

A compound that produces hydroxide ions (OH^-) when dissolved in water

Characteristics of Bases

If you have ever gotten soap in your mouth, you are familiar with a characteristic of bases—their bitter taste. Soap solutions, which are mildly basic, demonstrate another characteristic of bases—a slippery feel. Characteristics of bases are illustrated in the photos on the next page.

The Formation of Bases

A **base** can be defined as a compound that produces *hydroxide* ions (OH^-) when dissolved in water. Solutions that form when bases dissolve are called *basic.* The word *alkaline* is also frequently used to refer to basic solutions.

The formulas for some familiar bases, as well as their common names, are given in the following table. When these bases dissolve in water, they produce hydroxide ions.

Some Common Bases

Chemical name	Formula	Common name
ammonia	NH_3	household ammonia
calcium hydroxide	$Ca(OH)_2$	slaked lime
magnesium hydroxide	$Mg(OH)_2$	milk of magnesia
sodium hydroxide	NaOH	lye, caustic soda

Never touch or taste a concentrated solution of a strong base.

Although most bases, such as NaOH and $Ca(OH)_2$, already contain hydroxide ions, some bases do not. Ammonia (NH_3), for example, produces hydroxide ions only when it dissolves in water.

$$NH_3 + H_2O \rightarrow NH_4^+ + OH^-$$

Because ammonia forms OH^- ions in solution, it is considered a base. Notice how the hydroxide ion forms. In this case, the base (NH_3) takes a hydrogen ion (H^+) from a water molecule (H_2O). As a result, dissolving ammonia in water produces NH_4^+ and OH^- ions, and the solution is therefore basic.

DID YOU KNOW . . .

that bases feel slippery because they dissolve some of the fat molecules in your skin?

The ability to dissolve fats makes bases good cleaning agents.

The Strength of Bases The stronger the base is, the greater the number of hydroxide ions it produces when dissolved in water. Strong bases are just as dangerous as strong acids. They can cause serious burns and must be handled with care.

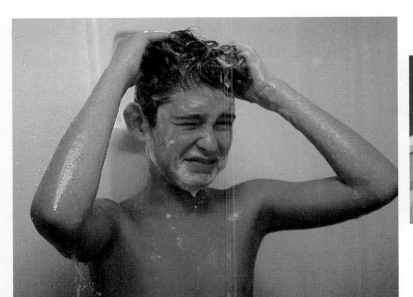

What characteristics of bases are indicated by the images shown here?

Strength of Some Common Bases

Strong bases	Weak bases
calcium hydroxide—$Ca(OH)_2$	ammonia—NH_3
potassium hydroxide—KOH	aniline—$C_6H_5NH_2$
sodium hydroxide—NaOH	sodium carbonate—Na_2CO_3

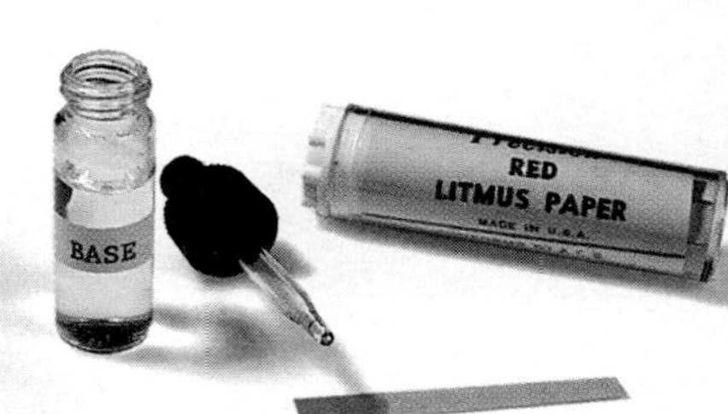
BASE
RED
LITMUS PAPER

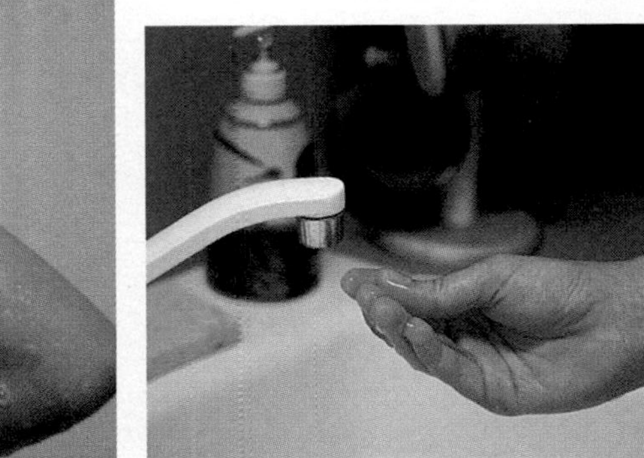

S63

Uses of Bases

Bases are used for a variety of things, from washing yourself and your clothes to unclogging sinks and even making cement. One example of a commonly used base is household ammonia. This compound is added to many household cleaners because of its ability to dissolve grease.

Sodium hydroxide is the most commercially important base. It is used by many industries, such as the paper industry. When paper is made, sodium hydroxide is used to remove pulp fibers from wood. Like sulfuric acid, sodium hydroxide is an *economic indicator.* That is, the amount of sodium hydroxide that a country produces for use in industry indicates how well the economy of the country is doing.

Many other bases are found in common products or are used to make products. For example, magnesium hydroxide is sold as an antacid. More than 35 billion pounds of calcium hydroxide are produced annually in the United States. This base is used to make cement, mortar, and plaster.

Bases Neutralize Acids Acids and bases react in a process called *neutralization.* Neutralization occurs when equal amounts of hydronium ions and hydroxide ions are in solution. These ions produce water, as is shown in the following equation.

$$H_3O^+ + OH^- \rightarrow 2H_2O$$

Once the water forms, the solution is no longer acidic or basic; it is neutral. This characteristic reaction between acids and bases occurs under many circumstances—even in the human body. When a person has an acid stomach or heartburn, he or she may take an antacid tablet. The antacid tablet is a weak base that dissolves in the stomach and combines with the excess acid to neutralize it.

The paper-making process, which starts with wood chips (top), requires great amounts of sodium hydroxide to turn the wood chips into pulp (middle). Pulp is then made into rolls of finished paper (bottom).

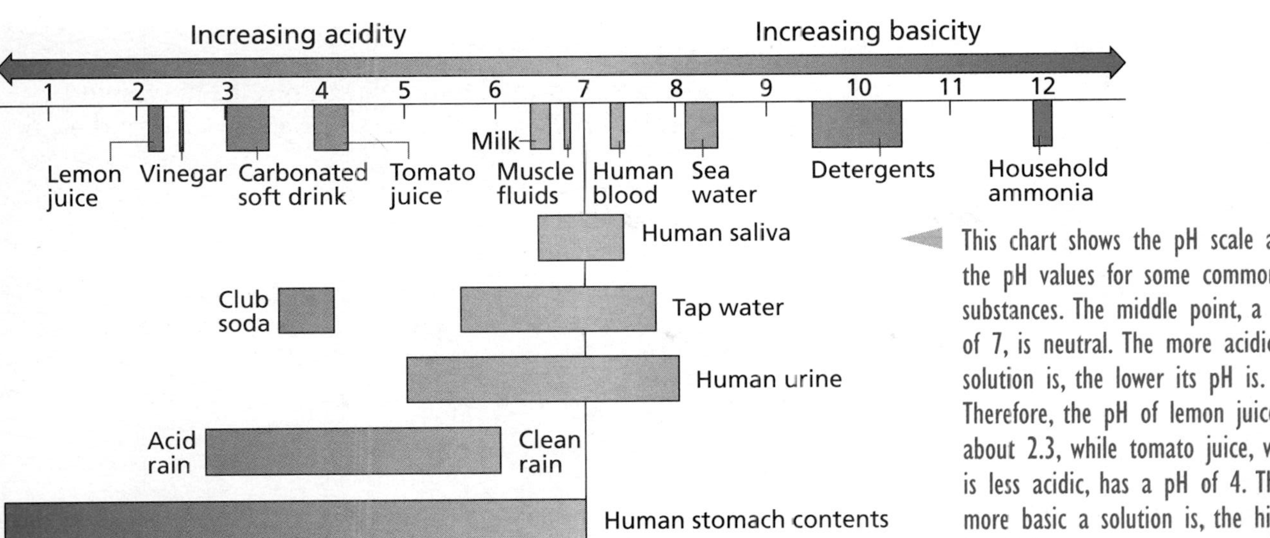

This chart shows the pH scale and the pH values for some common substances. The middle point, a pH of 7, is neutral. The more acidic a solution is, the lower its pH is. Therefore, the pH of lemon juice is about 2.3, while tomato juice, which is less acidic, has a pH of 4. The more basic a solution is, the higher its pH is.

The pH Scale

Since you cannot see H_3O^+ or OH^- ions in solution, how can you identify a solution as acidic or basic? To determine how acidic or basic a solution is, scientists use the pH scale. The **pH** of a solution is a measure of the hydronium-ion concentration in the solution. The pH of most solutions falls between 0 and 14. Acids have a pH of less than 7. Bases have a pH higher than 7. A solution with a pH of 7 is neither acidic nor basic, but is *neutral.*

pH Indicators There are several ways to determine whether a solution is acidic, basic, or neutral. The precise pH of a solution can be measured with an instrument called a *pH meter*. A pH meter has electrodes that are placed in the solution. A voltage is created and measured by the meter. The pH is read from a dial or display.

Other, less expensive methods can be used to estimate the pH of common substances. For example, garden stores sell kits to help people determine the pH of their soil. The pH of swimming pools, which is important if the water is to be kept clean and free of bacteria, can also be determined by using a pH kit. Both of these kits work on the basis of color changes. The kits contain special solutions called *indicators.* An indicator is a substance that changes color in an acid or a base. For an indicator to be useful, it must show a very distinct color change within a narrow range of pH.

Some indicators show only two colors—one for acids and another for bases. Litmus paper is widely used to test for acids and bases. This paper is dyed with an indicator called litmus. Strips of litmus paper come in two colors: red and blue. Blue litmus paper turns red in an acid. Red litmus paper turns blue in a base. Phenolphthalein is another common indicator. It is colorless in an acid and red in a base.

pH

A measure of the hydronium-ion concentration of a solution

This scientist is using a pH meter to test lake water for traces of acid-rain contamination.

DID YOU KNOW...

that litmus is made from lichens? Lichens are composed of two types of organisms that live together in a symbiotic relationship. Many lichens contain brightly colored pigments.

When pH paper touches an acidic or basic solution, it changes color. The pH of the solution may be estimated by comparing the color produced with the scale provided.

This helicopter is spreading lime, which is a base, over a lake polluted by acid rain. Adding a base to the water may help to neutralize the acid and save the organisms that live in the lake.

Other indicators produce a variety of colors depending on the pH of the substance. Some indicator paper is dyed with what is called a *universal indicator.* Universal indicator is a mixture of several different indicators combined so that the paper gives a different color for each pH. Paper dyed with universal indicator is called pH paper. The advantage of pH paper is that it indicates the actual pH of a solution rather than just showing whether the solution is an acid or a base.

Nature and pH Many plants and animals are very sensitive to changes in pH. Some fish and other aquatic life have died in lakes that became more acidic than normal due to pollution. Lake water usually has a pH between 6 and 7. However, some lakes in northern states have a pH as low as 3. Many scientists believe that this change in pH is caused by acid rain, which is produced by the mixing of rain and pollutants from factory smokestacks and automobile exhaust.

Scientists are experimenting with methods to bring the pH of these acidic lakes back to normal levels. One method involves dumping large quantities of a weak base such as ground limestone (calcium carbonate) into the lakes from the air. Compounds that react with water to form strong bases, such as lime (calcium oxide), have also been added to acidic lakes. Scientists hope that the addition of these bases will neutralize the excess acid. The results of these efforts, however, will only be temporary if the pollution that causes acid rain is not stopped.

DID YOU KNOW...

that the human body is very sensitive to changes in the pH of its bloodstream?

The normal pH of human blood is between 7.38 and 7.42. If the pH is above 7.8 or below 7, the body cannot function normally. If a sudden pH change is not corrected rapidly, it can be fatal.

Characteristics of Salts

You are used to calling sodium chloride "salt." Scientists, however, use the term *salt* in a more general way. A **salt** is any ionic compound formed when the negative ion from an acid combines with the positive ion from a base. You may recall that when acids and bases are mixed together, the H_3O^+ and OH^- ions form water. The other two ions that were part of the acid and base form a salt. All neutralization reactions produce both water and a salt. Look at the following neutralization reaction:

$$HBr + KOH \rightarrow H_2O + KBr$$

In this equation, hydrogen bromide is an acid and potassium hydroxide is a base. When solutions of these compounds are mixed, the potassium and bromide ions in these solutions combine to form the ionic compound potassium bromide (KBr).

Not all salts form by neutralization reactions, however. Some form by the reaction of metals with acids. For instance, when pieces of magnesium metal are dropped into hydrochloric acid, hydrogen gas is released and a white solid forms. This white solid is magnesium chloride ($MgCl_2$), a salt. The following equation describes this reaction:

$$Mg + 2HCl \rightarrow MgCl_2 + H_2$$

Sodium chloride, potassium bromide, and magnesium chloride are all white. However, not all salts are white. The color of a salt depends on the type of ions that it contains. The photo below shows several examples of brightly colored salts.

Salt

Any ionic compound formed when the negative ion from an acid combines with the positive ion from a base or by the reaction of an acid with a metal

The color of a salt depends on the elements that compose it. Pictured here are four different salts, each a different color—copper sulfate (blue), nickel chloride (green), potassium chromate (yellow), and potassium dichromate (red).

Uses of Salts

Salts are very common compounds. You probably use them every day without even knowing it. For example, most canned and processed foods contain salts. These salts either improve the flavor of the food or help to preserve it.

Huge amounts of many different salts are dissolved in Earth's oceans. These salts can be obtained by the evaporation of sea water. Of the salts taken from the sea, sodium chloride, or table salt, is the most common. In some parts of the world, drinking water is obtained from sea water by removing its dissolved salts. The process of removing dissolved salts from sea water is called *desalination.* Approximately 40 million metric tons of sodium chloride are collected each year by desalination. Sodium chloride is an important part of our diet. It is also used in many industrial processes. Lye, chlorine, hydrochloric acid, and baking soda are some of the chemicals that are made from sodium chloride.

As you know, salt is able to lower the temperature at which water freezes. People often take advantage of this property of salts. For example, sodium chloride is spread on wet streets in the winter to prevent ice from forming at 0°C. Salty water freezes at a much lower temperature than does pure water. Other salts have a similar effect on water, but because sodium chloride is so abundant, it is the most economical to use.

Sodium chloride can be collected from the sea or mined from the Earth's crust. Solid sodium chloride (inset) is often found in the form of large halite crystals, or rock salt.

SUMMARY

Acids are compounds that produce H^+ ions when dissolved in water. Bases are compounds that produce OH^- ions when dissolved in water. The strengths of acidic and basic solutions are measured on the pH scale. Acids have a pH less than 7, while bases have a pH greater than 7. A solution with a pH of 7 is neutral. Indicators such as litmus are used to indicate the pH of solutions. When acidic and basic solutions are mixed, the H^+ and OH^- ions combine to form water in a neutralization reaction. In addition to water, a salt is produced in a neutralization reaction when the negative ion of an acid combines with the positive ion of a base. Salts are also made by reactions of an acid and a metal.

Unit CheckUp

Concept Mapping

The concept map shown here can be used to illustrate major ideas in this unit. Complete the map by placing terms from the list in the appropriate position. Then extend your map by answering the additional question below. Write your answers in your ScienceLog. **Do not write in this textbook.**

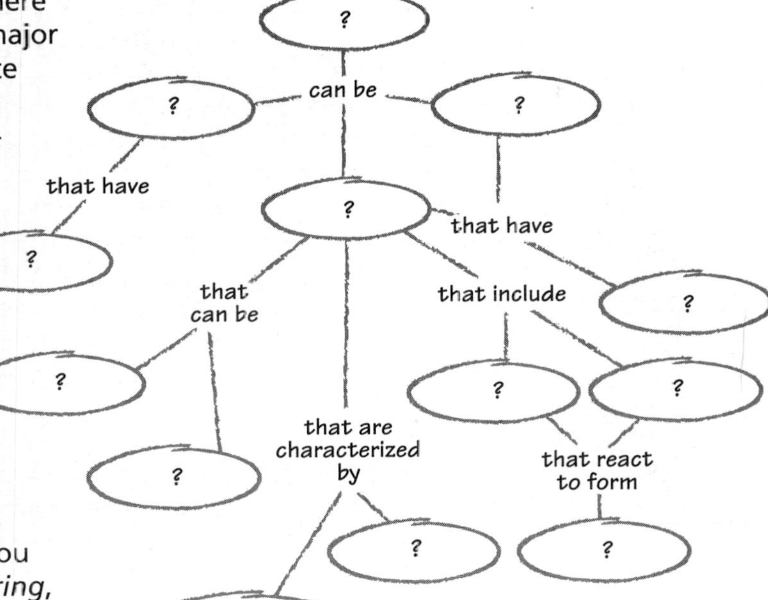

acids
bases
boiling-point elevation
colloids
concentrated
dilute
freezing-point depression
invisible particles
mixtures
salts
solutions
suspensions
visible particles

Where and how would you connect the terms *scattering*, *solvent*, and *sodium hydroxide*?

Checking Your Understanding

Select the choice that most completely and correctly answers each of the following questions.

1. A concentrated solution has a
  a. small amount of solvent dissolved in a large amount of solute.
  b. small amount of solute dissolved in a large amount of solvent.
  c. large amount of solute dissolved in a small amount of solvent.
  d. large amount of solvent dissolved in a small amount of solute.

2. Which of the following does NOT affect the solubility of a particular solid?
  a. type of solvent
  b. amount of solvent
  c. amount of solute
  d. temperature

3. The difference between the temperatures at which pure water and a saltwater solution boil is called the
  a. boiling-point elevation.
  b. freezing-point elevation.
  c. boiling-point depression.
  d. freezing-point depression.

4. Which of the following is a characteristic of the pH scale?
  a. As acidity increases, pH increases.
  b. The more basic a solution is, the higher its pH is.
  c. Basic solutions have a pH between 0 and 7.
  d. Weak acids have a lower pH than strong acids.

5. Which of the following does not indicate pH with a color change?
  a. litmus paper
  b. phenolphthalein indicator
  c. pH paper
  d. pH meter

Interpreting Photos

Look at the photo of the sunset shown here. Based on evidence visible in this photo, what type of mixture does the atmosphere represent? Explain.

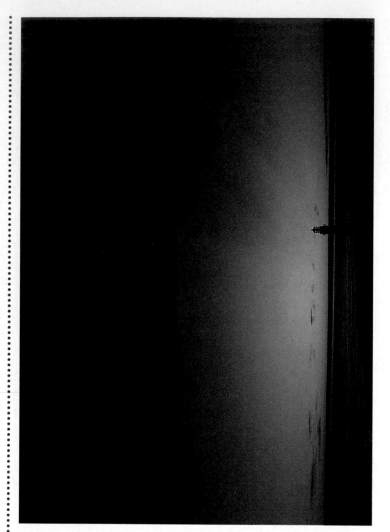

Critical Thinking

Carefully consider the following questions, and write a response in your ScienceLog that indicates your understanding of science.

1. Suppose your younger brother or sister asks how you can tell the difference between solutions and other types of mixtures. What would you tell him or her?
2. Which would be more affected by gravity, a colloid or a suspension? Explain.
3. What salt will form when hydrochloric acid (HCl) reacts with magnesium hydroxide ($Mg(OH)_2$)? Use a chemical equation to determine your answer.
4. What do you think would be the easiest method to use to accurately determine the pH of water in lakes and streams? Justify your answer.
5. Calcium oxide, CaO, reacts with water to form calcium hydroxide, $Ca(OH)_2$. Suppose calcium oxide is added to a lake that has been polluted by acid rain containing sulfuric acid, H_2SO_4. Use your knowledge of neutralization reactions to explain how this event might affect the lake. How will any chemical reactions that occur affect the water's pH and the organisms in the lake? What salts would remain in the lake?

Portfolio Idea

Make a Solutions Scrapbook that includes as many common solutions as you can find. Use magazine pictures, your own photos, or even drawings to illustrate the solutions. Also include labels of commercial products that contain solutions. For each solution, identify the solvent and solute and whether the solution is acidic, basic, or neutral.

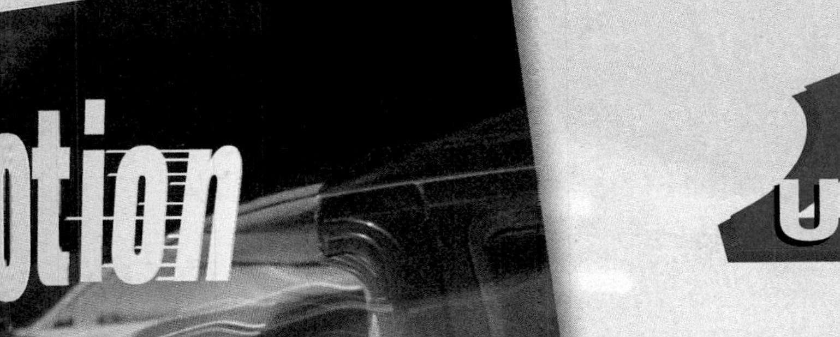

IN THIS UNIT

Now that you have been introduced to forces and motion, consider these questions.

1. How do we observe and describe the motion of objects?
2. In what ways can motion be described by physical laws?
3. How are motion, force, acceleration, and gravity related?

In this unit, you will take a closer look at how we describe motion and how motion relates to force.

What Is Motion?

The world around us is full of motion. People walk from place to place, cars and trucks move along city streets, and leaves fall from trees. Maybe you walked to school today, or maybe you rode the bus. In either case, you were in motion. You see motion everywhere. But what exactly *is* motion?

This balloon is in motion. What reference points do you see in these pictures? Can you see that during the time it took to take these two photos, the balloon moved over the trees? Now assume that the time interval between the two photographs was 20 seconds. If the balloon moved a distance of 100 m during that time, the speed of the balloon was 5 m/s.

Observing Motion

To determine if an object is in motion, we must compare the object with another object that appears to stay in place. The object that stays in place is called a *reference point*. When an object changes its position with respect to a reference point, it is in **motion**. On Earth, we usually compare moving objects with objects on the ground that appear to be stationary. For example, if you look at clouds in the sky, you can tell if they are moving by comparing them with a stationary tree. If the clouds change position with respect to the tree, they are in motion. If there is no change, you know they are still.

Speed Every moving object covers a certain distance in a certain period of time. By dividing this distance by the time of travel, you can find an object's rate of motion. The rate at which an object moves from one place to another is called its **speed**. Speed depends on the distance traveled and the time taken to travel that distance.

Motion

A change in position of an object when compared with a reference point

Speed

The rate at which an object moves, as measured by the distance it covers in a unit of time

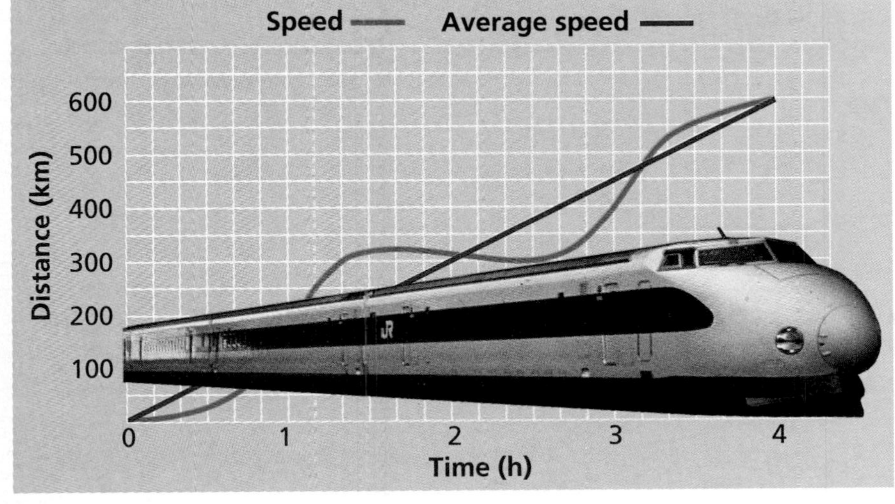

The speed of a train changes as it travels from town to town. Because this train traveled 600 km in 4 hours, its average speed for this journey was 150 km/h.

Most of the time, an object's speed is not constant. A train, for example, moves more slowly as it leaves a station than it does when it is in the country. Hills can slow it down or speed it up. Therefore, an object's speed over a long distance is best expressed as an average. To calculate *average speed*, use the following formula:

$$\text{average speed} = \frac{\text{total distance}}{\text{total time}}, \quad \text{or} \quad v = \frac{d}{t}$$

Velocity Speed simply refers to how fast an object is moving. But objects can move either *toward* or *away* from a reference point. To completely describe an object's motion, you must specify two things—the speed *and* the direction of the motion. The speed of an object *in a particular direction* is called **velocity**.

Velocity

The speed and direction of a moving object

People often use the words *speed* and *velocity* as if they have the same meaning. But there is a difference. For example, 60 km/h is a speed, while 60 km/h west is a velocity. As you can see, velocity has a direction, while speed does not. To calculate an object's velocity, you can use the same formula that is used to calculate speed: $v = d/t$. Then you must specify its direction. For an object to have a constant velocity, *both* its speed and the direction of its motion must remain the same. If either its speed or direction changes, so does its velocity.

Although the cars on these intersecting roads are moving at about the same speed, their velocities are different because they are moving in different directions.

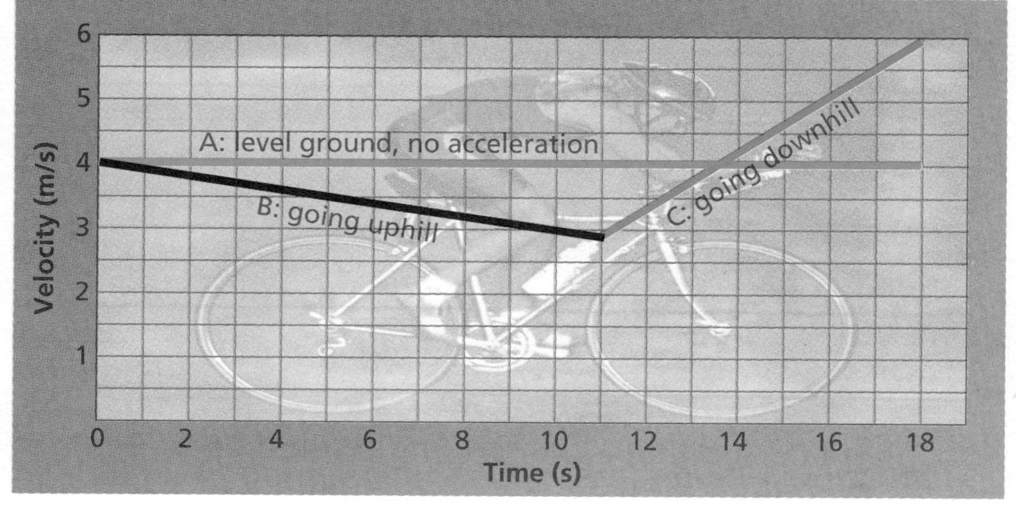

The motion of a bicycle can be graphed as the relationship between velocity and time. The blue line shows constant velocity, or no acceleration. The green line shows a positive change in velocity, or acceleration. The red line shows a negative change in velocity, or deceleration.

Acceleration Most moving objects do not move at a constant velocity. When you ride a bicycle, for example, your speed increases as you go downhill and decreases as you go uphill. In addition, the direction of your motion changes as you turn corners or go around curves. In each of these cases, the bicycle's velocity changes. Any time an object's velocity changes, we say that it *accelerates*.

The rate at which the velocity of an object changes is called **acceleration**. To determine an object's acceleration, you must know how its velocity changes and the time it takes to make that change. You can calculate acceleration by using the following formula.

$$\text{acceleration} = \frac{\text{change in velocity}}{\text{time}}, \quad \text{or} \quad a = \frac{v_f - v_i}{t}$$

In this equation, v_f is the final velocity, and v_i is the initial velocity. But remember, you must add a direction to your answer. What happens when something slows down? In this case, calculating acceleration gives a negative number. This type of acceleration is called negative acceleration, or *deceleration*.

Since velocity refers to speed *and* direction, acceleration occurs whenever an object, such as a car, changes it s speed or direction, or both. The only time that a moving object is not accelerating is when both its speed and direction are constant. As the diagrams on this page show, the direction of an object's acceleration is not always the same as the direction of its motion.

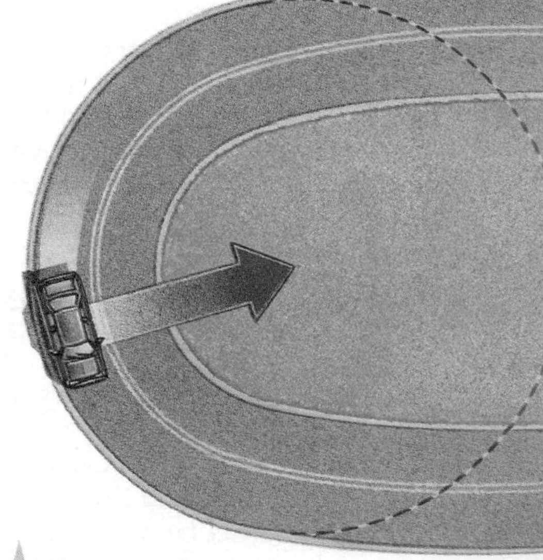

When a car goes around a curve at a constant speed, the direction of its acceleration is toward the center of a circle that could be made by continuing the curve.

Acceleration

A change in the speed or direction of motion

As a car slows down, the direction of its acceleration is opposite to the direction of its motion.

As a car speeds up, the direction of its acceleration and motion are the same.

Considering the Earth's motion—as it rotates, revolves, and moves with our galaxy—just how fast are you going when you are "standing still"?

Frames of Reference

If you look around, you will notice that many objects do not move with respect to each other, and thus they seem to be stationary. A set of stationary surroundings, such as the walls of a room, defines a **frame of reference**. Frames of reference are used to specify the positions and relative motions of objects. We describe the motion of an object by comparing it with a reference point within a given frame of reference.

For many of the motions we observe, the Earth's surface is our frame of reference. Stationary objects on the ground serve as reference points for describing motion. However, the Earth itself is in motion. Although you appear to be sitting still as you read these words, you are hurtling through space at an incredible speed.

We can also describe motion that occurs within other, smaller frames of reference. Imagine that you are riding in a train. The walls, ceiling, and floor of the car you are in are stationary with respect to each other and thus form a frame of reference. Inside this *moving frame of reference*, objects appear to behave as they would if you were standing on the Earth's surface. You can toss a coin into the air and catch it—never mind that the coin is actually going forward at 100 km/h, the same speed as the train. As long as the train's velocity is constant, the laws of physics act as though the train were standing still. In fact, if there were no windows in the train, you might not even notice you were moving.

Frame of reference

A system for specifying the precise location and motion of objects in space

How can you tell that the train in this photograph is moving?

Friction and Motion

The ancient-Greek philosopher Aristotle once said that a moving object would stop moving unless it were pushed or pulled. Do you think this is true? Think about a wagon. As long as you pull it, it will follow you. If you let go of the handle, it will slow down and stop. This simple experiment seems to indicate that Aristotle was right. But was he?

Any push or pull is called a **force**. Aristotle thought that a force was needed to keep an object moving, but he didn't know why. Aristotle was not aware of a very important aspect of nature—friction.

Force

Any push or pull

In science, great discoveries are often made when someone asks the right question. Aristotle asked, "Why do objects move?" The early scientists who investigated motion asked the same question. Their experiments indicated that a force is needed to cause motion and that objects with no force acting on them do not move. However, because they only investigated what *causes* motion, they did not discover the unseen forces that also affect motion. In the seventeenth century, Isaac Newton saw a different problem. He asked, "Why do objects *stop* moving?"

Friction

The force that opposes motion when two objects are touching

Newton proposed that moving objects are often stopped by a force we call **friction**. As you know, friction results when two objects rub against each other. The force of friction causes a moving object to slow down and, if the force acts long enough, to stop. Friction makes it harder to move some objects. But friction is also a very useful force that enables certain types of motion to occur. For example, if there were no friction, you would not be able to walk. As you may have discovered, smooth surfaces have less friction than rough surfaces and thus are more difficult to walk on.

Have you ever slipped on a polished floor? Friction between your shoes and the floor enables you to apply the horizontal force that moves you forward.

Types of Motion

You have just learned that whenever an unbalanced force is applied to an object, the object's motion changes. If the object is at rest, it moves; if it is moving, its velocity changes. You also know that if a force does not change an object's motion, the force must have been *balanced* by another force. As the diagram on this page shows, only *unbalanced* forces are able to change the motion of, or accelerate, objects.

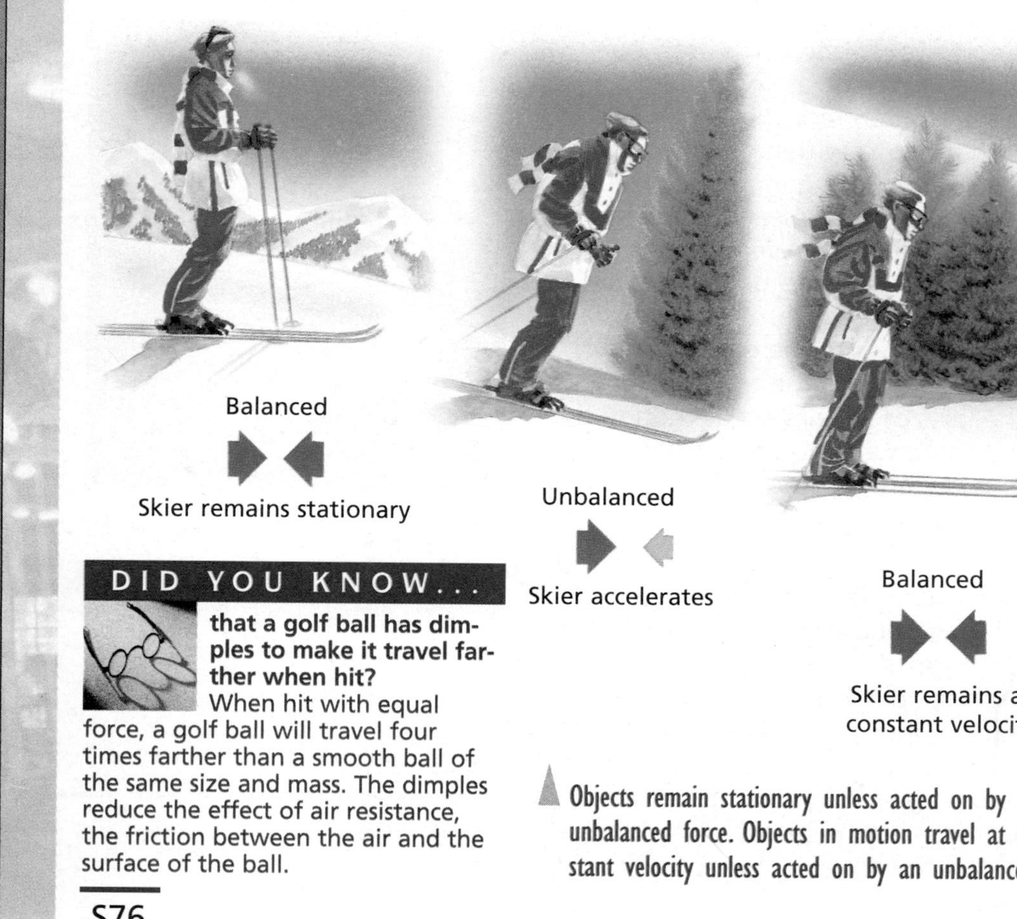

Objects remain stationary unless acted on by an unbalanced force. Objects in motion travel at a constant velocity unless acted on by an unbalanced force.

DID YOU KNOW...

that a golf ball has dimples to make it travel farther when hit? When hit with equal force, a golf ball will travel four times farther than a smooth ball of the same size and mass. The dimples reduce the effect of air resistance, the friction between the air and the surface of the ball.

S76

The motion of an object is the result of *all* the forces that act on it. On Earth, the motion of all objects is affected by gravity and friction. By analyzing all the forces that act on a moving object, we can better understand how some familiar types of motion occur.

Free Fall A freely falling object is a good example of accelerated motion in one direction. If there were no friction, an object falling toward the Earth would accelerate downward at 9.8 m/s^2 because of the pull of gravity. This means that the object's speed would increase by 9.8 meters per second for each second that it falls.

Actually, a falling object accelerates downward at 9.8 m/s^2 only if it falls in a vacuum. An object falling through air accelerates only for a short time before its acceleration is stopped by friction with the air, or *air resistance*. The greatest velocity that an object can reach while falling through air is called its *terminal velocity*. For example, if you were to jump from an airplane, you would accelerate until you reached a terminal velocity of about 200 km/h downward. After opening your parachute, your terminal velocity would be reduced to a much slower rate by the additional air resistance of the parachute.

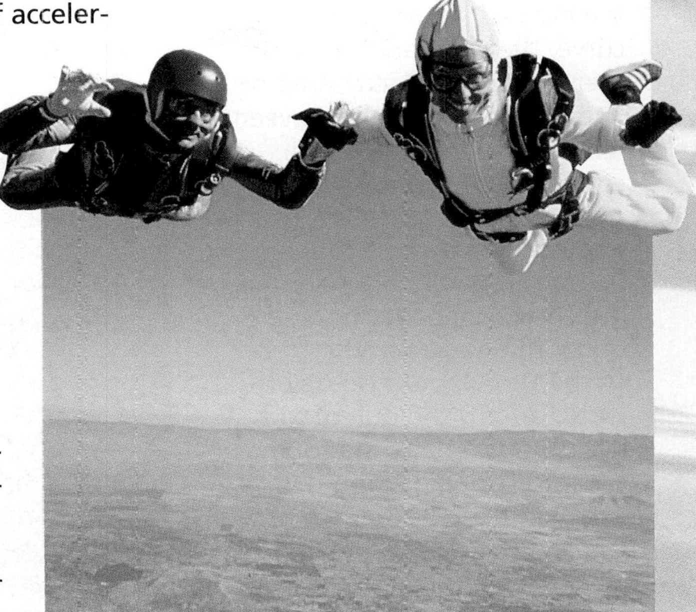

All objects falling through the air reach a terminal velocity once the force of air resistance balances the force of gravity. The upward force of air resistance on these sky divers works against the downward force of gravity.

Circular Motion A race car moving around a circular track, a child on a merry-go-round, and clothes being spun in a washing machine are all examples of objects in *uniform circular motion*. Uniform circular motion refers to motion at constant speed in a circular path. When an object travels in a circle, its direction constantly changes, and it is therefore accelerating. Consider a race car speeding around a curve on the track. The car must constantly change direction to follow the track. An unbalanced force causes the car to accelerate toward the center of a circle that would be made by completing the curve.

The force that causes objects to move in a circular path is called **centripetal force**. The word *centripetal* means "toward the center."

Centripetal force

A force that causes objects to move in a circular path

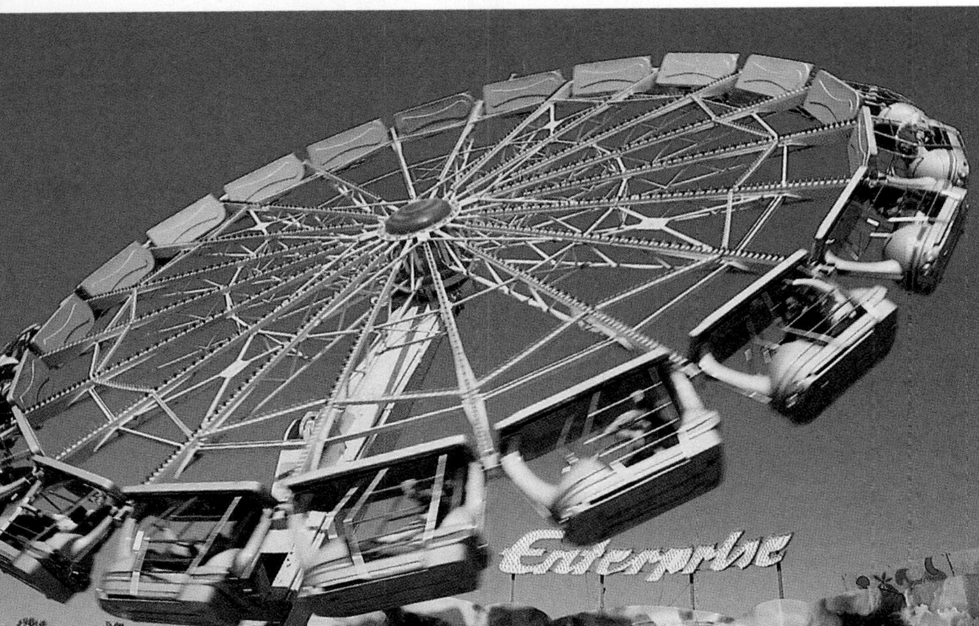

As this amusement-park ride spins, its frame provides the centripetal force necessary to keep the riders moving in a circular path.

As a car travels around a corner, traction (friction between the road and the car's tires) provides the centripetal force. If the tires were to lose traction as the car was moving around the curve, the car would immediately start sliding in a straight line. Without friction between its tires and the road, a car could not be turned.

Projectile Motion When a ball is thrown in a horizontal direction, it moves not only horizontally, due to the force of the throw, but also downward, due to the pull of gravity. Thus, the ball's motion has two directional components. Such an object is called a *projectile*. Projectiles, once they have been thrown, travel horizontally while they accelerate vertically downward. These horizontal and vertical components are independent—that is, they have no effect on each other.

The curved path that a projectile takes is called its *trajectory*. One interesting thing about trajectories is that an object thrown sideways will reach the ground at the same time as an identical object that falls the same distance straight down. This is because gravity's acceleration acts equally on all objects, regardless of their horizontal motion.

When you are in a car that is moving around a curve, it feels as if you are applying an outward force to the seat belt of the car. However, the seat belt actually transfers the centripetal force that causes the change in the car's path to you and changes your direction of travel.

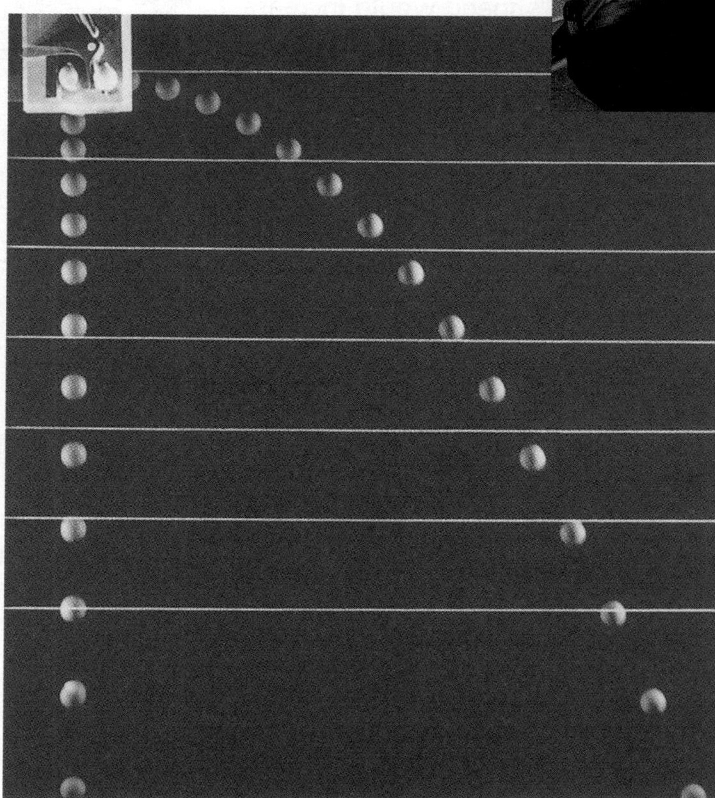

Both balls have the same vertical acceleration due to gravity, but the ball on the right also has horizontal motion.

SUMMARY

The motion of an object is described by comparing it with a reference point, which is an object that is considered stationary. Speed is determined by dividing the distance traveled from a reference point by the amount of time it takes to cover that distance. Velocity is the speed of an object in a particular direction. Acceleration is a change in the velocity of an object. Frames of reference are used to specify the positions and relative motions of objects. Forces cause changes in motion. Friction is a force that opposes motion. The motion of an object is determined by all the forces that act on it.

LAWS OF MOTION

Our understanding of motion is due largely to a book published by Isaac Newton in 1687. Newton's book, *Principia*, describes his findings about moving objects. The basic principles he wrote about are now known as Newton's laws of motion. These and other laws describe the motion we observe and explain the forces behind this motion.

Inertia

The resistance of a mass to any change in its motion

Newton's First Law of Motion

Newton thought that a moving object with no friction or other force acting on it would continue to move in the same direction and at the same speed forever. He decided that a force had to be present for an object's speed or direction to change, as seen below.

A certain amount of force must be applied to an object with a particular mass in order to change the object's motion. If force is applied, the object *resists* any change in its motion. This resistance to change in motion is a property of matter called **inertia**. A moving object, in the absence of an outside force, thus continues to move because of its inertia. Similarly, a stationary object needs a force to start it moving. The amount of inertia that an object has depends on the mass of the object.

If an object is moving at a constant velocity, no additional force is needed to continue its motion.

To change the object's direction, a force must be applied from the side.

To make the object go faster, a force must be applied in the direction the object is going.

To make the object slow down, a force must be applied in the opposite direction.

The effect of inertia on motion is summarized in what is now called ***Newton's first law of motion***. It states: *Every object remains at rest or moves with a constant speed in a straight line unless acted on by some unbalanced force*. Think back to the example of pulling a wagon. Do you see that the force of friction slows down a wagon when you stop pulling on it? If there were no friction, the wagon would continue to move forever. If a wagon is at rest, it will remain that way until a force—such as your push or pull—puts it in motion.

In space, there is almost no friction acting against moving objects. For example, the *Pioneer 10* spacecraft, launched from Earth in the early 1970s, has now left our solar system. Once it left Earth's orbit, its rocket engines were no longer needed to keep it in motion. It has since traveled billions of kilometers with no additional push. Until some outside force affects it differently, *Pioneer 10* will continue moving out into our galaxy indefinitely.

In the absence of friction, *Pioneer 10* will remain in motion in the same direction, even without additional rocket thrusts, until another force acts on it.

The first law of motion explains why you should always wear a seat belt when riding in a car. If the car stops suddenly, you will keep moving forward. If you are not wearing a seat belt, you might hit the windshield or dashboard. A seat belt makes it less likely that you will be hurt, since it supplies the force needed to hold you firmly in your seat.

Newton's Second Law of Motion

Newton's first law of motion states that a force is needed to change the motion of an object. The way that force affects an object's motion is described by ***Newton's second law of motion***, which states: *The change in motion of an object corresponds to the amount of force exerted on it and takes place in the same direction as that force*. The acceleration of the object depends on the size of the force acting on it and on the object's mass. The greater the mass of the object, the greater the amount of force needed to cause a given acceleration. To see how Newton's second law of motion can be applied, look at the illustrations on this page.

The relationship of acceleration (a) to force (F) and mass (m) is summarized by the equation $a = F/m$. All forces act in a particular direction and, of course, may vary in strength. Scientists measure the size of forces with a unit called the *newton* (N). For example, you would need to apply an upward force of about 10 N to lift an object with a mass of 1 kg off the ground. Another way to state Newton's second law of motion is as follows:

$$\text{Force} = \text{mass} \times \text{acceleration}$$

or

$$F = ma$$

DID YOU KNOW...

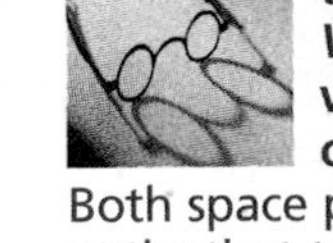

that the *Voyager 1* and *Voyager 2* space probes were literally slung out of our solar system?

Both space probes were sent on paths that took them close to Jupiter. The force of Jupiter's gravity accelerated the space probes, which swung around the huge planet and headed out into the galaxy at far greater speeds.

A professional baseball pitcher can apply more force to a baseball than a Little League pitcher. Therefore, the professional's fastball accelerates more than the Little Leaguer's.

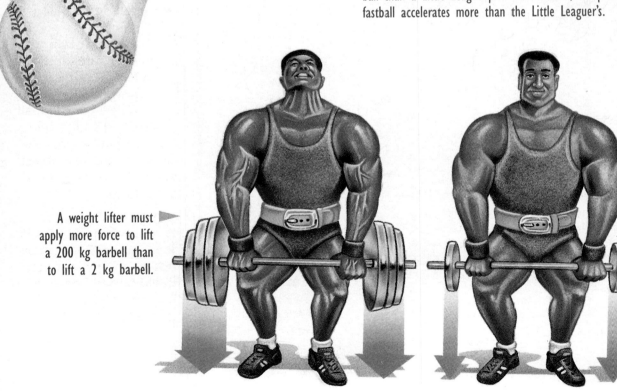

A weight lifter must apply more force to lift a 200 kg barbell than to lift a 2 kg barbell.

Newton's Third Law of Motion

When two objects interact, each exerts a force on the other. The force applied by one object causes a change in the motion of the other object, and vice versa. These changes in motion, or *actions*, are described by ***Newton's third law of motion***, which states: *For every action there is an equal and opposite reaction.* In other words, the force exerted by one object on a second object is equal to but acts in the opposite direction of the force exerted by the second object on the first. The equal and opposite forces that two objects exert on each other are called *action-reaction pairs* of forces. To picture these forces, look at the diagram shown here.

Pushing a grocery cart around a store involves several pairs of action-reaction forces.

1 When you push the grocery cart, the cart pushes back against you.

2 As you walk forward, your feet push horizontally on the floor, and friction with the floor pushes back against your feet.

3 As the wheels of the cart roll forward, they push horizontally on the floor, and friction with the floor pushes back against the wheels.

Examples of Newton's third law are all around you. A book lying on a table pushes down on the table with a certain force. The table, in turn, pushes up on the book with an equal but opposite force. If you lean against the wall with your hand, the wall pushes back on your hand with the same amount of force. Imagine what would happen to you if the wall suddenly disappeared. Since it would no longer be pushing on your hand, you would fall.

Momentum

If you have ever watched American football on television, you have probably heard an announcer use the word *momentum*. The announcer may have said something like this: "Even though number 34 was hit at the 1-yard line, his momentum carried him into the end zone for a touchdown." In physics, momentum is a property of objects in motion. **Momentum** is determined by multiplying an object's mass by its velocity. This relationship is summarized by the following equation:

$$\text{momentum} = \text{mass} \times \text{velocity}$$
$$\text{or } p = mv$$

Because velocity is a factor in the momentum of an object, momentum, like velocity and acceleration, has a direction.

The momentum of an object depends solely on its mass and velocity. Therefore, an object with a small mass and a high velocity can have more momentum than an object with a large mass and a low velocity.

For example, a 1000 kg car moving at 50 km/h has more momentum (p = 1000 kg × 50 km/h = 50,000 kg·km/h) than a 2000 kg truck moving at 20 km/h (p = 2000 kg × 20 km/h = 40,000 kg·km/h).

What happens to the momentum of a moving object when it collides with another object? Some of this momentum can be transferred to the other object. Imagine, for example, that you are skating along when another skater accidentally runs into you from behind. The push from the other skater causes you to move at a higher velocity, and thus your momentum increases. The other skater receives an equal but opposite force from you (Newton's third law of motion) that causes him or her to move at a slower velocity, decreasing momentum. The momentum you gained is equal to the momentum lost by the other skater. This example illustrates the ***law of conservation of momentum,*** which states: *Momentum can be transferred from one object to another, but the total amount of momentum in a closed system cannot change.* In other words, when two or more objects collide, the total momentum of the objects combined is the same after the collision as it was before the collision.

Momentum

The mass of an object multiplied by its velocity

If hit just right, a cue ball will transfer all of its momentum to a rack of billiard balls and come to a dead stop. The total amount of momentum in the billiard balls after the collision equals the amount of momentum that the cue ball had before the collision. In other words, the momentum of this system has been conserved.

SUMMARY

Isaac Newton formulated three laws that describe motion and its related forces. The first law deals with the cause of motion. The second law accounts for changes in motion. The third law describes pairs of forces that act on different objects. Momentum, the product of mass and velocity, is conserved in collisions.

GRAVITATION

What do you think would happen if two stones of different masses were dropped from the same height? Would the larger, heavier one reach the ground first? Or would they both reach the ground at the same time? In 1590, this was the topic of an intense debate among the scientists of Europe. Most scientists believed that the heavier stone would reach the ground first.

Development of Gravitational Theory

Nearly 2000 years earlier, Aristotle had proposed that heavier objects would fall faster. He believed that all matter was made of four elements, each with its own natural position in the world. From lightest to heaviest, these elements were fire, air, water, and earth. If, for example, fire were placed below its natural position, it would tend to rise above air. Similarly, if a stone were dropped, it would pass through fire, air, and water until it came to rest on the ground, its natural place. Aristotle thought that the movement of an object depended on its mixture of the four elements. Therefore, he expected a heavier stone to drop faster than a lighter stone because it contained more of the earth element. Aristotle's physics seemed to fit everyday observations. For instance, you know that if you drop a feather and a stone at the same time, the stone will reach the ground first. What did Aristotle not realize?

By 1589, a young scientist named Galileo had confronted this view of the world. Appointed professor of mathematics at the university in Pisa, Italy, at the age of 25, he soon began to challenge the opinions of the older professors. As it turned out, Galileo's ideas signaled the beginning of the end for Aristotle's worldview.

According to Aristotle, a large mass would hit the ground before a smaller mass if both were dropped from the same height. Galileo, however, predicted that both masses would hit the ground at the same time. We now know that Galileo was right.

In Aristotle's worldview, a 100 kg mass would reach the ground in the time that a 10 kg mass would fall only 1 m if both were dropped from a height of 10 m.

DID YOU KNOW...

that in the United States, the force of Earth's gravity is greater in the eastern states than in the western states?
As a result, you would weigh about 0.3 N less in California than you would in New York. This slight difference in gravitational force is caused by differences in the thickness and density of the Earth's crust.

S84

Legend has it that Galileo dropped stones from the Leaning Tower of Pisa, as shown in this illustration, to test his hypothesis that objects of different mass fall at the same rate. Although he probably did not do this particular experiment, Galileo did do experiments with objects rolling down an inclined plane. Nevertheless, his results showed that objects with different masses accelerate downward at the same rate.

Galileo's legendary experiment

Newton's Law of Gravity

According to Newton's first law of motion, an unbalanced force must act on an object to accelerate it. **Gravity** is the force that causes all things near the Earth to accelerate toward the center of the Earth. Newton believed that gravity was a force of attraction between any mass, such as the Earth, and the matter near it.

Through his observations and calculations, Newton showed that the Earth and all other objects in the universe exert pulling forces on each other. For example, the Earth and the moon exert a gravitational pull on each other. Earth's gravity keeps the moon in orbit about the Earth. Ocean tides are the most obvious effect of the moon's gravitational pull on the Earth. The sun's gravity keeps all the planets orbiting in the solar system. It would be too small to measure, but all the objects around you exert a slight gravitational pull on everything else around them. Another aspect of gravitational pull is that it decreases as the distance between two objects increases. For example, as a spacecraft moves farther from Earth, the strength of Earth's pull on it decreases.

The relationship between gravitational pull and mass is stated in ***Newton's law of gravitation***: *The gravitational attraction between two objects depends on their masses and the distance between them.* For example, a large mass exerts a stronger pull than does a smaller mass at the same distance.

So how can two objects with different masses accelerate and hit the ground at the same time? Look at the photo to the right. If the effects of air resistance could be removed, everything on Earth would fall at the same rate.

Gravity

The force that causes two masses to accelerate toward each other

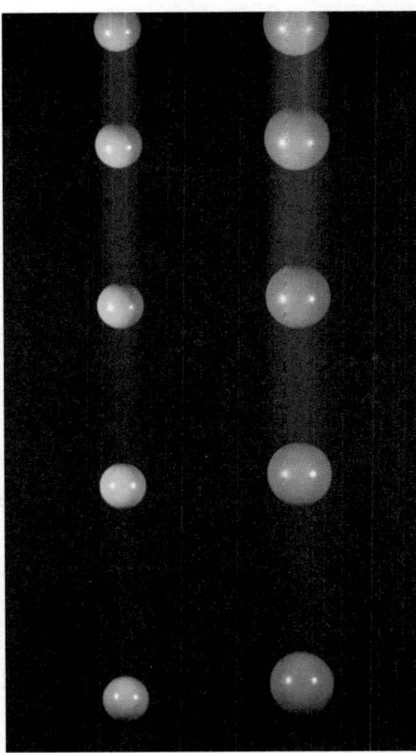

These two objects have different masses, yet they accelerate and fall at the same rate. More force is needed to accelerate a large mass. And because of its greater mass, the pull of gravity acting on the large mass is greater than the pull of gravity acting on the small mass.

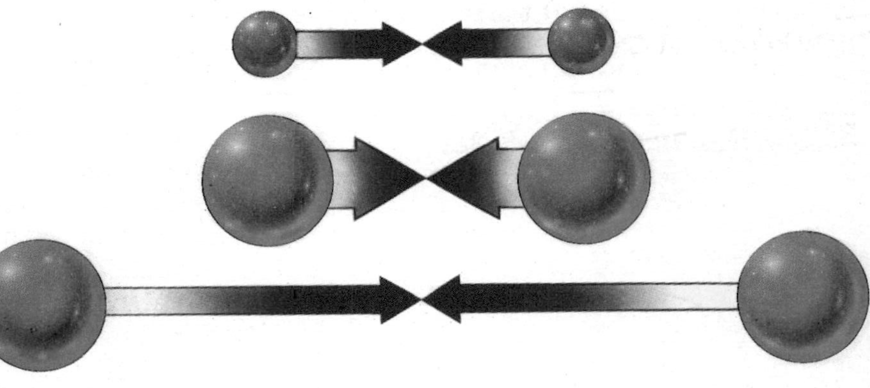

Gravitational pull, represented by the width of the arrows, varies according to the mass of the objects and their distance from one another.

Like all other scientific laws, the law of gravitation describes how some parts of the natural world work. It does not explain what *causes* gravity. Instead, it is a scientific tool that helps *explain* the effects of gravitational attraction. For example, Newton's law predicts (and astronauts have demonstrated) that the gravitational pull on objects on the moon's surface is about one-sixth of that on objects on Earth's surface. This is because the moon's mass and size are both much smaller than those of Earth. How do you think this would affect you if you lived on the moon?

Mass and Weight

The terms *mass* and *weight* are often used as if they mean the same thing. But they have different meanings. As you know, mass is the *amount* of matter in an object. When gravity pulls on the mass in an object, it causes the object to exert a force. **Weight** is the *force* exerted by an object because gravity is pulling on its mass. In science classes, you normally use a *balance* to measure mass and a *force meter* to measure weight.

Since mass and weight are not the same thing, scientists are always careful to distinguish between them. Mass, as you know, is measured in kilograms. But because weight is a force, it is measured in newtons. On Earth, a person with a mass of 50 kg weighs 490 N. On the surface of the moon, the same person would weigh about 80 N. This does not mean, however, that the person's mass has changed. The weight of an object changes as the size of the gravitational pull on it changes, but the mass (amount of matter) in the object does not change. As long as two objects are in the same place, the object with more mass will have more weight than the object with less mass.

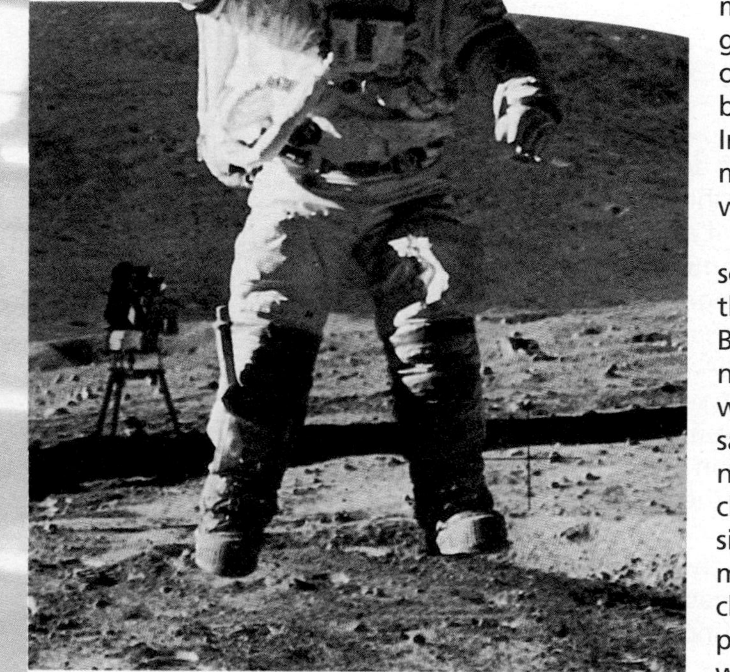

Astronauts on the moon can jump much higher than they can on Earth because their weight is only one-sixth what it is on Earth.

Weight

The amount of force an object exerts because of the pull of gravity acting on the object

SUMMARY

Gravity is described as the force that causes all objects to accelerate toward the center of a mass, such as the Earth. The gravitational attraction between two objects increases as the masses of the objects increase, and the attraction decreases as the distance between the objects increases. Weight is the amount of force an object exerts because of the pull of gravity on it.

S86

Unit CheckUp

Concept Mapping

The concept map below illustrates major ideas in this unit. Complete the map by supplying the words needed to connect the terms. Then extend your map by answering the additional question below. Write your answers in your ScienceLog. **Do not write in this textbook.**

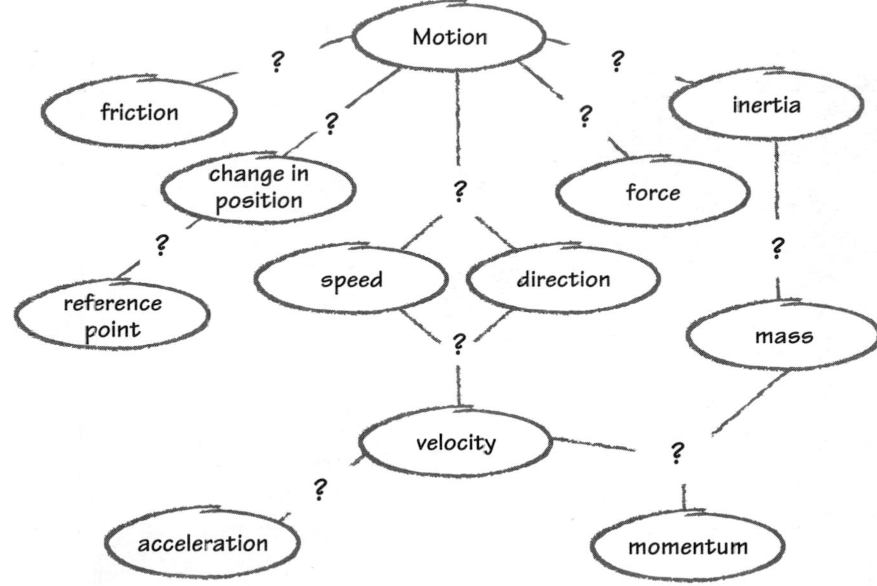

Where and how would you connect the terms *deceleration*, *gravity*, and *Newton's first law?*

Checking Your Understanding

Select the choice that most completely and correctly answers each of the following questions.

1. If a car were to travel 1600 km from Texas to Nebraska in 16 hours, its average speed would be
  a. 100 km/h north.
  b. 256 km/h south.
  c. 100 km/h.
  d. 256 km/h.

2. A frame of reference
  a. can only be stationary.
  b. has parts that are stationary with respect to each other.
  c. defines an area where objects would *not* behave as they would elsewhere.
  d. is the same as a reference point.

3. The force that always opposes motion is called
  a. inertia.
  b. centripetal force.
  c. gravitational force.
  d. friction.

4. According to Newton's second law of motion, the correct relationship of force (F), mass (m), and acceleration (a) is
  a. $F = ma$.
  b. $m = Fa$.
  c. $a = m/F$.
  d. $F = a/m$.

5. The gravitational force between two objects
  a. increases as the objects get farther apart.
  b. increases as the masses of the objects increase.
  c. decreases as the masses of the objects increase.
  d. decreases as the objects get closer together.

Interpreting Photos

Newton's third law of motion explains why it is possible to hang a mobile from a ceiling. Identify all of the action-reaction pairs of forces that exist in the mobile photographed here.

Critical Thinking

Carefully consider the following questions, and write a response in your ScienceLog that indicates your understanding of science.

1. Considering that everything in the universe is in motion, is there such a thing as a stationary frame of reference? Explain.
2. How are each of Newton's three laws of motion demonstrated by shooting a basketball?
3. If an arrow is shot straight forward, it will fall to Earth in the same amount of time as an arrow that is simply dropped. Explain why this happens.
4. The current trend in football is to recruit linemen (the players that line up directly across from each other) that are both heavy and fast. How can the law of conservation of momentum be used to justify this trend?

S88

5. When the starships of popular science fiction develop engine trouble, they tend to come to a standstill in the depths of space. On other occasions, a starship might use a continuous tractor beam to tow a disabled shuttle. Using your understanding of inertia, explain how the activities of these starships are scientifically unrealistic.

Portfolio Idea

Write a poem or song lyric, or produce a videotape, that uses the concepts of force, speed, velocity, acceleration, and Newton's laws of motion and gravitation to explain the motion of objects that you see daily.

Structures and Design

In This Unit

Now that you have learned something about the construction of buildings, bridges, and other structures, consider these questions.

1. What kinds of changes can cause structural failure?

2. How do molecular forces contribute to the durability and strength of structures?

3. How might geography, climate, religion, and culture affect the architectural styles of a region?

4. Why might different cultures solve design problems in different ways?

In this unit, you'll take a closer look at certain aspects of structure and design. You'll then take a tour of some different kinds of architecture from around the world, both past and present.

The Right Stuff

As you know, the materials that make up structures such as buildings and bridges must be able to withstand assaults of many kinds. They must endure not only environmental effects, but also forces produced by the various parts of the structures themselves. Buildings are designed to last a long time, and they must be constructed to withstand many negative influences for many years.

The constant force of gravity pulls at a building and its parts, and the effects of wind, rain, and temperature changes can weaken it as well. Unexpected natural phenomena, such as earthquakes and tornadoes, must also be taken into account. When a building is good at resisting these various effects, we say that it is *durable.* The materials used in a building, and the way those materials are used, determine its durability.

Will this skyline be the same in 100 years? in 1000 years?

Metal stairways, building supports, and frames can rust. Rust can reduce the thickness of unprotected metal by approximately 0.01 mm per year.

Wooden beams and rafters can deteriorate and weaken when they are exposed to water. The rotting of wood is another example of a chemical change.

Changes in Materials

Structures sometimes fail because the design was flawed. But even well-designed structures cannot last forever. While unforeseen factors like extreme weather conditions may cause a structure to fail quickly, other agents work slowly, almost invisibly, causing changes over time that can weaken a structure from the inside out.

Chemical Changes Sometimes changes occur within a structure's materials that decrease the overall strength of the structure. *Chemical changes,* for example, occur when different substances combine to produce completely new substances. These new substances do not have the same properties as the original substances. One example of this is the combination of iron with oxygen from the air to produce rust.

In constructing a building, great care must be taken to minimize the effects of chemical change. This includes selecting materials that resist such changes or using protective measures to delay the onset of the changes. Can you name one way to protect wood or metal?

Physical Changes A *physical change* can also weaken a structure. In physical changes, no new substances are produced, but the original materials are broken down mechanically. You may have broken a length of wire, such as a paper clip, by bending it back and forth many times. This is one type of physical change that can cause structural failure. The result of bending such as this is called **metal fatigue.** The flexing of pieces of metal over and over again can cause parts of a structure to break. Engineers and designers must know how long certain parts will last before metal fatigue occurs. This is particularly important for the structural parts of airplanes and other vehicles.

Metal fatigue

The wearing out of metal parts due to repeated bending

Properties of Materials

The physical and chemical properties of the materials used in a structure have a tremendous effect on strength and durability. Recall that matter can be either gaseous, liquid, or solid. Solids are more useful for construction because they generally maintain their shape and volume. But what is it about solids that determines their utility?

Structure of Solids Solids, like liquids and gases, are composed of atoms and molecules. But in most solids, these microscopic particles are held together in fixed geometric patterns. The way in which these orderly patterns are organized is called *crystal structure.* For example, the atoms in steel are ordered in regular columns and rows. This gives steel its strength and durability. In marble, a rock composed of calcium carbonate, the particles are held in another crystal shape. You know that marble is a strong and durable material that has been used by many cultures throughout history.

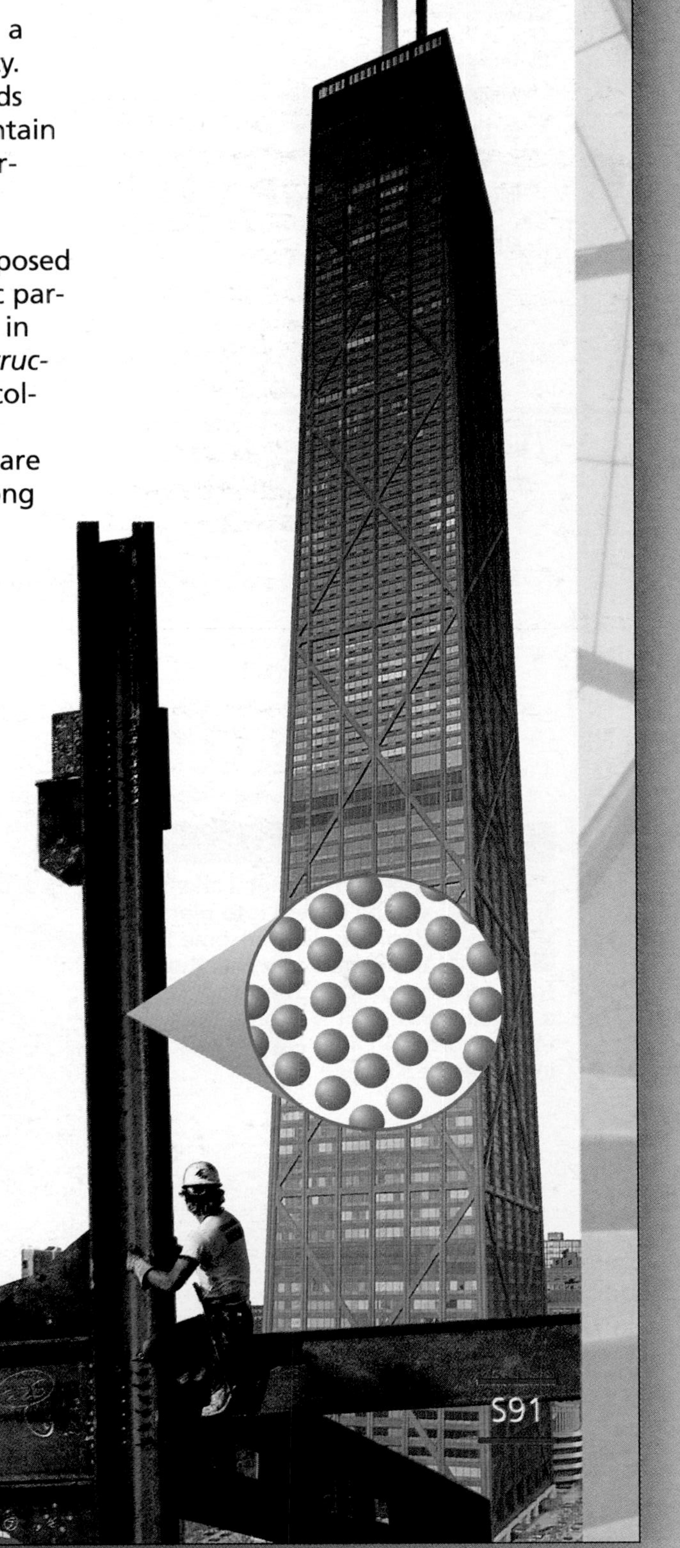

The particles of a crystalline solid, such as steel or marble, are arranged in specific crystal patterns.

The glass in this ancient window has slowly settled to the bottom as its molecules gradually flowed downward due to gravity.

Some solids, such as glass, do not have regular crystal patterns. Their atoms and molecules are arranged more randomly throughout the material without any consistent pattern. Because there is no crystal structure, these materials may not hold their shape as well over time. In fact, glass is actually considered a very slow-flowing liquid, as shown to the left.

The type of chemical bonding and crystal arrangement determines the strength and flexibility of a particular material. In diamond, the hardest natural substance known, each atom forms strong bonds with four neighboring atoms in a tetrahedral arrangement. But carbon atoms can also be arranged in layers that easily slide past each other, forming graphite, the soft substance you leave on paper when you use a pencil.

Sticking Together In addition to molecular arrangement, the forces between the particles in a solid help determine a material's physical properties. You have seen that some materials handle stretching (*tension*) and squeezing (*compression*) better than others. We can look at the effects of tension and compression of a material on a microscopic level as well. On the molecular level, the force that keeps molecules together is called **cohesion.** Cohesion refers to the attraction of *similar* particles to each other. Without cohesion, solids would disintegrate. When the particles of a solid substance have stronger cohesion, it takes more tensile force to separate them from each other. This affects our ability to change the shape of the material or break it apart.

Cohesion

The force of attraction between the particles of like substances

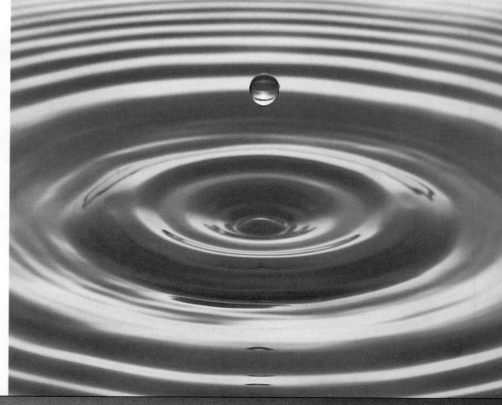

Cohesion is also responsible for the shape of raindrops. Without the cohesion of the water molecules attracting each other, the rounded shape of drops would not form.

DID YOU KNOW...

that certain metal alloys can be turned into glass? Rapid cooling of some liquid metals prevents them from forming crystals. This produces a metallic glass that is strong, resistant to corrosion, and easy to mold into delicate shapes.

To Bend but Not Break You have investigated several types of forces that affect structures, including tension, compression, and shear forces. When these forces are exerted on a structure, the structure is put under stress. *Stress* is any force that, if great enough, would cause a change in the shape of a material. The amount of change, or deformation, caused by a stress is called *strain.* Designers need to know how materials are deformed, or strained, by different stresses.

When stress is applied to a solid object, its size and shape will be changed, sometimes permanently. For example, a spring is considered very elastic because it can be stretched a long way and still return to its original shape. The elasticity of a spring is very high. That of a piano wire, on the other hand, is much lower. All materials, even concrete and stone, have a small amount of elasticity.

Yet even a spring can be stretched too far. If it is stretched beyond the point where it will return to its original shape, it is said to have exceeded its **elastic limit.** Like cohesion, you can understand elastic limits in terms of chemical structure. Beyond the elastic limit of any solid material, the forces of cohesion are overcome to such an extent that the atoms or molecules of which the solid is made slide past each other. They are then realigned into a new pattern, and the original shape of the solid is permanently changed. Elastic limits can be exceeded by tensile, compressive, or shear stress. The amount of stress required to exceed the elastic limit can be different for each type of stress.

Elastic limit

The point beyond which a material cannot return to its original size or shape when external, deforming forces are removed

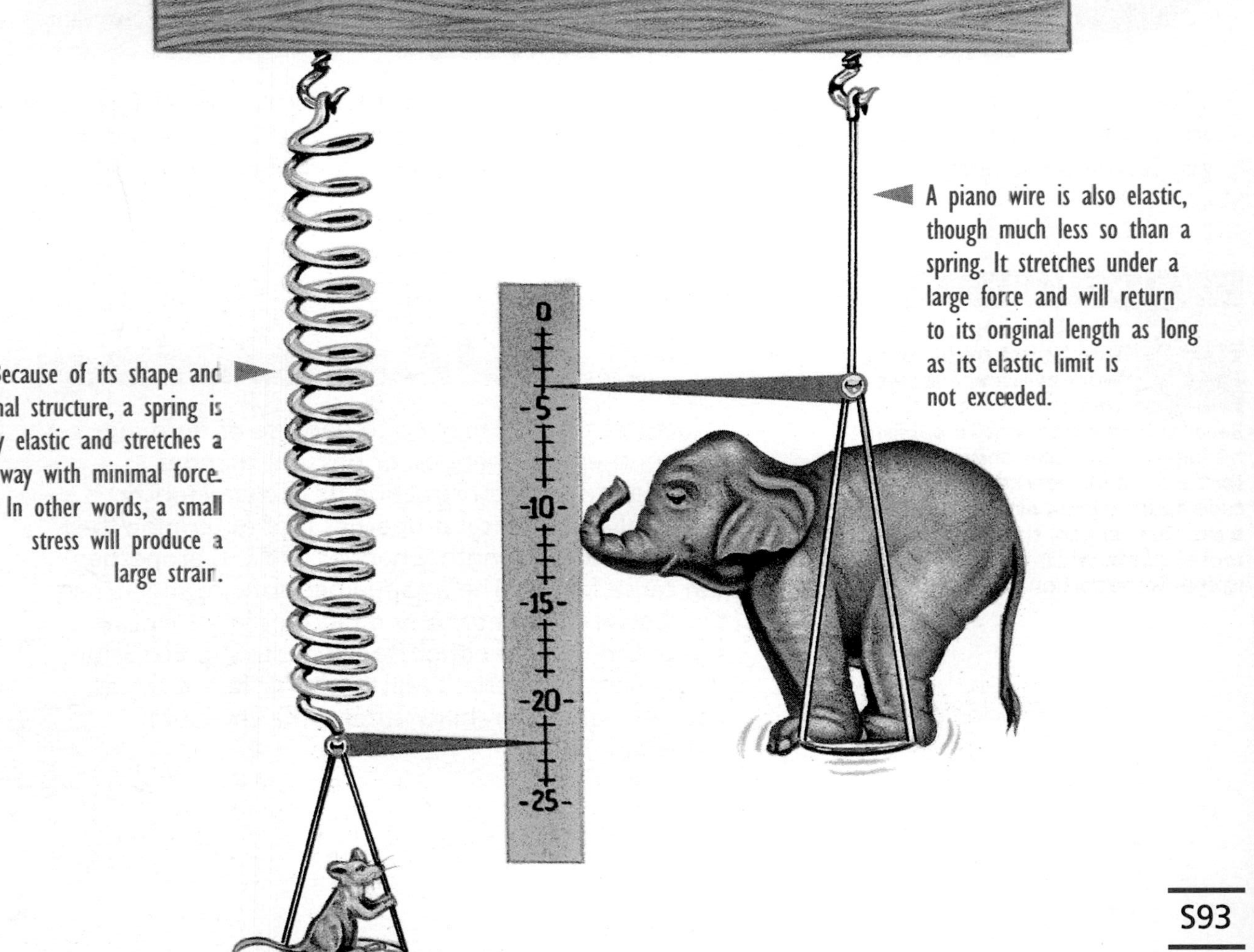

Because of its shape and internal structure, a spring is very elastic and stretches a long way with minimal force. In other words, a small stress will produce a large strain.

A piano wire is also elastic, though much less so than a spring. It stretches under a large force and will return to its original length as long as its elastic limit is not exceeded.

Robert Hooke, the scientist who discovered cells with the aid of his microscope, also spent time measuring the elasticity of many solids. He found that strain is directly proportional to the stress on the material, at least within certain limits. This relationship is now called ***Hooke's law***. The effects of stress and strain on different materials can be compared by the ratio of stress to strain, called the *elastic modulus.* If the elastic modulus is low, the material is not very elastic; if it is high, it can withstand more stress before reaching its elastic limit. Since the elastic modulus is unique for every solid substance, the materials that are most appropriate for use in construction can be readily determined.

Most materials behave differently under tension than under compression. Concrete, for example, has a higher elastic limit for compression than it does for tension. In other words, a column of concrete can support a heavy load, but a beam made of concrete will sag and eventually crack under a similar load. To avoid this problem, concrete is often reinforced or prestressed, as shown in the diagram below.

Steel, on the other hand, has a high modulus for both tension and compression, each of which are much higher than concrete. Why, then, would any designer decide to use concrete? One answer is in the cost. Steel would be much too expensive to use exclusively for all buildings and bridges. Concrete, on the other hand, is relatively inexpensive. It can also be poured into shapes at the job site and thus provides an efficient way to build a variety of structures. Can concrete be used in all situations? No. It takes a much larger column of concrete to support the same weight as a column of steel. In many cases, saving space is more important that cutting the cost of the material.

A beam made of concrete alone will sag under a heavy load and could break. Steel reinforcement rods help reduce the sagging. Concrete can be strengthened even more by holding steel rods under tension as the concrete is poured. Once the concrete has hardened, the external forces are released and the concrete is permanently compressed by the tension in the steel.

DID YOU KNOW...

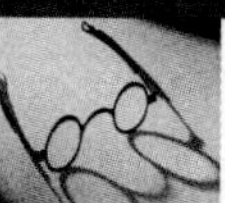

that Robert Hooke made sure he got credit for his work by using a secret code?

Because he wasn't ready to publish his ideas on the relationship of stress to strain, he included the code word *ceiiinosssttuv* in an article on another topic. Rearranged, the letters spell *ut tensio, sic vis,* which is Latin for "the stretch is proportional to the force."

SUMMARY

Structures that fail may do so because of unsound design, chemical changes, or physical changes. Materials used in construction have characteristic chemical and physical properties that determine their durability and strength. Changes in these properties can cause failure. The arrangement of and the attraction between the atoms or molecules in a solid substance can make it either fragile or strong. Elasticity determines a material's reaction to different forces. Cost, as well as durability, affects the choice of materials.

Architecture in Other Cultures

The development of design as seen in the structures of ancient Greece and Rome, Medieval and Renaissance Europe, and the Modern era represents a history of Western architecture. In other areas of the world, however, different architectural styles were developed by various cultures along separate lines.

While the scientific principles that govern the strength and durability of structures are the same in all locations and for every culture, a tremendous variety of designs have been used by different peoples through the ages. We will take a short tour of some of the architectural achievements from around the world.

The Byzantine Empire

In A.D. 330, Constantine the Great designated Byzantium as capital of the Eastern Roman Empire and renamed it Constantinople. This capital (now Istanbul, Turkey) became the center of the great Byzantine Empire. Although the Western Roman Empire collapsed in A.D. 476, the empire in the east lasted until the Turkish conquest of 1453. At its height, the Byzantine Empire included parts of Turkey, Italy, the Middle East, and North Africa.

Interior of Hagia Sophia. The Turkish government now maintains the church as a museum.

Architecture was the greatest form of Byzantine art. One of the architectural masterpieces of the ancient world is the church of Hagia Sophia, built in the sixth century A.D. The building was built in the form of a cross, measuring 72 m wide by 76 m long. Though the outside was composed of plain brick and mortar, the interior surface was once covered by intricate murals, mosaics, stone carvings, and metalwork. Insets of ivory, silver, and jewels were used to adorn the pulpit. A dome 56 m high and 33 m across dominates the former cathedral. Resting on massive columns instead of walls, the dome illustrates the talent of Byzantine architects who first solved the difficult problem of placing a round dome over a rectangular building. Byzantine architectural design was later adopted in Armenia, Italy, Russia, and Serbia, where it was modified to fit the various climates and building materials that were available.

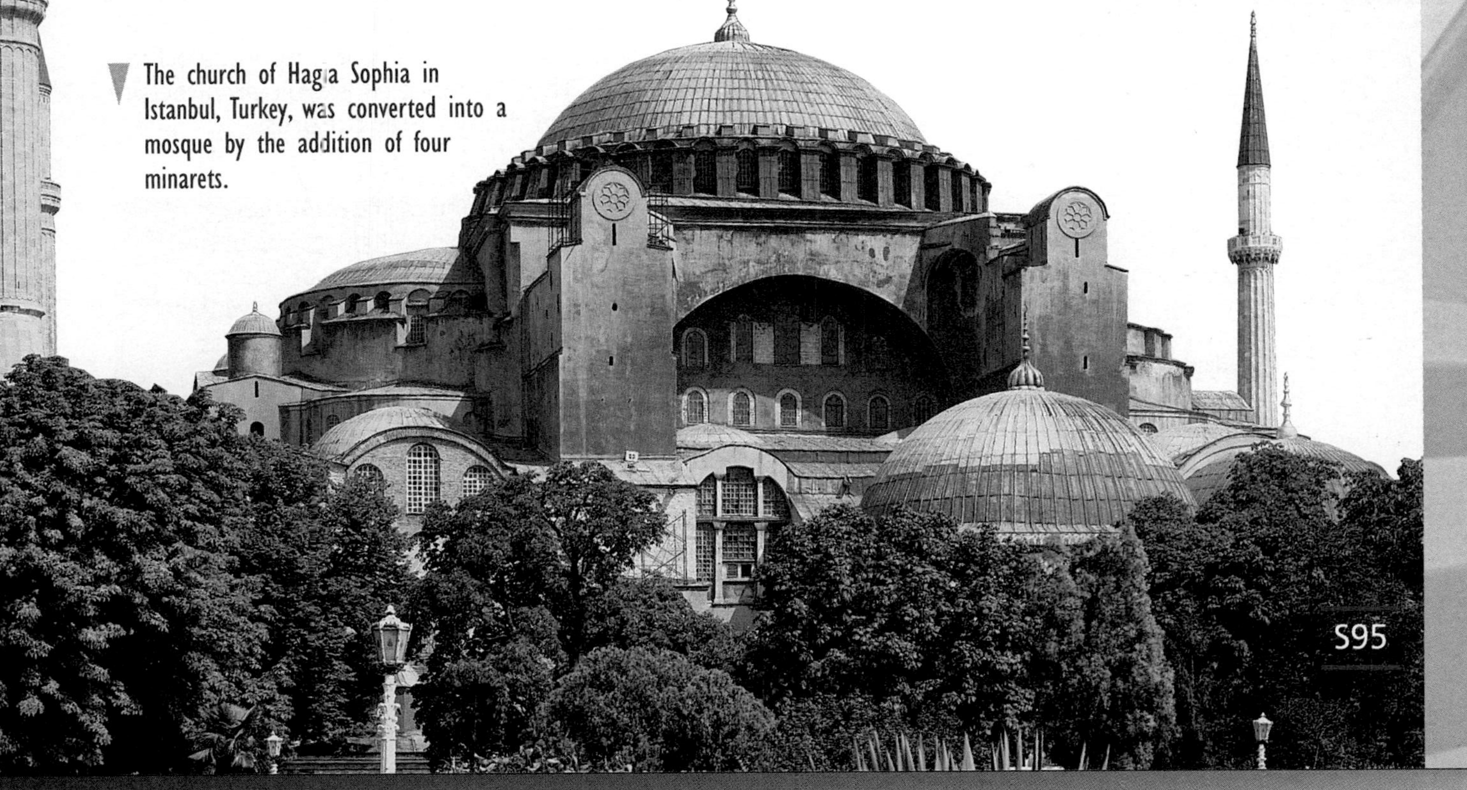

The church of Hagia Sophia in Istanbul, Turkey, was converted into a mosque by the addition of four minarets.

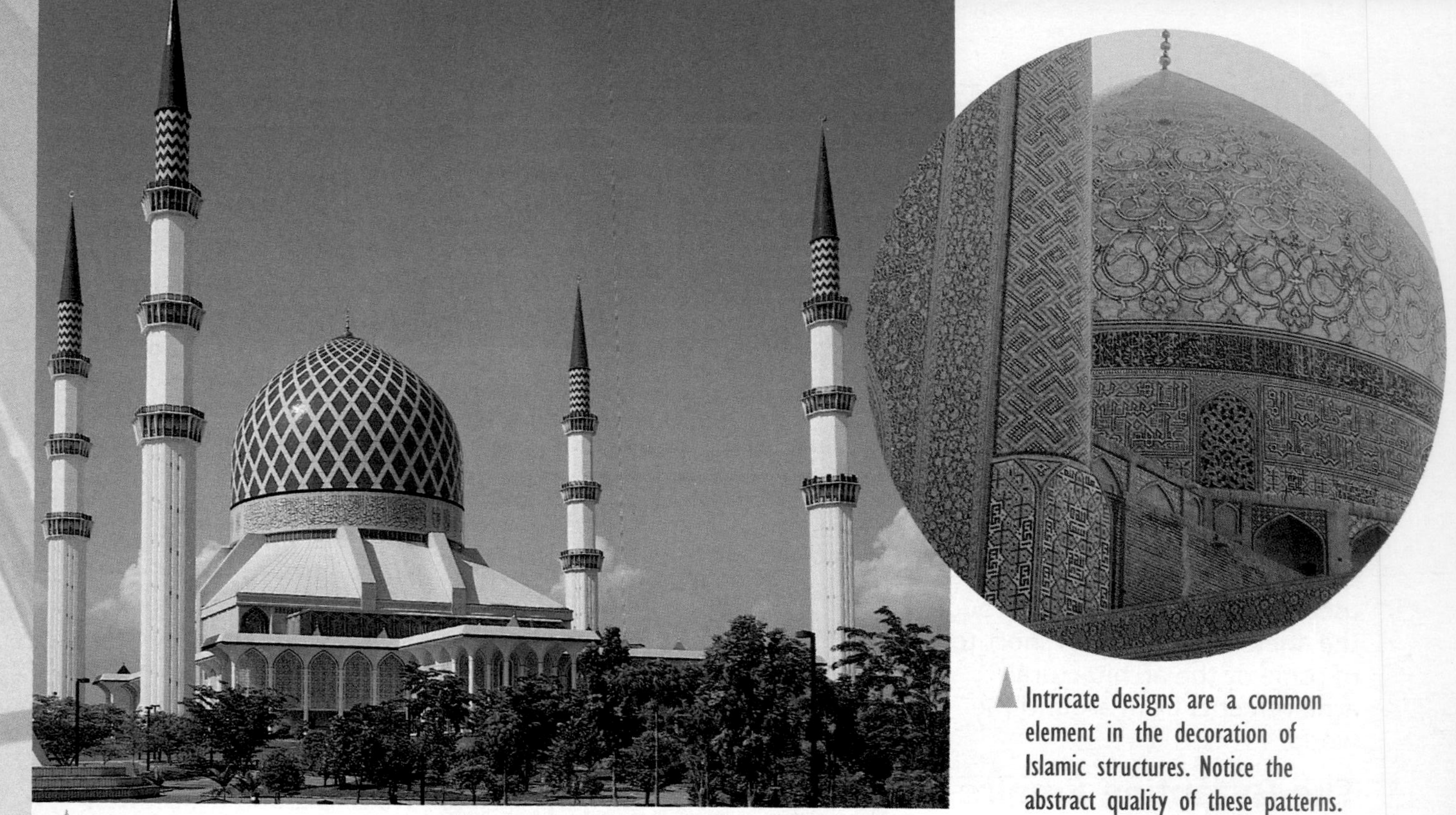

This beautiful mosque exhibits the typical dome and minarets that are characteristic of Islamic meeting places the world over.

Intricate designs are a common element in the decoration of Islamic structures. Notice the abstract quality of these patterns.

The Influence of Islam

Religion has influenced the development of architecture in the Islamic world, much as it has influenced architecture in Western culture. During the Middle Ages, the Islamic world stretched from Spain in the west to Samarkand and India in the east. The most important buildings to be designed were meeting places for prayer. These buildings are called **mosques.** Domes and *minarets* helped give the mosques a distinctive appearance. The minaret is a tall and slender tower from which the call to prayer is sung.

Islam forbids the use of statues and paintings of humans or animals as decorations in the mosques. Instead, mosques are decorated with designs based on flowers, plants, leaves, geometric figures, and Arabic inscriptions. The result is an unparalleled richness of abstract decoration. The influence of Islamic architecture can be seen in the structures of many other cultures.

Mosque

An Islamic place of worship and prayer

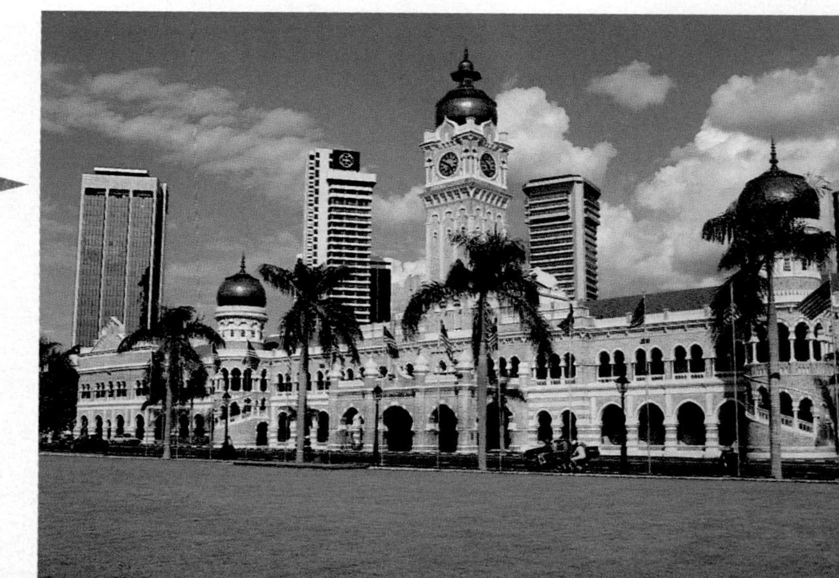

Nestled among modern skyscrapers, this government building in Kuala Lumpur, demonstrates the influence of Islam on the architecture of Malaysia.

India: A Contrast of Styles

India is the original home of two of the world's major religions—Buddhism and Hinduism. Much of Indian architecture and art reflect this religious influence. For example, **stupas,** or dome-shaped shrines, were built to hold artifacts and objects associated with Buddhism. The Indian stupa is a large mound-shaped structure covered by brick or stone and surrounded by a fence with elaborate stone gates.

A Hindu temple, on the other hand, is typically composed of a square building with heavy walls. The outer walls of the temple are covered with a multitude of colorful statues and carvings, and the ornamentation is typically more elaborate than that of Buddhist structures.

In 1506, the Mogul Empire rose to power after years of war and conflict in India. The most famous creation of the Mogul period was the work of the emperor Shah Jahan. In the mid-1600s, he built the Taj Mahal, a magnificent tomb for himself and his favorite wife. It is built of white marble on a platform of red sandstone and is studded with gems. Although it is an Islamic building, the Taj Mahal, as well as other Islamic structures in India, incorporated many architectural elements of Hindu temples. These elements include pinnacles, pavilions, and elaborate ornamentation.

Exterior of a Hindu temple in southern India, showing figures of various Hindu deities

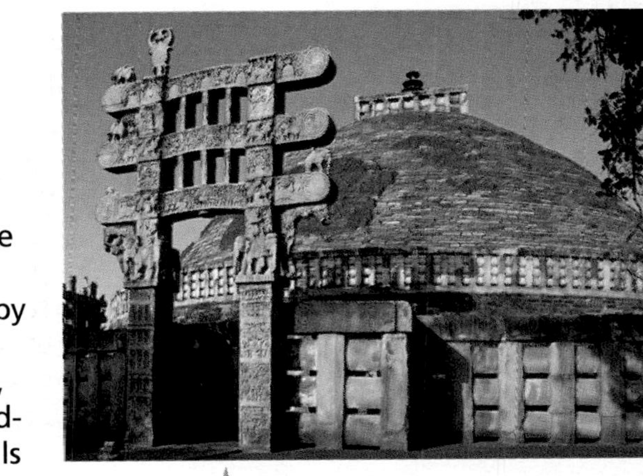

A stupa in central India. What do you think the carvings represent?

Stupa

A dome-shaped Buddhist shrine

DID YOU KNOW...

it took 20,000 workers more than 20 years to complete the Taj Mahal? Considered an example of perfect proportion, the main building is 57 m square with a dome 21 m in diameter and 37 m high.

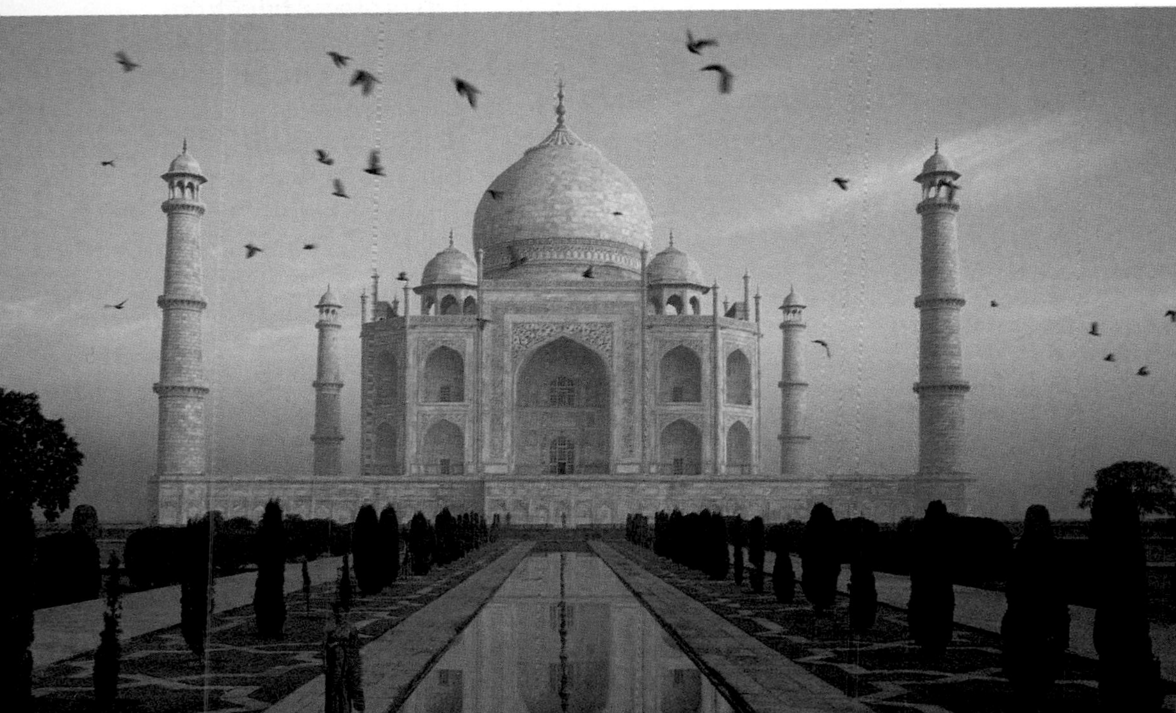

The Taj Mahal's exterior design typifies the blending of Islamic and Hindu cultures.

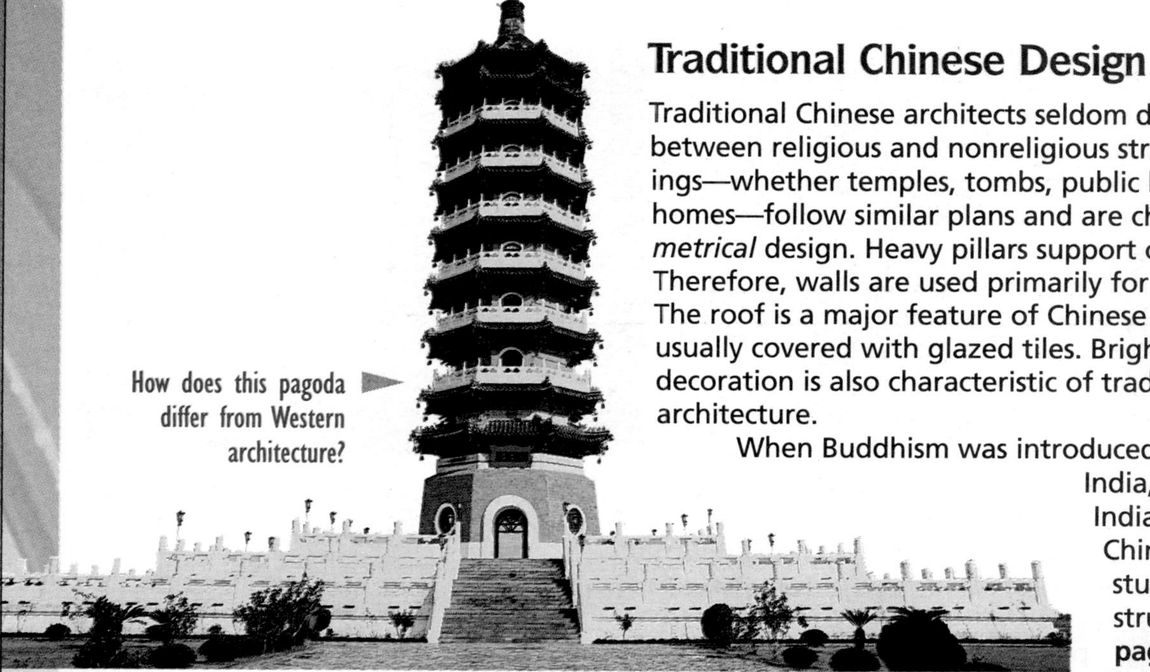

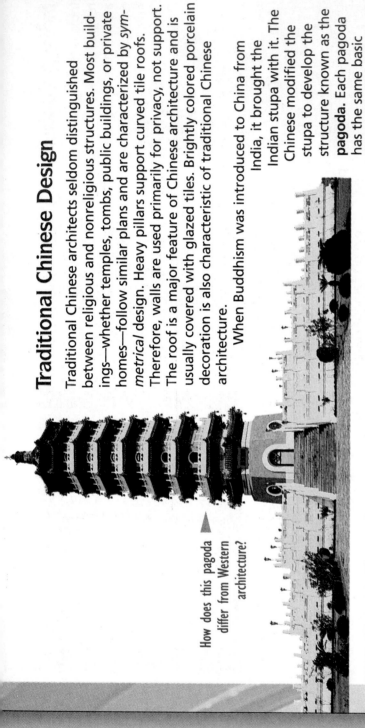

How does this pagoda differ from Western architecture?

Traditional Chinese Design

Traditional Chinese architects seldom distinguished between religious and nonreligious structures. Most buildings—whether temples, tombs, public buildings, or private homes—follow similar plans and are characterized by *symmetrical* design. Heavy pillars support curved tile roofs. Therefore, walls are used primarily for privacy, not support. The roof is a major feature of Chinese architecture and is usually covered with glazed tiles. Brightly colored porcelain decoration is also characteristic of traditional Chinese architecture.

When Buddhism was introduced to China from India, it brought the Indian stupa with it. The Chinese modified the stupa to develop the structure known as the **pagoda.** Each pagoda has the same basic design. It usually has eight sides and many levels, with each level being smaller than the one below it. Each level of the pagoda also has its own roof with long, upward-curving corners. Pagodas in China are usually constructed of wood, masonry, and highly glazed tile or porcelain.

Pagoda

A temple that is several stories high with floors of different sizes

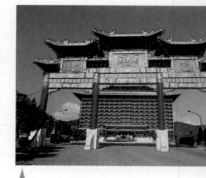

Chinese gateways, known as *pai-lous,* are built as memorials to great individuals. They usually have one or three openings and are constructed of the same materials as are pagodas. Pai-lous and pagodas are the most typical examples of traditional Chinese architecture.

A typical Chinese pai-lou

One of the most impressive feats of construction ever completed is the Great Wall of China. For several hundred years, roughly one-third of all ablebodied Chinese men were forced to help build the original wall to protect China from invaders from the north. With a foundation of granite blocks and sides of stone or brick, the inside is filled with earth. The wall is a massive structure that is nearly 8 m high and 5 m wide, with towers over 12 m tall. The longest structure ever built, it stretches for 2400 km east and west across northern China and is one of the few structures made by humans that can be seen from space.

The Great Wall of China took almost 2000 years to build. Started by the Emperor Shi Huangdi (221–206 B.C.), construction continued off-and-on until the 1600s. Even today, portions are kept in repair as a tourist attraction.

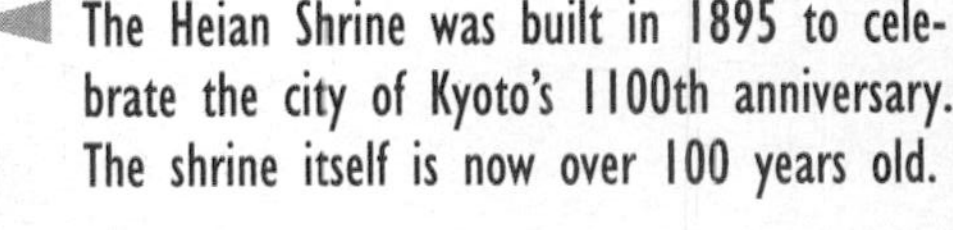

The Heian Shrine was built in 1895 to celebrate the city of Kyoto's 1100th anniversary. The shrine itself is now over 100 years old.

Yoyogi Stadium in Tokyo, built for the 1964 Olympics, retains the curving rooflines of traditional Japanese architecture.

Japan: A Natural Approach

The architecture of Japan is strikingly different from that of Western cultures. Much of Japanese architecture has its roots in Chinese designs. Temples and shrines were constructed in a style similar to that of Chinese structures, with curving roofs and supporting columns. Traditional Japanese houses are made of wood and often have roofs of thatch. However, the use of tile roofs is also common. The gently curved corners of Japanese roofs add a sense of graceful proportion that is often imitated in modern Japanese buildings, as shown in the photograph to the right.

The entrances to shrines are often marked by a structure called a **torii.** It is a ceremonial gate that consists of columns with a curved beam across the top. Though similar to the Chinese pai-lou, the torii is typically simpler in design. The torii in the photograph below marks the entrance to the Meiji Shrine in Tokyo. This shrine is dedicated to Emperor Mutsuhito, who opened Japan to the West and its new ideas.

Torii

A Japanese ceremonial gate

DID YOU KNOW...

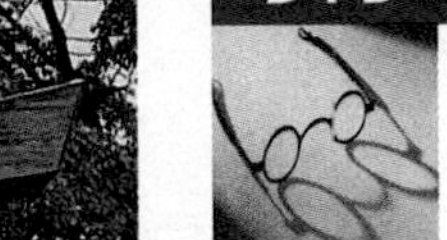

that from 1630 to 1853 Japan was virtually isolated from the rest of the world?

Japanese were forbidden to leave the country and foreigners were not allowed in. The only contact maintained was through visits by one Dutch ship per year.

The torii—gateway to Meiji Shrine in Tokyo, Japan

A Japanese garden

In addition to their religious shrines, Japanese designers also created beautiful gardens designed to imitate nature. These gardens are often miniature representations of the world, with rocks representing mountains, ponds representing oceans, and sand and gravel representing lakes and rivers.

Japanese temples, shrines, and houses are designed to blend in with their natural surroundings and thereby emphasize harmony between the buildings and nature. The Japanese broke from traditional Chinese architecture by using more asymmetrical designs and a less formal approach that more closely suggest the unusual formations found in nature.

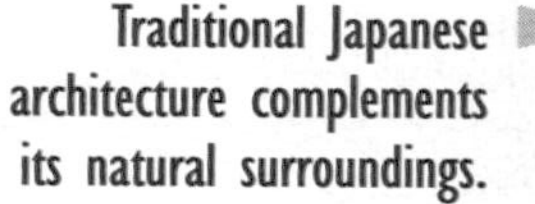

Traditional Japanese architecture complements its natural surroundings.

The Americas Before Columbus

The ancient civilizations of Central and South America developed some unique architectural styles. Although there were several distinctive cultures in these areas, three stand out for their achievements in what is called *pre-Columbian* architecture.

A model of the Aztec temple area at Tenochtitlán

Aztecs The Aztecs, who dominated the area of central Mexico at the time of European exploration, built great cities, pyramid temples, marketplaces, and palaces. Altars and plazas accompanied the pyramids as centers of Aztec religious ceremony.

When the Spanish conquistador Hernando Cortés discovered Tenochtitlán in 1519, it was one of the largest cities in the world. Tenochtitlán was a regularly planned city with straight streets. Built on an island in the middle of a lake, it was linked to the mainland by stone-lined causeways and bridges. Its buildings were constructed of rough stone that was plastered with polished and painted *stucco*. The original location of this grand city is now completely surrounded by Mexico City.

The temple of Quetzalcoatl at Teotihuacán

Near Mexico City, in Teotihuacán, an earlier culture built one of the greatest temple complexes of central Mexico. The people of Teotihuacán built the Pyramid of the Sun around A.D. 100. The base of this pyramid is larger than that of the Great Pyramid at Giza in Egypt. The Pyramid of the Sun was built with a core of clay, rock, and mud and was faced with stone slabs that were plastered and brightly colored. Other buildings in this complex were covered with ornate carvings.

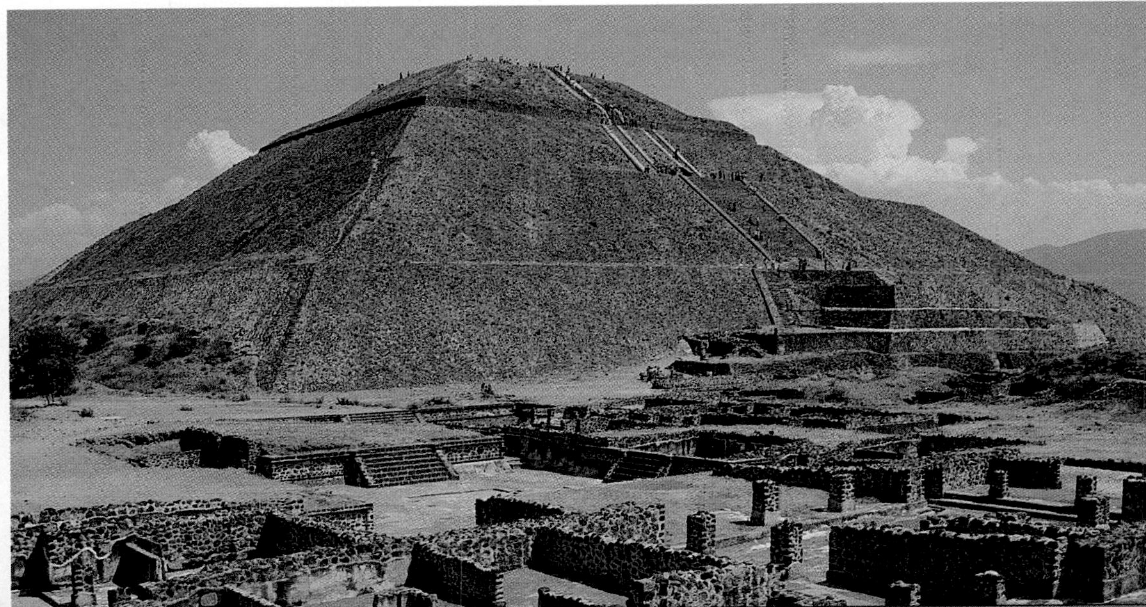

Pyramid of the Sun at Teotihuacán

Maya In southern Mexico and northern Guatemala, the Mayan culture built complex cities that included temples and other large public buildings. At the peak of their civilization, about A.D. 500, the Maya built great cities centered around tall pyramids of limestone that supported small temples at the top, as shown to the left. In addition to their architectural achievements, the Maya also devised a highly accurate calendar, a form of mathematics, and an intricate writing system.

A stone pyramid built by the Maya at Tikal, Guatemala

Incas In South America, the Incan empire covered the mountainous areas of what are now the nations of Peru, Ecuador, Bolivia, and Chile. The Incas built fortresses, irrigation systems, and paved roads. They used huge, cut stones to build the fortresses and palaces, many of which were located high in the mountains. Homes were made of stone and had thatched roofs.

One of the earliest examples of architecture in the area is the Gateway of the Sun at Tihuanaco. This archway, shown to the right, is made of two immense upright slabs of stone that support a 3 m carved stone *lintel,* or crosspiece. The lintel alone has a mass of approximately 10,000 kg. Perhaps the best known example of Incan architecture, though, is the mountaintop city of Machu Picchu.

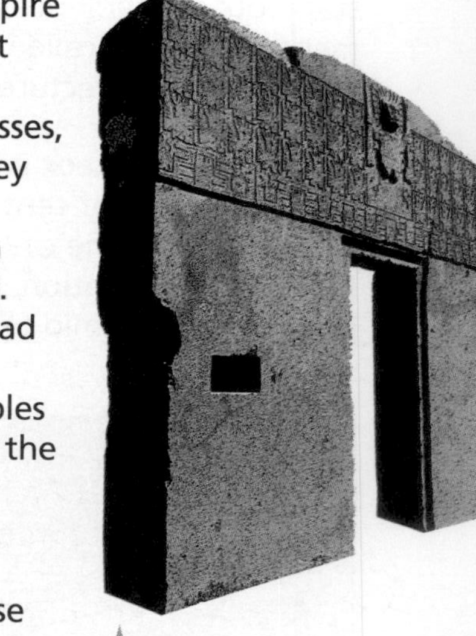

Gateway of the Sun, at Tiahuanaco, Bolivia

DID YOU KNOW...

that the Incas were among the world's finest stonemasons? They were able to cut stones so accurately that large blocks, weighing many tons, fit together without the need for concrete or mortar.

Machu Picchu, Peru. The ruins of this Incan city were not discovered until 1910.

S102 SOURCEBOOK UNIT 5 • STRUCTURES AND DESIGN

Grass-covered mounds (left) are all that remain of a North American culture whose architecture is illustrated in the drawing (right).

North American Tribes The influence of the Aztec and other Central American cultures also spread to North America. Tribes living along the Ohio and Mississippi Rivers built large cities and huge ceremonial mounds. The largest of these mounds was 30 m high and covered an area of about 16 acres (an area larger than 12 football fields). These mounds were used for burials as well as support for temples. Some mounds were built in the shape of birds, bears, mountain lions, and other animals.

In other parts of North America, various cultures with different kinds of architecture flourished. Many of these differences related to the climate and the availability of materials in these areas. In the hot, dry Southwest, Pueblo tribes used stone or **adobe,** a sun-dried brick, to build flat-roofed communal houses. Many of these houses were several stories high, and some were located under cliffs for protection.

How are these Pueblo buildings suited to their environment?

Since many North American tribes were nomadic, their architecture was developed around a need for mobility. Plains tribes, such as the Sioux, used buffalo skins to build cone-shaped *tepees.* These portable tents allowed the tribes to migrate with the buffalo herds. Tribes living in the eastern end of the continent used forest materials, such as leaves, reed mats, and bark, to cover a frame of poles. The *wigwam* of the Chippewa tribe, shown here, was such a structure. The Iroquois used similar methods to build rectangular long houses, some of which were 30 m long.

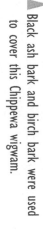

Black ash bark and birch bark were used to cover this Chippewa wigwam.

Adobe

Sun-dried brick made of clay used to build houses

The remains of the city of Kilwa, Africa

The Cities of Africa

The architecture of African civilizations is reflected in the development of urban centers. Dominated by walls and passageways, early cities—like Kilwa in Kenya and Kano in Nigeria—had marketplaces, palaces, wide streets, and other structures. The Hausa people, who lived in what is now Nigeria, built cities surrounded by high walls, a common feature in the architecture of Africa. As a town grew, its residents built walls around it. The first, outer wall probably served as a defense against attackers. A second wall often enclosed an area for food storage.

Ancient walled cities are found in what is now Zimbabwe.

One of the greatest fortresses was built in Zimbabwe in the 1300s. Shown in the photo to the right, it had 10 m high walls that were made of cut stone and laid in a variety of patterns. The wall is over 240 m long and consists of some 900,000 large granite blocks.

The development of cities initiated the smaller-scale development of homesteads, such as those of the Kuria along the Kenya-Tanzania border. Here, circular houses with cone-shaped thatched roofs were clustered together and connected by a tall fence that formed a secure yard. The number of houses in the homestead depended on the size of the family.

The use of urban space was as important to Africans as their use of walls. They tried to preserve something of a rural setting, even in the largest cities. Animals were allowed to wander about, and trees shaded streets and plazas. In some towns, people even raised crops along walkways or on small plots of land.

SUMMARY

Many kinds of architecture have developed in different countries around the world. The buildings and cities built by early cultures in Africa, Asia, India, and the Americas represent the skill and creativity of their designers. The architecture of each of these civilizations differs from traditional Western architecture, yet it fits each culture in function and form.

Unit CheckUp

Concept Mapping

Starting with the terms supplied here, construct a concept map that illustrates major ideas from this unit. Arrange the terms in an appropriate manner and connect them with linking words. Then extend the concept map by adding as many additional terms from the unit as you can. Use your ScienceLog. **Do not write in this textbook.**

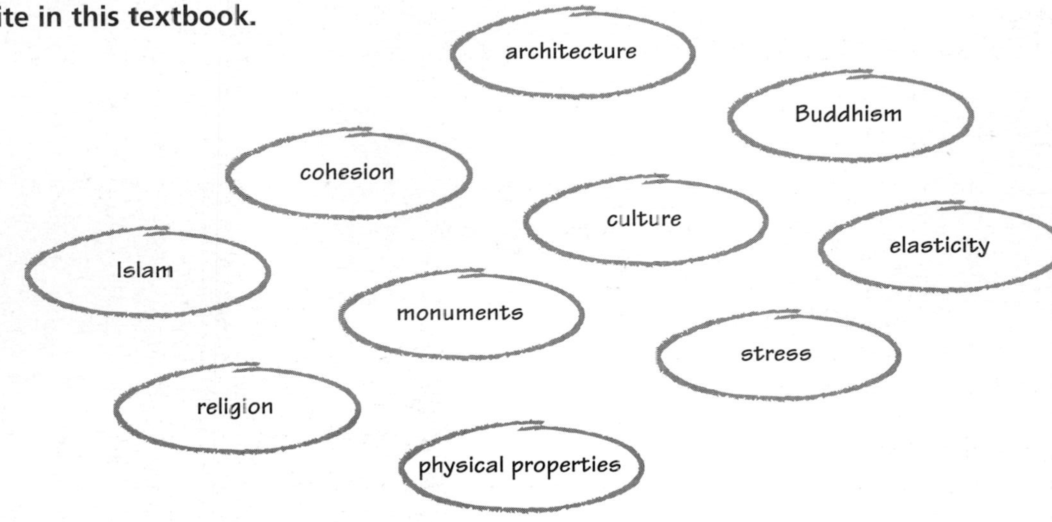

Checking Your Understanding

Select the choice that most completely and correctly answers each of the following questions.

1. Which of the following would NOT be considered a chemical change?
a. the oxidation of a steel arch
b. the rusting of a metal walkway
c. the cracking of a stained-glass window
d. the rotting of a wooden beam

2. Metal fatigue could be considered a failure of
a. cohesion.
b. adhesion.
c. tension.
d. Hooke's law.

3. The elastic limit of a solid material refers to
a. its ability to return to its normal shape.
b. the point beyond which it will permanently deform.
c. its ability to bend or twist under stress.
d. the way it handles tension and compression.

4. The Japanese *torii* was influenced by the
a. Indian stupa.
b. Chinese pai-lou.
c. Islamic minaret.
d. Incan archway.

5. The Taj Mahal represents a blending of which two architectural traditions?
a. Byzantine and Islamic
b. Chinese and Japanese
c. Indian and Islamic
d. Chinese and Indian

UNIT CHECKUP S105

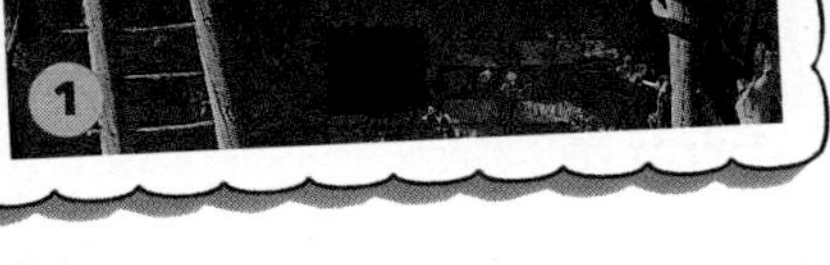

Interpreting Photos

Look at each of these photos. What architectural influences do you see in these buildings? In what countries might you find each of them?

Critical Thinking

Carefully consider the following questions, and write a response in your ScienceLog that indicates your understanding of science.

1. Exposure to oxygen in the atmosphere is a concern when using iron and steel to construct a building or bridge. Explain why this is so.
2. Diamond is the hardest known natural substance. Although it looks like glass, it is much more durable. How might the difference between these two materials be explained in terms of crystal arrangement?
3. If a certain material had a high elastic modulus for compression but a low elastic modulus for tension, would it be more suited for use as a cable or a column? Explain.
4. In what way does Japanese design more closely reflect nature than does Chinese design?
5. Byzantine architecture involved intricate design work on the inside of buildings while often leaving the outside very plain. How might this tendency be related to the culture of the people and the types of buildings that were usually constructed in this style?

S106

Portfolio Idea

Choose a structure from history, either from a Western architectural tradition or from that of another place in the world. Find out when the structure was built, where it is located, and how it was constructed. Then write an account of its design and construction from the viewpoint of someone involved—a laborer, architect, engineer, etc. Your account should include relevant factual information in the form of an interesting fictional story.

The Restless Earth

In This Unit

Now that you have been introduced to the movements of the Earth's crust and the rocks of which the crust is made, consider these questions.

1. What do earthquakes, volcanoes, and mid-ocean ridges have in common?

2. What is the rock cycle?

3. How are minerals related to rocks, and how can you identify them?

4. In what ways do rocks record the history of the Earth?

In this unit, you will take a closer look at the movements of the crust, the makeup of its rocks, and the kinds of information these rocks can give us.

S107

EARTH'S MOVING CRUST

If you take a look at a map of the world, you might notice that the shapes of the continents seem to resemble pieces of a puzzle. If you could move the pieces around, you might find that some of them almost fit together. A similar thought occurred to the German meteorologist Alfred Wegener in the early 1900s. He proposed that the continents were long ago part of one great landmass surrounded by a single gigantic ocean. Then, over time, the great landmass broke into separate pieces. These pieces eventually drifted apart to become the continents we know today.

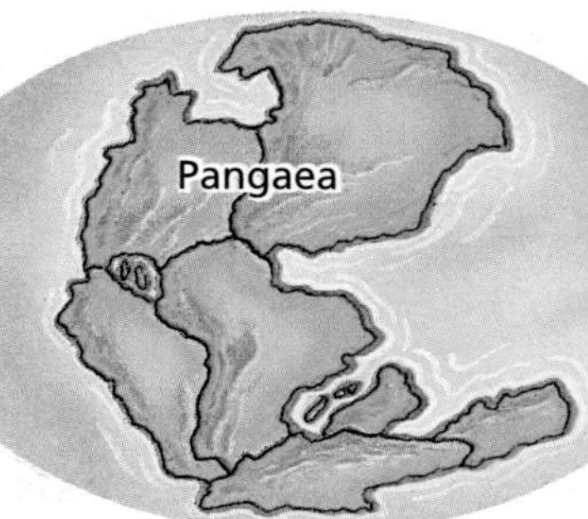

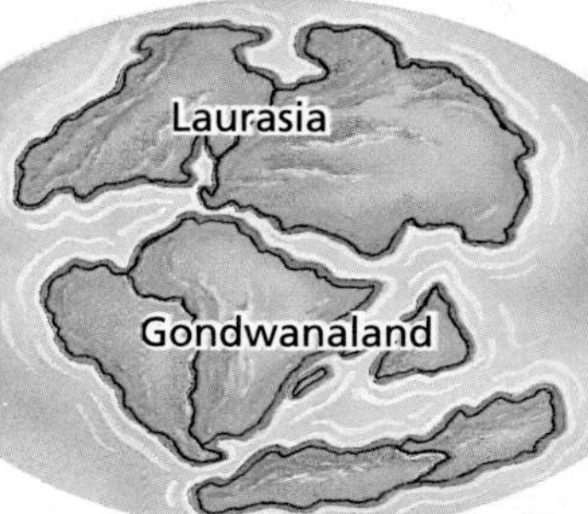

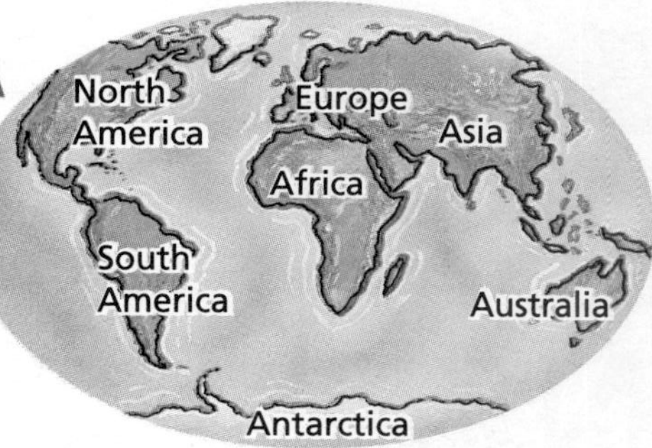

About 200 million years ago, Pangaea began to break apart. The modern continents then separated and moved toward their current positions.

A Supercontinent Breaks Up

Wegener called that great landmass *Pangaea* (pan-JEE-uh), Greek for "all Earth." He estimated that this vast supercontinent began to break up about 200 million years ago into two smaller continents, called Laurasia and Gondwanaland. As these landmasses moved apart, a new sea filled the area between them. Eventually, Laurasia separated into the continents of North America and Eurasia, while Gondwanaland split apart to become South America, Africa, Australia, Antarctica, and the subcontinent of India.

Wegener was able to provide three major pieces of evidence in support of his ideas. Not only did the shapes of the continents seem to "fit," but fossils of plants and animals discovered on either side of the Atlantic Ocean were very similar. He also noted evidence of glacial activity in locations that were much too warm for permanent ice. Therefore, he concluded, these areas must have been located in regions of the globe that experienced much colder temperatures than their current positions would allow.

When Wegener first proposed his ideas in 1912, many scientists were skeptical. It seemed ridiculous that huge masses of solid rock could somehow drift across the globe. Because of widespread opposition in the scientific community, Wegener's theory was discredited and all but ignored. Then in the mid-1960s, J. Tuzo Wilson of Canada came up with the idea that it wasn't the continents themselves that were moving; instead, he theorized, huge pieces of the Earth's crust were driven back and forth by forces within the planet. The theory by which the movement of these plates is explained is called **plate tectonics,** and the pieces of crust are named *tectonic plates*.

DID YOU KNOW...

that as far back as 1620 it was noticed that the continents bordering the Atlantic Ocean looked as though they might fit together? The English philosopher Sir Francis Bacon recorded such an observation almost 300 years before Wegener proposed his ideas.

Plate Tectonics

Scientists now realize that Wegener was on the right track. They know that great rifts under the oceans are spreading apart, producing new rock at their boundaries. Where plates come together, tremendous forces can cause mountains to form in the crust. For example, the Himalayas are a result of two plates—the Indian and the Asian—converging. Because the density of each plate is about the same, neither can sink beneath the other. The rock has nowhere to go but up, creating a mountain chain.

With the help of sophisticated equipment, scientists have confirmed that the Earth's surface is indeed moving. Using special sensors linked to satellites orbiting overhead, scientists can plot the exact location of two or more points in relation to one another. From the data they collect, it is evident that the Atlantic Ocean is growing wider by an average of about 2 cm per year as the plates on either side of the Atlantic move apart.

Two centimeters per year may not sound like much, but it adds up over time. A hundred million years from now, the continents will have a much different arrangement. The Atlantic, not the Pacific, may be the largest ocean. Asia and the Americas will be much closer together. Perhaps one day another supercontinent will form as the continents come together yet again.

Plate tectonics

The theory that describes the Earth's crust as divided into numerous rigid yet mobile plates

DID YOU KNOW...

that dinosaur fossils were recently discovered on Antarctica?
Now frigid and nearly lifeless, Antarctica must once have had a much warmer climate to have supported these giant reptiles.

▼ The Earth's crust is broken into several tectonic plates. Heat within the mantle creates convection currents that carry warmer rock up and cooler rock down. These currents form *convection cells* that may move the tectonic plates apart or together.

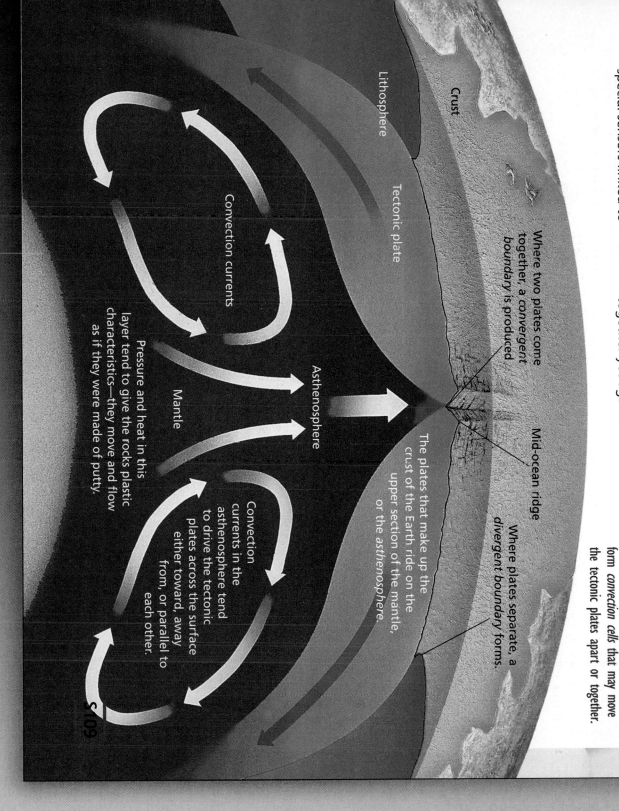

Where two plates come together, a *convergent boundary* is produced.

Mid-ocean ridge

Where plates separate, a *divergent boundary* forms.

Crust

Tectonic plate

Lithosphere

The plates that make up the crust of the Earth ride on the upper section of the mantle, or the *asthenosphere*.

Asthenosphere

Convection currents

Convection currents in the asthenosphere tend to drive the tectonic plates across the surface either toward, away from, or parallel to each other.

Mantle

Pressure and heat in this layer tend to give the rocks plastic characteristics—they move and flow as if they were made of putty.

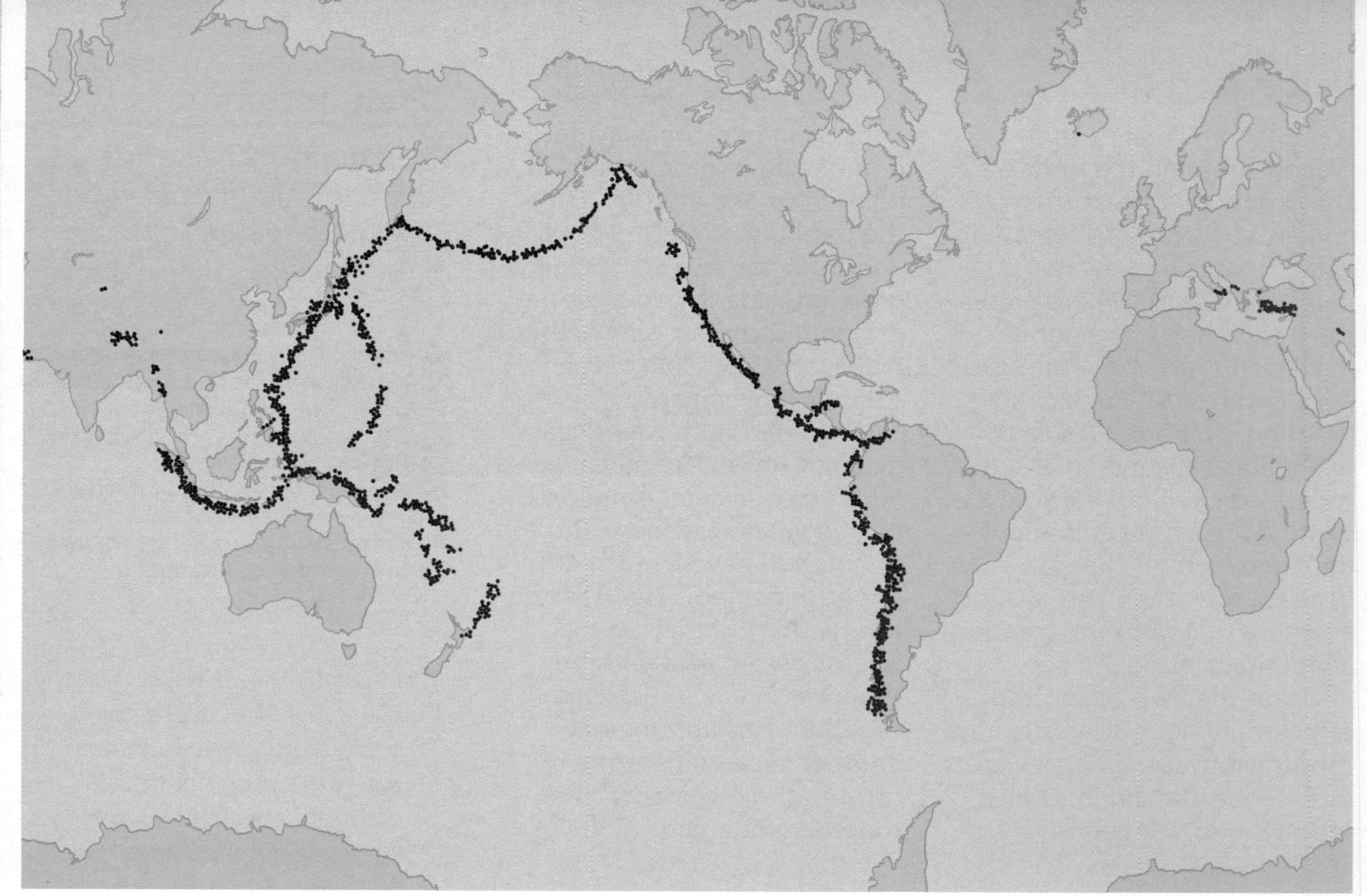

This map shows the locations of earthquake epicenters over a 10-year span. The locations of these earthquakes mirror the locations of the edges of tectonic plates.

DID YOU KNOW . . .

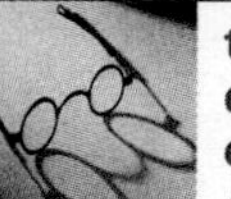

that over 2200 earthquakes occur each and every day?
As the crust of the Earth continually adjusts, tectonic plates bump and rub, causing earthquakes, most of which cannot be felt by humans.

Earthquakes

Few natural events are as destructive as earthquakes. In just a few seconds, a powerful earthquake can reduce a city to rubble and take tens of thousands of human lives. Earthquakes occur almost anywhere, but they seem to occur more often in certain areas. Why is this?

You know that most earthquakes are caused by the release of tension between two plates that are moving against each other. That is why most earthquakes take place along the boundaries of tectonic plates. Notice on the map above that the locations of these earthquakes follow the outlines of the plates you saw in the map on page 358 of your text.

Waves in Rock Earthquakes produce several types of waves. The fastest waves are called *primary waves*, or P waves. These waves are longitudinal waves, which move the ground back and forth like pulses traveling through a spring. They are actually ultralow-frequency sound waves that can travel through both solid and liquid parts of the Earth's interior.

Secondary waves, or S waves, are transverse waves that move the ground up and down or side to side. However, S waves can travel only through the solid parts of the Earth and therefore are stopped by the molten outer core. Because S waves travel slightly slower than P waves, they take longer to travel from the focus (origin) of an earthquake to any other location.

Seismologists determine how far away an earthquake was from their monitoring station by measuring the difference in arrival times between P and S waves. However, just knowing the distance does not locate the earthquake. It could be anywhere on the circumference of a circle with a radius equal to that distance. Even the information from two seismographs is not enough. It takes data from three stations to pinpoint the epicenter of an earthquake. Once three circles are plotted, the epicenter of a quake can be determined. The epicenter is the spot on the surface of the Earth directly above the focus. The earthquake originated below the spot at which all three circles intersect at a single point.

When P and S waves meet at the surface, they form another type of wave called a *surface wave*. Surface waves cause the ground to heave and roll like the sea in a storm. These are the waves that usually do the most damage to buildings and other structures on the surface.

Earthquake Warnings

Predicting earthquakes is a tricky business. Although we know that certain areas of the crust are under more stress than others, scientists have not been able to determine the exact point where this stress will be released or when. But as seismologists gather more and more information, the science of earthquake prediction continues to improve. Someday, seismologists hope to be able to issue earthquake warnings just as meteorologists issue severe weather warnings today.

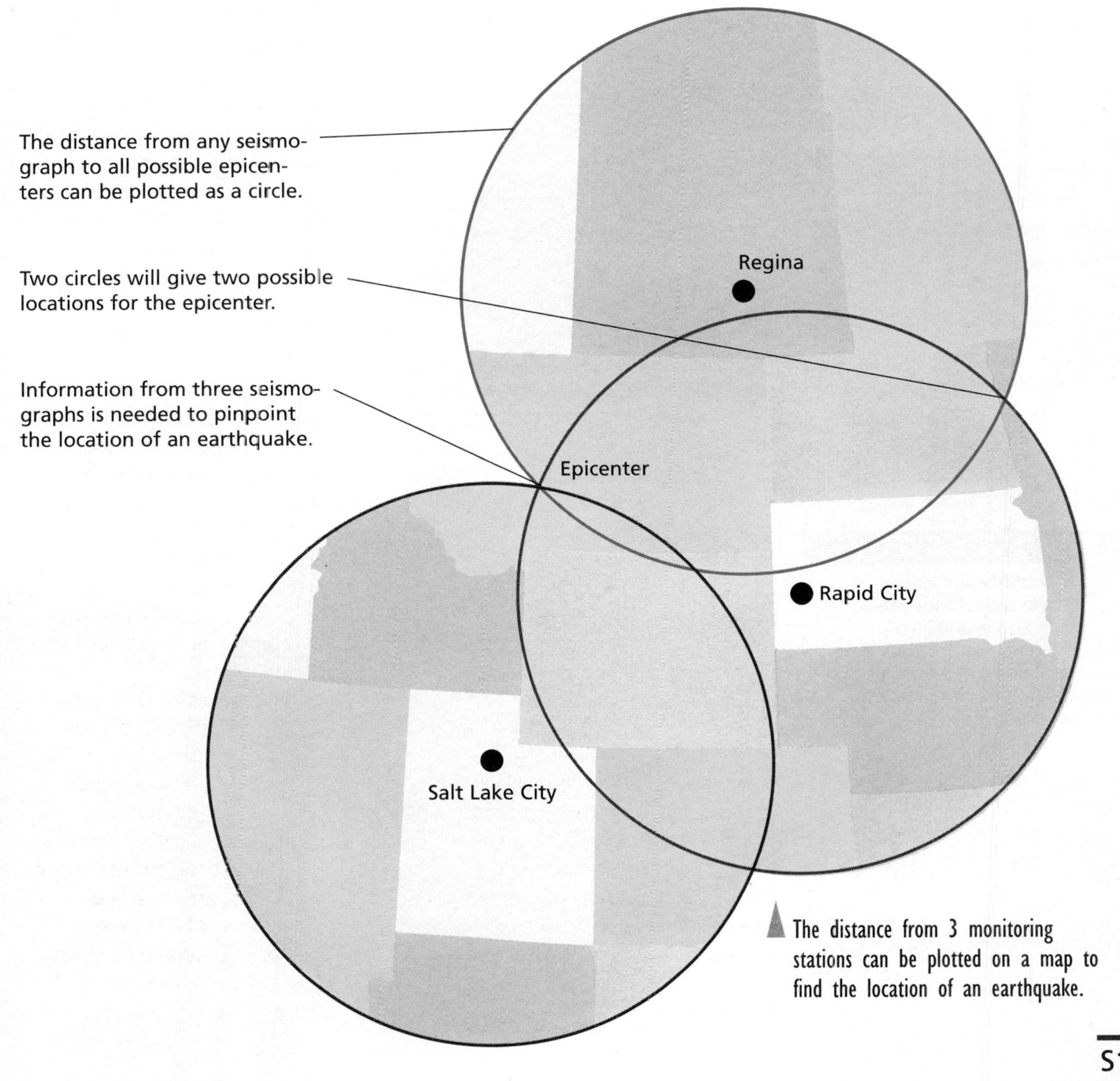
The distance from any seismograph to all possible epicenters can be plotted as a circle.
Two circles will give two possible locations for the epicenter.
Information from three seismographs is needed to pinpoint the location of an earthquake.
Regina
Epicenter
Rapid City
Salt Lake City

The distance from 3 monitoring stations can be plotted on a map to find the location of an earthquake.

Belching Fire

About 3500 years ago, an eruption of almost unimaginable violence ripped apart the island of Thera in the Mediterranean Sea, wiping out every living thing on this and nearby islands. Large rocks rained down on the island of Crete 125 km away. Untold thousands of people were swept away by huge sea waves generated by the eruption. The explosion, which may have been the loudest noise in recorded history, was heard thousands of kilometers away. This cataclysm is thought to have been responsible for the destruction of the Minoan civilization and is probably the source of ancient legends about the lost continent of Atlantis.

Volcano

An opening in the Earth's surface through which lava and other materials erupt

What causes volcanoes to erupt? Scientists have determined that volcanoes are associated with the movement of tectonic plates. In fact, about 300 volcanoes encircle the Pacific Ocean along converging plate boundaries in what has been called the "Ring of Fire." As plates make contact, molten rock, or *magma*, forms as one plate moves under the other. Masses of magma may then rise to the surface and erupt.

Types of Volcanoes In some volcanic eruptions, lava simply pours from an opening in the crust. Hardened lava gradually builds up, forming a mountain with gently sloping sides. This type of volcano is called a *shield volcano* because its shape is flat and round, like a shield. Shield volcanoes form from lavas that flow very easily because they have a comparatively runny consistency.

Not all eruptions are as gentle as those that form shield volcanoes. If the magma happens to be thick, pressure from trapped gases within the magma may build up, causing an explosive eruption to take place.

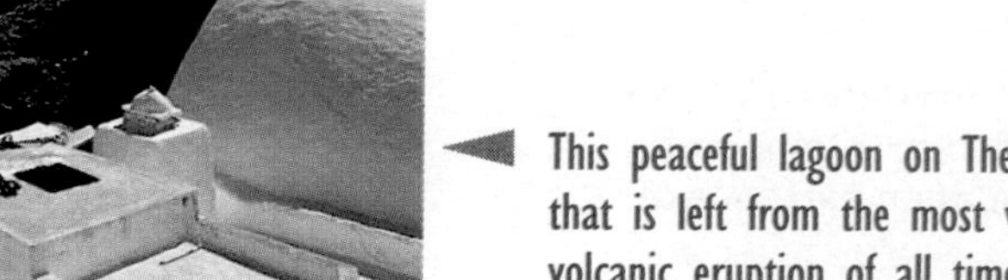

◄ This peaceful lagoon on Thera is all that is left from the most violent volcanic eruption of all time.

DID YOU KNOW...

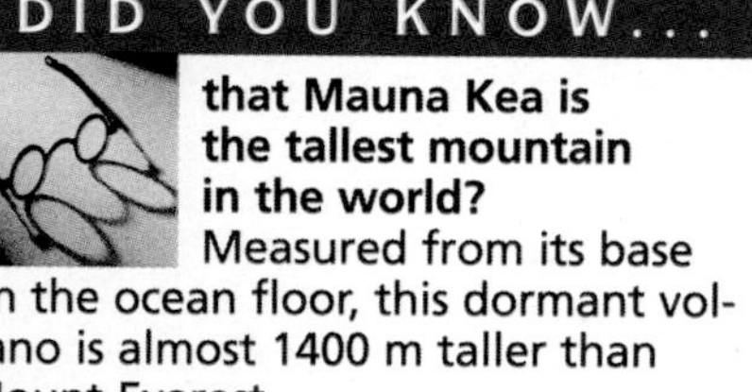

that Mauna Kea is the tallest mountain in the world? Measured from its base on the ocean floor, this dormant volcano is almost 1400 m taller than Mount Everest.

Mount Fuji, the highest mountain in Japan, is a dormant composite volcano. It last erupted in 1707.

Such an explosion can spew lava and *tephra*—volcanic rock, cinders, and ash—several kilometers into the air. This type of eruption creates a *cinder cone*—a volcano with steep sloping sides and a narrow base. Once the pressure is released, this type of volcano usually becomes inactive. Made of loose materials, cinder-cone volcanoes erode easily and, geologically speaking, do not last very long.

Some volcanoes, called *composite volcanoes*, are formed of alternating layers of lava and tephra. Many well-known volcanoes, such as Vesuvius and Mount St. Helens, are composite volcanoes. They usually have steep slopes and broad bases. Because of their layered construction, composite volcanoes are sometimes referred to as *stratovolcanoes*.

Mid-Plate Volcanoes

Occasionally, volcanoes can be found far from the edge of a tectonic plate. For example, the famous volcanoes on the island of Hawaii are located in the middle of the Pacific plate. What causes these volcanoes?

Paricutín, in Mexico, is a cinder cone that grew to a height of 100 m in its first five days of eruption in 1943.

Mauna Kea, a shield volcano, rises 10,203 m from the sea floor to its summit.

S113

Scientists hypothesize that convection currents within the asthenosphere create hot spots, where columns of rising magma can erupt through the crust.

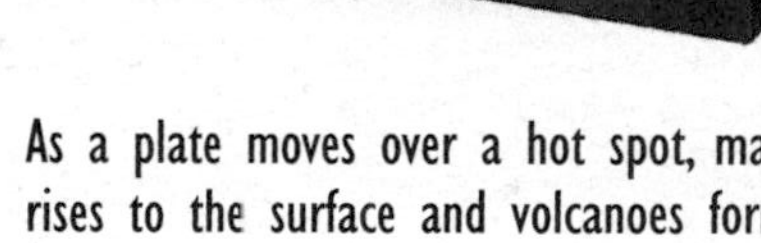

As a plate moves over a hot spot, magma rises to the surface and volcanoes form.

Hot spots stay fixed in place while the plates move slowly over them. Because of this, hot spots produce a series of volcanoes.

One clue is that the Hawaiian Islands get older from southeast to northwest. The youngest, and only volcanically active, island is Hawaii. It is also the largest island in the chain. The oldest islands in the Hawaiian chain are submerged seamounts far to the northwest, near Alaska. Geologists reason that molten material rising out of the Earth's interior is causing the Hawaiian volcanoes to form. Geologists call the rising plume of molten material a *hot spot*. As the Pacific plate moves over the hot spot, volcanoes continuously form. These mid-plate volcanoes, therefore, are also referred to as hot-spot volcanoes. A given volcano is active for only a short time, geologically speaking, before it is carried away from the hot spot and becomes extinct.

Hot spots can also occur under continents. When this happens, the result can be seen in the form of geysers, hot springs, and other thermal phenomena. Yellowstone National Park, with its famous geysers, is located over such a hot spot. Since the North American plate moves northwesterly, can you predict where new geysers would form?

DID YOU KNOW...

that a new volcano is currently forming southeast of the island of Hawaii? This undersea volcano, named Loihi, will become the next island in the chain once it reaches the surface of the ocean.

SUMMARY

Alfred Wegener proposed that all of the continents were once part of a great supercontinent that broke into separate pieces and drifted apart. Further research revealed that the surface of the Earth is divided into a number of tectonic plates that move across the Earth's surface, driven by convection currents within the mantle. The interaction of these plates gives rise to earthquakes and volcanic eruptions. Three types of volcanoes—shield, cinder cone, and composite—are formed, depending on the type of volcanic material involved. Hot spots give rise to mid-plate volcanoes and geysers.

Rocks and the Rock Cycle

Have you ever picked up a rock because of its interesting shape, appearance, or color? Perhaps you have a rock collection. If so, you already know something about the variety of rocks that can be found on Earth. You have learned by now that rocks are the materials that compose the Earth's solid crust. Sometimes these rocks form spectacular natural structures, such as the one pictured in the photo below. In addition to being interesting, rocks are important to us both scientifically and economically. They not only reflect the history of the Earth, but also contain valuable resources for our increasingly technological society.

DID YOU KNOW...

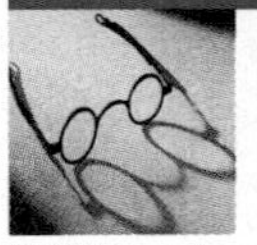

that a certain rock in Saudi Arabia commands the daily attention of millions of people? Within the *Ka'bah*—the most holy shrine of Islam, which worshipers pray toward five times a day—is the Black Stone of Mecca (al *Hajar al-Aswad*), an igneous rock of great religious significance.

The Formation of Rocks

Scientific studies indicate that the Earth was probably once made mostly of molten materials. As the Earth cooled, the materials at the Earth's surface solidified into the first rocks. These rocks were then exposed to forces that caused them to change. Other rocks developed as a result of these changes. By studying rocks and the forces that change them, geologists have identified three basic processes by which rocks form. The activity on the following page simulates these processes.

Devil's Tower in eastern Wyoming

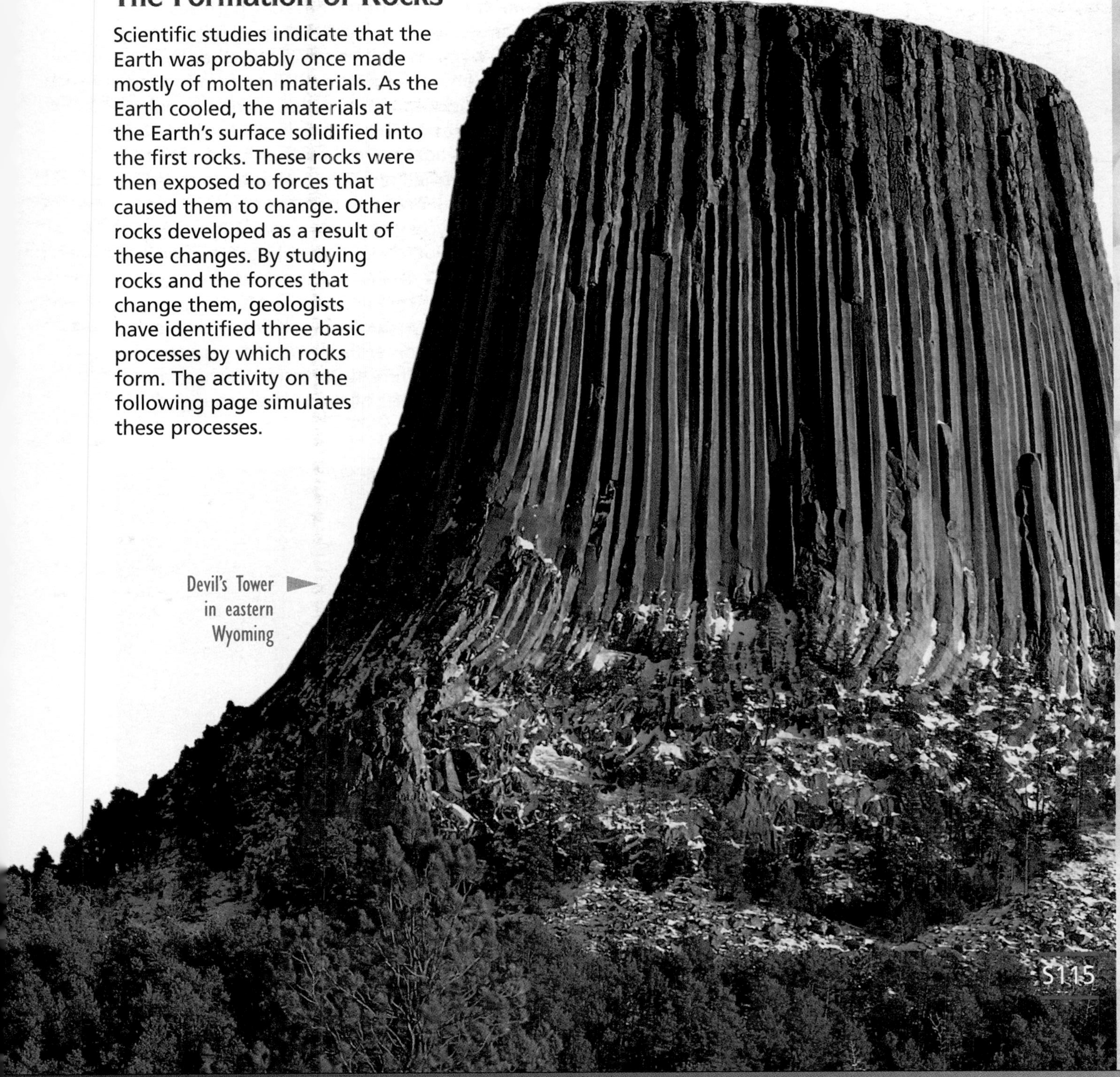

S115

The dark igneous rocks of the Pacific islands are formed of solidified magma.

Imagine using candle wax to simulate the process of rock formation. Take some old candles and melt them in a pan. If you pour the melted wax onto paper or into a mold and allow it to cool, the liquid wax will gradually harden into a solid mass. In a similar manner, rocks solidify from magma. Rocks that form when molten rock cools are called *igneous rocks*. Devil's Tower, shown on the previous page, is made of igneous rock.

Now imagine breaking off some small pieces of the solid wax. If you place a pile of these pieces under a heavy object, such as a book, they should eventually stick together, forming a single piece. In the same way, most *sedimentary rocks* form from rock particles that were broken up by weathering processes. Deposits of these particles stick together by cementation or compaction. When forming sedimentary rocks, larger particles, such as pebbles, sand, and silt normally are "glued" together by a cementing agent. On the other hand, if the particles are small enough, pressure alone can cause them to *lithify*, or become stone.

Finally, if you warm the remaining solid wax slightly, you should be able to bend and squeeze it. In a similar fashion, the heat and pressure inside the Earth change both igneous and sedimentary rocks into *metamorphic rocks*. Sometimes the minerals in the rocks are rearranged into a different pattern. In other cases, the mineral content of the rocks is changed.

The mud of this dry lake bed may eventually become sedimentary rock. The mud cracks here may also be preserved in the rock that forms.

The Rock Cycle

Igneous, sedimentary, and metamorphic rocks are formed through individual steps of an endless process that has been going on for billions of years. Old rocks are constantly made into new rocks by a continuous process called the **rock cycle**. A simplified version of the rock cycle is shown below. As you examine the diagram, consider that any rock you see has been through at least one pathway in this cycle.

Rock cycle

The continuous process of change in which new rocks are formed from old rock material

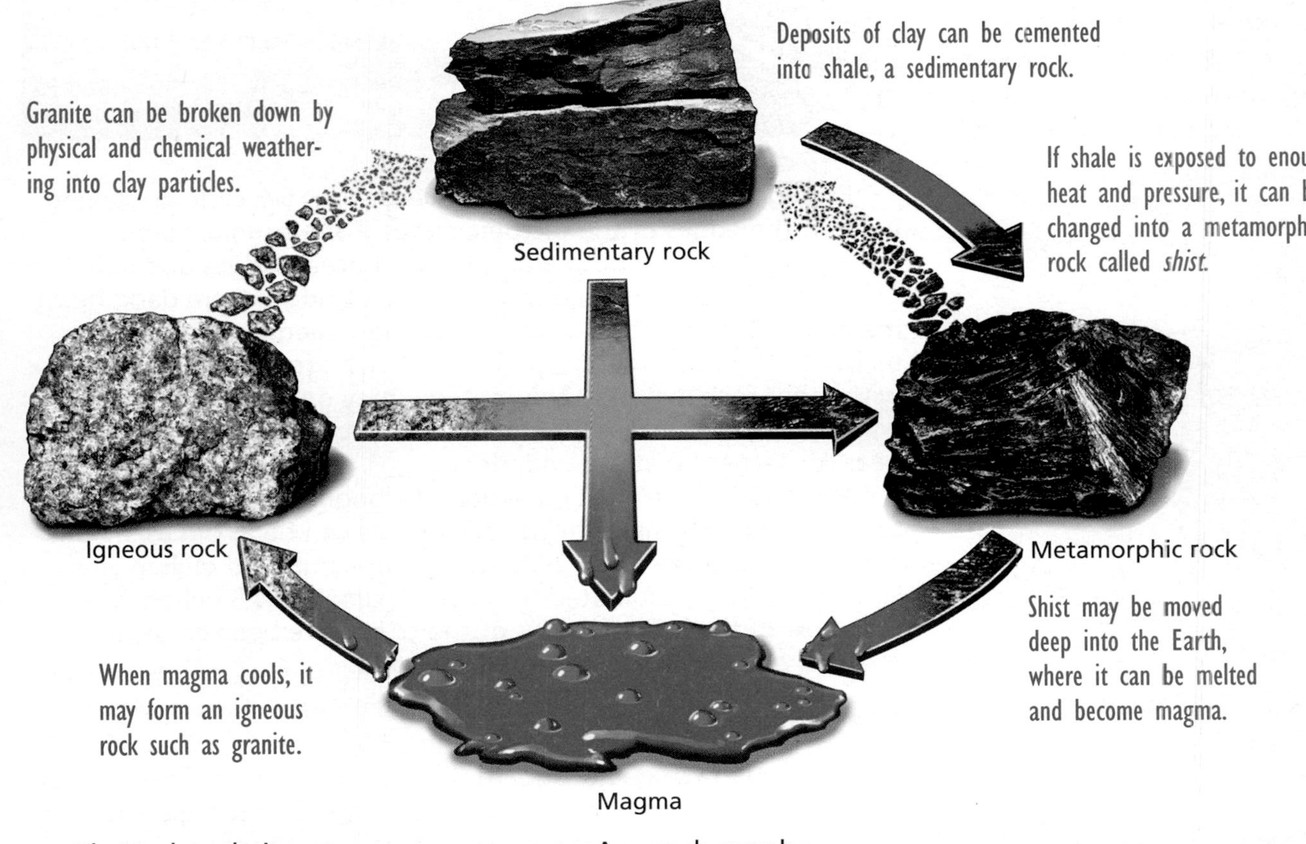

Sedimentary rock
Deposits of clay can be cemented into shale, a sedimentary rock.
Granite can be broken down by physical and chemical weathering into clay particles.
If shale is exposed to enough heat and pressure, it can be changed into a metamorphic rock called *shist.*
Igneous rock
Metamorphic rock
Shist may be moved deep into the Earth, where it can be melted and become magma.
When magma cools, it may form an igneous rock such as granite.
Magma

The rock cycle is not a one-way process. Any rock may be changed into any other type of rock by going through the cycle in various ways. But what other pathways are there? Suppose that granite is not broken down into sand but is exposed to high temperature and extreme pressure instead. In this case, the granite could become a metamorphic rock called *gneiss*, an example of which is shown to the right. The granite itself may also be remelted to become magma.

Gneiss is a metamorphic rock that is formed when certain types of rock are exposed to high temperature and pressure.

Though new igneous rock is continually being formed on Iceland (left), a volcanic island located along the mid-Atlantic ridge, the Canadian Shield (right) is composed of bedrock that is millions of years old.

Evidence suggests that major parts of the rock cycle are taking place at the edges of the tectonic plates. For example, magma comes to the surface both along the mid-ocean ridges and near ocean trenches. New igneous rocks form at ridges from dark, heavy magma. In the trenches, old rock is being pulled down into the mantle. There, the lighter rocks of the crust are melted and mixed with other rock material. This material may once again make its way back to the surface at a mid-ocean ridge or in volcanic belts near convergent plate boundaries.

The rocks of the crust have gone through the rock cycle many times. While rocks near the Earth's surface can be as old as 3.5 billion years, most are rarely older than a few hundred million years. Since the Earth is estimated to be approximately 4.5 billion years old, most of the original rocks of the crust have been changed by the rock cycle since they were first formed.

Classifying Rocks

Although these rocks look different, they belong to the same rock family.

You have many characteristics that distinguish you as a person. For example, you have two arms and two legs, you have hair, you have a personality, and you have the ability to speak, read, and write. But even though you share these traits with other humans, you are still different from every other person on Earth. You are unique!

Rocks also have definite characteristics. For example, look at the photographs on the left. You can tell that the objects in these photographs are rocks. But what is it about these objects that makes them rocks? Just as all humans share certain characteristics, all rocks share certain characteristics as well.

Like living organisms, rocks can be classified by their similarities and differences. One method of classifying rocks is by the way they are formed. On this basis, rocks can be classified as igneous, sedimentary, and metamorphic. In addition, within each of these groups, rocks can be further classified and named on the basis of their origin, texture (which refers to crystal or particle size), and mineral content.

Igneous Rocks As you know, there are two basic types of **igneous** rocks—*intrusive* igneous rocks and *extrusive* igneous rocks. These names indicate *where* the rocks are formed. Intrusive igneous rocks form deep underground in bodies of magma that squeeze in between, or *intrude* into, layers of other rocks. Long periods of time are required for intrusive rocks to solidify because the heat of the magma dissipates very slowly underground. Extrusive igneous rocks, on the other hand, are formed from *lava*. Lava forms when molten rock exits, or *extrudes*, from inside the Earth through openings such as volcanoes. Above ground, lava cools quickly to form solid rock.

Although intrusive and extrusive rocks are made of the same types of minerals, they can be easily distinguished by their texture. These two types of rock have very different appearances because of their different textures. For example, look at the two rocks shown here.

Mineral content affects the color and density of igneous rocks. Basalt, for example, is dark and heavy because it is composed of minerals containing large amounts of the elements iron and magnesium. Gabbro, which is also dark gray, is made of the same minerals as basalt. Basalt and gabbro are both igneous rocks formed from magma that comes from deep inside the Earth. Basalt and other dark extrusive rocks form the crust on the ocean bottom. Lightcolored rocks, such as granite and rhyolite, contain elements such as calcium and sodium that make them less dense than basalt and gabbro. Minerals containing these elements are very common in rocks of the continental crust.

Magma from deep within the Earth often comes to the surface as lava. As the lava cools, igneous rock is formed.

If you look closely, you can see that granite, an intrusive igneous rock, is composed of large visible crystals. The slow cooling of intrusive rocks allows time for large crystals to grow. Granite is therefore said to have *coarse texture.*

On the other hand, if an igneous rock cooled quickly, it feels fairly smooth and may not have any visible crystals. Geologists say that such rocks, like the rhyolite shown here, have *fine texture.*

Notice the coarse-grained texture of this sample of gabbro, an intrusive igneous rock.

Igneous rocks

Rocks produced by the cooling of molten rock either at or below the Earth's surface

S119

Sedimentary rocks

Rock produced by cemented or compacted rock particles that are deposited by wind, water, ice, or chemical reactions

Sedimentary Rocks As you know, most **sedimentary rocks** are made from sediments, which are pieces of other rocks that have been deposited on the Earth's surface by wind, water, or ice. Sedimentary rocks are classified by the type of sediments they contain. There are three different kinds of sediments that can become sedimentary rocks—clastic, chemical, and organic. Therefore, there are three major types of sedimentary rock.

Clastic sedimentary rocks are made of broken pieces of rock. These pieces can be large and rounded pebbles, sharp-edged chunks, or small particles such as sand, silt, or clay. Clastic rocks are often found in layers because of the way they are formed. When agents of erosion, such as water and wind, carry rock particles and other products of weathering away from their source, these materials become sorted according to their size. For example, a swift-flowing mountain stream may carry large pebbles and even boulders as well as sand, silt, and clay. As the water slows down, it drops the largest particles first and then gradually drops the smaller and smaller pieces. By the time the stream becomes a slow-moving river, only the smallest particles are left. This sorting produces layers of particles that are about the same size. Sorting also happens in the ocean when sediments are deposited there. Thus, clastic sedimentary rocks can be further classified according to particle size, as shown in the photos on this page.

Chemical sedimentary rocks form when minerals that were dissolved by weathering processes separate from the water and crystallize. In some cases, the minerals separate during the process of *evaporation*. As water evaporates, the dissolved minerals left behind form crystals and eventually rocks.

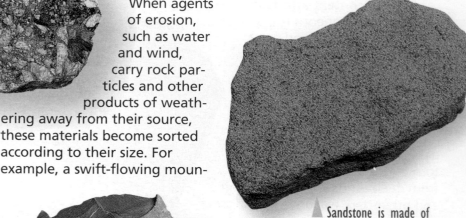

Conglomerates are made of a mixture of large and small pebbles.

Breccias contain angular pieces, unlike the rounded pebbles found in conglomerates.

Sandstone is made of small rock particles.

Shale is composed of clay—the smallest of rock particles.

DID YOU KNOW...

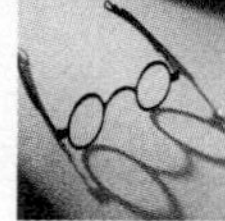

that geologists who study sedimentary rocks are sometimes called "soft-rockers" by other geologists?

This is because sedimentary rocks are generally softer than igneous or metamorphic rocks. Conversely, geologists who study igneous and metamorphic rocks are called "hard-rockers."

Rock salt and gypsum are two examples of chemical sedimentary rocks that form when water containing dissolved material evaporates. Other dissolved materials simply fall out of solution under proper conditions in a process called *precipitation*. Limestone often forms in this way. Another familiar rock that forms by precipitation is chert, also called flint, a dense rock made of tiny quartz crystals. Precipitates such as calcium carbonate and quartz are two of the materials that also cement the particles of clastic rocks together. These minerals fill in the spaces between rock particles much like mortar fills the spaces between bricks or stones in a wall.

Organic sedimentary rocks are a type of sedimentary rock that comes from the remains of once-living organisms. For example, limestone (which can also be classified as a chemical sedimentary rock) can be formed from marine organisms that remove calcium carbonate from the water to build their shells. When these organisms die and their shells settle to the bottom of the ocean, layers of limestone can build up. Chalk is a type of limestone that is formed in this way.

Coal is also considered to be a type of organic sedimentary rock. As you know, coal is formed from masses of dead plants that are compressed for millions of years. As this organic material is gradually compressed, the water and gases it contains are driven off, and it becomes denser and harder. There are three types of coal, but only two are considered to be organic sedimentary rocks: *lignite*, a very soft, dark brown coal, and the slightly harder, black *bituminous coal*. The third type of coal is classified as a metamorphic rock because of the conditions under which it forms.

Rock salt is a chemical sedimentary rock.

Chert is precipitated from sea water.

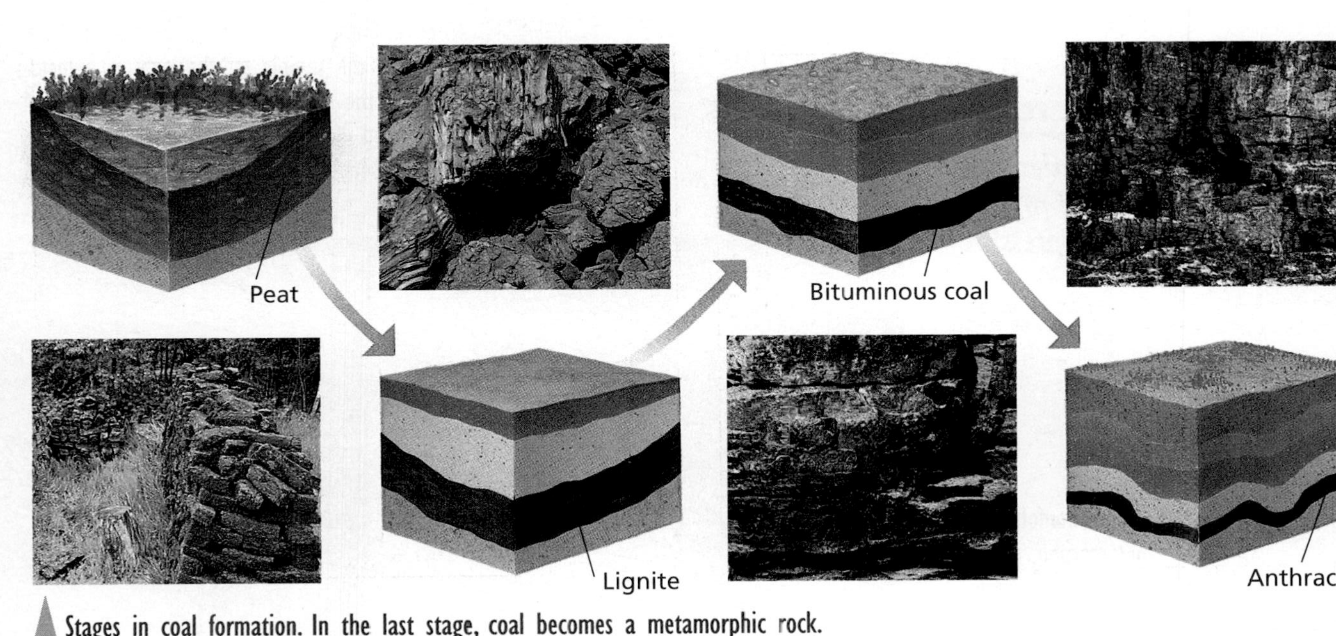

Stages in coal formation. In the last stage, coal becomes a metamorphic rock.

Metamorphic Rocks Just as living things go through physical changes during their lives, many rocks change physically as well. **Metamorphic rocks** form from other rocks when they are exposed to heat and pressure. Some of these changes are structural changes that result in the rearrangement of the crystals in a rock. In other cases, chemical changes occur, recombining the original minerals of the rock into new ones. Note the pairs of rocks in the photos shown here.

Geologists divide metamorphic rocks into two groups based on internal structure. *Foliated rocks* have obvious bands or zones of similar materials that indicate they were formed by pressure from one direction. These bands often appear to be shiny, because of the melting and reforming of mineral crystals. *Nonfoliated rocks*, like the marble in the photograph below, have no visible alignment.

Sandstone

Quartzite

Limestone

Marble

Granite

Gneiss

Three examples of rock pairs are shown here. In each case, the sample on the right is a metamorphic rock, and the sample on the left is the parent rock from which it formed. Notice that both sedimentary rocks and igneous rocks have been changed into metamorphic rocks.

Metamorphic rocks

Rocks that have undergone change as a result of intense heat and pressure

Marble, a nonfoliated metamorphic rock, is a valuable building material.

S122

The amount of foliation, or banding, is used to classify metamorphic rocks. For example, slate, schist, and gneiss are types of foliated rock. Slate, with the least amount of foliation and no visible crystals, shows the least amount of alteration by the processes of metamorphism. As you can see in the photos on this page, gneiss has larger crystals and more definite bands and therefore was exposed to higher temperatures and pressures than the schist.

When bituminous coal, an organic sedimentary rock, is put under intense heat and pressure, it becomes a harder, more compact form of coal called *anthracite*. Anthracite, an organic metamorphic rock, is very hard and has few of the impurities found in bituminous coal.

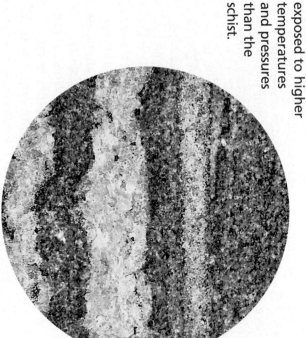

The crystal structure of gneiss (left) is much more defined than that of schist (above).

SUMMARY

As the Earth cooled, some of its molten materials solidified into the first rocks. Rocks that solidify from magma or lava are called igneous rocks. Sedimentary rocks form from pieces of other rocks that were broken up by weathering or from organic and chemically derived materials. Heat and pressure inside the Earth change both igneous and sedimentary rocks into metamorphic rocks. The rock cycle is a continuous process in which old rocks are made into new rocks. Rocks are classified on the basis of their origin, texture, and mineral content. Igneous rocks are classified as either intrusive or extrusive. There are three different kinds of sedimentary rock: clastic, chemical, and organic. Metamorphic rocks are classified as either foliated or nonfoliated.

ROCK-FORMING MINERALS

Rocks, as you have learned, are made of minerals. As magma and lava cool, the chemical compounds that make up the molten rock link together to form crystals of certain minerals. Most minerals are made up of at least two kinds of atoms. That is, minerals are usually made up of two or more chemical elements.

Kinds of Minerals

The Earth's crust contains over 3000 different minerals, but only about 20 are common. These 20 minerals are the basic rock-forming minerals of the Earth's crust. Whether or not they are common, all minerals can be classified into two major categories—silicates or nonsilicates—based on their chemical composition.

Silicates Silicon and oxygen are the two most common elements in the Earth's crust, as illustrated in the graph below. Therefore, it is not surprising that many minerals contain these two elements. Minerals that contain silicon and oxygen are called *silicates.* Silicate minerals make up more than 90 percent of the mass of the Earth's crust and are found in igneous, sedimentary, and metamorphic rocks.

One common silicate mineral that might be familiar to you is quartz, shown below. A crystal of quartz is composed of single silicon atoms, each of which is bonded to four oxygen atoms. These atoms form a network in the shape of a *tetrahedron,* which is a three-sided pyramid. Notice in the diagram on the next page that the oxygen atoms are located at the corners and the silicon atom is in the center. When silicon-oxygen tetrahedrons bond together with others to form this framework, a very hard mineral is formed.

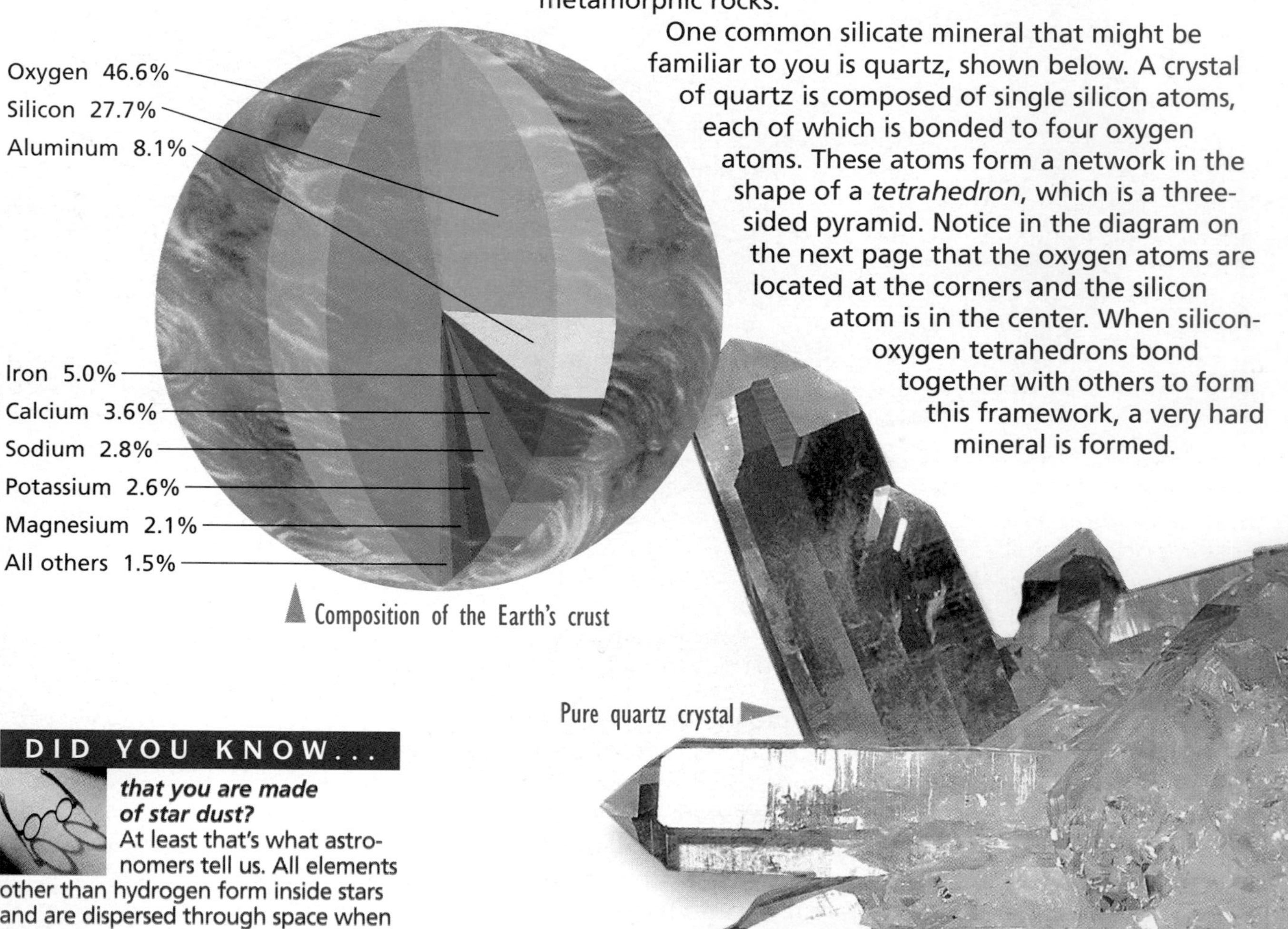

Composition of the Earth's crust

Pure quartz crystal

DID YOU KNOW...

that you are made of star dust? At least that's what astronomers tell us. All elements other than hydrogen form inside stars and are dispersed through space when the stars explode as supernovae. In other words, the matter on Earth, including the molecules in your own body, originally came from the stars.

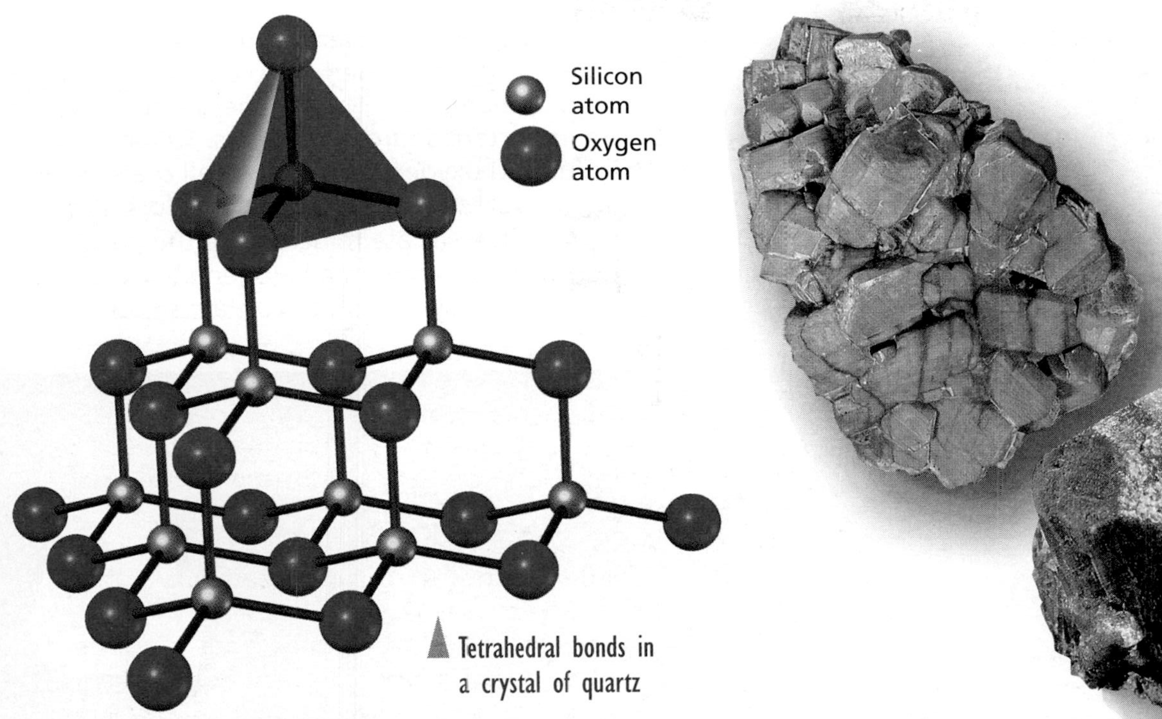

Tetrahedral bonds in a crystal of quartz

Feldspars are the most abundant silicate minerals. In addition to silicon and oxygen, they contain aluminum and either calcium, potassium, or sodium atoms in their crystals. The varied chemical composition and the presence of impurities, such as iron or flecks of the iron-containing mineral hematite, produce many form and color variations in the feldspar group. Three varieties of feldspar are shown to the right. Quartz and the feldspars together make up more than 50 percent of the minerals in the Earth's crust.

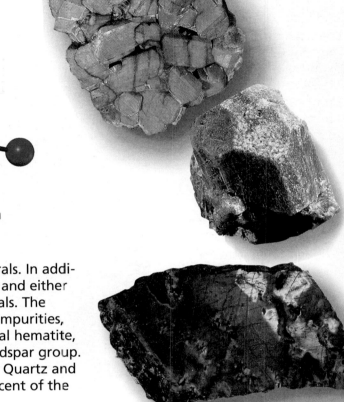

Feldspars, which take three basic forms, are among the most common minerals in the Earth's crust.

Nonsilicates Less than 10 percent of the Earth's crust is composed of *nonsilicate* minerals. Still, based on chemical composition, there are several different families of these minerals. One of the most important families of nonsilicate minerals is the *carbonates*. For example, calcite is the mineral name for calcium carbonate, the substance that makes up limestone, which is shown to the right. Dolomite is another example of a carbonate mineral. Carbonates are common in sedimentary rocks.

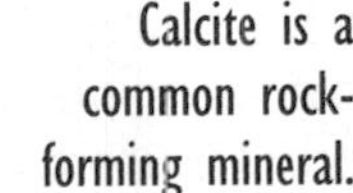

Calcite is a common rock-forming mineral.

Oxides, such as those shown above, are commercial sources of iron, aluminum, and other metals.

Other nonsilicate families are characterized by different chemical groups. *Oxides* contain oxygen bonded to an element other than silicon, such as iron or aluminum. These minerals, such as magnetite (iron oxide), are important sources of metals. *Sulfates*, such as gypsum, contain sulfur and oxygen. You may recall that gypsum is also the name for a chemical sedimentary rock that forms by evaporation. Other types of nonsilicate minerals are the *halides*, such as halite and fluorite, which contain a halogen element; and the *sulfides*, such as galena and pyrite, which contain sulfur.

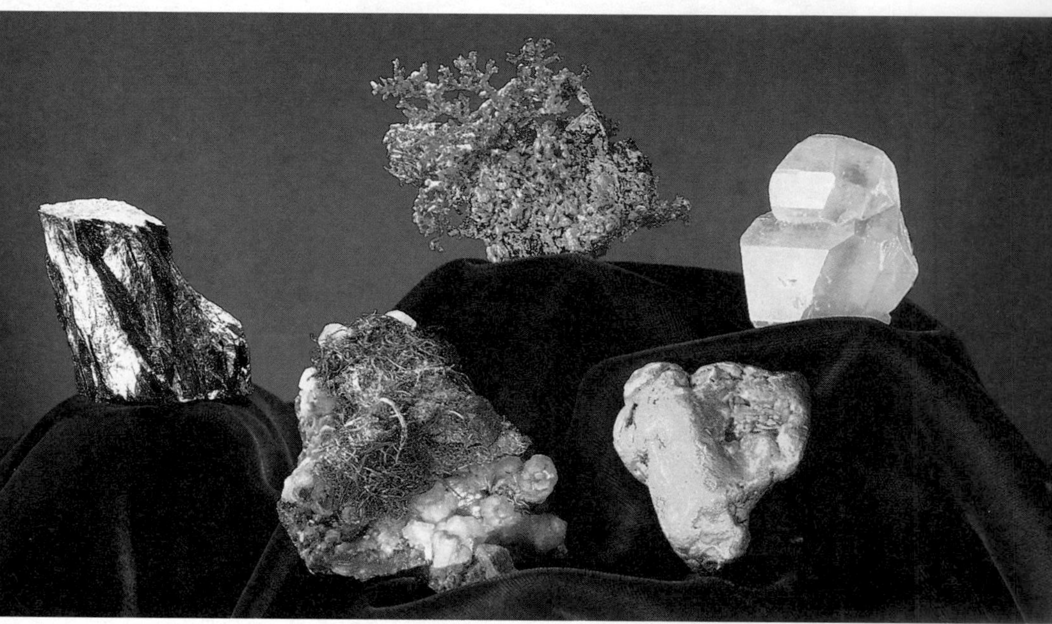

A few elements occur naturally in native (uncombined) form.

Some nonsilicate minerals contain only one kind of element. These rare minerals are called *native elements*. Many of the minerals in this group are valuable metals. Some examples of native elements are copper, sulfur, gold, silver, and lead, shown clockwise from the top in the photo above. Nickel and iron may also appear as native elements, but they are very rare in the Earth's crust. However, much of the Earth's interior is believed to consist of uncombined nickel and iron. Diamond and graphite are also native elements because they are made of pure carbon. They differ only in the arrangement of their carbon atoms.

Quartz is found in many colors due to mineral impurities.

Identifying Minerals

But just how are all of these types of minerals identified? Though detailed analysis of minerals can be done only in a laboratory with special equipment, there are some basic physical properties that you can use to help identify a mineral sample.

Color Many minerals have bright colors. For example, cinnabar is red, azurite is deep blue, serpentine is green, and sulfur is a bright yellow. Although most obvious, color is one of the least dependable characteristics you can use to identify a mineral. The color of many minerals can be affected by small amounts of impurities. For example, pure quartz is colorless, but with impurities it can be pink, tan, red, purple, or even black.

Luster Have you ever admired the shine of a newly waxed car? What you are admiring is the *luster* of the car's surface, or the way it reflects light. Similarly, the luster of a mineral refers to the way its surface appears when it reflects light. Each mineral has a characteristic luster. Some of the terms used to describe luster—metallic, glassy, waxy, pearly, greasy, or earthy—are related to familiar objects. How would you describe the luster of each of the minerals shown to the right?

▲ Metals, such as gold and silver, have a luster that you can easily recognize. The luster of other minerals is compared with those of common substances.

Streak If you rub a piece of chalk on a sidewalk, it leaves a white mark on the concrete. Some minerals also leave marks when rubbed on a rough surface. The color of the powder in these marks is called the mineral's *streak*. Sometimes the color of the streak is different from the color of the mineral sample. For instance, the streak of gold-colored pyrite is greenish black, and the streak of silver-colored hematite is brick red. Notice the streak produced by the samples shown below.

Streak is a good characteristic for identifying minerals because it is not influenced by such things as tarnishing and impurities.

DID YOU KNOW...

that sapphires and rubies have the same chemical formula?

Tiny amounts of impurities make the difference between these gemstones. If a crystal of pure corundum (Al_2O_3, or aluminum oxide), a normally clear and colorless mineral, contains a trace of titanium, it becomes a blue sapphire. If chromium is present, it is a red ruby.

S127

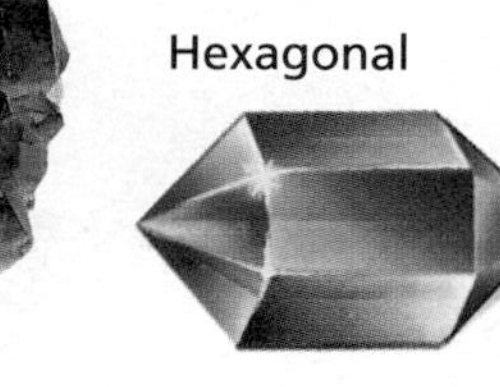

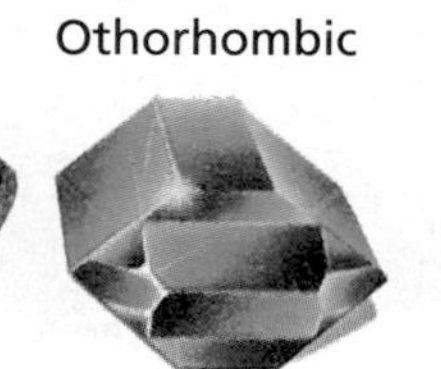

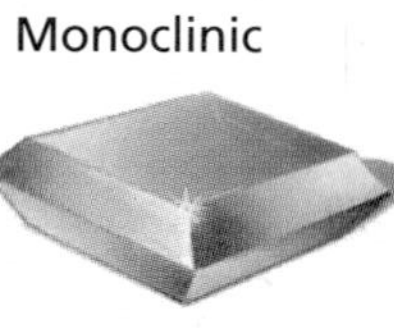

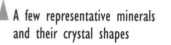

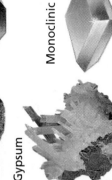

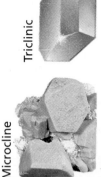

A few representative minerals and their crystal shapes

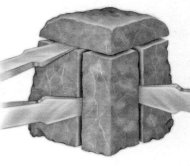

Crystal Shape Each mineral also has a characteristic crystal shape that can be useful in identifying the mineral. This shape is due to the orderly arrangement of the molecules that make up the mineral. The diagram on the left shows some of the different crystal shapes that minerals can have.

Hardness Another characteristic of minerals is hardness, or resistance to being scratched. *Mohs' scale of mineral hardness* consists of 10 standard minerals. Because harder minerals can scratch softer minerals, these 10 minerals can be used to compare the hardness of all other minerals.

Mohs' Scale of Mineral Hardness

Mineral	Hardness	Common test
Talc	1	Easily scratched by fingernail
Gypsum	2	Can be scratched by fingernail
Calcite	3	Barely scratched by copper penny
Fluorite	4	Easily scratched by a steel knife blade
Apatite	5	Can be scratched by a steel knife blade
Feldspar	6	Easily scratches glass
Quartz	7	Scratches both glass and steel
Topaz	8	Scratches quartz
Corundum	9	No simple test
Diamond	10	No simple test

You probably do not have a sample of each of the minerals on Mohs' scale, but you can still determine the hardness of an unknown mineral by using some common objects. For example, if the mineral in question can be scratched by a penny, its hardness is 3 or less. If a steel nail can scratch it, you know its hardness is less than 5. What is the only mineral that can scratch corundum? Although diamond is one step higher on the hardness scale than corundum, diamond is actually about four times harder.

Cleavage and Fracture These terms describe how minerals break. Minerals that always break along flat surfaces have a property called *cleavage*. Cleavage reflects the arrangement of a mineral's molecules. The cleavage of a particular mineral always occurs in the same way and produces faces that meet at characteristic angles. Halite, for example, breaks into cubes, calcite pieces break diagonally, and mica peels off in thin sheets.

The diagram on the left demonstrates cleavage in one direction, which is characteristic of the mineral mica. The diagram on the right demonstrates cleavage in three directions, as would be found in a sample of galena.

Some minerals do not display cleavage but instead break into irregular pieces, a property called *fracture*. Minerals usually fracture in a characteristic way. Some minerals fracture along a curved surface that resembles the inside of a clam shell. For example, quartz and obsidian (seen in the photo to the left) break in this way. Fracturing may also produce jagged, uneven surfaces.

Obsidian exhibits fracture, not cleavage.

Special Properties Minerals show a variety of other interesting properties, some of which are unique to certain minerals. Some minerals are *radioactive*. Pitchblende, for example, contains the radioactive element uranium. Many minerals are *fluorescent*—they glow brightly when exposed to ultraviolet light. Other minerals are *phosphorescent*—they glow after exposure to bright light.

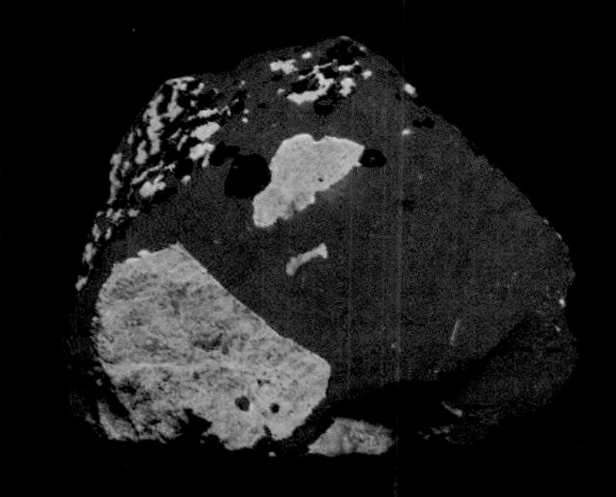

Many minerals *fluoresce* when exposed to ultraviolet (UV) light. The same mineral is shown here in normal light (top) and in UV light (bottom).

Magnetite is a type of iron oxide that is magnetic. Rocks that contain magnetite attract iron objects just as a magnet does.

Calcite is known for the property of *double refraction*, which is caused by the way light passes through the crystal.

SUMMARY

Minerals can be classified into two major categories based on their chemical composition. Those that contain silicon and oxygen are called silicates; those that do not are called nonsilicates. Certain physical properties, such as color, luster, and streak, can be used to identify minerals.

ROCK-FORMING MINERALS S129

Stories in Rocks

Imagine that you find a family photo album from 100 years ago. As you lift it, the binding breaks and the pages fall out in bunches. Several bunches fall together, landing in a pile on the floor. Other bunches scatter away from the pile. How are you going to put the album together again? How will you decide the order in which the pages belong? Is it possible that some pages may be missing?

In a similar way, the Earth's history is written in layers of rock. Trying to piece this history together is like putting the album mentioned above back together in the proper order. However, piecing together Earth's geologic history is more difficult because many of the "pages" have been widely scattered and even destroyed. For example, some rock formations have been split apart by processes such as plate tectonics and mountain-building. Others are lost through erosion or remelting. Still others are out of order because they were rearranged or distorted by movements of the crust. Although the story is incomplete, clues in the rocks can be used to piece together a fairly complete picture of the Earth's past.

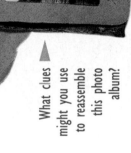

What clues might you use to reassemble this photo album?

Uniformity

Have you ever noticed how uniform or consistent some things are? For instance, next time you cut open a loaf of bread, notice how uniform the texture of the loaf is. A slice of bread from a different loaf of the same brand of bread will show the same uniformity of color and texture. The bread-making process ensures this uniformity. When the same process is used, even at different times, the results are similar.

Uniformity is an underlying principle of the basic geological processes. In the same way that a bread-making process makes uniform loaves of bread, the geological processes that form rocks and mountains today formed similar rocks and mountains in the distant past.

This idea was formalized as the *principle of uniformitarianism* in 1788 by James Hutton, a Scottish physician and geologist. Simply put, this principle states that similar processes produce similar results, wherever and whenever they happen. Observations support Hutton's theory. For example, sediments may be carried away by water or wind and then deposited in new places, but they are always deposited in similar ways regardless of where or when. Over long periods of time, these sediments are compressed and cemented to form sedimentary rocks.

If you slice open just about any loaf of bread, it will look much like any other loaf of bread.

Records of Environmental Change

Would you think to look for seashells on top of a mountain? Probably not. Nevertheless, fossil seashells are found in the rocks of mountains throughout the world. For this to have happened, either the rocks must have moved or the environments in which they formed must have changed.

Fossils such as brachiopods indicate that this rock formed in an underwater environment.

Like a mystery story, certain rocks contain clues about what happened in the past. Geologists can "read" these clues to figure out the mystery of their history. With these clues, they can figure out what the environment was like when a rock was formed and how those conditions have changed. To use our earlier example, what does finding fossil seashells in the rocks on the top of a mountain tell you? You could reasonably infer that the rocks first formed in an ocean and were later uplifted to become a mountain.

The kinds of fossils contained in rocks can tell us where the rocks were formed—in an ocean, in a stream, or on land. This is possible because the organisms that live in each place are different from one another. Fossilized sedimentary structures, such as ripple marks and mud cracks, also indicate what the environment was like when and where the rocks were formed.

The shape of mineral grains may also help geologists determine what sort of environment existed when rocks were formed. For instance, if a piece of sandstone contains rounded quartz grains, it probably formed on a beach. Geologists know this because the action of waves rounds the sand grains and because sand-sized sediments are often deposited near shore.

The size of the fragments in sedimentary rocks is another clue to where rocks form. Large fragments are deposited by rapidly moving water; smaller fragments are deposited by slower currents. Waves and currents also separate fragments by size. Large fragments are deposited close to shore to form sandstone, whereas smaller particles are carried out into deeper water. Ocean currents may carry clay particles far from land. In calm water, away from the beach, clay settles to the bottom to form shale.

DID YOU KNOW . . .

that most Earth scientists used to support a theory called catastrophism? Before Hutton presented the principle of uniformitarianism, it was widely believed that the Earth's landforms were caused by a series of catastrophic events.

Notice how the sediment spreads out once it reaches the ocean. Which size of particles would be found farthest from shore?

Relative Dating One clue to the order of the pages in Earth's history is the *law of superposition,* which simply states that younger rock layers lie on top of older rock layers. As layers of sedimentary and extrusive igneous rock form, the oldest layers are deposited first. Successive layers form one at a time, one on top of the other. Thus, when you look at the layers of rock in a cliff, like those shown to the right, the law of superposition tells you that the upper layers were deposited last. As you go down the cliff, the layers get older, as long as these rock layers have been undisturbed. Using the law of superposition, geologists can determine the **relative age** of rock formations by studying the order in which they were laid down.

Relative age

The age of a rock (older or younger) in comparison with other rocks

Another geologic principle states that layers of sedimentary and extrusive igneous rock are laid down in flat, level beds. Once this happens, however, these level beds may become folded, tilted, or offset by Earth processes.

Look at the figure below. Can you list the rock layers from oldest to youngest? If you said that **G** is the oldest layer and **A** is the youngest layer, you are correct. However, this is not the only information that can be determined from this diagram.

Geologists can study the exposed rocks of the Grand Canyon to learn about the geologic history of the region.

It is important to remember that the relative age of a rock cannot be given in years; it can only state that one rock is older or younger than another rock. You might not think that this information is very useful, but it is very important if you are trying to put together the geologic history of an area. By determining the relative age of rock layers and studying the way they are tilted or broken, a sequence of geologic events can be reconstructed.

1. Notice that layer **G** was deposited first, followed by **F** and then **E**.
2. After the lower layers were put down, they were broken by a fault and tilted during a time of mountain-building.
3. Layer **D** was deposited next, followed by an igneous intrusion, or *dike,* that cut across the 4 lower layers.

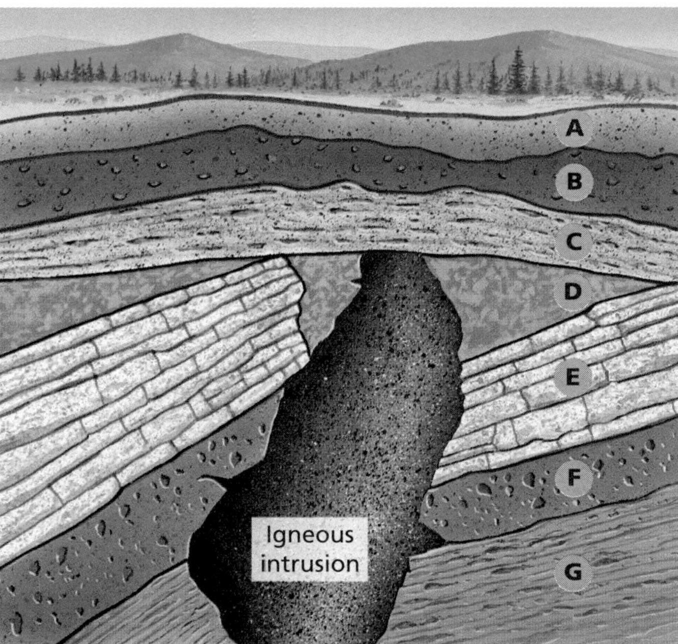

4. Then, a period of erosion wore down the landscape to a nearly level plain.
5. The boundary between layers **D** and **C** suggests that pages are missing from the story in this area. Such a boundary is called an *unconformity.*
6. Finally, the process of deposition began again, laying down layers **C**, **B**, and finally **A**.

Numerical Dating You may recall that some elements are unstable and therefore undergo *radioactive decay*. By giving off atomic particles, these radioactive elements change into more stable elements. For example, uranium decays over time to eventually form lead—a stable element that does not decay further.

Some radioactive isotopes of elements are very unstable and decay rapidly; others decay more slowly. The time it takes for half of the atoms of a radioactive isotope to decay is called its **half-life**. After each half-life, the number of unstable atoms is reduced by one-half as they change into atoms of a more stable element. Some isotopes have a half-life shorter than one second; others have a half-life of billions of years. Because radioactive isotopes decay at known rates, they act as an internal clock to mark the passage of time.

Numerical dating (also known as absolute dating) is the process of estimating the age in years of a sample using the radioactive isotopes in the sample. Rocks, minerals, fossils, water, and ice are dated using this method. The isotopes most commonly used in numerical dating are listed in the table above.

Isotope	Half-life (years)	Datable material
Tritium	12	Ground water, sea water, ice
Carbon-14	5730	Wood, bones, shells
Potassium-40	1,300,000,000	Rocks, minerals
Uranium-238	4,500,000,000	Rocks, minerals

Scientists date rocks by comparing the amount of the parent (unstable) element they contain with the amount of the daughter (stable) element present. The ratio of daughter to parent element gives the approximate age of the sample. The older the mineral or rock, the more daughter element there is.

However, because sedimentary or metamorphic rocks are made of pieces of other, much older rocks, they cannot usually be numerically dated. In sedimentary rock, measuring radioactive elements and their decay products gives the age of these older rock pieces, not the age of the younger sedimentary rock itself. Furthermore, as a metamorphic rock forms, heat and pressure alter the rock's original minerals and partially "reset" the numerical clock, affecting the amounts of both the original isotopes and their decay products.

The approximate age of sedimentary and metamorphic rocks can only be determined in relation to other rocks. For example, a nearby igneous formation may be dated numerically, and then the age of the neighboring sedimentary or metamorphic formations can be estimated by their position relative to the igneous rock.

Half-life

The time it takes for half of the atoms of a radioactive isotope to decay

DID YOU KNOW...

that carbon-14 can only be used to date materials that are less than 50,000 years old?

Because dinosaur fossils are millions of years old, they must be dated by other methods.

Fossil Correlation Sometimes rocks are moved far away from their places of origin, for example, when a continent breaks apart and its parts become separated by an ocean. When this happens, it becomes difficult to compare layers of rock between the two sections. Still, by using the fossil remains of plants and animals, the age of rock layers can often be determined.

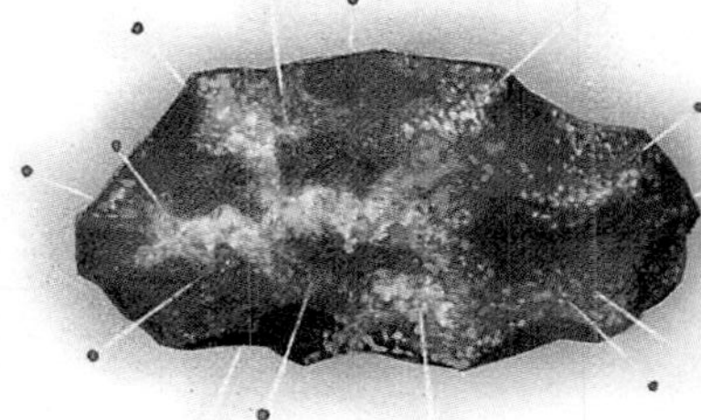

Radioactive elements can be used to find the approximate age of a material. For example, after 4.5 billion years, half of all the uranium atoms in the origial sample of uranium (left) will have changed into atoms of lead.

Geologic column

An arrangement that shows rock layers in the order in which they were formed

On a geologic time scale, most plant and animal species exist for only a limited time. Because life-forms change over time, some fossils can be used as time indicators. Such fossils are called *index fossils*. Different index fossils appear in sedimentary rock of different ages. By dating the fossils, or by dating similar fossils that occur elsewhere, the approximate age of sedimentary rock can be determined.

Fossil correlation is the use of index fossils to determine the relative age of widely separated rock layers. If similar fossils are found in rock layers separated by an ocean, the two rock layers would be the same age. By comparing index fossils with the rock layers in which they are found, geologists have been able to combine their observations to make a **geologic column**. In this way, fossils and the rock layers they are found in can be used to sort out the pages of the Earth's "family album," as shown below.

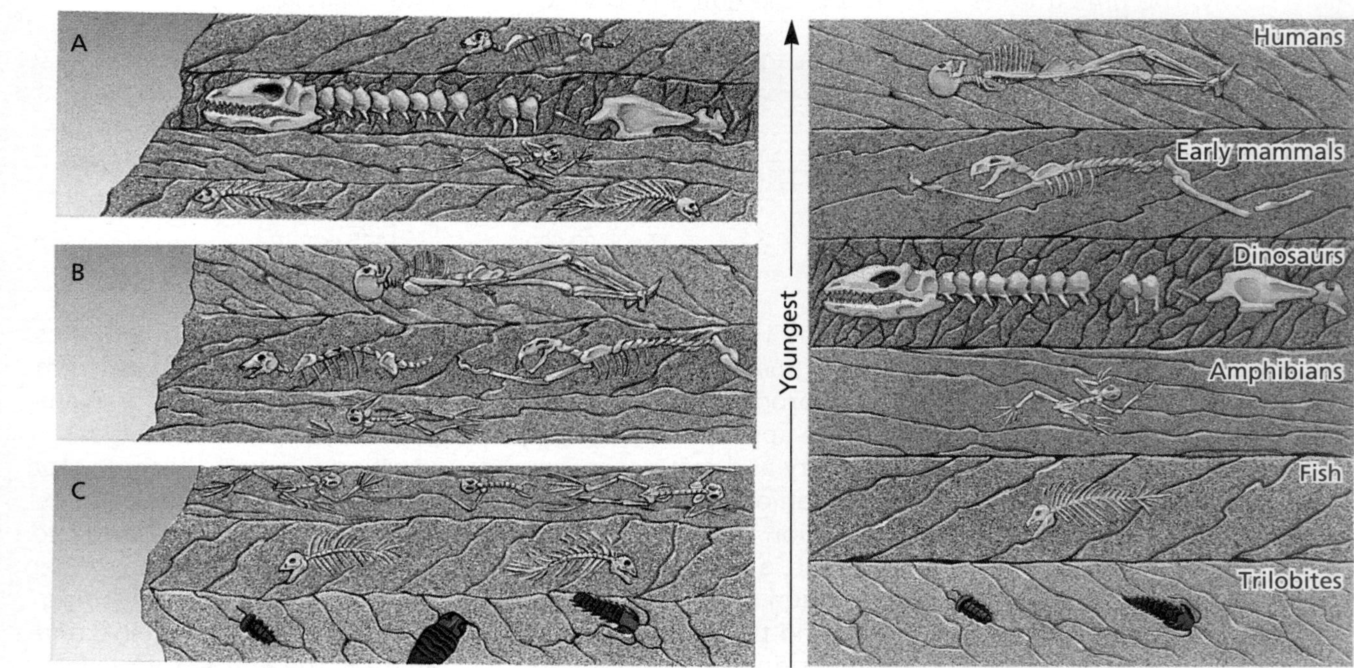

Layers that have the same rocks and fossils are the same geologic age no matter where they are found. A combined geologic column (right) can be constructed by arranging the 3 sets of rock layers on the left.

SUMMARY

Earth's history is written in layers of rock. Geologists piece this history together using the underlying principle of uniformitarianism. Certain rocks contain clues that geologists can use to determine what the environment was like when a rock was formed and how the conditions have changed. Using the law of superposition, geologists can estimate the relative age of rocks. Radioactive isotopes can be used to find the approximate numerical age of geologic material. By using index fossils, the age of widely separated rock layers can be correlated. The geologic history of an area can be pieced together by observing its rocks and the fossils contained in them.

Unit CheckUp

Concept Mapping

The concept map shown here illustrates major ideas in this unit. Complete the map by supplying the missing terms. Then extend your map by answering the additional question below. Write your answers in your ScienceLog. **Do not write in this textbook.**

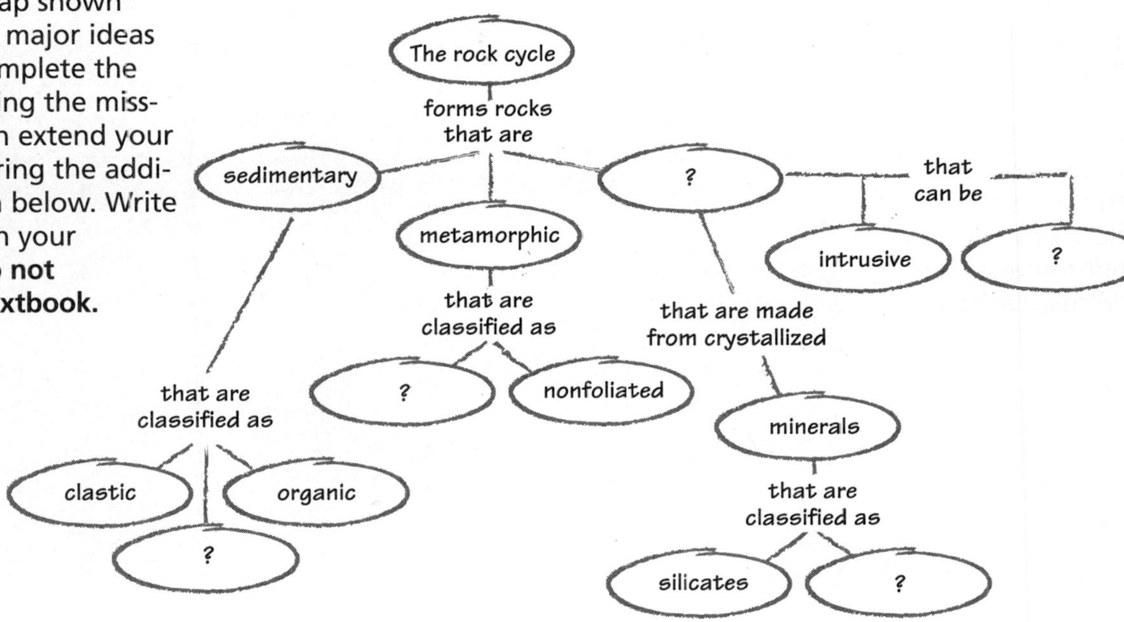

Where and how would you connect the terms *feldspar*, *breccia*, and *marble*?

Checking Your Understanding

Select the choice that most completely and correctly answers each of the following questions.

1. Explosive volcanic eruptions tend to form
a. cinder cones.
b. composite volcanoes.
c. shield volcanoes.
d. stratovolcanoes.

2. The Hawaiian Islands are a chain of volcanoes that formed over
a. a mid-ocean ridge.
b. a tectonic plate boundary.
c. an earthquake zone.
d. a hot spot.

3. A rock made of large angular particles of igneous rock cemented together would most likely be considered
a. a nonfoliated metamorphic rock.
b. a clastic sedimentary rock.
c. a chemical sedimentary rock.
d. an extrusive igneous rock.

4. While rock can be made of more than one mineral, a mineral grain cannot be made of
a. more than one rock.
b. less than three elements.
c. only one type of element.
d. inorganic material.

5. The law of superposition supports which of the following dating methods most?
a. numerical dating
b. absolute dating
c. relative dating
d. Dutch dating

Interpreting Illustrations

Label the diagram shown here according to the law of superposition. Number the seven layers from youngest to oldest. Then write a brief geological history of the area, based on your understanding of the principle of uniformitarianism and relative age.

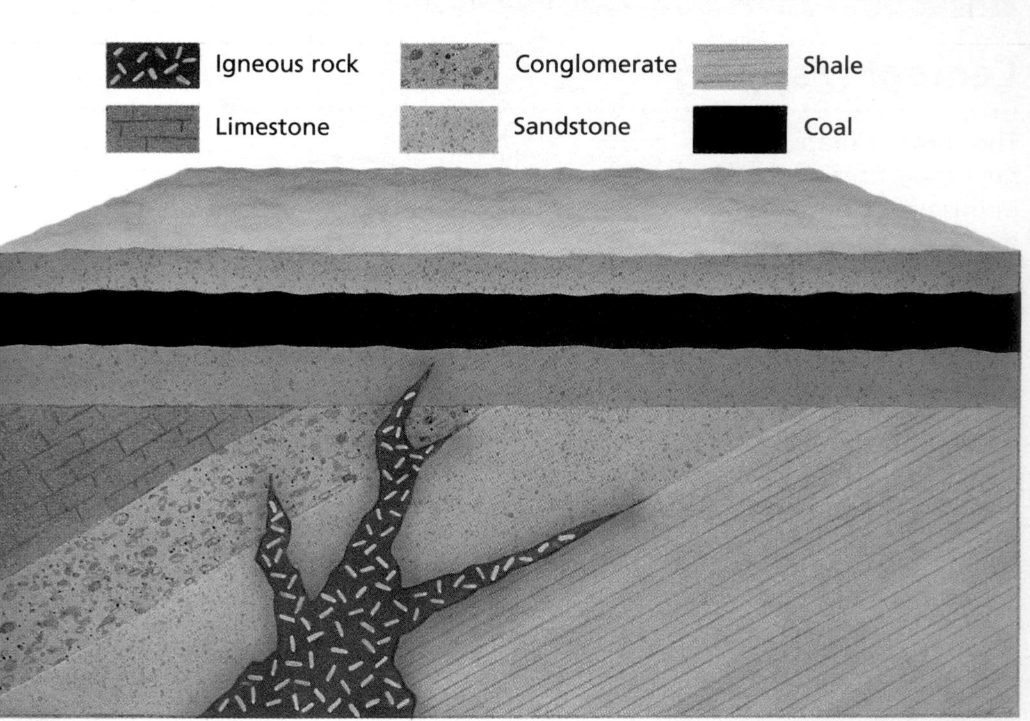

Critical Thinking

Carefully consider the following questions, and write a response in your ScienceLog that indicates your understanding of science.

1. How does plate-tectonic theory incorporate mountain-building, volcanism, and the occurrence of earthquakes into one explanation of observed geologic processes?
2. Because of the Earth's great age, it is likely that every part of the crust has been through the rock cycle at least once. If some of the original rock still existed, what type would it be? Why?
3. The slower a magma cools, the larger are the crystals in the resulting rock. An igneous rock sample was found to consist of microscopic crystals. How did this sample most likely form?
4. A geologist submitted a sample of gneiss to the radioisotope laboratory of a nearby university for dating. The lab returned the sample along with a letter. What did the letter likely say and why?

Portfolio Idea

Prepare a display on the rock cycle. Use photographs, diagrams, and captions as well as actual samples of different rocks to illustrate the interactions between and among the rock types, their sources, and the forces that make them change. Your display should include as many ideas from this chapter as possible. Make sure there are connections between the ideas presented. You may want to work with one or more classmates.

Toward the Stars

In This Unit

Now that you have been introduced to astronomy and to the exploration of the universe, consider these questions.

1. What are the differences and similarities among the various bodies of the solar system?
2. How do astronomers get information from the stars?
3. In what way is our knowledge of the universe changing?

In this unit, you'll take a closer look at the members of our solar system and the stars beyond and at how scientists are furthering their investigation of space.

UPWARD AND OUTWARD

A view of the Earth from the first Apollo mission to the Moon in 1969

For most of history, people have looked into the blackness of space and wondered at the marvels they saw. Without modern technology, the ability to collect information was limited. Recently, however, we have been able to get a better look at our cosmic companions, as well as at the Earth itself. With the help of space-age technology, we can look back at the Earth from orbit and view it as an entire planet. From this vantage point, we are able to get firsthand information about our own planet and its relationship to other celestial bodies.

The Moon, Our Nearest Neighbor

It would be difficult to ignore the most visible object in the night sky—the Moon. It has long been the object not only of scientific speculation, but of poetry and song as well. In some ways, the Moon resembles the Earth. They are both spherical and have a crust and mantle made of rock. The Moon may also have an iron core like the Earth, although it is probably much smaller.

The astronauts who visited the Moon, however, were more impressed by the differences between the Moon and the Earth. For one thing, they found the Moon an inhospitable environment with no evidence of any living things. The Moon has no atmosphere to shield it from the Sun's radiant energy or to help hold in heat during the night. This causes wide fluctuations in temperature. During the day, the temperature on the Moon's surface rises to above 100°C; during the night, it drops below –170°C. There is also no water on the Moon. Its surface features, therefore, have not been changed by wind or water as the Earth's have been. Most of the features that can be seen on the Moon today—craters and lava fields—have been there since its early history.

The Moon is not free from all change, however. The lack of an atmosphere allows the surface to be hit constantly by small particles from space. This bombardment breaks up some of the rocks on the surface. Since the small pieces of rock are not washed or blown away, a fine dust covers almost all of the Moon's surface.

DID YOU KNOW . . .

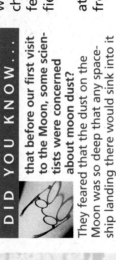

that before our first visit to the Moon, some scientists were concerned about moon dust?
They feared that the dust on the Moon was so deep that any spaceship landing there would sink into it and disappear!

James Irwin leaves his footprints on the Moon.

Lunar Origins It is unusual for a planet to have a moon so big compared to its own size. The diameter of the Moon is about one-fourth the diameter of Earth. Various models have been proposed to explain how an object as large as the Moon formed and began orbiting the Earth. If it were not already the Earth's only natural satellite, the Moon is large enough to qualify as a small planet itself.

1 In fact, some astronomers thought that the Moon actually was, at one time, a small planet that was captured by the attraction of Earth's gravity and pulled into orbit.

2 Some scientists thought the Moon was formed by material from Earth's mantle that broke off of the swiftly spinning planet, leaving the depression now filled by the Pacific Ocean.

3 Others claimed that the Earth and the Moon were formed at the same time from the same cloud of dust and gas that had surrounded the early Sun.

4 Recently, some astronomers have theorized that debris orbiting the Earth eventually came together because of gravity.

5 Still other astronomers speculate that a body the size of Mars hit the Earth and knocked material out into orbit around the planet. This material later came together to form the Moon.

There are problems with each of these ideas, however. If the Moon was captured by the Earth, gravity should have given the Moon a different orbit than it has today. It is also unlikely that the Moon was once part of the Earth because too many differences have been found in the chemical makeup of the two bodies. The theory that the Earth and Moon were both formed together has a similar problem.

These models are still being investigated. The information gained so far has not settled the question of the Moon's origin. What do you think?

The Sun: Source of Life

The Sun is the source of most of the energy used on the Earth. It not only heats our atmosphere and warms our oceans, but also is used by plants to make food—food that ends up on our tables. Without the Sun, there could be no life on Earth. But what is this radiant sphere in the sky? And why does it shine?

Composition of the Sun No one has ever been able to see into the interior of the Sun. No space probes from Earth have ever penetrated the Sun's interior. Yet, by using mathematical models, scientists have been able to learn about the inner structure of the Sun.

The Sun is thought to be made up of several layers that blend into one another. Because all of the layers are gases, there are no sharp dividing lines between them. The Sun is so hot that these gases may well be a mixture of atomic nuclei and free electrons—a form of matter called *plasma.* As shown in the diagram below, the Sun can be divided into six layers: three that form the Sun itself and three that make up the Sun's atmosphere.

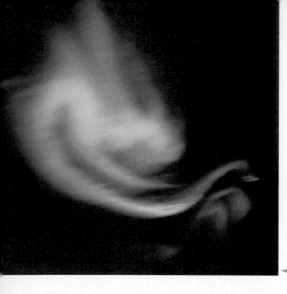

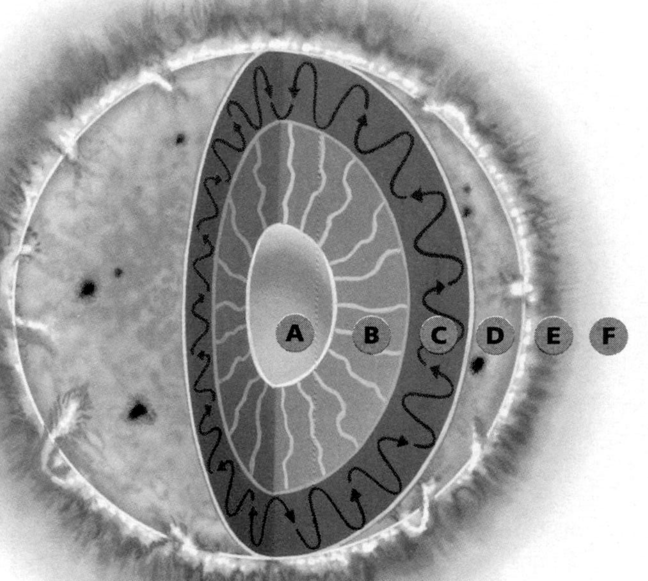

A The reactions that produce the Sun's huge supply of energy take place in its center, or *core.* Scientists calculate its temperature to be at least 15,000,000°C.

B Energy coming from the core is carried outward through the *radiative* zone, which is warmed in the same way that a room is warmed by sunlight through a window.

C The energy then enters the *convection* zone, where it is transferred by convection to the Sun's surface in the same way that water carries energy from the bottom to the surface of a pot of boiling water.

D Energy that reaches the surface causes a thin layer of gases to give off a brilliant light. This glowing layer, called the *photosphere,* produces the visible light that comes from the Sun. The photosphere, the part of the Sun that we see, has a temperature of about 6000°C.

E The *chromosphere* gives off a faint red light that cannot be seen against the bright background of the photosphere except during eclipses.

F The chromosphere blends into a much less dense layer, called the *corona,* that surrounds the Sun like a halo. It has no outer boundary, and some of the Sun's material escapes into space in the form of a *solar wind* that flows past the Earth.

The northern lights—*aurora borealis*—as seen from Canada.

Fortunately for living organisms on Earth, the Earth's magnetic field pushes the solar wind aside as it passes. This causes the wind, which consists of particles moving at speeds of 300 to 700 km/s, to flow around the planet like a river flowing around an island. Some of the charged particles in the solar wind, however, become trapped by the Earth's magnetic field. When these particles collide with gas molecules in the Earth's upper atmosphere, they often form eerie curtains of light that stretch across the sky in polar regions. These greenish white and blue lights are known as *auroras,* or northern lights and southern lights.

Energy Production by the Sun Each second, the Sun gives off an amount of energy equal to 200 billion hydrogen bombs. We are fortunate that this solar furnace is 150,000,000 kilometers away!

How the Sun works, however, remains an intriguing question for astronomers. It was first thought that the Sun burned fuel, much like the fuel we know on Earth—wood, coal, and oil. But if the Sun were burning one of these fuels, the rate at which it produces energy would cause it to burn out in only a few thousand years. Since geologists have estimated that the Earth is about 4.5 billion years old, the Sun cannot be using any common fuel. Throughout the nineteenth century, scientists proposed other explanations for the Sun's energy output; yet, one by one, each had to be ruled out.

Then in 1905, a young physicist named Albert Einstein came up with his famous formula, $E = mc^2$. By establishing this relationship between mass and energy, Einstein paved the way for the discovery of a new source of energy—fusion. During the process of fusion, hydrogen is converted into helium, releasing enormous amounts of radiant energy. Fusion is thus able to account for the Sun's tremendous output of energy. Even though 4 million tons of hydrogen are consumed each second, the Sun contains so much hydrogen that it should continue to burn for at least another 5 billion years.

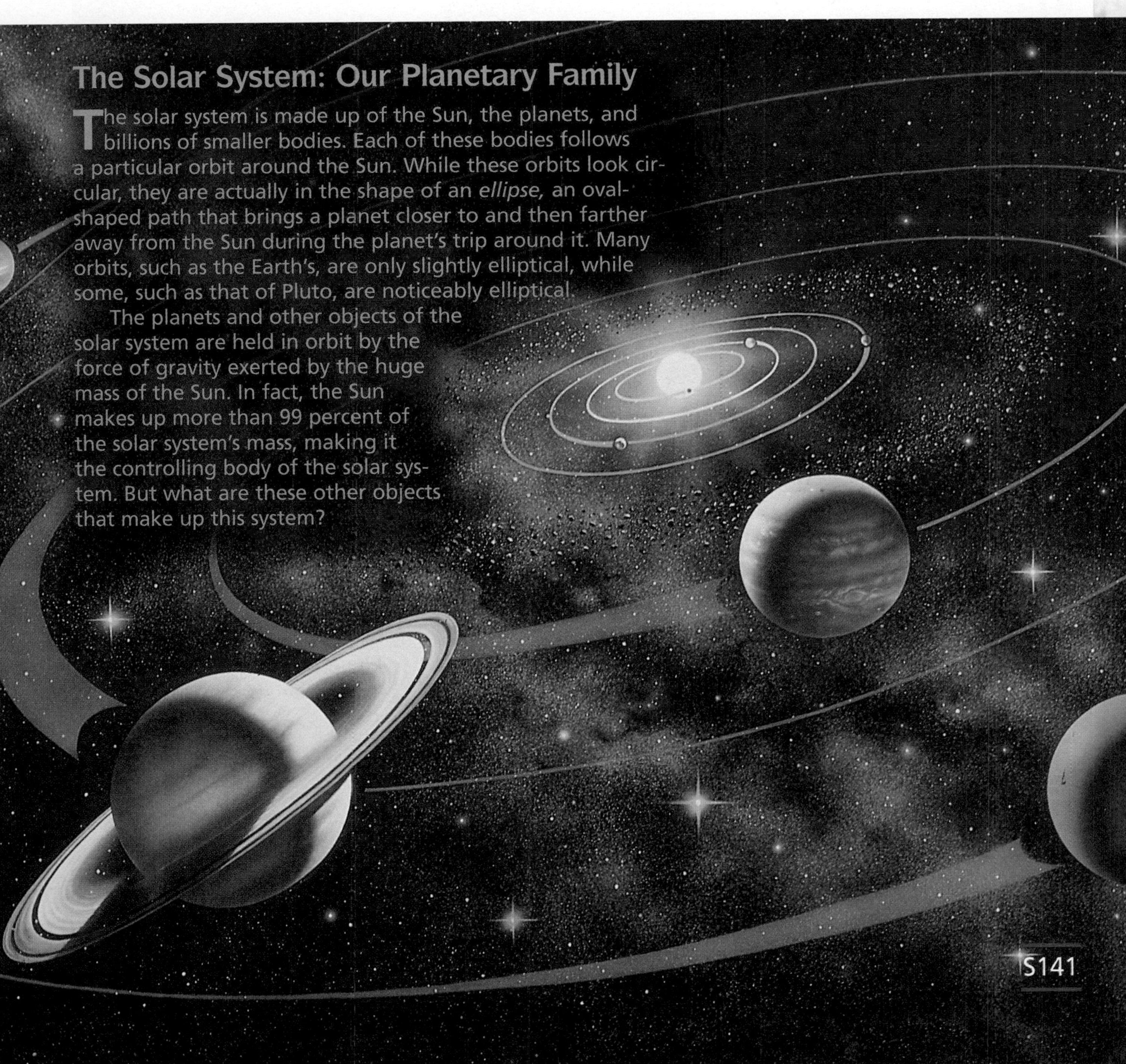

The Solar System: Our Planetary Family

The solar system is made up of the Sun, the planets, and billions of smaller bodies. Each of these bodies follows a particular orbit around the Sun. While these orbits look circular, they are actually in the shape of an *ellipse,* an oval-shaped path that brings a planet closer to and then farther away from the Sun during the planet's trip around it. Many orbits, such as the Earth's, are only slightly elliptical, while some, such as that of Pluto, are noticeably elliptical.

The planets and other objects of the solar system are held in orbit by the force of gravity exerted by the huge mass of the Sun. In fact, the Sun makes up more than 99 percent of the solar system's mass, making it the controlling body of the solar system. But what are these other objects that make up this system?

The Planets The best known objects in the solar system are the nine planets. Most of the planets can be classified according to two major groups—the inner planets and the outer planets. The inner planets include Mercury, Venus, Earth, and Mars. These planets are small and consist of solid rock with metal cores. The four large, outer planets include Jupiter, Saturn, Uranus, and Neptune. The giants Jupiter and Saturn are made mostly of hydrogen and helium. They have gaseous atmospheres that turn liquid inside and have small solid cores of hot rock and ice. Uranus and Neptune are similar except that they have much more rock and ice. The remaining planet, Pluto, is very small and seems to be different from all of the other planets. It is usually the farthest planet from the Sun, but its orbit is such that it was actually closer to the Sun than Neptune between 1989 and 1999. The table below compares some of the major characteristics of the planets.

DID YOU KNOW...

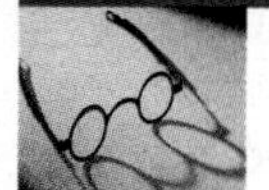

there's a clever way to remember the order of the planets? Just use the first letter of each word in the following phrase: My Very Energetic Mother Just Served Us Nine Pizzas.

Characteristics of the Planets

Planet	Distance from Sun (AU)	Revolution (years)	Radius at equator (km)	Rotation period (days)	Mass compared to Earth's	Density (g/cm³)	Surface gravity (Earth = 1)	Surface temperature (°C)
Mercury	0.39	0.24	2,439	58.6	0.055	5.43	0.38	–173 to 427
Venus	0.72	0.62	6,052	–243.0	0.82	5.24	0.91	453
Earth	1.00	1.00	6,378	0.997	1.00	5.52	1.00	–13 to 87
Mars	1.52	1.88	3,397	1.026	0.11	3.9	0.38	–83 to –33
Jupiter	5.20	11.86	71,398	0.41	317.8	1.3	2.53	–153
Saturn	9.54	29.46	60,000	0.43	95.1	0.7	1.07	–185
Uranus	19.19	84.07	26,200	–0.65	14.6	1.3	0.92	–214
Neptune	30.06	164.82	25,225	0.72	17.2	1.5	1.18	–225
Pluto	39.53	248.6	1,150	–6.39	0.0022	2.0	0.09	–236

The differences between the inner and the outer planets may have been determined by the way in which the solar system formed. Scientists theorize that the inner planets formed close to the Sun, where lighter materials could not accumulate because of the Sun's intense radiation, leaving mostly solid rock and metals. The outer planets formed farther away from the Sun's radiation, where they were able to hold on to the lighter materials. These materials, in the form of ices and gases, make up most of the mass of the outer planets.

The matter from which the planets formed began as a cloud of gas surrounding the early Sun. Over time, gravity drew together particles from the cloud to form the planets we know today.

Smaller Members of the Solar System

There are other bodies in the solar system besides the nine major planets. These objects, while generally much smaller than the planets, far outnumber them. You have already learned about meteors and comets. There are two other important members of the solar system—satellites and asteroids.

Jupiter's four largest satellites—Io, Europa, Ganymede, and Callisto—were discovered by Galileo in 1610, using his primitive telescope. More recently, active volcanoes have been observed on Io.

Satellites As you know, the Earth has a very familiar satellite—the Moon. Other planets have satellites, or moons, of their own. Mars has two small satellites, Phobos and Deimos, while the giant planets have many satellites. Jupiter has 16, Saturn has at least 18, and Uranus and Neptune have 15 and 8, respectively. Mercury and Venus, on the other hand, do not have any natural satellites.

One of the moons of Saturn, named Titan, is the only satellite in our solar system known to have a substantial atmosphere. Neptune's largest moon, Triton, rotates around its planet in a direction opposite that of the planet's rotation. This has led some astronomers to believe Triton was captured by Neptune's gravitational pull. All but two of Neptune's moons remained undiscovered until the visit by *Voyager 2* in 1989. In 1995, several more moonlike objects were discovered around Saturn when its thin ring was edge-on to Earth, essentially making it disappear from view.

Combined photos of Saturn and its moons, with Dione in the foreground

Pluto has one moon, named Charon, that is almost half the diameter of Pluto itself. Thus, they form a double-planet system, each one orbiting the other in a period of about six days.

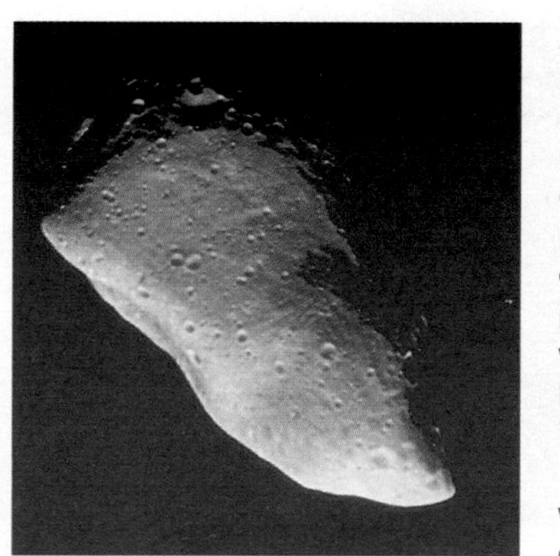

Asteroids are fragments of rock that orbit the Sun along with the planets. This photo of Gaspra was the first picture ever taken of an asteroid from space.

Asteroids Between the orbits of Mars and Jupiter is a wide region in which a large number of rocky bodies, called *asteroids,* are found. This region is known as the asteroid belt. The total number of asteroids is not known, but at least 6000 have been discovered already. Ceres, the largest asteroid, makes up about 30 percent of the total mass of all known asteroids. In 1991, the first close-up photographs of an asteroid were taken by a passing space probe. This asteroid, named Gaspra, is only 13 km long and 5 km wide. Another asteroid, named Ida, was photographed in 1993. From those photos, it was later discovered that Ida has a satellite of its own, which was given the name Dactyl.

At one time, scientists thought that the asteroids might be the wreckage of a small planet that was torn apart by Jupiter's tremendous gravity. However, scientists now think that asteroids are the parts of a planet that never formed. Some asteroids follow almost circular orbits around the Sun, while others have very elliptical orbits that often cross the paths of the inner planets, including that of Earth.

This meteorite was found in Australia. While small meteors seem to be associated with the dust trails of comets, many are fragments of asteroids that have wandered out of their usual place in the solar system.

DID YOU KNOW...

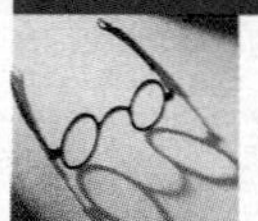

that some scientists blame an asteroid for the extinction of the dinosaurs?

A large crater, found in the Yucatan peninsula, is thought to be the remains of an impact that devastated Earth's climate about 65 million years ago.

SUMMARY

With modern technology, Earth can be viewed as one member of a family of celestial objects. The Moon, Earth's closest neighbor, still presents a puzzle regarding its formation. The Sun, made up of several layers of gases, dominates a solar system that includes nine planets. The inner planets are composed of solid rock and metals, while the outer planets are made mostly of gases and liquids. Most of the planets of our solar system have satellites, or moons. The most numerous of the smaller objects in the solar system are asteroids and comets.

STARS: A UNIVERSE OF SUNS

There are more stars in space than there are drops of water in all the oceans of the world or grains of sand on all the beaches of the Earth. Yet each star you see in the night sky is a sun, similar to our own. Whether or not those stars have planets of their own is not yet confirmed. However, new evidence indicates that many stars have characteristics that would allow for planetary systems. With such a vast number of stars in space, it is quite probable that other planetary systems do exist. Many scientists speculate that other planetary systems will eventually be found. What has already been found, however, is quite intriguing in its own right.

Except for the Sun, stars are so distant from the Earth that even the largest and most powerful telescopes show them only as pinpoints of light. Since stars are so far away, it might seem difficult to learn about them. But information about stars is carried in their light. By interpreting this information, scientists have learned what stars are made of, how large they are, and how they are moving in space.

How We Measure Stars

How far away is a star? When looking at the night sky, you can see that not all stars have the same brightness. Actually, some of the brightest objects are planets, not stars. Venus, Mars, Jupiter, and Saturn can be seen easily without a telescope. Planets, as you know, can be distinguished from stars by observing their positions from night to night. Since stars are much farther away from the Earth than the planets, individual stars appear to keep the same position in relation to other stars, while the planets seem to change position noticeably.

DID YOU KNOW...

that the brightest star you can see is one of the closest to Earth? The closest star that is visible from the continental United States is Sirius, which is also the brightest in the night sky. It is almost 9 light-years away, which means that the light you see tonight left the star before you started kindergarten!

The stars visible from Earth are just a small fraction of those that actually exist. These four photos were taken of the same area of the sky, with increasing magnification. Without a telescope, none of these stars would be visible.

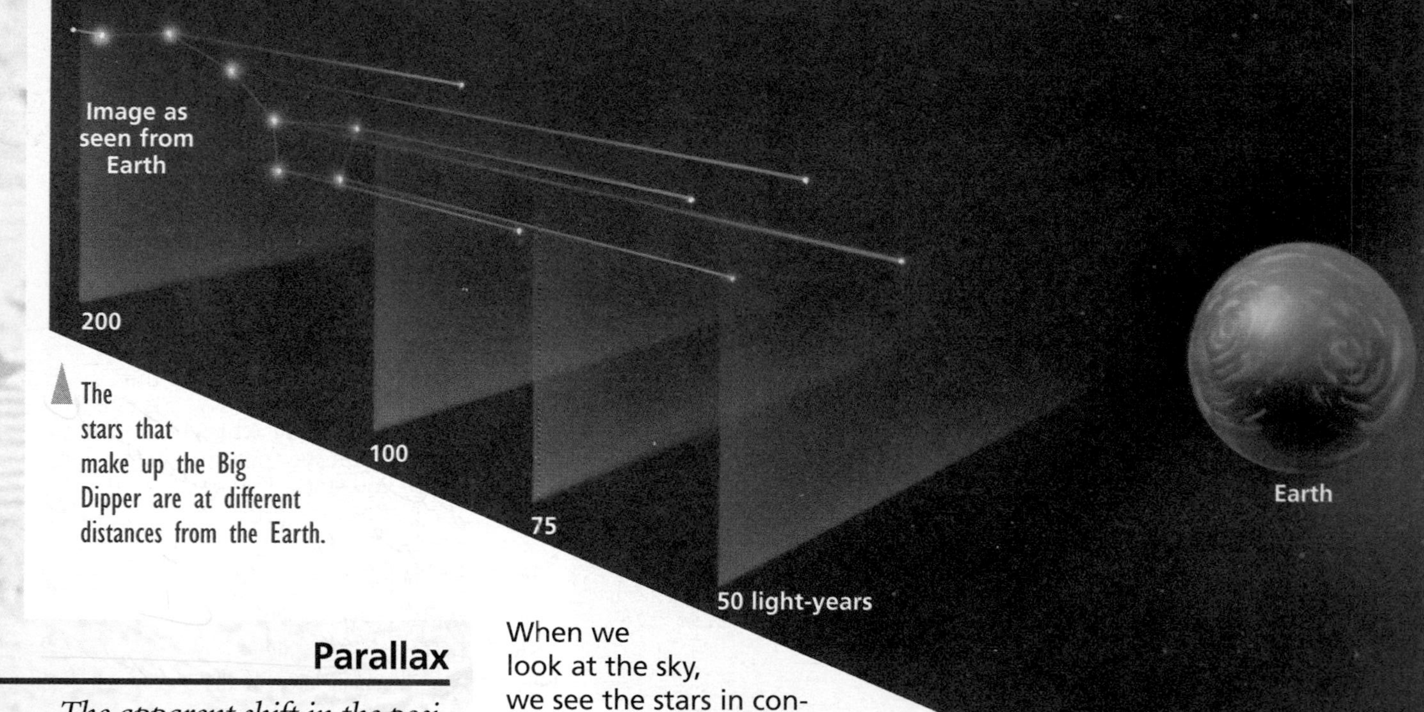

The stars that make up the Big Dipper are at different distances from the Earth.

Parallax

The apparent shift in the position of an object (such as a star) when seen from different locations or angles

When we look at the sky, we see the stars in constellations as if they were located on a kind of ceiling over our planet. For example, the stars in the Big Dipper all seem to be grouped together. But the stars are actually scattered and at different distances from the Earth.

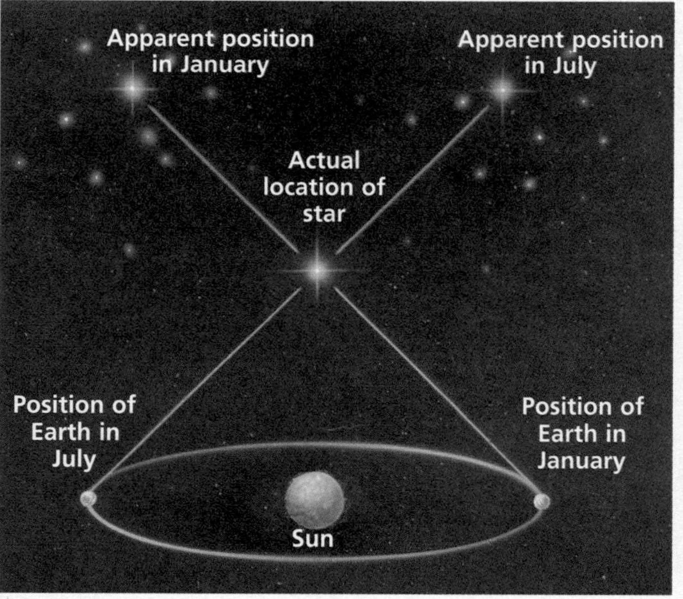

The movement of the Earth causes nearby stars to appear to move relative to those that are much farther away.

Some stars *do* seem to move very slightly when observed over a 6-month span. As the Earth moves in its orbit around the Sun, these stars seem to shift their positions in relation to more distant stars. This is similar to the way that a nearby tree appears to move against the distant background when viewed from a moving car. Trees that are farther away appear to move less against a distant background than do nearby trees. This apparent motion, called **parallax,** can be used to estimate how far away the trees are. By using parallax, the distances to nearby stars can also be measured, as shown in the diagram to the left.

Scientists use other methods to measure the distances to stars that are very far away. One method is based on the brightness of the star. Have you ever judged the distance to a street light by how bright it looked? If you have, you were estimating distance using brightness. The actual amount of light that a star sends out is called its **luminosity.** Stars have different luminosities; some are 10,000 times brighter than the Sun, while others are only a tiny fraction as bright. Once the luminosity of a star is known, its distance can be found by comparing its luminosity with its apparent brightness as actually seen from the Earth. For example, if a star that has a high luminosity appears dim, it means that the star is probably very far away. You might judge the distance to a street light by making the same kind of comparison.

Luminosity

A measure of the actual brightness of a star

Types of Stars

The information carried by light has shown scientists that there are many kinds of stars. Some stars would not fit within the bounds of Earth's orbit around the Sun, while others are smaller than the Earth itself. There are stars that are made of gas thinner than air and others that are harder than diamond. One kind of star even expands like a balloon, shrinks, and then expands again over and over.

Red and Yellow, Black and White One difference among stars is their color. You already know that the color of a star's light is related to the temperature of its surface. The coolest stars have a reddish color and surface temperatures of about 3000°C, while the hottest stars have a bluish color and temperatures of over 40,000°C.

Classification of Stars

Color	Surface temperature (°C)	Example stars
Blue	Above 30,000	10 Lacertae
Blue-white	10,000 to 30,000	Rigel, Spica
Blue-white	7500 to 10,000	Vega, Sirius
Yellow-white	6000 to 7500	Canopus, Procyon
Yellow	5000 to 6000	Sun, Capella
Orange	3500 to 5000	Arcturus, Aldebaran
Red	Less than 3500	Betelgeuse, Antares

Astronomers have discovered that there is also a relationship between the star's surface temperature and its brightness. Generally, the higher the temperature of the star, the brighter it is. When stars are plotted on a chart according to their surface temperature and brightness, they fall into three main groups. This chart is known as the Hertzsprung-Russell diagram, named after the Danish and American astronomers who first designed it.

Stars of several colors can be seen in the familiar constellation Orion.

DID YOU KNOW . . .

that stars vary greatly in size?
Betelgeuse is so large that if it were placed at the center of our solar system, its surface would reach well past the orbit of Mars. Some white dwarfs, on the other hand, are only the size of Earth.

Most stars fall into a narrow band on the chart, with hot stars at the upper left and cool stars at the lower right. Since most known stars fall into this group, they are called the *main-sequence* stars.

Another group of stars appers in the upper right corner, even though their red color indicates low surface temperatures. These are the red giants.

A third group of stars appears at the bottom of the chart. These tiny, dim stars are called white dwarfs, though some are very hot and produce intense white light.

Red giants
Betelgeuse
Anteres
Aldebaran
Capella
Arcturus
Rigel
Canopus
Spica
Vega
Sirius
Procyon
Main sequence
Sun
Tau Ceti
Companion to Sirius
White dwarfs
Barnard's star
Proxima Centauri
Luminosity (x Sun)
1,000,000
10,000
100
1
0.01
0.0001
30,000
10,000
7500
6000
5000
3500
Temperature (°C)

Binary Stars Most of the stars you see in the night sky are not single stars, as is our own Sun. In fact, about 60 percent of all stars are actually double-star systems, or *binary stars.* In binary star systems, two stars revolve around each other. Some star systems may even contain pairs of binary stars. The nearest star that is visible from the continental United States is Sirius, which you already know is a binary star.

These three photographs (covering a period of about 12 years) show the mutual revolution of the components of the nearby double star Kruger 60.

Variable Stars Stars that change in brightness or luminosity are called *variable stars.* One type of variable star is called a pulsating variable. Energy from below the surface of such a star heats the gases of the visible, shining surface. As the surface becomes hotter and brighter, it expands. Once the surface expands, however, the gases cool and the star becomes dimmer. The cooler gases then contract, causing them to once again get hotter, and a new cycle begins.

The first pulsating variable to be discovered was Mira, whose luminosity changes every 331 days. The time it takes to go from bright to dim and back to bright is the star's period. Most known pulsating variables have periods of 100 days or less. Mira has an unusually long period. The North Star, in contrast, is a pulsating variable that has a period of only 4 days.

The luminosities of other stars vary for different reasons. One type of variable star, called an *eclipsing binary,* is actually a pair of stars that move around each other. Their brightness appears to vary as one star periodically moves in front of the other, causing an eclipse that blocks its light. If, for example, a pair of stars happen to revolve around each other in line with the Earth, one will periodically block our view of the light from the other. As this happens, they will appear to vary in brightness.

Shown here, Proxima Centauri brightens and dims in only one day due to an X-ray flare.

Novas occur when stars explode.

This supernova, photographed in 1987, appeared in the sky over the Southern Hemisphere. It was the first supernova visible to the unaided eye in nearly four centuries.

Novas, such as the one shown in the photo above, are variable stars that give off bursts of energy that make them appear many thousands of times brighter for days or even years. The term *nova* comes from the Latin word for "new." They were given this name because in earlier times they were thought to be new stars. A nova eventually returns to its original brightness. A *supernova,* on the other hand, gives off such a huge burst of energy that the star blows itself completely apart.

Galaxies: Collections of Stars

Galaxies consist of billions of stars. There are many galaxies in the universe. However, the one we know most about is our own galaxy.

The Milky Way galaxy is a flat spiral galaxy of over 100 billion stars, stretching 100,000 light-years across. Have you ever noticed, on a very clear night, a band of stars so dense that it looks like a starlit cloud? This band of stars, called the Milky Way, is actually a portion of our galaxy, viewed from our position in one of its spiral arms.

A galaxy is held together by gravity, which tends to pull the stars toward one another. The Milky Way galaxy is shaped like a disk with a bulge at its center. Our solar system is located closer to its edge than to its center. Thus, by looking in the direction of the galaxy's center, we see many more stars than we would by looking toward the edge. However, the Milky way is not the only galaxy you can see. If you look at the constellation Andromeda, it is possible to see a small, fuzzy patch of light among the stars. A telescope reveals that this is a galaxy about 2.2 million light-years away. Our own Milky Way would look much the same if we could see it from far out in space.

If you look up at the sky on a very dark, clear night, you can see the Milky Way.

The spiral galaxy M31 in the constellation Andromeda (top left) may look very much like our own galaxy. Other types of galaxies are the barred spiral (top right), the elliptical (bottom left), and the irregular (bottom right).

Types of Galaxies
Telescopes that allow us to look deep into space show us that the universe contains billions of other galaxies. And just as stars differ from each other, so do galaxies. In addition to the spiral shape of the Milky Way and Andromeda galaxies, some galaxies are shaped like footballs or like spheres. Other galaxies have no regular shape. The major types of galaxies are spiral, barred spiral, elliptical, and irregular.

Except for a few galaxies in our immediate vicinity, all galaxies that can be observed have one thing in common: they are all moving away from each other at very high speeds. Scientists can tell that the galaxies are moving apart because of the information that they get from light. The colors in the spectra of the stars in these galaxies are shifted toward the red end. This *red shift* in the light from the galaxies, a type of Doppler effect, is caused when the light waves are spread out as the light source moves away. You may have heard the same effect in the changing pitch of sound waves that come from a moving source such as a train whistle or car horn.

DID YOU KNOW...

that extraterrestrials were thought to have contacted Earth?
The first recorded pulsars emitted such regular bursts of energy that some astronomers thought the pulses were being transmitted by intelligent beings. In fact, pulsars were originally called LGMs—an acronym for "little green men."

Strange Objects in Space

As if red and blue stars, novas and supernovas, and spiral and elliptical galaxies were not enough, even stranger objects can be found in the universe. Scientists have only begun to examine such objects, some of which are discussed below.

Pulsars While searching the skies, scientists detected rapid pulses of radio waves coming from certain stars. These radio pulses occurred at very regular intervals. These stars, or *pulsars,* gave off continuous pulses of radio waves, light, and X rays. As a pulsar spins on its axis, it sends out a stream of radiation in a pattern similar to that of the rotating light in a lighthouse. When the Earth is in the path of one of these streams of radiation, radio telescopes can pick up the signals as rapid pulses. A pulsar at the center of the Crab nebula, for example, rotates 30 times a second. Most pulsars are remnants of supernova explosions.

An artist's conception of a rotating pulsar

Quasars One of the most puzzling of all objects detected by astronomers is the quasi-stellar radio source, or *quasar* for short. Quasars emit vast amounts of radio waves and light. They appear to be several billions of light-years away and may be the most luminous objects in the universe, producing more energy than is produced by hundreds of galaxies combined. Since they seem to be so far away, quasars may also be among the oldest objects in the universe. Scientists still do not fully understand the nature of quasars.

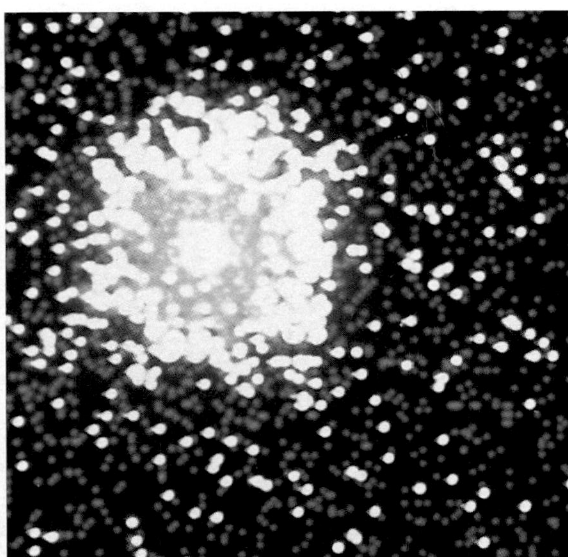

Quasar 3C 275, shown here in this X-ray image, was the first quasar to be discovered. It is the brightest and nearest known quasar—over 2 billion light-years away—and is moving away from Earth at 50,000 km/s.

Black Holes Astronomers have been searching for ways to explain the huge amounts of energy produced by quasars and other objects. Some have theorized that very large stars may collapse with such force that they become a "black hole." As you know, a black hole is an object so dense that its gravity will allow nothing—not even light—to escape from it. Because no energy of any kind can escape from a black hole, there is no way to observe it directly. However, some astronomers hypothesize that they may be able to locate black holes by the effect they have on nearby objects. For example, a black hole close to a star would strip matter from its neighbor. The material lost from the star would then give off energy such as X rays as it disappeared into the black hole. This process could explain some of the powerful energy sources now detected in space. In 1995, the Hubble Space Telescope produced compelling evidence that a black hole exists at the heart of galaxy M87. One is also thought to be at the center of the Milky Way. Yet black holes remain unobserved theoretical objects. Their existence can be inferred only by the effects they are assumed to produce.

Massive black holes are thought to exist at the center of many galaxies, including the Milky Way. This photo shows the central region of M87, a galaxy whose core structure could be explained by the presence of a black hole.

SUMMARY

There are billions and billions of stars in the universe. Astronomers use parallax and luminosity to estimate the distance to nearby stars. Astronomers estimate the surface temperature of a star by its color. Most stars are part of a binary system, and some stars are variables. Galaxies are huge collections of stars and are found in a variety of shapes and sizes. Some of the most unusual objects in space are pulsars, quasars, and the elusive black holes.

One Small Step

Stonehenge is the ruins of an ancient monument in southern England. It consists of two rings of rectangular stone columns. Other stone rectangles lie on top of some of these columns. The tallest column is 7.3 m high. Each stone weighs 30–60 tons. Through the center of the rings is a broad avenue. At the end of the avenue is a stone 5.4 m high called the Heel Stone. Why did people build this structure? Why did they carry these heavy stones from as far as 380 km away?

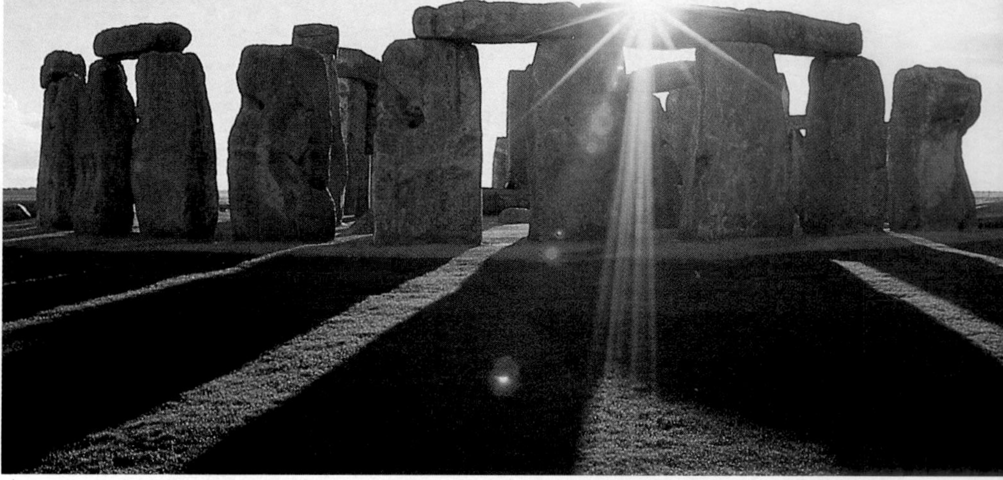

Stonehenge, begun around 2,600 B.C., is thought to have been used by early inhabitants of the British Isles to plot the motions of the Sun and Moon.

How We See the Universe

For centuries, astronomers have watched the evening skies. Using observatories such as Stonehenge, they noticed that the stars kept the same relative positions. However, five points of light, the visible planets, were observed to slowly move in relation to the stars. Ancient people could only wonder at the nature of these "wanderers."

But once the telescope was developed, these wanderers could be seen as planets illuminated by the Sun. While some surface features were visible through the first telescopes, it was not until space probes were launched that we got our first truly clear images of the planets. By 1989, eight of the nine known planets had been visited by probes sent from Earth.

The Telescope Is Born The actual invention of the telescope may have occurred several times. By the 1600s, it was being used as a field glass for viewing distant objects on Earth. In 1609, however, Galileo made his own telescope and pointed it toward the sky. Although his instrument was primitive, Galileo made many discoveries, such as the craters and mountains on the Moon, the phases of Venus, and the four large satellites of Jupiter.

Galileo's telescopes used lenses to focus light from distant objects.

The ancient Maya were accomplished astronomers for their level of technology. Built in A.D. 370, El Caracol in Chichén Itzá, Mexico, is the oldest known observatory in the Americas.

Isaac Newton constructed the first reflecting telescope. Reflecting telescopes use mirrors instead of lenses. Concave mirrors have three advantages over lenses. First, they reflect light instead of refracting it and therefore produce much less distortion. Second, a good mirror absorbs less light than even the most transparent lens, so the image is brighter. Third, a mirror is not as heavy as a lens of the same size. Therefore, a mirror is easier to support, and much larger telescopes can be constructed. The largest optical telescopes used today are reflecting telescopes. The new Keck telescope on top of Mauna Kea in Hawaii is the largest reflecting telescope in the world—twice the size of the famous Hale telescope at Mount Palomar. The images that this giant telescope can produce rival those of the orbiting Hubble Space Telescope.

A replica of the reflecting telescope made by Isaac Newton

Other Types of Telescopes Visible light is not the only radiation emitted by stars and galaxies. Radio waves, X rays, and ultraviolet waves are also produced by these objects. For example, stars, large planets, quasars, and the cores of many galaxies emit radio waves. Since radio waves are not absorbed by dust in space, as are visible light waves, radio waves may cross the entire universe with little loss of energy, giving us a clearer view.

Astronomers can capture radio waves with a special instrument called a radio telescope. Though some radio telescopes are like large television antennas, most are shaped like metallic dishes. Radio waves are reflected by metallic surfaces just as light is reflected by a mirror. At the focus of a radio telescope is a radio receiver. Radio astronomers use a tuner to select the radio frequency that they wish to study.

DID YOU KNOW...

that the Hubble Space Telescope has seen deeper into space than any other optical telescope?

Since it was repaired, the HST works extremely well. In fact, if it were placed in New York City, it could detect a firefly as far away as Tokyo!

The Keck telescope is the world's largest reflecting telescope. Its mirror consists of 36 individually adjusting panels that form a mirror 10 m in diameter.

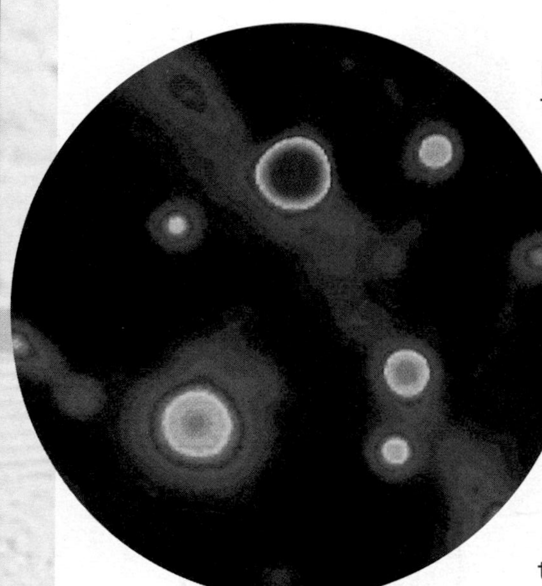

This false-color radio image of the center of M31, the Andromeda galaxy, shows what the top-left photo on page S150 looks like with radiowaves. The image was taken by a group of radio telescopes such as the one shown to the right.

A single radio telescope provides an image of the sky that is less precise than the image provided by an optical telescope. Radio telescopes can be linked together, however, so that the image received by one can be compared and mixed with the image received by another. Combining images from several radio telescopes can produce a much clearer radio image of the sky. By linking together radio telescopes from around the world, scientists have been able to create an instrument that is essentially the size of planet Earth!

Other telescopes, some of which are located on orbiting platforms, collect information from different parts of the spectrum. Ultraviolet and infrared telescopes, as well as X-ray and gamma-ray collectors, now give astronomers a much broader look at the universe around us.

A radio telescope provides images of the universe using radiowaves rather than visible light.

The Exploration of Space Begins

Suppose that observers from outer space have been watching the Earth since it was formed. For more than 4 billion years, they would not have seen anything leave the Earth. Then suddenly, starting about 40 years ago, they would have noticed that the people of Earth had begun to send tiny craft into space. At first these spacecraft only flew around the Earth. Then they landed on the Moon. Later the observers would have seen other spacecraft fly from the Earth to its neighboring planets. The observers might conclude that the people of Earth had begun a new age of exploration.

Rockets and Early Satellites

Even though in 1865 Jules Verne had written about a fictitious trip to the Moon, the first person to scientifically study the use of rockets for space travel was Konstantin Tsiolkovsky. In the early 1900s, this Russian inventor designed a spaceship with a bullet-shaped passenger cabin in the nose and fuel tanks filled with liquid hydrogen and liquid oxygen in the tail of his multistage rocket. Although Tsiolkovsky never launched a rocket himself, his designs were fundamental to the development of rocket flight.

DID YOU KNOW...

that the Chinese were actually the first to experiment with rocket propulsion?

They first developed rockets to launch fireworks. By the 1200s, the Chinese were using rockets to propel "arrows of flying fire" in warfare.

At about the same time, an American physicist named Robert Goddard was actually experimenting with rockets. In 1926 Goddard launched the world's first liquid-fueled rocket. During World War II, the Germans constructed many rockets to deliver bombs. At the end of the war, scientists in the United States used captured German rockets to explore the upper atmosphere. An American two-stage rocket, with a captured German rocket as the first stage, reached an altitude of 393 km in 1949.

The development and use of rockets proceeded slowly until October 4, 1957, when the Soviet Union surprised the world by placing into orbit an artificial satellite called *Sputnik 1.* This 80 kg satellite carried instruments to measure the density and temperature of the upper atmosphere. A month later, on November 3, 1957, the Soviet Union put another satellite into orbit. This second satellite carried a passenger—a dog named Laika. The purpose for sending the animal into orbit was to study the effects of space travel on a living organism.

The first American satellite, *Explorer 1,* was launched on January 31, 1958. It carried a scientific package for measuring temperature, cosmic rays, and meteors. *Explorer 1* discovered the Van Allen radiation belts that surround the Earth.

Sputnik 1, the first artificial satellite to orbit the Earth

Explorer 1, the first American satellite in space

Robert Goddard launched a gasoline-fueled rocket in 1926. The rocket was 1.2 m long, and it rose to a height of 56 m at an average speed of 103 km/h.

Human Exploration

On April 12, 1961, Yuri Gagarin, a 27-year-old cosmonaut, was launched into space. He orbited the Earth once, reaching an altitude of 327 km. On February 20, 1962, American astronaut John Glenn was launched into space and made three orbits around the Earth. The first woman to go into space was Valentina Tereshkova, who was launched in June of 1963. In 1965, Alexei Leonov became the first human to "walk" in space. Dressed in a protective spacesuit, he left his pressurized cabin and floated for 10 minutes in space. All of these space firsts were accomplished within 10 years of the launch of *Sputnik 1;* the next 10 years were even more dramatic.

Valentina Tereshkova was the first woman to venture into space.

As a challenge to the Soviet Union's leadership in space, the United States was determined to place an astronaut on the Moon before the end of the 1960s. In preparation for this journey to the Moon, many additional Earth-orbiting missions were launched. During the Mercury and Gemini missions, American astronauts learned to live and work in space. On July 16, 1969, *Apollo 11* was launched from Cape Canaveral, Florida, toward the Moon. The *Apollo 11* lunar module landed on the surface of the Moon on July 20 at 8:17 P.M. Neil Armstrong and Edwin "Buzz" Aldrin became the first humans to set foot on another world.

Neil Armstrong and "Buzz" Aldrin spent more than 2 hours walking on the Moon.

The space shuttle, a vehicle that can lift off like a rocket and return to Earth like an airplane, is now the United States' major vehicle for space travel. It was originally designed to be used in the construction of a space station, but many other missions involving the delivery of military, scientific, and telecommunications satellites as well as orbiting telescopes have been performed.

In anticipation of its role in building the planned International Space Station Alpha, the space shuttle *Atlantis* docked with the Russian space station *Mir* on June 29, 1995, marking the United States' 100th human space mission. Norm Thagard, an American astronaut and physician who had spent 105 days aboard the 9-year-old space station, was given a ride home, and a fresh crew was delivered to the station. During this historic meeting in space, a total of six American astronauts and four Russian cosmonauts were gathered in orbit 400 km above the Earth.

By landing on a runway, the space shuttle can be used again and again.

The Russian space station, *Mir*, has been in Earth orbit since 1986.

DID YOU KNOW...

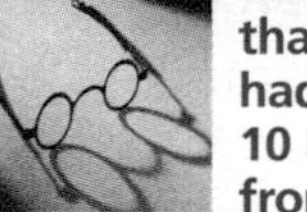

that by 1995, *Voyager 1* had traveled over 10 billion kilometers from Earth?

Launched in 1977, *Voyager 1* is traveling at a speed of 63,000 km/h and is still sending back information about space beyond our planetary system.

Interplanetary Probes Interplanetary probes, such as *Pioneer 10, Voyagers 1* and *2,* and *Magellan,* have sent back much more information about our neighbors in the solar system than could ever have been gained by looking through telescopes. Another space probe, *Galileo,* recently arrived on a historic mission to Jupiter. Having passed Mars and the asteroid belt, it finally reached Jupiter in December of 1995 on a planned 23-month mission. *Galileo* also released a smaller probe into the Jovian atmosphere. This probe was designed to send valuable information back to *Galileo* for about 75 minutes before being crushed by Jupiter's enormous atmospheric pressure.

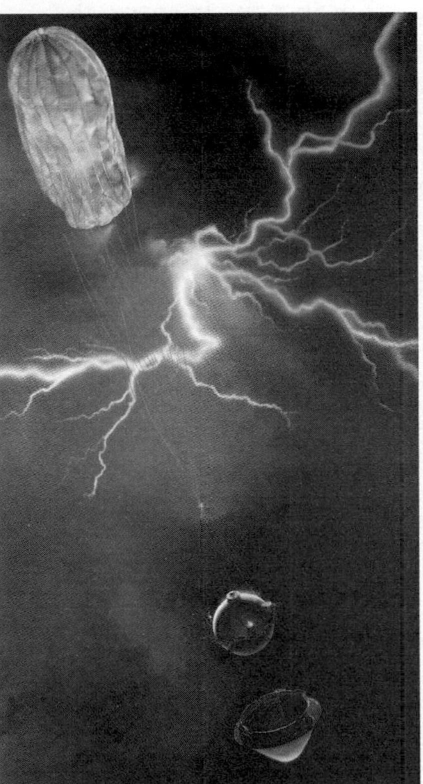

Artist's impression of *Galileo's* entry probe descending through Jupiter's atmosphere

The space probe *Ulysses* has recently made a historic rendezvous with the Sun. Traveling at 116,000 km/h, it is the first probe ever to journey outside the ecliptic plane of the planetary orbits. This means that rather than orbiting the Sun like the planets do, it is traveling at right angles to their paths. It crossed over the Sun's south pole in June 1994 and discovered that the solar wind was moving almost twice as fast as it does when going past Earth. *Ulysses* passed over the Sun's north pole in June of 1995, giving us our first view of the top of our own star, the Sun.

Ulysses has traveled around the Sun, gathering information from a new perspective.

Cosmology

The study of the structure, origin, and evolution of the universe

Theories of Cosmology

Since light takes time to travel, the farther we look out into space, the further back in time we can see. This ability may help us answer some very fundamental questions about our universe. **Cosmology** is the study of how the universe began and how it may eventually end. There is no way to witness the actual events that occurred billions of years ago, but scientists have proposed several theories to account for the universe as we know it today.

The Big Bang The observation that the universe is expanding has led scientists to develop the *big bang theory.* This theory states that at one time, all matter in the entire universe was concentrated in an extremely small volume. Then, about 10 to 20 billion years ago, there was a sudden event, the big bang, that caused the universe to begin expanding. About 1 billion years after the big bang, huge clouds of hydrogen and helium began to form. These clouds were the beginnings of galaxies. In time, stars started to develop in the galaxies. Within about 10 billion years after the big bang, all of the galaxies that can now be seen had formed. Today we see the results of the big bang as a rapidly expanding universe with galaxies moving away from each other.

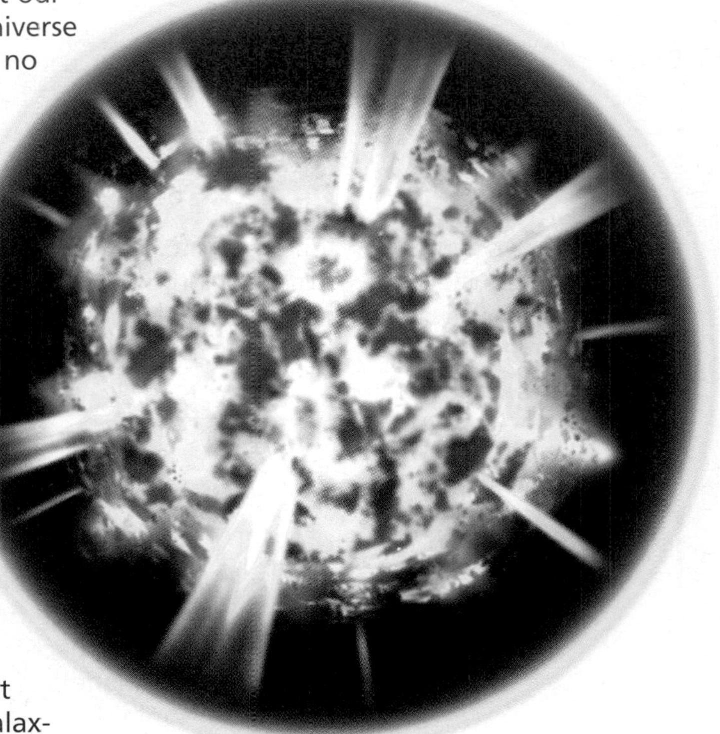

The big bang is one explanation of how our universe may have begun.

As you know, scientists have found that all space is filled with weak microwave radiation. One explanation of this cosmic background radiation is that it is evidence left by the big bang. Space was thought to have a uniform temperature of 3 K (–270°C), but faint readings by the *Cosmic Background Explorer,* an orbiting observatory, indicate that some areas of space are ever so slightly hotter than others. This indicates that the matter released during the big bang is not evenly distributed. The force of gravity in these denser regions could have drawn matter together to form the stars, galaxies, and other bodies in the universe. Yet many questions remain regarding just how this could have happened and whether or not it happens over and over again.

Pulsating Universe If there is enough matter in the universe, gravity should eventually stop the outward expansion of the galaxies caused by the big bang. A variation of the big bang theory, called the *pulsating* or *oscillating* universe theory, states that the universe will eventually contract, bringing the galaxies back together into another hot mass, until another big bang sends matter flying out again. This is repeated over and over, so the universe will continue to expand and shrink forever. However, the estimated density of the universe is such that scientists cannot yet determine whether the expansion will stop or if the universe will keep expanding forever.

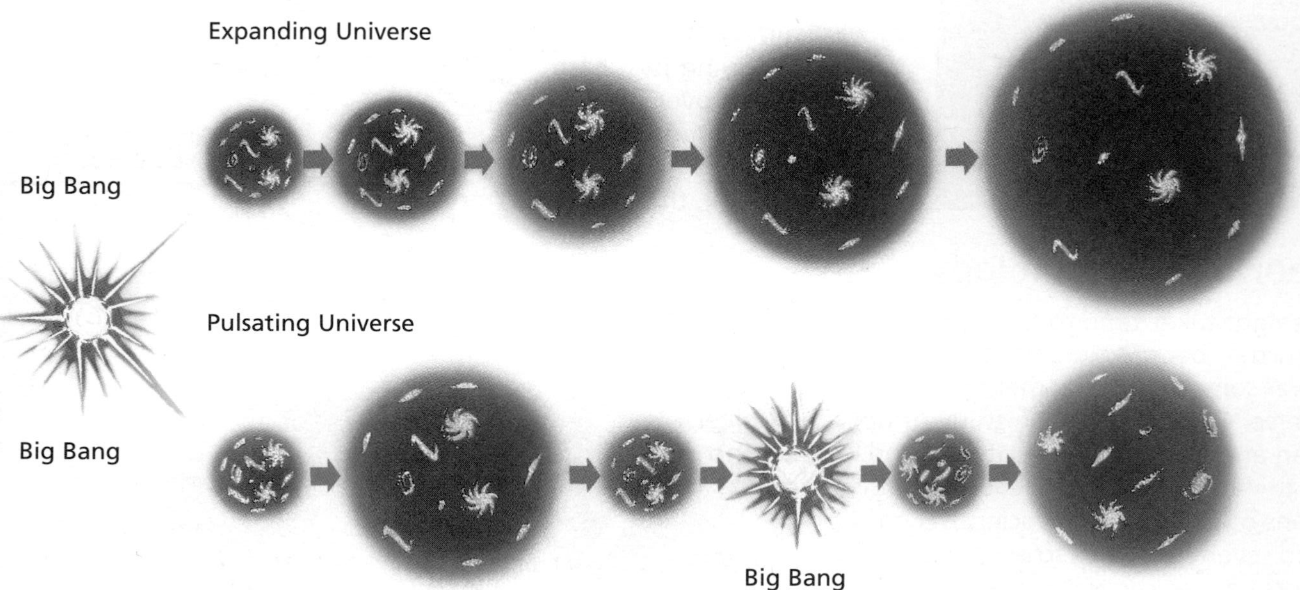

Many astronomers think that the universe will continue expanding forever. Others think there is enough matter in the universe to make it stop expanding and collapse back on itself, leading to another big bang and an endlessly repeating cycle.

SUMMARY

A variety of telescopes have enabled humans to observe the universe from Earth. With the development of the rocket, space probes have been sent on missions to view the planets of our solar system up close. The universe contains billions of other galaxies that are apparently moving apart as the universe expands. How the universe began and how it will end are two of the many questions that astronomers are still trying to answer.

Unit CheckUp

Concept Mapping

Starting with the terms shown here, construct a concept map that illustrates major ideas from this unit. Arrange the terms in an appropriate manner and connect them with linking words. Then extend your concept map by adding as many additional terms from the unit as you can. Use your ScienceLog. **Do not write in this textbook.**

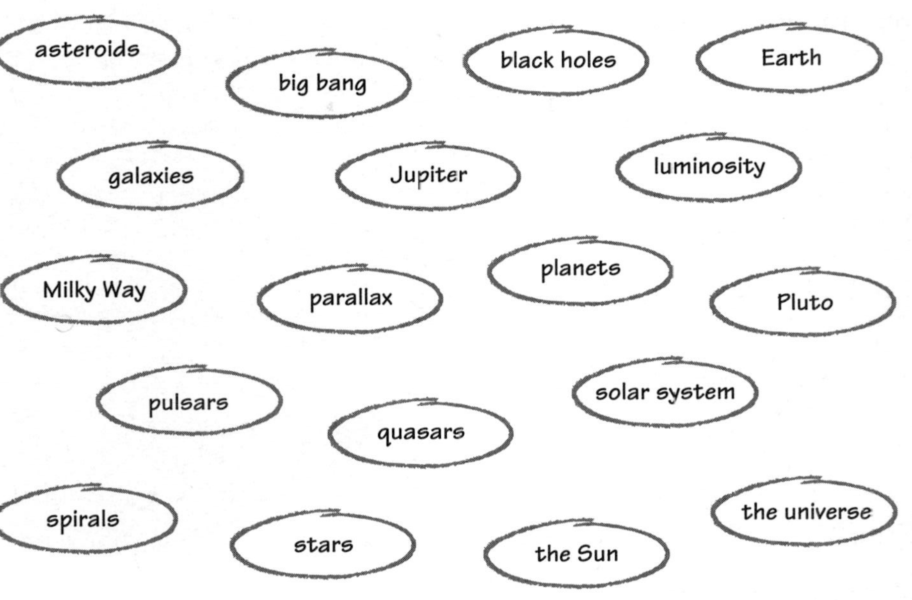

Checking Your Understanding

Select the choice that most completely and correctly answers each of the following questions.

1. The reactions inside the Sun most likely take place in which of the following layers?
a. the photosphere
b. the core
c. the radiative layer
d. the corona

2. The aurora borealis, or northern lights, are a result of the interaction between the Earth's magnetic field and
a. the solar wind.
b. the sun's radiative zone.
c. light coming from the photosphere.
d. polar thunderstorms.

3. If *Voyager 1* were to return to Earth, it would pass the orbits of the planets in the following order:
a. Neptune, Uranus, Jupiter, Saturn, Mars.
b. Neptune, Saturn, Uranus, Mars, Jupiter.
c. Uranus, Neptune, Saturn, Jupiter, Mars.
d. Neptune, Uranus, Saturn, Jupiter, Mars.

4. Of the following stars, which has the highest surface temperature?
a. Rigel, a blue star
b. Capella, a yellow star
c. Betelgeuse, a red star
d. Aldebaran, an orange star

5. One of the most efficient aspects of a space shuttle is that
a. it can carry both military and scientific equipment.
b. it can be reused over and over again.
c. it can be used to construct a space station.
d. it looks like an airplane.

Interpreting Illustrations

How many stars and constellations can you identify on the star chart shown here? Write their names in your ScienceLog. Then check your list with a classmate.

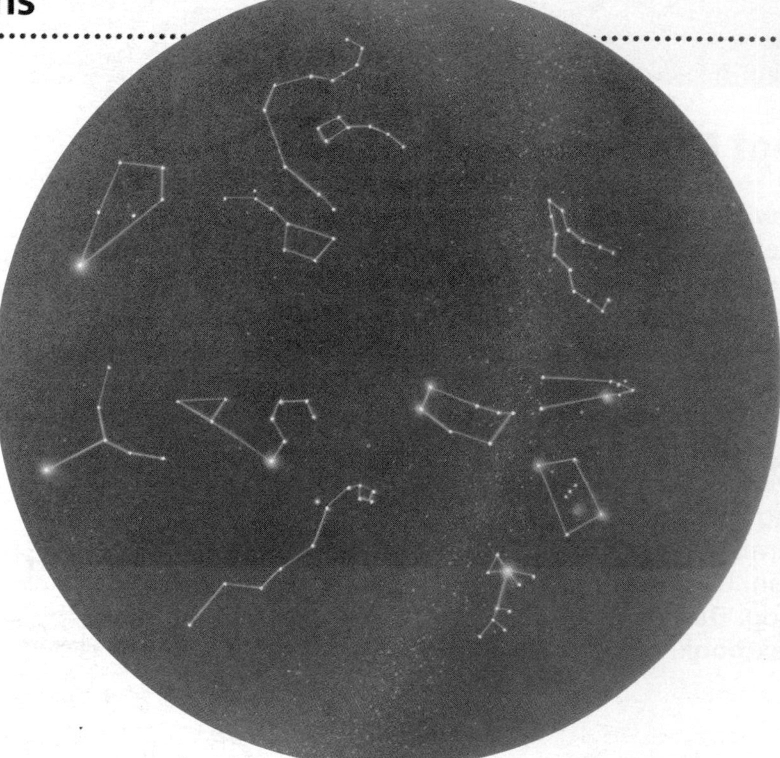

Critical Thinking

Carefully consider the following questions, and write a response in your ScienceLog that indicates your understanding of science.

1. Once the hydrogen supply in the Sun is used up, it cannot be replaced. Yet solar energy, or the energy we receive from the Sun and use to heat our water and homes, is considered a renewable resource. Explain.
2. Uranus is tilted almost on its side, with its axis lying near the orbital plane (ecliptic) and its south pole, now nearly pointing toward the Sun. Would this have more of an effect on the length of the planet's year, day, or seasons? Explain.
3. Why does the method of using parallax to calculate the distance to stars work only for relatively close stars?
4. The Apollo program was part of America's "race to the Moon" against the Soviet Union. How did that competition help or hinder the exploration of space? How might the present age of cooperation among different nations affect future developments?
5. Recent observations indicate that the universe could be only 8 to 12 billion years old. Yet some stars are thought to be older than that. What problems might these observations create for theories of cosmology?

Portfolio Idea

To get an idea of the difference in sizes of the Sun and the planets, make the following model of the solar sytstem: Let Earth's radius (6378 km) equal 3 cm. Calculate the relative radii in centimeters of the other planets and the Sun in comparison to that of Earth. (Refer to the table on page S142 for data.) Then, using a compass, draw circles on a large piece of paper to represent the size of each planet. (You may need to use a thumbtack, string, and a pencil to draw circles for the larger planets.) Finally, draw a circle to represent the Sun, which has a radius of 695,000 km.

GROWING PLANTS

In This Unit

Now that you have learned what plants need in order to grow, consider these questions.

1. How do the leaves, stems, and roots work together in a plant?
2. How have flowers become "specialized" to carry out their function?
3. In what ways do humans depend on plants?

In this unit, you will take a closer look at the structure and function of stems, roots, leaves, and flowers. You will also learn about some of the ways that we use plants.

S161

THE PLANT BODY

Have you ever wondered why a plant's body is so different from yours? These differences are related to the different ways that plants and animals live. Being an animal, your way of life involves moving around and eating food. Food gives you the energy and nutrients that your body needs in order to grow. Having arms, legs, eyes, and a mouth, not to mention all your other body parts, allows you to accomplish these necessary tasks.

A plant's way of life is very different from yours. Plants stay in one place and gather carbon dioxide and water from their surroundings. Then, by using energy from the sun, they convert these nutrients into *sugar*, which they use for food.

Most of the plants you see every day, such as trees and shrubs, are *vascular plants*. Vascular plants have become adapted to living on land by developing a body that consists of leaves, stems, and roots. Inside the leaves, stems, and roots of a vascular plant are *vascular tissues*. These tissues form tubes that transport water, minerals, and food throughout the plant. Together, all the vascular tissues form a vascular system that extends throughout the entire plant, similar to the way that your system of blood vessels extends through all the parts of your body. To understand how plants function, let's examine the basic parts of a plant's body.

Leaves

Leaves are the food-making organs of plants. Leaves use light, carbon dioxide, and water during *photosynthesis*. During this process, plants use energy from the sun to convert carbon dioxide and water into sugar and oxygen.

S162

Structure of Leaves Most leaves are thin and flat, which exposes a maximum number of leaf cells to sunlight. This shape also allows the maximum amount of a leaf's surface to contact the air, from which leaves take in carbon dioxide.

The veins of a leaf are made of vascular tissues that bring water and minerals into the leaf. These tissues are made of long, tubular cells that resemble small water pipes. Veins also transport sugar made by the leaf to other parts of the plant. Although leaves might appear to be simple in structure externally, inside they are quite complex.

The cells of a typical leaf blade are organized into several specialized layers. A layer of cells, called the *epidermis*, covers both the upper and lower surfaces of the leaf and prevents the loss of water from inner cells. Inside the leaf are two kinds of food-making cells that contain special organelles called **chloroplasts**. The chloroplasts contain chlorophyll, which absorbs light for photosynthesis.

Chloroplasts

Organelles of plant cells that contain chlorophyll and are the sites of photosynthesis

Stomata

Small pores in leaves that allow gases to enter and leave plants

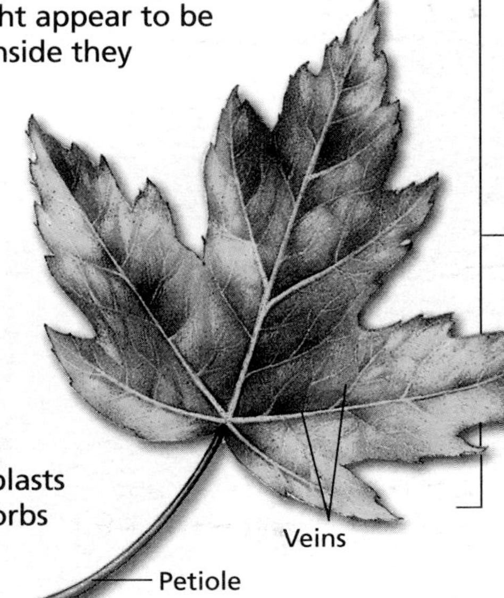

◄ Leaf veins are part of a plant's vascular system, which extends all the way down into the plant's roots. The main vein in a leaf divides many times into smaller and smaller veins. As a result, water and food molecules do not have far to travel between a vein and any cell in the leaf.

1 The tightly packed upper layer of food-making cells performs most of the photosynthesis in the leaf. The food-making cells on the lower side of the leaf are loosely packed and have many air spaces between them. These air spaces are connected to the outside by small openings called **stomata.**

2 Each stoma is surrounded by two *guard cells* that open and close the stomata by changing shape. The lower surface of a leaf may have thousands of stomata that allow carbon dioxide, oxygen, and water vapor to go into and out of the leaf. The top surface of a leaf has very few stomata and is coated with a waxy material that prevents water from evaporating.

3 The main vein in a leaf divides many times into smaller and smaller veins. As a result, water and food molecules do not have far to travel between a vein and any cell in the leaf.

4 The epidermis is tough, which gives strength to a leaf, and its cells are clear, which allows light to pass through.

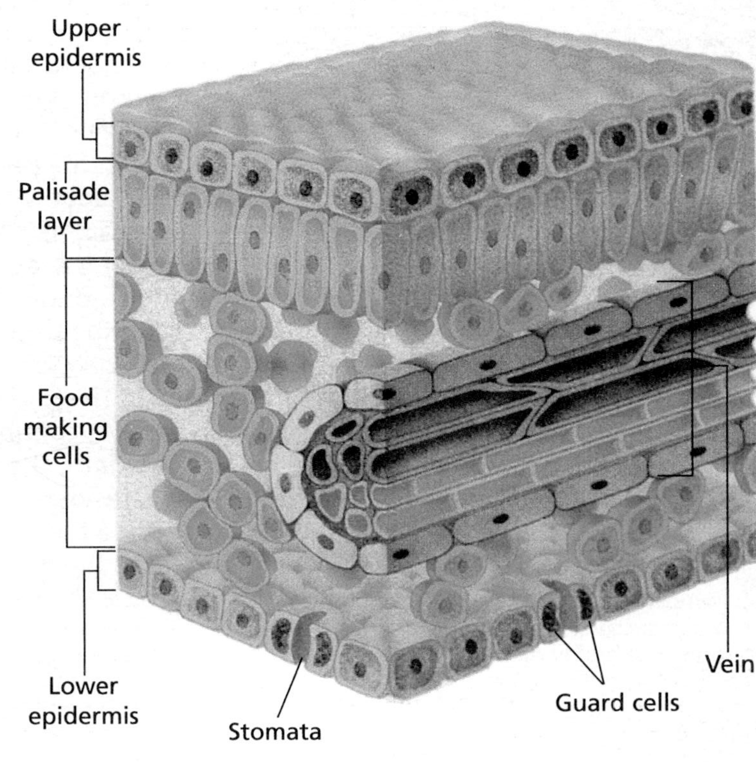

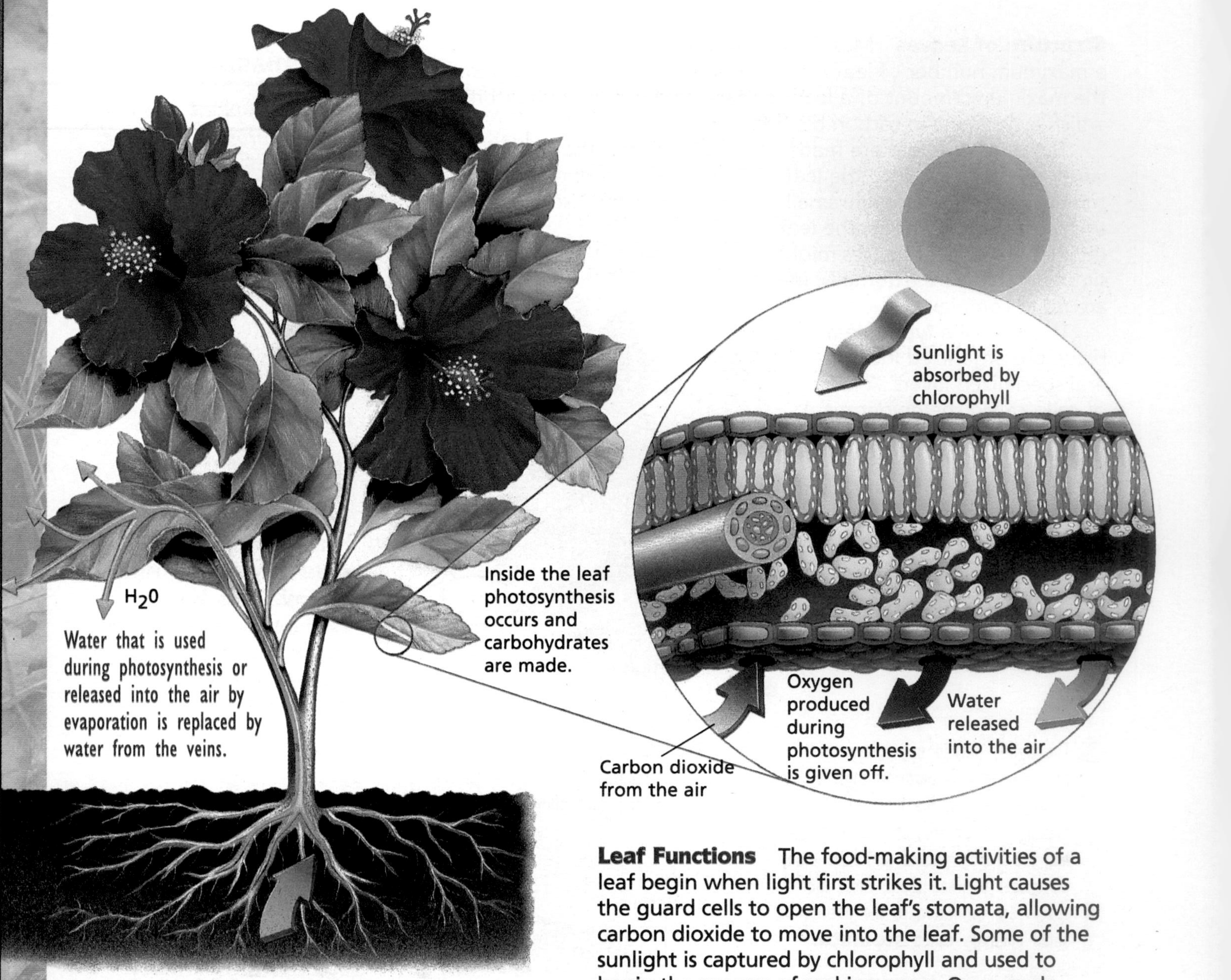

Sunlight is absorbed by chlorophyll

Inside the leaf photosynthesis occurs and carbohydrates are made.

Oxygen produced during photosynthesis is given off.

Water released into the air

Carbon dioxide from the air

H_2O

Water that is used during photosynthesis or released into the air by evaporation is replaced by water from the veins.

If the soil around the roots becomes too dry, the leaves will not receive enough water. What do you think this does to the food-making activities of the leaf?

The roots of a plant take up more water when it is used or lost by the leaves. As long as there is enough water available, it continues to move through a plant during daylight hours.

DID YOU KNOW...

that some leaves are not photosynthetic? In most cacti, it is the green succulent stems that make food. The highly modified leaves of cacti, often called spines, serve only to protect the plants from predators.

S164

Leaf Functions The food-making activities of a leaf begin when light first strikes it. Light causes the guard cells to open the leaf's stomata, allowing carbon dioxide to move into the leaf. Some of the sunlight is captured by chlorophyll and used to begin the process of making sugar. Once made, the sugar molecules move into the veins of the leaf and are carried to other parts of the plant. The sugar is used for energy or stored as *starch*. Oxygen made during photosynthesis moves out of the leaf into the surrounding air. At the same time, more carbon dioxide enters the leaf through the stomata. This exchange of gases, shown in the diagram above, continues throughout the daylight hours. At night, the stomata close.

Another function of the leaf is to release water inside the plant into the atmosphere. Water vapor exits the plant when water evaporates from cells in the leaf. The water vapor enters the outside air by passing through the stomata on the undersurface of the leaf. When water is used or lost by the leaves, the roots of the plant take up more water to replace it.

Stems

What do a cactus pad, a potato, a stalk of asparagus, and a tree trunk have in common? They are all stems of plants. The stem is the midsection of a plant's body that, in most plants, grows above the ground. In many plants, stems have specialized functions. The white potato plant, for example, stores food in specialized stems (the potatoes). In the cactus, the stem stores water. However, plant stems do have two basic functions—they support the leaves and transport food and water between the leaves and the roots.

Vascular bundles

Groups of water-carrying and food-carrying cells

Structure of Stems You may have noticed that some plant stems are soft and flexible while others are stiff and hard. Plants that have soft and flexible stems, such as beans, tomatoes, grasses, and petunias, are called *herbaceous* plants. Plants whose stems are stiff and hard, such as trees and shrubs, are called *woody* plants.

Both herbaceous and woody plants contain two kinds of vascular tissue. One kind of vascular tissue carries water and minerals upward from the roots to the leaves, while the other kind generally carries food away from the leaves to the stems, roots, and any developing fruits and seeds. Both kinds of tissue occur together in long strands, called **vascular bundles**.

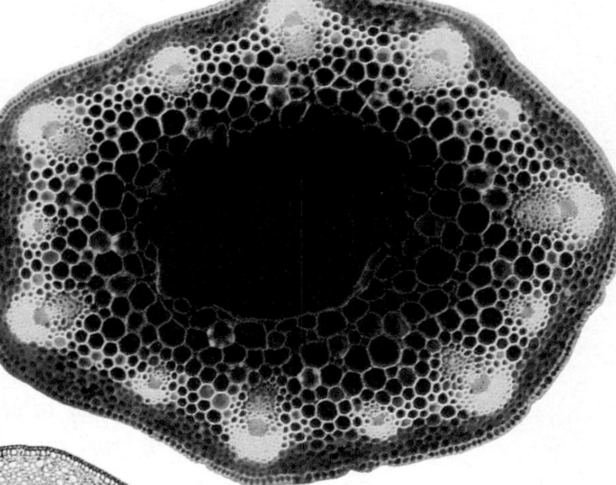

In some herbaceous stems, the vascular bundles are scattered throughout the stem (left) and in others, they are arranged in a ring (above). The larger, thick-walled cells in each vascular bundle are the water-carrying cells. The smaller, thin-walled cells are the food-carrying cells.

Woody plants can grow much taller and larger than herbaceous plants because the woody stems provide more support.

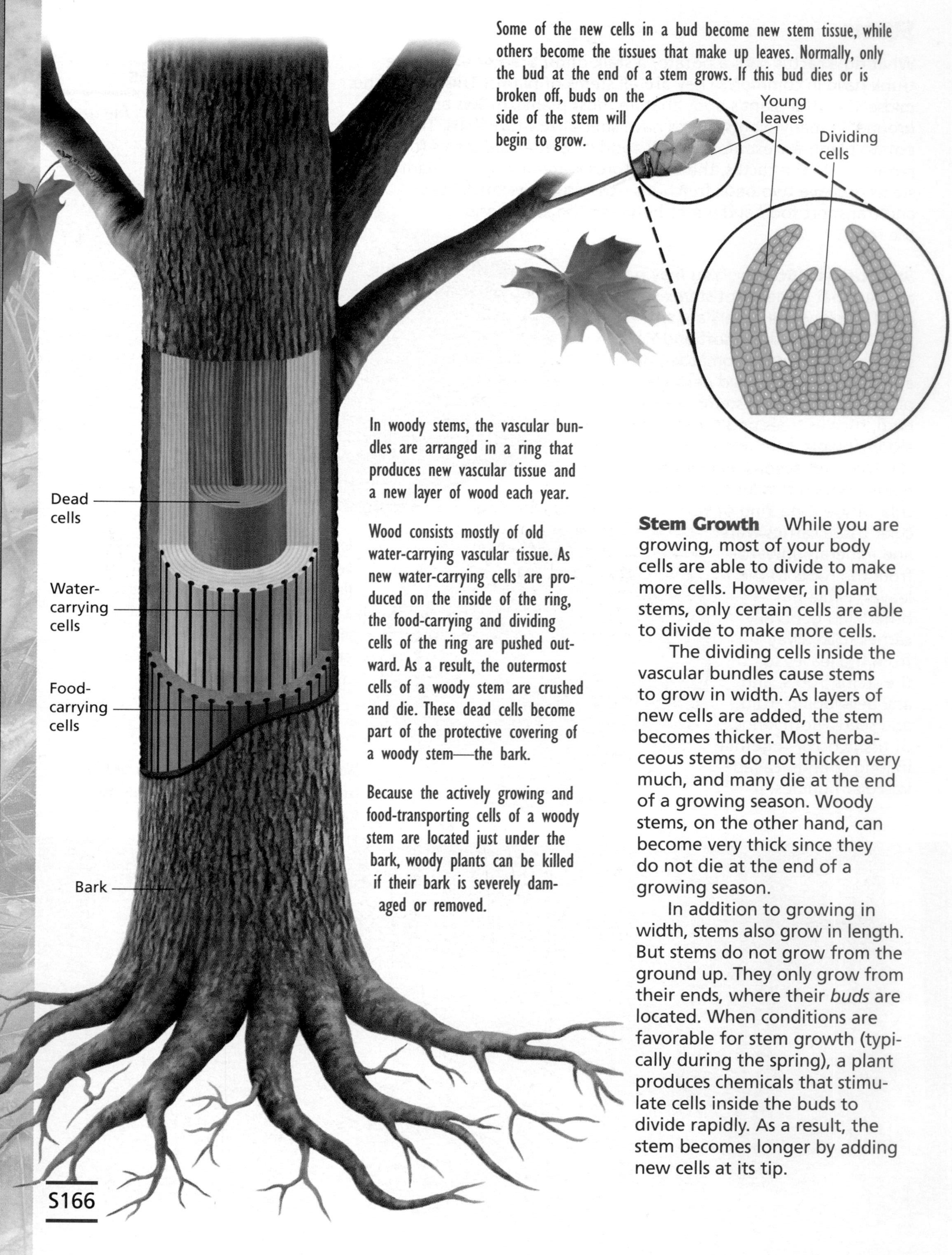

Some of the new cells in a bud become new stem tissue, while others become the tissues that make up leaves. Normally, only the bud at the end of a stem grows. If this bud dies or is broken off, buds on the side of the stem will begin to grow.

In woody stems, the vascular bundles are arranged in a ring that produces new vascular tissue and a new layer of wood each year.

Wood consists mostly of old water-carrying vascular tissue. As new water-carrying cells are produced on the inside of the ring, the food-carrying and dividing cells of the ring are pushed outward. As a result, the outermost cells of a woody stem are crushed and die. These dead cells become part of the protective covering of a woody stem—the bark.

Because the actively growing and food-transporting cells of a woody stem are located just under the bark, woody plants can be killed if their bark is severely damaged or removed.

Stem Growth While you are growing, most of your body cells are able to divide to make more cells. However, in plant stems, only certain cells are able to divide to make more cells.

The dividing cells inside the vascular bundles cause stems to grow in width. As layers of new cells are added, the stem becomes thicker. Most herbaceous stems do not thicken very much, and many die at the end of a growing season. Woody stems, on the other hand, can become very thick since they do not die at the end of a growing season.

In addition to growing in width, stems also grow in length. But stems do not grow from the ground up. They only grow from their ends, where their *buds* are located. When conditions are favorable for stem growth (typically during the spring), a plant produces chemicals that stimulate cells inside the buds to divide rapidly. As a result, the stem becomes longer by adding new cells at its tip.

Roots

The roots are the part of a plant's body that, in most cases, grows underground. Just like stems and leaves, roots also have several functions. One function of the roots is to anchor the plant in the ground, making it less likely that the plant will be blown away by wind or washed away by rain. Some roots, such as carrots, sweet potatoes, and turnips, also store food. However, the most important function of roots is to absorb water and minerals for the plant.

Some plants have roots that grow from aboveground parts, like their stems and leaves. The roots growing out from the stem of this corn plant are commonly called prop roots because they help stabilize the plant in the soil.

Types of Root Systems If you have ever pulled weeds in a garden or a lawn, you have probably noticed that some weeds come up easily, whereas others are held so firmly that the top of the plant usually breaks off before the roots come out of the ground. One reason for this is that there are two basic types of root systems in plants.

Taproots are the large, central roots that grow almost straight down. The main taproot of some trees may reach down hundreds of feet into the ground. They can be as long as a plant is tall, anchoring it firmly into the ground. Taproots can be long and slender, thick like carrots, or bulb-shaped like radishes. However, the smooth, unbranched structure of carrots and radishes is not typical of most taproots. These shapes have been developed through years of selective breeding.

Fibrous root systems, on the other hand, consist of many thin roots that are all about the same size. Fibrous roots often form a dense network near the surface of the soil. This network holds soil particles together and helps to keep them from being carried away by water.

Most taproots branch, but the branches are much smaller than the main root.

Grasses have fibrous roots that branch extensively and have no single main root.

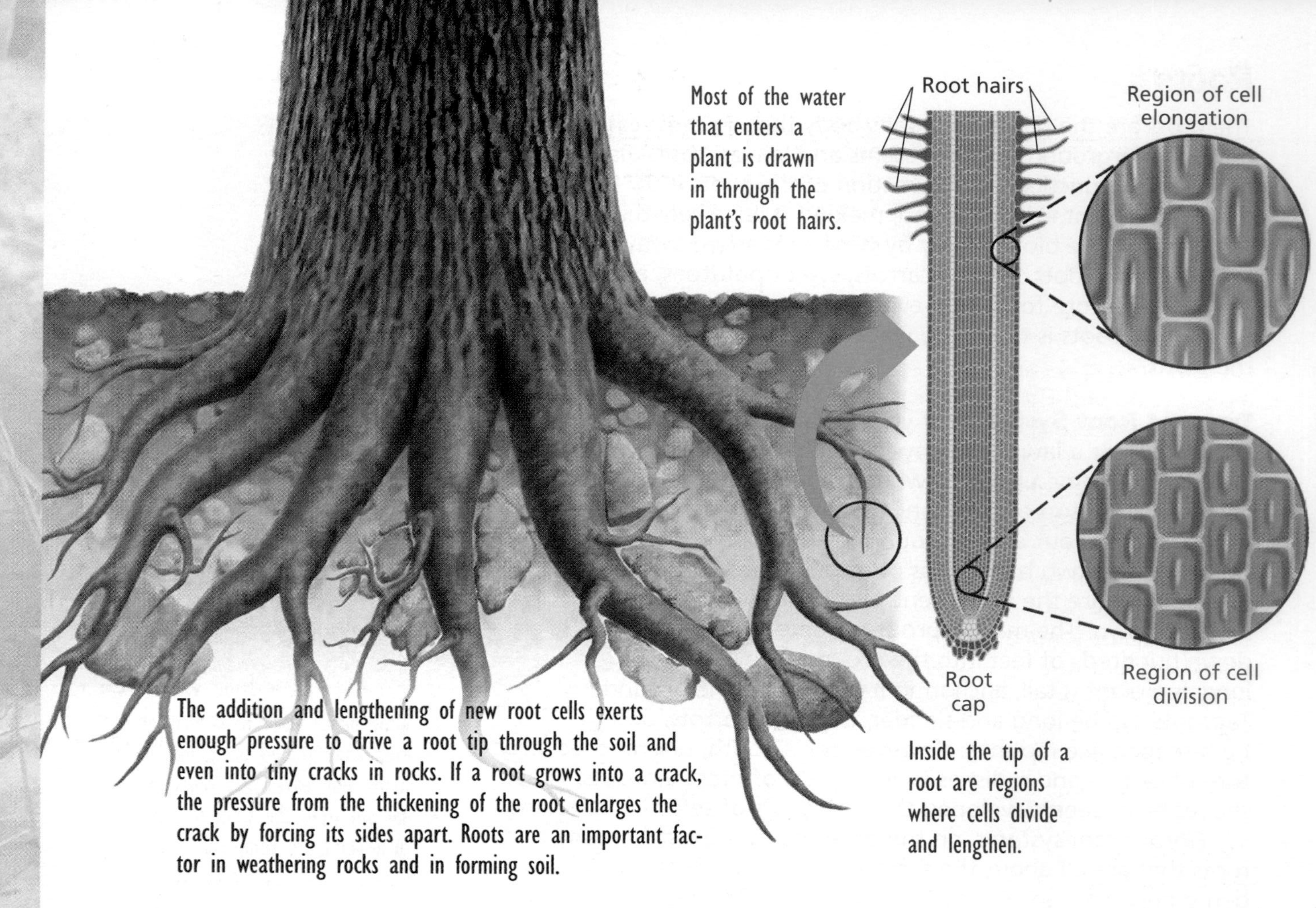

Most of the water that enters a plant is drawn in through the plant's root hairs.

The addition and lengthening of new root cells exerts enough pressure to drive a root tip through the soil and even into tiny cracks in rocks. If a root grows into a crack, the pressure from the thickening of the root enlarges the crack by forcing its sides apart. Roots are an important factor in weathering rocks and in forming soil.

Inside the tip of a root are regions where cells divide and lengthen.

Osmosis

The movement of water from an area of greater water concentration to an area of lesser water concentration through a membrane

DID YOU KNOW...

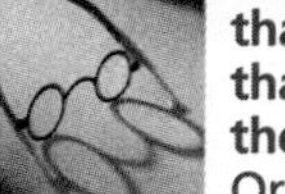

that there are plants that can make food in their roots?

Orchids are plants that usually live high above the ground, attached to the branches of trees. Amazingly, some species have evolved roots that contain photosynthetic tissue.

Root Growth Roots grow in length by making new cells at the ends of the roots, or *root tips*. The dividing cells are located just inside the very end of the root tip, which is protected by the *root cap*. After they are formed, new root cells lengthen, forcing the end of the root down into the soil. Cells that are more than 1 cm from the end of a root tip do not get larger. Thus, for a root to continue to grow in length, new cells must constantly be produced at the root tip.

Vascular tissues are found in the center of the root, while support and storage tissues develop around the vascular tissues. Some of the cells in the center of a root have the ability to divide. These cells cause a root to thicken by producing more vascular and support cells. The outermost cells become the epidermis, which protects the root.

Water and minerals are taken up by the actively growing tips of new root branches. Most of the water and minerals taken in by the roots are absorbed through *root hairs*, which are extensions of the epidermal cells near the root tips. The root system of a single plant may have billions of these tiny, threadlike structures. As long as there is enough water in the soil, water is drawn into the root hairs by the process of **osmosis**. This water continues to move through or between the root cells until it reaches the water-conducting cells in the center of the root branches. Once the water enters these cells, it can travel up the root, into the stem, and to the leaves where it will be used during photosynthesis.

Flowers

Although there are many different kinds of vascular plants, most are *angiosperms,* or flowering plants. Flowers are the reproductive organs of an angiosperm. Often, the appearance or fragrance of a flower attracts animals. Although animals usually visit flowers to feed on their sweet nectar, pollination—an important step in the life cycle of a plant—usually occurs in the process.

Structure of Flowers Most plants grow *perfect flowers,* which contain both male and female parts. A lily is a common example of a perfect flower. The male part of a flower produces pollen grains, which give rise to sperm cells. The female part, often located in the center of a flower, produces egg cells. *Petals* surround and protect the male and female reproductive structures. During reproduction, the sperm and egg cells of flowers fuse and cause the formation of *seeds.* Some plants, such as corn, develop *imperfect flowers,* in which each flower has either male or female structures, but not both. Is the flower shown to the right a perfect flower or an imperfect flower?

Flowers and Reproduction Plant reproduction begins with **pollination,** which is the transfer of pollen from a stamen to a pistil. When pollen is carried from the anther of one flower to the stigma of a flower on another plant, *cross-pollination* takes place. When a stigma receives pollen from the anther of the same flower, *self-pollination* occurs.

Pollination occurs by means of the wind or with the help of a variety of animals, such as insects, birds, and even mammals. Flowers have unique shapes, colors, and fragrances that make them visible and attractive to certain pollinators. For example, organ-pipe cacti grow long, tubular flowers that attract the long-nosed bat. The slender snout of this species of bat makes a near-perfect fit into the flower as it feeds on nectar produced at the bottom of the flower.

Pollination

The transfer of pollen from a stamen to a pistil

The *male part* of the flower is called the *stamen.* The stemlike part of a stamen is called the *filament.* At the top of a stamen is a structure called the *anther,* which contains the pollen grains. Most flowers have more than one stamen.

The *female part* of the flower is called the *pistil.* The base of the pistil, called the *ovary,* contains *ovules.* Each ovule produces one egg. The slender part of the pistil is the *style.* Found at the tip of the style is the *stigma,* which makes a sticky material that traps pollen grains.

Flowers that have red and yellow parts often attract birds. Here, a cactus wren feeds on the blooms of a saguaro cactus. At night, bats are also attracted to these flowers.

Life Cycle of a Plant

1. After a pollen grain lands on the stigma of a flower, the pollen grain begins to change in appearance.
2. A *pollen tube* grows from the pollen grain, through the stigma and style, and into an opening in one of the flower's ovules. Two sperm cells then move down the pollen tube and enter the ovule.
3. Fertilization occurs when one sperm combines with the egg inside the ovule to form a zygote, or fertilized egg. The zygote eventually develops into a tiny plant embryo.
4. The other sperm combines with two other cells inside the ovule. This fusion results in the development of an energy-rich food supply for the embryo.
5. Eventually, a tough protective "skin" forms around the outside of the ovule. Together, the embryo, its food reserve, and the surrounding coat are called a *seed.*
6. At the same time that the ovules are developing into seeds, the ovary is developing into a fruit.
7. Animals often help disperse the seeds inside fruit.
8. A new generation of plants eventually grows from the seeds released from the fruit.

SUMMARY

Most plants have leaves, stems, and roots and produce flowers. Leaves produce food by photosynthesis. Stems hold up the leaves and transport materials through continuous strands of vascular tissue. Roots anchor a plant in the ground and absorb water and minerals through root hairs. Both stems and roots grow in length by producing new cells at their tips. Flowers are structures used in reproduction. The colors of some flowers attract different animal pollinators, and the internal structures of flowers produce seeds and fruit.

S170

Uses of Plants

Plants are an important part of our lives—more important to you than you might think. Plants are used in a multitude of ways in modern society. They provide us with a variety of foods to eat, materials from which to make clothing for our bodies, and medicines that help to keep us healthy. Let's take a closer look at how we use plants.

Plants as Food

Have you ever eaten a flower? If you have eaten broccoli or cauliflower, you have. In fact, many different plant parts make excellent food. For example, an onion bulb is actually a bud, and celery is the petiole of a leaf. When you eat lettuce, spinach, or cabbage, you are eating whole, mature leaves. The tea in a tea bag is composed of leaves, as are seasonings such as parsley, chives, basil, rosemary, mint, and sage. Some herbal teas contain flower parts as well.

DID YOU KNOW...

that some seeds prefer to sprout on pottery? Chia seeds are from a type of wild sage plant found in Mexico. The seeds have a very thick protein coat and do not sprout easily in soil. But if they are placed on specially made hollow earthenware vessels, the seeds will soon produce a green blanket of protein-rich chia sprouts.

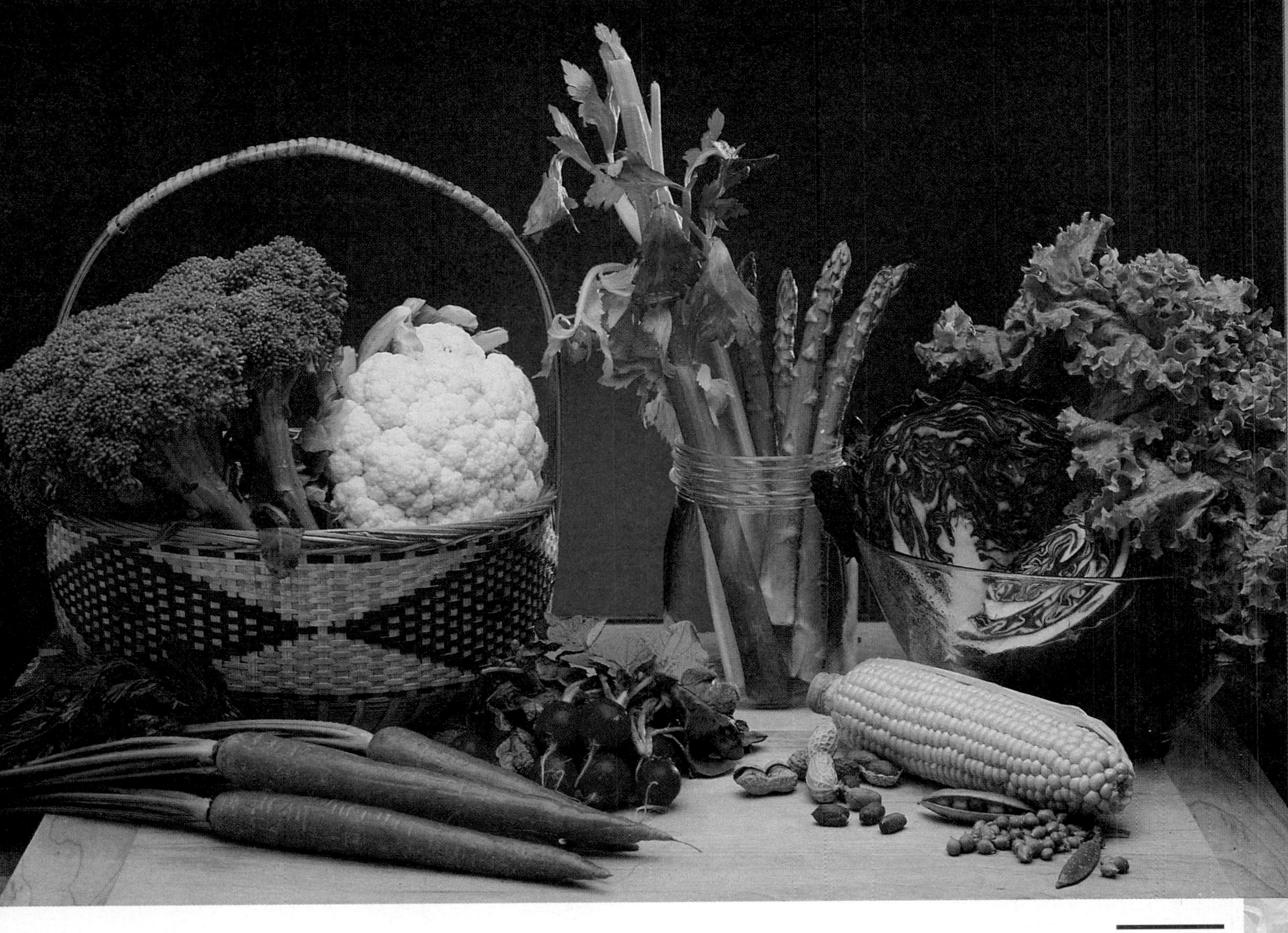

How many different plant parts are represented in this photograph?

Tapioca, which we make into a pudding, comes from the roots of the cassava plant. In the tropics, the cassava root is the main food for millions of people.

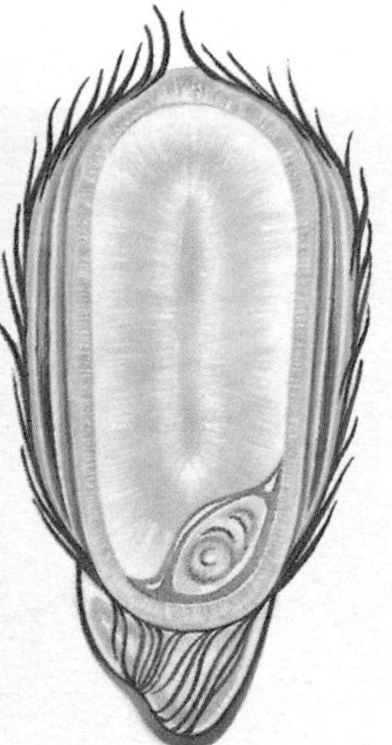

A rice grain is actually a fruit. The nutritious part of a cereal grain, such as rice, is the single, large seed contained in each fruit.

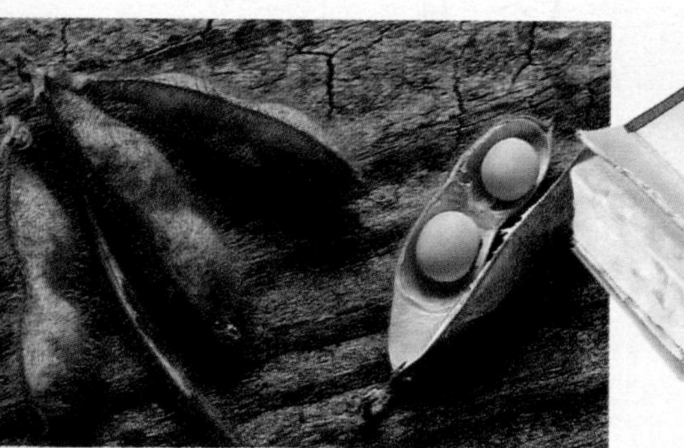

The soybean, which is high in protein, is sometimes used in place of meat. Tofu, a cheeselike food made from soybeans, is used in soups and various oriental dishes.

Our most important sources of food, however, are stems, roots, fruits, and seeds. Of course, you already know that potatoes and asparagus are stems. Potatoes, however, grow underground, while asparagus grows above ground, which is why asparagus is green and potatoes are not. Also, about half of the sugar we eat comes from the stems of a large grass called *sugar cane*. The rest of the sugar we eat comes from the roots of sugar beets. Carrots, turnips, radishes, and sweet potatoes are also plant roots.

You already know that bananas, apples, oranges, and grapes are fruits. But what about tomatoes, eggplant, squash, peppers, cucumbers, and green beans? Or for that matter, what about cereal grains such as rice, oats, corn, wheat, and barley? As a matter of fact, all of these foods—usually called "vegetables"—are considered fruits because they develop from the ovary and contain seeds.

Nutritious seeds, such as beans, peas, and peanuts, are found in the fruits of a group of plants called *legumes*. These plants produce seeds that are rich in protein. Peas and beans were also among the first types of plants to be cultivated, over 11,000 years ago.

Plant Fibers

You have probably heard the term *fiber* used in commercials for breakfast cereals. But what is fiber, anyway? The fiber you eat is the material found in the cell walls of plant cells. This material, called **cellulose**, cannot be digested by your body. However, cellulose does help to keep food moving along at a healthy rate as it passes through your digestive system. That is why it is healthy to include fiber in your diet. Some plants make very long strands of cellulose fibers. These fibers are used to make things such as cloth, rope, and paper.

DID YOU KNOW...

that cotton plants are used for more than just making clothes? The seeds of the cotton plant are used as a source of oil. After the oil is extracted, the left-over seed parts are used to make feed for livestock.

Cotton is the world's most important plant fiber. It has been grown by humans for over 5000 years. The thread used to make cotton cloth is spun from long fibers that grow from the seeds of the cotton plant. Cotton cloth is used to make clothing, bed sheets, towels, rugs, and diapers. Cotton fibers are also used to make bandages, stuffing for cushions, and paper.

Linen fibers come from the stems of the flax plant. Like cotton, linen is very strong and absorbent. The ancient Egyptians used linen cloth to wrap mummies. Now it is used to make cool summer clothing as well as tablecloths, napkins, handkerchiefs, and paper.

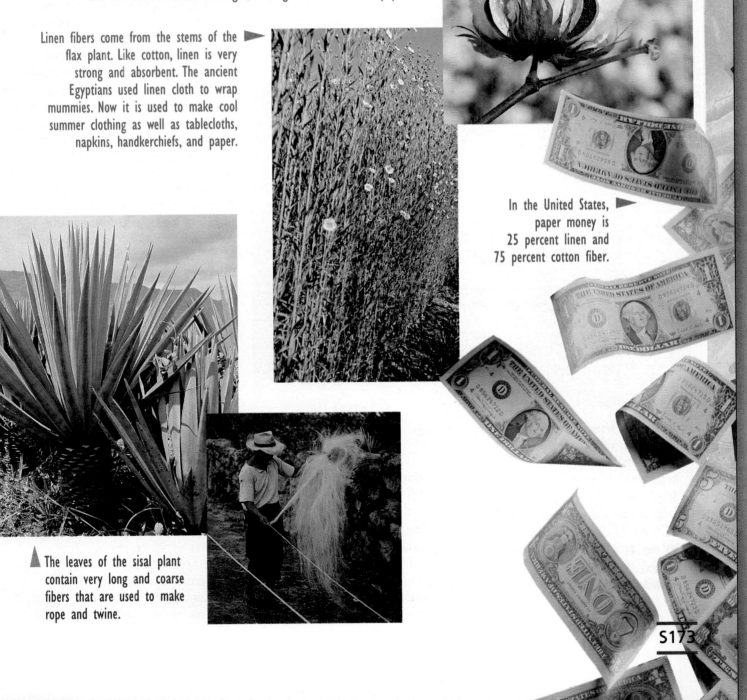

In the United States, paper money is 25 percent linen and 75 percent cotton fiber.

The leaves of the sisal plant contain very long and coarse fibers that are used to make rope and twine.

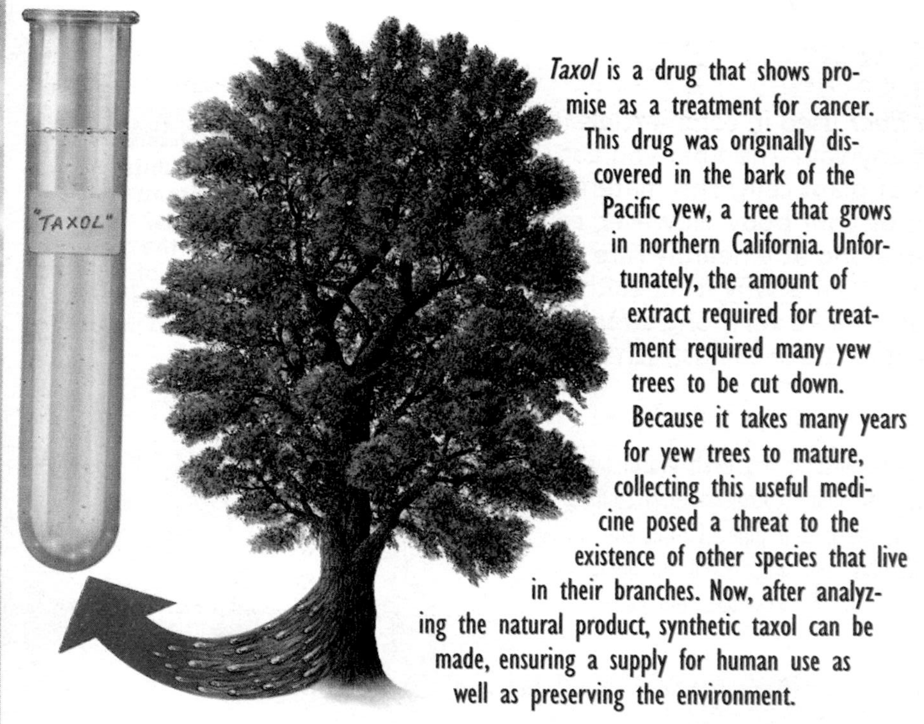

Taxol is a drug that shows promise as a treatment for cancer. This drug was originally discovered in the bark of the Pacific yew, a tree that grows in northern California. Unfortunately, the amount of extract required for treatment required many yew trees to be cut down.

Because it takes many years for yew trees to mature, collecting this useful medicine posed a threat to the existence of other species that live in their branches. Now, after analyzing the natural product, synthetic taxol can be made, ensuring a supply for human use as well as preserving the environment.

Medicine From Plants

Plants make a variety of chemicals in addition to the sugars and starches that we consume as foods. Some of the chemicals that plants make can be used as medicines. In fact, for most of the time humans have been on Earth, they have obtained the medicines they needed directly from plants. Although many modern medicines are now made synthetically, most of them were first derived from plants, and some are still extracted from those plants.

Aspirin is one medicine you have probably taken. It contains *salicylic acid*, which originally came from the bark of willow trees. Aspirin is an effective pain reliever and also controls fever. *Ephedrine*, a drug you may have taken for a stuffy nose, comes from a rare plant called ephedra. Periwinkles, plants you may have in your garden or neighborhood, are another source of important medicines. The drugs made from periwinkles help to relieve the symptoms of two types of cancer, Hodgkin's disease and leukemia. *Quinine*, the drug used to treat malaria, comes from the bark of the tropical cinchona tree and makes life in the tropics bearable for millions of people who suffer from this disease.

Much of the research into identifying plants with medicinal value is concentrated on tropical rain forests. These forests contain the richest variety of plant species on Earth. But tropical rain forests are being cut down at an alarming rate. As a result, untold numbers of plant species are becoming extinct before they can be properly studied.

SUMMARY

Plants are an important source of food for humans. Edible plant parts include roots, stems, leaves, flowers, fruits, and seeds. Plants also contain fibers that are used to produce cloth and other valuable products, such as rope and paper. Many plants produce chemicals that can be used as medicines. Most modern medicines were originally derived from plants. The destruction of the rain forests, however, may prevent scientists from discovering many new medicines.

Unit CheckUp

Concept Mapping

The concept map below illustrates major ideas in this unit. Complete the map by supplying the missing terms. Then extend your map by answering the additional question below. Write your answers in your ScienceLog. **Do not write in this textbook.**

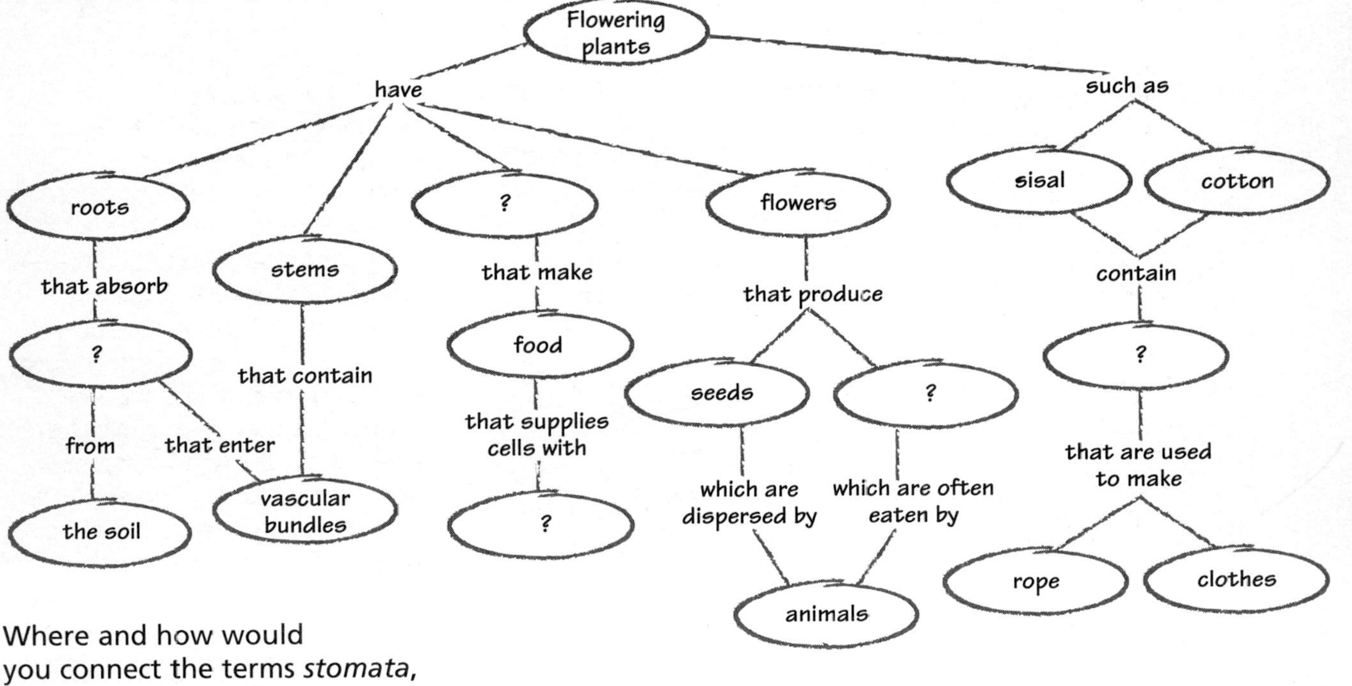

Where and how would you connect the terms *stomata*, *root hairs*, and *ovules*?

Checking Your Understanding

Select the choice that most completely and correctly answers each of the following questions.

1. Most of the plants on Earth are
  a. mosses.
  b. angiosperms.
  c. flowering plants.
  d. both *b* and *c*.

2. Leaves
  a. absorb sunlight.
  b. absorb water.
  c. use oxygen during photosynthesis.
  d. lack vascular tissue.

3. Wood is made mostly of
  a. chlorophyll.
  b. epidermis.
  c. old water-carrying cells.
  d. food-making cells.

4. Which statement is NOT true about roots?
  a. They anchor a plant.
  b. They grow in width but not length.
  c. They have "hairs."
  d. They contain vascular tissue.

5. Cellulose is
  a. digestible.
  b. part of a plant cell's nucleus.
  c. found only in cotton plants.
  d. also known as "fiber."

Interpreting Photos

What type of root system would you expect most of the plants in this photo to have? Why?

Critical Thinking

Carefully consider the following questions, and write a response in your ScienceLog that indicates your understanding of science.

1. Why would coating the undersides of all the leaves of a plant with petroleum jelly probably cause the plant to die?
2. A man hammers a nail at eye level into a tree that is 5 m tall. Twenty years later he returns to the same tree and observes that the tree is now 10 m tall, yet the nail is still level with his eye. Explain why the position of the nail has not changed.
3. Hay fever, a swelling of the membranes inside the nose, is often caused by pollen grains. Considering how pollination occurs, why do you think most hay fever is caused by pollen from grasses and trees?
4. Why would mixing a lot of salt into the soil around the roots of a plant probably cause the plant to die?
5. A certain person refuses to eat *any* vegetables because he believes they are low in protein and energy content. Convince the person that what he thinks about vegetables is not true.

Portfolio Idea

Imagine that you are a water molecule that is just entering the root hair of a plant. Your destiny within the plant is to participate in "the process of photosynthesis." Write a detailed story that describes your journey through the plant. Be creative, but make sure your story includes the scientific facts and concepts concerning plant structure and function that you learned about in this unit.

GLOSSARY

A

Abiotic nonliving, and having no living origin **(8)**

Acceleration a change in the speed of motion **(S74)**

Acid a chemical that has a sour taste, turns blue litmus paper red, reacts with a base to form a salt, and has a pH value below 7 **(60, S59)**

Acid rain highly acidic rain, sleet, or snow that results when gas, oil, and coal are burned and the compounds that result combine with water vapor in the air **(60)**

Action in physics, the act of doing something, or the state of being in motion **(281)**

Adaptation an inherited feature that arises over time and that better enables an organism to survive in a given environment **(97)**

Adobe sun-dried brick that is made of clay and is used to build houses **(S103)**

Aerobic occurring in the presence of free oxygen **(291)**

Alkaline having a pH value greater than 7; a basic solution **(60)**

Appendage a part of an organism that is attached to the main body, such as a tail or a finger **(112)**

Arch a curved structure, often constructed of stone, that supports the weight of the material above an open space **(317)**

Asteroids enormous rocks or boulders that revolve around the Sun, usually between the orbits of Mars and Jupiter **(457)**

Astronomy the science that studies the stars, planets, and other celestial objects **(424)**

Atmosphere the air surrounding the Earth to a height of about 1000 km **(197)**

Axis a straight line through an object, on which the object rotates **(446)**

B

Bacteria (singular, bacterium) a large class of single-celled organisms, considered neither plant nor animal, belonging to kingdom Monera. Their actions cause a number of diseases, as well as processes such as decay, fermentation, and soil enrichment. **(31)**

Base a chemical that has a bitter taste, turns red litmus paper blue, reacts with an acid to form a salt, and has a pH value greater than 7 **(60, S62)**

Beam a long, rigid piece of stone, wood, or metal used for horizontal support in ceilings, roofs, and other structures **(314)**

Big bang a theorized event believed to have occurred 15 to 20 billion years ago that resulted in the formation of the universe **(487)**

Biomes large areas with a similar climate and similar life-forms **(S14)**

Biosphere all of the environments on Earth that contain organisms **(501, S2)**

Biotic living or having a living origin **(8)**

Black hole a small and dense point in space that is formed from a massive collapsing star. Its gravitational forces are so strong that nothing, not even light, can escape. **(478, 495)**

Boiling the process in which a liquid rapidly changes into a gas at its boiling point (the maximum temperature at which the liquid can exist as a liquid) **(168)**

Boiling-point elevation the amount by which the boiling point of a solvent is raised when a solute is dissolved in it **(S56)**

Buoyant force the directional force exerted upward by a liquid on a submerged body **(225)**

Buttress a structure built perpendicular to a wall to reinforce or support it **(318)**

C

Camouflage the method by which an organism disguises itself or blends with its surroundings **(98)**

Cantilever a beam or structure that is fixed only at one end **(316)**

Carnivore a consumer of animals; a meat eater **(30)**

Cementation the process in which the spaces between loose particles are filled with a hardening or bonding agent. This is one form of lithification. **(395)**

Centripetal force a force that causes objects to move toward the center of rotation **(S77)**

Chloroplasts organelles of plant cells that contain chlorophyll and are the sites of photosynthesis **(S163)**

Classify to sort into groups **(114)**

Climax community the final stage in the succession of communities **(S13)**

Cohesion the force of attraction between the particles of like substances **(S92)**

Colloid a mixture containing particles that are not dissolved but are too small to settle **(S57)**

Comets small frozen masses of dust and gases that travel a definite path through the solar system. As they pass near the Sun, they are vaporized, leaving trails that resemble hair. **(457)**

Commensalism a relationship between organisms in which one organism benefits from the relationship, while the other neither benefits from nor is harmed by the relationship **(12)**

Community all of the populations of organisms that live in an area **(S5)**

Compaction the process in which overlying particles compress the spaces between underlying particles. This is one form of lithification. **(395)**

Compressive force a force that causes material to become more compact and pressed together **(305)**

Concentrated solution a solution that is made by dissolving a large amount of solute in a small amount of solvent **(188, S53)**

Concentration the strength of a solution, determined by the amount of solute dissolved in the solvent **(188)**

Conservation the protection and wise use of natural resources **(S22)**

Constellations groups of stars that are visible in the night sky **(440)**

Consumer an organism that depends on other organisms as food sources **(29)**

Cosmology the study of the structure, origin, and evolution of the universe **(S157)**

Crystallization a process in which the chemicals in a solution form solids. This is one form of lithification. **(395)**

D

Decomposer a consumer that breaks down dead food sources into substances that enrich the soil **(30)**

Deflection the amount of bending **(308)**

Desalination the removal of salt from a solution, especially sea water **(172, 210)**

Dike igneous rock that solidifies in a vertical crack in previously existing rock **(397)**

Dilute solution describes a solution that is made by dissolving a small amount of solute in a large amount of solvent **(188, S53)**

Dissolving the spreading of the particles of a solute evenly through a solvent **(154)**

Distillation a two-step process involving (1) the heating of a solution to change one part of it into a gas or vapor and (2) cooling the vapor to liquid form and collecting it **(172, 210)**

Diversity differences or variety among living things **(78)**

Dome a hemispherical roof **(317)**

Drag friction between a moving object and the fluid (air or water) through which it is moving **(289)**

E

Ecosystem a group of organisms and their physical environment **(S2)**

Ectotherm an organism whose body temperature changes according to the temperature of its environment **(S37)**

Elastic force the force that is exerted by a material when it is stretched **(225)**

Elastic limit the point beyond which a material cannot return to its original size or shape when external, deforming forces are removed **(S93)**

Elastic rebound theory a theory about the cause of earthquakes which states that rocks strained past a certain point will fracture and spring back to their original shape **(364)**

Elastic response the tendency of a material to return to its original shape or size after it has been stretched **(307)**

Electrical force the force that causes oppositely charged materials to attract each other and similarly charged materials to repel each other **(225)**

Endangered species a species whose numbers have fallen so low that it is likely to become extinct in the near future **(110, S22)**

Endoskeleton an internal skeletal system that supports the bodies of vertebrates **(S37)**

Endotherm an organism whose body temperature remains constant despite temperature changes in its environment **(S38)**

Energy pyramid a diagram that shows how energy decreases at each feeding level in a food chain **(S7)**

Environment the physical surroundings of an organism, including all of the biotic and abiotic conditions and circumstances that affect its development **(5)**

Epicenter the point on the surface of the Earth directly above the focus, or hypocenter, of an earthquake **(367)**

Epoch a subdivision of a geologic period on the geologic time scale **(409)**

Era the largest division of the geologic time scale, denoting an interval of time when certain rocks were formed **(409)**

Evaporation the process by which a liquid changes into a gas or vapor **(168)**

Event horizon the outer boundary of a black hole **(495)**

Exoskeleton a hard outer covering for protection that is typical of arthropods **(S36)**

Extinction the irreversible disappearance of a species **(108, S19)**

Extrusive rock rock produced by molten lava that flows onto the surface of the Earth; also called volcanic rock **(390)**

F

Fault the boundary between two rock sections that have been displaced relative to each other **(360)**

Focus the point where a rupture starts an earthquake; also called the hypocenter **(365)**

Food chain a chainlike diagram that shows the relationship between various organisms and their sources of food energy **(33)**

Food web a weblike diagram that shows relationships between various organisms and their sources of food energy **(34)**

Force any push or pull **(S75)**

Fossil the remains or imprint of a prehistoric plant or animal **(404, S39)**

Frame of reference a system for specifying the precise location and motion of objects in space **(S75)**

Freezing-point depression the amount by which the freezing point of a solvent is lowered when a solute is dissolved in it **(S56)**

Friction the rubbing of objects against one another, which produces heat and causes surfaces to wear away **(257, S76)**

Frictional force the force that resists the motion of objects rubbing against each other **(225)**

Fruit a ripened ovary that contains the seeds of a flowering plant **(S33)**

G

Gene therapy the process of replacing defective genes in cells by injecting healthy genes into the cells **(141)**

Genetic disorder an ailment caused by defective genes in the chromosomes of an organism **(141)**

Geologic column a model that shows rock layers in the order in which they were formed **(S134)**

Geologic time scale the sequence of events of the Earth's geologic history, as told in sedimentary rock **(409)**

Germinate to sprout; to begin to grow from a seed **(503)**

Gravitational force the mutual force of attraction exerted by particles of matter **(225, S85)**

H

Habitat the place where an organism lives **(10)**

Half-life the time it takes for half of the atoms in a sample of a radioactive isotope to decay **(S133)**

Hard water water that contains certain dissolved chemicals **(160)**

Herbivore a consumer of plants; a plant eater **(30)**

Humus a brown or black component of soil, resulting from the decay of once-living material **(508)**

Hydroponics the process of growing plants without soil by using solutions to provide the necessary nutrients for growth **(526)**

Hypocenter the point where a rupture starts an earthquake; also called the focus **(365)**

Hypothesis a possible explanation for why things happen the way they do. In an experiment, a hypothesis is tested. **(176)**

I

Igneous rock rock produced by the cooling and solidification of magma, either on or below the Earth's surface **(385, S119)**

Inertia the tendency of objects at rest to stay at rest and of objects in motion to stay in motion **(275, S79)**

Inference a conclusion, drawn from observations, that attempts to explain or to make sense of the observations **(176)**

Insoluble unable to dissolve or pass into solution **(162)**

Intrusion the pushing up of magma into cracks in existing rock **(396)**

Intrusive rock rock produced by the cooling of magma that pushes up into the Earth's crust; also called plutonic rock **(391)**

Invertebrate an animal without a backbone **(120)**

L

Lava magma that emerges onto the surface of the Earth **(378)**

Lichen a plantlike organism that consists of a fungus and an alga existing together to the mutual benefit of each other **(12)**

Light-year the distance that light travels through space in one year (9.46 trillion km) **(482)**

Limiting factors environmental elements that stabilize population size and keep species from producing too many offspring **(S4)**

Lithification the process by which sediments become rock **(395)**

Lithosphere the outer crust of the Earth, which is broken into seven large rigid plates and several smaller ones **(358)**

Loam a soil mixture composed of sand, clay, and organic matter **(509)**

Local Group a family of galaxies, including the Milky Way, that is held together by mutual gravitational forces **(485)**

Luminosity a measure of the actual brightness of a star **(S146)**

M

Magma melted rock deep within the Earth **(378)**

Magnetic force the attraction or repulsion that one magnet exerts on another, or the attraction between a magnet and certain metals **(225)**

Magnitude the strength, or measured size, of an earthquake **(366)**

Mass a measure of the amount of matter in an object **(229)**

Metal fatigue the wearing out of metal parts due to repeated bending **(S91)**

Metamorphic rock rock that has undergone change as a result of intense heat and pressure **(385, S122)**

Metamorphosis a change; a transformation of the nature of something **(396)**

Meteor a lump of a rock or metal that enters the Earth's atmosphere and burns **(455)**

Meteorite a meteor that reaches the ground without burning up completely in the Earth's atmosphere **(455)**

Milky Way galaxy a spiral-shaped system, of which our solar system is a part, containing billions of stars, huge clouds of dust particles, and gases **(484)**

Mimicry the resemblance in behavior, look, sound, or smell of an organism to another organism or object in its surroundings. Mimicry helps to protect organisms from predators **(101)**

Momentum the mass of an object multiplied by its velocity **(S83)**

Motion a change in position of an object relative to a reference point **(S72)**

Mutation a change in the DNA of a gene **(S42)**

Mutualism a relationship between organisms in which all of the organisms benefit from the relationship **(12)**

N

Natural resources all natural substances that humans can remove from the environment for their own use **(S20)**

Natural selection the process by which the organisms that are best suited to their environment survive and pass their traits on to their offspring **(92, S41)**

Neutron star a collapsed star composed mostly of neutrons, compressed to the size of an asteroid, and having the mass of many Suns **(477)**

Newton the international metric (SI) unit used to measure force. Abbreviated N, one newton (1 N) is approximately equal to the weight of a 100 g mass. **(237)**

Niche an organism's role in its community, including its behavior and its place in the food chain **(10)**

Noncontact force a force exerted by an agent that does not touch the receiver **(222)**

Nutrient a chemical that is needed for the functioning and growth of living things **(509)**

O

Omnivore an eater of both plants and animals **(30)**

Organism a living thing with organs or parts that work together **(78)**

Osmosis the movement of water across a semipermeable membrane from areas where water particles are more concentrated to areas where they are less concentrated **(519, S168)**

P

Paleobotanist a scientist who studies fossils of plants **(557)**

Paleontology the study of fossils, ancient life-forms, and their evolution **(407)**

Parallax the apparent shift in the position of an object (such as a star) when seen from different locations or angles **(S146)**

Parasitism the relationship between two organisms in which one organism, called a parasite, lives to the detriment or harm of the other, called the host **(13)**

Percolation rate the time it takes for a certain volume of water to pass through a sample of soil **(511)**

Period a subdivision of a geologic era on the geologic time scale **(409)**

Perspiration a solution that the body produces to help maintain body temperature and to rid the body of waste products **(203)**

pH a measure of the hydronium-ion concentration of a solution **(S65)**

pH scale a scale, ranging from 0 to 14, used to measure the alkalinity or acidity of a substance **(60)**

Photosynthesis the process by which green plants and plantlike organisms use energy from sunlight to convert water and carbon dioxide into sugars that can be used for food **(28)**

Plane of the ecliptic the plane of the Earth's orbit around the Sun. All of the other planets' orbits lie roughly in this plane. **(442)**

Plankton very small plants and animals that float near the surface of the ocean **(S18)**

Plate tectonics the theory that large crustal plates move on the Earth's surface **(358, S109)**

Plutonic rock rock produced by the cooling of magma within the Earth; also called intrusive rock **(391)**

Pollination the transfer of pollen from the stamen of one plant to the pistil of the same or another plant of the same species in order to produce seeds **(12, S169)**

Pollution contamination of the environment **(S20)**

Population a group of organisms of the same species living in the same area **(S3)**

Population density the number of individuals of the same species living in a given area **(S4)**

Postmodernist movement an architectural style that includes decorative elements that may have no practical purpose, but that make the exterior of a building look more interesting **(338)**

Predator a consumer that hunts or captures live animals as food sources **(30)**

Prey an animal that is hunted or captured and eaten by a predator **(30)**

Producer an organism that produces food for itself and others **(29)**

Q

Quasars bright energy sources, about the size of the solar system, that are billions of light-years from Earth **(487)**

R

Reaction in physics, the action that opposes a given action **(281)**

Recycle to reuse the remains of things **(52, S23)**

Red supergiant a star of great brightness, volume, and relatively large diameter; emits red light **(477)**

Regeneration the ability of some organisms to grow new body parts **(S36)**

Relative age the age of a rock (older or younger) in comparison with the age of other rocks **(S132)**

Relative time scale an organization of events based only on their sequence. No dates are included. **(408)**

Retrograde motion as seen from Earth, the apparent backward motion of a planet that occurs before it continues its forward motion **(434)**

Reverse osmosis a method of purifying water by forcing it through a series of semipermeable membranes that filter out unwanted particles **(210)**

Revolution the motion of one body around another **(442)**

Richter scale of magnitude a scale of measurement in which the number expressing the magnitude, or strength, of an earthquake is calculated by means of a formula **(366)**

Rock cycle the continuous process of change in which new rocks are formed from old rocks **(S117)**

Rotation the spinning motion of a planet or other object on its axis **(442)**

Runoff water from rain or other sources that does not soak into the ground, but rather flows over the ground to low areas **(211)**

S

Saliva the watery solution released by the salivary glands; it serves as an aid to swallowing and begins the process of digestion by softening food **(201)**

Salt any ionic compound formed when the negative ion from an acid combines with the positive ion from a base or when an acid reacts with a metal **(S67)**

Saturated describes a solution that contains all the dissolved solute that it can hold under existing conditions **(194, S54)**

Scavenger a consumer of dead food sources **(30)**

Scientific name the unique two-word name (corresponding to an organism's genus and species names) that is assigned by scientists to each kind of organism on Earth **(133)**

Sedimentary rock rock produced by cemented mineral particles deposited by wind, water, ice, or chemical reactions **(385, S120)**

Seismic wave an intense vibration in the Earth that occurs during an earthquake; also called a shock wave **(365)**

Seismograph the instrument used to record the vibrations of the Earth during an earthquake **(366)**

Seismologist a person who studies earthquakes **(366)**

Selective breeding the process of making hereditary improvements in animals and plants by breeding organisms that have the desired traits **(S47)**

Shear force a force that tends to cause a material to tear or be cut apart **(305)**

Shock wave the release of stored energy in the form of an intense vibration in the Earth; also called a seismic wave **(365)**

Sill igneous rock that solidifies in a horizontal crack in existing rock **(397)**

Singularity the point of incredible density at the center of a black hole **(495)**

Soft water water that is mostly free of dissolved chemicals **(160)**

Solubility the amount of substance able to pass into solution under certain conditions; the measure of the amount of solute needed to saturate a solution **(195, S54)**

Soluble able to dissolve or pass into solution **(162)**

Solute a substance that is dissolved in a solvent to make a solution **(155)**

Solution a mixture in which the solute particles are spread evenly throughout the solvent particles **(149, S52)**

Solvent a substance that dissolves a solute to make a solution **(155)**

Species a group of organisms that are able to reproduce together and that resemble each other in appearance, behavior, and internal structure **(40, 133)**

Speed the rate at which an object moves, as measured by the distance covered per unit of time **(S72)**

Stomata (singular, stoma) microscopic openings on the lower surface of leaves that regulate the passage of certain substances into and out of a plant **(523, S163)**

Succession the process in which a new group of organisms, which are better suited to the changes that have taken place in a particular area, replaces an old group of organisms **(48, S11)**

Supergiant a star of very great brightness and enormous size, at least 100 times larger and brighter than the Sun **(477)**

Supernova a very large and very bright exploding star **(477)**

Suspension a mixture in which particles are scattered throughout, but not dissolved in, a substance **(205, S57)**

Sustainable development the development of land, air, and water resources in such a way that the needs of other organisms and of future generations of humans are considered **(57)**

T

Tensile force the force that causes a material to stretch **(305)**

Texture a characteristic of soil based on particle size, roughness, and the way it feels to the touch **(508)**

Tolerance the ability to endure a set of conditions **(16)**

Transpiration the movement of water out of plants through the stomata in the leaves **(521)**

Truss a framework of connected beams or bars that is built to add strength and support to a bridge, roof, or other structure **(314)**

Tyndall effect a test for determining true solutions. True solutions are clear while non-solutions contain particles large enough to scatter or reflect light. **(149)**

V

Variable a factor in an experiment that may change or have different values and that affects the results of the experiment **(162)**

Vascular bundles groups of water- and food-carrying cells in plants **(S165)**

Velocity the speed and direction of a moving object **(S73)**

Vertebrate an animal with a backbone **(120)**

VLA (Very Large Array) a large radio telescope, consisting of 27 signal collection dishes, located in New Mexico **(496)**

Volcanic rock rock produced by molten lava that flows onto the surface of the Earth; also called extrusive rock **(390)**

Volcano an opening in the Earth's surface through which lava and other materials erupt **(370, S112)**

Vortices (singular, vortex) whirls or eddies of air or water; also called whirlpools **(288)**

W

Water cycle the process, involving evaporation and condensation, by which the Earth's water is constantly purified **(174)**

Weight the measure of the gravitational force acting on an object **(227, S86)**

White dwarf a small, dim, but extremely dense star **(478)**

Z

Zodiac in astronomy, 12 constellations, named mostly for animals, that lie close to the plane of the ecliptic (Note: These constellations do not correspond to the "birth signs" you may have read about.) **(440)**

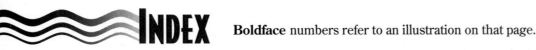

INDEX

Boldface numbers refer to an illustration on that page.

A

B

C

D

E

F

G

H

PHOTO CREDITS

Abbreviated as follows: (t) top; (b) bottom; (l) left; (r) right; (c) center.

All unlisted photos credited to Sam Dudgeon, HRW.

FRONT COVER: (tl), Martin Wendler/Okapia/Photo Researchers; (c), Hale Observatories/Science Source/Photo Researchers; (br), Peter Trusler, courtesy of Australian Post;

BACK COVER: (tl), G. Brad Lewis/Tony Stone Images; (c), Jeff Smith/FotoSmith;

TITLE PAGE: (l), Page Overtures; (b), Jeff Smith/FotoSmith;

Design and remaining illustrations by Randy Gates/Morgan Cain & Assoc.; Assorted textures by Page Overtures; Special thanks to Reptile Solutions of Tucson;

TABLE OF CONTENTS: Page iv(tr), Michael Salas/Michael Salas Photography; iv(cr) Peter Cade/Tony Stone Images; iv(cb), ZEFA Germany/The Stock Market; iv(bl), Suzanne & Joseph Collins/Photo Researchers, Inc.; v(tl), Tim Davis/Tony Stone Images; v(tr), Tim Davis/Photo Researchers, Inc.; v(r), George Bryce/Animals Animals; v(c), John Warden/Tony Stone Images; v(br), Breck P. Kent; v(bc), Ric Ergenbright Photography; v(bl), HRW Photo by Daniel Schaefer; vi(tl), (tc), (cr), (bl), Carlos Austin; vi(tr)Jeff Foott/Bruce Coleman, Inc.; vi(br), Warren Bolster/Tony Stone Images; vii(tl) (tr) (br), HRW Photo by Michelle Bridwell; vii(cr), Phil Degginger/Color-Pic, Inc.; viii(t)Larry Fisher Photographt/Masterfile; (tr), Lazio/Adros Studios; viii(br), Barry Sweet/AP/Wide World Photos; viii(b), Rob Talbot/Tony Stone Images; ix(tl), James L. Amos/Photo Researchers, Inc.; ix(cr), Chris Butler/Astrostock; ix(br) (bl), x(t) (tr) (br), HRW photo by Daniel Schaefer; x(bl), M.P. Gadomski/Bruce Coleman, Inc.; xi(tl), AP/Wide World Photos; xi(cr), E.R. Degginger/Color-Pic, Inc.; xi(lc), NASA/HRW Library; xi(b), NASA/Science Photo Library/Photo Researchers, Inc.; xii(l)HRW Photo by Patrick R. Dunn; xiii(tl), Carlos Austin; xv(tl), HRW Photo by Michelle Bridwell; ; xvi(cr), HRW Photo by Jana Birchum; xvi(bl), HRW Photo by Michelle Bridwell; xvii(b), HRW Photo by Daniel Schaefer;

UNIT 1: Page 2,3,4,25,26,27,44,45,46,64,65,66,67(bckgd), M. Thonig/H. Armstrong Roberts; Page 2(tl), G. Ryan & S. Beyer/Tony Stone Images; 2(bc), Ron Sanford/Tony Stone Images; 2-3, Joel Sartore/National Geographic Image Collection; 5(tr), Paul Barton/The Stock Market; 5(tr-inset), Tim Davis/Photo Researchers, Inc.; 5(cr), Phil Degginger/Color Pic; 5(cr-inset), Anthony Mercieca/Photo Researchers, Inc.; 5(br), Tom McHugh/Photo Researchers, Inc.; 5(tl), Chuck Savage/The Stock Market; 5(tl-inset), ZEFA/H. Armstrong Roberts; 5(cl), Peter Pearson/Tony Stone Images; 5(bl), David R. Frazier Photolibrary; 6-7, HRW Photo by Daniel Schaefer; 9(t), David R. Frazier Photolibrary; 9(cl), Anthony Mercienca/Photo Researchers, Inc.; 9(bl), John M. Roberts/The Stock Market; 9(br), Darrell Gulin/Tony Stone Images; 9(b), Nuridsany et Perennou/Photo Researchers, Inc.; 10(bl), Phil Degginger/Color-Pic; 10(br), Will & Deni McIntyre/Photo Researchers, Inc.; 11(cl), E.R. Degginger/Color-Pic; 11(cr), David R. Frazier Photolibrary; 11(b), Tom Bean/The Stock Market; 12(tr), Gregory K. Scott/Photo Researchers, Inc.; 12(bl), H. Armstrong Roberts; 12(bc), E.R. Degginger/Color-Pic; 12(br), Michael P. Gadomski/Photo Researchers, Inc.; 13(tl), Jeff Gnass; 13(tc), Breck P. Kent; 13(tr), William E. Ferguson; 13(bl), Bruce Iverson; 14(tl), Dwight R. Kuhn; 14(tr), David R. Frazier Photolibrary; 14(bl), William E. Ferguson; 14(bl-inset), Michael Abbey/Photo Researchers, Inc.; 14(br), Michael Long/Envision; 14(br-inset), Edward Grey/Science Photo Library/Photo Researchers, Inc.; 14(cr), Patti Murray/Animals Animals; 14(cr-inset), Dembinsky Photo Association; 15(tl), Gilbert S. Grant/Photo Researchers, Inc.; 15(tc), David Barron/Animals Animals; 15(tr), Donald Specker/Animals Animals; 15(cl), Vince Steano/The Stock Market; 15(cl-inset), William E. Ferguson; 15(cr), Viviane Holbrooke/The Stock Market; 15(bl), William E. Ferguson; 15(br), David Young-Wolff/Photo Edit; 15(br-inset), A.B. Dowsett/Photo Science Library/Photo Researchers, Inc.; 18, HRW Photo by Daniel Schaefer; 19(cl), Thomas Ives/The Stock Market; 19(cl-inset), Ric Ergenbright; 19(cr), Craig Tuttle/The Stock Market; 19(cr-inset), David Muench; 19(bl), Renee Lynn/Photo Researchers, Inc.; 19(bl-inset), Rosemary Calvert/Tony Stone Images; 20(t), HRW Photo by Daniel Schaefer; 20(b), David Muench/H. Armstrong Roberts; 22(tl), Ric Ergenbright; 22(tr), D. Katzenstein/Photonica; 22(cl), G. Ryan & S. Beyer/Tony Stone Images; 22(cr), Sarah J. Frankling/Tony Stone Images; 22(bl) (br), E.R. Degginger/Earth Scenes; 23, HRW Photo by Daniel Schaefer; 24(all), Jan Halaska/Photo Researchers, Inc.; 25, T. Ulrich/H. Armstrong Roberts; 27(tl), E.R. Degginger/Color-Pic; 27(c), Jose L. Pelaez/The Stock Market; 27(cr), S. Purdy Matthews/Tony Stone Images; 27(bl), Peter Cade/Tony Stone Images; 27(bc), M. Thonig/H. Armstrong Roberts; 28a, Tom Edwards/Animals Animals; 28b, ZEFA/The Stock Market; 28c, David R. Frazier Photolibrary; 28d, E.R. Degginger/Color-Pic; 28e, Mark M. Lawrence/The Stock Market; 28f, HRW Photo by Eric Beggs; 29g, Gary Holscher/Tony Stone Images; 29h, Tom Walker/Tony Stone Images; 29i, Norbert Wu/Tony Stone Images; 29j, Peter Parks/Animals Animals; 29k, Tom & Pat Leeson/Photo Researchers, Inc.; 29l, Suzanne & Joseph Collins/Photo Researchers, Inc.; 29m, Roy Morsch/The Stock Market; 29n, Len Rue Jr./Photo Researchers, Inc.; 30(tc), Breck P. Kent; 30(tr), Robert A. Lubeck/Animals Animals; 30(c), Chuck Davis/Tony Stone Images; 30(b), Leonard Lee Rue III/H. Armstrong Roberts; 31(l) (c), E.R. Degginger/Color-Pic; 31(r), Bruce Iverson; 32(t), Gay Bumgammer/Tony Stone Images; 32(cl), William E. Ferguson; 32(cr), Robert A. Lubeck/Animals Animals; 32(bl), Zig Leszczynski/Animals Animals; 32(br), Breck P. Kent; 34, HRW Photo by Daniel Schaefer; 35(t), ZEFA/UK/The Stock Market; 35(br), Larry Fisher/Masterfile; 36-7(bckgd), Great Smoky Mountains Natural History Association; 36(t), Tony Stone Images; 36(clt), Jeanne Drake/Tony Stone Images; 36(cl), JH Robinson/Earth Scenes;36(clb), E.R. Degginger/Color-Pic; 36(crt), ZEFA/UK/The Stock Market; 36(crb), Jeff Lepore/Photo Researchers, Inc.; 36(b), Leonard Lee Rue III/H. Armstrong Roberts; 37(tl), Robert Maler/Animals Animals; 37(tr), JH Robinson/Photo Researchers, Inc.; 37(tr-inset), William E. Ferguson; 37(clt), Stephen J. Krasemann/Photo Researchers, Inc.; 37(clb), William E. Ferguson; 37(cr), Maslowski/Photo Researchers, Inc.; 37(bl), Phil A. Dotson/ Photo Researchers, Inc.; 37(bc), E.R. Degginger/Animals Animals; 37(br), E.R. Degginger/Color-Pic; 38, HRW Photo by Daniel Shaefer; 39(tc), Terry Donnelly/Tony Stone Images; 39(cl), Bill Binzen/The Stock Market; 39(cr), William E. Ferguson; 39(br), Richard Packwood/Animals Animals; 40(bckgd), Pat Bates/Tony Stone Images; 40(c), Rod Planck/Dembinsky Photo Association; 41(tl), Jeff Lepore/Photo Researchers, Inc.; 41(r), Ric Ergenbright Photography; 41(cl), William E. Ferguson; 41(c), Paul Freed/Animals Animals; 41(bl), David Muench; 44(t), Stan Osolinski/Dembinsky Photo Association; 44(l), John Gerlach/Dembinsky Photo Association; 44(c)(r), Stan Osolinski/Dembinsky Photo Association; 45(r), HRW Photo by Daniel Schaefer; 46(tl), Scott Camazine/Photo Researchers, Inc.; 46(bl), Peter Beck/The Stock Market; 46(r), Francios Gohier/Photo Researchers, Inc.; 47, Tom & Pat Leeson/Photo Researchers, Inc.; 48(t), Stephanie S. Ferguson/William E. Ferguson; 48(c) (b), William E. Ferguson; 50, HRW Photo by Daniel Schaefer; 51(cl), William E. Ferguson; 51(br), HRW Photo by Daniel Schaefer; 52(bl), H. Thonig/H. Armstrong Roberts; 52(br), Tony Freeman/Photo Edit; 54(t), E.R. Degginger/Color-Pic; 54(bl), Ron Sanford/Tony Stone Images; 55(tr), John James Audubon, National Audubon Society/Photo Researchers, Inc.; 55(cl), Tom McHugh/Photo Researchers, Inc.; 55(br), HRW Photo by Daniel Schaefer; 56(tl), The Denver Public Library, Western History Department; 56(bl), Will & Deni McIntyre/Photo Researchers, Inc.; 56(br), Joan Teasdale/The Stock Market; 57, NASA; 58, Claus Mayer/Tyba/Black Star; 61(t), James Gritz/Photonica; 61(l), Jeremy Walker/Tony Stone Images; 61(b), R. Krubner/H. Armstrong Roberts; 62(tr), Will & Deni McIntyre/Photo Researchers, Inc.; 62(bl) (bc) (br), HRW Photo by Daniel Schaefer; 63(tr), Carson Baldwin/Earth Scenes; 63(br), Christopher Morris/Black Star; 64(bl), Mark Lewis/Tony Stone Images; 66(tr), Joel Sartore/National Geographic Image Collection; 66(cl), Frink/Waterhouse/H. Armstrong Roberts; 66(bl), Carolina Biological Supply Company/HRW Photo; 66(br), Sarah J. Frankling/Tony Stone Worldwide; 67(cr), M. Schnelders/H. Armstrong Roberts; 67(b), S. Purdy Mathews/Tony Stone Images; 68(all), Dr. Ronald H. Cohn/The Gorilla Foundation; 69(tl), Tony Stone Images; 69(tr), Sanford D. Porter, U.S. Dept. of Agriculture; 70(tr), Josh Blumenfeld, National Wildflower Research Center; 70(bl), Solar Survival Architecture; 71(tr), Ron Kimball; 71(tr), E.R. Degginger/Color Pic; 71(bl), c.1995 Discover Magazine;

UNIT 2: Page 72,73,74,86,87,88,111,113,136,137,138,139,(bckgd), Rod Planck/Tony Stone Images; 72(tl), Susan Middleton/Tony Stone Images; 72(tc), Terry Donnelly/Tony Stone Images, M.I. Walker/Photo Researchers, Inc., Mark Lewis/Tony Stone Images, Inc, M. Thonig/H. Armstrong Roberts,Ray Coleman/Photo Researchers, Inc., David Waters/Envision; 72-3, Merlin D. Tuttle/Bat Conservation International; 74, (l), Daniel J. Cox/Tony Stone Images; 74(c), VJ. Anderson/Animals Animals; 74(b), Mark D. Phillips/Photo Researchers, Inc.; 75, David Hardy/Science Photo Library/Photo Researchers, Inc.; 76-7(bckgd), E.R. Degginger/Color-Pic; 76(t), Greg Vaughn/Tony Stone Images; 76(b), Laura Riley/Bruce Coleman, Inc.; 77(br), Breck P. Kent; 78-9(c),

Suzanne L. Collins & Joseph T. Collins/Photo Researchers, Inc.; 78(t), E. R. Degginger/Color-Pic; 78(bl), Stephen P. Parker/Photo Researchers, Inc.; 79(bc), Phil A. Dotson/Photo Researchers, Inc.; 82(cl), Mueller/ZEFA/H. Armstrong Roberts; 82(c), Breck P. Kent; 82(cr), Chris Collins/The Stock Market; 82(br), T. Ulrich/H. Armstrong Roberts; 83(tl), Richard Kolar/Animals Animals; 83(tc), Suzanne L. Collins & Joseph T. Collins/Photo Researchers, Inc.; 83(tr), John Warden/Tony Stone Images; 83(cl), R. Kord/H. Armstrong Roberts; 83(bl), HRW Photo by Daniel Schaefer; 83(br), Rod Planck/Photo Researchers, Inc.; 84(t), Manoj Shah/Tony Stone Images; 84(cl), HRW Photo by Daniel Schaefer; 84(cr), Fritz Prenzel/Tony Stone Images; 84(bl), Peter Parks/Animals Animals; 84(br), ZEFA/Germany/The Stock Market; 85(tl), Stephen Dalton/Photo Researchers, Inc.; 85(tr), E. R. Degginger/Color-Pic; 85(clt), Bonnie Sue/Photo Researchers, Inc.; 85(clb), Jeff Lepore/Photo Researchers, Inc.; 85(crt), Tui De Roy/Oxford Scientific Films/Animals Animals; 85(crb), Tom McHugh/Photo Researchers, Inc.; 87(tl), Pat Lynch/Photo Researchers, Inc.; 87(tr), Max Gibbs/Animals Animals; 87(c), Tom McHugh/Photo Researchers, Inc.; 88. Art Wolfe/Tony Stone Images; 89(tr)(cr), E.R. Degginger/Color-Pic; 89(c), Sinclair Stammers/Photo Researchers, Inc.; 89(bl)(br), Breck P. Kent; 90(tl), Patti Murray/Animals Animals; 90(tc), Brian Stablyk/Tony Stone Images; 90(tr)(bl), Susan Middleton/Tony Stone Images; 91(tl), Susan Middleton/Tony Stone Images; 91(tr), Richard Shiell/Animals Animals; 92(l), Joe McDonald/Animals Animals; 82(br), Breck P. Kent/Animals Animals; 93(t), Breck P. Kent; 93(b), Breck P. Kent/Animals Animals; 94(bl), Michael Habicht/Animals Animals; 94-5(b), Tim Davis/Photo Researchers, Inc.; 95(br), HRW Photo by Daniel Schaefer; 96(tc), Breck P. Kent; 96(cr), Mickey Gibson/Animals Animals; 96(bl), E.R. Degginger/Photo Researchers, Inc.; 96(br), Larry Ulrich/Tony Stone Images; 97(tl), G. Ahrens/H. Armstrong Roberts; 97(tr), Rod Planck/Tony Stone Images; 97(c), Nuridsany et Perennou/Photo Researchers, Inc.; 97(bl), Ian Murphy/Tony Stone Images; 97(br), Anthony Mercieca/Photo Researchers, Inc.; 98(t), Stephen Krasemann/Tony Stone Images; 98(c), E.R. Degginger/Color-Pic; 98(bl), Kathy Bushue/Tony Stone Images; 99(tl), Leonard Lee Rue III/Animals Animals; 99(tr), Stephen Krasemann/Tony Stone Images; 99(c), Art Wolfe/Tony Stone Images; 99(bl), ZEFA/Germany/The Stock Market; 99(br), Ken Brate/Photo Researchers, Inc.; 100(tl), Breck P. Kent; 100(tr), Robert Maier/Animals Animals; 100(c), Nancy Sefton/Photo Researchers, Inc.; 100(cr), JH Robinson/Animals Animals; 100(bl), John Kapriellan/Photo Researchers, Inc.; 100(br), Stephen J. Krasemann/Photo Researchers, Inc.; 101(tr), S.L. & J.T. Collins/Photo Researchers, Inc.; 101(cl), G. Heilman/H. Armstrong Roberts; 101(cr), E.R. Degginger/Color-Pic; 101(bl), Breck P. Kent; 106, Richard Parker/Photo Researchers, Inc.; 107(t), Jeff Lepore/Photo Researchers, Inc.; 107(b), Andrew J. Martinez/Photo Researchers, Inc.; 110(t), E.R. Degginger/Color-Pic; 110(b), Mickey Gibson/Animals Animals; 113(tl), Breck P. Kent; 113(tr), D. Petku/H. Armstrong Roberts; 113(cl), M. H. Sharp/Photo Researchers, Inc.; 113(b), Tim Davis/Photo Researchers, Inc.; 115, HRW Photo by Daniel Schaefer; 117a, Dan Routh/The Stock Market; 117b, Richard Steedman/The Stock Market; 117c, Andrew Syred/Photo Researchers, Inc.; 117d, Luis Villota/The Stock Market; 117e, Breck P. Kent; 117f, James Barnett/The Stock Market; 117g, Terry Donnelly/Tony Stone Images; 117h, Michael Lustbader/Photo Researchers, Inc.; 118(tl), Christoph Burki/Tony Stone Images; 118(tr), Doug Wechsler/Animals Animals; 118(cl)(bl), E.R. Degginger/Color-Pic; 118(crt), Mickey Gibson/Animals Animals; 118(crb), Robert Maler/Animals Animals; 118(br), Reinhard/H. Armstrong Roberts; 118(icon-a), David Waters; 118(icon-b), M.I. Walker/Photo Researchers, Inc.; 118(icon-c), Ric Ergenbright; 119(tl), Norbert Wu/Tony Stone Images; 119(tr), Breck P. Kent; 119(clt), Kjell B. Sandved/Photo Researchers, Inc.; 119(cl), G. Soury/Jacana/Photo Researchers, Inc.; 119(clb), Tim Davis/Photo Researchers, Inc.; 119(c), Breck P. Kent; 119(crt), Mitch Reardon/Tony Stone Images; 119(crb), K. Scholz/H. Armstrong Roberts; 119(bl), Gregory G. Dimijian/Photo Researchers, Inc.; 119(br), Keith Gillett/Animals Animals; 120(tl), Robert C. Hermes/Photo Researchers, Inc.; 120(tr), Michael Lustbader/Photo Researchers, Inc.; 120(clt), E.R. Degginger/Color-Pic; 120(clb), Bruce Watkins/Animals Animals; 120(crt), E.R. Degginger/Color-Pic; 120(cr), Stephen Dalton/Photo Researchers, Inc.; 120(crb), E.R. Degginger/Color-Pic; 120(bl), Tom McHugh/Photo Researchers, Inc.; 120(br), Robert Dunne/Photo Researchers, Inc.; 121(tl), Zig Leszczynski/Animals Animals; 121(tcl), Phil Degginger/Color-Pic; 121(tcr), Patti Murray/Animals Animals; 121(tr), C.C. Lockwood/Animals Animals; 121(clt), William Curtsinger/Photo Researchers, Inc.; 121(cl), Fred McConnaughey/Photo Researchers, Inc.; 121(clb), Leonard Lee Rue III/Tony Stone Images; 121(c), John Walsh/Photo Researchers, Inc.; 121(crt), E.R. Degginger/Color-Pic; 121(cr), Ray Coleman/Photo Researchers, Inc.; 121(crb), J. Howard/Photo Researchers, Inc.; 121(bl), Frink/Waterhouse/H. Armstrong Roberts; 121(cb), Andrew Martinez/Photo Researchers, Inc.; 121(br), Robert Dunne/Photo Researchers, Inc.; 122(l), Leonard Lee Rue III/Photo Researchers, Inc.; 122(tr), Zig Leszczynski/Animals Animals; 122(br), Andrew G. Wood/Photo Researchers, Inc.; 123, HRW Photo by Daniel Schaefer; 124(tl), Tom & Pat Leeson/Photo Researchers, Inc.; 124(tr), Art Wolfe/Tony Stone Images; 124(cl), Tui De Roy/Oxford Scientific Films/Animals Animals; 124(bl), Tim Davis/Tony Stone Images; 124(br), Fred Morris/The Stock Market; 125(tl), Gregory Dimijian/Photo Researchers, Inc.; 125(tr), HRW Photo by John Langford; 125(clt), M.H. Sharp/Photo Researchers, Inc.; 125(cl), Suzanne L. & Joseph T. Collins/Photo Researchers, Inc.; 125(clb), N. Orabona/H. Armstrong Roberts; 125(crt), Marc Chamberlain/Tony Stone Images; 125(cr), Tom McHugh/Steinhart Aquarium/Photo Researchers, Inc.; 125(crb), M. Barrett/H. Armstrong Roberts; 125(bl), T. Ulrich/H. Armstrong Roberts; 125(br), ZEFA/H. Armstrong Roberts; 129(tr), Bruce Coleman, Inc.; 126(cl), Tim Davis/Photo Researchers, Inc.; 126(c), JH Robinson/Photo Researchers, Inc.; 126(bl), Stan Osolinski/The Stock Market; 126(br), Douglas Faulkner/Photo Researchers, Inc.; 127(tr), George Bryce/Animals Animals; 127(cl), E.R. Degginger/Color-Pic; 127(cr), Sinclair Stammers/Science Photo Library/Photo Researchers, Inc. 127(b), M. Barrett/H. Armstrong Roberts; 128(tl), Tony Stone Images; 128(tr), Mary Beth Angelo/Photo Researchers, Inc.; 128(crt), Jany Sauvanet/Photo Researchers, Inc.; 128(crb), Andrew G. Wood/Photo Researchers, Inc.; 128(bl), Art Wolfe/Tony Stone Images; 128(br), Michele Burgess/The Stock Market; 129(tl), ZEFA/Reinhard/The Stock Market; 129(tr), J.C. Carton/Bruce Coleman, Inc.; 129(cl), John M. Burnley/Photo Researchers, Inc.; 129(crt), Greg Ochocki/Photo Researchers, Inc.; 129(cr), Craig Tuttle/The Stock Market; 129(crb), Jim W. Grace/Photo Researchers, Inc.; 129(bc), Biophoto Associates/Science Source/Photo Researchers, Inc.; 130(tr), Chris Mihulka/The Stock Market; 130(br), Tom Bean/The Stock Market; 131(tl), Ted Levin/Earth Scenes; 131(tr), Leonard Lee Rue III/Earth Scenes; 131(cr), Ric Ergenbright; 131(bl), Jack Rosen/Photo Researchers, Inc.; 132, H. Armstrong Roberts; 133, Michael Newman/Photo Edit; 134(tr)(b), HRW Photo by John Langford; 134(cl), HRW Photo by Rodney Jones; 135(t), HRW Photo by John Langford; 135(bl), HRW Photo by Richard Haynes; 135(br), Robert E. Daemmrich/Tony Stone Images; 136(cl), L. West/Photo Researchers, Inc.; 136(c), Rod Planck/Photo Researchers, Inc.; 136(cr), Robert Maier/Animals Animals; 136(bl), Bruce Coleman, Inc.; 136(bcl), Doug Armand/Tony Stone Images; 136(bcr), Jim Zipp/Photo Researchers, Inc.; 136(br), Ralph A. Reinhold/Animals Animals; 136(cr-inset), John Warden/Tony Stone Images; 137(t), M.H. Sharp/Photo Researchers, Inc.; 137(cl), Henley & Savage/The Stock Market; 137(c), F. Gohier/Photo Researchers, Inc.; 137(cr), Tony Dawson/Tony Stone Images; 138(tl), HRW Photo by Daniel Schaefer; 138(tr), Merlin D. Tuttle/Bat Conservation International; 138(bl)HRW Photo; 139(1a), Tom McHugh/Steinhart Aquarium/Photo Researchers, Inc.; 139(1b), Donn McLaughlin/The Stock Market; 139(1c), J. Patton/H. Armstrong Roberts; 139(1d), ZEFA/Germany/The Stock Market; 139(2a), Gary Holscher/Tony Stone Images; 139(2b), Robert Maler/Animals Animals; 139(2c), Al Assid/The Stock Market; 139(2d), John Gerlach/Tony Stone Images; 139(2e), Pat Lynch/Photo Researchers, Inc.; 139(3a), Charles Krebs/The Stock Market; 139(3b), Otto Rogge/The Stock Market; 139(3c), Len Rue, Jr./H. Armstrong Roberts; 139(3d), Y. Lanceau Jacana/Photo Researchers, Inc.; 139(4a), Profy/H. Armstrong Roberts; 139(4b), Tony Stone Images; 139(4c), G. Heilman/H. Armstrong Roberts; 139(br), Tim Davis/Photo Researchers, Inc.; 140(c), Jana Birchum Photography; 141(tr), Dr. Thomas Broker/Phototake; 141(br), Fran Heyl Associates; 142(tl), Tony Stone Images; 142(tr), Zoological Society of San Diego; 142(bl), Ron Garrison/Zoological Society of San Diego; 143(tl), Howard Sochurek/The Stock Market; 143(tr), Svein P. Hardeburg/World Health Organization; 143(cl), World Health Organization; 143(bl), Dr. Seim/World Health Organization;

UNIT 3: Page 144,145,146,165,166,167,183,184,185,206,207,208, 209(bkgrd), HRW Photo by Letraset/Phototone; 144(tl)(b), Carlos Austin; 144-5, Marc Chamberlain/Tony Stone Images; 146(all),148,149, Carlos Austin; 152(both),153(all),154,155, Carlos Austin; 156(bl)(tr), W. Gregory Brown/Animals Animals; 156(br), Rudie H. Kuiter/Animals Animals; 156(cl),157(both),158, Carlos Austin; 159(tl), S. Nielsen/Bruce Coleman, Inc.; 159(cr), Al Grillo/Alaska Stock Images; 159(b), Randy Brandon/Alaska Stock Images; 160, Gareth Trevor/Tony Stone Images; 161,162,163,164,165,166, Carlos Austin; 167(tl), Warren Bolster/Tony Stone Images; 167(tr), Larry Ulrich/Tony Stone Images; 167(bl), Mike McQueen/Tony Stone Images; 167(br), Henley & Savage/The Stock Market; 168,170,171, Carlos Austin; 172(t), Roy Morsch/The Stock Market; 172(c), Warren Bolster/Tony Stone Images; 172(b), Porterfield-Chickering/Photo Researchers, Inc.; 177,179, Carlos Austin; 180(tl), Leonard Lee Rue III/Bruce Coleman, Inc.; 180(tc), Carlos Austin; 182(t), David R. Frazier Photolibrary; 182(b), Jeff Foott/Bruce Coleman, Inc.; 183, HRW Photo by Daniel Schaefer; 185(l), 185(r),187,189,190, Carlos Austin; 191, HRW Photo by John Langford; 192, Carl Purcell/Photo Researchers, Inc.; 193,196, Carlos Austin; 197, World Perspectives/Tony Stone Images; 198, Jeffrey L. Rotman Photography; 199(l), Warren Bolster/Tony Stone Images; 199(tr), Jeremy Walker/Tony Stone Images; 199(br), Carson Baldwin/Earth Scenes; 199(b), Nasa/HRW; 201, Carlos Austin; 202, Tom McCarthy Photos/Photo Edit; 203(tr), David Young-Wolf/Photo Edit; 203(bl), Carlos Austin; 204(tr), Norman Owen Tomalin/Bruce Coleman, Inc.; 204(cr), Prof. W. Villiger/Science Photo

Library/Photo Researchers, Inc.; 205(tr), Charles D. Winters/Photo Researchers, Inc.; 205(cr), Larry Mulvehill/Science Source/Photo Researchers, Inc.; 205(br), Carlos Austin; 207(tr), Warren Bolster/Tony Stone Images; 208(tr), Marc Chamberlain/Tony Stone Images; 208(cl), Gary Milburn/Tom Stack & Associates; 209(bl), Jeff Foott/Bruce Coleman, Inc.; 210(tl), Rogge/The Stock Market; 210(bl), Mary Van de Ven/Pacific Stock; 211(tl), Tony Stone Images; 211(tr), Tony Stone Images; 211(bl), Andy Sacks/Tony Stone Images; 212(bl), Francisco Erize/Bruce Coleman, Inc.; 212(bc), Richard Steedman/The Stock Market; 213(tl), Tony Stone Images; 213(tr), Elvia E. Niebla/National Forest Service;

UNIT 4: Page 214,215,216,231,232,233,248,249,250,263,264,284,285,286, 287(bckgrd), Vince Streano/The Stock Market; 214-5, Romilly Lockyer/The Image Bank; 216(t), Peter Gridley/FPG International; 216(br), Michael Salas; 217, Tom Martin/The Stock Market; 218(t), David Madison/Bruce Coleman, Inc.; 218(cl) (bl), S. Lissau/H. Armstrong Roberts; 221(tl), Tom Raymond/Tony Stone Images; 221(tr), Michael Salas; 221(bl), Michael P. Gadomski/Bruce Coleman, Inc.; 221(br), Linda McQueeney/Sports Photo Masters; 220(tr), Alan Becker/The Image Bank; 220(bl), Lily Solmssen/Photo Researchers, Inc.; 222, J. Irwin/H. Armstrong Roberts; 223,224(tl)(b), HRW Photo by Michelle Bridwell; 224(tr),225(both),226(kids), HRW Photo by Daniel Schaefer; 230(br), The Bettmann Archive; 232(tr), David Madison/Bruce Coleman, Inc.; 233(t), Mark E. Gibson; 233(cr), Tony Freeman/Photo Edit; 233(bl), Michael Salas; 238,239(both), HRW Photo by Michelle Bridwell; 240, HRW photo by Michelle Bridwell; 241, HRW Photo by Michelle Bridwell; 242(all but br), HRW Photo by Michelle Bridwell, 242(br),243(tl) (tc), HRW Photo by Daniel Schaefer; 243(tr),244, 245(t) (bl), HRW Photo by Michelle Bridwell; 246(both), HRW Photo by Daniel Schaefer; 246-7,NASA; 247, Tony Stone Images; 248, HRW Photo by Michelle Bridwell; 250(all), Michael Salas; 253,254, HRW Photo by Daniel Shaefer; 257(t) (cr), HRW Photo by Michelle Bridwell; 257(br), David M. Phillips/Visuals Unlimited; 258(tr), B. Paulson/H. Armstrong Roberts; 258(inset), Phil Degginger/Color-Pic; 261(tr), Kim Taylor/Bruce Coleman, Inc.; 261(cl), David R. Frazier Photolibrary; 261(cr), R. Shuman/Bruce Coleman, Inc.; 261(bl), Michael Salas; 261(br), T. del Amo/H. Armstrong Roberts; 262, HRW Photo by Michelle Bridwell; 263(tr), L. Nelson/H. Armstrong Roberts; 264(tl)Michael Newman/Photo Edit; 264(cr), Frank E. Meitz/FPG International; 265, NASA; 267, Daniel Shaefer; 270(car-icon), Thomas Zimmermann/FPG International; 280(all), HRW Photo by Daniel Schaefer; 282(both),283(r), HRW Photo by Michelle Bridwell; 285(l), The Flying Redpaths/The Royal Hanneford Circus/Superstock; 285(tr), Tony Freeman/Photo Edit; 285(cr), Lommis Dean/Life Magazine/Time, Inc.; 286(tl), Stephen Dunn/Allsport USA; 286(bl), HRW Photo by Steven Gottlier; 286(r), Romilly Lockyer/The Image Bank; 287(br), Michael Salas; 288(tl), Ron Kimball; 288((tr), Reprinted with permission. Scientific American. All rights reserved.; 288(bl), Courtesy of T.J. Mueller, from An Album of Fluid Motion, Stanford University; 289(tl), Rogge/The Stock Market; 290(tl), Argonne National Laboratory Photo; 290(tc), U.S. Department of Energy/Science Photo Library/Photo Researchers, Inc.; 291(tl), Howard Sochurek/The Stock Market; 291(bl), Garolina Biological Supply Company/Phototake;

UNIT 5: Page 292,293,294,324,325,326,344,345,346,347(bckgrd), In House blue print; 292(tl), Phillip Habib/Tony Stone Images; 292-3, Mark E. Gibson/The Stock Market; 292(tr), HRW Photo by Letraset/Phototone; 294(tl), Linda Albrizio/The Stock Market; 294(tr), J. Hesserschmidt/H. Armstrong Roberts; 294(c), Paul Chauncey/The Stock Market; 294(bl), Larry Fisher/Masterfile; 298(tr) (bl), AP/Wide World Photo; 299(tl), Department of Clinical Radiology, Salisbury District Hospital/Science Photo Library/Photo Researchers, Inc.; 299(r), Shahn Kemani/Gamma-Liaison; 299(cr), Gen Nishino/Photonica; 300(tr), C. Campbell-Boulanger/FPG International; 300(cr), Pete Saloutos/Tony Stone Images; 300(bl), Sarah Stone/Tony Stone Images; 300(br), Merlin D. Tuttle/Bat Conservation International/Photo Researchers, Inc.; 301(tr), Perry Mastrovito/Reflexion; 301(c), Albert Moldvay/Eriako Associates; 301(br), P. Wallick/H. Armstrong Roberts; 302(t), Jeanne Drake/Tony Stone Images; 302(cr), H. Abernathy/H. Armstrong Roberts; 302(bl), ZEFA Germany Rossenbach/The Stock Market; 302(br), L. Nelson/H. Armstrong Roberts; 314(tr), Mitchell Reidel/Sports Photo Masters, Inc.; 314(b), R. Rizzo/FPG International; 317(t), M. Thonig/H. Armstrong Roberts; 317(c), Oddo & Sinibaldi/The Stock Market; 317(b), Yvette Cardozo/Tony Stone Images; 322(tl), Ed Cooper/H. Armstrong Roberts; 322(tr), Waldkrich/ZEFA/H. Armstrong Roberts; 322(cl), Paul Thompson/International Stock Photo; 322(bl), B.P. Maiti/Dinodia Picture Agency; 322(br), Robert Harding Picture Library, London; 323, Mike Penney/David R. Frazier Photolibrary; 324(tl), Terry Qing/FPG International; 324(tr), Peter Pearson/Tony Stone Images; 324(bl), Superstock; 324(br), Jeffrey W. Myers/The Stock Market; 326(tr), Bob Burch/Reflexion; 326(b), K. Scholz/H. Armstrong Roberts; 328(t), Thomas Gilcrease Institute, Tulsa, Oklahoma; 328(b), Paolo Koch/Photo Researchers, Inc.; 329(t), Lawrence Migdale; 329(b), Larry Fisher/Masterfile; 330(b), John Lamb/Tony Stone Images; 331(t), Mauritius-Vidler/Reflexion; 331(b), M. Grimoldi/A & F Archivio; 332(t), Al Michaud/FPG International; 332(b), G. Martin-Guillou/Explorer/Photo Researchers, Inc.; 333(tl), Lazio/Adros Studio; 333(tr), Andrea Palladio/Art Resource, NY; 333(b), Adros Studio; 334(tr), A. Tovy/H. Armstrong Roberts; 334(cl), Fred Salaff/FPG International; 334(b), Larry Fisher/Masterfile; 335(br), A. Higgott/Architectural Association; 336(bl), Peter Aaron/Esto Photographics; 337(tr), Andrea Sperling/FPG International; 337(c), Scott Frances/Esto Photographics; 337(bl), Photo by Samuel Stewart/Property of Fallingwater, Mill Run, PA/Courtesy of Western Pennsylvania Conservancy; 338(tr), Dennis Crompton/Architectural Association; 338(cl), Peter Aaron/Esto Photographics; 339(tr), John M. Roberts/The Stock Market; 339(bl), Jeffrey Aaronson/Network Aspen; 340(t), Susan McCartney/Photo Researchers, Inc.; 340(b), Eugen Gebhardt/FPG International; 341(tr), J.P. Karsenty/Publiphoto; 341(bl), David Hundley/The Stock Market; 342, C. Vergara/Photo Researchers, Inc.; 344(tr), Ezra Stoller/Esto Photographics; 344(c), J. Paul Kennedy/The Stock Market; 344(bl), John Zoiner; 344(br), Chris Hamilton/The Stock Market; 345, Comstock; 346(tl), Telegraph Colour Library; 346(bl), HBJ Photo; 346(tr), Mark E. Gibson/The Stock Market; 347(t), Neg/Transp no. K13813, Courtesy Department of Library Services, American Museum of Natural History; 348(tl), Ron Kimball; 348(tr) (cl), Desmond Murray; 349(tl), Rogge/The Stock Market; 349(tr), Max Aguilera-Hellweg; 349(bl), Courtesy Lorraine Lin, Department of Zoology, UK; 350(tl), David Bookstaver/AP/Worlwide; 351(tr), Art Resource, NY;

UNIT 6: Page 352,353,354,379,380,381,400,401,402,412,413,414, 415(bckgrd), M. Angelo/Westlight; 352-3, P.F. Bentley/Time Magazine, C. Time, Inc.; 354(bl), Greg Vaughn/Tony Stone Images; 355, Dieter & Mary Plage/Bruce Coleman, Ltd.; 356(tr), Archive Photos; 356(cl), The Bettmann Archive; 356(br), T. Algire/H. Armstrong Roberts; 357, HRW Photo by Daniel Schaefer; 358(tr), UPI/The Bettmann Archive; 360(tr), Van Bucher/Photo Researchers, Inc.; 360(cl) (bl), E.R. Degginger/Color-Pic; 360(br), Phil Degginger/Color-Pic; 361(l), HRW Photo by Daniel Schaefer; 361(tr), Tom Bean/Tony Stone Images; 362, U.S. Geological Survey/Photo Researchers, Inc.; 363(all), HRW Photo by Daniel Schaefer; 364(l), Comstock; 364(br), HRW Photo by Daniel Schaefer; 366(t), Andrew Rafkind/Tony Stone Worldwide; 366(bl), Tom McHugh/Photo Researchers, Inc.; 366(br), HRW Photo by Michelle Bridwell; 367(bl), Itauo Inouye/AP/Wide World Photos; 367(br), NASA/E.R. Degginger/Color-Pic; 368(t), U.S. Geological Survey/Science Photo Library/Photo Researchers, Inc.; 370-1,372(cr), R. Lamb/H. Armstrong Roberts; 372(tr), David Muench/Tony Stone Images; 372(bl), Phil Degginger/Color-Pic; 372(br), FPG International; 373(bckgrd), E. Cooper/H. Armstrong Roberts; 373(bc), Gary Braasch; 374(tr), The Bettmann Archive; 374(bl), AP/Wide World Photos; 374(br), Wayne & Karen Brown; 375(t), UPI/The Bettmann Archive; 375(bl), Dennis Oda/Photo Resource Hawaii; 375(bc), Barry Sweet/AP/Wide World Photos; 376(tr), Icelandic Tourist Office/HRW Library; 376(cl), Robert M. Carey, NOAA/Science Photo Library/Photo Researchers, Inc.; 377, HRW Photo by Michelle Bridwell; 378(tr), Chlaus Lotscher/Peter Arnold, Inc.; 378(bl), Brad Lewis/Photo Resource Hawaii; 380(rug), Courtesy of McCall's Needlework; 380(tr), Fred George/Tony Stone Images; 380(bl)FPG International; 380(bl), Eugen Gebhardt/FPG International; 380(bl), James Porto/FPG International; 381(tr), Phil Degginger/Color-Pic; 381(bl), Jo McBride/Tony Stone Images; 382(tr), Rob Talbot/Tony Stone Worldwide; 382(cl), R. Richardson/H. Armstrong Roberts; 382-3(b), Mike Surowiak/Tony Stone Images; 383(tl), H. Armstrong Roberts; 384(bl), HRW Photo by Daniel Schaefer; 389(bl), HRW Photo by Daniel Schaefer; 390(tr), The Bettmann Archive; 390(bl), Salmoiraghi/Photo Resource Hawaii; 390(br), G. Brad Lewis/Tony Stone Worldwide; 391(tr), H.R. Bramaz/Peter Arnold, Inc.; 393(b), HRW Photo by Michelle Bridwell; 394(l), HRW Photo by Daniel Schaefer; 395(t), HRW Photo by Michelle Bridwell; 396(tr), Science Photo Library/Photo Researchers, Inc.; 397(bl), James Cowlin/Image Enterprises; 402(c), Louis Psihoyos/Matrix; 403(bckgrd), Courtesy Department Library Services American Museum of Natural History; 404-5(bckgrd) (t), Francois Gohier/Photo Researchers, Inc.; 404(bc), Phil Degginger/Color-Pic; 404(bl), Martin Land/Science Photo Library/Photo Researchers, Inc.; 405(tl), James L. Amos/Photo Researchers, Inc.; 405(cl), Walter Hodge/Peter Arnold, Inc.; 405, Jane Burton/Bruce Coleman, Ltd.; 408(tl), Doug Plummer; 408(tr), NASA; 408(cr), The Bettmann Archive; 408(bl), Fred Ward/Black Star; 408(br), Tim Davis/Photo Researchers, Inc.; 409(b), H. Abernathy/H. Armstrong Roberts; 410(tr), Phil Degginger/Color-Pic; 410(br), Science Photo Library/Photo Researchers, Inc.; 412(t), Michelle Bridwell/Frontera Photos; 413(t), Sinclair Stammers/Science Photo Library/Photo Researchers, Inc.; 414(tl), HRW Photo by Michelle Bridwell; 414(bl), HRW Photo by Letraset/Phototone; 414(tr), P.F. Bentley/Time Magazine, C. Time, Inc.; 415(bl), Brad Lewis/Photo Resource Hawaii; 416(tl) (tr), Louis Psihoyos/Matrix International; 416(bl), E.R.

Degginger/Color-Pic; 417(r), Superstock; 417(bl), E.R. Degginger/Color-Pic; 418(tl), Tony Stone Images; 418(tr), Superstock; 418(cr), E.R. Degginger/Color-Pic; 419(tr), Jerry Jacka; 419(bl), DTM Corporation;

UNIT 7: Page 420,421,422,436,437,438,452,453,454,470,471,472,490,491, 492,493(bckgrd), Kazuaki Iwasaki/The Stock Market; 420(tc), Royal Observatory, Edinburgh/AATB/Science Photo Library/Phot Researchers; 420(tl), NASA/Peter Arnold, Inc.; 420-1, Roger Ressmeyer/Corbis; 422(tl), Chris Butler/Astrostock; 422(r), HRW Photo by Daniel Schaefer; 422(bl), c. Mary Evans Picture Library/Photo Researchers, Inc.; 423, Keith Gunnar/Bruce Coleman, Inc.; 425(tl), David A. Hardy/Science Photo Library/Photo Researchers, Inc.; 425(cl), Jet Propulsion Labs/HRW Library; 425(c), Photri/The Stock Market; 425(cr), NASA/Astrostock*Sanford; 425(bl), NASA/Corbis; 425(br), NASA; 426(t), HRW Photo by Daniel Schaefer; 426(b), NASA; 427, Malcom S. Kirk/Peter Arnold, Inc.; 428(l), HRW Photo by Daniel Schaefer; 428(r), Charles G. Summers/Bruce Coleman, Inc.; 429, HRW Photo by Jana Birchum; 430, Astrostock*Sanford; 431(tr), Comstock; 431(bl), NASA/Peter Arnold, Inc.; 431(br), Scala/Art Resource, NY; 432(tr), Erich Lessing/Magnum Photos; 432(b), HRW Photo by Daniel Schaefer; 433(tr), Telescope Board/HRW Photo by David F. Malin; 433(b), Histotisches Bildarchiv, Handke-Bad Berneck/Omikron Collection/Photo Researchers, Inc.; 434(t), Dr. Jeremy Burgess/Science Photo Library/Photo Researchers, Inc.; 434(b), George Loui; 435, Steve Solum/Bruce Coleman, Inc.; 436(tr), Image Select/Ann Ronan Picture Library; 438(cl) (cr), Gay Bumgamer/Tony Stone Images; 439(t), HRW Photo by Jana Birchum; 439(b),440(bckgrd) Erich Lessing/Art Resource, NY; 440(tr), HRW Photo by Daniel Schaefer; 442(cr), HRW Photo by Eric Beggs; 442(bl), Eunice Harris/Photo Researchers, Inc.; 443(tr), Phil Degginger/Color-Pic; 446(bl),450(both),451, HRW Photo by Daniel Schaefer; 452, Astronomical Society of the Pacific/HRW Library; 453(tr), Dr. Fred Espenak/Science Photo Library/Photo Researchers, Inc.; 454(bl), HRW Photo by Daniel Schaefer; 455(t), Paolo Koch/Photo Researchers, Inc.; 455(b), K. Iwasaki/The Stock Market; 456, Astrostock*Sanford; 457(tr), Richard J. Wainscoat/Peter Arnold, Inc.; 457(c), Giraudon/Art Resource, NY; 457(bl), HRW Photo by Daniel Schaefer; 458(bl), Mary Evans Picture Library; 459(b), Dennis Di Cicco/Peter Arnold, Inc.; 460(b), Astronomical Society of the Pacific/HRW Library; 461(tl) (tc), NASA/JPL/Astrostock*Sanford; 461(tr), Phil Degginger/Color-Pic; 461(cl), David A. Hardy/Photo Researchers, Inc.; 461(cr),462, NASA/Science Photo Library/Photo Researchers, Inc.; 462(tr) (c) (cr), Jet Propulsion Labs/HRW Library; 462(br), NASA; 462(bc), NASA/Astrostock*Sanford; 463(tr), Jet Propulsion Labs/HRW Library; 463(cr), NASA/Phil Degginger/Color-Pic; 463(br), Sovfoto/Eastfoto; 464(cl), NASA/Corbis; 464(cr), NASA/Mark Marten/Science Source/Photo Researchers, Inc.; 464(br), Astronomical Society of the Pacific/HRW Library; 465(b), Chromosohm/SohmMCMXCII/Photo Researchers, Inc.; 466, NASA; 468(c), NASA; 468(bl), Jet Propulsion Labs/HRW Library; 469, National Geographic Society Image Collection; 470, NASA; 471(t), Mary Evans Picture Library; 471(c), Science Photo Library/Photo Researchers, Inc.; 471(b), NASA/Astrostock*Sanford; 472(bckgrd), Roger Ressmeyer/Corbis; 472(c), HRW Photo by Daniel Schaefer; 473(t), Naval Research Laboratory/Hansen Planetarium/HRW Library; 473(b), NASA; 474(l), Astrostock*Sanford; 475, HRW Photo by Daniel Schaefer; 476, Dennis Di Cicco/Peter Arnold, Inc.; 477(bckgrd), Kitt Peak National Observatory/Cerro Tololo Inter-American Observatory/HRW Library; 477(tr), Astrostock*Sanford; 478(tr), Naval Research Laboratory/Hansen Planetarium/HRW Library; 478(bl), Robin Scagell/Science Photo Library/Photo Researchers, Inc.; 480(bl), HRW Photo by Daniel Schaefer; 480(br), NASA; 483(b), Comstock; 485(b), Hale Observatories/HRW Library; 486(bckgrd), Roger Ressmeyer/Corbis; 486(t), Jean Miele/The Stock Market; 486(b), HRW Photo by Daniel Schaefer; 487(b), Dana Berry/HRW Library; 488(t), NASA; 488(b), Chris Butler/Photo Researchers, Inc.; 489(all), HRW Photo by Daniel Schaefer; 490(t) (bl), Science Photo Library/Photo Researchers, Inc.; 490(br), Compliments of Antony Hewish/University of Cambridge; 492(tl), NASA; 492(tr), Roger Ressmeyer/Corbis; 492(bl), Roger Ressmeyer/Corbis; 492(br), HRW Photo by Daniel Schaefer; 493, NASA; 494(tl), Rogge/The Stock Market; 494(tr), NASA; 494(c), sovfoto/eastfoto; 495(tl), Ron Kimball; 495(tr), NASA; 495(cr), NASA/Corbis; 496(tl), Rogge/The Stock Market; 496(tr), Pete Saloutos/The Stock Market; 496(bl), Max-Planck-Institute for Radioastronomy/Science Photo Library/Photo Resarchers, Inc.; 497(tl), Tony Stone Images; 497(tr), HRW Photo; 497(bl), NASA;

UNIT 8: Page 498,499,500,515,516,517,538,539,540,554,555(bckgrd), Randy Wells/Tony Stone Images; 498(tc), Andy Sacks/Tony Stone Images; 498(tl), HRW Photo by John Langford; 498-9, Frans Lanting/Minden Pictures; 500(tr), Adam Hart-Davis/Science Photo Library/Photo Researchers, Inc.; 500(bl), HRW Photo by Jana Birchum; 501, NASA; 504(t), HRW Photo by Jana Birchum; 505,506(both), HRW Photo by Daniel Schaefer; 506(both), 507(tr), Andy Sacks/Tony Stone Images; 507(c), Peter Pearson/Tony Stone Images; 507(cr), Phil Degginger/Color-Pic; 507(br), Joan Menschenfreund/The Stock Market; 508,510(bl), HRW Photo by Daniel Schaefer; 510(c), HRW Photo by Jana Birchum; 511(b), John Eastcott/Yva Momatium/The Image Works; 514(all), HRW Photo by Jana Birchum; 515(tr), HRW Photo by Daniel Schaefer; 515(bl), Andy Sacks/Tony Stone Images; 516(t), Letraset/Phototone; 516(tc), Peter D'Angelo/Tony Stone Images; 516(bc), Cosmo Condina/Tony Stone Images; 516(b), Tim Thompson/Tony Stone Images; 517(tl), HRW Photo by Jana Birchum; 518(bckgrd), Jonathan A. Meyers; 518(c), F. Stuart Westmorland/Photo Researchers, Inc.; 519,520(all),521, HRW Photo by Jana Birchum; 522(l), Topham Picture Point; 522-3, HRW Photo by Jana Birchum; 523(cr), Topham Picture Point; 525(cr), HRW Photo by Jana Birchum; 525(bl), Tom Dietrich/Tony Stony Images; 527(br), Fujifotos/The Image Works; 528(tr), Anne Heimann/The Stock Market; 528(bl), Superstock; 528(br), M. P. Gadomski/ Bruce Coleman, Inc.; 529(tr) (bl), Roy Morsch/The Stock Market; 529(cl), W. Weisser/Bruce Coleman, Inc.; 529(br), David Cavagnaro/Peter Arnold, Inc.; 532(t), Topham Picture Point; 532(bl) (br), Ilex/Muriel Orans/Horticulture Photography; 533(br), Rollin Sears/Kansas State University; 535, HRW Photo by Daniel Schaefer; 536, Dr. Jeremy Burgess/Science Photo Library/Photo Researchers, Inc.; 537(t), Patrick Lynch/Science Source/Photo Researchers, Inc.; 537(br), HRW Photo by Daniel Schaefer; 540(t), S. Weiss/Photo Researchers, Inc.; 540(cr), J. H. Robinson/Photo Researchers, Inc.; 541,543, HRW Photo by Daniel Schaefer; 544(t), Jerome Wexler/Photo Researchers, Inc.; 544(cl), Topham Picture Point; 544(cr), Tom McHugh/Photo Researchers, Inc.; 544(b), Betty Derig/PhotoResearchers, Inc.; 547, HRW Photo by Daniel Schaefer; 548(bl), Douglas Faulkner/Photo Researchers, Inc.; 548(br), Tom McHugh/Photo Researchers, Inc.; 550(b), HRW Photo by Daniel Schaefer; 553(t), c. Richard B. Peterson/Environmental and Conservation Services Department of Austin; 553(b), R. Perron; 554(tl), HRW Photo by Daniel Schaefer; 554(bl), David Muench; 554(tr), Frans Lanting/Minden Pictures; 556(tl), Howard Sochurek/The Stock Market; 556(tr) (bl), Phillip-Lorca DiCorcia/c. 1993 The Walt Disney Co. Reprinted with permission of Discover Magazine; 557(tl), Tony Stone Images; 557(tr) (bl), Courtesy Bonnie Fine Jacobs/Southern Methodist University; 558(tl), Rogge/The Stock Market; 558(c), Dann Blackwood/c. 1991 The Walt Disney Co. Reprinted with permission of Discover Magazine; 559(tl), Tony Stone Images; 559(tr) (br), Cary S. Wolinsky/Natural Cotton Colours, Inc.; 559(cr), Alan Pitcaim/Grant Heilman Photography, Inc.;

SOURCEBOOK:

UNIT 1: All glasses icons throughout Sourcebook, Courtesy of the Franklin Institute Science Museum, Philadelphia, PA./HRW Photo; Page S1-S26(bckgrd-strip), Carolina Biological Supply Company/HRW Photo; S1(tl), Joel Sartore/National Geographic Society Photographic Library; S3(tr), Alissa Crandall/Alaska Stock; S3(bl), Wide World Photos; S5(tl), Johnny Johnson/DRK Photo; S5(bl), John Hendrickson; S6(b), Jeffrey L. Rotman Photography; S9(tc), John Cancalosi/Peter Arnold, Inc.; S11(tr), S.J. Krasemann/Peter Arnold, Inc.; S11(bl), Steve McCutcheon/Alaska Pictorial Service; S11(bc), Bill Ross/Tony Stone Images, Inc.; S13(r), Daryl Benson/Masterfile; S14(c), Steve McCutcheon/Alaska Pictorial Service;S15(tr), Tom Ulrich/Tony Stone Worldwide; S15(bl), Larry Ulrich; S15(inset-bc), John Cancalosi/Peter Arnold, Inc.; S16(tl), Tom Bean/The Stock Market; S16(br), John Foster/Masterfile; S17(tl), Charles Kreb/The Stock Market; S17(cr), Breck P. Kent; S19(c), National Audubon Society/Photo Researchers, Inc.; S20(c), Superstock; S21(tr), Jeffrey Sylvester/FPG International; S21(cl), Ray Pfortner/Peter Arnold, Inc.; S21(br), Carr Clifton; S22(t), Dave Watts/Tom Stack & Associates; S22(c), Barbara Von Hoffmann/Tom Stack & Associates; S22(br), John Zoiner/International Stock; S23(tr), Gabe Palmer/Mugshots 1992/The Stock Market; S23(bl), HRW Photo by Daniel Schaefer; S26(tr), E.R. Degginger/H. Armstrong Roberts; S24(tl), Visual Horizons/FPG International.

UNIT 2: Page S27-S50(bckgrd-strip), HRW Photo; S27(tl), Merlin D. Tuttle/Bat Conservation International; S28(tl), Bruce Iverson; S28(cr) (cl) (br) (bc), E.R. Degginger/Color-Pic; S29(t), Runk/Schoenberger/Grant Heilman Photography, Inc.; S29(cr), Michael Abbey/Science Source/Photo Researchers, Inc.; S29(br), Randy Morse/Tom Stack & Associates; S29(bc) , Manfred Kage/Peter Arnold, Inc.; S29(bl), Tom E. Adams/Peter Arnold, Inc.; S30(cr), E.R. Degginger/Color-Pic S30(br), M.P. Kahl/Photo Researchers, Inc.; S31(tr), Runk/Schoenberger/Grant Heilman Photography, Inc.; S31(b), Wayne Aldridge/International Stock; S32(tr), Bruce Iverson; S32(cl), Michael P. Gadomski/Photo Researchers, Inc.; S32(b), Larry LaFever/Grant Heilman Photography, Inc.; S32(inset-cr), Biophoto Associates/Photo Researchers, Inc.; S33(tc), Fritz Polking/Peter Arnold, Inc.; S33(inset-tc), E.R. Degginger/Color-Pic; S33(inset-cr), Gilbert S. Grant/Photo Researchers, Inc.; S33(cl), Doug Anderson/Natural Selection; S33(br), Arthur Beck/Photo Researchers, Inc.; S34(tl), Mike Bacon/Tom Stack &

Associates; S34(cl), Nancy Sefton/Photo Researchers, Inc.; S34(tc)Brian Parker/Tom Stack & Associates; S34(br), Kim Taylor/Bruce Coleman, Inc.; S36(tr), Fred Bavendam/Peter Arnold, Inc.; S36(cl), E.R. Degginger/Photo Researchers, Inc.; S36(br), Runk/Schoenberger/Grant Heilman Photography, Inc.; S37(tl), Gary Milburn/Tom Stack & Associates; S37(cr), J. Christian/Leo De Wys, Inc.; S37(cl), Stuart Westmorland/Natural Selection; S37(b), R. Andrew Odum/Peter Arnold, Inc.; S38(tl), Rod Planck/ Dembinsky Photo Associates; S38(cl), John Bova/Photo Researchers, Inc.; S38(cr), S. Purdy Matthews/Tony Stone Images; S39(inset-cl), Soames Summerhays/Photo Researchers, Inc.; S40(c), HRW Photo by Bob Tucek; S41(cr), Breck P. Kent; S42(tl), M. Austerman/Animals Animals; S42(bl), S43(tl), Superstock; S43(tc)Michael Tweedie/Photo Researchers, Inc.; S44(tl), Superstock; S46(tl), Joyce Butler/Photo Researchers, Inc.; S47(tc), Nigel Cattlin/Photo Researchers, Inc.; S47(tr), D. Cavagnaro/Peter Arnold, Inc.; S47(cl), Alan Pitcairn/Grant Heilman Photography, Inc.; S48(cl), Bruce Iverson.

UNIT 3: Page S51(tl), Jeffrey L. Rotman Photography; S52(tl), HRW Photo by Daniel Schaefer; S52(br), Runk/Schoenberger/Grand Heilman Photography, Inc.; S53(cl)(bl), E.R. Degginger/Color-Pic; S54(all), HRW Photo by Jack Newkirk; S57(c), Charles D. Winters/Photo Researchers, Inc.; S57(bl), David York/The Stock Shop; S57(bc), Maratea/International Stock; S58(cr), HRW Photo by Richard Haynes; S59(tl), HRW Photo by Jack Newkirk; S59(tr), Runk/Schoenberger/Grant Heilman Photography, Inc.; S59(cr), E.R. Degginger/Color-Pic; S61(tl), David Pollack/The Stock Market; S62(tc), Brownie Harris/The Stock Market; S62(l), Michael A. Keller/The Stock Market; S62(cr), Palmer/Kane, Inc./The Stock Market; S63(cr), E.R. Degginger/Color-Pic; S63(bl)(br), HRW Photo by Daniel Schaefer; S64(tl)(cl), David R. Frazier Photolibrary; S64(bl), Telegraph Colour Library/FPG International; S64(inset-tc), Tom Tracy/FPG International; S65(bc), Grapes/Michaud/Photo Researchers, Inc.; S66(tr), Runk/Schoenberger/Grant Heilman Photography, Inc.; S66(cl), BIOS(M. Edwards)/Peter Arnold, Inc.; S67(bc), HRW Photo by Jack Newkirk; S68(tl), Kevin Schafer/Peter Arnold, Inc.; S68(cl), Superstock; S70(t), Bryan F. Peterson/The Stock Market;

UNIT 4: Page S71-S88(bckgrd-strip), HRW Photo by Steven Gottlier; S71(tl), Romilly Lockyer/The Image Bank; S71(both), Superstock; S73(tl), Randall Hyman; S73(br), Baron Wolman/Tony Stone Images; S74(tl), Lori Adamski Peek/Tony Stone Images; S75(tl), Gordon Garradd/Science Photo Library/Photo Researchers, Inc.; S75(br), Thomas Braise/The Stock Market; S76(tc), Movie Still Archives; S77(tr), M. Herker/The Image Bank; S77(bl), Peter G. Aitken/Photo Researchers, Inc.; S78(tr), Park Street Photography; S83(cl)(cr), HRW Photo by Scott Van Osdel; S85(tc), The Bettmann Archive; S85(tr), Peticolas/Megna/Fundamental Photographs; S86(tl), NASA;

UNIT 5: Page S89-S106(bckgrd-strip), HRW Photo by Bob Mates; S89(tl), Mark E. Gibson/The Stock Market; S90(tc), Tony Stone Images; S90(tr), Casimir/Leo de Wys; S90(c), Johnny Stockshooter/International Stock; S90(bc), HRW Photo by Bob Tucek; S90(inset-cl), Cameron Davidson/Comstock; S90(inset-bl), HRW Photo by Bob Tucek; S91(cl), Pete Saloutos/The Stock Market; S91(cr), Chuck Mason/International Stock; S91(r), Ezra Stoller/Esto Photographics; S91(bl), Superstock; S92(tl)(br),S95(b), Superstock; S95(c), Paul Merideth/Tony Stone Images; S96(tl), Superstock; S96(tr), H. Armstrong Roberts; S96(br), Miwako Ikeda/International Stock; S97(tr), Adam Woolfitt/Woodfin Camp & Assoc, Inc.; S97(cl), Archive Photos; S97(bc), Hilarie Kavanagh/Tony Stone Images; S98(tl), Richard Gorbun/Leo de Wys; S98(cr), Cameramann International; S98(bl), Tony Stone Images; S99(tl), Alain Evrard/Photo Researchers, Inc.; S99(tr), Cameramann International; S99(bl), Superstock; S100(t), Tony Stone Images; S100(br), HRW Photo by Bob Tucek; S101(tr), Kenneth Garrett/FPG International; S101(cl), James K. Hackett/Leo de Wys; S101(bl), Superstock; S102(tr), Robert Frerck/The Stock Market; S102(tl), Van Phillips/Leo de Wys; S102(br), Olaf Soot/Tony Stone Images; S103(tl)(tr), Cahokia Mounds State Historic Site; S103(cr), E.R. Degginger/Color-Pic; S103(bl), Smithsonian Institute/The Bettmann Archive; S104(t), M & E Bernheim/Woodfin Camp & Assoc; S104(cr), E. Streichan/Superstock; S106(tl), David Muench; S106(tr), Steve Vidler/Leo de Wys; S106(cl), Josef Beck/FPG International; S106(c), Fridmar Samm/Leo de Wys; S106(cr), Louis Goldman/FPG International;

UNIT 6: Page S107-S136(bckgrd-strip), HRW Photo by Letraset-Phototone; S107(tl), P.F. Bentley/Time Magazine, C. Time Inc.; S112(cl), Roland Weber/Masterfile; S112-3(b), David Cornwell/Pacific Stock; S113(tl), ZEFA/H. Armstrong Roberts; S113(cr), Frafft/Explorer/Science Source/Photo Researchers, Inc.; S115(cr), T. Dietrich/H. Armstrong Roberts; S116(tl), Breck P. Kent; S116(bc), E.R. Degginger/Color-Pic; S117(cl)(tc), Breck P. Kent; S117(cr)(br), E.R. Degginger/Color-Pic; S118(tr), Kevin Schafer/Peter Arnold, Inc.; S118(tc), Terry Domico/Earth Images; S118(cl), E.R. Degginger/Color-Pic; S118(bl), Breck P. Kent; S119(tr), Giannil Tortoli/Photo Researchers, Inc.; S119(cl), Pat Lanzafield/Bruce Coleman, Inc.; S119(cr)(bc), E.R. Degginger/Color-Pic;S120(tl), Breck P. Kent; S120(cl)(br)(bc), S121(t)(tr) E.R. Degginger/Color-Pic; S121(bl), Horst Schafer/Peter Arnold, Inc.; S121(cl), Paolo Koch/Photo Researchers, Inc.; S121(br), Brian Parker/Tom Stack & Assoc.; S121(cr), C. Kuhn/The Image Bank; S122(tr)(tl)(cl)(cr)(bl), E. R. Degginger/Color-Pic; S122(br), Breck P. Kent; S122(b), Don Spiro/Tony Stone Images, Inc.; S123(tr)(cl), Breck P. Kent; S124-5(b), S125(t)(cl), E.R. Degginger/Color-Pic; S125(tr), Mary A. Root/Root Resources; S125(br), Breck P. Kent; S126(tl)(bl), Paul Silverman/Fundamental Photographs; S126(c),#3067(2), J. Beckett, Courtesy Department of Library Services, American Museum of Natural History; S127(tr), Paul Silverman/Fundamental Photographs; S127(l), Fundamental Photographs; S128(all except chalcopyrite), E.R. Degginger; S128(other), Breck P. Kent/Earth Scenes; S129(tl)(c), Breck P. Kent; S129(tr)(cr), E.R. Degginger/Color-Pic; S130(tr), HRW Photo by Eric Beggs; S130(bc), Park Street Photography; S131(tl), Breck P. Kent; S131(br), Terry Domico/Earth Images; S132(tr), Breck P. Kent;

UNIT 7: Page S137-S160(bckgrd-strip), NASA/HRW Library; S137(tl), Roger Ressmeyer/Corbis; S138(tl), Frank P. Rossotto/The Stock Market; S138(inset), NASA/Corbis; S140(tr), N. Rosing(Sunset)/Peter Arnold, Inc.; S141(b), David A. Hardy/ Science Photo Library/ Photo Researchers, Inc.; S143(tc), NASA/Peter Arnold, Inc.; S143(cr), Jet Propulsion Laboratory/HRW; S143(bl), NASA; S144(tl), Phil Degginger/ Color-Pic, Inc.; S144(c), E.R. Degginger/Color-Pic, Inc.; S145(c), Courtesy of Larousse Encyclopedia of Astronomy/copyright Prometheus Press, New York 1967; S147(tr), John Sanford/Science Photo Library/ Photo Researchers, Inc.; S148(tl)(tr)(c), Yerkes Observatory; S148(bl)(br), Bernhard Haisch, Lockheed Martin Solar and Astrophysics Laboratory; S149(tl), John Sanford/Photo Researchers, Inc.; S149(tr), Royal Observatory, Edinburgh/Science Photo Library/Photo Researchers, Inc. ; S149(bl), Allan Morton/Dennis Milon/Science Photo Library/Photo Researchers, Inc.; S150(tl), Tony Hallas/Science Photo Library/Photo Researchers, Inc.; S150(tc)(cl)(cr), Anglo-Australian Observatory 1991; S150(bl), Chris Butler/Astrostock; S151(tc), Mark Marten/NASA; S151(cr), Space Telescope Science Institute/NASA/Science Photo Library/Photo Researchers, Inc.; S152(tr), John Serafin/Peter Arnold, Inc.; S152(bl), Douglas Waugh/Peter Arnold, Inc.; S152(c), Scala 1984/Art Resource; S153(tr), The Granger Collection; S153(bc), Roger Ressmeyer/Corbis ; S154(tr), Richard Megna/Fundamental Photographs; S154(tl), NRAO/AUI/Science Photo Library/Photo Researchers, Inc.; S155(tr), CulverPictures, Inc.; S155(cl), NASA/Photri, Inc.; S155(l), NASA/Photri Inc.; S155(br), Novosti Press Agency/Science Photo Library/Photo Researchers, Inc.; S156(tl)World Perspectives/Explorer/Photo Researchers, Inc ; S156(cl), Frank Rossotto/The Stock Market; S156(cr), NASA; S157(tl), NASA/HRW Photo;

UNIT 8: S161-S176(bckgrd-strips), David Muench ; S161(tl), Frans Lanting/Minden Pictures; S162, Wayne Eastep/Tony Stone Images; S165(tr), P. Dayanandan/Photo Researchers, Inc.; S165(c), E.R. Degginger/Color-Pic, Inc.; S165(b), Chad Ehlers/International Stock; S167(tr), Mark C. Burrell/Photo Researchers, Inc.; S169(bc), Merlin D. Tuttle/Bat Conservation International; S169(inset), Francois Gohier/Photo researchers, Inc.; S171(c), HRW Photo by Richard Haynes; S172(tl), G. Buttner/Naturbild/OKAPIA/Photo Researchers, Inc.; S172(cl), Nigel Cattlin/Holt Studios Int./Photo Researchers, Inc.; S172(bl), Dan Routh/The Stock Market; S173(tr), Lance Nelson/The Stock Market; S173(c), Nigel Cattlin/Holt Studios International; S173(br), Caffee/ International Stock; S173(bl), Nigel Cattlin/Holt Studios International; S173(bc), W.H. Hodge/Peter Arnold, Inc. S174(cl), Art Wolfe/Tony Stone Images; S176(t), James Randklev/Tony Stone Images.

ART CREDITS

Abbreviated as follows: (t) top; (b) bottom; (l) left; (r) right; (c) center.

All work, unless otherwise noted, is contributed by Holt, Rinehart and Winston.

UNIT 1: Page 4, Steven Adler/Jeff Lavaty Artist Agent; 8, Bruce Metherd; 10(tr), Doug Henry/American Artists Rep., Inc.; 16, Donna Kae Nelson/Sharon Langley Artist Representatives; 17, John Francis; 18, Jim Sutton/Renée Kalish; 21(tr), Gary Yealdhall/American Artists Rep., Inc.; 23(tr), Larry McEntire/Fran Seigel Representing Illustrators; 25(tc),Glasgow & Associates; 33(b), Bruce Bowles; 42-43, Robin Bouttell/ Morgan Cain & Associates; 48-49(b), Sarah Woodward/Morgan Cain & Associates; 52(tr), David Merrell/Suzanne Craig Represents; 53, June Workman; 54(c), Bruce Metherd; 59, Jim Auckland/Liz Sanders Agency; 60(t), Bruce Metherd; 60(b), Glasgow & Associates; 63(br), Steven Schudlich/Ceci Bartels Associates; 64(c), Glasgow & Associates; 67(tr), Blake Thornton/Rita Marie and Friends; 69, Precision Graphics.

UNIT 2: Page 74(t), Steve Roberts/Morgan Cain & Associates; 80-81, Sarah Woodward/Morgan Cain & Associates; 86(c),Truc An Lam/Garden Studio Illustrators; 87(bl), Rhoda Grossman; 90-91(br), Chuck Passarelli/American Artists Rep., Inc.; 91(cr), Simon Turvey/Morgan Cain & Associates; 94, Peter Van Gulik; 102-105, Pond & Giles/Morgan Cain & Associates; 108-109, Paul Blakey/Susan Wells & Associates, Inc.; 111, Donna Kae Nelson/Sharon Langley Artist Representatives; 112, Amy L. Wasserman; 114, Pamela Hamilton/American Artists Rep., Inc.; 116, Robin Bouttell/ Morgan Cain & Associates; 130-131(bkgd), Richard Pembroke/American Artists Rep., Inc.; 132, Blake Thornton/Rita Marie & Friends; 136(t) Bruce Metherd.

UNIT 3: Page 147, Keith Locke/Suzanne Craig Represents; 148(tl), Keith Locke/ Suzanne Craig Represents; 149(l), Keith Locke/Suzanne Craig Represents; 149(r), Glasgow & Associates, 150, Uhl Studio; 151, Uhl Studio; 152(t), Uhl Studio; 155(t), Uhl Studio; 160, Mark Oliver/Garden Studio Illustrators; 161(l), Rainey Kirk/The Neis Group, 163(c), Karen Snave/Renée Kalish; 163(b), Karen Snave/Renée Kalish; 164(t), Karen Snave/Renée Kalish; 166(b), Richard Wehrman; 168-169, Holly Cooper; 170-175(b), Johanna E. Kimball; 173, Brian Harrold/American Artists Rep., Inc.; 174-175(b), Robin Carter/Mogan Cain & Associates; 175(t), Don Sullivan; 176(t) Tim Ladwig/Suzanne Craig Represents; 178, Donna Kae Nelson/Sharon Langley Artist Representatives; 180-181, Paul E. Kimball; 183(t), Karen Snave/Renée Kalish; 184, Rhoda Grossman; 186, Victoria Bruck; 187(t), Joel Iskowitz; 188-189(t), Uhl Studio; 189(b), Mark J. Persyn; 190(t), Michael Koester/Woody Coleman Presents Inc.; 191, Rainey Kirk/The Neis Group; 193, Heather Collins; 194-195, David Merrell/Suzanne Craig Represents; 198(t), Bruce Metherd; 198(c), Uhl Studio; 200, Vesna Krstanovich; 202(t), Gary Yealdhall/American Artists Rep., Inc.; 203(b), 204(tl), 205(tl), Patti Bonham/Washington-Artists' Representative, Inc.; 206(t), Uhl Studio; 206(b), Valerie Marsella; 207(b), David Merrell/Suzanne Craig Represents; 209(tr) Mark Oliver/Garden Studio Illustrators; 210, James E. Pfeffer, Jr.

UNIT 4: Page 219, Lane Dupont/John Brewster Creative Services; 223(trl), John Francis; 226, Uhl Studio; 227, Valerie Marsella; 228-229(l), Don Weller/Jae Wagoner, Artists' Representative; 228(br) Bruce Metherd; 229(br) Blake Thornton/Rita Marie & Friends; 230(t), Donna Kae Nelson/Sharon Langley Artist Representatives; 231, Bob Dorsey; 232, Don Weller/Jae Wagoner, Artists' Representative; 234-237, Uhl Studio; 241, David Reed; 247, Bill Geisler; 249, Keith Locke/Suzanne Craig Represents; 251(t), Reggie Holladay; 251(b), John Francis; 252, Donna Kae Nelson/Sharon Langley Artist Representatives; 255-256(t), Lane Dupont/John Brewster Creative Services; 256(bl), Tom Shephard/Woody Coleman Presents; 259-260, John Francis; 262, Sandy Kossin/ Creative Freelancers; 263, Jean Calder; 264, Valerie Marsella; 266, Bruce Metherd; 267(t), Jim Theodore/Illustrated Alaskan Moose Studio; 267(b)-274(b), Bruce Metherd; 268(t), Eric Joyner/Freda Scott, Inc.; 269(c), David Reed; 271(t)-274(t), Tim Ladwig/ Suzanne Craig Represents; 275, Tom Shephard/Woody Coleman Presents; 276, Mr. Stobbs/Renée Kalish; 279, Brad Gaber; 281(tr), David Reed; 281(b), Jim Kopp/Carol Chislovsky Design Inc.; 282, Jim Kopp/Carol Chislovsky Design Inc.; 284(bl), David Puckett Design; 285, Jim Kopp/Carol Chislovsky Design Inc.; 287(t), Tom Shephard/ Woody Coleman Presents; 289, Jim Effler/American Artists Rep., Inc.; 291, Jean Calder.

UNIT 5: Page 295, Dean Kennedy/Carol Guenzi Agents; 296-297, Keith Kohler/ Susan Wells and Associates, Inc.; 303, Bruce Metherd; 304, Emmanuel Lopez; 305, Joann Daley/Cliff Knecht Artist Representative; 306(t), John Francis; 306(b), Bruce Metherd; 307, Bruce Metherd; 308, Jim Kopp/Carol Chislovsky Design Inc.; 310, Gary Locke/Suzanne Craig Represents; 311, Bruce Metherd; 313(tl), Bruce Metherd; 315, Rainey Kirk/The Neis Group; 316, Randy Nelson/Munro Goodman, Ltd., Artist Representatives; 318, Bruce Metherd; 319-321, Academy Artworks, Inc.; 325, Reggie Holladay; 326, Bob Lange/Diann Roche Represents; 327, Don Weller/ Jae Wagoner, Artists' Representatives; 329(c), Randy Nelson/Munro Goodman, Ltd., Artist Representatives; 330, Dennis Doheny/Jae Wagoner, Artists' Representative; 345, Bob Lange/Diann Roche Represents; 347, Dennis Doheny/Jae Wagoner, Artists' Representatives; 351, Joe Scrofani/American Artists Rep., Inc.

UNIT 6: Page 354, Chris Forsey/Morgan Cain & Associates; 357(l), Frank Demes; 358, GeoSystems Global Corporation; 359(t), GeoSystems Global Corporation; 359(b), Bill Geisler; 365, David Fischer/Morgan Cain & Associates; 367(t), Michael Koester/Woody Coleman Presents Inc.; 369, David Puckett Design; 379(b), Bruce Metherd; 380, Aletha Reppel/Suzanne Craig Represents; 385, Richard Wehrman; 387-388(l), John Francis; 388(br), Richard Wehrman; 391, David Fischer; 392(t), Richard Wehrman; 392(b), David Fischer; 394, Margaret DeNeergaard/Wilson Zumbo Illustration Group, Inc.; 396(l), Gareth Llewhellin/Morgan Cain & Associates; 397(tl), David Fischer; 399, David Fischer/Morgan Cain & Associates; 400(tr), Ken Smith; 401, Tim Ladwig/Suzanne Craig Represents; 407, Uhl Studio; 410(cl), Mark J. Persyn; 411, Steve Roberts/Morgan Cain & Associates; 412(cl), GeoSystems Global Corporation; 415(tr), Richard Wehrman.

UNIT 7: Page 429, John Rowe; 430, John Rowe; 435, David Puckett Design; 436, David Puckett Design; 441(b), Bruce Metherd; 444, Bruce Metherd; 445, Daniel Vasconcellos; 449, Victoria Bruck; 451, Tim Ladwig/Suzanne Craig Represents; 453, Truc An Lam/Garden Studio Illustrators; 454(tr), David Merrell/Suzanne Craig Represents; 454(b), Joe Scrofani/American Artists Rep., Inc.; 458(t), Joe Scrofani/ American Artists Rep., Inc.; 461(b)-462(b), Joe Scrofani/American Artists Rep., Inc.; 467, Dennis Jones; 479, Uhl Studio; 484, Donna Kae Nelson/Sharon Langley Artist Representatives.

UNIT 8: Page 502, Dennis Doheny/Jae Wagoner, Artists' Representative; 503, Douglas Schneider; 509(cl), Uhl Studio; 512(b), Uhl Studio; 513, Robert Bergin/Planet Rep.; 517(tr), Doug Henry/American Artists Rep., Inc.; 519(cr), Bruce Metherd; 520(bl), Don Sullivan; 524, Richard Pembroke/American Artists Rep., Inc.; 526(cr), Uhl Studio; 527(cl), Uhl Studio; 530-531, Pond & Giles/Morgan Cain & Associates; 532(l), Pond & Giles/Morgan Cain & Associates; 533(l), Pond & Giles/Morgan Cain & Associates; 534, Douglas Schneider; 538, Pond & Giles/Morgan Cain & Associates, 539, Joe McDermott/Koralik Associates; 540(bl), Mark Oliver/Garden Studio Illustrators; 542(bc, cr), Pond & Giles/Morgan Cain & Associates; 545(cl), David Griffin; 546, Keith Kohler/Susan Wells and Associates, Inc.; 551, Don Collins; 552, Robert Bergin/ Planet Rep.; 555(tr), Rebecca Merrilees; 555(bl), Joe McDermott/Koralik Associates.

SOURCEBOOK ILLUSTRATION CREDITS

UNIT 1: Page S2, Craig Attebery/Jeff Lavaty Artist Agent; S4(l), Don Sullivan; S4(r), Don Sullivan; S6(t), Pedro Julio Gonzalez/Melissa Turk & The Artist Network; S7, Jerry Werner/Christine Prapas Artist Representative; S8, Pedro Julio Gonzalez/ Melissa Turk & The Artist Network; S9(b), Pedro Julio Gonzalez/Melissa Turk & The Artist Network; S10(t), Pedro Julio Gonzalez/Melissa Turk & The Artist Network; S12-13, Craig Attebery/Jeff Lavaty Artist Agent; S14(tr), GeoSystems Global Corportation; S18, Jerry Werner/Christine Prapas Artist Representative.

UNIT 2: Page S30(tl), Todd Lockwood/Carol Guenzi Agents; S30(bl), Todd Lockwood/Carol Guenzi Agents; S33(b), Lori Anzalone/Jeff Lavaty Artist Agent; S35, Walter Stuart/Richard W. Salzman Artist Representative; S40(b), Glasgow & Associates; S42(br), Lori Anzalone/Jeff Lavaty Artist Agent; S44(b), Lori Anzalone/Jeff Lavaty Artist Agent; S44(b), Todd Lockwood/Carol Guenzi Agents; S45(t), Walter Stuart/ Richard W. Salzman Artist Representative; S45(bc), Walter Stuart/Richard W. Salzman Artist Representative; S46(b), Martens & Kiefer/Carol Chislovsky Design Inc.; S48(tl), Don Brautigam/Bill Erlacher Artists Associates; S50(t), Glasgow & Associates.

UNIT 3: Page S55(t), Joe McDermott/Koralik Associates; S55(c), Glasgow & Associates; S56(bc), Uhl Studio; S56(br), Uhl Studio; S60(tl), George Kelvin Science Graphics; S60(b), David Fischer; S61(b), Wendy Smith-Griswold/Melissa Turk & The Artist Network; S65(t), Glasgow & Associates.

UNIT 4: Page S74(cr), Joe McDermott/Koralik Associates; S74(br), Joe McDermott/Koralik Associates; S76(b), Tim Ladwig/Suzanne Craig Represents; S79, Richard Wehrman; S80, Tony Randazzo/American Artists Rep., Inc.; S81(c), Joe McDermott/Koralik Associates; S81(b), Joe McDermott/Koralik Associates; S82(c), Winson Trang/Creative Freelancers Management Inc.; S84, Jack Graham; S85(b), Don Brautigam/Bill Erlacher Artists Associates.

UNIT 5: Page S91(bl), Graber Graphics; S91(br), Graber Graphics; S92(tr), Graber Graphics; S92(cl), George Kelvin Science Graphics; S93(b), Joe McDermott/Koralik Associates; S94(c), Glasgow & Associates; S98(br), George Kelvin Science Graphics.

UNIT 6: Page S108(c), Greg Harris/Cornell & McCarthy Artist Representatives; S109(b), George Kelvin Science Graphics; S110(t), GeoSystems Global Corportation; S111(b), Susan Johnston Carlson/ Melissa Turk & The Artist Network; S114(t), George Kelvin Science Graphics; S117(tc), Mark J. Persyn; S121(bl), David Fischer; S121(bc), David Fischer; S121(bc), David Fischer; S121(br), David Fischer; S124(cl), Don Brautigam/Bill Erlacher Artists Associates; S125(tl), Graber Graphics; S128(lt-b), Mark Mille/Sharon Langley Artist Representatives; S128(b), Sarah Woodward/ Morgan Cain & Associates; S132(b), Greg Harris/Cornell & McCarthy Artist Representatives; S133(b), David Fischer; S134(c), Uhl Studio; S136(t), Mark J. Persyn.

UNIT 7: Page S139, Craig Attebery/Jeff Lavaty Artist Agent; S140(c), Tony Randazzo/American Artists Rep., Inc.; S142(b), George Kelvin Science Graphics; S146(t), John Francis; S146(b), John Francis; S147(b), Stephen Durke/Washington-Artists' Representative, Inc.; S149(br), Tony Randazzo/American Artists Rep., Inc.; S157(tr), NASA; S157(br), Stephen Durke/Washington-Artists' Representative, Inc.; S158(c), Tony Randazzo/American Artists Rep., Inc.; S160(t), John Francis.

UNIT 8: Page S163(t), Lori Anzalone/Jeff Lavaty Artist Agent; S163(b), James Dowdalls; S164(t), Lori Anzalone/Jeff Lavaty Artist Agent; S166, Darrel Tank; 167(b), Darrel Tank; S168(t), Darrel Tank; S169(c), Joani Pakula; S170, Lori Anzalone/Jeff Lavaty Artist Agent; S172(tc), Joani Pakula; S172(c), Joani Pakula; S172(bc), Joani Pakula; S174(tl), Don Brautigam/Bill Erlacher Artists Associates.

Acknowledgments

For permission to reprint copyrighted material grateful acknowledgment is made to the following sources:

Mary Jarrell, Executrix of Estate of Randall Jarrell: "Bats" from *The Bat Poet* by Randall Jarrell. Copyright © 1963, 1964 by Randall Jarrell.

Canadian Heritage: From an article on citizens' group participation in conservation of old buildings and communities by Phyllis Lambert from *Canadian Heritage,* August 1980. Published by Heritage Canada, Ottawa, ON.

Larousse PLC: Quote by Boutros Boutros-Ghali from *Rescue Mission-Planet Earth.* Copyright © 1994 by Peace Child Charitable Trust. Published by Kingfisher.

Life Picture Sales: From "The Red Planet May be the Next Giant Step for Mankind" by Brad Darrach, Steve Petranek and Anne Hollister from Life , vol. 14, no. 5. May 1991. Copyright © 1991 by The Time Inc. Magazine Company.

Ray Lincoln Literary Agency, Elkins Park House, 107-B, Elkins Park, PA: From "Search and Rescue" from *Mount St. Helens: A Sleeping Volcano Awakes* by Marian T. Place. Copyright © 1981 by Marian T. Place.

Scott Meredith Literary Agency: From *Cosmos* by Carl Sagan. Copyright 1980 by Carl Sagan. All rights reserved.

National Geographic Society: From *"Mars"* from *National Geographic Picture Atlas of Our Universe* by Roy A. Gallant. Copyright © 1980 by National Geographic Society. Illustration of the Martian waterseeker by Michael Whelan.

Pantheon Books, a Division of Ramson House, Inc.: Adaptation of "How the Crow Came to Be Black" as told by Good White Buffalo and adaptation of "Why the Owl Has Big Eyes" from *American Indian Myths and Legends,* selected and edited by Richard Erdoes and Alfonso Ortiz. Copyright © 1984 by Richard Erdoes and Alfonso Ortiz.

ATE CREDITS

Abbreviations used: (t) top, (c) center, (b) bottom, (l) left, (r) right

PHOTO CREDITS

OWNER'S MANUAL: Page T17, T18, T22, T24, T25(b), T28, T29, T36, T41, T44(t), T45, T46, T55(r), T56, T58, HRW photo by Sam Dudgeon; T19, T42, T44(b), T49, HRW photo by Daniel Shaefer; T16, T50, T51, HRW photo by Tomas Pantin; T25, T26, T33, T35, T38, T39, T47, T48, HRW photo by John Langford; T55(l), HRW photo by Jack Newkirk; T20, T23, T30(t), HRW photo by Scott Van Osdal; T15, T31, David Young Wolff/Photo Edit; T43, Frontera Fotos/Michelle Bridwell

UNIT 1: Page 1A, H. Armstrong Roberts; 1D, HRW photo by Michelle Bridwell; 68, Ronald H. Cohn/The Gorilla Foundation; 69, Tony Stone Images; 70, HRW photo by Sam Dudgeon; 71, Ron Kimball

UNIT 2: Page 71A, Phil A. Dotson/Photo Researchers, Inc.; 71D, HRW photo by Daniel Schaefer; 141, HRW photo by Sam Dudgeon; 143, Howard Sochurek/The Stock Market

UNIT 3: Page 143A, David R. Frazier Photography; 143B(l), W. Gregory Brown/Animals Animals/Earth Scenes; 143B(r), Rudie H. Kuiter/ Animals Animals/Earth Scenes; 143D, HRW photo by Daniel Schaefer; 211, Tony Stone Images

UNIT 4: Page 213D, HRW photo by Michelle Bridwell; 288, Ron Kimball; 289, Rogge/The Stock Market; 290, 291, Howard Sochurek/The Stock Market

UNIT 5: Page 291A, Jeanne Drake/Tony Stone Images, Inc; 291B(l), Paul Thompson/International Stock Photo; 291B(r), Robert Harding Picture Library, London; 291D, HRW photo by Daniel Shaefer; 348, Ron Kimball; 349, Rogge/The Stock Market; 350, David Bookstaver/AP/ Wideworld

UNIT 6: Page 351D, HRW photo by Michelle Bridwell; 418, Tony Stone Images; 419, HRW photo by Sam Dudgeon; 416, Louis Psihoyos/Matrix International

UNIT 7: Page 419A, NASA; 419C(bl), Harrison Schmidt/NASA JSC/ Starlight/Corbis; 419C(t), Photri/The Stock Market; 419C(br), NASA; 419D, HRW photo by Daniel Schaefer; 494, HRW photo by Sam Dudgeon; 495, Ron Kimball; 496, Rogge/The Stock Market; 497, Tony Stone Images

UNIT 8: Page 497A, Roy Morsch/The Stock Market; 497D, HRW photo by Daniel Schaefer; 556, Howard Sochurek/The Stock Market; 558, Rogge/The Stock Market; 559, Tony Stone Images

ART CREDITS

All work, unless otherwise noted, is contributed by Holt, Rinehart and Winston.

Page 1C, Robin Boutell/Morgan Cain & Associates; 71B, Blake Thornton/ Rita Marie & Friends; 71C, Sarah Woodward/Morgan Cain & Associates; 213A, John Francis; 213C, John Francis; 291C, Don Weller/Jae Wagoner, Artists' Representative; 351C, David Fischer; S200, John Francis; S204(b), David Fischer; S205(t), David Fischer; S209, The Quarasan Group, Inc.

S193

Answers to SourceBook Unit CheckUp: Unit 1, page S25

Concept Mapping
Sample concept map:

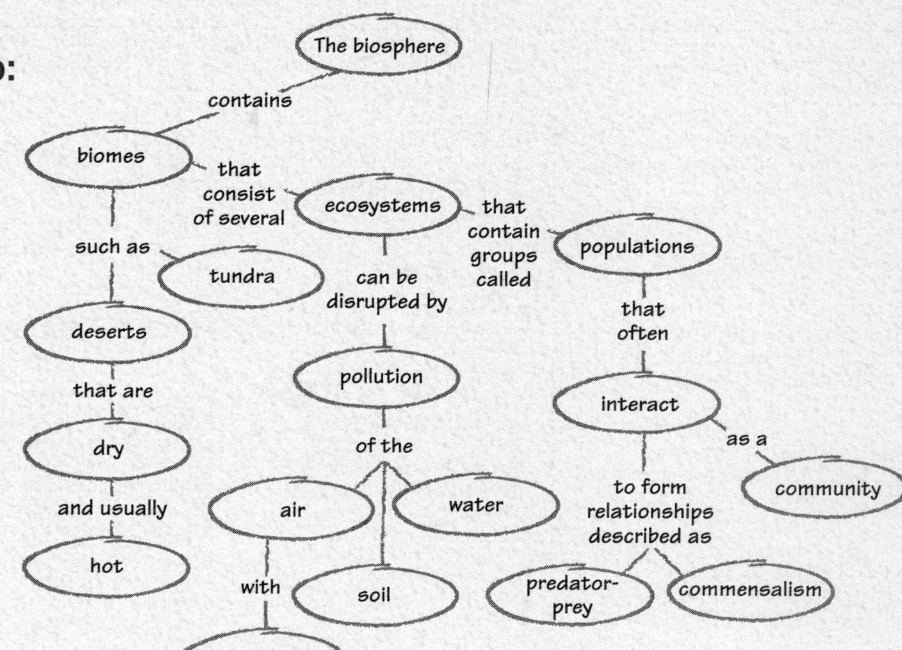

Checking Your Understanding

1. c. the sun
2. b. community.
3. a. evaporation
4. d. lichens.
5. b. pollution.

Interpreting Photos

Temperate deciduous forest. These ecosystems are generally found about 40 degrees north of the equator.

Critical Thinking

1. Space exploration could lead to the colonization of moons and other planets by humans. The building of undersea cities would provide humans with still another place to live. Providing new places for people to live could keep the population density of the human species on the surface of the Earth at a sustainable level.

2. The carbon dioxide–oxygen cycle provides your body with oxygen. Your body uses this oxygen to release energy from food during cellular respiration. You can then use the energy to carry out many processes, including running, talking, and reading.

3. The plastic cover prevented seeds carried by the wind or by animals from falling onto the plowed soil, so new plant growth could not begin. The cover also prevented sunlight from reaching seedlings that grew from seeds present in the soil when it was overturned. Without sunlight, the seedlings could not live.

4. Many animals cannot be active during the extremely hot daytime temperatures that can occur in deserts. Tropical rain forests generally have cooler daytime temperatures. Also, tropical rain forests contain more vegetation than deserts do. This vegetation provides shade and many places where animals can hide. The sparse vegetation of deserts provides little cover for animals.

5. Without sanitary landfills, garbage was simply left to accumulate on the ground during the Middle Ages. This exposed garbage became food for rats. With more than enough food to eat, the rat population exploded and the plague quickly spread.

Portfolio Idea

Answers will vary, but a report may contain information such as the following:

Investigative methods—testing the water for phosphates (algae and water plants grow out of control in a high-phosphate environment)

Findings—Phosphate levels in the water are abnormally high. Washing-machine water from newly built homes has been entering the lake for the past year.

Solutions—Suggest building an adequate drainage system for nearby houses that doesn't empty into the lake and fining residents who continue to empty their washing-machine water into the lake.

Answers to SourceBook Unit CheckUp: Unit 2, page S49

Concept Mapping
Sample concept map:

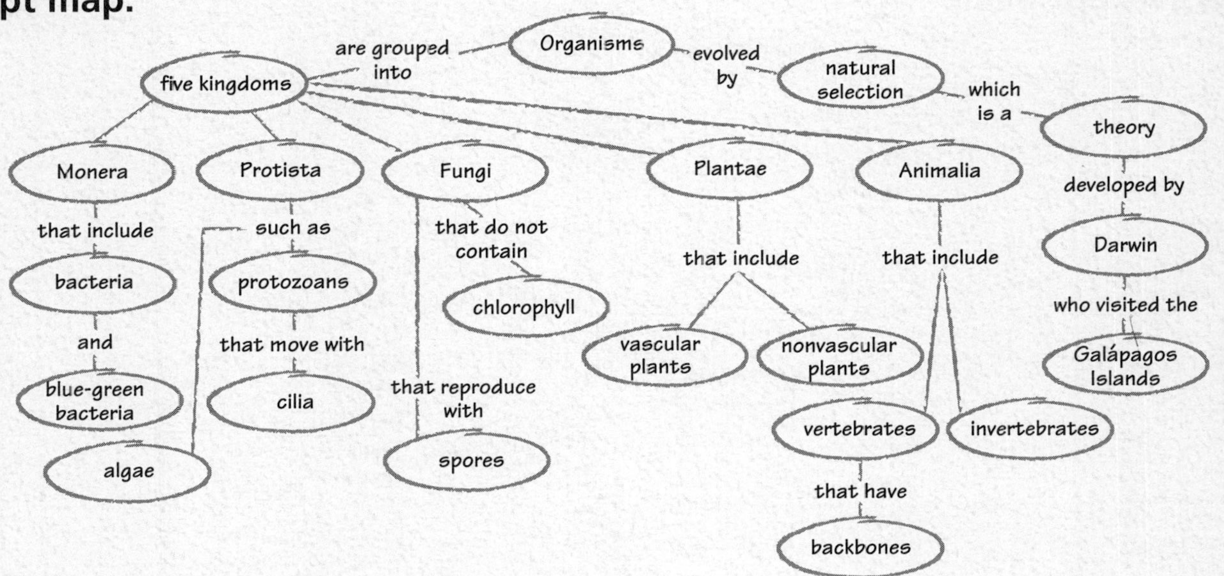

Checking Your Understanding
1. c. Protozoa
2. d. a vertebrate.
3. b. hyphae
4. b. Precambrian time
5. a. homologous structures.

Interpreting Graphs
After day two, the size of the population will probably stabilize for a short while. But because there is a limited food supply, the struggle for existence will soon cause the population to start decreasing. Eventually the food will be used up and the entire population will die out.

Critical Thinking
1. Looking at the stem under a microscope will reveal if the stem contains vascular tissue. Nonvascular plants do not grow tall and they must live in moist environments. It makes sense to assume that the plants in question are nonvascular—they are short and grow near a swampy area.

2. A planarian is a flatworm. Flatworms have a simple nervous system. Earthworms are segmented worms that have a complex nervous system consisting of a simple brain and a nerve cord. The more advanced nervous system of the earthworm allows this animal to crawl faster than the planarian can.

3. South America, because the Galápagos islands are near South America. The other two places are much farther away.

4. Look at the protozoan under a microscope and see if it moves around. If it is able to move, then try to determine *how* it moves (for example, by flowing, like an amoeba, or with cilia, like a paramecium). Determining how it moves will indicate which protozoan group it belongs to. If it doesn't move, it may be some sort of parasitic protozoan. To further pinpoint the classification of this organism, you should conduct a DNA analysis, and compare its DNA to the DNA of other, known protozoans.

5. They are not homologous because their structure is different.

Portfolio Idea
The design of the table will vary, but tables should contain information such as the following:
Kingdom names—(1) Monera, (2) Protista, (3) Fungi, (4) Plantae, (5) Animalia
Characteristics—(1) the simplest organisms, single-celled, no nuclei or organelles, (2) single-celled or multicellular, cells contain nuclei, no tissues or organs, contains plantlike and animal-like organisms, (3) mostly multicellular, have cell walls, cells nucleated, digest food outside their bodies by releasing digestive enzymes, (4) contain chlorophyll, make their own food, cells nucleated, all multicellular, have cell walls, have tissues and organs, (5) cells nucleated, all multicellular, have complex tissues and organ systems, must eat food, digest food inside their bodies

Answers to SourceBook Unit CheckUp: Unit 3, page S69

Concept Mapping
Sample concept map:

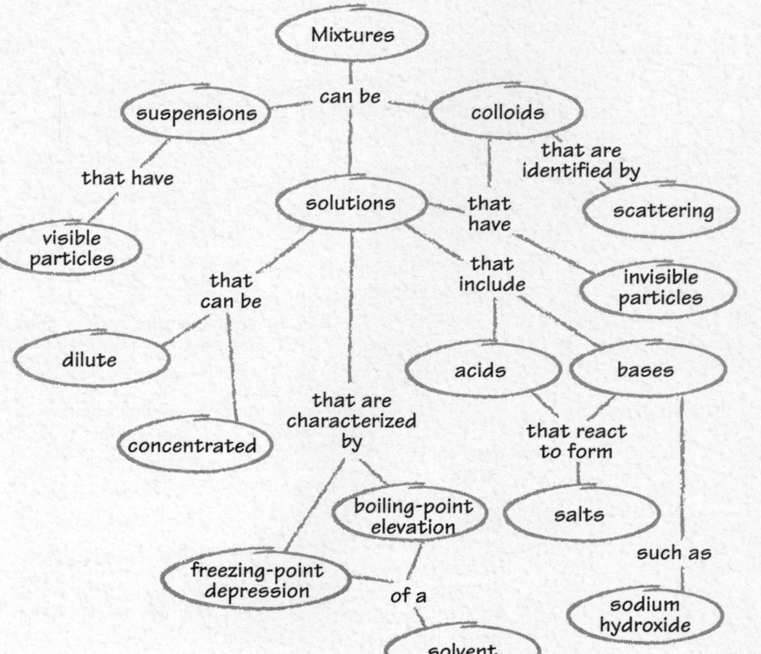

Checking Your Understanding

1. c. large amount of solute dissolved in a small amount of solvent.
2. c. amount of solute
3. a. boiling-point elevation.
4. b. The more basic a solution is, the higher its pH is.
5. d. pH meter

Interpreting Photos

Streaks of light in a clear sky at sunrise or sunset indicate that there are particles suspended in the air. Thus, the air in that area is a suspension of gases and small particles of a solid or liquid and is not just a solution of several gases.

Critical Thinking

1. Answers will vary but should mention that solutions look like a single substance and do not exhibit the Tyndall effect. Other mixtures either exhibit the Tyndall effect or obviously have two or more distinct parts that can be separated by physically sorting, centrifuging, or allowing them to settle.
2. A suspension is more affected by gravity because its particles are large enough to settle in normal gravity. To separate the parts of a colloid, which has particles too small to settle in normal gravity, a centrifuge must be used to enhance gravity.
3. Magnesium chloride ($MgCl_2$);
$2HCl + Mg(OH)_2 \rightarrow MgCl_2 + 2H_2O$
4. The easiest method of accurately determining the pH of water in lakes and streams would be to use pH paper, which gives an accurate pH value and comes in a dispenser that is smaller and easier to carry than a pH meter.
5. If calcium oxide were added to a lake polluted with acid rain containing sulfuric acid, the water would become less acidic, as the following equations indicate:
$CaO + H_2O \rightarrow Ca^+ + 2OH^-$
$H_2SO_4 + 2H_2O \rightarrow 2H_3O^+ + SO_4^-$
$H_3O^+ + OH^- \rightarrow 2H_2O$
The salt calcium sulfate would also form in the lake, as the following equation shows:
$Ca^+ + SO_4^- \rightarrow CaSO_4$
If the calcium sulfate is not toxic, organisms that would be killed by water with an acidic pH would be more likely to survive with this treatment than without it.

Portfolio Idea

Solutions Scrapbooks may contain original photographs, pictures cut from magazines, labels from commercial products, or drawings of products or other applicable solutions. All mixtures included should be solutions and should have labels that identify the solvent and the solute and that indicate whether the solution is acidic, basic, or neutral.

Answers to SourceBook Unit CheckUp: Unit 4, page S87

Concept Mapping
Sample concept map:

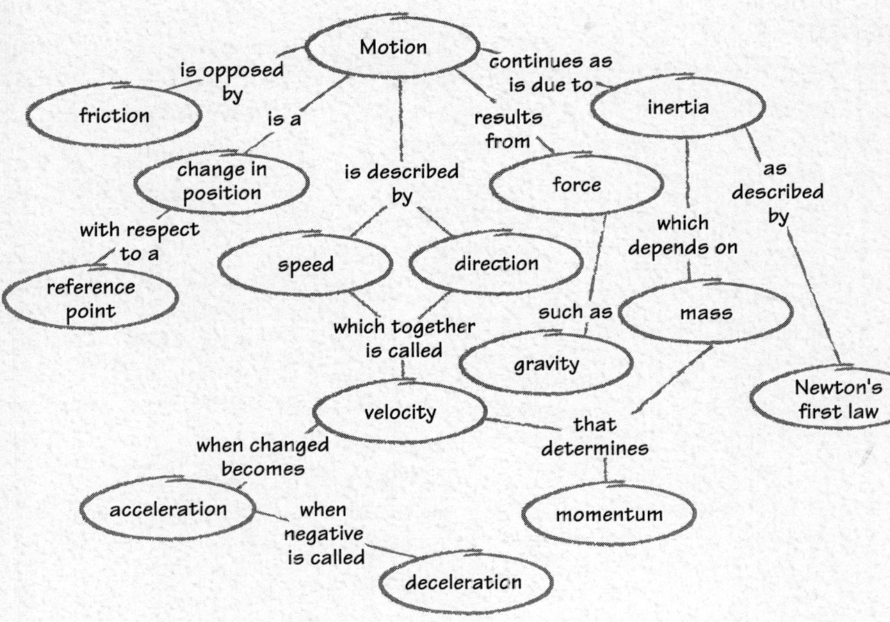

Checking Your Understanding

1. c. 100 km/h.
2. b. has parts that are stationary with respect to each other.
3. d. friction.
4. a. $F = ma$.
5. b. increases as the masses of the objects increase.

Interpreting Photos

There are action-reaction force pairs between the ceiling and the chain suspending the mobile, between the mobile itself and the chain suspending it, and between the Earth and the mobile, as well as between each suspended item in the mobile and the support holding each item.

Critical Thinking

1. Answers will vary. If students answer no, they should point out that even the Earth, which we normally consider to be stationary, is spinning on its axis, orbiting the sun, and moving through space with the Milky Way galaxy. If students answer yes, they should provide some reason to justify their response (e.g., perhaps the universe itself is stationary even though everything in it is moving).

2. Answers will vary but should demonstrate an understanding of Newton's three laws of motion. First law: Unless a force is exerted on the ball, it will not leave the shooter's hand; it will remain still. Second Law: To shoot the ball, the shooter applies the force necessary to move the ball toward the hoop. Third Law: As the shooter pushes the ball, he or she feels resistance, or the force of the ball pushing back on his or her hand.

3. An arrow that is shot straight forward and one that is dropped from the same height will both hit the ground at the same time because their downward acceleration, which is due to the pull of gravity, will be the same.

4. The amount of momentum that a moving object has is determined by its mass times its velocity. This means that momentum increases with both mass and speed. Linemen that are both heavier and faster than other linemen have more momentum and thus are more likely to be able to push opposing linemen backward when they collide.

5. Students' answers should indicate their understanding of inertia. When an object is in motion, it will remain in motion unless an opposing force, such as the gravity of a nearby star or planet, acts upon it. Once a starship or shuttle is moving, it no longer needs engines or tractor beams to keep it moving.

Portfolio Idea

Accept all reasonable answers. Students should demonstrate an understanding of the concepts of force, speed, velocity, acceleration, and Newton's laws of motion and gravitation.

Answers to SourceBook Unit CheckUp: Unit 5, page S105

Concept Mapping
Sample concept map:

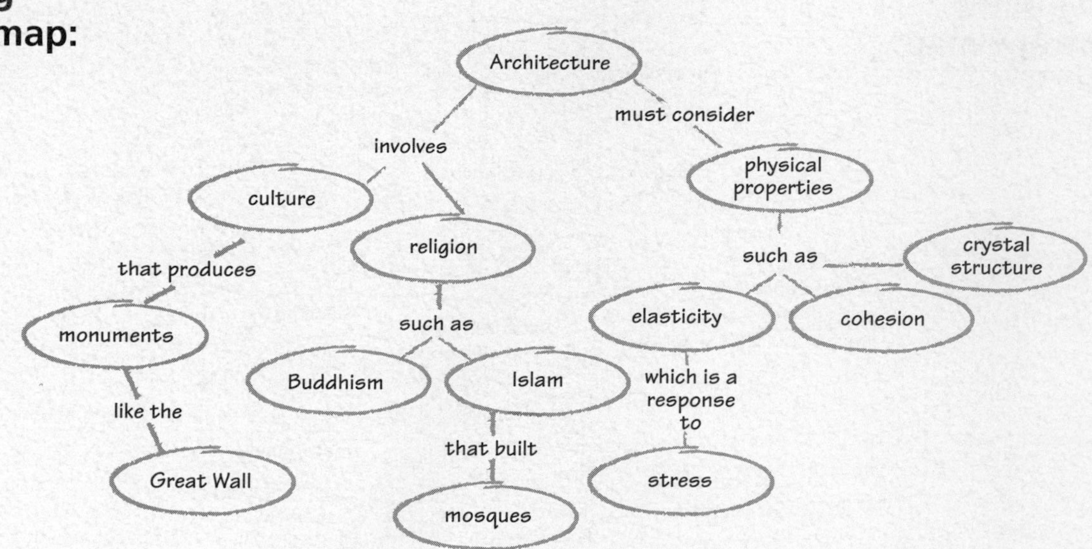

Checking Your Understanding

1. c. the cracking of a stained-glass window
2. c. tension.
3. b. the point beyond which it will permanently deform.
4. b. Chinese pai-lou.
5. c. Indian and Islamic

Interpreting Photos

1. This architecture was influenced by pre-Columbian Pueblo cultures. This is a photograph of the Acoma Pueblo in New Mexico, United States. Accept all reasonable answers.
2. This architecture was influenced by Hindu and Buddhist cultures. This is a photograph of the Angkor Wat in Cambodia. Accept all reasonable answers.
3. This architecture was influenced by the Chinese culture. This is a photograph of the Lama Temple in the Inner City, Beijing, China. Accept all reasonable answers.
4. This architecture was influenced by Islamic and Byzantine cultures. This is a photograph of the Blue Mosque in Istanbul, Turkey. Accept all reasonable answers.
5. This architecture was influenced by pre-Columbian Mayan cultures. This is a photograph of the Castle in Chichén Itzá, Yucatán, Mexico. Accept all reasonable answers.

Critical Thinking

1. Oxygen combines with iron to form iron oxide or rust. Rust affects the properties that make iron a good building material—strength and elasticity.
2. Diamond has a crystalline arrangement in which the carbon atoms are held in a rigid pattern. The molecules in glass are randomly arranged, which allows for more movement between the molecules of glass, thereby reducing its hardness.
3. It would be more suitable for use as a column because its high elastic modulus for compression would allow it to support heavy weights without breaking under the strain. If it were used as a cable, its low elastic modulus for tension would make it likely to snap when stretched.
4. Japanese design involves a more asymmetric approach that resembles the unusual shapes found in nature, such as rocks and trees.
5. Byzantine architecture (especially the design of churches) represented the Christian view of the human. While a human may have a rough exterior, it is what is on the inside that counts.

Portfolio Idea

Accounts will vary depending on the structure selected by the student. Written accounts should be accurate and should reflect the viewpoint of the writer chosen.

Answers to SourceBook Unit CheckUp: Unit 6, page S135

Concept Mapping
Sample concept map:

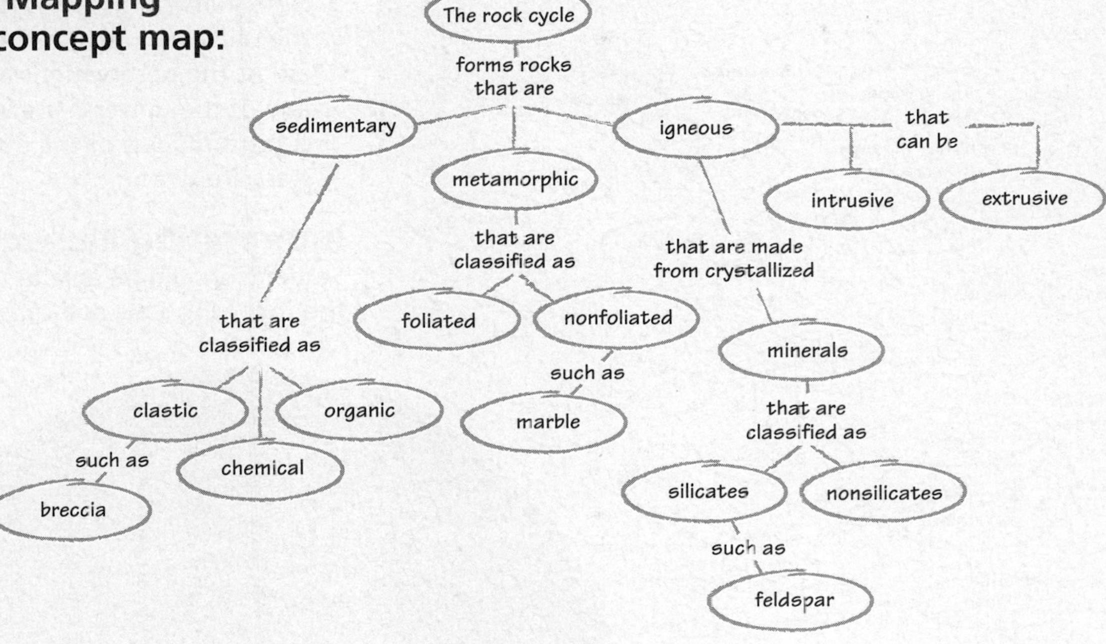

Checking Your Understanding
1. a. cinder cones.
2. d. a hot spot.
3. b. a clastic sedimentary rock.
4. a. more than one rock.
5. c. relative dating

Interpreting Illustrations
The shale, older sandstone, conglomerate, and limestone were deposited and then tilted. The younger sandstone was then deposited. The igneous rock intruded after deposition of the younger sandstone. The formation of the coal bed may or may not have taken place prior to the intrusion.

Critical Thinking
1. The movement of tectonic plates, on which continents are located, explains their apparent drift across the oceans and provides a mechanism for sea-floor spreading. Volcanism occurs at both divergent boundaries and subduction zones. Earthquakes occur along the plate boundaries.

2. Any original rock would have to be igneous, formed from the cooling magma. Both sedimentary and metamorphic rock are produced from preexisting rock.

3. It was formed during a volcanic eruption in which lava cooled rapidly. (The sample is obsidian, a volcanic glass with microscopic crystals.)

4. Since this is a metamorphic rock, the radioactive elements have been affected by the heat and pressure, and we cannot determine the original date.

Portfolio Idea
Student displays can vary, but they should all represent accurate concepts from the unit.

Answers to SourceBook Unit CheckUp: Unit 7, page S159

Concept Mapping
Sample concept map:

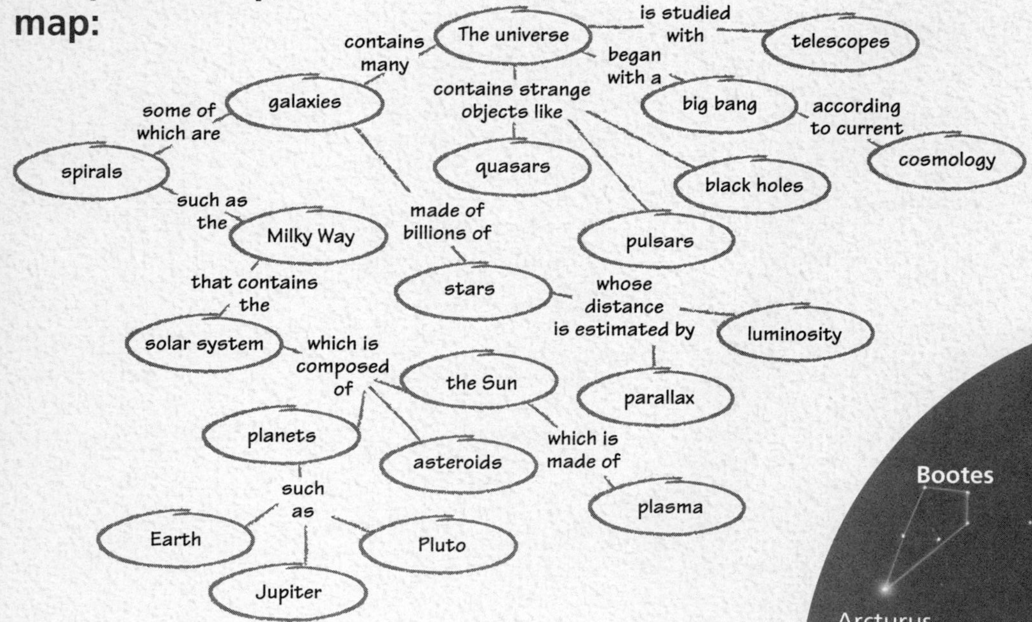

Checking Your Understanding

1. b. the core
2. a. the solar wind.
3. d. Neptune, Uranus, Saturn, Jupiter, Mars.
4. a. Rigel, a blue star
5. b. it can be reused over and over again.

Critical Thinking

1. The amount of hydrogen present in the Sun is so great that it will take billions of years for all of it to be used up. For all practical purposes, it will last as long as there is life on Earth.
2. A planet's year is the amount of time it takes the planet to orbit the Sun. A day is the time it takes the planet to rotate on its axis. Neither of these would be affected by the inclination of the axis. The seasons would be affected because, over most of the planet, the greater the tilt, the longer the daytime is in the summer and spring, and the longer the nighttime is in the winter and autumn.
3. If a star is too far away, the angle formed between it and the Earth's position on opposite sides of its orbit becomes too small to measure with accuracy.
4. Accept all reasonable answers. Sample answer: Competition between the United States and the Soviet Union provided a great deal of motivation for the advancement of space technology. However, because the competition made cooperation impossible, each country was unable to benefit from the other's success. In the present age of cooperation, countries will probably be able to pool their resources and develop technologies that would be difficult or impossible to develop separately.
5. Since stars cannot be older than the universe in which they exist, either the observations about the age of the universe are wrong or current models of the evolution of stars are wrong.

Interpreting Illustrations

Students should be able to identify the following stars and constellations:

Draco
Ursa Minor (Little Dipper)
Polaris (North Star)
Bootes
Arcturus
Ursa Major (Big Dipper)
Perseus
Castor
Taurus
Leo
Pollux
Aldebaran
Virgo
Betelgeuse
Gemini
Regulus
Orion
Spica
Rigel
Sirius
Hydra
Canis Major

Portfolio Idea

The planets in the students' models of the solar system should have the following radii: Mercury, 1.15 cm; Venus, 2.85 cm; Earth, 3.0 cm; Mars, 1.60 cm; Jupiter, 33.6 cm; Saturn, 28.2 cm; Uranus, 12.3 cm; Neptune, 11.9 cm; Pluto, 0.52 cm. The Sun's radius should be 3.27 m, which is too large to fit on one piece of paper.

Answers to SourceBook Unit CheckUp: Unit 8, page S175

Concept Mapping
Sample concept map:

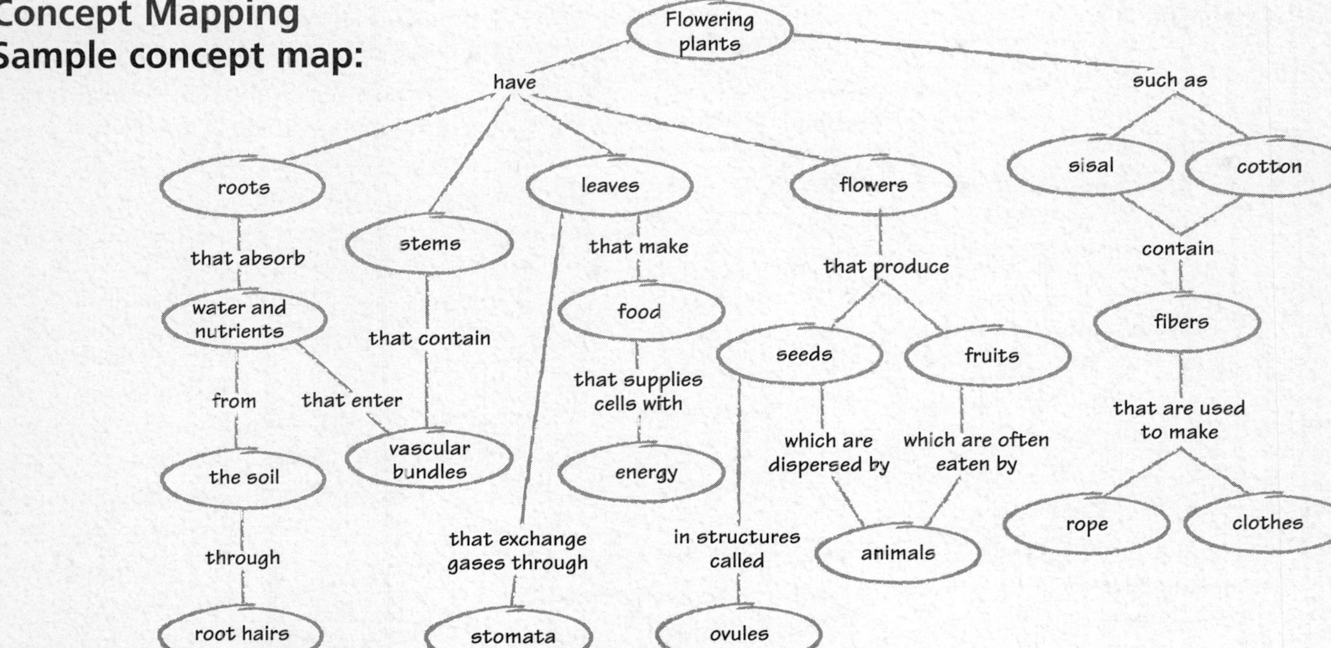

Checking Your Understanding

1. d. both b and c.
2. a. absorb sunlight.
3. c. old water-carrying cells.
4. b. They grow in width but not length.
5. d. also known as "fiber".

Interpreting Photos

Since the soil is so dry, you would expect many of the plants in this environment to have taproot systems.

Critical Thinking

1. Coating the undersides of the leaves would prevent oxygen and carbon dioxide from entering the plant. Without these gases, the plant could not carry out cellular respiration or photosynthesis. Eventually, the plant would die from a lack of energy and food.
2. Plants grow in length by adding new tissue at the tips of their stems. They don't grow from the ground up. Therefore, a nail placed at a certain level above the ground will remain at that same level, no matter how tall the plant grows.
3. The pollen of many grasses and trees is distributed by the wind, so the air can contain a lot of pollen from these kinds of plants.
4. Mixing salt with the soil would cause water to move out of the plant's roots instead of into the roots. Without water, the plant will eventually die.
5. Some plants, such as the legumes (peas, beans, peanuts), are rich in proteins. All plants contain a great deal of energy. Plants trap energy from the sun during photosynthesis to make glucose molecules. The glucose molecules contain some of the energy from the sun.

Portfolio Idea

Answers will vary, but the stories should include the following points: the water molecule moves into the root hair by osmosis; the molecule enters the water-conducting tissue in the middle of the root; the molecule moves up the roots, up the stem, through the leaf petiole, and into the vein of a leaf; the molecule exits the vein and enters a food-making cell in the leaf; the water molecule, light, and carbon dioxide are used during photosynthesis to make sugar and oxygen.

Answers to Tables and Diagrams

Unit 1, Chapter 2, page 35
Exploration 1

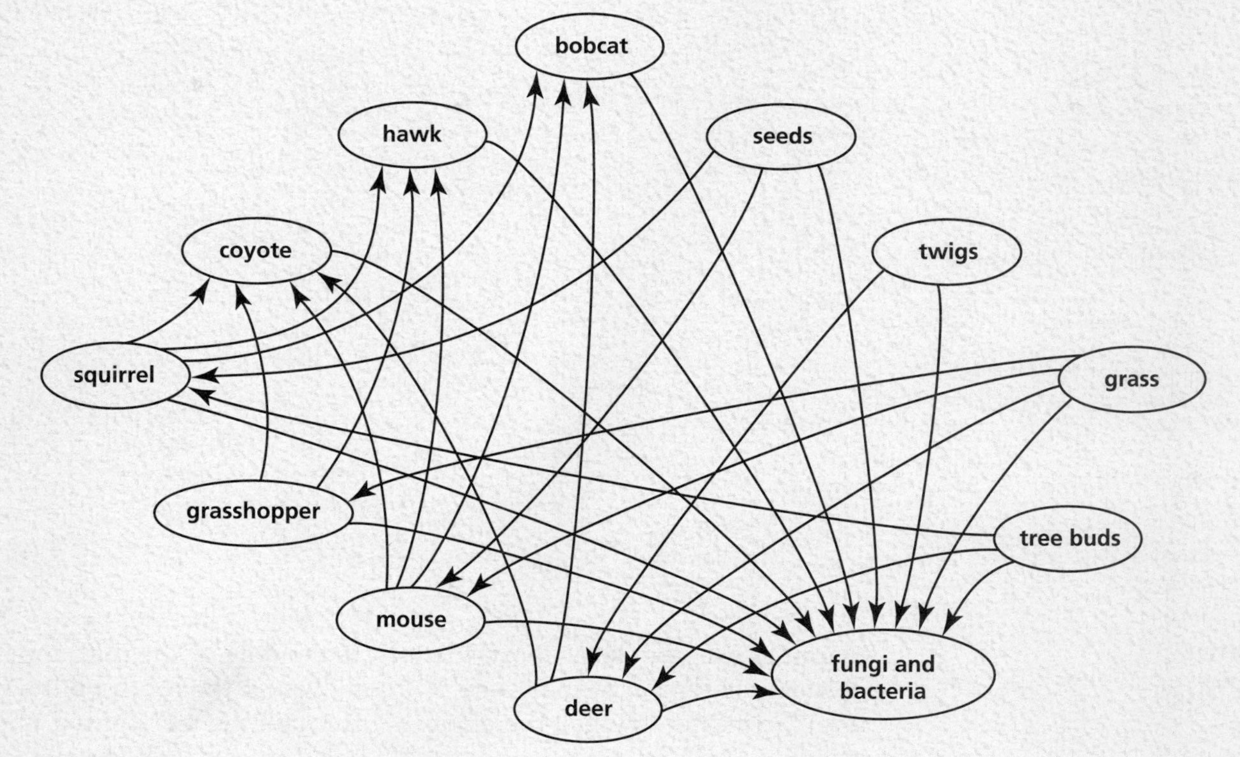

Unit 2 Making Connections, page 139
Checking Your Understanding, question 2

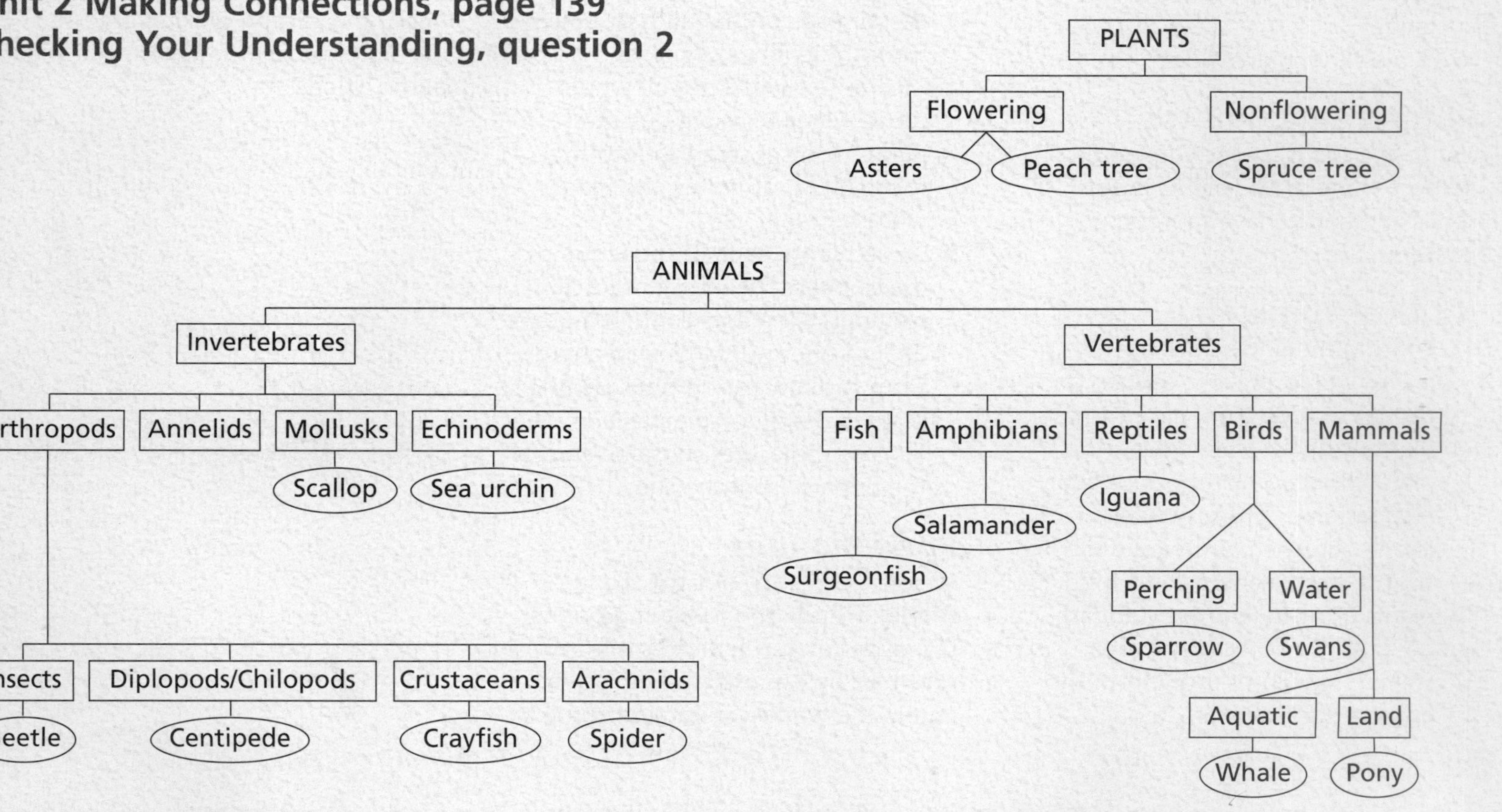

Unit 2 Making Connections, page 139
Checking Your Understanding, question 3

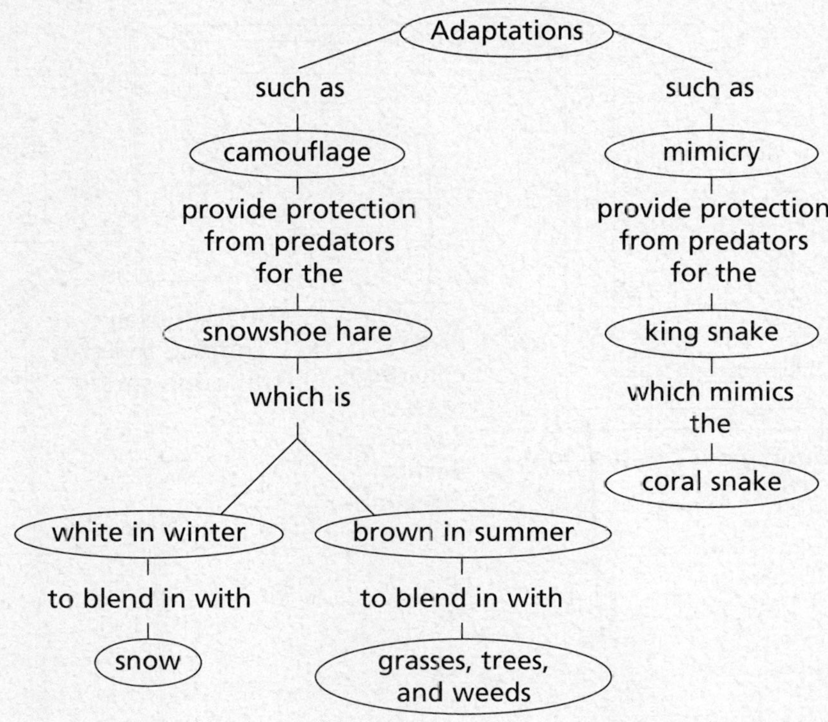

Unit 4 Making Connections, page 286
Checking Your Understanding, question 1

Type of force	Agent	Receiver	Effect
a. frictional	the boat and the sand	the boat and the sand	The two objects act on each other, resisting movement.
	the boat and the water	the boat and the water	same as above
b. gravitational	the Earth	the boat	weight of the boat (floating)
	the Earth	Kathleen	causes Kathleen to fall into the water
c. buoyant	the water	the boat	keeps the boat from sinking
	the water	Kathleen	keeps Kathleen from sinking due to the force of gravity

Type of force	Example	Effect
d. balanced	The force of the wind on the sails = the frictional force between the water and the boat.	The boat moves along at a steady speed.
	The downward force of gravity on Kathleen and Kimiko = the upward force of the boat seats.	Kathleen and Kimiko are at rest in the boat.
e. unbalanced	The force of Kathleen and Kimiko pulling the boat is greater than the frictional force between the boat and the sand.	The boat slides into the water.
	The force of the wind on the sails is greater than the friction between the boat and the water.	The boat speeds up.
	The force of friction between the boat and the water is greater than the force of the wind on the boat.	The boat slows down.

Answers to Tables and Diagrams

Unit 6, Chapter 16, page 366
Measuring the Strength of Earthquakes

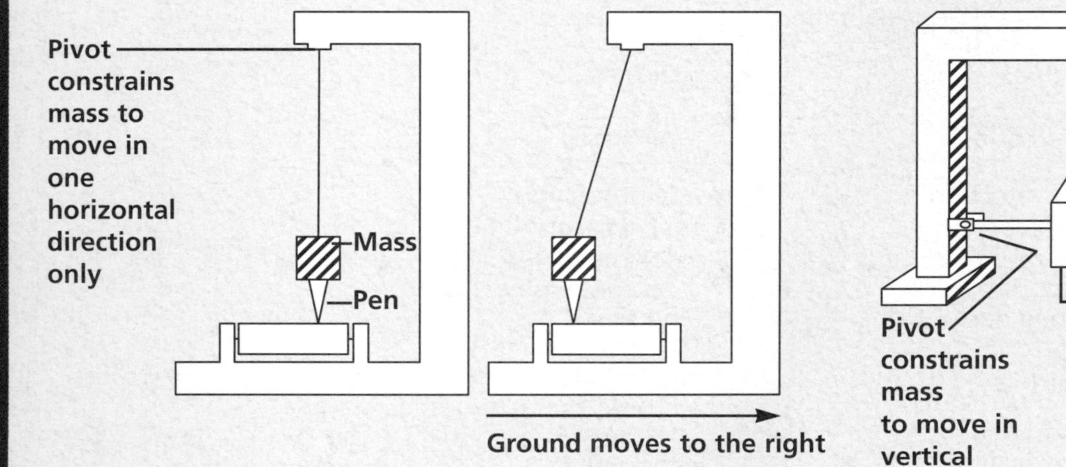

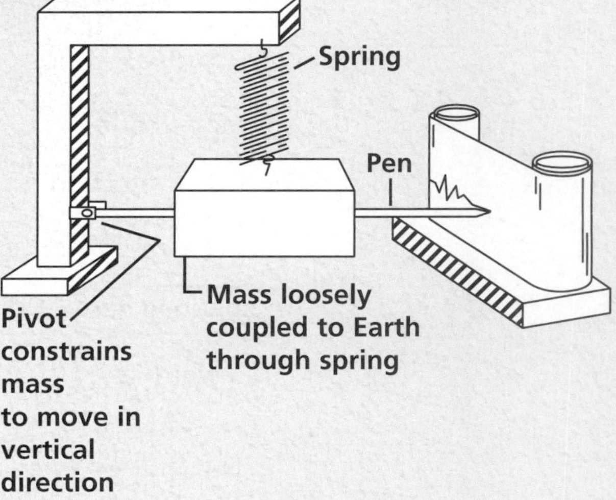

Unit 6, Chapter 17, page 391
The Formation of Igneous Rocks

a. extrusive (volcanic) igneous rocks
b. rocks formed from the slow cooling of magma
c. rocks formed from the fast cooling of lava
d. rocks with large crystals
e. rocks with small crystals
f. intrusive (plutonic) igneous rocks

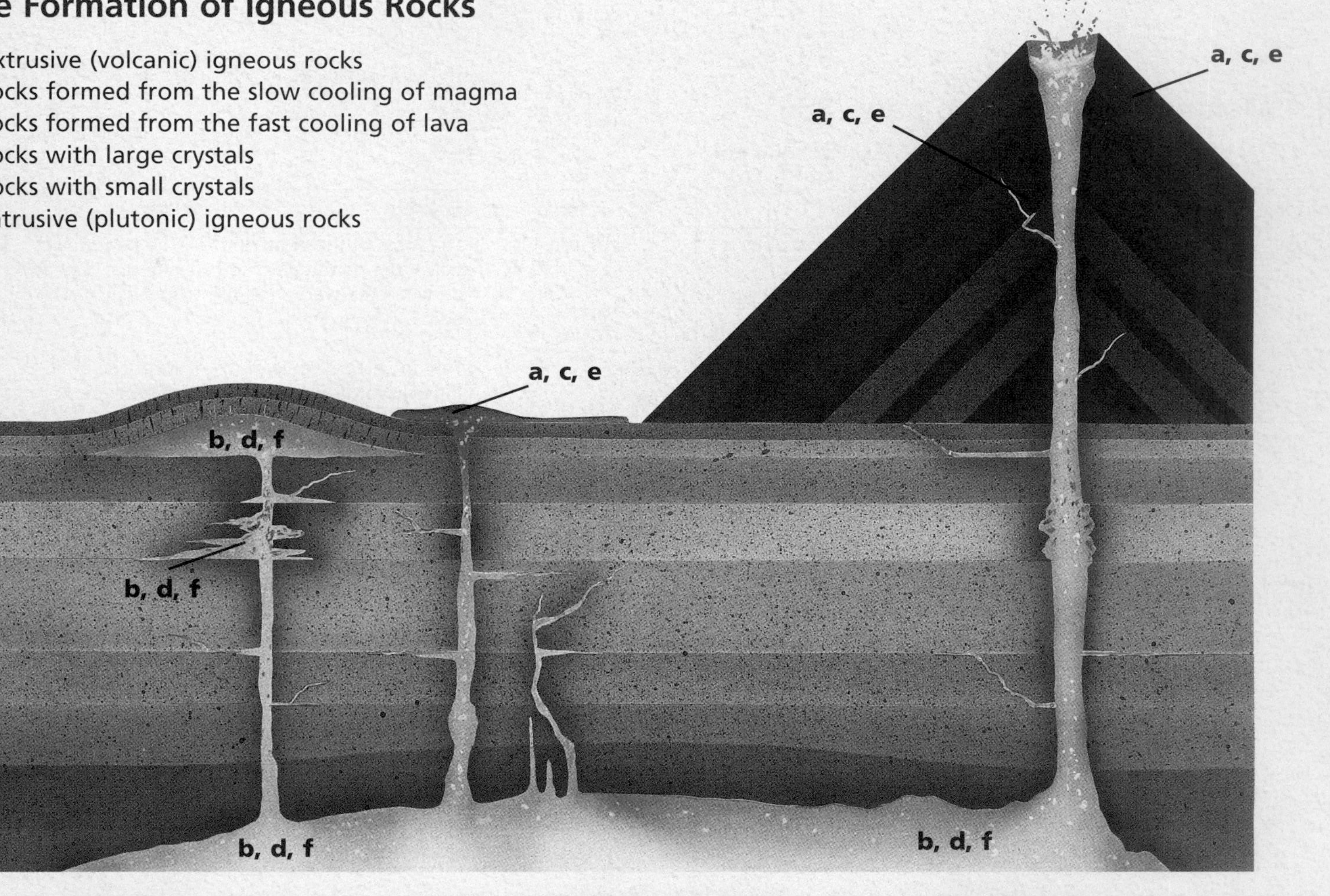

Unit 6, Chapter 17, page 399
Those Plates Again!!

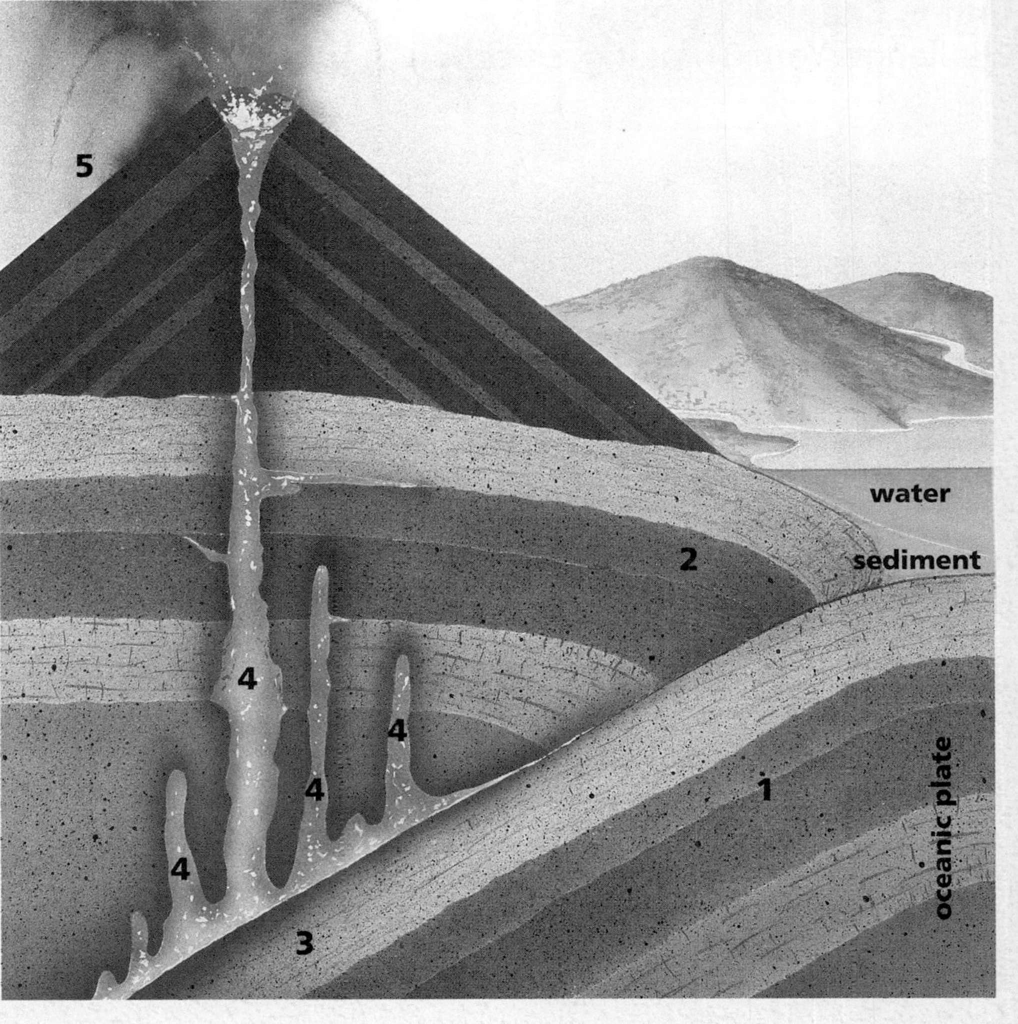

Unit 6, Chapter 18, page 413
Challenge Your Thinking, question 5, part a

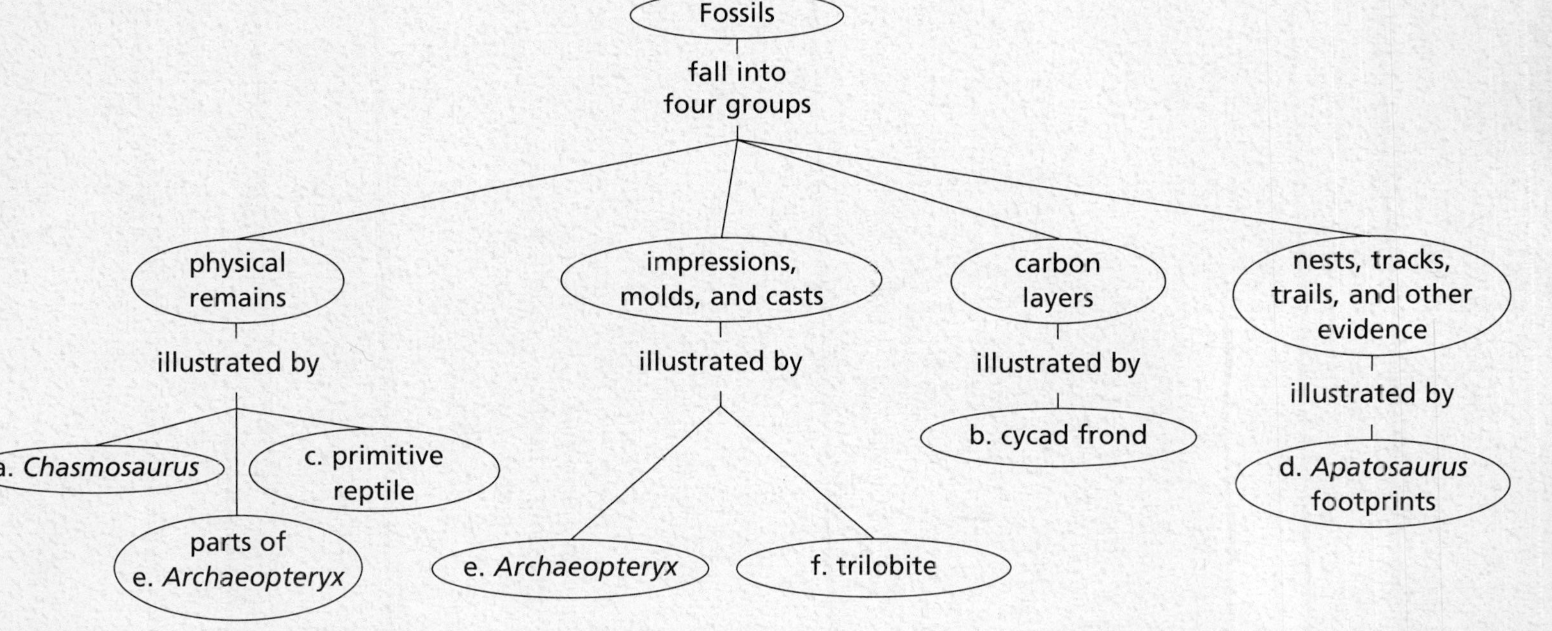

Unit 6, Chapter 18, page 413
Challenge Your Thinking, question 5, part b

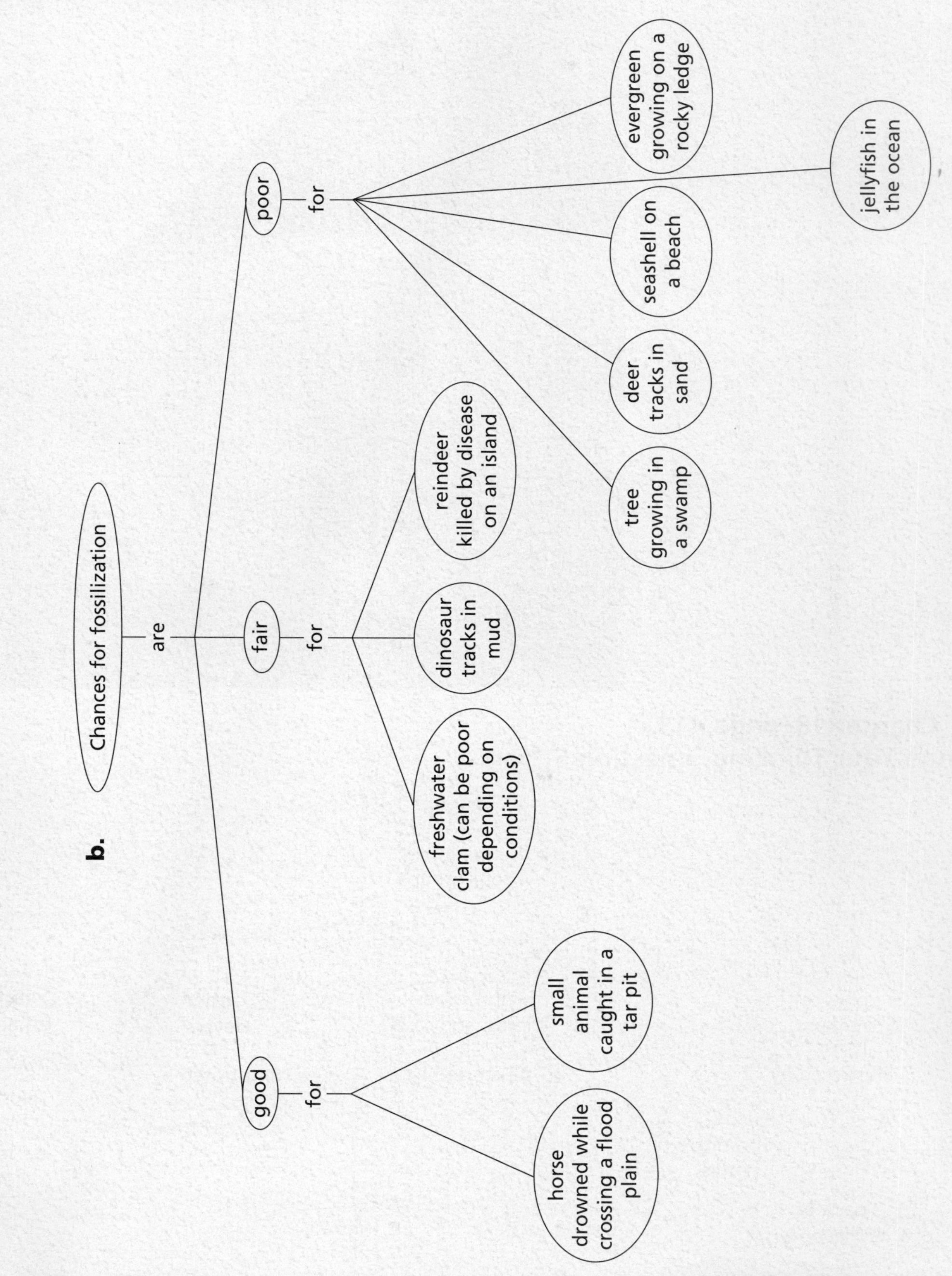

Unit 6, Chapter 18, page 413
Challenge Your Thinking, question 5, part c

c.

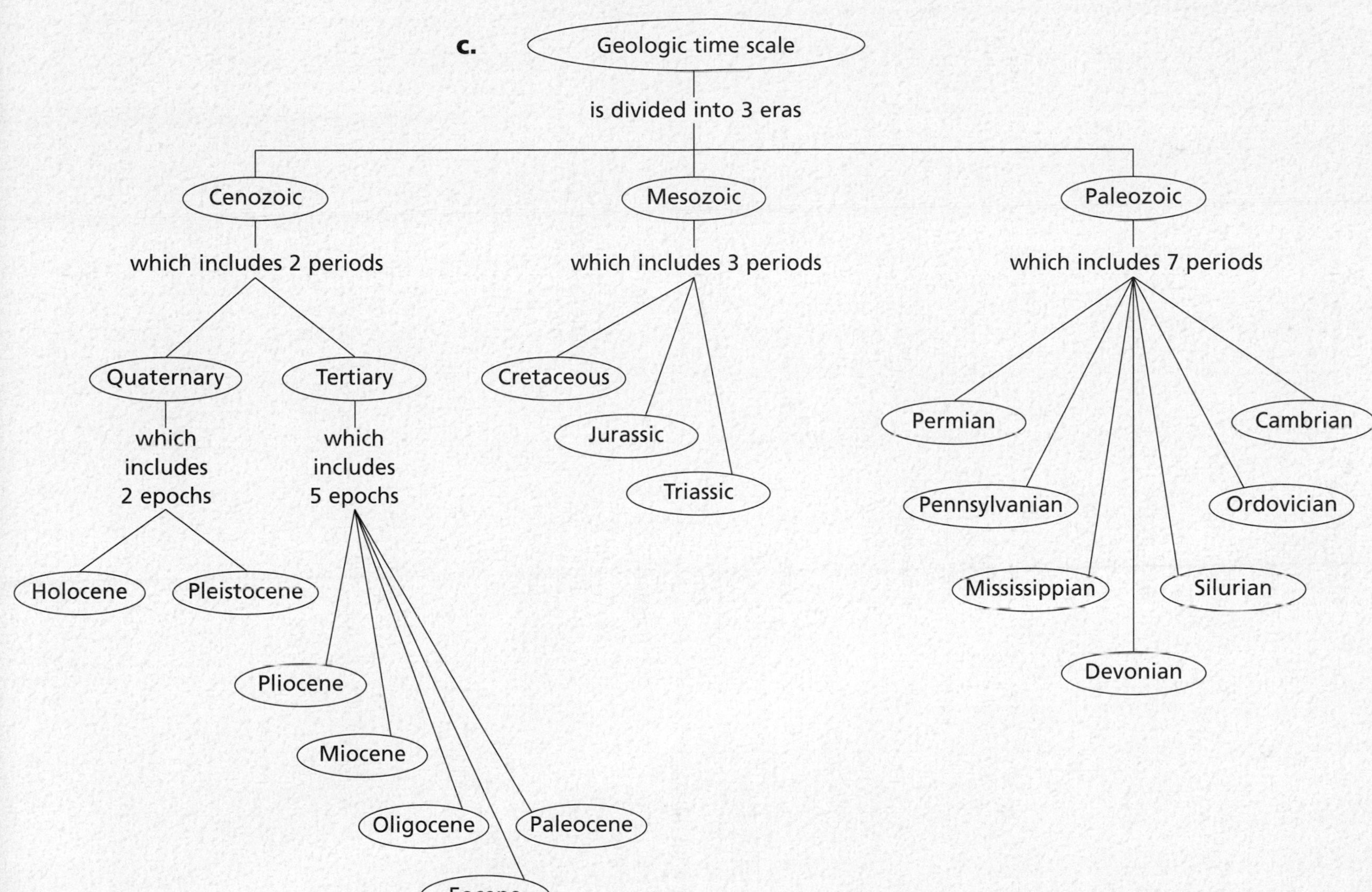

Answers to Tables and Diagrams

Unit 7 Making Connections, page 493
Checking Your Understanding, question 4

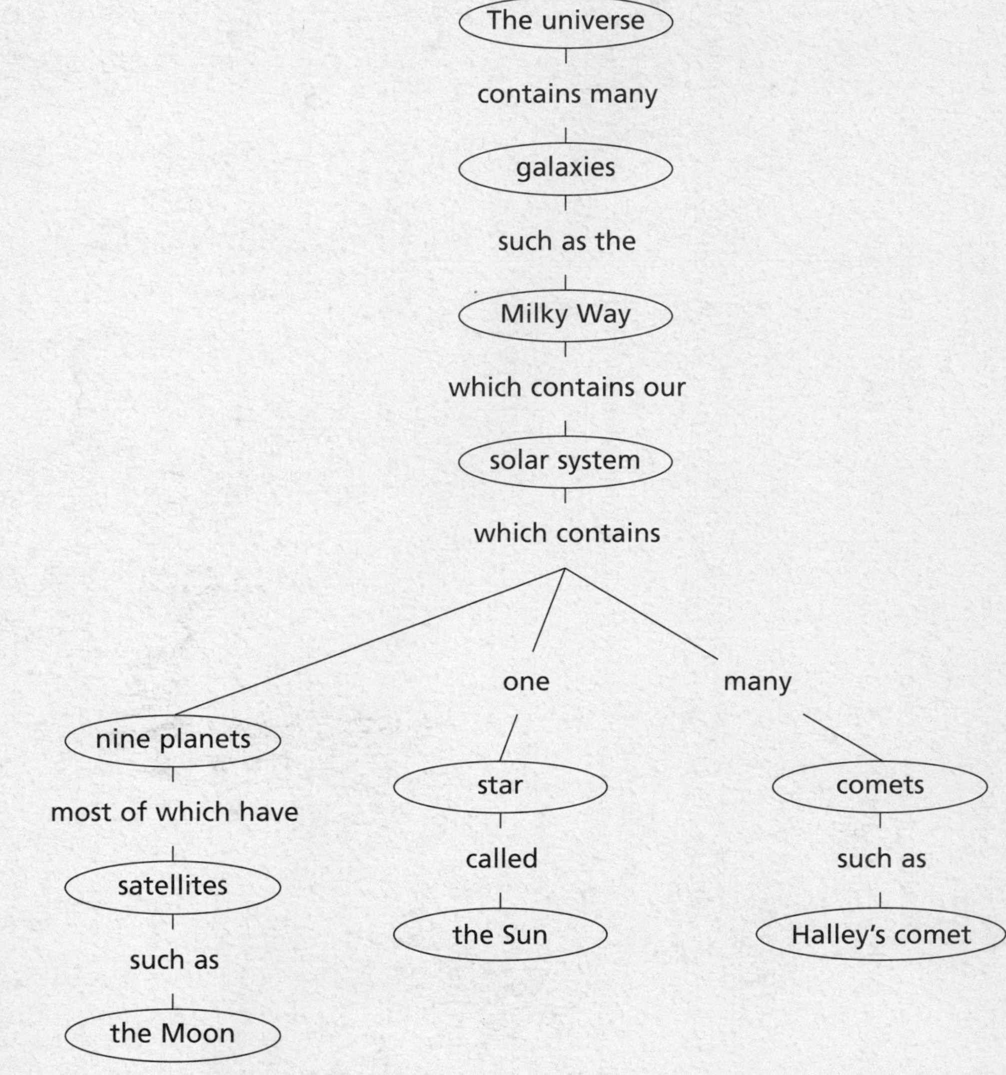

Unit 8, Chapter 24age 538
Challenge Your Tking, question 2, parts a and b

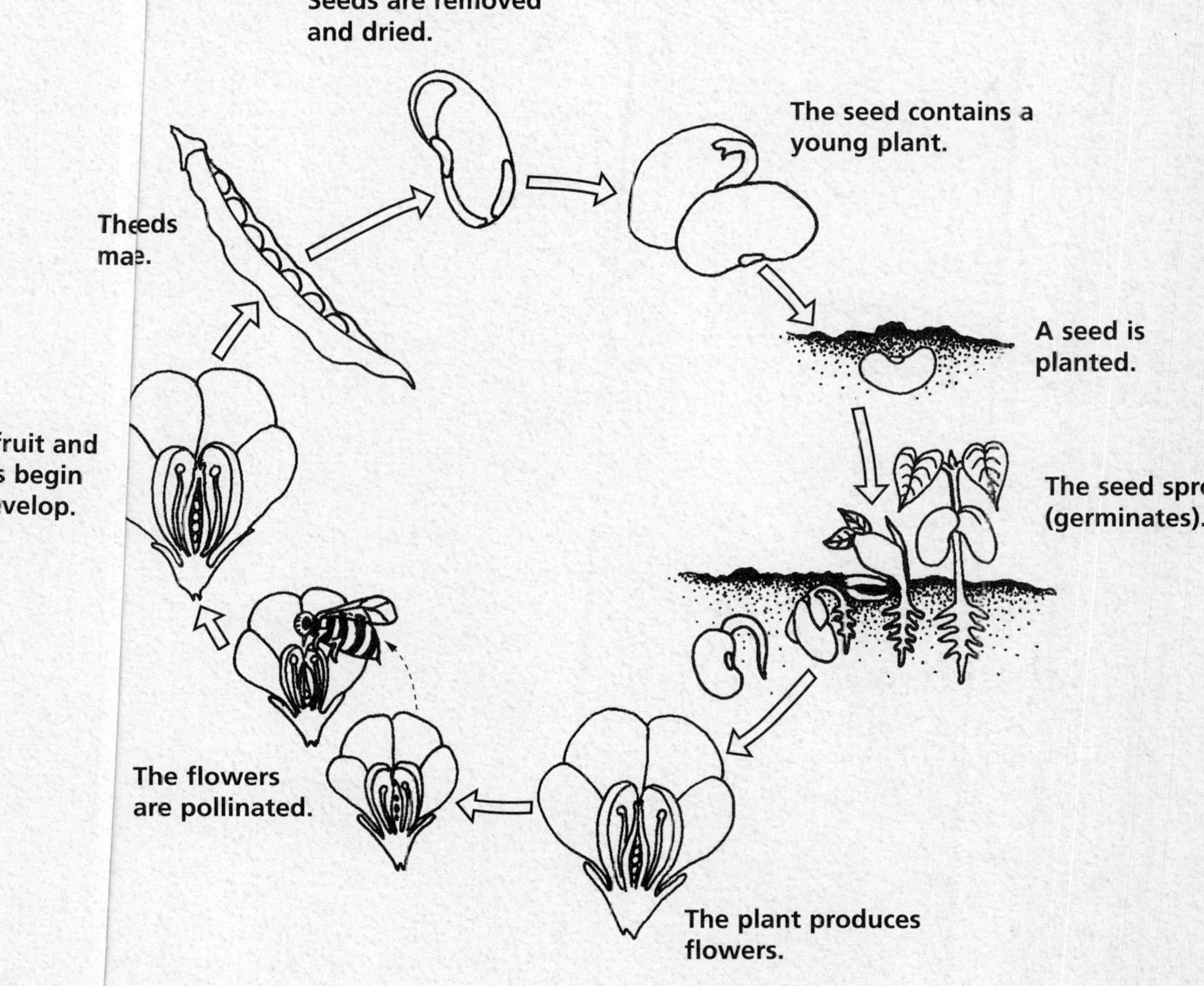

Seeds are removed and dried.
The seed contains a young plant.
A seed is planted.
The seed sprouts (germinates).
The plant produces flowers.
The flowers are pollinated.
The fruit and seeds begin to develop.
Theeds mae.